CONVERSIONS BETWEEN U.S. CUSTOMARY UNITS AND SI UNITS (Continued)

U.S. Customary unit		Times conversion factor		Equals SI unit	
		Accurate	**Practical**		
Moment of inertia (area)					
inch to fourth power	in.^{4}	416,231	416,000	millimeter to fourth power	mm^4
inch to fourth power	in.^{4}	0.416231×10^{-6}	0.416×10^{-6}	meter to fourth power	m^4
Moment of inertia (mass)					
slug foot squared	slug-ft^2	1.35582	1.36	kilogram meter squared	kg·m^2
Power					
foot-pound per second	ft-lb/s	1.35582	1.36	watt (J/s or N·m/s)	W
foot-pound per minute	ft-lb/min	0.0225970	0.0226	watt	W
horsepower (550 ft-lb/s)	hp	745.701	746	watt	W
Pressure; stress					
pound per square foot	psf	47.8803	47.9	pascal (N/m^2)	Pa
pound per square inch	psi	6894.76	6890	pascal	Pa
kip per square foot	ksf	47.8803	47.9	kilopascal	kPa
kip per square inch	ksi	6.89476	6.89	megapascal	MPa
Section modulus					
inch to third power	in.^{3}	16,387.1	16,400	millimeter to third power	mm^3
inch to third power	in.^{3}	16.3871×10^{-6}	16.4×10^{-6}	meter to third power	m^3
Velocity (linear)					
foot per second	ft/s	0.3048*	0.305	meter per second	m/s
inch per second	in./s	0.0254*	0.0254	meter per second	m/s
mile per hour	mph	0.44704*	0.447	meter per second	m/s
mile per hour	mph	1.609344*	1.61	kilometer per hour	km/h
Volume					
cubic foot	ft^3	0.0283168	0.0283	cubic meter	m^3
cubic inch	in.^{3}	16.3871×10^{-6}	16.4×10^{-6}	cubic meter	m^3
cubic inch	in.^{3}	16.3871	16.4	cubic centimeter (cc)	cm^3
gallon (231 in.^{3})	gal.	3.78541	3.79	liter	L
gallon (231 in.^{3})	gal.	0.00378541	0.00379	cubic meter	m^3

*An asterisk denotes an *exact* conversion factor

Note: **To convert from SI units to USCS units, *divide* by the conversion factor**

Temperature Conversion Formulas

$$T(°C) = \frac{5}{9}[T(°F) - 32] = T(K) - 273.15$$

$$T(K) = \frac{5}{9}[T(°F) - 32] + 273.15 = T(°C) + 273.15$$

$$T(°F) = \frac{9}{5}T(°C) + 32 = \frac{9}{5}T(K) - 459.67$$

Fundamentals of Chemical Engineering Thermodynamics

Kevin D. Dahm
Rowan University

Donald P. Visco Jr.
University of Akron

SI Edition prepared by

Jayant Singh
Indian Institute of Technology Kanpur

CENGAGE
Learning®

Australia · Brazil · Japan · Korea · Mexico · Singapore · Spain · United Kingdom · United States

Fundamentals of Chemical Engineering Thermodynamics, SI Edition
Kevin D. Dahm
Donald P. Visco, Jr.
SI Edition by: Jayant Singh

Publisher: Timothy Anderson

Senior Developmental Editor:
Mona Zeftel

Senior Editorial Assistant:
Tanya Altieri

Senior Content Project Manager:
Kim Kusnerak

Production Director:
Sharon Smith

Media Assistant:
Ashley Kaupert

Rights Acquisition Director:
Audrey Pettengill

Rights Acquisition Specialist, Text and
Image: Amber Hosea

Text and Image Researcher:
Kristiina Paul

Manufacturing Planner:
Doug Wilke

Copyeditor: Fred Dahl

Proofreader: Patricia Daly

Indexer: Shelly Gerger-Knechtl

Compositor: MPS Limited

Senior Art Director:
Michelle Kunkler

Internal Designer: MPS Limited

Cover Designer: Rose Alcorn

Cover Illustration: © Rob Schuster

Library of Congress Control Number: 2013948648

ISBN-13: 978-1-111-58071-1

ISBN-10: 1-111-58071-5

Cengage Learning
200 First Stamford Place, Suite 400
Stamford, CT 06902
USA

Cengage Learning is a leading provider of customized learning solutions with office locations around the globe, including Singapore, the United Kingdom, Australia, Mexico, Brazil, and Japan. Locate your local office at:
international.cengage.com/region

Cengage Learning products are represented in Canada by Nelson Education Ltd.

For your course and learning solutions, visit
www.cengage.com/engineering

Purchase any of our products at your local college store or at our preferred online store **www.cengagebrain.com**

Unless otherwise noted, all items © Cengage Learning.

Printed in the United States
3 4 5 17 16 15 14

To Robin, my family, and all of my students and my colleagues at Rowan.

Kevin D. Dahm

To my past, present and future students as well as my family, especially Tracey, Mary, Matthew, and Lucy.

Donald P. Visco, Jr.

Preface to the SI Edition

This edition of *Fundamentals of Chemical Engineering Thermodynamics* has been adapted to incorporate the International System of Units (*Le Système International d'Unités* or SI) throughout the book.

Le Système International d'Unités

The United States Customary System (USCS) of units uses FPS (foot–pound–second) units (also called English or Imperial units). SI units are primarily the units of the MKS (meter–kilogram–second) system. However, CGS (centimeter–gram–second) units are often accepted as SI units, especially in textbooks.

Using SI Units in this Book

In this book, we have used both MKS and CGS units. USCS units or FPS units used in the US Edition of the book have been converted to SI units throughout the text and problems. However, in case of data sourced from handbooks, government standards, and product manuals, it is not only extremely difficult to convert all values to SI, it also encroaches upon the intellectual property of the source. Some data in figures, tables, and references, therefore, remains in FPS units. For readers unfamiliar with the relationship between the FPS and the SI systems, a conversion table has been provided inside the front cover.

To solve problems that require the use of sourced data, the sourced values can be converted from FPS units to SI units just before they are to be used in a calculation. To obtain standardized quantities and manufacturers' data in SI units, the readers may contact the appropriate government agencies or authorities in their countries/regions.

Instructor Resources

The Instructors' Solution Manual in SI units is available through your Sales Representative or online through the book website at www.login.cengage.com. A digital version of the ISM and PowerPoint slides of figures, tables, and examples and equations from the SI text are available for instructors registering on the book website.

Feedback from users of this SI Edition will be greatly appreciated and will help us improve subsequent editions.

Cengage Learning

Preface

"Thermodynamics is a funny subject. The first time you go through it, you don't understand it at all. The second time you go through it, you think you understand it, except for one or two small points. The third time you go through it, you know you don't understand it, but by that time you are so used to it, so it doesn't bother you any more."

Arnold Sommerfeld

Most likely, just about anyone who's studied thermodynamics or taught it can relate to the above quote. Though we were undergraduate students a generation ago, we still remember how we, and many of our classmates, perceived the subject of Chemical Engineering Thermodynamics when we first encountered it: complex and abstract, with tons of different equations and terms like "entropy" and "fugacity" that were often hard to connect to anything that seemed *real* (not to mention the symbols, with an array of carats, overbars, subscripts and superscripts).

As teachers of the subject, we can't shy away from its complexity—we have to tackle it head on. What we can do is frame the subject in ways that make it more *accessible*. The range of thermodynamics concepts and the long lists of equations may always seem intimidating at first, but they needn't seem arbitrary. Our goal with this book is to provide a practical and relatable introduction for students who are encountering Chemical Engineering Thermodynamics for the first time. This is an important distinction, since several of the most popular books on the subject for this course have also been used at the graduate level as well. By contrast, this book is truly aimed at providing the "fundamentals" of chemical engineering thermodynamics for the undergraduate student. Once complete, the student will have the proper background for follow-on undergraduate courses that rely on a solid foundation in this field of study or for advanced courses in thermodynamics. In an effort to provide this solid foundation in chemical engineering thermodynamics, we have incorporated several features into the book that are intended to make it more accessible to a wide variety of learners:

Motivational Examples

Each chapter begins with a "Motivational Example" that introduces the topic of the chapter and illustrates its importance. This is intended to benefit all learners, but particularly global learners who require big picture insights to connect information, and technical learners who require a practical application for everything.

Worked Examples

The book makes extensive use of examples in which the thought process behind the solution is explained, step-by-step, and the practical significance of the material is underscored. For some problems, an expanded version of the solution is available in the students' electronic supplements.

Detailed Derivations

We have made a big effort not to skip steps in the derivation of key concepts and fundamental equations, instead taking an extra step or two such that the student (who is new to the field) can follow the approach.

Margin Notes

The book makes extensive use of margin notes. These are intended to serve as a "voice over the reader's shoulder" guiding them through the book. Placing these notes in the margins avoids interrupting the flow of the main text. The notes include three recurring themes:

- Margin Notes: These should be interpreted as an aside to the reader, providing an interesting fact about the concept being presented or a quick digression on the scientist or engineer associated with the development of that concept.

- Margin Notes: Pitfall Prevention: This special type of margin note calls out to the reader where, from our experience, common errors (both conceptual and from a calculation standpoint) will normally occur.

- Margin Notes: Food for Thought: These are special points that the reader might consider in a deeper way related to the concept being presented. Some are simpler while others are more challenging. The student supplemental materials include feedback on the Food for Thought questions.

Exercises and Problems

Each chapter ends with problems suitable for homework, which are divided into "Exercises" and "Problems." The Exercises are very focused and comparatively short, and the answers are included in the students' supplemental materials. Each exercise will test the student's ability to apply one specific concept or perform one specific type of calculation, and the student can obtain immediate feedback on whether he/she did it correctly. The Problems are longer and require synthesis of more concepts. Solutions to problems are available to the instructor both electronically and in a printed solution manual. For many problems, the technology used to arrive at the solutions (such as an Excel sheet) is provided for the instructor.

Organization

Each chapter is organized to be helpful to students with a variety of learning styles.

- The introduction to each chapter includes a list of instructional objectives.
- Each chapter closes with a bulleted summary that includes definitions of key terms and highlights key concepts.
- The book uses both inductive and deductive reasoning. In some places, key equations and/or concepts are developed in the context of an example, and after the example there is a discussion of how the concept generalizes. In other places a more traditional deductive approach is used.

The book comes with a number of additional resources for both students and instructors.

Instructor Resources

The supplemental material available to instructors includes:

- An Instructor's Solutions Manual, available in both print and electronic form.
- Electronic resources used in the solutions of problems, such as an Excel sheet or a PolyMath worksheet.
- A Test Bank of suggested exam problems
- Lecture Builder PowerPoint slides for each chapter

Student Resources

The electronic resources for students include:

- Answers to Exercises
- Feedback on Food for Thought questions
- Links to experiential learning activities

MindTap Online Course and Reader

In addition to the print version, this textbook is also available online through Mind-Tap, a personalized learning program. Students who purchase the MindTap version will have access to the book's MindTap Reader and will be able to complete homework and assessment material online, through their desktop, laptop, or iPad. If your class is using a Learning Management System (such as Blackboard, Moodle, or Angel) for tracking course content, assignments, and grading, you can seamlessly access the MindTap suite of content and assessments for this course.

In MindTap, instructors can:

- Personalize the Learning Path to match the course syllabus by rearranging content, hiding sections, or appending original material to the textbook content
- Connect a Learning Management System portal to the online course and Reader
- Customize online assessments and assignments
- Track student progress and comprehension with the Progress app
- Promote student engagement through interactivity and exercises

Additionally, students can listen to the text through ReadSpeaker, take notes and highlight content for easy reference, and check their understanding of the material.

Acknowledgments

The manuscript was peer-reviewed at three different stages of completion. The authors know this valuable feedback led to significant improvements in the book and gratefully acknowledge the contributions of the following individuals who served as reviewers:

D. Eric Aston	University of Idaho
Daniel H. Chen	Lamar University
Michael Mark Domach	Carnegie Mellon University
Isabel C. Escobar	The University of Toledo
Jeanette M. Garr	Youngstown State University
Douglas Ludlow	Missouri University of Science and Technology
Edward Maginn	University of Notre Dame
Sohail Murad	University of Illinois at Chicago
William E. Mustain	University of Connecticut
Steven Nartker	Kettering University
Athanassios Panagiotopoulos	Princeton University
Ajit Sandana	University of Mississippi
Rafael Tadmor	Lamar University
John C. Telotte	Florida State University
Reginald P. T. Tomkins	New Jersey Institute of Technology

We'd also like to recognize a number of current and former students who made substantial contributions to the book: Jason Giacomelli, Andrew Garrison, David T. Hitchcock, William John Hoffman, Marc A. Izquierdo, Michele L. Marandola, Juan Riveros, Christopher N. Robinson Zavala, Heather Malino, and Zachary Hinton all participated through the Rowan University Junior/Senior Engineering Clinic, Sarah E. Gettings and Chris Gies as summer interns at Rowan, Kate Clark through the Williamstown High School Engineering Academy, and Keith McIver simply volunteered. These people reviewed the manuscript and provided insightful feedback and drafted many of the figures, and the Engineering Clinic students contributed much to the solutions manual. The book wouldn't be what it is without their time and effort, and we greatly appreciate their contributions. We'd also like to thank Pamela Kubinski, who collected the data used in Example 1-1, and Gina Tierno, who took the photos in Figure 1-26. Ryan Pavlovsky, a student at Tennessee Tech University, performed some research on systems that were important for Chapter 9, and we gratefully acknowledge his efforts. Keith McIver contributed to the writing of questions and answers to the online tests for MindTap.

I (KD) wish I could list all of the hundreds of people who have contributed to my perspective on thermodynamics: authors, classmates and teachers, current and former faculty colleagues. I'd also like to acknowledge the many teachers who have helped me to become an effective writer. Of course, I have to thank all of the students who have shared a classroom with me over the years. Every semester as a teacher I see new ideas, hear questions I've never heard before, and gain new insights. The book wouldn't be what it is without all of these people. But most of all I want to recognize my very first teachers, Arlene and Donald Dahm, and my wife Robin for all of the help and support she's given me throughout this project.

I (DV) would like acknowledge the efforts of those who have been important to my personal education on the topic of thermodynamics, be it conversations with colleagues, or textbook and journal authors that I have learned from. In particular I would like to thank David Kofke, whose humility and insight have been inspirational. Finally, I would like to thank all of my past, current, and future students who ask the question "why", whether in class, after class, during office hours, or in emails. When you ask "why" about a topic in thermodynamics, it challenges me to think more deeply on the concept and, in turn, my comprehension of the subject becomes that much deeper. I often tell students that thermodynamics is a beautiful and powerful subject, whose study helps us to a better understanding of why our world is the way it is. And last, but certainly not least, I would like to thank my wife, Tracey, and my children, Mary, Matthew, and Lucy. Many weekends, nights, and early mornings were spent "working on the book", and I have always appreciated their patience.

Contents

About the Authors

Kevin D. Dahm   joined the Rowan University Chemical Engineering department in 1999, and was promoted from Associate Professor to Professor in 2013. He received his B.S. in Chemical Engineering from Worcester Polytechnic Institute in 1992 and his Ph.D. in Chemical Engineering from Massachusetts Institute of Technology in 1998 He has published over 30 journal articles, many of which are in the area of engineering pedagogy, on topics such as instilling metacognition in engineering students, pedagogically sound uses for process simulation, and assessment of student learning. He has received four national awards from the American Society for Engineering Education: the 2002 ASEE PIC-lll Award, the 2003 Joseph J. Martin Award, the 2004 Raymond Fahien Award, and the 2005 Corcoran Award. In addition, he and his father Donald Dahm authored the book *Interpreting Diffuse Reflectance and Transmittance: A Theoretical Introduction to Absorption Spectroscopy of Scattering Materials*. Prior to joining Rowan University, he was a postdoctoral researcher at UC Berkeley and an adjunct professor at North Carolina A&T State University.

Donald P. Visco, Jr.   is the Associate Dean for Undergraduate Studies and a Professor of Chemical & Biomolecular Engineering in the College of Engineering at the University of Akron. He was previously employed at Tennessee Technological University. Prof. Visco's research work focuses on molecular design and thermodynamic modeling. He has won several awards for his research and educational activities, including both the Dept. of Energy PECASE and the ASEE National Outstanding Teaching Award. He has served as Chair of both the ASEE Chemical Engineering Division as well as the Education Division of AIChE. Prof. Visco received both his B.S. and Ph. D. degrees in Chemical Engineering from the University at Buffalo, State University of New York.

Fundamentals of
Chemical Engineering
Thermodynamics

Introduction

1

Thermodynamics is the study of energy, including the conversion of energy from one form into another and the effects that adding or removing energy have on a system. Thermodynamics is essential for the practice of chemical engineering. The principles of thermodynamics have a fundamental role in how chemical processes are understood, analyzed, and designed. This book is intended for readers who are being introduced to this crucial subject for the first time.

The first chapter gives an overview of how and why thermodynamics is important and introduces some fundamental concepts. In particular, two abilities that are foundational in chemical engineering thermodynamics are

1. Recognizing the forms in which energy can be stored and transferred.
2. Identifying and analyzing systems.

Every chapter of this book opens with a list of objectives like the one above. Each chapter also features a "motivational example," where an examination of an engineering application underscores the significance of the topics presented. Normally, the motivational example immediately follows the chapter instructional objectives, but in this chapter, the motivational example is in Section 1. 2. First, we briefly explore why the field of thermodynamics *as a whole* is essential to the practice of chemical engineering. ■

1.1 The Role of Thermodynamics in Chemical Engineering

The number and variety of chemical products is staggering. Walking through the aisles of a drug store or a supermarket, you will see hundreds of different products, where chemical engineers played some role in the production of almost all of them. Consider these examples:

- The *2011–2013 Alfa Aesar® catalog* contains a 2210-page alphabetical list of the company's chemical products, from Abietic Acid to Zirconium(IV) 1,1,1-trifluoro-2,4-pentanedionate.

- The *2010 Physicians' Desk Reference* lists over 2400 prescription drugs sold in the United States.

- In a trip to a local supermarket, one of the authors counted over 200 different household cleaning products, over 200 different hair care products, over 50 kinds of toothpaste, and over 40 different insect repellents—not to mention the multitude of processed food products.

- Many of these household products contain 10 or more ingredients, each of which can itself be considered a chemical product.

> Chemical engineers are also concerned with a wide variety of by-products and waste materials. The EPA's 2010 Toxic Release Inventory lists 593 individual chemicals and 30 categories that encompass dozens of additional chemicals. (This information is available at: http://www.epa.gov/tri/trichemicals/index.htm)

These various products are manufactured in factories and chemical plants from a wide variety of raw materials. Chemical engineers are responsible for designing processes to manufacture them efficiently, economically, reliably, and (in particular) *safely*.

Chemical engineering plays an important role in the manufacture of products in a wide range of industries. Even across all of these different kinds of applications, all chemical manufacturing processes basically follow a universal rule of thumb: Chemical reactions convert raw materials (chemicals) into desired products (other chemicals). From a business standpoint, the product must have a greater monetary value than the raw materials being consumed (see Figure 1-1).

> Figure 1-1 shows a single chemical reaction, but many chemical manufacturing processes require multistep reactions carried out in series of reactors.

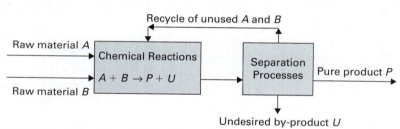

FIGURE 1-1 A schematic of a typical chemical process.

As illustrated in Figure 1-1, chemical reactions almost never lead to products that are pure. The mixture leaving a reactor normally contains not only the desired product but also by-products and/or unused raw materials. Consequently, some chemical and/or physical process is needed to separate the substances that leave the reactor.

But what do the boxes labeled "Chemical Reactions" and "Separation Processes" in Figure 1-1 actually represent? The specific processes depend on the answers to questions such as

> Figure 1-1 shows two raw materials (*A* and *B*), one product (*P*), and one by-product (*U*). Some chemical processes will have significantly larger numbers of raw materials and by-products, especially when multiple distinct reaction steps are required to make the product.

- How many different chemical reactions are necessary?

- Can all reactions be carried out in a single reactor, or is a separate piece of equipment needed for each reaction?

- Figure 1-1 shows pure products leaving the Separation processes. How close to 'pure' is required?

■ How are the separations to be carried out?

■ Can the separations all be completed with a single piece of equipment, or are several distinct separation steps needed?

■ At what temperature and pressure does each piece of equipment operate?

Typically, chemical engineering programs include a course on chemical reactor design and at least one course on separations.

Answering questions like these is fundamental to the practice of chemical engineering, and the process designer will find that every process presents unique challenges and opportunities. Indeed, some products have more than one plausible route to making them, each of which could have completely different answers to these questions.

Thermodynamics plays a vital role in the design of processes. The chemical engineer must consider thermodynamic properties when addressing questions such as

■ How much raw material and energy will it take to make 10 million kg of this product annually?

■ What methods can be used to separate this product from any by-products and unused raw materials?

■ How much energy does it take to heat this process stream to the required temperature of 422 K?

■ How can the reactor conditions be optimized for the maximum production of the desired product while minimizing the production of undesired by-products?

Distillation is the most commonly used separation technique, though there are many others.

Minimizing the production of by-products (U in Figure 1-1) is a major goal of green engineering.

This book introduces and illustrates the principles of chemical engineering thermodynamics by exploring the use of thermodynamics in the solution of engineering problems. Although focusing on examples of particular interest to chemical engineers, the text also illustrates the extreme breadth of systems for which thermodynamics is applicable. Consequently, this book emphasizes examples drawn from the design or components of chemical manufacturing processes, while also examining a broad range of engineering systems and problems. The next section considers an extremely broad and practical problem in our society: the generation of electricity.

1.2 MOTIVATIONAL EXAMPLE: The Conversion of Fuel into Electricity

You might recognize the generation of electricity as an important engineering problem, but not realize that electricity generation is also a *chemical* engineering problem. And it's true that the design of the electric generator itself falls more within the sphere of electrical and mechanical engineering. Historically, however, most electricity has come from burning fuels. A more global look at the problem of obtaining electricity from fuel reveals a complex engineering challenge that is multidisciplinary in nature and in which chemical engineers have a clear role.

The cover of this book shows an example of a process that converts fuel into electricity.

1.2.1 Generation of Electricity

If you've taken a physics course that covered electricity and magnetism, you've learned about induction of an electric current using a magnetic field. Consider the conducting wire shown in Figure 1-2. The area surrounded by the wire is a flat, round surface. Any closed circuit can be considered the boundary of a surface, although the surface may not have a regular shape like that in Figure 1-2. If the magnetic flux through the surface bound by the circuit changes, an electric current is induced in the circuit. One strategy for inducing current is to create a magnetic field with stationary magnets and to rotate the conductor within the field, as illustrated in Figure 1-3.

The use of electro-
magnetic principles
to generate electric-
ity originated in the
1830s. (Cantor, 1996)

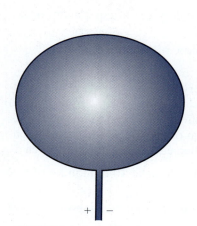

FIGURE 1-2 A conducting wire defining a circular surface.

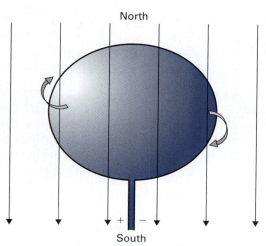

FIGURE 1-3 A conducting wire rotated in a stationary magnetic field induces a current.

You may have seen
a children's museum
exhibit in which you
powered a light bulb
by pedaling a bicycle,
and the bulb burned
brighter as you ped-
aled harder. Much of
the world's electricity
is generated in exactly
this way, although
using more sophisti-
cated equipment.

The details of how an electric generator is designed are outside the scope of this book. For our purpose, it is enough to know that rotary motion in the form of a spinning axle or *shaft* can be used to generate electricity. However, we remain faced with the challenge: How do we turn the shaft?

1.2.2 Forms of Energy and Energy Conversion

Take a moment to make a list of sources of energy that are used in our society. Note this doesn't mean "batteries" and "wall sockets" but the *root* sources of the energy.

Your list might look something like this:

- Coal
- Oil
- Natural gas
- Sun
- Wind
- Running water
- Lumber/plants
- Nuclear sources
- Geothermal sources

Herbert Hoover,
the 31st President
of the United States,
graduated with a
Geology degree from
Stanford University
and worked for many
years as a mining
engineer.

On this list, wind and water are of particular interest, because they are naturally occurring fluids in motion and readily can be used to turn a shaft. Figure 1-4 shows a variety of devices that harness wind and water power. They span the range from a pinwheel, which is a simple child's toy, to the Hoover Dam, which is one of the most massive engineering projects ever undertaken. While widely different in scale, these devices have much in common. They all have some type of fins or blades designed to "catch" the motion of the wind or water. These blades are connected to a central axle, which turns. In the language of thermodynamics, *kinetic energy is converted into shaft work*.

A pinwheel

Hoover Dam (outside)

Hoover Dam (inside)

Turbines in an offshore wind farm

A waterwheel

A windmill

FIGURE 1-4 Devices that harness wind and water power.

Kinetic energy and *work* are probably familiar concepts from physics. Briefly, kinetic energy is energy of motion. Any object in motion has kinetic energy, which is quantified by

$$\text{K. E.} = \frac{1}{2} M v^2 \tag{1.1}$$

where M is the mass and v is the velocity of the object. Kinetic energy is discussed in Section 1.4.4.

Work is a measure of energy transferred from one object to another. It is commonly quantified by

$$W = \int F \, dl \tag{1.2}$$

where l represents the linear distance travelled and F represents the force opposing the motion. In devices like wind turbines, the motion is rotary—not linear—so *shaft work* (W_s) is quantified by

$$W_s = \int \tau \, d\omega \tag{1.3}$$

where ω represents the angle through which the shaft turns and τ represents torque. In Section 1.4.3, three different manifestations of work—flow work, expansion work, and shaft work—are discussed.

Wind turbines and hydroelectric dams illustrate a crucial theme in thermodynamics: energy conversion.

> In nuclear processes, matter can be converted into energy. This conversion is described by Einstein's famous $E = mc^2$ equation. Such processes are not considered in this book. Thus, we model energy as a "conserved quantity."

Energy can be modeled as a "conserved quantity"; it can be converted from one form into another but cannot be created or destroyed. *This observation is known as the **first law of thermodynamics**.*

In thermodynamics, we are concerned with how energy is *stored* and how it is *transferred*. Substances store energy in one of three forms: internal energy, potential energy, and kinetic energy.

- *Kinetic energy* is energy of motion.
- *Potential energy* is energy of position. A stationary object 200 meters above the ground is said to have "potential" energy because it can gain kinetic energy as Earth's gravity works on it.
- *Internal energy* is energy stored within matter on a microscopic level. It is by far the most important of the three in chemical engineering thermodynamics, but it is less tangible than potential and kinetic energy. Temperature is a crucial metric for quantifying how much internal energy a substance *has*, but one has to go down to the molecular level to build a physical picture of what internal energy actually *is*.

Stored energy can be transferred in the sense that material can be moved from location to location. Thus, if a tank is filled with water, all of the energy stored by the water is now inside the tank. Energy can also be transferred in two additional forms: work and heat.

> Two fundamental aspects of green engineering are obtaining more energy from renewable sources and using energy (regardless of its source) more efficiently.

- *Work* is energy used to impart motion. It takes work to move a box, compress a gas, or turn a shaft. Motion occurs because the forces on an object are imbalanced.
- *Heat* is energy transferred without macroscopic motion. Heat transfer occurs because of differences in temperature.

Section 1.4 discusses the various forms of energy in more detail. A major aspect of engineering is designing machines and processes that harness the energy in natural resources and then convert that energy (as efficiently as possible) into the form needed to accomplish desired tasks. The devices shown in Figure 1-4 effectively convert kinetic energy into shaft work. The spinning shaft can be used in turn to power electric generators or any other task that requires shaft work; a classic application for windmills and waterwheels is grinding grain into flour.

Return to the list of energy sources at the start of this section. Wind and running water represent naturally occurring kinetic energy that can be directly converted into work. Most of the energy sources on the list are fuels—coal, oil, gas, and lumber—substances that have large amounts of internal energy. We know from experience that this internal energy is readily converted into heat simply by burning the fuel. The next section examines the challenge of converting this heat into work.

The same strategies can be applied when the heat comes from other sources, such as sunlight or nuclear reactions.

1.2.3 The Rankine Cycle

The Rankine cycle converts heat into work. A schematic of the four steps of the Rankine cycle—boiler, turbine, condenser, and pump—is shown in Figure 1-5. This section explains the roles of these steps.

The Rankine cycle is named after William John Macquorn Rankine, a 19th-century engineering professor at the University of Glasgow, for whom the Rankine temperature scale is also named.

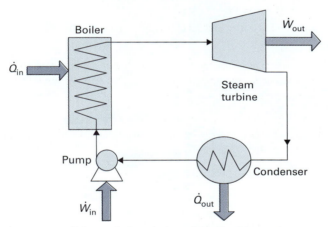

FIGURE 1-5 Schematic description of the Rankine cycle.

When burning fuel is used as the source of our energy, there is no naturally occurring motion—we have to convert energy from some other form into kinetic energy. A *nozzle* is a device that accelerates a fluid. Nozzles are used in rockets to propel gases downward at a high velocity, causing the rocket to move upward due to the conservation of momentum. As a more modest example, garden hoses have nozzle attachments that produce faster-moving streams of water. (You may have used your thumb as a sort of nozzle to influence the velocity of water coming from a hose.) The kinetic energy in the exiting stream has to "come from" someplace, and in fact, it comes from the fluid itself.

Figure 1-6 shows a schematic of a nozzle, where steam enters at $P = 0.5$ MPa and $T = 723.15$ K with negligible velocity and leaves at $P = 0.1$ MPa and $T = 501.15$ K with a velocity of 669 m/s. The decreased temperature and increased velocity of the exiting fluid represent a conversion of the steam's internal energy into kinetic energy. (Example 4-6 demonstrates why the exiting stream has $T = 501.15$ K and $v = 669$ m/s, specifically.)

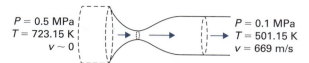

FIGURE 1-6 A nozzle converts internal energy of a fluid into kinetic energy; here steam is accelerated from a low velocity to 669 meters per second.

Example 4-8 examines the compression of a gas and reveals that the work required is orders of magnitude larger than in the pump in Example 3-8.

A turbine (Figure 1-7) is an integrated device that in effect combines nozzles with windmills. Consequently, *a turbine is a machine that converts internal energy into shaft work*. The details of how nozzles and turbine blades are designed are beyond the scope of this book; however, it must be noted that significant pressure drops occur in a turbine, so the entering vapor or gas must be at a high pressure. Furthermore, internal energy increases with increasing temperature, and at a given

(a) A steam turbine

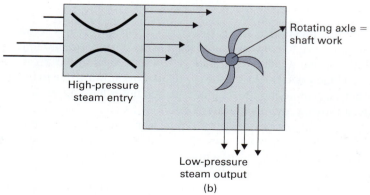

(b)

FIGURE 1-7 (a) Photograph of a large steam turbine, taken during repair. (b) A schematic of a turbine.

set of conditions, vapors are higher in internal energy than liquids. Thus, in the Rankine cycle, the stream entering the turbine is a high-temperature, high-pressure vapor. In principle, the vapor can be any compound, but it is typically steam, which is what we will assume here.

Where does the high-temperature, high-pressure steam come from? It takes a great deal of shaft work to compress gases to higher pressures, because the changes in volume accompanying pressure changes are so large. It would be self-defeating to use up a lot of work to compress steam to a high pressure just so this steam could be fed to a turbine to produce work. However, the volume of a liquid is much smaller than the volume of an equivalent mass of vapor. As a result, compressing a liquid requires significantly less work (often 2 to 3 orders of magnitude less) than compressing an equivalent mass of a vapor to the same pressure. Consequently, in the Rankine cycle, a pump is used to compress liquid water to high pressure. This liquid water then enters a *boiler*.

Figure 1-8 shows a schematic of a boiler in which fuel is burned in a furnace. The liquid water enters, travels through coils inside the furnace, and emerges as water vapor. In effect:

- The internal energy in the fuel is converted into heat through a combustion reaction.

- The heat is transferred to the water and increases the internal energy of the water.

- The increase in internal energy causes the water to increase in temperature, boil into vapor, and possibly increase in temperature further after the phase change is complete.

When describing a pure compound, a "saturated" liquid or vapor is a pure liquid or vapor at its boiling point. This and other terms describing the state of a fluid are defined in Section 2.2.1.

Coils, when compared to straight pipes, provide more surface area and produce more effective heat transfer. The typical chemical engineering curriculum includes a heat transfer course that explores this in detail.

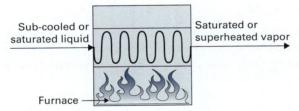

FIGURE 1-8 Schematic of a boiler. In this case, the heat required to boil the liquid comes from burning fuel.

Thus, low-pressure liquid water enters a series of three steps (pump, boiler, and turbine, see Figure 1-5) and leaves as low-pressure steam. This device can be run by continuously taking in fresh water and expelling steam, but that is inefficient and impractical. For example, consider one of the early applications of the Rankine cycle: the steam engine for trains. One cannot reasonably stop a train every ten miles to take on fresh water. Instead, the steam leaving the turbine goes to another unit where it gives off heat to the surroundings, condenses into liquid, and then goes to the pump. This allows the Rankine cycle to operate at a *steady state*. A steady-state process is a continuous process in which all parameters are constant with respect to time. For example, if the material leaving the condenser is at $T = 373$ K, then a minute, an hour, or a day from now, it will still be at $T = 373$ K if the process is operating at a steady state.

Thus, the Rankine cycle is a continuously operating, closed loop in which water sequentially circulates through the four steps—pump, boiler, turbine, condenser—thereby converting heat into shaft work. We can now make several observations

The condenser in a Rankine cycle is a heat exchanger that operates much like the radiator of a car or the coils on the back of a refrigerator.

Steady state and equilibrium are discussed in Section 1.3.2.

about chemical engineering and thermodynamics, and how these observations are expressed in the Rankine cycle:

- Energy can be regarded as a *conserved* quantity. It cannot be created or destroyed, but it can be converted from one form into another. This rule of experience is called the first law of thermodynamics and is discussed in a mathematical sense in Section 3.3.

- An integral part of engineering practice is the design of processes that accomplish an objective. In the Rankine cycle, the objective is the conversion of heat into work.

- Burning fuels is essentially conversion of the internal energy of the fuel into heat. The heat can subsequently be converted into work using the Rankine cycle.

- A chemical process typically is comprised of several steps called *unit operations*—each of which accomplishes a specific task. The Rankine cycle includes four unit operations: a pump, a turbine, and two heat exchangers (a boiler and a condenser). Other examples of unit operations familiar to chemical engineers include reactors, compressors, and various separation processes such as flash drums (in which liquid is partially vaporized due to a decrease in pressure), distillation columns (which separate compounds through differences in their boiling points), absorbers (which use liquid solvents to dissolve gases), extractors (which use two immiscible liquid phases where solutes are purified through their attraction to one or the other of these phases), and strippers (in which gas is bubbled through liquids, evaporates volatile components of the liquid, and leaves non-volatiles behind).

- Many chemical processes are designed to operate continuously—at a steady state. Electrical generation, for example, needs to operate continuously, as there is constant demand.

This section discussed the engineering challenge of generating electricity by burning fuel, primarily in qualitative terms. Section 1.4 expands upon the concepts of energy and energy conversion, laying a foundation for quantitative approaches to thermodynamics problems. We will find that quantitative analyses cannot be carried out without a very clear understanding of what specifically is being analyzed. Consequently, Section 1.3 introduces the concepts of a *system* and a *process*.

1.3 Systems and Processes

Briefly, a system is a specific object or location that we are analyzing, and a process is a specific time period or event that we are analyzing. We begin with an example. The concept of *work* is most likely familiar from introductory physics courses but is reviewed in Section 1.4.3. As stated mathematically in Equation 1.2, the work done is equal to the distance moved multiplied by the force opposing the motion. Consider the following problem:

While billiard balls are not a system of practical interest to chemical engineers, the concept of work is fundamental in analyzing chemical processes. Section 1.4.3 outlines three forms of work that are of primary significance in chemical engineering.

Two billiard balls, a cue ball and an 8-ball, are initially at rest as shown in Figure 1-9a. The cue ball is tapped, rolls two meters and strikes the 8-ball (Figure 1-9b). The collision also causes the cue ball to slow down and change direction; it rolls another meter and stops. The 8-ball moves two meters and stops. The final position of both balls is shown in Figure 1-9c. The collisions are essentially frictionless, but the frictional force of the felt surface on the rolling balls is 0.89 N. Find the work done.

We know some distances and the force of friction, so it seems like we should be able to calculate work. The difficulty is "find the work done" isn't a clear question.

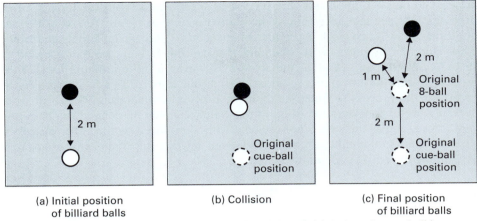

(a) Initial position
of billiard balls

(b) Collision

(c) Final position
of billiard balls

FIGURE 1-9 Position of cue ball (white) and 8-ball (black) (a) before the cue ball is tapped, (b) at the moment of collision, and (c) after both come to rest.

Work was done on the cue ball when it was tapped. Work was done by the cue ball on the 8-ball when they collided. Work occurred continuously as the balls rolled. We could calculate any or all of these values of work. But which of these are we actually interested in? Which is meant by "find the work done?"

A clearer question would be "*How much work was done BY the 8-ball, BETWEEN the time it was struck by the cue ball and the time it came to rest?*" This is something we can calculate immediately. The ball moved two meter, and the friction force opposing the motion was a constant 0.89 N, so the work is (2 m)(0.89 N) = 1.78 J.

In the language of thermodynamics, when we chose to examine the 8-ball, it became our *system*. Furthermore, we defined a *process*, which has a clear beginning (the moment the cue ball struck the 8-ball) and a clear end (the moment the ball stopped rolling.) The location of the 8-ball in Figure 1-9a is the *initial state* of the system, and the location of the ball in Figure 1-9c is the *final state* of the system.

A **system** is the *specific portion of the universe that we are modeling.* We will find that the solution of practical problems is often much easier if we start by identifying a specific system. In the Rankine cycle described in Figure 1-5, there are many different systems one might define:

■ The entire cycle.

■ Any single piece of equipment: the turbine, the pump, the boiler or the condenser.

■ Two sequential pieces of equipment, such as the pump and the boiler.

■ The furnace that supplies the heat to the boiler.

Anything that has clear, unambiguous boundaries can be considered a system—even if the boundaries are moving.

■ "The rocket and its contents" could be a system. Here the system would have constant volume but would be moving as the rocket flies.

■ "The water in the pool" could be a system. The amount of water may be constantly changing through processes like rain and evaporation, but at any specific time, the system has clear boundaries and a measurable amount of water in it.

In some applications, carefully defining a particular system might seem an unnecessary step; it might appear "obvious" what the system is. However, subtle errors

In a typical heat exchanger, there are at least three different systems: the side that supplies the heat (e.g., the furnace that contains the burning fuel), the side that receives the heat (e.g., the pipe that carries the water that is being boiled), or the entire exchanger. This is discussed further in Section 3.6.

are easy to make if one is not very precise or specific in defining an approach to a problem. For example, referring to the Rankine cycle in Figure 1-10, consider the question, "How much work is produced per second?" Recognizing that the turbine is the step that produces work, one might look at the turbine, see that it produces 100 kJ/s of work, and call that the answer. However, a more in-depth look at the process reveals that the pump requires work. If one must add 1 kJ/s of work to the pump for the process to function, then the process is generating 99 kJ/s of work overall. Careful consideration of how "the system" should be defined can help prevent errors such as this one. Since the "net work" produced by the entire process is a more meaningful number than the work from any one step, it makes sense to start by defining the entire cycle as the system.

Many of the figures in this book use dashed lines to show the boundaries of the chosen system.

The *sign* of work is negative when energy leaves the system in the form of work (as in the turbine) and positive when energy is added to the system in the form of work (as in the pump). The sign convention is discussed in Section 1.4.3.

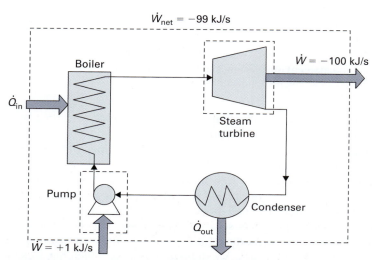

FIGURE 1-10   A sample Rankine cycle. If "the system" is the turbine, then $\dot{W} = -100$ kJ/s. If "the system" is the pump, then $\dot{W} = +1$ kJ/s. If "the system" is the entire cycle, then $\dot{W}_{net} = -99$ kJ/s.

Examples 3-6, 3-7, and 5-9 are additional problems in which there is more than one way the system could reasonably be defined. Thus, choosing the most meaningful system definition is a step in solving the problems.

A process in which the state of the system is constant with respect to time is called a "steady-state" process. Steady state and equilibrium are discussed in detail in Section 1.3.2.

Notice the use of dashed lines in Figure 1-10 to indicate the boundaries of three possible systems; one contains the pump, one contains the turbine, and one contains the entire heat engine. The **control volume** is a term used to describe the region of space inside the boundaries of the system. Our primary interest is often in the material and energy *crossing* the boundaries of the system. The turbine is a good example. In designing a heat engine, we are concerned with how much work is produced (this is energy leaving the system), and we need to know how the turbine affects the material flowing through it, which can be understood by examining how the material entering the system is different from the material leaving the system.

The *specific events* or *period of time* are called the **process**. Just as we need to be clear and specific in defining the system, we also need to have a clear understanding of when and how the process begins and ends. Example 1-6 will concern an object that falls from a height of 15 m and asks "What is the velocity when it hits the ground?" Thus, to be useful, the process is defined as ending at the exact instant the object strikes the ground. The language used throughout this book states

- The *system* is the object itself.
- The *initial state* of the system is that it is located at a height of 15 m above the ground and is stationary.

■ The *process* is everything that happens to the system as it transitions from the initial state to the final state. Here, the process is very simple: The force of gravity acts upon the object, causing it to fall. In more complex processes, there could be several different events that affect the system, occurring either consecutively or concurrently.

■ In the *final state*, the object is at a height of zero feet and has a non-zero downward velocity, which is the unknown we are trying to determine.

1.3.1 Fundamental Definitions for Describing Systems

This section introduces some terminology that is used throughout the book. We will see, particularly in Chapters 3 and 4, that these are not merely semantic definitions. The fact that a system is "open," "closed," or "adiabatic" has important implications in how we build mathematical models of the system.

*An **open system** can have both matter and energy entering and/or leaving the system.* A lake is an open system (see Figure 1-11); material can enter in the form of rain and leave in the form of evaporation. In addition, rivers or streams might flow into and out of the lake. Sunlight is an example of energy transfer, where it causes the lake water to get warmer. Thus, the system can be affected even though no water enters or leaves the system.

*A **closed system** has no matter entering or leaving the system, although energy transfer can still occur.* Thus, a sealed aluminum can is a closed system (see Figure 1-12); material cannot enter or leave the can, although its contents can be heated or cooled.

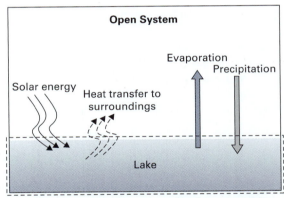

FIGURE 1-11 A lake is an example of an open system.

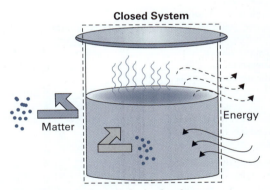

FIGURE 1-12 A sealed can that has no insulation is an example of a closed system.

The Rankine cycle described in Figures 1-5 and 1-10 provides examples of both closed and open systems. If the turbine is defined as the system, it is an open system, since there is steam continuously flowing both into and out of it. Similarly, the pump, evaporator, and condenser are individual open systems. However, if one defines the system to include all four unit operations as in Figure 1-10, then it is a closed system. Water, as liquid or vapor, circulates continuously inside the system, but no material crosses the boundaries of the system, and that fact makes it a closed system.

An **adiabatic** system has *no heat* entering or leaving the system. Note that it's possible for either closed or open systems to be adiabatic, as illustrated in Figure 1-13. A section of pipe in a house may have water continuously flowing into

The Greek word "adiabatos" means "pathless" or "not to be crossed." In an adiabatic system, heat does not cross the boundaries of the system; it takes no "path" in or out.

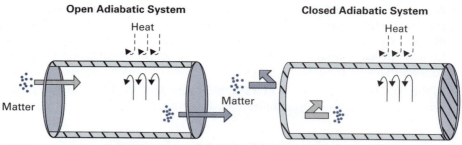

FIGURE 1-13 An adiabatic system can be either an open or a closed system.

and out of it, making it an open system, but it also may be insulated enough to be considered adiabatic. Note that if a system is described as adiabatic, this doesn't exclude the possibility that *work* is added to or removed from the system.

An **isolated** system has *neither matter nor energy entering or leaving* the system. If a cooler or thermos (shown in Figure 1-14) is closed and is so well insulated that it can be modeled as perfectly insulated, then it is be considered an isolated system. A system that is both closed and adiabatic can have work added or removed (e.g., the "expansion work" described in Example 1-2), but an isolated system is not affected by the surroundings in *any* way.

The prefix "iso" means "equal," and the Greek word "khoros" means "space." An isochoric system has constant volume—it always contains the same amount of space.

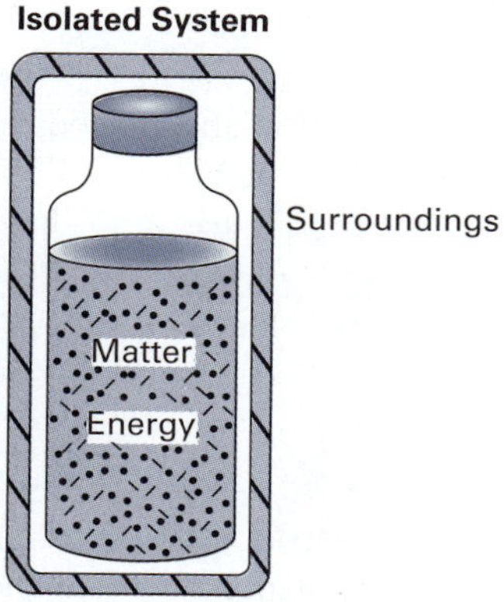

An **isochoric** system has a *constant volume*. Consider the water tank shown in Figure 1-15. The tank has rigid walls, so if the tank is the system, it is isochoric. However, if the water in the tank is the system, its volume changes as the tank fills or empties, so it is not isochoric.

Isolated System

FIGURE 1-14 A schematic of an isolated system.

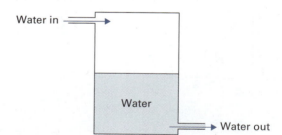

FIGURE 1-15 Schematic of a water tank. "The tank" is an isochoric system, but "the water" is not necessarily isochoric.

An **isothermal** system has a *constant temperature* throughout the system. The inside of a small refrigerator typically is regarded as an isothermal system. There is probably little difference between the temperatures at different locations inside the refrigerator. A three-story house usually can't be regarded as isothermal, because the top floor is often noticeably warmer than the bottom floor.

An **isobaric** system has a *constant pressure* throughout the system. Gaseous phases can normally be modeled as isobaric systems, while liquids often cannot. Consider the swimming pool shown in Figure 1-16. The surface of the water is in contact with the air, which exerts 1 atmosphere of pressure. At the bottom of the pool, the pressure is higher, because in addition to the atmospheric pressure, the force of gravity on the water is also pushing downward. You've probably felt this effect while swimming or diving—the pressure increases as the depth of the water increases. If the system were a glass of water, the pressure difference between the top and bottom of the glass is small enough that it can be modeled as isobaric. For larger liquid systems, pressure is not assumed to be uniform.

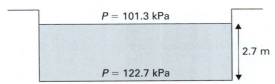

$P = 101.3$ kPa

2.7 m

$P = 122.7$ kPa

FIGURE 1-16 Pressure is a function of depth in a swimming pool because of the force of gravity pulling the water down.

The Greek word "barrios" means "weight." Weight, force, and pressure are related (but not interchangeable) concepts.

Note that the terms "isothermal" and "isobaric" can cause confusion if not used carefully, because they could reasonably be applied to either processes or systems. The distinction is given here.

- In an *isothermal/isobaric system*, the temperature/pressure is constant with respect to *position*.
- In an *isothermal/isobaric process*, the temperature/pressure is constant with respect to *time*.

For example, if a can of soda is removed from a refrigerator and placed on a counter, it gradually warms to room temperature, as illustrated in Figure 1-17. The process is not isothermal, because the temperature of the soda changes with time. But it is likely reasonable to say that the temperatures at all points in the soda can are identical to each other at any particular time, so one can describe the system as isothermal. In this book, in order to avoid such confusion, we only use the words "isothermal" or "isobaric" when temperature or pressure is uniform with respect to

$T = 278$ K

$T = 278$ K

$T = 278$ K

time passes

$T = 294$ K

$T = 294$ K

$T = 294$ K

Temperature at different locations at a given time

FIGURE 1-17 A cold soda can is removed from a refrigerator and warms to room temperature. Although the process is not isothermal (temperature changes with time), the system has uniform temperature at any given time.

both position and time; that is, if *both* the system and the process are isothermal or isobaric.

1.3.2 Equilibrium and Steady State

Equilibrium and steady state are two fundamental concepts that can be applied to systems and processes. They are distinct from each other but are discussed together, because they are frequently confused with each other. This section defines steady state and equilibrium and uses an example to illustrate the distinctions between them.

> *A system is at **steady state** when all properties of the system are constant with respect to time.*

> *A system is at **equilibrium** when there is no driving force for any change to its state properties.*

There are three possible "driving forces" that are important in chemical engineering thermodynamics.

1. **Mechanical driving force:** If the mechanical forces on an object are not balanced, there is a driving force for acceleration. A falling object is at neither steady state nor equilibrium. It's not at steady state because the velocity and height are changing with time. It's not in *mechanical equilibrium* because the downward force (gravity) is stronger than the upward force (air resistance).

2. **Thermal driving force:** If two objects are at different temperatures, there is a driving force for heat transfer, which leads to changes in the temperatures of both objects. When cold water is mixed with hot water, heat flows from the hot water to the cold water. The water is not at *thermal equilibrium* until it reaches a uniform temperature.

3. **Chemical driving force:** Chemical reactions and phase changes are both chemical processes that are of interest to chemical engineers, and a system is not in *chemical equilibrium* if there are driving forces for either or both processes present. When pure nitrogen and pure hydrogen are brought together, there is a chemical driving force for nitrogen and hydrogen to combine, forming the more stable compound ammonia. If water at atmospheric pressure is placed in an environment where the temperature is 268 K, there is a chemical driving force for it to freeze, because the solid phase is the more stable phase at this temperature and pressure.

Thermal equilibrium exists when temperature is balanced throughout a system, and mechanical equilibrium exists when force (or pressure) is balanced throughout a system. Force, pressure, and temperature are described in more detail in Section 1.4. The analogous property that we use to model chemical equilibrium is the *fugacity*. Fugacity is an abstract, intangible property. It cannot be measured directly like temperature and pressure can; it only can be calculated using models that are introduced in Chapter 8. We will revisit fugacity at that point.

It is important to recognize two facts:

1. A system can be at steady state without being at equilibrium.

2. A system can have state variables that change with position and still be at steady state if the variables at all positions are constant with respect to time.

These points are illustrated by the following example.

FOOD FOR THOUGHT 1-3

Look at the swimming pool in Figure 1-16. The pressure at the bottom is higher than the pressure at the top. Is the pool in mechanical equilibrium?

A mixture of nitrogen and hydrogen isn't in equilibrium at ambient T and P, but the reaction forming ammonia is so slow it could be mistaken for an equilibrium. While a "chemical driving force" for reaction does indeed exist at ambient temperature, in practice high temperatures are needed to produce significant reaction rates.

| A POT OF WATER ON A STOVE | EXAMPLE 1-1 |

A covered pot of water is placed on a stove and the heat is turned on. The temperature is monitored using thermocouples placed as shown in Figure 1-18. Predict what the temperature data will look like before reading on.

Defining the pot and its contents as the system, is the system at steady state? At equilibrium?

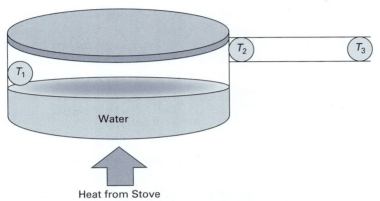

FIGURE 1-18   A covered pot of water, with three temperature sensors, T_1, T_2, and T_3.

SOLUTION: The specifics will vary based upon the size of the pot, the material from which the pot is made, the amount of water, etc., but the behavior should be qualitatively similar with any pot.

Initially, the temperature climbs at all three locations, as shown in Figure 1-19:

■ T_1, the temperature inside the pot just above the liquid surface, climbs rapidly to almost 373 K.

■ T_2, the temperature at the base of the handle, climbs less rapidly and never exceeds 350 K.

■ T_3, the temperature at the end of the handle, increases only fractionally.

During this time, the system (the pot and its contents) is clearly not at a steady state.

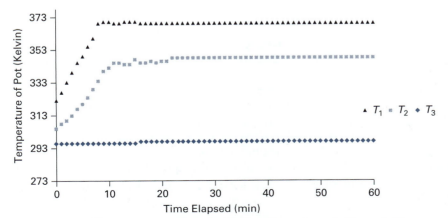

FIGURE 1-19   Temperature measured at the locations shown in Figure 1-18.

FOOD FOR THOUGHT 1-4

How would you expect these results to be different if the pot lid was sealed? If the pot was well insulated? If it was both sealed and insulated?

After approximately 15 minutes, the temperatures level off. The system is now at steady state, because while the temperature is different at different locations, none of these temperatures are changing with time, nor are any other properties changing with time. The system, however, is not at equilibrium. There are temperature gradients and therefore driving forces for heat transfer throughout the system. However, because the heat transferred *into* each location is equal to the heat transferred *out*, the temperature at each location is constant and there is a steady state.

While the system, as a whole, is not at equilibrium, some aspects of equilibrium are evident in this system. The temperature of the water levels off at ~373 K. At $P = 101.3$ kPa, the boiling point, $T = 373$ K, is the only temperature at which pure liquid water and pure water vapor can exist in equilibrium with each other. Here, the vapor phase doesn't actually contain pure water vapor; it contains a mixture of air and water vapor. The topic of transitions between phases, and how they relate to equilibrium, is introduced in Chapter 2 for pure compounds. Discussion of mixtures begins in Chapter 9. For now, we will simply note that if "the system" had been defined as a small region of space containing the surface of the liquid, we could likely say this system was at equilibrium:

■ The water at the surface and the air immediately above the surface are both at $P = 101.3$ kPa.

■ The water at the surface and the air immediately above the surface are both at $T = 373$ K.

■ The water at the surface and the water vapor in the air immediately above the surface are in chemical equilibrium; though we won't be discussing how to model chemical equilibrium mathematically until Chapter 8.

But when we define the system as "the pot and its contents," we see significant temperature gradients in this larger system; the entire system is not at equilibrium.

Note that if the pot didn't have a lid, the system would be at neither steady state nor equilibrium. Vapor is continuously formed from the boiling liquid. Without the lid, this water vapor would escape. Thus, the amount of water in the pot would decrease with time, and if any property of the system is changing with time, then the system is not at steady state. However, with the lid in place, the contents of the pot are a *closed system*; liquid water turns into vapor, and vapor turns back into liquid, but the total amount of water in the system is constant.

Figure 1-19 shows the top of the pot was at about 350 K. Consequently the water boiled, rose to the top as vapor, condensed on the lid, and fell back down.

Chemical processes that are designed to operate at steady state can run continuously for weeks or months at a time.

Example 1-1 examined a real system—a pot of water on a stove—for a period of one hour. Figure 1-19 illustrates a period of time during which the system was not at steady state, and a point after which steady state was maintained for the rest of the hour, but there was never an equilibrium state.

We said in Section 1.3 that a thermodynamic *process* needs to have a clear beginning and end. If a system is at steady state, identifying a beginning or end to the process isn't necessary, since a steady state can last indefinitely. Consider the time period from 20 to 60 minutes in Figure 1-19; temperatures T_1, T_2, and T_3 were constant during this time and would have remained constant until some change disturbed the system, such as the burner being turned off, or the lid being removed from the pot.

1.4 The Forms of Energy

Here we formally state the first law of thermodynamics, which was alluded to previously:

> The **first law of thermodynamics** states that energy is a conserved quantity. Energy can be converted from one form into another, and it can be transferred from one location to another, but it cannot be created or destroyed.

This fundamental principle can be applied to a huge range of practical engineering problems, but it begs the question, "what precisely is energy?"

This section describes three forms in which energy can be stored in a material: internal energy, potential energy, and kinetic energy. It also describes two forms in which energy can be transferred from one location to another: heat and work. We start with work because it allows us to create a framework for all of the others. Briefly, when an object is moved, "work" can be regarded as the amount of energy required to move it. Work is defined formally and mathematically in Section 1.4.3. The potential to do work is what makes the other forms of energy also recognizable as energy.

We begin with a discussion of force and pressure, which are not forms of energy but are foundational principles in the analysis of work.

1.4.1 Force

Newton's second law of motion quantifies the **force** (F) acting on an object as the product of the mass (M) of the object times the acceleration (a) resulting from the force:

$$F = Ma \tag{1.4}$$

Don't be confused if one of your textbooks says $F = ma$ and another book says $F = ma/g_c$. All books agree that 1 Newton equals 1 kg · m/s²; it doesn't much matter whether we give this unit conversion factor a name (g_c) or not.

Gravity is the most familiar force in our everyday lives, and an object's acceleration due to gravity is given the symbol g. On the surface of the Earth, $g = 9.8066$ m/s² (also 32.174 ft/s²).

When SI units are used in Equation 1.4, the right-hand side will have units of kg · m/s². A *Newton* is defined as the force required to accelerate 1 kg by 1 m/s²; thus, 1 N = 1 kg · m/s². This conversion factor is used so routinely that in some texts it is written into Equation 1.5 explicitly, as

$$F = \frac{1}{g_c} ma \tag{1.5}$$

Recall from physics that acceleration is the first derivative of velocity, and velocity is the first derivative of position. Both derivatives are taken with respect to time.

where $g_c = 1$ kg · m · s⁻² · N⁻¹.

In English units, a *pound-force* (lb$_f$) is the force required to accelerate 1 pound-mass (lb$_m$) by 32.174 ft/s². Thus, a pound-force is equivalent to the force exerted by the earth's gravity on a 1 pound-mass object or, numerically, as 1 lb$_f$ = 32.174 lb$_m$ · ft · s⁻².

1.4.2 Pressure

Pressure (P) is defined as the force (F) acting on a surface, divided by the area (A) of the surface upon which the force is acting:

$$P = \frac{F}{A} \tag{1.6}$$

FOOD FOR THOUGHT 1-6

Can you think of a situation in which a system expands, but there is no force opposing the motion?

The SI unit of pressure is the Pascal, defined as the pressure resulting from one Newton of force acting on an area of 1 m². Another commonly used pressure unit is psi, which stands for pounds-force per square inch. At sea level, atmospheric pressure is 101.325 kPa (14.696 psi). The bar is another unit of pressure, defined as 100 kPa.

In thermodynamics, one physical phenomenon we encounter frequently is expansion (and contraction). When something expands, there is generally a force opposing the expansion.

Figure 1-20, in Example 1-2, shows a type of system considered many times in this book: a gas confined in a piston-cylinder device. A *piston* is a moving seal that confines a gas in a cylinder while allowing it to expand or contract. In a typical automobile engine, the piston is attached to a crankshaft. The release of heat when fuel is combusted leads to rapid expansion of gases, and the piston and crankshaft transfer the work from this expansion into power to move the vehicle. The piston in Example 1-2 isn't attached to anything; it has no purpose other than to confine the gas while allowing the volume to change. This isn't a particularly practical system, but it is used routinely in thermodynamics and throughout this book because it is a simple system that is relatively easy to picture, and it provides a useful way to illustrate concepts.

In many engines, cylinders have openings that are covered or uncovered depending upon the position of the piston. Thus the piston can also be made to act like a valve.

EXAMPLE 1-2	**GAS IN A PISTON-CYLINDER DEVICE**

The force opposing the motion is used in calculating work. This is discussed in Section 1.4.3.

A gas is confined in a cylinder that is 0.05 m in diameter with a movable piston at the top, as in Figure 1-20. The piston itself weighs 0.1 kilograms. If the gas is heated and expands, what is the force *opposing* the expansion? Assume friction is negligible.

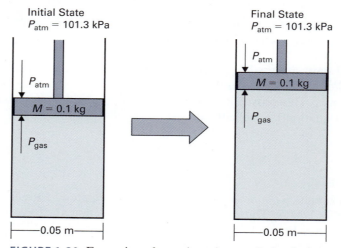

FIGURE 1-20 Expansion of a gas in a piston-cylinder device.

FOOD FOR THOUGHT 1-7

How would the problem be affected if the cylinder were oriented sideways instead of vertically?

SOLUTION:

Step 1 *Define the system*
Define the gas as the system. Why choose the gas rather than the piston? Because we are trying to find the force opposing the expansion of the gas; if we define the gas as the system, then the force we're trying to find is an external force that is acting on the system.

Step 2 *Identify forces opposing the motion*
The piston moves upward as the gas expands, so the force of gravity on the piston is opposing the motion. In addition, the atmospheric pressure, which is pushing down on the top of the piston, is also opposing the motion. Conceptually, friction would be an additional force opposing the motion, but we are told friction is negligible in this case.

$$F_{total} = F_{piston} + F_{atmosphere} \qquad (1.7)$$

Step 3 *Quantify the force of gravity on the piston*
Force due to gravity is calculated as

$$F = Mg \qquad (1.8)$$

So for a 0.1 kg piston, this force is

$$F_{piston} = (0.1\,\text{kg})\left(9.8\,\frac{\text{m}}{\text{s}^2}\right)\left(\frac{1\,\text{N}}{1\,\frac{\text{kg}\cdot\text{m}}{\text{s}^2}}\right) = \mathbf{0.98\,N} \qquad (1.9)$$

Step 4 *Quantify the force of the atmosphere on the piston*
Pressure was previously defined as

$$P = \frac{F}{A} \qquad (1.10)$$

The force *F* exerted by the atmosphere is what we want to determine. Thus,

$$F_{atmosphere} = PA \qquad (1.11)$$

Since nothing unusual is stated about the location of the cylinder, we can assume the atmospheric pressure is $P = 101.3$ kPa. *A* can be found from the known piston diameter of 0.05 m:

$$F_{atmosphere} = (101.3\,\text{kPa})\left(\frac{\pi}{4}\right)(0.05\text{ m})^2 \qquad (1.12)$$

Recall that 1 Pascal is defined as 1 N/m²; introducing this definition and converting units gives

$$F_{atmosphere} = (101.3\text{ kPa})\left(\frac{1000\,\text{Pa}}{\text{kPa}}\right)\left(\frac{1\,\frac{\text{N}}{\text{m}^2}}{\text{Pa}}\right)\left(\frac{\pi}{4}\right)(0.05\,\text{m})^2 = \mathbf{198.9\,N}$$

Step 5 *Find total pressure*

$$F_{total} = F_{piston} + F_{atmosphere} = 0.98\text{ N} + 198.9\text{ N} = \mathbf{199.9\,N} \qquad (1.13)$$

Note that the piston mass is almost irrelevant, and it is common when solving this sort of problem to assume the mass of the piston is negligible.

Pressure is often quantified using a gauge, which generally compares the pressure being measured to the pressure of the surrounding atmosphere. For example, if you measure the pressure of a car tire with a gauge and it reads 206.8 kPa, this probably means 206.8 kPa above ambient pressure. The pressure inside the tire in absolute terms is 308.1 kPa (206.8 + 101.3) if the gauge is operating around sea level.

Sidebar notes:

If the piston were attached to something, like a crankshaft, then that could contribute additional forces opposing the motion.

The area of a circle is $\pi d^2/4$.

"Gage" is sometimes used as a secondary spelling of the word "gauge."

NIST is the National Institute of Standards and Technology.

In this book, *pressure is always expressed on an absolute basis, unless explicitly stated otherwise.*

1.4.3 Work

Work (W) can be quantified as the product of the distance moved (l) times the magnitude of the force opposing the motion (F). In many applications, the force opposing the motion is not constant, so it's necessary to express the definition in differential form as

$$dW = F\, dl \qquad (1.14)$$

The SI unit for work (and more fundamentally, for energy) is the Joule, which is defined as the work required to move an object one meter against an opposing force of one Newton (1 J = 1 N · m).

In some applications, it is convenient to quantify the **power**, which is the *rate* at which work is done:

$$\dot{W} = Fv \qquad (1.15)$$

> Throughout this book, a dot over a symbol signifies a rate. Thus W represents work, and $\dot{W}$ represents work per unit time.

In this equation, $\dot{W}$ stands for the power, and v is the velocity. You can make the connection between Equations 1.14 and 1.15 clearer by considering that velocity is equivalent to the derivative of position or dl/dt. The SI unit for power is the Watt, which is defined as 1 J/s.

The following example applies Equation 1.14 to a familiar everyday application.

EXAMPLE 1-3	**LIFTING A BOX**

A mover lifts a 23 kg box off the ground and places it on a truck (Figure 1-21). If the floor of the truck is 1.2 m off the ground, how much work was required to lift the box?

FIGURE 1-21 A 23 kg box is lifted off the ground and placed on the flatbed of a truck.

SOLUTION:

Step 1 *Define a system*
Define the box as the system.

Step 2 *Apply definition of work*
The system is moving upward. The force opposing the motion is gravity, and is thus given by

$$F = Mg \qquad (1.16)$$

Consequently, work is given by

$$dW = F\,dl = Mg\,dl \tag{1.17}$$

In this case, M and g are constant, and integration of Equation 1.17 is

$$W = \int_{l=0}^{l=1.2\,m} Mg\,dl = (23\text{ kg})\left(9.81\,\frac{m}{s}\right)(1.2\text{ m} - 0)\left(\frac{N}{kg\,\frac{m^2}{s}}\right) = 270.8\text{ N} \tag{1.18}$$

$$W = 270.8\text{ N}$$

Note that this problem only examined the work required to lift the box up 1.2 m; horizontal motion was not considered. Horizontal motion requires work when there is friction opposing the motion (e.g., to push the box along the flatbed of the truck one would have to do work to overcome friction).

Example 1-3 reviews material that should be familiar to most readers from introductory physics. However, one point worth examining is *the sign of l (length)*. It probably seems natural that dl was positive here since the box was moving up. However, in many cases it might seem arbitrary which direction is positive and which is negative. We wish to establish a sign convention where $W > 0$ is when energy is transferred TO the system in the form of work and $W < 0$ is when energy is transferred FROM the system to the surroundings in the form of work. To accomplish this, the distance moved (dl) is defined as positive when the motion is in the same direction as the external force acting on the system and negative when the motion is in the opposite direction of the external applied force.

Consider this definition as it applies to two possible systems (the box and the mover):

When the box was the system, there were two external forces—the force applied by the mover, which was acting upward, and the force of gravity, acting downward. The force applied by the mover was larger; otherwise the box wouldn't have moved up. The motion was upward, and the NET external force was also upward. Thus the displacement was POSITIVE 1.2 m, as illustrated in Equation 1.18, reflecting that energy was added to the system by the mover.

If we defined the mover as the system, then the force of gravity on the box would be an external force acting on the system. In this case there is no other external applied force; the mover is producing the upward force internally. The motion would be upward, but the external force would be acting downward. Consequently in this case we would say the displacement is NEGATIVE 1.2 m, and the work is −270.8 N, reflecting the fact that the system (the mover) is doing work on the surroundings (the box).

Example 1-3 illustrates the basic concept of work but is not really a chemical engineering application. There are three categories of work that are of great significance throughout this book.

1. Work of expansion/contraction
2. Flow work
3. Shaft work

These are examined through the following examples.

$W > 0$ for work done ON system

$W < 0$ for work done BY system

Be aware that some books and references use the opposite sign convention.

EXAMPLE 1-4	**EXPANSION DUE TO HEATING**

A gas is confined in a piston-cylinder device and has an initial volume of 1 dm³. It is heated and expands to a volume of 2 dm³ (Figure 1-22). Assuming the external pressure from the atmosphere is $P = 0.1$ MPa and the piston has negligible mass and negligible friction, how much work was done by the gas?

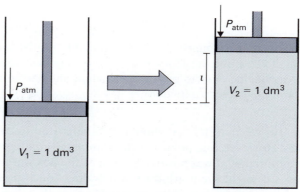

FIGURE 1-22 A gas expands from $V = 1$ dm³ to $V = 2$ dm³ against a constant pressure.

SOLUTION:

Step 1 *Define a system*
At a quick glance, this problem might look hopeless: The known pressure (0.1 MPa) represents force *per unit area*. Without knowing the area or diameter of the cylinder, how can we determine the force opposing the motion?

However, if we approach the problem methodically, by defining the gas inside the cylinder as the system and applying definitions, we will find the problem can be solved without calculating force explicitly.

Step 2 *Apply definitions of force and pressure*
By the definition in Equation 1.14, work is

$$dW = F\,dl$$

PITFALL PREVENTION

Keep in mind that a physical dimension of a system, like length (l), area (A) or volume (V), is always positive, but a *change* in a dimension (dl, dA, or dV) can be positive or negative.

Force is unknown, but pressure is known, so it makes sense to apply the relationship between F and P: pressure (P) is defined as force (F) divided by cross-sectional area (A). Consequently,

$$dW = PA\,dl \qquad (1.19)$$

Step 3 *Relate length to volume*
Area (A) times length (l) is volume—thus the term $A\,dl$ in Equation 1.19 is equivalent to the change in volume of the gas.

A common mistake at this point would be to overlook the importance of the sign of dl. Here the system is the gas; thus the system is driving the motion upward, and the external force is the atmospheric pressure opposing the motion. Consequently, the change in l is negative (the direction of the motion is the opposite of the direction of the external force), but the change in volume (V) is positive, so $dV = -A\,dl$. Introducing this observation into Equation 1.20 gives

$$dW = -P\,dV \qquad (1.20)$$

Step 4 *Solve the differential equation*

The pressure opposing the motion is the external pressure of the atmosphere and is a constant 0.1 MPa. Consequently, Equation 1.20 integrates to

$$W = -\int_{V=0.001\ m^3}^{V=0.002\ m^3} P\,dV \tag{1.21}$$

$$= -(10^5\ \text{Pa})(0.002\ \text{m}^3 - 0.001\ \text{m}^3)\left(\frac{1\ \dfrac{\text{N}}{\text{m}^2}}{\text{Pa}}\right)\left(\frac{\text{J}}{\text{N}\cdot\text{m}}\right) = -100\ \text{J}$$

FOOD FOR THOUGHT 1-8

How would the math in this example be different if, instead of expanding, the gas were being compressed?

Reviewing Example 1-4, we see that in order to find the work, we needed to know the pressure opposing the motion, but it was not necessary to know any dimensions of the cylinder other than its total volume. Indeed, the fact that the system was specifically a cylinder full of gas was irrelevant; the mathematics would apply to *any* system.

> Consequently, when the total volume of a system changes, work is done and is quantified by
>
> $$W_{EC} = -\int P\,dV \tag{1.22}$$
>
> where W_{EC} stands for **work of expansion or compression**, V is the total volume of the system, and P is the force *opposing* the motion.

Expansion/contraction work is sometimes called "PV work" in reference to Equation 1.22. However we don't recommend the term "PV work," since it could plausibly refer either to expansion/compression work or to flow work.

The next example explores the concept of flow work.

| PIPE FLOW | EXAMPLE 1-5 |

Pure water flows through a horizontal section of pipe that is 0.3 m long and 3 centimeters in diameter. The water entering the pipe has a flow rate of 2 kg/s with a temperature of 373 K and pressure of 0.2 MPa. The exiting water also has a flow rate of 2 kg/s and temperature of 373 K, but the pressure is 0.195 MPa (Figure 1-23). The density of water at these conditions is 961.5 kg/m³. Determine the rate at which work is done, both at the pipe entrance and at the pipe exit.

Water is widely used in chemical processes as a solvent and as a coolant, so water flowing through a pipe is a simple, yet common and practical system to consider.

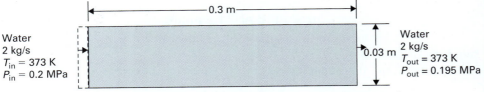

FIGURE 1-23 Water flowing through a pipe experiences a gradual pressure drop.

In design settings, "given" information is often on a mass basis (such as 2 kg/s in this problem), because process design specifications are typically on a mass basis (e.g., "We need to make 50 million kilograms per year of product.").

SOLUTION:

Step 1 *Define a system*

We first want to find the rate at which work is done at the pipe entrance. If we define the fluid in the pipe as the system, then the work we're trying to calculate is identical to the external work done on the system.

We are given no information about where the water entering the pipe is coming from (a tank? a river? another section of pipe?) and the analysis here is not dependent upon knowing that.

Step 2 *Apply definitions*

As specified by Equation 1.15, the rate at which work is done, or power, is given by

$$\dot{W} = Fv$$

Step 3 *Examine the system to evaluate F*

Figure 1-23 shows a small volume of water that is about to enter the pipe section. In order to enter, it must overcome the pressure exerted by the water at the pipe entrance, which opposes the motion.

The problem statement indicates that pressure gradually decreases along the length of the pipe. However, as the section of pipe under consideration gets shorter, the change in pressure along the length becomes smaller. The volume of liquid shown in Figure 1-23 (immediately outside the pipe entrance) is infinitesimally small, so it's valid to model the pressure as constant in this "slice" of water.

Consequently, Equation 1.15 can be expressed as

$$\dot{W} = PAv \tag{1.23}$$

When we look at large volumes of fluid (like the entire pipe) the pressure is not uniform, but the pressure *at any particular point* in a fluid is exerted equally in all directions.

where A is the cross-sectional area of the pipe, v is the velocity of the fluid, and P is the pressure of the water at the exact point that it enters the system. What about the sign of $\dot{W}$? There are two ways we can rationalize the fact that this sign is positive.

1. Physically, energy is being transferred to the system from the surroundings — the system is the fluid inside the pipe, and the fluid immediately outside the system is pushing the system.

The velocity (v) is the derivative of length (l), or $v = dl/dt$. So, what we said previously about a sign convention for l can also be regarded as true for v.

2. The external force and the motion are both acting in the same direction — in Figure 1-23, they both act from left to right. In the discussion following Example 1-3, we established that l (length travelled) is positive if the direction of travel and the direction of the applied force are the same.

Step 4 *Relate unknown velocity to known mass flow rate*

The linear velocity of the liquid times the cross-sectional area of the pipe is equal to the volumetric flow rate ($\dot{V}$) of the liquid.

$$\dot{W} = P\dot{V} \tag{1.24}$$

The specific volume in this example is $\hat{V} = 1/\rho = 1/961.5 = 0.00104$ m³/kg. Specific volume is discussed further in Section 2.2.3.

Here, volumetric flow rate is not given but mass flow rate ($\dot{m}$) is. You're likely accustomed to relating to mass through the *density, ρ* (defined as mass per unit volume), which was given in this example as $\rho = 961.5$ kg/m³. In this book, we will more commonly use the *specific volume* ($\hat{V}$), which is defined as volume per unit mass. Density and specific volume are reciprocals of each other, and either can be used here.

The steam tables, located in Appendix A, give specific volume and other properties of liquid water and steam at many conditions.

$$\dot{W} = P\dot{V} = P\left(\frac{\dot{m}}{\rho}\right) = P(\dot{m}\hat{V}) \tag{1.25}$$

Step 5 *Solve equation*

Inserting the given value of density gives

$$\dot{W} = (0.2 \text{ MPa})\left(\frac{10^6 \, Pa}{MPa}\right)\left(1\frac{\frac{N}{m^2}}{Pa}\right)\left(\frac{2\frac{kg}{s}}{961.6\frac{kg}{m^3}}\right) = \mathbf{416} \, \frac{\mathbf{J}}{\mathbf{s}} \tag{1.26}$$

Step 6 *Calculate work at exit*

In order to leave the pipe, the water must overcome the pressure exerted by whatever is immediately downstream of the pipe. If we go through steps 1 through 5, we see that the derivation of Equation 1.25 still applies to the pipe exit, but there are two differences:

1. The pressure at the outlet is 0.195 MPa, not 0.2 MPa.
2. The sign of the work is now negative. The motion is from left to right in Figure 1-23, but the *external* applied force is acting from right to left. Because the directions are opposite, the sign of v, and consequently of W, is negative. Another rationale is $W < 0$, because the system (the fluid in the pipe) is doing work on to the surroundings (the fluid immediately outside the pipe).

Consequently the work done at the pipe outlet is:

$$\dot{W} = P\dot{V} = P\left(\frac{\dot{m}}{\rho}\right) = (0.195 \text{ MPa})\left(\frac{10^6 \, Pa}{MPa}\right)\left(\frac{1 \frac{N}{m^2}}{Pa}\right)\left(\frac{-2 \frac{kg}{s}}{961.6 \frac{kg}{m^3}}\right) = -\mathbf{405.6} \, \frac{\mathbf{J}}{\mathbf{s}} \qquad (1.27)$$

An important observation about Example 1-5 is that the expression for flow work turned out to be independent of pipe area, or any other aspect of the geometry of the system. Consequently, Equation 1.28 is applicable to any material stream.

> **Flow work is given by**
>
> $$\dot{W} = P\dot{V} = \dot{m}(P\hat{V}) \qquad (1.28)$$
>
> where $\dot{W}$ is the rate at which work is done, P is pressure, $\dot{V}$ is volumetric flow rate, $\dot{m}$ is mass flow rate, and $\hat{V}$ is specific volume. Pressure and flow rate are evaluated for the fluid at the exact point that it enters or leaves the system.

SHAFT WORK

Shaft work is characterized by rotary, rather than linear, motion. Work is as always equal to the distance travelled times the force opposing the motion, but "distance" is measured by revolutions, as indicated in Equation 1.3. Many chemical processes involve shaft work being either added to, or produced by, a moving fluid; the pump and turbine in the Rankine cycle are prime examples. The moving parts in pumps and compressors are powered by shaft work. Conversely, the kinetic energy in moving liquids or gases can be converted into shaft work by devices like windmills and turbines. *Shaft work can immediately be recognized as equal to zero in any system that has no moving parts.*

Equations 1.22 and 1.28 relate expansion work and flow work to pressure and volume. There is an analogous equation for shaft work but its derivation is more complex and requires the concept of energy balances. This derivation is shown in Example 3-8.

Chapter 3 introduces an energy balance equation that we will use extensively. In this equation, flow work is accounted for using the enthalpy (H), which is formally defined in Section 2.3.1.

Pressure times volume is a quantity that occurs frequently in thermodynamics. Volume can be expressed in absolute terms (e.g., m³), a molar basis (m³/mol) or a specific basis (m³/kg), as discussed further in Section 2.2.3.

Throughout this book, the hat signifies a property expressed on a mass basis (e.g., specific volume) and the underline signifies a property expressed on a molar basis (e.g., molar volume $\underline{V}$).

1.4.4 Kinetic Energy

Kinetic energy (K.E.) is stored energy that takes the form of *motion*. It is quantified by

$$\text{K.E.} = \frac{1}{2}Mv^2 \tag{1.29}$$

In this expression, M is the mass of the object, and v is the velocity of the object.

The connection between kinetic energy and the potential to do work is straightforward; if an object is in motion, and there is a force opposing the motion, then work is being done. For example, consider a ball at the base of a hill (see Figure 1-24). It takes work to move the ball up the hill against the opposing force of gravity, and a person could supply this work by pushing the ball with his or her hand or foot. But if the ball, instead of being stationary at the bottom of the hill, is rolling along a flat surface toward the hill, then the ball rolls at least partway up the hill without any external work required; the kinetic energy of the ball is equivalent to the work done by a person's hand. Kinetic energy also can be transferred to other objects in the form of work, as when the moving billiard ball strikes the stationary one in Figure 1-9.

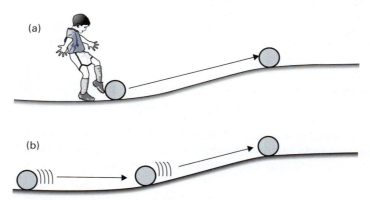

FIGURE 1-24 (a) A ball is moved up a hill by adding work. (b) A ball's stored energy, in the form of kinetic energy, allows it to climb the hill without addition of external work.

Taking the example a step further, we can use Equation 1.14 to determine how much work was done when the ball rolled up the hill, but where did the work "go?" The first law of thermodynamics states that energy cannot be created or destroyed, and (neglecting friction) the ball did not transfer the energy to any other object, so the energy must be stored inside the object in some form. As the ball rolls up the hill, it gains potential energy, which is discussed in the next section.

In theory, Equation 1.29 is applicable to a particle of any size, but in this book, we draw a distinction between *macroscopic* and *microscopic* kinetic energy. Consider a glass of water. We know that the individual water molecules are constantly in motion. Indeed individual water molecules can move with extremely high velocity, and they are constantly colliding, accelerating, decelerating, and changing directions. In addition to translational motion (motion of the entire molecule through space), molecules with more than one atom also exhibit vibrational motion (e.g., bonds stretching and compressing) and rotational motion (e.g., the free rotation of single bonds.)

However, all of this molecular motion does not contribute to macroscopic motion; the water (as a whole) appears stationary. If we pour out the water, the water as it is moving has a measurable macroscopic velocity, and the macroscopic kinetic energy of the water can be calculated using Equation 1.29.

If we knew the velocity of a molecule, we could compute its kinetic energy using Equation 1.29. However, practically speaking, it's unrealistic to quantify the kinetic energy of each individual molecule in a glass. Instead, all of the energy stored as kinetic energy of individual molecules is lumped into a macroscopic property called internal energy, described in Section 1.4.6. Thus, even though molecular motion does, strictly speaking, represent kinetic energy, throughout the rest of this book we will use the term "kinetic energy" to describe only the *macroscopic* motion of an entire object or system.

> An 250 cm³ glass of water contains approximately 8×10^{24} individual water molecules.

1.4.5 Potential Energy

Potential energy is stored energy associated with an object's *position* in a force field. For example, consider two objects that are initially both stationary: They have no kinetic energy, so if they start moving toward each other with no outside intervention, it would appear that kinetic energy is being spontaneously created. However, the motion is not arbitrary; the objects move toward each other because they are attracted to each other by gravitational or electromagnetic forces.

Consequently, objects that are located within significant electromagnetic or gravitational force fields are said to have potential energy; the attractive forces acting upon them have the potential to add kinetic energy to the objects, and in turn, to do work. The waterwheel is a practical example of a system that converts potential energy into work. Water at the top of a slope has potential energy. If the water was allowed to flow downhill in a natural channel the potential energy might be converted into kinetic energy—the water would flow faster. Instead the stream is re-routed to the top of the waterwheel. This water is captured in buckets, and as the force of gravity lowers the bucket, the axle turns. Historically, waterwheels were largely used to grind grain into flour, but conceptually, the shaft work produced by a waterwheel can be used to power any device that requires rotary motion.

Like kinetic energy, potential energy exists on both a macroscopic and microscopic level. Intermolecular attractions and repulsions give rise to potential energy on a microscopic level, but for our purposes the microscopic potential and kinetic energy of individual molecules are lumped together into internal energy, described in the next section.

Potential energy resulting from electromagnetic fields, while of great importance in physics and some engineering disciplines, is not typically a significant consideration in chemical engineering thermodynamics. In this book we will primarily be concerned with macroscopic potential energy resulting from an object's position in Earth's gravitation force field..

> **FOOD FOR THOUGHT 1-9**
>
> The river at the top of Niagara Falls appears to flow much faster than the river leading away from the falls. How can the water at the top of the falls have BOTH more potential energy and more kinetic energy than the water at the bottom?

Potential energy (P.E.) resulting from gravity is quantified by

$$\text{P.E.} = Mgh \tag{1.30}$$

where M is the mass of the object, g the acceleration due to gravity, and h the height.

In using Equation 1.30, "height" is usually regarded as equal to 0 at the ground, as in Example 1-6 below. Conceptually, however, height is measured on a relative scale, and any convenient point (e.g., the top of a table or the roof of a building) can be defined as $h = 0$.

EXAMPLE 1-6	**OBJECT IN FREE FALL**

An object is dropped from a height of 15 m off the ground. Initially it has no velocity (Figure 1-25). Assuming no forces act on the object other than gravity, what is its velocity when it hits the ground?

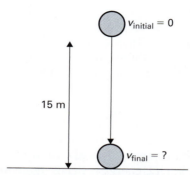

$v_{initial} = 0$

15 m

$v_{final} = ?$

FIGURE 1-25 An object experiences free fall from a height of 15 m.

SOLUTION:

Step 1 *Define a system*

The object is the system. The process ends at the moment the object strikes the ground.

Step 2 *Apply the concept of conservation of energy*

The first law of thermodynamics states that energy can be converted from one form into another, but it cannot be created or destroyed. Initially, the object has zero kinetic energy ($v_{init} = 0$) but has potential energy as given by Equation 1.30. At the end of the process the object has zero potential energy ($h_{final} = 0$) and kinetic energy as given by Equation 1.29. Since there are no other forces acting upon the object, we can assume the initial potential energy equals the kinetic energy at the end of the process

$$Mgh_{init} = \frac{1}{2} Mv_{final}^2 \tag{1.31}$$

Step 3 *Solve the equation*

The mass of the object cancels. Inserting known values into the left-hand side gives

$$\left(\frac{9.81 \text{ m}}{\text{s}^2}\right)(15 \text{ m}) = \frac{1}{2} v_{final}^2 \tag{1.32}$$

This is a workable equation in that the units of the left-hand side are m²/s², so when we multiply through by 2 and take the square root of both sides, we will be left with a velocity in m/s.

$$v_{final} = 17.39 \, \frac{\text{m}}{\text{s}} \tag{1.33}$$

You've likely solved problems like this one in introductory physics, using a different approach that involves applying the facts that acceleration is the first derivative of velocity and velocity is the first derivative of position. This approach is shown in an expanded version of the problem located in the electronic supplements. It's instructive to see that both approaches produce equivalent answers.

1.4.6 Internal Energy

The **internal energy** of a substance consists of the energy stored by individual molecules. As mentioned in Sections 1.4.4 and 1.4.5, materials store large quantities of energy in the form of *microscopic potential and kinetic energy*; these are the components of internal energy. The "potential to do work" that accompanies internal energy is more subtle than it is for macroscopic potential and kinetic energy, but it does exist. The Rankine cycle described in Section 1.2.3 is essentially a machine used to convert the internal energy of a fuel (e.g., coal or oil) into shaft work. Conversely, shaft work can be converted into internal energy: pumps and compressors are unit operations that are used to increase the pressure of a fluid; this requires the addition of work, and the work added leads directly to an increase in the internal energy of the fluid.

In principle, one might envision measuring internal energy on an absolute scale:

- If the position and velocity of all molecules were known, internal energy can be quantified using Equations 1.29 and 1.30.
- A material in which there was no molecular motion, and no intermolecular attractions and repulsions between particles, would have zero internal energy.
- There would be no such thing as negative internal energy.

Practically speaking, one cannot imagine implementing the above strategy, due to the number of molecules and the complexity of knowing their exact velocity and position at any given time. Consequently, though internal energy is understood at the molecular level, it is normally quantified with macroscopic measurement.

- We define the property specific internal energy ($\hat{U}$) as energy per unit mass (e.g., kJ/kg).
- A "reference state" is chosen for which $\hat{U} = 0$.
- Internal energy at other conditions is determined by measuring the energy added or removed for a particular change in state (e.g., how much energy did it take to raise the temperature 5 degrees above the temperature of the reference state, 10 degrees above, etc.).

Because internal energy is measured relative to a reference state, negative values of $\hat{U}$ are possible, even though in an absolute sense all materials have positive amounts of stored energy. The specific choice of reference state is not fundamentally important; what is crucial is that a single reference state is used within each calculation. Reference states are explored in the following example.

INTERNAL ENERGY OF LIQUID WATER AND STEAM	EXAMPLE 1-7

Your team is designing a process that has six streams containing liquid water or steam at different temperatures and pressures, and you need to know the specific internal energy for each. Your teammate has given you the information in Table 1-1, but you need to fill in the missing data. Using the steam tables in Appendix A, determine the values that should go in the two empty cells.

TABLE 1-1 Specific internal energy of liquid water and steam at various conditions.

Stream	Temperature (K)	Pressure (MPa)	Phase	$\hat{U}$ (kJ/kg)
1	298.15	0.1	Liquid	0 (reference state)
2	323.15	0.1	Liquid	104.5
3	348.15	0.1	Liquid	209.2
4	373.15	0.1	Vapor	2401.4
5	373.15	0.5	Liquid	
6	473.15	0.5	Vapor	

PITFALL PREVENTION

We can't use data from the steam tables directly. They have a different reference state than the data given in this example.

The "triple point" of a compound is discussed in Section 2.2.1.

Physically, $\hat{U}_4 - \hat{U}_5$ represents the amount of internal energy that must be added to a kilogram of water in order to convert it from liquid at $T = 373$ K and $P = 0.5$ MPa to vapor at $T = 373$ K and $P = 0.1$ MPa.

In step 2, the calculations use the same reference state as the steam tables. In step 3, the calculations use the same reference state as Table 1-1.

SOLUTION:

Step 1 *Collect data*

According to Appendix A-4, for stream 5 (liquid water at $P = 0.5$ MPa and $T = 373.15$ K) the specific internal energy is $\hat{U}_5 = 418.9$ kJ/kg, and according to Appendix A-3, for stream 6 (steam at $P = 0.5$ MPa and $T = 473.15$ K) the specific internal energy is $\hat{U}_6 = 2643.3$ kJ/kg. The conditions for the liquid water in streams 1, 2, and 3 are not shown in the steam tables. However, stream 4 (steam at $P = 0.1$ MPa and $T = 373.15$ K) is shown, and its value is $\hat{U}_4 = 2506.2$ kJ/kg. This is different from the value in your teammate's table.

Step 2 *Compute changes in specific internal energy*

The steam tables use liquid water at the triple point ($T = 273.16$ K, $P = 611.7$ Pa) as a reference state, while your teammate's table uses liquid water at approximately ambient conditions ($T = 298$ K and $P = 0.1$ MPa) as a reference state. The data we fill into the two empty cells must use the same reference state as the rest of the table, so that the team can use this table for accurate, self-consistent design calculations. How can we convert the steam table data to your teammate's alternative reference state?

Engineers are primarily concerned with *changes* in the conditions of materials—how much energy does it take to heat water from $T = 373$ K to $T = 473$ K, or compress it from $P = 0.1$ MPa to $P = 0.5$ MPa? The answers to questions like these do not depend upon what we choose as a reference state. Thus, while the values of $\hat{U}_4$ and $\hat{U}_5$ depend upon the reference state, the difference between them does not. Using the data from the steam tables,

$$\hat{U}_4 - \hat{U}_5 = 2506.2 - 418.9 \, \frac{kJ}{kg} = 2087.3 \, \frac{kJ}{kg} \tag{1.34}$$

and

$$\hat{U}_4 - \hat{U}_6 = 2506.2 - 2643.3 \, \frac{kJ}{kg} = -137.1 \, \frac{kJ}{kg} \tag{1.35}$$

Step 3 *Determine $\hat{U}_5$ and $\hat{U}_6$ in the team's reference state*

According to Table 1-1, the specific internal energy of stream 4 is $\hat{U}_4 = 2401.4$ kJ/kg. The value of $\hat{U}_4 - \hat{U}_5$ was calculated using a different reference state, but is the same regardless of the reference state. We can combine these to find the value of $\hat{U}_5$ for the reference state used in Table 1-1.

$$\hat{U}_4 - \hat{U}_5 = 2087.3 \frac{kJ}{kg}$$

$$2401.4 \frac{kJ}{kg} - \hat{U}_5 = 2087.3 \frac{kJ}{kg} \qquad (1.36)$$

$$\hat{U}_5 = \mathbf{314.1} \frac{\mathbf{kJ}}{\mathbf{kg}}$$

Applying a similar procedure to stream 6:

$$\hat{U}_4 - \hat{U}_6 = -137.1 \frac{kJ}{kg}$$

$$2401.4 \frac{kJ}{kg} - \hat{U}_6 = -137.1 \frac{kJ}{kg} \qquad (1.37)$$

$$\hat{U}_6 = \mathbf{2538.5} \frac{\mathbf{kJ}}{\mathbf{kg}}$$

The steam tables, which were used in Example 1-7, give data for $\hat{U}$ and other properties of water, in the liquid and vapor phases at a broad range of conditions. Chapter 6 discusses how changes in internal energy can be estimated in circumstances where experimental data does not exist.

> The tables in Appendix A are typically called "steam tables" even though they contain properties of liquid water as well as steam.

1.4.7 Heat

Heat is a form of energy transfer that is distinct from mechanical work in that the energy transfer occurs without accompanying macroscopic motion. Motion, and consequently work, occur because of an imbalance in force. Heat transfer occurs because of an imbalance in temperature. We use the symbol Q to represent heat energy entering the system.

> Like W, we define Q to be positive when energy enters the system and negative when energy leaves the system.

The phenomenon of heat transfer can be understood, on a molecular level, as an extension of the previous discussion on internal energy. Recall that internal energy is a macroscopic property that encompasses microscopic potential and kinetic energy. The molecules in liquid water move more quickly than the molecules in an ice cube. If ice cubes are placed in water (Figure 1-26a), collisions between slow-moving (ice) molecules and fast-moving (liquid water) molecules result. Since one gram of water

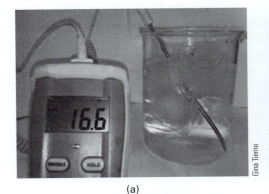
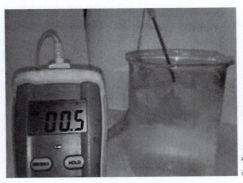

(a)   (b)

FIGURE 1-26 (a) A glass of water immediately after ice cubes are dropped in and (b) the glass of water after it reaches equilibrium. Temperatures are displayed in degrees Celsius.

contains $>3 \times 10^{22}$ molecules, the number of individual collisions is almost unimaginably large. The net effect of all these collisions is to "even out" the internal energy:

- Macroscopically, the ice cube melts and the liquid water loses heat and decreases temperature.
- On a molecular level (Figure 1-27), the ice molecules gain energy and the liquid water molecules lose energy.

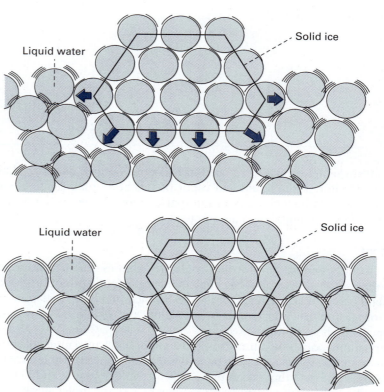

FIGURE 1-27 Molecular-scale imagining of the melting of an ice cube in liquid water.

If a single ice cube were dropped into a glass of hot water, the ice cube would melt completely. At this point, the molecules that were originally part of the ice and those that were originally part of the water are indistinguishable. The molecules are now all liquid water and their internal energy, on average, falls somewhere between the hot water's original internal energy and the ice's original internal energy. Their temperature, too, falls between the original temperatures of the ice and liquid water.

In Figure 1-26a, there is a large amount of ice and the liquid water is cool (16.6 Celsius, or 289.8 K) to begin with. Here again, internal energy is transferred (through molecular collisions) from the liquid water to the ice, melting the ice and cooling the liquid water. But in this experiment, the liquid water reached its freezing point before all the ice had melted, producing an ice-water mixture in which ice and water are both at the freezing point (measured as 0.5 Celsius, or 273.7 K in Figure 1-26b). This final state provides an example of thermal equilibrium; equilibrium was introduced in Section 1.3.2.

This example illustrates a familiar everyday phenomenon; that when high-energy "hot" objects and low-energy "cold" objects are brought into contact, energy

FOOD FOR THOUGHT 1-11

According to the thermocouple, the temperature of the ice-water equilibrium mixture in Figure 1-26b is 0.5 degrees Celsius. Shouldn't it be exactly 0?

flows from the "hot" object to the "cold" one. Temperature is the property that is used to quantify how "hot" something is, allowing us to predict whether heat transfer will occur and quantify the physical effects of the heat transfer. While temperature is a familiar quantity in an everyday sense, formal definitions for temperature are presented in Section 1.4.8.

Throughout this section we have observed that the various forms of stored energy (kinetic, potential and internal) are recognizable as energy because they can be converted into work. Heat, in turn, is recognizable as energy because it is a transfer of stored energy from one location to another.

While heat transfer is a process distinct from work, the two routinely occur simultaneously. For example, heating something generally causes it to expand, and expansion implies work unless there is no pressure opposing the expansion.

1.4.8 Temperature

Internal energy was described in Section 1.4.6 as *microscopic* kinetic and potential energy; the energy of individual molecules. A difficulty is that microscopic potential and kinetic energy is not something that can be measured directly. The virtue of **temperature** is that it is a readily measurable property that allows us to benchmark the internal energy of a system.

The Celsius temperature scale was originally defined as

- The freezing point of water at atmospheric pressure is 0°C.
- The boiling point of water at atmospheric pressure is 100°C.

The phenomenon of heat transfer is what makes a temperature scale meaningful. Two objects are said to be in "thermal equilibrium" if no heat is transferred between them, even though they are in contact such that heat transfer could occur. By definition, materials in thermal equilibrium are said to be at the same temperature. Thermal equilibrium has been observed to be transitive; that is, if material A is in thermal equilibrium with material B, then any other material that is in thermal equilibrium with material A will also be in thermal equilibrium with material B.

Thus, any material that is in thermal equilibrium with water at its normal freezing point is said to have a temperature of 273.15 K. Any material that would transfer heat TO water at its normal freezing point has a temperature higher than 273.15 K, and any material to which heat would transfer FROM water at its normal freezing point has a temperature lower than 273.15 K. The definition of two points permits construction of a linear scale for temperature in between these two points. The transitive nature of thermal equilibrium allows us to calibrate instruments that measure temperature (e.g., we know that any two objects that provide the same reading on a thermometer will be in thermal equilibrium with each other).

While temperature is related to internal energy, it is not a simple or linear relationship. For example, liquid water at its freezing point and ice at its melting point are in thermal equilibrium with each other, and therefore at the same temperature, but the liquid water has far more internal energy per unit mass than does the ice. Temperature is best understood not as a "measurement of internal energy" but as a measurement of the propensity of a material to transfer heat to or from other materials.

The Celsius temperature scale was first devised before the concept of an absolute temperature scale existed. We can imagine a state, termed **absolute zero**, in which there is zero molecular motion and zero stored energy. A material that has zero stored energy has no ability to transfer heat to any other material, and therefore has, in an absolute sense, zero temperature.

The Celsius temperature scale was originally called "Centigrade," because it had 100 "gradations" between the freezing point and boiling point of water. In 1948, at the Ninth International Conference on Weights and Measures, it was formally renamed "Celsius," in honor of Anders Celsius. (Olson, 1998)

This definition of Celsius has formally been replaced, though the new definition was crafted to be practically consistent with this one.

The freezing point and boiling point of a pure substance measured at atmospheric pressure are also termed the "normal freezing point" and "normal boiling point."

In Section 2.3.2, the relationship between internal energy and temperature is discussed further.

The Kelvin temperature scale is named after Sir William Thompson, whose title was Baron Kelvin of Largs, and is consequently often referred to as "Lord Kelvin." (Olson, 1998)

The **Kelvin** temperature scale is defined as

- Absolute zero corresponds to 0 K.
- The temperature at the triple point of water is defined as 273.16 K.

"One Kelvin" is defined as 1/273.16 of the temperature difference between these points. Temperature in degrees Celsius is now formally defined as temperature in Kelvin minus 273.15. This definition aligns almost perfectly with the original "centigrade" definition. By the original scale, the normal boiling point of water was *exactly* 100°C by definition. When measured on the temperature scale as currently defined, the normal boiling point of water has been reported at various values ranging from 373.15 to 373.20 K, which corresponds to 100.00 to 100.05°C (Arce, 1998; Fandary, 1999; Rajendran, 1989). For common use, it remains accurate to say that at atmospheric pressure, water freezes at 0°C and boils at 100°C.

The "triple point" of a compound is discussed in Section 2.2.1.

The Rankine scale, like the Kelvin, measures temperature on an absolute basis; 0°R represents absolute zero. Figure 1-28 illustrates the relationships between the Kelvin, Celsius, Fahrenheit, and Rankine temperature scales, and the following equations summarize the conversions between them.

$$T_{\text{Celsius}} = T_{\text{Kelvin}} - 273.15 \tag{1.38}$$

$$T_{\text{Rankine}} = \frac{9}{5} T_{\text{Kelvin}} \tag{1.39}$$

$$T_{\text{Fahrenheit}} = T_{\text{Rankine}} - 459.67 \tag{1.40}$$

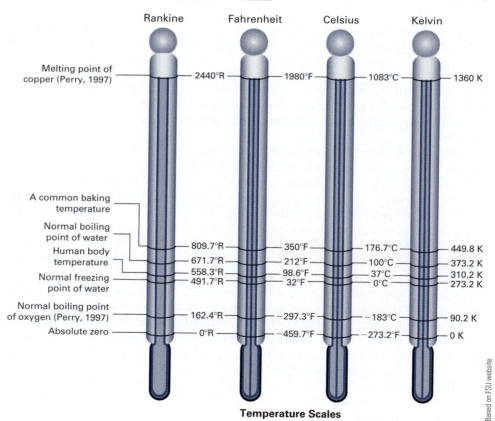

FIGURE 1-28 Comparison of Fahrenheit, Celsius, Kelvin, and Rankine temperature scales.

The concept of an absolute temperature has broad utility in chemical engineering thermodynamics. The ideal gas law, discussed in Section 2.3.3, relates temperature, pressure and molar volume to each other, but requires that all three quantities (T, P, and $\underline{V}$) are measured on an absolute scale. Many additional thermodynamics properties (e.g., entropy, Gibbs energy) developed later in this book also depend upon an absolute temperature. A simple point, but one that often causes confusion, is addressed in the following example.

| A CHANGE IN TEMPERATURE | EXAMPLE 1-8 |

On a stove, water is heated from room temperature $T = 25°C$ to boiling temperature $T = 100°C$. Determine ΔT in degrees Celsius and in degrees Kelvin.

SOLUTION:

In Celsius:

$$\Delta T = T_{final} - T_{initial} = 100 - 25°C = \mathbf{75°C} \tag{1.41}$$

In Kelvin:

$$\Delta T = T_{final} - T_{initial} = (100 + 273.15 \text{ K}) - (25 + 273.15 \text{ K}) \tag{1.42}$$

$$= \mathbf{75 \text{ K}}$$

As mentioned above, in many thermodynamics applications, temperature (T) must be expressed on an absolute scale. A danger is becoming so accustomed to adding 273.15 to any Celsius temperature that one carelessly does so for a *change* in temperature. For example "the water temperature increased by 75°C," does not convert to "the water temperature increased by 348.15 K." Example 1-8 illustrates that a change in temperature (ΔT) is identical on the Celsius and Kelvin scales, because the conversion factor cancels. A similar demonstration could be made for the Fahrenheit and Rankine scales.

1.4.9 Overview of the Forms of Energy

Conceptually, the crucial points in this section include:

- Work occurs because of an imbalance in force, and is recognizable when motion occurs against an opposing force.
- Other forms of energy are recognizable and quantifiable through their potential to be converted to and from work.
- Kinetic energy and potential energy are the forms in which energy can be stored.
- Internal energy is stored energy in the form of *microscopic* kinetic and potential energy.
- Heat is energy transferred due to an imbalance in temperature.

Quantitatively, the most important mathematical relationships in this section are

$$dW = F\,dl \tag{1.43}$$

$$dW_{EC} = -P\,dV \tag{1.44}$$

$$\text{K.E.} = \frac{1}{2}Mv^2 \tag{1.45}$$

$$\text{P.E.} = Mgh \tag{1.46}$$

1.5 SUMMARY OF CHAPTER ONE

- **The first law of thermodynamics** states that **energy** cannot be created or destroyed, but it can be converted from one form into another.

- Engineers design products and processes that accomplish specific tasks or meet specific needs. Converting and using energy as efficiently as possible is a fundamental concern in design.

- Energy can be **stored** in the forms of **internal energy, kinetic energy and potential energy**.

- Energy can be **transferred** in the forms of **heat (Q) and work (W)**.

- **Work** is the energy required to move something through a **distance** against an opposing **force**.

- Work occurs when there is an imbalance in force, and heat transfer occurs when there is an imbalance in **temperature**.

- For our purposes, three forms of work are significant: **flow work, shaft work,** and **expansion/contraction work**.

- Solving problems is facilitated by defining a **system**. A system must have clear, unambiguous boundaries, such that we can recognize when mass or energy enters or leaves the system.

- W and Q are considered positive when energy is added to the system and negative when energy is removed from the system.

- Solving problems is often facilitated by defining a **process,** which has a clear beginning and a clear end.

- The **initial state** is a description of the system at the beginning of the process, and the **final state** is a description of the system at the end of the process.

- A **closed system** is one in which no matter enters or leaves the system, and an **open system** is one in which matter can cross the boundaries of the system.

- In an **isolated system,** neither matter nor energy crosses the boundaries of the system.

- An **adiabatic** process is one in which no heat is added or removed.

- An **isothermal** process is one in which temperature is constant with respect to time, and an isothermal system is one in which the temperature is uniform throughout the system.

- An **isobaric** process is one in which pressure is constant with respect to time, and an isobaric system is one in which the pressure is uniform throughout the system.

- An **isochoric** system has a constant volume.

- A system is at **steady state** if ALL properties of the system remain constant with respect to time.

- A system is at **equilibrium** when there is no driving force present that will cause the properties of the system to change.

- The **driving forces** for change that we are concerned with in chemical engineering thermodynamics are differences in **temperature, pressure,** and **fugacity**.

- It is possible for a system to be at steady state without being at equilibrium.

1.6 EXERCISES

1-1. Several systems are described below. For each system, go through each of the following adjectives, and determine which ones apply: open, closed, adiabatic, isolated, isochoric, isothermal, or isobaric.
   A. An ice cube that has just been dropped into a glass of warm water.
   B. An aluminum can full of soda that has been fully submerged in ice water for some time.
   C. The inside of a closed, perfectly insulated cooler that contains ice, water, and cans of soda.
   D. A closed, reasonably well insulated cooler that contains ice, water, and cans of soda.
   E. Your body, over the last 24 hours.
   F. Your body, over the last five years.
   G. The Atlantic Ocean, over the last 24 hours.
   H. The Atlantic Ocean, over the last million years.

1-2. Suppose a car, and everything inside it, is defined as the system. A car is not necessarily an open or closed system; it could be either, depending upon the specific process and circumstances under consideration.
   A. While a car is being driven with the windows rolled up, is the system open or closed? Which, if any, of these apply: isolated, isochoric, isothermal, isobaric, adiabatic?
   B. While a car is sitting in a garage, with no passengers inside, all windows rolled up and the engine not running, is the system open or closed? Which, if any, of these apply: isolated, isochoric, isothermal, isobaric, adiabatic?
   C. Describe a process (different from either A or B) in which the car can be considered a closed system.
   D. Describe a process (different from either A or B) in which the car can be considered an open system.

1-3. Several systems and processes are described below. For each of them, determine whether the system is at steady state, at equilibrium, or neither. Determine also whether any of these adjectives are applicable to the system: isothermal, isobaric, isochoric, adiabatic, closed, or open.

	System	Process
A	An ice cube	The ice cube is inside a freezer, where is has already been for some time before the process begins. The process lasts 24 hours, during which the freezer door is never opened.
B	An ice cube	The ice cube is inside a freezer. The freezer door is opened, allowing room temperature air in. The door is open for one minute, and the process ends 10 minutes after the door is closed.

C	The air inside a hot air balloon	The balloon rises from the ground to a height of 100 m.
D	The air inside a hot air balloon	The balloon floats, stationary, 100 m above the ground. The air in the balloon is maintained at a constant temperature.
E	A wood furnace and its contents	Wood is piled inside the furnace. When the process begins, the wood is already burning and the furnace and its contents are at a uniform T = 420 K. At the end of the process, 10 minutes later, the temperature is still a uniform T = 420 K.
F	A section of a water pipe	Water is flowing through the pipe at a constant flow rate and a constant, uniform temperature.
G	The water in a sink	The process begins immediately after the faucet is turned off. The water is initially at 305 K, the ambient air is at 294 K. The process continues for 24 hours, during which evaporation is negligible and no water is added or removed.
H	Your body	You are stationary and immersed in bathwater at 310 K throughout the process.

1-4. Ray Bradbury's novel title *Fahrenheit 451* is based upon the temperature at which paper, or more specifically books, burn. Express 451.0°F in Celsius, Kelvin, and Rankine.

1-5. Convert $P = 5.00$ atm into Pa, bar, and psia.

1-6. Convert 10,000 ft-lb$_f$ of energy into BTU, Joules, and kilojoules.

1-7. Convert 5 megawatts of power into BTU/hr, ft-lbs/s, and kJ/hr.

1-8. Use the data in the steam tables (Appendix A) to answer the following:
   A. Find the change in internal energy when 50 kg of steam at constant pressure $P = 0.5$ MPa has its temperature reduced from 773.15 K to 573.15 K.
   B. Find the change in internal energy when 200 kg of liquid water at constant pressure $P = 10$ MPa has its temperature increased from 513.15 K to 573.15 K.

1-9. Use the data in the steam tables to answer the following:
- **A.** Find the change in volume when 50 kg of steam, initially at $T = 573.15$ K and $P = 0.2$ MPa, is heated and compressed into steam at $T = 623.15$ K and $P = 0.5$ MPa.
- **B.** Find the change in volume when 100 kg of liquid water at $P = 100$ MPa and $T = 573.15$ K is converted into 100 kg of steam at $P = 5$ MPa and $T = 573.15$.

1-10. An object has a mass of 50 kg.
- **A.** Find the change in potential energy when it is initially 10 meters above the ground and is raised to 25 meters above the ground.

- **B.** Find the change in kinetic energy when it is initially stationary and is accelerated to 10 m/s.

1-11. Find the expansion work done when the volume of a gas is increased from 0.2 m³ to 1.5 m³ against a constant pressure of 50 kPa.

1-12. Find the flow work added to a system by a fluid that enters with a volumetric flow rate of 8 dm³/s at a pressure of 0.3 MPa. Express the answer in Watts.

1.7 PROBLEMS

1-13. The value g = 9.81 m/s² is specific to the force of gravity on the surface of the Earth. The universal formula for the force of gravitational attraction is

$$F = G\frac{m_1 m_2}{r^2}$$

where m_1 and m_2 are the masses of the two objects, r is the distance between the centers of the two objects, and G is the universal gravitation constant, $G = 6.674 \times 10^{-11} \text{N(m/kg)}^2$.
- **A.** Research the diameters and masses of the Earth and Jupiter.
- **B.** Demonstrate that $F = m(9.81 \text{ m/s}^2)$ is a valid relationship on the surface of the Earth.
- **C.** Determine the force of gravity acting on a 1000 kg satellite that is 3200 km above the surface of the Earth.
- **D.** One of the authors of this book has a mass of 90 kg. If he was on the surface of Jupiter, what gravitational force in N would be acting on him?

1-14. A gas at $T = 300$ K and $P = 100$ kPa is contained in a rigid, rectangular vessel that is 2 meters long, 1 meter wide, and 1 meter deep. How much force does the gas exert on the walls of the container?

1-15. A car weighs 1360 kg and is travelling 96 km/h when it has to make an emergency stop. The car comes to a stop 5 seconds after the brakes are applied.
- **A.** Assuming the rate of deceleration is constant, what force is required?
- **B.** Assuming the rate of deceleration is constant, how much distance is covered before the car comes to a stop?

1-16. Solar panels are installed on a rectangular flat roof. The roof is 4 m by 9 m, and the mass of the panels and framing is 408 kg.
- **A.** Assuming the weight of the panels is evenly distributed over the roof, how much pressure does the solar panel array place on the roof?
- **B.** The density of fallen snow varies; here assume its ~30% of the density of liquid water. Estimate the total pressure on the roof if 0.1 m of snow fall on top of the solar panels.

1-17. A box has a mass of 20 kg, and a building has a height of 15 meters.
- **A.** Find the force of gravity acting on the box.
- **B.** Find the work required to lift the box from the ground to the roof of the building.
- **C.** Find the potential energy of the box when it is on the roof of the building.
- **D.** If the box is dropped off the roof of the building, find the kinetic energy and velocity of the box when it hits the ground.

1-18. 100 kg of steam is enclosed in a piston-cylinder device, initially at 573.15 K and 0.5 MPa. It expands and cools to 473.15 K and 0.1 MPa.
- **A.** What is the change in internal energy of the steam in this process?
- **B.** If the external pressure is constant at 0.1 MPa, how much work was done by the steam on the surroundings?
- **C.** Research and briefly describe at least two examples of machines, either historical or currently in use, that harness the energy in steam and convert it into work. Any form of work is acceptable; you needn't confine your research to expansion work (which was examined in parts A and B).

1-19. A. An object is dropped from a height of 6 m off the ground. What is its velocity when it hits the ground?

B. Instead of being dropped, the object is thrown down, such that when it is 6 m off the ground, it already has an initial velocity of 6 m/s straight down. What is its velocity when it hits the ground?

C. What did you assume in answering parts A and B? Give at least three examples of objects for which your assumptions are very good, and at least one example of an object for which your assumptions would fail badly.

1-20. An airplane is 6000 m above the ground when a 100 kg object is dropped from it. If there were no such thing as air resistance, what would the vertical velocity and kinetic energy of the dropped object be when it hits the ground?

1-21. A filtration system continuously removes water from a swimming pool, passes the water through filters, and then returns it to the pool. Both pipes are located near the surface of the water. The flow rate is 1 dm³ per second. The water entering the pump is at a gauge pressure of 0 Pa, and the water leaving the pump at a gauge pressure of 70 kPa.

A. The diameter of the pipe that leaves the pump is 0.02 m. How much flow work is done by the water as it leaves the pump and enters the pipe?

B. The water returns to the pool through an opening that is 0.04 m in diameter, located at the surface of the water, where the pressure is 101.325 kPa. How much work is done by the water as it leaves the pipe and enters the pool?

C. "The system" consists of the water in the pump and in the pipes that transport water between the pump and the pool. Is the system at steady state, equilibrium, both, or neither?

1-22. The Reaumur temperature scale, while now obscure, was once in common use in some parts of the world. The normal freezing point of water is defined as 0 degrees Reaumur and the normal boiling point of water is defined as 80 degrees Reaumur. Tolstoy's *War and Peace* mentions the temperature "minus 20 degrees Reaumur." What is this in Celsius, Fahrenheit, Kelvin and Rankine?

1-23. Use the data in the steam tables to answer the following:

A. Find the change in internal energy when 100 kg of steam at constant pressure $P = 0.1$ MPa has its temperature reduced from 573.15 K to 373.15 K.

B. Find the change in internal energy when 100 kg of liquid water at constant pressure $P = 20$ MPa has its temperature increased from 513.15 K to 573.15 K.

C. Energy was transferred from the system in part A, and energy was transferred to the system in part B. What form would you expect these energy transfers took?

D. Your answers to parts A and B should be similar in magnitude, though different in sign. Would it be possible to accomplish both of the processes in parts A and B simultaneously, by taking most of the energy that was removed from the steam described in part A and transferring it to the liquid water described in part B?

1-24. A balloon is inflated from a negligible initial volume to a final volume of 200 cm³. How much work is done by the balloon on the surroundings if the pressure opposing the expansion is

A. $P = 0.1$ MPa
B. $P = 0.05$ MPa
C. $P = 0$ MPa
D. $P = 0.3$ MPa
E. Can you think of locations where each of the "surroundings" pressures given in parts A–D would be realistic?

1-25. "The system" is a large ship and everything inside it. The ship turns off its engines, and is floating in the ocean. No material enters or leaves the ship during the process. Some descriptions of this situation are given below. For each, indicate whether it is true or false, and explain why.

A. The system is not at equilibrium, because equilibrium implies no external forces are acting on the system and the ship has forces acting upon it. The system is at a steady state, however, because the forces acting upon it are balanced.

B. The system is not at equilibrium, because the force of gravity is acting upon the ship but there is no upward force balancing the downward force of gravity. However, the system is at a steady state, because it is not moving.

C. The system is both at equilibrium and at a steady state, because the ship is not moving, and there is no driving force for motion: the forces acting upon it balance each other and there is no driving force for change.

D. The system is neither at equilibrium nor at a steady state, because no object in the ocean is perfectly motionless. The ship bobs up and down with the waves, and likely drifts in a horizontal direction due to currents. If the position of the system is changing, it can't be at steady state or equilibrium.

1-26. "The system" is a large ship and its contents. The inside of the ship and the air outside the ship are at the same temperature. The ship is sailing north at a constant speed of 20 kilometers per hour. The engines are powered by burning liquid fuel, and the gaseous by-products (primarily carbon dioxide and water) are vented to the atmosphere, but nothing else enters or leaves the system. Some descriptions of this situation are given below. For each, indicate whether it is true or false, and explain why.

A. The system is at a steady state, because its velocity is constant. However, it is not at equilibrium-the fact that the ship is moving indicates that the forces are not balanced.

B. The system is not a at steady state, because the amount of fuel inside the system is changing. However, the system is at equilibrium, because it is at the same temperature as the surroundings; there is no driving force for heat transfer.

C. The system is neither at equilibrium nor at steady state.

D. The system is adiabatic, because there is no temperature driving force that would cause heat transfer to occur.

E. The system is an isolated system, because nothing is entering it and there is no heat transfer.

1-27. You are collecting data from the literature on a compound, for which you need to know the specific internal energy at a number of different states. You've found some data from three different sources, but they each use different reference states and the units aren't uniform either. The data is shown in the table below. Fill in all of the empty cells in the table, so that you have correct values of $\hat{U}$ for all seven conditions (A–G) at all three reference states.

State	Phase	T (K)	P (MPa)	$\hat{U}$, Source 1	$\hat{U}$, Source 2	$\hat{U}$, Source 3
A	Solid	290.15	0.1		0	
B	Liquid	290.15	0.1	−18.0 J/g	20,470 ft · lb$_f$/lb$_m$	
C	Liquid	298.15	0.1	0		
D	Liquid	355.15	0.1		102,970 ft · lb$_f$/lb$_m$	0
E	Vapor	355.15	0.1			67.0 J/g
F	Vapor	373.15	0.133			136.0 J/g
G	Liquid	373.15	0.133			37.0 J/g

1.8 GLOSSARY OF SYMBOLS

A	area	$\dot{m}$	mass flow rate	v	velocity
a	acceleration	P	pressure	W	work
F	force	P.E.	potential energy	$\dot{W}$	power
g	gravitational constant	Q	heat	W_{EC}	work of expansion/contraction
h	height	T	temperature	W_S	shaft work
K.E.	kinetic energy	t	time	ρ	density
l	linear distance, or length	V	volume	ω	angular distance
M	mass	$\dot{V}$	volumetric flow rate		

1.9 REFERENCES

Arce, A., Martinez-Ageitos, J., Soto, A., *Vapor-Liquid Equilibria at 101.32 kPa of the Ternary Systems 2-Methoxy-2-methylpropane + Methanol + Water and 2-Methoxy-2-methylpropane + Ethanol + Water, J. Chem. Eng. Data*, 1998, 43, 708–13.

Cantor, G. N., Gooding, D., James, F. A. J., *Michael Faraday*, Humanities Press, Atlantic Highlands, NJ, 1996.

Fandary, M. S. H., Aljima, A. S., Al-Kandary, J. A., *Liquid-Liquid Equilibria for the System Water + Ethanol + Ethyl tert-Butyl Ether, J. Chem. Eng. Data,* 1999, 44, 1129–31.

Olson, R. (editor), *Biographical Encyclopedia of Scientists,* Marshall Cavendish, Tarrytown, NY 1998.

Perry, R. H., Green, D. W., *Perry's Chemical Engineers' Handbook,* 7th ed., McGraw-Hill, New York, NY, 1997.

Rajendran, M., Renganarayanan, S., Madhavan, P. R., Srinivasan, D., *Effect of Dissolved Salts on the Heat of Mixing of Three Binary Systems*, J. Chem. Eng. Data, 1989, 34, 375.

The Physical Properties of Pure Compounds

2

LEARNING OBJECTIVES

This chapter is intended to help you learn how to:

- Envision and qualitatively predict the interrelationships among pressure, temperature, and molar volume of a pure material
- Define four states of matter: *solid, liquid, vapor,* and *supercritical fluid*
- Recognize qualitatively the conditions (primarily temperature and pressure) at which solids, liquids, vapors, and supercritical fluids exist and at which transitions between the phases occur
- Appreciate the physical significance of the *critical point* and the *triple point*
- Distinguish between *intensive* and *extensive* properties of a material
- Recognize *state properties* of a material, and appreciate the significance of state properties in solving problems
- Define *enthalpy* in relation to other state properties
- Quantify changes in enthalpy and internal energy using *heat capacity*
- List the three attributes of an *ideal gas*, and recognize situations in which real vapors are reasonably modeled as ideal gases
- Define *equation of state*
- Quantify P, $\underline{V}$, and T using the ideal gas law or van der Waals equation of state

The first chapter presented an overview of the forms of energy and introduced the concept that energy cannot be created or destroyed but can be converted from one form into another. Converting a chemical's internal energy into a different form (e.g., heat, work, or, as in a nozzle, kinetic energy) is one of the recurring themes in the examples throughout this book. A challenge in modeling or designing such processes comes from the fact that internal energy cannot be measured directly. Three properties that are measurable—pressure, temperature, and volume—are the primary focus of this chapter. We discuss the interrelationships among P, V, and T that are observed in pure compounds, and consider how more abstract properties like internal energy can be related to measurable properties. ∎

2.1   MOTIVATIONAL EXAMPLE: Vapor Pressure of Water and Its Effect on the Rankine Cycle

Section 1.2.3 described the purpose of the Rankine cycle—conversion of heat into shaft work—and the role of each of the four steps: boiler, turbine, condenser, and pump. However, the discussion in Chapter 1 was qualitative. Before a working Rankine cycle actually can be designed and built, a number of specific, quantitative design questions need to be answered, such as the following:

- What are the pressures of the water entering and leaving the pump?
- What is the temperature of the boiler, and what is the source of its heat?
- What is the temperature of the condenser, and how is the cooling accomplished?
- How much shaft work must the engine produce?
- What flow rate of water/steam is needed to attain the necessary shaft work?

The answers to these questions are rooted in the *physical properties of water* or whatever fluid is circulating in the heat engine. To illustrate this, consider the operation of the turbine. The entire purpose of the Rankine cycle is to produce shaft work. Mechanically, this shaft work comes from the force of the water vapor acting on the blades of the turbine—the higher the pressure of the entering vapor, the faster the blades will spin. However, as designers, we cannot simply assign an arbitrary high pressure (e.g., 10 MPa) to the steam entering the turbine—we are constrained by the boiling point of water.

For a given temperature, there is a unique pressure above which water will be liquid and below which it will be vapor. This pressure is called the *vapor pressure*. Conversely, at any pressure, there is a unique boiling temperature above which water will be vapor and below which it will be liquid. At $P = 10$ MPa, this boiling temperature is 584.15 K.

In the Rankine cycle, the steam entering the turbine comes from a boiler. Its temperature is constrained by the source of the heat used in the boiler. If the heat source is at 573 K, then the water/steam in the boiler must be below 573 K, or there will be no driving force for heat transfer to occur. But if the liquid water entering the "boiler" is at $P = 10$ MPa and it isn't heated to at least 584.15 K, then it won't boil at all. There also may be practical and safety barriers to operating a turbine at such a high pressure, but the vapor pressure represents a fundamental constraint.

Similarly, the temperature of the condenser is constrained by the available heat sink. If the coolant in the condenser is simply air or water at ambient temperature (e.g., $T = 298$ K), the condenser must operate above ambient temperature. At $T = 313.15$ K, for example, the vapor pressure of water is 7.38 kPa.

If the steam leaving the turbine is at a pressure lower than this, it won't condense at 313.15 K. At pressures below 611.7 Pa, the liquid state doesn't exist at all; water vapor that is cooled at these pressures will eventually reach a temperature at which it converts directly into ice. Thus, knowing vapor pressure and its relationship to temperature is of fundamental importance in designing a Rankine heat engine or any other process that involves liquid-vapor transitions.

The aforementioned vapor pressures are available in the "steam tables," which are located in Appendix A. The steam tables also contain several other physical properties relevant in modeling processes like the Rankine Cycle. For example, Appendix A-3 indicates that for steam at $T = 523.15$ K (250°C) and $P = 2.5$ MPa, the *specific volume* is 0.0871 m³/kg, the *specific internal energy* is 2663.3 kJ/kg, the *specific enthalpy* is 2880.9 kJ/kg and the *specific entropy* is 6.4107 kJ/kg · K. Internal energy was discussed in Section 1.4.6, while enthalpy is introduced in Section 2.3.1 and

Though we often think of water only as a liquid (and use the word in that way), water is the name for the compound H_2O, whether it is solid, liquid or vapor.

The relationships among pressure, temperature, and phase for pure compounds are discussed further in Section 2.2.1.

entropy is defined in Chapter 4. Volume, internal energy, enthalpy and entropy are all *state properties*; their values depend only upon the current state of the substance, and are independent of how the current state was reached. This fact is what makes tabulation of data in Appendix A possible; a kilogram of steam at $T = 523.15$ K and $P = 2.5$ MPa always has the same volume and internal energy, regardless of how that temperature and pressure were reached. State properties are discussed in Section 2.2.3. Note, too, that all of these properties are expressed "per kilogram" of steam. The volume of steam (in m^3 for example) depends upon how much steam is present. However, specific volume (m^3/kg) is the same at $T = 523.15$ K and $P = 2.5$ MPa, regardless of the amount of steam present. Specific volume is an example of an *intensive property*, discussed in detail in Section 2.2.4.

In Section 5.2, we will return to the Rankine cycle a third time. We will apply energy balances (covered in Chapter 3) and entropy balances (covered in Chapter 4) to make quantitative determinations regarding the Rankine cycle: the rate at which work is produced in the turbine, the state of the material leaving the turbine, etc. All of these calculations rely upon the data in the steam tables. Unfortunately, most chemical compounds have not been studied as extensively as water. Indeed, one thing chemical engineers do is invent new compounds! Consequently, a major thrust in thermodynamics is building models that allow one to estimate the physical properties of chemical compounds from limited data.

- Section 2.3 provides an introduction to simple thermodynamic models of pure compounds.

- Chapters 6 through 8 describe modeling strategies that are significantly more complex — but more broadly applicable. Again, the focus is on pure compounds.

- Chapters 9 through 13 describe modeling strategies that are applicable to mixtures of compounds.

2.2 Physical Properties of Pure Chemical Compounds

Pressure, temperature and volume are three physical properties of matter that are familiar to us from everyday experience, as are the physical states of matter (e.g., solid, liquid, or vapor). Section 2.1 provided one example of the significance of physical properties, where effective operation of the Rankine cycle is dependent upon H_2O changing between the liquid and vapor phases, and these phase transitions are governed by temperature and pressure. Throughout this book, we will explore many other properties (e.g., internal energy, enthalpy, entropy, Gibbs energy, and fugacity) which are less tangible than pressure, temperature, and volume but are just as important in the modeling and designing of chemical processes. A unifying theme in chemical engineering thermodynamics is

> The physical properties of chemical compounds are inherently related to each other. The practice of thermodynamics consists of exploring and quantifying these interrelationships and applying the insights gained to answer questions, make predictions, and solve problems.

2.2.1 The *P-V-T* Behavior of Real Compounds

To illustrate the interrelationships of physical properties of compounds, we consider the most commonly observed states of matter. Figure 2-1 displays how the state of matter of a pure compound relates to pressure and temperature. As temperature increases at a constant pressure (e.g., P_1 in the diagram), the compound undergoes

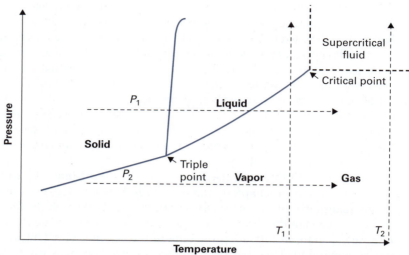

FIGURE 2-1 Typical phase transition diagram for pure compounds.

the familiar transition from solid to liquid to vapor. At lower pressures (e.g., P_2), no liquid phase occurs. Instead, at a certain temperature, the solid converts directly into gas, which is a process called sublimation. This is observed when dry ice (which is solid CO_2) sublimates at ambient conditions. Key definitions and observations illustrated in Figure 2-1 are given here.

- For a particular pressure, the temperature at which the solid–liquid transition occurs is called the **melting point** or **freezing point**, the temperature at which the liquid–vapor transition occurs is called the **boiling point**, and the temperature at which the solid–vapor transition occurs is called the **sublimation point**.

- The melting point, boiling point, and sublimation point all increase with increasing pressure.

- The melting point and boiling point at atmospheric pressure specifically are called the **normal melting point** and **normal boiling point**.

- The pressure at which the liquid–vapor transition or solid–vapor transition occurs is called the **vapor pressure**. Vapor pressure increases with increasing temperature.

- There is a unique temperature and pressure at which the solid, liquid, and vapor states can all exist in equilibrium with each other. This is called the **triple point**.

- There is a unique temperature above which no liquid phase occurs regardless of pressure. This is called the **critical temperature (T_C)**.

- There is a unique pressure above which no vapor phase occurs regardless of temperature. This is called the **critical pressure (P_C)**.

- A material that is both above the critical pressure and above the critical temperature is called a **supercritical fluid**.

The triple point temperatures and pressures for different compounds vary tremendously, as illustrated in Table 2-1. For each individual compound, however, the triple point is unique. Consequently, the triple point is useful for providing repeatable temperatures for calibrations. For example, the triple point temperature of ethylene carbonate is 309.5 K, which is a useful reference point for the calibration

Section 2.1 illustrated the significance of the vapor pressure in designing a Rankine heat engine.

TABLE 2-1 Comparison of conditions at which phase changes occur for various compounds.

Compound	Normal Melting Point (K)[1]	Normal Boiling Point (K)[1]	Triple Point T (K)[2]	Triple Point P (kPa)[2]	Critical Temperature T_c (K)[1]	Critical Pressure P_c (MPa)[1]
Water	273.15	373.2	273.16	0.61	647	22.06
Oxygen	54.8	90.2	54.36	0.152	155	5.04
Nitrogen	63.3	77.4	63.18	12.6	126.2	3.40
Carbon Dioxide	N/A	N/A	216.55	517	304.18	7.38
Methane	85.7	111	90.68	117	190.6	4.61
Ammonia	195	238	195.4	6.076	405.4	11.30

Based on data from NIST Webbook (Brown, 2012); Cengel, et. al., 2012

FOOD FOR THOUGHT 2-1

Why are the normal boiling and melting points for carbon dioxide listed as "not applicable?"

of clinical thermometers because it's close to human body temperature (Machin, 2005). As mentioned in Section 1.3.8, the triple point of water is used as a reference point in defining the Kelvin temperature scale.

The supercritical state is also worth a bit more discussion. Typically, supercritical fluids are transparent and look like gases, but they have properties that are more commonly associated with liquids, such as high densities. Supercritical fluid extraction is an important application for supercritical fluids. In this process, compounds are separated by dissolving one or more of them in a supercritical fluid. Carbon dioxide is the most commonly used solvent. The high densities allow supercritical fluids to act as solvents, but the process is made easier by the fact that supercritical fluids generally have transport properties (such as high diffusivities) that are more typical of gases.

As a note on nomenclature, the term "fluids" includes liquids, vapors, gases, and supercritical fluids. All of these states of matter have flexible macroscopic structures that take on the shape of their container, rather than retaining their shape as solids do. This "catch-all" term is useful because of similarities in how liquids, gases, and supercritical fluids are handled. For example, all fluids are easily transported through pipes, but solids are not. Another note on terminology is that the words "gas" and "vapor" are often used interchangeably. In this book, the term "gas" is used to describe a compound that is above its critical temperature but below its critical pressure, and the term "vapor" is used for a compound that is below its critical temperature but above its boiling temperature, as illustrated in Figure 2-1. Thus, a vapor condenses if it is isothermally compressed to its vapor pressure (see T_1 in Figure 2-1), while a gas does not condense—no matter how much it is isothermally compressed (see T_2 in Figure 2-1). This distinction is primarily semantic, as our methods of modeling gases and vapors are substantially the same.

The temperatures and pressures at which phase transitions occur vary tremendously from one compound to another, as illustrated in Table 2-1. Qualitatively, however, almost all compounds exhibit the behavior illustrated in Figure 2-1. Rare exceptions occur—such as in compounds that have no measurable boiling point because they decompose at elevated temperatures. In this case the process of heating converts them into other chemical compounds more readily than it converts them into vapors. Note that Figure 2-1 specifically describes a pure compound. The phase behavior of mixtures is much more complex and will be addressed in Chapter 9.

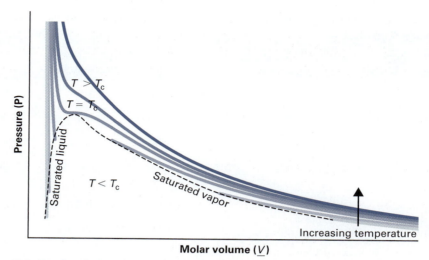

FIGURE 2-2 Typical relationships between P, $\underline{V}$ and T for liquid, vapor, and supercritical phases in a pure compound. Each line shows the pressure (P) versus the molar volume ($\underline{V}$) for a constant temperature (T).

Beyond the question of *which* phase will be present at a particular temperature and pressure, the physical properties of a phase are also dependent upon temperature and pressure. Figure 2-2 illustrates typical interrelationships among pressure, temperature, and molar volume for the vapor, liquid, and supercritical phases. Combined, Figures 2-1 and 2-2 give a good overview of the interrelationships among P, $\underline{V}$, and T for a typical compound.

Notable observations and definitions related to Figure 2-2 include the following statements.

- Vapor molar volume increases with increasing temperature and decreases with increasing pressure.

- Liquid molar volumes also increase with increasing temperature and decrease with increasing pressure, but the effects of temperature and pressure on liquid molar volume are generally much less significant than on vapor molar volume.

- A vapor that is at a boiling point is called a **saturated vapor**, and a liquid that is at a boiling point is called a **saturated liquid**.

- A vapor that is at a temperature above its boiling point is called a **superheated vapor**.

- A liquid that is at a temperature below its boiling point is called a **subcooled liquid**.

- A liquid that is at a pressure above its vapor pressure is called a **compressed liquid**.

- As pressure increases, the molar volumes of the saturated liquids and saturated vapors at a given pressure become closer together, as illustrated in Figure 2-2. At the critical point, the liquid and vapor phases become indistinguishable from each other.

- At pressures above the critical pressure, there is no longer a boiling point or a liquid–vapor transition, but supercritical fluids do follow the trend of decreasing volume with increasing pressure and increasing volume with increasing temperature.

The motivational example (Section 2.1) mentioned several other physical properties of chemical compounds, including internal energy, enthalpy, and entropy (which will be discussed in Chapter 4). Quantifying the interdependence of the physical properties of chemical compounds allows us to solve practical problems like those given here:

- Knowledge of both the enthalpy and entropy of the steam entering the turbine in the Rankine cycle is essential for predicting the work produced, as described in Section 5.2. Enthalpy is defined in Section 2.3.1 and entropy in Section 4.3.

- Many chemical reactions are best carried out at high pressure and/or high temperature, so the reactants need to be compressed and/or heated prior to entering the reactor. Knowing how internal energy and enthalpy relate to temperature and pressure allows the designer to predict how much heat and work are required for these processes.

Sections 2.2.3 and 2.2.4 provide a context for defining and comparing physical properties of compounds. Before closing this section, the question "How many states of matter are there?" is worth mentioning briefly. Solids, liquids, and gases are familiar in our everyday lives and are sometimes termed "the three states of matter." Introductory chemistry and physics books often indicate that there are four states of matter: solid, liquid, gas, and plasma. Plasma is a highly ionized gaseous state that occurs at high temperatures. Much of the material in stars is in the plasma state. Here on Earth, naturally occurring plasma is comparatively rare, although it is produced by lightning bolts and flames. Plasmas have many practical engineering applications (e.g., lighting, conduction, and surface cleaning) but do not play a significant role in mainstream *chemical* engineering practice.

There is no clear consensus on exactly how many distinct states of matter are known. While most introductory chemistry and physics books do not describe the supercritical fluid as a distinct 'state of matter,' a plausible argument can be made that it should be regarded as one. Other esoteric states such as superglass, supersolid, and Bose-Einstein condensates exist at extreme conditions and are sometimes described as additional "states of matter."

Regardless of the semantic issue of whether it is or is not a state of matter, the supercritical fluid is certainly important in chemical engineering thermodynamics. The application of supercritical fluid extraction was mentioned previously. In addition, the critical point plays a prominent role in our efforts to build mathematical models of real compounds, as illustrated in Chapter 7. By contrast, plasmas and the various esoteric states of matter mentioned here are not of central importance to chemical engineering thermodynamics, and detailed discussion of them is beyond the scope of this introductory book. Consequently, through the remainder of this book, **solid**, **liquid**, **vapor/gas**, and **supercritical fluid** are the only **"states of matter"** to be discussed.

> **FOOD FOR THOUGHT 2-3**
>
> You can recognize solids, liquids, and gases from your everyday experience, but can you give concise definitions of "solid," "liquid," and "gas" that a grade school student would understand? Try it.

2.2.2 Forms and Sources of Physical Property Data

The previous section described the pressure-volume-temperature (or *P-V-T*) and phase behavior of real compounds and underscored the importance of being able to quantify the relationships between the properties of compounds.

For some compounds, there exists very complete data regarding phase transitions, *P-V-T* behavior, and other physical properties (e.g., internal energy) that are of interest. Appendix A contains various steam tables that provide detailed data for water.

- The saturated steam tables show the properties of saturated liquid water and saturated water vapor at a broad range of temperatures and pressures. Table A-1 is organized by pressure and Table A-2 is organized by temperature.

- Table A-3 gives properties of superheated steam and is organized by pressure. For each pressure, the physical properties of steam at a variety of temperatures are shown.

- Table A-4 is organized similarly to the superheated-steam tables and shows the properties of compressed liquid water.

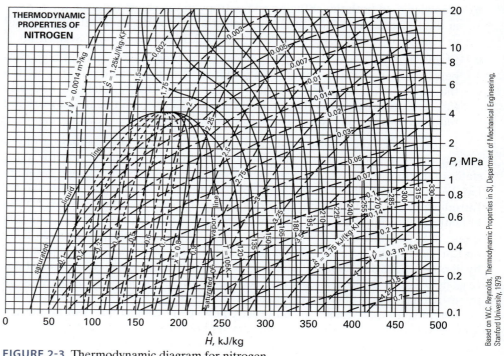

FIGURE 2-3 Thermodynamic diagram for nitrogen.

Based on W.C. Reynolds, Thermodynamic Properties in SI, Department of Mechanical Engineering, Stanford University, 1979

■ Appendix B-1 reviews interpolation, which can be used to estimate data when the exact number needed is not in the tables (e.g., $\underline{V}$ is needed at $T = 533$ K and is known at $T = 523$ K and $T = 548$ K).

"Specific" properties (e.g., specific enthalpy and specific volume) are defined in Section 2.2.3.

Data analogous to that given in the steam tables is often consolidated into a figure. Figure 2-3, for example, shows physical property data for nitrogen. The axes of Figure 2-3 are pressure versus specific enthalpy, but the figure also contains temperature, specific entropy, and specific volume. If any two of these five properties are known, the other three can be found. Comparable diagrams for several other compounds are provided in Appendix F. More extensive data is available in sources such as *Perry's Chemical Engineers Handbook* (Green, 2007), *The Properties of Gases and Liquids* (Poling, 2001), and the NIST Chemistry WebBook (available at time of writing at http://webbook.nist.gov/chemistry/).

Overall, relatively few chemical compounds have been studied in the same level of detail as water or nitrogen. Chemical engineers are routinely required to estimate physical properties of compounds and/or mixtures when the relevant data is unavailable. Much of this book is devoted to constructing mathematical models that are used to produce such estimates.

2.2.3 State Properties and Path-Dependent Properties

A **state property** is one that describes the condition of a material or system at a particular time and is independent of how the system or material reached that condition. Without the idea of a state property, it would be impossible to make something like the steam tables or Figure 2-3; we could not tabulate a value for the specific internal energy of water at 373.15 K and 1 MPa unless it was always the same for water at 373.15 K and 1 MPa.

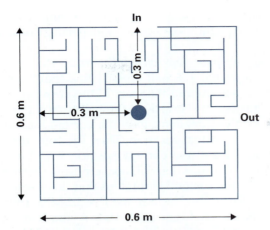

FIGURE 2-4 The current location of the ball is "the exact center of the maze," and this is true regardless of the path taken to reach the location.

Consider the ball in the maze in Figure 2-4. We can describe the ball's position in different ways.

- It is in the exact center of the maze—equidistant from the four corners.
- It is 0.3 meters below the entrance.
- It is 0.3 meters to the left of the exit.

There are many paths by which it could have reached this point. Those paths have different lengths, and it would take the ball different amounts of time to follow those paths from the entrance. However, these bulleted facts are true regardless of the path taken.

Similarly, the current **state** of a material can be described in many different ways: There is a specific amount of the material (typically quantified through mass or moles), and it has a specific pressure, temperature, and volume. These properties are independent of the path taken to reach the current state. Consider the ice/water mixture in Figure 2-5. Assuming the mixture is well mixed and at atmospheric pressure, its temperature must be 273.15 K; this is the only temperature at which water can exist in both liquid and solid forms in equilibrium at atmospheric pressure. Five minutes ago, it may have been all ice at 270 K, all liquid water at 275 K, or any one of innumerable other possibilities, but the current temperature of 273.15 K is independent of the path taken to reach the current state.

The significance of state properties in thermodynamics comes from the fact that they are intrinsically related to each other. Some useful relationships between state properties include:

- The data provided in the steam tables, and similar tables and charts for other compounds.
- The formula of Mass = Density × Volume ($M = \rho V$).
- The relationship between height and potential energy (Equation 1.30).
- The relationship between velocity and kinetic energy (Equation 1.29).
- The ideal gas law and other equations of state (discussed in Section 2.3.3).
- Numerous additional expressions to be developed later in the book, particularly in Chapter 6.

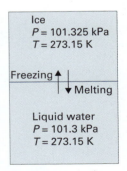

FIGURE 2-5 An ice-water mixture in equilibrium at $P = 101.325$ kPa *must* have a temperature of 273.15 K.

Examples 4-2 and 4-3 will give an illustration of two very different paths from the same initial state to the same final state.

Like the properties they relate, these equations and relationships can be applied to the state of a system at any time—independent of how that state came about.

We can also imagine *path-dependent* properties. Returning to Figure 2-4, if asked "How many times has the ball crossed this spot?" or "How far has the ball travelled since it first entered the maze?" the answer clearly depends upon the path, while the answer to "How far is the ball from the left-hand edge of the maze right now?" is only dependent upon the current location of the ball—not on the path taken to reach that location. Thus, the total distance travelled is a path-dependent property and the current position is a state property. The primary path-dependent quantities that will concern us in thermodynamics are *heat* and *work*. For example, the steam tables tell us that the specific internal energy of steam is $\hat{U} = 2663.3$ kJ/kg at $T = 523.15$ K and $P = 2.5$ MPa. There are many ways in which this steam can be expanded and cooled into a saturated liquid at $T = 313.15$ K (40°C). The specific internal energy of the water is $\hat{U} = 167.5$ kJ/kg at the end of the process—regardless of the path. $\hat{U}$ is a state function. Thus, the steam experienced a change in internal energy of -2495.8 kJ/kg regardless of the path. But how much of this energy was transferred in the form of heat, and how much in the form of work? Many different answers to this question exist, but heat and work depend upon the specific path.

Specific internal energy is an example of an intensive property, which is discussed further in the next section.

The symbol Δ represents the change in a state property. Using volume as an example, the change in volume of a system during a process (ΔV) is

$$\Delta V = V_{final} - V_{initial} \tag{2.1}$$

An analogous definition can be made for any state property. (Example 1-8 used ΔT to represent the change in the temperature of water.)

A common careless error is inverting the order of the terms (e.g., computing ΔP as $P_{initial} - P_{final}$). To help avoid such confusion, in this book we use the symbol Δ in discussions and in posing questions (e.g., "find ΔU"), but in performing the actual calculations we generally avoid this shorthand (e.g., "$U_{final} - U_{initial}$" or "$U_2 - U_1$").

2.2.4 Intensive and Extensive Material Properties

In chemical engineering thermodynamics, much of our interest is in predicting how one state property (such as volume) of a material will be influenced by changes in another state property (such as pressure). In modeling the many physical properties of a material, there is an important distinction between intensive and extensive properties.

An **intensive property** of a system is not dependent upon either the size of the system or the amount of the material present in the system.

An **extensive property** of a system is a property that is proportional to the amount of material present in the system.

The word "extensive" shares a root with the word "extend." You might say that as the amount of a material *extends* (gets larger) the *extensive* properties change, but the *intensive* properties do not.

For example, the density of liquid water at 293 K and 0.1 MPa is 998 kg/m³ (Table 2-2 shows the density of water at a variety of conditions). This value of the density is valid whether examining a drop of water or a reservoir full of water: thus, density is an example of an intensive property. Conversely, the mass and volume of water are examples of extensive properties; they are directly proportional to the number of moles of water present.

TABLE 2-2 Density of liquid water in kg/m³ at various temperatures and pressures.

	P = 0.1 MPa	P = 1 MPa	P = 10 MPa	P = 100 MPa
T = 20°C	998	999	1003	1040
T = 40°C	992	993	997	1032
T = 60°C	983	984	987	1023
T = 80°C	972	972	976	1012
T = 100°C	958*	959	963	1000

*This is actually the density of liquid water at 99.6°C, which is the boiling point at P = 0.1 MPa. At T = 100°C and P = 0.1 MPa, pure water is a vapor.

The distinction between intensive and extensive properties is important, because intensive properties are fundamental and repeatable properties of a material, while extensive properties only can be used to describe a particular sample of a material. For example, if told "the mass of the water is 1 kg," the value of 1 kg is an extensive property that is applicable only to that particular sample of water. But "the density of water at $T = 323.15$ K and $P = 0.1$ MPa is 988 kg/m³" is an intensive property and is valid for any system or process involving water at that T and P.

The relationship among the temperature, pressure, and density of pure liquid water is illustrated in Table 2-2. Notice that, if any two of these properties are known, there is only one value the third can have, and it can be determined from the table. This table illustrates a general principle:

> For a pure substance in a single homogeneous phase, if two intensive properties have been fixed, all other intensive properties must have a unique value.

If the substance is pure but is in two phases in equilibrium (e.g., vapor–liquid equilibrium), you can only "set" one intensive variable.

Because temperature and pressure are relatively easy to measure and control, all intensive properties are generally thought of as being related to temperature and pressure. We often refer to the density, specific volume, specific internal energy, or specific enthalpy of a compound at a particular T and P. Conceptually, however, we can set any two intensive properties. If the specific volume of water is $\hat{V} = 0.001043$ m³/kg and the specific internal energy is $\hat{U} = 418.8$ kJ/kg, there is no freedom for also setting the temperature and pressure; they must be $T = 373.15$ K and $P = 1$ MPa. This is one of the data points given in Appendix A-4.

Because of these fundamental relationships between intensive properties, it is usually convenient or necessary to focus on intensive properties, so it is useful to define intensive counterparts for all extensive properties. For example, volume is an extensive property, but one frequently refers to the specific volume or the molar volume. The **specific volume** is the volume of a material per unit mass, and the **molar volume** is the volume per mole of material, as summarized in Table 2-3. Table 2-4 provides a number of prominent examples of intensive and extensive properties.

Throughout this book, the hat (circumflex) signifies a property expressed on a mass basis (e.g., specific internal energy $\hat{U}$) and the underline signifies a property expressed on a molar basis (e.g., molar internal energy $\underline{U}$).

TABLE 2-3 Common units for volume, specific volume, and molar volume.

Property	Symbol	Typical Units
Volume	V	m³, L, cm³, ft³
Specific Volume	$\hat{V}$	m³/kg, L/kg, ft³/lb$_m$
Molar Volume	$\underline{V}$	m³/mol, cm³/mol, ft³/lb-mole

TABLE 2-4 Examples of common intensive and extensive properties.

Intensive Property: Symbol and SI Units		Extensive Property: Symbol and SI Units	
Specific Volume = $\hat{V}$	m³/kg	Volume = V	m³
Density = ρ	kg/m³	Mass = M	kg
Specific Enthalpy = $\hat{H}$	J/kg	Enthalpy = H	J
Specific Internal Energy = $\hat{U}$	J/kg	Internal Energy = U	J
Specific Entropy = $\hat{S}$	J/kg·K	Entropy = S	J/K
Temperature = T	K	Kinetic Energy = K.E.	J
Pressure = P	Pa	Potential energy = P.E.	J

The fact that the water is at its boiling point doesn't guarantee that two phases are present. Water at 373.15 K and 101.3 kPa can exist as saturated liquid, as saturated vapor, or as a vapor–liquid equilibrium (VLE) mixture.

2.2.5 State Properties of Multiphase Mixtures

Section 2.2.3 introduced the concept of a state property and Section 2.2.4 discussed intensive and extensive properties. This section examines applying these concepts to systems that, like the ice-water mixture in Figure 2-5, contain more than one phase. In Section 2.2.4, we noted that if two intensive properties have been specified for a pure compound in a single, homogeneous phase, all others are constrained to unique values.

A more general statement of the interrelationship between intensive properties is given by the **Gibbs phase rule**:

$$F = C - \pi + 2 \tag{2.2}$$

where F is the number of *degrees of freedom* in the system, C is the number of distinct chemical *compounds* in the system, and π is the number of distinct *phases* in the system.

The Gibbs phase rule was proposed by Josiah Willard Gibbs in 1878. (Olson, 1998)

FOOD FOR THOUGHT 2-4

If $C = 1$ and $\pi = 3$, then $F = 0$. How is this fact reflected in Figure 2-1?

The number of degrees of freedom (F) is the number of intensive variables that can be established before all others are constrained to unique values. Degrees of freedom are important in design applications; you might think of the number of degrees of freedom as the number of independent decisions you are able to make regarding particular process conditions. For example, with two degrees of freedom you can decide what the temperature and pressure of a gas will be, but you can't choose a third intensive variable, such as molar volume, as only one molar volume is physically attainable at a particular temperature and pressure.

The number of distinct chemical *compounds* in the system is C. The first several chapters of this book focus on pure compounds ($C = 1$), but starting in Chapter 9, we will examine mixtures. Adding compounds introduces more degrees of freedom, and the mole fraction of a particular compound is an intensive property that is commonly used to satisfy these degrees of freedom.

The number of distinct *phases* in the system is π. Pure liquid water is a simple system: There is a single compound ($C = 1$) and a single phase ($\pi = 1$). Therefore, there are two degrees of freedom, as illustrated in Table 2-2. However, if pure liquid water and ice are present *in equilibrium*, as in Figure 2-5, then $C = 1$, but there are two phases ($\pi = 2$). The simultaneous presence of liquid and solid means that the mixture must be at the melting point, a fact that accounts for one of the degrees of freedom and leaves only one degree remaining ($F = 1$).

In examining multiphase systems, we must recognize that each individual phase can be described by a set of state properties, and in addition, there are state properties that describe the system as a whole. Consider water at its boiling point at $P = 0.1$ MPa. According to the steam tables, saturated liquid water at $P = 0.1$ MPa has a specific internal energy of 417.4 kJ/kg. This is a state property; it is always true for saturated liquid water at $P = 0.1$ MPa—regardless of how the water arrived at that state, of the quantity of liquid water present, and of whether there is or is not a vapor phase present in equilibrium with the liquid. Similarly, saturated water vapor at $P = 0.1$ MPa has a specific internal energy of 2505.6 kJ/kg—regardless of whether the saturated vapor is a single homogeneous phase or is in equilibrium with a liquid phase. We can also compute the specific internal energy of a liquid–vapor mixture as a whole, as illustrated in Example 2-1.

A VAPOR–LIQUID EQUILIBRIUM (VLE) MIXTURE IN A RIGID CONTAINER	**EXAMPLE 2-1**

Five kilograms of water is contained in a rigid, 3 m³ vessel at $P = 0.1$ MPa.

A. What is the temperature and physical state (liquid, vapor, or both liquid and vapor) of the water?

B. What is the specific internal energy of the water in the vessel?

SOLUTION A:

Step 1 *Identify intensive properties*

We know from the Gibbs phase rule that it takes two intensive properties to specify the state of a pure, homogeneous phase. Here one intensive property is given as $P = 0.1$ MPa. A second intensive property can be inferred from the known mass and volume:

> According to the Gibbs phase rule, $F = 2$ when $P = 1$ and $C = 1$.

$$\hat{V} = \frac{V}{M} = \frac{3 \text{ m}^3}{5 \text{ kg}} = 0.6 \, \frac{\text{m}^3}{\text{kg}} \tag{2.3}$$

Our first instinct is to consult the steam tables and find the temperature where saturated liquid, saturated vapor or superheated vapor has a specific volume of 0.6 m³/kg. But there is no such temperature. Saturated vapor has $\hat{V}^V = 1.6919$ m³/kg, and superheated steam at $P = 0.1$ MPa has even higher specific volumes. There is also no circumstance in which liquid water at $P = 0.1$ MPa has $\hat{V} = 0.6$ m³/kg; the specific volume of saturated liquid water is only $\hat{V}^L = 0.001043$ m³/kg. The only state in which water can have a pressure $P = 0.1$ MPa and a specific volume 0.6 m³/kg is when it is in vapor–liquid equilibrium.

> The container is a closed system; no material is entering or leaving. Individually, the liquid or vapor phase is an open system, since material can move between them.

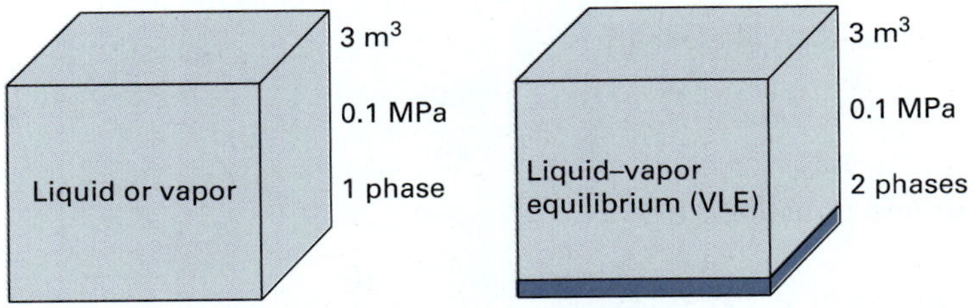

FIGURE 2-6 Possible states for water in a 3 m³ vessel at $P = 0.1$ MPa. In Example 2-1, we determine which is observed in reality.

According to the Gibbs phase rule, there is only one degree of freedom ($F = 1$) for a single component ($C = 1$) in a two-phase mixture ($\pi = 2$). The pressure $P = 0.1$ MPa fills this one degree. The only possible temperature is the boiling point $T = \textbf{372.78 K}$. Similarly, all intensive properties of the saturated liquid and vapor phases individually ($\hat{V}$, $\hat{U}$, etc.) have fixed values, which also can be obtained from the steam tables.

We have partially answered the question by determining the temperature and demonstrating that there is a vapor–liquid equilibrium mixture. To fully characterize the state of the material, we will quantify the amounts of liquid and vapor.

Step 2 *Quantify phases within VLE mixture*

The total volume of the mixture is the sum of the liquid and vapor volumes:

$$V = V^L + V^V$$

$$V = M^L \hat{V}^L + M^V \hat{V}^V \tag{2.4}$$

This is an isochoric system; the volume is constant. Notice the problem statement indicated the vessel was "rigid."

The total volume V is a constant and is given. While the specific volumes of the liquid and vapor phases are known from step 1, this is one equation in two unknowns (M^L and M^V). The known total mass is the second constraint:

$$M = M^L + M^V$$

$$M^L = M - M^V \tag{2.5}$$

Substituting Equation 2.5 into Equation 2.4 gives

$$V = (M - M^V)\hat{V}^L + M^V \hat{V}^V \tag{2.6}$$

$$3\,\text{m}^3 = (5\,\text{kg} - M^V)\left(0.001043\,\frac{\text{m}^3}{\text{kg}}\right) + M^V\left(1.6919\,\frac{\text{m}^3}{\text{kg}}\right)$$

$$M^V = 1.77\,\text{kg}$$

Solving Equation 2.6 gives $M^L = \textbf{3.23 kg}$.

SOLUTION B:

Step 3 *Determine specific internal energy of system*

The total internal energy of the system is the sum of the internal energies of the individual phases.

$$U = U^L + U^V = M^L \hat{U}^L + M^V \hat{U}^V \tag{2.7}$$

Some books define "quality" as the liquid fraction (rather than the vapor fraction) in a VLE mixture.

The specific internal energy of a system comprised of two phases is defined the same way as the specific internal energy of a single phase. This is the internal energy divided by the mass:

$$\hat{U} = \frac{M^L \hat{U}^L + M^V \hat{U}^V}{M} \tag{2.8}$$

$M^L/M = (M - M^V)/M$
$= 1 - q$

For a vapor–liquid system, the quality (q) is the mass or mole fraction of a system that is in the vapor phase. Here the quality is

$$q = \frac{M^V}{M} = \frac{1.77\,\text{kg}}{5\,\text{kg}} = 0.354 \tag{2.9}$$

q represents both the mass fraction of vapor and the mole fraction of vapor. Since the liquid and vapor are the same compound they have the same molecular mass.

Inserting this into Equation 2.8 gives

$$\hat{U} = (1 - q)\hat{U}^L + q\hat{U}^V \tag{2.10}$$

$$\hat{U} = (1 - 0.354)\left(417.4\,\frac{\text{kJ}}{\text{kg}}\right) + (0.354)\left(2505.6\,\frac{\text{kJ}}{\text{kg}}\right) = \textbf{1156.6}\,\frac{\textbf{kJ}}{\textbf{kg}}$$

Example 2-1 introduced the **_quality_(_q_)** of a system in vapor–liquid equilibrium, which is the _fraction of mass that is in the vapor phase_. It also illustrated how intensive properties representing a multiphase equilibrium system can be computed using a weighted average of the intensive properties of the individual phases. If a system consists of liquid and vapor in equilibrium, an intensive property of the overall system can be computed as

$$M = (1 - q)M^L + qM^V \qquad (2.11)$$

where M can represent any _intensive_ property (e.g., specific enthalpy or molar entropy) and q represents the quality.

2.3 Thermodynamic Models of Physical Properties

Pressure, volume, and temperature are easily measured properties of materials that have great practical significance in both engineering and thermodynamics. Section 1.4.6 described internal energy, which cannot be measured directly but is a fundamental quantity in chemical engineering thermodynamics. Sections 2.2.1 and 2.2.2 outlined the importance of quantifying the interrelationships between various physical properties of compounds, but they also noted that the relevant data is not always available. This section begins the process of developing mathematical models that quantify the interrelationships among the key thermodynamic properties of materials.

2.3.1 Enthalpy

Enthalpy is another state property that quantifies energy. Like internal energy, it cannot be measured directly and cannot be known in an absolute sense, but it is quantified relative to a reference state. This section introduces and explores how enthalpy is useful in solving problems.

Recall that the rate of flow work is equal to

$$\dot{W} = P\dot{V} = \dot{m}P\hat{V} \qquad (2.12)$$

A chemical plant is shown in Figure 2-7. One immediate observation is that the chemical plant contains hundreds of pipes, so liquids and gases are continuously moving throughout the plant. When material flows through a pipe, flow work is always occurring. The material flowing through the pipe also has stored energy. Potential and kinetic energy may or may not be significant, depending upon height of the pipe and velocity of the fluid, but internal energy is always significant. Consequently, in accounting for the energy in any process that involves moving streams, the sum $U + PV$ consistently occurs, where U represents the internal energy in a stream and PV quantifies the flow work done by the stream.

For phases in equilibrium, temperature and pressure of the phases—the "liquid T," "vapor T," and "system T"—are all identical. Other intensive properties of the system like specific volume, specific internal energy, specific enthalpy, and specific entropy can be determined from an equation analogous to Equation 2.11.

Flow work per kilogram of a fluid is equal to $P\hat{V}$, as demonstrated in Section 1.4.3. Thus, $\hat{U} + P\hat{V}$ can be considered "internal energy plus flow work" for a stream of material that is flowing.

Physically, internal energy is always "significant" in that all chemical streams contain a significant amount of internal energy. Mathematically, however, internal energy is sometimes unchanged in a process and therefore may not be "significant" in the solution of a particular problem.

Marafona/Shutterstock.com

FIGURE 2-7 A chemical plant.

The Greek word "thalpein" means "to warm."

The sum is defined as **enthalpy**:

$$H = U + PV \tag{2.13}$$

where H is enthalpy, U is internal energy, P is pressure, and V is volume.

Internal energy is quantified relative to a reference state, as discussed in Section 1.4.6. When converting between U and H using Equation 2.13, take care that both are expressed relative to the same reference conditions.

Note that the U and PV terms both have units of energy (e.g., Joules). Thus, enthalpy also has units of energy. Furthermore, note that U, P, and V are all state properties. Since H is computed exclusively from state properties, it also must be a state property. This illustrates a general principle used several times throughout this book: *Any property that is defined as a mathematical function calculated entirely from state properties is itself a state property.*

For example, mass is a state property. Dividing both sides of Equation 2.13 by the mass gives

$$\hat{H} = \hat{U} + P\hat{V} \tag{2.14}$$

$\hat{H}$ is the **specific enthalpy** and is also a state property. Unlike H, $\hat{H}$ is also an intensive property, because $\hat{U}$, P, and $\hat{V}$ are all intensive properties. We can divide both sides of Equation 2.13 by the number of moles, giving the **molar enthalpy** as

$$\underline{H} = \underline{U} + P\underline{V} \tag{2.15}$$

Again, molar enthalpy is an intensive property as well as a state property.

2.3.2 Heat Capacity

The number 4.18 kJ/kg · K is specific to liquid water and cannot be applied to ice or steam. Furthermore, it is only an approximation for liquid water at most conditions, as heat capacity is not a constant with respect to temperature and pressure.

Internal energy, heat, and temperature were described in Sections 1.3.6 through 1.3.8. It was noted that temperature is not precisely a measurement of internal energy, but temperature and internal energy are related to each other. Heat capacity is the property used to quantify that relationship. Heat capacity is described as the amount of energy required to increase the temperature of a unit of material by one degree (any T scale can be used). The unit of material can be either moles or a mass unit. For example, as taught in introductory chemistry or physics, the heat capacity of liquid water is about 4.18 kJ/kg · K. That is, it takes 4.18 kiloJoule of heat to raise the temperature of one kilogram of water by one degree Kelvin (or one degree Celsius). However, this definition of heat capacity isn't very clear. To explore why, consider this question: *How would you experimentally measure the heat capacity of helium gas?*

One possibility is shown in Figure 2-8. If a vessel is filled with a known mass of helium (e.g., 0.1 kg), a known amount of heat is added (e.g., 300 Joules), and the

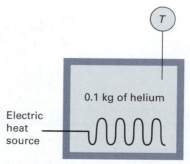

FIGURE 2-8 Schematic of a rigid (fixed volume) calorimeter system for measuring heat capacity of helium.

temperature change is accurately measured (e.g., initially 293.15 K, at the end 294.12 K), the heat capacity is

$$\text{Heat Capacity} = \frac{0.3 \text{ kJ}}{(0.1 \text{ kg})(294.12 \text{ K} - 293.15 \text{ K})} = 3.1 \frac{\text{kJ}}{\text{kg} \cdot \text{K}} \qquad (2.16)$$

However, heating a substance in a sealed, rigid container is generally regarded as unsafe because of the potential for pressure build-up, resulting in rupture or explosion. Instead, picture a system like the one in Figure 2-9, where a piston is used to maintain a constant applied pressure while allowing the volume to change. At first glance, this looks like a slightly different way to measure exactly the same property. But it produces a different answer. This time, the same 300 Joules of heat are added to the same 0.1 kg of helium, but the temperature only increases to 293.73 K. Thus,

$$\text{Heat Capacity} = \frac{0.3 \text{ kJ}}{(0.1 \text{ kg})(293.73 \text{ K} - 293.15 \text{ K})} = 5.2 \frac{\text{kJ}}{\text{kg} \cdot \text{K}} \qquad (2.17)$$

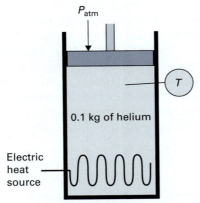

FIGURE 2-9 Schematic of a constant-pressure, variable volume calorimeter system for measuring the heat capacity of helium.

Why are the answers in Equations 2.16 and 2.17 different? Because there is more going on in these processes than just changes in temperature. When helium was heated in a closed, rigid vessel, both the temperature and the pressure increased (though only the temperature change was monitored). When the same gas was heated in a constant-pressure vessel, the temperature and volume both increased.

Which is the "right" heat capacity? Both are based on real data obtained from well-designed experiments and both are "right." The problem is that "the heat required to raise the temperature by one degree" is not a satisfactory definition of heat capacity; the answer depends upon how the experiment is conducted. The challenge arises from the fact that temperature, pressure, volume, and internal energy are all interrelated in ways that are often complex and subtle.

Consequently, we need formal mathematical definitions of heat capacity—definitions that explicitly account for the fact that internal energy is not only a function of temperature. The two heat capacities in common use are given here.

This result, to two significant digits, is the correct *ideal gas heat capacity* C_V of helium gas. C_V is defined in this section, and the concept of an ideal gas heat capacity is discussed in Section 2.3.3.

FOOD FOR THOUGHT 2-5

Can you propose an explanation for why the constant-pressure heat-capacity measurement is larger than the constant-volume heat-capacity measurement?

Remember that "Q" is not a state function, but is path-dependent. We have to specify how we added the heat (constant molar volume or constant pressure).

The **constant-volume heat capacity (C_v)** is the partial derivative of molar internal energy ($\underline{U}$) with respect to temperature (T) at constant molar volume ($\underline{V}$). Thus,

$$C_v = \left(\frac{\partial \underline{U}}{\partial T}\right)_{\underline{V}} \tag{2.18}$$

The **constant-pressure heat capacity (C_p)** is the partial derivative of molar enthalpy ($\underline{H}$) with respect to temperature (T) at constant pressure (P). So,

$$C_p = \left(\frac{\partial \underline{H}}{\partial T}\right)_{P} \tag{2.19}$$

Heat capacity is defined here on a molar basis, but it also can be defined on a mass basis as

$$\hat{C}_v = \left(\frac{\partial \hat{U}}{\partial T}\right)_{\hat{V}}$$

Example 2-2 helps illustrate the rationale for these mathematical definitions and how the heat capacity can be used. The question of "why do different compounds have different heat capacities?" is discussed here. When 300 Joules of energy is added to 0.1 kg of helium, the helium increases in temperature by almost a degree. (Remember, the actual temperature increase depends on how we added the energy.) But what form does this added energy take? In general, molecules can store kinetic energy in the forms of translational, vibrational, or rotational motion (see Figure 2-10). Helium is a monatomic gas. Because there are no chemical bonds, there is no freedom for vibrational or rotational motion. Thus, the entire 300 Joules takes the form of added translational kinetic energy, which is the motion of the entire molecule. Larger molecules with more vibrational and/or rotational degrees of freedom have higher heat capacities, because they can store energy in other forms in addition to translation. Thus, the more non-translational modes in a compound, the larger the heat capacity and the less the temperature of the compound will increase for a given amount of a heat added.

Some chemical engineering programs include a physical chemistry course as either an elective or a requirement. A course like this provides a deeper insight into the relationships between molecular structure and physical properties.

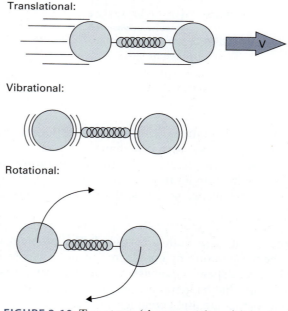

Translational:

Vibrational:

Rotational:

FIGURE 2-10 Two atoms (shown as spheres) joined by a single bond, and the three forms their molecular kinetic energy can take.

| LIQUID HEAT CAPACITY | EXAMPLE 2-2 |

A liquid has $\hat{C}_V$ of 2.15 kJ/kg · K. What is the change in specific internal energy when this liquid is heated from 295 to 300 K?

$$
\boxed{T_1 = 295\ \text{K}} \xrightarrow{\ \hat{U}_2 - \hat{U}_1 = ?\ } \boxed{T_2 = 300\ \text{K}}
$$

FIGURE 2-11 Schematic of process examined in Example 2-2.

SOLUTION: Intuitively, you might look at this question and say "It takes 2.15 kJ to increase the temperature by one degree, so the answer is 2.15 × 5." For this example, that is a satisfactory answer. However, the following is a more rigorous mathematical solution with a discussion of the potential shortcomings of the "intuitive" approach.

Step 1 *Relate internal energy to specific volume and temperature*
For a *pure, single-phase compound*, the Gibbs phase rule (introduced in Section 2.2.5) indicates that there are two degrees of freedom. This means that two intensive variables are enough to completely describe the state of the material (e.g., if the temperature and specific volume are known, there is only one value the specific internal energy can have). Furthermore, changes in specific internal energy can be related to changes in temperature and specific volume using partial differential calculus:

$$
d\hat{U} = \left(\frac{\partial \hat{U}}{\partial T}\right)_{\hat{V}} dT + \left(\frac{\partial \hat{U}}{\partial \hat{V}}\right)_{T} d\hat{V} \tag{2.20}
$$

While many students initially find partial derivative expressions like this one intimidating, they are readily placed in a physical context.

- ▪ dT represents a small change in the temperature, and $\left(\dfrac{\partial \hat{U}}{\partial T}\right)_{\hat{V}}$ or $\hat{C}_V$ quantifies the effect of this small change in temperature on the specific internal energy.

- ▪ $d\hat{V}$ represents a small change in specific volume, and $\left(\dfrac{\partial \hat{U}}{\partial \hat{V}}\right)_{T}$ quantifies the effect of this small change in specific volume on the internal energy.

- ▪ Summing the small changes (integrating) yields the total change in specific internal energy for a process.

Step 2 *Simplifying assumption*
Liquids and solids typically expand with increasing temperature, but the change in volume is often comparatively small, as illustrated in Table 2-1. Since this example examines a small temperature interval, it is logical to assume that the volume of the liquid is essentially unchanged, and the $d\hat{V}$ term is negligible. Stated mathematically, if volume is unchanged, the limits of integration on the $d\hat{V}$ term are identical, which means the integral equals zero regardless of what $\left(\dfrac{\partial \hat{U}}{\partial \hat{V}}\right)_{T}$ is. With this simplification, Equation 2.20 becomes

$$
d\hat{U} = \left(\frac{\partial \hat{U}}{\partial T}\right)_{\hat{V}} dT = \hat{C}_V dT \tag{2.21}
$$

While finding $\hat{U}_2 - \hat{U}_1$ by itself may not seem like a particularly interesting calculation, Chapter 3 covers energy balances, which quantify relationships among internal energy, work, and heat. Work and heat have more tangible significance than $\hat{U}$; they are resources that have monetary value.

Chapter 6 focuses on writing and solving partial derivative expressions like this one.

FOOD FOR THOUGHT 2-6

You'd like to quantify the expansion of water vapor with increasing temperature and compare it to the expansion of liquid water with increasing temperature. What's the best way to make this comparison using data from the steam tables?

Heat capacity expressions account for the changes in energy that accompany changes in temperature. If the liquid boiled between 295 and 300 K, these equations wouldn't be adequate—we also would have to account for the change in specific internal energy on vaporization.

Step 3 *Integration*

Because $\hat{C}_V$ is constant, Equation 2.21 integrates to

$$\hat{U}_2 - \hat{U}_1 = \hat{C}_V(T_2 - T_1) = \left(2.15\,\frac{kJ}{kg \cdot K}\right)(300 - 295\,K) = \mathbf{10.75}\,\frac{\mathbf{kJ}}{\mathbf{kg}}$$

DISCUSSION: This is the same answer obtained from the "intuitive" approach, but having analyzed the situation mathematically, we recognize that this answer relies upon the *assumption of specific volume being constant.*

Example 2-2 demonstrates the use of the constant volume heat capacity. A key result is Equation 2.21:

$$d\hat{U} = \hat{C}_V\,dT$$

But the example reveals that the derivation of Equation 2.21 requires assuming that specific volume is constant. Many scenarios where such an assumption would not be valid can be imagined.

- What if the problem had concerned a gas, rather than a liquid? Then we certainly couldn't assume volume was constant unless the gas was confined in a rigid vessel.

- Thermal expansion of liquids tends to become more significant near the critical point. Since this example did not specify a pressure or even identify the liquid, we don't know if it is near the critical point.

- What if the temperature increase had been 100 K instead of 5 K? Would it still be reasonable to assume the change in volume was negligible over this much larger temperature range?

If neglecting the change in volume isn't realistic, we cannot complete the calculation unless we are able to evaluate $\left(\frac{\partial \hat{U}}{\partial \hat{V}}\right)_T$. Chapter 6 discusses methods of evaluating partial derivative expressions like this one to solve such problems.

Note that the constant pressure heat capacity can be applied using an equation analogous to 2.21:

$$d\hat{H} = \hat{C}_P\,dT \tag{2.22}$$

But the derivation of Equation 2.22 requires assuming that pressure is constant.

2.3.3 Ideal Gases

An **ideal gas** is a hypothetical gas in which the molecules have no intermolecular interactions and no volume. This section explores the utility of this hypothetical gas state.

The **ideal gas law** is undoubtedly familiar to most readers. It is expressed as

$$P\underline{V} = RT \tag{2.23}$$

where P is pressure, $\underline{V}$ is molar volume, T is temperature, and R is the gas constant. P, $\underline{V}$, and T must be expressed on an *absolute scale*. The ideal gas law is also frequently written in terms of the total volume:

$$PV = NRT \tag{2.24}$$

where N is the number of moles of gas and V is the total volume.

In practice volume is always expressed on an absolute scale, but pressure and temperature are not.

It can be proved (K. J. Laidler, 2003) that this expression holds for any gas that has the following identifying characteristics.

1. The gas molecules have zero volume.
2. The particles are in constant motion, and collisions are perfectly elastic.
3. There are no intermolecular attractions or repulsions between the molecules.

The derivation of the ideal gas model from a starting point of these three assumptions has been called the "kinetic theory of the ideal gas."

We started this section by calling an ideal gas "hypothetical," because descriptors like "molecules have zero volume" and "no intermolecular attractions" are never literally true for real gases. However, there are many practical situations where the ideal gas law provides an accurate and useful model. Knowing the three characteristics of the ideal gas listed above will help us identify situations in which the ideal gas model is a reasonable approximation. The following example compares properties computed for steam using the ideal gas law with data from the steam tables.

ASSESSING ACCURACY OF IDEAL GAS LAW	EXAMPLE 2-3

Using the ideal gas law, estimate the specific volume of steam (in m^3/kg) at each of the following temperatures and pressures, and compare to the specific volume in the steam tables.

A. $T = 100°C, P = 0.01$ MPa
B. $T = 150°C, P = 0.1$ MPa
C. $T = 200°C, P = 1$ MPa
D. $T = 300°C, P = 5$ MPa

SOLUTION:

Step 1 *Determine molar volume from ideal gas law*
The ideal gas law, as written in Equation 2.23, can be solved for molar volume as

$$\underline{V} = \frac{RT}{P} \tag{2.25}$$

Substitute the known values for case A, recognizing that T must be expressed on an absolute scale.

$$\underline{V} = \frac{\left(8.314 \dfrac{m^3 Pa}{mol\ K}\right)(373.15\ K)}{0.01\ MPa}\left(\frac{MPa}{10^6\ Pa}\right) = 0.310 \frac{m^3}{mol} \tag{2.26}$$

Step 2 *Convert to specific volume*
Converting the answer from step 1 into specific volume allows direct comparison to the data in the superheated steam tables (Appendix A-3). Thus,

$$\hat{V} = \left(0.310 \frac{m^3}{mol}\right)\left(\frac{1\ mol}{18.015\ g}\right)\left(\frac{1000\ g}{kg}\right) = 17.221 \frac{m^3}{kg} \tag{2.27}$$

Cases B, C, and D can be solved in a similar manner. The results are summarized in Table 2-5.

PITFALL PREVENTION

Forgetting to convert from Celsius to Kelvin is a common error in using the ideal gas law. There are many equations throughout this book in which temperature MUST be expressed on an absolute scale.

TABLE 2-5 Comparison of actual $\hat{V}$ for water vapor to those obtained from the ideal gas law.

Temperature	Pressure	m³/kg (Ideal Gas Law)	m³/kg (Steam Tables)	% Difference
100°C	0.01 MPa	17.221 m³/kg	17.196 m³/kg	0.15%
150°C	0.1 MPa	1.953 m³/kg	1.936 m³/kg	0.87%
200°C	1 MPa	0.2184 m³/kg	0.206 m³/kg	6.0%
300°C	5 MPa	0.0529 m³/kg	0.04532 m³/kg	16.7%

Example 2-3 illustrates that the departures from ideal gas behavior tend to increase with increasing pressure. This observation can be rationalized by considering the three characteristics of ideal gases mentioned below:

1. *The gas molecules have zero volume.* All molecules have finite volume, but in gases at low pressure, the volume of the molecules is not significant compared to the volumes of the spaces between the molecules. As pressure increases, volume decreases, and Figure 2-12 illustrates the increasingly large fraction of the total volume occupied by actual particles.

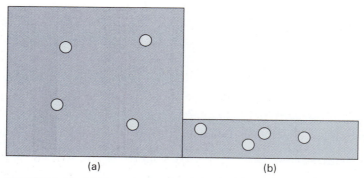

(a)   (b)

FIGURE 2-12 (a) Four gas "molecules" in a large volume. (b) The same four "molecules" compressed into a smaller volume.

2. *Collisions are perfectly elastic.* Physically, pressure (P) reflects collisions between gas molecules and the walls of the container. The kinetic theory of the ideal gas assumes these collisions are perfectly elastic. Real gas molecules collide not only with the walls but also with each other. Intermolecular collisions aren't necessarily elastic. In real gases, collisions can lead directly to associations and chemical reactions, which in turn affect the pressure and other measurable properties of the gas. However, as volume increases, whether or not these intermolecular collisions are elastic becomes less and less important simply because they become less common.

3. *There are no intermolecular attractions or repulsions between the molecules.* Water is an extremely polar compound. Consequently, this sounds like quite an unreasonable description for water. There are definitely strong intermolecular attractions and repulsions between the dipoles in water molecules. However, the example implies that water is well approximated by the ideal gas law for pressures below atmospheric. Why should this be?

Coulomb's law states that the force between two charged particles is directly proportional to the product of their charges and inversely proportional to the square of the distance between them. Thus,

$$F = k_e \frac{q_1 q_2}{r^2} \qquad (2.28)$$

where k_e is a constant, q_1 and q_2 are the charges of the particles, and r is the distance between the particles. Consequently, regardless of how strong the charges, the force approaches zero as the intermolecular distance gets very large. This helps explain why even the strongly polar compound of water behaves almost like an ideal gas at low pressures (and correspondingly at high volumes).

These characteristics are summarized here.

- At low pressure, the total volume is large, and the volumes of the particles themselves are insignificant compared to the total volume.

- At low pressure, it does not matter much if the collisions are perfectly elastic; high volumes mean relatively infrequent collisions.

- At low pressure, it does not matter much if the particles are attracted to each other or not; they are generally too far apart for the attractions (or repulsions) to matter.

In thermodynamics, the concept of "ideal gas behavior" has implications far beyond relating pressures, temperatures, and volumes to each other using the ideal gas law. For example, consider the molar internal energy of a gas ($\underline{U}$). We saw that $\underline{U}$ increases with increasing T, and the constant-volume heat capacity C_V can be used to quantify this effect. In general, $\underline{U}$ is also a function of volume (or, conversely, pressure.) However, *for an ideal gas, $\underline{U}$ is dependent upon temperature—and only temperature.* This statement is demonstrated mathematically with techniques covered in Chapter 6. For now, it will be rationalized as given here.

- Internal energy is comprised of microscopic potential and kinetic energy—the potential and kinetic energy of individual molecules.

- Ideal gases have no intermolecular interactions. Therefore, moving the molecules farther apart or closer together has no effect on the energy of each individual molecule.

- Consequently, changing the pressure or volume of an ideal gas would not be expected to affect its molar internal energy ($\underline{U}$).

Recall that C_V is defined as

$$C_V = \left(\frac{\partial \underline{U}}{\partial T} \right)_V$$

Example 2-2 showed that, for constant-volume processes, we can equate

$$d\underline{U} = C_V dT \qquad (2.29)$$

However, in an ideal gas, volume does not influence $\underline{U}$ at all, so it does not matter whether volume is constant. *Equation 2.29 is always valid for ideal gases.* Note that C_v can be expressed on either a mass or a molar basis. If it is expressed on a mass basis, Equation 2.29 would read as

$$d\hat{U} = \hat{C}_V dT \qquad (2.30)$$

Recall that the molar enthalpy is given by

$$\underline{H} = \underline{U} + P\underline{V}$$

Coulomb's Law is named after the French physicist Charles Augustin de Coulomb. (Olson, 1998)

Coulomb's Law models the charged particles as "point charges," which is a good model if the particle size is small compared to the distance between the particles. Section 7.4.4 describes a model for molecular interaction that is useful when the molecules are very close together.

Intensive properties such as molar internal energy $\underline{U}$ were introduced in Section 2.2.3.

Equations 2.29 and 2.30 are valid for constant-volume processes and for any process involving an ideal gas.

For an ideal gas, however, the $P\underline{V}$ term can be equated to RT. So

$$\underline{H} = \underline{U} + RT \tag{2.31}$$

Note the right-hand side, where we have established that $\underline{U}$ is a function of temperature—and only temperature. R is a constant, so RT is also a function of only temperature.

Equation 2.32 is valid for any constant-pressure process and for any process involving an ideal gas.

"Example 6-9 revisited" gives a mathematical proof of the fact that Equation 2.32 is always valid for ideal gases, even if P isn't constant.

Consequently, ideal gas molar enthalpy is also only a function of temperature, and the equation

$$d\underline{H} = C_p dT \tag{2.32}$$

is valid for any constant-pressure process and for any ideal gas process, regardless of whether the pressure is constant.

When a vapor is not modeled as an ideal gas, these simplifications do not occur; $\underline{U}$ and $\underline{H}$ for real gases are (or at least, can be) dependent upon pressure and volume as well as temperature. Consequently, throughout the rest of this book, symbols C_P^* and C_V^* are used for heat capacity values that specifically describe the ideal gas state to distinguish them from values that are valid in general.

EXAMPLE 2-4	A HELIUM BALLOON

At time of writing, the three Goodyear blimps each hold 5740 m³ of helium, weigh 5824.18 kg with no lifting gas, and typically fly at heights of 305 to 457 meter (obtained from the Goodyear Blimp official web site in October 2013).

The balloon portion of a dirigible is filled with 5000 moles of helium (Figure 2-13). Initially, the helium is at $P = 101.325$ kPa and $T = 298$ K. As the dirigible gains altitude, the pressure drops to $P = 96.3$ kPa, and the temperature drops to $T = 288$ K. Find the changes in volume, internal energy, and enthalpy for the helium in this process, assuming helium has a constant ideal gas heat capacity of $C_V^* = (1.5)R$.

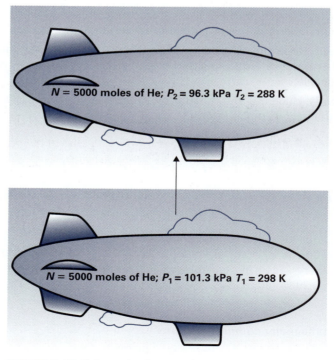

FIGURE 2-13 Schematic of system in Example 2-4.

SOLUTION:

Step 1 *Define system*
The helium inside the balloon is the system.

Step 2 *Obtain volume from ideal gas law*
The ideal gas law can be solved for volume where

$$V = \frac{NRT}{P} \tag{2.33}$$

$$V_1 = \frac{(5000 \text{ mol})\left(8.314 \frac{\text{m}^3 \text{ Pa}}{\text{mol K}}\right)(298 \text{ K})}{101{,}300 \text{ Pa}} = 122.3 \text{ m}^3 \tag{2.34}$$

$$V_2 = \frac{(5000 \text{ mol})\left(8.314 \frac{\text{m}^3 \text{ Pa}}{\text{mol K}}\right)(288 \text{ K})}{96{,}300 \text{ Pa}} = 124.5 \text{ m}^3 \tag{2.35}$$

Therefore, the volume expands by 2.2 m³, as the drop in pressure is more significant than the drop in temperature.

Step 3 *Evaluate physical situation*
If we assume helium is an ideal gas, the changes in volume and pressure have no effect on the internal energy and enthalpy. Is such an assumption reasonable? Example 2-3 suggested that the ideal gas model is a reasonable approximation for water vapor at pressures up to 1 atm. Since one of the ideal gas criteria is "no intermolecular interactions," we'd expect a non-polar compound like helium to be closer to ideal gas behavior than water vapor at the same temperature and pressure. Since pressure is at most one atmosphere in this example, we can assume (quite logically) that helium acts like an ideal gas.

Step 4 *Apply heat capacity C_V to find change in internal energy*
Equation 2.29 is valid for any ideal gas process:

$$d\underline{U} = C_V^* dT$$

Because the given C_V^* is constant, step 4 simply integrates to

$$\underline{U}_2 - \underline{U}_1 = \frac{3}{2}R(T_2 - T_1) \tag{2.36}$$

Since the total number of moles is known, multiply both sides of the equation by N to give the total change in internal energy, rather than the change in molar internal energy. Thus,

$$U_2 - U_1 = (5000 \text{ mol})\left(\frac{3}{2}\right)\left(8.314 \frac{\text{J}}{\text{mol} \cdot \text{K}}\right)(288 - 298 \text{ K}) = -\mathbf{623{,}550\,J} \tag{2.37}$$

Step 5 *Relate enthalpy to internal energy*
The change in enthalpy is not as straightforward to determine. We know that, for an ideal gas, Equation 2.32 is valid, regardless of whether the pressure is constant. Thus,

$$d\underline{H} = C_P^* dT \tag{2.38}$$

However, no value of C_P^* is given. What other approach to computing the change in enthalpy is possible? Since we've been successful at quantifying the change in internal energy, it makes sense to apply the definition of enthalpy, which relates H to U as

$$H = U + PV \tag{2.39}$$

FOOD FOR THOUGHT 2-7

This is a closed system; no helium is entering or leaving. Are any of the other system descriptors covered in Section 1.3.1 correct for this system?

or dividing both sides by the number of moles gives

$$\underline{H} = \underline{U} + P\underline{V} \qquad (2.40)$$

This definition can be inserted directly into Equation 2.38 to give

$$d(\underline{U} + P\underline{V}) = C_P^* dT \qquad (2.41)$$

Step 6 *Apply ideal gas law*
For an ideal gas, $P\underline{V} = RT$, so Equation 2.41 becomes

$$d(\underline{U} + RT) = C_P^* dT \qquad (2.42)$$

$$d\underline{U} + R dT = C_P^* dT$$

Introducing Equation 2.29 to relate internal energy to temperature gives

$$C_V^* dT + R dT = C_P^* dT \qquad (2.43)$$

Comparing the left- and right-hand sides yields an important property of ideal gases:

$$C_P^* = C_V^* + R \qquad (2.44)$$

Step 7 *Calculate enthalpy*
In this problem, Equation 2.44 is useful because C_V^* is known to be $(1.5)R$; thus, $C_P^* = (2.5)R$. Now that C_P^* is known to be a constant, it is possible to integrate Equation 2.38 to

$$\underline{H}_2 - \underline{H}_1 = \frac{5}{2}R(T_2 - T_1) \qquad (2.45)$$

Multiplying both sides by the number of moles to get total enthalpy, rather than molar enthalpy, gives

$$H_2 - H_1 = (5000 \text{ mol})\frac{5}{2}\left(8.314\frac{\text{J}}{\text{mol} \cdot \text{K}}\right)(288 - 298 \text{ K}) = -\mathbf{1{,}039{,}250 \text{ J}} \qquad (2.46)$$

What if we were doing an example for which using the ideal gas model is NOT a reasonable approximation (e.g., because the pressure was higher)? Chapter 6 explores how to quantify the effects of pressure and/or volume on U and H when the ideal gas approximation does not apply.

Example 2-4 illustrates two significant properties of ideal gases:

1. $C_P^* = C_V^* + R$ *for all ideal gases.* In Example 2-4, C_V^* was a constant, while in Example 2-5 it is a function of temperature, but in all cases $C_P^* = C_V^* + R$.

2. The given value of $C_V^* = (1.5)R$ is not specific to helium; one result of the Kinetic Theory of the Ideal Gas is that for *any monatomic ideal gas*, $C_V^* = (1.5)R$.

Modeling C_P^* as constant with respect to temperature is a simplifying approximation that is quite reasonable over small temperature intervals ($\Delta T = -10$ K in Example 2-4) but becomes suspect over larger temperature intervals. Example 2-5 illustrates a more realistic modeling of C_P^*.

EXAMPLE 2-5	INTERNAL ENERGY OF NITROGEN

The ideal gas heat capacity of nitrogen varies with temperature; it is $C_P^* = 29.42 - (2.170 \times 10^{-3})T + (0.0582 \times 10^{-5})T^2 - (1.305 \times 10^{-8})T^3 - (0.823 \times 10^{-11})T^4$, with T equal to the temperature in Kelvin and C_P^* is in J/mol $\cdot$ K.

A. How much internal energy (per mole) must be added to nitrogen to increase its temperature from 450 K to 500 K at low pressure?

B. Repeat part A for an initial temperature of 273 K and a final temperature of 1073 K.

C. The Ideal gas heat capacity of nitrogen is sometimes roughly modeled as $C_P^* = (7/2)R$. How different would the answers be if you used that constant value for parts A and B?

> The heat capacity data in Appendix D is tabulated in the form "$C_P^*/R = $," so that you can readily express C_P in preferred units by selecting a value for R. Here the expression in Appendix D was multiplied by $R = 8.314$ J/mol · K.

SOLUTION A:

Step 1 *Assess physical situation*

The pressure is not known exactly, but "at low pressure" implies that we can *assume nitrogen acts like an ideal gas*. Recall that for an ideal gas, changes in internal energy are given by Equation 2.29:

$$d\underline{U} = C_V^* dT$$

Step 2 *Determine C_V^**

In this case, C_V^* is not given, but C_P^* is. It was proved in Example 2-4 that for an ideal gas $C_P^* = C_V^* + R$.

$$C_V^* = C_P^* - R = 29.42 - (0.00217)T \tag{2.47}$$
$$+ (5.82 \times 10^{-7})T^2 - (1.305 \times 10^{-8})T^3$$
$$- (8.23 \times 10^{-12})T^4 - 8.314$$

$$C_V^* = 21.106 - (0.00217)T + (5.82 \times 10^{-7})T^2$$
$$- (1.305 \times 10^{-8})T^3 - (8.23 \times 10^{-12})T^4$$

Step 3 *Integrate*

Inserting the value of C_V^* into Equation 2.29 gives

$$\underline{U}_2 - \underline{U}_1 = \int_{T=450\text{ K}}^{T=500\text{ K}} 21.106 - (0.00217)T \tag{2.48}$$
$$+ (5.82 \times 10^{-7})T^2 - (1.305 \times 10^{-8})T^3$$
$$- (8.23 \times 10^{-12})T^4 dT$$

$$\underline{U}_2 - \underline{U}_1 = 21.106T - (0.00217)\frac{T^2}{2}$$
$$+ (5.82 \times 10^{-7})\frac{T^3}{3} - (1.305 \times 10^{-8})\frac{T^4}{4} - (8.23 \times 10^{-12})\frac{T^5}{5} \Big|_{T=450\text{ K}}^{T=500\text{ K}}$$

$$\underline{U}_2 - \underline{U}_1 = \mathbf{1059}\ \frac{\mathbf{J}}{\mathbf{mol}}$$

SOLUTION B:

Step 4 *Compare to part A*

The only difference between part B and part A is the limits of integration for Equation 2.48 are now $T_1 = 273$ K and $T_2 = 1073$ K. The result is

$$\underline{U}_2 - \underline{U}_1 = \Big[21.106T - (0.00217)\frac{T^2}{2} + (5.82 \times 10^{-7})\frac{T^3}{3} - (1.305 \times 10^{-8})\frac{T^4}{4}$$
$$- (8.23 \times 10^{-12})\frac{T^5}{5} \Big]_{T=273\text{ K}}^{T=1073\text{ K}}$$

$$\underline{U}_2 - \underline{U}_1 = \mathbf{17,920}\ \frac{\mathbf{J}}{\mathbf{mol}}$$

> If a chemical reaction is highly exothermic, the released heat can lead to unsafe temperature rises. Diluting the reactants with a carrier gas such as nitrogen is one way of controlling the resulting temperature increase. In this example, we calculate the amount of energy required to produce a particular temperature increase.

SOLUTION C:

Step 5 *Compare to part A*

As it was in part A, our starting point for calculating the change in $\underline{U}$ is

$$d\underline{U} = C_V^* \, dT$$

This is a property of ideal gases that is valid regardless of the value of C_p^* or C_V^*. Also, $C_p^* = C_V^* + R$ is a property of ideal gases, which is valid whether C_p^* is constant or not. Thus, if $C_p^* = (7/2)R$, then

$$C_V^* = C_P^* - R = \frac{7}{2}R - R = \frac{5}{2}R \tag{2.49}$$

Step 6 *Integrate*

Because C_V here is considered a constant, the integration of Equation 2.29 simplifies to

$$\underline{U}_2 - \underline{U}_1 = \int_{T=450\,\text{K}}^{T=500\,\text{K}} \frac{5}{2}R \, dT \tag{2.50}$$

$$\underline{U}_2 - \underline{U}_1 = \frac{5}{2}R(T_2 - T_1) = \frac{5}{2}\left(8.314 \frac{\text{J}}{\text{mol}\cdot\text{K}}\right)(500 - 450\,\text{K}) = \mathbf{1039} \frac{\textbf{J}}{\textbf{mol}}$$

or, for the conditions in part B,

$$\underline{U}_2 - \underline{U}_1 = \frac{5}{2}R(T_2 - T_1) = \frac{5}{2}\left(8.314 \frac{\text{J}}{\text{mol}\cdot\text{K}}\right)(1073 - 273\,\text{K}) = \mathbf{16{,}628} \frac{\textbf{J}}{\textbf{mol}}$$

For the relatively small temperature interval of 50 K, the answer is only ~2% different when a constant heat capacity is assumed. For the larger temperature interval of 800 K, the answers are 7.2% different.

Example 2-5 illustrates that, while internal energy of an ideal gas is solely a function of temperature, it's not usually a linear function. The heat capacity is itself dependent on temperature. There is no theoretical reason to expect that C_p^* will follow the form $C_p^* = A + BT + CT^2 + DT^3 + ET^4$; it is simply an empirical formula that works well. Appendix D gives values for C_p^* in this form for a number of compounds.

2.3.4 Equations of State

They are called "equations of state" because they relate state functions to each other.

FOOD FOR THOUGHT 2-9

"Data trumps theory" is a popular expression in science and engineering. Do you agree with this statement?

Example 2-3 used H_2O to illustrate the fact that real gases behave like ideal gases at low pressure, and suggested that $P \leq 101.3$ kPa is not a bad general guideline for what constitutes "low." The example likely seems contrived in that there's no obvious practical need to use the ideal gas approximation for water; there is data readily available and one can simply look up the volume, internal energy, and enthalpy at most any conditions. However, chemical engineers routinely work with thousands of different compounds, very few of which have been studied as extensively as water. Consequently, while an engineer should certainly make use of reliable data when available, it is extremely valuable to have available strategies for the accurate prediction of unknown properties for chemical compounds. Equations of state are at the heart of such strategies.

Formally, an **equation of state** is a *relationship among temperature, pressure, and molar volume, such that if any two are known, the third can be calculated.*

The ideal gas law:

$$P = \frac{RT}{\underline{V}}$$

fits the definition. If any two of P, V, and T are known, the third can be found, as was done in Example 2-3. Thus the ideal gas law is an equation of state that has practical value at low pressures. However, at elevated pressures, real gases exhibit substantial departure from ideal gas behavior. Many more complex equations of state have been proposed to quantify P, V, and T at a broader range of conditions.

Among the oldest of these is the van der Waals equation:

$$P = \frac{RT}{V - b} - \frac{a}{V^2} \tag{2.51}$$

where a and b are constants that have unique values for each compound and R is the gas constant. Temperature (T), pressure (P), and molar volume (V) must all be expressed on an absolute scale.

Physical significances of a and b can be considered as

- "b" is a parameter that has the units of specific volume and accounts for the volume of the particles. According to the ideal gas law, as P approaches infinity, V approaches zero. But $V = 0$ is an unrealistic limit because real particles have finite volume. If we picture molecules as a collection of rigid spheres, increasing pressure moves the spheres closer together as illustrated in Figure 2-12, but once the spheres are packed together a minimum volume has been reached. The parameter b represents this minimum volume; according to the van der Waals equation, as pressure approaches infinity, V approaches not 0 but b.

> Formally, b is called the "excluded volume," because if a particle is occupying space it "excludes" other particles from that space.

- "a" is a parameter that accounts for the effect of intermolecular attractions on pressure. Consider that if two molecules are attracted to each other, they might "stick together" and act like one particle. This consolidation of particles would tend to decrease the pressure exerted by the gas, and thus the sign of the second term is negative. This term is proportional to $1/V^2$ because as V gets smaller, the particles are closer together and the effect of intermolecular interactions ("a") becomes more significant.

> **FOOD FOR THOUGHT 2-10**
>
> If a rigid sphere had a volume of 1 cm³, would you guess its "excluded volume" was 1 cm³, or would it be some other number?

While qualitative, these explanations of "a" and "b" allow us to rationalize the van der Waals equation compared to the ideal gas law.

- The ideal gas model assumes particles have zero volume; the van der Waals parameter "b" accounts for real gas departures from this assumption.
- The ideal gas model assumes particles have no intermolecular interactions; the van der Waals parameter "a" accounts for the fact that they do interact.

> Section 7.2.5 discusses how numerical values of a and b can be assigned for specific compounds.

- When V is very large, the a/V^2 term will become vanishingly small, and b will be insignificant compared to V. Thus, as V becomes larger, the van der Waals equation becomes closer and closer to $P = RT/V$, which is the ideal gas law. This observation is logical in that high volume, and correspondingly low pressure, is the exact condition under which real gases are expected to behave like ideal gases.

At this point, we introduce the **compressibility factor (Z)**

$$Z = \frac{PV}{RT} \tag{2.52}$$

> Note that if $a = b = 0$, the van der Waals equation simplifies to the ideal gas law:
> $$P = \frac{RT}{V}$$

Notice that $Z = 1$ for an ideal gas ($PV = RT$). Equations of state are frequently expressed in terms of Z. Doing this makes comparisons to ideal gas behavior easier;

if a gas has $\underline{V} = 0.02$ m³/mol this number might not hold much insight for you, but if $Z = 0.65$ you know immediately that the gas departs significantly from ideal gas behavior. The van der Waals EOS, written in terms of Z, is

$$Z = \frac{\underline{V}}{\underline{V} - b} - \frac{a}{RT\underline{V}} \tag{2.53}$$

We close this section with an example illustrating the accuracy of the van der Waals equation compared to the ideal gas law.

EXAMPLE 2-6	**COMPARING THE VAN DER WAALS EQUATION TO REAL DATA**

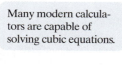
Example 7-4 illustrates the process by which these values of the parameters a and b were calculated.

For H_2O, the following parameters have been proposed for the van der Waals equation of state: $a = 0.5542$ Pa · m⁶/mol² and $b = 3.051 \times 10^{-5}$ m³/mol. Find the specific volume of water at the same temperatures and pressures given in Example 2-3.

SOLUTION:

Step 1 *Manipulate van der Waals equation*
A complication in solving the van der Waals equation is that it gives the pressure explicitly as

$$P = \frac{RT}{\underline{V} - b} - \frac{a}{\underline{V}^2}$$

As a result, it is simple to solve when $\underline{V}$ and T are given and P is unknown, but less straightforward when $\underline{V}$ is unknown. Furthermore, if we multiply through the denominators, we have

$$P(\underline{V} - b)(\underline{V}^2) = RT\underline{V}^2 - a(\underline{V} - b) \tag{2.54}$$

Many modern calculators are capable of solving cubic equations.

We can now see more clearly that this is a *cubic equation;* the left-hand side is of third degree in $\underline{V}$. Appendix B outlines methods for solving cubic equations, so here we will simply give numerical answers. A source of confusion, however, is that cubic equations can have three real solutions.

Step 2 *Physical interpretation of result*
For the first case, $T = 100°C$ and $P = 10$ kPa, and when the three solutions for $\underline{V}$ are converted to specific volume, the results are $\hat{V} = 2.17 \times 10^{-3}$ m³/kg, $\hat{V} = 7.76 \times 10^{-3}$ m³/kg, and $\hat{V} = 17.213$ m³/kg. Which one is "right?" *Mathematically*, all three are "right," but *physically*, there can only be one value of $\hat{V}$ for steam at a particular T and P. If you look at the ideal gas calculations in Example 2-3, you can probably deduce that $\hat{V} = 17.213$ m³/kg is the physically realistic value; the others are much too small for the specific volume of a low-pressure

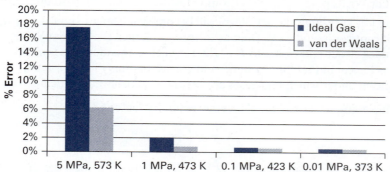

FIGURE 2-14 Accuracy of $\underline{V}$ predicted by ideal gas law versus van der Waals EOS for steam.

gas. This would be correct; in using a cubic equation of state, when there are three solutions for $\underline{V}$, the largest value corresponds to the molar volume of the vapor. Cubic equations of state can be developed such that the lowest value of $\underline{V}$ models the liquid molar volume. This is discussed further in Section 7.2. For some values of T and P, there is only one real solution for $\hat{V}$; the other two roots of the equation are imaginary numbers. These are discussed further in Section 7.2.

The $\hat{V}$ (van der Waals) shown in Table 2-6 is the largest of the solutions for $\hat{V}$ at each temperature and pressure. Notice that at $P = 0.1$ MPa, use of the ideal gas law is justified; it is accurate to within 1%, and the more complex van der Waals equation is only fractionally better. However, as pressure increases, we find that the van der Waals equation models the actual data for steam much more closely than does the ideal gas law. Chapter 7 covers modern equations of state, which can be used to model a wide range of compounds with a still greater degree of accuracy than that shown for the van der Waals equation.

TABLE 2-6 Comparison of actual specific volumes of water vapor to calculations using ideal gas and van der Waals equations of state.

Temp. (°C)	Pressure (MPa)	$\hat{V}$ (steam tables) m³/kg	$\hat{V}$ (van der Waals) (m³/kg)	% error, van der Waals	$\hat{V}$ (ideal gas law) (m³/kg)	% error, ideal gas law
100	0.01	17.196	17.213	0.10	17.221	0.15
150	0.1	1.936	1.946	0.52	1.953	0.87
200	1	0.206	0.2121	2.96	0.2184	6.00
300	5	0.04532	0.0477	5.25	0.0529	16.70

2.3.5 Simple Models of Liquids and Solids

Up to this point only modeling of the vapor phase has been discussed. Chapter 7 discusses the use of equations of state to describe liquids; indeed, it shows how a single equation of state can be used to quantify the molar volumes of a substance in both the liquid and the vapor phases. Generally, however, the volumes of liquids and solids are much less sensitive to changes in pressure and temperatures than the volumes of vapors and gases. Because of this observation, both liquids and solids are frequently modeled as having constant volume (especially over small ranges of temperature and/or pressure).

$$\underline{V}_L \approx \text{constant} \tag{2.55}$$

$$\underline{V}_S \approx \text{constant} \tag{2.56}$$

If a material (liquid or solid) is modeled as having constant volume, internal energy can be calculated as shown in Example 2-2, where

$$d\underline{U} = C_V dT \tag{2.57}$$

Further, because $\underline{V}$ is much smaller in solids and liquids than in vapors and gases, at low pressures $P\underline{V}$ is usually very small compared to $\underline{U}$. Consequently, for solids and liquids at low pressures, it is often reasonable to approximate enthalpy and internal energy as equal:

$$\underline{H} = \underline{U} + P\underline{V} \approx \underline{U} \tag{2.58}$$

This means that the constant pressure and constant volume heat capacities are approximately equal:

$$C_V \approx C_P \tag{2.59}$$

FOOD FOR THOUGHT 2-11

Are Equations 2.55 and 2.56 equations of state?

2.3.6 Summary of Simple Thermodynamic Models for Materials

Table 2-7 summarizes the properties of ideal gases that are of particular utility in solving thermodynamics problems. Section 2.3.5 summarizes some ways to model thermodynamic properties for liquids and solids, based on the assumption that volume is constant with respect to both pressure and temperature. While simplified, these models are adequate for many practical problems. Chapters 3 through 5 will focus on thermodynamic analysis of systems for which either comprehensive data (e.g., the steam tables) are available, or else these simple models are sufficient. Chapter 6 outlines the limitations of the simple models, discusses more complex models capable of describing real compounds at a broader range of conditions, and illustrates how accurate equations of state play a pivotal role in such models. A more complete discussion of equations of state is then presented in Chapter 7.

TABLE 2-7 Properties of ideal gases.

Property	Equation	Comment
$P, \underline{V}$, and T	$P\underline{V} = RT$	The ideal gas law
$\underline{U}$	$d\underline{U} = C_V^* dT$	Internal energy for an ideal gas is dependent only upon temperature
$\underline{H}$	$d\underline{H} = C_P^* dT$	Enthalpy for an ideal gas is dependent only upon temperature
Heat capacity	$C_P^* = C_V^* + R$	Valid whether C_V^* is modeled as constant or not

2.4 SUMMARY OF CHAPTER TWO

- A **state property** is a property that depends upon the state of the system, as described by other state properties, but does not depend upon the path taken to arrive at the current state.

- **Temperature**, **pressure**, **volume**, and **internal energy** are all state properties.

- **Heat** and **work** are NOT state properties.

- Any property such as **enthalpy**, which is defined as a mathematical function of other state properties ($H = U + PV$), is also a state property.

- State properties of materials are related to each other. **Degrees of freedom** are the number of independent intensive state properties that must be determined before all other intensive state properties have unique values.

- Degrees of freedom can be quantified using the **Gibbs phase rule**:

$$F = C - \pi + 2$$

- A **pure material in a single phase** has two degrees of freedom.

- An **intensive property** of a material does not depend upon the quantity of the material, while an **extensive property** does depend upon the quantity of the material present.

- An **equation of state** is an equation that quantifies the interrelationships among pressure, temperature, and molar volume of a material.

- The **ideal gas law** is the simplest equation of state for gases.
- Real vapors and gases behave like ideal gases at low pressures, and Table 2-7 summarizes several useful properties of ideal gases.
- The **heat capacity** of a material is the amount of energy required to raise the temperature of a set amount of the material (e.g., one mole, one gram) by one degree.
- **Heat capacity** can be defined through either a constant-volume heat capacity (C_V) or a constant-pressure heat capacity (C_P).

2.5 EXERCISES

2-1. Consider the *P-T* diagram in Figure 2-15 for a pure substance:

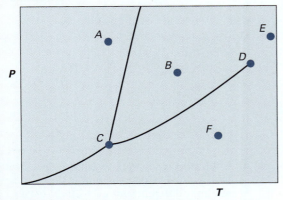

FIGURE 2-15 Phase diagram for a typical compound.

A. What phase is represented by point *A*? According to the Gibbs phase rule, how many degrees of freedom exist at this point?

B. What phase is point *B*? How many degrees of freedom exist at this point?

C. How many phases are in equilibrium at point *C*? What is the name of this point? How many degrees of freedom exist at this point?

D. How many phases are in equilibrium at point *D*? What is the name of this point? How many degrees of freedom exist at this point?

E. What phase is point *E*?

F. What phase is point *F*?

G. What is the physical significance of the curve that exists between points *C* and *D*?

2-2. A pure compound is initially in the vapor phase, at its vapor pressure. The pressure is increased isothermally. Which phase will exist at the end of the process? For each of the options listed below, determine whether or not it is possible.

A. Solid

B. Liquid

C. Solid–liquid equilibrium mixture

D. Supercritical fluid

2-3. A pure compound is initially in the liquid phase. Each of the following processes ends with the compound in a single, pure phase that is not liquid. Identify the phase that will exist at the end of each process. If there is more than one possibility, give all the possibilities.

A. The pressure is increased isothermally.

B. The pressure is decreased isothermally.

C. The temperature is increased isobarically.

D. The temperature is decreased isobarically.

E. Temperature and pressure are increased simultaneously.

F. Temperature and pressure are decreased simultaneously.

G. Temperature is increased while pressure is decreased.

H. Temperature is decreased while pressure is increased.

2-4. Read the following paragraph.

A plane was scheduled to fly a round trip from an airport to a rural landing strip. The mass of the plane, when empty, was 45,000 kg. It was loaded with 38 m³ gallons of fuel. The fuel weighed 719 kg/m³. The plane took off from its home airport, and made a detour to take aerial photographs, so it flew 354 km before landing at the landing strip, which is 241 km from the airport. There were 17.4 m³ of fuel remaining in the plane when it landed at the landing strip.

A. For every number in the paragraph, indicate whether or not the number represents a state property.

B. For every number in the paragraph, excluding the distances, indicate whether the number represents an intensive or extensive property.

2-5. Read the following paragraph.

A room has a volume of 34 m³. The initial temperature was 292.6 K. The room contained 40 kilograms of air, and 20.9 mol% of it oxygen. A space heater was accidentally turned on. The space heater added 1000 kJ of heat to the room, increasing the temperature by 15.0 K.

A. For every number in the paragraph, indicate whether the number represents a state property.

B. For every state property, indicate whether the number represents an intensive or extensive property.

2-6. Most (though not all) of the people who climb Mt. Everest do so with the aid of oxygen masks and tanks. This is because the barometric pressure at the peak of Mt. Everest is about one third that at sea level. Estimate the number of moles of oxygen in one cubic meter of air, assuming air is an ideal gas that is 21 mol% oxygen:

A. $P = 0.1$ MPa and $T = 298$ K

B. $P = 35$ kPa and $T = 263$ K

2-7. A 0.5 m³ container holds ideal gas, initially at $T = 323$ K and $P = 0.1$ MPa. The ideal gas has $C_V^* = 2.5R$.

A. How many moles of ideal gas are in the container?

B. What is the change in internal energy of the gas when it is heated from 323 to 348 K?

C. What is the change in enthalpy of the gas when it is heated from 323 to 348 K?

2-8. If the heat capacity of an ideal gas is $C_P^* = 30 + 0.05T$, with C_P^* in J/mol · K and T representing the temperature in K, what is the change in internal energy and the change in enthalpy for five moles of this gas when its heated from $P = 0.1$ MPa and $T = 300$ K to $P = 0.5$ MPa and $T = 600$ K?

2-9. A quantity of carbon dioxide is confined in a sealed container. For each of the following cases, estimate the pressure of the carbon dioxide, using both the ideal gas law and the van der Waals equation of state. The parameters used to model carbon dioxide using the van der Waals equation of state are $a = 0.3658$ Pa · m⁶/mol² and $b = 42.86 \times 10^{-6}$ m³/mol.

A. $T = 500$ K, $V = 50$ dm³, $N = 1$ mol

B. $T = 500$ K, $V = 50$ dm³, $N = 5$ mol

C. $T = 500$ K, $V = 50$ dm³, $M = 1$ kg

D. $T = 750$ K, $V = 50$ dm³, $M = 1$ kg

E. $T = 750$ K, $V = 15$ dm³, $N = 1$ mol

F. $T = 750$ K, $V = 1$ dm³, $N = 1$ mol

2.6 PROBLEMS

2-10. The boiler is an important unit operation in the Rankine cycle. This problem further explores the phenomenon of "boiling."

A. When you are heating water on your stove, before the water reaches 373 K, you see little bubbles of gas forming. What is that, and why does that happen?

B. Is it possible to make water boil at below 373 K? If so, how?

C. What is the difference between "evaporation" and "boiling?"

2-11. 10 mol/s of gas flow through a turbine. Find the change in enthalpy that the gas experiences:

A. The gas is steam, with an inlet temperature and pressure $T = 873.15$ K and $P = 1$ MPa, and an outlet temperature and pressure $T = 673.15$ K and $P = 0.1$ MPa. Use the steam tables.

B. The gas is steam, with the same inlet and outlet conditions as in part A. Model the steam as an ideal gas using the value of C_P^* given in Appendix D.

C. The gas is nitrogen, with an inlet temperature and pressure of $T = 300$ K and $P = 1$ MPa, and an outlet temperature and pressure $T = 200$ K and $P = 0.1$ MPa. Use Figure 2-3.

D. The gas is nitrogen with the same inlet and outlet conditions as in part C. Model the nitrogen as an ideal gas using the value of C_P^* given in Appendix D.

E. Compare the answers to parts A and B and to parts C and D. Comment on whether they are significantly different from each other, and if so, why.

2-12. Using data from the steam tables in Appendix A, estimate the constant pressure heat capacity of superheated steam at 350 kPa, 473.15 K and at 700 kPa, 473.15 K. Are the answers very different from each other? (NOTE: You may need to use the "limit" definition of the derivative to help you get started.)

2-13. The specific enthalpy of liquid water at typical ambient conditions, like $T = 298.15$ K and $P = 0.1$ MPa, is not given in the steam tables. However, the specific enthalpy of saturated liquid at $P = 0.1$ MPa is given.

A. Using the approximation that $\hat{C}_P$ for liquid water is constant at 4.19 kJ/kg · K, estimate the specific enthalpy of liquid water at $T = 298.15$ K and $P = 0.1$ MPa.

B. Compare the answer you obtained in part A to the specific enthalpy of *saturated* liquid water at $T = 298.15$ K.

2-14. This problem is an expansion of Example 2-3. The table below lists 10 sets of conditions—five temperatures at a constant P, and five pressures at a constant T. For each T and P, find:

▪ The specific volume of steam, from the steam tables

- The specific volume of steam, from the ideal gas law

Comment on the results. Under what circumstances does departure from ideal gas behavior increase?

Temperature	Pressure
473.15 K	0.5 MPa
573.15 K	0.5 MPa
673.15 K	0.5 MPa
773.15 K	0.5 MPa
1273.15 K	0.5 MPa
523.15 K	0.01 MPa
523.15 K	0.1 MPa
523.15 K	0.5 MPa
523.15 K	1 MPa
523.15 K	2.5 MPa

2-15. A refrigeration process includes a compressor, as explained in detail in Chapter 5, because it is necessary to change the boiling point of the refrigerant, which is done by controlling the pressure. Chapter 3 shows that the work required for compression is well approximated as equal to the change in enthalpy. Use Appendix F to find the change in specific enthalpy for each of the scenarios A through C.
  A. Freon 22 enters the compressor as saturated vapor at $P = 0.05$ MPa and exits the compressor at $P = 0.2$ MPa and $T = 293$ K.
  B. Freon 22 enters the compressor as saturated vapor at $P = 0.2$ MPa and exits the compressor at $P = 0.8$ MPa and $T = 333$ K.
  C. Freon 22 enters the compressor as saturated vapor at $P = 0.5$ MPa and exits the compressor at $P = 2$ MPa and $T = 353$ K.
  D. In these three compressors, the inlet and outlet pressures varied considerably, but the "compression ratio" P_{out}/P_{in} was always 4. What do you notice about the changes in enthalpy for the three cases?

2-16. The classic way to synthesize ammonia is through the gas phase chemical reaction:

$$N_2 + 3H_2 \rightarrow 2\,NH_3$$

This reaction is carried out at high pressures, most often using an iron catalyst.
  A. Use Equation 2.32 and C_p^* from Appendix D to determine the change in molar enthalpy when nitrogen is compressed from $T = 300$ K and $P = 0.1$ MPa to $T = 700$ K and $P = 20$ MPa.
  B. Repeat part A for hydrogen.
  C. What assumptions or approximations were made in step B? Comment on how valid you think the approximations are.

2-17. Liquid water enters a steady state heat exchanger at $P = 1$ MPa and $T = 353.15$ K, and exits as saturated water vapor at $P = 1$ MPa.
  A. Using the steam tables, find the change in specific enthalpy for this process.
  B. Using the approximation that $\hat{C}_p = 4.19$ kJ/kg · K for liquid water, find the change in enthalpy when liquid water is heated from 353.15 K to its boiling point at $P = 1$ MPa.
  C. How do the answers to parts A and B compare to each other? What is the reason for the difference between them?

2-18. One mole of ideal gas is confined in a piston-cylinder device, which is 0.30 m in diameter. The piston can be assumed weightless and frictionless. The internal and external pressures are both initially 101.3 kPa. An additional weight of 4.54 kg is placed on top of the piston, and the piston drops until the gas pressure balances the force pushing the piston downward. The temperature of the gas is maintained at a constant temperature of 300 K throughout the process.
  A. What is the final pressure of the gas?
  B. What is the final volume of the gas?
  C. How much work was done on the gas during the process?

2-19. Two moles of an ideal gas are confined in a piston-cylinder arrangement. Initially, the temperature is 300 K and the pressure is 1 MPa. If the gas is compressed isothermally to 5 MPa, how much work is done on the gas?

2-20. A gas is stored in an isochoric, refrigerated tank that has $V = 5$ m³. Initially, the gas inside the tank has $T = 288$ K and $P = 0.5$ MPa, while the ambient surroundings are at 298 K and atmospheric pressure. The refrigeration system fails, and the gas inside the tank gradually warms to 298 K.
  A. Find the final pressure of the gas, assuming it is an ideal gas.
  B. Find the final pressure of the gas, assuming the gas is described by the van der Waals equation of state with $a = 0.08$ m⁶ Pa/mol² and $b = 1 \times 10^{-4}$ m³/mol.
  C. For the cases in parts A and B, how much work was done by the gas on the surroundings?

2-21. Biosphere II is an experimental structure that was designed to be an isolated ecological space, intended for conducting experiments on managed, self-contained ecosystems. Biosphere II (so named because the earth itself was regarded as the first "Biosphere") is covered by a rigid dome. One of the engineering challenges in designing Biosphere II stemmed from temperature fluctuations of the air inside the dome. To prevent this from resulting in pressure fluctuations that could rupture the dome, flexible diaphragms called "lungs" were built into the structure. These "lungs" expanded and contracted so that changes in air volume could be accommodated while pressure was maintained constant.

For the purposes of this problem, assume that air is an ideal gas, with heat capacity equal to a weighted average of the ideal gas heat capacities of nitrogen and oxygen: $C^*_{P,\text{air}} = 0.79 C^*_{P,N_2} + 0.21 C^*_{P,O_2}$.

A. Suppose the volume of air contained within the Biosphere II was 125,000 m³, there were no lungs, and the pressure of the air was 100 kPa when the temperature was 294 K. What air pressure would occur at a temperature of 255 K?

B. Repeat part A, calculating the pressure resulting if the temperature were 314 K rather than 255 K.

For parts C–E assume the Biosphere II has lungs, and that the total volume of air inside the dome is 125,000 m³ *when the lungs are fully collapsed.*

C. If the lungs are designed to maintain a constant pressure of 0.1 MPa at all temperatures between 255 and 314 K, what volume of air must the lungs hold when fully expanded?

D. If the temperature inside the dome increases from 255 K to 314 K at a constant P = 0.1 MPa, what change in internal energy does the air undergo?

E. If the temperature inside the dome increases from 255 K to 314 K at a constant P = 0.1 MPa, and the lungs are sized as calculated in part C, give your best estimate of the work done by the lungs on the surroundings.

Use the following data for Problems 2-22 through 2-26.

A compound has

▨ A molecular weight of 50 kg/kmol.

▨ A normal melting point of 300 K.

▨ A normal boiling point of 350 K.

▨ A constant heat capacity in the solid phase of $C_p = 3000$ J/kg · K.

▨ A constant heat capacity in the liquid phase of $C_p = 3600$ J/kg · K.

▨ A constant density in the solid phase of $\rho = 640$ kg/m³.

▨ A constant density in the liquid phase of $\rho = 480$ kg/m³.

▨ An ideal gas heat capacity of $C^*_p = 4186.8 - 6.28\,T + 13.56 \times 10^{-3}\,T^2$ J/kg · K with T representing the temperature in Kelvin.

▨ An enthalpy of fusion $\Delta \hat{H}_{\text{fus}} = 175$ kJ/kg at atmospheric pressure.

▨ An enthalpy of vaporization $\Delta \hat{H}_{\text{vap}} = 300$ kJ/kg at atmospheric pressure.

2-22. Estimate the change in $\hat{U}$ when this compound melts at atmospheric pressure.

2-23. Estimate the change in $\hat{U}$ when this compound boils at atmospheric pressure.

2-24. When 1 kmol is heated from 269 K to 366 K at a constant pressure of P = 101.325 kPa:
A. Give your best estimate of the change in enthalpy
B. Give your best estimate of the change in internal energy

2-25. The compound is initially at 278 K. If the pressure is constant at P = 101.325 kPa throughout the process, and the specific enthalpy of the compound is increased by 815 kJ/kg, what is the final temperature?

2-26. The compound is initially at 370 K. If the pressure is constant at P = 101.325 kPa throughout the process, and the specific internal energy of the compound is decreased by 350 kJ/kg, what is the final temperature?

2-27. This question involves using the steam tables in Appendix A, but answer questions A through C before looking at the steam tables.
A. The density of liquid water at ambient conditions is about 1 g/cm³. Convert this into a specific volume, expressed in m³/kg—the units used in the steam tables.
B. You're asked to look up the largest and smallest values of $\hat{V}$ found anyplace in the steam tables. Recall that Appendix A includes saturated steam tables, superheated steam tables, and compressed liquid tables, and that the conditions in the tables range from 0.01–100 MPa and 273.15 K–1273.15 K. Where will you look for the largest and smallest values?
C. Before looking up the largest and smallest values of $\hat{V}$, guess or estimate what they will be.
D. Look up the largest and smallest values of $\hat{V}$ found anyplace in the steam tables. How do they compare to the guess/estimates you made in part C? Are they located in the place you predicted in part B?

2-28 Model water using the van der Waals equation of state with $a = 5.53 \times 10^6$ bar · cm⁶/mol² and $b = 30.48$ cm³/mol. (The values of the van der Waals a and b were determined using the method illustrated in Exam-ple 7-4).
A. Make a graph of P vs. $\hat{V}$ for water at T = 100°C. Use the data from the steam tables.

B. Using the van der Waals equation of state, plot a graph of P versus $\hat{V}$ for water, holding temperature constant at $T = 100°C$. Compare this to the data from part A and comment on the quality of the predictions produced by the van der Waals equation.

C. Repeat parts A and B for temperatures of 200°C, 300°C, 400°C, and 500°C. When you compare the five temperatures to each other, what differences in behavior do you notice, for both the real data and the equation of state?

2-29. A compound has a molecular mass of 120 kg/kmol, and the information in the table below is the only other data available for a compound. Fill in all of the empty cells with your best estimate of the value. Explain any assumptions or approximations you make.

	Phase	T (K)	P (MPa)	$\hat{U}$ (kJ/kg)	$\hat{H}$ (kJ/kg)	$\hat{V}$ (m³/kg)
A	Solid	273	0.5	−42.6		8.0 × 10⁻⁴
B	Solid	298	0.5	0		
C	Liquid	298	0.5	63.0	63.6	
D	Liquid	333	0.5			
E	Liquid	358	0.5		223.2	
F	Vapor	358	0.5	977.2	1014.2	
G	Vapor	358	0.1	961.2		

2.7 GLOSSARY OF SYMBOLS

A	area
a	parameter in van der Waals EOS, accounting for intermolecular interactions
b	parameter in van der Waals EOS, accounting for excluded volume
C	number of chemical compounds
$\hat{C}_P$	constant-pressure heat capacity (mass basis)
C_P	constant-pressure heat capacity (molar basis)
C_P^*	constant-pressure heat capacity for ideal gas (molar basis)
$\hat{C}_V$	constant-volume heat capacity (mass basis)
C_V	constant-volume heat capacity (molar basis)
C_V^*	constant-volume heat capacity for ideal gas (molar basis)
F	degrees of freedom (in Gibbs phase rule)
H	enthalpy
$\hat{H}$	specific enthalpy
$\underline{H}$	molar enthalpy
$\dot{m}$	mass flow rate
N	number of moles
P	pressure
Q	heat
$\dot{Q}$	rate of heat addition
R	gas constant
T	temperature
U	internal energy
$\hat{U}$	specific internal energy
$\underline{U}$	molar internal energy
V	volume
$\hat{V}$	specific volume
$\underline{V}$	molar volume
$\dot{V}$	volumetric flow rate
$\dot{W}$	power
π	number of phases

2.8 REFERENCES

Brown, R. L., Stein, S. E. "Boiling Point Data," *NIST Chemistry WebBook, NIST Standard Reference Database Number 69*, Eds. P. J. Linstrom and W. G. Mallard, National Institute of Standards and Technology, Gaithersburg MD, 20899, http://webbook.nist.gov (retrieved July 6, 2012).

Cengel, Y. A., Cimbala, J. M., Turner, R. H. *Fundamentals of Thermal-Fluid Sciences*, 4th ed., McGraw-Hill, 2012.

Green, D. W. (editor) and Perry, R. H. (late editor), *Perry's Chemical Engineers' Handbook*, 8th edition, McGraw-Hill, 2008.

Laidler, K. J., Meiser, J. H., Sanctuary, B. C., *Physical Chemistry*, Cengage Learning, Inc., 2003.

Machin, G. "System for Calibrating Thermometers," U.S. Patent #6,939,035, issued Sept. 6, 2005.

Olson, R. (editor), *Biographical Encyclopedia of Scientists*, Marshall Cavendish, Tarrytown, NY 1998.

Poling, B. E., Prausnitz, J. M., O'Connell, J. P., *The Properties of Gases and Liquids*. 5th edition, McGraw-Hill, New York, 2001.

Reynolds, W. C., *Thermodynamic Properties in SI*, Department of Mechanical Engineering, Stanford University, 1979.

Material and Energy Balances

3

LEARNING OBJECTIVES

This chapter is intended to help you learn how to:

- Define *systems* and *processes* that are well chosen for solution of a problem
- Write *material* and *energy balances* that describe specific systems and processes
- Recognize the implications of system and process descriptors such as *adiabatic, steady state, closed,* or *open,* and how they are reflected in material and energy balance equations
- Identify *physical properties* that are necessary for solution of energy balance equations
- Quantify relevant physical properties using simple *thermodynamic models*
- Solve realistic engineering problems using material and energy balances

Various forms that energy can take and the fact that energy can be converted from one form into another were discussed in Chapter 1. Indeed, to say energy "can be" converted from one form into another isn't going far enough; in the vast array of engineering applications, energy conversion represents a major unifying theme. Almost all of the products and systems you work with as a practicing engineer will use energy in some form, and using energy *efficiently* is a major priority for modern engineers of all disciplines, including chemical engineers. In this chapter, we examine quantitative analysis of energy conversions. ■

3.1 MOTIVATIONAL EXAMPLE: Rockets

The first law of thermodynamics states that energy is conserved; it can be converted from one form into another, but cannot be created or destroyed. In the first chapter we examined the forms energy can take. Here we explore how to apply the first law of thermodynamics to a broad range of problems. We've previously invoked the principle of energy conservation to solve a problem. Example 1-6 examined an object in free fall. The first law of thermodynamics was relatively straightforward to apply in this case, because energy in one form, potential energy, was completely converted into one (and only one) other form, kinetic energy. Many practical engineering systems and processes are more complex than this, with several forms of energy relevant. Consider, for example, a rocket.

Conventional rockets operate through the combustion of liquid or solid fuel. As the fuel burns, gases are produced. These gases are expelled out of the back of the

Robert Goddard, the "Father of Modern Rocketry," is credited with launching the world's first liquid-fueled rocket in 1926. It rose to a height of 12.5 m and landed 56 m from the launch point. (Kluger, 1999)

rocket through nozzles, at extremely high velocity. As the gases move downward, the rocket is propelled upward through conservation of momentum. In this process, internal energy, kinetic energy and potential energy are all important; and there are two distinct velocities (that of the rocket and that of the exhaust) to consider in quantifying kinetic energy. Accounting for the energy conversion is much more complicated than in the simple case of a falling object, and it would be very easy to make an error. One such error is illustrated below.

In the early days of rocketry, many were doubtful that the technology could be used to reach outer space. Indeed, here is a "proof," attributed to A. W. Bickerton (Clarke, 1962), in which the first law of thermodynamics was used to demonstrate that space flight is impossible:

1. The velocity needed to escape the Earth's gravity—the escape velocity—is 7 miles per second.

2. The kinetic energy of a gram at this velocity is 15,180 calories.

3. The energy of the most violent explosive—nitroglycerin—is less than 1500 calories per gram. (The calorie was a commonly used unit for energy in the 1920s; a calorie is equal to 4.184 J.)

The argument is that even if the fuel had nothing to carry, it has less than one-tenth the energy required to escape the Earth's gravity, and that space flight is, therefore, impossible.

This proof is clearly an attempt to apply the principle of conservation of energy, but it must be mistaken in some respect, since we now know space flight is indeed possible. Before you read further, can you spot the error in the "proof"? The math is not the problem-the numbers are all accurate.

The crucial point is: *the rocket fuel doesn't have to get into outer space.* If you've ever seen a rocket launch, you know that a great deal of fuel is burned and released very close to the ground. It is the payload (the rocket itself, equipment, astronauts, etc.) that needs to get into space. In Bickerton's time, kerosene was widely used as a fuel for lighting and heating. Ten grams of a kerosene/oxygen mixture releases over 20,000 calories when burned. If 100% of this energy is transferred to the payload in the form of kinetic energy, then ten grams of fuel is more than enough to propel one gram of payload into outer space. While modern rockets use better fuels than kerosene, we have now refuted the above "proof" using information that was available in the 1920s.

How can we avoid making errors such as the one illustrated above? One principle that is helpful is that of a *system*, introduced in Section 1.3. One cannot answer questions or draw conclusions about a system or process without a clear understanding of what that system or process *is*. In this case, "the rocket and all of its contents" would be a logical definition of the system. Next, we can account for all the energy present in the system, which is illustrated in Figure 3-1. At the start of the process, the system is at rest (kinetic energy = 0) on the ground (potential energy = 0), but contains both payload and rocket fuel (the fuel has a large internal energy). At a later point, the system contains no fuel, but has significant potential and kinetic energy that can be quantified if height and velocity are known.

Figure 3-1 and the above discussion might help us rationalize why Bickerton's "proof" is wrong, but it leaves significant questions unanswered. How much energy is actually released when the fuel burns? What form does this energy take, specifically? What about the combustion products-presumably primarily carbon dioxide and

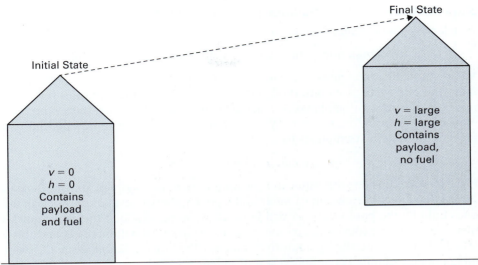

FIGURE 3-1 In a rocket launch, the internal energy of the fuel is transferred to the rocket in the forms of kinetic energy ($Mv^2/2$) and potential energy (Mgh).

water? These gases do not simply disappear; they have stored energy (internal, potential and kinetic) that they take with them when they leave the system. How does this phenomenon influence the velocity and height attained by the rocket? What about air resistance? Rockets move quite quickly, and the faster an object moves, the more air resistance it encounters. Since the air represents a force opposing the rocket's motion, this falls into the category of work.

There is no single, simple answer to the questions raised in the last paragraph. In analyzing a system that involves energy conversion, one must consider all of the forms in which energy can be stored or transferred, and for each, either determine it is negligible for the case at hand or else account for it quantitatively. Section 3-3 presents a mathematical expression of the first law of thermodynamics, which is called the *generalized energy balance,* that makes such an analysis easier.

The focus of this chapter is energy balances, but energy balances are often applied in combination with material balances. For example, in analyzing a rocket, one cannot quantify the stored energy in the exhaust gases without knowing how much gas is actually released. A material balance allows one to relate the amounts of carbon dioxide and water produced to the amount of fuel burned. Consequently, this chapter examines material and energy balances, applied to physical systems both individually and in combination with each other. We begin with material balances.

> Remember, the fuels we burn are predominantly hydrocarbons, and carbon dioxide and water are the products when a hydrocarbon is burned to completion.

> The space shuttle *Endeavour*, which was retired in 2011, had a mass of ~78,000 kg when empty and usually carried about 24,000 kg of additional payload. Over 1,700,000 kg of fuel were used to launch the orbiter. (http://www.nasa.gov, accessed October 2013.)

3.2 Material Balances

This section discusses how to apply and solve material balances. Section 3.2.2 contains three worked examples, but first we develop the material balance equation in a form that is applicable to any system.

> Material balances are also frequently called "mass balances."

FOOD FOR THOUGHT 3-2

What would the material balance look like for the Motivational Example process illustrated in Figure 3-1?

3.2.1 Mathematical Formulation of the Material Balance

A balance equation can be envisioned as

$$\text{Accumulation} = \text{In} - \text{Out} + \text{Generation} - \text{Consumption}$$

We are modeling total mass as a conserved quantity. Nuclear processes exist where mass and energy are not conserved, but these are not considered in this book. The principle of conservation of mass is reflected mathematically in that there is no "generation" or "consumption" term. When our material balance represents the *total* mass of the system, it simplifies to

$$\text{Accumulation} = \text{In} - \text{Out}$$

FOOD FOR THOUGHT 3-3

What if a chemical reaction occurred, and water were converted into hydrogen and oxygen? How would this be reflected in the material balance?

For example, consider the water in a swimming pool as a system. We can imagine events that affect the amount of water in the pool: water can evaporate, rain can fall, one could fill the pool with a garden hose, etc. If water is added (in) more quickly than it is removed (out), the amount of water in the pool will increase, and this is reflected mathematically by a positive value for the 'accumulation' term. (A swimming pool is examined quantitatively in Problem 3-8.)

Using the nomenclature of this book, the **balance equation for total mass** is

$$\frac{dM}{dt} = \sum_{j=1}^{j=J} \dot{m}_{j,\text{in}} - \sum_{k=1}^{k=K} \dot{m}_{k,\text{out}} \tag{3.1}$$

where M is the total mass of the system, $\dot{m}$ is the mass flow rate of a stream entering or leaving the system, t is time. J is the total number of entering streams, K is the total number of exiting streams, and j and k are subscripts used to distinguish entering and exiting streams from each other.

FOOD FOR THOUGHT 3-4

A swimming pool normally has a circulating filtration system that removes water, filters it and returns it to the pool. How would this be included in a material balance?

Since M represents total mass of the system, the left-hand side dM/dt represents "accumulation;" the rate at which the total mass is changing. The right-hand side represents material flows entering or leaving the system. The symbol $\dot{m}$ represents a mass flow rate, and has the dimension mass/time; common units include kg/s. The summations (Σ) account for the fact that there may be several different sources of mass entering or leaving the system. For example, Figure 3-2 shows a swimming

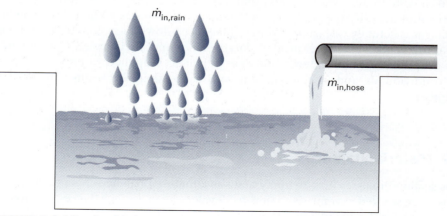

FIGURE 3-2 Mass is added to a swimming pool from two different sources, or "streams": rain and a garden hose.

pool that is being filled with a garden hose while rain is also falling. Consequently, if the pool were the system, there would be two separate "in" terms, which might be called $\dot{m}_{1,in}$ and $\dot{m}_{2,in}$, or $\dot{m}_{in,hose}$ and $\dot{m}_{in,rain}$. Distinct sources of mass entering or leaving the system are typically called "streams."

The next section illustrates the use of this equation through examples.

3.2.2 Examples of Material Balances

The material balance equation is applied using three examples.

■ Example 3-1 examines steam in a piston-cylinder device; an unsteady-state process carried out in a closed system.

■ Example 3-2 examines a bathroom sink with the water left on. We use an open-system material balance to determine when the sink will overflow.

■ Example 3-3 examines the growth of an oak tree, and requires three simultaneous material balances to solve.

| COMPRESSION OF STEAM | EXAMPLE 3-1 |

A vessel initially contains 5.0 m³ of superheated steam at $T = 573.15$ K and $P = 0.1$ MPa. The vessel is sealed but is equipped with a piston-cylinder arrangement so that the volume can be changed (Figure 3-3). If the steam is compressed isothermally to $P = 0.5$ MPa, what will be the new total volume?

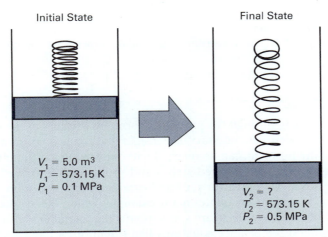

Initial State   Final State

$V_1 = 5.0\ m^3$
$T_1 = 573.15\ K$
$P_1 = 0.1\ MPa$

$V_2 = ?$
$T_2 = 573.15\ K$
$P_2 = 0.5\ MPa$

FIGURE 3-3 Initial and final states of the steam in the vessel described in Example 3-1.

SOLUTION:

Step 1 *Define the system*
We're trying to predict the final volume of steam, so define the system as the steam.

Step 2 *Gather data*
We saw in Chapter 2 that, if two *intensive* properties of a pure compound in a single phase are known, all others will have unique values. Here T and P are known for both

In a steady-state process, ALL parameters of the system are constant with respect to time. Here, temperature is constant, but pressure changes with time, so there is no steady state.

PITFALL PREVENTION

If you are careless with units, you could confuse the specific volume (m³/kg) with the total volume (m³).

the initial and final states. Thus, specific volumes can be obtained from the superheated steam tables in Appendix A-3:

$$\hat{V} = 0.5226 \frac{m^3}{kg} \text{ at } P = 0.5 \text{ MPa and } T = 573.15 \text{ K}$$

$$\hat{V} = 2.639 \frac{m^3}{kg} \text{ at } P = 0.1 \text{ MPa and } T = 573.15 \text{ K}$$

Step 3 *Material Balance*

We know the specific volume at the end of the process, but we also need to know the total mass in order to find the total volume. Thus, doing a material balance is a logical step:

$$\frac{dM}{dt} = \sum_{j=1}^{j=J} \dot{m}_{j,\text{in}} - \sum_{k=1}^{k=K} \dot{m}_{k,\text{out}} \tag{3.2}$$

Here the vessel is described as "sealed;" no mass enters or leaves. Thus, the mass balance simplifies to:

Equation 3.3 is a material balance equation that could be used to describe any closed system.

$$\frac{dM}{dt} = 0 \tag{3.3}$$

Consequently, M is constant; the initial and final masses of the system are equal.

Step 4 *Find initial value of M*

At the initial state, total volume and specific volume are known, so the total mass can be determined:

$$M = \frac{V}{\hat{V}} = \frac{5.0 \, m^3}{2.639 \dfrac{m^3}{kg}} = 1.89 \, kg \tag{3.4}$$

FOOD FOR THOUGHT 3-5

Could you have solved this problem using the ideal gas law? How would the answer have compared?

Step 5 *Calculate final volume*

We proved in step 3 that mass is constant; thus, the unknown final volume is

$$V = M\hat{V} = (1.89 \, kg)\left(0.5226 \frac{m^3}{kg}\right) = \mathbf{0.99 \, m^3} \tag{3.5}$$

Equation 3.34 in Section 3.3 is a generalized energy balance, analogous to the material balance in Equation 3.1.

In Example 3-1, going through the process of formally writing out the mass balance equation probably seems unnecessary. The insight that "mass is constant," while necessary to solve the problem, can be reached without writing a differential equation. One could, therefore, have skipped step 3 and probably still arrived at a correct answer. However, as we move forward and examine more and more complex examples, you will inevitably encounter many examples where the needed insights are not intuitively obvious and there are subtle points that could easily be overlooked. The value of generalized balance equations, like Equations 3.1 and 3.34, is that they are always valid; applying them helps ensure you don't overlook anything and they give you a starting point when the solution strategy isn't obvious.

EXAMPLE 3-2	**OVERFLOWING BATHROOM SINK**

A 4-dm³ sink is initially half-full of water. The faucet is turned on, and 1 kg/min of water begins pouring into the sink. At the same moment that the faucet is turned on, the drain is opened, and 10 cm³/s of water flows out of the sink (Figure 3-4). How long after the faucet is turned on will the sink overflow?

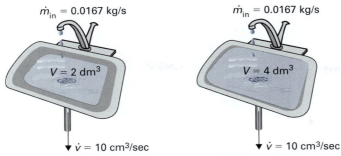

FIGURE 3-4 Initial and final states of the bathroom sink described in Example 3-2.

> The water in the sink is an example of an open system, and is not at steady state, because the mass of the system is changing.

SOLUTION:

Step 1 *Define the system and process*
The sink will overflow when the volume of water exceeds 4 dm³. Consequently, we define the water in the sink as the system, the start of the process as the moment the faucet is turned on, and the end of the process as the moment at which the volume of water equals 4 dm³.

Step 2 *Material balance*
Start with the general mass balance in Equation 3.1:

$$\frac{dM}{dt} = \sum_{j=1}^{j=J} \dot{m}_{j,\text{in}} - \sum_{k=1}^{k=K} \dot{m}_{k,\text{out}}$$

Since there is only one stream entering the system, and one stream leaving, this equation simplifies to

$$\frac{dM}{dt} = \dot{m}_{\text{in}} - \dot{m}_{\text{out}} \tag{3.6}$$

Step 3 *An assumption*
Some of the given information is on a mass basis, and some on a volume basis. We can't solve the mass balance equation until everything is expressed on a mass basis. Since no information about temperature and pressure are given, we assume the density of water is the standard 1000 kg/m³, or 1 kg/dm³.

> **PITFALL PREVENTION** ⚠
>
> Volume is not a conserved quantity; we cannot in general write "volume balances."

Step 4 *Unit conversions*
The initial volume V_0 of the water is 2 dm³, so its initial mass M_0 is

$$M_0 = V_0\rho = (2\,\text{dm}^3)\left(\frac{1\,\text{kg}}{\text{dm}^3}\right) = 2\,\text{kg} \tag{3.7}$$

Analogously we can determine that the sink will be full when $M = 4$ kg.
   The rates at which water leaves and enters the system are

$$\dot{m}_{\text{out}} = \left(10^{-5}\frac{\text{m}^3}{\text{s}}\right)\left(\frac{1000\,\text{kg}}{\text{m}^3}\right) = 0.01\,\frac{\text{kg}}{\text{s}} \tag{3.8}$$

$$\dot{m}_{\text{in}} = \left(1\frac{\text{kg}}{\text{min}}\right)\left(\frac{1\,\text{min}}{60\,\text{s}}\right) = 0.0167\,\frac{\text{kg}}{\text{s}} \tag{3.9}$$

> **FOOD FOR THOUGHT 3-6**
>
> How would the problem be different if $\dot{m}_{\text{out}}$ turned out to be bigger when $\dot{m}_{\text{in}}$ and $\dot{m}_{\text{out}}$ were converted into the same units?

Step 5 *Solve material balance*

Since $\dot{m}_{in}$ and $\dot{m}_{out}$ are both constant, the material balance is readily solved by separation of variables:

$$\frac{dM}{dt} = \dot{m}_{in} - \dot{m}_{out} \tag{3.10}$$

$$\int_{M=M_0}^{M=M} dM = \int_{t=0}^{t=t} (\dot{m}_{in} - \dot{m}_{out})\, dt$$

$$M - M_0 = (\dot{m}_{in} - \dot{m}_{out})t$$

Substituting in known values:

$$M - 2\ \text{kg} = \left(0.0167 - 0.01\,\frac{\text{kg}}{\text{s}}\right)t \tag{3.11}$$

When $M = 4$ kg, $t = 300$ sec.

The next example involves using multiple material balances in conjunction. It also illustrates a common feature of thermodynamics problems in that it describes a process (the growth of a tree) in which the final state (the way the tree looks right now) and the initial state (when there was no tree at all) are well understood, but the *duration* of the process is unknown.

EXAMPLE 3-3	GROWTH OF A TREE

This problem ignores the leaves, branches and roots of the tree, which are also made of hydrocarbons, so the real CO_2 required to grow a 9-m-tall tree is even larger than this problem illustrates.

An oak tree is 9.0 m tall and 0.6 m in diameter (Figure 3-5). How many pounds of CO_2 and H_2O were consumed to form the tree's trunk? How many pounds of O_2 gas were released? Assume the density of oak wood is 720.83 kg/m³ and that the trunk is pure cellulose ($C_6H_{10}O_5$) and is perfectly cylindrical in shape.

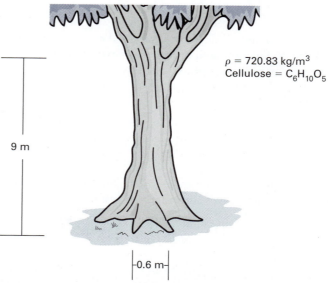

$\rho = 720.83$ kg/m³
Cellulose $= C_6H_{10}O_5$

9 m

0.6 m

FIGURE 3-5 The tree described in Example 3-3.

SOLUTION:

Step 1 *Define the system and process*
Define the tree trunk as the system, and define the process to encompass the entire growth of the tree.

Step 2 *Write overall material balance*
The tree takes in water and carbon dioxide and generates oxygen. Thus, the overall mass balance is

$$\frac{dM}{dt} = \dot{m}_{CO_2} + \dot{m}_{H_2O} - \dot{m}_{O_2} \qquad (3.12)$$

Initially there is no tree, so $M = 0$, and the final mass can be determined from the given information, but this still leaves three unknowns on the right hand side and only one equation.

Step 3 *Recognize additional material balances*
We can write a balance equation analogous to Equation 3.12 not only for total mass, but for the quantity of any element. Instead of M for mass we will use N for moles.

$$\frac{dN}{dt} = \sum_{j=1}^{j=J} \dot{n}_{j,in} - \sum_{k=1}^{k=K} \dot{n}_{k,out} \qquad (3.13)$$

Here we will write material balances for carbon and hydrogen atoms.

Step 4 *Apply simplifying assumptions*
We assume that ALL of the carbon, hydrogen, and oxygen for the tree trunk came from CO_2 and H_2O; no other compounds. This may not be literally true, but it's a logical approximation and one that is implied by the wording of the question. The balance equations are

$$\frac{dN_H}{dt} = 2\dot{n}_{H_2O} \qquad (3.14)$$

$$\frac{dN_C}{dt} = \dot{n}_{CO_2} \qquad (3.15)$$

In which N_H and N_C represent the numbers of moles of hydrogen and carbon atoms in the tree, and $\dot{n}_{H_2O}$ and $\dot{n}_{CO_2}$ are the molar flow rates at which water and carbon dioxide are added to the system.

Step 5 *Examine the degrees of freedom*
We're asked to solve for three unknowns: water in, carbon dioxide in, and oxygen out. Conceptually, we need three independent equations to solve for three unknowns. We now have three equations, but they contain additional variables that must be computed: M, N_H, N_C, and t.

Step 6 *Calculate quantities describing final state of system*
The volume of a cylinder is $\pi r^2 h$, so the FINAL mass (M) of the tree is

$$M = \rho V = \rho\pi r^2 h = \left(720.83\ \frac{kg}{m^3}\right)(3.1416)\,(0.3m)^2\,(9m) = 1834.3\ kg \qquad (3.16)$$

We're told cellulose has the empirical structure $C_6H_{10}O_5$, which means a molecular mass of 162. Thus,

$$N_{cellulose} = \frac{M}{MW} = \frac{1834.3\ kg}{162\ \dfrac{kg}{kmol}} = 11.32\ kmol \qquad (3.17)$$

Recall that "the process" as we defined it began when there was no tree trunk at all.

FOOD FOR THOUGHT 3-8

If you plotted the rate of tree growth with respect to time, what would you expect it to look like?

This definition of m can be applied even if (as in this example) we have no quantitative information on what t_{final} and $\dot{m}$ are.

PITFALL PREVENTION ⚠️

It's easy to confuse values of mass and mole in a complex problem. In our nomenclature, capital M or N refers to mass or moles of the system, and lowercase m or n refers to mass or moles entering or leaving the system.

Thus, at the END of the tree's growth, the moles of carbon and hydrogen are

$$N_C = 6N_{cellulose} = 6\,(11.32 \text{ kmol}) = 67.92 \text{ kmol} \tag{3.18}$$

$$N_H = 10N_{cellulose} = 113.2 \text{ kmol} \tag{3.19}$$

Thus, we know the exact composition of the tree at the end of the process, and at the beginning M, N_C, and N_H are all zero. But how do we handle the time-dependence of the material balances?

Step 7 *Integration of differential equations*
We have no information regarding how long the growth took or the *rate* at which water and carbon dioxide were added. In such a case, time can be eliminated from the material balance equations by integrating with respect to dt. Multiplying both sides of Equation 3.12 by dt gives

$$dM = (\dot{m}_{CO_2} + \dot{m}_{H_2O} - \dot{m}_{O_2})dt \tag{3.20}$$

The rate at which mass is added ($\dot{m}$) is not necessarily constant with respect to time. However, if $\dot{m}$ is integrated with respect to time from the beginning of the process to the end, the result is the total mass added (or removed). This quantity m is

$$m = \int_{t=0}^{t=t_{final}} \dot{m}\,dt \tag{3.21}$$

If we apply this definition to the CO_2 and H_2O added, and the O_2 removed, Equation 3.20 integrates to

$$M_{final} - M_{initial} = m_{CO_2} + m_{H_2O} - m_{O_2} \tag{3.22}$$

Analogous to Equation 3.21, we define n as the total number of moles added (or removed) to the system over the entire process:

$$n = \int_{t=0}^{t=t_{final}} \dot{n}\,dt \tag{3.23}$$

And integrating Equations 3.14 and 3.15 with respect to dt, just as we did with Equation 3.22, gives

$$N_{H,final} - N_{H,initial} = 2n_{H_2O} \tag{3.24}$$

$$N_{C,final} - N_{C,initial} = n_{CO_2} \tag{3.25}$$

Step 8 *Plugging in known values*
Inserting values from step 5 into the integrated equations from step 6 gives

$$n_{H_2O} = \frac{113.2 \text{ kmol}}{2} = 56.6 \text{ kmol} \tag{3.26}$$

$$n_{CO_2} = 67.92 \text{ kmol} \tag{3.27}$$

Converting to mass gives

The growth of this tree's trunk took approximately 3000 kilograms of carbon dioxide out of the atmosphere.

$$m_{H_2O} = (56.6 \text{ kmol})\left(18\,\frac{\text{kg}}{\text{kmol}}\right) = \mathbf{1018.8 \text{ kg}} \tag{3.28}$$

$$m_{CO_2} = (67.92 \text{ kmol})\left(44\,\frac{\text{kg}}{\text{kmol}}\right) = \mathbf{2988.5 \text{ kg}} \tag{3.29}$$

Substituting these values into Equation 3.22 gives

$$1834.3 \text{ kg} - 0 = 1018.8 \text{ kg} + 2988.5 \text{ kg} - m_{O_2} \tag{3.30}$$

$$m_{O_2} = \mathbf{2173\,kg}$$

Example 3-3 illustrates some ideas that will be recurring themes in the study of thermodynamics. First, while the most general form of the material balance is the time-dependent Equation 3.1:

$$\frac{dM}{dt} = \sum_{j=1}^{j=J} \dot{m}_{j,\text{in}} - \sum_{k=1}^{k=K} \dot{m}_{k,\text{out}}$$

There are many examples, such as the growth of the tree, in which no information is available about the duration or rate of the process. Furthermore, no such information is needed: We know the initial state (no tree) and the final state (the current tree), and we can answer the given questions without knowing how old the tree is or what the tree looked like at any particular intermediate time.

For such examples, where no information on time is available or needed, we can use *a time-independent mass balance*:

$$M_{\text{final}} - M_{\text{initial}} = \sum_{j=1}^{j=J} m_{j,\text{in}} - \sum_{k=1}^{k=K} m_{k,\text{out}} \tag{3.31}$$

where *m* represents the total mass added from, or removed by, one stream throughout the entire process, and M_{initial} and M_{final} represent the masses of the system at the beginning and end of the process.

Equation 3.31 can be derived from Equation 3-1 by multiplying through by *dt* and integrating, as in step 6 of Example 3-3.

FOOD FOR THOUGHT 3-9

To model the rocket process illustrated in Figure 3-1, would you use Equation 3.1 or Equation 3.31?

Another principle illustrated in Examples 3.1 through 3.3 is that we can write material balances on total mass, and also on the amount of any element. We can also write material balances for compounds (e.g., a "water balance" rather than a "hydrogen balance"), but the amount of a compound is not always conserved quantity. A material balance for water, or any other compound, must include a term accounting for generation and/or consumption, at least for cases where chemical reactions occur.

Chapters 1 through 8 of this book focus on applications in which the system contains only a pure compound, so in these chapters we won't typically have need of material balance equations beyond the "total mass" balance. Mole balances on specific chemical species are of particular importance in Chapters 14 and 15, which cover chemical reactions.

3.3 Mathematical Expression of the First Law of Thermodynamics

As indicated in Section 1.4, the first law of thermodynamics states that energy can be converted from one form into another, but cannot be created or destroyed. This fundamental principle is applicable to an enormous range of engineering problems. "How much work (or heat) would it take to do this job?" is one example of a straightforward, practical type of engineering question that can be answered using the principle of conservation of energy.

In Chapter 1, the first law was invoked in the solution of Example 1.6, which examined an object in free fall. In that example, it was assumed that as the object fell, potential energy was converted into kinetic energy, and no other forms of energy

where

- W_{EC} is the total expansion/contraction work added throughout the process.
- W_s is the total shaft work added throughout the process.
- Q is the total heat added throughout the process.
- $m_{j,in}$ and $m_{k,out}$ are individual quantities of mass added to and removed from the process, respectively, and the summations are carried out over all such quantities.

Equations 3.34 and 3.35 are both completely general, and therefore valid for any system. Which one is more convenient to use for a particular problem depends on whether or not time shows up explicitly in either the known information (e.g., "Heat is added at a rate of 200 J/s") or in the questions to be answered (e.g., "How long will it take for the water to reach boiling temperature?")

The next two sections are devoted to applying the energy balance—sometimes in combination with the mass balance—to a variety of examples. These examples are also used to illustrate some of the key engineering unit operations that are of interest throughout the book.

3.4 Applications of the Generalized Energy Balance Equation

The first example shows a nozzle, a device that is used to accelerate a flowing fluid to high velocity. A nozzle has a tapered neck, as shown in Figure 3-6, and the decreasing diameter creates a large pressure drop and forces a high velocity. Typically, no energy is transferred to the fluid from external sources.

EXAMPLE 3-4	**NOZZLE**

The term "adiabatic" was first defined in Section 1.3 along with several other essential definitions used to describe thermodynamic systems and processes.

If steam flows through an adiabatic nozzle at steady state, entering the nozzle with a pressure of 0.5 MPa and a temperature of 400°C and exiting at 0.1 MPa and 350°C (Figure 3-6), what is its exiting velocity?

T_{in} = 400°C
P_{in} = 0.5 MPa

T_{out} = 350°C
P_{out} = 0.1 MPa

FIGURE 3-6 An adiabatic nozzle at steady state described in Example 3-4.

FOOD FOR THOUGHT 3-11

This example is in the energy balance section, so of course there's an energy balance. But suppose your boss asked you to calculate the velocity coming out of a nozzle. How would you recognize this as an "energy balance" problem?

SOLUTION:

Step 1 *Define the system*

Defining the nozzle as the system allows us to make use of the described properties of the nozzle (e.g., it's adiabatic, it operates at steady state).

Step 2 *Apply and simplify the energy balance*

At steady state, the accumulation term is 0, and the time-independent energy balance simplifies to

$$0 = \sum_{j=1}^{j=J} \left\{ m_{j,in} \left(\hat{H}_j + \frac{v_j^2}{2} + gh_j \right) \right\} - \sum_{k=1}^{k=K} \left\{ m_{k,out} \left(\hat{H}_k + \frac{v_k^2}{2} + gh_k \right) \right\} + W_{EC} + W_s + Q \quad (3.36)$$

Examining the right-hand side of the equation, note:

- There is only one stream entering (m_{in}) and only one stream leaving (m_{out}).
- Because this is a steady-state process with only one entering stream and only one exiting stream, the material balance (Equation 3.31) simplifies to $m_{in} = m_{out}$.
- $Q = 0$ since the problem statement says the nozzle is adiabatic.
- $W_{EC} = 0$ because the boundaries of the system are not moving.
- $W_S = 0$ because the system contains no moving parts or any mechanism for shaft work.

Thus, the energy balance further simplifies to:

$$0 = m_{in}\left(\hat{H}_{in} + \frac{v_{in}^2}{2} + gh_{in}\right) - m_{out}\left(\hat{H}_{out} + \frac{v_{out}^2}{2} + gh_{out}\right) \tag{3.37}$$

Because $m_{in} = m_{out}$, we can simply divide both sides by this mass:

$$0 = \left(\hat{H}_{in} + \frac{v_{in}^2}{2} + gh_{in}\right) - \left(\hat{H}_{out} + \frac{v_{out}^2}{2} + gh_{out}\right) \tag{3.38}$$

According to Figure 3-6 the nozzle is oriented horizontally, so there is no difference in height between the entering and leaving streams. If $h_{in} = h_{out}$, the potential energy terms cancel each other out.

$$0 = \left(\hat{H}_{in} + \frac{v_{in}^2}{2}\right) - \left(\hat{H}_{out} + \frac{v_{out}^2}{2}\right) \tag{3.39}$$

Step 3 *Simplifying assumption*
The exiting velocity (v_{out}) is what we are trying to determine. There is no information in the problem statement that would enable us to determine the inlet velocity (v_{in}), so we will *assume it is small enough that the entering kinetic energy is negligible.* Thus,

$$0 = (\hat{H}_{in}) - \left(\hat{H}_{out} + \frac{v_{out}^2}{2}\right) \tag{3.40}$$

which can be rearranged as

$$\frac{v_{out}^2}{2} = \hat{H}_{in} - \hat{H}_{out} \tag{3.41}$$

This is a typical energy balance for nozzles. The result is somewhat intuitive in that the purpose of the nozzle is to accelerate the fluid to higher velocity, and Equation 3.41 shows that a portion of the entering fluid's enthalpy is converted into kinetic energy.

Step 4 *Plug in numerical values*
The temperatures and pressures of the entering and leaving streams are known so the enthalpies can be found from the superheated steam tables (Appendix A-3) as

$$\hat{H}_{in}(T = 400°C, P = 0.5 \text{ MPa}) = 3272.3 \text{ kJ/kg}$$

$$\hat{H}_{out}(T = 350°C, P = 0.1 \text{ MPa}) = 3175.8 \text{ kJ/kg}$$

Inserting these into Equation 3.41 and performing the necessary unit conversions gives

$$\frac{v_{out}^2}{2} = \left(3272.3 - 3175.8\frac{\text{kJ}}{\text{kg}}\right)\left(1000\frac{\text{J}}{\text{kJ}}\right)\left(1\frac{\text{N}\cdot\text{m}}{\text{J}}\right)\left(\frac{\frac{\text{kg}\cdot\text{m}}{\text{s}^2}}{1\,\text{N}}\right) \tag{3.42}$$

$$v_{out} = \mathbf{439.5}\,\frac{\mathbf{m}}{\mathbf{s}}$$

FOOD FOR THOUGHT 3-12

Would it have been ok to use the time-dependent energy balance instead?

FOOD FOR THOUGHT 3-13

If the nozzle were oriented vertically, would potential energy be significant?

Recall from Section 1.2.3 that a nozzle is an integral component of a turbine. This example shows internal energy of steam being converted to kinetic energy.

439 meters per second is 1680 kilometers per hour.

The speed of sound moving through air at ambient temperature and pressure is about 343 m/s, so the velocity of 439.5 m/s obtained in Example 3-4 is a high velocity, but not physically unrealistic. The next example considers velocities that are more commonly found in chemical process equipment.

EXAMPLE 3-5	STEADY-STATE BOILER

In this process, the exiting steam is superheated; it is at a temperature above its boiling temperature.

A boiler operates at steady state. The entering water is saturated liquid at $P = 0.5$ MPa and has a flow rate of 2.78 kg/s. The exiting steam is also at $P = 0.5$ MPa and has $T = 400°C$. The pipe entering the boiler is 0.08 m in diameter, and the pipe leaving the boiler is 0.3 m in diameter (Figure 3-7). What is the rate at which heat is added in the boiler?

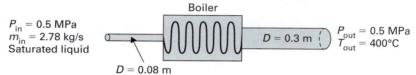

$P_{in} = 0.5$ MPa
$m_{in} = 2.78$ kg/s
Saturated liquid

$D = 0.3$ m

$P_{out} = 0.5$ MPa
$T_{out} = 400°C$

$D = 0.08$ m

FIGURE 3-7 A boiler operating at steady state, as described in Figure 3-5.

FOOD FOR THOUGHT 3-14

This boiler is producing steam at the same conditions that entered the nozzle in Example 3-4. Combined, the boiler and the nozzle effectively convert heat into kinetic energy. How is this like what goes on in a rocket? How is it different?

The boiler is an example of an open system.

SOLUTION:

Step 1 *Define the system*
Define the contents of the boiler as the system.

Step 2 *Apply and simplify the energy balance*
We use the time-dependent form of the energy balance, because we are trying to determine the *rate* at which heat is added. Because the system operates at steady state, the accumulation term of the energy balance is zero. We can neglect all work terms, because the boundaries of the system do not move, and there isn't normally any shaft work in a boiler (or any other conventional heat exchanger). Finally, there is only one stream entering the system and only one stream leaving. Thus, the energy balance simplifies to

$$0 = \dot{m}_{in}\left(\hat{H}_{in} + \frac{v_{in}^2}{2} + gh_{in}\right) - \dot{m}_{out}\left(\hat{H}_{out} + \frac{v_{out}^2}{2} + gh_{out}\right) + \dot{Q} \quad (3.43)$$

Because this is a steady-state process, the mass flow rates entering ($\dot{m}_{in}$) and leaving ($\dot{m}_{out}$) the process have to be identical; if they were not, the mass would be accumulating in the system. Going forward, we simply set both equal to $\dot{m}$.

The diagram implies that the boiler is oriented horizontally. As noted in Example 3-4, if the entering and exiting streams are at the same height ($h_{in} = h_{out}$) and their mass flow rates are identical, the potential energy terms cancel each other out. (See Example 3-6 for a system in which potential energy is significant.) The energy balance becomes

$$0 = \dot{m}\left(\hat{H}_{in} + \frac{v_{in}^2}{2} - \hat{H}_{out} - \frac{v_{out}^2}{2}\right) + \dot{Q} \quad (3.44)$$

Step 3 *Obtain data*
From the steam tables, $\hat{H} = 640.1$ kJ/kg for saturated liquid water at 0.5 MPa and $\hat{H} = 3272.3$ kJ/kg for steam at 0.5 MPa and 400°C.

Step 4 *Calculate unknown velocities*
The velocities of the entering and exiting streams can also be found, because the sizes of the pipes are known. First, the mass flow rate can be related to the volumetric flow rate

through the specific volume, which is also available in the steam tables:

$$\dot{V}_{in} = \dot{m}\,\hat{V}_{in} = \left(2.78\,\frac{kg}{s}\right)\left(0.001093\,\frac{m^3}{kg}\right) = 3.036 \times 10^{-3}\,\frac{m^3}{s} \tag{3.45}$$

$$\dot{V}_{out} = \dot{m}\,\hat{V}_{out} = \left(2.78\,\frac{kg}{s}\right)\left(0.6173\,\frac{m^3}{kg}\right) = 1.715\,\frac{m^3}{s} \tag{3.46}$$

The cross-sectional areas (A) of the pipes are

$$A_{in} = \pi r^2 = \pi\left(\frac{0.08\,m}{2}\right)^2 = 50.27 \times 10^{-4}\,m^2 \tag{3.47}$$

$$A_{out} = \pi r^2 = \pi\left(\frac{0.3\,m}{2}\right)^2 = 706.9 \times 10^{-4}\,m^2 \tag{3.48}$$

And the linear velocity (v) is equal to the volumetric flow rate divided by the cross-sectional area:

> Typical vapor velocities are much larger than typical liquid velocities.

$$v_{in} = \frac{\dot{V}_{in}}{A_{in}} = \frac{3.036 \times 10^{-3}\,\dfrac{m^3}{s}}{50.27 \times 10^{-4}\,m^2} = 0.604\,\frac{m}{s} \tag{3.49}$$

$$v_{out} = \frac{\dot{V}_{out}}{A_{out}} = \frac{1.715\,\dfrac{m^3}{s}}{706.9 \times 10^{-4}\,m^2} = 24.26\,\frac{m}{s} \tag{3.50}$$

Step 5 *Solve energy balance*
The energy balance expression can be rearranged as

$$0 = \dot{m}\left[(\hat{H}_{in} - \hat{H}_{out}) + \left(\frac{v_{in}^2}{2} - \frac{v_{out}^2}{2}\right)\right] + \dot{Q} \tag{3.51}$$

Inserting known values gives

$$0 = \left(2.78\,\frac{kg}{s}\right)\left[\left(640.1\,\frac{kJ}{kg} - 3272.3\,\frac{kJ}{kg}\right) + \left(\frac{\left(0.604\,\dfrac{m}{s}\right)^2}{2} - \frac{\left(24.26\,\dfrac{m}{s}\right)^2}{2}\right)\right.$$

$$\left.\times \left(\frac{1\,N}{\dfrac{kg \cdot m}{s^2}}\right)\left(\frac{1\,J}{N \cdot m}\right)\left(\frac{1\,kJ}{1000\,J}\right)\right] + \dot{Q}$$

$$0 = \left(2.78\,\frac{kg}{s}\right)\left[\left(-2632.2\,\frac{kJ}{kg}\right) + \left(-0.3\,\frac{kJ}{kg}\right)\right] + \dot{Q}$$

$$\dot{Q} = 7318.63\,\frac{kJ}{s} \tag{3.52}$$

> Even though the vapor velocity is orders of magnitude larger than the liquid velocity, the exiting stream kinetic energy is only 0.3 kJ/kg larger than that of the incoming stream. By contrast the change in specific enthalpy is over 2000 kJ/kg.

In Example 3-5, two distinct phenomena were identified and quantified:

1. The increase in enthalpy from the entering stream to the exiting stream.
2. The increase in kinetic energy from the entering stream to the exiting stream.

Physically, heat (Q) is added to the system, and this accounts for the higher energy of the exiting stream. However, the kinetic energy term was insignificant compared to the enthalpy term. This result is common in chemical process equipment. Going forward, when we are examining standard chemical process equipment (e.g., heat exchangers, pumps, turbines, compressors) we will assume kinetic energy of flowing streams is negligible unless we have a specific reason to think otherwise. Example 3-4 provides an example of a scenario in which kinetic energy cannot realistically be neglected: in a nozzle, converting internal energy into kinetic energy is the entire purpose of the process.

The significance of potential energy is examined in the next example.

EXAMPLE 3-6	GEOTHERMALLY HEATED STEAM

An underground reservoir 3 km below the surface contains geothermally heated steam at $T = 523.15$ K and $P = 0.5$ MPa. A vertical shaft allows this steam to flow upward, through the shaft, and to the surface, where it enters a turbine. The effluent leaving the turbine is saturated steam at $P = 0.1$ MPa (Figure 3-8).

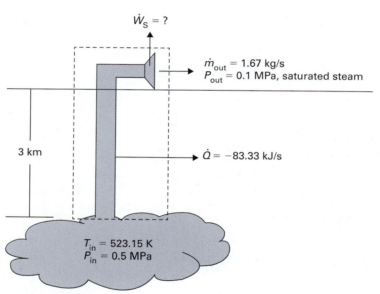

FIGURE 3-8 Schematic of turbine/shaft described in Example 3-6. The dashed line represents the boundary of the system as defined in steps 2 and 3.

Determine the rate at which shaft work is produced by the turbine. Assume

■ The process operates continuously at steady state.

■ The flow rate of steam is 1.67 kg/s.

■ The turbine itself is adiabatic, but the steam loses 83.33 kJ/s of heat to the surroundings as it travels upward through the shaft.

■ The steam velocity is negligible at the turbine exit and the shaft entrance.

SOLUTION:

Step 1 *Define a system*
In many problems, defining the system doesn't seem like much of a decision, but in this problem, there are several possible systems. At first glance, it seems natural to define the turbine as the system, since the question asks how much work the turbine is producing.

However, what happens when we do this? This is a steady-state system, so the energy balance (Equation 3.34) immediately simplifies to

$$0 = \sum_{j=1}^{j=J}\left\{ \dot{m}_{j,\text{in}}\left(\hat{H}_j + \frac{v_j^2}{2} + gh_j \right)\right\} - \sum_{k=1}^{k=K}\left\{ \dot{m}_{k,\text{out}}\left(\hat{H}_k + \frac{v_k^2}{2} + gh_k \right)\right\} + \dot{W}_{\text{S}} + \dot{W}_{\text{EC}} + \dot{Q} \quad (3.53)$$

The turbine is adiabatic ($\dot{Q} = 0$), and the system boundaries are not moving ($\dot{W}_{\text{EC}} = 0$). There is only one stream entering and one leaving, and at steady state, these must have equal flow rates ($\dot{m}_{\text{in}} = \dot{m}_{\text{out}}$). The energy balance further simplifies to

$$0 = \dot{m}\left(\hat{H}_{\text{in}} + \frac{v_{\text{in}}^2}{2} + gh_{\text{in}} \right) - \dot{m}\left(\hat{H}_{\text{out}} + \frac{v_{\text{out}}^2}{2} + gh_{\text{out}} \right) + \dot{W}_{\text{S}} \quad (3.54)$$

The exiting velocity is explicitly stated in the problem statement as negligible, and based on what we saw in Example 3-5, it would be logical to assume the same is true at the turbine entrance. Figure 3-8 suggests the turbine is oriented horizontally, so we infer the entering and exiting streams have equal potential energy ($h_{\text{in}} = h_{\text{out}}$). However, the enthalpy of the stream entering the turbine ($\hat{H}_{\text{in}}$) is unknown, so even if we make these various simplifying assumptions, we cannot solve for the shaft work $\dot{W}_{\text{S}}$.

To determine $\hat{H}_{\text{in}}$, we could do a second energy balance around the vertical shaft, knowing that the steam leaving the top of the shaft is the same as the steam entering the turbine. However, this solution strategy is unnecessarily complicated. While no information is given about the steam that enters the turbine, it is also the case that none is needed. In defining a system, our purpose is to relate what we *want to know* ($\dot{W}_{\text{S}}$ in the turbine) to what we *do know*. The material entering the bottom of the shaft and the material leaving the turbine are both fully described. Consequently, the most efficient solution strategy is to draw our system boundary around *both* the shaft and the turbine, as shown in Figure 3-8. This way, we have complete information about all of the material entering and leaving the system.

Step 2 *Apply and simplify the energy balance*
We use the time-dependent form of the energy balance because the given information is in the form of rates. Since this is a steady-state process, the accumulation term is zero:

$$0 = \sum_{j=1}^{j=J}\left\{ \dot{m}_{j,\text{in}}\left(\hat{H}_j + \frac{v_j^2}{2} + gh_j \right)\right\} - \sum_{k=1}^{k=K}\left\{ \dot{m}_{k,\text{out}}\left(\hat{H}_k + \frac{v_k^2}{2} + gh_k \right)\right\} + \dot{W}_{\text{S}} + \dot{W}_{\text{EC}} + \dot{Q} \quad (3.55)$$

There is only one stream entering and leaving, and both have negligible velocity and identical flow rates ($\dot{m}$) of 1.67 kg/s. Finally, there is no expansion/contraction work; the boundaries of the system are not moving. With these simplifications, the energy balance becomes

$$0 = \dot{m}(\hat{H}_{\text{in}} + gh_{\text{in}}) - \dot{m}(\hat{H}_{\text{out}} + gh_{\text{out}}) + \dot{W}_{\text{S}} + \dot{Q} \quad (3.56)$$

Step 3 *Solve the energy balance*
The enthalpies of both streams are available in the steam tables. $\dot{Q}$ is given and $\dot{W}_{\text{S}}$ is what we are trying to determine. The only remaining unknowns are the values of h. We define the surface of the Earth as $h = 0$, so the steam exiting the turbine is at $h = 0$ and the steam entering the shaft is at $h = -3$ km.

Inserting all known values produces

$$0 = \left(1.67\frac{\text{kg}}{\text{s}} \right)\left[2961.0\frac{\text{kJ}}{\text{kg}} + \frac{\left(9.8\,\dfrac{\text{m}}{\text{s}^2}\right)(-3000\,\text{m})\left(\dfrac{1\,\text{J}}{\text{N}\cdot\text{m}}\right)\left(\dfrac{1\,\text{kJ}}{1000\,\text{J}}\right)}{1\dfrac{\text{kg}\cdot\dfrac{\text{m}}{\text{s}^2}}{\text{N}}} \right] \quad (3.57)$$

This is the energy balance when the *turbine* is defined as the system.

PITFALL PREVENTION

Suppose we decided to solve the problem as described in this paragraph, but the kinetic energy of the steam entering the turbine in reality wasn't negligible. Would our final answer be wrong?

The velocity of the steam at some places inside the turbine is undoubtedly quite high, but that material never crosses the boundaries of the system.

This is the energy balance when the system is defined as in Figure 3-8.

Defining the reservoir to be at $h = 0$ and the turbine to be at $h = +3$ km would have been equally logical and produced the same answer.

$$- \left(1.67\frac{\text{kg}}{\text{s}}\right)\left[2675.0\frac{\text{kJ}}{\text{kg}} + 0\right] + \dot{W}_s - 83.33\frac{\text{kJ}}{\text{s}}$$

$$0 = \left(1.67\frac{\text{kg}}{\text{s}}\right)\left[2961.0\frac{\text{kJ}}{\text{kg}} + \left(-29.4\frac{\text{kJ}}{\text{kg}}\right)\right] - \left(1.67\frac{\text{kg}}{\text{s}}\right)\left[2675.0\frac{\text{kJ}}{\text{kg}} + 0\right] + \dot{W}_s - 83.33\frac{\text{kJ}}{\text{s}}$$

$$-345.2\,\frac{\text{kJ}}{\text{s}} = \dot{W}_s$$

This example is somewhat contrived in that it makes harvesting geothermal energy look much simpler than it really is, but it illustrates a strategy for defining a system: choose a system that will allow you to use the known information and relate it to the information needed.

Example 3-6 illustrates a worthwhile point about potential energy. The process had one material stream entering and one stream leaving, and the heights of these two streams differed by 3 km. With $\Delta h = 3$ km, the potential energy term of the energy balance accounted for ~30 kJ of energy per kilogram of steam. However, it is not common for a chemical production process to involve heights of this magnitude. If the difference in height between the entering and exiting streams had been 10 m (i.e., slightly more than the height of a typical three story house) instead of 3 km, the potential energy term would have been ~0.1 kJ/kg, which can typically be considered negligible compared to $\hat{H}$ terms that are on the order of hundreds or thousands of kJ/kg.

Consequently, going forward, we will treat the potential energy of entering and exiting streams as negligible unless our system involves changes or differences in height that are considerably larger than 10 m.

Examples 3-4 through 3-6 were steady-state, open-system processes. The next example illustrates a closed system that is not at steady state.

| **EXAMPLE 3-7** | **COOLING OF A TWO-PHASE MIXTURE** |

"Rigid" is a common and useful key word. You know that the system is isochoric if the boundaries of the system cannot move.

Conceptually, you could define "the liquid" or "the vapor" as the system, but if you did that, you'd have to consider it an open system, because material could evaporate or condense during this process.

A rigid storage tank has a total volume of 5.00 m³ and is sealed throughout the process. Initially, the storage tank contains 0.50 m³ of saturated liquid water at $P = 0.5$ MPa, while the rest of the tank is filled with saturated steam at the same pressure. The tank gradually cools until the liquid–vapor system is at $P = 0.4$ MPa (Figure 3-9). How much heat was lost to the surroundings during this process, and what are the final volumes of liquid and vapor?

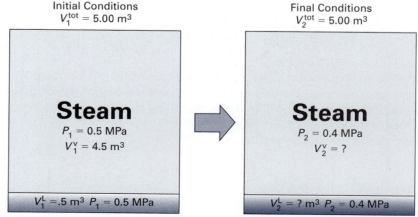

FIGURE 3-9 Schematic of initial and final conditions of tank described in Example 3-7.

SOLUTION:

Step 1 *Define a system*

Define the contents of the tank (both liquid and vapor) as the system. The tank is sealed, so it is a closed system. While the total mass of the system must remain constant, the masses of liquid and vapor phases can change.

Step 2 *Apply and simplify the energy balance*

We use the time-independent energy balance since no information regarding duration of the process is known. This is not a steady-state process; we know the pressure is changing and must expect that other properties (like U) will also be changing. There are, however, several immediate simplifications to the energy balance:

- There is no mass entering or leaving the system (it is a closed system).
- Potential and kinetic energy portions of the "accumulation" term can be ignored; the tank is not moving.
- The tank is rigid, so there can be no work of expansion or contraction.
- There is no mention of a mixer or any other mechanism for shaft work in the tank.

Thus, the energy balance simplifies to

$$\Delta(M\hat{U}) = Q \tag{3.58}$$

Or, equivalently:

$$U_2 - U_1 = Q \tag{3.59}$$

Heat is the only form in which energy is added to or removed from the system. We expect Q to be negative, since it is stated that the tank "cools"; this implies energy is leaving the system.

Step 3 *Quantify initial liquid and vapor masses*

Step 2 shows we cannot find Q without determining the initial (U_1) and final (U_2) values of internal energy, which requires determining how much liquid and vapor are present at the beginning and end of the process.

According to the steam tables, the specific volumes of saturated liquid and vapor at $P = 0.5$ MPa are $\hat{V}_L = 0.001093$ m³/kg and $\hat{V}_V = 0.3748$ m³/kg. Consequently, the initial masses of liquid and vapor can be found as

$$M_1^L = \frac{V_1^L}{\hat{V}_1^L} = \frac{0.50\,\text{m}^3}{0.001093\dfrac{\text{m}^3}{\text{kg}}} = 457.5\,\text{kg} \tag{3.60}$$

$$M_1^V = \frac{V_1^V}{\hat{V}_1^V} = \frac{4.5\,\text{m}^3}{0.3748\dfrac{\text{m}^3}{\text{kg}}} = 12.0\,\text{kg} \tag{3.61}$$

Thus, the total mass is $457.5 + 12.0 = 469.5$ kg.

Step 4 *Quantify final liquid and vapor masses*

The amounts of liquid and vapor at the end of the process (M_2^L and M_2^V) are unknown, so we need two equations to determine them. One equation comes from the fact that this is a closed system; total mass is constant:

$$M_2^L + M_2^V = 469.5\,\text{kg} \tag{3.62}$$

The other equation comes from the fact that the container is rigid, so the total volume must remain 5.0 m³.

$$M_2^L \hat{V}_2^L + M_2^V \hat{V}_2^V = 5.0\,\text{m}^3 \tag{3.63}$$

$U = M\hat{U}$ by the definition of specific internal energy

Recall the Gibbs phase rule: $F = C - \pi + 2$. With one component and two phases, there is only one degree of freedom. When pressure is set ($P = 0.5$ MPa), there are no degrees of freedom remaining; any other intensive property of the system (e.g., specific volume) must have a unique value.

Saturated liquid and vapor volumes at 0.4 MPa are again available in the steam tables. Thus,

$$M_2^L \left(0.001084 \frac{m^3}{kg}\right) + M_2^V \left(0.4624 \frac{m^3}{kg}\right) = 5.0\,m^3 \tag{3.64}$$

Solving these two equations simultaneously gives M_2^L = **459.8 kg** and M_2^V = **9.7 kg**. The final volume of liquid is 0.498 m³. Recall that the original volume was given with two significant figures, 0.50 m³, so to this level of accuracy, the liquid and vapor volumes are unchanged. This result is plausible, because approximately 20% of the original vapor has condensed, but the remaining vapor is significantly less dense than the original vapor due to the decrease in pressure.

Step 5 *Solve energy balance for Q*
Returning to the energy balance, we know $\Delta U = Q$ and that the total internal energy at either the beginning or end of the process is the sum of the liquid and vapor internal energy. Thus,

$$Q = U_2 - U_1 = (M_2^L \hat{U}_2^L + M_2^V \hat{U}_2^V) - (M_1^L \hat{U}_1^L + M_1^V \hat{U}_1^V) \tag{3.65}$$

The masses have all been determined and the internal energy of saturated liquid and vapor at P = 0.5 MPa and P = 0.4 MPa can be obtained from the steam tables. Thus,

$$Q = \left[(459.8\,kg)\left(604.2\frac{kJ}{kg}\right) + (9.7\,kg)\left(2553.1\frac{kJ}{kg}\right)\right]$$
$$- \left[(457.5\,kg)\left(639.5\frac{kJ}{kg}\right) + (12.0\,kg)\left(2560.7\frac{kJ}{kg}\right)\right] \tag{3.66}$$
$$Q = -20,723\,kJ$$

We have now examined examples of open and closed systems, as well as examples of steady-state and unsteady-state processes. This section closes with an example where the shaft work required in a pump is determined. The following example combines the results of an open system, steady-state energy balance with the results of a closed system, unsteady-state energy balance.

EXAMPLE 3-8	**SHAFT WORK IN A WATER PUMP**

An adiabatic pump operates at steady state. Water enters as saturated liquid at P = 0.02 MPa and is compressed to P = 1 MPa (Figure 3-10). What is the rate at which shaft work is added in the pump, per kilogram of entering water?

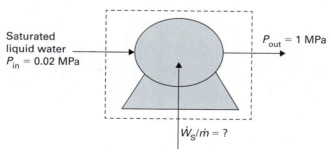

Saturated liquid water
P_{in} = 0.02 MPa

P_{out} = 1 MPa

$\dot{W}_S/\dot{m}$ = ?

FIGURE 3-10 An adiabatic pump operating at steady state, described in Example 3-8.

SOLUTION:

Step 1 *Define a system*

The system is defined as a pump with the system boundaries shown in Figure 3-10.

Step 2 *Apply and simplify the energy balance*

Because the pump operates at a steady state, the left-hand side of the energy balance is 0. Thus,

$$0 = \sum_{j=1}^{j=J}\left\{\dot{m}_{j,\text{in}}\left(\hat{H}_j + \frac{v_j^2}{2} + gh_j\right)\right\} - \sum_{k=1}^{k=K}\left\{\dot{m}_{k,\text{out}}\left(\hat{H}_k + \frac{v_k^2}{2} + gh_k\right)\right\} + \dot{W}_S + \dot{W}_{\text{EC}} + \dot{Q}$$

There is only one stream entering and one stream leaving, the boundaries of the system are not moving ($\dot{W}_{\text{EC}} = 0$), and the pump is described as adiabatic ($\dot{Q} = 0$). Neglecting potential and kinetic energy leaves the energy balance:

$$0 = \dot{m}_{\text{in}}\hat{H}_{\text{in}} - \dot{m}_{\text{out}}\hat{H}_{\text{out}} + \dot{W}_S \qquad (3.67)$$

The mass flow rates entering and leaving are equal, so the "in" and "out" subscripts can be eliminated from the mass flow rates. It is convenient to rearrange the equation so that it is explicit in "work produced per kilogram of water," since that is what we are trying to determine. Thus,

$$\frac{\dot{W}_S}{\dot{m}} = \hat{H}_{\text{out}} - \hat{H}_{\text{in}} \qquad (3.68)$$

At this point, however, there is no clear way to progress with Equation 3.68. We know exactly what is entering the pump—saturated liquid at $P = 0.02$ MPa—and can look up $\hat{H}_{\text{in}}$ in the steam tables. We know the exiting fluid is at $P = 1$ MPa, but as the Gibbs phase rule states, it takes two intensive properties to fully specify the state of a pure fluid, and we have no second piece of information. We can gain some deeper insight into what's happening in this process through a second energy balance.

Step 3 *Define a second system*

The only plausible systems we could define are the pump, examined in step 2, and the water itself. We can define a small "slug" of liquid as the system, with the process beginning right before the slug enters the pump and ending when the liquid leaves the pump. We will model this "slug" of liquid as a closed system; thus we assume it does not mix with the water surrounding it. This system is illustrated in Figure 3-11.

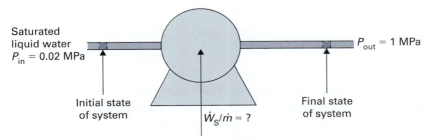

Saturated
liquid water
$P_{\text{in}} = 0.02$ MPa

$P_{\text{out}} = 1$ MPa

Initial state
of system

Final state
of system

$\dot{W}_S/\dot{m} = ?$

FIGURE 3-11 Alternative approach to modeling the pump, where the system is a slug of fluid.

Step 4 *Apply and simplify the energy balance*

Since we don't know how long it takes a slug of liquid to pass through the pump, we apply the time-independent energy balance, which in its most general form is

---

In steps 1 and 2, the system is the pump.

PITFALL PREVENTION

The given information says the liquid entering the pump is saturated, but there is no reason to expect that the liquid exiting the pump is also saturated.

In steps 3 through 6, the system is a small slug of liquid water, and the process is its passage through the pump.

$$\Delta\left\{M\left(\hat{U} + \frac{v^2}{2} + gh\right)\right\} = \sum_{j=1}^{j=J}\left\{m_{j,\text{in}}\left(\hat{H}_j + \frac{v_j^2}{2} + gh_j\right)\right\}$$

$$- \sum_{k=1}^{k=K}\left\{m_{k,\text{out}}\left(\hat{H}_k + \frac{v_k^2}{2} + gh_k\right)\right\} + W_{\text{EC}} + W_{\text{S}} + Q$$

Since we know that at least one property of the system (pressure) is changing, this is not a steady-state process; ΔU on the left-hand side cannot be neglected. But we will neglect potential and kinetic energy for reasons that were explored in Examples 3-5 and 3-6.

On the right-hand side, the terms associated with material flow can be eliminated, since the slug is a closed system. The pump as a whole is adiabatic, so we expect there is also no heat transfer into this small slug of liquid, thus, $Q = 0$. In addition, by this definition of the system, shaft work is zero. This is because:

■ When the pump, and everything inside the pump, is the system, the machinery is working inside the boundaries of the system and transfers energy in the form of shaft work.

■ When a small slug of liquid is the system, there are no moving parts inside the system. The shaft is outside the system, but it acts on the liquid, compressing it. Thus, by this definition of the system, the work done is work of expansion/contraction.

The energy balance thus simplifies to

$$\Delta(M\hat{U}) = W_{\text{EC}} \tag{3.69}$$

The mass of the slug of liquid is constant, because this is a closed system. The specific internal energy is not constant. While we don't know exactly how the pump works, we know the initial and final values of $\hat{U}$ are those of the liquid entering and leaving the pump: the process begins when the fluid enters the pump and ends when the liquid leaves the pump.

$$M(\hat{U}_{\text{out}} - \hat{U}_{\text{in}}) = W_{\text{EC}} \tag{3.70}$$

Step 5 *Relate W_{EC} to intensive properties of liquid*
We were unable to solve the problem in step 2 because we knew only one intensive property (P) of the liquid exiting the pump. To be useful, Equation 3.70 must also be related to intensive properties of the liquid. We learned in Section 1.4.3 that expansion work is equal to $-P\,dV$. Thus,

By definition, $\hat{V} = \dfrac{V}{M}$

$$M(\hat{U}_{\text{out}} - \hat{U}_{\text{in}}) = \int_{V_{\text{in}}}^{V_{\text{out}}} - P\,dV \tag{3.71}$$

Dividing both sides by the system mass M gives

$$(\hat{U}_{\text{out}} - \hat{U}_{\text{in}}) = \int_{\hat{V}_{\text{in}}}^{\hat{V}_{\text{out}}} - P\,d\hat{V} \tag{3.72}$$

At this point, we appear to have a similar barrier to further progress as we had in step 2. The initial state of the water is fully specified and we can look up $\hat{V}_{\text{in}}$ in the steam tables, but we don't have any information about the liquid leaving the pump that would help us resolve the integral. However, the connection between Equations 3.72 and 3.68 becomes more apparent when we use integration by parts.

Step 6 *Integration by parts*
Integration by parts yields

$$(\hat{U}_{\text{out}} - \hat{U}_{\text{in}}) = -P\hat{V}\Big|_{\text{in}}^{\text{out}} + \int_{P_{\text{in}}}^{P_{\text{out}}} \hat{V}\,dP \tag{3.73}$$

$$(\hat{U}_{out} - \hat{U}_{in}) = P_{in}\hat{V}_{in} - P_{out}\hat{V}_{out} + \int_{P_{in}}^{P_{out}} \hat{V}\,dP \tag{3.74}$$

But we can combine the U and PV terms, applying the definition of enthalpy for

$$(\hat{U}_{out} + P_{out}\hat{V}_{out}) - (\hat{U}_{in} + P_{in}\hat{V}_{in}) = \int_{P_{in}}^{P_{out}} \hat{V}\,dP$$

$$\hat{H}_{out} - \hat{H}_{in} = \int_{P_{in}}^{P_{out}} \hat{V}\,dP \tag{3.75}$$

Step 7 *Combine energy balance equations*
Equations 3.68 and 3.75 were derived through two different energy balances on two different systems, but both relate the change in enthalpy experienced by the liquid as it travels through the pump: $\hat{H}_{out} - \hat{H}_{in}$. Combining these expressions gives

$$\frac{\dot{W}_S}{\dot{m}} = \int_{P_{in}}^{P_{out}} \hat{V}\,dP \tag{3.76}$$

Step 8 *Solve for pump work*
There are two key points about Equation 3.76 that make it solvable. First, the integral is with respect to dP rather than $d\hat{V}$, and P is the one property that is known for both the entering and exiting water. Second, the specific volume of a liquid (under most conditions) is reasonably approximated as constant. Applying this gives

$$\frac{\dot{W}_S}{\dot{m}} \approx \hat{V}(P_{out} - P_{in}) \tag{3.77}$$

$$\frac{\dot{W}_S}{\dot{m}} \approx \left(0.001017\frac{m^3}{kg}\right)(1 - 0.02\,MPa)\left(\frac{10^6\,Pa}{MPa}\right)\left(1\frac{N}{m^2}{Pa}\right)\left(\frac{1\,J}{N\cdot m}\right) = \mathbf{996.7}\frac{\mathbf{J}}{\mathbf{kg}}$$

In this example, it only takes about 1 kJ of work to increase the pressure of 1 kg of water by a factor of 50.

A key result from Example 3-8 is Equation 3.76, which relates the shaft work in a pump to the volume and pressure of the fluid flowing through the pump. Multiplying both sides of Equation 3.76 by mass flow rate gives

$$\dot{W}_S = \int_{P=P_{in}}^{P=P_{out}} \dot{V}\,dP \tag{3.78}$$

The derivation of this equation is not specific to a pump; in principle it could be applied to a compressor, a turbine, or any other process involving shaft work. In practice, it is very difficult to apply Equation 3.78 to a process in which the fluid is a vapor (or liquid–vapor system.) A vapor in a turbine or compressor undergoes large changes in both pressure and temperature. One would need a very detailed model of the process in order to relate the volumetric flow rate $\dot{V}$ to pressure P, and without this relationship one cannot integrate Equation 3.78. Turbines and compressors are commonly modelled with strategies introduced in Chapter 4.

Equation 3.78 is quite useful for pumps, however, because in many applications, liquid volumes are reasonably modeled as constant with respect to T and P, as was done in Equation 3.77 in Example 3-8.

3.5 Combining the Energy Balance with Simple Thermodynamic Models

While the physical systems in the examples found in Section 3.4 were varied, the only chemical compound present in any of them was H_2O. Application of the energy balance was therefore simple in one respect: $\hat{U}$, $\hat{H}$, and $\hat{V}$ were known at all of the conditions of interest. In this section, we will examine the application of the energy balance in conjunction with the simple thermodynamic models developed in Section 2-3, which is crucial in situations where physical property data is less extensive or not available at all.

EXAMPLE 3-9	**HELIUM-FILLED BALLOON**

The balloon portion of a dirigible contains 5000 moles of helium. Initially, the helium is at $T = 288$ K and $P = 96.26$ kPa, and is 500 m off the ground (where the atmospheric pressure is also 96.26 kPa). The dirigible flies south at a constant height and a slow velocity, and gradually warms to $T = 298$ K (though the pressure remains constant) (Figure 3-12). How much heat was added to the balloon during the process?

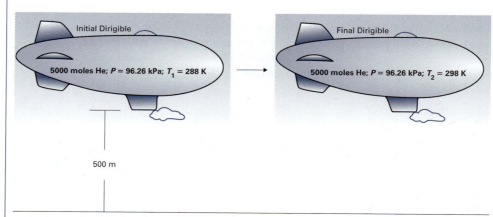

FIGURE 3-12 Schematic of dirigible flight described in Example 3-9.

SOLUTION:

Step 1 *Define the system*
The given information is all in the form of properties of the helium so define the helium as the system.

Step 2 *Apply and simplify the energy balance*
We use the time-independent energy balance because there is no information on how long the flight took.

   This is a closed system; no mass enters or leaves. Thus, $m_{in} = m_{out} = 0$, and the energy balance simplifies to

$$\Delta\left\{M\left(\hat{U} + \frac{v^2}{2} + gh\right)\right\} = W_{EC} + W_s + Q \tag{3.79}$$

We know Q is significant because it is what we are trying to find. One likely error, however, would be crossing out both of the work terms. Because producing work is not the apparent purpose of the process, it would be easy to fall into the trap of assuming that no work occurs. There are no moving parts inside of a balloon, so no shaft work is produced. However, the gas is being heated at constant pressure, so it must expand, and anytime the system is expanding, work is being done on the surroundings (as explained in Section 1.4.3).

FOOD FOR THOUGHT 3-15

Try simplifying the energy balance equation yourself before reading step 2.

PITFALL PREVENTION

The fact that pressure is constant does not indicate the process is at steady state. Steady state means ALL properties of the system are constant with respect to time, and in this case, the temperature of the system is changing.

On the left-hand side, kinetic energy can be neglected, since velocity (v) is described as "slow." The height h is constant at 500 m, so the potential energy relative to the ground is not 0, but it is constant. Thus, the energy balance simplifies to

$$M_2\hat{U}_2 - M_1\hat{U}_1 = W_{EC} + Q \tag{3.80}$$

Because the number of moles of helium is given, it's more convenient to express this on a molar basis:

$$N_2\underline{U}_2 - N_1\underline{U}_1 = W_{EC} + Q \tag{3.81}$$

Thus, if we can evaluate any two of heat, work and molar internal energy, we can compute the third with the energy balance. Note the amount of helium doesn't change, so $N_2 = N_1 = N$.

Step 3 *Find expansion work*
As derived in Section 1.4.3, work of expansion/contraction is given by

$$dW_{EC} = -P\,dV \tag{3.82}$$

where P is the pressure opposing the motion. In this case, the external pressure is from the atmosphere. However, because of the elevation, the "external pressure" is 96.26 kPa, not 1 atm. As the external pressure is constant, Equation 3.82 integrates to

$$W_{EC} = -(96.26\,\text{kPa})(V_2 - V_1) \tag{3.83}$$

Since the gas pressure is below atmospheric, we will assume ideal gas behavior. V_2 and V_1 can be found with the ideal gas law as

$$W_{EC} = -(96.26\,\text{kPa})\left(\frac{NRT_2}{P_2} - \frac{NRT_1}{P_1}\right) \tag{3.84}$$

In this case $P_1 = P_2 = 96.26$ kPa, so the equation simplifies. (The fact that the gas pressure and external pressure are equal is discussed further at the end of the example.)

$$W_{EC} = -NR(T_2 - T_1) \tag{3.85}$$

$$W_{EC} = -(5000\,\text{moles})\left(8.314\frac{\text{J}}{\text{mol} \cdot \text{K}}\right)(298 - 288\,\text{K})\left(\frac{1\,\text{kJ}}{1000\,\text{J}}\right) = -415.7\,\text{kJ}$$

Step 4 *Calculate change in molar internal energy*
In Section 2.3.3, we showed that for an ideal gas

$$d\underline{U} = C_V^*\,dT \tag{3.86}$$

Furthermore, helium is a *monatomic* gas. As discussed in Example 2-4, we can assume $C_V^* = (1.5)R$ for any monatomic ideal gas. Since C_V^* is constant, Equation 3.86 integrates to

$$\underline{U}_2 - \underline{U}_1 = C_V^*(T_2 - T_1) = \frac{3}{2}\left(8.314\frac{\text{J}}{\text{mol} \cdot \text{K}}\right)(298 - 288\,\text{K}) = 124.7\frac{\text{J}}{\text{mol}} \tag{3.87}$$

Step 5 *Solve energy balance for Q*
Inserting the results of steps 3 and 4 into the energy balance gives

$$(5000\,\text{moles})\left(124.7\frac{\text{J}}{\text{mol}}\right)\left(\frac{1\,\text{kJ}}{1000\,\text{J}}\right) = -415.7\,\text{kJ} + Q \tag{3.88}$$

$$\mathbf{Q = 1039.2\ kJ}$$

The "accumulation" term quantifies changes in the energy of the system. If height is constant ($h_2 = h_1$) then there is no change in potential energy.

Equation 3.80 represents a common special case: a simple, closed system with no moving parts and no significant changes in kinetic or potential energy.

$U = M\hat{U} = N\underline{U}$ by the definitions of specific and molar internal energy

Example 2-3 showed that water vapor is reasonably approximated as an ideal gas at pressures up to $P = 101.3$ kPa. The ideal gas approximation is even more realistic for helium than for H_2O at these low pressures, because the noble gases are non-polar.

A possible source of confusion

According to the given information, the atmosphere and the helium have constant $P = 96.26$ kPa. You may ask, if the inside and outside pressures are identical, how can the expansion occur in the first place?

The key word in the problem statement is "gradually." The heat is added very slowly, such that the pressure inside the balloon never climbs significantly higher than the external pressure. Since the pressure difference driving the expansion is very small, the expansion is very slow. Consequently, even though we know the helium pressure must be fractionally higher than the external pressure (at least part of the time) in order for the expansion to occur, in building a mathematical model of the process, we can treat these pressures as identical. This is actually an example of a *reversible process*, a phenomenon discussed in detail in Chapter 4.

The next example again examines a gas confined in a piston-cylinder device. It is instructive to note that, while the physical system looks very different from the dirigible in Example 3-9, when we go through the exercise of identifying relevant forms of energy entering, leaving, or accumulating in the system, the energy balance equation ends up looking substantially the same.

EXAMPLE 3-10	IDEAL GAS IN A PISTON-CYLINDER DEVICE

An ideal gas has $C_P^* = (7/2)R$. One mole of this gas is confined in a piston-cylinder device. Initially, the gas is at $T = 300$ K and $P = 0.1$ MPa (Figure 3-13). If the gas is compressed isothermally to $P = 0.5$ MPa, find the amounts of work and heat associated with the process.

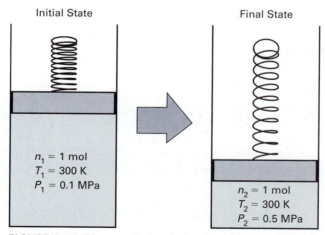

Initial State   Final State

$n_1 = 1$ mol
$T_1 = 300$ K
$P_1 = 0.1$ MPa

$n_2 = 1$ mol
$T_2 = 300$ K
$P_2 = 0.5$ MPa

FIGURE 3-13 Piston-cylinder device described in Example 3-10.

SOLUTION:

Step 1 *Define the system*
Define the system as the gas inside the cylinder.

Step 2 *Apply and simplify the energy balance*
There is no information given or needed on how long the process takes, so we use the time-independent form of the energy balance. Note the following simplifications.

- This is not a steady-state process; the accumulation term as a whole can't be neglected. However, the cylinder, as a unit, is not moving through space, and changes in the kinetic or potential energy of the system *can* be considered negligible.

- There is no material entering or leaving the system, $m_{in} = m_{out} = 0$.

- While a piston may have moving/rotating parts, these are outside the system. There are no moving parts inside the cylinder so $W_S = 0$.

- The system gets smaller so W_{EC} is not negligible.

- There is no reason to assume Q is negligible.

Consequently, the energy balance simplifies, as it did in Example 3-9, to

$$N_2 \underline{U}_2 - N_1 \underline{U}_1 = W_{EC} + Q \tag{3.89}$$

The internal energy here is expressed on a molar (rather than a mass) basis.

Step 3 *Determine work of compression*

The work can be determined using Equation 1.22:

$$dW_{EC} = -P\,dV$$

where P is the pressure opposing the motion—in this case, the gas pressure. Solving the ideal gas law for P gives

$$P = \frac{NRT}{V} \tag{3.90}$$

Introducing the ideal gas law into Equation 1.23 gives

$$dW_{EC} = -\frac{NRT}{V}\,dV \tag{3.91}$$

This is stated as an isothermal compression, and the system is closed. Therefore, T and N are both constant in this process, and R is always a constant. Consequently, Equation 3.91 can be integrated as

$$W_{EC} = -NRT \int_{V=V_1}^{V=V_2} \frac{1}{V}\,dV = -NRT(\ln V_2 - \ln V_1) = -NRT \ln\left(\frac{V_2}{V_1}\right) \tag{3.92}$$

N, R, and T are all known. Because the initial and final pressures and temperatures are given, we could certainly determine V_1 and V_2, but it is not actually necessary to do so. We can instead use the ideal gas law to express V_1 and V_2 in terms of the known temperatures and pressures:

$$W_{EC} = -NRT \ln\left(\frac{\dfrac{NRT_2}{P_2}}{\dfrac{NRT_1}{P_1}}\right) \tag{3.93}$$

Because N, R, and T are all constant ($T_1 = T_2$ and $N_1 = N_2$), Equation 3.93 simplifies to:

$$W_{EC} = -NRT \ln\left(\frac{P_1}{P_2}\right) = -(1\,\text{mol})\left(8.314\frac{\text{J}}{\text{mol}\cdot\text{K}}\right)(300\,\text{K}) \ln\left(\frac{0.1\,\text{MPa}}{0.5\,\text{MPa}}\right) \tag{3.94}$$

$$= \mathbf{4014\,J}$$

Side notes:

A piston-cylinder device is a very different looking piece of equipment than a balloon, but they have the same energy balance equation; the forms in which energy enters, leaves, and accumulates are the same in both systems.

Whenever a system is expanding or contracting, work occurs, and can be quantified using Equation 1.22.

A crucial difference between this example and Example 3-9 is that P is not constant. In order to integrate with respect to V, we must therefore express P as a mathematical function of V.

It makes sense that W is positive in this case. Work is being done on the system by the surroundings, which means energy is added to the system.

Step 4 *Determine change in molar internal energy*

There is no direct way to compute Q, but now that we know W_{EC}, if we can quantify the change in internal energy, we can compute Q from the energy balance. Section 2.3.3 indicated that for an ideal gas:

$$d\underline{U} = C_V^* \, dT$$

This is an isothermal process ($dT = 0$). Thus, even though this is not a steady-state process, the change in internal energy is 0. Consequently the energy balance simplifies to

$$C_V \, dT = 0 = W_{EC} + Q = 4014 \, J + Q \tag{3.95}$$

$$\mathbf{-4014 \, J = Q}$$

Note several points illustrated by the previous example:

▪ A process is at steady state if ALL state variables remain constant with time. In the example, the temperature and molar internal energy of the system were both constant, but the pressure was changing, so the system was not at steady state.

▪ For an ideal gas, internal energy is only a function of temperature. It would have been reasonable, at the start of the problem, to recognize that internal energy does not change because temperature does not change, and cross out the U term in the generalized energy balance for that reason. However, the solution demonstrated that even if one had missed this physical insight at the start of the problem, Equation 3.95 would still lead to the correct answer.

The expression $C_P^* = C_V^* + R$ is specifically valid *for ideal gases* and is not valid in general.

▪ C_V^* appears in Equation 3.95, but its value doesn't matter because $dT = 0$. Note, however, that C_P^* was given, and it was shown in Section 2.3.3 that for an ideal gas, $C_V^* + R = C_P^*$. Therefore, in this case, had C_V^* been needed, one could have computed it:

$$C_V^* = C_P^* - R = \frac{7}{2}R - R = \frac{5}{2}R$$

The previous two examples concerned ideal gases. The next problem involves a liquid, which will be modeled using the assumptions discussed in Section 2.3.5. Example 3-11 also requires use of the time-dependent energy balance.

EXAMPLE 3-11	**A STORAGE TANK**

A tank has a volume of 0.5 m³ and is initially full of a liquid that is at 300 K, has a density of 800 kg/m³, and a constant heat capacity of $\hat{C}_V = 3 \, kJ/kg \cdot K$. The tank can be modeled as perfectly insulated, and its contents can be assumed to be well mixed, although the shaft work added by the mixer can be considered negligible.

A system can be adiabatic either because (1) it is so well insulated that there is essentially no heat transfer, as in this problem or (2) because the system and surroundings are at the same temperature throughout the process; there is no driving force for heat transfer.

The inlet and outlet valves are both opened at the same time. The flow rates of the entering and leaving liquids are both 4.17×10^{-4} m³/s. The entering liquid is at $T = 400$ K, and the temperature of the exiting liquid at any time is identical to the temperature of the liquid in the tank at that particular time (Figure 3-14). After five minutes, what is the temperature of the liquid in the tank?

SOLUTION:

Step 1 *Define a system*

Define the contents of the tank as the system.

Step 2 *Apply the material balance*

There is no information in the problem statement that would let us quantify how much the liquid expands with increasing temperature—we don't even know what the liquid *is*,

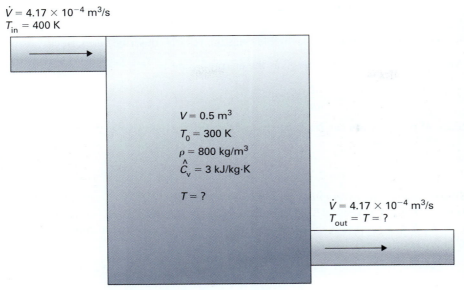

$\dot{V} = 4.17 \times 10^{-4}$ m³/s
$T_{in} = 400$ K

$V = 0.5$ m³
$T_0 = 300$ K
$\rho = 800$ kg/m³
$\hat{C}_v = 3$ kJ/kg·K

$T = ?$

$\dot{V} = 4.17 \times 10^{-4}$ m³/s
$T_{out} = T = ?$

FIGURE 3-14 Tank described in Example 3-11.

FOOD FOR THOUGHT 3-16

What would happen to a *real* tank filled with liquid if it was heated and there was no space above the liquid to accommodate the thermal expansion?

specifically. We do know the density of liquids (and solids) is frequently, and to a reasonable approximation, constant with respect to temperature, so we will assume the density 800 kg/m³ (given at $T = 300$ K) is valid at all temperatures.

Consequently, the mass of liquid entering and leaving the tank is

$$\dot{m}_{in} = \dot{m}_{out} = \left(4.17 \times 10^{-4} \frac{m^3}{s}\right)\left(800 \frac{kg}{m^3}\right) = 0.334 \frac{kg}{s} \qquad (3.96)$$

Because the entering and exiting mass flow rates are identical, the mass balance simplifies to

$$dM/dt = 0 \qquad (3.97)$$

Thus, the mass of the system (M) is constant with respect to time and can be computed as

$$M = (0.5\,m^3)\left(800 \frac{kg}{m^3}\right) = 400\,kg \qquad (3.98)$$

Step 3 *Apply and simplify the energy balance*

$$\frac{d}{dt}\left\{M\left(\hat{U} + \frac{v^2}{2} + gh\right)\right\} = \sum_{j=1}^{j=J}\left\{\dot{m}_{j,in}\left(\hat{H}_j + \frac{v_j^2}{2} + gh_j\right)\right\}$$
$$- \sum_{k=1}^{k=K}\left\{\dot{m}_{k,out}\left(\hat{H}_k + \frac{v_k^2}{2} + gh_k\right)\right\} + \dot{W}_S + \dot{W}_{EC} + \dot{Q}$$

Examining the terms from left to right:

- $\hat{U}$ is not constant, since we expect the tank temperature to increase. However, the system is a stationary tank, so there is no change in kinetic or potential energy.

- There is one stream entering and one leaving the process. The kinetic and potential energies of these streams can be neglected (see Examples 3-5 and 3-6).

- Shaft work is explicitly stated as negligible.

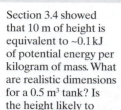

Section 3.4 showed that 10 m of height is equivalent to ~0.1 kJ of potential energy per kilogram of mass. What are realistic dimensions for a 0.5 m³ tank? Is the height likely to be more or less than 10 m?

- There is no work of expansion or contraction since the system is a constant-volume vessel.

- Heat can be considered negligible since the vessel is described as perfectly insulated.

Applying these simplifications gives the energy balance:

$$\frac{d}{dt}(M\hat{U}) = \dot{m}_{in}\hat{H}_{in} - \dot{m}_{out}\hat{H}_{out} \tag{3.99}$$

Since M is constant it can be pulled out of the differential. Also, $\dot{m}_{in} = \dot{m}_{out} = \dot{m}$, so

$$M\frac{d\hat{U}}{dt} = \dot{m}(\hat{H}_{in} - \hat{H}_{out}) \tag{3.100}$$

Step 4 *Simplifying assumption regarding enthalpy*

It was observed in Section 2.3.5 that for most liquids and solids, $PV << U$, so $U \approx H$ and $C_V \approx C_P$. Since we have no better way to progress using the given information, we assume $U = H$ (and $\hat{U} = \hat{H}$) here.

The steam tables show that for saturated liquid water at 298.15 K,

$\hat{U} = 104.8\ \dfrac{kJ}{kg}$, and

$\hat{H} = 104.8\ \dfrac{kJ}{kg}$; these are identical to four significant figures. The saturated liquid values start to diverge as temperature and pressure increase, so at $T = 448.15$ K and $P = 0.8926$ MPa, $\hat{U}$ and $\hat{H}$ differ by 1.0 kJ/kg.

$$M\frac{d\hat{U}}{dt} = \dot{m}(\hat{U}_{in} - \hat{U}_{out}) \tag{3.101}$$

Step 5 *Relate specific internal energy to known heat capacity*

This is a constant-volume process (see Section 2.3.2), and for a constant-volume process:

$$d\hat{U} = \hat{C}_V\, dT \tag{3.102}$$

We can relate the term $\hat{U}_{in} - \hat{U}_{out}$ to temperature by integrating Equation 3.102. Since $\hat{C}_V$ is constant,

"The given information" is an artificial constraint in textbook problems. In engineering practice, you need to consider whether the uncertainty introduced by your assumptions is acceptable, and you always have the option of seeking additional information.

$$\int_{\hat{U}=\hat{U}_{in}}^{\hat{U}=\hat{U}_{out}} d\hat{U} = \int_{T=T_{in}}^{T=T_{out}} \hat{C}_V\, dT \tag{3.103}$$

$$\hat{U}_{out} - \hat{U}_{in} = \hat{C}_V(T_{out} - T_{in})$$

Substituting this and Equation 3.102 into Equation 3.101 gives

$$M\hat{C}_V\frac{dT}{dt} = -\dot{m}\hat{C}_V(T_{out} - T_{in}) \tag{3.104}$$

Step 6 *Solve differential equation*

Because the tank is well mixed, the exiting liquid temperature (T_{out}) is the same as the temperature of the system (T). Making this substitution, and also cancelling $\hat{C}_V$, gives an equation we can integrate:

Heat capacity can be measured on either a molar or mass basis. Here $\hat{C}_V = 3\ \text{KJ/kg} \cdot \text{K}$, so we write $d\hat{U} = \hat{C}_V\, dT$. In Example 3-10, C_V was given on a molar basis, so we wrote $d\underline{U} = C_V\, dT$.

$$M\frac{dT}{dt} = \dot{m}(T_{in} - T) \tag{3.105}$$

$$\frac{dT}{T_{in} - T} = \frac{\dot{m}}{M}\, dt \tag{3.106}$$

$$\int_{T=300\,K}^{T=T} \frac{dT}{T_{in} - T} = \int_{t=0}^{t=t} \frac{\dot{m}}{M}\, dt$$

$$-\ln\left(\frac{T_{in} - T}{T_{in} - 300\,K}\right) = \frac{\dot{m}}{M}t$$

Substituting in known values:

$$-\ln\left(\frac{400\,\text{K} - T}{400\,\text{K} - 300\,\text{K}}\right) = \frac{0.334\dfrac{\text{kg}}{\text{s}}}{400\,\text{kg}}t$$

$$T = 400\,\text{K} - (100\,\text{K})e^{-(0.00083\,\text{s}^{-1})t} \tag{3.107}$$

$T = 322.1\,\text{K}$ at a time of 5 minutes, and a full graph of T versus t is shown in Figure 3-15.

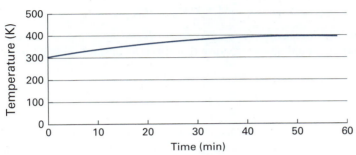

FIGURE 3-15 Temperature versus time for the water in the tank in Example 3-11.

Up to this point, the examples in this section all involved a single phase (gas in Examples 3-9 and 3-10, liquid in Example 3-11). The next example illustrates a steady-state process that involves both a phase change and a temperature change.

| STEADY-STATE BOILER | **EXAMPLE 3-12** |

16.67 mol/s of pure ethanol enters a steady-state boiler as liquid at $P = 101.3$ kPa and $T = 298$ K, and leaves the boiler as vapor at $P = 101.3$ kPa and $T = 373$ K as shown in Figure 3-16. The normal boiling point of ethanol is 351.55 K. At what rate is heat is added to the boiler?

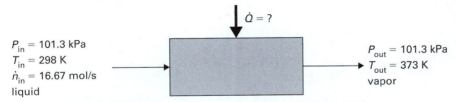

$\dot{Q} = ?$

$P_{\text{in}} = 101.3$ kPa
$T_{\text{in}} = 298$ K
$\dot{n}_{\text{in}} = 16.67$ mol/s
liquid

$P_{\text{out}} = 101.3$ kPa
$T_{\text{out}} = 373$ K
vapor

FIGURE 3-16 Schematic of ethanol boiler modeled in Example 3-12.

SOLUTION:

Step 1 *Apply and simplify the energy balance*
The system is the boiler and its contents. Since this is a steady-state process, the accumulation term of the energy balance is zero:

$$0 = \dot{m}_{\text{in}}\left(\hat{H}_{\text{in}} + \frac{v_{\text{in}}^2}{2} + gh_{\text{in}}\right) - \dot{m}_{\text{out}}\left(\hat{H}_{\text{out}} + \frac{v_{\text{out}}^2}{2} + gh_{\text{out}}\right) + \dot{Q} + \dot{W}_{\text{S}} + \dot{W}_{\text{EC}}$$

There is no mechanism for work in a typical heat exchanger, and neglecting potential and kinetic energy changes produces:

$$0 = \dot{m}_{in}(\hat{H}_{in}) - \dot{m}_{out}(\hat{H}_{out}) + \dot{Q}$$

the entering and exiting mass flow rates are equal so:

$$0 = \dot{m}(\hat{H}_{in} - \hat{H}_{out}) + \dot{Q}$$

The given information is on a molar basis so we switch to this ($m\hat{H} = n\underline{H}$) and solve for rate of heat addition:

$$\dot{Q} = \dot{n}(\underline{H}_{out} - \underline{H}_{in}) \tag{3.108}$$

Step 2 *Formulate strategy for computing $\underline{H}_{out} - \underline{H}_{in}$*

A common error would be to recognize that this is a constant-pressure process and simply apply C_P:

$$\underline{H}_{out} - \underline{H}_{in} = \int_{T_{in} = 298 \text{ K}}^{T_{out} = 373 \text{ K}} C_P \, dT$$

and look up the heat capacity of ethanol in Appendix D-1 or D-2. But a closer look at Appendices D-1 and D-2 reveals that ethanol liquid and ethanol vapor have different heat capacities. More fundamentally, $d\underline{H} = C_P \, dT$ accounts only for temperature changes and does not account for the change in molar enthalpy that accompanies the phase change.

Instead, we find $\underline{H}_{out} - \underline{H}_{in}$ by applying the fact that molar enthalpy is a state property: The change in molar enthalpy is independent of path. We envision a three-step path, illustrated in Figure 3-17; step 1 is the heating of the liquid to its normal boiling point, step 2 is the phase change, and step 3 is the heating of saturated vapor to 373 K. The molar enthalpy changes for these individual steps sum to the molar enthalpy change for the whole path:

$$\underline{H}_{out} - \underline{H}_{in} = \Delta\underline{H}_1 + \Delta\underline{H}_2 + \Delta\underline{H}_3 \tag{3.109}$$

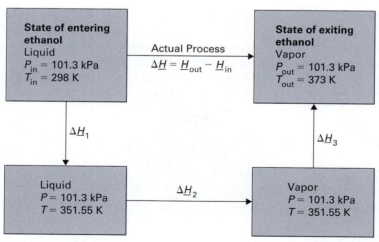

FIGURE 3-17 Calculation of change in molar enthalpy for ethanol in Example 3-12.

Step 3 *Calculate $\Delta\underline{H}$ for individual steps*

Step 2 is boiling at atmospheric pressure, so $\Delta\underline{H}_2$ is simply $\Delta\underline{H}^{vap}$ for ethanol at atmospheric pressure, which is available in Appendix C.

$$\Delta\underline{H}_2 = 38,560 \, \frac{\text{J}}{\text{mol}}$$

Steps 1 and 3 are both temperature changes at constant pressure, so $d\underline{H}=C_P dT$ applies to each, as discussed in Section 2.3.2.

The integration of a temperature-dependent expression for C_P was illustrated in Example 2-5.

$$\Delta\underline{H}_1 = \int_{T=298\,K}^{T=351.55\,K} C_{P,liq}\, dT$$

$$\Delta\underline{H}_3 = \int_{T=351.55\,K}^{T=373\,K} C_P^*\, dT$$

Note that C_P^* is the ideal gas heat capacity of ethanol, which is available in Appendix C-1. Because this process is carried out at a low pressure (101.3 kPa), it is reasonable to model ethanol vapor as an ideal gas, as discussed in Section 2.3.3. The value of C_P for liquid ethanol is available in Appendix C-2. Both heat capacities are functions of T. The full integration of $C_P\, dT$ is shown in the more detailed solution provided in the electronic appendices.

$$\Delta\underline{H}_1 = 6253\,\text{J/mol} \qquad\qquad \Delta\underline{H}_3 = 1621\,\text{J/mol}$$

Step 4 *Calculate Q*
Returning to Equation 3.108 to solve for the heat addition:

$$\dot{Q} = \dot{n}(\underline{H}_{out} - \underline{H}_{in})$$

$$\dot{Q} = \left(16.67\,\frac{\text{mol}}{\text{s}}\right)\left(6253 + 38{,}560 + 1621\,\frac{\text{J}}{\text{mol}}\right)\left(\frac{1\,\text{kJ}}{1000\,\text{J}}\right) = \mathbf{773.9}\,\frac{\text{kJ}}{\text{s}}$$

Notice the molar enthalpy of vaporization is much larger in magnitude than the changes in molar enthalpy resulting from the temperature changes.

The previous examples illustrate the process of applying and simplifying the generalized energy balance equation to a variety of physical systems. We close this section by considering a rocket launching, which revisits the motivational example in a more quantitative way.

LAUNCH OF A ROCKET | **EXAMPLE 3-13**

A rocket's payload (everything EXCEPT fuel) has a mass of 10,000 kg and $\hat{C}_V = 2.5$ kJ/kg · K. Initially, the rocket is at rest at ground level, is at ambient temperature ($T = 298$ K) and pressure ($P = 0.1$ MPa), and contains the payload plus 100,000 kg of rocket fuel. The fuel has a specific enthalpy of formation $\hat{H} = 1000$ kJ/kg, using the same reference state that is used for the data in Appendix C.

Rocket fuel is here modeled as if it were a single compound. Real rocket fuel would more likely be a mixture of one or more fuels plus an oxidizer to support their combustion.

The fuel burns to completion, and the emitted exhaust consists of 30,000 kg water vapor and 70,000 kg carbon dioxide. The exhaust leaves at $T = 298$ K and $P = 0.1$ MPa and has a velocity of 3 km/s. When the last of the fuel is consumed, the rocket is 5 km above the Earth, and the payload has $T = 523$ K. What is its velocity at this point? Exhaust is released continuously as the rocket climbs, but a large fraction of the fuel burns close to the ground. Assume the *average* height at which the exhaust is released is 0.5 km. Assume further that the rocket is adiabatic and that it neither produces nor uses shaft work.

FOOD FOR THOUGHT 3-18

Is it plausible that the rocket itself is at 523 K but the fuel exhaust is at 298 K?

SOLUTION:

Step 1 *Define a system*
The rocket, and everything inside it, will be the system.
Step 2 *Apply and simplify the energy balance*
This example describes the initial and final states of the rocket, but doesn't indicate the time required to climb to this height, so we use the time-independent form of the energy balance.

No immediate simplification of the accumulation term is possible; the system experiences significant changes in mass, internal energy, kinetic energy and potential energy. On the right-hand side, there is no Q or W_S. We assume there is also no W_{EC}, because the size of the rocket doesn't change. There is no matter entering the system, but the exhaust gases leave. The energy balance simplifies to

$$\Delta\left\{M\left(\hat{U} + \frac{v^2}{2} + gh\right)\right\} = -\sum_{k=1}^{k=K}\left\{m_{k,out}\left(\hat{H}_k + \frac{v_k^2}{2} + gh_k\right)\right\} \tag{3.110}$$

While the CO_2 and H_2O leave the system as a single exhaust stream, we haven't yet developed how to model mixtures, so we will *model* them as two separate exiting streams:

$$\Delta\left\{M\left(\hat{U} + \frac{v^2}{2} + gh\right)\right\} = -m_{CO_2,out}\left(\hat{H}_{CO_2} + \frac{v_{CO_2}^2}{2} + gh_{CO_2}\right) \tag{3.111}$$

$$- m_{H_2O,out}\left(\hat{H}_{H_2O} + \frac{v_{H_2O}^2}{2} + gh_{H_2O}\right)$$

Writing out the initial and final energy of the system explicitly gives

$$\left\{M\left(\hat{U} + \frac{v^2}{2} + gh\right)\right\}_{final} - \left\{M\left(\hat{U} + \frac{v^2}{2} + gh\right)\right\}_{initial} \tag{3.112}$$

$$= -m_{CO_2,out}\left(\hat{H}_{CO_2} + \frac{v_{CO_2}^2}{2} + gh_{CO_2}\right) - m_{H_2O,out}\left(\hat{H}_{H_2O} + \frac{v_{H_2O}^2}{2} + gh_{H_2O}\right)$$

The initial height and velocity are both zero. However, initially, the mass of the system includes payload and fuel; at the end of the process there is only payload:

$$\left\{M_{payload}\left(\hat{U}_{payload} + \frac{v^2}{2} + gh\right)\right\}_{final} - \left\{M_{payload}\hat{U}_{payload} + M_{fuel}\hat{U}_{fuel}\right\}_{initial} \tag{3.113}$$

$$= -m_{CO_2,out}\left(\hat{H}_{CO_2} + \frac{v_{CO_2}^2}{2} + gh_{CO_2}\right) - m_{H_2O,out}\left(\hat{H}_{H_2O} + \frac{v_{H_2O}^2}{2} + gh_{H_2O}\right)$$

Step 3 *Insert known values and identify what is unknown*

v_{final} is the unknown we seek to calculate. Most of the other information in Equation 3.113 is given. The only unknowns we need to resolve are $\hat{U}_{payload,\,final}$ and $\hat{U}_{payload,\,initial}$, and $\hat{H}_{CO_2}$ and $\hat{H}_{H_2O}$.

Step 4 *Collect data*

Appendix C contains enthalpy of formation data for CO_2 and H_2O. Recall that enthalpy is only known relative to a reference state. The "enthalpy of formation" is the molar enthalpy relative to a reference state in which all elements have $\underline{H} = 0$ at $P = 0.1$ MPa and $T = 25°C$. Can we relate these to $\hat{H}_{CO_2}$ and $\hat{H}_{H_2O}$ in Equation 3.113? Let us consider this more closely. The exiting stream is a mixture of carbon dioxide and water vapor. We haven't yet learned how to calculate $\hat{H}$ for a mixture, so we are *imagining* that there are two separate exiting streams. What are the pressures of these two streams? Since the total pressure of the exiting gas is $P = 0.1$ MPa, then CO_2 and H_2O each must have a partial pressure that is less than 0.1 MPa. The value of $\underline{H} = -393.5$ kJ/mol in Appendix C is valid for CO_2 at $P = 0.1$ MPa and $T = 298$ K; is it applicable to CO_2 at 298 K and a lower pressure? We will say "yes" because again, real gases act like ideal gases at low pressure, and for an ideal gas, enthalpy is not a function of pressure. Converting the enthalpy of formation data from Appendix C to mass basis using the molecular weight gives

---

This is the only example in the first eight chapters of the book that involves mixtures of gases or liquids: the exhaust is a mixture of CO_2 and H_2O. A complete treatment of the mathematical modeling of mixtures begins in Chapter 9.

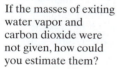

FOOD FOR THOUGHT 3-19

If the masses of exiting water vapor and carbon dioxide were not given, how could you estimate them?

PITFALL PREVENTION

Don't confuse the different v terms and h terms. The left-hand side represents the system (rocket) and the right-hand side represents material (exhaust gases) entering or leaving the system. Thus, the v on the left-hand side is the unknown rocket velocity, and the v terms on the right-hand side are 3 km/s, the exhaust velocity.

In Chapter 9, the properties of a mixture for ideal gases are found simply by summing the properties of the individual pure gases that make up the mixture multiplied by how much of each gas is present.

$$\hat{H}_{CO_2} = \left(-393.5\frac{kJ}{mol}\right)\left(\frac{1\,mol}{44\,g}\right)\left(\frac{1000\,g}{kg}\right) = -8943\frac{kJ}{kg} \tag{3.114}$$

$$\hat{H}_{H_2O} = \left(-241.8\frac{kJ}{mol}\right)\left(\frac{1\,mol}{18\,g}\right)\left(\frac{1000\,g}{kg}\right) = -13,430\frac{kJ}{kg}$$

Meanwhile, the internal energy of the rocket fuel, $\hat{U}_{fuel,\,initial}$, is unknown, but the enthalpy of formation of the fuel at the correct temperature (298 K) and pressure ($P = 0.1$ MPa) is given, $\hat{H}_{fuel,\,initial} = 1000\frac{kJ}{kg}$. Using the general rule stated in Section 2.3.5 that for liquids and solids, $\hat{U} \sim \hat{H}$. So

$$\hat{U}_{fuel,initial} \sim 1000\frac{kJ}{kg} \tag{3.115}$$

Step 5 *Account for change in internal energy*

The left-hand side of Equation 3.114 can be rearranged so that the $\hat{U}_{payload}$ terms are grouped together:

$$\left\{M_{payload}\left(\frac{v_{final}^2}{2} + gh_{final}\right)\right\} + \{M_{payload}(\hat{U}_{payload,final} - \hat{U}_{payload,initial}) - M_{fuel}\hat{U}_{fuel,initial}\} \tag{3.116}$$

$$= -m_{CO_2,out}\left(\hat{H}_{CO_2} + \frac{v_{CO_2}^2}{2} + gh_{CO_2}\right) - m_{H_2O,out}\left(\hat{H}_{H_2O} + \frac{v_{H_2O}^2}{2} + gh_{H_2O}\right)$$

Assuming the rocket is a constant-volume system, the change in internal energy can be related to $\hat{C}_V$:

$$M_{payload}(\hat{U}_{payload,final} - \hat{U}_{payload,initial}) = M_{payload}(\hat{C}_{V,payload})(T_{payload,final} - T_{payload,initial}) \tag{3.117}$$

Step 6 *Calculate velocity*

Inserting Equation 3.117 and all of the known values identified in steps 3 and 4 into Equation 3.116 gives an equation in which the final velocity is unknown. The required algebra is shown in the more detailed solution given in the supplemental material. The solution is:

$$v_{final} = 11,600\,\frac{m}{s}$$

In terms of analyzing a real rocket, Example 3-13 is over-simplified in several ways: it treated "rocket fuel" as a single compound, assumed the rocket fuel was at ambient conditions initially (real rocket fuel would probably be pressurized), and used an "average" height at which the fuel was released. A rigorous treatment would include air resistance, model the rocket's flight continuously with differential equations, and combine energy balances with momentum balances.

While simplistic, this example illustrates a system where all three forms of stored energy—internal, kinetic, and potential—are important and demonstrates the conversion of chemical energy into kinetic and potential energy. Furthermore, it illustrates mathematically how a rocket can indeed achieve escape velocity without violating the first law of thermodynamics.

3.6 Energy Balances for Common Chemical Process Equipment

The focus of this chapter has been on developing a systematic approach to writing and solving energy balances that is applicable to any physical system. However, some specific cases that are of particular interest to chemical engineers are explored here.

FOOD FOR THOUGHT 3-20

Using only the information in the appendix of this book, how could you find $\hat{H}_{CO_2}$ and $\hat{H}_{H_2O}$ if the exhaust were at $T = 373$ K, rather than at 298 K?

PITFALL PREVENTION

You can't use the data in the steam tables for $\hat{H}_{H_2O}$. The data for rocket fuel, carbon dioxide, and water vapor must all have the same reference state. The data in the steam tables and Appendix C use different reference states.

When $\hat{C}_V$ is constant, the equation $d\hat{U} = \hat{C}_V\,dT$ simply integrates to $\hat{U}_2 - \hat{U}_1 = \hat{C}_V(T_2 - T_1)$.

This example is oversimplified in that modern rockets are multi stage; meaning that portions of the rocket are fuel tanks that are jettisoned when empty. Thus, $M_{payload}$ wouldn't be constant.

Recall that the Motivational Example stated escape velocity is ~7 miles/s. Here the final answer, 11,600 m/s, is approximately 7.2 miles/s.

TABLE 3-1 Steady state energy balance equations for common chemical processes.

Unit Operation	Symbols	Typical Steady State Energy Balance	Assumptions/Simplifications in Energy Balance
Valve		$\hat{H}_{in} = \hat{H}_{out}$	Adiabatic, potential, and kinetic energy negligible
Nozzle		$\dfrac{v_{out}^2}{2} = \hat{H}_{in} - \hat{H}_{out}$	Adiabatic and potential energy negligible, velocity of *entering* stream negligible
Pump, Turbine or Compressor		$\dfrac{\dot{W}_S}{\dot{m}} = \hat{H}_{out} - \hat{H}_{in}$	Adiabatic, potential, and kinetic energy negligible
Heat Exchanger (single stream)	$\dot{Q}$	$\dfrac{\dot{Q}}{\dot{m}} = \hat{H}_{out} - \hat{H}_{in}$	Potential and kinetic energy negligible
Heat Exchanger (both hot and cold streams)		$\begin{aligned} 0 &= \dot{m}_1(\hat{H}_{1,in} - \hat{H}_{1,out}) \\ &+ \dot{m}_2(\hat{H}_{2,in} - \hat{H}_{2,out}) \end{aligned}$	Potential and kinetic energy negligible, no heat lost to surroundings

In Example 3-6 we defined a system that included a turbine, but was larger than just the turbine, so it would have been wrong to apply the turbine equation from Table 3-1 to that system.

Most large-scale chemical processes are designed to operate at steady state. Table 3-1 summarizes steady-state energy balance equations for several unit operations commonly found in chemical processes. *In all cases, the single piece of equipment is the system.*

The equations in Table 3-1 are broadly useful, but care must be taken to apply them only to steady-state processes in which the noted assumptions and simplifications are valid. Notice there is no W_{EC} in any of the equations, but the words "expansion work negligible" don't even appear in the "Assumptions/Simplifications" column of the table. If a process is operating at steady state, the volume of the system cannot be changing, so no expansion/contraction work can occur. Thus, once we've established that a process is steady state, "expansion work negligible" is *not* an assumption; it's inherently true.

Reactors and separators are arguably the most fundamental unit operations in chemical engineering. These, however, are not included in Table 3-1 for two reasons.

1. Reactors and separators both inherently involve mixtures of compounds. Thus, while Example 3-13 did involve a chemical reaction, we defer any in-depth discussion of reactions and separations until the second half of the book. In Chapters 9–13, when we develop methods of modeling the thermodynamic properties of mixtures, many of the examples examine the thermodynamics governing separation processes. Chapter 14 covers chemical reactions.

2. There are many different types of reactors and separators; it isn't realistic to propose a single "typical energy balance equation" for these operations.

The following sections discuss the unit operations summarized in Table 3-1.

3.6.1 Valves

Valves can be used to control the flow rate or pressure of a stream. Some pictures of common types of valves are given in Figure 3-18.

The primary physical effect of a valve on the fluid flowing through it is a decrease in pressure. This decrease is a result of the constriction in diameter.

FIGURE 3-18 Photos of valves.

While it takes work to open and close a valve, the opening or closing is a transient process. During steady-state operation we expect the state of the valve (i.e., how far open) to be static. Consequently, there is no mechanism for transfer of shaft work to or from the fluid. Valves are commonly assumed adiabatic: if the contents of a pipe are at high temperature, one might need to account for heat lost in a long section of pipe, but can safely ignore heat lost in the small space occupied by the valve. Thus, if potential and kinetic energy are considered negligible, the energy balance simplifies to:

$$0 = \dot{m}_{in}\hat{H}_{in} - \dot{m}_{out}\hat{H}_{out} \tag{3.118}$$

But because the entering and leaving mass flow rates are equal at steady state, the mass flow rates can be divided out, providing the energy balance shown in Table 3-1:

$$0 = \hat{H}_{in} - \hat{H}_{out} \quad \text{or equivalently} \quad \hat{H}_{in} = \hat{H}_{out} \tag{3.119}$$

3.6.2 Nozzles

A nozzle is designed to accelerate a fluid to high velocity, converting some of the fluid's internal energy into kinetic energy. Rockets operate on the principle of conservation of momentum; the exhaust is expelled through nozzles, giving it a very high downward velocity, which results in an upward velocity for the rocket. A more familiar everyday example of nozzles is an attachment on a garden hose. In principle the water leaving the nozzle must be slightly colder than the water in the hose, but the velocities achieved by these nozzles are modest enough that the effect likely isn't noticeable to your bare hands.

The typical energy balance for a nozzle was derived in the context of Example 3-4.

3.6.3 Pumps, Compressors, and Turbines

While pumps, compressors, and turbines are distinct unit operations, they are here discussed simultaneously because in steady-state operation they typically have identical energy balance equations, and there is also some overlap in the physical phenomena at work in these three processes.

Turbines were discussed in Section 1.2.3 and Example 3-6. Their purpose (in effect) is to convert internal energy into shaft work. A turbine can be pictured as two steps: (1) a nozzle is used to accelerate the fluid and (2) the fluid moves over the blades of a windmill. The nozzle converts internal energy into kinetic energy, and the windmill converts kinetic energy into rotary motion or shaft work. The turbine encompasses both steps. The fluid leaving the nozzle is at high velocity, but this fluid never crosses the boundaries of the system. This allows us to consider kinetic energy as negligible in the energy balance around a turbine; the fluid entering the nozzle and the fluid leaving the windmill are both moving at comparatively low velocity.

Turbines normally are modeled as adiabatic, meaning that the only significant terms of the energy balance are the enthalpy of the entering and leaving streams and the shaft work produced.

$$0 = \dot{m}_{in}\hat{H}_{in} - \dot{m}_{out}\hat{H}_{out} + \dot{W}_S \tag{3.120}$$

> The energy balance derived here for a turbine is identical to that derived in Example 3-8 for a pump.

The mass flow rates entering and leaving are equal, so the "in" and "out" subscripts can be eliminated. It is convenient to rearrange the equation so that it is explicit in "work produced per mass of fluid":

$$\frac{\dot{W}_S}{\dot{m}} = \hat{H}_{out} - \hat{H}_{in} \tag{3.121}$$

The turbine always produces a substantial drop in pressure. The converse unit operations are the pump and compressor, both of which are designed to increase the pressure of a fluid, which is accomplished through the addition of work. While the intent of pumps and compressors are identical, the physical equipment is very different, as illustrated in Figure 3-19. Pumps are used to compress liquids and compressors are used for gases. Both pumps and compressors can be damaged if liquid-vapor mixtures are fed into them.

Jan de Wild / Shutterstock.com

FIGURE 3-19 Photo of a compressor.

Since pumps, compressors, and turbines are all well modeled as adiabatic, the typical steady-state energy balance equations for pumps and compressors are identical to that for turbines. For pumps and compressors, work should be positive (energy is added to the system in the form of work, $\hat{H}_{out} > \hat{H}_{in}$) while for turbines, work is negative ($\hat{H}_{out} < \hat{H}_{in}$, energy is transferred to the surroundings.) For our purposes, applying the energy balance is typically the only way to determine the shaft work for a turbine or compressor. For a pump, however, the shaft work can frequently be estimated using Equation 3.78:

$$\dot{W}_S = \int_{P=P_{in}}^{P=P_{out}} \dot{V}\, dP$$

with the assumption that volume is constant for the liquid.

3.6.4 Heat Exchangers

Heat exchangers are unit operations intended to add or remove heat from streams; among the many applications are changing the temperature of the stream to a desired value or changing the phase of the stream (as in the condenser and boiler in the Rankine cycle described in Section 2-1).

The classical design for a heat exchanger is the shell-and-tube heat exchanger shown in Figure 3-20. A large number of small tubes provide more surface area than a single large tube, thereby allowing more efficient heat transfer. One fluid (the "tube-side" fluid) travels through the narrow tubes, and the other phase (the "shell-side" fluid) surrounds the tubes. Thus, heat is exchanged between the two fluids without the fluids coming into direct contact with each other.

> Heat exchangers are frequently abbreviated HE or HX in drawings and schematics.

> **FOOD FOR THOUGHT 3-21**
>
> Is the bulb of a mercury thermometer a heat exchanger?

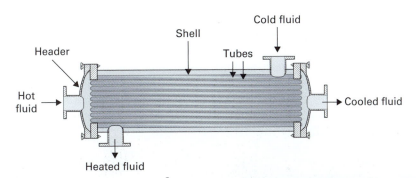

FIGURE 3-20 Schematic and photos of shell-and-tube heat exchanger.

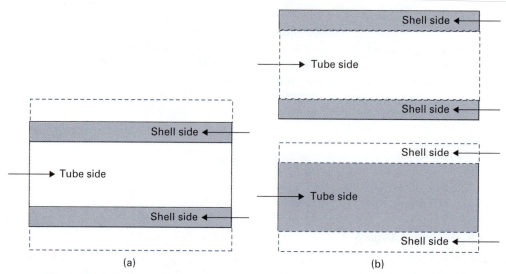

FIGURE 3-21 Possible system definitions: (a) the entire heat exchanger, (b) one side of the heat exchanger.

If the system is an entire shell-and-tube heat exchanger, the heat transfer occurs inside the boundaries of the system, as illustrated in Figure 3-21a. Thus, assuming no heat is lost to the surroundings, the Q and W terms of the energy balance are zero. However, in this case there are two streams entering and two streams leaving. Using the subscripts 1 and 2 to distinguish the two fluids from each other gives

$$0 = \dot{m}_1(\hat{H}_{1,in} - \hat{H}_{1,out}) + \dot{m}_2(\hat{H}_{2,in} - \hat{H}_{2,out}) \qquad (3.122)$$

While Equation 3.122 is a correct energy balance for complete shell-and-tube type heat exchangers, one could also define one side or the other of the heat exchanger as the system, as illustrated in Figure 3-21b. Here, there is only one material stream entering and leaving the system, and Q is not equal to zero; the heat transfer *does* cross the boundaries of the system when the system is defined this way. The energy balance for "one side" of a heat exchanger, rearranged to mirror Equation 3.121, is thus:

$$\frac{\dot{Q}}{\dot{m}} = \hat{H}_{out} - \hat{H}_{in} \qquad (3.123)$$

There are several reasons why one might use Equation 3.123 rather than Equation 3.122. Two examples are given here.

■ Some heat exchangers don't have two distinct fluids. If the heat is supplied by electricity, then the heat exchanger has material flow on only one "side"; Equation 3.122 would not apply.

■ In a design setting, especially early in the design process, one frequently wants to know how much heat would be required to accomplish a specific task, but doesn't yet know what the source of the heat will be. Indeed, the decision regarding the source of the heat might itself be dependent upon the value of Q!

3.7 SUMMARY OF CHAPTER THREE

▪ **Mass** and **energy** can both be treated as **conserved quantities**; they cannot be created or destroyed.

▪ The principles of mass conservation and energy conservation can be expressed mathematically through **material balances** (Equation 3.1) and **energy balances** (Equation 3.34).

▪ The energy balance equation is a mathematical statement of the **first law of thermodynamics.**

▪ The generalized material and energy balances can be applied to **any clearly defined system.**

▪ Any more simplified version of a material or energy balance, such as the ones in Table 3-1, can only be applied when the **assumptions** used to derive the equation are understood and are valid for the system of interest.

▪ A process that has a known **initial state** and a known **final state**, but an unknown duration, can be analyzed using **time-independent** material and energy balances (Equations 3.31 and 3.35).

▪ Material and energy balances both take the form accumulation = in − out.

▪ For a steady-state process, accumulation = 0.

▪ Kinetic and potential energy are often negligible in typical chemical processes.

3.8 EXERCISES

3-1. 10 m³ of saturated steam at T = 423.15 K is mixed with 0.1 m³ of saturated liquid water at T = 423.15 K How many total kilograms of H_2O does the mixture contain?

3-2. A 1 m³ vessel contains a mixture of saturated liquid water and saturated steam at T = 473.15 K. If the vessel contains 15 kg in total, find the mass of liquid water and the mass of steam.

3-3. Two moles of an ideal gas with $C_v^* = 3R$ are confined in a piston-cylinder arrangement. The piston is frictionless and the cylinder contains no mechanism for shaft work. Initially, the temperature is 300 K and the pressure is 0.1 MPa. The external atmospheric pressure is also 0.1 MPa. Find the heat, work and change in internal energy of the gas if:
  A. It is heated to 500 K at constant volume.
  B. It is heated at constant pressure to 500 K.
  C. It is compressed isothermally to 0.5 MPa.

3-4. A mole of gas in a closed system undergoes a four-step, cyclic process that returns it to the original state 1. The following table gives some data on the steps of the process. Fill in the blanks.

Step	ΔU (J)	Q (J)	W(J)
$1 \rightarrow 2$	−200		−6000
$2 \rightarrow 3$		−3800	
$3 \rightarrow 4$		−800	300
$4 \rightarrow 1$	4700		
$1 \rightarrow 2 \rightarrow 3 \rightarrow 4 \rightarrow 1$			−1400

3-5. 1.5 kg of water is placed in a closed container. The vessel has a frictionless piston that maintains a constant pressure. This water is initially all saturated liquid at 353.15 K, and is boiled into saturated vapor. The pressure opposing expansion can be assumed equal to the pressure inside the cylinder.
  A. At what pressure does the water boil?
  B. What are the initial and final volumes of the cylinder?
  C. What is the work done in this process?
  D. What is the change in internal energy of the water?
  E. What is the heat added for this process?

3-6. Liquid water enters a steady-state heat exchanger with a flow rate of 3.33 kg/s at 15 MPa and 608.15 K. It exits the boiler as superheated steam at 15 MPa and 873.15 K. There is no mechanism for shaft work in the heat exchanger.

A. Find the difference in specific enthalpy between the entering and exiting streams.

B. Find the rate at which heat is added to the steam in the heat exchanger.

3-7. Steam enters an adiabatic, steady-state turbine at $T = 573.15$ K and $P = 0.5$ MPa. The effluent from the turbine has $P = 0.1$ MPa, and is a liquid–vapor mixture with $q = 0.9$.

A. Find the difference between $\hat{H}$ of the exiting stream and $\hat{H}$ of the entering stream.

B. In order to produce 1.67 MJ/s of work, what is the required flow rate of the entering stream?

3.9 PROBLEMS

3-8. A swimming pool is 3.6 m wide, 9.1 m long, and 3.0 m deep, and the pool is initially 90% filled with water. A hose is placed in the pool, and water flows into the pool through the hose at a constant rate of 0.45 kg every 8 seconds, until the pool is completely filled. During this time, an inch of rain falls, and a 0.11 m³ rain barrel full of water is emptied into the pool. How long, from the moment the hose is turned on, does it take to fill the pool? Assume the density of water is constant at 1000 kg/m³.

3-9. A 4 m³ storage tank is filled with steam and maintained at a constant $T = 523.15$ K. The initial pressure is 2 MPa. A valve is opened and 1 kg/min of steam leaves the tank.

A. What is the initial mass of the steam in the tank?

B. How much time does it take for the pressure to decrease to 0.5 MPa?

3-10. Many chemical products are synthesized by biological mechanisms, in which living cells take in "reactants" as food, and emit "products." A typical fermentation process involves at least two stages— in the first, cells multiply until the desired concentration of cell mass is reached, and in the second, the bulk of the product is synthesized. A cell is a complex living organism, not a single chemical compound. However, in modeling fermentation processes, it is sometimes useful to assign an empirical chemical formula to the cells. Here we will assume the cells, or the "biomass," can be represented by the chemical formula $C_4H_7O_2N$.

The "growth phase" of a fermentation process produces a broth that has a volume of 5 m³ and contains 100 kg/m³ of biomass.

A. If all of the nitrogen in the biomass comes from ammonia, what mass of ammonia was required to produce the biomass?

B. If all of the carbon in the biomass comes from glucose ($C_6H_{12}O_6$), what mass of glucose was required to produce the biomass?

C. If the ammonia from part A and the glucose from part B are added to a batch fermentation reactor, and are completely consumed forming biomass, how much by-product is formed? Can you propose a balanced chemical reaction that represents the process?

3-11. A liquid enters a steady-state flash chamber with a flow rate of 2 mol/s. One liquid stream and one vapor stream, each at $T = 323$ K and $P = 0.1$ MPa, exit. The exiting vapor is an ideal gas with a flow rate of 1.8 m³/min. What are the molar flow rates of the liquid and vapor product streams?

3-12. Water in a creek is at $T = 298$ K and is flowing at 2 m/s when it reaches the top of a waterfall. It falls 50 meters into a pool. The velocity leaving the pool is negligible. The creek, waterfall, and pool can be modeled as at steady state and adiabatic.

A. What is the velocity of the water at the point that it hits the surface of the pool?

B. Assuming the heat capacity of liquid water is constant at 4.19 kJ/kg · K, and neglecting any external heat source, what is the temperature of the water in the pool?

3-13. One kilogram of liquid water is placed in a piston-cylinder device, initially at $P = 0.1$ MPa and $T = 278$ K. At these conditions, the density of water is 1000.4 kg/m³. The water is heated at constant pressure to $T = 368$ K, at which point the density is 962.3 kg/m³. The external pressure is also $P = 0.1$ MPa.

A. Find the work and heat for this process. Use the temperature-dependent value of C_P given in Appendix D. Apply the relationship $H = U + PV$ if conversion between H and U is necessary.

B. Find the heat for this process. This time, assume the density is constant at 1000 kg/m³, that C_P is constant at 4.19 KJ/kg·K, that $C_V \sim C_P$, and that $U \sim H$ for a liquid.

C. Compare your answers to parts A and B. The simplifying approximations in part B are often used for liquids (and solids); how well do they work here?

3-14. One kilogram of steam is placed in a piston-cylinder device, initially at $P = 0.1$ MPa and $T = 373.15$ K. The steam is heated at constant pressure to $T = 473.15$ K.

  A. Find the heat and work for this process. Obtain all needed data from the steam tables.

  B. Find the heat and work for this process. Use the ideal gas law and the ideal gas heat capacity for steam, located in Appendix D. How far off are your answers compared to part A?

  C. Suppose in part A you had overlooked the fact that the steam expands when heated, and assumed that work was negligible. How far off would your answers for Q have been?

3-15. An adiabatic valve operates at steady state. Saturated liquid water at $P = 0.5$ MPa enters. The exiting stream is a saturated liquid–vapor mixture at $P = 0.1$ MPa. What is the quality of the exiting stream?

3-16. In a large chemical plant, steam is used as a heat source in several different processes. The steam condenses in heat exchangers, and the liquid water is recycled to a boiler, which converts it back into steam. Because the liquid water streams are coming from different processes, they are at different conditions, but there is one uniform exit stream which is at $P = 0.8$ MPa.

  The boiler operates at steady state and has the following streams entering:

  ▦ 1.67 kg/s of saturated liquid at $T = 423.15$ K.

  ▦ 3.33 kg/s of saturated liquid at $T = 453.15$ K.

  ▦ 1.25 kg/s of saturated liquid at $P = 0.6$ MPa.

  A. What rate of heat must be added to the boiler if the exiting steam is to be at $T = 573.15$ K?

  B. If the rate at which heat is added is 13,333 kJ/s, what is the temperature of the exiting steam?

3-17. A lake initially contains 20,000 kg of water and has a uniform temperature of 293 K. The ambient pressure is $P = 0.1$ MPa. During the course of a day, the following events affecting the lake occur:

  ▦ 200 kg of rain falls into the lake. The rainwater has a temperature of 298 K.

  ▦ Some water evaporates from the lake.

  ▦ Heat is both added to (sunlight) and removed from (convection into the cooler air) the lake.

  If you need the heat capacity of LIQUID water, use the approximation $C_V \sim C_P \sim 4.184$ KJ/kg · K.

  A. Find a reasonable estimate of the velocity of a raindrop and determine the kinetic energy of the rain when it strikes the lake. Is the kinetic energy significant compared to the enthalpy?

  B. The surface of the lake is at $T = 293$ K and $P = 0.1$ MPa. The enthalpy of water vapor at $T = 293$ K and $P = 0.1$ MPa is not in the steam tables. You could model the evaporating water

as saturated vapor at $T = 293.15$ K or saturated vapor at $P = 0.1$ MPa. Which is a better estimate?

  C. If 200 kg of water evaporated, and at the end of the day, the temperature of the lake is again a uniform 293 K, calculate the NET amount of heat that was added to or removed from the lake.

  D. Calculate the NET heat if the final temperature was a uniform 294 K and the amount of water evaporating was 200 kg.

  E. Calculate the NET heat if the final temperature was a uniform 294 K and the amount of water evaporating was 50 kg.

3-18. 10 mol/s of gas enters a steady state, adiabatic nozzle at $T = 573$ K and $P = 0.5$ MPa and leaves at $T = 373$ K and $P = 0.1$ MPa. Find the exiting velocity of the gas if it is:

  A. Steam

  B. Nitrogen, modeled as an ideal gas

  C. An ideal gas with $C_p = (7/2)R$ and a molecular mass of 28 kg/kmol

3-19. This problem examines the effect of mixing water at two different temperatures and/or two different phases, such as dropping ice cubes into warm water. Assume that $\Delta \hat{H}_{fus} = 333.55$ KJ/kg, that C_p for ice is constant at 2.11 KJ/kg·K, and that C_p for liquid water is constant at 4.19 KJ/kg·K. For each the following cases, assume the mixing is adiabatic and carried out at atmospheric pressure. For each case, determine the temperature and phase of the water when the mixing is complete. If there are two phases in equilibrium, determine the mass of each phase.

  A. 0.1 kg of liquid water at 50°C is mixed with 0.05 kg of liquid water at 0°C.

  B. 0.1 kg of liquid water at 50°C is mixed with 0.05 kg of ice at 0°C.

  C. 0.05 kg of liquid water at 0°C is mixed with 0.05 kg of ice at 0°C.

  D. 0.05 kg of liquid water at 20°C is mixed with 0.05 kg of ice at –20°C.

3-20. Superheated steam enters a nozzle that has an inlet diameter that measures 0.025 m and an outlet diameter that measures 0.01 m. Its outlet velocity has been measured at 30 m/s. The steam that enters the nozzle is at 923.15 K and 1350 kPa and the steam that exits the nozzle is at 673.15 kPa and 100 kPa.

  A. Determine the velocity of the steam upon entering the nozzle.

  B. Determine the rate at which this process gains or loses heat.

  C. Example 3-4 and Section 3.6.2 indicate that in modeling a nozzle, it is common to assume the nozzle is adiabatic and that the velocity entering the nozzle is negligible. Do your answers in this case seem consistent or inconsistent with those common assumptions?

3-21. A rigid, 10 m³ vessel initially contains 100 kg of water/steam at $P = 0.2$ MPa. 10 kilograms of steam at $P = 0.5$ MPa and $T = 573.15$ K is gradually added to the vessel. During the same time period, heat exchange occurs with the surroundings. The final water/steam mixture is at $P = 0.3$ MPa.
  A. What fraction of the initial contents of the vessel were in the vapor phase?
  B. What fraction of the final contents of the vessel are in the vapor phase?
  C. How much heat was added or removed?

3-22. A vessel initially contains 1000 kg of saturated liquid water at $T = 303.15$ K, which needs to be cooled. The vessel is adiabatic but has a piston that allows pressure and volume to be changed and a valve that can be opened to allow vapor to escape through the top. So, a small portion of the water is allowed to evaporate and escape, resulting in the cooling of the remaining water. Calculate each of the following, and state and justify any assumptions that you make.
  A. The temperature of the water remaining in the vessel if 1 kg of water evaporated.
  B. The amount of water that would have to evaporate in order to cool the water remaining in the vessel to 298.15 K.
  C. The amount of water that would have to evaporate in order to cool the water remaining in the vessel to 293.15 K.

3-23. Water at $P = 2.5$ MPa and $T = 473.15$ K enters a steady-state flash chamber with a flow rate of 5 kg/s. The liquid and vapor streams exiting the flash chamber are both at $P = 0.1$ MPa.
  A. If the flash chamber is adiabatic, what fraction of the entering water leaves the flash as vapor?
  B. How much heat must be added to or removed from the flash chamber in order to vaporize half of the entering water?

3-24. This problem presents a comparison of the work required in adiabatic pumps and compressors.
  A. Saturated water vapor enters a steady-state compressor at $P = 0.1$ MPa. Estimate the work required to compress this vapor up to $P = 1.0$ MPa and $T = 673.15$ K.
  B. Saturated liquid water at $P = 0.1$ MPa enters a pump. Estimate the work required to pump the liquid up to $P = 1$ MPa, and estimate the temperature of the exiting liquid.

3-25. A two-step, steady-state process is used to compress 100 kg/min of nitrogen from $P = 0.1$ MPa and $T = 250$ K to $P = 1$ MPa and $T = 250$ K. First, an adiabatic compressor is used to convert the nitrogen from $P = 0.1$ MPa and $T = 250$ K to $P = 1$ MPa and $T = 300$ K. Then a heat exchanger is used to cool the nitrogen to $T = 250$ K. Find the work added in the compressor, and the heat removed in the heat exchanger.
  A. Use an ideal gas model.
  B. Use Figure 2-3 as much as possible. If you need an equation of state, use the van der Waals equation, with $a = 0.137$ Pa · m⁶/mol² and $b = 3.86 \times 10^{-5}$ m³/mol.

3-26. 1 kg of nitrogen is contained in a piston-cylinder device. The nitrogen is isothermally compressed from 0.1 MPa to 1 MPa at $T = 300$ K. Find the initial volume, final volume, and work and heat added to or removed from the nitrogen.
  A. Assume nitrogen behaves as an ideal gas.
  B. Use Figure 2-3 as much as possible. If you need an equation of state, use the van der Waals equation, with $a = 0.137$ Pa · m⁶/mol² and $b = 3.86 \times 10^{-5}$ m³/mol.

Problems 3-27 through 3-29 involve the same compound that was considered in Chapter 2, Problems 2-22 through 2-26:

■ A molecular weight of 50 kg/kmol

■ A normal melting point of 300 K

■ A normal boiling point of 350 K

■ A constant heat capacity in the solid phase of $C_P = 3$ KJ/kg·K

■ A constant heat capacity in the liquid phase of $C_P = 3.6$ KJ/kg·K

■ A constant density in the solid phase of $\rho = 640$ kg/m³

■ A constant density in the liquid phase of $\rho = 480$ kg/m³

■ An ideal gas heat capacity of $C_P^* = 4186.8 - 6.28T + 0.01356T^2$ J/kg·K, where T is the temperature in K

■ An enthalpy of fusion $\Delta \hat{H}_{fus} = 175$ KJ/kg at atmospheric pressure

■ An enthalpy of vaporization $\Delta \hat{H}_{vap} = 300$ KJ/kg at atmospheric pressure

3-27. 0.15 kg-mol/s of the compound enters a steady-state boiler as saturated liquid at $P = 101.325$ kPa. Find the rate at which heat is added if the exiting stream is:
  A. Saturated vapor at $P = 101.325$ kPa
  B. Vapor at $P = 101.325$ kPa and $T = 375$ K
  C. Vapor at $P = 70.927$ kPa and $T = 375$ K

3-28. Initially, one kmol of the compound is placed in a piston-cylinder device at $T = 270$ K and $P = 101.325$ kPa. The compound is heated at constant pressure. Find the heat (Q) added if the final state is:
  A. 50% solid and 50% liquid at $T = 300$ K
  B. Liquid at 325 K
  C. 25% vapor and 75% liquid at $T = 350$ K
  D. Vapor at 366 K

3-29. A liquid stream contains 1 kg/s of the compound at $T = 311$ K and $P = 101.325$ kPa. It needs to be boiled and heated to $P = 101.325$ kPa and 353 K, as that is the temperature at which it must enter a chemical reactor. This heating/boiling process is to be carried out in a shell- and-tube heat exchanger with steam at $P = 0.1$ MPa and $T = 573.15$ K to be used as the source of the heat. Find the flow rate of entering steam required if

A. The steam exits the heat exchanger as saturated vapor at $P = 0.1$ MPa.

B. The steam condenses completely, and exits the heat exchanger at saturated liquid at $P = 0.1$ MPa.

3-30. A small water tower has a cylindrical chamber 3 m in diameter and 4 m tall. The bottom of the cylinder is 15 m above the ground. Water is pumped from an underground reservoir into the chamber, via a pipe that enters through the bottom of the chamber. The surface of the reservoir is 5 m below the ground. Assume that the tower, reservoir, and pipes are adiabatic, and that all water is at a uniform temperature, and a uniform density of 1000 kg/m³.

A. When the water tower is completely empty, what is the minimum pump work required to move 1 kg of water into the tower?

B. When the water tower is almost full, what is the minimum pump work required to move 1 kg of water into the tower?

C. A home has a second floor faucet is 3 m above the ground. The ground in the area is level, and there is nothing powering the water except the force of gravity. When the faucet it turned on, what is the maximum velocity the water can have?

D. Suppose the water tower has these same dimensions but is located on top of a hill. Repeat part C, assuming the altitude of the water faucet is 5 m *below* the altitude at the top of the hill.

3-31. A steady-state distillation column is designed to separate benzene from toluene. The separation is nearly enough complete that, for the purposes of designing the reboiler and the condenser, we can *model* the material in the reboiler as pure toluene and the material in the condenser as pure benzene.

▪ 10 mol/s of toluene enters the reboiler as saturated liquid at $P = 0.11$ MPa. 80% of the entering toluene is boiled and returned to the column as saturated vapor at $P = 0.11$ MPa; the rest is removed as the bottom product.

▪ 8 mol/s of benzene enters the condenser as saturated vapor at $P = 0.09$ MPa. All of the entering vapor is condensed into saturated liquid at $P = 0.09$ MPa. 75% of the liquid benzene is returned to the column and the rest is removed as the top product.

▪ The heat for the reboiler is to be provided by saturated steam, which is available at $P = 0.1$ MPa, $P = 0.3$ MPa, or $P = 0.5$ MPa.

▪ Either air or water can be used as the coolant in the condenser. Either way, the coolant enters at $P = 0.1$ MPa and $T = 298$ K, and leaves at $P = 0.1$ MPa and a temperature 10 K lower than the temperature of the benzene in the condenser.

You will probably have to look beyond this book to find physical properties of benzene, toluene, and water that are needed to solve the problem. Indicate your sources. If you can't find the exact data you need, state and justify any assumptions that you make in solving the problem.

A. Determine the rate at which heat is added to the toluene in the reboiler.

B. Determine the flow rate of steam entering the reboiler, for each of the three possible inlet pressures. In each case, assume the steam is completely condensed and leaves the reboiler as saturated liquid, with the same pressure at which it entered.

C. Comment on the results of part B. Are any of the three steam pressures obviously better or worse options than the others? Knowing that the higher-pressure steam is more expensive than the lower-pressure steam, is there any rationale for using the higher-pressure steam?

D. Determine the flow rate of water, if water is used as the coolant.

E. Determine the flow rate of air, if air is used as the coolant. Assume air at $P = 0.1$ MPa acts like an ideal gas, and that $C^*_{P,\text{air}} = 0.79\, C^*_{P,N_2} + 0.21\, C^*_{P,O_2}$

F. The air for the condenser is at the local ambient conditions and can be considered free. The water for the condenser is not free. Do you see any rationale for using water instead of air?

3.10 GLOSSARY OF SYMBOLS

A	area	C_V	constant-volume heat capacity (molar basis)
C	number of chemical compounds	C^*_V	constant-volume heat capacity (molar basis) for ideal gas
C_P	constant-pressure heat capacity (molar basis)		
C^*_P	constant-pressure heat capacity (molar basis) for ideal gas	$\hat{C}_V$	constant-volume heat capacity (mass basis)

$\hat{C}_P$	constant-pressure heat capacity (mass basis)	n	number of moles added to or removed from system	v	velocity
F	degrees of freedom	$\dot{n}$	molar flow rate	V	volume
g	gravitational constant	P	pressure	$\hat{V}$	specific volume
H	enthalpy	P.E.	potential energy	$\underline{V}$	molar volume
$\hat{H}$	specific enthalpy	$\dot{Q}$	rate of heat addition	W_S	shaft work
$\underline{H}$	molar enthalpy	Q	heat	W_{EC}	expansion/contraction work
h	height	R	gas constant	$\dot{W}_S$	rate of shaft work
K.E.	kinetic energy	r	radius	$\dot{W}_{EC}$	rate of expansion/contraction work
M	mass of system	T	temperature	$\Delta \underline{H}^0_f$	enthalpy of formation
m	mass added to or removed from system	t	time	ρ	density
$\dot{m}$	mass flow rate	U	internal energy	π	number of distinct phases
N	number of moles in system	$\hat{U}$	specific internal energy		
		$\underline{U}$	molar internal energy		

3.11 REFERENCES

A. C. Clarke, *Profiles of the Future*, Harper & Row, New York, 1962.

J. Kluger, "Rocket Scientist: Robert Goddard," *Time,* 153, 12, March 29, 1999.

Entropy

4

LEARNING OBJECTIVES

This chapter is intended to help you learn how to:

- Define a *reversible* process and recognize when processes are reasonably modeled as reversible
- Define *entropy* and predict qualitatively whether a physical change (melting, boiling, temperature increase, volume increase, etc.) should cause entropy to increase or decrease
- Recognize the *irreversible* features of real processes that lead to *generation of entropy*
- Quantify changes in the entropy of a substance using the fundamental definition of entropy
- Write *entropy balances* that describe a variety of physical systems
- Compute the maximum work that can be produced by a process, or the minimum work required to accomplish a specific task
- Use *efficiencies* to compare real processes to idealized processes
- Apply entropy balances in combination with energy balances to solve engineering problems

The conservation of mass and energy are foundational engineering principles that can be applied to a vast array of engineering problems. Conversion of energy from one form to another is a common theme in engineering applications throughout the first three chapters, and the generalized energy balance is a powerful fundamental tool in modeling these applications. However, by itself, the energy balance might imply that it is possible to convert energy from any form into any other, at any time, without restrictions or limitations of any kind. This is not true, and one role of the concept of *entropy* in thermodynamics is revealing the limitations on energy conversion. This chapter examines entropy and its use. ■

FOOD FOR THOUGHT 4-1

Suppose ice cubes were placed in a warm glass of water, and the ice cubes got colder while the water got warmer. Is the first law of thermodynamics violated?

4.1 MOTIVATIONAL EXAMPLE: Turbines

This chapter introduces a new state property called *entropy*. We first illustrate how entropy is a useful property through a closer examination of turbines. The introductions to Chapters 1 and 2 discussed the Rankine cycle, how the turbine is the unit operation in the cycle that actually produces work, and how the fluid experiences large pressure drops in the turbine. But in previous examples, the inlet and outlet conditions of the steam were simply given. How can we relate the work produced to the pressure drop in a more fundamental way? We will see that this is a question for which the energy balance, while useful, does not tell the whole story.

EXAMPLE 4-1	**FOUR POSSIBLE TURBINE CONFIGURATIONS**

Steam at 823.15 K and 3 MPa enters a steady-state, adiabatic turbine (Figure 4-1). The exiting stream is at $P = 0.1$ MPa. Find the work produced per kilogram of entering steam if the exit stream is

A. Saturated liquid
B. Saturated vapor
C. Superheated steam at 473.15 K
D. A 50-50 mixture (by mass) of saturated liquid and saturated vapor

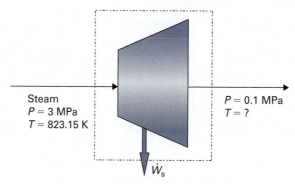

Steam
$P = 3$ MPa
$T = 823.15$ K

$P = 0.1$ MPa
$T = ?$

$\dot{W}_s$

FIGURE 4-1 Turbine described in Example 4-1.

SOLUTION:

Step 1 *Define the system and write the energy balance*
In Section 3.6.3, we applied the energy balance to a steady-state, adiabatic turbine. Using the turbine as the system, the energy balance simplifies to:

$$\frac{\dot{W}_s}{\dot{m}} = \hat{H}_{out} - \hat{H}_{in}$$

> $\dot{W}_s$ is the *rate* at which work is added. The time dimensions of $\dot{W}_s$ and $\dot{m}$ cancel, giving dimensions of energy per unit mass.

The left-hand side is exactly what we're trying to find: work per unit mass of entering steam.

Step 2 *Collect data*
The right-hand side can be evaluated using the steam tables. The entering stream is the same in all four variations of the problem:

$$\hat{H}_{in} = 3569.7 \text{ kJ/kg} \qquad (\text{steam at } T = 823.15 \text{ K and } P = 3 \text{ MPa})$$

In parts A through C, the exiting stream is a single phase, so $\hat{H}_{out}$ is also obtained directly from the steam tables.

A. $\hat{H}_{out} = 417.5$ kJ/kg   (saturated liquid at 0.1 MPa)
B. $\hat{H}_{out} = 2674.9$ kJ/kg   (saturated vapor at 0.1 MPa)
C. $\hat{H}_{out} = 2875.5$ kJ/kg   (superheated vapor at 0.1 MPa)

Part D requires an additional step.

Step 3 *Compute properties of a 50-50 mixture*
In Example 2-1, an intensive property of a system that contains both saturated liquid and saturated vapor was found as a weighted average of the properties of the individual phases. For specific enthalpy, the expression is

$$\hat{H} = (1 - q)\hat{H}_L + q\hat{H}_V \tag{4.1}$$

In part D, the exiting stream is a mixture of liquid and vapor, so $\hat{H}_{out}$ can be found using equation 4.1. The vapor fraction or quality (q) is given as 0.5 (50-50 mixture), and $\hat{H}_L$ and $\hat{H}_V$ are the specific enthalpies of the saturated liquid and vapor at 0.1 MPa, which we found for parts A and B.

$$\hat{H}_{out} = (1 - 0.5)\left(417.5\frac{kJ}{kg}\right) + (0.5)\left(2674.9\frac{kJ}{kg}\right) = 1546.2\frac{kJ}{kg} \tag{4.2}$$

Step 4 *Plug values into energy balance*
The results are summarized as follows:

Part	$\hat{H}_{in}$ (kJ/kg)	$\hat{H}_{out}$ (kJ/kg)	$\dot{W}_S/\dot{m} = \hat{H}_{out} - \hat{H}_{in}$ (kJ/kg)
A	3569.7	417.5	−3152.2
B	3569.7	2674.9	−894.8
C	3569.7	2875.5	−694.2
D	3569.7	1546.2	−2023.5

Example 4-1 proposes four possible exit streams for a turbine. Based on the energy balance, all of the scenarios presented look plausible and we would clearly say that example A was the "best": it produces the most work. But can we simply "choose" outlet stream A? No, we cannot. Designers do not have infinite freedom to specify process variables. For example, in practice one normally controls the temperature of a piece of equipment by adding or removing heat. Consequently, as the designer, if you define a piece of equipment as adiabatic, you can't realistically also choose a specific outlet temperature. So when we feed real steam into a real turbine, what actually comes out? The energy balance can be applied to many possible outlet streams, but it can't tell us which one is "right."

> In reality, parts A, B, and D in Example 4-1 are all physically impossible. The energy balance would never have told us this, but the concept of entropy does.

Looking, for example, at part A, we see that the calculated work is $\dot{W}_S/\dot{m} = -3152.2$ kJ/kg. And it's true that −3152.2 kJ/kg of energy must be removed in order to convert one kilogram of steam at 823 K and 3 MPa into one kilogram of saturated liquid water at $P = 0.1$ MPa. However, there is simply no way to remove *all* of this energy in the form of work. There is a theoretical limit to how much work can be obtained when the steam expands to a pressure of 0.1 MPa. In Example 4-7, we will use the concept of entropy to prove that this maximum is −887.4 kJ for each kilogram of steam. Parts A, B, and D are physically impossible; the maximum work is exceeded. Part C is physically possible, but the work obtained is 193.2 kJ/kg less than the maximum; you might call this 193.2 kJ/kg of "lost work." In this chapter, entropy allows us to quantify maximum work and lost work.

It might seem strange that we are talking about the "maximum" work being a negative number. Work produced by the system is energy *leaving* the system, and therefore negative. The negative $\dot{W}_S$ with the largest absolute value corresponds to the maximum work produced.

In applying the concept of entropy, two points are helpful.

1. *Entropy is a state property.*
2. *The entropy of the universe cannot decrease.*

Knowing these two facts actually allows us to show quantitatively that parts A, B, and D in Example 4-1 are impossible. If we define the turbine as our system, then the system is at a steady state, meaning that the entropy of the system (like all state properties) must be constant. Since the entropy of the universe cannot decrease, and no entropy is accumulating in the system, the rate at which entropy leaves the system must be greater than or equal to the rate at which entropy enters the system.

We will learn in Section 4.3.1 that heat transfer affects the entropy of a system, but here the system is adiabatic. The only way entropy enters or leaves this system is through material streams, quantified by the mass flow rate times the specific entropy:

Equation 4.3 quantifies the rate at which entropy is transported in a single stream.

$$\dot{S} = \dot{m}\hat{S} \tag{4.3}$$

The mass flow rates of the entering and exiting streams are equal in this steady-state process. Consequently, based on the previous paragraph, we can say that *the specific entropy of the stream leaving the turbine can't be smaller than the specific entropy of the entering stream.*

Section 4-4 presents the entropy balance, which we can use to prove the italicized statement in a rigorous mathematical way.

Because entropy is a state property, there is only one value the specific entropy of superheated steam at $T = 823.15$ K and $P = 3$ MPa can have. That value, $\hat{S} = 7.3768$ kJ/kg·K, is available in the steam tables. According to the steam tables, the entropy of saturated steam at $P = 0.1$ MPa is 7.3588 kJ/kg·K. The entropies of saturated liquid or of liquid–vapor mixtures at this pressure are even smaller. All of these values are less than 7.3768 kJ/kg·K and therefore are unrealistic. However, the specific entropy of steam at $P = 0.1$ MPa and $T = 473.15$ K is 7.8356 kJ/kg·K; this at least is a physically realistic value based upon what we have said about entropy so far.

At this point, even if we have relatively little insight into what entropy actually is, we can see that the concept has practical value. The energy balance is very useful in solving engineering problems, but by itself the energy balance reflects no limitations or restrictions on conversion of energy from any form into any other form. As a simple example, consider that heat transfer from a colder object to a hotter object does not violate the first law of thermodynamics, but we know from experience that such a process does not occur. Entropy is a property that allows us to incorporate realistic limitations like this one into our mathematical models of thermodynamic systems and processes.

We begin our examination of entropy by introducing the concept of the reversible process, and how it relates to the notions of "maximum work" and "minimum work."

4.2 Reversible Processes

The attributes that characterize an ideal gas, listed in Section 2.3.3, are never literally true for real gases. Similarly, these four statements will never be literally true for any real process. But the "ideal gas" and the "reversible process" are often *useful models.*

This section defines what a reversible process is, and explains why the concept is useful. Briefly, **a reversible process** is identified as follows:

1. A process in which all driving forces for change are infinitesimally small.
2. A process that progresses through a series of equilibrium states.
3. A process that can be returned to its original state with no NET addition of heat or work.
4. A process that has no friction.

Reversible and irreversible processes are illustrated in Examples 4-2 and 4-3, where these attributes of a reversible process are explained in detail.

Consider a gas in a piston-cylinder device. Initially, the gas is at $T = 300$ K and $P = 0.5$ MPa, while the surrounding atmosphere is at $T = 300$ K and $P = 0.1$ MPa. Because the gas is at higher pressure than the surroundings, the gas has the potential to produce mechanical work.

Suppose the gas expands until it is in equilibrium with the surroundings, reaching a final state of $T = 300$ K and $P = 0.1$ MPa. There are many ways this expansion could occur. Two of the possibilities are illustrated below. Example 4-2 is an *irreversible expansion* and Example 4-3 is a *reversible expansion*. We will see that the reversible path proves to produce more work.

| IRREVERSIBLE EXPANSION | EXAMPLE 4-2 |

Two moles of an ideal gas are confined in a piston-cylinder device, initially at $P = 0.5$ MPa and $T = 300$ K. The surroundings are also at $T = 300$ K. The atmosphere is at $P = 0.1$ MPa but clamps hold the piston in place, as shown in Figure 4-2. When the clamps are removed, the gas expands rapidly—so rapidly that heat transfer is negligible during this step—until the pressure of the gas is $P = 0.1$ MPa. Heat is then transferred to the gas from the surroundings until the gas reaches a final state of $P = 0.1$ MPa and $T = 300$ K. Find the total work done by the gas on the surroundings, and the total heat transferred to the gas.

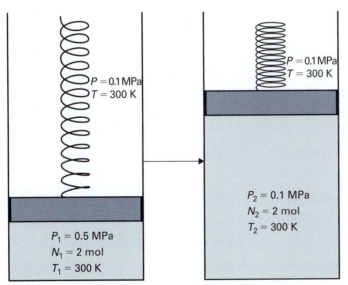

FIGURE 4-2 Irreversible expansion of a gas as discussed in Example 4-2.

SOLUTION: Before beginning, we note that this is computationally a fairly simple problem, but if the system and process aren't defined carefully, one could easily make it more complicated than it needs to be. We are asked to find Q and W. The problem statement describes a two-step process:

1. An adiabatic expansion from $P = 0.5$ MPa to $P = 0.1$ MPa, during which a temperature drop occurs.

2. Isobaric heating of the gas to 300 K.

Your first instinct might be to calculate Q and W for each step, and add them together to find the total Q and W for the process. But we will see that this two-step approach is neither necessary nor efficient.

Step 1 *Define a system*

Q and W for the gas is what we're asked to compute, so we define the gas as the system.

Step 2 *Simplify the energy balance*

We use the time-independent energy balance (no information on duration of process). The gas is a closed system with no moving parts (no shaft work) and changes in kinetic and potential energy are negligible. Thus, as in Examples 3-9 and 3-10, the energy balance simplifies to:

$$M_2 \hat{U}_2 - M_1 \hat{U}_1 = W_{EC} + Q \tag{4.4}$$

Step 3 *Envisioning the process*

Examining the first step of the process only, we know $Q = 0$ and we know work is done by the system on the surroundings. Thus we expect W_{EC} to be negative, which means that U_2 is smaller than U_1, which in turn means that the temperature decreases. However, the heat capacity of the gas is not given, so there is currently no way to quantify this temperature change.

Instead, we can apply the energy balance to the *entire* process, looking only at the initial and final states. This is a closed system, so the mass is constant ($M_2 = M_1 = M$) and can be factored out:

$$M(\hat{U}_2 - \hat{U}_1) = W_{EC} + Q \tag{4.5}$$

The initial and final temperatures are both 300 K. For an ideal gas, the specific internal energy ($\hat{U}$) is a function of temperature only. Thus, while $\hat{U}$ is not constant throughout the process, it is the same at the initial and final states ($\hat{U}_1 = \hat{U}_2$). Consequently the energy balance *for the entire process* simplifies to:

$$W_{EC} = -Q \tag{4.6}$$

Step 4 *Calculating expansion work*

It was established in Section 1.4.3 that expansion/contraction work is given by:

$$W_{EC} = -\int P \, dV$$

P is the pressure opposing the motion, while V is the volume of the system. Here, the gas is expanding, and the pressure opposing the motion is constant at $P = 0.1$ Mpa. Consequently the expansion work is

$$W_{EC} = -(0.1 \text{ MPa})(V_2 - V_1) = (0.1 \text{ MPa})(V_1 - V_2) \tag{4.7}$$

The ideal gas law is used to evaluate V_2 and V_1:

$$W_{EC} = (0.1 \text{ MPa}) \left(\frac{N_1 R T_1}{P_1} - \frac{N_2 R T_2}{P_2} \right) \tag{4.8}$$

$$W_{EC} = (0.1 \text{ MPa}) \left[\frac{(2 \text{ mol})\left(8.314 \dfrac{\text{J}}{\text{mol} \cdot \text{K}}\right)(300 \text{ K})}{0.5 \text{ MPa}} - \frac{(2 \text{ mol})\left(8.314 \dfrac{\text{J}}{\text{mol} \cdot \text{K}}\right)(300 \text{ K})}{0.1 \text{ MPa}} \right] = \mathbf{-3991\,J}$$

This is physically reasonable in that W should be negative; the system is doing work on the surroundings.

Step 5 *Solve energy balance*

Substituting the result of step 4 into the energy balance (Equation 4.6) gives $Q = \mathbf{3991\ J}$.

Example 4-3 shows that an expansion with the same initial and final states as Example 4-2 can produce a very different result for Q and W.

Internal energy is a state property. Therefore, the internal energy of the final state is determined only by the temperature of the final state and is independent of the temperature at any previous time. While temperature isn't uniform throughout the entire process, the initial and final temperatures are identical.

FOOD FOR THOUGHT 4-3

Suppose we made a plot of the internal energy of the system vs. time. Sketch what this plot would look like.

A real gas wouldn't necessarily be well described by the ideal gas law at $P = 0.5$ MPa. The ideal gas model is used for simplicity in Examples 4-2 and 4-3 to illustrate reversible and irreversible processes.

| REVERSIBLE EXPANSION | **EXAMPLE 4-3** |

Two moles of an ideal gas are confined in a piston-cylinder device, initially at $P = 0.5$ MPa and $T = 300$ K. The atmosphere is at $P = 0.1$ MPa but there is a pile of sand on top of the piston, as shown in Figure 4-3. The mass of the sand is sufficient to make the total downward pressure on the piston 0.5 MPa. The grains of sand are removed one at a time, leading to a gradual decrease in pressure and a gradual expansion of the gas, until the gas is at $P = 0.1$ MPa. The temperature of the gas is $T = 300$ K throughout.

FOOD FOR THOUGHT 4-4

List the ways this process is different from the one in Example 4-2, and consider how your list relates to the four properties of reversible processes given at the beginning of this section.

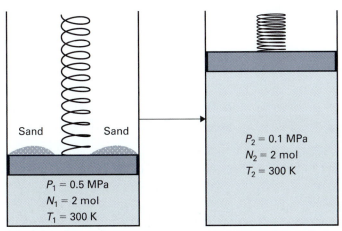

FIGURE 4-3 Reversible expansion process described in Example 4-3.

SOLUTION:

Step 1 *Compare/contrast with Example 4-2*

As in Example 4-2, we can define the gas as the system. This is a closed system with no moving parts, and changes in potential and kinetic energy can be neglected, so again the energy balance is

$$M_2 \hat{U}_2 - M_1 \hat{U}_1 = W_{EC} + Q$$

As in Example 4-2, the gas is an ideal gas, so $\hat{U}$ is only a function of temperature. Therefore $\hat{U}$ is the same at the start and end of the process ($T_1 = T_2 = 300$ K). Thus, $M_2 \hat{U}_2 - M_1 \hat{U}_1 = 0$, which in turn means

$$W_{EC} = -Q$$

Step 2 *Evaluating the external pressure*

The work of expansion/contraction is given by:

$$W_{EC} = -\int P \, dV \qquad (4.9)$$

P is the pressure opposing the motion, which here means the external pressure pushing down on the piston. Unlike Example 4-2, this pressure is not constant; it starts at $P = 0.5$ MPa and gradually decreases to $P = 0.1$ MPa.

In order to integrate Equation 4.9, we must express P as a function of V. V represents the volume of the system, which is the gas. If P is the gas pressure, it could be related to V through the ideal gas law. But P represents the external pressure.

The key insight comes from the description of the process as *gradual*. The grains of sand are being removed one at a time. When a grain is removed, the external pressure

"Slow" and "gradual" are key words that might lead you to *consider* modeling a process as reversible. However, the definitive way to decide whether it *is* reasonable to model a process as reversible is comparing the process to the criteria listed at the beginning of this section.

Work and heat are not state properties, so we cannot assume these are the same as they were in Example 4-2 based on the initial and final states being the same.

The energy balance is $Q = -W_{EC}$. Every time a grain of sand is removed, a tiny amount of work is done (dW_{EC}) and an equivalent amount of energy is added to the system as heat (dQ). Q is the sum of all these tiny heat transfers ($Q = \int dQ$).

is *incrementally* smaller than the gas pressure, and the gas expands *incrementally*, equalizing the pressure. This expansion (dV) leads to an incremental decrease in the gas temperature. Since the surroundings are now incrementally warmer than the gas, a very small amount of heat (dQ) is added to the gas, equalizing the temperatures. Then the next grain is removed. As grains are removed, the external and internal pressures both gradually decrease from 0.5 MPa to 0.1 MPa. However, at any specific instant in time, the external and internal pressures are essentially identical; the mass of a single grain of sand accounts for the entire difference between them. Consequently, in modeling the process it is reasonable to equate the external pressure to the gas pressure, which means we can relate P to V through the ideal gas law:

$$W_{EC} = -\int \frac{NRT}{V} dV \tag{4.10}$$

Step 3 *Calculate expansion work*

In this process, N, R, and T are all constant, so the integration is

$$W_{EC} = -\int_{V=V_1}^{V=V_2} \frac{NRT}{V} dV = -NRT \ln\left(\frac{V_2}{V_1}\right) \tag{4.11}$$

One of the descriptors for a reversible process is that it progresses "through a series of equilibrium states." The discussion in Step 2 illustrates an example of what that means.

While we can calculate V_1 and V_2 using the ideal gas law, it actually isn't necessary. Since we know P and T at both the beginning and end of the process, we can substitute for V_1 and V_2 for

$$W_{EC} = -NRT \ln\left(\frac{\dfrac{NRT_2}{P_2}}{\dfrac{NRT_1}{P_1}}\right) = -NRT \ln\left(\frac{P_1}{P_2}\right) \tag{4.12}$$

$$W_{EC} = -(2\,\text{mol})\left(8.314 \frac{\text{J}}{\text{mol} \cdot \text{K}}\right)(300\,\text{K}) \ln\left(\frac{0.5\,\text{MPa}}{0.1\,\text{MPa}}\right) = -8029\,\text{J}$$

Step 4 *Close energy balance*

Substituting W_{EC} into the energy balance from step 1, we find $Q = 8029\,\text{J}$ of heat was added during the course of the process.

FOOD FOR THOUGHT 4-5

The discussion so far suggests that a reversible process is "better" than an irreversible one; it produces more work. Do you see any advantages for the irreversible process (Example 4-2) compared to the reversible one (Example 4-3)?

In both processes described in Examples 4-2 and 4-3, the initial and final states of the process were the same. We showed mathematically that more work was actually done in the reversible expansion. Physically, we can rationalize this result by comparing the physical effects of the processes. In both processes, the piston moved upward and displaced the atmosphere. In both cases, the air exerted a downward pressure of 0.1 MPa on the piston, the gas expanded to the same final volume, and the work required to "push the atmosphere up" was the same. But in Example 4-2, pushing the atmosphere up was the only work done. In Example 4-3, more work was done: in addition to pushing up the atmosphere, the piston also lifted the grains of sand.

At the start of this section we listed four attributes of reversible processes. We now discuss each in more detail, with reference to Example 4-2 and Example 4-3.

1. **A reversible process is a process in which all driving forces for change are infinitesimally small.**

Example 4-3 illustrates this aspect of a reversible process through two physical phenomena: expansion and heat transfer.

■ The driving force for expansion is a difference in the upward and downward pressures acting on the piston, but the difference in pressure at any given time is very small—the weight of a single grain of sand accounts for the difference. Therefore, in building a mathematical model of the process, it is valid to assume $P_{external} = P_{internal}$ at any particular time.

■ The driving force for heat transfer is a difference in temperature between the gas and the surroundings, but the difference in temperature is very small; it is valid to assume that the gas and the surroundings are both at a temperature of 300 K throughout the entire process.

2. **A reversible process is a process that progresses through a series of equilibrium states.**

The FINAL state in both Example 4-2 and Example 4-3 is easy to recognize as an equilibrium state; both the gas and the surroundings are at $P = 0.1$ MPa and $T = 300$ K. But the gas in Example 4-3 can be regarded as "at equilibrium" for the entire process.

The definition of a system at equilibrium requires that there be no driving force for change. It might seem counter-intuitive to describe a system that undergoes a pressure change from 0.5 MPa to 0.1 MPa as "at equilibrium." But the external and internal pressures acting on the gas were never significantly different from each other. Thus, in a reversible process, the system can be modeled as in equilibrium at any particular moment. The system progresses very slowly from one "equilibrium state" to another.

By contrast, an *irreversible* process is one in which there is at least one driving force for change that is significant in magnitude. In Example 4-2, when the blocks holding the piston in place were withdrawn, the pressure pushing upward on the piston was 0.5 MPa while the pressure pushing downward was 0.1 MPa. This was a large driving force that led to rapid, irreversible expansion, and the only times this process can be modeled as "at equilibrium" are the beginning—before the blocks were removed—and the end—when the gas reached the same T and P as the surroundings.

3. **A reversible process is a process that can be returned to its original state with no NET addition of heat or work.**

The word "reversible" is descriptive in that it only takes a fractional change in driving force to cause the process to reverse direction. In Example 4-3, we imagined causing a gas to expand by removing grains of sand from the piston one at a time, but we could at any point have caused the gas to compress by replacing the grains of sand one at a time.

Comparing Examples 4-2 and 4-3 demonstrates that the starting and ending conditions of the gas were identical, but the reversible pathway produced more work. Indeed, one can demonstrate that it is always true that reversible expansions produce maximum work. Recall that work of expansion/contraction is given by

$$W_{EC} = -\int_{V_1}^{V_2} P \, dV$$

where P is the pressure opposing the motion. So in an expansion, the larger the external pressure, the larger the work. External pressure cannot be greater than the system pressure. If it were, compression would occur instead, and the work

Examples 4-2 and 4-3 illustrate that Q and W are path-dependent properties, but the final equilibrium state ($T_2 = 300$ K, $P_2 = 0.1$MPa, $U_2 = U_1$) was the same regardless of the path taken to reach that state.

FOOD FOR THOUGHT 4-6

Here we have illustrated maximum work produced by an expansion. What kind of example could we use to illustrate the minimum work required for a compression?

would be positive (work done on system) rather than negative (work done by system). Since external pressure cannot *exceed* internal pressure, work is maximized when external pressure is *equal to* internal pressure, which is the case in a reversible expansion.

These observations generalize to processes that produce shaft work as well.

- *For any process that produces work, the maximum work is obtained from a reversible pathway*

- *For any process that requires work, the minimum work is required by a reversible pathway*

Examples 4-2 and 4-3 had different values of Q and different values of W, though $Q + W$ was equal to zero for both examples. The sum $Q + W$ had to be the same for both because it was equal to ΔU, which is a state function.

Let us now further consider reversing the processes described in Examples 4-2 and 4-3. Because the internal and external pressures are essentially the same at any given time on a reversible path, the work required to reversibly re-compress the gas to its original pressure would be equal in magnitude and opposite in sign to the work produced when it expanded. In Example 4-3, we computed $W = -8029$ J and $Q = +8029$ J for an isothermal reversible expansion of the gas from $P = 0.5$ MPa to $P = 0.1$ MPa. If the gas were compressed from $P = 0.1$ MPa to $P = 0.5$ MPa, again by an isothermal reversible path, the work required would be $W = +8029$ J and the heat released would be $Q = -8029$. The system would be restored to its original state, and the NET work and heat exchanged with the surroundings would be 0. By contrast, the irreversible process in Example 4-2 did only 3991 Joules of work on the surroundings ($W = -3991$ J), but at least 8029 Joules of work would be required to return the gas to its original state. In sum:

- In a reversible process, the system and surroundings can BOTH be returned to their original state.

- In an irreversible process, the system and surroundings cannot both be returned to their original state, because returning the system to its original state would require a net addition of heat and/or work.

FOOD FOR THOUGHT 4-7

Thinking back to the Motivational Example, consider a reversible turbine. What would need to be true for it to be considered reversible? Does it seem plausible that a reversible turbine could actually be built?

4. A reversible process is frictionless.

Throughout this book, we have assumed that the piston in a piston-cylinder apparatus is frictionless. Significant friction between the piston and the cylinder would serve as an additional force opposing motion in either direction. Suppose, for example, that the force of friction was equivalent to 10 kPa of pressure acting on the piston. As Figure 4-4 illustrates, the final pressure of the system would be 0.11 MPa rather than 0.1 MPa: at this point, the driving force for change still exists but is no longer sufficient to overcome friction. Thus, there is no further motion of the piston beyond this point. In a frictionless process the gas would continue expanding until its pressure was 0.1 MPa, thus doing more work. Figure 4-4 further illustrates that in order to get the process to reverse direction, one would need to change the applied force by at least 20 kPa. Recall that in a reversible process, an infinitesimal change in driving force is sufficient to change the direction of the process.

FOOD FOR THOUGHT 4-8

Force and pressure are two different things. How can a frictional "force" equal 10 kPa of pressure?

Friction is a force that opposes motion of the piston in either direction.

This section started with four identifying attributes of a reversible process. We have now considered the significance of each of these in the context of reversible and irreversible processes. The concept of the reversible process is fundamentally necessary for defining, understanding, and quantifying the property of entropy, which we introduce formally in the next section.

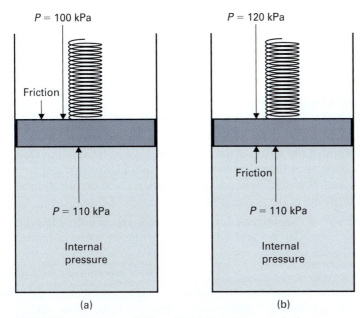

FIGURE 4-4 A piston cylinder with frictional force equivalent to 10 kPa of pressure. (a) The internal pressure is 10 kPa greater than the external pressure, but this isn't enough of a driving force to overcome the force of friction, so no further motion occurs. (b) In order to compress the gas, the external pressure must be increased until it exceeds the internal pressure by more than 10 kPa.

4.3 Defining and Describing Entropy

Examples 4-2 and 4-3 described a gas confined in a piston-cylinder arrangement. Because the gas pressure ($P = 100$ kPa) was initially significantly greater than the external pressure, one might say the gas had the *potential* to do mechanical work. Because the final temperature and pressure of the gas were in equilibrium with the surroundings, the gas at this point had no more potential to do mechanical work, unless there was first some kind of outside intervention (adding heat, adding work, moving the cylinder to surroundings where the external pressure was lower, etc.).

Example 4-3 demonstrated that the gas was capable of doing 8029 Joules of work before it reached an equilibrium state with the surroundings. In Example 4-2, the initial and final states were the same as in Example 4-3, so the potential to do work was the same; however, in reality the work produced (3991 J) was less than half of the potential work. One might therefore say there were over 4000 Joules of "lost work." Entropy is a property that allows one to quantify "lost work."

> The **second law of thermodynamics** can be stated as follows. The entropy of the universe cannot decrease. The entropy of the universe is unchanged by any reversible process and increased by any irreversible process.

Like the first law of thermodynamics, this "law" is reflected consistently in observation and experience. The second law won't seem particularly helpful until you develop more insight into what entropy actually is. Therefore, we present several essential facts about entropy here.

In Examples 4-2 and 4-3, the entropy change of the gas was the same; the gas experienced the same decrease in *potential* to do work. In the reversible case the potential work was 100% converted; in the irreversible case less than 50% of the potential work was harnessed.

In 1865, Rudolf Clausius famously summarized the first two laws of thermodynamics as follows: "The energy of the world is constant. The entropy of the world tends toward a maximum." (Olson, 1998)

At some point, you may have heard a definition like "entropy is a measure of disorder." This will be addressed in Section 4.3.5.

■ **Entropy is a state property.** The change in entropy experienced by the gas in Example 4-2 and Example 4-3 must therefore be the same, since the initial and final states are the same.

■ **The entropy of the universe is unchanged by a reversible process.** Entropy can be viewed as a property that quantifies "lost work." There is no lost work in a reversible process: the maximum work possible for a particular initial and final state is obtained if the process produces work, and the minimum possible work is used by a process that requires work.

■ **The entropy of the universe never decreases.** Qualitatively this observation can be rationalized in that "lost work" can only be zero or positive; actual work can't exceed potential work. A more mathematical demonstration of the fact that entropy never decreases is given in Example 4-5.

■ **The entropy of the universe increases during any irreversible process,** as illustrated in Examples 4-5 and 4-9. The larger the increase in entropy, the more potential work was lost.

While the entropy change *of the gas* was identical in Examples 4-2 and 4-3, Example 4-4 will show that the entropy *change of the universe* was 0 in the reversible case and positive in the irreversible case. This reflects the fact that, in the reversible case, there was no "lost work."

Notice that several of the above statements refer to the entropy "of the universe." When we are doing a problem, "the universe" essentially means "the system AND the surroundings." When thinking about entropy it is important to distinguish among "the system," "the surroundings" and "the universe." The change in entropy *for a system* can be negative, zero, or positive, but if it's negative, then according to the second law of thermodynamics there must be a corresponding increase in the entropy of the surroundings. Consider, for example, a cup of water left outside on a cold winter night. In the morning the water will be ice, and ice has lower specific entropy than liquid water at the same pressure. Thus the entropy of the water in the cup has decreased, and the next section introduces the mathematics that would allow you to quantify this decrease. But the entropy of the universe has not decreased—it has increased through the heat transfer from the water to the surrounding air. Example 4-5 and Example 4-9 illustrate how heat transfer processes lead to an increase in the entropy of the universe.

A "homogeneous" system is uniform in temperature, pressure, and composition. A system that contains distinct components or regions, such as an ice–water mixture, is not homogeneous.

4.3.1 Mathematical Definition of Entropy

Quantitatively, the *change in entropy for a closed, homogeneous system* is defined as

$$dS = \frac{dQ_{rev}}{T} \tag{4.13}$$

In integrated form, it is

$$S_{final} - S_{initial} = \int_{initial}^{final} \frac{dQ_{rev}}{T} \tag{4.14}$$

where

■ S represents the total entropy of the system.

■ dS represents a differential change in the entropy of the system.

■ Q_{rev} is the heat energy added to the system *on a reversible path.*

■ T is the absolute temperature of the system at the boundary where the heat transfer occurs.

FOOD FOR THOUGHT 4-9

Entropy was introduced as a property that allows us to quantify "lost work," yet work doesn't appear in Equation 4.13 or 4.14. Is there a contradiction here?

Because Equation 4.14 involves heat, which is a path-dependent property, it is not obvious from the definition that entropy is indeed a state property. The crucial point is that dQ_{rev} in Equation 4.14 represents the *reversible* heat; the heat transferred on a reversible path. It can be proved that while Q is not a state function, $\int dQ_{rev}/T$ is a state function. A full proof will not be shown here, but it can be done by demonstrating that $\int dQ_{rev}/T = 0$ for any cyclic process (Gutman, 1988) (Vemulapalli, 1986). If S always returns to its original value at the end of any cyclic process, then its value must be dependent only upon the original (and final) state of the system, and not upon the path.

Equation 4.14 can only be applied directly to reversible paths. Thus, in order to use this expression to calculate the change in entropy resulting from an irreversible process, one must take advantage of the fact that entropy is a state property: the change in entropy is a function only of the initial and final states, and is not dependent upon the path. Consequently, one can envision a *hypothetical reversible path* between the initial and final states of interest, and apply Equation 4.14 to this hypothetical path.

Common conceptual errors that can occur when one uses Equation 4.14 include:

- One cannot integrate the equation directly to $\Delta S = Q/T$ unless the system temperature is constant. If temperature is not constant, or if one does not know whether it is constant, one must find a mathematical relationship between Q and T before one can integrate. This relationship typically comes from the energy balance, as illustrated below in Example 4-4.

- For an irreversible process, one can fall into the trap of using the actual Q from the real path in Equation 4.14, rather than constructing a hypothetical reversible process. Example 4-4 demonstrates the potential consequences of this error.

- For a reversible process, one can fall into the trap of not applying Equation 4.14 at all, instead saying to oneself "This is a reversible process so I know the change in entropy is zero." The change in entropy of the universe is indeed zero for a reversible process, as illustrated in Example 4-4. But the change in entropy of a system can be positive, zero, or negative.

CHANGE IN ENTROPY FOR AN EXPANSION	**EXAMPLE 4-4**

Revisit the two processes described in Examples 4-2 and 4-3. Applying Equation 4.14, calculate the changes in entropy for the gas, the surroundings, and the universe for both processes.

SOLUTION: Table 4-1 summarizes the processes described in Examples 4-2 and 4-3, which involved gases confined in a piston-cylinder device.

Step 1 *Apply definition of entropy to the gas in the reversible expansion*
The mathematical definition of entropy is:

$$S_2 - S_1 = \int_{state\,1}^{state\,2} \frac{dQ_{rev}}{T} \tag{4.15}$$

The process in Example 4-3 is a reversible process, so the value of Q computed in Example 4-3 can be used directly in Equation 4.15. Further, T was constant in this process,

TABLE 4-1 Comparison of reversible and irreversible expansions of ideal gas.

	Reversible (Example 4-3)	Irreversible (Example 4-2)
Initial gas pressure (P_1)	0.5 kPa	0.5 MPa
Final gas pressure (P_2)	0.1 MPa	0.1 MPa
Initial gas temperature (T_1)	300 K	300 K
Final gas temperature (T_2)	300 K	300 K
Work (W)	−8029 J	−3991 J
Heat (Q)	+8029 J	+3991 J
Change in entropy, system (ΔS_{sys})	26.8 J/K	26.8 J/K
Change in entropy, surroundings (ΔS_{surr})	−26.8 J/K	−13.3 J/K
Change in entropy, universe (ΔS_{univ})	0	13.5 J/K

so Equation 4.15 integrates to:

$$S_2 - S_1 = \frac{Q_{rev}}{T} = \frac{8029\,\text{J}}{300\,\text{K}} = \mathbf{26.8}\,\frac{\mathbf{J}}{\mathbf{K}} \tag{4.16}$$

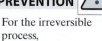

PITFALL PREVENTION

For the irreversible process,
$\frac{Q}{T} = \frac{3991\,\text{J}}{300\,\text{K}} = 13.3\,\frac{\text{J}}{\text{K}}$
However, this is not the correct change in entropy for the gas, because Q is not calculated for a reversible path.

Step 2 *Apply definition of entropy to gas in irreversible expansion*
What of the irreversible process illustrated in Example 4-2? Since this is not a reversible process, we cannot introduce the Q value calculated in Example 4-2 directly into Equation 4.14. In order to apply Equation 4.14, we can construct a hypothetical reversible path that converts the system from the initial state to the final state, and then evaluate Q on this hypothetical path. Because the initial and final states of the gas are the same in Examples 4-2 and 4-3, our reversible process in Example 4-3 IS a path we could apply to Example 4-2. Thus, the change in entropy of the gas is the same in both examples. We could have arrived at the same conclusion by noting that entropy is a state function.

Step 3 *Apply definition of entropy to surroundings in reversible expansion*
First, consider how the processes affected the surroundings. Some work was transferred from the gas to the surroundings, but work does not appear in Equation 4.14. Some heat was transferred from the surroundings to the gas, though this heat transfer was not significant enough to have a measurable impact on the temperature of the surroundings. Therefore, Q for the surroundings is identical in magnitude, and opposite in sign, to Q for the gas. The temperature of the surroundings is constant at 300 K. For the reversible process in Example 4-3, Equation 4.14 applies directly:

$$S_2 - S_1 = \frac{Q_{rev}}{T} = \frac{-8029\,\text{J}}{300\,\text{K}} = \mathbf{-26.8}\,\frac{\mathbf{J}}{\mathbf{K}} \tag{4.17}$$

Step 4 *Apply definition of entropy to surroundings in irreversible expansion*
On the irreversible path, $Q = -3991\,\text{J}$, but the expansion and heat transfer were not reversible. The temperature of the surroundings was 300 K throughout the process. The temperature of the gas started and ended the process at 300 K, but we know it was lower in between (though we never computed any exact values for it). Similarly, the expansion

was irreversible because the surroundings were uniformly at $P = 0.1$ MPa while the gas pressure varied between 0.1 MPa and 0.5 MPa. However, the effect *on the surroundings* would have been the same if the heat transfer was reversible. We pretend the system *was* at 0.1 MPa and 300 K throughout the process, as shown in Figure 4-5, and this gives us a "hypothetical reversible path" of the process- from the perspective of the surroundings.

$$S_2 - S_1 = \frac{Q_{rev}}{T} = \frac{-3991\,\text{J}}{300\,\text{K}} = -\mathbf{13.3}\,\frac{\mathbf{J}}{\mathbf{K}} \qquad (4.18)$$

If you calculate a negative change in entropy for the universe, then either you've made an error, or the process you are modeling is unrealistic.

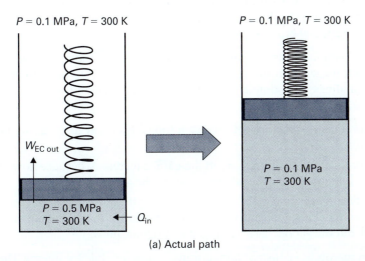

(a) Actual path

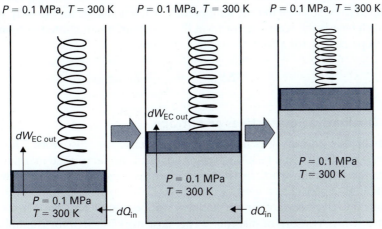

(b) Hypothetical reversible path

FIGURE 4-5 The (a) actual process, and (b) a hypothetical reversible process that has the same effect on the surroundings. The amount of work done on the surroundings and the *total* heat transferred from the surroundings is the same whether the process is sudden (as in a) or gradual (as in b).

Step 5 *Calculate change in entropy of the universe*

The change in entropy of the universe is simply the sum of the changes in entropy for the system and surroundings: $\Delta S_{universe} = \Delta S_{system} + \Delta S_{surroundings}$. Thus, $\Delta S_{universe}$ is 0 for the reversible process and $+13.5$ J/K for the irreversible process.

In perusing Example 4-4, you might be confused about one thing: On the irreversible path, $Q_{reversible}$ was the same as the actual Q when we modeled the surroundings, but the reversible and actual values of Q were different when we modeled the gas. This apparent contradiction arises because the gas was substantially affected by the process, while the surroundings were not.

■ The gas experienced a dramatic pressure change, and this change had to be modeled reversibly in order to apply Equation 4.14.

■ The surroundings remained at atmospheric pressure and 300 K throughout the process. Since the surroundings were substantially unaffected by the process, the hypothetical reversible path for the surroundings is not meaningfully different from the actual process.

This example illustrates a common situation: the surroundings are often modeled as so much bigger than the system that the process has no measureable effect on the surroundings. As a simple analogy, consider a person inflating a bicycle tire in a closed garage:

■ Theoretically, as the tire gets bigger, it transfers energy to the air in the garage in the form of work.

■ Theoretically, the air pressure in the garage increases, because now that the tire takes up more space, the volume of air in the garage is smaller.

■ Practically speaking, neither of the above effects is measurable or significant. The bike tire is significantly different at the end of the process than it was at the start, but the garage hasn't changed in a significant way.

FOOD FOR THOUGHT 4-10

Is the water in a car's radiator a "heat sink," and can you model it as a "heat reservoir"?

A **heat reservoir** is a constant temperature source, or sink, for energy. The reservoir is assumed to be so large that its temperature will not change measurably, no matter how much heat is transferred to or from the reservoir. As illustrated in Example 4-4, the entropy change for a heat reservoir is given by

$$\Delta S_{reservoir} = \frac{Q_{reservoir}}{T_{reservoir}} \tag{4.19}$$

4.3.2 Qualitative Perspectives on Entropy

In this chapter, entropy was introduced as a metric for "lost work." Increases in entropy occur (a) when a task that *requires* work is accomplished in a manner that exceeds the minimum work, or (b) when a process *produces* less work than it had the potential to produce.

You may have heard other nutshell descriptions and definitions of entropy, such as "entropy is a measure of disorder." The "lost work" descriptor is emphasized in this chapter for several reasons.

In Example 4-1, processes A, B, and D involved decreasing the entropy of the universe, meaning the "lost work" was negative; the work exceeded what was physically possible.

■ The concept of "work" has recognizable practical importance to engineers, and by identifying and quantifying "lost work" we can compute efficiencies.

■ The notion that entropy is a metric for quantifying "lost work" is logical from a historical perspective; entropy was first proposed as a property in the development of quantitative models for heat engines.

■ The idea of "lost work" helps emphasize the complementary nature of energy and entropy. Recall the motivational example in Section 4-1, which examined a turbine that took in superheated steam and expelled liquid water, producing enormous amounts of work without violating the energy balance. The entropy balance allows us to recognize such a process as impossible, benchmark how much potential work exists, and quantify how much of this potential is realized.

However, the following sections examine the concepts of entropy and reversibility from additional perspectives in order to provide deeper insights into entropy and also to help the reader reconcile any apparent contradictions between the discussion in this chapter and anything he/she has heard before.

4.3.3 Relating Entropy to Spontaneity and Directionality

It has been observed that the entropy of the universe is constantly increasing. Thus, the concept of entropy is useful in that it permits us to predict whether proposed processes will or will not occur *spontaneously*. We illustrate the concepts of spontaneity and directionality with a simple example.

HEAT TRANSFER BETWEEN HOT AND COLD WATER	**EXAMPLE 4-5**

"Vessel A" contains 0.1 kg of water initially at 353 K and "Vessel B" contains 0.05 kg of water initially at 293 K. They are brought into contact as shown in Figure 4-6, and allowed to reach equilibrium. Find the final temperature of the water and the change in entropy of the universe. Assume the heat capacity and specific volume of water are constant with respect to temperature (use $\hat{C}_P \sim \hat{C}_V \sim 4.2$ kJ/kg · K), and that no heat exchange with the surroundings occurs.

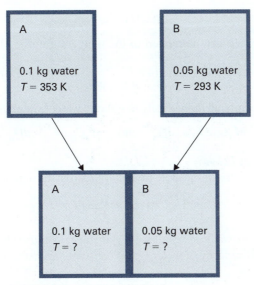

FIGURE 4-6 Two containers of water are brought into contact and heat transfer occurs as examined in Example 4-5.

SOLUTION:

Step 1 *Define the system*
A possible first thought is to define BOTH containers of water as the system. For this system, there is no heat entering or leaving ($Q = 0$), so on a first glance, Equation 4.14 might appear to imply that there was no change in entropy. However, a closer look reveals two problems with this analysis:

Neither $\hat{C}_P$ nor $\hat{V}$ for liquid water is precisely constant over the temperature range from 293 K to 353 K, but here it is convenient to use these simplifying approximations for illustrative purposes.

1. Equation 4.14 applies to homogeneous closed systems. A system can't be considered homogeneous if there are distinct regions with two different temperatures inside the system.

2. Equation 4.14 applies only to reversible paths. If we define both containers as the system, the process is not reversible—there is an irreversible heat transfer process occurring inside the boundaries of the system.

We will find it much more useful to define ONE container of water as the system; now we do have a homogeneous closed system. Further, the other container is outside the system and there is heat (Q) entering or leaving the system. Thus, we will be able to apply Equation 4.14 to each container.

Consequently, we define two distinct systems, A and B, each consisting of the water in one vessel.

Step 2 *Apply and simplify energy balance*
The energy balance equation and mathematical analysis for each system will be essentially identical, with only the initial temperature being different.

The containers of water are closed, stationary systems, so the energy balance for one container of water, in differential form, is

$$d(M\hat{U}) = dQ + dW_{EC} + dW_S \tag{4.20}$$

Since the volume of water is assumed constant, neither system will experience expansion or contraction ($W_{EC} = 0$). There are also no moving parts ($W_S = 0$), so the energy balance further simplifies to

$$d(M\hat{U}) = dQ \tag{4.21}$$

Mass is constant, and $d\hat{U}$ can be related to the temperature using the known, constant heat capacity:

$$M\hat{C}_V dT = dQ \tag{4.22}$$

This equation can be applied individually to either vessel of water, A or B:

$$M_A\hat{C}_V dT_A = dQ_A \quad \text{and} \quad M_B\hat{C}_V dT_B = dQ_B \tag{4.23}$$

Step 3 *Integrate energy balance*
To find the final temperature, we integrate Equation 4.23 from the start of the process to its end:

$$\int_{T_{A,\text{initial}}}^{T_{A,\text{final}}} M_A\hat{C}_V dT_A = Q_A \tag{4.24}$$

Because the mass of the water in vessel A (M_A) is constant and we have assumed for this example that the heat capacity of water is also constant, this becomes

$$M_A\hat{C}_V(T_{A,\text{final}} - T_{A,\text{initial}}) = Q_A \tag{4.25}$$

The analogous treatment of vessel B gives

$$M_B\hat{C}_V(T_{B,\text{final}} - T_{B,\text{initial}}) = Q_B \tag{4.26}$$

Step 4 *Relate the systems to each other to find final temperature*
The problem statement says that no heat transfer occurs with the surroundings. Thus, any heat leaving vessel A must go to vessel B; there is nowhere else for it to go, and there is no other source of heat added to vessel B. Consequently:

$$Q_A = -Q_B \tag{4.27}$$

$$M_A\hat{C}_V(T_{A,\text{final}} - T_{A,\text{init}}) = -M_B\hat{C}_V(T_{B,\text{final}} - T_{B,\text{init}})$$

FOOD FOR THOUGHT 4-11

Give both a physical and a mathematical explanation for how Equation 4.20 can be derived from Equation 3.35.

$d\hat{U} = \hat{C}_V dT$ because this is being modeled as a constant-volume process.

PITFALL PREVENTION

Be careful with signs. Q_A and Q_B represent exactly the same heat transfer, so they are equal in magnitude, but they are opposite in sign.

At equilibrium there is no driving force for more heat transfer; the temperatures of the vessels are the same ($T_{A,\text{final}} = T_{B,\text{final}} = T_{\text{final}}$). Applying this fact to Equation 4.27 gives

$$M_A \hat{C}_V (T_{\text{final}} - T_{A,\text{init}}) = -M_B \hat{C}_V (T_{\text{final}} - T_{B,\text{init}}) \tag{4.28}$$

$$(0.1\,\text{kg})\left(4.2\,\frac{\text{kJ}}{\text{kg} \cdot \text{K}}\right)(T_{\text{final}} - 353\text{ K}) = (0.05\,\text{kg})\left(4.2\,\frac{\text{kJ}}{\text{kg} \cdot \text{K}}\right)(T_{\text{final}} - 293\text{ K})$$

The final temperature is $T_{\text{final}} = \mathbf{333}$ **K**.

Step 5 *Apply definition of entropy*
The definition of entropy is

$$dS = \frac{dQ_{\text{rev}}}{T}$$

Since this heat transfer is irreversible (initially, the temperature gradient between the two containers is 60 degrees), we must construct a hypothetical reversible process in order to apply Equation 4.14.

Figure 4-7 focuses on Vessel B and illustrates the actual irreversible process and a hypothetical reversible process. In the real process, the source of the heat is initially at 353 K, and this temperature decreases until equilibrium is reached at 333 K. Because the heat source is outside the boundaries of the system, we can construct a hypothetical process in which the same amount of heat comes from a different source. In the hypothetical process, the temperature of the heat source is initially 293 K; this means there exists a very small driving force for heat transfer to the system, which is initially at 293 K. In the hypothetical process, the temperature of the heat source gradually increases along with the system temperature, so that the heat source temperature is always *fractionally* larger than the system temperature. Notice the energy balance in Equation 4.22 is not dependent upon the temperature of the heat source; it's the same for the reversible process as the irreversible one. The hypothetical process is not very realistic, but it has the same effect on container B as the real process has.

> The physical effect of heat transfer (e.g., temperature change, phase change, etc.) is the same, whether the heat is transferred reversibly or irreversibly.

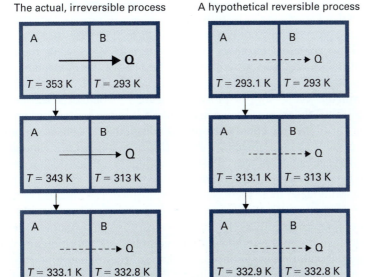

The actual, irreversible process   A hypothetical reversible process

FIGURE 4-7 An illustration of the actual process described in Example 4-5, and of a hypothetical reversible process that has the same effect on the water in container B.

> There's nothing special about the 0.1 degree temperature difference shown in Figure 4-7. We could've said the heat source was always 0.01 degrees warmer than the system, or 0.001. The crucial point is that the temperature difference must be *small enough that it's reasonable to consider it negligible*; otherwise we can't reasonably model the process as reversible.

Introducing Equation 4.22, for Q, into the definition of entropy

$$dS = \frac{M\hat{C}_V\,dT}{T} \tag{4.29}$$

This equation, like the energy balance in Equation 4.22, can be applied individually to either system.

Step 6 *Solve for changes in entropy*

Since mass and heat capacity are both constant, Equation 4.29 integrates to

$$S_2 - S_1 = M\hat{C}_V \ln\left(\frac{T_2}{T_1}\right) \tag{4.30}$$

Vessel B has an initial temperature of 293.15 K and a final temperature of 333.15 K. Thus,

$$S_{2,B} - S_{1,B} = (0.05\,\text{kg})\left(4.2\frac{\text{kJ}}{\text{kg}\cdot\text{K}}\right)\ln\left(\frac{333\,\text{K}}{293\,\text{K}}\right) = 26.9\frac{\text{J}}{\text{K}} \tag{4.31}$$

For Vessel A, the initial temperature at 353.15 K gives

$$S_{2,A} - S_{1,A} = (0.1\,\text{kg})\left(4.2\frac{\text{kJ}}{\text{kg}\cdot\text{K}}\right)\ln\left(\frac{333\,\text{K}}{353\,\text{K}}\right) = -24.5\frac{\text{J}}{\text{K}} \tag{4.32}$$

Since the process has no interaction with anything outside these two containers, the change in entropy of the universe is the sum of these two entropy changes: $\Delta S_{\text{univ}} = 26.9 - 24.5 = $ **2.4 J/K**.

Example 4-5 illustrates how irreversible heat transfer always leads to an increase in the entropy of the universe. Heat always flows from a high-temperature location to a lower-temperature location. Thus, Q is negative for the system with the larger initial T, and Q is positive for the system with the smaller initial T. So when the values of Q/T are summed, the result is positive.

When an object at 353 K is in contact with an object at 293 K, no action is required to make heat transfer begin; it begins predictably and spontaneously. Thus, the process illustrated in Figure 4-6 can be termed a *spontaneous process*. One can also imagine the reverse process shown in Figure 4-8, in which all of the water is at $T = 333$ K initially, and one container increases in temperature to 353 K while the other drops to 293 K. Such a process is consistent with the first law of thermodynamics (energy is conserved in Figure 4-8 whether the process runs "forward" or "backward") but is not consistent with the second law (the process in Figure 4-8 would decrease the entropy of the universe). Thus, the process shown in Figure 4-8 is *not* a spontaneous process.

We don't need the concept of entropy to know that heat transfers spontaneously from a hot object to a cold object. We know this from experience. Similarly, we can prove that the mechanical process illustrated in Figure 4-9, in which a gas increases pressure to 0.5 MPa even though there is no applied force acting on it besides the atmosphere, would lead to an increase in the entropy of the universe (Problem 4-15 asks you to do this proof). Perhaps it seems like a waste of time to do this mathematical proof; we know intuitively that such a process isn't spontaneous. However, many physical situations are more subtle or complex, and intuition and experience aren't enough. Recall that in Example 4-1, three of the four turbine configurations are shown to be physically impossible using the concept of entropy. These turbines did not violate the laws of nature in any way that most people would necessarily call "intuitively obvious." As another example, consider the gas phase chemical reaction:

$$C_2H_6 \leftrightarrow C_2H_4 + H_2$$

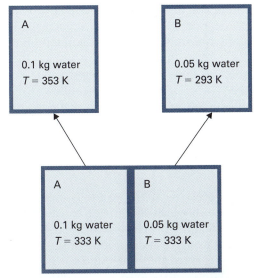

FIGURE 4-8 The reverse of the heat transfer process illustrated in Example 4-5. This process is physically unrealistic, though energy is conserved.

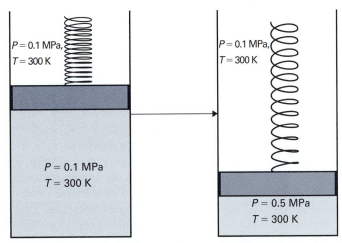

FIGURE 4-9 A mechanical process that would not occur spontaneously.

If 1.0 mole of ethane, 0.5 mole of ethylene, and 0.5 mole of hydrogen were placed in a container, would the reaction spontaneously run "forward" or "backward"? The summary answer is "It depends upon the temperature and pressure," but we can't answer the question quantitatively without the concepts of both enthalpy and entropy. Most likely, the reader has solved problems of this kind in chemistry courses, by calculating equilibrium constants (K) using the expression:

$$\ln K = \frac{-\Delta \underline{G}^0}{RT} \tag{4.33}$$

in which $\Delta \underline{G}^0$ is the standard change in **molar Gibbs free energy** from the reaction. This very chemical reaction is considered in the motivational example for Chapter 14, in which we examine chemical reaction equilibrium and learn the theoretical basis for Equation 4.33. For now we simply note that Gibbs energy (G) is a fundamental property for modeling equilibrium, and is defined as follows:

$$G = H - TS \tag{4.34}$$

Equation 4.34 shows that entropy is embedded into G. Gibbs energy in turn is the fundamental property that determines which direction (and how far) the reaction will run spontaneously.

Phase changes are another crucial physical phenomenon that can be addressed using Gibbs energy. Chapters 8 through 13 discuss how one can predict and model the temperatures and pressures at which phase transitions occur spontaneously, and the role entropy and Gibbs energy play in building these models.

Why do we refer to the process in Figure 4-6 as "spontaneous," and those in Figures 4-8 and 4-9 as "not spontaneous"? Why don't we simply call them "possible" and "impossible"? Because one of the things engineers do is design systems that allow non-spontaneous processes to happen! For example, heat would not spontaneously flow from a location at ~276 K to a location at 297 K, but a standard

The compound C_2H_4 is called both "ethene" and "ethylene." Ethene is the name by IUPAC convention, but ethylene is a name that predates the IUPAC system of nomenclature and is still in common use.

kitchen refrigerator is a machine that is designed to make this exact non-spontaneous process occur. Section 5-3 describes how refrigeration cycles actually work. For now we note that part of the refrigeration process involves the addition of work. The addition of work is an essential feature of carrying out processes that wouldn't occur spontaneously. A ball will roll downhill spontaneously, but can only be made to roll back up the hill if work is applied. Similarly, heat flows from high temperature to low temperature spontaneously, but heat can be made to flow "uphill" to higher temperatures through a well-designed process that involves the addition of work.

4.3.4 Spontaneity and Reversibility

Like the ideal gas, the reversible process is a useful theoretical model, which in some circumstances closely approximates reality.

Section 4.3.3 notes that in any *spontaneous* process, the entropy of the universe increases. A reversible process is a limiting case in which the driving force for the process is infinitesimally small. It was illustrated through Examples 4-2 and 4-3 that reversible processes provide the maximum work when work is produced. There is no "lost" work in a reversible process. It would be easy to infer from this discussion that reversible processes are the "best" processes. However, we must also be conscious of the *rate* at which processes occur. We consider the physical process of heat transfer as an example.

Q cannot equal $A\Delta T$; the units are different. A constant of proportionality called the heat transfer coefficient is needed to make the equality.

The rate of heat transfer by convection or conduction is proportional to the *area* (A) available for heat transfer and also proportional to the *temperature gradient* (ΔT) driving the heat transfer:

$$\dot{Q} \propto A\,\Delta T \qquad (4.35)$$

A approaches infinity as ΔT approaches 0.

Typically, undergraduate curricula in chemical engineering contain an entire course on heat transfer, which covers other factors influencing the rate of heat transfer and considerations in equipment design. However, Equation 4.35 is sufficient to illustrate the problem with reversible processes: in a reversible process, ΔT by definition approaches zero, so the heat transfer area (A) required to obtain any desired finite rate of heat transfer (Q) must approach infinity. A reversible heat transfer is optimal from the perspective of efficient use of energy, but equipment cost is minimized with a low heat transfer area (A), which requires a highly irreversible, highly spontaneous process. Such trade-offs are critical in engineering design and engineering practice.

Minimizing the cost of a piece of equipment and minimizing the costs (such as energy) associated with operating that equipment are often competing, contradictory goals. Typically, chemical engineering curricula include design courses that include examination of how to optimize this trade-off.

In sum, while the lost work that accompanies an irreversible, spontaneous process might at first glance seem wasteful compared to a reversible process, one might be better served by regarding some lost work as "the cost of doing business."

4.3.5 Relating Entropy to Disorder

You have likely heard that entropy is "a measure of disorder." The description of entropy as a metric for "lost work" is emphasized in this book largely because "work" is a physical quantity that has a formal mathematical definition. Thus, "maximum work" and "lost work" can be quantified, while a "measure of disorder" is an ambiguous notion. However, the description of entropy as a "measure of disorder" is justifiable, at least in a qualitative sense. Consider the following general observations about entropy:

- Specific entropy is higher in vapors than in liquids.
- Specific entropy is higher in liquids than in solids.

- Entropy of a multi-component mixture is higher than the sum of the entropies of the individual components of the mixture.

- Entropy tends to increase with increasing temperature.

- Entropy tends to increase with decreasing pressure.

Even without a formal definition for "order," it probably seems obvious that solids are more "ordered" than liquids, liquids are more ordered than gases or vapors, and pure substances are more ordered than mixtures. Furthermore, increasing temperature and decreasing pressure both typically lead to increases in volume; most people would probably agree that something becomes less "ordered" when it is spread out through a larger volume. So, you can probably see why "entropy is a measure of disorder" became a fashionable description: it is a concise summary of several broad observations regarding entropy.

When looking at more complex situations, the qualitative notions of "order" versus "disorder" become difficult to apply in any meaningful way. Consider, for example, the two systems shown in Figure 4-10: (a) superheated steam at 1 MPa and 200°C, and (b) a system comprised of both saturated liquid and saturated vapor at 0.1 MPa. Offhand, would you say system (b) is more "ordered," because it is partially liquid and liquids are more ordered than vapors? Or would you say the superheated system is more "ordered" because it's homogeneous while the liquid–vapor system is heterogeneous? Also, in measuring "order," is a phase change more or less important than a change in pressure or temperature?

In fact, *the two systems in Figure 4-10 have equal specific entropy.* If the superheated steam in Figure 4-10a is fed into an adiabatic reversible turbine and the exit pressure is 0.1 MPa, the effluent from the turbine would be the liquid–vapor mixture described in Figure 4-10b. In this scenario, there is no lost work. The reversible turbine produces the maximum possible work attainable for this particular change in state; thus there is no change in entropy. All of these facts can be demonstrated by applying the mathematical definition of entropy in Equation 4.14.

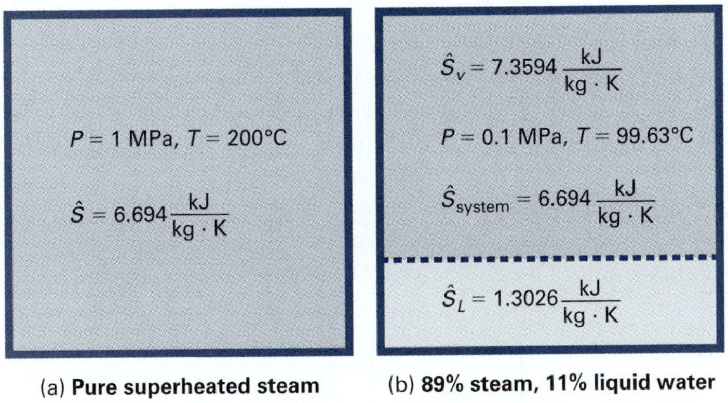

> Don't overextend generalizations such as "specific entropy is higher in liquids than in solids." For example, while liquid water has higher specific entropy than ice, you can't assume it has higher specific entropy than ALL solids.

$$P = 1 \text{ MPa}, \, T = 200°C$$

$$\hat{S} = 6.694 \, \frac{kJ}{kg \cdot K}$$

$$\hat{S}_v = 7.3594 \, \frac{kJ}{kg \cdot K}$$

$$P = 0.1 \text{ MPa}, \, T = 99.63°C$$

$$\hat{S}_{system} = 6.694 \, \frac{kJ}{kg \cdot K}$$

$$\hat{S}_L = 1.3026 \, \frac{kJ}{kg \cdot K}$$

(a) **Pure superheated steam**   (b) **89% steam, 11% liquid water**

FIGURE 4-10 (a) Superheated steam and (b) a liquid–vapor mixture.

Thus, entropy is well described as a property that can be used to quantify potential work and lost work. There exist other descriptors like "entropy is a measure of disorder" that you might find helpful in understanding or predicting changes in entropy, at least qualitatively. But the only definition of entropy that you actually *need* is the mathematical one, which is explored further in Section 4-4.

4.3.6 The Microscopic Definition of Entropy

The mathematical definition of entropy in Section 4.3.1 relies upon macroscopic properties—specifically temperature and reversible heat. This section overviews the fact that entropy can also be defined, and understood, at the microscopic level—the level of individual molecules. For simplicity, here we model molecules as spheres. The dark outer square in Figure 4-11 represents a vessel containing three "gas molecules," labeled 1, 2, and 3. There are any number of arrangements for these molecules, just one of which is shown in Figure 4-11. How might we describe the positions of the molecules?

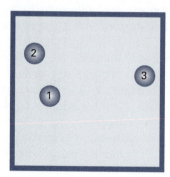

FIGURE 4-11 Three gas "molecules" in a "vessel" that has a fixed volume.

There is nothing "special" about our decision to divide the space into four regions—we can divide the space into as few or as many regions as we wish, as necessary for whatever model we are trying to build.

Let's divide the "vessel" up into four regions with equal volumes I, II, III, and IV. There are $4 \times 4 \times 4 = 64$ different ways to distribute the molecules into the regions, since there are four possible locations for molecule 1, and for each of these four possibilities, there are four possible locations for molecule 2, etc. We call each of these 64 distributions a "microstate."

Three "microstates" are shown in Figure 4-12. Notice that in all three of these "microstates," there are exactly two molecules in region I, and one in region II. Consider this question: "Are these three really different from each other?" They look different to us, because we've put numbers on the molecules, making them distinguishable. But real molecules within a pure compound are not distinguishable from each other, so the three "microstates" in Figure 4-12 are functionally identical. "Two in region I, one in region II" is an example of a *macrostate*—a specific, complete description of one possible state for the system. In this simple system of 3 molecules and 4 regions, there are 20 different macrostates, summarized in Table 4-2. However, these macrostates are not all equally likely to occur. There is, for example, only one

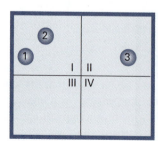

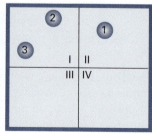

FIGURE 4-12 A "vessel" containing three "molecules." The vessel is divided up into four distinct regions, I, II, III, and IV.

TABLE 4-2 List of the 20 macrostates for the three-molecule, four-region system illustrated in Figure 4-12. The # of microstates sum to 64.

Region I	Region II	Region III	Region IV	# Microstates
3	0	0	0	1
0	3	0	0	1
0	0	3	0	1
0	0	0	3	1
2	1	0	0	3
2	0	1	0	3
2	0	0	1	3
1	2	0	0	3
0	2	1	0	3
0	2	0	1	3
1	0	2	0	3
0	1	2	0	3
0	0	2	1	3
1	0	0	2	3
0	1	0	2	3
0	0	1	2	3
1	1	1	0	6
1	1	0	1	6
1	0	1	1	6
0	1	1	1	6

microstate that has three molecules in region III; there is only one way to put all of the molecules in the same region. However, we saw in Figure 4-12 that there are 3 different ways to put "Two in region I and one in region II"; this macrostate has 3 different microstates. While real molecules within a pure compound are indistinguishable from each other, the thought exercise of making them distinguishable allowed us to count the number of microstates in each macrostate, and so demonstrate that some macrostates are more likely to occur by random chance than others.

Table 4-2 shows that there are four different macrostates in which the molecules are all in different regions. Each of these has six corresponding microstates. Therefore, if the "molecules" are being positioned randomly, these macrostates will occur most frequently. Each of these macrostates will occur $6/64 = 9.4\%$ of the time, while each macrostate in which all three molecules are in a single region will occur $1/64 = 1.6\%$ of the time.

This discussion doesn't account for the possibility that the molecules are attracted to each other, or the possibility that a region could get so full there's no room for more molecules. Section 7-4 discusses these issues.

Consider now adding a fourth molecule. There are now $4 \times 4 \times 4 \times 4 = 256$ different microstates. There is still only one in which *all* of the molecules are in region I, only one in which all of the molecules are in region II, etc., but there are 24 different ways to put exactly one molecule in each of the four regions. What if we had 8 molecules, or 12? As the number of molecules increases, the probability of *all* the molecules occupying a single region becomes infinitely small, as shown in Table 4-3.

TABLE 4-3 The probability of all gas molecules occupying the same quarter of the available space by random chance.

# of Molecules	# of Microstates	# of Microstates in which all molecules are in the same region as each other	Fraction of microstates in which all molecules are in the same region as each other
4	256	4	0.015625
8	65,536	4	0.000061
12	16,777,216	4	2.34×10^{-7}
20	1.0995×10^{12}	4	3.64×10^{-12}
100	1.6069×10^{60}	4	2.49×10^{-60}

Here, our discussion of microstates and macrostates connects to the idea of "spontaneity" discussed in Section 4.3.3. You can perhaps imagine all of the air molecules in a room, by coincidence, moving into one quarter of the room at the same time, leaving the other three-quarters of the room evacuated. Such an event has a small but not insignificant frequency of 1.6% when there are only three molecules, but no plausible frequency when there are $\approx 10^{25}$ molecules. The only "macrostates" that have realistic probabilities with this many molecules are ones in which the molecules are approximately evenly distributed across the four regions. Thus, we can use a microscopic (molecular level) model to demonstrate that the air in a room will not spontaneously compress itself into one-quarter of the room's volume. But we can also analyze this same situation macroscopically, using principles discussed earlier in this chapter. We learned through Example 4-4 that isothermally increasing the volume of a gas increases its entropy; conversely isothermally compressing the gas into a lower volume decreases its entropy. A gas can be compressed through a process that involves addition of work, such that the entropy of the system decreases but the entropy of the universe increases. However, if the volume of gas in a system spontaneously decreases, and *nothing else* happens, this process would violate the second law of thermodynamics.

The preceding paragraph gives a simple example that illustrates the consistency between the microscopic and macroscopic definitions of entropy. Mathematically, the microscopic description of entropy is that entropy (S) of a system is proportional to the number of microstates (Ω) corresponding to the macrostate that describes the system:

$$S = k_b \ln \Omega \qquad (4.36)$$

We learned in Section 4.3.3 that the entropy of a system can be decreased if we add work—here that could mean moving the walls so that the vessel is only one fourth its original volume.

Ludwig Boltzmann (1844-1906) was a physicist who held positions at several different universities in Austria. Equation 4.36 is engraved on his tombstone.

The macroscopic definition of entropy described in Section 4.3.1 was proposed first. Ludwig Boltzmann formulated the microscopic view of entropy (Sandler, 2011), and demonstrated its mathematical consistency with the macroscopic definition. Consequently, the constant of proportionality (k_b) in Equation 4.36 is called **Boltzmann's constant**.

At the introductory level we will not explore how to use Equation 4.36 to compute changes in entropy for real systems and processes. The microscopic view of entropy is introduced here to help build an understanding of entropy. It also lays the foundation for measuring entropy on an absolute scale:

$S = 0$ for a perfect crystal at absolute zero. This is sometimes called the **third law of thermodynamics**.

A perfect crystal is one that has no defects—no impurities, no "missing atoms," etc. Why should such a crystal have zero entropy? Consider the $10 \times 10 \times 10$ lattice pictured in Figure 4-13. If there are 1000 indistinguishable atoms in a crystal, then there

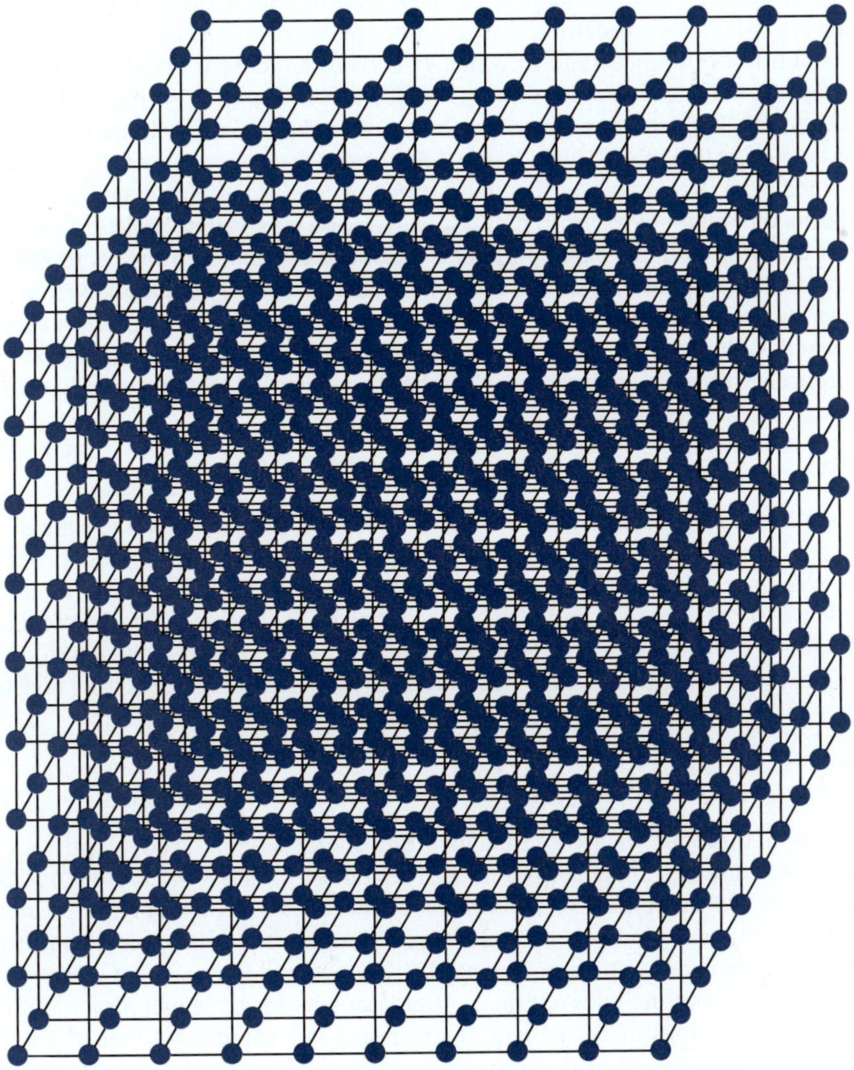

The Boltzmann constant that appears in equation 4.36 is different from the "Stefan-Boltzmann constant," which is important in modeling radiative heat transfer.

The equation $PV = nRT$ was derived empirically, from data, before the kinetic theory of the ideal gas "proved" it. Similarly, the macroscopic view of entropy predates the microscopic view, but both are now understood to be consistent with each other.

FIGURE 4-13 A simple crystal lattice containing 1000 atoms ($10 \times 10 \times 10$).

The fact that defects in solids lead directly to increased entropy is likely another reason why "measure of disorder" was at one time a fashionable description of entropy.

is only one microstate—every lattice point is identical. Equation 4.36 reveals that if the macrostate that describes the system contains only one microstate, then entropy is zero ($S = k_b \ln \Omega = k_b \ln 1 = 0$). However, if a macrostate has 999 indistinguishable atoms and 1 defect, then there are 1000 microstates corresponding to 1000 different locations in which the defect can be placed. More microstates mean greater entropy, as indicated by Equation 4.36.

Next, you may ask—why must the crystal be at absolute zero to have zero entropy? Doesn't the crystal in Figure 4-13 still have only one microstate, even if $T = 1$ K, or $T = 300$ K? The thing to remember is that atoms and molecules are not in reality stationary, even in a solid. While Figure 4-13 may accurately describe the location of atoms in a crystal *on average*, it doesn't capture the fact that the atoms are constantly vibrating. Only at absolute zero, a hypothetical state where there is zero molecular motion, could the simple picture shown in Figure 4-13 be valid all the time.

Because of the third law of thermodynamics, values of standard entropy in reference books can be expressed on an absolute scale (e.g., $\underline{S}^0$) rather than related to a reference state (e.g., $\Delta \underline{H}_f^0$). In an absolute sense, elements have positive internal energy, pressure, and volume and therefore have positive molar enthalpy ($\underline{H} = \underline{U} + P\underline{V}$) at any realistic conditions. But since we cannot measure energy on an absolute scale, we establish a reference state in which $\underline{H}$ is considered zero for all elements at a standard set of conditions (conventionally $T = 298.15$ K and $P = $ either 101.325 kPa or 0.1 MPa). The "standard molar enthalpy of formation" of a compound, $\Delta \underline{H}_f^0$, is then defined as the change in enthalpy that occurs when one mole of the compound is formed from its elements at this standard set of conditions. However, the third law of thermodynamics allows us to measure entropy on an absolute scale. If the change in molar entropy when a pure compound is isobarically heated from 0 K to 298.15 K (a process that may or may not include phase changes) were 300 J/mol·K, that compound has $\underline{S}^0 = 300$ J/mol·K in an absolute sense at 298.15 K, since it has zero entropy at 0 K. While it isn't realistic to make such a measurement experimentally, the theoretical models we are building in this book can be used to quantify values for $\underline{S}^0$.

4.4 The Entropy Balance

When we treat mass and energy as conserved quantities, we are ignoring nuclear reactions that convert mass into energy.

The mathematical definition of entropy, as given in Equation 4.14, is specified for a closed, homogeneous system and is only directly applicable to reversible processes. However, the examples in this book illustrate that engineers must be able to analyze a wide variety of processes and systems, including both open and closed systems, and including irreversible processes. Chapter 3 introduced generalized mass and energy balances that are applicable to this multitude of systems. In this section we construct the analogous entropy balance that it is applicable to *any* system.

As mentioned in Chapter 3, a balance equation can be written as

$$\text{Accumulation} = \text{In} - \text{Out} + \text{Generation} - \text{Consumption}$$

Throughout this book we are treating mass and energy as conserved quantities. Consequently, in our mass and energy balance equations (3.1 and 3.34), there is no "Generation" or "Consumption" term; our models need not account for the possibility that mass or energy can be created or destroyed. By contrast Example 4-5 illustrates that entropy is constantly being "created" by irreversible processes. Consequently, the entropy balance includes a term $\dot{S}_{gen}$ that accounts for the generation of entropy inside the boundaries of the system. It was noted in Example 4-4 that in a reversible process, the entropy of the universe is unchanged. The way this observation is reflected mathematically is that in the entropy balance, the $\dot{S}_{gen}$ *term is* 0 *for any reversible process.*

Regarding the "in" and "out" terms, entropy can be added to, or removed from, a system through material flows. Entropy is also affected by heat transfer; if heat is added to the system, then entropy is added to the system.

In the entropy balance, Q represents the actual heat transfer, not a reversible heat transfer as in Equation 4.14.

Thus, a general expression for an entropy balance equation is

$$\frac{d(M\hat{S})}{dt} = \sum_{j=1}^{j=J} \dot{m}_{j,in}\hat{S}_j - \sum_{k=1}^{k=K} \dot{m}_{k,out}\hat{S}_k + \sum_{n=1}^{n=N} \frac{\dot{Q}_n}{T_n} + \dot{S}_{gen} \qquad (4.37)$$

where

- t is the time.
- M is the mass of the system.
- $\hat{S}$ is the specific entropy of the system.
- $\dot{m}_{j,in}$ and $\dot{m}_{k,out}$ are the mass flow rates of individual streams entering and leaving the system, and the summations are carried out over all such streams.
- $\hat{S}_j$ and $\hat{S}_k$ are the specific entropies of streams entering and leaving the system.
- $\dot{Q}_n$ is the *actual* rate at which heat is added to or removed from the system at one particular location, and the summation is carried out over all locations where heat is added or removed.
- T_n is the temperature of the system at the boundary where the heat transfer labeled n occurs. Temperature must be expressed on an absolute scale (Kelvin or Rankine).
- $\dot{S}_{gen}$ is the rate at which entropy is generated within the boundaries of the system.

Equations 4.37 and 4.38 assume that heat transfer is occurring at a system boundary that has a constant temperature. Use Equation 4.39 for the case of temperature that changes with time.

In Chapter 3, the mass and energy balances were also expressed in time-independent forms, for situations in which no information about the duration of the process was known or needed.

FOOD FOR THOUGHT 4–12

If the entropy balance is valid for ANY system, why do we also need the "definition of entropy" in Equation 4.14? What role can Equation 4.14 play in solving an entropy balance?

The analogous time-independent entropy balance equation is

$$(M_2\hat{S}_2) - (M_1\hat{S}_1) = \sum_{j=1}^{j=J} m_{j,in}\hat{S}_j - \sum_{k=1}^{k=K} m_{k,out}\hat{S}_k + \sum_{n=1}^{n=N} \frac{Q_n}{T_n} + S_{gen} \qquad (4.38)$$

where

- $m_{j,in}$ and $m_{k,out}$ are individual quantities of mass added to or removed from the process, and the summations are carried out over all such quantities.
- Q_n is an individual quantity of heat added to or removed from the system during the process.
- S_{gen} is the total entropy generated inside the boundaries of the system during the process.

Any individual $\dot{Q}$ or Q can be either positive or negative. If heat is removed from a system, then entropy is also removed from the system, and this phenomenon is reflected by the negative sign of $\dot{Q}$ or Q. Note that if the temperature T_n is not constant with respect to time, then the heat added in each increment of time (dt) must be treated as a separate heat transfer, and the summation becomes an integral:

$$(M_2 \hat{S}_2) - (M_1 \hat{S}_1) = \sum_{j=1}^{j=J} m_{j,\text{in}} \hat{S}_j - \sum_{k=1}^{k=K} m_{k,\text{out}} \hat{S}_k + \int_{t_{\text{initial}}}^{t_{\text{final}}} \frac{\dot{Q}}{T} \, dt + S_{\text{gen}} \qquad (4.39)$$

> $\dot{Q}$ in Equation 4.39 is the real $\dot{Q}$ for the real process. It does not have to be $\dot{Q}$ for a reversible process.

It should be emphasized that $\dot{Q}$ or Q, in any entropy balance, is the actual heat added or removed. There is no requirement that the heat transfer be reversible, as there is in the mathematical definition of entropy (Equation 4.14). The rationale for this apparent contradiction is as follows: As illustrated in Figure 4-7, heat transfer can be either reversible or irreversible. The physical effect of the heat on the system (e.g., a temperature change and/or a phase change) is determined by the *amount* of heat added, and is the same whether the heat transfer was reversible or irreversible. Thus, when we are examining the process of heat transfer by itself, we can imagine the heat transfer was reversible (even if it wasn't) and apply Equation 4.14 to compute the change in entropy. This problem-solving strategy was illustrated previously in Example 4-5.

Next we address the $\dot{S}_{\text{gen}}$ (or S_{gen}) term. If all other terms of Equations 4.37, 4.38, or 4.39 can be evaluated, one can solve the entropy balance to determine $\dot{S}_{\text{gen}}$, much like we frequently solve an energy balance to determine Q or W. However, aside from solving the entropy balance, there is no fundamental way to calculate $\dot{S}_{\text{gen}}$. Consequently, in this book $\dot{S}_{\text{gen}}$ either will be computed using the entropy balance or will be set equal to 0 in processes that reasonably can be modeled as reversible.

The following sections provide several examples in which the entropy balance is applied.

4.4.1 Applying the Entropy Balance to a Reversible Nozzle

Example 4-6 illustrates the application of the entropy balance to a fairly simple physical system, demonstrating the use of both data and the ideal gas model to solve a problem. Subsequent examples involve more complex entropy balances.

EXAMPLE 4-6	A REVERSIBLE NOZZLE

> In Example 3-4, the exiting temperature was specified, but here it is unknown. The entropy balance is an additional equation that allows us to find this additional unknown.

A gas at $T = 450°C$ and $P = 0.5$ MPa flows at steady state through an adiabatic and reversible nozzle. The exiting pressure is 0.1 MPa. Find the velocity and temperature of the exiting stream if

A.   The gas is steam.

B.   The gas is an ideal gas with $C_V^* = 3R$ and molecular weight $= 75$ kg/kmol.

$T_{\text{in}} = 450°C$
$P_{\text{in}} = 0.5$ MPa

$T_{\text{out}} = ?$
$P_{\text{out}} = 0.1$ MPa
$v = ?$

FIGURE 4-14 Nozzle described in Example 4-6.

SOLUTION A:

Step 1 *Define a system and write the energy balance*
First, define the fluid in the nozzle as the system. The energy balance as listed in Table 3-1 is

$$\frac{v_{out}^2}{2} = \hat{H}_{in} - \hat{H}_{out}$$

There is no information in the problem statement that isn't consistent with the assumptions behind this equation (adiabatic, potential energy negligible, entering velocity negligible).

Step 2 *Degree of freedom analysis*
The specific enthalpy of the entering steam is found in the steam tables, $\hat{H}_{in} = 3377.7$ kJ/kg. In Example 3-4, which also involved a nozzle, the energy balance could be solved immediately because the outgoing temperature and pressure were both given. However, an adiabatic nozzle has no mechanism for controlling temperature, so the current example, in which the exiting temperature and velocity are both unknown, is a more realistic problem. With only one intensive property of the outlet stream known, we can't just look up $\hat{H}_{out}$. The energy balance is thus one equation in two unknowns ($\hat{H}_{out}$ and v_{out}). The entropy balance provides the needed second equation.

The Gibbs phase rule, introduced in Section 2.2.3, says that a single component in a single phase has two degrees of freedom. Here only one intensive property of the outlet stream is known.

Step 3 *Apply and simplify the entropy balance*
The entropy balance for this system simplifies greatly:

▪ It is a steady-state process, so the accumulation term is 0.

▪ It is adiabatic, so the $\dot{Q}/T$ term is 0.

▪ It is a reversible process, so there is no generation of entropy inside the system.

▪ There is only one stream entering and only one stream leaving.

Consequently, the entropy balance (Equation 4.37) simplifies to

$$0 = \dot{m}_{in}\hat{S}_{in} - \dot{m}_{out}\hat{S}_{out}$$

At steady state, the entering and leaving mass flow rates must be equal to each other. Dividing through by this mass flow rate gives

$$0 = \hat{S}_{in} - \hat{S}_{out} \tag{4.40}$$

Step 4 *Calculate terms in entropy balance*
From Appendix A, at 450°C and $P = 0.5$ MPa, $\hat{S}_{in} = 7.9465$ kJ/kg·K. Equation 4.40 shows that in this process the entropy of the steam is unchanged; thus $\hat{S}_{out} = 7.9465$ kJ/kg·K. From Appendix A, at $P = 0.1$ MPa,

Notice that specific entropy has the same units as heat capacity.

$$\hat{S} = 7.8356 \text{ kJ/kg·K at } 200°C$$

$$\hat{S} = 8.0346 \text{ kJ/kg·K at } 250°C$$

Linear interpolation is reviewed in Appendix B-1.

Interpolating between these values gives

$$T_{out} = 200°C + \left(\frac{7.9465 - 7.8356 \dfrac{\text{kJ}}{\text{kg} \cdot \text{K}}}{8.0346 - 7.8356 \dfrac{\text{kJ}}{\text{kg} \cdot \text{K}}} \right)(250°C - 200°C) = 227.6°C$$

Knowing the outlet temperature allows us to find $\hat{H}_{out}$. Interpolating from the steam tables gives

For a pure compound in a single phase, if two intensive properties are known, all others have unique values. P_{out} was given. Finding the second property T_{out} allows us to look up other properties in the steam tables.

$$\hat{H}_{out} = 2875.5 \frac{\text{kJ}}{\text{kg}} + \left(\frac{227.6°C - 200°C}{250°C - 200°C} \right)\left(2974.5 - 2875.5 \frac{\text{kJ}}{\text{kg}} \right) = 2930.2 \frac{\text{kJ}}{\text{kg}} \tag{4.41}$$

Step 5 *Solve energy balance*
Introducing this result into the energy balance gives

$$\frac{v_{out}^2}{2} = \hat{H}_{in} - \hat{H}_{out} = \left(3377.7 - 2930.2 \frac{kJ}{kg}\right)\left(1000 \frac{J}{kJ}\right)\left(1\frac{N \cdot m}{J}\right)\left(1\frac{\frac{kg \cdot m}{s^2}}{N}\right) \quad (4.42)$$

$$v_{out} = 946.0 \frac{m}{s}$$

SOLUTION B:

Step 6 *Compare/contrast with part A*
Now we consider solution of the problem using the ideal gas model, rather than data. The nozzle is adiabatic, reversible, and steady state, as it was in part A, so the energy and entropy balances from steps 1 and 3 again apply:

$$\frac{v_{out}^2}{2} = \hat{H}_{in} - \hat{H}_{out}$$

$$0 = \hat{S}_{in} - \hat{S}_{out}$$

For an ideal gas, enthalpy can be related to temperature through the heat capacity. Thus, just as in part A, we can resolve the energy balance if we can determine T_{out} from the entropy balance. However, unlike part A, there is no data available that relates entropy to temperature.

Step 7 *Apply definition of entropy*
Besides using experimental data, how else can entropy and temperature be related to each other? Through the definition of entropy:

$$dS = \frac{dQ_{rev}}{T}$$

which can be applied to any homogeneous, closed system. The nozzle is not a closed system, but we can utilize the fact that entropy is a state function.

This use of the concept of a state property illustrates a fundamental problem-solving strategy in thermodynamics, as discussed after the example.

■ We can imagine placing the ideal gas in a simple, closed system, like the piston-cylinder device from Example 4-3.

■ We can imagine that the gas in this closed system experiences the same change in temperature and pressure as the gas travelling through the nozzle ($T_1 = T_{in}$, $T_2 = T_{out}$, etc.).

■ We can imagine the changes in the closed system are reversible.

■ Because entropy is a state function, the change in $\underline{S}$ or $\hat{S}$ for a particular set of starting (T_1, P_1) and ending (T_2, P_2) conditions is always the same, regardless of the path.

Step 8 *Evaluate heat (dQ_{rev}) for hypothetical, closed system*
We know from Example 4-3 that, for our closed system, the energy balance is

The energy balance is converted into differential form because we ultimately need to integrate dQ/T

$$U_2 - U_1 = W_{EC} + Q$$

Or in differential form, it is

$$dU = dW_{EC} + dQ \quad (4.43)$$

For any ideal gas process, U can be related to T through C_V. Furthermore, we know dW_{EC} is given by $-P\,dV$.

$$dU = NC_V\,dT = -P\,dV + dQ \quad (4.44)$$

We know from Section 1.4.3 that in this expression, P is the pressure opposing the motion.

■ When a gas contracts, the pressure of the gas itself is the pressure opposing the motion.

■ When a gas expands, the pressure opposing the motion is an external pressure.

■ But for a reversible process, the pressure opposing the motion and the pressure driving the motion must be essentially identical. Consequently, in Equation 4.44, P can be the pressure of the ideal gas itself, regardless of whether the process involves expansion or contraction.

Thus for *any* reversible ideal gas process occurring in a homogeneous, closed system,

$$dQ_{rev} = NC_V^* \, dT + P \, dV \tag{4.45}$$

Step 9 *Evaluate change in entropy for hypothetical closed system*
Substituting Equation 4.45 into the definition of entropy (Equation 4.14) gives

$$dS = N\frac{C_V^*}{T} dT + \frac{P}{T} dV \tag{4.46}$$

Since this is a closed system, the number of moles (N) is constant. Thus, we can divide through by N, which has the effect of converting S and V into the molar volume and the molar entropy.

$$d\underline{S} = \frac{C_V^*}{T} dT + \frac{P}{T} d\underline{V}$$

The last term cannot be integrated unless P and T are expressed as functions of $\underline{V}$, which can be done through the ideal gas law: $P\underline{V} = RT$. Thus,

$$d\underline{S} = \frac{C_V^*}{T} dT + \frac{R}{\underline{V}} d\underline{V} \tag{4.47}$$

Step 10 *Solve differential equation for change in entropy*
For this example, C_V^* is a constant equal to $3R$, so the integration of Equation 4.47 is

$$\underline{S}_2 - \underline{S}_1 = C_V^* \ln\frac{T_2}{T_1} + R \ln\frac{\underline{V}_2}{\underline{V}_1} \tag{4.48}$$

The given information is in terms of T and P, not $\underline{V}$. So it is convenient to substitute for the molar volumes, again using the ideal gas law, for

$$\underline{S}_2 - \underline{S}_1 = C_V^* \ln\left(\frac{T_2}{T_1}\right) + R \ln\left(\frac{\frac{RT_2}{P_2}}{\frac{RT_1}{P_1}}\right) = C_V^* \ln\left(\frac{T_2}{T_1}\right) + R \ln\left(\frac{T_2 P_1}{T_1 P_2}\right) \tag{4.49}$$

Equation 4.49 consists entirely of state properties (molar entropy, temperature, and pressure), and is therefore path-independent; it can be applied to *any* ideal gas process in which the heat capacity (C_V^*) is constant. Apply it to the nozzle by equating $T_{in} = T_1$, $P_{in} = P_1$, $T_{out} = T_2$, and $P_{out} = P_2$ for

$$\underline{S}_{out} - \underline{S}_{in} = 3R \ln\left(\frac{T_{out}}{T_{in}}\right) + R \ln\left(\frac{T_{out} P_{in}}{T_{in} P_{out}}\right) \tag{4.50}$$

Step 11 *Find exiting temperature*
We determined in step 6 that $\underline{S}_{out} - \underline{S}_{in} = 0$. Combining this fact with Equation 4.50 gives

In steps 7 through 9, we have assumed nothing other than ideal gas behavior. Consequently, Equation 4.47 can be applied to *any* ideal gas process, as discussed further after the example. In step 10, the assumption is that C_V^* is constant, so subsequent equations are not valid for ideal gases in general.

S and $\hat{S}$ differ from each other only by a multiplier, which is the molecular weight. Thus, if $\hat{S}_{in} - \hat{S}_{out} = 0$, then $\underline{S}_{in} - \underline{S}_{out} = 0$.

$$0 = 3R \ln\left(\frac{T_{out}}{450 + 273.15\,\text{K}}\right) + R \ln\left[\frac{(T_{out})(0.5\,\text{MPa})}{(450 + 273.15\,\text{K})(0.1\,\text{MPa})}\right] \tag{4.51}$$

The solution to equation 4.51 is $T_{out} = \textbf{483.60 K}$. This is close to the answer from part A, which was 500.75 K. We did not expect the answers to be identical; high pressure steam is not an ideal gas.

Step 12 *Close energy balance*

For any ideal gas process, $d\underline{H} = C_P^* \, dT$. Here C_P^* is not given, but we can apply the ideal gas property $C_P^* = C_V^* + R$, so $C_P^* = 4R$. Thus,

$$\underline{H}_{out} - \underline{H}_{in} = \int_{T_{in}}^{T_{out}} C_P^* \, dT = 4R(T_{out} - T_{in}) \tag{4.52}$$

$$\underline{H}_{out} - \underline{H}_{in} = 4\left(8.314\,\frac{\text{J}}{\text{mol} \cdot \text{K}}\right)(483.6\,\text{K} - 723.15\,\text{K}) = -\textbf{7966.5}\,\frac{\textbf{J}}{\textbf{mol}}$$

We complete the problem by inserting this value into the energy balance:

$$\frac{v_{out}^2}{2} = \left(7966.5\,\frac{\text{J}}{\text{mol}}\right)\left(\frac{1\,\text{mol}}{75\,\text{g}}\right)\left(\frac{1000\,\text{g}}{1\,\text{kg}}\right)\left(\frac{1\,\text{N} \cdot \text{m}}{1\,\text{J}}\right)\left(\frac{\text{kg} \cdot \text{m}}{1\,\frac{\text{s}^2}{\text{N}}}\right) \tag{4.53}$$

The exiting velocity is $v_{out} = \textbf{460.9 m/s}$.

Steps 7 through 9 in Example 4-6 illustrate both a common problem-solving approach in thermodynamics and a useful property of ideal gases.

The problem-solving approach used the concept of a state property. The Gibbs phase rule shows it takes two intensive state properties (e.g., T and P) to fully define the state of a pure compound. Once the state is defined, all other intensive properties (e.g., $\underline{S}$) have unique values. Consequently, if one defines two states (T_1, P_1 and T_2, P_2), one can evaluate the change in a state property (S in the example) along *any* path between these two states, and know that the result is valid for all other paths. This problem-solving approach is of particular importance for entropy because the definition of entropy (Equation 4.14) only applies directly to reversible processes carried out in closed systems.

In steps 7–9 of Example 4-6, we modeled the change in entropy of an ideal gas. Since no assumptions (beyond ideal gas behavior) were made in these steps, the result is correct for ideal gases in general.

The change in molar entropy for any ideal gas is given by

$$d\underline{S} = \frac{C_V^*}{T}\,dT + \frac{R}{\underline{V}}\,d\underline{V} \tag{4.54}$$

One can carry out an analogous derivation on a mass basis and show that

$$d\hat{S} = \frac{\hat{C}_V^*}{T}\,dT + \frac{R}{\hat{V}}\,d\hat{V} \tag{4.55}$$

A common conceptual error is proceeding directly from Equation 4.54 to the integrated form of

$$\underline{S}_2 - \underline{S}_1 = C_V^* \ln\frac{T_2}{T_1} + R \ln\frac{\underline{V}_2}{\underline{V}_1} \tag{4.56}$$

The integration of Equation 4.54 only yields Equation 4.56 when C_V^* is constant with respect to temperature. If C_V^* is a function of temperature, as is generally true, a more complex integration results. Example 4-8 examines such a case.

Another thematic problem-solving strategy illustrated by Example 4-6 is the use of the ideal gas law to convert between pressure (P) and molar volume ($\underline{V}$). For example, substituting the ideal gas law into Equation 4.56 shows

$$\underline{S}_2 - \underline{S}_1 = C_V^* \ln\left(\frac{T_2}{T_1}\right) + R \ln\left(\frac{\dfrac{RT_2}{P_2}}{\dfrac{RT_1}{P_1}}\right) \tag{4.57}$$

which can be rearranged using the property $\log(xy) = \log(x) + \log(y)$ as

$$\underline{S}_2 - \underline{S}_1 = C_V^* \ln\left(\frac{T_2}{T_1}\right) + R \ln\left(\frac{T_2}{T_1}\right) + R \ln\left(\frac{P_1}{P_2}\right) \tag{4.58}$$

Since for an ideal gas, $C_P^* = C_V^* + R$, Equation 4.58 can be consolidated as

$$\underline{S}_2 - \underline{S}_1 = C_P^* \ln\left(\frac{T_2}{T_1}\right) + R \ln\left(\frac{P_1}{P_2}\right) \tag{4.59}$$

Equation 4.59, like Equation 4.56, is only valid for ideal gas processes in which C_P^* can be modeled as constant with respect to temperature.

> Equation 4.59 also can be written as
>
> $$\underline{S}_2 - \underline{S}_1 = C_P^* \ln\left(\frac{T_2}{T_1}\right)$$
>
> $$- R \ln\left(\frac{P_2}{P_1}\right)$$

4.4.2 Entropy and Efficiency

We have established that reversible processes are idealized processes that produce the maximum work, or use the minimum work. This section introduces the use of an efficiency to quantify how nearly real processes approximate these idealized, reversible processes. We begin with a closer look at the motivational example.

| MOTIVATIONAL EXAMPLE REVISITED | **EXAMPLE 4-7** |

Consider the turbine in part C of Example 4-1, in which steam entered at $T = 823.15$ K and $P = 3$ MPa and exited at $T = 473.15$ K and $P = 0.1$ MPa.

A. How much entropy is generated for each kilogram of steam that goes through this turbine?

B. What is the maximum work that can be obtained (per kilogram of entering steam) from a turbine with this inlet stream and an outlet pressure $P = 0.1$ MPa?

SOLUTION A:

Step 1 *Define a system and write an energy balance*
As in Example 4-1, the energy balance for the turbine is

$$\frac{\dot{W}_S}{\dot{m}} = \hat{H}_{out} - \hat{H}_{in}$$

Step 2 *Apply and simplify entropy balance*
To find the rate of entropy generation, an entropy balance is the logical way to begin. Starting with the time-dependent entropy balance (Equation 4.37):

$$\frac{d(M\hat{S})}{dt} = \sum_{j=1}^{j=J} \dot{m}_{j,in}\hat{S}_j - \sum_{k=1}^{k=K} \dot{m}_{k,out}\hat{S}_k + \sum_{n=1}^{n=N} \frac{\dot{Q}_n}{T_n} + \dot{S}_{gen}$$

This problem can be solved starting from either the time-dependent or the time-independent balance, since we seek answers on a basis of "per kilogram of steam." If we wanted to find the *rate* of entropy generation (e.g., in kJ/K·s) we would definitely use the time-dependent equation.

Like the nozzle in Example 4-6, this is an adiabatic, steady-state process, so the accumulation and rate of heat transfer ($\dot{Q}$) terms are 0. Consequently the entropy balance is

$$0 = \dot{m}_{in} \hat{S}_{in} - \dot{m}_{out} \hat{S}_{out} + \dot{S}_{gen} \tag{4.60}$$

For this steady-state process the entering and exiting mass flowrates are identical ($\dot{m}_{in} = \dot{m}_{out} = \dot{m}$). Solving for the generation of entropy:

$$\frac{\dot{S}_{gen}}{\dot{m}} = (\hat{S}_{out} - \hat{S}_{in}) \tag{4.61}$$

The left-hand side is now exactly what we wish to find: the entropy generated per kilogram of steam.

Step 3 *Introduce data*
Because temperature and pressure are both known for the entering and exiting streams, entropy of each can be found in the steam tables.

$$\frac{\dot{S}_{gen}}{\dot{m}} = \left(7.8356 - 7.3768 \frac{kJ}{kg \cdot K}\right) = 0.4588 \frac{kJ}{kg \cdot K} \tag{4.62}$$

> You can show that for the other possible outlet streams listed in the Motivational Example, the value of $\dot{S}_{gen}$ calculated in Step 3 would be negative.

SOLUTION B:

Step 4 *Apply energy balance*
As in part A, the energy balance is

$$\frac{\dot{W}_S}{\dot{m}} = \hat{H}_{out} - \hat{H}_{in}$$

> As in Example 4-6, the energy balance here has two unknowns, and the entropy balance is a second equation.

In part A, the conditions of the outlet stream were exactly known, so $\hat{H}_{out}$ could be obtained immediately from the steam tables. Here, only pressure is known for the outlet stream—we don't know the outlet temperature that corresponds to "maximum work." We need a second property of the outlet stream before we can look up $\hat{H}_{out}$, and we can find it by applying the concept of the reversible process.

Step 5 *Entropy balance for reversible turbine*
As illustrated in Examples 4-2 and 4-3, the maximum work is obtained from a reversible process. Thus, in part B, we are analyzing a reversible turbine that has an outlet pressure of 0.1 MPa. As stated in Section 4.3.1, no entropy is generated in a reversible process. Consequently the entropy balance (Equation 4.61) simplifies to:

$$0 = (\hat{S}_{out} - \hat{S}_{in}) \tag{4.63}$$

Since the inlet stream is the same as it was in part A, for the reversible turbine:

$$\hat{S}_{in} = \hat{S}_{out} = 7.3768 \frac{kJ}{kg \cdot K} \tag{4.64}$$

Step 6 *Find outlet temperature*
According to the steam tables, at $P = 0.1$ MPa, $\hat{S} = 7.3614$ kJ/kg $\cdot$ K for steam at 100°C and $\hat{S} = 7.6134$ kJ/kg $\cdot$ K for steam at 150°C. Interpolating between these two values gives:

$$T_{out} = 100°C + (50°C) \left(\frac{7.3768 - 7.3614 \dfrac{kJ}{kg \cdot K}}{7.6134 - 7.3614 \dfrac{kJ}{kg \cdot K}} \right) = 103.1°C \tag{4.65}$$

Step 7 *Find specific enthalpy of outlet stream*
Interpolating a second time to find the specific enthalpy of steam at $P = 0.1$ MPa and $T = 103.1°C$:

$$\hat{H}_{out} = \hat{H}_{100} + \left(\frac{103.1 - 100°C}{150 - 100°C}\right)(\hat{H}_{150} = \hat{H}_{100}) \tag{4.66}$$

$$\hat{H}_{out} = 2676.2 \frac{kJ}{kg} + \left(\frac{103.1 - 100°C}{150 - 100°C}\right)\left(2776.4 - 2676.2 \frac{kJ}{kg}\right) = 2682.3 \frac{kJ}{kg}$$

Step 8 *Close energy balance*
The maximum work, which is equivalent to the reversible work, is found by inserting the result of part 7 into the energy balance:

$$\frac{\dot{W}_S}{\dot{m}} = \hat{H}_{out} - \hat{H}_{in} = 2682.3 - 3569.7 = \mathbf{-887.4} \frac{kJ}{kg} \tag{4.67}$$

Example 4-7 mirrors the motivational example given in Section 4-1. The motivational example described four possible sets of outlet conditions for a steady-state turbine and asserted that three of them were physically impossible, because the entropies of the possible outlet streams were smaller than the entropy of the inlet stream. The entropy balances presented in steps 1 through 3 of Example 4-7 illustrate the same principle in a more formal mathematical way; the process is physically impossible if the calculated $\dot{S}_{gen}$ is negative.

Example 4-7 also illustrates the application of the principle of reversibility to calculate maximum work. A reversible process is a hypothetical limiting case that represents the maximum work produced, or the minimum work required. We can define an *efficiency* that compares the work produced by a real turbine to the theoretical maximum represented by a reversible process, as

$$\eta_{turbine} = \frac{\dot{W}_{S,actual}}{\dot{W}_{S,reversible}} \tag{4.68}$$

For the turbine in Example 4-7, the real work is -694.2 kJ/kg and the reversible work is -887.4 kJ/kg, so the efficiency is 0.782 or 78.2%.

A similar definition can be applied to a pump or compressor:

$$\eta_{compressor} = \frac{\dot{W}_{S,reversible}}{\dot{W}_{S,actual}} \tag{4.69}$$

Note that a compressor is a process that requires work, and the reversible compressor represents the minimum work. Thus, for a compressor, the reversible work is smaller than the actual work. Equations 4.68 and 4.69 are both defined with the larger quantity in the denominator, so that efficiency will always be a fraction ($0 - 1$ or $0 - 100\%$). Equation 4.69 is applied in the next example.

> Our sign convention is that work done by the system on the surroundings is negative (energy leaving the system). Some books use the opposite sign convention. Reversible work and actual work will have the same sign as each other, so efficiency is positive regardless of the sign convention used.

COMPRESSION OF AN IDEAL GAS	**EXAMPLE 4-8**

Ammonia enters an adiabatic, steady-state compressor at $P = 20$ kPa and $T = 300$ K and leaves at $P = 0.1$ MPa (Figure 4-15). Ammonia can be assumed to act as an ideal gas at these conditions. The compressor efficiency is 75%. Find:

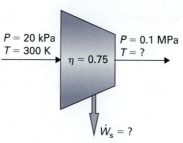

FIGURE 4-15 The compression of ammonia, which is examined in Example 4-8.

A. The work required for each mole of gas compressed.

B. The temperature of the outgoing gas.

SOLUTION:

Step 1 *Define a system*
The compressor is the system.

Step 2 *Apply and simplify the energy balance*
Section 3.6.3 presented the energy balance for an adiabatic, steady-state compressor:

$$\frac{\dot{W}_S}{\dot{m}} = \hat{H}_{out} - \hat{H}_{in} \tag{4.70}$$

Here the given information is on a molar basis, not a mass basis, so the equation is better written as

$$\frac{\dot{W}_S}{\dot{n}} = \underline{H}_{out} - \underline{H}_{in} \tag{4.71}$$

Step 3 *Relate enthalpy to temperature*
It was established in Section 2.3.3 that, for an ideal gas,

$$d\underline{H} = C_P^* \, dT$$

Appendix D gives ideal gas heat-capacity data for various compounds in the form of

$$C_P^* = A + BT + CT^2 + DT^3 + ET^4 \tag{4.72}$$

Thus, the change in molar enthalpy is given by

$$\underline{H}_{out} - \underline{H}_{in} = \int_{T_{in}}^{T_{out}} A + BT + CT^2 + DT^3 + ET^4 \, dT \tag{4.73}$$

$T_{in} = 300$ K, but T_{out} is unknown, which prevents us from solving the problem immediately.

Step 4 *Apply entropy balance to compressor*
As in Example 4-7, the entropy balance for an adiabatic, steady-state compressor simplifies to

$$0 = \dot{n}_{in} \underline{S}_{in} - \dot{n}_{out} \underline{S}_{out} + \dot{S}_{gen} \tag{4.74}$$

The entering and exiting flow rates are identical in this steady-state process:

$$0 = \dot{n}(\underline{S}_{in} - \underline{S}_{out}) + \dot{S}_{gen} \tag{4.75}$$

A compressor is an integral component of a refrigerator—the compressor is what actually makes the noise you hear when the refrigerator mechanism is operating. Refrigeration is discussed in detail in Section 5-3.

You can prove the equivalence of Equations 4.70 and 4.71 rigorously by applying the fact that $H = N\underline{H} = M\hat{H}$.

At the time of writing, a Wikipedia data page (http://en.wikipedia.org/wiki/Ammonia (data_page) said gaseous ammonia had $C_p = 35.06$ J/mol·K at $0°$C. What if we used a simpler approximation and assumed $C_p^* = 35.06$ J/mol·K was a constant value and correct at all temperatures?

If we were assuming C_p^* was constant, the integral in Equation 4.73 would be $\underline{H}_{out} - \underline{H}_{in} = C_P^*(T_{out} - T_{in})$.

Here, $\dot{S}_{gen}$ is an unknown that makes Equation 4.75 impossible to solve directly for the real turbine.

Step 5 *Apply entropy balance to a reversible compressor*
We now consider a hypothetical reversible compressor, with the same inlet stream and outlet pressure as the actual compressor. Since $\dot{S}_{gen}$ is zero in a reversible process, the entropy balance is now

$$0 = \dot{n}(\underline{S}_{in} - \underline{S}_{out,rev}) \tag{4.76}$$

$$0 = (\underline{S}_{in} - \underline{S}_{out,rev}) \tag{4.77}$$

Step 6 *Relate entropy change to temperature*
Equation 4.54 is valid for any ideal gas process:

$$d\underline{S} = \frac{C_V^*}{T}\,dT + \frac{R}{\underline{V}}\,d\underline{V}$$

C_V^* isn't known directly, but C_P^* is known, and we can apply the property of ideal gases $C_V^* + R = C_P^*$ for

$$C_V^* = (A - R) + BT + CT^2 + DT^3 + ET^4 \tag{4.78}$$

Step 7 *Integrate differential equation to find outlet temperature*
Applying Equation 4.54 to the reversible compressor yields

$$\underline{S}_{out,rev} - \underline{S}_{in} = \int_{T_{in}}^{T_{out,rev}} \frac{(A - R) + BT + CT^2 + DT^3 + ET^4}{T}\,dT + \frac{R}{\underline{V}}\,d\underline{V} \tag{4.79}$$

$$\underline{S}_{out,rev} - \underline{S}_{in} = (A - R)\ln\frac{T_{out,rev}}{T_{in}} + B(T_{out,rev} - T_{in}) + \frac{C}{2}(T^2_{out,rev} - T^2_{in})$$
$$+ \frac{D}{3}(T^3_{out,rev} - T^3_{in}) + \frac{E}{4}(T^4_{out,rev} - T^4_{in}) + R\ln\frac{\underline{V}_{out,rev}}{\underline{V}_{in}}$$

We showed in step 5 that the entropy change for the reversible process is 0. In addition, since the given information is in terms of P and T, rather than $\underline{V}$, we can substitute for $\underline{V}$ using the ideal gas law for

$$0 = (A - R)\ln\frac{T_{out,rev}}{T_{in}} + B(T_{out,rev} - T_{in}) + \frac{C}{2}(T^2_{out,rev} - T^2_{in}) \tag{4.80}$$

$$+ \frac{D}{3}(T^3_{out,rev} - T^3_{in}) + \frac{E}{4}(T^4_{out,rev} - T^4_{in}) + R\ln\frac{\dfrac{RT_{out,rev}}{P_{out}}}{\dfrac{RT_{in}}{P_{in}}}$$

The parameters A, B, C, D, and E are available in Appendix D. For ammonia, $A = 25.235$, $B = -35.044 \times 10^{-3}$, $C = 16.969 \times 10^{-5}$, $D = -17.676 \times 10^{-8}$, and $E = 6.327 \times 10^{-11}$ for T expressed in K and C_P^* expressed in J/mol·K. Consequently, while this is a messy equation, the only unknown in it is the outlet temperature from the reversible turbine, $T_{out,rev}$. It can be solved to determine $T_{out,rev} = 428.7$ K.

If we were assuming C_P^* was constant at 35.06 J/mol·K, then we would say $C_V^* = C_P^* - R = 35.06 - 8.314 = 26.75$ J/mol·K.

PITFALL PREVENTION

The temperature found in step 7 is not the answer to part B. Here we are calculating the temperature leaving a hypothetical reversible compressor, NOT the real compressor.

If we were assuming C_P^* was constant, Equation 4.80 would be

$$0 = (C_P^* - R)\ln\frac{T_{out,rev}}{T_{in}}$$
$$+ R\ln\frac{\dfrac{RT_{out,rev}}{P_{out}}}{\dfrac{RT_{in}}{P_{in}}}$$

and its solution would be $T_{out,rev} = 444.4$ K.

There is no easy analytical solution to Equation 4.80, but many modern calculators can be used to solve such equations numerically, as can software like Excel, MATLAB, and POLYMATH.

If we were assuming C_p^* was constant, this equation would be

$$\frac{\dot{W}_{S,rev}}{\dot{n}} = \underline{H}_{out} - \underline{H}_{in}$$

$$= C_p^*(T_{out,rev} - T_{in})$$

and the answer would be

$$\frac{\dot{W}_{S,rev}}{\dot{n}} = 5063 \frac{J}{mol}$$

In computing efficiency, work can be expressed in absolute terms (W_S), as a rate ($\dot{W}_S$), or on a mass or molar $\left(\dfrac{\dot{W}_S}{\dot{n}}\right)$ basis. But reversible and actual work must be expressed on the same basis.

If we were assuming C_p^* was constant, this would be

$$\frac{\dot{W}_{S,actual}}{\dot{n}} = \frac{5063 \dfrac{J}{mol}}{0.75}$$

$$= 6750 \frac{J}{mol},$$

which is ~5% larger than the real answer.

If we were assuming C_p^* was constant, this would be

$$\frac{\dot{W}_{S,act}}{\dot{n}} = C_p^*(T_{out,act} - T_{in})$$

and the solution would be $T_{out,act} = 492.5$ K.

Step 8 *Find reversible work*

The energy balance in Equation 4.71 can now be solved:

$$\frac{\dot{W}_{S,rev}}{\dot{n}} = \underline{H}_{out} - \underline{H}_{in} = \int_{T=300\,K}^{T=428.7\,K} A + BT + CT^2 + DT^3 + ET^4\, dT \qquad (4.81)$$

$$\frac{\dot{W}_{S,rev}}{\dot{n}} = A(T_{out,rev} - T_{in}) + \frac{B}{2}(T_{out,rev}^2 - T_{in}^2) + \frac{C}{3}(T_{out,rev}^3 - T_{in}^3)$$

$$+ \frac{D}{4}(T_{out,rev}^4 - T_{in}^4) + \frac{E}{5}(T_{out,rev}^5 - T_{in}^5)$$

$$\frac{\dot{W}_{S,rev}}{\dot{n}} = 4839 \frac{J}{mol}$$

Step 9 *Apply efficiency*

The definition of efficiency for a compressor is

$$\eta_{compressor} = \frac{\dot{W}_{S,reversible}}{\dot{W}_{S,actual}}$$

Here the efficiency is 0.75, so

$$0.75 = \frac{4839 \dfrac{J}{mol}}{\dfrac{\dot{W}_{S,actual}}{\dot{n}}} \qquad (4.82)$$

$$\frac{\dot{W}_{S,actual}}{\dot{n}} = \mathbf{6452\, \frac{J}{mol}}$$

Step 10 *Determine actual outlet temperature*

In step 8, we knew the outlet temperature for the reversible turbine and used Equation 4.81 to find the reversible work. Here we are faced essentially with the inverse problem: we know the actual work, and wish to find the actual outlet temperature. The integration in Equation 4.81, however, is still valid:

$$\frac{\dot{W}_{S,act}}{\dot{n}} = A(T_{out,act} - T_{in}) + \frac{B}{2}(T_{out,act}^2 - T_{in}^2) + \frac{C}{3}(T_{out,act}^3 - T_{in}^3) \qquad (4.83)$$

$$+ \frac{D}{4}(T_{out,rev}^4 - T_{in}^4) + \frac{E}{5}(T_{out,rev}^5 - T_{in}^5)$$

The actual outlet temperature is **468.8 K**.

4.4.3 Unsteady-State Entropy Balances

Examples 4-6 through 4-8 examined steady-state, adiabatic systems. Consequently, the entropy balance equations were comparatively simple. This section examines the application of the entropy balance to systems that are not at steady state.

| COOLING OF A SOLID | EXAMPLE 4-9 |

A solid sculpture (Figure 4-16) with M = 1 kg and $\hat{C}_p$ = 2.1 kJ/kg·K is heat-cured in an oven at 394 K and ambient pressure. It is removed from the oven, placed on a table, and allowed to cool to room temperature, which is 294 K. The room is large enough that the heat released by the sculpture has no significant effect on the room temperature. Find the amount of heat released, the change in entropy of the sculpture, and the change in entropy of the universe.

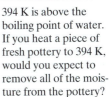

FOOD FOR THOUGHT 4-13

394 K is above the boiling point of water. If you heat a piece of fresh pottery to 394 K, would you expect to remove all of the moisture from the pottery?

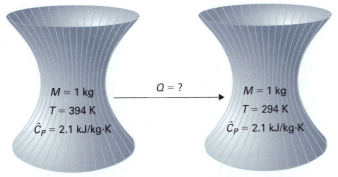

FIGURE 4-16 Cooling process examined in Example 4-9.

SOLUTION:

Step 1 *Define the system and process*
First, define the sculpture as the system. The process begins when the sculpture is first placed on the table (T_1 = 394 K) and ends when the sculpture reaches room temperature (T_2 = 294 K).

Step 2 *Simplifying assumption*
The surface of the object will reach room temperature sooner than the inside. However, we will simplify the process by modeling the object as at a uniform temperature.

Step 3 *Apply and simplify entropy balance*
There is no information concerning rate or duration of the process, so we use the time-independent entropy balance. The solid object is a closed system, so our first inclination might be to simplify the time-independent entropy balance (Equation 4.38) to

$$(M_2\hat{S}_2) - (M_1\hat{S}_1) = \frac{Q}{T_{sculpture}} + S_{gen} \qquad (4.84)$$

The left-hand side of the equation (the accumulation term) represents the change in entropy of the system, which is what we were asked to determine. Q can be computed using an energy balance; however we must recognize that $T_{sculpture}$, the temperature of the system, is changing throughout the process. Consequently we begin with Equation 4.39 rather than Equation 4.37, and the entropy balance is

$$S_2 - S_1 = \int \frac{dQ}{T_{sculpture}} + S_{gen} \qquad (4.85)$$

The integral reflects that each increment of heat is added at a different temperature, and these increments all need to be summed.

Step 4 *Consider generation of entropy*
A likely source of confusion concerns the S_{gen} term. We know that S_{gen} = 0 in reversible processes. This is clearly not a reversible process; there is initially a 100 K temperature

FOOD FOR THOUGHT 4-14

Is this assumption actually necessary to solve the problem?

A typical chemical engineering curriculum includes at least one course on heat transfer, covering the principles needed to model the non-uniform temperature of an object.

By definition $S = M\hat{S}$

PITFALL PREVENTION

T in the entropy balance is the temperature of the system at the boundary where the heat transfer takes place.

A process that has irreversible heat transfer at the system boundary, but no other irreversible feature, is called "internally reversible." By our definition $S_{gen} = 0$ for *reversible* and also for *internally reversible* processes.

difference between the sculpture and the surroundings. The crucial insight is that S_{gen} represents entropy generated *inside* the boundaries of the system. The irreversible feature of the process is heat transfer, which occurs *at* the boundaries of the system. The system is a solid object that is being modeled as constant temperature, and thus there is no particular process occurring *inside* the system that would generate additional entropy.

Consequently, though the process overall is not reversible, we assume $S_{gen} = 0$. Equation 4.85 becomes:

$$S_2 - S_1 = \int \frac{dQ}{T_{sculpture}} \tag{4.86}$$

When we compute the change in entropy of the universe we will see that the irreversible heat transfer is indeed accounted for, even with S_{gen} set equal to 0.

It is not possible to make further progress with Equation 4.86 unless Q and T are related to each other.

As in Example 4-5, this example involves irreversible heat transfer. We again note that the physical effect of heat transfer (here, lowering of the temperature of the sculpture) is the same whether the heat transfer is reversible or irreversible.

Step 5 *Apply and simplify energy balance*

The sculpture is stationary, so there is no accumulation of kinetic or potential energy. A solid object with no moving parts would not produce shaft work. Consequently the energy balance simplifies to:

$$U_2 - U_1 = Q + W_{EC} \tag{4.87}$$

Step 6 *Evaluate work*

As always $dW_{EC} = -P\,dV$. We don't know whether the volume of the object changes significantly as T changes, but we do know the object and room are at constant, ambient pressure, so W_{EC} integrates to:

$$W_{EC} = -P(V_2 - V_1) \tag{4.88}$$

Step 7 *Quantify heat*

Introducing the result of step 6 into the energy balance and solving for Q gives

$$Q = (U_2 - U_1) + P(V_2 - V_1) \tag{4.89}$$

Since $H = U + PV$, we can equate the right-hand side to

$$Q = (H_2 - H_1) \tag{4.90}$$

PITFALL PREVENTION

A common conceptual error at this point is inserting the value $Q = -210\,kJ$ directly into the Q/T term of the entropy balance, forgetting that T isn't constant.

Because this is a constant-pressure process, we know $dH = MC_p\,dT$. Here $\hat{C}_P$ is a constant:

$$Q = (H_2 - H_1) = \int_{T_1}^{T_2} M\hat{C}_P\,dT \tag{4.91}$$

$$Q = M\hat{C}_P(T_2 - T_1) = (1\ kg)\left(2.1\,\frac{kJ}{kg \cdot K}\right)(294\ K - 394\ K) = -210\ kJ$$

Step 8 *Find entropy change of sculpture*

Since T isn't constant, we need a relationship between Q and T that would allow us to integrate Equation 4.86. Thus, we can take the derivative of Equation 4.91, producing:

$$dQ = dH = M\hat{C}_P\,dT \tag{4.92}$$

Inserting expression (4.92) into expression (4.86) gives:

$$S_2 - S_1 = \int_{T_1}^{T_2} \frac{M\hat{C}_P}{T}\,dT \tag{4.93}$$

$$S_2 - S_1 = M\hat{C}_p \ln\left(\frac{T_2}{T_1}\right) = (1 \text{ kg})\left(2.1 \frac{\text{kJ}}{\text{kg} \cdot \text{K}}\right) \ln\left(\frac{294}{394}\right)$$

$$S_2 - S_1 = -0.615 \frac{\text{kJ}}{\text{K}}$$

This value represents the change in entropy of the sculpture. Recall that it is physically impossible for the change in entropy of the universe to be negative, or for entropy generation (S_{gen}) to be negative, but there is nothing physically unrealistic about a *system* experiencing a negative change in entropy.

Step 9 *Apply entropy balance to room*
We were also asked to calculate the change in entropy of the universe. Aside from the sculpture, the only other portion of the universe affected by the process is the room, so we now define a new system as "the entire room *excluding* the sculpture." For this system, there is no mass entering or leaving the system, and this time, the temperature of the system is constant, so the entropy balance is

$$S_{2,\text{room}} - S_{1,\text{room}} = \frac{Q}{T_{\text{room}}} + S_{gen} \tag{4.94}$$

As we did when we were analyzing the sculpture, we recognize that the heat transfer is irreversible, but because the sculpture is not part of the system, the heat transfer occurs at the system boundary. We again assume that no entropy is generated inside the system boundaries and that S_{gen} is again 0.

Step 10 *Calculate change in entropy of universe*
When the room is the system, the total heat transferred is +210 kJ, since the heat given off by the sculpture was added to the room.

$$S_{2,\text{room}} - S_{1,\text{room}} = \frac{Q}{T_{\text{room}}} = \frac{210 \text{ kJ}}{294 \text{ K}} = 0.714 \frac{\text{kJ}}{\text{K}} \tag{4.95}$$

Thus, the change in entropy of just the room is 0.714 kJ/K. The change in entropy of the universe is $\Delta S_{\text{room}} + \Delta S_{\text{sculpture}}$, or $0.714 - 0.615 = \mathbf{0.099}$ **kJ/K**.

Example 4-9 shows what some might consider an odd, counter-intuitive result: When we analyzed the system, we set $S_{gen} = 0$, and when we analyzed the surroundings, we again set $S_{gen} = 0$, yet the final result was a positive change in entropy of the universe, showing that entropy was indeed generated. The entropy generation occurred at the boundary between the system and the surroundings, in the form of an irreversible heat transfer. S_{gen} represents entropy generated inside the boundaries of the system.

The next example re-visits Example 3-11, in which a hot liquid is continuously mixed into a cooler liquid. It is both an open system and an unsteady-state process.

AN UNSTEADY-STATE, OPEN SYSTEM	EXAMPLE 4-10

Recall the 0.5 m³ adiabatic tank that was analyzed in Example 3-11. It was full of liquid initially at $T = 300$ K. Valves were opened, allowing 0.33 kg/s of liquid at $T = 400$ K to enter while 0.33 kg/s of liquid was also removed, as illustrated in Figure 4-17. After 300 s, the liquid in the tank had increased in temperature to 322.1 K. How much entropy is generated within the tank during these 300 s?

SOLUTION:

Step 1 *Define the system*
We define the tank as the system, since our goal is to find the entropy generation in the tank.

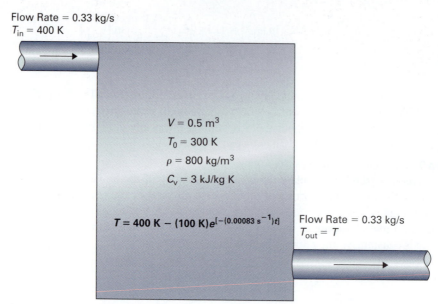

FIGURE 4-17 Schematic of a tank with liquid continuously entering and exiting. The temperature of the tank, as a function of time, was determined in Example 3-11.

Step 2 *Apply and simplify the entropy balance*

Accounting for time is clearly important in part B, so we start with the time-dependent entropy balance:

$$\frac{d(M\hat{S})}{dt} = \sum_{j=1}^{j=J} \dot{m}_{j,\text{in}}\hat{S}_j - \sum_{k=1}^{k=K} \dot{m}_{k,\text{out}}\hat{S}_k + \sum_{n=1}^{n=N} \frac{\dot{Q}_n}{T_n} + \dot{S}_{\text{gen}} \tag{4.96}$$

This is an adiabatic process, and there is only one stream entering and one stream leaving, so the entropy balance simplifies to

$$\frac{d(M\hat{S})}{dt} = \dot{m}_{\text{in}}\hat{S}_{\text{in}} - \dot{m}_{\text{out}}\hat{S}_{\text{out}} + \dot{S}_{\text{gen}} \tag{4.97}$$

Three further simplifications can be noted:

1. The entering mass flow rate is equal to the exiting mass flow rate ($\dot{m}_{\text{in}} = \dot{m}_{\text{out}} = \dot{m}$).
2. The mass of the system (M) is constant, and can be pulled out of the differential.
3. The system is well mixed, so the properties of the exiting stream are identical to the properties of the liquid in the tank. This means $\hat{S}$, which is the specific entropy of the liquid in the tank, is identical to $\hat{S}_{\text{out}}$, which is the specific entropy of the exiting liquid.

Applying these simplifications gives

$$M\frac{d(\hat{S})}{dt} = \dot{m}(\hat{S}_{\text{in}} - \hat{S}) + \dot{S}_{\text{gen}} \tag{4.98}$$

Step 3 *Define S_{gen}*

We were asked to determine the quantity of entropy, S_{gen}, that is generated in 5 minutes. This quantity is mathematically defined as:

$$S_{\text{gen}} = \int_{t=0}^{t=300\text{ s}} \dot{S}_{\text{gen}}\, dt \tag{4.99}$$

PITFALL PREVENTION

$d(M\hat{S})$ is the change in the *total entropy* of the system, and $\dot{S}_{\text{gen}}$ is the rate at which entropy is *generated* in the system. These are related in that entropy generation is one of the phenomena that affect the total system entropy, but they are two different things.

$\dot{S}_{\text{gen}}$ is a *rate* of entropy generation. To find the total quantity of entropy generated, we must integrate with respect to *dt*. The total, S_{gen}, is what we are actually asked to calculate.

Step 4 *Evaluate differential equation*

Multiplying Equation 4.98 through by dt and integrating allows us to introduce S_{gen}, as defined in step 3.

$$M\,d\hat{S} = \dot{m}(\hat{S}_{in} - \hat{S})dt + \dot{S}_{gen}\,dt$$

$$M(\hat{S}_{final} - \hat{S}_{initial}) = \int_{t=0}^{t=5\,min} \dot{m}(\hat{S}_{in} - \hat{S})dt + S_{gen} \qquad (4.100)$$

Notice the left-hand side was easy to integrate, but $\hat{S}_{final} - \hat{S}_{initial}$ is not yet known. The $\dot{m}(\hat{S}_{in} - \hat{S})$ term can't currently be integrated with respect to dt, because we don't have a mathematical relationship between $\hat{S}$ and t. From Example 3-11, we do know a relationship between temperature (T) of the tank and time (t). What we still need is a relationship between $\hat{S}$ and T.

Step 5 *Construct a hypothetical reversible process for computing entropy*

Consider the quantity $\hat{S}_{final} - \hat{S}_{initial}$; we know that the system contains 400 kg of liquid, that the initial system temperature is $T = 300$ K, and the final system temperature is $T = 322.1$ K. If we imagine that the 400 kg of liquid was a closed, homogeneous system and that it was heated reversibly from $T = 300$ K to $T = 322.1$ K, we can find Q_{rev} and apply the definition of entropy (Equation 4.14).

Step 6 *Apply and simplify energy balance to closed, reversible system*

The energy balance for a closed system with no shaft work can be written as

$$dU = dQ_{rev} + dW_{EC} \qquad (4.101)$$

The density of the liquid is constant (0.8 kg/L, as given in Example 3-11). This means the system (liquid) volume V will remain constant as the temperature changes. This fact is significant in two respects: first, there is no work of expansion/contraction, and second, dU can be equated to $MC_V\,dT$:

$$M\hat{C}_V\,dT = dQ_{rev} \qquad (4.102)$$

Step 7 *Apply definition of entropy*

Introducing Equation 4.102 into the definition of entropy:

$$dS = \frac{dQ_{rev}}{T} = \frac{M\hat{C}_V\,dT}{T} \qquad (4.103)$$

We divide through by M, since Equation 4.100, which is what we are ultimately trying to solve, is written in terms of specific entropy:

$$d\hat{S} = \frac{\hat{C}_V\,dT}{T} \qquad (4.104)$$

Step 8 *Evaluate $\hat{S}_{final} - \hat{S}_{initial}$*

Physically, $\hat{S}_{final} - \hat{S}_{initial}$ represents the change in specific entropy as the liquid increases temperature from $T_{initial} = 300$ K to $T_{final} = 322.1$ K, which can be found by integrating Equation 4.104 for

$$\hat{S}_{final} - \hat{S}_{initial} = \int_{T=300\,K}^{T=322.1\,K} \frac{\hat{C}_V\,dT}{T} = \hat{C}_V \ln\left(\frac{322.1\,K}{300\,K}\right) \qquad (4.105)$$

$$\hat{S}_{final} - \hat{S}_{initial} = \left(3\frac{kJ}{kg \cdot K}\right) \ln\left(\frac{322.1\,K}{300\,K}\right) = 0.213\frac{kJ}{kg \cdot K}$$

$\hat{S}_{in}$ is a constant, because the entering stream has a constant $T = 400$ K. However, $\hat{S}$ is not constant; it represents the specific entropy of the liquid in the tank, which is changing as the tank temperature changes.

FOOD FOR THOUGHT 4-16

Why are we using a closed system in this step, when the tank is obviously an open system and the entropy balance in step 2 was for an open system?

Step 9 *Relate $\hat{S}_{in} - \hat{S}$ to t*

The entropy balance also includes the integral of $(\hat{S}_{in} - \hat{S})\,dt$, which represents the difference in specific entropy between the entering and exiting streams. The entering stream is at a constant temperature of $T = 400$ K, but we cannot assign a specific numerical value to the exiting temperature, because it changes with time. Integrating Equation 4.104 from the variable outlet temperature T to $T = 400$ K:

$$\hat{S}_{in} - \hat{S} = \int_{T=T}^{T=400K} \frac{\hat{C}_V\, dT}{T} = \hat{C}_V \ln\left(\frac{400\,K}{T}\right) \tag{4.106}$$

We now substitute the known relationship between T and t into Equation 4.106, so that we can integrate with respect to dt:

$$\hat{S}_{in} - \hat{S} = \hat{C}_V \ln\left(\frac{400\,K}{400\,K - (100\,K)\exp(-(0.00083\,s^{-1})t)}\right) \tag{4.107}$$

Step 10 *Solve entropy balance*

We can now evaluate Equation 4.100

> The mass flow rate $\dot{m}$ and heat capacity $\hat{C}_V$ are both constant with respect to t and can be pulled out of the integral.

$$M(\hat{S}_{final} - \hat{S}_{initial}) = \int_{t=0}^{t=300\,s} \dot{m}(\hat{S}_{in} - \hat{S})\,dt + S_{gen}$$

$$M(\hat{S}_{final} - \hat{S}_{initial}) = \int_{t=0}^{t=300\,s} \dot{m}\,C_V \ln\left[\frac{400\,K}{400\,K - (100\,K)e^{-.00083t}}\right] dt + S_{gen} \tag{4.108}$$

$$(400\,kg)\left(0.213\,\frac{kJ}{kg\cdot K}\right) = \left(0.33\,\frac{kg}{s}\right)\left(3\,\frac{kJ}{kg\cdot K}\right)\bigg|_{t=0}^{t=300\,s} = \ln\left[\frac{400\,K}{400\,K - (100\,K)e^{-.00083}}\right] dt + s_{gen}$$

The integral can be evaluated numerically, and is equal to 75.06 s

> The integral was evaluated using POLY-MATH, a software package that includes a numerical ODE solver.

$$(400\,kg)\left(0.213\,\frac{kJ}{kg\cdot K}\right) = \left(0.33\,\frac{kg}{s}\right)\left(3\,\frac{kJ}{kg\cdot K}\right)(75.06\,s) + S_{gen} \tag{4.109}$$

$$S_{gen} = 10.14\,\frac{kJ}{K}$$

Example 4-10 shows that when hot and cool liquid are mixed together, entropy is generated, and it shows how the entropy generation can be quantified. Example 4-9 showed that entropy is generated when a hot object cools. This chapter has proposed that entropy can be regarded as a metric for lost work. Since entropy is being generated in Example 4-9 and Example 4-10, these examples must involve "lost work" in some form. Where is the "lost work" in these heat transfer processes?

The motivational examples in Chapters 1 and 2 presented the Rankine cycle as an important system that illustrates the significance of several different thermodynamics concepts. These sections illustrated a way to convert heat into work. Thus, any hot substance has some *potential to do work*. Example 4-10 involved mixing liquid at 400 K into a tank of cooler liquid. As the liquids mixed, the entering liquid cooled, giving it less potential to do work, while the liquid in the tank warmed, giving it more potential to do work. Thus at first glance it is unclear that a *net* loss of potential work has occurred. The next section demonstrates how "potential to do work" is inherently related to temperature.

4.5 The Carnot Heat Engine

Throughout this chapter, we have considered systems that were single items: single pieces of equipment, a cylinder containing a gas, a sculpture, etc. Like the energy balance, the entropy balance can also be applied to larger systems, such as complete processes. Here, we consider application of the entropy balance to a heat engine. We define a **heat engine** as a process that is designed to convert heat into work.

> The Rankine cycle described in Sections 1.2.3 and 2.1 is a heat engine, but it is not the only possible design of a heat engine.

A REVERSIBLE HEAT ENGINE	EXAMPLE 4-11

This problem investigates a heat engine that operates as shown in Figure 4-18.

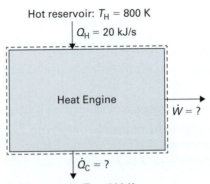

FIGURE 4-18 Heat engine described in Example 4-11.

> Many processes involve two distinct heat reservoirs: one for heating and one for cooling. We use T_H to represent the temperature of the "hot" reservoir and T_C for the temperature of the "cold" reservoir. These symbols are used extensively in Chapter 5.

■ The engine is a closed system that operates continuously at a steady state.

■ 20 kJ/s of heat are added to the engine from a high-temperature reservoir at $T_H = 800$ K.

■ Heat is emitted from the engine to a low-temperature reservoir at $T_C = 300$ K.

■ Shaft work is produced.

Assuming that the entire process is reversible, determine the rate at which work is produced and the rate at which heat is emitted.

SOLUTION: Before beginning, we note that we know nothing specific about the unit operations used in the heat engine. We know only that the process is reversible and operating at a steady state. We will see here that these facts are sufficient to solve the problem. There are two unknowns, the shaft work produced, $\dot{W}_S$, and the heat emitted to the low-temperature reservoir, which we will call $\dot{Q}_C$. Consequently, we need two equations: an energy balance and an entropy balance.

Step 1 *Define a system*
The system is the entire heat engine, as shown in Figure 4-18. Heat transfer between the engine and the reservoirs is crossing the boundaries of the system, as the reservoirs are outside the system.

> The Rankine cycle described in the introductions to Chapters 1 and 2 had four steps—each with a well-defined purpose. Here, we know nothing about the steps—we know what the process *does*, but we don't know *how*.

> Heat that is removed from a system without being converted into work or recycled in any useful way is often called "spent" or "rejected" heat.

Step 2 *Apply and simplify energy balance*

Because this is a steady-state process, the accumulation term in the energy balance is 0. The heat engine is also a closed system ($\dot{m}_{in} = \dot{m}_{out} = 0$), so the only terms present in the energy balance are heat and work. We include two $\dot{Q}$ terms to represent the heat exchanged with each of the reservoirs.

$$0 = \dot{Q}_H + \dot{Q}_C + \dot{W}_S \tag{4.110}$$

$\dot{Q}_H$ is the rate of heat transfer to the engine and is known to be $+20$ kJ/s. From our sign conventions, $\dot{Q}_C$ and $\dot{W}_S$ are both negative (heat transferred out of system, work done by system on surroundings).

Step 3 *Apply and simplify entropy balance*

Again, the system is at steady state, so the accumulation term of the entropy balance is 0. There is no material entering or leaving the system, so the entropy balance is

$$0 = \sum_{n=1}^{n=N} \frac{\dot{Q}_n}{T_n} + \dot{S}_{gen} \tag{4.111}$$

Here there will be two different heat terms, $\dot{Q}_H$ and $\dot{Q}_C$.

The fact that the entire process is reversible is significant in this process two different ways:

- In a reversible process, no entropy is generated inside the boundaries of the system.

- In order for heat transfer to be reversible, the driving forces for heat transfer must be very small. Consequently, the system temperature at the boundary where the heat transfer takes places has to be essentially identical to the temperature of the reservoir itself.

Consequently, the entropy balance becomes

$$0 = \frac{\dot{Q}_H}{T_H} + \frac{\dot{Q}_C}{T_C} \tag{4.112}$$

Step 4 *Solve entropy balance*

By definition, the temperature of a heat reservoir is assumed constant. Since T_C, T_H, and $\dot{Q}_H$ are all given, the only unknown in Equation 4.112 is $\dot{Q}_C$, which is the heat transferred to the cold reservoir.

$$0 = \frac{20\,\frac{kJ}{s}}{800\,K} + \frac{\dot{Q}_C}{300\,K} \tag{4.113}$$

$$\dot{Q}_C = 7.5\,\frac{kJ}{s}$$

Step 5 *Solve energy balance*

The result of step 4 allows us to solve the energy balance for the work.

$$0 = \dot{Q}_H + \dot{Q}_C + \dot{W}_S$$

$$0 = +20\,\frac{kJ}{s} - 7.5\,\frac{kJ}{s} + \dot{W}_S \tag{4.114}$$

$$\dot{W}_S = 12.5\,\frac{kJ}{s}$$

Sidebar (left margin):

$\dot{Q}_H$ represents heat transferred to/from the "hot" reservoir and $\dot{Q}_C$ represents heat transferred to/from the "cold" reservoir.

Entropy was first proposed as a property because Q/T was recognized as a useful quantity in analyzing heat engines.

Notice that in Example 4-11, we were able to determine the rate at which the heat engine produced work even though we knew nothing about how, specifically, the engine functions, other than that it was reversible. Because we assumed nothing specific, the analysis is valid for *any* reversible heat engine.

The energy balance for a closed system heat engine (derived in step 2 of Example 4-11) can be written:

$$0 = \dot{Q}_H + \dot{Q}_C + \dot{W}_{net} \tag{4.115}$$

> This equation was derived in Example 4-11 for a steady-state process, but Example 4-12 will show that substantially the same equation can be derived for unsteady-state processes that are *cyclic*.

- $\dot{Q}_H$ represents heat added to the heat engine from a high-temperature reservoir.
- $\dot{Q}_C$ represents heat removed from the heat engine into a low-temperature reservoir.
- $\dot{W}_{net}$ represents the *total rate* at which the engine produces work.

There are typically at least two steps in a heat engine that produce or use work, and the work contributions from these individual steps are combined into the $\dot{W}_{net}$ term. You might wonder, if all the individual contributions to work are lumped together into $\dot{W}_{net}$, why aren't $\dot{Q}_H$ and $\dot{Q}_C$ combined into a $\dot{Q}_{net}$? One reason is that in the entropy balance, $\dot{Q}_H$ and $\dot{Q}_C$ must be kept distinct, because the temperatures at which they enter or leave the system are different. Work by contrast does not appear in the entropy balance at all. Another point is that all heat is not equally useful. Consider the results from Example 4-11.

- $\dot{Q}_H$ represents 20 kJ/s of heat available at 800 K, which is valuable and versatile; it can be used for any heating application in which the desired final temperature is below 800 K. This heat was likely generated by burning a consumable resource such as fossil fuels.

- $\dot{Q}_C$ represents 7.5 kJ/s of heat available at 300 K. This heat simply has little or no practical value. Because 300 K is barely above typical ambient temperatures, it isn't realistic to expect there will be many opportunities to transfer this heat elsewhere in a useful way.

> The efficiency of the heat engine in Example 4-11 is 62.5%. 20 kJ/s of heat is added, of which 62.5% is converted into work.

- Thus, combining $\dot{Q}_H$ and $\dot{Q}_C$ into a single "net heat" term ($\dot{Q}_{net} = 12.5$ kJ/s) wouldn't be wrong (at least, it wouldn't be wrong in the energy balance), but it would be misleading. We are consuming 20 kJ/s of a valuable resource—not 12.5 kJ/s.

The purpose of a heat engine is to convert heat energy into work. The **efficiency** of a heat engine is to quantify how effectively that is done. Thus,

$$\eta_{H.E.} = \frac{-\dot{W}_{net}}{\dot{Q}_{added}} \tag{4.116}$$

where

- η is the symbol conventionally used for efficiency. The subscript H.E. is used to emphasize that it stands for the efficiency of a heat engine (to distinguish it from, for example, a turbine or compressor efficiency).

- $\dot{W}_{net}$ is the **net work produced** by the heat engine and is typically computed by summing the individual work contributions of several steps. The negative sign reflects the fact that W is negative when work is produced by the engine, while the efficiency is conventionally a positive number.

(Continued)

PITFALL PREVENTION

Equation 4.116 defines the efficiency of a heat engine. Equations 4.68 and 4.69 defined efficiencies for turbines, pumps and compressors, which are often components of heat engines. Consequently, in Chapter 5 we will often solve problems that involve more than one value of efficiency, and must be careful to keep them straight.

■ $\dot{Q}_{added}$ is the **total heat added** to the heat engine. In Example 4-11, there was only one step in which heat was added, and it came from a high-temperature reservoir. In this common case, $\dot{Q}_{added}$ is simply equal to $\dot{Q}_H$.

■ $\dot{W}_{net}$ and $\dot{Q}_{added}$ are here expressed as rates (e.g., J/sec), but could also be expressed as absolute quantities (e.g., kJ), or on the basis of "energy per quantity of operating fluid" (e.g., kJ/kg). $\dot{W}_{net}$ and $\dot{Q}_{added}$ must be expressed on the same basis as each other since efficiency is a dimensionless quantity.

Note that Equation 4.116 does not indicate the form of the work. The following two examples show that *conceptually*, heat engines can be used to produce either expansion work (Example 4-12) or shaft work (Example 4-13). In practice, what we need is more commonly shaft work.

EXAMPLE 4-12	THE CARNOT CYCLE

The idea of "isothermal heating" sounds counter-intuitive, but if a gas expands as it's heated, it transfers the energy it receives as heat to the surroundings in the form of work. Maintaining a constant temperature is therefore possible.

A gas is confined in a piston-cylinder arrangement. The gas undergoes a series of four *reversible* steps, detailed below, which together constitute the **Carnot heat engine**, or the **Carnot cycle**. The steps are illustrated in Figure 4-19 and the state of the gas at each is shown in Figure 4-20.

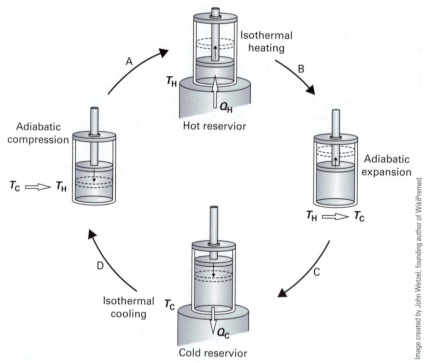

Section 5.3.3 illustrates that when the four steps described here are implemented in reverse sequence, they constitute a **Carnot refrigerator**.

Image created by John Wetzel, founding author of WikiPremed.

FIGURE 4-19 Schematic of the four steps in the Carnot cycle.

■ The gas, initially at a high pressure and high temperature (illustrated as point A on Figure 4-20), is *isothermally heated at a constant temperature* T_H. The gas expands as it's heated. The final state is represented by point B on Figure 4-20.

■ The gas is *adiabatically expanded*, from state B to state C. In this process, the temperature changes from T_H to T_C.

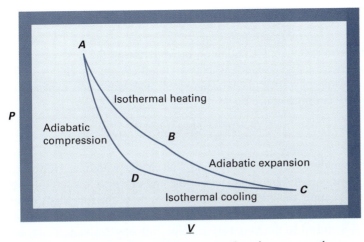

FIGURE 4-20 The four steps of a Carnot cycle, using a gas as the operating fluid, illustrated on a P-$\underline{V}$ diagram.

■ The gas is *isothermally cooled* from state C to state D. In order to maintain a constant temperature, the gas must be compressed while heat is removed.

■ The gas is *adiabatically compressed* back to state A.

A. Find a relationship between the net work produced by this process and the measureable properties (P, $\underline{V}$, and/or T) of the gas.

B. Find a relationship between the *efficiency* of the heat engine and the measureable properties (P, $\underline{V}$, and/or T) of the gas.

SOLUTION: No specific conditions are given (e.g., specific values of P or T); we are looking for mathematical expressions that are true in general for a heat engine of this kind.

SOLUTION A:

Step 1 *Define the system*
Define the gas in the cylinder as the system.

Step 2 *Assess work*
There is no shaft work; only expansion/contraction work. Figure 4-20 demonstrates that the gas volume changes in all four steps, and therefore work is done either on or by the system in each step. Recall that expansion/compression work is given by:

$$W_{EC} = -\int_{\text{initial state}}^{\text{final state}} P\, dV$$

P is the pressure opposing the motion, which in some steps is the gas pressure and in others is the external pressure. However, in a reversible process, the driving force for any change is negligible, so external and gas pressures are essentially equal; P can be equated to the gas pressure for all four steps.

Thus the *net work* for the complete cycle is:

$$W_{EC,\text{net}} = -\int_A^B P\, dV - \int_B^C P\, dV - \int_C^D P\, dV - \int_D^A P\, dV \qquad (4.117)$$

Graphically, the work in a particular step is given by the area under the curve representing that step in the P-V diagram, and the total work is the area bounded by the curves representing the four steps.

We use a P-V plot rather than (for example) P-T, because this plot permits a simple graphical visualization of work, as explained in step 2.

The Carnot cycle is named after Nicolas Leonard Sadi Carnot, a French physicist and military engineer who published the first theoretical description of heat engines in 1824. (Carnot, 1890)

SOLUTION B:

Step 3 *Apply and simplify energy balance*
Since there is no specific information on the duration of the process, we start with the time-independent energy balance. This is a closed system, so energy only enters and leaves in the forms of work and heat. There is no shaft work; only expansion/contraction work. Thus, the energy balance is:

$$U_2 - U_1 = Q_H + Q_C + W_{EC,net} \tag{4.118}$$

Step 4 *Define the process*
This is not a steady-state process; P, V, and T all change with time. However, the process is cyclic; the gas goes through four steps at the end of which it is returned to its original state. We could apply the energy balance to any one of the four steps, or to more than one consecutive step. However, if we define the process as *one complete cycle*, then the final state is identical to the initial state. Since enthalpy is a state function, this means $U_2 = U_1$. The energy balance thus further simplifies to

$$0 = Q_H + Q_C + W_{EC,net} \tag{4.119}$$

or solving for the net work gives

$$-W_{EC,net} = (Q_H + Q_C) \tag{4.120}$$

Step 5 *Apply definition of efficiency*
Dividing both sides of Equation 4.120 by Q_H and rearranging gives us an expression for the efficiency of the heat engine:

$$\eta_{H.E.} = \frac{-W_{EC,net}}{Q_H} = \frac{Q_H + Q_C}{Q_H} = 1 + \frac{Q_C}{Q_H} \tag{4.121}$$

Q_C and Q_H are opposite in sign, so Q_C/Q_H is negative, making the efficiency less than one.

Step 6 *Apply and simplify the entropy balance*
We start with the time-independent entropy balance. This is a closed system, so $m_{in} = m_{out} = 0$, and the process is reversible, so $S_{gen} = 0$. Thus the entropy balance simplifies to

$$S_2 - S_1 = \sum_{n=1}^{n=N} \frac{Q_n}{T_n} \tag{4.122}$$

The entropy balance further simplifies because entropy, like enthalpy, is a state function. Because the process is one complete cycle, $S_1 = S_2$.

This is essentially identical to the entropy balance obtained for the steady-state process in Example 4-11. Here it is only valid if the process is defined as *one complete cycle*.

Two of the four steps are adiabatic. The other two involve heat transfer, but the system is isothermal during these steps. Thus, the summation becomes

$$0 = \frac{Q_H}{T_H} + \frac{Q_C}{T_C} \tag{4.123}$$

Solving for Q_C produces

$$Q_C = \frac{-Q_H T_C}{T_H} \tag{4.124}$$

Step 7 *Relate efficiency to temperature*
Substituting the expression for Q_C in Equation 4.124 into Equation 4.121 gives us an expression for the efficiency of the cycle in terms of the temperatures of the heat reservoirs. We call this the Carnot efficiency

$$\eta_{H.E.} = \frac{-W_{EC,net}}{Q_H} = 1 + \frac{\left(\frac{-Q_H T_C}{T_H}\right)}{Q_H} = 1 - \frac{T_C}{T_H} \tag{4.125}$$

Example 4-12 described a series of four reversible steps that are known as the Carnot cycle and shows expressions for work produced and efficiency when this sequence is implemented using a gas in a piston-cylinder arrangement. There is no requirement that the operating fluid be an ideal gas, or even a gas. The next example outlines a method of implementing a Carnot heat engine that uses water as the operating fluid.

| CONTINUOUS CARNOT CYCLE | EXAMPLE 4-13 |

A Carnot heat engine operates continuously and reversibly at steady state, using the following cycle (illustrated in Figure 4-21):

- Saturated liquid water at temperature T_H (state A) enters a boiler, and emerges as saturated vapor (state B). Because the process is a phase change occurring at constant pressure, this is an *isothermal heating*.

- The saturated steam enters a reversible turbine, in which it *expands adiabatically* to state C. In this step, some of the steam condenses into liquid.

- The saturated liquid/vapor mixture enters a heat exchanger and is *isothermally cooled* to state D.

- The saturated liquid/vapor mixture enters a reversible pump and an *adiabatic compression* returns it to state A.

A. What is the physical state of the material (liquid, vapor, mixture of liquid and vapor, etc.) at conditions C and D?

B. What is the efficiency of the heat engine?

As mentioned in Section 1.2.3, continuous recirculation of the operating fluid in a heat engine (regardless of what that fluid is) is logical. Any other design would involve the continuous expense of adding fresh material and disposing of waste material.

Recall that T_H and T_C are defined as the temperatures of the hot and cold heat reservoirs, not of the system. However, in a reversible heat transfer, the reservoir and system temperatures are essentially identical.

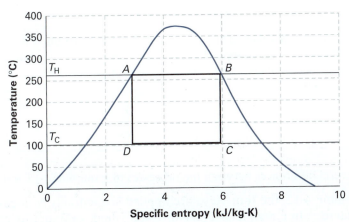

FIGURE 4-21 The four steps in a Carnot cycle, using H_2O as the operating fluid, illustrated on a T-$\hat{S}$ diagram.

SOLUTION A:

Step 1 *Apply entropy balance to single steps*
Figure 4-21 shows the process sketched on a temperature-entropy diagram. States A and B are saturated liquid and saturated vapor, respectively, both at temperature T_H. Because the transition between state C and state D is an isothermal cooling, we know that states C and D must both be at the same temperature as each other, specifically T_C.

The transition between states B and C is carried out in an adiabatic, reversible, steady-state process. However, it is an open system; saturated steam continuously enters and continuously leaves as a liquid–vapor mixture. Consequently, as in the nozzle in Example 4-6, the entropy balance simplifies to

$$0 = \dot{m}_{in}\hat{S}_{in} - \dot{m}_{out}\hat{S}_{out} \tag{4.126}$$

Since the mass flow rates entering and leaving must be the same at steady state, equation 4.126 demonstrates that the specific entropy must be equal for the entering and exiting material. Thus, the transition from state B to C must be represented by a vertical line on Figure 4-21. Identical reasoning applies to the transition from state D to A. Examining Figure 4-21, states C and D are both liquid–vapor mixtures with the same specific entropies as states B and A, respectively.

SOLUTION B:

Step 2 *Apply entropy balance to entire process*

In step 1 we established that a single component of the heat engine is an open system. However, the heat engine as a whole is a closed system; the H_2O continuously recirculates inside the boundaries of the system but no material enters or leaves. The process is also steady state (accumulation = 0) and reversible ($S_{gen} = 0$). Consequently, for the whole process, the entropy balance is

$$0 = \frac{\dot{Q}_H}{T_H} + \frac{\dot{Q}_C}{T_C} \tag{4.127}$$

Step 3 *Apply energy balance to entire process*

For a steady-state process in a closed system, the energy balance consists of only work and heat:

$$0 = \dot{Q}_H + \dot{Q}_C + \dot{W}_{S,net} \tag{4.128}$$

The energy and entropy balances are identical to the ones in Example 4-12, aside from the fact that the process in Example 4-12 produced expansion work while this one produces shaft work. Consequently, the subsequent derivation of the efficiency of the heat engine is also identical.

$$\eta = \frac{-\dot{W}_{S,net}}{\dot{Q}_H} = 1 - \frac{T_C}{T_H}$$

Examples 4-11 through 4-13 consider three different reversible heat engines. This entropy balance has been valid for all of them, but only when specific definitions of "the system" were employed.

We note several differences between the processes in Examples 4-12 and 4-13.

■ In Example 4-12, the four steps are applied sequentially to a gas confined in a chamber. In Example 4-13, the operating fluid continuously recirculates through the four processes, which are carried out continuously at steady state.

■ In the piston-cylinder process, the operating fluid is always a gas, and the circulating system uses liquid water and water vapor at different points.

■ The piston-cylinder maintains the constant temperature in the isothermal steps by adding or removing work, thus balancing the heat that is removed or added. In the circulating system, the heat energy added or removed produces a phase change, which is an isothermal process. There is no need to add or remove work.

■ The piston-cylinder process yields expansion work; the circulating process yields shaft work.

Despite these differences, the four steps in both cycles mirror each other. *Both processes are composed of an isothermal heating, adiabatic expansion, isothermal cooling, and adiabatic compression.* Any reversible cycle comprised of these four steps is a **Carnot heat engine**.

The efficiency expression derived in Example 4-12 was also valid for Example 4-13 and in fact applies to any Carnot engine.

$$\eta_{Carnot} = 1 - \frac{T_C}{T_H} = \frac{T_H - T_C}{T_H} \tag{4.129}$$

The efficiency of a Carnot heat engine depends only upon the temperatures of the heat reservoirs, and not upon the working fluid or the details of how the process is carried out.

This **Carnot efficiency** represents the highest possible efficiency for a heat engine operating between the the temperatures T_H and T_C. The Carnot cycle is a fully reversible process, and is thus an idealized limit which cannot be fully attained in practice. Section 5-2 explores the design of real heat engines, and how they can most closely approximate the Carnot cycle.

4.6 SUMMARY OF CHAPTER FOUR

- A **reversible** process is one in which all **driving forces** for changes are negligible.
- **Entropy** is a state property that can be regarded as a metric for lost work. It is mathematically defined as

$$dS = \frac{dQ_{rev}}{T}$$

- Specific entropy has dimensions of energy per unit mass per temperature (e.g., kJ/kg·K).
- The entropy of the universe in **unchanged** by any **reversible** process, and is **increased** by any **irreversible** process.
- Irreversible processes lead to **lost work**, and entropy can be regarded as a property that quantifies the loss of potential to do work.
- A **macrostate** is a specific description of the state of a system.
- A **microstate** is a specific configuration for the components that make up the system.
- Entropy of a system can be computed from the number of distinct microstates that each correspond to the macrostate that describes the system, through Equation 4.36.
- The **entropy balance** takes the form Accumulation = In – Out + Generation.
- The **generalized entropy balance** (Equations 4.37, 4.38, or 4.39) can be applied to any system, and can be used to establish limitations on what is possible within a process.
- **Accumulation** of entropy is zero in any steady-state process.
- **Generation** of entropy is zero in any **reversible** process.
- Because entropy is a state property, the change in entropy of a system can be determined by calculating the change in entropy along a **hypothetical reversible path** from the initial state to the final state. This is valid whether the real path is reversible or not.
- The **efficiency** of a real process can be defined by comparing it to a hypothetical reversible process.

4.7 EXERCISES

4-1. Several processes are described. Indicate whether each is reasonably modeled as reversible, and if not, indicate what aspect of the process makes it irreversible.

A. The inside of a refrigerator is at exactly 273.1 K. A glass of room temperature water is placed inside the refrigerator and left there until it is frozen.

B. The inside of a refrigerator is at exactly 273.1 K. A glass containing an equilibrium mixture of ice and water is placed inside the refrigerator and left there until it is frozen.

C. The inside of a refrigerator is at exactly 273.2 K. A glass containing an ice–water mixture is placed inside the refrigerator and left there until the ice is all melted.

D. A diver is 304 m below the surface of the ocean, where the pressure is 3.1 MPa. She rapidly ascends to the surface, where the pressure is 101.325 kPa.

E. A diver is 304 m below the surface of the ocean, where the pressure is 3.1 MPa. Over a period of half an hour she gradually ascends to the surface, where the pressure is 101.325 kPa.

4-2. Find the change in entropy for the system in each of the following processes.

A. The system is 5 kg of initially saturated steam at $P = 0.1$ MPa, and it is heated isobarically to 573.15K.

B. The system is 5 kg of steam, initially at $P = 0.5$ MPa and $T = 573.15$ K, and it is cooled and condensed into saturated liquid at $P = 0.5$ MPa.

C. The system is 10 moles of an ideal gas with $C_V^* = 1.5\,R$, and it is compressed from $T = 400$ K and $P = 0.1$ MPa to $T = 500$ K and $P = 0.3$ MPa.

D. The system is 4.5 kmol of an ideal gas with $C_P^* = 3.5\,R$, and it is expanded from $T = 533$ K and $P = 405.3$ kPa to $T = 311$ K and $P = 101.325$ kPa.

4-3. A heat reservoir is at 300 K. Find the change in entropy of the heat reservoir when

A. 1 kJ of heat is added.

B. 1 kJ of heat is removed.

C. Heat is added at a rate of 100 W for a total of one hour.

4-4. The system is 0.1 kg of a solid that has $\hat{C}_P = 0.3\,\text{kJ/kg K}$. Find the change in entropy when

A. It is heated reversibly from 300 K to 500 K.

B. It is cooled reversibly from 323 K to 298 K.

4-5. The system is a 5 kg object that has $\hat{C}_P = 2.5$ kJ/kg K. Find the change in entropy when:

A. It is heated reversibly from 500 K to 1000 K.

B. It is cooled reversibly from 200°C to 50°C.

C. It is heated reversibly from –5°C to 100°C.

D. It is cooled reversibly from 50°C to 0°C.

4-6. The system is 1 kg of nitrogen, initially at $T = 300$ K and $P = 0.4$ MPa. For each of the following cases, find the change in entropy of the system using Figure 2-3, then find the change in entropy again by assuming nitrogen is an ideal gas and using the data in Appendix D.

A. The nitrogen is isobarically heated to 330 K.

B. The nitrogen is isobarically cooled to 225 K.

C. The nitrogen is isothermally compressed to $P = 1$ MPa.

D. The nitrogen is compressed to $T = 330$ K and $P = 1$ MPa.

4-7. Find the efficiency of each of the following Carnot heat engines.

A. $T_\text{H} = 500$ K, $T_\text{C} = 300$ K

B. $T_\text{H} = 500°C$, $T_\text{C} = 300°C$

4-8. A system is operating at a steady state. Find the rate of at which entropy is generated inside the system for each of the following cases.

A. The system is adiabatic. 1.5 kg/s of saturated steam enters at 0.5 MPa and 1.5 kg/s of saturated steam leaves at 0.1 MPa.

B. The system is adiabatic. 1.5 kg/s of saturated liquid water enters at 0.1 MPa and 1.5 kg/s of liquid water leaves at $T = 393.15$ K and $P = 1$ MPa.

C. The system is closed. 1000 kJ/s enter the system at a boundary where $T = 1000$ K, and 750 kJ/s exit the system at a boundary where $T = 500$ K.

D. The system is closed. 1000 kJ/s enter the system at a boundary where $T = 1000$ K, and 500 kJ/s exit the system at a boundary where $T = 750$ K.

4.8 PROBLEMS

4-9. Steam enters a turbine at 1 MPa. The effluent pressure is 0.1 MPa and the efficiency of the turbine is 80%. Determine the state of the turbine effluent (if pure liquid or vapor, find the temperature, and if a mixture, find the quality) and compute the work produced per kg of entering steam, when

A. The entering steam has $T = 523.15$ K

B. The entering steam has $T = 598.15$ K.

4-10. A series of two turbines and a heat exchanger are used to obtain shaft work from steam in a steady-state process. The steam enters the first turbine at 1 MPa and 523.15 K, and exits the turbine at 0.3 MPa. In the heat exchanger, the steam is heated back up to 523.15 K, while the pressure remains 0.3 MPa. This steam enters the second turbine in which it is expanded to 0.1 MPa. Each turbine has an efficiency of 80%. For each kilogram of steam entering the process, find the amount of work produced in each turbine, and the amount of heat added in the heat exchanger.

4-11. Compare the results of Problems 4-9 and 4-10.
  A. Without the heat exchanger, the pair of turbines in Problem 4-10 would function identically to the turbine in Problem 4-9A. Comment on the effects the heat exchanger has on the process.
  B. If the heat exchanger in Problem 4-10 was moved in front of the first turbine, and the two turbines were combined, the result would essentially be the process examined in Problem 4-9B. The process in Problem 4-9B looks simpler: the same heat is added, it's just being added sooner, so any benefits you noted in part A of this problem are still obtained, and there's no need for two separate turbines. Can you see any rationale for using the apparently more complicated process described in Problem 4-10?

4-12. A 0.28 kg glass of water is initially at 288 K. It is left outside overnight in a location where the air temperature is 268 K. By morning the glass is in equilibrium with the surroundings. Find the change in entropy of the water, the surroundings, and the universe for this process. Assume for liquid water $\hat{C}_P \sim \hat{C}_V \sim 4.19$ kJ/kg K and for ice $\hat{C}_P \sim \hat{C}_V \sim 2.11$ kJ/kg K.

4-13. An inventor claims to have built a machine that operates as follows:

  ■ 1.67 kg/s of steam at $T = 673.15$ and $P = 0.5$ MPa enters the machine.

  ■ 1.67 kg/s of saturated liquid water at 313.15 K leaves the machine.

  ■ 800 kJ/s of shaft work is produced by the system. There is no heat exchange with surroundings.

  ■ An undisclosed amount of nitrogen enters the process at 315 K and 0.1 MPa, and the same amount of nitrogen leaves at 330 K and 0.1 MPa.

  ■ Kinetic and potential energy of all entering and leaving streams is negligible.

  ■ This is a steady-state process that can continue as described above indefinitely.

Without knowing anything about how the machine works, what can you say about the feasibility of the inventor's claim?

4-14. Steam at $T = 673.15$ K and $P = 0.5$ MPa enters an adiabatic, steady-state nozzle. The exiting pressure is $P = 0.1$ MPa.
  A. If the nozzle is reversible, find the temperature and velocity of the exiting steam.
  B. If the exiting temperature is $T = 623.15$ K, then the nozzle is identical to the one that was examined in Example 3-4. Find the amount of entropy generated in this nozzle for each kilogram of entering steam.

4-15. Prove that the process shown in Figure 4-9 is impossible if the cylinder contains a monatomic ideal gas.

4-16. Ten moles of a gas are placed in a rigid container. Initially the gas is at $P = 0.05$ MPa and $T = 300$ K and the container is also at $T = 300$ K. The container is placed in a furnace, where its surroundings are at a constant $T = 600$ K. The container is left in the furnace until both the container and the gas inside it reach thermal equilibrium with the surroundings. The gas can be modeled as an ideal gas at the pressures attained throughout this process, and has a constant heat capacity of $C_V^* = 2.5R$. The container itself has a mass of 10 kg (not including the mass of the gas inside) and a heat capacity of $\hat{C}_V = 1.5$ kJ/kg K.
  A. Find the heat added to the gas.
  B. Find the heat added to the container.
  C. Find the change in entropy of the gas.
  D. Find the change in entropy of the universe.
  E. What aspect of this process is irreversible?

4-17. This problem re-examines the dirigible in Example 2-4. The balloon contains 5000 moles of helium, which can be modeled as an ideal gas with $C_P^* = (5/2)R$ at the conditions in this problem. The dirigible is initially at $T = 298$ K and $P = 101.325$ kPa. The dirigible flies slowly enough that all changes to the helium are reversible. Find Q, W, ΔU, ΔH, and ΔS for the helium if its final state is:
  A. $T = 288$ K and $P = 101.325$ kPa. Pressure is uniform throughout the process.
  B. $T = 298$ K and $P = 91.2$ kPa. Temperature is uniform throughout the process.

4-18. Problem 3-19 described four different adiabatic mixing processes, and the relevant data presented was that the enthalpy of fusion of water is 333.55 kJ/kg, the heat capacity of ice is constant at 2.11 kJ/kg K, and the heat capacity of liquid water is constant at 4.19 kJ/kg K. Find the change in entropy of the universe resulting from each of these processes, and comment on whether each result is consistent with whether the process appears to be reversible or irreversible.
  A. 0.1 kg of liquid water at 323 K is mixed with 0.05 kg of liquid water at 273 K
  B. 0.1 kg of liquid water at 323 K is mixed with 0.05 kg of ice at 273 K.

C. 0.05 kg of liquid water at 273 K is mixed with 0.05 kg of ice at 273 K

D. 0.05 kg of liquid water at 298 K is mixed with 0.05 kg of ice at 253 K.

4-19. A 10 kg copper block has an initial temperature of 800 K. It is placed in a well-insulated vessel containing 100 kg of water initially at 290 K. The process is isobaric at atmospheric pressure. Assume copper and liquid water have constant heat capacity of $C_p = 4.184$ kJ/kg K for water and 0.398 kJ/kg K for copper.

A. When the system reaches equilibrium will be the temperature of the block?

B. What is the change in entropy of the universe resulting from the process?

4-20. A stream of liquid nitrogen enters an adiabatic, steady-state valve as a saturated liquid at $P = 2$ MPa. The material leaves the valve at $P = 0.6$ MPa. Use the data in Figure 2-3 to determine the following.

A. The temperature of the nitrogen leaving the valve.

B. The physical state of the nitrogen leaving the valve (if VLE mixture, indicate quality).

C. The rate at which entropy is generated in the valve (per kg of entering nitrogen.)

D. The irreversible feature of the process, if any, that explains the result of part C.

E. Repeat parts A, B, and C for methane instead of nitrogen. Use Figure 7-1.

4-21. One of the steps in a typical refrigeration process is a boiler. Freon® 22 enters a steady-state boiler as a VLE mixture with a pressure $P = 0.05$ MPa and quality of $q = 0.2$, and leaves the boiler as saturated vapor at $P = 0.05$ MPa. Use the data in Appendix F to determine the following.

A. How much heat is added to the boiler per kilogram of entering Freon® 22.

B. The flow rate of Freon through the boiler if the rate of heat addition is $\dot{Q} = 3.33$ kJ/s.

4-22. One of the steps in a typical refrigeration process is a compressor. Freon® 22 enters a steady-state compressor as saturated vapor at $P = 0.05$ MPa, and leaves at $P = 0.3$ MPa. Use the data in Appendix F for the following.

A. Determine the work added to the compressor, per kilogram of entering Freon, if the compressor is reversible.

B. Determine the work added to the compressor, and the temperature of the exiting Freon, if the compressor has an efficiency of 75%.

4-23. One of the steps in a typical refrigeration process is a condenser. Freon® 22 enters a condenser at steady state as superheated vapor at $P = 0.3$ MPa and $T = 303$ K. It leaves the condenser as saturated liquid at $P = 0.3$ MPa. Use the data in Appendix F for the following.

A. Find the heat removed from the condenser, per kilogram of entering Freon.

B. Find the flow rate of Freon in the condenser, if the rate at which heat is expelled from the condenser is $\dot{Q} = 4.17$ kJ/s.

C. Determine the change in entropy of the universe resulting from the process in part B, assuming the coolant used for the condenser is a heat reservoir at a temperature 5 K lower than the temperature of the Freon exiting the condenser.

4-24. Four eggs labeled A, B, C, and D are to be placed in three bowls labeled 1, 2, and 3.

A. How many distinct "microstates" are there?

B. How many distinct "macrostates" are there? List the macrostates, and indicate how many microstates there are which produce each macrostate.

4-25. You have four molecules labeled 1, 2, 3, and 4 and two bins labeled A and B. Any number of molecules can be placed in each bin.

A. List all of the possible microstates. How many are there?

B. List all of the macrostates, and indicate which microstates correspond to each if the four molecules are all the same compound.

C. List all of the macrostates, and indicate which microstates correspond to each, if molecules 1, 2, and 3 are the same as each other but 4 is a different compound.

4-26. A high-temperature reservoir at $T = 648$ K is to be used as the heat source for a steady-state heat engine, and a low-temperature reservoir at $T = 298$ K is to be used as the heat sink. The boiler takes in liquid water at $P = 0.8$ MPa and $T = 313$ K and expels steam at $P = 0.8$ MPa and $T = 623.15$ K. The condenser takes in steam at $T = 373.15$ MPa and $P = 0.01$ MPa and expels saturated liquid water at $P = 0.01$ MPa. The mass flow rate of the steam throughout the process is 5 kg/s. Find the rate at which the entropy of the universe is increased by the processes in the boiler and the condenser.

4-27. A high-temperature reservoir at $T = 533$ K is to be used as the heat source for a steady-state heat engine, and a low-temperature reservoir at $T = 297$ K is to be used as the heat sink. The boiler takes in a liquid organic compound at $P = 500$ kPa and $T = 311$ K and expels the organic compound as a vapor at $P = 500$ kPa and $T = 516$ K. The condenser takes in the organic compound as a saturated vapor at $T = 311$ K and expels it as a saturated liquid at $T = 311$ K. The mass flow rate of the organic compound throughout the process is 3.78 kg/s. Relevant data for the organic compound include:

■ Boiling point at $P = 500$ KPa is 422 K.

■ $\Delta\hat{H}_f = 1163$ kJ/kg at $T = 311$ K and $\Delta\hat{H}_f = 1050$ kJ/kg at $T = 422$ K.

■ Liquid phase $\hat{C}_p = 2.8$ kJ/kgK.

■ Vapor phase heat capacity at $P = 0.5$ MPa is $\hat{C}_p = 6.276 + 0.0105T$ with T expressed in Kelvin and C_p expressed in kJ/kgK.

A. Find the efficiency of a Carnot heat engine operating between 533 K and 297 K.

B. Find the efficiency of a Carnot heat engine operating between $T = 516$ K and $T = 311$ K.

C. Find the rate at which the entropy of the universe is increased by the processes in the boiler and the condenser.

4-28. Two chambers are separated by a partition. One of the chambers is evacuated, and the other has a volume of 1 m³ and contains steam at $T = 773.15$ K and $P = 1$ MPa. The partition is removed, allowing the gas to expand into the evacuated chamber. No heat transfer occurs with the surroundings. The final steam pressure is $P = 0.4$ MPa. Find

A. The mass of steam.

B. The volume of the chamber that was initially evacuated.

C. The change in entropy of the universe resulting from the process.

D. Re-do parts A–C, but this time, instead of steam, assume the 1 m³ chamber was filled with an ideal gas with $C_p^* = (7/2)R$.

4-29. Throughout Chapters 3 and 4, we have assumed that potential and kinetic energy are negligible in standard chemical process equipment. This problem tests how valid these approximations are for a turbine. Steam at 1.4 MPa and 573.15 K enters a turbine through a 0.05 m diameter pipe at a velocity of 2.5 m/s. The exhaust exits via a 0.2 m diameter pipe that is located 1.5 m below the inlet pipe. The exhaust is saturated steam at 0.09 MPa. Assume steady-state, adiabatic operation, but assume *nothing else* about various contributions to the energy balance.

A. Determine the mass flow rate entering the turbine.

B. Determine the power output from the turbine.

C. Determine the efficiency of the turbine.

4-30. Saturated liquid water enters a steady-state, adiabatic throttling valve at $T = 373.15$ K. The pressure of the fluid leaving the valve is $P = 0.05$ MPa. It has been suggested that perhaps it would make more sense to replace the valve with a turbine—the desired pressure drop will still occur, but some work would be obtained from the turbine.

A. Determine the physical state of the fluid leaving the valve.

B. Determine the rate at which entropy is generated in the valve per kilogram of entering water.

C. Determine the maximum work that could be produced by a turbine per kilogram of entering water if the stream entering the turbine was identical to the stream entering the valve, and the pressure leaving the turbine was $P = 0.05$ MPa.

D. Compare the physical state of the fluid leaving the "idealized" turbine described in part C to the physical state of the fluid leaving the valve as described in part A. Are they identical? Can you rationalize why they are, or are not, identical?

4-31. Superheated steam enters a steady-state, adiabatic throttling valve at $T = 473.15$ K and $P = 0.1$ MPa. The pressure of the fluid leaving the valve is $P = 0.05$ MPa. It has been suggested that perhaps it would make more sense to replace the valve with a turbine—the desired pressure drop will still occur, but some work would be obtained from the turbine.

A. Determine the physical state of the fluid leaving the valve.

B. Determine the rate at which entropy is generated in the valve, per kilogram of entering steam.

C. Determine the maximum work that could be produced by a turbine, per kilogram of entering steam, if the steam entering the turbine was identical to the steam entering the valve, and the pressure leaving the turbine was $P = 0.05$ MPa.

D. Compare the physical state of the fluid leaving the "idealized" turbine described in part C to the physical state of the fluid leaving the valve as described in part A. Are they identical? Can you rationalize why they are or are not identical?

E. Compare your answers to this problem to your answers to Problem 4-30. How is the outcome for steam different from the outcome of a similar process involving liquid water?

4-32. The thermostat in a house is set at 294 K. Consequently, the inside of the house is always at 294 K, regardless of the season.

A. One day, the outside temperature is 283 K, and the house loses 100 kJ of heat to the surroundings. What is the change of entropy for the universe resulting from this process?

B. Besides the heat transfer noted in part A, is there any other source of entropy generation associated with maintaining this house at 294 K?

C. If a Carnot engine were designed to operate with the house as the heat source and the outdoors as the heat sink, what would its efficiency be on this day? How much work would be produced if the 100 kJ was sent to this Carnot heat engine, rather than simply being lost to the outdoors?

4-33. A nuclear power plant generates 750 MW of power. The heat engine uses a nuclear reactor operating at 588 K as the source of heat. A river is available (at 293 K) which has a volumetric flow rate of 165 m³/s. If you use the river as a heat sink, estimate the temperature rise in the river at the point where the heat is dumped. Assume the actual efficiency of the plant is 60% of the Carnot efficiency.

4-34. You are designing a chemical process that involves a relatively unstudied compound, and you need to know $\Delta \underline{H}^{vap}$, $\Delta \underline{U}^{vap}$ and $\Delta \underline{S}^{vap}$, and C_P^* for the compound at atmospheric pressure. You place 1 mole of the compound (in the liquid state) in a piston-cylinder device and heat it very slowly while maintaining the pressure at 0.1 MPa. You can't actually see what's going on inside the cylinder, but you can monitor the heat addition, temperature, and volume. Initially, the temperature increases steadily while the volume increases very little. At a temperature of $T = 309.0$ K, the volume of the cylinder is 8.9×10^{-5} m³. At $T = 309.4$ K, the temperature levels off, and the volume begins to expand more rapidly. 89 kJ of heat are added between the time the temperature levels off and the time the temperature begins to increase again. Unfortunately, you don't have an accurate recording of the volume at the exact time the temperature started increasing. An additional 1.3 kJ of heat are added during the time the temperature climbs from 309.4 K to 311 K, at which time you stop the experiment. Give your best estimates of $\Delta \underline{H}^{vap}$, $\Delta \underline{U}^{vap}$, and $\Delta \underline{S}^{vap}$ and C_P^*.

4.9 GLOSSARY OF SYMBOLS

C_P	constant pressure heat capacity (mole basis)	$\dot{n}$	molar flow rate	T_H	temperature of a high-temperature heat reservoir
C_P^*	constant pressure heat capacity (mole basis) for ideal gas	P	pressure	t	time
		Q	heat	U	internal energy
C_V	constant volume heat capacity (mole basis)	Q_C	heat exchanged with a low-temperature heat reservoir	$\hat{U}$	specific internal energy
C_V^*	constant volume heat capacity (mole basis) for ideal gas	Q_H	heat exchanged with a high-temperature heat reservoir	$\underline{U}$	molar internal energy
				V	volume
$\hat{C}_V$	constant-volume heat capacity (mass basis)	$\dot{Q}$	rate of heat addition	$\hat{V}$	specific volume
$\hat{C}_p$	constant-pressure heat capacity (mass basis)	$\dot{Q}_C$	rate of heat exchange with low temperature reservoir	$\underline{V}$	molar volume
		$\dot{Q}_H$	rate of heat exchange with low temperature reservoir	v	velocity
H	enthalpy	R	gas constant	W	work
$\hat{H}$	specific enthalpy	S	entropy	$\dot{W}$	power
$\underline{H}$	molar enthalpy	$\hat{S}$	specific entropy	W_{EC}	work of expansion/contraction
M	mass of system	$\underline{S}$	molar entropy	W_S	shaft work
m	mass added to or removed from system	$\dot{S}_{gen}$	rate of entropy generation	η	efficiency
$\dot{m}$	mass flow rate	T	temperature	$\eta_{H.E.}$	overall efficiency of a heat engine
N	number of moles in system	T_C	temperature of a low-temperature heat reservoir	η_{carnot}	overall efficiency of a completely reversible heat engine
n	number of moles added to or removed from system				

4.10 REFERENCES

Carnot, S., Thurston R. (editor and translator), *Reflections on the Motive Power of Heat and on Machines Fitted to Develop That Power*, John Wiley & Sons, 1890.

Gutman, P., Djurdjevic, I., "A simple method for showing entropy is a function of state," *Journal of Chemical Education*, 65, 1988.

Sandler, S. *An Introduction to Applied Statistical Thermodynamics*, John Wiley & Sons, 2011.

Vemulapalli, G. K., "A Simple Method for Showing Entropy is a Function of State," *Journal of Chemical Education*, 63, 1986.

Thermodynamic Processes and Cycles

5

LEARNING OBJECTIVES

This chapter is intended to help you learn to:

- Apply entropy and energy balances to larger chemical processes with *multiple unit operations*
- Solve energy and entropy balances to fully characterize such processes
- Recognize the purposes of *Rankine heat engines, refrigerators,* and *Linde liquefaction* processes and appreciate the role of each unit operation in these processes
- Propose preliminary designs of Rankine heat engines, refrigerators, and Linde liquefaction processes that meet given specifications and operate within given constraints
- Compare and evaluate process designs, using appropriate benchmarks such as *efficiency* for heat engines and *coefficient of performance* for refrigerators

Chapters 3 and 4 examined the application of material, energy, and entropy balances to a variety of systems. Most of the examples focused on comparatively simple systems, such as single pieces of equipment. This chapter examines combining individual unit operations into complete processes, and analyzing these processes using material, energy, and entropy balances. ∎

5.1 MOTIVATIONAL EXAMPLE: Chemical Process Design

Chemical engineers can and do work in a great variety of industries and in a great variety of roles. However, one cross-cutting aspect of the practice of chemical engineering is the design and analysis of complete chemical processes. As an example, we examine the problem of synthesizing ammonia. Imagine you were tasked with designing a continuously operating, steady-state chemical process that makes 30 million kg/year of ammonia. Imagine further that all you really knew about ammonia synthesis to begin with was the information given in Problem 2-16, where nitrogen and hydrogen can be combined at high pressure over an iron catalyst. Your first step might be to draw a basic block diagram like that shown in Figure 5-1.

Figure 5-1 describes ammonia synthesis and is quite similar to Figure 1-1, which described chemical processes in general.

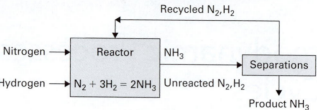

FIGURE 5-1 Preliminary schematic of process for ammonia synthesis.

Ultimately, we need to identify and design each of the individual unit operations in the process, but right now we have nothing more than boxes labeled "Reactor" and "Separations." Developing a complete design will require a great many engineering decisions. Let's start at the beginning: where does the nitrogen come from? A natural answer is "air contains almost 80% nitrogen and is free." This simple answer affects our vision of the process in at least two ways.

1. If we use air as a feedstock, then oxygen is entering the process, along with other impurities.

2. Air enters the process at ambient T and P, but the reaction is carried out at "high" pressure.

Thus, we could revise our block diagram into the one shown in Figure 5-2, in which the nitrogen is purified and compressed before entering the reactor. Figure 5-3 shows another alternative: rather than designing a separation process to purify the nitrogen, send compressed air directly to the reactor, and allow the oxygen to circulate through the process with the nitrogen. The alternative in Figure 5-3 looks like a simpler process requiring less equipment, so offhand we might expect it's cheaper to build. But which would be cheaper to operate on a day-to-day basis? On the one hand, shaft work is expensive, and the process in Figure 5-3 involves compressing not only the nitrogen and hydrogen, but also oxygen. On the other hand, the "purification" step in Figure 5-2 might require work of its own. Finally, we don't yet know whether the presence of oxygen would have any negative effects on the safety or operation of the reactor.

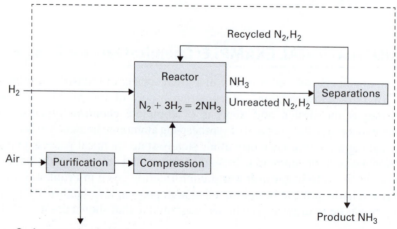

O$_2$, Ar, any other impurities

FIGURE 5-2 Schematic of the ammonia synthesis process in which nitrogen is purified before being fed to the reactor.

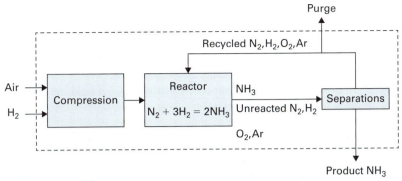

FIGURE 5-3 Schematic of the ammonia synthesis process in which compressed air is fed, without separation, to the reactor.

The foregoing discussion illustrates the significant challenge of designing a complete process. In considering the question "where does the nitrogen come from," we have raised several new questions, and are already wrestling with two fundamentally different approaches, between which we must somehow choose. The next question could be "Where does the hydrogen come from?" but we will not explore that at this time. Our goal with this discussion is not to reach any final conclusions about ammonia synthesis but to consider *What is the role of thermodynamics in the development of complete processes*? Before reading further, try putting some thought into how what we've learned so far applies to the design of this ammonia synthesis process.

Up to this point, we have primarily applied the principles of material, energy, and entropy balances developed in Chapters 3 and 4 to simple systems, such as single pieces of equipment. However, we can apply these same balance equations to any system—including large systems like an entire chemical plant or the entire world. Notice the system boundaries drawn in Figure 5-2. Even without making any decisions about how the process actually works, we can make significant determinations by applying material balances to this system.

- It takes 24.7 million kg/year of nitrogen and 5.3 million kg/year of hydrogen to make 30 million kg/year of ammonia.

- Modeling air as 78 mol% nitrogen, 21 mol% oxygen, and 1 mol% argon, we can show that the "air entering" stream must contain 7.6 million kg/year of oxygen and ~200,000 kg/year of argon, along with 24.7 million kg/year of nitrogen.

Writing and solving material balances for the system shown in Figure 5-3 is much more complicated due to the presence of the purge stream, but another perspective is that material balances are why we know the purge stream is necessary to begin with. If the purge stream were not present, writing an oxygen balance for this system would reveal that oxygen enters the system but has no way to leave. Achieving a steady state would therefore be impossible; oxygen and the other inerts would accumulate in the system, producing unsafe pressure increases.

In summary, useful insights can be gained from applying balance equations to complete processes. Looking into the design of individual steps in the process, the principles of thermodynamics are again evident.

- Some combination of compressors and heat exchangers will presumably be used to deliver the feedstock gases at the desired temperature and pressure to enter the reactor. We have examined compressors and heat exchangers individually in Chapters 3 and 4. In this chapter Example 5-7 studies the design of a series of heat exchangers and compressors.

In Example 4-11, we applied energy and entropy balances to a heat engine, without knowing how it actually worked. Similarly, here we apply material balances to an open-system chemical synthesis process, even though we know nothing yet about the inner workings of the process.

FOOD FOR THOUGHT 5-1

Do for yourself the calculations that produce these numbers.

■ Ammonia has a much higher boiling point than oxygen, nitrogen, or hydrogen (some of the relevant data was shown in Table 2-1). Ammonia is also highly soluble in water, while oxygen, nitrogen, and hydrogen gases are minimally soluble in water. Either of these facts can logically be considered as the basis for designing the "separation" step. In either case, the principles of multi-component phase equilibrium covered in Chapters 10 through 12 would be crucial for understanding and designing the process.

Chapters 1 through 4 present a basic "toolbox" of thermodynamic principles. The foregoing discussion outlines how these "tools," and additional tools we will develop in subsequent chapters, can be applied to exactly the types of problems chemical engineers are tasked with solving. In this chapter, we will begin to examine complete, integrated processes that are comprised of multiple unit operations. We are not yet ready to further our analysis of the ammonia synthesis process specifically—we have only covered modeling of pure compounds, and the reaction and separation processes involve mixtures. We will re-visit the ammonia synthesis challenge in Chapter 15, at which point we will have covered both chemical reaction equilibrium and multi-component equilibrium. For now, we begin our study of process analysis and design by revisiting the Rankine heat engine, which was introduced in Sections 1.2.3 and 2.1.

5.2 Real Heat Engines

The Carnot cycle is analogous to the "hypothetical reversible paths" in Examples 4-4 and 4-5, in that while it exists only in our imagination, we can learn things from it that are useful in solving real problems.

Section 4-5 introduced the Carnot cycle—a completely reversible heat engine. Unfortunately, realistically speaking, it is not possible to build a reversible heat engine. For example, if the addition of heat were truly reversible, this would mean the driving force for heat transfer is infinitesimally small, which in turn means the heat transfer is infinitesimally slow. The Carnot heat engine is thus a hypothetical, idealized case that can't be implemented in the real world.

However, the Carnot cycle is a useful theoretical construct because it represents the maximum conversion of heat into work that is possible for a heat engine operating between hot and cold reservoirs of two particular temperatures. Thus, it makes sense to design real heat engines so that they approximate Carnot heat engines as closely as is realistically possible.

Consequently, Section 5.2.1 examines real heat engines that are designed using the Rankine cycle and how design decisions can be understood through consideration of, and comparison to, the Carnot cycle.

5.2.1 Comparing and Contrasting the Rankine Cycle with the Carnot Cycle

The Rankine cycle was discussed in Chapters 1 and 2 because it provided a good framework for introducing a number of thermodynamic concepts. At this point, we are ready to analyze a Rankine heat engine in a more thorough and quantitative way.

The Rankine cycle was introduced in Section 1.2. It operates as given here.

■ High-pressure liquid enters a boiler, absorbs energy from a heat source, and emerges as either saturated vapor or superheated vapor.

■ The high-pressure vapor enters a turbine, producing work and emerging as low-pressure vapor, possibly with a small amount of liquid entrained in the vapor.

■ The low-pressure vapor enters a condenser, emits energy to a heat sink, and emerges as saturated liquid.

■ The low-pressure liquid enters a pump, is compressed into high-pressure liquid, and is returned to the boiler.

■ The entire process is typically designed to operate at steady state.

■ Water is typically used as the operating fluid, though the logic of the four steps is equally valid for other operating fluids.

Figure 5-4 shows the steps of the Carnot cycle, and Figure 5-5 the steps of the Rankine cycle, each sketched on a T vs. Ŝ diagram for water. Here, we will review the Carnot cycle as it was described in Example 4-13, and consider what aspects of the Carnot cycle are, and are not, realistic to implement in real heat engines.

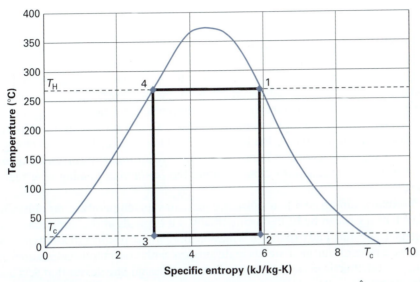

FIGURE 5-4 An example Carnot heat engine cycle, illustrated on the T-$\hat{S}$ plot for water. State 1 is the steam after isothermal heating, 2 is the liquid/vapor mixture after adiabatic expansion, 3 is the liquid/vapor mixture after isothermal cooling, and 4 is the liquid after adiabatic compression. Recall that the entire cycle is reversible; consequently the adiabatic steps are isentropic, and there is no temperature difference between the heat reservoir and the working fluid in the heating/cooling steps.

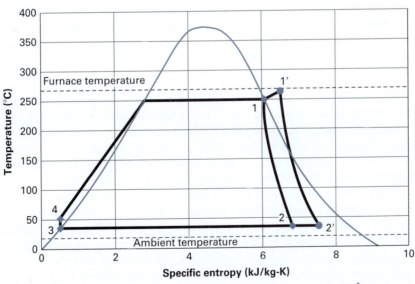

The temperature difference between states 3 and 4 in Figure 5-5 is exaggerated to make it clear that the points are distinct. The temperature change in a pump is typically closer to 1 to 2 degrees, rather than 15 to 20 degrees as the figure implies.

FIGURE 5-5 An example Rankine heat engine cycle, illustrated on the T-$\hat{S}$ plot for water. State 1 is the steam leaving the boiler, 2 is the liquid/vapor mixture leaving the turbine, 3 is the liquid leaving the condenser, and 4 is the liquid leaving the pump. States 1′ and 2′ represent an alternative cycle in which the steam leaving the boiler is superheated rather than saturated.

FOOD FOR THOUGHT 5-2

If the direction of the Carnot cycle were reversed, progressing through the states in the sequence 4-3-2-1-4 instead of 1-2-3-4-1, what would it do?

In this discussion we assume the fuel burns at a specific, fixed temperature, and thus we treat $T = 543$ K for the high-temperature reservoir as a design constraint.

In their chapter on "experience-based principles" for design, Turton et al. (Turton, 2009) recommend a heat exchanger use a *minimum* approach temperature (ΔT) of 10 K for fluids above ambient temperature, and a minimum ΔT of 5 K for heat exchangers that use refrigeration.

Problem 5-6 asks you to calculate values of the heat and work in each step of the Rankine engine described by Figure 5-5.

Enthalpy of vaporization is defined as the molar enthalpy of saturated vapor minus the molar enthalpy of saturated liquid at the same T and P: $\Delta \underline{H}^{vap} = \underline{H}^V - \underline{H}^L$

OVERALL PROCESS: CONTINUOUS, STEADY-STATE, AND CLOSED SYSTEM

Two prominent applications for heat engines are powering vehicles and generating electricity. In such applications the engine is required to operate continuously for significant periods of time, so steady state is the most practical mode of operation. The cycle is a closed system and the operating fluid is continuously circulated through the following four steps.

Carnot Step 1: Isothermal Heating A reversible heat engine requires that the heating of the operating fluid is carried out isothermally, and at essentially the same temperature as the high-temperature reservoir, as shown in Figure 5-4. Practically speaking, there must be a significant ΔT between the reservoir and the operating fluid in order for the heat transfer to be accomplished in a heat exchanger of manageable size. Thus, if the high-temperature reservoir is a furnace at $T_H = 543$ K (as in Figure 5-5), the operating fluid would most likely leave the heat exchanger at ≈ 523–533 K (523 K is used in Figure 5-5). This observation still leaves the question: how can a real process most closely mimic the *isothermal* heating of a real heat engine?

Normally, adding heat to a substance increases its temperature. As illustrated in Figure 4-19, a gas can be heated isothermally *if* it expands at the same time. This simultaneous heating and expansion is feasible in a closed, piston-cylinder type system. Here, we are considering a steady-state process with continuous circulation of the working fluid. Practically speaking, it's difficult to design one device that acts both as a heat exchanger and as a pressure controller in a continuous, steady-state operation.

Instead, the most straightforward way to heat something isothermally is through a phase change, as in Example 4-13. A heat exchanger that took in saturated liquid and emitted saturated vapor would be isothermal (assuming the pressure drop across the heat exchanger was negligible).

Realistically, the boiler in a real Rankine heat engine doesn't truly operate isothermally: the entering liquid is not saturated liquid and the exiting vapor is sometimes superheated, for reasons that are explained in the discussion of the remaining steps. However, water has a large enthalpy of vaporization, so a boiler is certainly a logical operation for adding a large amount of heat to the circulating liquid without exceeding the limit imposed by the temperature of the heat source.

Carnot Step 2: Adiabatic, Reversible Expansion Since the goal of the design is to convert heat into shaft work, the expansion is carried out in a turbine. In a Carnot cycle this step is adiabatic and reversible. Real turbines are well approximated as adiabatic but are not reversible, as discussed in Example 4-7. Since a detailed discussion of the design and construction of turbines is beyond the scope of this book, here we simply note that the most effective heat engine will use turbines that are as close to reversible as possible. "Close to reversible" for a turbine can be quantified through the efficiency, as discussed in Section 4.4.2.

For the moment, assume the material entering the turbine is saturated vapor, as illustrated by path 1-2 in Figure 5-5. The material exiting the turbine in such a case is typically a liquid–vapor mixture, composed primarily of vapor with a small amount of liquid entrained in it. If the liquid fraction in a turbine becomes too large, the equipment is vulnerable to damage from erosion. Consequently, in designing a heat engine, a maximum acceptable liquid fraction leaving the turbine would likely be specified as a design constraint. If the predicted liquid fraction is unacceptably large, one way to alter the design is to send superheated vapor, rather than saturated vapor, into the turbine. This alternative is illustrated by path 1′ to 2′ in Figure 5-5; note that state 2′ is nearer to the saturated vapor curve than state 2, meaning the vapor fraction (q) is higher. If the entering steam were superheated to an even higher temperature, the exiting steam could contain no liquid at all.

> The **quality** (q) of a liquid–vapor mixture was defined in Section 2.2.5 as the fraction of mass in the vapor phase.

Carnot Step 3: Isothermal Cooling The cooling step is carried out in a condenser. A real condenser operates at a higher temperature (313.15 K in Figure 5-5) than the low-temperature reservoir (293.15 K), for the same reason that the boiler operates at a lower temperature than the high-temperature reservoir: a significant temperature difference is required to allow the desired heat transfer to occur in a heat exchanger of manageable volume.

Figure 5-4 shows that, if a heat engine were truly reversible and water were the operating fluid, the material exiting the condenser (state 3 in Figure 5-4) would be a liquid–vapor mixture. This is necessary for the following reasons:

- If any entropy change occurred in the pump, the pump wouldn't be reversible.
- Low-pressure saturated liquid has a lower specific entropy than high-pressure saturated liquid.
- Therefore, in order to have the same entropy as the high-pressure liquid, the material entering the compression step cannot be saturated liquid; it must be a liquid–vapor mixture.

However, it is easier and cheaper to design a pump to compress a pure liquid rather than compressing a liquid–vapor mixture. Consequently, in the Rankine cycle, the condenser is designed to produce a pure liquid, as practical and safety considerations are prioritized over the fractionally larger efficiency that might result from compressing a liquid–vapor mixture. There is no practical advantage to further cooling the water once it has all condensed, so the condenser is designed to produce saturated liquid.

> Because vapors typically have a specific volume far larger than liquids, it typically takes far more work to compress a vapor than to compress an equivalent mass of liquid. See Example 3-9 which examines a pump.

Carnot Step 4: Adiabatic Compression The compression is carried out in a pump. The work required to compress a pure liquid is comparatively small, and it is therefore normal that a large pressure increase is accompanied by only a small temperature increase. Consequently, the liquid leaving the pump is subcooled liquid, as illustrated by path 3 to 4 in Figure 5-5.

Thus, in general terms, the Rankine cycle illustrated in Figure 5-5 can be rationalized as "the closest thing to a Carnot heat engine that we can build after taking real-world considerations into account."

The next section examines the detailed quantitative design of a Rankine heat engine.

5.2.2 Complete Design of a Rankine Heat Engine

This section illustrates the design and analysis of a Rankine heat engine.

COMPLETE QUANTITATIVE DESCRIPTION OF A RANKINE CYCLE	EXAMPLE 5-1

A Rankine heat engine is to be designed, with the following specifications and constraints:

- The boiler will operate at a maximum $T = 200°C$.
- The condenser will operate at a minimum $T = 100°C$.
- The turbine has an efficiency of 75%.

■ The liquid fraction exiting the turbine cannot exceed 10%.

■ Water, in liquid and/or vapor form, is the operating fluid.

Determine the state of the water entering each of the four steps of the process in Figure 5-6, the efficiency of the heat engine, and the mass flow rate of water necessary for the engine to produce 10 MW of net work.

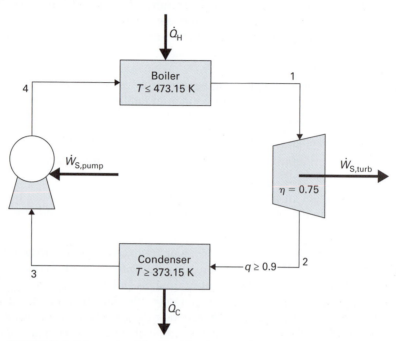

FIGURE 5-6 Schematic of the Rankine heat engine examined in Example 5-1.

SOLUTION: This example is not fully specified; it requires us to make some decisions and assumptions.

Step 1 *Preliminary design decisions—boiler*

Section 4.4.4 demonstrated that efficiency in a Carnot engine is given by:

$$\eta_{Carnot} = 1 - \frac{T_C}{T_H} \tag{5.1}$$

While the exact equation does not hold for real, irreversible heat engines, it provides a physical insight—that efficiency is maximized when T_C is minimized and T_H is maximized—which *is* true for real heat engines. Consequently, we will assign the steam exiting the boiler to be at $T = 473.15$ K, which is the specified maximum temperature. We cannot set the pressure of the steam exiting the boiler above the vapor pressure at 473.15 K (1.555 MPa), or the liquid will not boil. We could set the pressure lower than 1.555 MPa, but we note that turbine work will be maximized if the pressure difference between the entering and exiting steam is maximized. Consequently, we begin by assigning:

The steam exiting the boiler and entering the turbine (1) will be saturated steam at $T = 473.15$ K.

Step 2 *Preliminary design decisions—condenser*

We wish the turbine outlet pressure to be as low as possible, so that the pressure drop across the turbine is as large as possible. The lowest temperature at which the condenser can operate is 373.15 K. The vapor pressure at this temperature, $P = 101.3$ kPa, is the lowest pressure at which the vapor will condense. Consequently:

The water exiting the condenser and entering the pump (3) will be saturated liquid at $T = 373.15$ K.

Step 3 *Simplifying assumptions*
We were asked to find the state of the fluid leaving each of the four steps of the process. We've already determined two. We can assume, as we did in our discussions of the processes in Figures 5-4 and 5-5, that the pressure of the fluid does not change in the boiler or the condenser. Therefore, the steam exiting the turbine is at $P = 101.3$ kPa and the water exiting the pump is at $P = 1.555$ MPa. We can apply entropy and energy balances to determine more properties of these streams.

Step 4 *Apply entropy and energy balances to reversible turbine*
To determine the work produced in the turbine, we must apply the known turbine efficiency of 75%. First we consider a reversible turbine, which has the following entropy balance (see Example 4-7):

$$\hat{S}_{out} - \hat{S}_{in} = \hat{S}_{2,rev} - \hat{S}_1 = 0 \tag{5.2}$$

And any turbine, reversible or irreversible, has the following energy balance (see Section 3.6.3):

$$\frac{\dot{W}_{S,turbine}}{\dot{m}} = \hat{H}_{out} - \hat{H}_{in} \tag{5.3}$$

Step 5 *Collect data*
From Appendix A, $\hat{S}_1 = \hat{S}_{2,rev} = 6.4302 \frac{kJ}{kg \cdot K}$. At $T = 373.15$ K, saturated liquid and saturated vapor have $\hat{S}^L = 1.3072 \frac{kJ}{kg \cdot K}$ and $\hat{S}^V = 7.3541 \frac{kJ}{kg \cdot K}$. Consequently, the material leaving the reversible turbine must be a liquid–vapor mixture; its specific entropy falls between $\hat{S}^L$ and $\hat{S}^V$.

Step 6 *Compute q for reversible turbine*
The mixture leaving the reversible turbine is 84.7% vapor, as

$$\hat{S}_{2,rev} = (1 - q_{rev})\hat{S}^L + q_{rev}\hat{S}^V \tag{5.4}$$

$$6.4302 \frac{kJ}{kg \cdot K} = (1 - q_{rev})\left(1.3072 \frac{kJ}{kg \cdot K}\right) + (q_{rev})\left(7.3541 \frac{kJ}{kg \cdot K}\right)$$

$$q_{rev} = \mathbf{0.847}$$

Step 7 *Compute specific enthalpy of stream leaving reversible turbine.*
The specific enthalpy of the exiting stream can be determined as

$$\hat{H}_{2,rev} = (1 - q_{rev})\hat{H}^L + (q_{rev})\hat{H}^V \tag{5.5}$$

$$\hat{H}_{2,rev} = (1 - 0.847)\left(419.2 \frac{kJ}{kg}\right) + (0.847)\left(2675.6 \frac{kJ}{kg}\right) = 2330.8 \frac{kJ}{kg}$$

Step 8 *Solve energy balance for reversible turbine*
The energy balance (Equation 5.3) can now be solved to find the reversible work:

$$\frac{\dot{W}_{S,turbine,rev}}{\dot{m}} = \hat{H}_{2,rev} - \hat{H}_1 = 2330.8 \frac{kJ}{kg} - 2792.0 \frac{kJ}{kg} = -461.2 \frac{kJ}{kg} \tag{5.6}$$

FOOD FOR THOUGHT 5-3

Throughout this chapter we assume the properties of a fluid do not change as the fluid is transported from one unit operation to the next. Is this true?

The state of a pure, homogeneous phase has two degrees of freedom, as shown by the Gibbs phase rule. Thus, to describe the "state" of a material, one intensive property (here, pressure) is not a complete answer — we need two.

In steps 4 through 8, the system is a reversible turbine. We account for the non-idealities in the turbine using the turbine efficiency.

PITFALL PREVENTION

The symbol q_{rev} is used to emphasize that this number is specific to the reversible turbine. The problem specified that the turbine effluent cannot exceed 10% liquid. The result of 84.7% vapor and 15.3% liquid in step 6 does not violate that constraint because the reversible turbine is purely hypothetical.

Step 9 *Apply turbine efficiency*

The turbine is known to be 75% efficient. The turbine efficiency is defined as

$$\eta = \frac{\dot{W}_{S,\text{turbine,act}}}{\dot{W}_{S,\text{turbine,rev}}} \tag{5.7}$$

and can be solved for the actual work:

$$\frac{\dot{W}_{S,\text{turbine,act}}}{\dot{m}} = \frac{\eta(\dot{W}_{S,\text{turbine,rev}})}{\dot{m}} = (0.75)\left(-461.2\frac{\text{kJ}}{\text{kg}}\right) = -345.9\frac{\text{kJ}}{\text{kg}} \tag{5.8}$$

Step 10 *Determine q for fluid leaving real turbine*

Now that we've analyzed the turbine, it seems natural to progress to the next step: the condenser. However, we cannot forget the specification that the stream leaving the turbine can contain no more than 10% liquid; we need to verify that this is true. Applying the energy balance to the actual turbine gives

$$\frac{\dot{W}_{S,\text{turbine,act}}}{\dot{m}} = \hat{H}_{2,\text{act}} - \hat{H}_1 \tag{5.9}$$

Find the specific enthalpy of the stream leaving the actual turbine:

$$\hat{H}_{2,\text{act}} = \frac{\dot{W}_{S,\text{turbine,act}}}{\dot{m}} + \hat{H}_1 = -345.9\frac{\text{kJ}}{\text{kg}} + 2792.0\frac{\text{kJ}}{\text{kg}} = 2446.1\frac{\text{kJ}}{\text{kg}} \tag{5.10}$$

Now determine the actual liquid fraction of the material leaving the actual turbine:

$$\hat{H}_{2,\text{act}} = (1 - q_{\text{act}})\hat{H}^L + (q_{\text{act}})\hat{H}^V \tag{5.11}$$

$$2446.1\frac{\text{kJ}}{\text{kg}} = (1 - q_{\text{act}})\left(419.2\frac{\text{kJ}}{\text{kg}}\right) + (q_{\text{act}})\left(2675.6\frac{\text{kJ}}{\text{kg}}\right)$$

$$q_{\text{act}} = 0.90$$

Thus, the material leaving the turbine (2) is a mixture of 10% liquid and 90% vapor at $T = 373.15$ K, and has a specific enthalpy of 2446.1 kJ/kg.

The fraction of liquid in the turbine effluent is essentially at its maximum allowable value, so we will assume the decisions we've made up to this point are acceptable and continue with the design process.

Step 11 *Find heat removed from condenser*

The energy balance for one side of a heat exchanger, as derived in Section 3.6.4, is

$$\frac{\dot{Q}}{\dot{m}} = \hat{H}_{\text{out}} - \hat{H}_{\text{in}} = \hat{H}_3 - \hat{H}_2 \tag{5.12}$$

For this condenser, the leaving stream is saturated liquid at 100°C, and the entering stream is the stream that exits the turbine, the specific enthalpy of which was found in step 10.

$$\frac{\dot{Q}_C}{\dot{m}} = \hat{H}_3 - \hat{H}_2 = 419.2\frac{\text{kJ}}{\text{kg}} - 2446.1\frac{\text{kJ}}{\text{kg}} = -2026.9\frac{\text{kJ}}{\text{kg}} \tag{5.13}$$

Step 12 *Apply energy balance to entire heat engine*

The *entire* heat engine is a closed system operating at steady state, so the energy balance is

$$0 = \dot{Q} + \dot{W} \tag{5.14}$$

But there are two distinct steps involving heat and two distinct steps involving work:

$$0 = \dot{Q}_C + \dot{Q}_H + \dot{W}_{S,\text{turbine}} + \dot{W}_{S,\text{pump}} \tag{5.15}$$

At this point $\dot{Q}_C$ and $\dot{W}_{S,turbine}$ are known. If either the pump work or the boiler heat is determined, the other can be found by closing this energy balance. We cannot calculate either $\dot{Q}_H$ or $\dot{W}_{S,pump}$ from an energy balance because we don't have enough information to determine the specific enthalpy of stream 4, so what alternatives are available?

Step 13 *Calculate pump work*
Refer to Example 3-8, which demonstrated that pump work is well approximated by

$$\dot{W}_{S,pump} = \int \dot{V}\, dP \approx \dot{V}(P_{out} - P_{in}) \tag{5.16}$$

in which liquid volume is modeled as constant. Here, stream 4 is leaving the pump and stream 3 is entering, and since we don't know the mass flow rate, we divide both sides by it to get

$$\frac{\dot{W}_{S,pump}}{\dot{m}} = \hat{V}(P_4 - P_3) \tag{5.17}$$

$$\frac{\dot{W}_{S,pump}}{\dot{m}} = \left(0.001043\,\frac{m^3}{kg}\right)(1.555\ \text{MPa} - 0.1013\ \text{MPa})$$

$$\times \left(10^6\,\frac{Pa}{MPa}\right)\left(1\,\frac{\frac{N}{m^2}}{Pa}\right)\left(\frac{1\ J}{1\ N\cdot m}\right)\left(\frac{1\ kJ}{1000\ J}\right) = 1.52\,\frac{kJ}{kg}$$

Notice that only 1.52 kJ/kg of work is required to compress the liquid from 0.1013 MPa to 1.555 MPa. While the addition of energy in a pump results in a temperature rise, it is only an incremental increase.

Step 14 *Apply energy balance to pump*
The energy balance for a pump mirrors the energy balance for a turbine:

$$\frac{\dot{W}_{S,pump}}{\dot{m}} = \hat{H}_{out} - \hat{H}_{in} = \hat{H}_4 - \hat{H}_3 \tag{5.18}$$

Thus, the specific enthalpy of the water leaving the pump can be determined by

$$1.52\,\frac{kJ}{kg} = \hat{H}_4 - 419.2\,\frac{kJ}{kg} \tag{5.19}$$

$$\hat{H}_4 = 420.7\,\frac{kJ}{kg}$$

Step 15 *Describing the state of stream 4*
We now have two physical properties of the pure water leaving the pump: $P = 1.555$ MPa and $\hat{H} = 420.7$ kJ/kg. This means (according to Gibbs phase rule) all other intensive properties of the stream must have unique values. If we were required to find the temperature, we could estimate it from data in Appendix A.

Step 16 *Find heat added in boiler*
We can now solve the energy balance for the entire heat engine, Equation 5.15 from step 12:

$$0 = \dot{Q}_C + \dot{Q}_H + \dot{W}_{S,turbine} + \dot{W}_{S,pump}$$

$$0 = \frac{\dot{Q}_C}{\dot{m}} + \frac{\dot{Q}_H}{\dot{m}} + \frac{\dot{W}_{S,turbine}}{\dot{m}} + \frac{\dot{W}_{S,pump}}{\dot{m}}$$

$$0 = \left(-2026.9\,\frac{kJ}{kg}\right) + \frac{\dot{Q}_H}{\dot{m}} + \left(-345.9\,\frac{kJ}{kg}\right) + 1.52\,\frac{kJ}{kg} \tag{5.20}$$

$$\frac{\dot{Q}_H}{\dot{m}} = 2372.7\,\frac{kJ}{kg}$$

PITFALL PREVENTION

A common error is modeling the liquid *leaving* the pump as saturated liquid. There is no reason to expect this liquid is saturated; it is done just because the properties of saturated liquids are available. But we will see below how bad this assumption is.

By definition, $V = M\hat{V}$.

We use the specific volume of stream 3 in solving Equation 5.17, because it is available in the steam tables. The specific volume of stream 4 is currently unknown (but assumed to be approximately the same).

In step 14, the system is the pump.

PITFALL PREVENTION

The specific enthalpy of saturated liquid at 473.15 K is 852.3 kJ/kg. Had we assumed the material leaving the pump was saturated liquid, in effect, we would have been assuming the work added in the pump was over 430 kJ/kg (852.3−419.04), when in reality it is less than 2 kJ/kg.

The compressed liquid table in Appendix A contains data at $P = 1$MPa and $P = 5$MPa. The compressed liquid leaving the pump is at $P = 1.555$ MPa.

Step 17 *Find overall efficiency of heat engine*

The overall efficiency is:

$$\eta_{\text{H.E}} = \frac{\dfrac{-\dot{W}_{\text{S,net}}}{\dot{m}}}{\dfrac{\dot{Q}_{\text{H}}}{\dot{m}}} = \frac{\dfrac{-(\dot{W}_{\text{S,turbine}} + \dot{W}_{\text{S,pump}})}{\dot{m}}}{\dfrac{\dot{Q}_{\text{H}}}{\dot{m}}} = \frac{-\left(-345.9\dfrac{\text{kJ}}{\text{kg}} + 1.52\dfrac{\text{kJ}}{\text{kg}}\right)}{2372.7\dfrac{\text{kJ}}{\text{kg}}} \tag{5.21}$$

$$\eta_{\text{H.E.}} = \mathbf{0.145}$$

Step 18 *Calculate mass flow rate*

Finally, we see from step 17 that the net work produced is 344.4 kJ per kilogram of steam, so the mass flow rate required to produce 10 MW is calculated as follows:

$$10\ \text{MW} = \dot{m}\left(344.4\dfrac{\text{kJ}}{\text{kg}}\right) \tag{5.22}$$

$$\dot{m} = \left(\dfrac{10\ \text{MW}}{344.4\dfrac{\text{kJ}}{\text{kg}}}\right)\left(10^6\dfrac{\text{W}}{\text{MW}}\right)\left(\dfrac{1\dfrac{\text{J}}{\text{s}}}{1\,\text{W}}\right)\left(\dfrac{1\ \text{kJ}}{1000\ \text{J}}\right) = \mathbf{28.9}\dfrac{\textbf{kg}}{\textbf{s}}$$

Here "MW" is the unit megawatts, though the symbol MW is also frequently used for molecular weight.

In Example 5-1 we determined the physical state of the material leaving each piece of equipment, the heat and work duties in each piece of equipment, the mass flow rate of H_2O circulating through the process, and the overall efficiency of the heat engine. This is as far as we can go using thermodynamics. The next step in designing a real heat engine would probably be detailed sizing of the individual pieces of equipment, which is addressed in other courses in a typical chemical engineering curriculum.

The preliminary design decisions made in Example 5-1 are not definitive. Changing the operating temperature in the condenser or boiler by a few degrees can have a measurable impact on the economics of the process, since these temperatures are design parameters that influence both the efficiency of the process and the cost of the equipment. Problems 5-17 and 5-20 examine the economics of a Rankine heat engine.

5.2.3 Design Variations in the Rankine Heat Engine

Section 5.2.2 illustrated a complete quantitative analysis of a Rankine heat engine, and illustrated some design decisions (e.g., assigning outlet pressures to pump and turbine). This section looks deeper into design considerations in the Rankine heat engine.

Section 5.2.1 noted that excessive amounts of liquid entrained in the vapor in a turbine can damage the equipment. In Example 5-1, the fraction of liquid in the stream exiting the turbine $(1 - q)$ was 0.1, which was the exact maximum liquid fraction stated for that turbine. Suppose this liquid fraction had been 0.13. What aspect of the design could we have changed to fix the problem? Example 5-2 explores quantitatively the effect of superheating the liquid coming out of the boiler.

| SATURATED VERSUS SUPERHEATED STEAM ENTERING A TURBINE | EXAMPLE 5-2 |

A turbine (Figure 5-7) has an efficiency of 80% and an outlet pressure of $P = 0.05$ MPa. The inlet stream is at $T = 300°C = 573.15$ K, and can be either a saturated steam or a superheated steam at either $P = 0.3$ MPa, 1 MPa, or 3 MPa.

Recall we defined quality (q) as the vapor fraction, so a maximum liquid fraction of 0.1 corresponds to a minimum q of 0.9.

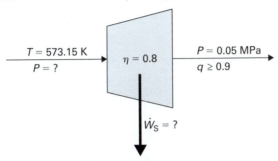

FIGURE 5-7 Turbine examined in Example 5-2.

A. Which of these inlet streams provides the most work *without* exceeding a maximum allowable liquid fraction of 10% in the steam exiting the turbine?

B. What is the efficiency of a Rankine heat engine operating using this turbine, with the inlet stream chosen in part A?

SOLUTION:

Step 1 *Apply entropy and energy balances to reversible turbine*
Define stream 1 as entering the turbine and stream 2 as exiting. As in Example 5-1, the entropy balance for a reversible turbine is

$$\hat{S}_{2,rev} - \hat{S}_1 = 0 \tag{5.23}$$

$$\hat{S}_{2,rev} = \hat{S}_1$$

and the energy balance for a turbine is

$$\frac{\dot{W}_{S,turbine,rev}}{\dot{m}} = \hat{H}_{2,rev} - \hat{H}_1 \tag{5.24}$$

Step 2 *Determine condition of stream leaving reversible turbine.*
Each of the five possible inlet streams has a different specific entropy, but they are all known, and summarized in Table 5-1. At the outlet pressure $P = 0.05$ MPa, the saturated liquid and vapor specific entropies are $\hat{S}^L = 1.0912\dfrac{kJ}{kg \cdot K}$ and $\hat{S}^V = 7.5930\dfrac{kJ}{kg \cdot K}$. Consequently, *if* the exiting material is a liquid–vapor mixture, the fraction of liquid in the material leaving the turbine can be found using

$$\hat{S}_{2,rev} = (1 - q_{rev})\hat{S}^L + (q_{rev})\hat{S}^V \tag{5.25}$$

with $\hat{S}_{2,rev}$ representing the specific entropy of the outlet stream, which is equal to the specific entropy of the inlet stream, as shown in Equation 5.23.

Step 3 *Solve for reversible work*
Once q_{rev} is known, the outlet specific enthalpy can be determined using

$$\hat{H}_{2,rev} = (1 - q_{rev})\hat{H}^L + (q_{rev})\hat{H}^V \tag{5.26}$$

and $\dot{W}_{S,rev,turbine}$ can be found from Equation 5.24. The results of steps 1 through 3 are summarized in Table 5-1.

TABLE 5-1 Effect of inlet pressure on shaft work produced by turbine, for uniform inlet temperature of 300°C.

Inlet T (K)	Inlet P (MPa)	$\hat{S}_{2,\,rev} = \hat{S}_1 \left(\dfrac{kJ}{kg \cdot K} \right)$	q_{rev}	$\hat{H}_{2,\,rev} \left(\dfrac{kJ}{kg} \right)$	$\hat{H}_1 \left(\dfrac{kJ}{kg} \right)$	$\dot{W}_{S,\,rev} \left(\dfrac{kJ}{kg} \right)$
573.15	0.3	7.7037	1.017*	2685.7	3069.6	−383.9
573.15	0.5	7.4614	0.980	2598.6	3064.6	−466.0
573.15	1.0	7.1246	0.928	2479.2	3051.6	−572.4
573.15	3.0	6.5412	0.838	2272.4	2994.3	−721.9
573.15	8.58 (sat'd)	5.7059	0.710	1976.3	2749.6	−773.3

*For $P = 0.3$ MPa, the result of $q_{rev} = 1.017$ is obtained from Equation 5.25. The "vapor fraction" being greater than 1 is physically unrealistic and simply tells that this is, in fact, NOT a liquid–vapor mixture. The fact that $\hat{S}_{2,\,rev}$ is not between $\hat{S}^L$ and $\hat{S}^V$ indicates that the stream is not a mixture of liquid and vapor, but even if we didn't notice this, we can come to the same conclusion mathematically when we obtain an unrealistic result from Equation 5.25.

As it turns out, the specific entropy $\left(7.7037 \dfrac{kJ}{kg \cdot K} \right)$ from the $P = 0.3$ MPa inlet stream is slightly larger than that of superheated steam at 0.05 MPa and 100°C $\left(7.6953 \dfrac{kJ}{kg \cdot K} \right)$, and interpolating between the values for $T = 373.15$ K and 423.15 K yields a temperature of 374.85 K. This temperature is used in the determination of $\hat{H}_{2,\,rev}$ and subsequent calculations. For the other four inlet streams, the outlet stream is indeed a VLE mixture and equations 5.25 and 5.26 do model them accurately.

Overall, Table 5-1 shows that the work increases as the inlet pressure increases, as expected. However, now we must determine which outlet streams exceed the allowable 10% liquid.

Step 4 *Determine condition of stream exiting turbine*
The actual work can be determined from the reversible work by applying the efficiency of 80% for

$$\eta = \frac{\dot{W}_{S,turbine,act}}{\dot{W}_{S,turbine,rev}} = 0.8 \tag{5.27}$$

The actual specific enthalpy of the outlet stream can be found by solving the energy balance:

$$\hat{H}_{2,act} = \frac{\dot{W}_{S,turbine,act}}{\dot{m}} + \hat{H}_1 \tag{5.28}$$

And the quality of the actual outlet stream is determined using Equation 5.26, as was done for the reversible outlet stream. These results are summarized in Table 5-2. Notice that in the case of the $P = 0.5$ MPa inlet stream, the actual outlet stream is actually a superheated steam, though it was a VLE mixture in the reversible case. For the higher inlet pressures, the actual outlet stream is a VLE mixture, but with higher quality than the corresponding reversible turbine. These observations can be understood by the fact that the actual turbine removes less energy than does the reversible turbine.

ASIDE: *This example presents a very plausible scenario.* The temperature of the steam leaving the boiler is constrained by the temperature of whatever is being used as the heat source, so all the variants we are examining have an inlet $T = 573.15$ K. Our goal is to obtain the maximum possible work for this boiler outlet temperature. Our calculations reveal that maximum work is obtained for a saturated steam ($P = 8.58$ MPa) leaving the

FOOD FOR THOUGHT 5-4

In the last three rows of Table 5-2, where the exit stream is a VLE mixture, we don't actually need to know its temperature to solve this problem—but how would we find T if we needed to?

TABLE 5-2 Effect of inlet pressure on physical state of outlet stream, for a turbine with fixed inlet temperature ($T = 573.15$ K and fixed outlet pressure ($P = 0.05$ MPa).

Inlet T (K)	Inlet P (MPa)	$\dot{W}_{S,\text{rev}}\left(\dfrac{kJ}{kg}\right)$	$\dot{W}_{S,\text{act}}\left(\dfrac{kJ}{kg}\right)$	$\hat{H}_{2,\text{act}}\left(\dfrac{kJ}{kg}\right)$	State of Exit Stream
573.15	0.3	−383.9	−307.1	2762.5	superheated vapor, $T = 414.05$ K
573.15	0.5	−466.0	−372.8	2691.8	Superheated vapor, $T = 377.95$ K
573.15	1.0	−572.4	−457.9	2593.7	VLE mixture, $q = 0.978$
573.15	3.0	−721.9	−577.5	2416.8	VLE mixture, $q = 0.901$
573.15	8.58 (sat'd)	−773.3	−618.6	2131	VLE mixture, $q = 0.777$

A design specification like "q must be greater than 0.9" typically has a safety margin built into it, since fluctuations in process conditions can occur due to process upsets or during the unsteady-state operation associated with startup and shutdown.

boiler, but that this gives us an unacceptable amount of condensation in the turbine. Setting the turbine inlet pressure to 3 MPa solves this problem; q is now almost exactly 0.9. Further reducing the turbine inlet pressure unnecessarily lowers the work produced by the turbine.

SOLUTION B:

Step 5 *Apply energy balance to condenser*

We now establish the specifications for a Rankine heat engine that incorporates the turbine selected in part A (inlet $P = 3$ MPa). The VLE mixture entering the condenser, as shown in Table 5-2, has $\hat{H}_{2,\text{act}} = 2416.8$ kJ/kg. Saturated liquid at $P = 0.05$ MPa has $\hat{H} = 340.5$ kJ/kg, so the heat removed in the condenser is

$$\frac{\dot{Q}_C}{\dot{m}} = \hat{H}_{\text{out}} - \hat{H}_{\text{in}} = \hat{H}_3 - \hat{H}_2 = 340.5\frac{kJ}{kg} - 2416.8\frac{kJ}{kg} = -2076.3\frac{kJ}{kg} \quad (5.29)$$

Step 6 *Determine pump work*

The work added in the pump can be estimated, as in Example 3-8, using Equation 5.30 for

$$\frac{\dot{W}_{S,\text{pump}}}{\dot{m}} = \int \hat{V}\,dP \approx \hat{V}(P_{\text{out}} - P_{\text{in}}) \quad (5.30)$$

$$\frac{\dot{W}_{S,\text{pump}}}{\dot{m}} = \left(0.00103\frac{m^3}{kg}\right)(3\text{ MPa} - 0.05\text{ MPa})\left(10^6\frac{Pa}{MPa}\right)\left(\frac{1\frac{N}{m^2}}{1\,Pa}\right)\left(1\frac{J}{N\cdot m}\right)\left(\frac{1\text{ kJ}}{1000\text{ J}}\right)$$

$$\frac{\dot{W}_{S,\text{pump}}}{\dot{m}} = 3.04\frac{kJ}{kg}$$

Step 6 *Apply energy balance to entire process*

The overall energy balance for the entire process is solved to find $\dfrac{\dot{Q}_H}{\dot{m}}$ using

$$0 = \frac{\dot{Q}_C}{\dot{m}} + \frac{\dot{Q}_H}{\dot{m}} + \frac{\dot{W}_{S,\text{turbine}}}{\dot{m}} + \frac{\dot{W}_{S,\text{pump}}}{\dot{m}} \quad (5.31)$$

$$0 = -2076.3\frac{kJ}{kg} + \frac{\dot{Q}_H}{\dot{m}} + \left(-577.5\frac{kJ}{kg}\right) + 3.04\frac{kJ}{kg}$$

$$\frac{\dot{Q}_H}{\dot{m}} = 2650.8\frac{kJ}{kg}$$

Figure 5-5 illustrates that compression of a saturated liquid doesn't inherently lead to a phase change, as does expansion of a saturated vapor. Consequently, while there exists a practical incentive to consider feeding superheated vapor to the turbine, there is no analogous incentive to consider subcooling the liquid entering the pump.

The specific volume used is for the inlet condition, saturated liquid at $P = 0.05$ MPa, because the condition of the outlet stream isn't fully specified.

Step 7 *Determine overall efficiency of complete process*

The overall efficiency of the entire cycle can be determined by

$$\eta_{H.E} = \frac{\dfrac{-\dot{W}_{S,net}}{\dot{m}}}{\dfrac{\dot{Q}_H}{\dot{m}}} = \frac{\dfrac{-(\dot{W}_{S,turbine} + \dot{W}_{S,pump})}{\dot{m}}}{\dfrac{\dot{Q}_H}{\dot{m}}} = \frac{-\left(-577.5\,\dfrac{kJ}{kg} + 3.04\,\dfrac{kJ}{kg}\right)}{2650.8\,\dfrac{kJ}{kg}} \tag{5.32}$$

$$\eta_{H.E.} = \mathbf{0.217}$$

Example 5-2 illustrates the cause and effect relationship between the pressure of the steam entering the turbine and the physical state of the stream exiting the turbine. It further demonstrates the motivation for using superheated steam entering the turbine, even though saturated steam may have the potential to produce more work. Two other design variants in the Rankine cycle are worth mentioning:

If the steam entering the turbine is to be superheated, a separate boiler and superheater can be employed. The boiler and the superheater are two separate heat exchangers; the first produces the steam and the second heats the steam to the desired turbine inlet temperature. The advantage of this approach is that separate heat sources can be used. In the example illustrated in Figure 5-8b, the boiler heat source can be ~473 K, while the superheater heat source must be above 673 K. Table 5-3 illustrates that heat becomes more expensive as the temperature at which it is delivered increases, so the scheme in Figure 5-8a, in which ALL of the heat is provided from a source operating above 673 K, would not be favorable. The energy balance is substantially the same whether the "boiler" is a single piece of equipment or two; the absolute amounts of heat added in Figures 5-8a and b are the same.

The single turbine can be replaced by multiple turbines with inter-stage heating. The purpose of superheating the steam entering a turbine is to reduce the fraction

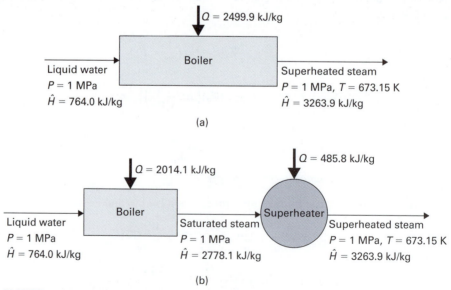

FIGURE 5-8 Formation of steam at $P = 1$ MPa and $T = 673.15$ K by (a) a single boiler and (b) a boiler/superheater sequence.

TABLE 5-3 Estimates of the cost of steam heating, as a function of temperature (Turton, 2009).

Heat Source	Steam Pressure (gage) (MPa)	Steam Temperature (K)	Utility Cost ($/GJ)
Low-pressure steam	0.5	433	14.05
Medium-pressure steam	1	457	14.83
High-pressure steam	4.1	527	17.70

Based on data from Turton, Bailie, Whiting and Shaewitz, Analysis, Synthesis and Design of Chemical Processes, 3rd ed. Prentice-Hall, 2009.

of steam that condenses in the turbine. It is possible to achieve the same effect by using two turbines with moderate pressure drops, rather than a single turbine with a large pressure drop, as illustrated in Figure 5-9. Figure 5-9a is the single turbine and requires superheated steam at 623.15 K. Figure 5-9b begins with superheated steam at a lower temperature of 523.15 K, but heat is added between the two turbines. Because of the temperature drop in the first turbine, it is possible for the inter-heater to operate at the same temperature (and use the same heat source) as the boiler, rather than operating at a higher temperature as does the super-heater in Figure 5-9. The two systems produce similar amounts of work, and both avoid significant amounts of liquid condensing in a turbine.

These design alternatives are further explored in Problem 5-15.

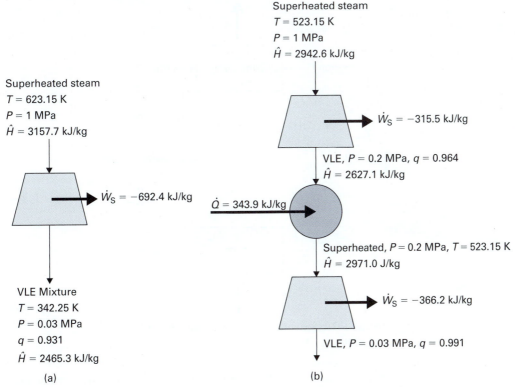

FIGURE 5-9 Comparison of (a) a single reversible turbine to (b) a sequence of two reversible turbines with inter-stage heating. The overall pressure drop in both cases is identical, and both designs maintain a liquid fraction $(1-q)$ below 10% in the turbines.

5.3 Refrigeration—The Vapor-Compression Cycle

This section discusses design and analysis of refrigeration cycles. Section 5-1 showed that the physical properties of water figure prominently in the design of a Rankine heat engine. We needed to know properties (e.g., values of $\hat{H}$ and $\hat{S}$) in order to perform the energy and entropy balances, but more fundamentally our design choices were guided by the properties of water (particularly the relationship between pressure and temperature). We will find the same is true in designing refrigeration cycles.

While industrial manufacturing processes often employ refrigeration on a large scale, we will begin by considering a household kitchen refrigerator, since that is more likely familiar to the reader.

5.3.1 A Household Refrigerator

Household refrigerators are normally set to maintain a temperature at or below 277.5 K. For simplicity, let's assume the interior temperature is 278.15 K, which is 5°C. The ambient temperature may vary throughout the day and year; for this example let's assume its 293.15 K. In effect, the refrigerator is a machine that makes a non-spontaneous process occur: heat is transferred from the cool inside (278.15 K) to the warmer surroundings (293.15 K, which is 20°C). This section describes the classic process by which refrigerators operate: the **vapor-compression cycle**.

Figure 5-10 shows a schematic of the vapor-compression cycle, in which a refrigerant is circulated through a continuous, steady-state process.

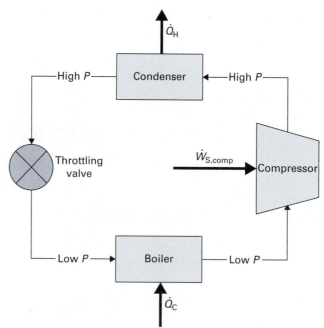

FIGURE 5-10 Schematic of the vapor-compression refrigeration cycle.

- In the *boiler*, heat is transferred to the refrigerant, causing the refrigerant to boil.
- The refrigerant, now a vapor, enters a *compressor*, which increases its pressure.
- The high-pressure vapor refrigerant enters a *condenser*, where heat is transferred to the surroundings, causing the refrigerant to condense to liquid.

- The high-pressure liquid refrigerant enters a *throttling valve*, where its pressure and temperature drop.
- The low-pressure refrigerant enters the boiler.

For this to work, the refrigerant must have a temperature below 278.15 K when it contacts the inside of the refrigerator (here we assign it to 273.15 K = 0°C) and must have a temperature above 293.15 K when it contacts the ambient air (here we will set it to 298.15 K = 25°C). Frequently, *the contents of the refrigerator, and the surroundings, can each be modeled as heat reservoirs*, and that is what we will do in this section. Note the following to analyze a refrigeration cycle.

- $\dot{Q}_C$ will represent the rate of heat transfer from the cold reservoir.
- $\dot{Q}_H$ will represent the rate of heat transfer to the hot reservoir.
- These definitions mirror the definitions used for heat engines, but in a refrigerator, $\dot{Q}_C$ will be positive and $\dot{Q}_H$ will be negative.

Like the Rankine heat engine, the vapor-compression cycle makes use of the physical phenomenon of a phase change to transfer significant quantities of heat. Thus, the refrigerant must boil at 273.15 K and then condense at 298.15 K. The only way to cause this dramatic change in boiling temperature is a pressure change. Consequently, the compressor and the valve are used to control the pressure.

One question that might arise is "Why not place a turbine, instead of a valve, after the condenser? That way some useful work could be obtained while the necessary pressure drop occurs." While this is a logical thought, it is instructive to recall Examples 3-8 and 4-8. These illustrate that the work required to compress a liquid is typically very small compared to the work required to compress an equivalent mass of vapor to the same pressure. The inverse is also true: expansion of a typical liquid in a turbine produces an insignificant amount of work compared to a vapor expanding from the same initial pressure to the same final pressure. Compared to a valve, a turbine is more expensive, more complex, and more likely to require maintenance and repair, and the work it would provide in a typical vapor-compression cycle is not enough for the turbine to be cost-effective.

| **VAPOR-COMPRESSION CYCLE** | **EXAMPLE 5-3** |

The contents of a refrigerator are at 278.15 K. The surroundings are at 293.15 K. Heat transfers through the walls to the inside of the refrigerator at a rate of 1.67 kJ/s. Consequently, the refrigeration process operates continuously at steady state, and removes 1.67 kJ/s (100 kJ/min) of heat, thus allowing the contents to be maintained at a constant temperature of 278.15 K (Figure 5-11).

The refrigerant is Freon® 22, and thermodynamic properties are provided in Appendix F. The material leaving the boiler is saturated vapor at 0°C (273.15 K) and the material leaving the condenser is saturated liquid at 25°C (298.15K). The compressor has an efficiency of 75%. Find the refrigerant flow rate required to remove 100 kJ/min of heat from the inside of the refrigerator, and the rate at which the compressor requires work.

In order to design and build a refrigerator, you'd need to determine the size of each piece of equipment, the diameter of the pipes, etc. While these are not directly "thermodynamics" questions, the flow rate of coolant is an essential parameter that informs these design decisions.

SOLUTION: The 1.67 kJ/s removed from the contents of the refrigerator by our definition is $\dot{Q}_C$.

Step 1 *Overall energy balance around entire process*
Like the Rankine heat engine, each individual step in the vapor-compression process is an open system, but the refrigeration process as a whole is a closed system that operates at a steady state. Thus, the energy balance simplifies to

$$0 = \dot{Q} + \dot{W} \qquad (5.33)$$

Freon® is a registered trademark of DuPont. Several different refrigerants (distinguished from each other by a following number) have been marketed under the trade name Freon.

In step 1, the system is the entire cycle.

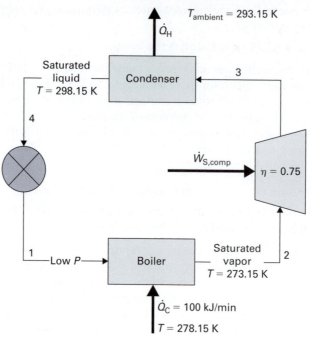

FIGURE 5-11 Vapor-compression cycle described in Example 5-3.

Heat transfer occurs in the evaporator and condenser, but no work occurs in these steps. Shaft work is added in the compressor, but typical compressors are reasonably modeled as adiabatic (see Section 3.6.3). The throttling valve changes the pressure of the refrigerant without transferring work or heat. Consequently, the energy balance can be written more specifically as:

$$0 = \dot{Q}_C + \dot{Q}_H + \dot{W}_S \tag{5.34}$$

Thus, if we can find any two of $\dot{Q}_C$, $\dot{Q}_H$, and $\dot{W}_S$, we can compute the other through Equation 5.34.

There's no immediate way to find $\dot{Q}_H$—we know exactly what's coming out of the condenser but not what's going in. Similarly, we know exactly what's coming out of the boiler (saturated vapor at 273.15 K), but we don't know exactly what's going in. However, we can determine the compressor work by applying the known efficiency.

Step 2 *Apply energy and entropy balances to reversible compressor*
Section 3.6.3 derived the energy balance for a compressor as

$$\frac{\dot{W}_{S,compressor}}{\dot{m}} = \hat{H}_{out} - \hat{H}_{in} = \hat{H}_3 - \hat{H}_2 \tag{5.35}$$

which is valid whether the compressor is reversible or not. Because we know the vapor entering the compressor is saturated, $\hat{H}_2$ can be found from Appendix F and is 405 kJ/kg. However, at this point, we don't have any information on the vapor leaving the compressor.

This is a steady-state process, so the accumulation term in the entropy balance is zero. Thus,

$$0 = \dot{m}_{in}\hat{S}_{in} - \dot{m}_{out}\hat{S}_{out} + \frac{\dot{Q}}{T} + \dot{S}_{gen} \tag{5.36}$$

FOOD FOR THOUGHT 5-6

Since we have saturated liquid entering the valve, and the valve adds neither heat nor work, why can't we assume the exiting stream is also saturated liquid?

In steps 2 through 4, the system is a hypothetical reversible compressor with exactly the same inlet stream, and the same outlet pressure, as the real compressor.

Compressors, like pumps and turbines, are reasonably modeled as adiabatic, so $\dot{Q} = 0$. For the particular case of a *reversible* compressor, $\dot{S}_{gen}$ is also zero, leaving

$$0 = \dot{m}_{in}\hat{S}_{in} - \dot{m}_{out}\hat{S}_{out,rev} = \dot{m}_2\hat{S}_2 - \dot{m}_3\hat{S}_{3,rev} \qquad (5.37)$$

At this point, $\dot{m}$ is unknown, so even though we have to find it eventually, it makes sense to divide it out:

$$\hat{S}_{3,rev} = \hat{S}_2 \qquad (5.38)$$

Step 3 *Determine specific enthalpy of vapor leaving reversible compressor*

The vapor leaving the boiler is saturated vapor at $T = 273.15\,\text{K}$; which, according to Appendix F, has $\hat{S} \approx 1.755\,\dfrac{\text{kJ}}{\text{kg} \cdot \text{K}}$. We know from Equation 5.38 that $\hat{S}$ for the vapor leaving the compressor is the same. We need a second property of the exiting stream in order to find $\hat{H}_3$ from Appendix F. We get this by considering the design of the remainder of the process. The condenser produces saturated liquid at $T = 298.15\,\text{K}$, which according to Appendix F must have $P \approx 10.5$ bar, which is 1.05 MPa. Assuming the pressure does not change significantly in the condenser, the vapor leaving the compressor also has $P = 1.05$ MPa.

According to Appendix F, vapor with $P = 1.05$ MPa and $\hat{S} = 1.755\,\dfrac{\text{kJ}}{\text{kg} \cdot \text{K}}$ has $\hat{H} = 425\,\dfrac{\text{kJ}}{\text{kg}}$. This is $\hat{H}_{3,rev}$, which is the specific enthalpy of the gas leaving our hypothetical reversible compressor.

Step 4 *Solve energy balance for reversible compressor*

We can now determine the shaft work required for the *reversible* compressor as

$$\frac{\dot{W}_{S,\,compressor,rev}}{\dot{m}} = \hat{H}_{3,rev} - \hat{H}_2 \qquad (5.39)$$

$$\frac{\dot{W}_{S,compressor,rev}}{\dot{m}} = 425\,\frac{\text{kJ}}{\text{kg}} - 405\,\frac{\text{kJ}}{\text{kg}} = 20\,\frac{\text{kJ}}{\text{kg}}$$

Step 5 *Calculate actual work required for compressor*

The definition of efficiency for a compressor is:

$$\eta_{compressor} = \frac{\dot{W}_{S,rev}}{\dot{W}_{S,act}} \qquad (5.40)$$

Applying the known efficiency of 75% and the result of step 4 gives

$$\frac{\dot{W}_{S,act}}{\dot{m}} = \frac{20\,\dfrac{\text{kJ}}{\text{kg}}}{0.75} = 26.7\,\frac{\text{kJ}}{\text{kg}} \qquad (5.41)$$

Step 6 *Determine specific enthalpy of ACTUAL vapor leaving compressor*

We now know the work required to run the compressor, but still need to know the actual specific enthalpy of the stream leaving the compressor, so that we can analyze the condenser. Thus, we now apply the energy balance to the ACTUAL compressor:

$$\frac{\dot{W}_{S,compressor,\,act}}{\dot{m}} = \hat{H}_{out,act} - \hat{H}_{in} = \hat{H}_{3,act} - \hat{H}_2 \qquad (5.42)$$

which can be solved using

$$\hat{H}_{3,act} = \hat{H}_2 + \frac{\dot{W}_{S,compressor,act}}{\dot{m}} = 405\,\frac{\text{kJ}}{\text{kg}} + 26.7\,\frac{\text{kJ}}{\text{kg}} = 431.7\,\frac{\text{kJ}}{\text{kg}} \qquad (5.43)$$

Because this is a steady-state process, and there is only one stream entering and one leaving, $\dot{m}_{in} = \dot{m}_{out}$.

Graphically, this step can be carried out by following the lines of constant entropy on the P-H diagram from $T = 0°C$ to $T = 25°C$.

PITFALL PREVENTION

A careless error at this point would be ignoring units and substituting $\dot{Q}_C = 100$ and $\dot{W}_S = 20$ into the overall energy balance (Equation 5.34) to solve for $\dot{Q}_H$. We cannot add or subtract numbers until we have them in the same units.

In step 6, the system is the ACTUAL compressor, not the reversible one we analyzed in steps 2-4.

In step 7, the system is the condenser; specifically the side of the heat exchanger that has the refrigerant flowing through it.

Step 7 *Apply energy balance to condenser*

The condenser is a heat exchanger, and the energy balance (see Section 3.6.4) is

$$\frac{\dot{Q}_H}{\dot{m}} = \hat{H}_{out} - \hat{H}_{in} = \hat{H}_4 - \hat{H}_3 \tag{5.44}$$

$\hat{H}_3$ is known from step 6, and the stream leaving the condenser (4) is saturated liquid, so its specific enthalpy can be found in Appendix F. Thus,

$$\frac{\dot{Q}_H}{\dot{m}} = \hat{H}_4 - \hat{H}_3 = 230\frac{kJ}{kg} - 431.7\frac{kJ}{kg} = -201.7\frac{kJ}{kg} \tag{5.45}$$

Step 8 *Solve energy balance to find $\dot{Q}_C$*

The work ($\dot{W}_S$) and the condenser heat ($\dot{Q}_H$) are both now known—not on an absolute basis but in energy per unit mass. Consequently, the energy balance for the whole system (Equation 5.34) can be solved if it is divided through by the mass flow rate:

$$0 = \frac{\dot{Q}_C}{\dot{m}} + \frac{\dot{Q}_H}{\dot{m}} + \frac{\dot{W}_S}{\dot{m}} \tag{5.46}$$

which can be solved for $\dot{Q}_C$ by

$$\frac{\dot{Q}_C}{\dot{m}} = \frac{-\dot{Q}_H}{\dot{m}} + \frac{-\dot{W}_S}{\dot{m}} = \left(201.7\frac{kJ}{kg}\right) + \left(-26.7\frac{kJ}{kg}\right) = 175\frac{kJ}{kg} \tag{5.47}$$

Step 9 *Find mass flow rate*

The heat added to the boiler was given as 100 kJ/min, so Equation 5.47 can be solved for $\dot{m}$ as

$$\frac{\dot{Q}_C}{\dot{m}} = 175\frac{kJ}{kg} \tag{5.48}$$

$$\dot{m} = \frac{1.67\frac{kJ}{s}}{175\frac{kJ}{kg}} = 0.0095 \text{ kg/s}$$

FOOD FOR THOUGHT 5-7

How would you characterize the overall effectiveness of this cycle with a single number, similar to the efficiency of the heat engine?

Step 10 *Calculate rate at which compressor uses work*

In step 5, we found the work required in the compressor for each kilogram of refrigerant. Now that we know the mass flow rate, we can find the rate at which work is added in absolute terms:

$$\frac{\dot{W}_{S, act}}{\dot{m}} = 26.7\frac{kJ}{kg} \tag{5.49}$$

$$\dot{W}_{S, act} = \dot{m}\left(26.7\frac{kJ}{kg}\right) = 0.0095 \text{ kg/s}\left(26.7\frac{kJ}{kg}\right) = 0.19\frac{kJ}{s}$$

Example 5-3 presents quantitative modeling of a vapor-compression cycle. Notice that there was very little mention of the throttling valve. The energy balance for the typical valve (see Section 3.6.1) is $0 = \hat{H}_{in} - \hat{H}_{out}$. Thus, in Example 5-3 (and Figure 5-11), $\hat{H}_1 = \hat{H}_4 = 230 \text{ kJ/kg}$, and the valve unit operation makes no explicit contribution to the overall energy balance (Equation 5.34) for the cycle. While it wasn't necessary to solve the problem, we can quantify the state of the fluid leaving the valve knowing that $\hat{H}_1 = 230\text{kJ/kg}$ and $P_1 = P_2 = 0.5$ MPa. From Appendix F, it is a liquid–vapor mixture with $q = 0.15$. The fluid entering the valve is saturated liquid. The pressure drop causes some of the liquid to evaporate, and the energy required for evaporation results in a decrease in the temperature of the fluid.

The coefficient of performance of a refrigeration cycle is analogous to the efficiency of a heat engine—they are both single numbers that benchmark the overall effectiveness of the process.

5.3.2 Coefficient of Performance

The purpose of refrigeration is to maintain a space below ambient temperature by removing heat from that space. The effectiveness of the cycle is quantified by the **coefficient of performance (C.O.P.)**:

$$\text{C.O.P.} = \frac{\dot{Q}_C}{\dot{W}_S} \tag{5.50}$$

The coefficient of performance for the refrigeration cycle in Example 5-3 was $(1.67 \text{ kJ/s})/(0.19 \text{ kJ/s})$ $= 8.78$.

where $\dot{Q}_C$ is the rate at which heat is removed from the contents of the refrigerator and $\dot{W}_S$ is the rate at which work is added. This definition reflects the fact that we wish to maximize $\dot{Q}_C$ (removing this heat is the whole point of the process) and we wish to minimize $\dot{W}_S$ (since this work represents the expense required for operating the process).

Why is it called the "coefficient of performance" rather than the "efficiency"? Because conventionally, "efficiency" is a fraction; its value ranges from 0 to 1 (or 0 to 100%). The efficiency of a heat engine is always a fraction: specifically the fraction of entering heat that is converted into useful work (see Equation 5.21 and Example 4-11). However, Examples 5-3 and 5-4 show that $\dot{Q}_C$ can be, and often is, larger than $\dot{W}_S$. Consequently, the word "efficiency" is not used, but the "coefficient of performance" serves exactly the same purpose as the "efficiency" of a heat engine: It summarizes the overall effectiveness of the entire process in a single numerical value.

$\dot{Q}_H$ does not appear in the coefficient of performance equation, since it represents energy expelled to the surroundings and likely has little cost or practical value.

5.3.3 The Carnot Refrigerator

Section 4.4.4 introduced the **Carnot heat engine**, which is an idealized heat engine that is 100% reversible and follows four steps.

- Isothermal heating
- Adiabatic expansion
- Isothermal cooling
- Adiabatic compression

Like the Carnot heat engine, the Carnot refrigerator is an idealized limit that can never be attained in the real world, but serves as a useful point of comparison for real refrigerators.

If these four steps were executed in the opposite order, they would mirror the steps in the vapor-compression cycle introduced in Section 5.3.1:

- Adiabatic compression occurs in the compressor
- Isothermal cooling occurs in the condenser
- Adiabatic expansion occurs in the valve
- Isothermal heating occurs in the boiler

A reversible cycle following these steps is called a **Carnot refrigerator**, as examined in Example 5-4.

EXAMPLE 5-4	A CARNOT REFRIGERATOR

A refrigeration process is designed to operate at steady state, removing 16.7 kJ/s (1000 kJ/min) of heat from a low-temperature reservoir at $T = 278.15$ K and expel the heat to a high-temperature reservoir at $T = 293.15$ K. If the cycle is reversible, what is its coefficient of performance?

SOLUTION: Note the question, as asked, gives no specific information about the process: what is used as a refrigerant, etc. Just like the reversible heat engine in Example 4-11, we will find we can analyze a reversible refrigerator knowing nothing other than the fact that it's reversible and the temperatures of the two heat reservoirs between which it operates.

In order to calculate the coefficient of performance, we need to find two things: $\dot{Q}_C$ and $\dot{W}_S$.

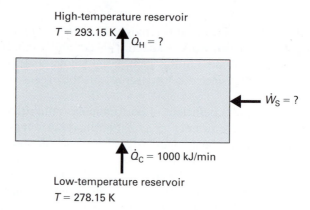

High-temperature reservoir
$T = 293.15$ K $\dot{Q}_H = ?$

$\dot{W}_S = ?$

$\dot{Q}_C = 1000$ kJ/min
Low-temperature reservoir
$T = 278.15$ K

FIGURE 5-12 Reversible refrigerator described in Example 5-4.

> A conventional refrigeration process only has one step involving work, but if there were more than one, $\dot{W}$ in this equation would simply represent the NET work.

FOOD FOR THOUGHT 5-8

Why didn't we do an overall entropy balance in Example 5-3?

Step 1 *Apply energy balance to entire process*
We are explicitly told the process is designed to operate at steady state. We are not explicitly told that the cycle is a closed system, but this is a reasonable assumption based on what we've learned about cycles up to this point. Thus, there is no accumulation of energy in the system, and the only terms that appear in the energy balance are heat and work:

$$0 = \dot{Q}_C + \dot{Q}_H + \dot{W}_S \qquad (5.51)$$

Again, we have two distinct terms, $\dot{Q}_C$ and $\dot{Q}_H$, representing heat exchange with each of the two heat reservoirs. $\dot{Q}_C$ is given, so we have two unknowns, and need one additional equation to solve for $\dot{W}_S$.

Step 2 *Apply entropy balance to entire process*
Again, the accumulation term is 0 for this steady-state process, and there is no material entering or leaving the system. The process is reversible, so $\dot{S}_{gen} = 0$. The entropy balance simplifies to:

$$0 = \frac{\dot{Q}_C}{T_C} + \frac{\dot{Q}_H}{T_H} \qquad (5.52)$$

> For heat transfer to be reversible, the temperature difference driving the heat transfer must be negligible. This means the temperature of the heat reservoir is essentially identical to the temperature of the system at the boundary where the heat transfer occurs.

Since $\dot{Q}_H$ does not appear in the coefficient of performance equation, we solve Equation 5.52 for $\dot{Q}_H$, so that we can eliminate it from the energy balance. Thus,

$$\dot{Q}_H = \frac{-\dot{Q}_C T_H}{T_C} \qquad (5.53)$$

Step 3 *Solve energy balance*
Substituting Equation 5.53 into Equation 5.51 gives

$$0 = \dot{Q}_C + \frac{-\dot{Q}_C T_H}{T_C} + \dot{W}_S$$

$$\dot{W}_S = \dot{Q}_C \left(\frac{T_H}{T_C} - 1 \right) \tag{5.54}$$

Step 4 *Evaluate coefficient of performance*
By definition the coefficient of performance is $\dot{Q}_C/\dot{W}_S$. Using Equation 5.54 allows us to relate coefficient of performance to the temperatures of the heat reservoirs:

$$\text{C.O.P.} = \frac{\dot{Q}_C}{\dot{W}_S} = \frac{\dot{Q}_C}{\dot{Q}_C \left(\dfrac{T_H}{T_C} - 1 \right)} = \frac{1}{\dfrac{T_H}{T_C} - 1} \tag{5.55}$$

$$\text{C.O.P.} = \left(\frac{1}{\dfrac{T_H}{T_C} - 1} \right) \left(\frac{T_C}{T_C} \right) = \frac{\boldsymbol{T_C}}{\boldsymbol{T_H - T_C}}$$

Step 5 *Introduce numerical values for T_C and T_H*

$$\text{C.O.P.} = \frac{278.15 \text{ K}}{(293.15 \text{ K} - 278.15 \text{ K})} = \textbf{18.5} \tag{5.56}$$

> The overall energy and entropy balances for a refrigeration cycle are not substantially different from the balances for a heat engine, but the signs of $\dot{Q}_C$, $\dot{Q}_H$, and $\dot{W}_S$ are opposites in the two processes.

> **FOOD FOR THOUGHT 5-9**
>
> Why is the C.O.P. of 6.57 for the real refrigerator in Example 5-3 so much lower than that of a Carnot refrigerator that operates between the same temperatures?

Equation 5.55 is valid for any *reversible* refrigeration process. Again, the reversible process is an idealized, unattainable limit, but Equation 5.55 provides some useful insight into the performance of real refrigerators. Thus,

$$\text{C.O.P.} = \frac{T_C}{T_H - T_C} \tag{5.57}$$

> Equation 5.57 is valid only for a Carnot refrigerator.

- The lower the refrigerator temperature (T_C), the lower the coefficient of performance.
- The bigger the temperature difference between the ambient surroundings (T_H) and the refrigerator temperature, the lower the coefficient of performance.

5.3.4 Analyzing a Refrigeration Cycle with Simple Models

Previous examples have examined water as the operating fluid in heat engines and Freon® 22 as the operating fluid in a refrigerator. Both of these are compounds for which complete data is available at the conditions of interest. However, we cannot assume such data will always be available. For example, in the 1970s, household refrigerators used chlorofluorocarbons (CFCs) like Freon® 22 as refrigerants. In the 1980s, it was determined that these compounds were contributing to the depletion of the atmospheric ozone layer, and in a relatively short amount of time, a new generation of refrigerants was developed and adopted, though this development and refinement process is on-going even today. When one is developing or working with new chemical products, one cannot expect that comprehensive data is going to be available. The motivational example for Chapter 8 discusses this further. In Example 5-5, we examine a problem in which data for enthalpy and entropy is not available, so we have to estimate needed properties using ideal gas models.

> The phasing out of CFCs and the development of new refrigerants is discussed again in the Motivational Example of Chapter 8.

EXAMPLE 5-5	AN IDEAL GAS REFRIGERANT

A refrigerator operates on the vapor-compression cycle as follows (see Figure 5-13).

- ■ The boiler produces saturated vapor at $T = 285$ K and $P = 0.02$ MPa.
- ■ The compressor has an efficiency of 70%.
- ■ The condenser produces saturated liquid at $T = 305$ K and $P = 0.09$ MPa.

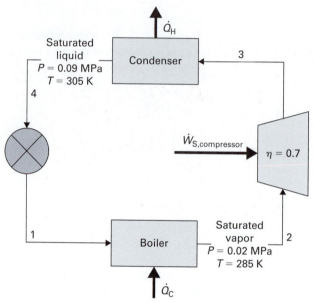

FIGURE 5-13 Refrigeration cycle examined in Example 5-5.

The refrigerant has MW = 100 kg/kmol, $\hat{C}_P^* = 2$ kJ/kg K, and $\Delta\hat{H}_{vap} = 250$ kJ/kg. Both heat capacity and enthalpy of vaporization can be assumed constant over the range of conditions in this process, and ideal gas behavior can be assumed at pressures below 0.1 MPa. Estimate the coefficient of performance.

SOLUTION:

The system in step 1 is the entire refrigeration cycle.

Step 1 *Overall energy balance for the entire process*
As in Example 5-3, the overall energy balance for the refrigerator is:

$$0 = \dot{Q}_C + \dot{Q}_H + \dot{W}_{S,compressor} \qquad (5.58)$$

So if we know any two of $\dot{Q}_C$, $\dot{Q}_H$, and $\dot{W}_{S,compressor}$, we can find the third. As in Example 5-3, we start with the compressor work, since the known efficiency gives us a starting point in analyzing the compressor.

The system in steps 2 and 3 is the compressor.

Step 2 *Apply energy balance to compressor*
As in Example 5-3, the energy balance for a compressor is

$$\frac{\dot{W}_{S,compressor}}{\dot{m}} = \hat{H}_{out} - \hat{H}_{in} = \hat{H}_3 - \hat{H}_2 \qquad (5.59)$$

Here we cannot look up the specific enthalpies. Instead we can calculate the change in enthalpy by applying the heat capacity.

This can also be written $dH = C_P^* dT$ when C_P^* is expressed on a molar basis.

Step 3 *Relate change in enthalpy to change in temperature*
As shown in Section 2.3.3, for *any* ideal gas process, we have

$$d\hat{H} = \hat{C}_P^* dT \qquad (5.60)$$

Because $\hat{C}_p^*$ is constant, Equation 5.60 here integrates to

$$\frac{\dot{W}_{S,\text{compressor}}}{\dot{m}} = \hat{H}_3 - \hat{H}_2 = \hat{C}_P^*(T_3 - T_2) \tag{5.61}$$

We cannot progress further without knowing the temperature change that occurs in the compressor. There is no immediate way to find this for the real compressor, but we can assess it using the entropy balance for the reversible compressor.

Step 4 *Apply entropy balance to a reversible compressor*
As in Example 5-3, the compressor is an adiabatic, steady-state process, so if it is assumed to also be reversible, the entropy balance simplifies to

$$\hat{S}_{3,\text{rev}} - \hat{S}_2 = 0 \tag{5.62}$$

This is an ideal gas that has a constant heat capacity; the exact situation for which Equation 4.59 was derived. Applying the equation to this process:

$$\hat{S}_{3,\text{rev}} - \hat{S}_2 = C_P^* \ln\left(\frac{T_{3,\text{rev}}}{T_2}\right) + R \ln\left(\frac{P_2}{P_{3,\text{rev}}}\right) = 0 \tag{5.63}$$

$$\ln\left(\frac{T_{3,\text{rev}}}{T_2}\right) = \frac{R}{C_P^*} \ln\left(\frac{P_{3,\text{rev}}}{P_2}\right)$$

Step 5 *Solve for outlet temperature from reversible compressor*
The only unknown in Equation 5.63 is $T_{3,\text{rev}}$, so it can be found: Note that heat capacity must be converted from mass to molar basis, so the units are consistent with R.

$$\ln\left(\frac{T_{3,\text{rev}}}{285\,\text{K}}\right) = \left(\frac{8.314\,\dfrac{\text{kJ}}{\text{kmol} \cdot \text{K}}}{2\,\dfrac{\text{kJ}}{\text{kg K}}}\right)\left(\frac{1}{100}\,\frac{\text{kmol}}{\text{kg}}\right) \times \ln\left(\frac{0.09\,\text{MPa}}{0.02\,\text{MPa}}\right) \tag{5.64}$$

$$T_{3,\text{rev}} = 303.4\,\text{K}$$

Step 6 *Find work for reversible compressor*
Applying Equation 5.61 to the reversible compressor gives

$$\frac{\dot{W}_{S,\text{comp,rev}}}{\dot{m}} = \hat{C}_P^*(T_{3,\text{rev}} - T_2) = \left(2\,\frac{\text{kJ}}{\text{kgK}}\right)(303.4\,\text{K} - 285\,\text{K}) = 36.8\,\frac{\text{kJ}}{\text{kg}} \tag{5.65}$$

Step 7 *Apply definition of efficiency of compressor*
Knowing that the compressor efficiency is 70%, we can find the actual compressor work to be

$$\eta_{\text{compressor}} = \frac{\dot{W}_{S,\text{rev}}}{\dot{W}_{S,\text{act}}} \tag{5.66}$$

$$\frac{\dot{W}_{S,\text{act}}}{\dot{m}} = \frac{36.8\,\dfrac{\text{kJ}}{\text{kg}}}{0.7} = 52.6\,\frac{\text{kJ}}{\text{kg}}$$

Step 8 *Find temperature of gas leaving compressor*
Having modeled the compressor, it is logical to move on to the condenser. However, first we'd like to know exactly what is entering the condenser. We already know the pressure of the gas is 0.09 MPa, and the temperature can be found by applying Equation 5.61 to the ACTUAL compressor for

---

Equation 5.61 can be applied to either the real compressor or a reversible compressor; we will need to do both.

In steps 4 through 6, the system is a hypothetical reversible compressor with the same inlet stream and outlet pressure as the real compressor.

FOOD FOR THOUGHT 5-10

Vapor pressure increases with increasing temperature, but the specific numerical values vary drastically from compound to compound. Based on Equation 5.63, what kind of vapor pressure vs. temperature relationship would make for a good refrigerant?

$$\frac{\dot{W}_{S,comp,act}}{\dot{m}} = \hat{C}_P^*(T_{3,act} - T_2) \tag{5.67}$$

$$52.6\,\frac{kJ}{kg} = \left(2\,\frac{kJ}{kg\cdot K}\right)(T_{3,act} - 285\,K)$$

$$T_{3,act} = 311.3\,K$$

Step 9 *Apply energy balance to condenser*
As in Example 5-3, the energy balance for the condenser is

$$\frac{\dot{Q}_H}{\dot{m}} = \hat{H}_{out} - \hat{H}_{in} = \hat{H}_4 - \hat{H}_3 \tag{5.68}$$

What is $\hat{H}_4 - \hat{H}_3$? Stream 3 is a gas at $P = 0.09$ MPa and $T = 311.3$ K; stream 4 is a saturated liquid at $T = 305$ K. Since $\hat{H}$ is a state property, we can evaluate the change in enthalpy for any path from state 3 to state 4 and use the result in Equation 5.68.

Step 10 *Construct two-step path*
Using the methodology illustrated in Figure 5-14, we define state 3* as saturated vapor at $T = 305$ K. We can introduce the specific enthalpy of this saturated vapor into Equation 5.68 as

$$\frac{\dot{Q}_H}{\dot{m}} = \hat{H}_4 - \hat{H}_3 = (\hat{H}_4 - \hat{H}_{3*}) + (\hat{H}_{3*} - \hat{H}_3) \tag{5.69}$$

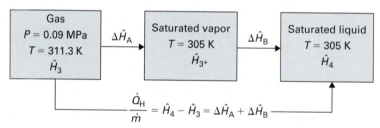

FIGURE 5-14 Decomposition of the condenser into two steps so that change in specific enthalpy can be determined.

The quantity $\hat{H}_4 - \hat{H}_{3*}$ is the change in enthalpy when saturated vapor turns to saturated liquid; this is the opposite of the enthalpy of vaporization.

$$\frac{\dot{Q}_H}{\dot{m}} = -\Delta\hat{H}_{vap} + (\hat{H}_{3*} - \hat{H}_3) \tag{5.70}$$

Meanwhile $\hat{H}_{3*} - \hat{H}_3$ represents the change in enthalpy when the ideal gas is cooled from 311.3 K to 305 K and can be related to temperature through Equation 2.33:

$$\frac{\dot{Q}_H}{\dot{m}} = -\Delta\hat{H}_{vap} + \hat{C}_P^*(T_{3*} - T_3) \tag{5.71}$$

Step 11 *Solve for heat removed in condenser*
Inserting known values into Equation 5.71 gives

$$\frac{\dot{Q}_H}{\dot{m}} = -250\,\frac{kJ}{kg} + \left(2\,\frac{kJ}{kg\cdot K}\right)(305\,K - 311.3\,K) = -262.6\,\frac{kJ}{kg} \tag{5.72}$$

Step 12 *Solve overall energy balance*

The mass flow rate is unknown but is uniform throughout the process, so we can divide the energy balance through by $\dot{m}$ for

$$0 = \frac{\dot{Q}_C}{\dot{m}} + \frac{\dot{Q}_H}{\dot{m}} + \frac{\dot{W}_S}{\dot{m}} \tag{5.73}$$

which allows us to find the heat in the boiler:

$$0 = \frac{\dot{Q}_C}{\dot{m}} + \left(-262.6\,\frac{kJ}{kg}\right) + 52.6\,\frac{kJ}{kg} \tag{5.74}$$

$$\frac{\dot{Q}_C}{\dot{m}} = 210.0\,\frac{kJ}{kg}$$

Step 13 *Evaluate coefficient of performance*

Applying the definition of coefficient of performance gives

$$\text{C.O.P.} = \frac{\dot{Q}_C}{\dot{W}_{S,comp,act}} = \frac{\dfrac{\dot{Q}_C}{\dot{m}}}{\dfrac{\dot{W}_{S,comp,act}}{\dot{m}}} = \frac{210\,\dfrac{kJ}{kg}}{52.6\,\dfrac{kJ}{kg}} = \mathbf{3.99} \tag{5.75}$$

> Efficiency and coefficient of performance are dimensionless quantities. In computing them, we can use work and heat on an absolute basis (J), rate (J/s), or a mass (J/kg) or molar (J/mol) basis, but the units must cancel.

5.4 Liquefaction

This chapter examines applications in which several individual steps are combined into a complete process; specifically to this point a heat engine or a refrigeration cycle. In all examples presented in Sections 5-2 and 5-3, the process as a whole was a closed system; the operating fluid was continuously re-circulated. In this section, we consider **liquefaction**, which is an open-system application for the purpose of producing a chemical product in the liquid phase.

When we think of compounds like N_2, O_2, CO_2, and CH_4, we think of them as gases. Nitrogen, for example, has a normal boiling point of 77.2 K, and thus exists only as a gas in our everyday experience. However, liquid nitrogen has practical applications as an extreme low-temperature coolant. For example, extreme cold can be used to destroy growths and tissues that are unsightly (warts) or potentially malignant (tumors). Liquids also have the merit of being much denser than gases; the Motivational Example in Chapter 7 discusses liquefying natural gas for shipping.

Consequently, in this section, we design processes that liquefy a compound that has an extremely low boiling point, primarily using nitrogen as an example. Before reading further, consider the question: *How many different ways can you think of to make liquid nitrogen?*

> Liquefaction is a very different application than a heat engine or a refrigeration process in terms of its purpose, but we will see it is similar in that understanding the physical and chemical properties of the compound is essential for making good design decisions.

> A common application of liquid nitrogen is its use to eliminate warts.

> The purpose of making liquid nitrogen is to use it as an ultra-low-temperature coolant; it would be self-defeating to use up an ultra-low-temperature coolant in order to make liquid nitrogen.

5.4.1 Simple Liquefaction of Nitrogen

The most straightforward way to condense a gas is to cool it below its boiling point. However, in the case of nitrogen at atmospheric pressure, this would require a coolant that had a temperature below 77 K (−196°C). We could also consider compressing the nitrogen beyond its vapor pressure, but nitrogen has a critical point of

$T = 126$ K and $P = 3.4$ MPa; thus, the liquid state is unattainable at temperatures above 126 K, regardless of pressure.

Examining the data in Figure 2-3 reveals another possibility. The enthalpy of supercritical nitrogen at $T = 135$ K and $P = 6.4$ MPa is about 170 kJ/kg. Meanwhile, at $P = 0.1$ MPa (approximately atmospheric pressure), saturated liquid nitrogen has $\hat{H} = 28$ kJ/kg and saturated vapor has $\hat{H} = 228$ kJ/kg. While the temperature of this equilibrium mixture is not shown explicitly in the figure, it is quite close to the normal boiling point of 77 K. Example 5-6 illustrates a straightforward approach for producing liquid nitrogen that is based upon these observations.

EXAMPLE 5-6	A SIMPLE LIQUEFACTION PROCESS

There is nothing "special" about the conditions $T = 135$ K and $P = 6.4$ MPa; they are simply convenient numbers to use for this example. Super-critical nitrogen with any initial state that is located directly above the two-phase region in Figure 2-3 can be expanded to produce liquid nitrogen.

100 kg/min (1.67 kg/s) of supercritical nitrogen at $T = 135$ K and $P = 6.4$ MPa flows into a chamber, as illustrated in Figure 5-15. The sudden increase in diameter leads to a large pressure drop; the pressure in the chamber is $P = 0.1$ MPa. The chamber has no moving parts, operates at steady state, and can be modeled as perfectly insulated. Find the flow rates of nitrogen liquid and vapor leaving the chamber.

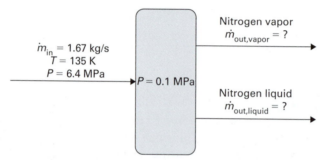

FIGURE 5-15 Schematic of simple liquefaction process.

SOLUTION:

Step 1 *Define a system*
The chamber is the system.

Step 2 *Apply mass balance to system*
This is a steady-state process, so the accumulation term is 0. There is one stream entering the system and two leaving (liquid and vapor), so the mass balance is

$$0 = \dot{m}_{in} - \dot{m}_{out,liq} - \dot{m}_{out,vap} \tag{5.76}$$

Step 3 *Apply energy balance to system*
The accumulation term is again 0. There are no moving parts and the system is perfectly insulated, so $\dot{Q} = \dot{W} = 0$. Thus, the only forms in which energy enters or leaves the system are in material flows:

$$0 = \dot{m}_{in} \hat{H}_{in} - \dot{m}_{out,liq} \hat{H}_{out,liq} - \dot{m}_{out,vap} \hat{H}_{out,vap} \tag{5.77}$$

Step 4 *Insert known information and data*
The entering mass flow rate is given, and according to Figure 2-3, the specific enthalpy of superheated nitrogen at $T = 135$ K and $P = 6.4$ MPa is 170 kJ/kg.

The exiting streams are at $P = 0.1$ MPa. At this pressure, saturated liquid has $\hat{H} = 28$ kJ/kg and saturated vapor has $\hat{H} = 228$ kJ/kg. Inserting this information into the energy balance (Equation 5.77) gives

$$0 = \left(1.67 \frac{kg}{s}\right)\left(170 \frac{kJ}{kg}\right) - \dot{m}_{out,liq}\left(28 \frac{kJ}{kg}\right) - \dot{m}_{out,vap}\left(228 \frac{kJ}{kg}\right) \tag{5.78}$$

PITFALL PREVENTION

In many examples, there is only one stream entering and one stream leaving the system. Here there are two exiting streams, so we can't simply equate $\dot{m}_{in} = \dot{m}_{out}$.

The importance of this process is that it allows us to produce liquid nitrogen at ≈ 77 K without needing to start with a coolant that is below $T = 77$ K.

And the mass balance (Equation 5.76) becomes

$$0 = 1.67 \frac{\text{kg}}{\text{s}} - \dot{m}_{\text{out,liq}} - \dot{m}_{\text{out,vap}} \tag{5.79}$$

Equations 5.78 and 5.79 are thus two equations in two unknowns; $\dot{m}_{\text{out,liq}}$ and $\dot{m}_{\text{out,vap}}$, which are exactly the two quantities we are trying to determine. The result is

$$\dot{m}_{\text{out,liq}} = 0.484 \frac{\text{kg}}{\text{s}}$$

$$\dot{m}_{\text{out,vap}} = 1.186 \frac{\text{kg}}{\text{s}}$$

Of the entering 100 kg/min, 29 kg/min is liquefied and 71 kg/min leaves the system as vapor.

Example 5-6 shows that we can produce pure liquid nitrogen at $T \approx 77$ K ($\approx -196°$C), if we first produce supercritical nitrogen at $P = 6.4$ MPa and $T = 135$ K. The example raises two questions.

1. What's the most efficient way to produce supercritical nitrogen at $P = 6.4$ MPa and $T = 135$ K?
2. The process in Example 5-6 only liquefies 29% of the entering nitrogen. What should be done with the 71 kg/min of vapor product? Is there a more efficient method that will liquefy a larger fraction of the entering nitrogen?

The following two sections examine these issues.

5.4.2 Energy-Efficient Compression Processes

The preceding section demonstrates that we can produce liquid nitrogen at atmospheric pressure ($T \approx 77$ K) if we start with a supercritical fluid that has a lower specific enthalpy than that of saturated nitrogen vapor. In Example 5-7, we examine the compression of an ideal gas to the same conditions seen in Example 5-6: $P = 6.4$ MPa and $T = 135$ K. It is clearly more accurate to use real data rather than modeling nitrogen as an ideal gas, and Example 5-8 does use data. However, the ideal gas model used in Example 5-7 will serve as a simpler way to illustrate some key physical phenomena.

COMPRESSION OF AN IDEAL GAS | **EXAMPLE 5-7**

An ideal gas has $C_P^* = (7/2)R$. We are designing a steady-state process to compress the gas from an initial state of $P = 0.1$ MPa and $T = 300$ K to a final state of $P = 6.4$ MPa and $T = 135$ K. Find the *work added* and *heat removed* per mole of gas for each of the following processes (summarized in Figure 5-16).

A. The gas is compressed to $P = 6.4$ MPa in an adiabatic, reversible compressor and then enters a heat exchanger in which it is cooled to $T = 135$ K.

B. The gas is cooled to $T = 135$ K in a heat exchanger, compressed to $P = 6.4$ MPa in an adiabatic reversible compressor, and in a second heat exchanger, is cooled to $T = 135$ K.

C. The gas is compressed in two separate adiabatic, reversible compressors. The first has inlet $P = 0.1$ MPa and outlet $P = 0.8$ MPa, and the second compresses the gas from $P = 0.8$ MPa to 6.4 MPa. Each compressor is preceded by a heat exchanger that cools the gas to $T = 135$ K, and a third heat exchanger cools the final product to $T = 135$ K.

It isn't realistic that a gas would behave like an ideal gas at $P = 6.4$ MPa, but this example is primarily intended to illustrate the differences between three approaches to designing the compression process.

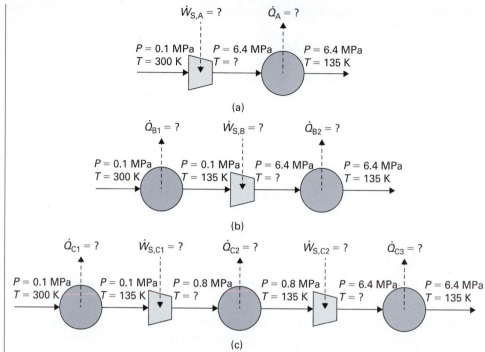

FIGURE 5-16 Three methods of compressing and cooling an ideal gas to $P = 6.4$ MPa and $T = 135$ K.

SOLUTION A:

Step 1 *Apply energy and entropy balances to compressor*

This is an adiabatic, steady-state compressor, so the energy balance we first saw in Section 3.6.3 applies. For this problem it is most convenient to write it on a molar basis:

$$\frac{\dot{W}_S}{\dot{n}} = \underline{H}_{out} - \underline{H}_{in} \qquad (5.80)$$

The entropy balance for the steady-state, adiabatic, reversible compressor that was established in Example 4-8 is

$$\underline{S}_{out} - \underline{S}_{in} = 0 \qquad (5.81)$$

> In steps 1 through 3, the system is the compressor. We cannot find $\dot{Q}$ in the heat exchanger without knowing the temperature of the stream entering it, which we will learn by modeling the compressor first.

Step 2 *Relate molar entropy to temperature and pressure*

The gas in this problem is an ideal gas with constant heat capacity, which is the exact situation for which Equation 4.59 was derived. Applying that equation to this compressor

$$\underline{S}_{out} - \underline{S}_{in} = 0 = C_P^* \ln\left(\frac{T_{out}}{T_{in}}\right) + R \ln\left(\frac{P_{in}}{P_{out}}\right) \qquad (5.82)$$

$$0 = \left(\frac{7}{2}R\right) \ln\left(\frac{T_{out}}{300\,\text{K}}\right) + R \ln\left(\frac{0.1\ \text{MPa}}{6.4\ \text{MPa}}\right) \qquad (5.83)$$

$$T_{out} = 984\ \text{K}$$

> The compressors in this problem are assumed to be reversible because this is a simplified example intended to illustrate specific physical phenomena. The compressors in Examples 5-8 and 5-9 are modeled more realistically.

Step 3 *Relate molar enthalpy to temperature*

For any ideal gas, $d\underline{H} = C_P^*\, dT$. Because C_P^* is constant, the change in molar enthalpy in Equation 5.80 becomes:

$$\frac{\dot{W}_{S,A}}{\dot{n}} = \underline{H}_{out} - \underline{H}_{in} = C_P^*(T_{out} - T_{in}) \tag{5.84}$$

$$\frac{\dot{W}_{S,A}}{\dot{n}} = \frac{7}{2}\left(8.314\frac{J}{mol \cdot K}\right)(984\,K - 300\,K)$$

$$\frac{\dot{W}_{S,A}}{\dot{n}} = \mathbf{19{,}915}\frac{\mathbf{J}}{\mathbf{mol}}$$

Step 4 *Apply energy balance to heat exchanger*
The energy balance for one side of a heat exchanger, as derived in Section 3.6.4, is

$$\frac{\dot{Q}}{\dot{n}} = \underline{H}_{out} - \underline{H}_{in} \tag{5.85}$$

Step 5 *Relate molar enthalpy to temperature*
As in step 3, the change in molar enthalpy can be related to temperature:

$$\frac{\dot{Q}}{\dot{n}} = \underline{H}_{out} - \underline{H}_{in} = C_P^*(T_{out} - T_{in}) \tag{5.86}$$

The temperature entering the heat exchanger is the same as the temperature leaving the compressor, and the heat exchanger is designed to reduce the temperature to 135 K.

$$\frac{\dot{Q}_A}{\dot{n}} = \left(\frac{7}{2}\right)\left(8.314\frac{J}{mol \cdot K}\right)(135\,K - 984\,K) = \mathbf{-24{,}717}\frac{\mathbf{J}}{\mathbf{mol}} \tag{5.87}$$

SOLUTION B: Now we turn to the case with one compressor and two heat exchangers. Since we know *exactly* what is entering and leaving the first heat exchanger, that is a simple place to start.

Step 6 *Model heat exchanger before compressor*
The ideal gas is cooled from 300 K to 135 K before entering the compressor. Applying Equation 5.86 to this cooling process:

$$\frac{\dot{Q}_{B1}}{\dot{n}} = C_P^*(T_{out} - T_{in}) = \left(\frac{7}{2}\right)\left(8.314\frac{J}{mol \cdot K}\right)(135\,K - 300\,K) = \mathbf{-4801}\frac{\mathbf{J}}{\mathbf{mol}} \tag{5.88}$$

Step 7 *Model compressor for part B*
The compressor in part B can be modeled in the same way as the one in part A, with the only difference being that T_{in} is now 135 K instead of 300 K. Applying Equation 5.82 gives

$$0 = \left(\frac{7}{2}R\right)\ln\left(\frac{T_{out}}{135\,K}\right) + R\ln\left(\frac{0.1\,MPa}{6.4\,MPa}\right) \tag{5.89}$$

$$T_{out} = \mathbf{443\,K}$$

and applying Equation 5.84 to the new inlet and outlet temperature gives

$$\frac{\dot{W}_{S,B}}{\dot{n}} = \frac{7}{2}\left(8.314\frac{J}{mol \cdot K}\right)(443\,K - 135\,K) = \mathbf{8962}\frac{\mathbf{J}}{\mathbf{mol}} \tag{5.90}$$

Step 8 *Model heat exchanger that follows compressor*
Applying Equation 5.86 to the second heat exchanger gives

$$\frac{\dot{Q}_{B2}}{\dot{n}} = \left(\frac{7}{2}\right)\left(8.314\frac{J}{mol \cdot K}\right)(135\,K - 443\,K) = \mathbf{-8962}\frac{\mathbf{J}}{\mathbf{mol}} \tag{5.91}$$

In steps 4 through 5, the system is the heat exchanger that follows the compressor.

PITFALL PREVENTION ⚠

$\dot{W}$ in step 7 and $\dot{Q}$ in step 8 are equal in magnitude and opposite in sign, because enthalpy is only a function of temperature for an ideal gas. The temperature changes in steps 7 and 8 also are equal in magnitude. For a *real* gas, $\dot{W}$ and $\dot{Q}$ would not be equal in magnitude, since the pressure change occurring in the compressor would also affect enthalpy.

COMPARE PARTS A AND B: The results reveal that far less work was required to compress the colder ($T = 135$ K) gas than the warmer gas ($T = 300$ K). *Physically*, the result can be understood because low-temperature gases are denser than high-temperature: work is related to the change in volume, and is therefore smaller when the initial and final volumes are both made smaller. *Mathematically*, the outcome can be understood by examination of Equations 5.82 and 5.84. In both processes, T_{out}/T_{in} is identical and approximately equal to 3.3, but when the values of T_{in} and T_{out} are decreased in magnitude, the difference $T_{out} - T_{in}$ also decreases.

Work is further decreased when the compression is accomplished in stages, as shown in part C.

SOLUTION C: Apply Equations 5.84 and 5.86 to two compressors and three heat exchangers.

The analysis of the individual compressors and heat exchangers in part C mirrors that in parts A and B and will not be shown in detail. Each compressor has a compression ratio $P_{out}/P_{in} = 8$ and an inlet temperature of 135 K, identical outlet temperatures of 245 K, and identical work required of 3188 J/mol. Table 5-4 presents a summary of the results and a more detailed solution is available in the electronic supplements.

> Equations 5.82 and 5.84 reveal that the work required in a compressor is primarily a function of the compression ratio P_{out}/P_{in}, rather than the absolute values of P_{out} and P_{in}. Consequently, staged compressors are normally designed with identical compression ratios: 8/1 and 64/8.

TABLE 5-4 Comparison of three different approaches to compression.

	Shaft Work Required (J/mol)	Heat Removal Required (J/mol)
Part A: Single compressor	19,915	−24,717
Part B: Single compressor with pre-cooling	8962	−4801 − 8962 = −13,763
Part C: Two-stage compressor with pre-cooling and inter-stage cooling	3188 + 3188 = 6376	−4801 − 3188 − 3188 = −11,177

> **FOOD FOR THOUGHT 5-11**
>
> Would a three- or four-stage compression sequence further reduce the work? Is there any disadvantage to adding more stages?

> Example 5-7 incorporates several heat exchangers that operate at $T = 135$ K. The coolant in these heat exchangers is not identified, but may well be produced by a separate liquefaction process.

While we can't expect a real compound to exhibit ideal gas behavior at $P = 6.4$ MPa, the solution to Example 5-7 reveals a strategy that informs our design of a complete process for liquefaction of nitrogen. Liquefaction requires formation of a supercritical fluid at a high pressure and low temperature. Specifically, in Example 5-6, we used nitrogen at 6.4 MPa and $T = 135$ K. Thus, the process will not work at all unless we have available a method of cooling the gas to 135 K. Rather than doing this once immediately before the expansion step, we minimize the needed work (and subsequent heat removal) by cooling the nitrogen to the lowest attainable temperature prior to compression. Example 5-8 examines the design of a complete liquefaction process for nitrogen that integrates a two-stage compressor analogous to the one in Example 5-7c.

5.4.3 Linde Liquefaction

A complete liquefaction process for nitrogen must combine the elements of Examples 5-6 and 5-7. First a series of compressors and heat exchangers is used to produce supercritical nitrogen at $P = 6.4$ MPa and $T = 135$ K. If this nitrogen is expanded to a pressure of 0.1 MPa, Example 5-6 reveals that ≈29% of the entering nitrogen is condensed. The remainder of the nitrogen leaves the expansion as vapor at $T \approx 77$ K and $P = 0.1$ MPa. It likely seems natural to recycle this nitrogen vapor to the beginning of the process, and this is done in Example 5-8.

| COMPLETE PROCESS FOR LIQUEFACTION OF NITROGEN | EXAMPLE 5-8 |

1.67 kg/s of liquid nitrogen is produced by the steady-state process shown in Figure 5-17.

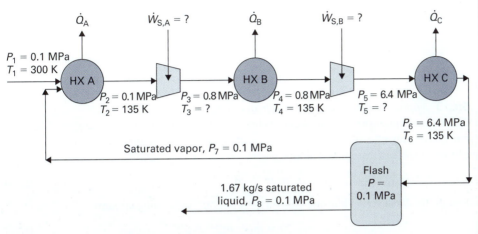

FIGURE 5-17 Schematic of liquefaction process for Example 5-8.

1. Nitrogen enters the process at $P = 0.1$ MPa and $T = 300$ K.

2. The nitrogen is cooled in a heat exchanger (A) to $T = 135$ K.

3. The nitrogen is compressed from $P = 0.1$ MPa to 0.8 MPa in one compressor (A), and from $P = 0.8$ MPa to $P = 6.4$ MPa in a second compressor (B). Each compression step is followed by a heat exchanger that cools the nitrogen to $T = 135$ K.

4. The nitrogen at $P = 6.4$ MPa and $T = 135$ K enters an adiabatic flash vessel in which $P = 0.1$ MPa.

5. The liquid leaving the vessel is removed as product: saturated liquid nitrogen at $P = 0.1$ MPa.

6. The vapor leaving the vessel is recycled to the heat exchanger described in step 2, where it mixes with the fresh nitrogen feed.

Assuming both compressors have an efficiency of 80% and no pressure drop occurs in any heat exchanger, find the total rate at which work is added to each of the two compressors. How much work is required for each kilogram of liquid nitrogen produced?

SOLUTION: The question focuses on the compressors, since work is a major expense in the process. So we will start by examining the compressors, but will quickly realize that a broader look at the process is needed.

Step 1 *Apply energy balance to a compressor.*
The energy balance here is substantially identical to Equation 5.80 in Example 5-7; but here we express it on a mass basis since Figure 2-3, the thermodynamic diagram for nitrogen, is on a mass basis.

$$\frac{\dot{W}_S}{\dot{m}} = \hat{H}_{out} - \hat{H}_{in} \qquad (5.92)$$

This energy balance equation can be applied to both compressors, and this is done explicitly in step 5. However, the equation immediately reveals a challenge: We cannot solve for shaft work unless the mass flow rate through the compressors is known. We know the product flow rate is 1.67 kg/s, but because of the presence of the recycle stream, there is more than 1.67 kg/s of material traveling through the compressors. Thus, we start by formulating a strategy to find all mass flow rates.

Step 2 *Apply material balance to entire process*
This is a steady-state process. If we draw our system around the entire process, there is only one stream entering (1) and only one stream leaving (8). Consequently, the material balance simplifies to

$$0 = \dot{m}_1 - \dot{m}_8 \tag{5.93}$$

$$\dot{m}_1 = \dot{m}_8 = 1.67 \frac{\text{kg}}{\text{s}}$$

Step 3 *Apply material balances to individual unit operations*
The two compressors, and the heat exchangers that follow them (exchangers B and C), each have only one inlet stream and one outlet stream, so the material balances for each of these pieces of equipment simplify to $m_{in} = m_{out}$. Consequently,

$$\dot{m}_2 = \dot{m}_3 = \dot{m}_4 = \dot{m}_5 = \dot{m}_6 \tag{5.94}$$

Finally, the flash vessel has one entering stream but two exiting streams:

$$0 = \dot{m}_6 - \dot{m}_7 - \dot{m}_8 \tag{5.95}$$

Or, applying the known value of the product flow rate gives

$$1.67 \frac{\text{kg}}{\text{s}} = \dot{m}_6 - \dot{m}_7 \tag{5.96}$$

Thus, Equation 5.96 is one equation in two unknowns. We need a second equation that relates $\dot{m}_6$ to $\dot{m}_7$, and can obtain it using an energy balance around the flash vessel.

Step 4 *Apply energy balance to flash vessel*
This is a steady-state process in which energy only enters and leaves through material streams:

$$0 = \dot{m}_6 \hat{H}_6 - \dot{m}_7 \hat{H}_7 - \dot{m}_8 \hat{H}_8 \tag{5.97}$$

Stream 6 is at $P = 6.4$ MPa and $T = 135$ K, and streams 7 and 8 are saturated vapor and liquid at $P = 0.1$ MPa. Thus, the state of each is fully known and specific enthalpies can be obtained from Figure 2-3:

$$0 = \dot{m}_6 \left(170 \frac{\text{kJ}}{\text{kg}}\right) - \dot{m}_7 \left(228 \frac{\text{kJ}}{\text{kg}}\right) - \left(\frac{100 \, \text{kg}}{\text{min}}\right)\left(28 \frac{\text{kJ}}{\text{kg}}\right) \tag{5.98}$$

Simultaneous solution of Equations 5.96 and 5.98 gives

$$\dot{m}_6 = 5.75 \frac{\text{kg}}{\text{s}} \qquad \dot{m}_7 = 4.08 \frac{\text{kg}}{\text{s}} \tag{5.99}$$

We can now analyze the compressors, knowing that the flow rate through each is 5.75 kg/s.

Step 5 *Apply energy balance equation to each compressor*
Applying Equation 5.92 to each of the two compressors and using the nomenclature of Figure 5-17 gives

$$\frac{\dot{W}_{S,A}}{\dot{m}_6} = \hat{H}_3 - \hat{H}_2 \tag{5.100}$$

$$\frac{\dot{W}_{S,B}}{\dot{m}_6} = \hat{H}_5 - \hat{H}_4 \tag{5.101}$$

The mass balance we would get from heat exchanger A is $0 = \dot{m}_1 + \dot{m}_7 - \dot{m}_2$, which is essentially identical to Equation 5.96, because $\dot{m}_2 = \dot{m}_6$.

PITFALL PREVENTION

The presence of a recycle stream has a huge effect on flow rates throughout a system. If we had used the product flow rate of 1.67 kg/s as the flow rate through each unit operation, our calculated W in step 7 would be <30% of the correct value.

The flow rate entering and leaving both compressors is equal to $\dot{m}_6$ as shown in Equation 5.94.

The state of the nitrogen entering each compressor is known. From Figure 2-3:

$$\hat{H}_2 = 288\frac{kJ}{kg}(P = 0.1\,\text{MPa}, T = 135\,\text{K}) \tag{5.102}$$

$$\hat{H}_4 = 282\frac{kJ}{kg}(P = 0.8\,\text{MPa}, T = 135\,\text{K}) \tag{5.103}$$

The conditions of the exit streams cannot be determined directly; we must model the compressors as reversible and then apply the known efficiency.

Step 6 *Apply entropy balance to hypothetical reversible compressor*
The entropy balance for a steady-state, adiabatic, reversible compressor, as established in Example 5-7, is

$$\hat{S}_{out,rev} - \hat{S}_{in} = 0 \tag{5.103}$$

Here we don't have an equation that relates entropy to T and P (as we did in Example 5-7); better yet, we have real data in graphical form. Equation 5.103 can be applied by locating the conditions of the entering stream on the graph and then following a line of constant entropy upward to the outlet pressure. Once this point has been located, the specific enthalpy (or any other property) can be obtained.

$$\hat{H}_{3,rev} = 410\frac{kJ}{kg}(P = 0.8\,\text{MPa}) \tag{5.104}$$

$$\hat{H}_{5,rev} = 390\frac{kJ}{kg}(P = 6.4\,\text{MPa})$$

Step 7 *Solve energy balances for reversible compressors*
Applying the specific enthalpy values that have been determined for compressor A gives

$$\frac{\dot{W}_{S,A,rev}}{\dot{m}_6} = \hat{H}_{3,rev} - \hat{H}_2 = 410 - 288\frac{kJ}{kg} \tag{5.105}$$

$$\dot{W}_{S,A,rev} = \left(5.75\frac{kg}{s}\right)\left(122\frac{kJ}{kg}\right) = \mathbf{701.5}\frac{kJ}{s}$$

The same process for compressor B gives:

$$\dot{W}_{S,B,rev} = \dot{m}_6(\hat{H}_{5,rev} - \hat{H}_4) = \left(5.75\frac{kg}{s}\right)\left(390\frac{kJ}{kg} - 282\frac{kJ}{kg}\right) = 621\frac{kJ}{s} \tag{5.106}$$

Step 8 *Apply known compressor efficiency*
By definition (see Section 4.4.2), the efficiency of a compressor is

$$\eta = \frac{\dot{W}_{S,rev}}{\dot{W}_{S,act}}$$

Thus, the actual work for each compressor can be obtained by applying the known efficiency:

$$\dot{W}_{S,A} = \frac{701.5\frac{kJ}{s}}{0.8} = 876.87\frac{kJ}{s} \tag{5.107}$$

$$\dot{W}_{S,B} = \frac{621\frac{kJ}{s}}{0.8} = 776.25\frac{kJ}{s}$$

In Example 5-7, with an ideal gas, the values of work in the two compression stages were identical. Here they are similar, but not identical, because in a real gas, enthalpy is a function of pressure in addition to temperature.

FOOD FOR THOUGHT 5-12

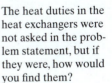

The heat duties in the heat exchangers were not asked in the problem statement, but if they were, how would you find them?

The total work required, per kilogram of liquid nitrogen produced, is

$$\frac{\dot{W}_{S,tot}}{\dot{m}_8} = \frac{\left(876.87 + 776.25\right)\frac{kJ}{s}}{1.67\frac{kg}{s}} = 990\frac{kJ}{kg} \tag{5.108}$$

Example 5-8 illustrates a complete liquefaction process that would work, but misses an opportunity for improved efficiency. See if you can spot it before reading further.

In chemical processes, there are frequently steps at which heat must be added or removed. Thus, when a stream is at either extremely high or extremely low temperature, we should think of it as a valuable resource. While producing nitrogen vapor at 77 K is not the purpose of the liquefaction process, this vapor is nonetheless a resource, and it is badly used in Example 5-8. Instead of simply recycling the cold vapor to the front of the process, we can use it to pre-cool the superheated nitrogen entering the flash vessel. This is the characteristic step of the **Linde liquefaction process**. Compared to Example 5-8, this requires only one additional piece of equipment, and Example 5-9 illustrates the benefits.

> This process is named after a German engineer named Carl von Linde, who also founded the company that is now known as The Linde Group.

EXAMPLE 5-9	**LINDE LIQUEFACTION OF NITROGEN**

> Example 5-9 makes good use of the "cold" vapor stream within the same process that produced it. Sometimes, when "in-process" recycle isn't possible, one can make use of "hot" or "cold" streams for heat exchange in other processes at the same facility.

1.67 kg/s of liquid nitrogen is produced by the following steady-state process shown in Figure 5-18.

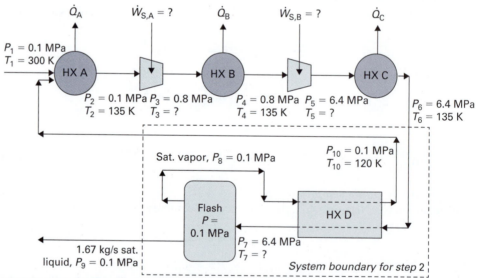

FIGURE 5-18 Schematic of the Linde liquefaction process employed in Example 5-9.

1. Nitrogen enters the process at $P = 0.1$ MPa and $T = 300$ K.

2. The nitrogen is cooled in a heat exchanger (A) to $T = 135$ K.

3. The nitrogen is compressed from $P = 0.1$ MPa to $P = 0.8$ MPa in one compressor (A), and from $P = 0.8$ MPa to $P = 6.4$ MPa in a second compressor (B). Each compression step is followed by a heat exchanger that cools the nitrogen to $T = 135$ K.

4. The superheated nitrogen at $P = 6.4$ MPa and $T = 135$ K enters another heat exchanger (D) and is cooled further.

5. The nitrogen from step 4 enters an adiabatic flash vessel that is maintained at $P = 0.1$ MPa.

6. The liquid leaving the flash vessel is removed as product: saturated liquid nitrogen at $P = 0.1$ MPa.

7. The vapor leaving the vessel is used as the coolant for the heat exchanger (D) described in step 4 and leaves this heat exchanger at $T = 120$ K.

8. The vapor is then recycled to the heat exchanger described in step 2, where it mixes with the fresh nitrogen feed.

Assuming both compressors have an efficiency of 80% and no pressure drop occurs in any heat exchanger, find the total rate at which work is added to each of the two compressors. How much work is required for each kilogram of liquid nitrogen produced?

SOLUTION:

Step 1 *Apply mass balances throughout process*
If we draw our system boundaries around the entire process, there is, like in Example 5-8, only one stream entering and one stream leaving. Thus,

$$0 = \dot{m}_1 - \dot{m}_9 \tag{5.109}$$

$$\dot{m}_1 = \dot{m}_9 = 1.67 \frac{\text{kg}}{\text{s}}$$

Also as in Example 5-8, the two compressors and heat exchangers B and C all have a single inlet and single outlet stream, so at steady state, these flow rates are all identical. Heat exchanger D has two inlet streams and two outlet streams, but the streams exchange heat without exchanging mass; the mass flow rates on both sides are unchanged. Consequently,

$$\dot{m}_2 = \dot{m}_3 = \dot{m}_4 = \dot{m}_5 = \dot{m}_6 = \dot{m}_7 \tag{5.110}$$

$$\dot{m}_8 = \dot{m}_{10} \tag{5.111}$$

Finally, applying a mass balance to the flash chamber gives

$$0 = \dot{m}_7 - \dot{m}_8 - \dot{m}_9 \tag{5.112}$$

Applying the known flow rate of stream 9, the product stream is

$$1.67 \frac{\text{kg}}{\text{s}} = \dot{m}_7 - \dot{m}_8 \tag{5.113}$$

As in Example 5-8, the material balance in Equation 5.113 reveals that we need one more equation, so an energy balance is the logical next step. But an energy balance for which unit? In Example 5-8, we used an energy balance around the flash vessel, but in that problem, the temperature of the stream entering the flash was known. Here T_7 is unknown, and as a result, an energy balance around the flash chamber would give us an additional equation but also an additional unknown ($\hat{H}_7$).

Step 2 *Energy balance around two unit operations*
If we define a system containing *both* the flash vessel and heat exchanger D, shown in Figure 5-18, the temperature and pressure of all entering and exiting streams are known,

and stream 7, for which T is unknown, is inside the boundaries of the system. The energy balance for this system is

$$0 = \dot{m}_6 \hat{H}_6 - \dot{m}_9 \hat{H}_9 - \dot{m}_{10} \hat{H}_{10} \tag{5.114}$$

Specific enthalpies for streams 6, 9, and 10 can be obtained from Figure 2-3, and the flow rate of stream 9 is also known. Thus,

$$0 = \dot{m}_6\left(170\,\frac{kJ}{kg}\right) - \left(1.67\,\frac{kg}{s}\right)\left(28\,\frac{kJ}{kg}\right) - \dot{m}_{10}\left(273\,\frac{kJ}{kg}\right) \tag{5.115}$$

This can be combined with Equation 5.113 to produce two equations in two unknowns, because in step 1, it was shown that $\dot{m}_6 = \dot{m}_7$ and $\dot{m}_8 = \dot{m}_{10}$. Thus,

$$1.67\,\frac{kg}{s} = \dot{m}_6 - \dot{m}_{10} \tag{5.116}$$

The simultaneous solution of Equations 5.113 and 5.116 is

$$\dot{m}_6 = 3.97\,\frac{kg}{s} \qquad \dot{m}_{10} = 2.3\,\frac{kg}{s} \tag{5.117}$$

> In Example 5-8, the fraction of nitrogen entering the flash step that actually condensed was 29%. Here, its $(1.67\ kg/s)/(3.97\ kg/s) = 42\%$, because the temperature of the nitrogen entering the chamber is lower.

Step 3 *Apply entropy balances to reversible compressors*
The compressors A and B are essentially identical in function to those in Example 5-8; though the flow rate of nitrogen is lower, the inlet temperatures and pressures are the same. Consequently, as in Example 5-8, the *reversible* compressors require 122 kJ/kg (A) and 108 kJ/kg (B). The actual work through each compressor is determined using the flow rates computed in step 2 and the 80% efficiency:

$$\dot{W}_{S,A} = \frac{\left(122\,\frac{kJ}{kg}\right)\left(3.97\,\frac{kg}{s}\right)}{0.8} = 605.43\,\frac{kJ}{s} \tag{5.118}$$

$$\dot{W}_{S,B} = \frac{\left(108\,\frac{kJ}{kg}\right)\left(3.97\,\frac{kg}{s}\right)}{0.8} = 535.95\,\frac{kJ}{s}$$

The total work required, per kilogram of liquid nitrogen produced, is thus:

> The total work required to make each kilogram of liquid nitrogen is 684 kJ in the Linde process, compared to 991 kJ in Example 5-8.

$$\frac{\dot{W}_{S,\,tot}}{\dot{m}_8} = \frac{605.43\,\frac{kJ}{s} + 535.95\,\frac{kJ}{s}}{1.67\,\frac{kg}{s}} = \mathbf{684}\,\frac{\mathbf{kJ}}{\mathbf{kg}} \tag{5.119}$$

Chapter 5 has examined application of thermodynamics principles to complete processes, but the examples presented were again predominantly reliant on the availability of comprehensive data (e.g., Figure 2-3, Appendix A). Chapter 6 considers what to do when such data isn't available.

5.5 SUMMARY OF CHAPTER FIVE

- A **heat engine** converts heat into work, most commonly shaft work.
- The **Rankine cycle** is commonly used in designing heat engines.
- The **efficiency** of a heat engine is the *net work produced* divided by the *total heat added*.

- A **refrigeration** process transfers heat from a low temperature location to a high-temperature location.
- The **vapor-compression cycle** is commonly used in designing refrigerators.
- The **coefficient of performance** of a refrigeration process is the *heat removed* from the low-temperature location divided by the *work required* to run the process.
- Both heat engines and refrigerators are typically modeled as operating between a high-temperature reservoir and a low-temperature reservoir.
- The efficiency of a **Carnot** heat engine, and the coefficient of performance of a **Carnot** refrigerator, are dependent only upon the temperatures of the high-and low-temperature reservoirs. These values are for reversible processes, and represent maximums that cannot be exceeded by real processes.
- The Linde liquefaction process can be used to condense low-boiling gases into the liquid phase.
- Phase changes, and the energy conversion associated with phase changes, play an integral role in the Rankine cycle, the vapor-compression cycle, and the Linde process. Consequently, we cannot design these processes effectively unless we know, or are able to predict with a model, the temperatures and pressures at which phase changes occur.

5.6 EXERCISES

5-1. The table below contains specifications for five different steady-state Rankine heat engines, A–E. Fill in all the missing data. NOTE: All numbers in the table are given as ABSOLUTE VALUES; determine whether they are positive or negative based upon what you know about the Rankine cycle.

Cycle	$\dot{Q}_H$ (kJ/s)	$\dot{W}_{S,turbine}$ (kJ/s)	$\dot{Q}_C$ (kJ/s)	$\dot{W}_{S,pump}$ (kJ/s)	η
A	10,000	3000		10	
B	5000			5	0.17
C		500		10	0.22
D	1000	300	725		
E		1500	6000	10	

5-2. The table below contains specifications for six different steady-state Rankine heat engines, A–F. Fill in all of the missing data.

	$\hat{H}$ Boiler Exit (kJ/kg)	$\hat{H}$ Turbine Exit (kJ/kg)	$\hat{H}$ Condenser Exit (kJ/kg)	$\hat{H}$ Pump Exit (kJ/kg)	Efficiency of Cycle	Flow Rate of Working Fluid (kg/s)	NET Power Produced (kW)
A	3000	2650	400	403		15	
B	2750		150	155	0.21		1000
C	1900	1550	250		0.20	5	
D	2750	2330			0.27	10	4100
E	3600	3000		500	0.18	20	
F		1900	398	400	0.22		20,000

5-3. The table below contains specifications for six different steady-state vapor-compression refrigerators, A–F. Fill in the missing data. NOTE: All numbers are given as ABSOLUTE VALUES. Determine whether they are positive or negative based upon what you know about the vapor-compression cycle.

Cycle	$\dot{Q}_H$ (kJ/s)	$\dot{W}_{S,compressor}$ (kJ/s)	$\dot{Q}_C$ (kJ/s)	C.O.P.
A	1500		1200	
B		400	800	
C	2000			2.6
D			100	3.3
E		10,000		0.8
F	750	375		

5-4. The table below contains specifications for six different steady-state vapor-compression refrigerators, A–F. Fill in all of the missing data.

	$\hat{H}$ Boiler Exit (kJ/kg)	$\hat{H}$ Compressor Exit (kJ/kg)	$\hat{H}$ Condenser Exit (kJ/kg)	$\hat{H}$ Valve Exit (kJ/kg)	Coefficient of Performance	Flow Rate of Refrigerant (kg/s)	Rate of Heat Removal from Refrigerated Space (kg/s)
A	1000	1350	100			40	
B	1500	1750		1000			1000
C		250	−200		3.5	10	
D	1200			500	0.9		15,000
E		1500			4	100	8000
F		1900		750		100	50,000

5-5. The table on the right contains specifications for four different liquefaction processes, A–D, all of which are designed using the basic flow sheet shown in Figure E5-5. Fill in the missing data.

	A	B	C	D
$\dot{n}_1$(mol/s)	200		100	
$\dot{n}_2$(mol/s)		100		
$\dot{n}_3$(mol/s)				250
$\dot{n}_4$(mol/s)		30		100
$\dot{n}_5$(mol/s)	300			
$\dot{n}_6$(mol/s)				
$\underline{H}_1$(J/mol)	3000	2000		2500
$\underline{H}_2$(J/mol)	1500	1800	1000	
$\underline{H}_3$(J/mol)	1300		750	700
$\underline{H}_4$(J/mol)	50			250
$\underline{H}_5$(J/mol)		1300	900	
$\underline{H}_6$(J/mol)		1600	1400	1500
$\dot{W}_{S,tot}$ (kJ/s)		35	30	55
$\dot{Q}_{tot}$ (kJ/s)	−650		−60	

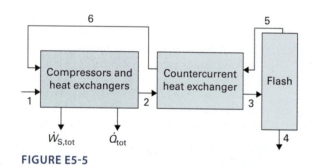

FIGURE E5-5

5.7 PROBLEMS

5-6. This problem examines the Rankine heat engine introduced in Figure 5-5. Saturated steam at $T = 523.15$ K enters the turbine and the condenser operates at $T = 313.15$ K.
   A. Assuming the turbine is reversible, give your best estimate of the efficiency of the cycle, and indicate the quality of the stream leaving the turbine.
   B. Assuming the turbine has an efficiency of 75%, give your best estimate of the efficiency of the cycle, and indicate the quality of the stream leaving the turbine.
   C. Find the flow rate of circulating water needed to produce a net power of 1 MW from the cycle with the turbine efficiency of 75%.

5-7. A refrigerator runs on the vapor-compression cycle. The boiler operates at $T = 265$ K and the condenser operates at 305 K. The compressor has an efficiency of 85%. Thermodynamic data for two different refrigerants is located in Appendix F.
   A. What is the maximum attainable coefficient of performance for Freon® 22?
   B. What is the maximum attainable coefficient of performance for refrigerant R-422A?
   C. Besides coefficient of performance, two considerations in choosing a refrigerant are price and safety. Research Freon® 22 and R-422A and comment on their suitability as refrigerants.

5-8. The engine on a steam ship runs on the Rankine cycle. The steam leaves the boiler at 2 MPa and 523.15 K. The turbine has an efficiency of 75% and an outlet pressure of 0.1 MPa. The pressure changes in the boiler and condenser can be considered negligible, and the liquid leaving the condenser is saturated.

  A. Determine the operating temperature of the condenser, and compute the efficiency of a Carnot cycle operating between the boiler and condenser temperatures.

  B. Determine the actual efficiency of this Rankine cycle and compare it to the Carnot efficiency.

  C. When the *Titanic* was sinking, the *Carpathia* received the S.O.S. and immediately set course to attempt a rescue. The captain of the *Carpathia* ordered the hot water turned off in the passengers' cabins (Lord, 1955). What effect do you expect this action had?

5-9. A schematic of a variation on the Rankine cycle is shown in Figure P5-9—not for steam but for an organic fluid. This process has been called the "organic Rankine cycle":

  A. Do some research and determine the major advantage of using a Rankine cycle with an organic as a working fluid as opposed to water.

  B. There are five unit operations in the process in Figure P5-9. Describe what is happening in each of those steps (for the organic working fluid).

  C. There is a valve between line 4 → 5. Why do you think that line exists and why do you think that valve is there?

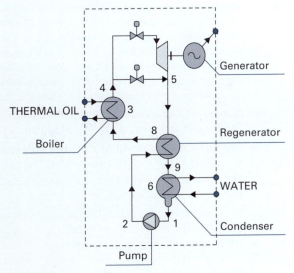

FIGURE P5-9

5-10. A refrigerator runs on the vapor compression cycle, using R-422A as a refrigerant. The boiler operates at

266 K. The effluent from the condenser is 5 K above ambient temperature. The compressor has an efficiency of 80%.
Find each of the following:

  ■ the temperature and pressure of the gas leaving the compressor

  ■ the fraction of vapor in the stream leaving the expansion valve

  ■ the coefficient of performance

  ■ the mass flow rate of refrigerant needed to attain 3 kJ/s of cooling

  A. The ambient temperature is 295 K.

  B. The ambient temperature is 310 K.

5-11. A heat engine operates on the Rankine cycle, with saturated steam at $T = 623.15$ K leaving the boiler, a condenser operating at $T = 373.15$ K, and a turbine efficiency of 80%.

  A. Find the liquid fraction leaving the turbine.

  B. Find the overall efficiency of the heat engine.

  C. A superheater is inserted into the cycle after the boiler, which increases the steam temperature to 723.15 K without changing its pressure. All other specifications remain the same. Find the liquid fraction leaving the turbine and the overall efficiency of the cycle, and compare to the answers from A and B.

  D. A heat engine operates with the same boiler, pump, and condenser specifications used in parts A and B (no superheater). Instead of a single turbine, there are two turbines, each with 80% efficiency. The steam leaving the first turbine has $P = 0.3$ MPa, is sent to a heat exchanger in which its temperature is increased to 473.15 K, and then continues to the second turbine. Find the liquid fractions in both turbine effluent streams and the overall efficiency of the heat engine, and compare to the answers in A, B, and C.

5-12. Water is the most common working fluid in a Rankine heat engine, but there is no fundamental reason why a Rankine heat engine couldn't be designed with other working fluids. Suppose a heat engine is to operate with the following specifications:

  ■ Quality (q) of fluid exiting turbine must be at least 0.9.

  ■ The fluid leaving the boiler is vapor at $T=373.15$ K. Its pressure can be specified as needed to meet the constraint that $q > 0.9$.

  ■ The fluid leaving the condenser is saturated liquid at $T = 313.15$ K

  ■ Turbine efficiency is $\eta = 0.8$

Find the state of the fluid exiting the turbine, the net work produced by the cycle per kilogram of operating fluid, and the efficiency of the cycle for each of the following operating fluids.

A. Water

B. Ammonia (Use Figure 6-2 for data)

5-13. A steady-state liquefaction process generates 7.6 kg/s of saturated liquid methane at $P = 0.07$ MPa. Fresh methane enters the process at $T = 297$ K and $P = 0.07$ MPa, and is mixed with the recycled methane to form the "methane feed" stream. This enters a heat exchanger (HX1) that cools the methane feed to $T = 200$ K and $P = 0.07$ MPa. The process has two compressors (both $\eta = 0.75$: the first compresses the methane to $P = 0.7$ MPa and the second compresses the methane to $P = 7$ MPa). Each compressor is followed by a heat exchanger (HX2 and HX3) that cools the methane to $T = 200$ K without changing the pressure. Next, the supercritical methane enters a counter-current heat exchanger (HX4) in which the methane vapor from the flash chamber is used as the coolant. The coolant leaves HX4 at $T = 195$ K and is recycled into the "methane feed" stream. The supercritical methane leaves HX4 and enters a flash chamber, where its pressure is reduced to $P = 0.07$ MPa.

A. Determine the flow rate of the supercritical methane entering the flash chamber.

B. Determine W for each of the two compressors.

C. Determine Q for each of the four heat exchangers.

5-14. The boiling point of a compound at $P = 0.1$ MPa is 150 K. The Linde liquefaction process will be used to produce saturated liquid at $P = 0.1$ MPa, which has a specific enthalpy of 20 kJ/kg. The Table P5-14 in the right-hand column contains some physical properties of the compound. The steady-state process works as follows.

Step 1. 1 kg/s enters the process at $T = 250$ K and $P = 0.1$ MPa.

Step 2. The feed enters a series of compressors and heat exchangers, and it leaves the last heat exchanger at $T = 200$ K and $P = 10$ MPa. The TOTAL work added by the compressors is 300 kilojoules per kilogram of feed.

Step 3. The stream leaving step 2 is cooled to $T = 175$ K in a heat exchanger.

Step 4. The stream leaving step 3 enters a flash chamber where it expands to 0.1 MPa and some of it condenses.

Step 5. The vapor from the flash chamber is used as the coolant for the heat exchanger in step 3.

Step 6. The vapor stream (at $P = 0.1$ MPa) from step 5 is NOT recycled; it exits the process as a by-product.

A. Find the flow rate of liquid product leaving the flash chamber.

B. Find the specific enthalpy (kJ/kg) of the vapor by-product described in step 6.

C. Find the total heat removed by the heat exchangers during step 2, in kJ/s.

D. What is the heat capacity of the compound, in kJ/kg · K, at ideal gas conditions?

Specific enthalpy in vapor or supercritical phase at various temperatures and pressures, in kJ/kg.

	150 K	175 K	200 K	225 K	250 K	275 K
0.1 MPa	200	215	230	245	260	275
1 MPa	192	202	215	229	241	258
10 MPa	184	190	200	215	228	241

5-15. A steady-state heat engine operates on the Rankine cycle.

■ The steam entering the turbine is 1 kg/s of steam at $P = 3.5$ MPa and $T = 523.15$ K.

■ The ACTUAL stream exiting the turbine is a mixture of 95% vapor and 5% liquid at $P = 50$ kPa.

■ The stream entering the pump is saturated liquid at $P = 50$ kPa.

A. Determine the efficiency of the turbine.

B. Estimate the work required by the pump.

C. Estimate the overall efficiency of the heat engine.

D. What is the efficiency of a Carnot heat engine that operates between a high temperature of 523.15 K and a low temperature that is the same as the temperature of the condenser in this problem?

5-16. A refrigeration process operates using the vapor-compression cycle, using a proprietary refrigerant that the inventor claims is better than R-422A. At low pressures, the refrigerant can be assumed to act as an ideal gas with constant $C_V = 14R$ and reportedly has a molecular weight of 400 kg/kmol. The steady-state refrigeration cycle reportedly works as follows:

■ A boiler produces saturated vapor at $P = 0.02$ MPa and $T = 278.15$ K, which has $\underline{H} = 50$ kJ/mol.

■ The vapor is compressed to $P = 0.06$ MPa and $T = 323.15$ K.

■ The vapor is then condensed to saturated liquid at $P = 0.06$ MPa, which has $\underline{H} = 20$ kJ/mol.

■ The liquid undergoes an isenthalpic expansion to $P = 0.02$ MPa, and enters the boiler.

A. What is the efficiency of the compressor?

B. If we wished to scale the process described above so that it provides 50 kJ/s of cooling, what is the required flow rate of refrigerant in kg/s?

C. What is the coefficient of performance of the refrigeration cycle?

D. Design a refrigeration cycle using R-422A as the refrigerant, in which the rate of cooling is 50 kJ/s. In designing this process, you should make all specifications and design decisions in such a way as to allow a meaningful comparison between R-422A and the proprietary refrigerant.

E. Based on the outcomes of questions A through D, comment on the inventor's claim that the proprietary refrigerant is better than R-422A.

5-17. A steady-state Rankine cycle currently in service operates as follows:

▧ Steam leaves the boiler at $P = 0.8$ MPa and 523.15 K.

▧ Steam leaves the turbine as saturated vapor at $P = 0.03$ MPa.

▧ Water leaves the condenser as saturated liquid at $P = 0.03$ MPa.

▧ Water leaving the condenser is pumped up to 0.8 MPa and returned to the boiler.

▧ The power output FROM THE TURBINE is 1 MW.

Your company has the opportunity to upgrade the turbine to one that is 85% efficient, at a cost of $1.2 million. This includes all costs associated with the replacement—the new equipment, installation, instru-mentation, etc. If the upgrade is done it will work this way:

▧ The flow rate, temperature, and pressure of the steam entering the turbine will all be unchanged, but the turbine is expected to produce more work.

▧ The pressure leaving the turbine will still be $P = 0.03$ MPa, but won't necessarily be saturated vapor, since the turbine is now removing more energy as work.

▧ The Q_C in the condenser will be adjusted such that the water leaving the turbine is saturated liquid, allowing the pump and boiler to operate EXACTLY the same in the upgraded cycle as they do in the current cycle.

▧ The current turbine produces 1 MW of power, 24 hours a day, for 350 days per year. The company values any "extra" work produced beyond this at $20/GJ.

A. Determine the flow rate at which water/steam circulates through the process.

B. Determine the efficiency of the turbine in the CURRENT cycle.

C. Determine the overall efficiency of the CURRENT cycle.

D. Determine the power produced by the turbine in the UPGRADED cycle.

E. Determine the overall efficiency of the UPGRADED cycle.

F. How long will the new turbine have to operate in order to pay for the $1.2 million cost of the upgrade?

5-18. You are designing a steady-state liquefaction process that will manufacture liquid methane. Part A of this problem will focus on two unit operations: the flash separation step itself, and the counter-current heat exchanger in which the vapor from the flash chamber is used to cool the feed entering the flash chamber. In answering parts B and C, however, consider the entire Linde process, not just the two unit operations for which you performed calculations.

Methane enters the counter-current heat exchanger as supercritical vapor at $P = 7$ MPa and $T = 200$ K. It is cooled and enters the flash chamber, where the pressure is reduced to $P = 70$ kPa. The flow rate of liquid methane product is 7.6 kg/s. The methane vapor from the flash is sent to the counter-current heat exchanger. The design parameter that is under your control is the temperature of this methane vapor stream when it leaves the counter-current heat exchanger—it can be 180 K, 185 K or 190 K.

A. Find the flow rate of supercritical methane entering the counter current heat exchanger for each of the three possible systems.

B. Discuss the factors that you would expect to affect the cost OF THE EQUIPMENT ITSELF for the Linde process, and what the results of part A suggest about these costs.

C. Discuss the factors that you would expect to affect the cost of OPERATION of the process, and what the results of part A suggest about these costs.

5-19. You are designing a refrigeration cycle, and have the option of using either a compressor with an efficiency of 70% or a compressor that is more expensive by $5000 but has an efficiency of 80%. The following specifications are valid regardless of which compressor is used:

▧ The refrigerant is R-422A.

▧ The liquid leaving the condenser is saturated liquid at $T = 323$ K.

▧ The boiler operates at $T = 273$ K.

▧ The vapor leaving the boiler is saturated vapor.

▧ The cycle must have $Q_C = 20$ kJ/s in the boiler.

A. Determine the flow rate of refrigerant. Is it the same or different in the two cycles?

B. Determine the compressor work for each of the two possible compressors.

C. Determine the coefficient of performance for the cycle with each of the two possible compressors.

D. The compressor runs on electricity, which is available for $0.10/kWhr. Assuming the refrigeration system runs constantly, how long would the system have to run in order for the higher-efficiency compressor to be cost effective?

5-20. You are designing a Rankine heat engine which must produce 200 kJ/s of NET shaft work. The heat source is at $T = 473$ K and the low-temperature reservoir at $T = 298$ K. The turbine efficiency is 75%, and there are no restrictions on the temperature, pressure, or quality of the water leaving the turbine. Your job is to design the most cost-effective Rankine engine possible. The boiler can be designed to operate at $T = 463.15$ K, $T = 458.15$ K, or $T = 453.15$ K, and the condenser can be designed to operate at $T = 308.15$ or $T = 313.15$ K. Costs of these heat exchangers can be determined from the formulas in the table on the right. Notice that the cost of the heat exchanger goes up as the heat duty goes up, and goes down as the ΔT between the heat reservoir and the fluid increases.

Exchanger	Operating Temp (K)	Formula
Boiler	453.15	$C = 10000 + \dot{Q}$
Boiler	458.15	$C = 12000 + 1.5\dot{Q}$
Boiler	463.15	$C = 15000 + 2\dot{Q}$
Condenser	308.15	$C = 10000 + 1.5\dot{Q}$
Condenser	313.15	$C = 7000 + \dot{Q}$

In all formulas, C represents the cost in dollars and $\dot{Q}$ represents the absolute value of the heat transferred in that exchanger in kJ/min.

The cost of the heat added to the boiler is $15/GJ. The heat removed in the condenser has neither cost nor value. Assume that the heat engine will be operating 24 hours per day, 350 days per year.

A. For each of the six possible Rankine cycles, determine $\dot{Q}_H$, $\dot{Q}_C$, the cost of the two heat exchangers, and the yearly cost of the heat.

B. Recommend which variation of the Rankine heat engine should be used if it's expected to be in service for 5 years.

5.8 GLOSSARY OF SYMBOLS

A	area	M	mass of system	$\hat{S}$	specific entropy
C.O.P.	coefficient of performance	$\dot{m}$	mass flow rate	$\underline{S}$	molar entropy
C_P	constant pressure heat capacity	$\dot{n}$	molar flow rate	$\dot{S}_{gen}$	rate of entropy generation
		P	pressure	T	temperature
C_P^*	constant pressure heat capacity for ideal gas	Q	heat	T_C	temperature of a low-temperature heat reservoir
C_V	constant volume heat capacity	Q_C	heat exchanged with a low-temperature heat reservoir	T_H	temperature of a high-temperature heat reservoir
C_V^*	constant volume heat capacity for ideal gas	Q_H	heat exchanged with a high-temperature heat reservoir	V	volume
$\eta_{H.E}$	overall efficiency of a heat engine	$\dot{Q}$	rate of heat addition	$\hat{V}$	specific volume
		$\dot{Q}_C$	rate of heat exchange with low temperature reservoir	W	work
η_{carnot}	overall efficiency of a completely reversible heat engine	$\dot{Q}_H$	rate of heat exchange with low temperature reservoir	$\dot{W}$	power
H	enthalpy			W_S	shaft work
$\hat{H}$	specific enthalpy	q	quality, or vapor fraction	$\Delta\hat{H}_{vap}$	enthalpy of vaporization
$\underline{H}$	molar enthalpy	R	gas constant	η	efficiency
		S	entropy		

5.9 REFERENCES

Lord, W. *A Night to Remember*. Henry Holt and Company, 1955.

Richard C. Bailie, Wallace B. Whiting, Joseph A. Shaeiwitz, and Richard Turton, *Analysis, Synthesis and Design of Chemical Processes*, 3rd ed. Prentice-Hall, 2009.

Thermodynamic Models of Real, Pure Compounds

6

LEARNING OBJECTIVES

This chapter is intended to help you learn to:

- Write *total derivative expressions* that relate unknown variables to known variables
- Manipulate partial derivatives using mathematical tools such as the *triple product rule* and *expansion rule*
- Apply the principles of calculus to express partial derivatives in terms of *measurable properties*
- Combine the mathematical techniques mentioned above with equations of state to solve problems in the absence of thermodynamic property data
- Use residual properties to solve problems involving real gases, when heat capacity is only known in the ideal gas state

The previous five chapters gave many examples of engineering problems that can be solved with energy and entropy balances. Solving these problems required knowledge of properties such as U, H, and S. In many cases we made use of experimental values for these properties, but we have noted several times (starting in Chapter 2) that we can't assume that such data will always be available. The ideal gas model has been used extensively in Chapters 3–5, but its use is limited to applications at very specific conditions. In modeling liquids and solids, we have assumed volume was constant with respect to both temperature and pressure, another simplifying approximation that is not always reasonable.

In this chapter, we will introduce mathematical modeling strategies that are applicable to pure compounds at a broad range of conditions. ∎

6.1 MOTIVATIONAL EXAMPLE: Joule-Thomson Expansion

We begin by posing a problem that should at this point appear comparatively routine: expansion in a throttling valve. This kind of example was first examined in Section 3.6.1, and it was shown that if the valve is adiabatic and operating at steady state, the energy balance simplified to:

$$\hat{H}_{in} = \hat{H}_{out} \quad \text{or} \quad \underline{H}_{in} = \underline{H}_{out} \tag{6.1}$$

Thus, the process is *isenthalpic*; the gas or liquid entering the valve experiences a sudden decrease in pressure but the enthalpy is unchanged. Isenthalpic expansion, which is also known as **Joule-Thomson expansion**, was an essential step in the refrigeration and liquefaction cycles examined in Chapter 5. In refrigeration, the

Section 5.3 (refrigeration) and Section 5.4 (liquefaction) both illustrate processes in which Joule-Thomson expansion plays a pivotal role.

expansion led to a phase change; a high-pressure saturated liquid entered the valve, and a portion of the liquid evaporated due to the pressure change. Because no work or heat was added to provide the enthalpy of vaporization, the energy came from the liquid itself, resulting in a large temperature drop. In the Linde liquefaction process, the feed to the Joule-Thomson expansion step was a supercritical fluid rather than a saturated liquid, but Examples 5-8 and 5-9 again showed a liquid–vapor mixture was produced and a significant temperature change occurred.

Example 6-1 illustrates the expansion of gaseous ammonia, and shows how Joule-Thomson expansion can have a significant effect on the temperature of a vapor, even in the absence of a phase change.

EXAMPLE 6-1	JOULE-THOMSON EXPANSION OF AMMONIA

Ammonia vapor enters an adiabatic, steady-state throttling valve at $T = 30°C$ and $P = 1$ MPa, illustrated in Figure 6-1. If it leaves the throttling valve at $P = 0.1$ MPa, what is the exiting temperature?

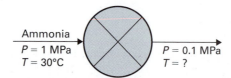

FIGURE 6-1 Joule-Thomson expansion examined in Example 6-1.

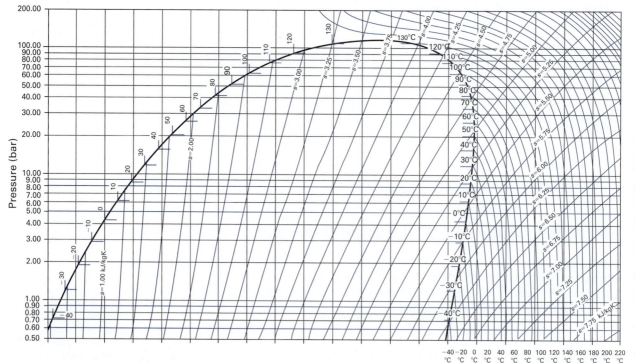

FIGURE 6-2 Pressure-enthalpy diagram for ammonia.

SOLUTION:

Step 1 *Apply energy balance*

If we define the valve as the system, then we have a steady-state, adiabatic throttling process; the exact situation described by Equation 6.1

$$\hat{H}_{in} = \hat{H}_{out}$$

Step 2 *Look up data*

Figure 6-2 shows that the specific enthalpy of ammonia at $P = 1$ MPa (10 bar) and $T = 30°C$ is 1500 kJ/kg; this is $\hat{H}_{in}$. The energy balance tells us that the exiting stream has $\hat{H}_{out} = 1500$ kJ/kg, and at $P = 0.1$ MPa (1 bar), this specific enthalpy falls between the $T = 0°C$ *and* $T = 10°C$ isotherms in Figure 6-2. $T_{out} \sim 3°C \sim 276$ **K**.

Example 6-1 illustrates the prediction of the temperature change that occurs when a gas expands. U.S. Patent #5,522,870 (Ben-Zion, 1994) gives an example of an application for this phenomenon. The patent describes a device that can either cool or heat a surface very rapidly, allowing one to produce large temperature changes almost instantly and with precise timing. The patent cites cryosurgery, surface curing, and sealing as applications. Thus, we can recognize Example 6-1 as a practical and important kind of problem, but we wouldn't have been able to solve the problem without Figure 6-2. This chapter, in a nutshell, addresses the question "what do you do when comprehensive data (such as Figure 6-2 in Example 6-1) is not available?"

> "Cryosurgery" is the use of extreme cold to destroy malignant or abnormal tissue. The ability to change the temperature of instruments quickly is helpful in localizing the effect of the treatment to the targeted tissue.

The chapter explores the construction of mathematical models that accurately describe the inter-relationship between physical and thermodynamic properties, including U, H, S, P, V, and T. Such models have two major purposes.

1. **Models allow us to solve problems in the absence of data.** It is not realistic to assume that the kind of data given in Figure 6-2 will be available for all compounds at all temperatures and pressures of interest; there's a need for accurate strategies for estimating quantities that have never been measured.

    The principles and techniques presented in Chapters 6–8 not only allow construction of models that describe the properties of pure compounds, but also lay the foundation for modeling of mixtures, which we cover beginning in Chapter 9.

2. **Models help us to gain deeper insights into the behavior of real chemicals, and the cause-effect relationships that occur between parameters in chemical processes.** Patent #5,522,870, mentioned above, describes a device that can rapidly cool a surgical probe to cryogenic temperatures, and just as rapidly return the device to room temperature. Interestingly, the heating and cooling are both achieved by Joule-Thomson expansions. We've now seen several examples (Examples 6-1, 5-6, and 5-4) in which Joule-Thomson expansion produced cooling; how can the same process also produce heating? Observation and data can tell us "most gases cool when they expand, but some gases actually increase in temperature," but the mathematical models developed in this chapter (see particularly Example 6-3) will give us a quantitative basis for rationalizing these apparently contradictory observations.

Example 6-1 essentially posed the question: If ammonia vapor is expanded isenthalpically from $P = 1$ MPa (10 bar) to $P = 0.1$ MPa (1 bar), what is the change in temperature? The next section begins to explore how to answer this kind of question without comprehensive data relating H to T and P.

6.1.1 The Total Derivative

We begin by noting a rule of partial differential calculus:

FOOD FOR THOUGHT 6-1

Why is X written as a function of two variables, Y and Z? Suppose we related X to three variables Y, Z, and W:

$$dX = \left(\frac{\partial X}{\partial Y}\right)_{W,Z} dY$$

$$+ \left(\frac{\partial X}{\partial Z}\right)_{W,Y} dZ$$

$$+ \left(\frac{\partial X}{\partial W}\right)_{Y,Z} dW$$

Isn't this just as valid as Equation 6.2?

If a variable X is a function of Y and Z, then changes in X can be related to changes in Y and Z through a **total derivative**:

$$dX = \left(\frac{\partial X}{\partial Y}\right)_Z dY + \left(\frac{\partial X}{\partial Z}\right)_Y dZ \qquad (6.2)$$

We can write such a total derivative expression for *any three intensive state properties* X, Y, and Z.

Thus, if we let X, Y, and Z represent, for example, $\underline{U}$, T, and $\underline{S}$, Equation 6.2 would become

$$d\underline{U} = \left(\frac{\partial \underline{U}}{\partial T}\right)_{\underline{S}} dT + \left(\frac{\partial \underline{U}}{\partial \underline{S}}\right)_T d\underline{S} \qquad (6.3)$$

Or if we let X, Y, and Z represent molar enthalpy, temperature, and pressure, Equation 6.2 becomes

$$d\underline{H} = \left(\frac{\partial \underline{H}}{\partial T}\right)_P dT + \left(\frac{\partial \underline{H}}{\partial P}\right)_T dP \qquad (6.4)$$

FOOD FOR THOUGHT 6-2

Here it is stressed that X, Y, and Z in Equation 6.2 can represent any *intensive* variable (e.g., $\underline{H}$). Why not an extensive variable?

Clearly, we could continue in this vein for some time, writing dozens more total derivative expressions. The question to ask ourselves is, which expression is the best starting point for the problem we are trying to solve? In Example 6-1, the goal was to find the change in temperature $(T_2 - T_1)$ resulting from a Joule-Thomson expansion. Consequently, if we wished to solve a similar problem using a model instead of data, we would do well to start with an equation that models changes in temperature (dT):

$$dT = \left(\frac{\partial T}{\partial \underline{H}}\right)_P d\underline{H} + \left(\frac{\partial T}{\partial P}\right)_{\underline{H}} dP \qquad (6.5)$$

Equations 6.3 through 6.5 are all correct, and we can imagine many more total derivative equations that could be written, relating various state properties to each other. However, in the case of Joule-Thomson expansion, Equation 6.3, while correct, is not particularly useful, since no information is given or asked about either internal energy or entropy. By contrast, Equation 6.5 is readily applicable to a Joule-Thomson expansion.

Equations 6.2 through 6.5 are the first of dozens of equations that will be presented in this chapter. Attempting to solve problems by drawing from this maze of equations in a haphazard or arbitrary way will not be productive. We will use a simple question to guide ourselves: how can I relate what I want to know to what I do know?

- The goal of Example 6-1 was to calculate the change in temperature of a vapor resulting from a process. Since "dT" represents precisely what we are trying to find, it makes sense to start by writing a total derivative expression for dT.

- In Example 6-1, the two properties we knew the most about were pressure (initial and final P are given) and enthalpy (we knew from the energy balance that the process was isenthalpic). Consequently, it is quite natural to let Y and Z in Equation 6.2 represent pressure and molar enthalpy; the resulting Equation 6.5 *relates what we want to know to what we do know.*

Example 6-2 demonstrates the application of a total derivative to solving a problem. It examines a Joule-Thomson process identical to that shown in Figure 6-1 and Example 6-1, except that instead of ammonia, the substance flowing through the valve is an ideal gas.

JOULE-THOMSON EXPANSION OF AN IDEAL GAS	EXAMPLE 6-2

An ideal gas with a constant $C_P^* = (7/2)R$ enters a steady-state throttling valve at $P = 1$ MPa and $T = 300$ K. If it leaves the valve at $P = 0.1$ MPa, what is the exiting temperature?

FIGURE 6-3 Joule-Thomson expansion of an ideal gas, examined in Example 6-2.

SOLUTION:

Step 1 *Define a system and apply energy balance*
Defining the system as the throttling valve, the energy balance, as derived in Section 3.6.1, is:

$$\underline{H}_{in} = \underline{H}_{out} \tag{6.6}$$

Notice that we do not have numerical values for $\underline{H}_{in}$ and $\underline{H}_{out}$, we simply know they are equal to each other, which proves to be sufficient to solve the problem.

Step 2 *Write a total derivative*
We wish to find the unknown outlet temperature of the gas. The two intensive state properties we know the most about are pressure (P) and molar enthalpy ($\underline{H}$). This is the exact situation for which Equation 6.5 was proposed:

$$dT = \left(\frac{\partial T}{\partial \underline{H}}\right)_P d\underline{H} + \left(\frac{\partial T}{\partial P}\right)_{\underline{H}} dP$$

This equation is not specific to our valve; T, P, and $\underline{H}$ are all state properties and this equation can be applied to any process involving a pure compound. Here, if we define the state of the entering gas as "state 1" ($P_1 = 1.0$ MPa, $T_1 = 300$ K) and the state of the exiting gas as "state 2," ($P_2 = 0.1$ MPa), we know from the energy balance that $\underline{H}_1 = \underline{H}_2$. Equation 6.5 can be solved by integrating from state 1 to state 2.

$$\int_{T_1}^{T_2} dT = \int_{\underline{H}_1}^{\underline{H}_2} \left(\frac{\partial T}{\partial \underline{H}}\right)_P d\underline{H} + \int_{P_1}^{P_2} \left(\frac{\partial T}{\partial P}\right)_{\underline{H}} dP \tag{6.7}$$

Step 3 *Simplify total derivative*
The first term on the right-hand side (the $d\underline{H}$ term) is zero. *Mathematically,* this is because $\underline{H}_1 = \underline{H}_2$; if the limits of integration are identical, the integral is zero, no matter what $(\partial T/\partial \underline{H})_P$ is. *Physically,* this can be understood as follows: $(\partial T/\partial \underline{H})_P$ quantifies the effect of a change in enthalpy on temperature, but in this case the enthalpy is not changing so the value of $(\partial T/\partial \underline{H})_P$ doesn't matter. Equation 6.7 simplifies to

$$\int_{T_1}^{T_2} dT = \int_{P_1}^{P_2} \left(\frac{\partial T}{\partial P}\right)_{\underline{H}} dP \tag{6.8}$$

In order to integrate $(\partial T/\partial P)_{\underline{H}}$ with respect to dP, we have to express $(\partial T/\partial P)_{\underline{H}}$ as a function of P.

Step 4 *Relate Joule-Thomson coefficient to measurable properties*
Using a rule of partial differential calculus called the *triple product rule*, which is

For this problem, it makes no real difference whether enthalpy is expressed on a molar or mass basis. However, equations of state are normally written on a molar basis, and we will use equations of state extensively throughout this chapter, so using molar properties is more convenient in general.

The manipulations in step 4 are likely hard to follow because they are unfamiliar. Sections 6.2.1 through 6.2.4 provide a detailed discussion of the mathematics used in this step.

explained in Section 6.2.1, we can express $(\partial T/\partial P)_H$ as

$$\left(\frac{\partial T}{\partial P}\right)_{\underline{H}} = -\frac{\left(\dfrac{\partial \underline{H}}{\partial P}\right)_T}{\left(\dfrac{\partial \underline{H}}{\partial T}\right)_P} \qquad (6.9)$$

By definition $(\partial \underline{H}/\partial T)_P$ is equal to C_P, so

$$\left(\frac{\partial T}{\partial P}\right)_{\underline{H}} = -\frac{\left(\dfrac{\partial \underline{H}}{\partial P}\right)_T}{C_P} \qquad (6.10)$$

Another rule of partial differential calculus called the expansion rule is presented in Section 6.2.3. We can use the expansion rule to prove:

$$\left(\frac{\partial \underline{H}}{\partial P}\right)_T = \underline{V} + T\left(\frac{\partial \underline{S}}{\partial P}\right)_T \qquad (6.11)$$

But an expression known as one of Maxwell's relations states that

$$\left(\frac{\partial \underline{V}}{\partial T}\right)_P = -\left(\frac{\partial \underline{S}}{\partial P}\right)_T \qquad (6.12)$$

> The differential expression $(\partial \underline{V}/\partial T)_P$ has a particular physical significance in thermodynamics that is discussed in Section 6.2.5; the **coefficient of thermal expansion** of a substance is defined as $(1/\underline{V})\,(\partial \underline{V}/\partial T)_P$.

Substituting Equations 6.11 and 6.12 into Equation 6.10 reveals that

$$\left(\frac{\partial T}{\partial P}\right)_{\underline{H}} = -\frac{\underline{V} - T\left(\dfrac{\partial \underline{V}}{\partial T}\right)_P}{C_P} \qquad (6.13)$$

Consequently, Equation 6.8 can be re-written as

$$\int_{T_1}^{T_2} dT = \int_{P_1}^{P_2} -\frac{\underline{V} - T\left(\dfrac{\partial \underline{V}}{\partial T}\right)_P}{C_P}\,dP \qquad (6.14)$$

These manipulations are discussed in detail in Section 6.2. For now, consider this: why is Equation 6.14 "better" than Equation 6.8? At first glance they both appear to have the same problem; neither can be integrated with respect to dP because the integrand is not expressed as a mathematical function of pressure. However, the expression $\underline{V} - T\left(\dfrac{\partial \underline{V}}{\partial T}\right)_P$ can be related to pressure through an equation of state. Here, the gas can be modeled using the ideal gas law.

Step 5 *Evaluate integral for ideal gas law*
According to the ideal gas law, molar volume can be expressed as

$$\underline{V} = \frac{RT}{P} \qquad (6.15)$$

> R is always a constant. We are evaluating the derivative with respect to T at constant P, so we treat P as a constant in evaluating this partial differential. Thus, the right hand side of Equation 6.15 is essentially a constant (R/P) times T, and the derivative of T with respect to T is 1.

Next, we can evaluate $(\partial \underline{V}/\partial T)_P$ by taking the derivative of both sides of Equation 6.15 with respect to T at constant P:

$$\left(\frac{\partial \underline{V}}{\partial T}\right)_P = \frac{R}{P} \qquad (6.16)$$

Plugging Equations 6.15 and 6.16 into Equation 6.14 gives

$$\int_{T_1}^{T_2} dT = \int_{P_1}^{P_2} -\frac{V - T\left(\frac{\partial V}{\partial T}\right)_P}{C_P}\, dP = \int_{P_1}^{P_2} -\frac{\frac{RT}{P} - \frac{RT}{P}}{C_P}\, dP \tag{6.17}$$

$$\int_{T_1}^{T_2} dT = \int_{P_1=1\text{ MPa}}^{P_2=0.1\text{ MPa}} 0\, dP \tag{6.18}$$

$$T_2 - T_1 = 0(P_2 - P_1)$$

$$T_2 = T_1 = \textbf{300 K}$$

Equation 6.18 shows that the temperature of an ideal gas is unchanged in Joule-Thomson expansion.

Example 6-2 demonstrated that modeling Joule-Thomson expansion requires evaluating the partial derivative $(\partial T/\partial P)_H$. Consequently this partial derivative is called the **Joule-Thomson coefficient**, μ_{JT}:

$$\mu_{JT} = \left(\frac{\partial T}{\partial P}\right)_H = -\frac{V - T\left(\frac{\partial V}{\partial T}\right)_P}{C_P} \tag{6.19}$$

The Joule-Thomson coefficient in Example 6-2 was 0, but this result was for an ideal gas specifically. The more general equation is

$$dT = \mu_{JT}\, dP = -\frac{V - T\left(\frac{\partial V}{\partial T}\right)_P}{C_P}\, dP \tag{6.20}$$

which is valid for any Joule-Thomson expansion. The next example uses the van der Waals equation of state.

The van der Waals equation of state was first introduced in Section 2.3.4.

JOULE-THOMSON EXPANSION OF A VAN DER WAALS GAS	EXAMPLE 6-3

A gas has $C_P = (7/2)R$ and is described by the van der Waals equation of state with $a = 0$ and $b = 1.5 \times 10^{-4}$ m³/mol. The gas enters a steady-state throttling valve at $P = 1$ MPa and $T = 300$ K and leaves the valve at $P = 0.1$ MPa. What is the exiting temperature?

Here, the heat capacity is modeled as constant for simplicity, as C_P is not the focus of the example. Section 6-3 presents more realistic methods of accounting for heat capacity of a gas.

SOLUTION:

Step 1 *Define a system and write an energy balance*
As in Example 6-2, if we define the system as the valve, the energy balance simplifies to Equation 6.1:

$$\underline{H}_{in} = \underline{H}_{out}$$

Step 2 *Compare and contrast with Example 6-2*
Our goal is to relate changes in temperature (T) to pressure (P) and specific enthalpy ($\underline{H}$), so we can apply known information regarding P and $\underline{H}$ to find the outlet temperature (T_2). Steps 2 through 4 of Example 6-2 did exactly this, writing and simplifying a total

derivative expression that related changes in temperature (dT) to changes in pressure (dP) and specific enthalpy ($d\underline{H}$) (Equation 6.14):

$$\int_{T_1}^{T_2} dT = \int_{P_1}^{P_2} -\frac{\underline{V} - T\left(\dfrac{\partial \underline{V}}{\partial T}\right)_P}{C_P}\, dP$$

Here, as in Example 6-2, we have an isenthalpic expansion, and Equation 6.14 is again valid. Now, however, instead of an ideal gas, we have a gas that is described by the van der Waals equation of state.

Step 3 *Evaluate* $\left(\dfrac{\partial \underline{V}}{\partial T}\right)_P$ *for this gas*

The van der Waals equation of state is Equation 2.53:

$$P = \frac{RT}{\underline{V} - b} - \frac{a}{\underline{V}^2}$$

However, for this gas, $a = 0$, so the equation simplifies to

$$P = \frac{RT}{\underline{V} - b} \tag{6.21}$$

The most straightforward way to evaluate $\left(\dfrac{\partial \underline{V}}{\partial T}\right)_P$ is to first solve the equation of state for $\underline{V}$:

$$\underline{V} = \frac{RT}{P} + b \tag{6.22}$$

and then to differentiate both sides with respect to T (treating P as a constant):

$$\left(\frac{\partial \underline{V}}{\partial T}\right)_P = \frac{R}{P} \tag{6.23}$$

Step 4 *Solve the integral*

In order to integrate Equation 6.14 with respect to dP, we need to express the integral as a function of P. So we use Equations 6.22 and 6.23 to express $\underline{V}$ and $\left(\dfrac{\partial \underline{V}}{\partial T}\right)_P$ in terms of P:

$$\int_{T_1}^{T_2} dT = \int_{P_1}^{P_2} -\frac{\underline{V} - T\left(\dfrac{\partial \underline{V}}{\partial T}\right)_P}{C_P}\, dP = \int_{P_1}^{P_2} -\frac{\dfrac{RT}{P} + b - T\left(\dfrac{R}{P}\right)}{C_P}\, dP \tag{6.24}$$

$$\int_{T_1}^{T_2} dT = \int_{P_1}^{P_2} -\frac{b}{C_P}\, dP$$

$$T_2 - T_1 = \left(-\frac{b}{C_P}\right)(P_2 - P_1)$$

$$T_2 - T_1 = \left[-\frac{1.5 \times 10^{-4} \dfrac{\text{m}^3}{\text{mol}}}{\dfrac{7}{2}\left(8.314 \dfrac{\text{Pa} \cdot \text{m}^3}{\text{mol} \cdot \text{K}}\right)}\right](0.1 - 1\,\text{MPa})(10^6\,\text{Pa}/1\,\text{MPa}) = 4.6\,\text{K} \tag{6.25}$$

$$\boxed{T_2 = 304.6\ \text{K}}$$

a and *b* are constants in the van der Waals equation, so the derivative of either with respect to *T* (or anything else) is 0. We will see in Chapter 7 that there are other equations of state that have parameters called *a* and *b*, but in which *a* is a function of temperature, rather than a constant.

Examples 6-1 through 6-3 all show a gas experiencing an adiabatic pressure drop.

- ■ In Example 6-1, the temperature decreased.
- ■ In Example 6-2, the temperature was unchanged.
- ■ In Example 6-3, the temperature increased.

Why the differences? The reason the ammonia in Example 6-1 decreased temperature on expansion was because of attractive intermolecular forces. A decrease in pressure means the molecules move farther apart. If the molecules are attracted to each other, then it takes energy to move them farther apart. In effect, the molecules are gaining microscopic potential energy, much like an object gains potential energy when it is moved farther from the Earth, to which its attracted by gravity. Where does this energy "come from"? With no external source of energy, it comes from the ammonia itself- the molecules gain microscopic potential energy and lose microscopic kinetic energy, observed as a decrease in temperature.

But in Example 6-3, we set $a = 0$. The van der Waals a represents the very attractive forces that the last paragraph described. If there are no attractive forces then there is no temperature drop. But why a temperature rise? The van der Waals b represents excluded volume, and the repulsive forces resulting when molecules that have a finite volume are compressed close together. Thus we see the opposite phenomenon: the intermolecular forces are repulsive not attractive, so when the molecules are spread apart (lower P) we see a temperature increase.

The ideal gas in Example 6-2, which had neither repulsive nor attractive forces, had no temperature change at all. Real gases, of course, have both attractive and repulsive intermolecular forces. The result of Example 6-1 shows that the attractive forces are more important for ammonia at 30°C and 0.1 MPa. This is typical- most gases at typical process conditions decrease in temperature when they expand. But gases at high temperatures increase temperature when they expand. The **inversion temperature** is the temperature at which the transition from cooling to heating takes place; it is defined as the temperature at which $(\partial T / \partial P)_H = 0$. Inversion temperature is different for every gas and is a function of pressure.

6.2 Mathematical Models of Thermodynamic Properties

The examples in Section 6-1 illustrate the need for building mathematical models that describe thermodynamic systems and quantify the relationships between physical properties. Our specific interest is expressing our mathematical model in terms of the *measurable properties* $(P, \underline{V}, T)$, so that a known equation of state can be applied. This section summarizes a number of mathematical tools and techniques that are useful for building thermodynamic models. These are presented, not in order of importance, but in the order in which they are employed in deriving Equation 6.19.

For now when we think of "measurable properties," we will include the heat capacity (C_P or C_V), and we will only consider applications in which the heat capacity is known. Section 6-3 looks at heat capacity, and situations in which it is and is not known, in more depth.

Equation 6.19 was useful for solving Example 6-2 and Example 6-3, but the derivation was only described briefly. Sections 6.2.1 through 6.2.4 each discuss one step in the derivation.

6.2.1 The Triple Product Rule

Equation 6.19 relates the Joule-Thomson coefficient to measurable properties and was useful for solution of Examples 6-2 and 6-3:

$$\mu_{JT} = \left(\frac{\partial T}{\partial P}\right)_H = -\frac{\underline{V} - T\left(\frac{\partial \underline{V}}{\partial T}\right)_P}{C_P}$$

Notice the constant-pressure heat capacity appears on the right-hand side. Consider the definitions of the Joule-Thomson coefficient and the constant pressure heat capacity:

$$\mu_{JT} = \left(\frac{\partial T}{\partial P}\right)_H$$

$$C_P = \left(\frac{\partial \underline{H}}{\partial T}\right)_P$$

Since they relate the same three quantities (T, P, and $\underline{H}$) it seems quite logical that there would be some mathematical relationship between them. The specific relationship can be understood through a useful property of partial differential calculus called the triple product rule.

FOOD FOR THOUGHT 6-4

Go back to Example 6-3. Suppose a had not been equal to 0. How would this have made evaluating $\left(\frac{\partial \underline{V}}{\partial T}\right)_P$ more difficult? How would the information in this section be helpful for evaluating $\left(\frac{\partial \underline{V}}{\partial T}\right)_P$?

The **triple product rule** can be expressed as

$$\left(\frac{\partial X}{\partial Y}\right)_Z \left(\frac{\partial Y}{\partial Z}\right)_X \left(\frac{\partial Z}{\partial X}\right)_Y = -1 \tag{6.26}$$

or equivalently as

$$\left(\frac{\partial X}{\partial Y}\right)_Z = \frac{-1}{\left(\frac{\partial Y}{\partial Z}\right)_X \left(\frac{\partial Z}{\partial X}\right)_Y} \tag{6.27}$$

Applying the triple product rule $X = T$, $Y = P$ and $Z = \underline{H}$ gives

$$\left(\frac{\partial T}{\partial P}\right)_{\underline{H}} = \frac{-1}{\left(\frac{\partial P}{\partial \underline{H}}\right)_T \left(\frac{\partial \underline{H}}{\partial T}\right)_P} \tag{6.28}$$

Another basic rule of partial differential calculus is

$$\left(\frac{\partial X}{\partial Y}\right)_Z = \frac{1}{\left(\frac{\partial Y}{\partial X}\right)_Z} \tag{6.29}$$

While heat capacity cannot be measured as simply and directly as pressure, temperature, and volume, it can be measured with calorimetry. In this section we regard C_P and C_v as "measurable properties."

Applying Equation 6.29 and the definition of heat capacity to Equation 6.28 gives

$$\left(\frac{\partial T}{\partial P}\right)_{\underline{H}} = -\frac{\left(\frac{\partial \underline{H}}{\partial P}\right)_T}{C_P} \tag{6.30}$$

Comparing Equation 6.30 to Equation 6.19 shows that, by using the triple product rule, we've now taken a significant step toward deriving Equation 6.19. To complete the process of relating μ_{JT} to measureable properties, we now need to evaluate $\left(\frac{\partial \underline{H}}{\partial P}\right)_T$.

This mathematical definition of C_p was first introduced in Section 2.3.2.

6.2.2 Fundamental Property Relationships

In this section we will derive a fundamental relationship between U and other state properties. We will do this by considering a gas confined in a simple piston-cylinder device. Since this is a closed system, the (time-independent) energy balance simplifies to

$$\Delta U = Q + W \tag{6.31}$$

In this case, an energy balance in differential form will be more convenient:

$$dU = dQ + dW \tag{6.32}$$

There is no mechanism for shaft work, so dW represents only work of expansion/contraction, which is given by $-P\,dV$, as first introduced in Equation 1.21.

$$dU = dQ - P\,dV \tag{6.33}$$

Furthermore, if the process is reversible and the gas is in a homogeneous phase, we can relate dQ to entropy through the definition of entropy (Equation 4.13):

$$dU = T\,dS - P\,dV \tag{6.34}$$

If we divide through by the number of moles of gas in the cylinder, we obtain

$$d\underline{U} = T\,d\underline{S} - P\,d\underline{V} \tag{6.35}$$

which is termed **the fundamental property relationship for molar internal energy**.

Examining Equation 6.35, we note that they consist entirely of *state functions*: molar internal energy, molar entropy, temperature, pressure, and molar volume. This observation has great significance. Even though Equation 6.35 was derived by considering a reversible process, the end result is path-independent. Equation 6.35 is thus valid for any process, whether it is reversible or irreversible. Further, it does not matter whether the process is carried out in a piston-cylinder device. Equation 6.35 can be used to describe the change in molar internal energy for any pure material in any process, and this is why it is known as a "fundamental" property relationship.

We can derive an analogous fundamental property relationship for the molar enthalpy by taking the derivative of the definition of enthalpy (first presented in Section 2.3.1). Thus,

$$d\underline{H} = d\underline{U} + d(P\underline{V}) \tag{6.36}$$

Substituting Equation 6.35 into Equation 6.36 gives

$$d\underline{H} = T\,d\underline{S} - P\,d\underline{V} + d(P\underline{V}) \tag{6.37}$$

Expanding the $d(P\underline{V})$ term gives

$$d\underline{H} = T\,d\underline{S} - P\,d\underline{V} + P\,d\underline{V} + \underline{V}\,dP$$

$$d\underline{H} = T\,d\underline{S} + \underline{V}\,dP \tag{6.38}$$

This is the **fundamental property relationship for molar enthalpy**.

The piston-cylinder device has been used as a system several times throughout the book, beginning with Examples 1-2 and 3-1. As usual we regard changes in the kinetic and potential energy of the system as negligible.

FOOD FOR THOUGHT 6-5

Can you give a mathematical and/or physical explanation for going from Equation 6.31 to Equation 6.32?

The fundamental property relationships are composed entirely of state functions, which means they are path-independent.

Here we are applying the product rule of calculus: $d(xy) = x\,dy + y\,dx$

Equations 6.35 and 6.38 are valid for any pure compounds. When we model a mixture of chemical compounds, there are more than two degrees of freedom; $\underline{H}$, for example, would depend upon composition as well as $\underline{S}$ and P. Discussion of modeling of mixtures begins in Chapter 9.

Recall from Section 6.1.1 that a total derivative can be used to relate any intensive property to any other two intensive properties. Expressing $\underline{H}$ as a function of $\underline{S}$ and P:

$$d\underline{H} = \left(\frac{\partial \underline{H}}{\partial \underline{S}}\right)_P d\underline{S} + \left(\frac{\partial \underline{H}}{\partial P}\right)_{\underline{S}} dP \tag{6.39}$$

Compare Equations 6.38 and 6.39.

- Both relate changes in molar enthalpy ($d\underline{H}$) to changes in molar entropy ($d\underline{S}$) and pressure ($d\underline{P}$).

- Equation 6.38 is a thermodynamic relationship between state properties; it is always valid for any pure substance.

- Equation 6.39 is a rule of partial differential calculus; it is always valid for any pure substance.

- The only way Equations 6.38 and 6.39 can both always be true is if the terms multiplying $d\underline{S}$ in both equations are equal to each other and the terms multiplying dP in both equations are also equal to each other. Thus,

$$\left(\frac{\partial \underline{H}}{\partial \underline{S}}\right)_P = T \tag{6.40}$$

$$\left(\frac{\partial \underline{H}}{\partial P}\right)_{\underline{S}} = \underline{V} \tag{6.41}$$

These equations are used in the derivation of Equation 6.19, which was needed to solve Examples 6-2 and 6-3.

In solving the motivational example, it was $(\partial \underline{H}/\partial P)_T$ we needed to evaluate, not $(\partial \underline{H}/\partial \underline{S})_P$ or $(\partial \underline{H}/\partial P)_{\underline{S}}$. However, we will see in Section 6.2.3 that Equations 6.40 and 6.41 can be applied directly to evaluating $(\partial \underline{H}/\partial P)_T$ using the expansion rule.

Everything we learned about $\underline{H}$ from Equations 6.38 through 6.41 has its counterpart for $\underline{U}$. We can write a total differential for $\underline{U}$ in terms of $\underline{S}$ and $\underline{V}$ as

$$d\underline{U} = \left(\frac{\partial \underline{U}}{\partial \underline{S}}\right)_{\underline{V}} d\underline{S} + \left(\frac{\partial \underline{U}}{\partial \underline{V}}\right)_{\underline{S}} d\underline{V} \tag{6.42}$$

By comparing Equations 6.35 and 6.42, we note that

$$\left(\frac{\partial \underline{U}}{\partial \underline{S}}\right)_{\underline{V}} = T \tag{6.43}$$

$$\left(\frac{\partial \underline{U}}{\partial \underline{V}}\right)_{\underline{S}} = -P \tag{6.44}$$

The potential usefulness of the fundamental property relationships for $\underline{U}$ and $\underline{H}$ is straightforward: because $\underline{U}$ and/or $\underline{H}$ routinely appear in energy balances, methods of relating them to known information are needed for the solution of real problems.

Two more state properties of interest in thermodynamics are given here. The **Gibbs free energy (G)** is:

$$G = H - TS \tag{6.45}$$

The **Helmholtz energy (A)** is

$$A = U - TS \tag{6.46}$$

Using derivations analogous to those in Equations 6.36 through 6.38, the **fundamental property relationships for the molar Gibbs free energy and molar Helmholtz energy** are

$$d\underline{G} = \underline{V}\,dP - \underline{S}\,dT \qquad (6.47)$$

$$d\underline{A} = -P\,d\underline{V} - \underline{S}\,dT \qquad (6.48)$$

The molar Gibbs free energy ($\underline{G}$) has not been used substantially up to this point, but has great importance in modeling phase equilibrium, as discussed starting in Chapter 8. The molar Helmholtz energy ($\underline{A}$) is not often used directly in thermodynamics at the introductory level, but it is used prominently in the field of statistical mechanics. For our purposes, the most immediate significance of Equation 6.48 is that it is used in the derivation of one of Maxwell's equations, which are discussed in Section 6.2.4.

> A brief description of statistical mechanics is given in Chapter 7.

6.2.3 The Expansion Rule

Another useful tool in partial differential calculus is the **expansion rule,** which says

$$\left(\frac{\partial X}{\partial Y}\right)_Z = \left(\frac{\partial X}{\partial K}\right)_L\left(\frac{\partial K}{\partial Y}\right)_Z + \left(\frac{\partial X}{\partial L}\right)_K\left(\frac{\partial L}{\partial Y}\right)_Z \qquad (6.49)$$

This is a rule of calculus; it can be applied to any set of variables X, Y, Z, K, and L.

Consider how Equation 6.49 could be applied to the solution of Example 6-2 and the derivation of Equation 6.19. We showed (see Equation 6.30) that solving the problem required evaluation of the partial derivative $(\partial\underline{H}/\partial P)_T$. In applying the expansion rule to this example, it is natural to let $X = \underline{H}$, $Y = P$, and $Z = T$, transforming the left-hand side into the exact partial derivative we wish to evaluate:

$$\left(\frac{\partial\underline{H}}{\partial P}\right)_T = \left(\frac{\partial\underline{H}}{\partial K}\right)_L\left(\frac{\partial K}{\partial P}\right)_T + \left(\frac{\partial\underline{H}}{\partial L}\right)_K\left(\frac{\partial L}{\partial P}\right)_T \qquad (6.50)$$

In effect, the expansion rule allows us to introduce two new variables, K and L, into the problem. Section 6.2.2 showed the derivation of known relationships among molar enthalpy, molar entropy, and pressure. Thus, if we assign K and L in the expansion rule to represent $\underline{S}$ and P, we can introduce this information into the problem we're trying to solve. Equation 6.50 becomes

$$\left(\frac{\partial\underline{H}}{\partial P}\right)_T = \left(\frac{\partial\underline{H}}{\partial P}\right)_S\left(\frac{\partial P}{\partial P}\right)_T + \left(\frac{\partial\underline{H}}{\partial\underline{S}}\right)_P\left(\frac{\partial\underline{S}}{\partial P}\right)_T \qquad (6.51)$$

> While X, Y, Z, K, and L can be *any* intensive variables, a common and powerful use of the expansion rule occurs when X is a measure of energy ($\underline{U}, \underline{H}, \underline{G}, \underline{A}$) and K and L are used to apply the fundamental property relationship for that measure of energy.

And in Section 6.2.2: we derived Equations 6.40 and 6.41, which expressed this known relationship among $\underline{H}$, $\underline{S}$, and P as

$$\left(\frac{\partial\underline{H}}{\partial\underline{S}}\right)_P = T$$

$$\left(\frac{\partial\underline{H}}{\partial P}\right)_{\underline{S}} = \underline{V}$$

Consequently, substituting Equations 6.40 and 6.41 into Equation 6.51 gives

$$\left(\frac{\partial \underline{H}}{\partial P}\right)_T = \underline{V}\left(\frac{\partial P}{\partial P}\right)_T + T\left(\frac{\partial \underline{S}}{\partial P}\right)_T \qquad (6.52)$$

Here we note two additional rules of partial differential calculus:

$$\left(\frac{\partial X}{\partial X}\right)_Y = 1 \qquad (6.53)$$

$$\left(\frac{\partial X}{\partial Y}\right)_X = 0 \qquad (6.54)$$

Rather than memorizing these, you can probably rationalize them: the change in X is always equal to the change in X, and the change in X is zero if X is held constant.

Notice that Equation 6.51 contains the expression $(\partial P/\partial P)_T$, which is a consequence of the fact that the variables X and K in the expansion rule were both set equal to the pressure P. Applying Equation 6.53 to Equation 6.51 leads to the simplification of

$$\left(\frac{\partial \underline{H}}{\partial P}\right)_T = \underline{V} + T\left(\frac{\partial \underline{S}}{\partial P}\right)_T \qquad (6.55)$$

One more step is required to complete the derivation of Equation 6.19.

Referring back to the solutions of Examples 6-2 and 6-3, our goal is to derive Equation 6.19, which related μ_{JT} to measureable properties. The molar entropy is the only thing on the right-hand side of Equation 6.55 that isn't a measurable property. It can be eliminated using one of Maxwell's equations.

6.2.4 Maxwell's Equations

Recall from Section 6.2.2 that the fundamental property relationship for internal energy is

$$d\underline{U} = T\,d\underline{S} - P\,d\underline{V}$$

One of the rules of partial differential calculus is that when one takes a mixed second derivative with respect to two variables, the order of differentiation does not matter. Stated mathematically, this rule gives

$$\left(\frac{\partial^2 X}{\partial Y\,\partial Z}\right) = \left(\frac{\partial^2 X}{\partial Z\,\partial Y}\right) \qquad (6.56)$$

This fact can be applied to the fundamental property relationship for $\underline{U}$ to learn a new thermodynamic identity. Equation 6.43 states that

$$\left(\frac{\partial \underline{U}}{\partial \underline{S}}\right)_V = T$$

We can differentiate both sides of Equation 6.43 with respect to $\underline{V}$ along a path that keeps $\underline{S}$ constant, producing

$$\frac{\partial}{\partial \underline{V}}\left[\left(\frac{\partial \underline{U}}{\partial \underline{S}}\right)_V\right]_{\underline{S}} = \left(\frac{\partial T}{\partial \underline{V}}\right)_{\underline{S}} \qquad (6.57)$$

The mixed second derivative of $\underline{U}$ with respect to $\underline{S}$ and $\underline{V}$ can also be found starting from Equation 6.44 as

$$\left(\frac{\partial \underline{U}}{\partial \underline{V}}\right)_S = -P$$

Taking the derivative with respect to $\underline{S}$ gives

$$\frac{\partial}{\partial \underline{S}}\left[\left(\frac{\partial \underline{U}}{\partial \underline{V}}\right)_S\right]_V = -\left(\frac{\partial P}{\partial \underline{S}}\right)_V \tag{6.58}$$

But because the order of differentiation doesn't matter, we have

$$\frac{\partial}{\partial \underline{V}}\left[\left(\frac{\partial \underline{U}}{\partial \underline{S}}\right)_V\right]_S = \frac{\partial}{\partial \underline{S}}\left[\left(\frac{\partial \underline{U}}{\partial \underline{V}}\right)_S\right]_V \tag{6.59}$$

Thus,

$$\left(\frac{\partial T}{\partial \underline{V}}\right)_S = -\left(\frac{\partial P}{\partial \underline{S}}\right)_V \tag{6.60}$$

While Equation 6.60 is not directly useful in solving the motivational examples, we note that the process of deriving it can be applied to the fundamental property relationships for $\underline{H}$, $\underline{G}$, and $\underline{A}$. Problem 6-3 asks you to perform the derivations.

The final results are

$$\left(\frac{\partial T}{\partial P}\right)_S = \left(\frac{\partial \underline{V}}{\partial \underline{S}}\right)_P \tag{6.61}$$

$$\left(\frac{\partial P}{\partial T}\right)_V = \left(\frac{\partial \underline{S}}{\partial \underline{V}}\right)_T \tag{6.62}$$

$$\left(\frac{\partial \underline{V}}{\partial T}\right)_P = -\left(\frac{\partial \underline{S}}{\partial P}\right)_T \tag{6.63}$$

Collectively, Equations 6.60 through 6.63 are known as **Maxwell's equations**.

Equation 6.63 figures directly into the solution of the motivational example. Using the expansion rule, we obtained Equation 6.55:

$$\left(\frac{\partial \underline{H}}{\partial P}\right)_T = \underline{V} + T\left(\frac{\partial \underline{S}}{\partial P}\right)_T$$

Applying Equation 6.63 to Equation 6.55 produces

$$\left(\frac{\partial \underline{H}}{\partial P}\right)_T = \underline{V} - T\left(\frac{\partial \underline{V}}{\partial T}\right)_P \tag{6.64}$$

Substituting this result into Equation 6.30 completes the process of deriving Equation 6.19. Thus,

$$\mu_{JT} = -\frac{\underline{V} - T\left(\frac{\partial \underline{V}}{\partial T}\right)_P}{C_P}$$

To review, the steps of the derivation are given here.

- Recognizing that μ_{JT} is defined from the same intensive variables (T, P, and $\underline{H}$) as C_P, we used the triple product rule to express μ_{JT} in terms of C_P. This step left the unknown $\left(\dfrac{\partial \underline{H}}{\partial P}\right)_T$ to resolve.

- We used the expansion rule to evaluate $\left(\dfrac{\partial \underline{H}}{\partial P}\right)_T$. The fundamental property relationship reveals that $\underline{H}$ is a natural function of $\underline{S}$ and P, so these were the two variables we introduced into the problem using the expansion rule.

- We used one of Maxwell's equations to eliminate $\underline{S}$ from the problem, leaving μ_{JT} as a function of measurable properties.

The next section applies Equation 6.19 to another example of a Joule-Thomson expansion—this time the fluid undergoing the expansion is a liquid.

6.2.5 Coefficient of Thermal Expansion and Isothermal Compressibility

We know from everyday experience that substances typically expand when heated (there do exist exceptions; liquids and solids that actually decrease in volume when temperature increases). This phenomenon has great practical importance in design and construction settings. If a product is expected to experience wide temperature fluctuations, it must be designed to withstand the resulting volume changes, especially where different materials that expand at different rates are attached to each other. Physically, this expansion is precisely what the partial derivative expression $(\partial \underline{V}/\partial T)_P$ measures—it tells us how much the molar volume of a substance changes for each degree the temperature is changed at constant pressure. Thus, we define the **coefficient of thermal expansion** as

$$\alpha_V = \frac{1}{\underline{V}}\left(\frac{\partial \underline{V}}{\partial T}\right)_P \tag{6.65}$$

Analogously, we can define a thermodynamic property that measures the effect of changing pressure on the molar volume of a material, which is the **isothermal compressibility,** as

$$\kappa_T = -\frac{1}{\underline{V}}\left(\frac{\partial \underline{V}}{\partial P}\right)_T \tag{6.66}$$

The definition of coefficient of thermal expansion can be introduced into Equation 6.19 for

$$\mu_{JT} = -\frac{\underline{V} - T\left(\dfrac{\partial \underline{V}}{\partial T}\right)_P}{C_P}$$

$$\mu_{JT} = -\frac{\underline{V} - T(\underline{V}\alpha_V)}{C_P}$$

$$\mu_{JT} = \left(\frac{\partial T}{\partial P}\right)_{\underline{H}} = \frac{\underline{V}}{C_P}(T\alpha_V - 1) \tag{6.67}$$

For liquids and solids, we have previously used the simple model that $\underline{V}$ is constant, which would imply that K_T and a_V are always zero. In Example 6-4, the coefficient of thermal expansion is modeled as a constant, but not equal to zero.

JOULE-THOMSON EXPANSION OF A LIQUID	EXAMPLE 6-4

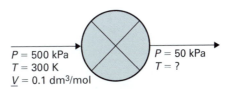

P = 500 kPa
T = 300 K
$\underline{V}$ = 0.1 dm³/mol

P = 50 kPa
T = ?

FIGURE 6-4 Valve examined in Example 6-4.

A liquid enters an adiabatic throttling valve, illustrated in Figure 6-4, at $T = 300$ K and $P = 500$ kPa, at which conditions it has $\underline{V} = 0.1$ dm³/mol. The exiting liquid has $P = 50$ kPa. For this steady-state process, what is the temperature of the exiting liquid? Assume constant heat capacity $C_P = 0.1$ kJ/mol K, constant coefficient of thermal expansion $\alpha_V = 0.001$ K⁻¹, and assume the isothermal compressibility is negligible.

SOLUTION:

Step 1 *Define a system and simplify the energy balance*
This is another Joule-Thomson expansion carried out at steady state; the exact situation described in Example 6-1 and Equation 6.1. The energy balance, using the valve as the system, is

$$\underline{H}_{in} = \underline{H}_{out} \qquad (6.68)$$

Step 2 *Write a total derivative that relates temperature to pressure and enthalpy*
We are trying to find the change in temperature (dT) as the liquid travels through the valve. The two intensive properties we know the most about are again enthalpy ($\underline{H}$) and pressure (P).

$$dT = \left(\frac{\partial T}{\partial \underline{H}}\right)_P d\underline{H} + \left(\frac{\partial T}{\partial P}\right)_{\underline{H}} dP \qquad (6.69)$$

The first term on the right-hand side is zero because the energy balance shows that enthalpy is constant. We can apply Equation 6.67, which relates the Joule-Thomson coefficient to measurable properties; for

$$dT = \left(\frac{\partial T}{\partial P}\right)_{\underline{H}} dP$$

$$dT = \frac{\underline{V}}{C_P}(T\alpha_V - 1)\, dP \qquad (6.70)$$

Step 3 *A simplifying assumption*
We can separate the variables by moving $(T\alpha_V - 1)$ over to the left-hand side for

$$\frac{dT}{T\alpha_V - 1} = \frac{\underline{V}}{C_P} dP \qquad (6.71)$$

The left-hand side is now strictly a function of T and constants (α_V being assumed constant). On the right-hand side, we are assuming C_P is constant, but what of the molar volume? Generally speaking (for liquids), liquid volume is not very sensitive to changes in temperature and pressure. For the moment, let's assume we can treat $\underline{V}$ as a constant with respect to both pressure and temperature, and we will re-evaluate that decision at the end of the problem.

Here the energy balance is written in terms of molar enthalpy because the heat capacity is given on a molar basis.

PITFALL PREVENTION

Temperature is expressed on an absolute scale in the definitions of entropy ($dS = dQ_{rev}/T$), Gibbs free energy ($G = H - TS$), and Helmholtz energy ($A = U - TS$). Temperature (T) must therefore be expressed on an absolute scale in Equation 6.71, which is derived from these.

Step 4 *Solve the integral*

The limits of integration for the right-hand side are the known pressures entering and leaving the valve. The temperature leaving the valve (T_{out}) is the unknown we are trying to find.

Integrating Equation 6.71, treating $\underline{V}$, C_P, and α_V all as constants, gives

$$\int_{T_{in}=300\,K}^{T_{out}=T_{out}} \frac{dT}{T\alpha_V - 1} = \int_{P_{in}=500\,kPa}^{P_{out}=50\,kPa} \frac{\underline{V}}{C_P} dP \tag{6.72}$$

$$\left(\frac{1}{\alpha_V}\right)\ln\left(\frac{T_{out}\,\alpha_V - 1}{T_{in}\,\alpha_V - 1}\right) = \left(\frac{\underline{V}}{C_P}\right)(P_{out} - P_{in})$$

Substituting in known values and applying necessary conversion factors gives

$$\left(\frac{1}{0.001\ K^{-1}}\right)\ln\left[\frac{T_{out}(0.001\ K^{-1}) - 1}{(300\ K)(0.001\ K^{-1}) - 1}\right] = \left(\frac{0.1\dfrac{dm^3}{mol}}{0.1\dfrac{kJ}{mol\cdot K}}\right)(50 - 500\ kPa) \tag{6.73}$$

$$\ln\left[\frac{T_{out}(0.001\ K^{-1}) - 1}{(300\ K)(0.001\ K^{-1}) - 1}\right] = (0.001\ K^{-1})\left(\frac{0.1\times10^{-3}\dfrac{m^3}{mol}}{0.1\dfrac{kJ}{mol\cdot K}}\right)(-450\ kPa)\left(\frac{1\ kJ}{1\ kPa\ m^3}\right)$$

$$\ln\left[\frac{T_{out}(0.001\ K^{-1}) - 1}{(300\ K)(0.001\ K^{-1}) - 1}\right] = -4.5\times10^{-4} \tag{6.74}$$

$$\frac{T_{out}(0.001\ K^{-1}) - 1}{(300\ K)(0.001\ K^{-1}) - 1} = 0.99957$$

The result is T_{out} = **300.3 K.**

Example 6-4 examined a liquid expanding from 500 kPa and 300 K to a final pressure of 50 kPa. According to this calculation, the liquid temperature only changes by 0.3 degrees as a result of the pressure drop. Examples 6-1 through 6-4 collectively illustrate that the effect of Joule-Thomson expansion on temperature is much more dramatic for vapors than liquids (unless the gas is ideal, in which case there is no temperature change at all). This can be understood mathematically in that the Joule-Thomson coefficient, according to Equation 6.67, is proportional to $\underline{V}$ and its derivatives, which are typically much larger for vapors than liquids. This also means that our decision to model $\underline{V}$ as a constant in step 3 is justified; the molar volume $\underline{V}$ is not going to change significantly when the temperature changes by just 0.3 K. Problem 6-10 asks you to estimate this quantitatively.

6.2.6 Additional Applications of Thermodynamic Partial Derivatives

Throughout the chapter up to this point, all examples have stemmed from the motivational example, and so involved Joule-Thomson expansion. The modeling techniques discussed throughout this chapter are, however, applicable to any physical process. To illustrate this, we now examine three different processes: the compression of a van der Waals gas in a closed (variable volume) system, the heating of a liquid in a rigid container, and a gas traveling through a reversible nozzle.

| ISOTHERMAL COMPRESSION OF A VAN DER WAALS GAS | EXAMPLE 6-5 |

Five moles of gas are confined in a piston-cylinder device (Figure 6-5). At the beginning of the process, the gas has $T = 300$ K and $V = 0.1$ m^3. If the gas is compressed isothermally to a final volume of 0.015 m^3, how much work is required, and how much heat is added or removed? Assume the heat capacity is constant at $C_V = 30$ J/mol · K and that the gas is modeled by the van der Waals equation of state

$$P = \frac{RT}{\underline{V} - b} - \frac{a}{\underline{V}^2}$$

with $a = 0.14$ m^6 Pa/mol^2 and $b = 3.8 \times 10^{-5}$ m^3/mol.

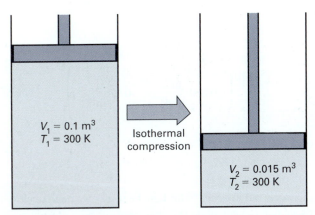

FIGURE 6-5 Compression of gas in a piston-cylinder device, modeled in Example 6-5.

An alternative to the Rankine heat engine is the Stirling cycle, in which the steps are isothermal compression (as in this example), isochoric heating, isothermal expansion, and isochoric cooling.

FOOD FOR THOUGHT 6-8

Published values (Weast, 1972) of van der Waals constants include $a = 0.1408$ m^3 Pa/mol^2 for nitrogen and $a = 0.1378$ m^3 Pa/mol^2 for oxygen. Is it reasonable to guess that air, which is ~79% nitrogen, would have $a \approx (0.79)(0.1408) + (0.21)(0.137^8) = 0.14$ m^3 Pa /mol^2?

SOLUTION:

Step 1 *Define a system and write an energy balance*
If we define the five moles of gas as the system, we have a closed system. Applying the time-independent energy balance and neglecting potential and kinetic energy gives

$$N(\underline{U}_2 - \underline{U}_1) = Q + W \tag{6.75}$$

Step 2 *Evaluate work*
The only kind of work in this process is expansion/contraction work, which is known (see Section 1.4.3) to be equal to $-P\,dV$. The initial and final volumes of the gas are known, so we can set up the integral:

$$W_{EC} = -\int_{V=0.1\text{ m}^3}^{V=0.015\text{ m}^3} P\,dV \tag{6.76}$$

Here, P is the pressure of the gas, which is given by the van der Waals equation of state:

$$W_{EC} = -\int_{V=0.1\text{ m}^3}^{V=0.015\text{ m}^3} \left[\frac{RT}{\underline{V} - b} - \frac{a}{\underline{V}^2}\right]dV \tag{6.77}$$

Recognizing that V and $\underline{V}$ are two different quantities, we relate them to each other as

$$V = N\underline{V} \tag{6.78}$$

Because N is constant in this case, we can write

$$dV = N\,d\underline{V} \tag{6.79}$$

P is the pressure opposing the motion. Here the gas is being compressed, so the pressure opposing the motion is the pressure of the gas itself.

PITFALL PREVENTION

Don't confuse V and $\underline{V}$. The limits of integration are either $V = 0.1$ m^3 to 0.015 m^3 if you integrate with respect to total volume, or $\underline{V} = 0.02$ m^3/mol to 0.003 m^3/mol if you integrate with respect to molar volume.

and insert this into Equation 6.77 for

$$W_{EC} = -N \int_{\underline{V}=0.02 \text{ m}^3/\text{mol}}^{\underline{V}=0.003 \text{ m}^3/\text{mol}} \left[\frac{RT}{\underline{V} - b} - \frac{a}{\underline{V}^2} \right] d\underline{V} \tag{6.80}$$

This is an isothermal compression. T is constant in this process, and can be moved outside the integral.

$$W_{EC} = -(NRT) \ln \left(\frac{\underline{V}_2 - b}{\underline{V}_1 - b} \right) + N \left(\frac{a}{\underline{V}_2} - \frac{a}{\underline{V}_1} \right)$$

Plugging in known values gives

$$W_{EC} = -(5 \text{ mol}) \left(8.314 \frac{\text{Pa} \cdot \text{m}^3}{\text{mol} \cdot \text{K}} \right) (300 \text{ K}) \ln \left(\frac{0.003 - 3.8 \times 10^{-5} \frac{\text{m}^3}{\text{mol}}}{0.02 - 3.8 \times 10^{-5} \frac{\text{m}^3}{\text{mol}}} \right) \tag{6.81}$$

$$+ (5 \text{ mol}) \left(\frac{0.14 \frac{\text{m}^6 \cdot \text{Pa}}{\text{mol}^2}}{0.003 \frac{\text{m}^3}{\text{mol}}} - \frac{0.14 \frac{\text{m}^6 \cdot \text{Pa}}{\text{mol}}}{0.02 \frac{\text{m}^3}{\text{mol}}} \right)$$

$$W_{EC} = (23792 \text{ Pa} \cdot \text{m}^3 + 198 \text{ Pa} \cdot \text{m}^3) = 23{,}990 \text{ J}$$

We now know the work, but in order to find the heat (Q), we need to close the energy balance. This means we need to calculate the change in specific internal energy ($\underline{U}_2 - \underline{U}_1$).

Step 3 *Write a total derivative that describes $\underline{U}$*
We can write a total derivative expression that relates $d\underline{U}$ to changes in any other two intensive properties. The two we know the most about are temperature and molar volume.

$$d\underline{U} = \left(\frac{\partial \underline{U}}{\partial \underline{V}} \right)_T d\underline{V} + \left(\frac{\partial \underline{U}}{\partial T} \right)_V dT \tag{6.82}$$

Because this is an isothermal process, the dT term can be eliminated ($dT = 0$) for

$$d\underline{U} = \left(\frac{\partial \underline{U}}{\partial \underline{V}} \right)_T d\underline{V} \tag{6.83}$$

Thus, solution of the problem requires us to relate the partial derivative $(\partial \underline{U}/\partial \underline{V})_T$ to known properties. This requires several steps, but can be done using a strategy analogous to the one illustrated in Sections 6.2.2 and 6.2.3. We can use the expansion rule to find an expression for $(\partial \underline{U}/\partial \underline{V})_T$, but what two parameters should we introduce into the problem (in other words, what should we choose as L and K in Equation 6.49)? We can answer this question by exploring the fundamental property relationship for $\underline{U}$.

Step 4 *Apply fundamental property relationship for $\underline{U}$*
Equation 6.35 gives the fundamental property relationship for $\underline{U}$:

$$d\underline{U} = T \, d\underline{S} - P \, d\underline{V}$$

We learned in Section 6.1.1 that a total derivative can be used to relate any intensive property to any other two intensive properties. Expressing $\underline{U}$ as a function of $\underline{S}$ and $\underline{V}$ gives

$$d\underline{U} = \left(\frac{\partial \underline{U}}{\partial \underline{S}} \right)_V d\underline{S} + \left(\frac{\partial \underline{U}}{\partial \underline{V}} \right)_S d\underline{V} \tag{6.84}$$

The only way Equations 6.35 and 6.84 can both be true is if the $d\underline{S}$ terms in both equations are equal to each other and the $d\underline{V}$ terms in both equations are also equal to each other:

$$\left(\frac{\partial \underline{U}}{\partial \underline{S}}\right)_V = T \tag{6.85}$$

$$\left(\frac{\partial \underline{U}}{\partial \underline{V}}\right)_S = -P \tag{6.86}$$

Step 5 *Apply expansion rule*
The expansion rule is

$$\left(\frac{\partial X}{\partial Y}\right)_Z = \left(\frac{\partial X}{\partial K}\right)_L\left(\frac{\partial K}{\partial Y}\right)_Z + \left(\frac{\partial X}{\partial L}\right)_K\left(\frac{\partial L}{\partial Y}\right)_Z \tag{6.87}$$

It is natural to define $X = \underline{U}$, $Y = \underline{V}$, and $Z = T$, since that transforms the left-hand side into the partial differential we seek to evaluate. If we define $K = \underline{S}$ and $L = \underline{V}$, we can make use of Equations 6.85 and 6.86 for

$$\left(\frac{\partial \underline{U}}{\partial \underline{V}}\right)_T = \left(\frac{\partial \underline{U}}{\partial \underline{S}}\right)_V\left(\frac{\partial \underline{S}}{\partial \underline{V}}\right)_T + \left(\frac{\partial \underline{U}}{\partial \underline{V}}\right)_S\left(\frac{\partial \underline{V}}{\partial \underline{V}}\right)_T \tag{6.88}$$

And introducing Equations 6.85 and 6.86 into Equation 6.88 gives

$$\left(\frac{\partial \underline{U}}{\partial \underline{V}}\right)_T = T\left(\frac{\partial \underline{S}}{\partial \underline{V}}\right)_T + (-P)\left(\frac{\partial \underline{V}}{\partial \underline{V}}\right)_T \tag{6.89}$$

We can eliminate the specific entropy from the right-hand side by applying one of Maxwell's equations. We also note that the derivative of $\underline{V}$ with respect to $\underline{V}$ is 1 (see Equation 6.53). Thus,

$$\left(\frac{\partial \underline{U}}{\partial \underline{V}}\right)_T = T\left(\frac{\partial P}{\partial T}\right)_V + (-P) \tag{6.90}$$

The right-hand side is now entirely composed of measurable properties, and can be evaluated.

Step 6 *Evaluate partial derivative of pressure for van der Waals equation*
A partial derivative expression that relates P, $\underline{V}$, and T to each other can be evaluated if we know an equation of state that describes the material well; in this case the van der Waals equation.

Differentiating the van der Waals equation with respect to T gives:

$$\left(\frac{\partial P}{\partial T}\right)_V = \frac{R}{\underline{V} - b} \tag{6.91}$$

Step 7 *Evaluate original total derivative*
The purpose of steps 4 through 6 was to evaluate Equation 6.83, which is the original total derivative relating $d\underline{U}$ to changes in temperature (dT) and specific volume $(d\underline{V})$. Applying Equation 6.90 to Equation 6.83 gives

$$d\underline{U} = \left(\frac{\partial \underline{U}}{\partial \underline{V}}\right)_T d\underline{V} = \left[T\left(\frac{\partial P}{\partial T}\right)_V - P\right]d\underline{V} \tag{6.92}$$

In order to integrate the right-hand side, we need to express everything as a function of $\underline{V}$. An expression for $(\partial P/\partial T)_V$ is known through Equation 6.91, and P can be found from the equation of state itself. Substituting these into Equation 6.92 gives

$$d\underline{U} = \left\{T\left(\frac{R}{\underline{V} - b}\right) - \left(\frac{RT}{\underline{V} - b} - \frac{a}{\underline{V}^2}\right)\right\}d\underline{V}$$

$$d\underline{U} = \frac{a}{\underline{V}^2}\,d\underline{V} \tag{6.93}$$

The simplification that occurs between Equations 6.88 and 6.89, in which $\underline{U}$ is eliminated from the right-hand side, is neither "luck" nor a fluke. We know $\underline{U}$ is related to $\underline{S}$ and $\underline{V}$ through the fundamental property relationship, and defining $K = \underline{S}$ and $L = \underline{V}$ introduces this knowledge into the solution of a problem.

In this differentiation, $\underline{V}$ is being held constant, so the term $a/\underline{V}^2$ is a constant, and its derivative is zero.

This shows mathematically a difference between ideal and real gases. For an ideal gas, specific internal energy is only a function of temperature; $\underline{U}_2 - \underline{U}_1$ would be 0 for an isothermal ideal gas process. Equation 6.93 reflects mathematically that compressing a real gas into a smaller volume—even if it is done isothermally—affects the internal energy due to intermolecular interactions.

Step 8 *Integrate the total derivative*
Integrating Equation 6.93 from the initial state to the final state gives

$$\int d\underline{U} = \int_{\underline{V}=0.020\ \text{m}^3/\text{mol}}^{\underline{V}=0.003\ \text{m}^3/\text{mol}} \frac{a}{\underline{V}^2}\, d\underline{V} \tag{6.94}$$

$$\underline{U}_2 - \underline{U}_1 = -\left(\frac{a}{\underline{V}_2} - \frac{a}{\underline{V}_1}\right) = -\left(\frac{0.14\ \dfrac{\text{m}^6 \cdot \text{Pa}}{\text{mol}^2}}{0.003\ \dfrac{\text{m}^3}{\text{mol}}} - \frac{0.14\ \dfrac{\text{m}^6 \cdot \text{Pa}}{\text{mol}^2}}{0.02\ \dfrac{\text{m}^3}{\text{mol}}}\right)$$

$$\underline{U}_2 - \underline{U}_1 = \left(-40\ \frac{\text{m}^3 \cdot \text{Pa}}{\text{mol}}\right) = -40\ \frac{\text{J}}{\text{mol}} \tag{6.95}$$

Step 9 *Close energy balance*
Inserting the known values for $\underline{U}$ and W into the energy balance gives

$$N(\underline{U}_2 - \underline{U}_1) = Q + W \tag{6.96}$$

$$(5\,\text{mol})\left(-40\ \frac{\text{J}}{\text{mol}}\right) = Q + 23{,}990\ \text{J}$$

$$Q = -24{,}190\ \text{J}$$

For an ideal gas, $\underline{U}$ is a function of temperature only, and an isothermal compression or expansion would have $\Delta \underline{U} = 0$ and $Q = -W$. We have noted previously that, for real gases, $\underline{U}$ is also a function of pressure and volume. In Example 6-5, we saw for the first time how pressure or volume dependence can be modeled.

The next example examines a case in which the isothermal compressibility and coefficient of thermal expansion of a liquid are both known and assumed to be constants.

| **EXAMPLE 6-6** | **HEATING A LIQUID IN A RIGID CONTAINER** |

The tank of a water heater (Figure 6-6) has $V = 0.1\ \text{m}^3$ and is initially sealed and full of water at $T = 328\ \text{K}$ and $P = 0.1\ \text{MPa}$. The water heater is designed to maintain the water at this temperature, but due to a fault in the temperature sensor, the heating element comes on and remains on. The temperature of the water increases, but there is no space into which the water can expand. The tank has a pressure relief value that is designed to open at $P = 1\ \text{MPa}$. At what water temperature will this occur?

For this example assume the isothermal compressibility and coefficient of thermal expansion of water are constant at $\alpha_V = 2.90 \times 10^{-4}\ \text{K}^{-1}$ and $\kappa_T = 4.5 \times 10^{-10}\ \text{Pa}^{-1}$.

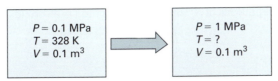

FIGURE 6-6 Isochoric heating of a water heater.

SOLUTION:

Step 1 *Define a system and process*
The system is the water in the tank. The description of the tank says that the water has "no room to expand," so we will model this as a rigid, constant-volume system. Since the tank is described as "sealed," this is also a closed system.

Since our goal is to find the temperature at which the relief valve will open, the process begins when the heating elements turns on, and ends when the pressure in the tank reaches exactly $P = 1$ MPa.

Step 2 *Write a total derivative that describes changes in temperature*
The problem asks us to find an unknown temperature. Changes in temperature (dT) can be related to changes in any other two intensive properties through a total derivative. In this case, the two variables we know the most about are volume (the container is rigid) and pressure (the initial and final pressure are both known). So we write the total derivative expression as

$$dT = \left(\frac{\partial T}{\partial \underline{V}}\right)_P d\underline{V} + \left(\frac{\partial T}{\partial P}\right)_{\underline{V}} dP \tag{6.97}$$

The total volume (V) is constant in this rigid container. The mass (M) is also constant; this is a closed system. Therefore, the specific volume ($\underline{V} = V/M$) must also be constant. Thus, the volume term of Equation 6.97 can be eliminated ($d\underline{V} = 0$).

$$dT = \left(\frac{\partial T}{\partial P}\right)_{\underline{V}} dP \tag{6.98}$$

Step 3 *Evaluate the partial differential $(\partial T/\partial P)_{\underline{V}}$*
To evaluate $(\partial T/\partial P)_{\underline{V}}$, our first thought might be to apply an equation of state, as we did when we found $(\partial P/\partial T)_{\underline{V}}$ for the gas in Example 6-5. We do not have an explicit equation of state for this liquid, but we do have two pieces of given information about the relationship among P, $\underline{V}$, and T. The isothermal compressibility and the coefficient of thermal expansion are both known. By definition,

$$\alpha_V = \frac{1}{\underline{V}}\left(\frac{\partial \underline{V}}{\partial T}\right)_P \quad \text{and} \quad \kappa_T = -\frac{1}{\underline{V}}\left(\frac{\partial \underline{V}}{\partial P}\right)_T$$

We can relate the known information to $(\partial T/\partial P)_{\underline{V}}$ through the triple product rule:

$$\left(\frac{\partial X}{\partial Y}\right)_Z \left(\frac{\partial Y}{\partial Z}\right)_X \left(\frac{\partial Z}{\partial X}\right)_Y = -1$$

which for T, P, and $\underline{V}$ becomes

$$\left(\frac{\partial T}{\partial P}\right)_{\underline{V}} \left(\frac{\partial P}{\partial \underline{V}}\right)_T \left(\frac{\partial \underline{V}}{\partial T}\right)_P = -1 \tag{6.99}$$

We apply Equation 6.29 to invert the $(\partial P/\partial \underline{V})_T$ term, so that it mirrors the definition of α_V.

$$\frac{\left(\dfrac{\partial T}{\partial P}\right)_{\underline{V}} \left(\dfrac{\partial \underline{V}}{\partial T}\right)_P}{\left(\dfrac{\partial \underline{V}}{\partial P}\right)_T} = -1 \tag{6.100}$$

We want to know $(\partial P/\partial T)_{\underline{V}}$. The fact that we know the value of a different partial derivative that relates the same three variables to each other is a hint that we might try applying the triple product rule.

Substituting the definitions of α_V and κ_T into Equation 6.100 yields

$$\frac{\left(\frac{\partial T}{\partial P}\right)_V (\underline{V}\alpha_V)}{(-\underline{V}\kappa_T)} = -1$$

$$\left(\frac{\partial T}{\partial P}\right)_V = \frac{\kappa_T}{\alpha_V} \quad\quad (6.101)$$

Equation 6.101 is not specific to this problem; we assumed nothing in its derivation so it's always true. Here, however, it's particularly useful, because the isothermal compressibility and coefficient of thermal expansion are both known and modeled as constant.

Thus, in this example, $(\partial T/\partial P)_V$ is equal to a constant.

Step 4 *Solve integral*

The purpose of step 3 was to convert $(\partial T/\partial P)_V$ into a form that can be integrated so that Equation 6.98 can be solved. Thus,

$$dT = \left(\frac{\partial T}{\partial P}\right)_V dP = \frac{\kappa_T}{\alpha_V} dP \quad\quad (6.102)$$

$$\int_{T_1 = 328\ K}^{T_2} dT = \int_{P_1 = 0.1\ MPa}^{P_2 = 1\ MPa} \frac{\kappa_T}{\alpha_V} dP$$

$$T_2 - T_1 = \frac{\kappa_T}{\alpha_V}(P_2 - P_1)$$

$$T_2 - 328\ K = \frac{4.5 \times 10^{-10}\ Pa^{-1}}{2.9 \times 10^{-4}\ K^{-1}}(1.0 - 0.1\ MPa) \times (10^6\ Pa/MPa) \quad\quad (6.103)$$

The result is $T_2 = 329.4\ K$.

You probably already knew that when liquids are heated in a confined space, there is the danger that if the liquid boils, a rapid, extreme pressure buildup can result. The previous example shows that even if the liquid does not boil, heating it in a rigid container can lead to a dramatic pressure increase. The volume of water is not very sensitive to temperature—heating water from 328 K to 329.4 K at a constant pressure of 0.1 MPa would only cause the volume to increase by about 0.04%. But if this fractional expansion is not permitted to occur, the pressure builds at a rather remarkable rate—in Example 6-6 a 1.4 K temperature increase accompanies an order-of-magnitude change in pressure. This example illustrates why a typical home water heater has a built-in expansion tank.

The final example looks at a gas traveling through a reversible nozzle, and in the process, derives another useful partial derivative expression, relating molar entropy to heat capacity.

EXAMPLE 6-7

GAS TRAVELING THROUGH A REVERSIBLE NOZZLE

A gas flows through an adiabatic nozzle (Figure 6-7) at steady state, entering at $P = 0.5\ MPa$ and $T = 675$ K, and leaving at $P = 0.1$ MPa. Assuming the nozzle is reversible, what is the temperature of the exiting gas? Assume the gas has a constant heat capacity of $C_P = 40$ J/mol · K and follows the equation of state:

FOOD FOR THOUGHT 6-10

Is this a realistic equation of state?

$$\underline{V} = \frac{RT}{P} + aTP \quad\quad \text{where } a = 10^{-11}\ \frac{m^3}{mol \cdot Pa \cdot K}$$

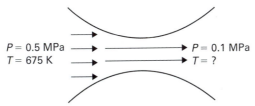

FIGURE 6-7 Nozzle modeled in Example 6-7.

SOLUTION:

Step 1 *Define a system and write an entropy balance*
The nozzle itself is the system. The inlet conditions and outlet pressure are identical to Example 3-4, but in that problem, the outlet temperature was also known, and we used an energy balance to find the unknown outlet velocity. Here, our first instinct might be to attempt to apply the energy balance, but the outlet temperature and outlet velocity are both unknown, so the energy balance would be one equation in two unknowns. Writing the entropy balance is more useful here because (1) the unknown velocity is not a factor in the entropy balance, and (2) we can directly apply the fact that the nozzle is well approximated as reversible.

For a steady-state, adiabatic, and reversible process, the entropy balance simplifies to

$$\underline{S}_{in} = \underline{S}_{out} \tag{6.104}$$

Step 2 *Write a total derivative expression relating molar entropy to measurable properties*
We know the inlet and outlet pressure, and the outlet temperature is what we're trying to find, so it's natural to relate $\underline{S}$ to P and T through the total derivative. Thus,

$$d\underline{S} = \left(\frac{\partial \underline{S}}{\partial P}\right)_T dP + \left(\frac{\partial \underline{S}}{\partial T}\right)_P dT \tag{6.105}$$

Step 3 *Convert Equation 6.105 into an expression that can be integrated*
The left-hand side is simply $d\underline{S}$, which could be integrated immediately. We can use one of Maxwell's relations (Equation 6.63) to re-express $(\partial \underline{S}/\partial P)_T$ as a function of measurable properties:

$$d\underline{S} = -\left(\frac{\partial \underline{V}}{\partial T}\right)_P dP + \left(\frac{\partial \underline{S}}{\partial T}\right)_P dT \tag{6.106}$$

While there was no clear way to evaluate $(\partial \underline{S}/\partial P)_T$, the expression $(\partial \underline{V}/\partial T)_P$ is readily evaluated using the known equation of state. Taking the derivative with respect to T (holding P constant) gives

$$\left(\frac{\partial \underline{V}}{\partial T}\right)_P = \frac{R}{P} + aP \tag{6.107}$$

And substituting this into Equation 6.106 gives us an expression that can be integrated as

$$d\underline{S} = -\left(\frac{R}{P} + aP\right)dP + \left(\frac{\partial \underline{S}}{\partial T}\right)_P dT \tag{6.108}$$

Thus, the $d\underline{S}$ and dP terms are now readily integrated. What of the dT term?

Step 4 *Relate $(\partial \underline{S}/\partial T)_P$ to the known heat capacity C_P*
While there is no simple way to evaluate $(\partial \underline{S}/\partial T)_P$ using the equation of state—as there was for $(\partial \underline{S}/\partial P)_T$—one can relate $(\partial \underline{S}/\partial T)_P$ to the heat capacity in a straightforward way. By definition, $C_P = (\partial \underline{H}/\partial T)_P$.

Here, we use the expansion rule with $X = H$, $Y = T$, and $Z = P$ (to make the left-hand side equal C_p), $K = \underline{S}$, and $L = P$ (which allows us to make use of the fundamental property relationship for $\underline{H}$).

Applying the expansion rule gives

$$\left(\frac{\partial \underline{H}}{\partial T}\right)_P = \left(\frac{\partial \underline{H}}{\partial \underline{S}}\right)_P \left(\frac{\partial \underline{S}}{\partial T}\right)_P + \left(\frac{\partial \underline{H}}{\partial P}\right)_{\underline{S}} \left(\frac{\partial P}{\partial T}\right)_P \quad (6.109)$$

The last term is 0 (Equation 6.54). We apply Equation 6.40 and the definition of heat capacity to give

$$\left(\frac{\partial \underline{H}}{\partial T}\right)_P = C_P = T\left(\frac{\partial \underline{S}}{\partial T}\right)_P \quad (6.110)$$

Applying this to Equation 6.108 gives

$$d\underline{S} = -\left(\frac{R}{P} + aP\right)dP + \frac{C_P}{T}dT \quad (6.111)$$

Step 5 *Integrate Equation 6.111*
Integrating from the state of the inlet gas to the state of the outlet gas gives

$$\int_{\underline{S}_{in}}^{\underline{S}_{out}} d\underline{S} = -\int_{P_{in}=0.5\,MPa}^{P_{out}=0.1\,MPa} \left(\frac{R}{P} + aP\right)dP + \int_{T_{in}=675\,K}^{T_{out}} \frac{C_P}{T}dT \quad (6.112)$$

$$\underline{S}_{out} - \underline{S}_{in} = -R\ln\left(\frac{P_{out}}{P_{in}}\right) - \frac{a}{2}(P_{out}^2 - P_{in}^2) + C_P\ln\left(\frac{T_{out}}{T_{in}}\right)$$

Step 6 *Substitute known values*
The entropy balance showed that $\underline{S}_{out} = \underline{S}_{in}$; thus, the left-hand side is 0. T_{out} is the only unknown on the right-hand side:

$$0 = -\left(8.314\,\frac{J}{mol \cdot K}\right)\ln\left(\frac{0.1\,MPa}{0.5\,MPa}\right) - \left(\frac{10^{-11}\,\frac{m^3}{Pa \cdot mol \cdot K}}{2}\right)(0.01 - 0.25) \times 10^{-12}\,Pa^2$$

$$+ \left(40\,\frac{J}{mol \cdot K}\right)\ln\left(\frac{T_{out}}{675\,K}\right)$$

$$0 = 13.38\,\frac{J}{mol \cdot K} + 1.20\,\frac{J}{mol \cdot K} + \left(40\,\frac{J}{mol \cdot K}\right)\ln\left(\frac{T_{out}}{675K}\right)$$

The result is $T_{out} = \mathbf{468.8\ K}$.

FOOD FOR THOUGHT 6-11

Would an irreversible nozzle produce higher or lower kinetic energy than a reversible one, for the same inlet stream and the same pressure drop?

Step 4 of Example 6-7 demonstrated that C_P, which is by definition a partial derivative of molar (or specific) enthalpy, also can be related to a partial derivative of molar (or specific) entropy:

$$\left(\frac{\partial \underline{S}}{\partial T}\right)_P = \frac{C_P}{T} \quad (6.113)$$

No assumptions were made in the derivation of Equation 6.113. Equations 6.113 and 6.114 are both completely general.

An analogous derivation can be used to arrive at the following equation for C_V:

$$\left(\frac{\partial \underline{S}}{\partial T}\right)_V = \frac{C_V}{T} \quad (6.114)$$

6.2.7 **Summary of Useful Partial Differential Expressions**

BASIC LAWS OF CALCULUS

The **total derivative** can be used to relate any INTENSIVE property of a pure, homogeneous phase to any other two INTENSIVE properties.

$$dX = \left(\frac{\partial X}{\partial Y}\right)_Z dY + \left(\frac{\partial X}{\partial Z}\right)_Y dZ$$

The **expansion rule** can be used to introduce two new properties (K and L) into a problem, helping one to express $(\partial X/\partial Y)_Z$ again as a function of other properties that are better known, or easier to find, in the problem at hand.

$$\left(\frac{\partial X}{\partial Y}\right)_Z = \left(\frac{\partial X}{\partial K}\right)_L \left(\frac{\partial K}{\partial Y}\right)_Z + \left(\frac{\partial X}{\partial L}\right)_K \left(\frac{\partial L}{\partial Y}\right)_Z$$

The **triple product rule** can be used to re-express $(\partial X/\partial Y)_Z$ in terms of other partial derivatives that relate X, Y, and Z.

$$\left(\frac{\partial X}{\partial Y}\right)_Z \left(\frac{\partial Y}{\partial Z}\right)_X \left(\frac{\partial Z}{\partial X}\right)_Y = -1$$

In modeling *extensive* properties of a pure substance, a three-term total derivative expression would be needed, because the quantity of mass is one additional degree of freedom beyond the two that come from intensive properties.

These **fundamental identities** are often helpful in simplifying partial differential equations:

$$\left(\frac{\partial X}{\partial Y}\right)_Z = \frac{1}{\left(\dfrac{\partial Y}{\partial X}\right)_Z}$$

$$\left(\frac{\partial X}{\partial X}\right)_Y = 1$$

$$\left(\frac{\partial X}{\partial Y}\right)_X = 0$$

FUNDAMENTAL PROPERTY RELATIONSHIPS

The molar internal energy, enthalpy, Helmholtz energy, and Gibbs free energy can be related to other intensive properties of any PURE compound through the fundamental property relations.

$$d\underline{U} = T\,d\underline{S} - P\,d\underline{V}$$

$$d\underline{H} = T\,d\underline{S} + \underline{V}\,dP$$

$$d\underline{A} = -P\,d\underline{V} - \underline{S}\,dT$$

$$d\underline{G} = \underline{V}\,dP - \underline{S}\,dT$$

DEFINITIONS

The **constant-pressure heat capacity** is defined as

When the heat capacity is valid at ideal gas conditions specifically, it is denoted with an asterisk, C_P^* or C_V^*.

$$C_P = \left(\frac{\partial \underline{H}}{\partial T}\right)_P = T\left(\frac{\partial \underline{S}}{\partial T}\right)_P$$

The **constant-volume heat capacity** is defined as

$$C_V = \left(\frac{\partial \underline{U}}{\partial T}\right)_{\underline{V}} = T\left(\frac{\partial \underline{S}}{\partial T}\right)_{\underline{V}}$$

The **Joule-Thomson coefficient** is defined as

$$\mu_{JT} = \left(\frac{\partial T}{\partial P}\right)_{\underline{H}} = -\frac{\underline{V} - T\left(\frac{\partial \underline{V}}{\partial T}\right)_P}{C_P}$$

The **coefficient of thermal expansion** is defined as

$$\alpha_V = \frac{1}{\underline{V}}\left(\frac{\partial \underline{V}}{\partial T}\right)_P$$

The **isothermal compressibility** is defined as

$$\kappa_T = -\frac{1}{\underline{V}}\left(\frac{\partial \underline{V}}{\partial P}\right)_T$$

MAXWELL'S EQUATIONS

These are often helpful in eliminating molar entropy ($\underline{S}$) from a problem, leaving it completely in terms of measurable properties ($P, \underline{V}, T$).

$$\left(\frac{\partial T}{\partial \underline{V}}\right)_{\underline{S}} = -\left(\frac{\partial P}{\partial \underline{S}}\right)_{\underline{V}}$$

$$\left(\frac{\partial T}{\partial P}\right)_{\underline{S}} = \left(\frac{\partial \underline{V}}{\partial \underline{S}}\right)_P$$

$$\left(\frac{\partial P}{\partial T}\right)_{\underline{V}} = \left(\frac{\partial \underline{S}}{\partial \underline{V}}\right)_T$$

$$\left(\frac{\partial \underline{V}}{\partial T}\right)_P = -\left(\frac{\partial \underline{S}}{\partial P}\right)_T$$

6.3 Heat Capacity and Residual Properties

Examples 6-2 through 6-7 illustrate that we can solve a lot of practical problems in the absence of data if we have two things: an equation of state that provides a good estimate of the properties of the substance, and an accurate measure of the heat capacity. This section takes a closer look at the heat capacity, specifically noting the following two things.

1. Heat capacity can itself be modeled using an equation of state.
2. Commonly, the heat capacity can be considered known at ideal gas conditions but not at higher pressures. We will show how "real gas" problems can be solved from this information.

6.3.1 Distinction between Real and Ideal Gas Heat Capacity

Section 2.3.2 introduced the heat capacity, and described how the constant-pressure heat capacity (C_P) can be measured over a range of temperatures from a single experiment. At this point we will consider the relationship between heat capacity and pressure, starting with an example that examines steam.

| ESTIMATING $\hat{C}_P$ OF STEAM FROM STEAM TABLES | EXAMPLE 6-8 |

Estimate the constant pressure heat capacity ($\hat{C}_P$) of steam at the various temperatures and pressures shown in Table 6-1, using only the data shown in the steam tables.

SOLUTION:

Step 1 *Compare data to definition of heat capacity*

By definition, $C_P = (\partial H/\partial T)_P$, or in this case, because the steam tables are on a mass basis, $\hat{C}_P = (\partial \hat{H}/\partial T)_P$. Taking $P = 0.01$ MPa and $T = 350°C$ as an example, we note that the steam tables contain data for $\hat{H}$ at $P = 0.01$ MPa and temperatures of $T = 300°C$ and $T = 400°C$. Calculating $\Delta\hat{H}/\Delta T$ between these points gives us a reasonable estimate of $(\partial \hat{H}/\partial T)_P$, since we are measuring the change in specific enthalpy with respect to the change in temperature, while holding the pressure constant at $P = 0.01$ MPa (in which we are interested). We would expect the estimate to become more accurate if based on data points that were closer to 350°C (e.g., 340°C and 360°C), but that data is not available in these steam tables.

Step 2 *Look up data*

$$\hat{C}_P \approx \frac{\Delta\hat{H}}{\Delta T} = \frac{\hat{H}_{400} - \hat{H}_{300}}{400 - 300} = \frac{3279.9 - 3076.7 \ \dfrac{\text{kJ}}{\text{kg}}}{400 - 300} = 2.032 \ \frac{\text{kJ}}{\text{kg K}} \qquad (6.115)$$

The remaining values shown in Table 6-1 were estimated using the same process.

TABLE 6-1 $\hat{C}_P$ for steam at various temperatures and pressures, as estimated using steam tables. All values are in kJ/kg · K.

	P = 0.01 MPa	P = 0.05 MPa	P = 1 MPa	P = 5 MPa
T = 350°C	2.032	2.036	2.128	2.710
T = 550°C	2.166	2.167	2.195	2.322
T = 750°C	2.307	2.308	2.320	2.374

The definition of a derivative $df(x)/dx$ is:

$$\lim_{\Delta x \to 0} \frac{f(x + \Delta x) - f(x)}{\Delta x}$$

In this numerical estimate of $(\partial \hat{H}/\partial T)_P$, you can think of 300°C as x and 100°C as Δx, which means $x + \Delta x = 400°C$. The specific enthalpy at 300°C, 3076.7 kJ/kg, is $f(x)$, and the specific enthalpy at 400°C, 3279.9 kJ/kg, is $f(x + \Delta x)$.

Plotting all of the data for $P = 0.01$ MPa and measuring the slope of the curve at $T = 350°C$ is a more time-consuming, but more rigorous, method of estimating $(\partial \hat{H}/\partial T)_P$.

Example 6-8 illustrates the estimation of heat capacity from discrete data points. The values in Table 6-1 demonstrate that the heat capacity of steam is a function of temperature; the relationship between C_P and T has also been explored previously (see Example 2-5). The table proves, however, that C_P is also dependent upon pressure.

Recall that real gases behave like ideal gases at low pressures, as first discussed in Section 2.3.3. A critical observation from Table 6-1 is that the estimates for $P = 0.01$ MPa and $P = 0.05$ MPa (low pressures) are practically identical, while the data for 1 MPa and 5 MPa (high pressures) are noticeably different.

Table 6-1 illustrates some facts that have been mentioned previously:

- For *ideal* gases, enthalpy is only a function of temperature, and consequently heat capacity is also only a function of temperature.
- For *real* gases, heat capacity depends upon both temperature and pressure.

Section 2.3.2 outlined how C_P could be measured experimentally in a constant-pressure vessel. By adding heat at a constant rate and monitoring the temperature of the material, one can measure C_P over a large range of temperatures in a single experiment. However, in general, the measured values of C_P would only be valid at the specific pressure for which the experiment was conducted.

At low pressures, however, real gases behave like ideal gases; C_P does not depend upon pressure. Consequently, if we measure C_P versus T at a low pressure, we can apply this value of C_P to any other low pressure. Thus, the numbers listed in the first two columns ($P = 0.01$ MPa and $P = 0.05$ MPa) of Table 6-1 can be considered estimates of the heat capacity of steam at ideal gas conditions. We have previously defined the ideal gas heat capacity (C_P^*) of a substance as the heat capacity under conditions at which the substance can be reasonably modeled as an ideal gas.

When one is solving problems or designing processes that involve high pressures, it is relatively common that an ideal gas heat capacity is known, but a value of the heat capacity at the exact pressure of interest is not available. Indeed, Appendix D of this book shows the ideal gas heat capacity for a number of common chemical compounds, as a function of temperature. The next sections explore how we can make use of a known ideal gas heat capacity, even at elevated pressures.

6.3.2 Motivation for the Residual Property

By now a turbine is a unit operation that is recognizable as a familiar and important physical system. Example 6-9 examines a turbine, and uses the common scenario that was posed in the previous section: the heat capacity is known only for ideal gas conditions. The example follows the problem-solving strategy outlined in Section 6-2.

EXAMPLE 6-9	**EXAMPLE IN WHICH C_P^* IS KNOWN**

Methane enters a turbine (Figure 6-8) at $T = 600$ K and $P = 1$ MPa and leaves at $T = 400$ K and $P = 0.2$ MPa. How much work is produced for each mole of gas? Use the following data: van der Waals constants for methane are $a = 0.2303$ Pa-m^6/mol^2 and $b = 4.306 \times 10^{-5}$ m^3/mol. C_P^* for methane is given in Appendix D.

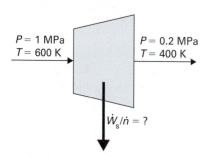

$P = 1$ MPa
$T = 600$ K

$P = 0.2$ MPa
$T = 400$ K

$\dot{W}_s/\dot{n} = ?$

FIGURE 6-8 Turbine described in Example 6-9.

SOLUTION:

Step 1 *Define a system and write an energy balance*
Lacking any contrary information, we apply the usual assumptions we've used for turbines (steady-state, adiabatic, negligible kinetic and potential energy), arriving at the equation given in Table 3-1:

$$\frac{\dot{W}_s}{\dot{m}} = \hat{H}_{out} - \hat{H}_{in}$$

For this example, however, it is more convenient to express the energy balance on a molar basis. If we call the state of the entering gas "1" and the exiting gas "2," the energy balance becomes

$$\frac{\dot{W}_S}{\dot{n}} = \underline{H}_2 - \underline{H}_1 \tag{6.116}$$

Step 2 *Write a total derivative expression for d$\underline{H}$*
The left-hand side of Equation 6.116 is exactly what we are trying to find—work produced per mole of gas—so solving the problem requires evaluating $\underline{H}_2 - \underline{H}_1$. The normal way to do this is to integrate $d\underline{H}$:

$$\int_{T_1, P_1}^{T_2, P_2} d\underline{H} = \underline{H}_2 - \underline{H}_1 \tag{6.117}$$

So as we've done throughout this chapter, we look to use a partial derivative expression to relate what we want to know ($d\underline{H}$) to the two intensive properties we know the most about:

$$d\underline{H} = \left(\frac{\partial \underline{H}}{\partial T}\right)_P dT + \left(\frac{\partial \underline{H}}{\partial P}\right)_T dP \tag{6.118}$$

Or, applying the definition of heat capacity:

$$d\underline{H} = C_P dT + \left(\frac{\partial \underline{H}}{\partial P}\right)_T dP \tag{6.119}$$

Step 3 *Relate $(\partial \underline{H}/\partial P)_T$ to measurable properties*
Applying the expansion rule with $X = \underline{H}, Y = P, Z = T, K = \underline{S}$, and $L = P$:

$$\left(\frac{\partial \underline{H}}{\partial P}\right)_T = \left(\frac{\partial \underline{H}}{\partial \underline{S}}\right)_P \left(\frac{\partial \underline{S}}{\partial P}\right)_T + \left(\frac{\partial \underline{H}}{\partial P}\right)_{\underline{S}} \left(\frac{\partial P}{\partial P}\right)_T \tag{6.120}$$

> *X, Y,* and *Z* are assigned to transform the left-hand side into the very partial derivative we are trying to evaluate. *K* and *L* are chosen because the fundamental property relationship expresses $\underline{H}$ as a function of $\underline{S}$ and *P*, so we introduce $\underline{S}$ and *P* into the problem.

Introducing Equations 6.40 and 6.41 into Equation 6.120 gives

$$\left(\frac{\partial \underline{H}}{\partial P}\right)_T = T\left(\frac{\partial \underline{S}}{\partial P}\right)_T + \underline{V}\left(\frac{\partial P}{\partial P}\right)_T \tag{6.121}$$

Entropy can be eliminated from the right-hand side by applying one of Maxwell's equations (6.63) and $(\partial P/\partial P)_T$ is simply equal to 1 (Equation 6.53):

$$\left(\frac{\partial \underline{H}}{\partial P}\right)_T = -T\left(\frac{\partial \underline{V}}{\partial T}\right)_P + \underline{V} \tag{6.122}$$

Thus, Equation 6.121 becomes

$$d\underline{H} = C_P dT + \left[\underline{V} - T\left(\frac{\partial \underline{V}}{\partial T}\right)_P\right] dP \tag{6.123}$$

> We will use Equation 6.123 in developing the *enthalpy residual function*, and will complete the example (see Example 6-9 revisited) using residual functions.

We assumed nothing in deriving Equation 6.123; it is a valid relationship among $\underline{H}, T$, and P for any process involving a pure compound. In Example 6-9, however, Equation 6.123 illustrates the barrier we face in solving the problem; the quantity $C_P dT$ cannot be integrated unless C_P is known as a function of temperature. It is indeed known for the ideal gas state, but assuming that methane behaves like an ideal gas is questionable at $P = 1$ MPa. A **residual property**, or **residual**, is a quantitative measure of the difference between the real material and an ideal gas at a given temperature and pressure. Residual properties would apply to Example 6-9 in that we can use the ideal gas heat capacity to calculate what the change in enthalpy *would* be *if* methane were an ideal gas at these conditions, and then use residual properties to quantify the difference between the real methane and an ideal gas.

Residual properties and their use are introduced in Section 6.3.3. First, we note that the previous discussion is intended to show a motivation for residual properties, but it shouldn't be misunderstood. We are not claiming Equation 6.123 is impossible to solve without them. Indeed, two other solution strategies are possible using what we've learned up to this point; you might try to identify at least one of them yourself before reading on. A brief consideration of these alternative solution methods will further illustrate why residual properties are useful.

> What we call "residual properties" are, in some books, called "departure functions," because they measure how much a real substance departs from ideal gas behavior.

One Possible Solution Strategy   We saw in Section 6.2 that, by applying principles of partial differential calculus, we can derive relationships between a host of intensive properties, such as U, H, S, P, V, and T. So it seems logical to look for a relationship between C_P and P and, indeed, it can be proved that

$$\left(\frac{\partial C_P}{\partial P}\right)_T = -T\left(\frac{\partial^2 V}{\partial T^2}\right)_P \tag{6.124}$$

Since the right-hand side is entirely a function of measurable properties, it can be evaluated using an equation of state. We could estimate the heat capacity of methane at 600 K and 1 MPa by finding the ideal gas heat capacity at 600 K, assuming this is valid at some low pressure (e.g., $P = 0.01$ MPa) and then use Equation 6.124 to quantify the change in C_P as the pressure is increased from $P = 0.01$ MPa to $P = 1$ MPa at constant temperature.

> The solution strategy described here sounds cumbersome, but we don't avoid it because it's difficult. We avoid it because there's a valid mathematical reason to expect residual functions will provide a more accurate answer.

While logical, this strategy has the drawback that Equation 6.124 contains a second derivative. One must always keep in mind that equations of state are *models* that provide estimates of real physical properties. While an equation of state might describe the *P-V-T* relationships for a compound quite accurately, uncertainties and errors tend to get magnified by operations such as differentiation, as illustrated in Figure 6-9 and Figure 6-10. Let us imagine that function A (solid line) in Figure 6-9 represents real data and function B (dashed line) represents a model that is used to estimate the data. At a glance, we would probably say the agreement is excellent, and

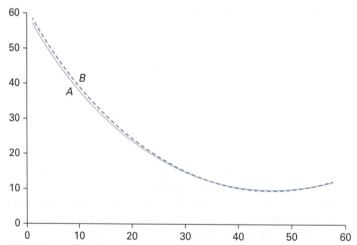

FIGURE 6-9 Two functions are shown. For any given value of x, the values of y for the two functions differ by less than 3% in magnitude.

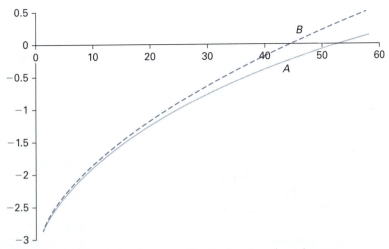

FIGURE 6-10 The derivative curves for the two functions shown in Figure 6-9.

a close inspection of the numerical values reveals that the "model value" of y (curve B) is always within 3% of the "real y" (curve A) for any given x. However, if we look at the slopes of curves A and B, we find values of dy/dx become noticeably different as x increases. Figure 6-10 shows that between $x = 46$ and $x = 52$ the derivatives are actually different in sign. Thus, the model fits "y" much more accurately than it fits "dy/dx." The uncertainty of the model would be compounded further if we took a second derivative with respect to x.

Residual properties, which are introduced in Section 6.3.3, allow us to solve problems like Example 6-9 and require only taking the first derivatives of equations of state—never a second derivative.

A Second Possible Strategy for Solving Example 6-9 Recognizing that $\underline{H}$ is a state property, you might have proposed the idea of constructing a hypothetical path for the process in which the temperature only changes at low pressure. For example, consider the path shown in Figure 6-11.

■ Steps 1 and 3 are isothermal, so the dT term in Equation 6.123 is 0, regardless of the value of C_P.

■ Step 2 is carried out at a low pressure of $P = 0.01$ MPa. At these conditions, the known ideal gas C_P^* can be assumed valid. Because it's an isobaric step, the dP term in Equation 6.123 is 0 here.

Equation 6.123 is a general relationship among $\underline{H}$, T, and P; we can apply it individually to find the change in molar enthalpy for each of the three steps in Figure 6-11.

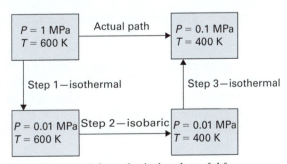

FIGURE 6-11 A hypothetical path useful for analyzing the process in Example 6-9.

This solution method is logical, and we will find that it is actually very similar to solving the problem using residual properties. However, Figure 6-11 is specific to the process in Example 6-9. Residual properties place this approach into a general framework that can be applied to any number of problems.

6.3.3 Definition of Residual Properties

The concept of a state property was first introduced in Section 2.2.4, and the Gibbs phase rule was introduced in Section 2.2.3.

Recall that a *state property* is dependent only upon the current state of the system and (as shown with the Gibbs phase rule) a pure, homogeneous phase has two degrees of freedom. If we specify two intensive properties, such as temperature and pressure, then all of the others—such as $\underline{H}$, $\underline{U}$, and $\underline{S}$—have a unique value that is always valid for a compound at that particular T and P.

Section 2.2.3 introduced the ideal gas model, and we have since established methods of quantifying how these same state properties ($\underline{H}$, $\underline{U}$, $\underline{S}$, P, T, and $\underline{V}$) are interrelated for an ideal gas. While the ideal gas model only describes real gases at low pressures, the equations can be applied at any conditions. Thus, even though the real world contains no such thing as an ideal gas at $P = 5$ MPa and $T = 500$ K, we can use the ideal gas model to evaluate what $\underline{H}$, $\underline{U}$, $\underline{S}$, or $\underline{V}$ *would* be *if* an ideal gas existed at those conditions. We denote such hypothetical ideal gas properties as $\underline{H}^{ig}$, $\underline{U}^{ig}$, $\underline{S}^{ig}$, and $\underline{V}^{ig}$.

This discussion focuses on $\underline{H}$, $\underline{U}$, and $\underline{S}$, because those are the properties that were important in the problems we've solved up to this point. In Chapter 8, the molar Gibbs free energy residual function $\underline{G} - \underline{G}^{ig}$ will have great importance.

> A **residual property** is defined as the difference between a real property (e.g., $\underline{H}$) for the real compound and the property for a hypothetical ideal gas that exists at the same temperature and pressure (e.g., $\underline{H}^{ig}$).

Thus, the residual property for molar enthalpy is written $\underline{H} - \underline{H}^{ig}$. It is termed a "residual" because it measures what remains after the value of the property in the ideal gas state is subtracted.

Example 6-9 in the previous section examined a turbine, and solution of the problem required us to find the difference in molar enthalpies of the inlet ($\underline{H}_1$) and outlet ($\underline{H}_2$) streams.

Like the "hypothetical reversible turbine" first introduced in Example 4-7, $\underline{H}_1^{ig}$ and $\underline{H}_2^{ig}$ are theoretical constructs that don't exist in the real world, but are useful in solving problems.

> Changes in molar enthalpy can be expressed, using residual properties, as
>
> $$\underline{H}_2 - \underline{H}_1 = (\underline{H}_2 - \underline{H}_2^{ig}) + (\underline{H}_2^{ig} - \underline{H}_1^{ig}) - (\underline{H}_1 - \underline{H}_1^{ig}) \qquad (6.125)$$
>
> In this expression:
>
> - $\underline{H}_2$ and $\underline{H}_1$ represent the molar enthalpy of a substance at two different states.
> - $\underline{H}_2^{ig}$ represents the molar enthalpy of a hypothetical ideal gas at state 2.
> - $\underline{H}_1^{ig}$ represents the molar enthalpy of a hypothetical ideal gas at state 1.
>
> An analogous expression can be written for any other intensive state property.

PITFALL PREVENTION

When using residual properties, it is easy to accidentally flip the sign of one of the terms. One way to avoid sign errors is to verify that the ideal gas properties cancel out, as they do on the right-hand side of Equation 6.125.

We can find the change in molar enthalpy for *any* process involving a pure compound using Equation 6.125. Practically speaking, Equation 6.125 is helpful when the heat capacity is unknown at the specific conditions of interest but the ideal gas heat capacity C_P^* is available, because the expression $\underline{H}_2^{ig} - \underline{H}_1^{ig}$ can be evaluated using C_P^*. Consider the application of Equation 6.125 to Example 6-9, which is also illustrated in Figure 6-12.

- Define states 1 and 2 to represent the gas entering the turbine (1) and leaving the turbine (2).

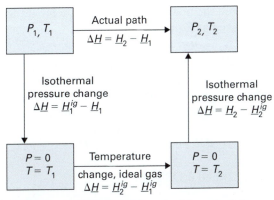

FIGURE 6-12 Summary of method for calculating change in enthalpy using the residual enthalpy.

- $H_2 - H_1$ represents the change of molar enthalpy for the gas as it travels through the turbine.

- $H_2^{ig} - H_1^{ig}$ represents the change in molar enthalpy for a hypothetical ideal gas experiencing the same change in temperature and pressure as the real gas.

- $H_2 - H_2^{ig}$ represents the residual molar enthalpy at state 2: How much the gas leaving the turbine departs from ideal gas behavior. Analogously, $H_1 - H_1^{ig}$ represents the residual molar enthalpy for the gas entering the turbine.

- Notice that in Equation 6.125 the left and right-hand sides both equal $H_2 - H_1$, as the properties of the hypothetical ideal gas cancel out on the right-hand side.

A comparable approach can be applied to molar entropy:

$$S_2 - S_1 = (S_2 - S_2^{ig}) + (S_2^{ig} - S_1^{ig}) - (S_1 - S_1^{ig}) \tag{6.126}$$

And analogous equations can be written for molar internal energy, molar Gibbs free energy, etc. In order to apply such an approach to solving problems, we must have a way to relate residual property expressions like $H - H^{ig}$ or $U - U^{ig}$ to measurable properties.

6.3.4 Mathematical Expressions for Residual Properties

How can we put a quantitative value on a residual property at a particular temperature and pressure? Recall that all gases approach ideal gas behavior as pressure decreases. In the limiting case of $P = 0$, all gases will be ideal gases. This in turn means that at this state the real gas properties are identical to the ideal gas properties; at $P = 0$, *all residual properties will be 0.*

We can use the $P = 0$ condition as a starting point in computing a residual property. For example, consider the state $T = 600$ K and $P = 1$ MPa found in Example 6-9. The value of the residual molar enthalpy is unknown at this T and P, but we know it is 0 at $T = 600$ K and $P = 0$.

$$H - H^{ig} = 0 \text{ at } T = 600 \text{ K}, P = 0$$

We also know that the ideal gas molar enthalpy is independent of pressure. Thus, as we isothermally increase the pressure, H changes, but H^{ig} remains constant. Consequently, the residual molar enthalpy is equal to the change in molar enthalpy for the gas as it is isothermally compressed from $P = 0$ MPa to $P = 1$ MPa. Thus,

$$(H - H^{ig})_{T=600\text{K}, P=1\text{ MPa}} = \int_{T=600\text{K}, P=0\text{ MPa}}^{T=600\text{K}, P=1\text{ MPa}} dH \tag{6.127}$$

Talking about a gas at $P = 0$ (e.g., methane at $P = 0$) may seem unrealistic, because for the pressure in a container to be literally 0, there would be no molecules. Mathematically, however, there is nothing wrong with modeling a gas at the limiting condition of $P = 0$.

In Example 6-9, we explored the relationship between molar enthalpy and temperature and pressure and arrived at Equation 6.123:

$$dH = C_p \, dT + \left[\underline{V} - T\left(\frac{\partial \underline{V}}{\partial T}\right)_P \right] dP$$

In assessing residual properties, we are considering a change in pressure while holding temperature constant, so Equation 6.123 simplifies to

$$d\underline{H} = \left[\underline{V} - T\left(\frac{\partial \underline{V}}{\partial T}\right)_P \right] dP \tag{6.128}$$

> Equation 6.128 is valid for any *isothermal* process.

Combining Equations 6.127 and 6.128 gives

$$(\underline{H} - \underline{H}^{ig})_{T=600\,\text{K},\,P=1\,\text{MPa}} = \int_{T=600\,\text{K},\,P=0\,\text{MPa}}^{T=600\,\text{K},\,P=1\,\text{MPa}} \left[\underline{V} - T\left(\frac{\partial \underline{V}}{\partial T}\right)_P \right] dP \tag{6.129}$$

Equation 6.129 can be integrated if an equation of state describing the gas is available. While Equation 6.129 is specific to the condition $T = 600$ K and $P = 1$ MPa, the logic behind the derivation is applicable to any temperature and pressure. Thus,

$$\underline{H} - \underline{H}^{ig} = \int_{T=T,\,P=0}^{T=T,\,P=P} \left[\underline{V} - T\left(\frac{\partial \underline{V}}{\partial T}\right)_P \right] dP \tag{6.130}$$

It is significant to note that the integral in Equation 6.130 is carried out along an isothermal path. This is a common feature of all residual properties; in evaluating this integral, T can be treated as a constant.

The residual molar entropy is slightly more complicated to derive, because $\underline{S}$ is a function of both temperature and pressure—even for an ideal gas. It is, however, still true that the residual is 0 at any temperature (T) when $P = 0$. Thus, the residual molar entropy at a higher pressure can be found by quantifying the changes in both $\underline{S}$ and $\underline{S}^{ig}$ as pressure increases.

$$\underline{S} - \underline{S}^{ig} = \int_{T=T,\,P=0}^{T=T,\,P=P} d\underline{S} - \int_{T=T,\,P=0}^{T=T,\,P=P} d\underline{S}^{ig} \tag{6.131}$$

Example 6-7 explored the relationship among $\underline{S}$, T, and P, leading to Equation 6.106. Thus,

$$d\underline{S} = -\left(\frac{\partial \underline{V}}{\partial T}\right)_P dP + \left(\frac{\partial \underline{S}}{\partial T}\right)_P dT$$

which, for an isothermal process, simplifies to

$$d\underline{S} = -\left(\frac{\partial \underline{V}}{\partial T}\right)_P dP \tag{6.132}$$

> As for enthalpy, the residual entropy is calculated on an isothermal path. The path begins at $P = 0$, because real and ideal gas properties are identical at that point, but there is no reason to vary from the actual temperature.

Because the integrals in Equation 6.131 are carried out on a constant-temperature path, we can combine Equations 6.131 and 6.132 for

$$\underline{S} - \underline{S}^{ig} = \int_{T=T,\,P=0}^{T=T,\,P=P} -\left(\frac{\partial \underline{V}}{\partial T}\right)_P dP - \int_{T=T,\,P=0}^{T=T,\,P=P} -\left(\frac{\partial \underline{V}^{ig}}{\partial T}\right)_P dP \tag{6.133}$$

The first integral requires an equation of state that describes the real gas. The second, however, can be evaluated further. For an ideal gas,

$$\underline{V}^{ig} = \frac{RT}{P}$$

and

$$\left(\frac{\partial \underline{V}^{ig}}{\partial T}\right)_P = \frac{R}{P}$$

This result can be inserted into Equation 6.133. Since both integrals are carried out with respect to dP and between the same limits, they can be combined as

$$\underline{S} - \underline{S}^{ig} = -\int_{T=T,P=0}^{T=T,P=P} \left[\left(\frac{\partial \underline{V}}{\partial T}\right)_P - \frac{R}{P}\right] dP \qquad (6.134)$$

The residual molar internal energy could be derived in an analogous manner, but it is instructive instead to recall the relationship between $\underline{U}$ and $\underline{H}$. By definition,

$$\underline{H} = \underline{U} + P\underline{V}$$

which means that

$$\underline{H} - \underline{H}^{ig} = (\underline{U} + P\underline{V}) - (\underline{U}^{ig} + P\underline{V}^{ig}) \qquad (6.135)$$

Rearranging, and recognizing that for an ideal gas $P\underline{V} = RT$:

$$\underline{H} - \underline{H}^{ig} = (\underline{U} - \underline{U}^{ig}) + (P\underline{V} - RT) \qquad (6.136)$$

Solving for $\underline{U} - \underline{U}^{ig}$ and also inserting the definition that $Z = P\underline{V}/RT$ gives

$$\underline{U} - \underline{U}^{ig} = RT(1 - Z) + (\underline{H} - \underline{H}^{ig})$$

$$\underline{U} - \underline{U}^{ig} = RT(1 - Z) + \int_{T=T,P=0}^{T=T,P=P} \left[\underline{V} - T\left(\frac{\partial \underline{V}}{\partial T}\right)_P\right] dP \qquad (6.137)$$

The compressibility Z was first defined in Section 2.3.4.

Alternative expressions of the residual properties can be derived by noting that, when $P = 0$, the molar volume is infinite. Thus, $\underline{V} = \infty$, like $P = 0$, is a condition at which all gases will show ideal gas behavior, and all residual properties are 0. When the integral is carried out with respect to $d\underline{V}$, the residual molar enthalpy becomes

$$\underline{H} - \underline{H}^{ig} = RT(Z - 1) + \int_{T=T,V=\infty}^{T=T,V=V} \left[T\left(\frac{\partial P}{\partial T}\right)_V - P\right] d\underline{V} \qquad (6.138)$$

Equations 6.130 and 6.138 are two different expressions for the residual molar enthalpy. They are both completely general, so it is never "right" or "wrong" to use either; we simply use whichever is more convenient for the case at hand. Note that Equation 6.130 involves a partial derivative of $\underline{V}$ and Equation 6.138 includes a partial derivative of P, whichever you use is guided by whether the equation of state being employed is more easily solved for P or $\underline{V}$.

6.3.5 Summary of Residual Properties

We close this section by summarizing the residual property expressions for molar internal energy, enthalpy, and entropy. We also, at this point, introduce commonly used shorthand: symbols of the form $\underline{M}^R$ are often used to represent residual properties of the form $\underline{M} - \underline{M}^{ig}$.

USAGE

The change in any intensive property $\underline{M}$ between states 1 and 2 can be quantified using residual properties as

$$\underline{M}_2 - \underline{M}_1 = \left(\underline{M}_2 - \underline{M}_2^{ig}\right) + \left(\underline{M}_2^{ig} - \underline{M}_1^{ig}\right) - \left(\underline{M}_1 - \underline{M}_1^{ig}\right)$$

or

$$\underline{M}_2 - \underline{M}_1 = \underline{M}_2^R + \left(\underline{M}_2^{ig} - \underline{M}_1^{ig}\right) - \underline{M}_1^R \tag{6.139}$$

INTEGRATING WITH RESPECT TO dP

$$\underline{U} - \underline{U}^{ig} = \underline{U}^R = RT(1 - Z) + \int_{T=T, P=0}^{T=T, P=P} \left[\underline{V} - T\left(\frac{\partial \underline{V}}{\partial T}\right)_P\right] dP \tag{6.140}$$

$$\underline{H} - \underline{H}^{ig} = \underline{H}^R = \int_{T=T, P=0}^{T=T, P=P} \left[\underline{V} - T\left(\frac{\partial \underline{V}}{\partial T}\right)_P\right] dP \tag{6.141}$$

$$\underline{S} - \underline{S}^{ig} = \underline{S}^R = -\int_{T=T, P=0}^{T=T, P=P} \left[\left(\frac{\partial \underline{V}}{\partial T}\right)_P - \frac{R}{P}\right] dP \tag{6.142}$$

INTEGRATING WITH RESPECT TO dV

$$\underline{U} - \underline{U}^{ig} = \underline{U}^R = \int_{T=T, \underline{V}=\infty}^{T=T, \underline{V}=\underline{V}} \left[T\left(\frac{\partial P}{\partial T}\right)_{\underline{V}} - P\right] d\underline{V} \tag{6.143}$$

$$\underline{H} - \underline{H}^{ig} = \underline{H}^R = RT(Z - 1) + \int_{T=T, \underline{V}=\infty}^{T=T, \underline{V}=\underline{V}} \left[T\left(\frac{\partial P}{\partial T}\right)_{\underline{V}} - P\right] d\underline{V} \tag{6.144}$$

$$\underline{S} - \underline{S}^{ig} = \underline{S}^R = R \ln(Z) + \int_{T=T, \underline{V}=\infty}^{T=T, \underline{V}=\underline{V}} \left[\left(\frac{\partial P}{\partial T}\right)_{\underline{V}} - \frac{R}{\underline{V}}\right] d\underline{V} \tag{6.145}$$

6.3.6 Application of Residual Properties

We begin this section by returning to Example 6-9, a turbine example posed in Section 6.3.2. It was used to illustrate the barriers that arise when C_P is known for the ideal gas case but not in general and, thus, motivate residual properties. Now that we have mathematical expressions for the residual molar enthalpy, we use them to complete the problem.

EXAMPLE 6-9	REVISITED: APPLYING RESIDUAL PROPERTIES

The van der Waals equation of state was first introduced in Section 2.3.4.

Methane enters a turbine at $T = 600$ K and $P = 1$ MPa and leaves at $T = 400$ K and $P = 0.2$ MPa. How much work is produced for each mole of gas? Use the following data: van der Waals constants for methane are $a = 0.2303$ Pa-m^6/mol^2 and $b = 4.306 \times 10^{-5}$ m^3/mol, and C_P^* for methane is given in Appendix D.

SOLUTION:

Step 1 *Define a system and write an energy balance*
Our previous inspection of this turbine process resulted in the energy balance:

$$\frac{\dot{W}_S}{\dot{n}} = \underline{H}_2 - \underline{H}_1 \tag{6.146}$$

Previously we looked at using a total derivative as a starting point for finding the change in molar enthalpy, and in the process established a motivation for residual properties. Now we will solve the problem using the residual molar enthalpy.

Step 2 *Apply residual molar enthalpy*
Equation 6.139 can always be used as a starting point when applying residual properties. For enthalpy we have

$$\underline{H}_2 - \underline{H}_1 = \underline{H}_2^R + \left(\underline{H}_2^{ig} - \underline{H}_1^{ig}\right) - \underline{H}_1^R$$

Step 3 *Find the ideal gas change in molar enthalpy*

Section 2.3.3 stated that for an ideal gas, $d\underline{H} = C_P^* \, dT$, and gave a physical rationalization for this equality. The equality will now be demonstrated mathematically. We can begin by writing the total derivative for enthalpy in terms of T and P for

$$d\underline{H} = \left(\frac{\partial \underline{H}}{\partial T}\right)_P dT + \left(\frac{\partial \underline{H}}{\partial P}\right)_T dP \tag{6.147}$$

In Section 6-2, we derived Equation 6.64:

$$\left(\frac{\partial \underline{H}}{\partial P}\right)_T = \underline{V} - T\left(\frac{\partial \underline{V}}{\partial T}\right)_P$$

This, along with the definition for C_P, can be introduced into Equation 6.147:

$$d\underline{H} = C_P \, dT + \left[\underline{V} - T\left(\frac{\partial \underline{V}}{\partial T}\right)_P\right] dP \tag{6.148}$$

For an ideal gas, $(\partial \underline{V}/\partial T)_P$ is determined by

$$\underline{V} = \frac{RT}{P}$$

$$\left(\frac{\partial \underline{V}}{\partial T}\right)_P = \frac{R}{P} \tag{6.149}$$

Introducing this into Equation 6.148 gives

$$d\underline{H} = C_P \, dT + \left[\underline{V} - T\left(\frac{R}{P}\right)\right] dP = C_P \, dT + 0 \, dP \tag{6.150}$$

$$d\underline{H} = C_P \, dT$$

Consequently, *for an ideal gas, $d\underline{H} = C_P^* \, dT$*, even if the pressure is not constant. According to Appendix D, the ideal gas heat capacity of methane is

$$\frac{C_P^*}{R} = 4.568 - (8.975 \times 10^{-3})T + (3.631 \times 10^{-5})T^2 \tag{6.151}$$

$$- (3.407 \times 10^{-8})T^3 + (1.091 \times 10^{-11})T^4$$

Here we've proved mathematically what was stated in Section 2.3.3; that molar enthalpy is only a function of temperature for an ideal gas.

Thus we know that if methane was an ideal gas at the conditions in this example, the change in molar enthalpy would be

$$\underline{H}_2^{ig} - \underline{H}_1^{ig} = \left(8.314 \, \frac{J}{mol \cdot K}\right) \int_{T=600\,K}^{T=400\,K} [4.568 - (8.975 \times 10^{-3})T \tag{6.152}$$

$$+ (3.631 \times 10^{-5})T^2 - (3.407 \times 10^{-8})T^3$$

$$+ (1.091 \times 10^{-11})T^4] \, dT$$

$$\underline{H}_2^{ig} - \underline{H}_1^{ig} = 37.98T - (0.07462)\frac{T^2}{2} + (3.019 \times 10^{-4})\frac{T^3}{3} - (2.833 \times 10^{-7})\frac{T^4}{4}$$

$$+ (3.019 \times 10^{-4})\frac{T^5}{5} \Big|_{T=600\,K}^{T=400\,K}$$

$$\underline{H}_2^{ig} - \underline{H}_1^{ig} = 14{,}035 \, \frac{J}{mol} - 23{,}324 \, \frac{J}{mol} = -9289 \, \frac{J}{mol}$$

Step 4 *Relate the residual molar enthalpy to temperature and pressure*

Equations 6.130 and 6.138 give expressions for the residual molar enthalpy. These are both completely general equations; we are free to use whichever one is convenient for the case at hand. Since the van der Waals equation gives P explicitly, and cannot be readily solved for $\underline{V}$, it is much easier to evaluate $(\partial P/\partial T)_{\underline{V}}$ than $(\partial \underline{V}/\partial T)_P$, so we will apply Equation 6.138

$$\underline{H}^R = RT(Z - 1) + \int_{T=T,\underline{V}=\infty}^{T=T,\underline{V}=\underline{V}} \left[T\left(\frac{\partial P}{\partial T}\right)_{\underline{V}} - P \right] d\underline{V}$$

The integral is with respect to $d\underline{V}$, so we need to relate everything to $\underline{V}$ in order to evaluate the integral. The van der Waals equation of state is

$$P = \frac{RT}{\underline{V} - b} - \frac{a}{\underline{V}^2} \tag{6.153}$$

Taking the derivative of both sides with respect to T at constant $\underline{V}$ gives

$$\left(\frac{\partial P}{\partial T}\right)_{\underline{V}} = \frac{R}{\underline{V} - b} \tag{6.154}$$

Because $\underline{V}$ is being held constant, the $a/\underline{V}^2$ term is a constant, and its derivative is zero. Substituting these expressions for P and $(\partial P/\partial T)_{\underline{V}}$ into the integral gives

$$\underline{H}^R = RT(Z - 1) + \int_{T=T,\underline{V}=\infty}^{T=T,\underline{V}=\underline{V}} \left[T\left(\frac{R}{\underline{V} - b}\right) - \left(\frac{RT}{\underline{V} - b} - \frac{a}{\underline{V}^2}\right) \right] d\underline{V}$$

$$\underline{H}^R = RT(Z - 1) + \int_{T=T,\underline{V}=\infty}^{T=T,\underline{V}=\underline{V}} \left[\frac{a}{\underline{V}^2} \right] d\underline{V} \tag{6.155}$$

which integrates to

$$\underline{H}^R = RT(Z - 1) + \frac{-a}{\underline{V}} \Big|_{\underline{V}=\infty}^{\underline{V}=\underline{V}} \tag{6.156}$$

Since $-a/\underline{V}$ is 0 at $\underline{V} = \infty$, the final expression is

$$\underline{H}^R = RT(Z - 1) - \frac{a}{\underline{V}} \tag{6.157}$$

Step 5 *Determine $\underline{V}_1$ and $\underline{V}_2$*

Equation 6.157 is an expression for the residual molar enthalpy for substances that are described by the van der Waals equation of state, valid at any set of conditions. To find a numerical value for this residual at the conditions of interest (P_1, T_1 and P_2, T_2), we need numerical values for $\underline{V}_1$ and $\underline{V}_2$. Conceptually, when we know two variables among P, $\underline{V}$, and T and we need the third, we should always look to apply an equation of state. In this case, finding a numerical value for $\underline{V}$ is somewhat complicated by the fact that the van der Waals equation of state is cubic in $\underline{V}$. Appendix B reviews methods of solving a cubic equation. In this case, the solutions are

For $P_1 = 1$ MPa and $T_1 = 600$ K, $\underline{V}_1 = 4.986 \times 10^{-3}$ m³/mol
For $P_2 = 0.2$ MPa and $T_2 = 400$ K, $\underline{V}_2 = 0.01659$ m³/mol

Cubic equations have three roots, but in this case, the other two solutions for the molar volume are complex numbers, which are physically meaningless solutions. Section 7.2.3 discusses the situation in which an equation of state has more than one real-number solution for volume.

The compressibility factor Z can now be found for both the inlet and outlet.

$$Z_1 = \frac{P_1 \underline{V}_1}{RT_1} = \frac{(1\,\text{MPa})\left(4.986 \times 10^{-3}\,\frac{\text{m}^3}{\text{mol}}\right)\left(10^6\,\frac{\text{Pa}}{\text{MPa}}\right)}{\left(8.314\,\frac{\text{Pa} \cdot \text{m}^3}{\text{mol} \cdot \text{K}}\right)(600\,\text{K})} = 0.9995 \qquad (6.158)$$

$$Z_2 = \frac{P_2 \underline{V}_2}{RT_2} = \frac{(0.2\,\text{MPa})\left(0.01659\,\frac{\text{m}^3}{\text{mol}}\right)\left(10^6\,\frac{\text{Pa}}{\text{MPa}}\right)}{\left(8.314\,\frac{\text{Pa} \cdot \text{m}^3}{\text{mol} \cdot \text{K}}\right)(400\,\text{K})} = 0.998 \qquad (6.159)$$

Applying Equation 6.157 to the conditions of the entering and exiting gas, we have

$$\underline{H}_1^R = RT(Z_1 - 1) - \frac{a}{\underline{V}_1}$$

$$\underline{H}_1^R = \left(8.314\,\frac{\text{J}}{\text{mol} \cdot \text{K}}\right)(600\,\text{K})(0.9995 - 1) - \left(\frac{0.2303\,\dfrac{\text{Pa} \cdot \text{m}^6}{\text{mol}^2}}{4.986 \times 10^{-3}\,\dfrac{\text{m}^3}{\text{mol}}}\right)\frac{1\,\text{J}}{\text{Pa} \cdot \text{m}^3} \qquad (6.160)$$

$$\underline{H}_1^R = -48.7\,\frac{\text{J}}{\text{mol}}$$

and analogously,

$$\underline{H}_2^R = -20.5\,\frac{\text{J}}{\text{mol}}$$

Step 6 *Combine calculated results into expression from step 2*
From Equation 6.139,

$$\underline{H}_2 - \underline{H}_1 = \underline{H}_2^R + \left(\underline{H}_2^{ig} - \underline{H}_1^{ig}\right) - \underline{H}_1^R$$

The terms on the right-hand side have now all been computed for

> The calculated values of Z are close to 1, which suggests the methane closely approximates ideal gas behavior. Thus it is logical that the calculated residuals are relatively small.

$$\underline{H}_2 - \underline{H}_1 = (-20.5)\,\frac{\text{J}}{\text{mol}} + (-9289)\,\frac{\text{J}}{\text{mol}} - (-48.7)\,\frac{\text{J}}{\text{mol}} = -9261\,\frac{\text{J}}{\text{mol}} \qquad (6.161)$$

The energy balance revealed that this change in molar enthalpy is identical to the work produced per mole, which is what we are trying to determine.

The problem-solving strategy established in Example 6-9, using the residual molar enthalpy, can be applied analogously to other intensive properties. Example 6-10 uses residuals to determine change in internal energy, and Example 6-11 applies the residual molar entropy.

PISTON-CYLINDER EXAMPLE USING RESIDUALS	**EXAMPLE 6-10**

One mole of gas is confined in a piston-cylinder apparatus (Figure 6-13) initially at $T = 300\,\text{K}$ and $P = 1\,\text{MPa}$. The gas is heated reversibly and at constant pressure to $T = 500\,\text{K}$. What are the values for Q and W as a result of this process?

The ideal gas heat capacity for this gas is constant at $C_V^* = 3R$, and the van der Waals parameters for the gas are $a = 0.250\,\text{Pa-m}^6/\text{mol}^2$ and $b = 3.00 \times 10^{-5}\,\text{m}^3/\text{mol}$.

In previous examples, the ideal gas constant pressure heat capacity C_P^* was given; here the constant-volume heat capacity C_V^* is known. Recall that, for an ideal gas, these can be related to each other,

$$C_V^* + R = C_P^*$$

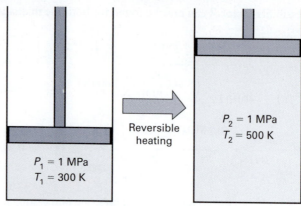

FIGURE 6-13 Piston-cylinder system described in Example 6-10.

SOLUTION:

Step 1 *Define a system and write an energy balance*
The gas inside the piston-cylinder device is the system. The work done by an expanding system is something we have calculated several times (first in Example 1-4), and it can be determined without an energy balance. But as usual, we need an energy balance to find Q, so we will start there. This is an unsteady-state process and also a closed system. Since the cylinder itself is stationary, we can neglect changes in potential and kinetic energy. The energy balance, therefore, is (see also Example 3-10 for a more detailed derivation)

$$\underline{U}_2 - \underline{U}_1 = Q + W_{EC} \tag{6.162}$$

Molar internal energy is a state function, and because we know two intensive properties (T, P) for the initial and final states, we should be able to compute $\underline{U}_2 - \underline{U}_1$.

Step 2 *Relate $\underline{U}_2 - \underline{U}_1$ to residual properties*
The ideal gas heat capacity is known, but the pressure is 1 MPa. While the methane in Example 6-9 had Z~1 at $P = 1$ MPa, many gases would not be well modeled as ideal at this pressure. Consequently we will not assume ideal gas behavior; we will calculate the residual properties and see whether they are significant or not.

To find the change in molar internal energy using residual properties, we begin by writing an equation for $\underline{U}$ that is analogous to Equation 6.139:

$$\underline{U}_2 - \underline{U}_1 = \underline{U}_2^R + \left(\underline{U}_2^{ig} - \underline{U}_1^{ig}\right) - \underline{U}_1^R \tag{6.163}$$

Step 3 *Evaluate ideal gas term of Equation 6.162*
The middle term of the right-hand side of Equation 6.163 is familiar: As first discussed in Section 2.3.3, internal energy for an ideal gas is a function of temperature only:

$$\left(\underline{U}_2^{ig} - \underline{U}_1^{ig}\right) = C_V^*(T_2 - T_1) = 3R(T_2 - T_1) \tag{6.164}$$

Step 4 *Evaluate residual molar internal energy for this equation of state*
Next we need to evaluate the residuals at the initial and final states. We can start with Equation 6.143, which gives the residual molar internal energy in its general form:

$$\underline{U} - \underline{U}^{ig} = \underline{U}^R = \int_{V=\infty}^{V=V} \left[T\left(\frac{\partial P}{\partial T}\right)_V - P\right] d\underline{V}$$

PITFALL PREVENTION

A common conceptual error is to say, "The $(\partial P/\partial T)_V$ term must be 0, because this is a constant-pressure process. However, *a residual is a state property*; its value is not path dependent. A more detailed discussion of this conceptual error is given at the end of the problem solution.

We need to express the term in brackets as a function of $\underline{V}$, so that we can integrate it with respect to $d\underline{V}$. The van der Waals equation of state is

$$P = \frac{RT}{\underline{V} - b} - \frac{a}{\underline{V}^2} \tag{6.165}$$

Taking the derivative of both sides with respect to T at constant V gives

$$\left(\frac{\partial P}{\partial T}\right)_{\underline{V}} = \frac{R}{\underline{V} - b} \tag{6.166}$$

Notice that because $\underline{V}$ is being held constant, the $a/\underline{V}^2$ term is a constant, and its derivative is zero.

Substituting these expressions for P and $(\partial P / \partial T)_{\underline{V}}$ into the integral gives

$$\underline{U}^R = \int_{\underline{V}=\infty}^{\underline{V}=\underline{V}} \left[T\left(\frac{R}{\underline{V} - b}\right) - \left(\frac{RT}{\underline{V} - b} - \frac{a}{\underline{V}^2}\right) \right] d\underline{V} \tag{6.167}$$

The first two terms within the integral cancel, leaving

$$\underline{U}^R = \int_{\underline{V}=\infty}^{\underline{V}=\underline{V}} \left[\frac{a}{\underline{V}^2} \right] d\underline{V}$$

The integral of $a/\underline{V}^2$ is $-a/\underline{V}$. Since $-a/\underline{V}$ is zero at the lower limit of integration ($\underline{V} = \infty$), the final result is

$$\underline{U}^R = \frac{-a}{\underline{V}} \tag{6.168}$$

Step 5 *Find numerical values for molar volumes $\underline{V}_1$ and $\underline{V}_2$*

The only thing we have assumed up to this point is that the gas follows the van der Waals equation of state, so the expression for the internal energy residual function is valid at any temperature and pressure. We need to evaluate $-a/\underline{V}$ at P_1, T_1 and at P_2, T_2. Solving the van der Waals equation for $\underline{V}$ gives

> The van der Waals equation is cubic in $\underline{V}$; it has three solutions for $\underline{V}$. Section 7.2.4 discusses the case in which more than one of the solutions are real.

At $P_1 = 1$ MPa and $T_1 = 300$ K, $\underline{V}_1 = .00242$ **m³/mol**

At $P_2 = 1$ MPa and $T_2 = 500$ K, $\underline{V}_2 = 0.00413$ **m³/mol**

As in Example 6-9, there is only one *real* solution for the cubic equation in this case; the other two solutions for $\underline{V}$ are complex numbers.

Step 6 *Solve for change in molar volume*

Recall that Equation 6.139 stated:

$$\underline{U}_2 - \underline{U}_1 = \underline{U}_2^R + \left(\underline{U}_2^{ig} - \underline{U}_1^{ig}\right) - \underline{U}_1^R$$

We can now express all three terms as functions of measurable properties as

$$\underline{U}_2 - \underline{U}_1 = \left(\frac{-a}{\underline{V}_2}\right) + 3R(T_2 - T_1) - \left(\frac{-a}{\underline{V}_1}\right) \tag{6.169}$$

$$\underline{U}_2 - \underline{U}_1 = \frac{-0.250\,\text{Pa}\,\dfrac{\text{m}^6}{\text{mol}^2}}{.00413\,\dfrac{\text{m}^3}{\text{mol}}} + 3\left(8.314\,\frac{\text{J}}{\text{mol K}}\right)(500\,\text{K} - 300\,\text{K}) + \frac{0.250\,\text{Pa}\,\dfrac{\text{m}^6}{\text{mol}^2}}{.00242\,\dfrac{\text{m}^3}{\text{mol}}}$$

The first and last term have units Pa · m³/mol, but a Pascal is a N/m², so this is really a N · m/mol, or equivalently J/mol. Thus, all three terms are in the same units and can be summed. Since a and b were given with three significant figures, we will round our answer (5031.7 J/mol) to three as well:

$$\underline{U}_2 - \underline{U}_1 = 5030 \, \frac{\text{J}}{\text{mol}}$$

Note that if we had assumed this was an ideal gas and used $\underline{U}_2 - \underline{U}_1 = C_V(T_2 - T_1)$, the result would have been $\underline{U}_2 - \underline{U}_1 = 4990$ J/mol. The residual functions are small in magnitude, but not insignificant, compared to the ideal gas change in internal energy: $\underline{U}_1^R = 103.2$ J/mol and $\underline{U}_2^R = -60.6$ J/mol. Because they are opposite in sign, significant cancellation occurs.

Step 7 *Calculate work*

The work done on, or by, an expanding system was first derived in Example 1-4 and is given by

$$W_{EC} = -\int_{V_1}^{V_2} P \, dV \tag{6.170}$$

where P is the pressure opposing the motion. In this case, because the process is reversible, the pressure of the gas and the external pressure are essentially the same, and so P can be considered constant at 1 MPa. Integrating gives

$$W_{EC} = -(1\,\text{MPa})\left(0.00413 - .00242 \, \frac{\text{m}^3}{\text{mol}}\right)\left(\frac{10^6 \, \frac{\text{N}}{\text{m}^2}}{1\,\text{MPa}}\right)\left(\frac{1\,\text{J}}{\text{N} \cdot \text{m}}\right)$$

$$W_{EC} = -1710 \, \frac{\text{J}}{\text{mol}}$$

Step 8 *Close energy balance to find heat*

Returning to the energy balance, we have

$$\underline{U}_2 - \underline{U}_1 = Q + W_{EC} \tag{6.171}$$

$$5030 \, \frac{\text{J}}{\text{mol}} = Q - 1710 \frac{\text{J}}{\text{mol}}$$

$$Q = 6740 \, \frac{\text{J}}{\text{mol}}$$

PITFALL PREVENTION: *A closer look at a common conceptual error*

Let us return to the possible conceptual error mentioned in step 4. This is a constant-pressure process, so why is the $(\partial P/\partial T)_V$ term not equal to zero when we are evaluating the residual? Aren't all derivatives of P equal to 0 if P is constant?

Conceptually, the thing to remember is that internal energy is a state function, and the residual internal energy is also a state function. If the initial pressure was 5 atm and the final pressure was 10 atm, the process would not be at constant pressure, and you would not be tempted to cross out the pressure derivative. But because a residual is a state function, it has a particular value at $P = 1$ MPa and $T = 300$ K, which is path independent. In other words, the quantity $(\underline{U}_1 - \underline{U}_1^{ig})$ depends only on P_1 and T_1; it does not depend on P_2 at all, so the fact that P_2 and P_1 happen to be equal to each other does not affect the residual property. $(\partial P/\partial T)_V$ can only be evaluated using an applicable equation of state, which quantifies the relationship between the measurable state properties P, $\underline{V}$, and T.

Mathematically, a residual property integral is not carried out over the process path; it is not integrated from P_1,T_1 to P_2,T_2. The residual compares a real substance to an ideal gas at a particular T and P. It has $P = 0$ (or its equivalent of $\underline{V} = \infty$) at the lower limit of integration, and the actual P at the upper limit. Temperature is always a constant when we are evaluating the integral to compute a residual property, but pressure is never constant.

Now, by contrast, suppose we had started the problem by writing a total derivative expression for $d\underline{U}$ as

$$d\underline{U} = \left(\frac{\partial \underline{U}}{\partial P}\right)_T dP + \left(\frac{\partial \underline{U}}{\partial T}\right)_P dT \tag{6.172}$$

This expression CAN be applied to the actual process path, so here it would be valid to say, "This is a constant pressure process, so $dP = 0$." This would, in fact, have been a logical start, but had we done this, we would have found $(\partial \underline{U}/\partial T)_P$ is dependent upon C_V. At this point, we would have recognized that we know C_V^* but not C_V, which would probably have convinced us to use residual properties.

The final example is similar to the nozzle problem examined in Example 6-7, with the only difference being what is stated about the heat capacity. Comparing the results of the two examples, which use two distinct solution strategies, is instructive.

APPLICATION OF ENTHALPY AND ENTROPY RESIDUAL FUNCTIONS TO A NOZZLE	EXAMPLE 6-11

A gas flows through an adiabatic nozzle at a steady state, entering at $P = 0.5$ MPa and $T = 675$ K and leaving at $P = 0.1$ MPa. Assuming the nozzle is reversible, what is the temperature of the exiting gas? Assume the gas has an ideal gas heat capacity of $C_P^* = 40$ J/mol · K and follows the equation of state:

$$\underline{V} = \frac{RT}{P} + aTP$$

where $a = 1.0 \times 10^{-11}$ m³/mol · Pa · K.

> Residual functions are used because C_P is given only for ideal gas conditions, but the pressure entering the nozzle is $P = 0.5$ MPa. While some gases may be well modeled as ideal gases at this pressure, we do not automatically assume that all gases are ideal at this pressure.

SOLUTION:

This problem is identical to Example 6-7 with one exception: The heat capacity was treated as a constant $C_P = 40$ J/mol · K in Example 6-7, but here this value is only considered valid at ideal gas conditions. Thus, the entropy balance in Example 6-7, which was simply $\underline{S}_{in} = \underline{S}_{out}$, is again valid, but here we will use residual properties to relate molar entropy to measurable properties.

Step 1 *Relate $\underline{S}_{in} - \underline{S}_{out}$ to temperature and pressure*
To calculate change in molar entropy from residual properties, start with Equation 6.139 as

$$\underline{S}_{out} - \underline{S}_{in} = \underline{S}_{out}^R + \left(\underline{S}_{out}^{ig} - \underline{S}_{in}^{ig}\right) - \underline{S}_{in}^R \tag{6.173}$$

From the entropy balance, we know the left-hand side is 0:

$$0 = \left(\underline{S}_{out} - \underline{S}_{out}^{ig}\right) + \left(\underline{S}_{out}^{ig} - \underline{S}_{in}^{ig}\right) - \left(\underline{S}_{in} - \underline{S}_{in}^{ig}\right) \tag{6.174}$$

The solution strategy is to express the right-hand side in terms of known quantities and the unknown temperature T_{out}.

Step 2 *Evaluate ideal gas term*
Section 4.4.1 examined the change in entropy for an ideal gas process. Here, the ideal gas heat capacity is constant, which means that Equation 4.58 can be applied for

> If C_P^* were not constant, we would apply the more general Equation 4.54 instead of Equation 4.59.

$$\underline{S}_{out}^{ig} - \underline{S}_{in}^{ig} = C_P^* \ln\left(\frac{T_{out}}{T_{in}}\right) + R \ln\left(\frac{P_{in}}{P_{out}}\right) \tag{6.175}$$

The entire right-hand side of Equation 6.175 is known except T_{out}, which is what we are trying to find.

Step 3 *Derive an expression for the residual molar entropy*

We can evaluate the residual molar entropy starting with either Equation 6.142 or Equation 6.145. Here, the equation of state is given in the form $\underline{V} = f(T, P)$, so it is straightforward to assess partial derivatives of $\underline{V}$. Consequently we choose to begin with Equation 6.142 as

$$\underline{S} - \underline{S}^{ig} = -\int_{T=T,P=0}^{T=T,P=P} \left[\left(\frac{\partial \underline{V}}{\partial T} \right)_P - \frac{R}{P} \right] dP$$

The partial derivative $(\partial \underline{V}/\partial T)_P$ is evaluated by differentiating both sides of the equation of state

$$\left(\frac{\partial \underline{V}}{\partial T} \right)_P = \frac{R}{P} + aP \tag{6.176}$$

Plugging this result into Equation 6.142 gives

$$\underline{S} - \underline{S}^{ig} = -\int_{T=T,P=0}^{T=T,P=P} \left[\left(\frac{R}{P} + aP \right) - \frac{R}{P} \right] dP$$

$$\underline{S} - \underline{S}^{ig} = -\int_{T=T,P=0}^{T=T,P=P} [aP] \, dP$$

$$\underline{S} - \underline{S}^{ig} = \frac{-aP^2}{2} \tag{6.177}$$

Step 4 *Apply results of steps 2 and 3 to Equation 6.174*

Combining Equations 6.175 and 6.177 into Equation 6.174 gives

$$0 = \left(\frac{-aP_{out}^2}{2} \right) + C_P^* \ln \left(\frac{T_{out}}{T_{in}} \right) + R \ln \left(\frac{P_{in}}{P_{out}} \right) - \left(\frac{-aP_{in}^2}{2} \right) \tag{6.178}$$

Inserting known values into Equation 6.178 gives

$$0 = -\frac{\left(10^{-11} \dfrac{m^3}{mol \cdot Pa \cdot K} \right)(0.1\,MPa)^2 \times \left(\dfrac{10^{12}\,Pa^2}{1\,MPa^2} \right)}{2} \tag{6.179}$$

$$+ \left(40 \frac{J}{mol \cdot K} \right) \ln \left(\frac{T_{out}}{675\,K} \right)$$

$$+ \left(8.314 \frac{Pa \cdot m^3}{mol \cdot K} \right) \ln \left(\frac{0.5\,MPa}{0.1\,MPa} \right) + \frac{\left(10^{-11} \dfrac{m^3}{mol \cdot Pa \cdot K} \right)(0.5\,MPa)^2 \times \left(\dfrac{10^{12}\,Pa^2}{1\,MPa^2} \right)}{2}$$

T_{out} is the only unknown in the equation, and can be solved $T_{out} = \textbf{468.8 K.}$

It would be reasonable, rather than computing a residual property for the outlet gas, to have assumed it's negligible, since the pressure is $P = 0.1$ MPa. This calculation reveals it to be very small compared to the residual at $P = 0.5$ MPa.

It is instructive to compare the results of Examples 6-7 and 6-11. In Example 6-7 we modeled the gas as having a constant $C_P = 40$ J/mol · K, with the assumption that this value was valid at ALL conditions. In Example 6-11, we used the same numerical value for C_P^*, but only assumed it was valid at ideal gas conditions. Why are the answers obtained from both examples identical? An explanation can be obtained from Equation 6.124, which provides a relationship between heat capacity and pressure:

$$\left(\frac{\partial C_P}{\partial P}\right)_T = -T\left(\frac{\partial^2 \underline{V}}{\partial T^2}\right)_P$$

We determined previously (Equation 6.176) that for this particular equation of state:

$$\left(\frac{\partial \underline{V}}{\partial T}\right)_P = \frac{R}{P} + aP$$

There is no temperature dependence on the right-hand side, which means that when we take the second derivative with respect to T:

$$\left(\frac{\partial C_P}{\partial P}\right)_T = -T\left(\frac{\partial^2 \underline{V}}{\partial T^2}\right)_P = 0 \tag{6.180}$$

This result reveals that, for *this* equation of state, heat capacity does not change with pressure; the distinction between "ideal gas heat capacity" and "actual heat capacity" does not matter here.

Another point is that, while residual properties had not yet been introduced in Section 6.2.6, there would have been nothing wrong with using them in Example 6-7. If the heat capacity can be modeled as constant at ALL conditions, it must certainly be constant at ideal gas conditions.

In sum, it makes sense that the two approaches taken in Example 6-7 (which started with a total derivative) and Example 6-11 (which started with residual properties) would produce the same answer; both are physically and mathematically valid approaches. The answer is only accurate to the degree that the assumptions are accurate: that $\underline{V} = RT/P + aTP$ and that $C_P = 40$ J/mol · K. This equation of state was contrived for illustrative purposes, and is not used in practice. Chapter 7 examines the equations of state that are in common use by chemical engineers.

6.4 SUMMARY OF CHAPTER SIX

- A **total derivative** can be written to relate changes in any *intensive* property to changes in any two other intensive properties.

- A common first step in solving problems is writing a total derivative that relates the unknown property to the two intensive properties about which the most is known.

- Total derivative expressions are most readily solved when expressed in terms of **measurable properties**: P, T, $\underline{V}$, and their derivatives, C_P and C_V.

- Among the useful tools for manipulating partial derivative expressions are the following, which are presented mathematically in Section 6.2.7.

  - **Fundamental property relations**, which express $\underline{U}$, $\underline{H}$, $\underline{G}$, and $\underline{A}$ as functions of $\underline{S}$, $\underline{V}$, T, and P.

  - The **expansion rule**, which allows one to introduce two new variables into a problem, and is often used in combination with fundamental property relations.

■ The **triple product rule**, which can be used to relate a partial derivative to other partial derivates involving the same three variables.

■ **Maxwell's equations**, which can often be used to convert a partial derivative involving $\underline{S}$ into a partial derivative that involves only measurable properties.

■ The mathematical definitions of **constant pressure heat capacity**, **constant volume heat capacity**, **isothermal compressibility**, and **coefficient of thermal expansion**.

■ **Equations of state** can be used to resolve partial derivative expressions involving pressure, temperature and molar volume into solvable ODEs.

■ **Residual properties** are useful for gas phase processes in which an **ideal gas heat capacity** is known, but the ideal gas model isn't a reasonable approximation.

■ Residual properties can be related to measurable properties through a known equation of state.

6.5 EXERCISES

6-1. Derive Equations 6.47 and 6.48.

6-2. Derive Equations 6.61, 6.62, and 6.63.

6-3. Find expressions for the isothermal compressibility and coefficient of thermal expansion for an ideal gas. Expressions should be in terms of measurable properties only.

6-4. Find expressions for the isothermal compressibility and the coefficient of thermal expansion for a fluid that is described by the van der Waals equation of state. Expressions should be in terms of measurable properties only.

6-5. Find expressions for the residual molar internal energy, residual molar enthalpy, and residual molar entropy for a fluid that is described by each of the following:
  A. The ideal gas law
  B. The equation of state $P = RT/\underline{V} + a/\underline{V}^2$, in which a is a constant.

C. The equation of state $Z = 1 + (CP^2)/(RT)$, in which C is a constant.

D. The equation of state $\underline{V} = \dfrac{RT}{P} + aTP^2$, in which a is a constant.

6-6. Equations 6.157 and 6.168 give the expressions for $\underline{H}^R$ and $\underline{U}^R$ for a fluid described by the van der Waals equation of state. For a compound that has $a = 1.8$ m^6 Pa/mol^2 and $b = 1.25 \times 10^{-4}$ m^3/mol, compute $\underline{H}^R$ and $\underline{U}^R$ at each of the following conditions. If there is more than one value for $\underline{V}$, find the $\underline{H}^R$ and $\underline{U}^R$ that correspond to each value.
  A. $T = 400$ K and $P = 0.1$ MPa
  B. $T = 400$ K and $P = 0.5$ MPa
  C. $T = 600$ K and $P = 0.3$ MPa
  D. $T = 800$ K and $P = 1.0$ MPa
  E. $T = 200°$C and $P = 0.2$ MPa
  F. $T = 375°$C and $P = 0.172$ MPa

6.6 PROBLEMS

6-7. Demonstrate that if a gas follows the ideal gas law, $(\partial \underline{U}/\partial P)_T$ and $(\partial \underline{U}/\partial \underline{V})_T$ are equal to 0.

6-8. Derive an expression for each of the following that is, as much as possible, in terms of measurable properties: P, $\underline{V}$, T, C_p, and C_V, and their partial derivatives with respect to each other. $\underline{S}$ can appear in the expression, but not derivatives of $\underline{S}$.
  A. $(\partial \underline{H}/\partial T)_V$
  B. $(\partial \underline{H}/\partial P)_T$
  C. $(\partial \underline{U}/\partial \underline{V})_P$

  D. $(\partial \underline{U}/\partial T)_P$
  E. $(\partial \underline{A}/\partial \underline{S})_P$
  F. $(\partial \underline{A}/\partial \underline{S})_T$
  G. $(\partial \underline{G}/\partial T)_P$
  H. $(\partial \underline{G}/\partial P)_S$

6-9. Prove that Equation 6.124 is valid. Hint: Start with $\left(\dfrac{\partial \underline{H}}{\partial T}\right)_P$ and $\left(\dfrac{\partial \underline{H}}{\partial P}\right)_T$ and use the fact that, in a mixed second partial derivative, the order of differentiation doesn't matter.

6-10. Example 6-4 examined a valve and a liquid that had a coefficient of thermal expansion $\alpha_V = 0.001\,K^{-1}$, an isothermal compressibility of 0, and a molar volume $V = 0.1$ dm³/mol at $T = 300$ K and $P = 500$ kPa.
  A. Estimate the molar volume of the liquid leaving the valve, which has $P = 50$ kPa and $T = 300.3$ K.
  B. Estimate the temperature at which the liquid would have a molar volume $V = 0.101$ dm³/mol.

6-11. This problem is an extension of Problem 2-19. Two moles of a gas are confined in a piston-cylinder. Initially, the temperature is 300 K and the pressure is 0.1 MPa. The gas is compressed isothermally to 0.5 MPa. The gas has $C_P^* = (7/2)R$. Find $Q, W, \Delta U, \Delta H,$ and ΔS for the gas if
  A. The gas is an ideal gas.
  B. The gas follows the van der Waals equation of state with $a = 0.5$ Pa m⁶/mol² and $b = 3 \times 10^{-5}$ m³/mol.

6-12. One mole of a gas is placed in a closed system with a 0.02 m³ vessel initially at $T = 300$ K. The vessel is then isothermally expanded to 0.04 m³. The gas follows the equation of state:
$$P = RT/V + a/V^2$$
where $a = 4.053$ m⁶ Pa/mol² and $R = 8.314$ J/mol K.
  A. Derive an expression relating $(dH/dV)_T$ to measurable properties.
  B. Find ΔH for the gas in this process.

6-13. A gas has an ideal gas heat capacity of $C_P^* = (7/2)R$ and is described by the equation of state:
$$Z = 1 + (CP^2)/(RT)$$
with $C = 10^{-9}$ m/Pa · mol.
  A. Find a general expression for the residual molar enthalpy for this gas.
  B. Find a general expression for the residual molar entropy for this gas.
  C. Find ΔH and ΔS for the gas if it is isothermally compressed from $P = 0.1$ MPa and $T = 400$ K to $P = 5$ MPa and $T = 400$ K.

Problems 14 through 17 involve a gas that follows the equation of state:
$$V = \frac{RT}{P} + aTP^2$$
with $a = 3 \times 10^{-17}$ m³/mol · Pa² · K.
The gas has a molecular mass of 120 g/mol and an ideal gas heat capacity $C_P^* = 40$ J/mol.

6-14. A gas, which follows the EOS given above, enters a turbine at $P = 1.5$ MPa and $T = 500$ K, and leaves with a pressure of $P = 0.05$ Mpa. If the turbine has an efficiency of 80%, find
  A. The work produced in the turbine.
  B. The actual temperature of the exiting stream.

6-15. **A.** Five moles of an ideal gas is confined in a piston-cylinder device, initially at $P_1 = 0.1$ MPa and $T_1 = 400$ K. First, the gas is compressed adiabatically to $P_2 = 1$ MPa and temperature T_2. Next the piston is locked in place, and the gas is cooled isochorically to $T_3 = 400$ K and pressure P_3. Determine $Q, W,$ and ΔU for the gas in each of the two steps of this process. Also find T_2 and P_3.
  B. Five moles of a gas that is described by the equation of state given above is confined in a piston-cylinder device. Calculate ΔU when this gas undergoes the same two changes in state described in part A, from T_1, P_1 to T_2, P_2 and from T_2, P_2 to T_3, P_3.
  For both parts A and B, assume the gas has $C_P^* = 40$ J/mol · K.

6-16. A gas with a flow rate of 5 mol/s enters a steady-state, adiabatic nozzle with negligible veloc-ity at $T = 500$ K and $P = 1$ MPa and leaves the nozzle at $P = 0.1$ MPa. The gas has $C_P^* = 40$ J/mol · K.
  A. Find the exiting temperature and exiting velocity if the gas is ideal and the nozzle is reversible.
  B. Find the exiting temperature and exiting velocity if the gas follows the equation of state above and the nozzle is reversible.
  C. The gas follows the equation of state above, but the nozzle is not reversible; it has a rate of entropy generation of $\dot{S}_{gen} = 25$ J/s. Find the exiting temperature and exiting velocity.

6-17. A gas stream with a flow rate of 16.67 mol/s leaves a reactor at $T = 450$ K and $P = 1$ MPa. For the next step, the temperature of the gas must be reduced to 300 K, but the pressure of the stream isn't very important. Consequently, a turbine is being placed after the reactor, to obtain some work, but the turbine must be designed to have an actual outlet temperature of 300 K. Find the turbine outlet pressure and the work produced for each of the following cases. In all cases, the gas has $C_P^* = 40$ J/mol · K.
  A. The gas is an ideal gas, and the turbine is reversible.
  B. The gas is an ideal gas, but the turbine efficiency is 80%. Find the turbine outlet pressure and the work produced.
  C. The turbine is reversible, and the gas follows the equation of state.
  D. The turbine efficiency is 80%, and the gas follows the equation of state.

6-18. One mole of a gas is compressed at a constant temperature of 400 K from $P = 0.1$ MPa to $P = 0.5$ MPa. The gas is known to follow the equation of state:
$$V = RT/P + aP^2$$
where $a = 0.01 \times 10^{-13}$ m³/Pa² mol.

The ideal gas heat capacity at $T = 400$ K is $C_p^* = 3.85R$.

A. Demonstrate that the heat capacity of this gas is independent of pressure.

B. Derive an expression for $(dU/dP)_T$ as a function of measurable properties.

C. Find ΔU for the gas in this process.

D. Find ΔS for the gas in this process.

6-19. A liquid has constant coefficient of thermal expansion and isothermal compressibility, $\alpha_V = 2.37 \times 10^{-3}$ K^{-1} and $\kappa_T = 3.54 \times 10^{-10}$ Pa^{-1}. The molar volume of the compound at $T = 300$ K and $P = 0.1$ MPa is $V = 0.1 \times 10^{-3}$ m^3/mol, and its heat capacity can be modeled as constant at $C_V = C_P = 27.5$ J/mol·K.

This liquid is stored in tank that has a "floating top." Thus, when the temperature changes, causing the liquid to expand or contract, the container expands or contracts. For each of the processes given, the tank initially has $V = 30$ m^3, $T = 300$ K, and $P = 0.1$ MPa.

A. The "floating top" works properly, maintaining the tank at a constant pressure. 5000 kJ of heat are added to the tank. Find the final temperature and volume.

B. The "floating top" malfunctions, such that the container is effectively rigid at $V = 30$ m^3. 5000 kJ of heat are added to the tank. Find the final temperature and pressure.

6-20. The liquid described in Problem 6-19 is compressed in a steady-state, adiabatic pump. 5 mol/s enter the pump at $T = 300$ K and $P = 0.1$ MPa and exit the pump at $P = 2$ MPa.

A. Using the approximation that V is a constant at 100 cm^3/mol, compute the rate at which work is done in the pump.

B. Assuming the value of work determined in part A is correct, find the temperature of the liquid leaving the pump. Use relevant physical properties given in Problem 6-19.

C. Assuming the temperature found in part B is correct, find the molar volume of the liquid leaving the pump. Use relevant physical properties given in Problem 6-19.

D. We assumed in part A that V was constant, but then in part C we computed the change in V for this process. Do you consider the answers obtained in parts A–C valid? How could you modify the solution approach to get a more accurate answer?

6-21. Using the compressed liquid tables, estimate the isothermal compressibility and coefficient of thermal expansion for liquid water at each of the following conditions.

A. $P = 10$ MPa and $T = 100°C$

B. $P = 20$ MPa and $T = 100°C$

C. $P = 10$ MPa and $T = 200°C$

D. $P = 20$ MPa and $T = 200°C$

E. Based upon the results of A–D, comment on how valid it would be to model the isothermal compressibility and coefficient of thermal expansion for liquids at high pressure as constants.

6-22. Re-do Problem 4-9. Instead of using the steam tables, assume that the material entering and leaving the turbine is all vapor, and that steam is accurately described by the van der Waals equation with $a = 0.5542$ Pa m^6/mol^2 and $b = 3.051 \times 10^{-5}$ m^3/mol. Comment on the accuracy of the van der Waals equation in this case.

6-23. This problem is a variation on Problem 4-16, in which pressure was in the range 0.05 to 0.1 MPa, and we used the ideal gas model. Here we explore a similar process with higher pressures and a real gas model.

Ten moles of gas are placed in a rigid container. Initially the gas is at $P = 0.5$ MPa and $T = 300$ K, and the container is also at $T = 300$ K. The container is placed in a furnace, where its surroundings are at a constant $T = 600$ K. The container is left in the furnace until both the container and the gas reach thermal equilibrium with the surroundings. The gas is described by the van der Waals equation of state, with $a = 0.08$ m^6 Pa/mol^2 and $b = 0.0001$ m^3/mol, and has an ideal gas heat capacity of $C_V^* = 2.5R$. The container itself has a mass of 10 kg (not including the mass of the gas inside) and a heat capacity of $\hat{C}_V = 1.5$ J/g·K.

A. Find the heat added to the gas.

B. Find the heat added to the container.

C. Find the change in entropy of the gas.

D. Find the change in entropy of the universe.

E. What aspect of this process is irreversible?

6-24. Methane enters a process at $T = 422$ K (300°F) and $P = 100$ kPa, and is heated and compressed to $T = 477.6$ K (400°F) and $P = 500$ kPa. Find the change in molar Gibbs free energy for the methane, using Figure 7-1.

6-25. One mole of a gas is compressed at a constant temperature of 400 K from $P = 0.01$ MPa to $P = 1$ MPa. The gas is known to follow the equation of state:

$$V = RT/P + aP^2$$

where $a = 2.5 \times 10^{-15}$ m^3/Pa^2 mol.

The ideal gas heat capacity at $T = 400$ K is $C_p^* = 3.85$ R.

A. Find an equation for the residual G for a gas that follows this equation of state.

B. Find the change in Gibbs free energy for this process.

6-26. Find the change in Gibbs free energy for the system and process described in Problem 6-15, part B.

The Helmholtz energy (A), which was introduced in this chapter, is rarely used in this introductory book, but has great utility in the field of statistical thermodynamics, so being able to model changes in A with respect to measurable properties is essential. The next three problems examine changes in A.

6-27. Steam is heated from an initial condition of saturated steam at $P = 0.15$ MPa to a final state of $P = 0.3$ MPa and $T = 573.15$ K. Use the steam tables to find the change in $\hat{A}$ for this process.

6-28. Find the change in Helmholtz energy for the process described in Problem 6-18.

6-29. A gas has $C_P^* = 35$ J/mol $\cdot$ K, and follows the equation of state:

$$\underline{V} = \frac{RT}{P} + aTP^2$$

with $a = 0.015 \times 10^{-15}$ m³/mol Pa² $\cdot$ K.
Find the change in $\underline{A}$ when the gas is compressed isothermally from $T = 300$ K and $P = 0.1$ MPa to $P = 1$ MPa.

6.7 GLOSSARY OF SYMBOLS

A	Helmholtz energy	$\underline{G}^{ig}$	molar Gibbs free energy for an ideal gas state	$\underline{S}^R$	residual molar entropy
$\underline{A}$	molar Helmholtz energy			T	temperature
a	parameter in van der Waals EOS, accounting for inter-molecular interactions	$\underline{G}^R$	residual molar Gibbs free energy	U	internal energy
		H	enthalpy	$\underline{U}$	molar internal energy
		$\hat{H}$	specific enthalpy	$\underline{U}^{ig}$	molar internal energy for an ideal gas state
b	parameter in van der Waals EOS, accounting for excluded volume	$\underline{H}$	molar enthalpy		
		$\underline{H}^{ig}$	molar enthalpy for an ideal gas state	$\underline{U}^R$	residual molar internal energy
C_P	constant-pressure heat capacity (molar basis)	$\underline{H}^R$	residual molar enthalpy	V	volume
C_P^*	constant-pressure heat capacity (molar basis) for ideal gas	M	mass of system	$\underline{V}$	molar volume
		$\dot{m}$	mass flow rate	$\underline{V}^{ig}$	molar volume for an ideal gas state
$\hat{C}_V$	constant-volume heat capacity (mass basis)	N	number of moles in system		
		$\dot{n}$	molar flow rate	W	work
$\hat{C}_P$	constant-pressure heat capacity (mass basis)	P	pressure	W_{EC}	work of expansion/contraction
C_V	constant-volume heat capacity (molar basis)	Q	heat	Z	compressibility factor
		R	gas constant	α_V	coefficient of thermal expansion
C_V^*	constant-volume heat capacity (molar basis) for ideal gas	S	entropy		
		$\underline{S}$	molar entropy	κ_T	isothermal compressibility
G	Gibbs free energy	$\underline{S}^{ig}$	molar entropy for an ideal gas state	μ_{JT}	Joule-Thomson coefficient
$\underline{G}$	molar Gibbs free energy				

6.8 REFERENCE

Ben-Zion, M., U.S. Patent #5,522,870.

Equations of State (EOS)

7

LEARNING OBJECTIVES

This chapter is intended to help you learn:

- How a *cubic equation of state* can be used to model a compound in the liquid, vapor, gas, and supercritical states
- To solve equations of state, and correctly interpret the results when multiple solutions exist
- How to compute values for the adjustable parameters in a cubic equation of state
- To recognize the analogies between different compounds that are revealed by the *principle of corresponding states,* and how they can be used to estimate unknown physical properties
- Estimation of physical properties using *group additivity*
- The basic elements of *microscopic modeling*

Chapter 2 introduced the need for reliable models that provide estimates of physical and thermochemical properties, for situations in which experimental data is not available. The models introduced at that time (e.g., ideal gas model) were simplistic, and valid only under specific conditions.

Chapter 6 introduced a broader range of modeling techniques that are applicable to any pure compound at any conditions, but using them often requires an equation of state. In Chapter 2, we defined an "equation of state" (frequently abbreviated EOS) as a mathematical relationship among P, $\underline{V}$, and T, such that if any two are specified, the third can be calculated. So far, the only equations of state we have introduced are the ideal gas law and van der Waals. The motivational examples take a closer look at these and demonstrate the need for more versatile equations of state. ■

> Chapter 6 contained some additional equations of state that were contrived by the authors for illustrative purposes but have never been used in engineering practice.

7.1 MOTIVATIONAL EXAMPLES: Transportation of Natural Gas

Natural gas is a widely used fuel source, composed primarily of methane. Within the continental United States, it is most commonly transported by pipeline. However, when natural gas is shipped overseas, or to remote locations, an alternative to building more pipelines is transporting it by ship. This involves *liquefying* the natural gas, so that it can be stored in small volumes for transport. Liquefied natural gas (LNG) is hazardous because it is flammable and also because of the extreme cold at which it must be maintained; typical temperatures would be in the vicinity of 110 K.

This chapter presents two examples in which we use the ideal gas law and the van der Waals equations of state to relate P, $\underline{V}$, and T to each other. In both examples, we use pure methane as a model for natural gas. Thermodynamic modeling of mixtures is covered beginning in Chapter 9.

EXAMPLE 7-1	SIZING A PIPELINE

We are designing a new process that uses natural gas, which will be delivered via pipeline at conditions of $P = 1000$ psia (6.89 MPa) and $T = 100°F$ (310.93 K). In order to size process equipment, we need to know the density of natural gas at this temperature and pressure. Use pure methane as a model for natural gas.

Examples 7-1 and 7-2 use English units for consistency with the data in Figure 7-1, in which density is reported in lb$_m$/ft³.

Find ρ in lb$_m$/ft³ for methane at $P = 6.89$ MPa and $T = 310.93$ K using:

A. The ideal gas law
B. The van der Waals EOS with the parameters $a = 8496$ ft⁶ psia/lb-mol² and $b = 0.685$ ft³/lb-mol

SOLUTION A:

Step 1 *Solve ideal gas law for molar volume*

Methane is a super-critical fluid at this temperature and pressure; its critical point is about $P = 4.67$ MPa and $T = 190.92$ K.

Temperature must always be expressed on an absolute scale when using the ideal gas law, and $T = 100°F$ is equivalent to $T = 559.6°R$. Since P and T are given, the ideal gas law can be solved immediately for $\underline{V}$:

$$\underline{V} = \frac{RT}{P} = \frac{\left(10.731 \dfrac{\text{psia} \cdot \text{ft}^3}{\text{lb-mol} \cdot °\text{R}}\right)(559.6°\text{R})}{1000 \text{ psia}} \tag{7.1}$$

$$\underline{V} = 6.005 \frac{\text{ft}^3}{\text{lb-mol}}$$

Step 2 *Convert to desired units*

One of the identifying characteristics of ideal gases is "no intermolecular interactions."

Density (lb$_m$/ft³) is the reciprocal of specific volume (ft³/lb$_m$), so we need to convert from moles to mass. The molecular weight of methane is 16.

$$\hat{V} = \left(6.005 \frac{\text{ft}^3}{\text{lb-mol}}\right)\left(\frac{1 \text{ lb-mol}}{16 \text{ lb}_m}\right) = 0.375 \frac{\text{ft}^3}{\text{lb}_m} \tag{7.2}$$

$$\rho = \frac{1}{\hat{V}} = \frac{1}{0.375 \dfrac{\text{ft}^3}{\text{lb}_m}} = \mathbf{2.66 \frac{lb_m}{ft^3}} \tag{7.3}$$

SOLUTION B:

2.66 lb$_m$/ft³ is equivalent to 42.7 kg/m³.

Step 3 *Apply van der Waals equation of state*

The van der Waals EOS is commonly written as

$$P = \frac{RT}{\underline{V} - b} - \frac{a}{\underline{V}^2}$$

but can be rearranged to show explicitly that it is a cubic equation in $\underline{V}$ for

$$P\underline{V}^3 - \underline{V}^2(Pb + RT) + a\underline{V} - ab = 0 \tag{7.4}$$

The two complex solutions to Equation 7.4 are 1.27 + 0.75i and 1.27 − 0.75i.

Here, P, T, a, and b are all known. The fact that Equation 7.4 is a cubic equation means that it has three solutions for $\underline{V}$. In this case, two of the solutions are imaginary. This is not a fluke—in this chapter we will learn a method for assigning parameters to cubic equations of state in a way that produces only one real solution for temperatures above the critical temperature. The real solution is $\underline{V} = \mathbf{5.293}$ **ft³/lb-mol**.

Converting this into density, by the same process shown in step 2, produces a result of $\rho = 3.02$ **lb$_m$/ft³**.

The two answers are about 13% different from each other. Which one is more accurate?

3.02 lb$_m$/ft³ is equivalent to 48.4 kg/m³.

Step 4 *Compare to data*

Examining Figure 7-1, we see the curve representing $\rho = 3$ lb$_m$/ft³ intersects the curve representing $T = 100°F$ almost exactly at the $P = 1000$ psia line, so to the limit of accuracy of this figure, we would say the answer is $\rho = 3$ lb$_m$/ft³. For a more accurate answer, one

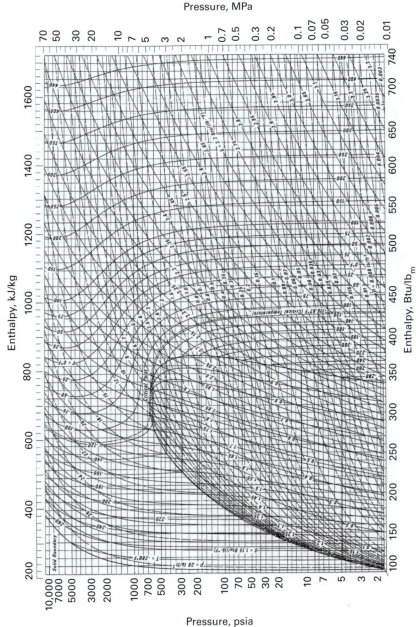

FIGURE 7-1 Thermodynamic properties of methane.

Based on NIST. Thermal Physics Division, Boulder, CO., as seen in Lira, J.R. Elliott and C. T., Introductory Chemical Engineering Thermodynamics, Prentice-Hall, 1999.

2.943 lb_m/ft^3 is equivalent to 47.2 kg/m^3.

good resource is the NIST WebBook, which gives a result of $\rho = 2.943 \ lb_m/ft^3$ (Lemmon, 2012). If we accept this value as the most accurate, then the ideal gas law is off by 9.6%, but the van der Waals equation is only off by 2.6%.

Example 7-1 illustrates a straightforward application for an accurate equation of state. In chemical processes, temperature and pressure are the most commonly specified parameters, but in order to size pipes and equipment, we often need to know the volumetric flow rate of a stream as well. This requires knowledge of the specific volume or density, which in Example 7-1 we obtained from an equation of state. This result, combined with a mass flow rate, would allow us to determine the volumetric flow rate. As another application, Example 7-2 will use equations of state in a safety analysis.

A question that often arises at this point is "Wouldn't real data be better?" and of course the answer is yes—reliable experimental data is preferred to any theoretical estimate. However, while extensive experimental data is available for compounds like water and methane, chemical engineers cannot count on having such data available for all chemicals they will ever work with. Furthermore, suppose your assignment was to collect experimental data of the kind shown in Figure 7-1. In the course of designing the experimental apparatus, you would have to size equipment and perform a safety analysis, requiring calculations analogous to those shown in Examples 7-1 and 7-2. Since your goal is to collect the data, it presumably doesn't exist yet—you would have to make estimates throughout your design process. Thus, accurate equations of state would be needed.

Real natural gas is not pure methane, but a mixture. Methane is the largest component of the mixture but the exact composition varies. Even if you have real data for each individual component in a mixture, you typically don't have data for every possible mixture composition. Thus, you need models, which rely on equations of state.

When the van der Waals EOS was introduced, Example 2-6 showed that for pressures above $P \approx 1$ atm, water vapor departed significantly from ideal gas behavior, and the van der Waals equation of state provided a much better model. H_2O is an extremely polar compound that can be expected to exhibit behavior that is not ideal. Example 7-1 illustrates that even methane, which is a spherical, non-polar compound, departs from ideal gas behavior at high pressures and demonstrates the improvement that results from the van der Waals EOS. In that example, the pressure was 6.89 MPa (1000 psia). Example 7-2 considers conditions supercritical methane at even higher pressure.

EXAMPLE 7-2 | A PRESSURIZED LNG STORAGE CONTAINER

This example is oversimplified in that LNG wouldn't be stored in a "rigid sealed" vessel. LNG storage vessels have a mechanism for venting natural gas that vaporizes. No vessel is perfectly insulated, so this "evaporative cooling" process sacrifices a small amount of the LNG in order to maintain the desired low temperature. Evaporative cooling was previously examined in Problem 3-22.

110 pounds (~50 kg) of liquefied natural gas (LNG) is stored in a rigid, sealed vessel with $V = 5$ ft³ (~0.14 m³). Due to a failure in the cooling/insulation system, the temperature increases to −100°F (199.82 K), which is above the critical temperature; thus the natural gas will no longer be in the liquid phase. What is the resulting pressure of the supercritical methane? If the vessel is rated as safe for pressures of up to 500 atmospheres (50.66 MPa), will the vessel rupture? In this problem, again model LNG as 100% methane.

Determine the answer using

A. the ideal gas law.
B. the van der Waals EOS.

The van der Waals parameters for methane are $a = 8496$ ft⁶ psia/lb-mol² and $b = 0.685$ ft³/lb-mol.

SOLUTION:

Step 1 *Determine number of moles of methane*
Pressure of the methane is ultimately what we're trying to determine. The temperature of the methane is known and the molar volume is readily found from the given

information. Whenever two of molar volume, temperature, and pressure of a material are known, the third can be determined if we have an equation of state that describes the material.

The molecular weight of methane is 16, so the number of moles of methane in the vessel is

$$n = \frac{M}{MW} = \frac{110 \text{ lb}_{m}}{16 \dfrac{\text{lb}_{m}}{\text{lb-mol}}} = 6.875 \text{ lb-mol} \tag{7.5}$$

6.875 lb-mol is equivalent to 3.125 kmol.

SOLUTION A:

Step 2 *Apply ideal gas law*
−100°F, on an absolute scale, is equivalent to 359.6°R. Solving the ideal gas law for pressure:

Recall that the ideal gas law can be written either $PV = NRT$ or $P\underline{V} = RT$, since $\underline{V} = V/N$.

$$P = \frac{NRT}{V} = \frac{(6.875 \text{ lb-mol})\left(10.731 \dfrac{\text{psia} \cdot \text{ft}^3}{\text{lb-mol} \cdot °\text{R}}\right)(359.6°\text{R})}{5 \text{ ft}^3} \tag{7.6}$$

$$P = 5306 \textbf{ psia}$$

The relationships between the Kelvin, Celsius, Fahrenheit and Rankine temperature scales were introduced in Section 1.4.8.

This is equivalent to approximately 361 times atmospheric pressure:

$$P = (5306 \text{ psia})\left(\frac{1 \text{ atm}}{14.696 \text{ psia}}\right) = 361.1 \text{ atm} \tag{7.7}$$

361.1 atm is equivalent to 36.6 MPa.

Thus the ideal gas law tells us the vessel is still safe if the methane temperature climbs to −100°F (199.82 K).

SOLUTION B:

Step 3 *Apply van der Waals EOS*
The molar volume of the methane is

$$\underline{V} = \frac{5 \text{ ft}^3}{6.875 \text{ lb-mol}} = 0.7273 \frac{\text{ft}^3}{\text{lb-mol}} \tag{7.8}$$

Inserting known information into the van der Waals equation gives

$$P = \frac{RT}{\underline{V} - b} - \frac{a}{\underline{V}^2} \tag{7.9}$$

$$P = \frac{\left(10.731 \dfrac{\text{psia} \cdot \text{ft}^3}{\text{lb-mol} \cdot °\text{R}}\right)(359.6°\text{R})}{0.7273 \dfrac{\text{ft}^3}{\text{lb-mol}} - 0.685 \dfrac{\text{ft}^3}{\text{lb-mol}}} - \frac{8496 \dfrac{\text{ft}^6 \cdot \text{psia}}{\text{lb-mol}^2}}{\left(0.7273 \dfrac{\text{ft}^3}{\text{lb-mol}}\right)^2}$$

$$P = \textbf{75,222 psia}$$

This is over 5100 times atmospheric pressure; ten times larger than the vessel's rated safe pressure. We have two equations of state that are leading us to two very different conclusions. Which is right?

FOOD FOR THOUGHT 7-1

Mathematically, the extremely large pressure can be understood as a consequence of $(\underline{V} - b)$ being very small. What does it mean if $\underline{V} = b$?

COMPARISON TO DATA

Figure 7-1 reveals that supercritical methane at $T = -100°\text{F}$ and $P \approx 5600$ psia has a density of $\rho = 22 \text{ lb}_m/\text{ft}^3$, which is exactly what we have in this example ($\rho = 110 \text{ lb}_m/5 \text{ ft}^3$). The correct answer is thus approximately 5600 psia, or approximately 381 atmospheres (38.6 MPa) of pressure.

Normally we expect a more complex method to be more rigorous and therefore more accurate. Here that is not the case.

The ideal gas estimate is off by about 5%. The van der Waals estimate is off by more than an order of magnitude and produces the wrong conclusion—it says the vessel will rupture while the real data says the vessel will still be safe at this temperature.

The outcome of Example 7-2 seems to contradict what we've learned so far.

We've learned the ideal gas law is only valid at low pressures; here the actual pressure is over 300 atmospheres. We've learned the van der Waals equation has parameters a and b, which are intended to take into account the intermolecular interactions between particles and the actual finite volumes of the particles. In other words, the parameters a and b are supposed to correct the exact limitations of the ideal gas model that cause it *not* to work at high pressures, so we would expect the van der Waals equation to be more accurate. In Example 7-2, however, the ideal gas law appears to be comparatively accurate, while the van der Waals equation appears to be worthless.

> Two properties of ideal gases are (1) no intermolecular interactions and (2) particles have zero volume.

The reasonable "accuracy" of the ideal gas law in this example is explained by Figure 7-2. While the figure is specific to methane at 200 K, the qualitative behavior shown in Figure 7-2 is typical for most compounds at most temperatures.

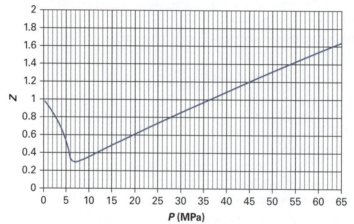

FIGURE 7-2 Compressibility (Z) of methane vs. pressure for a constant temperature $T = 200$ K.
(Produced using data obtained from Lemmon et al. 2012.)

Notice that $Z = 1$ (ideal gas behavior) at the low pressure limit, and Z falls below 1 as pressure increases. $Z < 1$ in this region because of attractive intermolecular forces that cause the molar volume to be smaller than it would be for an ideal gas. However, this trend of decreasing Z with increasing P cannot continue indefinitely. If molecules are too close together, repulsive forces will exceed attractive forces. At some point, the molecules are compressed so close together that increasing P has little further effect on $\underline{V}$. Since $Z = P\underline{V}/RT$, if P is increased while $\underline{V}$ remains nearly constant, Z must also increase. At $P \approx 36$ MPa, $Z = 1$, not because the attractive and repulsive phenomena are negligible (as they would be for an ideal gas), but because this happens to be the one pressure at which they cancel each other out. Example 7-2 was deliberately designed to fall near this pressure. Problem 7-8 looks at a variant on this example in which the ideal gas law prediction is far worse than it was in Example 7-2.

> While the ideal gas estimate of pressure was comparatively close, it's worth noting that the estimated pressure was 5% lower than the real pressure. In a safety analysis, we'd rather overestimate the pressure than underestimate it.

The "failure" of the van der Waals equation can be explained simply by saying the van der Waals equation isn't flexible enough to be used at all conditions. The numerical values of a and b that we used in Example 7-1 were determined by fitting the equation to critical point data for methane. This method is illustrated in detail in Example 7-4. If we had used other data (e.g., the data in Figure 7-2) as our starting point, we would have gotten different values for a and b. Alternative values of a and b might have worked better for this particular example, but there simply is no single set of parameters a and b that accurately model methane at all temperatures and pressures. Methane is a fairly simple compound: It is not polar, symmetrical, and contains only two different elements (C and H) and one kind of bond (the C—H single bond.) If the van der Waals equation doesn't describe methane satisfactorily, one can't expect it to be useful for larger, more complex compounds.

The van der Waals equation was first proposed in 1873, and at that time, it represented a breakthrough in the understanding of chemical compounds and the ability to build mathematical models. However, it has since been replaced by more accurate (and more complex) equations. Modern equations of state, and their use, are presented in this chapter.

7.2 Cubic Equations of State

The van der Waals equation is a cubic EOS—it is first degree in T and P but third degree in $\underline{V}$. Consequently, for a given T and P, there are three solutions for $\underline{V}$ (although in some circumstances such as Example 7-1 two solutions are imaginary). Although the van der Waals equation is no longer in common use, we will see in this chapter that there are equations of state in common use that are more complex variants of the van der Waals equation. This section outlines how values of adjustable parameters (such as a and b in the van der Waals equation) are determined, how cubic equations of state can be solved, and how they can be used in combination with the mathematical techniques described in the previous section. We begin by addressing the question: Why use *cubic* (third degree) equations, specifically?

7.2.1 The Rationale for Cubic Equations of State

Consider Figure 7-3, which shows a P-$\underline{V}$ diagram for H_2O. Imagine that we have a piston-cylinder device containing 1 mole of water vapor initially at $P = 100$ kPa and $T = 300°C$. We gradually lower the piston, compressing the vapor into a smaller volume. At the same time, we maintain a constant temperature of $T = 300°C$. The $T = 300°C$ isotherm in Figure 7-3 helps illustrate what is observed.

- At $T = 300°C$ and $P = 100$ kPa, H_2O is a vapor with a molar volume of $\underline{V} = 47{,}540$ cm³/mol (0.04754 m³/mol).

- As we decrease the volume at constant temperature, the pressure of the vapor increases, as illustrated by curve AB on Figure 7-3.

- This continues until we reach $P = 8.588$ MPa. At $P = 8.588$ MPa and $T = 300°C$, we have a saturated vapor with molar volume $\underline{V} = 391$ cm³/mol, as indicated as point B on Figure 7-3.

- If we continue to compress the saturated vapor into a smaller volume, the pressure will *not* increase. Instead, vapor will begin to condense into saturated liquid, which has a molar volume of 25.3 cm³/mol, as indicated by point C. Point C is also shown on Figure 7-4, which contains the same data shown in Figure 7-3 but focuses on the small molar volumes associated with liquids.

- As the volume of the cylinder decreases, the proportion of liquid will increase. For example, when the molar volume of the two-phase mixture is $\underline{V} = 208$ cm³/mol, it will be 50% vapor and 50% liquid: 208 cm³/mol = 0.5(391 cm³/mol) + 0.5(25.3 cm³/mol).

Phase changes and the theory that explains why they occur at specific temperatures and pressures are discussed in Chapter 8.

In this calculation the quality (q) of the VLE system is 0.5.

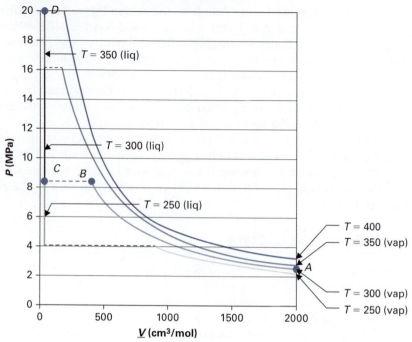

FIGURE 7-3 The P-$\underline{V}$ diagram for liquid and vapor H_2O at 250, 300, 350, and 400°Celsius.

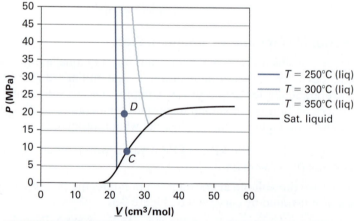

FIGURE 7-4 The P-$\underline{V}$ diagram for H_2O at various temperatures, focusing on the liquid phase.

FOOD FOR THOUGHT 7-2

This entire section on equations of state focuses on liquids and vapors. Why are solids not considered?

In Figure 7-3 the three curves representing liquid properties at 250, 300, and 350°C look like vertical lines that are indistinguishable from each other. Figure 7-4 focuses on small values of $\underline{V}$ to show the liquid phase behavior more clearly.

■ When the molar volume reaches $\underline{V} = 25.3$ cm³/mol, the last bubble of vapor will condense. We will then have a pure, saturated liquid at $T = 300°C$. All of the H_2O is now described by point C on Figures 7-3 and 7-4.

■ If we continue lowering the piston, the pressure will increase extremely sharply. For example, the pressure at point D ($P = 20$ MPa and $\underline{V} = 24.5$ cm³/mol) is more than double that of saturated liquid, but the molar volume of the liquid has changed by only ≈3%. Thus, on Figures 7-3 and 7-4, curve CD is very nearly a vertical line.

If you select a different temperature, say $T = 100°C$ or $T = 350°C$, and repeat this experiment, you will observe essentially the same sequence of events, although the

TABLE 7-1 Saturation conditions for H_2O at selected temperatures.

Temp. (°C)	Vapor Pressure (MPa)	Saturated Vapor Molar Volume $\underline{V}^V$ (cm³/mol)	Saturated Liquid Molar Volume $\underline{V}^L$ (cm³/mol)	$\underline{V}^V/\underline{V}^L$
100	0.1013	30,120	18.8	1600
200	1.555	2290	20.8	110
300	8.588	391	25.3	15.5
350	16.53	159	31.3	5.08
370	21.04	90.1	39.9	2.26
373.95	22.06	56.0	56.0	1

Table 7-1 illustrates that saturated liquid and saturated vapor properties converge and become identical at the critical point. At temperatures far below the critical point, $\underline{V}^L$ is insignificant compared to $\underline{V}^V$. This observation will form a basis for the Clausius-Clapeyron equation in Chapter 8.

numerical values of molar volume, saturation pressure, etc. might be very different, as illustrated in Table 7-1. However, if you choose a high enough temperature, say $T = 400°C$, the phase transition does not occur; the pressure simply increases continuously as molar volume decreases. This is because the critical temperature of H_2O is about $T = 374°C$. At temperatures higher than this, water vapor will not condense—regardless of pressure—so no two-phase equilibrium mixture is possible.

Figure 7-3, Figure 7-4, and Table 7-1 represent the actual, observed behavior of H_2O. While the specific numbers vary greatly from one compound to another, qualitatively, these figures are representative of P-$\underline{V}$-T behavior for most any chemical compound. Thus, they represent the kind of data we are trying to model when we propose an equation of state. Let us now consider what sort of equation would work.

We can easily imagine fitting an equation to curve AB in Figure 7-3, and thus modeling P vs. $\underline{V}$ for water vapor at $T = 300°C$. We can imagine building temperature dependence into the equation, so that it models the vapor phase at all temperatures—rather than just $T = 300°C$ specifically. We can also imagine fitting an equation to curve CD; thus, modeling P-$\underline{V}$ for water in the liquid phase. However, practically speaking, it would be useful to model both the liquid and the vapor with a single equation.

Figure 7-5 shows an attempt to fit a curve to the data from the $T = 300°C$ isotherm. For a P-$\underline{V}$ function to include both curves AB (vapor phase) and CD (liquid

The critical point was first described in Section 2.2.1.

Chapter 6 showed that we can use equations of state to model changes in enthalpy, entropy, internal energy, etc., but the theory in Chapter 6 is based on partial differential calculus. You may recall from calculus that you can't take a derivative of a function unless it is continuous.

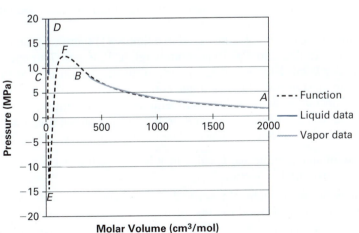

FIGURE 7-5 A continuous function that models the P vs. $\underline{V}$ data for liquid water and water vapor at $T = 300°C$ (curves AB and CD in Figure 7-3).

phase) and be continuous, the region in between must have a maximum and a minimum, as illustrated in Figure 7-5. Consequently, if a single, *continuous* equation of state is to be used to describe both the liquid and the vapor phases, it must be (at least) a third-degree equation in $\underline{V}$. While these data are for water at $T = 300°C$, the phenomenon is universal: In order to model P-$\underline{V}$-T for both the liquid and vapor phases with a single equation for any compound, it must be an equation of *at least* degree 3. Cubic equations have proven to work well and are in common use, as discussed throughout the remainder of Section 7-2.

7.2.2 Modern Cubic Equations of State

The van der Waals equation, introduced in Chapter 2, is the oldest cubic EOS. It can be written as in Equation 2.53:

$$P = \frac{RT}{\underline{V} - b} - \frac{a}{\underline{V}^2}$$

or in terms of the compressibility factor Z (Equation 2.55):

The compressibility factor (Z) was first defined in Section 2.3.4, where $Z = P\underline{V}/RT$

$$Z = \frac{\underline{V}}{\underline{V} - b} - \frac{a}{RT\underline{V}}$$

Peng and Robinson (Robinson, 1976) describe the two terms on the right-hand side of Equation 2.53 as the "repulsion pressure" and the "attraction pressure," so the van der Waals equation can be viewed as

$$P = P^{\text{rep}} + P^{\text{att}} \tag{7.10}$$

with

$$P^{\text{rep}} = \frac{RT}{\underline{V} - b} \quad \text{and} \quad P^{\text{att}} = -\frac{a}{\underline{V}^2} \tag{7.11}$$

The "repulsion pressure" is proportional to T, because as temperature increases, molecules move more quickly and collisions become more common. Because two particles cannot occupy the same space, a fluid has a minimum molar volume represented by b. As the actual molar volume $\underline{V}$ gets very close to the minimum possible volume b, the repulsion pressure goes toward infinity.

The "attraction pressure" is intended to model the effect of intermolecular attractive forces, which become less important when the molecules are father apart; hence, the term is proportional to $1/\underline{V}^2$.

The Peng-Robinson equation was used to fit the data in Figure 7-5.

The motivational example illustrated two things: (1) there's a practical need for equations of state that allow us to accurately relate P, $\underline{V}$, and T to each other and (2) the van der Waals EOS is not acceptably accurate, at least not for supercritical methane. The most useful EOS is a single equation that describes the liquid, vapor, and supercritical states at all realistic temperatures and pressures.

While the van der Waals equation is no longer in common use, it has proved to be a useful foundation for equations of state:

- The use of an equation that is cubic in $\underline{V}$ has stood the test of time for reasons that were explored in Section 7.2.1.

- The idea of an "attraction" term and a "repulsion" term with two adjustable parameters has also stood the test of time. In subsequent equations, these parameters have typically been called "a" and "b" for consistency with the van der Waals equation.

■   The van der Waals "repulsion pressure" term $RT/(V - b)$ reappears explicitly and unmodified in several subsequent equations of state.

The innovations that have been added since the van der Waals EOS was published in the 1870s are

■   The parameter a, rather than being a constant, is a function of temperature.

■   The parameter b, which represents the molar volume of a substance at maximum compression, appears in the "attraction" term as well as the "repulsion" term.

Modern cubic equations of state vary from each other primarily in the specific mathematical form of the "attraction" term. Two of the most prominent examples are the **Soave equation** (Soave, 1972) and the **Peng-Robinson equation** (Robinson, 1976), and these will be used as the primary cubic equations of state throughout the remainder of this book.

The Soave equation is

$$P = \frac{RT}{V - b} - \frac{a}{V(V + b)} \tag{7.12}$$

which, in terms of the compressibility factor, is

$$Z = \frac{V}{V - b} - \left(\frac{a}{RT}\right)\left(\frac{1}{V + b}\right) \tag{7.13}$$

The Peng-Robinson equation is

$$P = \frac{RT}{V - b} - \frac{a}{V(V + b) + b(V - b)} \tag{7.14}$$

which, when written in terms of the compressibility factor, is:

$$Z = \frac{V}{V - b} - \left(\frac{aV}{RT}\right)\left[\frac{1}{V(V + b) + b(V - b)}\right] \tag{7.15}$$

The development of these equations will not be discussed, as it was primarily empirical—many equations have been tried and these have proved to work particularly well for a broad range of compounds. *In both the Peng-Robinson and Soave equations, b is a constant, but a is a function of temperature.* In Section 7.2.3 and Section 7.2.4, we assume the parameters a and b are known, examine how to solve the equations, and interpret the solutions. In Section 7.2.5, we describe the temperature dependence of a and show how values for a and b can be determined when they are not known.

The "repulsion pressure" term is identical in the van der Waals, Peng-Robinson, and Soave equations. The "attraction pressure" terms are more complex in the Peng-Robinson and Soave equations than in the van der Waals equation, and these newer equations of state typically produce much better answers.

The Soave equation is also often referred to as the Soave-Redlich-Kwong, or Redlich-Kwong-Soave equation, because of its similarity to an equation Redlich and Kwong published in 1949.

Peng and Robinson (Robinson, 1976) used the word "semiempirical" to describe the van der Waals EOS and all of the cubic equations (including their own) stemming from it.

7.2.3 Solving Cubic Equations of State

An equation of state has been defined throughout this book as a mathematical relationship among P, V, and T, such that if any two are known, the third can be found. The van der Waals, Peng-Robinson, and Soave equations are conventionally written in the form $P = f(T, V)$ or $Z = f(T, V)$ and are readily solved if P is the unknown. Solving for T looks like a simple algebraic manipulation at first glance, since there is only one T explicitly shown in Equations 7.13 and 7.15, but keep in mind that in the Peng-Robinson and Soave equations, a is not a constant, it is a function of T.

This section, however, will focus on solving for $\underline{V}$, which is the variable for which the equations are cubic.

Cubic equations do not lend themselves to simple algebraic solutions, but they can be solved analytically, as described in Appendix B. However, modern computing means that engineers rarely need to implement this analytical solution approach "by hand." Indeed, solving cubic equations in practice is much simpler than it was even a decade or two ago, as many modern calculators are capable of solving cubic equations with no input required from the user beyond the equation itself. Mathematical software packages can also be used (Cutlip, 2007).

Example 7-3 illustrates a graphical solution approach that can be readily implemented using any conventional spreadsheet application, such as Microsoft Excel.

EXAMPLE 7-3	**DETERMINING MOLAR VOLUME OF PENTANE USING A CUBIC EQUATION OF STATE**

The value of a in this example is specific to the temperature $T = 100°C$. Section 7.2.4 discusses how the numerical values of a and b are determined for the van der Waals, Peng-Robinson, and Soave equations of state.

Find the molar volume of n-pentane at $T = 100°C$ and $P = 100$ kPa. The Peng-Robinson parameters for n-pentane at this temperature are: $a = 2.417 \times 10^{12}$ Pa · cm^6/mol^2 and $b = 90.18$ cm^3/mol.

SOLUTION:

Step 1 *Solve Peng-Robinson equation for one value of $\underline{V}$*
The Peng-Robinson equation can be expressed as

$$P = \frac{RT}{\underline{V} - b} - \frac{a}{\underline{V}(\underline{V} + b) + b(\underline{V} - b)}$$

Temperature must be expressed on an absolute scale, so $T = (100 + 273.15)$ K = 373.15 K. While there is no algebraic way to rearrange the equation so it is explicit in $\underline{V}$, we can easily find the P that corresponds to any $\underline{V}$. For example, if $\underline{V} = 100$ cm^3/mol,

$$P = \frac{\left(8.314 \times 10^6 \dfrac{\text{Pa} \cdot \text{cm}^3}{\text{mol} \cdot \text{K}}\right)(373.15\,\text{K})}{\left(100\dfrac{\text{cm}^3}{\text{mol}} - 90.18\dfrac{\text{cm}^3}{\text{mol}}\right)}$$

(7.16)

$$- \frac{2.417 \times 10^{12} \dfrac{\text{Pa} \cdot \text{cm}^6}{\text{mol}^2}}{\left(100\dfrac{\text{cm}^3}{\text{mol}}\right)\left(100\dfrac{\text{cm}^3}{\text{mol}} + 90.18\dfrac{\text{cm}^3}{\text{mol}}\right) + \left(90.18\dfrac{\text{cm}^3}{\text{mol}}\right)\left(100\dfrac{\text{cm}^3}{\text{mol}} - 90.18\dfrac{\text{cm}^3}{\text{mol}}\right)}$$

$$P = 194.5\,\text{MPa}$$

This is not a useful result by itself; we are trying to find the solution for $P = 0.1$ MPa. However, if we enter Equation 7.16 into a spreadsheet, solving for several hundred different values of $\underline{V}$ takes little more effort than solving for one.

Step 2 *Solve for P over a range of values of $\underline{V}$*
We solve for P at a wide range of values for $\underline{V}$ and identify the values that correspond to $P = 0.1$ MPa. Table 7-2 shows a small subset of the data, and Figure 7-6 shows plots.

TABLE 7-2 Solutions of Equation 7.16 for different values of $\underline{V}$.

$\underline{V}$ (cm³/mol)	P (MPa)
133	0.3996
134	−0.3055
135	−0.9574
136	−1.5603
465	−0.001328
470	0.0448748
475	0.0895148
480	0.1326498
30000	0.1010546
30200	0.1004008
30400	0.0997554
30600	0.0991182

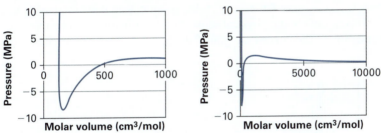

FIGURE 7-6 P-$\underline{V}$ plots for the $T = 100°C$ isotherm, as calculated for
n-pentane using the Peng-Robinson equation.

Step 3 *Identify the solutions that correspond to P = 0.1 MPa*
While step 2 produced a complete isotherm for $T = 100°C$, we were asked to find the
molar volume at $P = 0.1$ MPa. There are three molar volumes that correspond to a pressure of 0.1 MPa; Table 7-2 reveals that one of them falls between 133 to 134 cm³/mol, one
between 475 to 480 cm³/mol, and one between 30,200 to 30,400 cm³/mol. The exact values of $\underline{V}$ can be found using the "Solver" function of Microsoft Excel, or even by trial-
and-error until one finds values of $\underline{V}$ that yield $P = 0.1$ MPa to as many significant digits
as desired. In this case, the solutions are $\underline{V} = 133.42$ cm³/mol, $\underline{V} = 476.20$ cm³/mol, and
$\underline{V} = 30,324$ cm³/mol.

A graphical approach can be applied to any equation of state and any temperature. If
the temperature of interest is below the critical point, and the isotherm produced does
not qualitatively resemble Figure 7-6, most likely you either made an error in entering
the equation, or your range of $\underline{V}$ values is simply not sufficient to reveal all three roots.

While the solution illustrated in Example 7-3 is a "brute force" approach that may seem crude compared to an analytic solution or an efficient numerical solution algorithm, it can be implemented rapidly with modern spreadsheets, and it has instructional value. Rather than simply giving three answers as a calculator would do, the graphical approach allows one to examine the full behavior of the equation. Thus, using the graphical approach (at least some of the time) can help you develop an intuitive feel for equations of state. A graphical approach can also help resolve confusion. For example, if you are expecting three solutions and your calculator returns only one, there are many possibilities: Was your expectation wrong? Did you enter the equation wrong? Are the three roots identical? Did the calculator fail to detect two roots? Making a full graph of the equation can help you answer such questions.

7.2.4 Interpreting Solutions to Cubic Equations of State

Figure 7-7 illustrates solutions to a cubic EOS in the form of P vs. $\underline{V}$, for various values of T. For temperatures above the critical point, P simply decreases as $\underline{V}$ increases. For lower values of T, the isotherm has a maximum and a minimum. Consequently, over a large range of P values, there are three real values for $\underline{V}$ that each produce the same P. Section 7.2.3 discussed how to find these three real solutions. The natural next question is "If we have three solutions, which one is right?"

Examining Figure 7-5 in Section 7.2.1 reveals that a significant portion of the function doesn't correspond to any real data. Part of this function has negative values of pressure, which doesn't make physical sense. This is a commonly, but not always, observed feature of cubic equations of state. Another unrealistic feature of part of the function appears on curve EF, where molar volume is actually increasing as pressure increases. This doesn't make physical sense, but it is a feature of all cubic equations of state; the portion of the curve that "joins" the liquid and vapor data includes a region

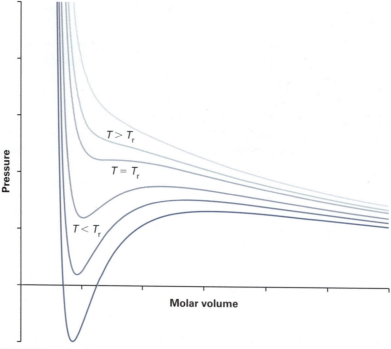

FIGURE 7-7 Isotherms generated using a cubic equation of state.

in which $\underline{V}$ increases with increasing P. The requirement that molar volume must decrease as pressure increases isothermally can be stated mathematically as

$$\left(\frac{\partial \underline{V}}{\partial P}\right)_T < 0 \qquad (7.17)$$

Equation 7.17 is often called the **stability criterion**; if the solution to an equation of state at a particular point doesn't meet this criterion, it doesn't describe a stable phase.

Revisiting Example 7-3, we determined that for n-pentane at $P = 0.1$ MPa and $T = 100°C$, the Peng-Robinson equation produces three solutions for molar volume: $\underline{V} = 133.42$ cm³/mol, $\underline{V} = 476.20$ cm³/mol, and $\underline{V} = 30,324$ cm³/mol. When we look at the form of the data we're actually modeling (e.g., Figure 7-3, Figure 7-5), these three solutions seem to lend themselves to a simple physical interpretation. The smallest $\underline{V}$ (133.42 cm³/mol) corresponds to a liquid $\underline{V}$, the largest (30,324 cm³/mol) corresponds to a vapor $\underline{V}$, and the intermediate one (476.20 cm³/mol) has no physical significance. As illustrated in Section 7.2.1, we are using a continuous function to model data that is not continuous, and the point $P = 0.1$ MPa, $\underline{V} = 476.20$ cm³/mol appears on the portion of the curve that violates the stability criterion.

However, Figure 7-5 also reveals that there are two sections of the curve (CE and FB) that meet the stability criterion but still don't represent the data. Curve CE represents the liquid phase, but at pressures below the vapor pressure at which liquid does not exist—at least not at equilibrium. Similarly, curve FB represents the vapor phase—but at pressures above the vapor pressure. The word **metastable** is used to describe the conditions represented by curves CE and FB.

According to Brown (2011), the experimental vapor pressure of n-pentane is about $P^{sat} = 0.593$ MPa at $T = 100°C$. Thus, at $P = 0.1$ MPa and $T = 100°C$, n-pentane exists in the vapor phase, and its molar volume is $\underline{V} = 30,324$ cm³/mol. The value $\underline{V} = 133.42$ cm³/mol is a metastable solution—for *pure* pentane at this temperature and pressure it isn't physically realistic either.

Thus, while "the smallest $\underline{V}$ is the liquid molar volume" sounds like a simple and logical interpretation, we need to be a little more specific and say that in the three-solution region:

■ *If* there is a stable liquid phase, its molar volume is represented by the smallest value of $\underline{V}$.

■ *If* there is a stable vapor phase, its molar volume is represented by the largest value of $\underline{V}$.

■ The vapor pressure is the only pressure at which the liquid and vapor values of $\underline{V}$ both represent stable phases.

■ The middle value of $\underline{V}$ never represents a stable phase.

In this analysis, we determined which of the three roots for $\underline{V}$ was metastable and which was "right" by comparing the actual pressure to the vapor pressure. What if the vapor pressure had been unknown? Chapter 8 addresses this question, discussing methods of estimating vapor pressure. Indeed, it is possible to compute vapor pressures directly from an equation of state, as illustrated in Example 8-5.

To close, we note that real compounds can be observed in metastable conditions in certain circumstances. For example, if steam at $P = 101.3$ kPa and $T = 100°C$ is either cooled by a fraction of a degree or compressed to a fractionally higher pressure, the steam should condense—liquid is the most stable state at the new T and P. However, condensation requires nucleation sites—locations where the liquid phase begins to form. Steam at 99.5°C and 101.3 kPa is thus an example of a "metastable" state.

Pure n-pentane is a vapor at $T = 100°C$ and $P = 0.1$ MPa, but n-pentane can exist in the liquid phase at $T = 100°C$ and $P = 0.1$ MPa if it is a component of a mixture. Chapter 9 discusses this distinction further.

You may recall from introductory chemistry that chemical reactions have an "activation energy" required to form the transition state. You can think of the nucleation site in a phase change as analogous to the transition state in a chemical reaction.

It would not exist at equilibrium, but it could exist temporarily until the nucleation process begins. In this introductory book, however, we are only concerned with states that are stable at equilibrium and will not substantially address "metastable" states or how this nucleation process occurs.

7.2.5 Fitting Parameters to Cubic Equations of State

This section has discussed the van der Waals, Soave, and Peng-Robinson equations and their solutions. Up to this point, the parameters *a* and *b* have simply been given. We now examine the question of how specific values are assigned to these parameters.

In Chapter 2, we introduced the physical uniqueness of the critical point. Below its critical temperature, a compound can exist in the liquid, solid, or vapor state; above its critical temperature, the compound does not condense into the liquid phase, regardless of pressure. Figure 7-7 illustrates the solutions to a cubic equation of state at various temperatures.

The curves in Figure 7-7 are isotherms; they represent P as a function of $\underline{V}$ with T held constant. Thus, the derivative of one of these curves is $(\partial P/\partial \underline{V})_T$.

Figure 7-7 shows the uniqueness of the critical point in the context of a mathematical model.

- For temperatures above the critical point, $(\partial P/\partial \underline{V})_T$ is always negative; pressure decreases continuously as molar volume increases.

- For temperatures below the critical point, each isotherm has a maximum and a minimum—two points at which $(\partial P/\partial \underline{V})_T = 0$. While these points occur on the portion of the curve that does not represent observed data (see Section 7.2.4), they are a recognizable aspect of our mathematical model.

> Recall from calculus that at a saddle point, the first and second derivatives (dy/dx and d^2y/dx^2) are both zero.

Mathematically, the transition between these two behaviors takes the form of a single temperature at which there is a saddle point. Notice in Figure 7-7 that as the temperature approaches the critical temperature, the maximum and minimum points on the isotherms get closer together; you can think of the saddle point as the point where the maximum and minimum "meet." Thus, *a cubic EOS has only one saddle point, and to be physically realistic, it must occur at the critical point.* This observation has tremendous significance in model-building. Mathematically, for a function $y = f(x)$, at a saddle point, the first and second derivatives dy/dx and d^2y/dx^2 both equal 0. Example 7-4 considers how this observation can be applied to the van der Waals EOS.

EXAMPLE 7-4	**FITTING PARAMETERS *a* AND *b* FOR THE VAN DER WAALS EQUATION TO CRITICAL POINT DATA**

The critical point of water is $T_c = 647.3$ K and $P_c = 22.12$ MPa. Determine the van der Waals parameters *a* and *b* for water, and also find the compressibility of water at its critical point (Z_c). Evaluate the results using data from the steam tables.

SOLUTION: As discussed previously, for a cubic equation of state in the form $P = f(T, \underline{V})$, the isotherm representing $T = T_c$ will have a saddle point located at the critical point. This means that at the critical point, $(\partial P/\partial \underline{V})_T$ and $(\partial^2 P/\partial \underline{V}^2)_T$ will both equal 0. To apply these facts, we must first evaluate the derivatives.

Step 1 *Differentiate the van der Waals equation with respect to $\underline{V}$*
Differentiating the van der Waals equation with respect to $\underline{V}$ while holding T constant produces

$$P = \frac{RT}{V - b} - \frac{a}{V^2}$$

$$\left(\frac{\partial P}{\partial V}\right)_T = \frac{-RT}{(V - b)^2} + \frac{2a}{V^3} \tag{7.18}$$

and

$$\left(\frac{\partial^2 P}{\partial V^2}\right)_T = \frac{2RT}{(V - b)^3} - \frac{6a}{V^4} \tag{7.19}$$

At the critical point $(T = T_c)$, as noted, these derivatives are both zero:

$$\left(\frac{\partial P}{\partial V}\right)_T = \frac{-RT_c}{(V_c - b)^2} + \frac{2a}{V_c^3} = 0 \tag{7.20}$$

and

$$\left(\frac{\partial^2 P}{\partial V^2}\right)_T = \frac{2RT_c}{(V_c - b)^3} - \frac{6a}{V_c^4} = 0 \tag{7.21}$$

Step 2 *Consider degrees of freedom*

Since T_c and P_c are both known (as is the constant R), Equations 7.20 and 7.21 contain three unknowns: the parameters a and b and the critical volume V_c. We can obtain a third equation by using the EOS itself. If we expect the van der Waals equation to be a reasonable model for liquids, vapors, gases and supercritical fluids at all temperatures and pressures, then we certainly expect it to be valid at the critical point. Thus, there are three equations and three unknowns: the values of a and b can be determined using no information beyond the critical pressure and temperature.

Step 3 *Express V in terms of b*

Equations 7.20 and 7.21 can be rearranged into

$$\frac{RT_c}{(V_c - b)^2} = \frac{2a}{V_c^3} \tag{7.22}$$

and

$$\frac{2RT_c}{(V_c - b)^3} = \frac{6a}{V_c^4} \tag{7.23}$$

These equations are greatly simplified when we divide Equation 7.23 by Equation 7.22 for

$$\frac{2}{(V_c - b)} = \frac{3}{V_c} \tag{7.24}$$

which simplifies to

$$V_c = 3b \tag{7.25}$$

Step 4 *Express a as a function of b*

Plugging the expression for V_c given in Equation 7.25 into Equation 7.22 gives

$$\frac{RT_c}{(3b - b)^2} = \frac{2a}{(3b)^3} \tag{7.26}$$

This can be solved for a, as

$$a = \frac{27}{8} RT_c b \tag{7.27}$$

Step 5 *Solve the equation of state for b*

Substituting the expressions for a and $\underline{V}$ (Equations 7.25 and 7.27) into the van der Waals EOS produces:

$$P_c = \frac{RT_c}{\underline{V}_c - b} - \frac{a}{\underline{V}_c^2}$$

$$P_c = \frac{RT_c}{3b - b} - \frac{\frac{27}{8}RTb}{(3b)^2} \tag{7.28}$$

$$P_c = \frac{RT_c}{2b} - \frac{3RT_c}{8b}$$

$$b = \frac{RT_c}{8P_c} \tag{7.29}$$

Step 6 *Solve for a and Z_c*

Back-substituting the expression for b into Equations 7.27 and 7.25 gives

$$a = \frac{27}{64} \frac{R^2 T_c^2}{P_c} \tag{7.30}$$

and

$$\underline{V}_c = \frac{3}{8} \frac{RT_c}{P_c} \tag{7.31}$$

We were asked to determine the compressibility factor at the critical point. Recall that by definition, the compressibility factor is $Z = P\underline{V}/RT$. The value of Z_c is obtained directly from Equation 7.31 as

$$Z_c = \frac{P_c \underline{V}_c}{RT_c} = \frac{P_c}{RT_c}\left(\frac{3}{8}\frac{RT_c}{P_c}\right) = 0.375 \tag{7.32}$$

Step 7 *Plug in known information*

The outcome of step 6 is that Z_c, which is the compressibility at the critical point, is not dependent upon the critical temperature and pressure; it is simply equal to 0.375. Inserting the known values of critical temperature and pressure into Equations 7.29 and 7.30 to find a and b for water gives

$$a = \frac{27}{64} \frac{R^2 T_c^2}{P_c}$$

$$a = \frac{27}{64} \frac{\left(8.314 \frac{\text{Pa} \cdot \text{m}^3}{\text{mol} \cdot \text{K}}\right)^2 (647.3\,\text{K})^2}{22.12\,\text{MPa} \times (10^6\,\text{Pa/1 MPa})} = \mathbf{0.552} \frac{\textbf{Pa} \cdot \textbf{m}^6}{\textbf{mol}^2} \tag{7.33}$$

$$b = \frac{RT_c}{8P_c}$$

$$b = \frac{\left(8.314 \frac{\text{Pa} \cdot \text{m}^3}{\text{mol} \cdot \text{K}}\right)(647.3\,\text{K})}{(22.12\,\text{MPa}) \times (10^6\,\text{Pa/1 MPa})} = \mathbf{30.4} \times \mathbf{10^{-6}} \frac{\textbf{m}^3}{\textbf{mol}} \tag{7.34}$$

Step 8 *Compare to published information*

The calculated Z_c in this example is not very good; we can determine from the steam tables that Z_c for water is only 0.233. This discrepancy is discussed further after the example.

We can evaluate the van der Waals constants, a and b, by comparing predictions computed using these values to data. For example, Appendix A shows that at $T = 100°C$, saturated liquid water has a specific volume $\hat{V}^L = 0.001043$ m³/kg, and saturated vapor has $\hat{V}^V = 1.6718$ m³/kg. Converting to a molar basis gives

$$\underline{V}^L = \left(0.001043\frac{m^3}{kg}\right)\left(\frac{1\,kg}{1000\,g}\right)\left(\frac{18.015\,g}{mol}\right) = 18.79 \times 10^{-6}\frac{m^3}{mol} \qquad (7.35)$$

$$\underline{V}^V = \left(1.6718\frac{m^3}{kg}\right)\left(\frac{1\,kg}{1000\,g}\right)\left(\frac{18.015\,g}{mol}\right) = 0.030117\frac{m^3}{mol} \qquad (7.36)$$

Using the van der Waals equation to compute pressure corresponding to these values of $\underline{V}$ gives

$$P = \frac{RT}{\underline{V} - b} - \frac{a}{\underline{V}^2}$$

$$P^L = \frac{\left(8.314\frac{Pa \cdot m^3}{mol \cdot K}\right)(373.15\,K)}{18.79 \times 10^{-6} - 30.4 \times 10^{-6}\frac{m^3}{mol}} - \frac{0.552\frac{Pa \cdot m^6}{mol^2}}{\left(18.79 \times 10^6\frac{m^3}{mol^2}\right)} = -1{,}830.7\,MPa \qquad (7.37)$$

$$P^V = \frac{\left(8.314\frac{Pa \cdot m^3}{mol \cdot K}\right)(373.15\,K)}{\left(0.030117 - 30.4 \times 10^63\right)\frac{m^3}{mol}} - \frac{0.552\frac{Pa \cdot m^6}{mol^2}}{\left(0.030117\frac{m^3}{mol^2}\right)} = 0.1025\,MPa \qquad (7.38)$$

The pressure P^L that was computed from $\underline{V}^L$ is a negative number; it's physically unrealistic. The source of the problem is our computed value of b. This parameter is intended to represent the molar volume of the molecule at "maximum compression," which is the lowest value $\underline{V}$ can possibly take. However, this value of b is larger than the experimental values of $\underline{V}^L$ that are observed at most any conditions.

Meanwhile, the pressure P^V that we computed from $\underline{V}^V$ (0.1025 MPa) is quite close to the familiar experimental value ($P^{sat} = 0.1013$ MPa at $T = 100°C$), but this is not really because the values of a and b are "good"—it's more that they have little effect for a vapor at this low pressure. The ideal gas law for this T and $\underline{V}$ produces $P = 103$ kPa; our answer is only fractionally different than it would have been if $a = b = 0$.

What we must remember is that real fluids do not "obey" our equations of state; rather, we build mathematical models that attempt to describe real behavior as closely as possible. Our calculations in steps 1 through 7 are correct if you start from the critical point, but the answer is only useful to the extent that our original assumption—that the van der Waals equation describes P-$\underline{V}$-T relationships for water—is correct. There simply is no single set of values for a and b that accurately model H_2O at all conditions.

However, with the more flexible Soave and Peng-Robinson equations, it is possible to more accurately model the P-$\underline{V}$-T behavior of many compounds at a wide range of conditions. Values of the parameters a and b can be found through examination of the critical point, as was done in this example.

ALL of the cubic equations of state examined in this section simplify to the ideal gas law when $a = b = 0$.

The data in Table 7-3 was not available when van der Waals originally proposed his equation of state in 1873. In fact, the concept of a "critical point" was itself very new at that time.

TABLE 7-3 Critical properties of various compounds.

Compound	T_c (K)	P_c (MPa)	Z_c
Water	647.3	22.12	0.233
Oxygen	154.6	5.043	0.288
Carbon dioxide	304.2	7.382	0.274
Freon-22	369.8	4.970	0.268
Ethylene	282.4	5.032	0.277
Propylene	364.8	4.613	0.275
1,3-Butadiene	425.4	4.330	0.270
Ethanol	516.4	6.384	0.248
Propanol	536.7	5.170	0.253
Isopropanol	508.3	4.764	0.248
Ethane	305.4	4.880	0.284
Propane	369.8	4.249	0.281
n-Butane	425.2	3.797	0.274
Isobutane	408.1	3.648	0.282

Based on data from Lira, J.R. Elliott and C. T., Introductory Chemical Engineering Thermodynamics, Prentice-Hall, 1999.

Notice the similarity between values of Z_c, especially among groups of similar compounds like ethane, propane, and butane, or ethanol, propanol, and isopropanol. This is a basis for the principle of corresponding state described in Section 7-3.

Ethylene, propylene, and butadiene are among the top chemicals in annual worldwide sales by mass, because they are commonly used as intermediates for synthesizing larger compounds. The double bonds provide an active site for either polymerization or adding new functional groups.

While Example 7-4 examined H_2O, no information specific to H_2O was used in steps 2–6. Thus, Equations 7.27, 7.29, and 7.32 are, mathematically speaking, valid for any compound. Equation 7.32 essentially says that, IF the vapor and liquid states of all real compounds were accurately described by the van der Waals EOS, then all real compounds would have a compressibility of 0.375 at their critical point. Table 7-3 illustrates that this outcome leaves much room for improvement, as Z_c is typically smaller than 0.375 and is clearly not identical for all compounds. This data is another illustration of why the van der Waals equation is no longer in common use. The data does, however, illustrate van der Waals' substantial improvement upon the ideal gas law, which says that $Z = 1$ at the critical point and at all other conditions.

The strategy of fitting parameters for an EOS by applying mathematical principles to the critical point can also be applied to the more modern cubic equations. Full derivations will not be shown, but the approach is analogous to that shown in Example 7-4 for the van der Waals equation.

In the Soave equation, the value of b is equal to

$$b = 0.08664R\frac{T_c}{P_c} \tag{7.39}$$

In the Soave equation, the parameter a is a function of temperature. However, the methodology outlined in Example 7-4 can be used to determine the value of a at the critical point (a_c):

Note that the forms of Equations 7.39 and 7.48 are identical to that of 7.29, although the constants relating b to P_c and T_c are different.

$$a_c = 0.42747R^2\frac{T_c^2}{P_c} \tag{7.40}$$

Soave used the variable α to quantify the temperature dependence of the parameter a; α is the ratio of a at the temperature of interest to a at the critical temperature:

$$\alpha = \frac{a}{a_c} \tag{7.41}$$

Soave found that α could be well expressed as a function of the reduced temperature (Soave, 1972), as

$$\alpha^{0.5} = 1 + m\left(1 - T_r^{0.5}\right) \tag{7.42}$$

where *the **reduced temperature** (T_r)* is defined as the ratio of the actual temperature to the critical temperature:

$$T_r = \frac{T}{T_c} \tag{7.43}$$

This is the first time we have encountered the reduced temperature. The analogous **reduced pressure (P_r)** is defined in Equation 7.44. We will find that T_r and P_r are useful in model building, particularly in Section 7.3 when we develop the principle of corresponding states.

$$P_r = \frac{P}{P_c} \tag{7.44}$$

Returning to Soave's EOS, Equation 7.42 is the expression for α as Soave originally published it. In practice, one usually calculates m and T_r first and then finds α from these, so we prefer a form that gives α explicitly.

$$\alpha = \left[1 + m\left(1 - T_r^{0.5}\right)\right]^2 \tag{7.45}$$

Finally, Soave related the value of m to the **acentric factor** as

$$m = 0.480 + 1.574\omega - 0.176\omega^2 \tag{7.46}$$

The acentric factor is

$$\omega = -1 - \left.\log_{10}\left(\frac{P^{\text{sat}}}{P_c}\right)\right|_{T_r = 0.7} \tag{7.47}$$

Thus, the acentric factor is derived from a single data point—if you know the vapor pressure of a compound at the reduced temperature of $T_r = 0.7$, you can determine the acentric factor using Equation 7.47. The rationale for this definition, and for its use in an EOS, is provided in Section 7.2.6. For now, the acentric factor is a state property that has been tabulated for many compounds. It is often published alongside critical properties (T_c and P_c) since these three properties are sufficient to allow one to model a compound in the liquid, vapor, and supercritical states using a cubic EOS.

Peng and Robinson (Robinson, 1976) developed their equation of state in a very similar way to Soave, although because the "attraction" terms in the two equations are different, the exact equations relating a_c and b to the critical temperature and pressure are slightly different. In the Peng-Robinson equation,

$$b = 0.07780R\frac{T_c}{P_c} \tag{7.48}$$

and

$$a_c = 0.45724R^2\frac{T_c^2}{P_c} \tag{7.49}$$

Recall that in algebra, a straight line is conventionally expressed in the form $y = mx + b$. Soave used the symbol m in Equation 7.42, presumably because he noted that the relationship between $\alpha^{0.5}$ and $T_r^{0.5}$ was in practice almost linear.

Kenneth Pitzer was the primary author of a pair of articles (Pitzer, 1955 and Pitzer et al., 1955) that introduced the acentric factor, which is sometimes called the "Pitzer acentric factor."

For Soave and Peng-Robinson, there are several equations and several steps involved in finding the value of the parameter a at a particular temperature, but there are only three pieces of data you need: critical temperature (T_c), critical pressure (P_c), and acentric factor (ω). These are tabulated in Appendix C-1.

Peng and Robinson's method of relating a to temperature was also substantially the same as Soave's:

$$\alpha = \frac{a}{a_c} \tag{7.50}$$

although they used the symbol κ where Soave used the symbol m for

$$\alpha = \left[1 + \kappa\left(1 - T_r^{0.5}\right)\right]^2 \tag{7.51}$$

While κ is related to the acentric factor in a manner analogous to Soave's m, the specific numerical values of the coefficients in Equation 7.52 are different from those in Equation 7.46. Thus,

$$\kappa = 0.37464 + 1.54226\omega - 0.269932\omega^2 \tag{7.52}$$

To summarize:

- Conceptually, the van der Waals constants a and b can be computed if the critical point (T_c and P_c) is known; no additional information is required.

- In the Soave and Peng-Robinson equations, because $a(T)$ is a function of temperature rather than a constant, one needs an additional piece of information beyond the critical point (T_c, P_c) in order to complete the equation of state. This additional parameter is the acentric factor ω, which is described further in the next section.

7.2.6 Vapor Pressure Curves and the Acentric Factor

This section has examined P-$\underline{V}$-T behavior of real compounds in the liquid and vapor phases. Figure 7-3 illustrated the P-$\underline{V}$-T data for water at a few temperatures. Section 7.2.2 through 7.2.4 discussed how to model P-$\underline{V}$-T behavior with cubic equations. To this point, we have examined single compounds—methane in Section 7.1, pentane in Example 7-3, and water in Section 7.2.1 and in Example 7-4. In this section, we begin to explore similarities and connections in the behaviors of a variety of compounds.

Qualitatively, essentially all chemical compounds exhibit the behavior illustrated in Figure 7-3 and Section 7.2.1, although the specific numbers (vapor pressures, liquid and vapor molar volumes, etc.) are very different from one compound to the next. Among the key thermodynamic properties of a chemical are the vapor pressures that correspond to each temperature; in Section 7.2.4, for example, we outlined the importance of vapor pressure in interpreting the three solutions of a cubic equation of state. It is instructive to plot the log of the reduced vapor pressure vs. the reciprocal of reduced temperature ($\log_{10} P_r^{sat}$ vs. $1/T_r$), as in Figure 7-8.

Notice that, for all of the compounds in Figure 7-8, the plot is essentially linear, although the slopes of the lines are different. The acentric factor, in effect, is used to model these vapor pressure curves as straight lines. We know from algebra that it takes two data points to define a straight line. However, all vapor pressure curves end at the critical point. Thus, ($1/T_r = 1$, $\log_{10} P_r^{sat} = 0$) is one point on all of the curves in Figure 7-8, and we only need one additional point to define each straight line.

The acentric factor essentially provides a second point on the line; it is derived from the value of the reduced vapor pressure that corresponds to $T_r = 0.7$. In fact, Figure 7-8 is adapted from the first of a pair of publications in which Pitzer and co-workers introduced the acentric factor (Pitzer, 1955).

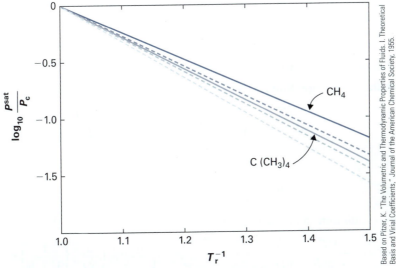

Based on Pitzer, K. "The Volumetric and Thermodynamic Properties of Fluids. I. Theoretical Basis and Virial Coefficients," Journal of the American Chemical Society, 1955.

FIGURE 7-8 $\log_{10}(P_r)$ vs. $1/T_r$ for five hydrocarbon compounds. The dashed curves represent propane, *n*-pentane, and *n*-heptane from top to bottom.

The value of $\log_{10}(P_r^{\text{sat}})$ at $T_r = 0.7$ falls between -1 and -2 for the vast majority of compounds, so the definition given in Section 7.2.4

$$\omega = -1 - \log_{10}\left(\frac{P^{\text{sat}}}{P_c}\right)\bigg|_{T_r = 0.7}$$

which is convenient because it produces values of ω that typically fall between 0 and 1. It is called the "acentric" factor because $\omega \sim 0$ for the spherical noble gases (argon, neon, xenon) and $\omega = 0.011$ for the spherically symmetrical methane, while nearly all compounds are spherically asymmetrical and have ω significantly greater than zero (Poling, 2001).

Why use $T_r = 0.7$ specifically in the definition of ω? This is somewhat arbitrary in that mathematically, if the $\log_{10}(P_r^{\text{sat}})$ vs. $1/T_r$ plot really is a straight line, any point on the straight line would define the line as well as any other point. That said, a point close to the critical point (say $T_r = 0.97$) would be a bad choice, since $\log_{10}(P_r^{\text{sat}})$ vs. $1/T_r$ for real compounds is not literally a straight line, and we can better model the entire vapor pressure curve by using two points that are farther apart ($T_r = 0.7$ and $T_r = 1$) to define a line. This reasoning perhaps suggests a lower reduced temperature like $T_r = 0.1$ would be even better. However, our goal with the acentric factor is to model the conditions at which *liquid–vapor equilibrium* occurs. If we choose a T_r that is too low, we fall below the triple point, and there is no liquid phase at all. Thus, when Pitzer and co-authors introduced the acentric factor, they chose $T_r = 0.7$, because it was well away from the critical point but, in almost all cases, above the triple point (Pitzer et al., 1955).

Chapter 6 showed that beyond P, T, and $\underline{V}$, other properties of a fluid such as $\underline{H}$, $\underline{S}$, $\underline{U}$, and $\underline{G}$ can be quantified through an equation of state that accurately models the fluid. Chapter 8 shows that vapor pressures and boiling temperatures can also be quantified using an equation of state. Recall that our goal is developing equations of state that are capable of modeling liquids, vapors, vapor–liquid equilibrium, and supercritical fluids—all with a single equation. Vapor pressures are, thus, a fundamental aspect of what we are trying to model, and this is what motivated the innovation of incorporating some vapor pressure data (in the form of the acentric factor) directly into the cubic equations of state.

7.2.7 Summary of Cubic Equations of State

The **van der Waals equation** can be written as

$$P = \frac{RT}{\underline{V} - b} - \frac{a}{\underline{V}^2} \quad \text{or} \quad Z = \frac{\underline{V}}{\underline{V} - b} - \frac{a}{RT\underline{V}}$$

with

$$a = \frac{27}{64}\frac{R^2 T_c^2}{P_c} \quad \text{and} \quad b = \frac{RT_c}{8 P_c}$$

The **Soave equation** can be written as

$$P = \frac{RT}{\underline{V} - b} - \frac{a}{\underline{V}(\underline{V} + b)} \quad \text{or} \quad Z = \frac{\underline{V}}{\underline{V} - b} - \left(\frac{a}{RT}\right)\left(\frac{1}{\underline{V} + b}\right)$$

with

$$a = a_c \alpha \qquad a_c = 0.42747 R^2 \frac{T_c^2}{P_c} \qquad \alpha = \left[1 + m\left(1 - T_r^{0.5}\right)\right]^2$$

$$m = 0.480 + 1.574\omega - 0.176\omega^2$$

$$b = 0.08664 R \frac{T_c}{P_c}$$

The **Peng-Robinson equation** can be written as

$$P = \frac{RT}{\underline{V} - b} - \frac{a}{\underline{V}(\underline{V} + b) + b(\underline{V} - b)} \quad \text{or} \quad Z = \frac{\underline{V}}{\underline{V} - b} - \left(\frac{a\underline{V}}{RT}\right)\left[\frac{1}{\underline{V}(\underline{V} + b) + b(\underline{V} - b)}\right]$$

with

$$a = a_c \alpha \qquad a_c = 0.45724 R^2 \frac{T_c^2}{P_c} \qquad \alpha = \left[1 + \kappa\left(1 - T_r^{0.5}\right)\right]^2$$

$$\kappa = 0.37464 + 1.54226\omega - 0.269932\omega^2$$

$$b = 0.07780 R \frac{T_c}{P_c}$$

7.2.8 Residual Properties from Cubic Equations of State

In Section 6.3.6, we solved problems involving residual functions computed using the van der Walls EOS. We derived Equation 6.157 (Example 6-9 Revisited), as an expression for the residual molar enthalpy:

$$\underline{H}^R = RT(Z - 1) - \frac{a}{\underline{V}}$$

and Equation 6.168 (Example 6-10), for the residual molar internal energy:

$$\underline{U}^R = \frac{-a}{\underline{V}}$$

An analogous expression can be found for entropy, beginning from the van der Waals equation:

$$P = \frac{RT}{\underline{V} - b} - \frac{a}{\underline{V}^2}$$

The partial derivative with respect to T is

$$\left(\frac{\partial P}{\partial T}\right)_{\underline{V}} = \frac{R}{\underline{V} - b}$$

The molar entropy departure function, in general, is described by Equation 6.145 as

$$\underline{S} - \underline{S}^{ig} = \underline{S}^R = R\ln(Z) + \int_{T=T,\underline{V}=\infty}^{T=T,\underline{V}=\underline{V}} \left[\left(\frac{\partial P}{\partial T}\right)_{\underline{V}} - \frac{R}{\underline{V}}\right] d\underline{V}$$

For the van der Waals equation, this becomes

$$\underline{S}^R = R\ln(Z) + \int_{T=T,\underline{V}=\infty}^{T=T,\underline{V}=\underline{V}} \left[\frac{R}{\underline{V} - b} - \frac{R}{\underline{V}}\right] d\underline{V} \tag{7.53}$$

$$\underline{S}^R = R\ln(Z) + R\int_{T=T,\underline{V}=\infty}^{T=T,\underline{V}=\underline{V}} \left[\frac{1}{\underline{V} - b} - \frac{1}{\underline{V}}\right] d\underline{V} \tag{7.54}$$

This integrates to

$$\underline{S}^R = R\ln(Z) + R[\ln(\underline{V} - b) - \ln(\underline{V})]\Big|_{\underline{V}=\infty}^{\underline{V}=\underline{V}} \tag{7.55}$$

$$\underline{S}^R = R\ln(Z) + R\left[\ln\left(\frac{\underline{V} - b}{\underline{V}}\right)\right]\Big|_{\underline{V}=\infty}^{\underline{V}=\underline{V}} \tag{7.56}$$

At the $\underline{V} = \infty$ limit, b is insignificant compared to $\underline{V}$, so the ln term is zero (ln 1). Thus, the result is

$$\underline{S}^R = R\ln(Z) + R\ln\left(\frac{\underline{V} - b}{\underline{V}}\right) \tag{7.57}$$

When $\underline{V} = \infty$, Z goes to 1, so $\ln(Z) = 0$

Analogous derivations can be performed to obtain residual properties for any equation of state, including the cubic equations introduced in this chapter. For the Peng-Robinson equation, we have

$$\frac{U^R}{RT} = -\left\{\left(\frac{A}{B\sqrt{8}}\right)\left(1 + \frac{\kappa\sqrt{T_r}}{\sqrt{\alpha}}\right)\ln\left[\frac{Z + (1 + \sqrt{2})B}{Z + (1 - \sqrt{2})B}\right]\right\} \tag{7.58}$$

$$\frac{H^R}{RT} = (Z - 1) - \left\{\left(\frac{A}{B\sqrt{8}}\right)\left(1 + \frac{\kappa\sqrt{T_r}}{\sqrt{\alpha}}\right)\ln\left[\frac{Z + (1 + \sqrt{2})B}{Z + (1 - \sqrt{2})B}\right]\right\} \tag{7.59}$$

$$\frac{S^R}{R} = \ln(Z - B) - \left\{\left(\frac{A}{B\sqrt{8}}\right)\left(\frac{\kappa\sqrt{T_r}}{\sqrt{\alpha}}\right)\ln\left[\frac{Z + (1 + \sqrt{2})B}{Z + (1 - \sqrt{2})B}\right]\right\} \tag{7.60}$$

in which:

$$A = \frac{aP}{R^2T^2} \tag{7.61}$$

$$B = \frac{bP}{RT} \tag{7.62}$$

The use of these expressions is illustrated in Example 7-5.

EXAMPLE 7-5	**COMPUTING RESIDUALS USING PENG-ROBINSON EOS**

Methane enters a turbine at $T = 600$ K and $P = 1$ MPa, and leaves at $T = 400$ K and $P = 0.2$ MPa. Use the Peng-Robinson equation to determine the work produced for each mole of gas.

SOLUTION:

Step 1 *Compare to Example 6-9*

This problem is identical to Example 6-9, except we are using the Peng-Robinson equation instead of the van der Waals equation. As in Example 6-9, the work will be calculated using the energy balance and residual properties where

$$\frac{\dot{W}_S}{\dot{n}} = \underline{H}_2 - \underline{H}_1 = \underline{H}_2^R + \left(\underline{H}_2^{ig} - \underline{H}_1^{ig}\right) - \underline{H}_1^R$$

The change in molar enthalpy for the ideal gas state does not depend upon the equation of state used to model the gas; it depends only upon the ideal gas heat capacity. Thus the value computed in step 3 of Example 6-9 is again valid as

$$\underline{H}_2^{ig} - \underline{H}_1^{ig} = -9289 \frac{J}{mol}$$

but the residual properties must be computed using the Peng-Robinson equation.

Step 2 *Fit parameters to the Peng-Robinson equation*

The Peng-Robinson b parameter is calculated using Equation 7.48:

$$b = 0.07780R\frac{T_c}{P_c} = (0.07780)\left(8.314 \frac{Pa \cdot m^3}{mol \cdot K}\right)\left(\frac{190.56\ K}{4.599\ MPa}\right) = 26.80 \times 10^{-6} \frac{m^3}{mol}$$

The parameter a is dependent upon temperature and is determined from Equations 7.49 through 7.52. Find a_c:

$$a_c = 0.45724R^2\frac{T_c^2}{P_c} = 0.45724\left(8.314 \frac{Pa \cdot m^3}{mol \cdot K}\right)^2 \frac{(190.56\ K)^2}{4.599\ MPa} \tag{7.63}$$

$$= 0.2496 \frac{Pa \cdot m^6}{mol^2}$$

The parameter κ is computed from the acentric factor ω where

$$\kappa = 0.37464 + 1.54226\omega - 0.269932\omega^2 = 0.37464 + 1.54226(0.011) - 0.269932(0.011)^2$$

$$\kappa = 0.3916$$

and is used to find α, which is temperature-dependent. At 600 K,

$$\alpha = \left[1 + \kappa\left(1 - T_r^{0.5}\right)\right]^2 = \left\{1 + (0.3916)\left[1 - \left(\frac{600\ K}{190.56\ K}\right)^{0.5}\right]\right\}^2 = 0.4855` \tag{7.64}$$

Finally, a is determined using the results of Equations 7.64 and 7.63 for

$$a = a_c\alpha = 0.1211\frac{Pa \cdot m^6}{mol^2} \tag{7.65}$$

Applying the same calculations at $T = 400$ K produces $a = 0.1695$ Pa m^6/mol.

Step 3 *Solve Peng-Robinson equation of state*

The Peng-Robinson equation is

$$P = \frac{RT}{\underline{V} - b} - \frac{a}{\underline{V}(\underline{V} + b) + b(\underline{V} - b)}$$

The critical point of methane ($T_c = 190.56$, $P_c = 4.599$ MPa) and the acentric factor ($\omega = 0.011$) are available in Appendix C.

and can now be solved for $\underline{V}$. Both the inlet temperature ($T_1 = 600$ K) and the outlet temperature ($T_2 = 400$ K) are above the critical temperature, so we only expect one real solution for $\underline{V}$. The results are summarized here.

	T (K)	P (MPa)	a (Pa m⁶/mol²)	b (m³/mol)	V (m³/mol)	Z
Inlet	600	1	0.1211	26.80×10^{-6}	4.99×10^{-3}	1.0006
Outlet	400	0.2	0.1695	26.80×10^{-6}	16.59×10^{-3}	0.999

The values of Z are both very close to 1 (even more so than when we were using the van der Waals equation in Example 6-9).

Step 4 *Calculate residual properties*

For the fluid entering the turbine, A is computed from Equation 7.61 as

$$A = \frac{aP}{R^2 T^2} = \frac{\left(0.1211 \dfrac{\text{Pa} \cdot \text{m}^6}{\text{mol}^2}\right)(1 \text{ MPa}) \times \left(\dfrac{10^6 \text{ Pa}}{1 \text{ MPa}}\right)}{\left(8.314 \dfrac{\text{Pa} \cdot \text{m}^3}{\text{mol} \cdot \text{K}}\right)^2 (600 \text{ K})^2} = 0.00487$$

From Equation 7.62, we have

$$B = \frac{bP}{RT} = \frac{\left(26.80 \times 10^{-6} \dfrac{\text{m}^3}{\text{mol}}\right)(1 \text{ MPa}) \times \left(\dfrac{10^6 \text{ Pa}}{1 \text{ MPa}}\right)}{\left(8.314 \dfrac{\text{Pa} \cdot \text{m}^3}{\text{mol} \cdot \text{K}}\right)(600 \text{ K})} = 0.00537$$

Everything required to compute the residual molar enthalpy from Equation 7.59 is now known

$$\frac{H_1^R}{RT} = (Z - 1) - \left\{\left(\frac{A}{B\sqrt{8}}\right)\left(1 + \frac{\kappa \sqrt{T_r}}{\sqrt{\alpha}}\right) \ln\left[\frac{Z + (1 + \sqrt{2})B}{Z + (1 - \sqrt{2})B}\right]\right\}$$

$$\frac{H_1^R}{RT} = (1.0006 - 1) - \left\{\left(\frac{0.00487}{0.00537\sqrt{8}}\right)\left(1 + \frac{0.3916\sqrt{\dfrac{600 \text{ K}}{190.56 \text{ K}}}}{\sqrt{0.4855}}\right)\right.$$

$$\left. \times \ln\left[\frac{1.0006 + (1 + \sqrt{2})(0.00537)}{1.0006 + (1 - \sqrt{2})(0.00537)}\right]\right\} = -0.000907$$

$$H_1^R = (-0.000907)\left(8.314 \frac{\text{J}}{\text{mol K}}\right)(600 \text{ K}) = -45.2 \frac{\text{J}}{\text{mol}}$$

And for the fluid exiting the turbine, we have

$$\frac{H_2^R}{RT} = (0.999 - 1) - \left\{\left(\frac{0.00309}{0.00162\sqrt{8}}\right)\left(1 + \frac{0.3916\sqrt{\dfrac{400 \text{ K}}{190.56 \text{ K}}}}{\sqrt{0.6794}}\right)\right.$$

$$\left. \times \ln\left[\frac{0.999 + (1 + \sqrt{2})(0.00162)}{0.999 + (1 - \sqrt{2})(0.00162)}\right]\right\} = -0.00521$$

$$H_{\underline{2}}^R = (-0.000521)\left(8.314\,\frac{J}{mol\,K}\right)(400\,K) = -20.7\,\frac{J}{mol}$$

$$\frac{\dot{W}_S}{\dot{n}} = H_{\underline{2}}^R + \left(H_{\underline{2}}^{ig} - H_{\underline{1}}^{ig}\right) - H_{\underline{1}}^R = \left(-20.7\,\frac{J}{mol}\right) + \left(-9289\,\frac{J}{mol}\right) - \left(-45.2\,\frac{J}{mol}\right) = \mathbf{-9265\,\frac{J}{mol}}$$

The residuals computed in Example 7-5 are very similar to those calculated in Example 6-9 and are small compared to the ideal gas change in enthalpy. Methane is a very light and spherically symmetrical compound, so even at 1 MPa, it is reasonably approximated by the ideal gas law in this case. The Motivational Examples in Section 7.1, however, showed that at higher pressures even methane departs dramatically from the ideal gas law.

7.3 The Principle of Corresponding States

In examining cubic equations of state, Section 7.2 also discussed the P-V-T behavior of real fluids—the data that equations of state are designed to model. Figure 7-9 shows a P-$\underline{V}$ diagram that encapsulates what we've learned about the interrelationships among P, $\underline{V}$, and T. Some of the key observations from Figure 7-9 include:

- $\underline{V}$ decreases as P increases at constant T, but the effect of P on $\underline{V}$ is much more dramatic for vapors and supercritical fluids than for liquids.
- Vapor pressure increases as temperature increases.
- The molar volume of saturated vapor decreases as temperature increases, but the molar volume of saturated liquid increases as temperature increases.

Qualitatively, Figure 7-9 describes most any compound in the liquid, vapor, and supercritical fluid states. Thus, we can say that P-$\underline{V}$-T relationships for the vast majority of compounds are similar. In this section, we further explore this "similarity" in the behavior of various compounds. The **principle of corresponding states** defines

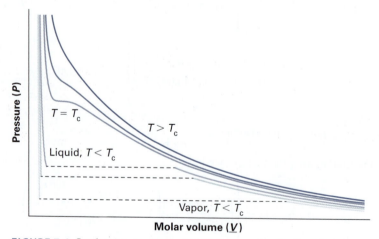

FIGURE 7-9 Isotherms on a typical P-$\underline{V}$ diagram for the liquid, vapor, and supercritical phases. The axes have no numbers because no specific chemical is described, but the P-$\underline{V}$ diagram for most any compound will have these features.

compounds as in "corresponding states" if they are at the same *reduced* temperature and *reduced* pressure as each other. Section 7.3.1 introduces the rationale behind this definition and demonstrates its use to make quantitative estimates and predictions.

7.3.1 Illustrations of the Principle of Corresponding States

This section focuses on two examples that illustrate what is meant by the "principle of corresponding states" and what kinds of calculations may be performed using the principle.

CORRESPONDING STATES AND COMPRESSIBILITY FACTORS	**EXAMPLE 7-6**

Compound A is an industrial solvent and is widely enough used that extensive data on it is available. Compound B is a recently invented compound that can serve the same purpose as compound A, and we now wish to determine whether compound B can be manufactured more cheaply than compound A. We have completed a flowsheet for the manufacturing process, but in order to design and size the needed equipment, we must be able to estimate the molar volume of compound B at a variety of temperatures and pressures. Here we will estimate the molar volumes of compound B:

A. In the liquid phase at $T = 900$ K and $P = 1.6$ MPa

B. In the vapor phase at $T = 900$ K and $P = 0.4$ MPa

C. In the supercritical state at $T = 1100$ K and $P = 2.4$ MPa

The critical points for both compounds are given here.

	T_c (K)	P_c (MPa)
Compound A	500	1.5
Compound B	1000	2

No further data is available for compound B, but the compressibility of compound A is known at a wide variety of conditions, some of which are shown in Figure 7-10.

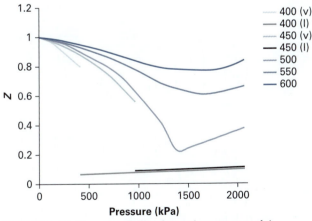

FIGURE 7-10 Compressibility factor for compound A for various pressures (kPa) and temperatures (K).

SOLUTION:

Step 1 *Explore the given information*

Because we know the critical point of compound B, we could use the van der Waals EOS to solve for $\underline{V}$ at any temperature and pressure we wish. However, we've now seen several examples (beginning with Example 7-2) that show the van der Waals EOS isn't very accurate. Why do we keep referring to the van der Waals EOS when it isn't very accurate? Because it is often useful as a relatively simple equation that illustrates principles. Here we will use it to illustrate how the known data for compound A can be used to make inferences and predictions regarding the behavior of compound B.

Step 2 *Compare different compounds through van der Waals equation*

The van der Waals equation is

$$P = \frac{RT}{\underline{V} - b} - \frac{a}{\underline{V}^2}$$

But we learned in Example 7-4 that, when we use the strategy of matching the critical point to assign values to the van der Waals parameters a and b, the results are

$$a = \frac{27}{64} \frac{R^2 T_c^2}{P_c} \quad \text{and} \quad b = \frac{RT_c}{8P_c}$$

Combining these expressions gives

$$P = \frac{RT}{\underline{V} - \dfrac{RT_c}{8P_c}} - \frac{\dfrac{27R^2 T_c^2}{64P_c}}{\underline{V}^2} \tag{7.66}$$

We now manipulate this expression so that it is all in terms of *reduced* properties. Dividing both sides by P_c gives us the exact definition of P_r on the left-hand side:

$$\left(\frac{P}{P_c}\right) = \frac{RT}{P_c\left(\underline{V} - \dfrac{RT_c}{8P_c}\right)} - \frac{\dfrac{27R^2 T_c^2}{64P_c^2}}{\underline{V}^2} \tag{7.67}$$

We can similarly convert T into T_r by dividing the first term on the right-hand side through by T_c:

$$\left(\frac{P}{P_c}\right) = \left(\frac{\dfrac{1}{T_c}}{\dfrac{1}{T}}\right) \frac{RT}{P_c\left(\underline{V} - \dfrac{RT_c}{8P_c}\right)} - \frac{\dfrac{27R^2 T_c^2}{64P_c^2}}{\underline{V}^2} \tag{7.68}$$

$$\left(\frac{P}{P_c}\right) = \frac{R\left(\dfrac{T}{T_c}\right)}{\dfrac{P_c}{T_c}\left(\underline{V} - \dfrac{RT_c}{8P_c}\right)} - \frac{\left(\dfrac{27R^2 T_c^2}{64P_c^2}\right)}{\underline{V}^2} \tag{7.69}$$

Similarly, we can convert $\underline{V}$ into $\underline{V}_r$ if we divide the right-hand terms through by $\underline{V}_c$ or $\underline{V}_c^2$ as needed:

$$\left(\frac{P}{P_c}\right) = \left(\frac{\frac{1}{V_c}}{\frac{1}{V_c}}\right)\left[\frac{R\left(\frac{T}{T_c}\right)}{\frac{P_c}{T_c}\left(V - \frac{RT_c}{8P_c}\right)}\right] - \left(\frac{\frac{1}{V_c}}{\frac{1}{V_c}}\right)^2\left[\frac{\left(\frac{27R^2T_c^2}{64P_c^2}\right)}{V^2}\right] \qquad (7.70)$$

$$\left(\frac{P}{P_c}\right) = \frac{R\left(\frac{T}{T_c}\right)}{\frac{V_c P_c}{T_c}\left[\left(\frac{V}{V_c}\right) - \frac{RT_c}{8V_c P_c}\right]} - \left[\frac{\left(\frac{27R^2T_c^2}{64V_c^2P_c^2}\right)}{\frac{V^2}{V_c^2}}\right] \qquad (7.71)$$

This looks like a messy expression, but it simplifies considerably when we recognize that $Z_c = P_c\underline{V}_c/RT_c$.

$$P_r = \frac{T_r}{Z_c\left[(\underline{V}_r) - \frac{1}{8Z_c}\right]} - \left[\frac{\frac{27}{64Z_c^2}}{\underline{V}_r^2}\right] \qquad (7.72)$$

But another result of Example 7-4 was that $Z_c = 0.375$, for *any* compound that follows the van der Waals equation. Notice that Equation 7.72 contains only the reduced properties T_r, P_r, $\underline{V}_r$, constants and Z_c, which is itself a constant for a fluid that follows the van der Waals equation. The conclusion is that, if the van der Waals equation is valid, then *all compounds will have the same $\underline{V}_r$ at a particular T_r and P_r.*

This derivation is an illustration of the "principle of corresponding states." If two compounds are at the same REDUCED temperature and REDUCED pressure as each other, they are said to be in "corresponding states." According to the van der Waals equation, any two compounds that are in corresponding states must have the same reduced molar volume ($\underline{V}_r$) as each other. We've shown in this chapter that the van der Waals equation can, in some circumstances, be very inaccurate. However, the insight that we have gained from the van der Waals equation has significance beyond the equation itself. The principle of corresponding states is discussed more broadly after this example. For now, let's assume the principle is 100% correct—that two compounds at the same T_r and P_r will also have the same $\underline{V}_r$ and see how we can use it to solve a problem.

Step 3 *Extend principle of corresponding states to the compressibility factor*
Since Figure 7-10 is in terms of the compressibility factor Z, rather than $\underline{V}_r$, let's introduce Z into the discussion. The compressibility factor in general is defined as

$$Z = \frac{P\underline{V}}{RT} \qquad (7.73)$$

And at the critical point, specifically for the VDW EOS, it is

$$Z_c = \frac{P_c\underline{V}_c}{RT_c} = 0.375 \qquad (7.74)$$

Dividing Equation 7.73 by Equation 7.74 gives

$$\frac{Z}{0.375} = \frac{P_r\underline{V}_r}{RT_r} \qquad (7.75)$$

Thus, *if two compounds have the same T_r, P_r, and $\underline{V}_r$, they will also have the same Z.* This is another way of expressing the principle of corresponding states. This result can be used with Figure 7-10.

We could do a similar derivation using the Soave or Peng-Robinson equation, but it would be more complicated and would require the assumption that the acentric factors of compounds A and B are identical.

You can think of reduced properties as measuring the "distance from the critical point." Two compounds are in corresponding states if they are both at $T_r = 0.5$ and $P_r = 0.5$; they are at different T and P but are each the same "distance" from their own critical point.

Note that the actual molar volumes of two compounds in corresponding states can be different. $\underline{V}_r$ is defined as $\underline{V}/\underline{V}_c$. Even if $\underline{V}_r$ is the same, $\underline{V}_c$, which is the molar volume at the critical point, can be very different.

This step, like step 2, assumes the van der Waals equation is valid; the result $Z_c = 0.375$ is specific to the van der Waals equation.

Step 4 *Compute reduced conditions for the states of interest*

We are asked to estimate $\underline{V}$ for compound B at three temperatures and pressures. These are expressed in *reduced* properties through the known critical point ($T_c = 1000$ K, $P_c = 2$ MPa) as follows.

T (K)	P (MPa)	T_r	P_r
900	1.6	0.9	0.8
900	0.4	0.9	0.2
1100	2.4	1.1	1.2

Step 5 *Find Z for compound A at states corresponding to the states of interest*

We are interested in predicting $\underline{V}$ for compound B at 900 K and 1.6 MPa. At this temperature and pressure, compound A is a supercritical fluid and compound B is a liquid (this was given), so they can't be compared to each other in any useful way at this T and P. But the principle of corresponding states tells us to compare them at the same T_r and P_r. For compound B, 900 K is equivalent to $T_r = 0.9$ and $P = 1.6$ MPa is equivalent to $P_r = 0.8$. The corresponding state for compound A is $T_r = 0.9$ (500) = 450 K and $P_r = 0.8(1.5) = 1.2$ MPa.

Now that we have a specific T and P, we can obtain Z from Figure 7-10. This process is summarized in Table 7-4.

TABLE 7-4 Results of Example 7-6.

T_r	P_r	T of Compound A (K)	P of Compound A (MPa)	Z of Compound A	T of Compound B (K)	P of Compound B (MPa)	$\underline{V}$ of Compound B (cm³/mol)
0.9	0.8	450	1.2	0.09	900	1.6	420.9
0.9	0.2	450	0.3	0.93	900	0.4	17,400
1.1	1.2	550	1.8	0.61	1100	2.4	2324

Step 6 *Apply assumption of corresponding states to solve for $\underline{V}$ of compound* B

Once Z is known at a state for compound A, we assume that compound B, at the corresponding reduced temperature and pressure, has the same value of Z. We can then compute $\underline{V}$ through the definition of Z as

$$\underline{V} = \frac{ZRT}{P} \tag{7.76}$$

These results, too, are shown in Table 7-4. Notice that in this final step we are using the actual P and T, not the reduced properties. The whole point of applying the principle of corresponding states was to obtain an estimate of the unknown Z. Once Z is "known," we can use Equation 7.76 like we would use any equation of state: If two of P, T, and $\underline{V}$ are known, the third can be calculated.

This calculation uses R = 8.314 Pa · m³/mol·K and 1 m³ = 10⁶ cm³.

Example 7-6 used the van der Waals equation to illustrate the principle of corresponding states. We define two compounds as being in "corresponding states" if they are at the same reduced temperature T_r and the same reduced pressure P_r, as

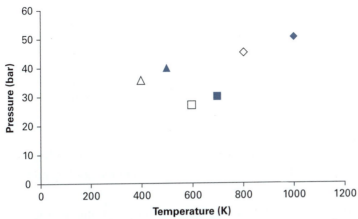

FIGURE 7-11 The three filled symbols represent three compounds at their critical points. The hollow symbols represent these same three compounds, each at $T_r = 0.8$ and $P_r = 0.9$. Two or more compounds at the same T_r and P_r as each other are in "corresponding states."

illustrated in Figure 7-11. *The **principle of corresponding states** indicates that if two compounds are in corresponding states, they will also have the same $\underline{V}_r$ and Z.* We now look at this principle more broadly.

First we note that, according to the Gibbs phase rule, a pure compound in a single phase has two degrees of freedom and, conceptually, we can use any two intensive variables to satisfy those two degrees of freedom. Consequently, we *could* have defined "corresponding states" as occurring when two compounds are at the same T_r and $\underline{V}_r$ or at the same P_r and $\underline{V}_r$. However, because $\underline{V}$ is extremely sensitive to T and P close to the critical point, $\underline{V}_c$ is more difficult to measure accurately than P_c or T_c. As a result, we will consistently use T_r and P_r as the two variables for characterizing "corresponding states."

Example 7-6 demonstrated that *if* real fluids followed the van der Waals equation, then the following would all be true.

1. All compounds have $Z = 0.375$ at the critical point.
2. If two compounds are at the same T_r and P_r, they will also have the same $\underline{V}_r$ and Z.
3. The vapor pressure curve, plotted as P_r^{sat} vs. T_r, is the same for all compounds.

Note that #3 was not explicitly mentioned in Example 7-6, but it follows from #2: If two compounds are to have the same Z at every value of T_r and P_r, then the transition from a liquid-phase value of Z to a vapor-phase value of Z must also occur at the same T_r and P_r. Example 7-7 illustrates how this aspect of the principle of corresponding states can be applied in the estimation of unknown vapor pressures. First, we address a natural and important question: Since we know the van der Waals equation isn't very accurate in general, *how valid is the principle of corresponding states, really?*

We begin by noting that the critical point is itself an example of a "corresponding state"; all compounds have $T_r = 1$ and $P_r = 1$ at the critical point. If the principle of corresponding states is not valid at the critical point, then it certainly cannot be valid in general, since all other states are compared to the critical state through reduced properties. It is thus instructive to begin with another look at Table 7-3, which listed critical properties of a number of compounds. Previously, we observed from this table that

$Z_c = 0.375$, which is the value that is obtained from the van der Waals equation, is routinely too high. However, it is perhaps not unreasonable to say "most compounds have roughly the same Z_c as each other." The values of Z_c in Table 7-3 range from 0.233 to 0.288. A broader range is seen when examining the larger list of compounds in Appendix C, but most values do fall in the range $0.23 < Z_c < 0.30$. What is perhaps more enlightening is examining groups of closely related compounds within Table 7-3.

- For the unsaturated hydrocarbons ethene, propene, and butadiene, Z_c ranges from 0.27 to 0.277.
- For the alcohols ethanol, propanol, and isopropanol, Z_c ranges from 0.248 to 0.253.
- For the saturated alkanes ethane, propane, butane, and isobutane, Z_c ranges from 0.274 to 0.284.

Thus, the principle of corresponding states is not as absolutely true as the van der Waals EOS predicts it should be. But the principle appears to have some credibility based upon this limited sample of data. Examining wider ranges of compounds leads to the same conclusion: *The principle of corresponding states is a useful model within groups of compounds that are chemically similar to each other.*

The principle is applied to estimation of vapor pressures in the next example.

EXAMPLE 7-7	**APPLYING THE PRINCIPLE OF CORRESPONDING STATES TO VAPOR PRESSURES**

The critical properties for compounds A and B, which are two compounds with similar molecular structures, are given here. No further information is available about compound B, but the vapor pressure of compound A as a function of temperature is given in Figure 7-12. Give the best possible estimates of the vapor pressures for compound B at $T = 360, 420, 480$ and 540 K.

	T_c (K)	P_c (kPa)
Compound A	444	1379
Compound B	600	1750

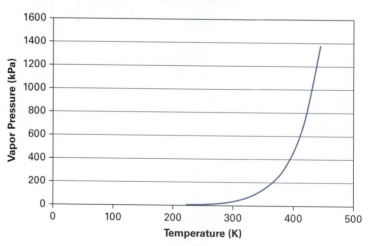

FIGURE 7-12 Vapor pressure vs. temperature for compound A.

SOLUTION:

Step 1 *Explore the given information*

We are asked to estimate vapor pressures for compound B. The only direct information we have about compound B is the critical point. We learned in Example 7-4 that we can fit the parameters for the van der Waals equation (*a* and *b*) if we know the critical point, but what we've learned so far gives us no way to find vapor pressures from the van der Waals equation.

Thus, we must use Figure 7-12, which describes compound A, to estimate vapor pressure data for compound B. Our first thought might be to assume compounds A and B have the same vapor pressures at each temperature, because they are said to have "similar molecular structures." However, this is not a realistic assumption. As an illustration, *n*-butane, *n*-pentane, *n*-hexane, *n*-heptane, and *n*-octane are all saturated linear hydrocarbons, so they certainly have "similar molecular structures." However, an assumption that they have similar vapor pressures at a particular temperature isn't reasonable. Table 7-5 illustrates that as hydrocarbons get heavier, they becomes less volatile; the vapor pressures at a given temperature decrease as molecular weight increases. Returning to our example, the thought that compounds A and B have identical vapor pressures is certainly impossible at 500 K, where A is above its critical temperature (no liquid phase can form at all) but B will still have a finite vapor pressure.

> Chapter 8 discusses how to estimate vapor pressures from an equation of state, but that information wouldn't necessarily help with this particular example. The van der Waals EOS generally predicts vapor pressures poorly, and the more accurate cubic equations of state require the acentric factor, which is not known here.

TABLE 7-5 Vapor pressures of some linear alkanes at $T = 20°C$.

Compound	Vapor Pressure (kPa)
n-butane	207.5
n-pentane	56.55
n-hexane	16.16
n-heptane	4.63
n-octane	1.28

Rather than comparing compounds A and B at the same temperatures, we will apply the principle of corresponding states and compare them at the same *reduced* temperatures.

Step 2 *Compute T_r and P_r for saturated A*

We are asked to find the vapor pressure of B at $T = 360$ K, 420 K, 480 K and 540 K. Since the critical temperature of B is 600 K, this means our specific interest is in reduced temperatures of 0.6, 0.7, 0.8, and 0.9. While we know nothing about compound B at these reduced temperatures, we can determine the vapor pressures of compound A at these reduced temperatures.

■ The actual temperature that corresponds to each reduced temperature can be found because the critical temperature is known ($T_c = 444$ K for compound A).

■ The vapor pressure at each temperature can be obtained from Figure 7-12.

■ Each vapor pressure can be converted into a reduced vapor pressure through the known critical pressure ($P_c = 1379$ kPa).

> By definition $T_r = T/T_c$ and $P_r = P/P_c$, as first introduced in Section 7.2.4.

Table 7-6 summarizes the results of this analysis.

TABLE 7-6 Saturation conditions for compound A.

Reduced Temperature T_r	Actual T (K)	Vapor Pressure P^{sat} (kPa)	Reduced Vapor Pressure P_r^{sat}
0.6	266.4	21	0.015
0.7	310.8	55	0.04
0.8	355.2	165	0.12
0.9	399.6	469	0.34

Step 3 *Apply principle of corresponding states*

Using $T_r = 0.9$ as an example, step 2 showed that compound A is at vapor–liquid equilibrium when $T = 400$ K and $P = 469$ kPa. These conditions are equivalent to $T_r = 0.9$ and $P_r = 0.34$. We assume that vapor–liquid equilibrium will also occur for a similar compound B that is in the corresponding state $T_r = 0.9$ and $P_r = 0.34$. Since $P_c = 1750$ kPa for compound B, our estimate of the vapor pressure at $T = 540$ K would be:

$$P^{sat} = P_r^{sat} P_c = (0.34)(1750 \text{ kPa}) = 595 \text{ kPa} \tag{7.77}$$

Table 7-7 summarizes the results when the same reasoning is applied to the other temperatures.

TABLE 7-7 Summary of estimated vapor pressures for compound B.

Actual T (K)	Reduced Temp. T_r	Reduced Vapor pressure P_r^{sat}	Vapor Pressure (kPa)
360	0.6	0.015	26
420	0.7	0.04	70
480	0.8	0.12	210
540	0.9	0.34	595

This section demonstrated through two examples what the principle of corresponding states says and overviewed essential physical properties—compressibility factors, molar volumes, and vapor pressures—that the principle can help us estimate. The following sections provide some practical strategies for applying the principle of corresponding states.

7.3.2 Generalized Correlations and Aggregated Data

Examples 7-6 and 7-7 were contrived to provide a compact illustration of the principle of corresponding states. However, in practice, the scenarios described—using data from one compound as the basis for making predictions for one other compound—aren't particularly likely to occur. Recognizing the usefulness of the principle of corresponding states, some authors have published aggregated data showing the relationship among Z, T_r, and P_r for a variety of compounds. Figure 7-13, for example, was first published in 1946, and its use is demonstrated in Example 7-8.

| ESTIMATING COMPRESSIBILITY FROM GENERALIZED CHARTS | EXAMPLE 7-8 |

Use Figure 7-13 to estimate the compressibility (Z) of supercritical toluene at $T = 651$ K and $P = 5.75$ MPa.

SOLUTION:

Step 1 *Compute reduced properties*
Applying the principle of corresponding states always involves the use of reduced properties. According to Appendix C, toluene has $T_c = 591.75$ K and $P_c = 4.108$ MPa. Thus, the reduced temperature and pressure for the state considered in this problem is

$$T_r = \frac{T}{T_c} = \frac{651\,\text{K}}{591.75\,\text{K}} = 1.1 \tag{7.78}$$

$$P_r = \frac{P}{P_c} = \frac{5.75\,\text{MPa}}{4.108\,\text{MPa}} = 1.4 \tag{7.79}$$

Step 2 *Obtain Z from Figure 7-13*
Figure 7-13 shows data for a number of compounds, but toluene isn't one. Consequently, our best strategy is to use the trendline that represents average data for hydrocarbons. According to the figure, $P_r = 1.4$ and $T_r = 1.1$ corresponds approximately to **Z = 0.55**.

Problem 7-9 involves a turbine in which the entering fluid is toluene at this temperature and pressure.

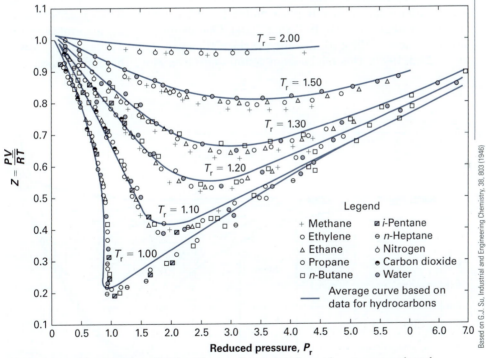

FIGURE 7-13 The Z-P_r plot, showing discrete data points for various compounds and average trendlines for hydrocarbons.

Mechanically, Example 7-8 is very similar to Examples 7-6 and 7-7; they all use the principle of corresponding states, and the mathematical manipulations needed to solve them are essentially the same. However, by using Figure 7-13 as the basis of our calculations, we make estimates of unknown properties that are based on aggregated data for a variety of "model" compounds, rather than basing our estimates on a single "model" compound as we did in Examples 7-6 and 7-7.

Figure 7-13 shows curves fit to real data for a handful of chosen compounds. The **Lee-Kesler equation** takes the development of generalized correlations a step further. The Lee-Kesler equation is built upon the hypothesis that the principle of corresponding states is applicable to compounds that have the same acentric factor; that is, if two compounds have the same ω and are in corresponding states (same T_r and P_r), then they should have the same Z. In principle, we could apply this idea by making many separate figures (each analogous to Figure 7-13) for many different values of ω. The Lee-Kesler equation in effect consolidates the approach into two figures. The compressibility Z is estimated as

$$Z = Z^0 + \omega Z^1 \tag{7.80}$$

Z^0 represents the compressibility of a compound with $\omega = 0$, and is obtained from Figure 7-14.

Z^1 represents the difference between Z^0 and the Z that would be expected for a hypothetical compound that had an acentric factor of 1.0, and it is obtained from Figure 7-15.

Almost all compounds have an acentric factor that falls between 0 and 1, so in effect, Equation 7.80 represents a simple linear interpolation between the values tabulated for $\omega = 0$ and $\omega = 1$.

The Lee-Kesler approach is not recommended for highly polar compounds, but for non-polar or weakly polar molecules, it typically provides results that are accurate within ~2% (B. E. Poling, 2001). One caution is that, at conditions that are quite close to the critical point, the correlation can give very inaccurate answers (C. T. Lira, 1999). This might seem counter-intuitive. Since the critical point is so

The information in Figure 7-14 and in Figure 7-15 was first published in tabular form. (K. Pitzer, 1955)

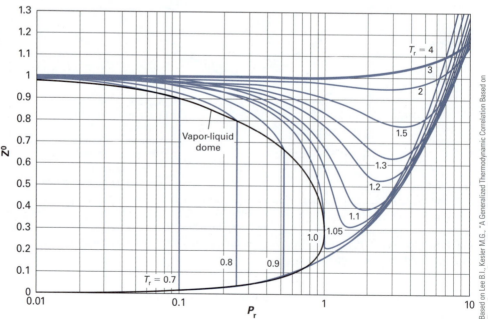

FIGURE 7-14 Compressibility factor for a compound with $\omega = 0$ for use in the Lee-Kesler equation.

Based on Lee B.I., Kesler M.G., "A Generalized Thermodynamic Correlation Based on Three-Parameter Corresponding States", AIChE J., 21(3), 510-527, 1975.

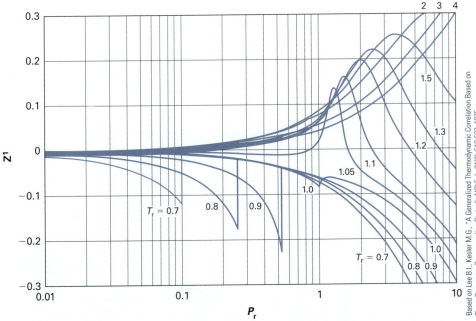

There are sharp vertical lines in Figures 7-14 and 7-15 because of the vapor–liquid transition; the phase change causes Z to be a non-continuous function of P at the vapor pressure.

Based on Lee B.I., Kesler M.G., "A Generalized Thermodynamic Correlation Based on Three-Parameter Corresponding States", AIChE J., 21(3), 510-527, 1975.

FIGURE 7-15 Compressibility correction factor for a compound with $\omega = 1$ for use in the Lee-Kesler equation.

fundamentally integrated into the construction of the model, shouldn't predictions be *better* close to the critical point? However, near the critical point, real compounds display behavior that is strikingly different from what is observed at most temperatures and pressures. For example, liquid molar volumes for most compounds are approximately constant with respect to T and P—except near the critical point, where they are extremely sensitive to changes in T and P. *Since behavior near the critical point is unique, it is unlikely to be modeled accurately with a generalized approach.*

Chapter 6 demonstrated that properties like enthalpy, entropy, and internal energy can be mathematically related to measurable properties. The mathematical relationships rely on partial derivatives of P, $\underline{V}$, and T. Changes in molar enthalpy, for example, can be quantified using Equation 6.123 by

$$d\underline{H} = C_p \, dT + \left[-T\left(\frac{\partial \underline{V}}{\partial T}\right)_P + \underline{V} \right] dP$$

In Example 6-9, $(\partial \underline{V}/\partial T)_P$ was computed using the van der Waals equation of state, but conceptually it can be determined using any equation of state. The information in Figures 7-14 and 7-15 is not in the form of an analytic equation, but it *does* constitute a relationship among P, $\underline{V}$, and T. Thus, it is possible to quantify $(\partial \underline{V}/\partial T)_P$ (or any other partial derivative involving P, $\underline{V}$, and T) using the Lee-Kesler equation, and this makes it possible to construct generalized figures analogous to Figures 7-14 and 7-15 for properties such as molar enthalpy.

This is done here; the molar enthalpy residual function can be estimated through

$$\frac{(\underline{H} - \underline{H}^{ig})}{RT_c} = \frac{(\underline{H} - \underline{H}^{ig})^0}{RT_c} + \omega \frac{(\underline{H} - \underline{H}^{ig})^1}{RT_c} \qquad (7.81)$$

FOOD FOR THOUGHT 7-6

Use the data in the steam tables to test the assertion that liquid molar volumes are approximately constant except near the critical point.

FOOD FOR THOUGHT 7-7

Figures 7-14 and 7-15 were generated from an analytical equation, as explained in Section 7.4.1, but pretend you didn't know this. How could you go about calculating $(\partial \underline{V}/\partial T)_P$ at a specific T and P, using only the figures themselves?

where the $(H - H^{ig})^0$ term can be obtained from Figure 7-16 and the $(H - H^{ig})^1$ term is determined from Figure 7-17. The analogous equation for the molar entropy residual function is

$$\frac{(S - S^{ig})}{R} = \frac{(S - S^{ig})^0}{R} + \omega\frac{(S - S^{ig})^1}{R} \tag{7.82}$$

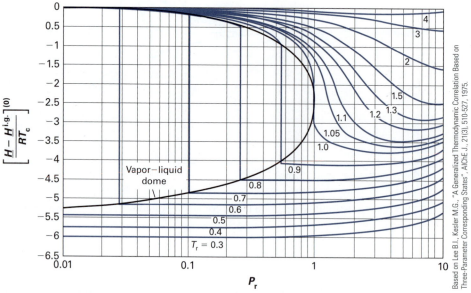

FIGURE 7-16 Molar enthalpy residual function for a compound with $\omega = 0$, as determined using the Lee-Kesler equation.

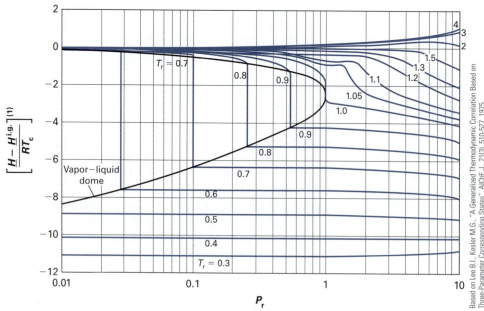

FIGURE 7-17 Correction to the molar enthalpy residual function for a compound with $\omega = 1$, as determined using the Lee-Kesler equation.

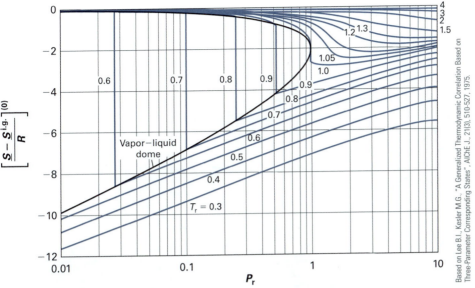

FIGURE 7-18 Molar entropy residual function for a compound with $\omega = 0$, as determined using the Lee-Kesler equation.

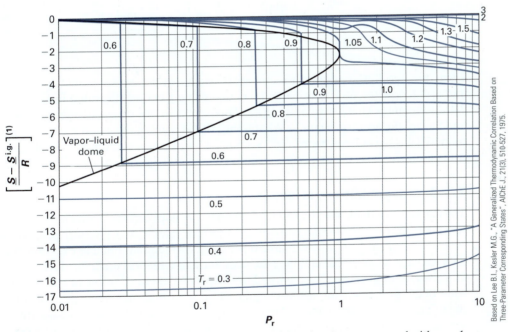

FIGURE 7-19 Correction to the molar entropy residual function for a compound with $\omega = 1$, as determined using the Lee-Kesler equation.

Equation 7.82 can be applied using Figures 7-18 and 7-19, as illustrated next in Example 7-9. In Section 6.3, it was noted that other residual functions (e.g., $\underline{U}, \underline{A}$, and $\underline{G}$) can be determined from the $\underline{H}$ and $\underline{S}$ residual functions by applying the definitions of H, A, and G. This is true whether the $\underline{H}$ and $\underline{S}$ residual functions are computed from the analytical equations found in Section 6.3 or from generalized correlations like the ones given in this section.

EXAMPLE 7-9	**APPLYING THE LEE-KESLER EQUATION**

Use the Lee-Kesler correlation to find the molar entropy residual for toluene at $T = 651$ K and $P = 5.75$ MPa.

Computing this residual function is a step in Problem 7-9, which examines a turbine.

SOLUTION:

Step 1 *Calculate reduced properties*
The T and P of toluene are the same as in Example 7-8, where we determined $P_r = 1.40$ and $T_r = 1.1$.

Step 2 *Obtain relevant data*
In the Lee-Kesler formulation, the molar entropy residual function is computed from Equation 7.82 as

PITFALL PREVENTION

The y axis in Figure 7-18 and Figure 7-19 is the dimensionless group $(\underline{S} - \underline{S}^{ig})/R$, not simply the residual function $(\underline{S} - \underline{S}^{ig})$.

$$\frac{(\underline{S} - \underline{S}^{ig})}{R} = \frac{(\underline{S} - \underline{S}^{ig})^0}{R} + \omega \frac{(\underline{S} - \underline{S}^{ig})^1}{R}$$

According to Appendix C, the acentric factor of toluene is 0.264. From Figure 7-18 $P_r = 1.4$ and $T_r = 1.1$, $(\underline{S} - \underline{S}^{ig})^0/R = -1.3$ and from Figure 7-19 $(\underline{S} - \underline{S}^{ig})^1/R = -0.6$.

$$\frac{(\underline{S} - \underline{S}^{ig})}{R} = -1.3 + (0.264)(-0.6) = -1.46 \qquad (7.83)$$

$$(\underline{S} - \underline{S}^{ig}) = (-1.46)\left(8.314 \ \frac{J}{mol \cdot K}\right) = \mathbf{-12.1} \frac{J}{mol \cdot K}$$

Using the Lee-Kesler approach requires knowledge of the critical point (T_c and P_c) and acentric factor, ω. Fitting parameters to a cubic EOS, as illustrated in Example 7-4, also requires knowledge of the critical point plus (in the cases of Peng-Robinson and Soave) the acentric factor ω. What if the critical point of the compound you wish to model has never been measured? Section 7.3.3 describes an estimation strategy—using group additivity—that is also inspired by the principle of corresponding states.

7.3.3 Group Additivity Methods

Throughout this book, we've examined engineering applications in which one needs to know the inter-relationships between thermodynamic and physical properties such as $P, \underline{V}, T, \underline{S}, \underline{U}, \underline{H}$, etc. We have repeatedly stressed that reliable experimental data isn't always available and thus the need for mathematical modeling techniques. However, most everything we learned in Chapter 6 depended upon an equation of state being available, and in this chapter we've seen that one needs at least some data (e.g., T_c, P_c, ω) to propose a realistic equation of state. Using group additivity schemes, it is possible to generate reasonable estimates of the properties of a compound even if one has no access to any experimental data at all.

Table 7-8 gives a simple illustration of the rationale for group additivity schemes. The table compares the normal boiling points of pairs of isomers—each pair has a linear hydrocarbon and another hydrocarbon with the same chemical formula but a methyl substitution at the #2 position in the chain. Observe that the boiling point of the branched compound is consistently 7 to 8 K lower than that of the linear compound—except in the C_4H_{10} row where the difference is 11 K. Observe too that the boiling points within each family (linear and branched) of compounds increase

TABLE 7-8 Comparison of normal boiling points for saturated hydrocarbon isomers.

Chemical Formula	Linear Compound	Linear Compound Normal Boiling Point (K)	Branched Compound	Branched Compound Normal Boiling Point (K)
C_4H_{10}	n-butane	272.95	Isobutane	261.95
C_5H_{12}	n-pentane	309.15	2-methyl butane	301.15
C_6H_{14}	n-hexane	341.85	2-methyl pentane	334.05
C_7H_{16}	n-heptane	371.45	2-methyl hexane	363.25
C_8H_{18}	n-octane	398.65	2-methyl heptane	390.75
C_9H_{20}	n-nonane	423.75	2-methyl octane	416.15
$C_{10}H_{22}$	n-decane	447.15	2-methyl nonane	440.15

(roughly) linearly with the number of carbon atoms—although the lightest hydrocarbons are again something of an outlier in this trend.

Observations like these lead to the hypothesis that functional groups within a chemical compound—groups like CH_3, CH_2, COOH, etc.—make contributions to the physical properties of a compound that are consistent—regardless of what those groups are attached to. Thus, one can estimate the chemical properties of a compound by summing the contributions of all of the individual groups in the compound. Table 7-8 implies that this approach might not work as well for the lightest compounds—ones that are too small to be meaningfully broken down into "groups." This is not such a drawback, since (practically speaking) it is unlikely we would wish to apply a group additivity scheme to such a small molecule. One point is that the number of possible compounds with four or fewer carbon atoms is comparatively small, and there is extensive experimental data available for many of these (e.g., methane, ethane, and ethylene). Another point is that thermodynamic properties of small compounds can be predicted using more advanced techniques involving microscopic modeling, which are introduced in Section 7.4. While quite useful, these methods require extensive calculations that become exponentially more time-consuming as the molecule being simulated gets larger. Thus, a virtue of group additivity schemes is that they can produce reasonable estimates of the properties of larger, more complex molecules for which microscopic modeling approaches are less practical.

Appendix G describes one group additivity scheme—the Joback (Reid, 1987) method—that can be used to estimate 11 different properties of a compound. Example 7-10 provides an illustration that focuses on estimating critical temperature and critical pressure, since the critical point has been of vital interest throughout this chapter.

> Using group additivity schemes, you can estimate the properties of a chemical compound that has never been synthesized in a lab or found in nature—as long as the chemical structure is plausible such that it *could* exist.

> This additivity scheme was introduced in a 1987 publication by Joback and Reid. It has been described both as the "Joback method" and as the "Joback and Reid method."

ESTIMATION OF CRITICAL TEMPERATURE AND CRITICAL PRESSURE USING THE JOBACK METHOD	EXAMPLE 7-10

The chemical compound of interest is CH_3–CHCl–CH_2–CHCl–CH_2–CH = CH_2

Estimate the critical temperature and critical pressure of this compound using group additivity.

SOLUTION:

Step 1 *Identify the groups within the compound*

The complete list of groups identified within the Joback method is in Appendix G. In this case, the compound contains

One $-CH_3$ group

Two $-CH_2-$ groups

Two $>CH-$ groups

Two $-Cl$ groups

One $=CH-$ group

One $=CH_2$ group

Step 2 *Identify relevant equations*

From Appendix G, the Joback equations for critical temperature and critical pressure are

$$T_c = T_b[0.584 + 0.965\Sigma - (\Sigma)^2]^{-1} \qquad (7.84)$$

$$P_c = (0.113 + 0.0032n_a - \Sigma)^{-2} \qquad (7.85)$$

where n_a is the number of atoms and Σ represents the sum of the contributions of individual groups to the property. The units are Kelvin for the critical temperature and bar for the critical pressure. The formula for T_c requires the normal boiling point T_b. We would use an experimental value of T_b if we had one, but as it is, we will estimate this too from the Joback method:

$$T_b = 198.2 + \Sigma \qquad (7.86)$$

Step 3 *Sum group contributions*

The contribution of each group to each property is obtained from Appendix G. The results are summarized in Table 7-9.

TABLE 7-9 Applying the Joback group additivity method to the compound $CH_3–CHCl–CH_2–CHCl–CH_2–CH=CH_2$.

Group	# Occurrences	T_b	T_c	P_c
$-CH_3$	1	23.58	0.0141	−0.0012
$-CH_2-$	2	2 × 22.88	2 × 0.0189	2 × 0
$>CH-$	2	2 × 21.74	2 × 0.0164	2 × 0.0020
$-Cl$	2	2 × 38.13	2 × 0.0105	2 × (−0.0049)
$=CH-$	1	24.96	0.0129	−0.0006
$=CH_2$	1	18.18	0.0113	−0.0028
Σ		232.22	0.1299	−0.0104

Step 4 *Compute properties*

Substituting the results from Table 7-9 into Equations 7.84 through 7.86 for normal boiling point gives

$$T_b = 198.2 + \Sigma = 198.2 + 232.22 = 430.42\,K \qquad (7.87)$$

For critical temperature, we have

$$T_c = T_b[0.584 + 0.965\Sigma - (\Sigma)^2]^{-1} \qquad (7.88)$$

$$T_c = (430.42\,K)[0.584 + 0.965(0.1299) - (0.1299)^2]^{-1} = \mathbf{621.6\,K}$$

For critical pressure, the number of atoms is $n_a = 21$ (7 C, 12 H, 2 Cl) and the estimated P_c is

$$P_c = (0.113 + 0.0032n_a - \Sigma)^{-2} \qquad (7.89)$$

$$P_c = [0.113 + 0.0032(21) - (-0.0104)]^{-2} = \mathbf{27.53\,bar = 2.753\,MPa}$$

Step 5 *Assessing Accuracy*

This compound is a di-chloro-heptene. It is not a prominent enough chemical product to appear in mainstream resources like the NIST Webbook, the CRC Handbook of Chemistry and Physics or Perry's Chemical Engineers' Handbook. The value of group additivity schemes in they allow us to generate an estimate of physical properties when no data is available. But how accurate is the estimate? One way to benchmark this is to apply the Joback method to the compound 1-heptene, the most similar compound for which data is readily available. The compound 1-heptene contains one "–CH$_3$" group, four "–CH2–" groups, one "=CH–" group and one "=CH$_2$" group. The Joback method for this compound predicts that T_c = **523.5 K** and P_c = **29.3 bar**. (The calculations are shown in an expanded solution available in the supplemental material.) The experimental values (Poling, 2001) are T_c = 537.3 K and P_c = 29.2 bar. Thus, for this case, the Joback method was accurate within ~2.5% for the critical temperature and ~0.5% for the critical pressure.

Example 7-10 illustrates a typical example of using a group additivity method—the Joback method—to estimate critical temperature and pressure. This estimation, in turn, could be used for fitting parameters to a cubic equation of state or in the application of the principle of corresponding states. It is possible to estimate or predict physical properties of a compound for which no experimental data exist.

Other group additivity schemes have been proposed by authors such as Benson (Benson, 1976) and Marrero and Pardillo (Pardillo-Fontdevila, 1999). The method of Constantinou and Gani (L. Constantinou, 1995) is of particular interest for use with equations of state, because it allows estimation of the critical point *and* the acentric factor, along with several other properties such as C_P, $\Delta \underline{H}_{vap}$, etc. Poling et al. (B. E. Poling, 2001) provide an excellent summary of these methods.

Note that you don't actually need the principle of corresponding states in order to understand and apply a group additivity scheme. Notice that Example 7-10 required only knowledge of the molecular structure and access to the Joback group property tables; the calculations do not "use" the principle of corresponding states in any explicit way. However, the principle does provide a rationale that explains why group additivity schemes can work: There are strong observed correlations in the relationships among P_r, T_r, and Z for groups of "similar" compounds, and a group additivity scheme allows us to attribute these correlations directly to the groups of atoms that make the compounds "similar."

7.4 Beyond the Cubic Equations of State

This chapter has discussed strategies for modeling the P-$\underline{V}$-T relationships observed in *fluids*: liquids, vapors, and supercritical fluids. Section 7.2 provided an overview of cubic equations of state, highlighting the van der Waals, Soave, and Peng-Robinson equations. These cubic equations of state are single equations that are capable of modeling the liquid, vapor, and supercritical phases—each with a single equation containing only two adjustable parameters. This section describes alternative approaches to developing equations of state, in which *macroscopic fluid properties are estimated from microscopic models of the behavior of individual molecules.*

At this point, you might be thinking, How is that different? Don't the cubic equations of state themselves model molecular-level behavior? The answer is "yes they do," but only in a limited way. It was noted in Section 7.2.2 that cubic equations of state can be understood as modeling the actual pressure as a sum of the "repulsion pressure" and the "attraction pressure" with

$$P = P^{rep} + P^{att}$$

Liquids, vapors, gases and supercritical fluids are collectively called "fluids" because they all take on the shape of their container.

Thus, in a broad way, the equations are indeed inspired by some understanding of molecular-level behavior. However, the cubic equations of state are, as Peng and Robinson described them, "semi-empirical" (Robinson, 1976). The cubic equations of state model the macroscopically observed behavior of fluids, and while they incorporate some consideration of molecular-level phenomenon, they do not directly incorporate any properties of *individual* molecules, such as their sizes or dipole moments. Consider the following statements.

- The rationale for using cubic equations, as described in Section 7.2.1, relies entirely on macroscopic observations (e.g., What happens to volume as pressure is increased? What happens to vapor pressure as temperature is increased? etc.).

- The intent of parameter b in these equations is to represent the "excluded volume" of the molecules. However, values of b are not conventionally obtained from measurements of the sizes of individual molecules.

- The intent of the parameter a in these equations is to represent "intermolecular attractive forces." However, values of a are not conventionally calculated from dipole moments or any other molecular-level property that is directly related to intermolecular forces.

- Rather, a and b are conventionally calculated as in Example 7-4 by finding the values that produce a saddle point at $T = T_c$ and $P = P_c$. While quite logical, this process relies entirely upon macroscopic properties.

This discussion is not intended to marginalize the use of cubic equations of state; they have tremendous practical value. Instead, our intent is to contrast two approaches to mathematical model-building that are both in common use by modern chemical engineers: the semi-empirical approach represented by cubic equations of state and the molecular-level approach, which we call the "microscopic modeling" of thermodynamic systems.

We now define two terms that help create a framework for microscopic modeling:

Quantum chemistry is the use of quantum mechanics to build detailed mathematical models of atoms and molecules.

The nucleus of each atom has a known mass and a known positive charge. Individual electrons do not have distinct positions within an atom or molecule. Rather, electrons are modeled by probability functions that describe the *electron density*, which is the fraction of time an electron spends in each location as it orbits a nucleus.

Statistical mechanics is the application of probability theory to modeling thermodynamic systems.

Some chemical engineering curricula include a course on physical chemistry, which would likely include an introduction to quantum chemistry and statistical mechanics.

Where quantum chemistry allows us to build a very detailed model of an individual molecule, statistical mechanics allows us to model the behavior of systems that contain huge numbers of individual molecules. We know molecules are constantly in motion and interacting with other molecules. Statistical mechanics can be used to quantify the frequency of different types of interactions and compute the properties of molecules *on average*. For example, you can think of the macroscopically observed internal energy as a reflection of the average internal energy of all molecules, and statistical mechanics allows us to compute this average from a microscopic model.

The cubic equations of state are comparatively straightforward relationships that are accessible to introductory students of chemical engineering thermodynamics. Quantum chemistry and statistical mechanics are both mathematically intense fields that can be regarded as graduate-level chemical engineering topics. While detailed coverage of microscopic modeling is beyond the scope of a first course in thermo-dynamics, this section is intended to give you a sense of what is possible through microscopic modeling. For example, process simulators allow the user to choose among many different methods for computing physical properties. Even if the user isn't knowledgeable enough to build a new microscopic model on his or her own, a general understanding of microscopic modeling can help the user make informed decisions on a modeling strategy when using simulation software.

The next section presents an equation of state—the only one in this chapter that is not a cubic equation—that can be understood on a microscopic level.

7.4.1 The Virial Equation of State

The virial equation of state expresses the compressibility factor Z as a simple power series as

$$Z = 1 + \frac{B}{\underline{V}} + \frac{C}{\underline{V}^2} + \frac{D}{\underline{V}^3} + \cdots \qquad (7.90)$$

where B, C, and D are the "second," "third," and "fourth" virial coefficients and are functions of temperature.

This equation has a simple form that can be reconciled with a microscopic model-ing approach in the following way. A microscopic model of a fluid must integrate an accurate accounting of the interactions between molecules. The term $\underline{V}$ represents the molar volume of a fluid, which we are accustomed to expressing in units like m^3/mol. The inverse $(1/\underline{V})$ is the *molar density*, which is the number of molecules in a specific volume of space. Thus, we might expect the frequency with which a typical molecule interacts with ONE other molecule to be directly proportional to $(1/\underline{V})$. A "three-body interaction" requires our typical molecule to be in proximity with two other molecules simultaneously: thus, its frequency would be proportional to $(1/\underline{V})^2$. It is theoretically possible to calculate values of B, C, D, etc. using statistical mechanics with $B/\underline{V}$ rep-resenting the effects of two-body interactions, $C/\underline{V}^2$ representing three-body interac-tions, etc. In theory, the power series can continue indefinitely with succeeding terms representing interactions between larger and larger numbers of molecules until all possible interactions are accounted for.

In practice, while much has been published about the virial equation, it is not widely used. While the discussion above implies a theoretical basis that is more fun-damental than that of the cubic equations of state, the equation in practice is not very accurate for dense fluids—meaning liquids, supercritical fluids, or even vapors at high reduced pressures (B. E. Poling, 2001). This observation can be rationalized by reviewing the form of the equation. $Z = 1$ is the ideal gas model, and you can view the other terms as correction factors that account for departures from ideal gas behavior. Thus, it's not surprising that the virial equation works best for vapors in which the departure from ideal gas behavior is modest and is less useful in describing dense fluids that are fundamentally different from ideal gases.

Consequently, the virial equation in practice is primarily applied to the *vapor phase only* in applications with pressure high enough for departures from ideal gas behavior to be significant—yet low enough to be well below the critical pressure. For vapors at lower pressures, $\underline{V}$ is large, so the $1/\underline{V}^2$ and higher-order terms are often negligible. In such cases, it is logical to truncate the virial equation after the second term. Also com-mon is expressing the equation as a function of P instead of $\underline{V}$, since P and T are in practice the variables that are most often used as design specifications. Consequently,

$$Z = 1 + \frac{B'P}{RT} \qquad (7.91)$$

"Interactions" is used as a catch-all term in this section. Dipole-dipole, induction, and dispersion are all examples of what can be meant by "interac-tions," as detailed in Section 7.4.4.

The virial equation is analogous to a Taylor series, which repre-sents a function as an infinite summation. The "1" that starts the series represents the fact that if there are no intermolecular interactions ($0 = B = C = D = \ldots$) then the fluid is an ideal gas ($Z = 1$).

Equations 7.90 and 7.91 are both called "the virial equation" in the open literature, which is somewhat mislead-ing, since they are not equivalent. You cannot derive Equation 7.91 directly from Equation 7.90, and the numerical values of B and B' in the two equations are not the same.

is sometimes called "the virial equation." In this expression, B' is again a function of temperature, and its numerical value in Equation 7.91 is not identical to the value of B in Equation 7.90.

Notice that Equation 7.91 implies a linear relationship between Z and P with the slope of the linear relationship being a function of temperature (B/RT). Inspection of Figures 7-13, 7-14, and 7-15 shows that modeling Z vs. P as a linear function is quite reasonable at lower pressures.

It is possible to strike a generalized correlation for values of B in Equation 7.91 through the principle of corresponding states. Equation 7.91 can be rewritten as a function of reduced properties:

$$Z = 1 + \frac{B'P}{RT} = 1 + \left(\frac{B'P_c}{RT_c}\right)\frac{P_r}{T_r} \tag{7.92}$$

And the factor $(B'P_c)/(RT_c)$, like the virial coefficients themselves, can be modeled as a function of temperature only. As reported by Smith et al. (J. M. Smith, 1996), the following equations well represent $(BP_c)/(RT_c)$ as a function of temperature and the acentric factor.

$$\left(\frac{B'P_c}{RT_c}\right) = B^0 + \omega B^1 \tag{7.93}$$

$$B^0 = 0.083 - \frac{0.422}{T_r^{1.6}} \tag{7.94}$$

$$B^1 = 0.139 - \frac{0.172}{T_r^{4.2}} \tag{7.95}$$

The Lee-Kesler equation was introduced in Section 7.3.2.

Next we note the linkage between the virial equation and the Lee-Kesler equation. The information in Figures 7-14 and 7-15 was originally published by Pitzer et al. (K. Pitzer, 1955) in the form of tables. At that time, computers were in their infancy and tables were the most straightforward and accurate way to make the information readily useable. Later, Lee and Kesler recognized the desirability of fitting the data to analytical equations that could be programmed into computers and published this equation in 1975 (Lee, 1975).

$$Z = 1 + \frac{B}{V_r} + \frac{C}{V_r^2} + \frac{D}{V_r^5} + \frac{E_{c_4}}{V_r^2 T_r^3}\left(E_\beta + \frac{E_\gamma}{V_r^2}\right)\exp\left(-\frac{E_\gamma}{V_r^2}\right) \tag{7.96}$$

in which B, C, D, E_{c_4}, E_β, and E_γ are all functions of temperature. While this should be regarded as a strictly empirical equation, its form is inspired in part by the virial equation.

As noted previously, there exist in practice large ranges of conditions in which Z is essentially linear with respect to P. At these conditions, the very simple Equation 7.91 provides results that are practically identical to the very complex Equation 7.96. Elliott and Lira (C. T. Lira, 1999) recommend the following heuristics for determining whether Z is in the linear region:

$$T_r > 0.686 + 0.439P_r \quad \text{or} \quad \underline{V}_r > 2.0 \tag{7.97}$$

The following example provides an illustration of the virial equation and the use of these heuristics.

APPLICATION OF THE VIRIAL EQUATION TO BENZENE	EXAMPLE 7-11

Use the $Z = f(T, P)$ virial equation to determine the compressibility (Z) of benzene in the vapor phase at

A. $T = 523.15$ K and $P = 0.4$ MPa

B. $T = 523.15$ K and $P = 2.8$ MPa

Equation 7.90 expresses Z as a function of T and V, while Equation 7.91 expresses Z as a function of T and P.

SOLUTION A:

Step 1 *Obtain data and determine reduced properties*
According to Appendix C, benzene has $T_c = 562.2$ K, $P_c = 4.898$ MPa, and $\omega = 0.211$. Thus,

$$T_r = \frac{523.15 \, \text{K}}{562.2 \, \text{K}} = 0.931 \tag{7.98}$$

$$P_r = \frac{0.4 \, \text{MPa}}{4.898 \, \text{MPa}} = 0.0817 \tag{7.99}$$

Step 2 *Solve virial equation*
Equations 7.94 and 7.95 are functions of only the reduced temperature for

$$B^0 = 0.083 - \frac{0.422}{T_r^{1.6}} = 0.083 - \frac{0.422}{(0.931)^{1.6}} = -0.390$$

$$B^1 = 0.139 - \frac{0.172}{T_r^{4.2}} = 0.139 - \frac{0.172}{(0.931)^{4.2}} = -0.093$$

Once these are known, Equation 7.93 can be solved as

$$\left(\frac{B' P_c}{R T_c} \right) = B^0 + \omega B^1 = -0.390 + (0.211)(-0.093) = -0.410$$

which allows solution of Equation 7.92 for Z to be

$$Z = 1 + \left(\frac{B' P_c}{R T_c} \right) \frac{P_r}{T_r} = 1 + (-0.410) \frac{0.0817}{0.931} = \mathbf{0.964}$$

SOLUTION B: Solve virial equation for higher pressure where in part B the reduced temperature and the critical point data for benzene are the same. The reduced pressure is now higher:

The vapor pressure of benzene at this temperature is ≈ 2.98 MPa.

$$P_r = \frac{2.8 \, \text{MPa}}{4.898 \, \text{MPa}} = 0.572$$

resulting in a lower value of Z for

$$Z = 1 + \left(\frac{B' P_c}{R T_c} \right) \frac{P_r}{T_r} = 1 + (-0.410) \frac{0.572}{0.931} = \mathbf{0.748}$$

Step 3 *Compare to data*
According to data obtained from the NIST WebBook (Lemmon, 2012), at $P = 0.4$ MPa and $T = 523.15$ K, $Z = 0.9644$. Note that the acentric factor, which we used in the calculations, is only reported to three significant digits, and the agreement between the virial equation and the data is perfect to three significant digits.

At $P = 2.8$ MPa and $T = 523.15$ K, the virial equation produces an estimate of 0.748, while the value obtained from the NIST WebBook (Lemmon, 2012) is 0.666, a difference of about 12%. Modern cubic equations of state are typically much more accurate than this.

Step 4 *Compare to heuristics*

The recommendation in Equation 7.97 is that this form of the virial equation only be used if $T_r > 0.686 + 0.439\,P_r$. The conditions in part A meet this criterion easily, but in part B, we have

$$T_r > 0.686 + 0.439 P_r$$

$$0.931 > 0.686 + 0.439(0.572)$$

$$0.931 \not> 0.947 \quad \text{FAILS} \tag{7.100}$$

The heuristic shows that the virial equation, at least in this form, should not have been applied to part B. The example is written this way to illustrate reasonable expectations for accuracy of the equations, but in a methodical problem-solving approach, applying the heuristic should be the first step, not the last.

In summary, the "virial equation" in its most general form (Equation 7.90) can be *understood* on a microscopic level. In practice, however, the most commonly used "virial equation" is actually a different, simpler equation (Equation 7.91). Furthermore, the coefficient B' is typically determined, not through a microscopic molecular model, but through the method illustrated in Example 7-11, which relies upon the principle of corresponding states.

The general form of the virial equation does, however, provide a framework for understanding some aspects of microscopic model-building. This section discussed how we expect the frequency of two-body interactions is proportional to $1/\underline{V}$, three-body interactions proportional to $1/\underline{V}^2$, etc. The following sections discuss aspects of microscopic model-building, and look more closely at what it would mean to model "interactions" between molecules.

7.4.2 Microstates and Macrostates

You may wish to review Section 4.3.6 before reading this one.

A starting point in building a microscopic model is shown in Figure 7-20. The dark outer square represents a "vessel" containing three spherical "gas molecules," labeled 1, 2, and 3. Figure 7-20 is identical to Figure 4-11, which was used in developing the microscopic view of entropy. That discussion involved accounting for the possible positions of molecules within the vessel. Here we extend that discussion by noting that molecules are not stationary; they are in constant motion. We could program a computer to model movement of these three spheres through the space. To do this, we would need to specify rules that guide the movement, such as "each

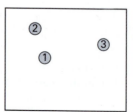

FIGURE 7-20 Three gas "molecules" in a "vessel" that has a fixed volume.

sphere moves in a straight line until it collides with either a wall or another sphere," and "collisions between spheres, or between spheres and the walls, are perfectly elastic." We would also have to specify an initial velocity for each molecule. Then we could simply run the program and watch the results.

In a sense, the computer program described in the previous paragraph is an equation of state. Consider that an EOS, by definition, is a relationship among P, $\underline{V}$, and T, such that if any two are specified, the third can be found. By specifying the size of the box in Figure 7-20 and deciding that there were three spheres, we effectively specified the molar volume, which is the number of "molecules" in a particular volume. By specifying the velocity of each sphere, we effectively specified the kinetic energy of each sphere, and internal energy and temperature are both macroscopic manifestations of the kinetic energy of individual molecules. Thus, T and $\underline{V}$ can be considered "known." Every time a gas molecule collides with the container wall, it exerts a force. Since pressure is force per unit area, our program could compute P by counting the number of collisions and computing the force of each. If we lower the molar volume (by making the box smaller) or raise the temperature (by making the spheres move faster), there will be more collisions between spheres and the walls; thus, our computer program will correctly show that lower $\underline{V}$ or higher T leads to higher P.

Thus, in principle, it's possible to create an equation of state by modeling the behavior of individual molecules, but Figure 7-20 represents a crude model consisting of only three individual molecules. Real gases, even at low density conditions like atmospheric pressure, consist of $\approx 10^{20}$ individual molecules per cubic centimeter. Therefore, it is not realistic to model the movements of each molecule individually, as Figure 7-20 and the previous discussion described. Rather, the essence of statistical mechanics is using mathematical functions based on the laws of probability to build models that are representative of the almost countless individual molecules. These functions allow one to answer questions like "How many times per second does the *average* molecule collide with the wall?" or "How fast is the *average* molecule moving?" The answers to questions like these can then be related to macroscopic properties.

How do we take the step from the described model in which the exact location of each molecule is known at any given time to a model that is based upon probability functions? Our discussion of the microscopic view of entropy in Section 4.3.6 represents the beginning of an answer. Recall the following.

- A microstate is one possible configuration for the components of a system.
- A macrostate is a specific state for a system.
- There can be many distinct microstates that each represent identical macrostates.

In Section 4.3.6, we established that the entropy of a macrostate can be computed from the number of corresponding microstates (Equation 4.32). This microscopic entropy reflects this probability: If we assume the gas molecules are located randomly, then all microstates are equally likely to occur. This would mean that the macrostates in which gas molecules are evenly distributed throughout the space, which have the most corresponding microstates, are the most likely to occur by random chance.

Thus, we can begin to envision what it means to represent the positions of molecules using probability functions. However, there are at least two limitations to the discussion up to this point:

We understand internal energy as microscopic potential and kinetic energy. This section mentions only kinetic energy of individual molecules. Section 7.4.4 addresses potential energy.

1. We haven't accounted for the fact that the molecules can be attracted to each other.

   The microscopic view of entropy suggests that molecules should spread out and fill all of the available space approximately evenly. But we know from experience that, at many temperatures and pressures, water can take the form of an ice cube or a glass of water rather than water vapor. The H_2O molecules remain in the compact solid or liquid phases—even when there is space available into which they could expand. Our discussion of microscopic entropy correctly rules out the possibility that H_2O molecules would all congregate together in a small space *by coincidence*, but we haven't yet accounted for the possibility that they congregate together *because they are attracted to each other*.

2. We haven't accounted for the fact that the molecules cannot occupy the same space.

   When we were picturing a computer program virtually moving three molecules around in the box, we acknowledged that molecules would collide, and we mentioned "perfectly elastic collisions" as a way of modeling these collisions. But our discussion of microstates and macrostates in Section 4.3.6 ignored this issue— we put no maximum on the number of molecules that could occupy one region.

These two aspects of microscopic modeling are discussed in the next two sections. Section 7.4.3 describes the distribution of molecules in space and shows that even molecules that experience no intermolecular forces can influence each other simply through excluded volume. Section 7.4.4 gives an overview of the modeling of intermolecular forces.

7.4.3 Radial Distribution Functions

In this section, we will imagine that molecules are *hard spheres*. We picture molecules as identical rigid spheres with radius R, and for the moment, we will assume there are no intermolecular attractive or repulsive forces. The significance of the fact that they are rigid is that if they collide they will not bend or deform—the shape and total volume of each hard sphere is constant. We now begin a thought experiment in which we imagine a chamber that is initially evacuated except for one sphere in the center, as shown in Figure 7-21. As we begin filling the chamber with more spheres, where will they be located relative to the original sphere?

Let's start with three spheres—the original one and two more. Figure 7-22 shows several possible configurations for the spheres—some with a sphere quite close to the original one and some with a large distance between them. Since the spheres are neither attracted to nor repelled by each other, no one of these configurations is any more likely to occur than any other. The only restriction on the spheres is that they cannot occupy the same space as each other. As we add more spheres, the number

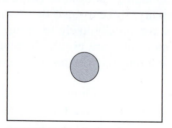

FIGURE 7-21 A chamber containing a single sphere.

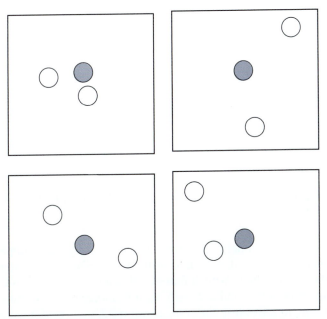

FIGURE 7-22 When the chamber contains only three spheres, there is very little restriction on where they can be placed.

of possible configurations increases exponentially. How can we summarize all this information into a single mathematical function?

A **radial distribution function** quantifies the density of molecules as a function of inter-atomic distance (r). Figure 7-23 helps illustrate what this means. It shows the space around the central sphere divided into concentric radial "shells" with each shell having the same thickness as the sphere's radius (R). Note that parts of the white sphere shown in Figure 7-23 are in three different shells, but we define a sphere's position by the location of its *center*, which is located in the shell that encompasses $2R < r < 3R$. The volume of the $2R < r < 3R$ shell is

$$V = \frac{4}{3}\pi(3R)^3 - \frac{4}{3}\pi(2R)^3 = 79.59R^3 \qquad (7.101)$$

The volume of a sphere is
$$V = \frac{4}{3}\pi r^3$$

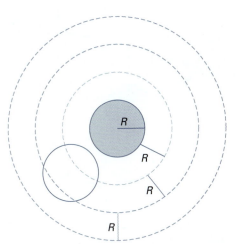

FIGURE 7-23 A series of concentric "shells" around the central sphere.

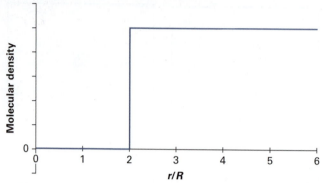

FIGURE 7-24 A radial distribution function for hard spheres at low density.

Dividing the number of spheres contained in the shell by the volume of the shell gives the (molar) density in that shell. If instead of shells with thickness R we have an infinite number of shells with infinitesimal thickness dr, we can formulate a continuous function of molecular density vs. intermolecular distance r. Figure 7-24 shows such a function. Because the spheres are rigid, their centers must be at least $2R$ apart; otherwise at least a portion of the spheres would overlap. This is reflected in the figure by a 0 density for all intermolecular distances less than $2R$. Beyond the requirement that the spheres do not occupy the same space, they do not interact at all—there are no attractive forces. Consequently, all other intermolecular distances are equally likely. This is reflected by uniform density when $r \geq 2R$.

Figure 7-22 represents low density, as the volume of the spheres themselves is very small compared to the volume of the space available. Figure 7-24 represents the radial distribution function that describes a fluid with low density; at low density, the only restriction on position is that two particles not occupy the same space. Offhand, you might think Figure 7-24 should be valid for hard spheres all the time—if the spheres don't interact, why should one interatomic distance be more or less likely than any other, regardless of how many or few molecules are present? Figure 7-25 shows why this offhand thought is not correct. If we pack spheres into the chamber

Figure 7-25 represents "simple packing" of spheres. The density of spheres would be slightly higher in "close packing."

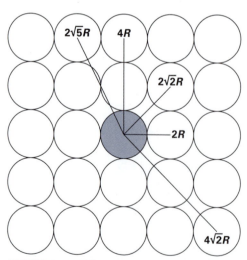

FIGURE 7-25 When the chamber is packed to capacity with spheres, there is no flexibility in where they can be located.

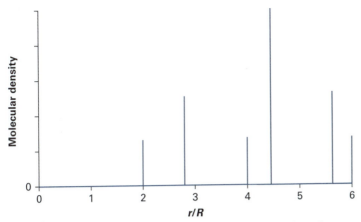

FIGURE 7-26 The radial distribution function that describes the spheres in Figure 7-25.

until it's completely filled, each sphere is inevitably in contact with several other spheres, so the density has a peak at $2R$. Furthermore, the ordered structure that occurs at maximum density ensures that our central sphere will have neighbors at regular, repeating intervals, some of which are shown in Figure 7-25. Consequently, the radial distribution function is as shown in Figure 7-26: The density is 0 except at specific, repeating interatomic distances.

Figure 7-24 is a logical radial distribution function for ideal gases that have low density, while Figure 7-26 is representative of solids that have repeating, ordered structures. A molecular state that is more representative of liquids or supercritical fluids is illustrated in Figure 7-27. Here, the density is high, but the spheres are not locked into a repeating structure. The radial distribution function here does not have distinct sharp peaks, but is not uniform either. In Figure 7-25, it is impossible to have $r = 2.05R$ or $r = 2.1R$; the spheres are in contact with each other, so $r = 2R$ exactly. In Figure 7-27, interatomic distances on the order of $r = 2.05R$ or $r = 2.1R$ are quite common, but an interatomic distance of $r = 2.5R$ is less common; there simply isn't usually this much space between a molecule and its nearest neighbor. In Figure 7-26, the shortest intermolecular distance beyond $2R$ is $2\sqrt{2}R$, and "next nearest neighbor" distances of this magnitude are again prevalent in Figure 7-27.

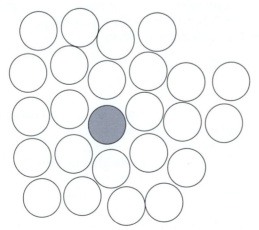

FIGURE 7-27 Sphere density is high, but not maximum, and there is no repeating ordered structure as in Figure 7-25.

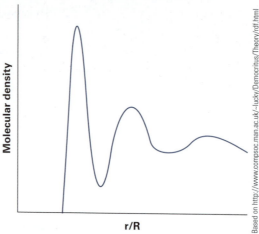

FIGURE 7-28 A radial distribution function that describes the spheres in Figure 7-27.

Liquids are sometimes described as having "short range order." Figure 7-28 and this paragraph illustrate what short range order means. By contrast you might say vapors have "no "order" and solids have "long range order."

As r increases, the density becomes more uniform. While we can recognize that high density makes, for example, $r = 2.1R$ more likely than $r = 2.5R$, the lack of an ordered structure means there aren't discrete repeatable distances as in Figure 7-26, and it's difficult to distinguish meaningfully between the probabilities of, for example, $r = 10.1R$ and $r = 10.5R$. Figure 7-28 gives a radial distribution function that represents the kind of distribution shown in Figure 7-27.

This section illustrates how mathematical functions can be used to describe the positions of molecules relative to each other. More significantly, it demonstrates that even if particles do not interact through attractive and repulsive forces, they still influence each other simply through excluded volume—the space occupied by one molecule isn't available to any other molecule. Of course, real molecules do have intermolecular attractive and repulsive forces, as examined in the next section, but the "hard sphere" effects discussed in this section also are present in real molecules.

Intermolecular forces, and how they can be modeled, are discussed in Section 7.4.4.

7.4.4 The Lennard-Jones Potential

Mathematically, a point describes a location in space that has zero volume. Thus a "point charge" is a charged particle that is modeled as having negligible volume.

Up to now we have used "intermolecular forces" or "intermolecular attractions" as rather broad terms, not really exploring the specific reasons why molecules can be attracted to, or repelled by, each other. The forces between two *point charges* can be quantified using Coulomb's Law, which is probably familiar from introductory physics or chemistry courses:

$$F \propto \frac{q_1 q_2}{r^2}$$

(7.102)

in which q_1 and q_2 are the charges, r is the distance between the two points. Unlike charges yield a negative value of F, representing an attractive force, and like charges yield a positive value of F, representing a repulsive force.

This section is concerned with the attractive forces between stable molecules. Ions, which have much stronger intermolecular forces, are not specifically considered here.

It would be straightforward to apply Equation 7.102 to individual protons and electrons, since they have known charges and are small enough that they can be regarded as points. But how does it apply to entire molecules? If we modeled the entire molecule as a "point," we would predict that there are no intermolecular forces at all—an individual molecule *as a whole* has no charge. However, many molecules have permanent dipoles; nonuniformities in electron density that cause part of the molecule to be positively charged and part to be negatively charged. Consequently, we must consider the following.

- **Dipole–dipole interactions.** As polar molecules move and rotate, they experience both attractive and repulsive interactions with each other, but the attractions and repulsions do not entirely cancel each other out. Statistical mechanics shows that the molecules (on average) spend more time aligned in attractive configurations than in repulsive ones. Thus, there is a net intermolecular attraction between permanent dipoles.

- **Induction forces.** When a molecule with a permanent dipole comes into proximity with a molecule that is not polar, the electron density of this molecule is influenced by the presence of the charged dipole. Thus an *induced dipole* is created in the normally non-polar molecule, and there is a net attractive interaction between the permanent and induced dipoles. The induced dipole effect can be modeled using quantum chemistry.

- **London dispersion forces.** Even non-polar molecules influence each other's electron densities and induce in each other weak dipoles that lead to attractive forces. For example, an individual noble gas atom is perfectly spherical and perfectly non-polar, but the nearby presence of another noble gas molecule can distort the spherical distribution of electrons, as the electron clouds shift to minimize overlap. Thus, even non-polar molecules induce dipoles in each other and experience weak attractive forces.

These attractive forces are the basis for *microscopic potential energy*, which we understand as a component of the macroscopic property internal energy. All objects are attracted to the Earth through gravity and have potential energy due to their distance from the Earth's center of gravity. Similarly, if charged objects that are attracted to each other are positioned a distance apart, they have potential energy; the attractive forces have the potential to move them toward each other.

Of the forces listed, dipole–dipole interactions are by far the strongest and dispersion forces the weakest. However, statistical mechanics reveals that the potential energy resulting from each of these kinds of forces can be modeled as proportional to $1/r^6$:

$$\Gamma \propto \frac{-1}{r^6} \tag{7.103}$$

with Γ representing intermolecular potential energy and r the distance between the *centers* of the molecules.

The discussion so far implies molecules are always attracted to each other, and become more strongly attracted as they move closer together (r decreases so $1/r^6$ increases). However, as two molecules move closer and closer together, at some point their electron clouds start to overlap. Looking back at Coulomb's Law (Equation 7.102), we can picture that if like-charged particles (electrons) are occupying practically the same space as each other ($r \approx 0$), then this repulsive force would overwhelm any attractive forces between the molecules (for which r is much larger). To continue the gravity analogy, consider an object 20 feet above a trampoline. The object has potential energy due to the Earth's gravity. As it falls, this potential energy gets converted into another form—kinetic energy. When the object hits the trampoline, the attractive force is still acting on it—the object would keep falling all the way to the center of the Earth if the force of gravity was the only relevant one. The object stops, because the trampoline is in the way. The trampoline isn't rigid—the object doesn't stop falling instantly—but the farther the object pushes into the trampoline,

Arunan et al. (Arunan 2011) have proposed a formal definition for "hydrogen bonding." In the context of this chapter, you can think of hydrogen bonds as strong dipole–dipole interactions, rather than as a separate kind of bond.

London dispersion forces are named after the physicist Fritz London, who first used quantum mechanics to explain how noble gas atoms can be attracted to each other. (Miller et al., 2002)

The negative sign in Equation 7.103 indicates that potential energy due to attractive intermolecular forces decreases (it becomes larger in magnitude but is negative) as r decreases, just like the potential energy of an object in the Earth's gravity decreases as the object gets closer to the Earth.

the harder the trampoline pushes back. When the object's velocity reaches zero, its kinetic energy has all been converted into potential energy, which now takes the form of the tension in the trampoline. This potential energy is converted back into kinetic energy when the object bounces back up. While the analogy isn't perfect, you can think of the intermolecular attractive forces as gravity and the repulsive forces from overlapping electron clouds as the trampoline.

How do these phenomena figure into intermolecular potential energy? Consider the following possibilities.

- If molecules are far apart, they have potential energy, because there are attractive forces pulling them towards each other.

- If the molecules are extremely close together, they have potential energy, because now there are repulsive forces pushing them apart.

- Somewhere in between, there is an intermolecular distance at which potential energy is minimized.

The **Lennard-Jones potential** is a model for intermolecular potential energy that integrates these aspects of potential energy as

$$\Gamma = 4\varepsilon \left[\left(\frac{\sigma}{r} \right)^{12} - \left(\frac{\sigma}{r} \right)^{6} \right]$$ (7.104)

> The Lennard-Jones potential is named after Sir John Edward Lennard-Jones, who was a professor of theoretical physics.

where

- Γ represents intermolecular potential energy.
- r represents the actual distance between the two molecules.
- σ represents the intermolecular distance at which $\Gamma = 0$.
- ε is the characteristic Lennard-Jones energy.

> This is sometimes called the "Lennard-Jones 12-6 potential" because of the exponents on the two terms.

σ and ε are parameters that have been tabulated for various compounds; see, for example, B. E. Polling (2001). Figure 7-29 shows these characteristics.

- At the lowest values of r, the term $(\sigma/r)^{12}$, which represents repulsive forces, dominates, and potential energy rapidly goes to infinity if r decreases below σ.
- At high values of r, both terms are negligible and potential energy is 0.
- At intermediate values, the term $(\sigma/r)^{6}$, which represents attractive forces, dominates and potential energy is negative.

> σ, which Poling (2001) calls the "characteristic Lennard-Jones length", can be regarded as a measure of molecular diameter.

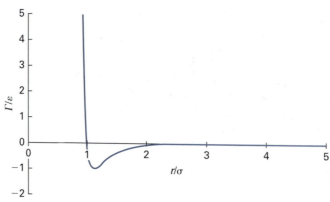

FIGURE 7-29 Lennard-Jones potential plotted in a non-dimensional form, Γ/ε vs. r/σ.

In Figure 7-29 the "potential energy well" appears to reach a minimum of $\Gamma \approx -\varepsilon$ when $r \approx 1.1\sigma$. By setting $d\Gamma/dr = 0$, one can prove that in fact the minimum occurs when $(\sigma/r)^6 = 0.5$ and that indeed $\Gamma = -\varepsilon$ at this point.

While the discussion in this chapter is not sufficient to let you build your own microscopic models, it should provide a sense of how a microscopic model can indeed be built. Two fundamental questions are (1) How frequently do specific intermolecular interactions occur? and (2) What is the effect of these interactions? Accounting for the relative positions of molecules, on average, was discussed in the previous section, and this allows us to model how frequently interactions will occur. A model of intermolecular potential like the Lennard-Jones allows us to account for the effects of the interactions.

> The minimum in the Lennard-Jones potential curve occurs when $(\sigma/r)^6 = 0.5$, which is approximately $r \approx 1.12\sigma$.

7.5 SUMMARY OF CHAPTER SEVEN

- A **cubic equation** can be used to model a compound in the vapor, liquid, and supercritical phases with a single set of parameters.

- The **van der Waals equation** was the original cubic EOS, although it is no longer in common use.

- The **Soave** and **Peng-Robinson** equations are modern improvements on the van der Waals EOS.

- The parameter **a** in the cubic equations of state represents intermolecular interactions.

- The parameter **b** in the cubic equations of state represents excluded volume of individual molecules.

- For temperatures below the critical temperature, a cubic equation of state generally has three real solutions for $\underline{V}$, the smallest of which represents a liquid $\underline{V}$ and the largest a vapor $\underline{V}$.

- Values of the parameters a and b are often assigned to fit a saddle point to the critical point.

- The **acentric factor** ω is defined from the reduced vapor pressure that corresponds to a reduced temperature of 0.7.

- Parameters for the Peng-Robinson and Soave equations can be assigned if the critical point and acentric factor are known.

- Two compounds are defined as being in **corresponding states** when they are at the same reduced temperature (T_r) and (P_r).

- The **principle of corresponding states** indicates that, if two compounds are in corresponding states, they will have approximately equal values of Z.

- **Generalized correlations** based upon the principle of corresponding states can be used to estimate compressibility factors (Z) and residual functions.

- The **Joback** method is a group additivity scheme, where a molecule's physical properties are estimated by summing the contributions of the groups that make up the molecule.

- The **virial equation** is a simple EOS that is primarily applicable to gases at moderate pressures.

- A **radial distribution function** quantifies molecular density as a function of inter-molecular distance.
- The **Lennard-Jones potential** quantifies microscopic potential energy as a function of intermolecular distance.

7.6 EXERCISES

7-1. Using data in Appendix C-1, determine the van der Waals parameters a and b for each of the following compounds.
  A. Methanol
  B. Carbon dioxide
  C. Ethanol
  D. Butane
  E. Octane

7-2. Using data in Appendix C-1, determine the Soave parameters a and b for each of the following compounds at the temperature $T = 300$ K.
  A. Argon
  B. Carbon monoxide
  C. 1-Propanol
  D. Pentane
  E. Decane

7-3. Using data in Appendix C-1, determine the Peng-Robinson parameters a and b for each of the following compounds at the temperature $T = 100°C$.
  A. Ethane
  B. Acetone
  C. Benzene
  D. Toluene
  E. Decane

7-4. Find the molar volume of methane at $P = 1.5$ MPa and $T = 473.15$ K (it is a gas at these conditions), using the following methods.

  A. The Soave equation of state
  B. The Peng-Robinson equation of state
  C. The virial equation of state
  D. The Lee-Kesler generalized correlation
  E. Figure 7-1

7-5. Find the molar volume of Freon® 22 at $P = 0.5$ MPa and $T = 20°C$ (it is a vapor at these conditions), using the following methods.
  A. The Soave equation of state
  B. The Peng-Robinson equation of state
  C. The virial equation of state
  D. The Lee-Kesler generalized correlation
  E. Appendix F

7-6. Find the molar volume of nitrogen at $P = 1$ MPa and $T = 330$ K, using the following methods.
  A. The Soave equation of state
  B. The Peng-Robinson equation of state
  C. The virial equation of state
  D. The Lee-Kesler generalized correlation
  E. Appendix F

7-7. Use the Joback method to estimate T_c and P_c of each of the following compounds.
  A. Butane
  B. 1-Hexanol
  C. 2-chloropentane
  D. 3-Hexene
  E. 1,3-butadiene

7.7 PROBLEMS

7-8. 55 kg of liquefied natural gas (LNG) are stored in a rigid, sealed 0.17 m³ vessel. In this problem, model LNG as 100% methane. Due to a failure in the cooling/insulation system, the temperature increases to 200 K, which is above the critical temperature; thus, the natural gas will no longer be in the liquid phase. Find the final pressure of the supercritical methane using the following methods.
  A. The ideal gas law
  B. The van der Waals equation
  C. The Soave equation
  D. The Lee-Kesler generalized correlation in Section 7.3.2
  E. Figure 7-1

7-9. 16.67 mol/s of toluene enters a turbine at $P = 5.75$ MPa and $T = 651$ K, and the exiting pressure is $P = 0.5$ MPa. If the efficiency of the turbine is 80%, determine the rate at which work is produced.
  A. Using the Lee-Kesler approach in Section 7.3.2
  B. Using the Peng-Robinson equation

7-10. Using data from the steam tables, make isotherm plots of P-$\underline{V}$ for both liquid water and steam at the temperatures $T = 50, 100, 200, 300$ and $400°C$. Then, make plots of P-$\underline{V}$, at the same temperatures and over the same range of values of $\underline{V}$, using the Peng-Robinson, Soave, and van der Waals equations. Comment on the accuracies of the equations of state and on the different features shown in the plots at the five different temperatures.

7-11. Imagine a compound has $T_c = 500$ K and $P_c = 2$ MPa. Use the Peng-Robinson equation throughout this problem.
  A. Plot P-$\underline{V}$ at $T = 400$ K, $T = 500$ K, and $T = 600$ K, assuming the compound has $\omega = 0$.
  B. Repeat part A for $\omega = 0.5$.
  C. Repeat part A for $\omega = 1.0$.
  D. What do the plots reveal about the inter-relationship between ω and the P-$\underline{V}$-T behavior of a compound?

7-12. Consider the chemical compound:

$$CH_3\text{--}CHCl\text{--}CH = CH\text{--}CHF\text{--}CH_2\text{--}CH_2OH$$

Estimate the normal boiling point, critical temperature, critical pressure, and standard enthalpy of formation of the compound using the Joback method.

7-13. Consider the chemical compound:

$$CH_3\text{--}CH_2\text{--}CHCl\text{--}CH = CH\text{--}CHF\text{--}CH_2\text{--}CH_3$$

  A. Estimate the critical temperature and critical pressure using the Joback method.
  B. Estimate the acentric factor of this compound by identifying structurally similar compounds in Appendix C and assuming similar compounds have similar acentric factors.
  C. Give your best estimate the molar volume at $P = 0.5$ MPa and $T = 500$ K.
  D. Give your best estimate of the molar volume at $P_c = 0.5$ MPa and $T = 800$ K.

7-14. A new and relatively unstudied compound is being investigated for its potential use as a refrigerant. The compound has a critical temperature $T_c = 500$ K and $P_c = 3$ MPa. The vapor pressure at $T = 350$ K is $P^{sat} = 0.24$ MPa. Estimate the molar volume of the compound at each of the following conditions.
  A. In the liquid phase at $T = 350$ K and $P = 0.6$ MPa.
  B. In the vapor phase at $T = 350$ K and $P = 0.2$ MPa.
  C. In the vapor phase at $T = 500$ K and $P = 0.6$ MPa.
  D. In the liquid phase at $T = 300$ K and $P = 0.3$ MPa.

7-15. You are a project engineer. You find a tank in your plant salvage yard which has a volume of 50 dm³. The specs on the tank indicate that it will burst at 3 MPa (absolute). Company policy is to never exceed half the bursting pressure of any tank. You would like to use the tank as a surge tank which will hold propane. The maximum amount of material to be stored will be 0.975 kg and the maximum temperature will be 425 K. Will you be in violation of company policy if you use this tank? Perform calculations using each of the following methods, and discuss the results.
  A. The ideal gas law
  B. The Lee-Kesler charts
  C. The virial equation

7-16. This problem re-visits Problem 2-16, which concerned the high pressure synthesis of ammonia by the reaction:

$$N_2 + 3H_2 \rightarrow 2\,NH_3$$

That problem asked you to estimate the change in molar enthalpy when hydrogen is compressed from $T = 300$ K and $P = 0.1$ MPa to $T = 700$ K and $P = 20$ MPa. At that time, the ideal gas model was the only method we had available to solve the problem. Now use the Peng-Robinson equation of state to solve this problem.
  A. Find the change in molar enthalpy when hydrogen is compressed from $T = 300$ K and $P = 0.1$ MPa to $T = 700$ K and $P = 20$ MPa.
  B. Find the work required to compress each mole of hydrogen from $T = 300$ K and $P = 0.1$ MPa to $T = 600$ K and $P = 1$ MPa in an adiabatic, steady-state compressor.
  C. Find the heat removed from each mole of hydrogen if the gas leaving the compressor in part B is cooled to $T = 300$ K and $P = 1$ MPa in a steady-state heat exchanger.
  D. Find the work required to compress each mole of hydrogen from $T = 300$ K and $P = 1$ MPa to $T = 800$ K and $P = 20$ MPa in an adiabetic, steady-state compressor.
  E. For each of the compressors in parts B and D, comment on whether their operation is physically realistic as described, and show calculations that support your conclusions.

7-17. A compound has a critical temperature of 500 K, a critical pressure of 3.039 MPa, and an acentric factor of $\omega = 0.45$. Estimate $\underline{V}$ for this compound in the vapor phase at $T = 450$ K and $P = 0.506$ MPa, using the following methods.
  A. The ideal gas law
  B. The van der Waals equation
  C. The Soave equation
  D. The Peng-Robinson equation
  E. The virial equation
  F. The Lee-Kesler equation

7-18. You are designing a process in which toluene is used as a solvent. In order to size process equipment, you need to know the properties at several conditions. Use the Peng-Robinson equation to estimate the following quantites.
  A. The molar volume at the critical point.
  B. The molar volume in the liquid phase at $T = 300$ K and $P = 0.1$ MPa.
  C. The molar volume in the vapor phase at $T = 500$ K and $P = 1$ MPa.
  D. The change in molar enthalpy when toluene is converted from $T = 300$ K and $P = 0.1$ MPa to $T = 500$ K and $P = 1$ MPa.

7-19. You are designing a process in which benzene is used as a solvent. In order to size process equipment, you need to know the properties at several conditions. Estimate the following quantites.
A. The molar volume at the critical point.
B. The molar volume in the liquid phase at $T = 278$ K and $P = 101.325$ kPa.
C. The molar volume in the vapor phase at $T = 444$ K and $P = 405.3$ kPa.
D. The change in molar enthalpy when benzene is heated and compressed from $T = 278$ K and $P = 101.325$ kPa to $T = 444$ K and $P = 405.3$ kPa.

7-20. A new and relatively unstudied compound is being investigated for its potential use as a refrigerant in a vapor-compression cycle that operates with $T = 268.15$ K in the boiler and $T = 308.15$ K in the condenser. The properties of the compound at these two conditions have been measured as follows:

- At $T = 268.15$ K, the vapor pressure is $P^{sat} = 0.06$ MPa, the enthalpy of vaporization is 35 kJ/mol.

- At $T = 308.15$ K, the vapor pressure is $P^{sat} = 0.42$ MPa and the enthalpy of vaporization is 30 kJ/mol.

However, there is very little data for the compound at other temperatures and pressures. Best estimates are the ideal gas heat capacity is $C_P^* = 110$ J/mol · K and the van der Waals parameters are $a = 1.5$ Pa m^6/mol^2 and $b = 4 \times 10^{-5}$ m^3/mol.

Give your best estimate of the coefficient of performance of a refrigeration cycle using this compound, if the efficiency of the compressor is 80%. State any assumptions that you make.

7-21. A new and relatively unstudied compound is being investigated for potential use as the working fluid in a Rankine heat engine. In the proposed heat engine, the boiler will operate at $T = 200°C$, with saturated vapor leaving the boiler, and the condenser will operate at $T = 40°C$, with saturated liquid leaving the condenser. You have access to the following data:

- At $T = 40°C$, the vapor pressure is $P^{sat} = 0.025$ MPa

- At $T = 200°C$, the vapor pressure is $P^{sat} = 1$ MPa

- The ideal gas heat capacity is approximately $C_P^* = 75$ J/mol · K.

- The liquid and vapor phases are described by the van der Waals EOS, with $a = 4$ Pa m^6/mol^2 and $b = 7.5 \times 10^{-5}$ m^3/mol.

Assuming the turbine has an efficiency of $\eta = 0.75$, give your best estimates of:
A. The work produced in the turbine, per mole
B. The work required in the pump, per mole
C. The efficiency of the cycle.

7-22. A compound has $T_c = 800$ K, $P_c = 4.5$ MPa, $C_P^* = 8R$, and $\omega = 0.25$ and is well modeled by the Peng-Robinson equation of state. 10 mol/s of the compound enters a turbine in the vapor state at $T = 600$ K and $P = 1$ MPa. It leaves at $T = 450$ K and $P = 0.075$ MPa, as a mixture with $q = 0.97$.
A. Determine the rate of which work is produced by the turbine.
B. Determine the rate of entropy generation in the turbine.
C. Determine the efficiency of the turbine.

7-23. This problem examines the generalization mentioned in Section 7-3, that the Lee-Kesler generalized approach should not be applied to highly polar compounds.
A. Choose three temperatures and three pressures that, for water, fall within the range $0.7 < T_r < 1.0$ and $0.1 < P_r < 0.9$. Combinations of these temperatures and pressures define nine data points.
B. For each of the nine data points identified in part A, estimate the molar volume of water from the Lee-Kesler approach, and compare to the data in the steam tables.
C. Repeat parts A and B for methane. Use the same *reduced* temperatures and pressures that you identified in part A for water, and compare the results of the Lee-Kesler approach to the data in Figure 7-1.
D. Comment of the results, recognizing that water is highly polar and methane is non-polar.
E. Choose your own compound that is more polar than methane, but less polar than water, and repeat parts A and B for this compound.

7-24. One mole of *n*-butane vapor is compressed at a constant temperature of 400 K from $P = 0.01$ MPa to $P = 1.5$ MPa. Using the Peng-Robinson equation of state:
A. Find ΔU for the gas in this process.
B. Find ΔS for the gas in this process.

7-25. Five moles per second of n-butane enter a turbine at $P = 1.5$ MPa and $T = 500$ K and leave the turbine at $P = 0.1$ MPa. The turbine has an efficiency of 80%. Find the rate at which work is done, using the following methods.
A. The Lee-Kesler approach
B. The Peng-Robinson equation of state

7-26. The Lennard-Jones parameters that describe intermolecular interactions for a compound are $\varepsilon = 1$ kJ/mol and $\sigma = 0.4$ nm.
A. Determine the intermolecular distance at which potential energy is minimized.
B. Determine the potential energy when the intermolecular distance is 0.5 nm.
C. Determine the potential energy when the intermolecular distance is 0.39 nm.
D. Make a complete plot of potential energy in kJ/mol vs. intermolecular distance in nm.

7.8 GLOSSARY OF SYMBOLS

a	parameter in van der Waals, Soave or Peng-Robinson EOS	m	fitting parameter used to find a for Soave EOS	$\underline{V}$	molar volume
		MW	molecular weight	$\underline{V}_c$	molar volume at critical point
a_c	value of Soave or Peng-Robinson parameter a at the critical point	N	number of moles in system	$\underline{V}_r$	reduced molar volume
		P	pressure	$\hat{V}$	specific volume
B	second virial coefficient	P_c	critical pressure	Z	compressibility factor
B'	coefficient in truncated, pressure-dependent form of virial equation	P_r	reduced pressure	Z_c	compressibility factor at critical point
		P^{sat}	vapor pressure		
		q	charge	α	parameter used in calculation of a for Soave or Peng-Robinson EOS
b	parameter in van der Waals, Soave, or Peng-Robinson EOS	R	gas constant		
		R	radius of a sphere	Γ	intermolecular potential energy
C	third virial coefficient	r	interatomic distance		
D	fourth virial coefficient	S	entropy	ε	Lennard-Jones energy
F	force	$\underline{S}$	molar entropy	κ	parameter used in calculation of a for Peng-Robinson EOS
H	enthalpy	$\underline{S}^{ig}$	molar entropy for an ideal gas state		
$\underline{H}$	molar enthalpy				
$\underline{H}^{ig}$	molar enthalpy for an ideal gas state	T	temperature	ρ	density
		T_c	critical temperature	σ	interatomic distance at which intermolecular potential energy is zero
M	mass of system	T_r	reduced temperature		
		V	volume	ω	acentric factor

7.9 REFERENCES

Arunan Elangannan, Desiraju, Gautam R., Klein Roger A., Sadlej Joanna, Scheiner Steve, Alkorta Ibon, Clary David C., Crabtree Robert H., Dannenberg Joseph J., Hobza Pavel, Kjaergaard Henrik G., Legon Anthony C., Mennucci Benedetta, Nesbitt David J., "Definition of the Hydrogen Bond (IUPAC Recommendations 2011), *Pure Appl. Chem.*, 83, 8 (2011).

Benson, S. W. *Thermochemical Kinetics: Methods for the Estimation of Thermochemical Data and Rate Parameters*, John Wiley & Sons, 1976.

Brown R. L., and Stein S. E., "Boiling Point Data" in *NIST Chemistry WebBook, NIST Standard Reference Database Number 69*, Eds. P. J. Linstrom and W. G. Mallard, National Institute of Standards and Technology, Gaithersburg MD, 20899, http://webbook.nist.gov (retrieved December 5, 2011)

Constantinou L., Gani R., and O'Connell J. P., *Fluid Phase Equilibrium*, 1995.

Cutlip, M. and Shacham, M., *Problem Solving in Chemical and Biochemical Engineering with POLYMATH, EXCEL and Matlab*, Prentice Hall, 2007.

Lee, B. I. and Kesler, M. G., "A Generalized Thermodynamic Correlation Based on Three-Parameter Corresponding States," *AIChE J.*, 21(3), 510-527, 1975.

Lemmon E. W., McLinden M. O., and Friend D. G., "Thermophysical Properties of Fluid Systems" in *NIST Chemistry WebBook, NIST Standard Reference Database Number 69*, Eds. P. J. Linstrom and W. G. Mallard, National Institute of Standards and Technology, Gaithersburg MD, 20899, http://webbook.nist.gov (retrieved May 31, 2012)

Lira, J. R., Elliott C. T., *Introductory Chemical Engineering Thermodynamics*, Prentice-Hall, 1999.

Miller, D., Miller, I., Miller, J., Miller, M., *The Cambridge Dictionary of Scientists*, 2nd ed., Cambridge University Press, 2002.

Pardillo-Fontdevila, J., Marrero-Marejon, E. *AICHE Journal*, 1999.

Pitzer, K. "The Volumetric and Thermodynamic Properties of Fluids. I. Theoretical Basis and Virial Coefficients," *Journal of the American Chemical Society*, 1955.

Pitzer, K., Lippmann, D., Curl, R., Huggins, C., Petersen, D., "The Volumetric and Thermodynamic Properties of Fluids. II. Compressibility Factor, Vapor Pressure and Enthalpy of Vaporization," *Journal of the American Chemical Society*, 1955.

Poling, B. E., Prausnitz J. M., O'Connell J. P., *The Properties of Gases and Liquids.* 5th edition, McGraw-Hill, New York, 2001.

Reid, K. J., Joback, R. C., "Estimation of Pure-Component Properties from Group Contributions," *Chemical Engineering Communications*, 1987.

Sandler, S. *An Introduction to Applied Statistical Thermodynamics*, John Wiley & Sons, 2011.

Smith J. M., Van Ness, H. C., Abbott, M. M., *Introduction to Chemical Engineering Thermodynamics*, McGraw-Hill, 1996.

Soave, G., "Equilibrium Constants from a Modified Redlich-Kwong Equation of State," *Chemical Engineering Science*, 27, 1972.

Robinson, D., Peng D., "A New Two-Constant Equation of State," *Ind. Eng. Chem. Fundamentals*, 15, 1, 1976.

Modeling Phase Equilibrium for Pure Components

8

LEARNING OBJECTIVES

This chapter is intended to help you learn how to:

- Identify the most stable phase for a compound at a particular T and P, using the molar Gibbs free energy $\underline{G}$
- Recognize and apply the equilibrium criterion: for phases in equilibrium, $\underline{G}$ is identical
- Estimate vapor pressure using the *Clausius-Clapeyron, shortcut,* and *Antoine equations*
- Recognize the approximations made in deriving the Clausius-Clapeyron, shortcut, and Antoine equations, and determine which equation is best suited for the case at hand
- Calculate the fugacity of a pure compound at any T and P using an equation of state
- Use fugacity to estimate vapor pressure
- Use the *Poynting correction* to estimate the fugacity of liquids and solids at elevated pressures

In the first five chapters of this book, all properties needed to solve problems were obtained either from data (e.g., the steam tables) or from simple models (e.g., the ideal gas law). Chapter 6 demonstrated that if one has an equation of state, one can quantify not only the properties explicitly present in the equation of state (P, $\underline{V}$, T), but a number of other essential thermodynamic properties as well (e.g., $\underline{H}$, $\underline{U}$, $\underline{S}$). Chapter 7 broadened our knowledge of equations of state and showed how, in many cases, the same equation can be used to describe both the vapor and liquid phases. Throughout these chapters, one fundamental property has been treated as "given," and that was the physical state of the material. Up to this point, we have solved problems in which the material was specified to be a solid, liquid, gas, vapor, or supercritical fluid. When phase changes were part of the process, the temperature or pressure at which the phase transition (e.g., boiling, melting, or sublimation) occurred was known. This chapter explores how phase transitions can be predicted or modeled for cases in which data is not available. ∎

8.1 MOTIVATIONAL EXAMPLE: VLE Curves for Refrigerants

Refrigeration was first discussed in Section 5-3. We examined the design of a refrigeration cycle (Example 5-3) in which Freon® 22 was used as the refrigerant. Freon® 22 is a *chlorofluorocarbon* (*CFC*), which is a class of compounds that was commonly used as refrigerants until the 1980s, when it was demonstrated that they (CFCs) were damaging the ozone layer. An international agreement called the Montreal Protocol

Chlorodifluoro-methane is known as HCFC-22, R-22, and Freon 22. Freon® is a registered trademark of DuPont.

was ratified, which mandated a specific rate at which production and use of CFCs would be reduced until completely phased out (Benedick, 1991). Manufacturers of refrigerants were thus faced with a significant engineering challenge: design a new generation of refrigerants that were not ozone depleting, but provided similar performance to CFCs. How might you approach such a task?

If you review Example 5-3, you will note that we started with a pressure vs. specific enthalpy diagram for Freon® 22, located in Appendix F. Reviewing the example, we see that important design decisions in the refrigeration cycle stemmed directly from this data:

Liquid–vapor phase changes also play an essential role in the Rankine heat engine and the Linde liquefaction process illustrated in Chapter 5. These also can be used as "motivational examples" for this chapter.

- The boiler operates at $P = 0.5$ MPa because that is the vapor pressure corresponding to the desired temperature $T = 273.15$ K.
- The compressor delivers an outlet pressure of $P = 1.05$ MPa because that pressure corresponds to a boiling point of $T = 298.15$ K.

Example 5-3 also introduced the coefficient of performance, a metric for the overall effectiveness of a refrigeration cycle. The coefficient of performance is defined as

$$\text{C.O.P.} = \frac{\dot{Q}_C}{\dot{W}_S}$$

with $\dot{Q}_C$ representing the rate at which heat is removed from the space being refrigerated and $\dot{W}_S$ as the work required to run the process. If you review the examples in Section 5-3, you'll see that $\dot{Q}_C$ and $\dot{W}_S$ are both strongly influenced by the physical properties of the refrigerant—the heat capacity, enthalpy of vaporization, the rate at which vapor pressure increases with increasing temperature, etc. One logical strategy for evaluating and comparing potential refrigerants would be to design refrigeration cycles and compare the resulting coefficients of performance. Doing this would require thermodynamic data—possibly in the form of a figure analogous to Appendix F-1—for each candidate refrigerant. This leads to the question, where did Appendix F-1 come from?

The straightforward answer that likely first occurred to you is "lots of experiments." One property, contained in Figure 8-1, is the vapor pressure at each temperature (or conversely, the boiling temperature at each pressure). We can imagine making a vapor pressure curve by conducting more and more vapor pressure experiments (see the progression from Figure 8-1a to Figure 8-1c) until we are satisfied by the level of accuracy our curve provides. Similarly, we can imagine measuring the other data in Appendix F-1 experimentally. For example, while $\underline{H}$ and $\underline{S}$ cannot be measured directly, we learned in Chapters 2 through 4 ways to relate $\underline{H}$ and $\underline{S}$ to properties like Q and W, which can be measured directly.

While we can imagine conducting all of these experiments, we can also see that generating Appendix F-1 experimentally would be an extremely time-consuming and expensive process. While such comprehensive data may be available for a widely used commodity chemical, the scenario we are considering is a search for new compounds to use as refrigerants; chemicals that probably haven't been studied so extensively, or perhaps haven't even been synthesized in a lab or found in nature yet.

However, all of the information shown in Appendix F-1 and Figure 8-1 can be estimated directly from an equation of state. The fact that we can do this is powerful—instead of comprehensive data describing a chemical, we need only enough data to fit parameters for an equation of state, and then we can produce estimates of that chemical's properties at most any conditions. What's more, using the tools and techniques we will present in Chapters 9 through 13, one can estimate the properties of mixtures of compounds. The Montreal Protocol illustrates one example of why such

FOOD FOR THOUGHT 8-1

Can you draw the structure of a chemical compound that could, in theory, exist, but hasn't actually been discovered or invented yet?

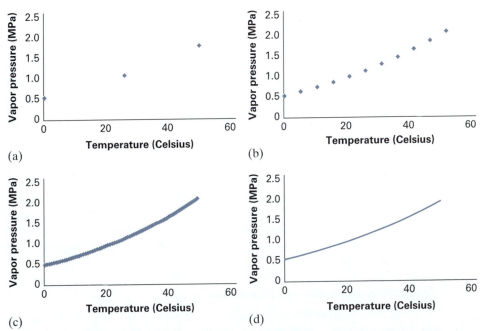

FIGURE 8-1 Vapor pressure vs. temperature for the refrigerant Freon® 22 with (a) data points every 25°C, (b) data points every 5°C, (c) data points every 1°C, and (d) smooth curve fit to the data.

models are useful. Equation of state models allow for a rapid method of evaluating the potential of a compound as a refrigerant. One can even use models to evaluate mixtures of compounds and to identify situations in which a mixture might have more desirable properties than any of the individual compounds that make up the mixture. In-depth experiments need only be completed for the compounds and/or mixtures that appear to be the most promising.

How is the generation of thermodynamic and physical property data from an EOS done? Consider that, in the examples throughout Chapters 1 through 7, properties that have been useful in solving problems have included $\underline{V}$, $\underline{U}$, $\underline{H}$, C_p, and $\underline{S}$. We can find molar volume ($\underline{V}$) at a given temperature and pressure from the EOS directly. Chapters 6 and 7 demonstrated how changes in internal energy, enthalpy, heat capacity, or entropy can be modeled using an EOS, and how a cubic equation of state can be used to model both the liquid and vapor phases. However, at this point, there is one piece missing—we don't yet know how to estimate the vapor pressure that corresponds to each temperature. To recognize the importance of vapor pressure it is instructive to recall the functioning of the vapor-compression refrigeration cycle:

- ■ Heat is transferred from the space being cooled to the refrigerant, causing the refrigerant to boil.

- ■ Heat is expelled to the surroundings, causing the refrigerant to condense.

- ■ A compressor is used to increase the pressure, so that the refrigerant, which boiled at a low temperature, can then condense at a temperature that is higher than that of the surroundings.

Thus, the relationship between vapor pressure and temperature is one of the determining factors in whether a compound is or is not a good refrigerant, and it certainly must be known, or estimated, in order to design a refrigeration cycle. We will see in Section 8.2.2 that the state of aggregation can be predicted through the Gibbs free energy.

Other relevant properties for selecting refrigerants include enthalpy of vaporization and heat capacity, as well as safety considerations like compatibility with materials of construction.

Solid, liquid, and vapor phases all have a molar Gibbs free energy ($\underline{G}^S$, $\underline{G}^L$, and $\underline{G}^V$) that is a function of temperature and pressure. For a specific T and P, the smallest $\underline{G}$ corresponds to the phase that will exist at equilibrium, which we will often call the "most stable phase." The vapor pressure is the pressure at which $\underline{G}^L = \underline{G}^V$ for a specific temperature. At the vapor pressure, liquid and vapor are equally stable and can exist in equilibrium. This chapter presents several methods of estimating vapor pressure, some of which make explicit use of equations of state (see Section 8.3.2).

8.2 Mathematical Models of Phase Equilibrium

This book contains several warnings not to apply the ideal gas law to high-pressure systems (or worse yet, to liquids.) Similarly, specific restrictions exist for each of the vapor pressure equations derived in Section 8.2.

This chapter examines some useful methods of predicting, or modeling, the conditions at which phase transitions occur in pure compounds. The discussion of each emphasizes the assumptions or approximations employed in the derivation. Examples of estimating boiling points, vapor pressures, etc. are shown in Sections 8.2.4 through 8.2.6. First, we lay a theoretical foundation for building mathematical models of phase equilibrium.

8.2.1 Qualitative Discussion of Phase Transitions

FOOD FOR THOUGHT 8-2

Why do some food recipes include alternative cooking instructions for high altitudes?

Some of the people reading this page likely have a glass or bottle of water nearby. We know from everyday experience that H_2O is a liquid at ambient temperature and pressure. Most readers have probably boiled water while cooking. When doing so, even if you never placed a thermometer in the water, you likely feel confident the temperature was quite close to 100°C. But how often do we think about this: why is H_2O a liquid at ambient temperature, and why should boiling occur spontaneously at the temperature we call 100°C, specifically?

At a given T and P, a compound has both higher H and higher S as a liquid than as a solid. Thus, our analysis of the process of liquids boiling into vapors applies analogously to the process of solids melting into liquids.

Chapter 4 introduced the property of entropy and gave us one metric for whether a process occurs spontaneously or not: The entropy of the universe is increased by any spontaneous process. The steam tables tell us that for water at $T = 100°C$, $\hat{S}^L = 1.3072$ kJ/kg·K, and $\hat{S}^V = 7.3541$ kJ/kg·K. This indicates that the change of specific entropy from vaporization is 6.0469 kJ/kg·K or, on a molar basis, 108.9 J/mol·K. While the specific numbers vary for different temperatures and different compounds, the principle is general: $\Delta \underline{S}^{vap}$ is positive, and is usually quite significant. So here's a question: If the entropy of the universe is always increasing, shouldn't *all* liquids spontaneously boil into vapors and never turn back? *Shouldn't all the matter in the universe be gaseous?* You would presumably answer "yes" if you looked at $\Delta \underline{S}^{vap}$ by itself. We need to consider, when a liquid boils into vapor, what *else* changes besides the entropy?

Problem 3-22 explored evaporative cooling, which is employed in cooling towers.

The vapor phase also has a higher molar enthalpy than the liquid phase ($\Delta \underline{H}^{vap} > 0$); energy must be added to water to make it boil. Where does this energy come from? It depends upon the setting. In a cooling tower, this energy comes from the water itself; a small portion of the liquid water evaporates, and the temperature of the water remaining in the liquid phase decreases. In our kitchen, the energy for vaporization is added to a pot of water in the form of heat from the stove. If heat (Q) is being added to the water ($Q_{water} > 0$), then this heat has to be taken from the surroundings ($Q_{surr} < 0$). This effect can be related to entropy; recall that the change in entropy from a process is by definition:

When your body is sweating, it is making use of evaporative cooling.

$$dS = \frac{dQ_{rev}}{T}$$

Thus when $Q < 0$, the change in entropy is negative. In summary when water goes from the liquid to the vapor phase, the entropy of *that water* increases, but the entropy of *the rest of the universe* decreases, because the energy required to vaporize the liquid had to come from someplace. And, of course, in order for the process to be spontaneous, the entropy decrease in the surroundings would have to be less than the entropy increase of the water, in accordance with the second law of thermodynamics.

Thus, we return to the original question: does a liquid spontaneously boil into vapor? The answer to this question lies in a trade-off between the effects of entropy and energy. How do we account for this trade-off quantitatively? The definition of Gibbs free energy incorporates both energy and entropy as

$$G = H - TS = U + PV - TS$$

Section 8.2.2 reveals that the Gibbs free energy can be used as a foundation in building quantitative models of phase equilibrium.

We could have also used internal energy (U) instead of enthalpy (H) to make this point. ΔU^{vap} is also positive, and in Chapter 3, we saw some specific circumstances where $dU = dQ$ and other cases in which $dH = dQ$.

8.2.2 Mathematical Expression of the Equilibrium Criterion

In this section, we begin to build mathematical models for systems in which phase equilibrium occurs. In practice, we will commonly use the property *fugacity* in the solution of specific problems, both for pure compounds in this chapter and for mixtures in subsequent chapters. This section demonstrates how the Gibbs energy is used to lay a foundation for equilibrium models. The fugacity is derived from the Gibbs energy, through definitions introduced in Section 8.3.2.

Consider the piston-cylinder device illustrated in Figure 8-2, containing liquid water and water vapor in equilibrium at $P = 101.325$ kPa and $T = 100°C$. Imagine that the pressure is being maintained constant at $P = 101.325$ kPa, and consider the outcomes we would observe if a small amount of heat was added.

Gibbs free energy was first defined in Section 4.3.3.

- The addition of the heat causes a small amount of liquid to boil. Thus, the **quality** of the VLE mixture would increase.

- Because saturated vapor has a larger molar volume than saturated liquid, the VLE mixture expands slightly. The piston allows the VLE mixture volume to change while maintaining both temperature and pressure as constants.

- The $\underline{S}$, $\underline{U}$, and $\underline{H}$ of the liquid phase is unchanged by the process, as are the $\underline{S}$, $\underline{U}$, and $\underline{H}$ of the vapor phase. These are all intensive properties and are not affected by the change in the amount of material in the respective phases. Said another

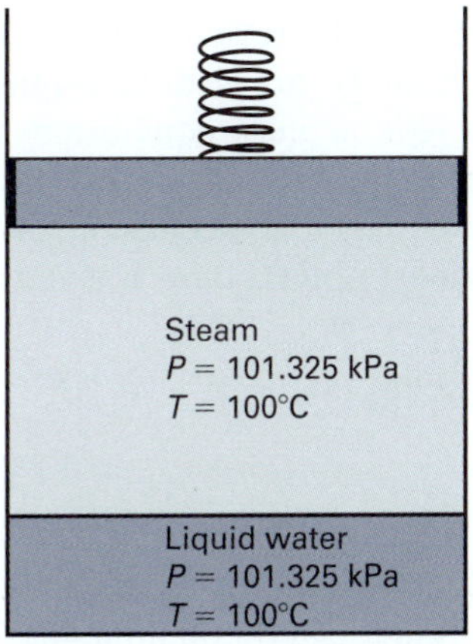

The vapor fraction of a VLE mixture is called the "quality," as introduced in Section 2.2.5.

Steam
$P = 101.325$ kPa
$T = 100°C$

Liquid water
$P = 101.325$ kPa
$T = 100°C$

FIGURE 8-2 Piston-cylinder device holding a steam–water VLE mixture at atmospheric pressure.

The liquid volume in this figure is exaggerated for illustrative purposes. If the figure is drawn to scale, then the VLE mixture is, on a mass basis, practically all liquid.

You have likely used the molar Gibbs free energy of formation ($\Delta \underline{G}_f$) in chemistry courses to calculate equilibrium constants for chemical reactions. In this book, we use $\underline{G}$ to analyze a broader range of equilibrium applications.

way, the Gibbs phase rule illustrates there are two degrees of freedom ($F = 2$) for a pure component in a single phase. If two intensive properties (T and P) both remain constant, no other intensive properties can change.

■ Because the vapor phase is higher in $\underline{S}$, $\underline{U}$, and $\underline{H}$ than the liquid phase, the conversion of liquid to vapor increases the *total* entropy, internal energy, and enthalpy of the system.

The Gibbs free energy is defined as $G = H - TS$. Notice that in the boiling process, the total enthalpy and entropy of the system are both increased. The effect of the boiling process on the Gibbs energy is, at first glance, unclear because of the negative sign; H increases but TS also increases. It is instructive to recall the fundamental property relations for $\underline{U}$, $\underline{H}$, $\underline{A}$, and $\underline{G}$, which were derived in Section 6.2.2:

$$d\underline{U} = T\,d\underline{S} - P\,d\underline{V}$$

$$d\underline{H} = T\,d\underline{S} + \underline{V}\,dP$$

$$d\underline{G} = \underline{V}\,dP - \underline{S}\,dT$$

$$d\underline{A} = -P\,d\underline{V} - \underline{S}\,dT$$

Notice that $\underline{G}$ is unique in that it is a natural function of pressure and temperature. When saturated liquid converts into saturated vapor, $\underline{V}$, $\underline{U}$, $\underline{H}$, and $\underline{S}$ all increase, but pressure and temperature do not change ($dP = 0$, $dT = 0$), and consequently the molar Gibbs free energy of the system as a whole ($G/n = \underline{G}^{\text{sys}}$) does not change ($d\underline{G}^{\text{sys}} = 0$). Thus, we can infer that the *saturated liquid and saturated vapor phases must have identical values of molar Gibbs energy* ($\underline{G}^{\text{L}} = \underline{G}^{\text{V}}$), since if they were different from each other, the molar Gibbs energy of the system overall ($\underline{G}^{\text{sys}}$) would be changed by the boiling of some liquid.

If you have 100% saturated vapor ($q = 1$) or 100% saturated liquid ($q = 0$), only one phase is present; there is no two-phase equilibrium. However, $\underline{G}$ is a state property. This means $\underline{G}^{\text{V}}$ of saturated steam at $T = 100°C$ and $P = 101.325$ kPa is the same—regardless of whether there is also some liquid present or not.

This observation generalizes to phase equilibrium of any kind. Thus, the criterion for phase equilibrium in a pure compound can be expressed as follows.

■ In vapor–liquid equilibrium, $\underline{G}^{\text{L}} = \underline{G}^{\text{V}}$.
■ In liquid–solid equilibrium, $\underline{G}^{\text{L}} = \underline{G}^{\text{S}}$.
■ In vapor–solid equilibrium, $\underline{G}^{\text{S}} = \underline{G}^{\text{V}}$.
■ The triple point of a compound is the unique temperature and pressure at which $\underline{G}^{\text{S}} = \underline{G}^{\text{L}} = \underline{G}^{\text{V}}$.

Up to this point we've considered only the scenario in which two or more phases occur in equilibrium. What if $\underline{G}^{\text{L}}$, $\underline{G}^{\text{S}}$, and $\underline{G}^{\text{V}}$ are all different from each other at a specific temperature and pressure? There is no phase equilibrium at these conditions, but which of the three phases will be present? We can address this question by considering a closed system that is being maintained at a constant, uniform temperature and pressure. The energy balance, in its most general form, is given in Equation 3.34. Thus,

$$\frac{d}{dt}\left\{ M\left(\hat{U} + \frac{v^2}{2} + gh \right) \right\} = \sum_{j=1}^{j=J}\left\{ \dot{m}_{j,\text{in}}\left(\hat{H}_j + \frac{v_j^2}{2} + gh_j \right) \right\}$$

$$- \sum_{k=1}^{k=K}\left\{ \dot{m}_{k,\text{out}}\left(\hat{H}_k + \frac{v_k^2}{2} + gh_k \right) \right\} + \dot{W}_{\text{S}} + \dot{W}_{\text{EC}} + \dot{Q}$$

If our closed system is stationary (kinetic and potential energy constant) and contains no moving parts (no shaft work), this simplifies to

$$\frac{d}{dt}(M\hat{U}) = \dot{W}_{\text{EC}} + \dot{Q} \tag{8.1}$$

And the entropy balance for a closed system is

$$\frac{d(M\hat{S})}{dt} = \sum_{n=1}^{n=N} \frac{\dot{Q}_n}{T_n} + \dot{S}_{gen} \tag{8.2}$$

There is no need for a summation of different Q/T terms, since the system temperature is uniform and constant. We can multiply through by this constant T for

$$\frac{d(M\hat{S}T)}{dt} = \dot{Q} + T\dot{S}_{gen} \tag{8.3}$$

Solving Equation 8.3 for $\dot{Q}$ and substituting into Equation 8.1:

$$\frac{d}{dt}(M\hat{U}) = \dot{W}_{EC} + \frac{d(M\hat{S}T)}{dt} - T\dot{S}_{gen} \tag{8.4}$$

Recall from Chapter 1 that expansion/contraction work is given as $dW_{EC} = -P\,dV$. Since $\dot{W}_{EC}$ represents the *rate* at which work is added, this can be expressed as

$$\frac{d}{dt}(M\hat{U}) = -P\frac{dV}{dt} + \frac{d(M\hat{S}T)}{dt} - T\dot{S}_{gen} \tag{8.5}$$

Since pressure is a constant in this system, we can move P inside the differential. In addition, we can apply the definition of specific properties, $M\hat{U} = U$ and $M\hat{S} = S$ for

$$\frac{d(U)}{dt} = -\frac{d(PV)}{dt} + \frac{d(ST)}{dt} - T\dot{S}_{gen} \tag{8.6}$$

By definition, $G = U + PV - TS$. Consequently, the three d/dt terms in Equation 8.6 can be combined as

$$\frac{d(U + PV - ST)}{dt} = -T\dot{S}_{gen}$$

$$\frac{dG}{dt} = -T\dot{S}_{gen} \tag{8.7}$$

> Neither T nor $\dot{S}_{gen}$ can be negative (when T is expressed on an absolute scale), so $-T\dot{S}_{gen} \le 0$.

dG/dt represents the rate at which the Gibbs free energy of the system changes. It was established in Chapter 4 that any spontaneous process increases the entropy of the universe ($\dot{S}_{gen} > 0$). Equation 8.7 thus reveals that any process occurring in a closed system at fixed T, P will decrease the Gibbs energy (G) of the system. Equilibrium will occur when G reaches a minimum, at which time dG/dt will equal zero. Thus, if $\underline{G}^L < \underline{G}^V$ at a given temperature and pressure, then according to Equation 8.7, vapor will spontaneously condense into liquid. However, if $\underline{G}^L = \underline{G}^V$, there is no driving force for liquid to vaporize or for vapor to condense; the two can exist in equilibrium.

A practical lesson from Equation 8.7 is this: If we use a model to calculate values of $\underline{G}^L$, $\underline{G}^S$, and $\underline{G}^V$ for a substance at a particular temperature and pressure, and they are all different, we know that **the most stable phase, at equilibrium, is the one that corresponds to the minimum value of $\underline{G}$.**

> In this chapter, we focus on pure compounds, but Equations 8.1 through 8.7 don't assume the system is a pure compound, so the derivation is also valid for mixtures. In Chapter 14, we apply the principle "G is minimized at equilibrium" to chemical reactions.

8.2.3 Chemical Potential

Here we define a new property that is useful in analyzing equilibrium. We have seen that Gibbs energy is a fundamental property for modeling equilibrium, so we need to be able to quantify Gibbs energy. The **chemical potential** (μ_i) of a compound i is defined as the partial derivative of Gibbs free energy with respect to the number of moles of compound i:

$$\mu_i = \left(\frac{\partial G}{\partial n_i}\right)_{T,P,n_{j\ne i}} \tag{8.8}$$

For a pure compound
$$G = n\underline{G}$$

Thus, chemical potential indicates the change in Gibbs energy of a system when a small amount of compound i is added, if everything else (temperature, pressure, and the amount of all compounds other than i) is held constant. For a pure compound, this is simply the molar Gibbs energy, as

$$\mu_i = \left(\frac{\partial G}{\partial n_i}\right)_{T,P,n_{j\neq i}} = \left(\frac{\partial n_i \underline{G}}{\partial n_i}\right)_{T,P,n_{j\neq i}} \tag{8.9}$$

Applying the chain rule gives

$$\mu_i = \underline{G}\left(\frac{\partial n_i}{\partial n_i}\right)_{T,P,n_{j\neq i}} + n_i\left(\frac{\partial \underline{G}}{\partial n_i}\right)_{T,P,n_{j\neq i}}$$

The requirement that number of moles of all other compounds ($n_{j\neq i}$) be held constant is part of the definition of chemical potential, but this requirement is meaningless in the case of a pure compound—there is no compound present other than i.

The derivative of any variable with respect to itself is simply one, so the first term on the right-hand side simplifies to $\underline{G}$. But $\underline{G}$ for a pure compound is an intensive property; its value is independent of the number of moles, so $(\partial \underline{G}/\partial n_i)$ is 0, and

$$\mu_i = \underline{G} \tag{8.10}$$

Thus, when modeling phase equilibrium for a *pure* compound, it matters little whether we express the equilibrium criterion in terms of molar Gibbs free energy (e.g., $\underline{G}^L = \underline{G}^V$) or chemical potential (e.g., $\mu_i^L = \mu_i^V$); they are identical. Consequently chemical potential is not mentioned again in this chapter. However, we will see in Chapter 10 that for mixtures, chemical potential is a function of composition as well as temperature and pressure. We will see that for mixtures, μ_i is essential for modeling phase equilibrium and is not necessarily equal to $\underline{G}_i$.

8.2.4 The Clapeyron Equation

Section 8.2.2 demonstrated that when a pure compound exists in two phases at equilibrium, the molar Gibbs energies of each of the phases are identical. Example 8-1 applies this fact to the estimation of unknown vapor pressures.

EXAMPLE 8-1	**ESTIMATING THE VAPOR PRESSURE OF WATER**

For liquid water at $T = 50°C$, the vapor pressure is $P^{sat} = 12.4$ kPa, and $\Delta \underline{H}^{vap} = 42.91$ kJ/mol.

A. Using only the given information, estimate the vapor pressures of water at $T = 55°C$, $T = 60°C$, $T = 75°C$, $T = 100°C$, and $T = 150°C$.

B. Compare the estimates obtained in part A to the data in the steam tables.

No engineer working on a real problem would estimate the vapor pressure of water using the method shown in this example—using the data in the steam tables is better. But this example gives us an idea of how accurately we can estimate vapor pressures when comprehensive data isn't available.

SOLUTION A:

Step 1 *Apply the equilibrium criterion*

Figure 8-3 provides a graphical interpretation of what this problem is asking. We know that, for every boiling temperature, there is a unique vapor pressure, and that vapor pressure increases with increasing temperature. However, Figure 8-3 is nothing more than a rough sketch—we only know one point on the graph and are trying to estimate the locations of other points from the given information.

We know that vapor–liquid equilibrium occurs for water at $T = 50°C$ and $P = 12.4$ kPa. Thus at this temperature and pressure, $\underline{G}^L = \underline{G}^V$. At $T = 55°C$ and an unknown higher pressure, liquid water and water vapor are again in vapor–liquid equilibrium (VLE). $\underline{G}^L$ and $\underline{G}^V$ at this new temperature and pressure will be different than they were at $T = 50°C$ and $P = 12.4$ kPa, but must again be the same as each other. The same is true for any point on the vapor pressure curve: $\underline{G}^L = \underline{G}^V$.

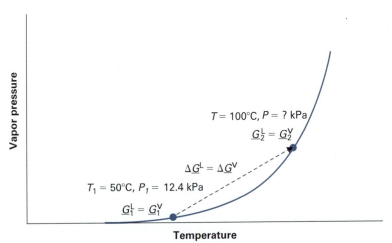

FIGURE 8-3 Sketch of vapor pressure vs. temperature.

Thus, as we move from one point to another along the vapor pressure curve, the change in $\underline{G}^L$ and the change in $\underline{G}^V$ must be equal to each other:

$$d\underline{G}^L = d\underline{G}^V \tag{8.11}$$

Our goal is to determine a relationship between vapor pressure and temperature, so we must next relate $\underline{G}$ to these.

Step 2 *Relate Gibbs energy to pressure and temperature*

We can use the fundamental property relationship to relate $d\underline{G}$ to temperature and pressure:

$$d\underline{G}^L = d\underline{G}^V$$

$$\underline{V}^L dP^{\text{sat}} - \underline{S}^L dT = \underline{V}^V dP^{\text{sat}} - \underline{S}^V dT \tag{8.12}$$

which can be rearranged to

$$(\underline{V}^V - \underline{V}^L) dP^{\text{sat}} = (\underline{S}^V - \underline{S}^L) dT$$

$$\frac{dP^{\text{sat}}}{dT} = \frac{(\underline{S}^V - \underline{S}^L)}{(\underline{V}^V - \underline{V}^L)} \tag{8.13}$$

Equation 8.13 looks useful—we are trying to locate several points on the vapor pressure curve (P^{sat} versus T), and Equation 8.13 gives us the slope of that curve (dP^{sat}/dT). However, $\Delta \underline{S}^{\text{vap}}$, which appears in the equation, is unknown, while $\Delta \underline{H}^{\text{vap}}$ is given. We can relate the two to each other by applying again the facts that on the vapor pressure curve $\underline{G}^L = \underline{G}^V$ and the definition of $\underline{G}$ is $\underline{G} = \underline{H} - T\underline{S}$:

$$\underline{G}^L = \underline{H}^L - T\underline{S}^L = \underline{H}^V - T\underline{S}^V = \underline{G}^V \tag{8.14}$$

Rearranging Equation 8.14 to parallel the structure of Equation 8.13 gives

$$T(\underline{S}^V - \underline{S}^L) = \underline{H}^V - \underline{H}^L \tag{8.15}$$

Equation 8.15 allows us to substitute molar enthalpy for molar entropy in Equation 8.13:

$$\frac{dP^{\text{sat}}}{dT} = \frac{(\underline{H}^V - \underline{H}^L)}{T(\underline{V}^V - \underline{V}^L)} = \frac{\Delta \underline{H}^{\text{vap}}}{T(\underline{V}^V - \underline{V}^L)} \tag{8.16}$$

The only assumption in the derivation of Equation 8.16 is that the two phases are in equilibrium, so this equation describes not only vapor–liquid equilibrium, but also solid–liquid and solid–vapor; we need only change the superscripts. This is discussed further after the example.

Table 7-1 showed how saturated vapor volumes are orders of magnitude larger than saturated liquid volumes at low temperatures, but the values of these volumes converge to each other near the critical point.

The values of V^L and V^V in Equation 8.17 are not part of the given information, but here they are only used to make a point. Neither of these numbers is actually used directly in solving the problem.

Step 3 *Make simplifying assumptions*

Equation 8.16 is called the Clapeyron equation, and is discussed further after the example. Here, our goal is to determine P^{sat} at various values of T. Equation 8.16 is a first-order ordinary differential equation (ODE) that relates P^{sat} to T. Conceptually, we can treat T as the independent variable, P^{sat} as the dependent variable, and the known data point ($P^{sat} = 12.4$ kPa when $T = 50°C$) as the initial value, and solve the ODE to get an equation for P^{sat} as a function of T, or $P^{sat} = f(T)$. However, if we do try to integrate Equation 8.16 using only the given information, we encounter two barriers:

1. $\Delta \underline{H}^{vap}$ is only known at one particular point on the vapor pressure curve.

2. The liquid and vapor molar volumes aren't known at all.

We address the first by assuming $\Delta \underline{H}^{vap}$ is a constant. A simple perusal of the steam tables shows that in reality $\Delta \underline{H}^{vap}$ gets smaller as boiling temperature and vapor pressure increase, and approaches zero at the critical point—the point at which vapor and liquid become indistinguishable. However, in order to solve our differential Equation 8.16, we need to express $\Delta \underline{H}^{vap}$ as a mathematical function of T or P, so that we can separate the variables and integrate. Since the given information provides no way to quantify a relationship among $\Delta \underline{H}^{vap}$, T, and P, we will assume $\Delta \underline{H}^{vap}$ is constant with respect to T and P. At the end of this example, we will evaluate this assumption.

Regarding the molar volumes of liquid and vapor, we know that liquid molar volumes are small compared to vapor molar volumes—except near the critical point—and we are not near the critical point here. For saturated H_2O at $T = 50°C$, for example:

$$\underline{V}^V - \underline{V}^{dm^3} = 216.66 \frac{dm^3}{mol} - 0.0182 \frac{dm^3}{mol} = 216.64 \frac{dm^3}{mol} \tag{8.17}$$

Thus, the quantity $\underline{V}^V - \underline{V}^L$ can be considered identical to $\underline{V}^V$, unless one requires five digits of accuracy. We will move forward with the approximation that $\underline{V}^L$ is insignificant compared to $\underline{V}^V$. Therefore,

$$\frac{dP^{sat}}{dT} = \frac{\Delta \underline{H}^{vap}}{T(\underline{V}^V - \underline{V}^L)} \approx \frac{\Delta \underline{H}^{vap}}{T\underline{V}^V} \tag{8.18}$$

In order to solve Equation 8.18, we must relate $\underline{V}$ to T and P, so that we can separate the variables and integrate with respect to dT and dP^{sat}. Our first instinct always should be to apply an equation of state when we need to relate $\underline{V}$, T, and P to each other. Here the known vapor pressure at $T = 50°C$ is $P = 12.4$ kPa, which is low enough to assume ideal gas behavior:

$$\frac{dP^{sat}}{dT} = \frac{\Delta \underline{H}^{vap}}{T\left(\dfrac{RT}{P^{sat}}\right)} \tag{8.19}$$

Separating the variables gives

$$\frac{dP^{sat}}{P^{sat}} = \frac{\Delta \underline{H}^{vap}}{RT^2} dT \tag{8.20}$$

Step 4 *Integrate*

We can solve our differential equation directly through integration of Equation 8.20 from the known data point ($T_1 = 50°C$, $P_1 = 12.4$ kPa) to a different, unknown point on the vapor pressure curve (T_2, P_2). Recognizing that R is a constant and treating $\Delta \underline{H}^{vap}$ as a constant gives

$$\int_{P_1}^{P_2} \frac{dP^{sat}}{P^{sat}} = \int_{T_1}^{T_2} \frac{\Delta \underline{H}^{vap}}{RT^2} dT$$

$$\ln\left(\frac{P_2^{sat}}{P_1^{sat}}\right) = \frac{-\Delta \underline{H}^{vap}}{R}\left(\frac{1}{T_2} - \frac{1}{T_1}\right) \qquad (8.21)$$

Equation 8.21 is called the Clausius-Clapeyron equation, named after the physicists Rudolf Clausius and Benoit Pierre Emile Clapeyron (Olson, 1998).

Step 5 *Plug in known values*

If $T_2 = 55°C$, the only unknown in Equation 8.21 is P_2^{sat}, which is exactly what we are trying to find. Thus,

$$\ln\left(\frac{P_2^{sat}}{12.4\ kPa}\right) = \left(\frac{-42,910\ \dfrac{J}{mol}}{8.314\ \dfrac{J}{mol\cdot K}}\right)\left[\frac{1}{(55+273.15\ K)} - \frac{1}{(50+273.15\ K)}\right] \qquad (8.22)$$

$$P_2^{sat} = \textbf{15.8 kPa}$$

Vapor pressures at the other unknown temperatures are computed analogously.

SOLUTION B:

Compare estimates obtained to steam table data

The approach taken in part A models the vapor pressure curve as a linear relationship between $\ln(P^{sat})$ and $(1/T)$ with the assumption that $\Delta\underline{H}^{vap}$ is constant (as seen in Equation 8.21). Table 8-1 illustrates that when the assumption of constant $\Delta\underline{H}^{vap}$ is applied over small temperature intervals, the accuracy of the estimate is excellent. When the method is applied over a 50°C temperature range (50 to 100°C), the error is $\approx 4\%$. As the actual vapor pressure P^{sat} climbs, the approximation that the vapor volume $(\underline{V}^V)$ follows the ideal gas law also becomes suspect; at $T = 150°C$ the error is more than 10%.

There are *no* approximations in the derivation of the Clapeyron equation. It can be applied to any phase equilibrium.

TABLE 8-1 Comparison of estimated vapor pressure of water to data from steam tables.

Temperature (°C)	Estimated P^{sat} (kPa)	P^{sat} from Steam Table (kPa)
55	15.8	15.8
60	20.0	19.9
75	39.0	38.6
100	105.4	101.3
150	540.3	476.2

Example 8-1 introduces two equations that are applicable to a variety of phase equilibrium problems.

The **Clapeyron equation** can be used to model the relationship between saturation pressure and saturation temperature for vapor–liquid equilibrium:

$$\frac{dP^{sat}}{dT} = \frac{\Delta\underline{H}^{vap}}{T(\underline{V}^V - \underline{V}^L)} \qquad (8.23)$$

where P^{sat} represents the vapor pressure.

While Example 8-1 and Equation 8.23 refer to vapor–liquid equilibrium specifically, the logic of the derivation is equally applicable to solid–vapor equilibrium:

$$\frac{dP^{\text{sat}}}{dT} = \frac{\Delta \underline{H}^{\text{sub}}}{T(\underline{V}^{\text{V}} - \underline{V}^{\text{S}})} \tag{8.24}$$

FOOD FOR THOUGHT 8-3

Would you expect the Clausius-Clapeyron equation to be valid for solid–vapor equilibrium? For solid–liquid equilibrium?

or to solid–liquid equilibrium:

$$\frac{dP^{\text{sat}}}{dT} = \frac{\Delta \underline{H}^{\text{fus}}}{T(\underline{V}^{\text{L}} - \underline{V}^{\text{S}})} \tag{8.25}$$

The Clapeyron equation is powerful in that it allows us to model the entire equilibrium P vs. T curve from a single data point. However, in order to solve the equation, we must relate the $\Delta \underline{H}$ and $\Delta \underline{V}$ of the phase change to pressure and temperature. This likely involves making approximations of some kind.

Example 8-1 illustrates one commonly used special case, the **Clausius-Clapeyron equation**:

$$\ln\left(\frac{P_2^{\text{sat}}}{P_1^{\text{sat}}}\right) = \frac{-\Delta \underline{H}^{\text{vap}}}{R}\left(\frac{1}{T_2} - \frac{1}{T_1}\right) \tag{8.26}$$

When using the Clausius-Clapeyron equation, we must recall the assumptions made in its derivation:
1. $\Delta \underline{H}^{\text{vap}}$ is constant with respect to temperature and pressure.
2. $\underline{V}^{\text{L}}$ is negligible in comparison to $\underline{V}^{\text{V}}$.
3. $\underline{V}^{\text{V}}$ is well modeled by the ideal gas law.

The second and third assumptions are most reasonable at low pressures. The first assumption is reasonable if T_1 and T_2 are close to each other and neither is near the critical point. In Example 8-1, the Clausius-Clapeyron equation produced excellent estimates of P^{sat} over a temperature interval of 25°C.

8.2.5 The Shortcut Equation

The Clausius-Clapeyron equation is very useful, but only applicable for modeling phase equilibriums that occur at low pressure. The next example examines vapor pressure vs. temperature at high pressures, again using water as an example.

EXAMPLE 8-2	**ESTIMATING THE HIGH-TEMPERATURE VAPOR PRESSURE OF WATER**

At $T = 179.88$°C, the vapor pressure of water is $P = 1$ MPa. At these conditions, $\Delta \hat{H}^{\text{vap}} = 2014.59$ kJ/kg, $\hat{V}^{\text{V}} = 0.1944$ m³/kg, and $\hat{V}^{\text{L}} = 0.001127$ m³/kg.

A. Using only this information, estimate the vapor pressure of water at 200°C, 250°C, and 300°C.

B. Compare the results obtained in part A to the values available in the steam tables.

The Clausius-Clapeyron equation assumed $\Delta \hat{H}^{\text{vap}}$ is constant, which is a poor assumption over a large temperature range.

SOLUTION A:

Step 1 *Compare physical situation to assumptions behind Clausius-Clapeyron equation*
This is similar to Example 8-1 in that we are asked to estimate several unknown vapor pressures from a single known data point, but here we need a different approach. In

the Clausius-Clapeyron equation, the vapor phase is modeled using the ideal gas law, which is valid at low pressures. Here, our known data point is a vapor pressure of 1 MPa, and the unknown vapor pressures will be even higher, since they occur at higher temperatures. Thus, the Clausius-Clapeyron equation cannot be expected to be valid at these pressures; we need a different modeling approach that does not use the ideal gas law.

However, the Clapeyron Equation (8.23) derived in Example 8-1 does not depend upon any assumptions at all, and so can be used as a starting point:

$$\frac{dP^{\text{sat}}}{dT} = \frac{\Delta \underline{H}^{\text{vap}}}{T(\underline{V}^{\text{V}} - \underline{V}^{\text{L}})}$$

> One approach to high-pressure problems we could imagine is using a different equation of state, rather than the ideal gas law, to model $\hat{V}^{\text{V}}$. Section 8.3 will present strategies for predicting P^{sat} vs. T from an equation of state.

Step 2 *Relate molar volume to temperature and pressure*

Our goal is to integrate Equation 8.23 into an expression relating P^{sat} to T, so that we can solve for unknown values of P^{sat}. The other quantities in Equation 8.23—$\Delta \underline{H}^{\text{vap}}$, $\underline{V}^{\text{V}}$, and $\underline{V}^{\text{L}}$—are known at one point, but there is not enough information to propose a mathematical relationship between these quantities and T and/or P, and it is not realistic to assume they are constant over this large temperature range. We can relate $\underline{V}^{\text{V}}$ and $\underline{V}^{\text{L}}$ to pressure and temperature through the compressibility factor ($Z = P\underline{V}/RT$):

> The given information is actually $\Delta \hat{H}^{\text{vap}}, \hat{V}^{\text{V}}$, and $\hat{V}^{\text{L}}$ not $\Delta \underline{H}^{\text{vap}}, \underline{V}^{\text{V}}$, and $\underline{V}^{\text{L}}$, but only the molecular weight (known) is needed to make conversions.

$$\frac{dP^{\text{sat}}}{dT} = \frac{\Delta \underline{H}^{\text{vap}}}{T\left(\dfrac{Z^{\text{V}}RT}{P^{\text{sat}}} - \dfrac{Z^{\text{L}}RT}{P^{\text{sat}}}\right)} \tag{8.27}$$

$$\frac{dP^{\text{sat}}}{dT} = \frac{P^{\text{sat}}\Delta \underline{H}^{\text{vap}}}{RT^2(\Delta Z^{\text{vap}})} \tag{8.28}$$

But without an equation of state, how can we progress further?

Step 3 *Make a simplifying assumption*

Consider what happens with VLE as pressure increases. Close to the critical point, $\Delta \underline{H}^{\text{vap}}$ gets very small. Similarly, as temperature and pressure approach the critical point, the compressibility factors (Z) for the saturated liquid and saturated vapor get closer together, as they both converge on Z_{c}. Thus, while $\Delta \underline{H}^{\text{vap}}$ and ΔZ^{vap} individually are not constant, it is logical to speculate that the ratio remains fairly constant, since both quantities will approach 0 at the same point.

$$\frac{\Delta \underline{H}^{\text{vap}}}{\Delta Z^{\text{vap}}} \approx \text{constant as } P \text{ approaches } P_{\text{C}}$$

As we did in Example 8-1, we will evaluate the simplifying assumption at the end of the example.

Step 4 *Integrate the Clapeyron equation*

Applying the simplifying assumption, the Clapeyron equation can be solved:

> R is a constant, and $\Delta \underline{H}^{\text{vap}}/\Delta Z^{\text{vap}}$ is being assumed constant, so these can be pulled outside the integral.

$$\frac{dP^{\text{sat}}}{P^{\text{sat}}} = \frac{\Delta \underline{H}^{\text{vap}}}{R\Delta Z^{\text{vap}}}\frac{dT}{T^2} \tag{8.29}$$

$$\int_{P_1}^{P_2} \frac{dP^{\text{sat}}}{P^{\text{sat}}} = \frac{\Delta \underline{H}^{\text{vap}}}{R\Delta Z^{\text{vap}}}\int_{T_1}^{T_2} \frac{dT}{T^2}$$

$$\ln\left(\frac{P_2^{\text{sat}}}{P_1^{\text{sat}}}\right) = \frac{-\Delta \underline{H}^{\text{vap}}}{R\Delta Z^{\text{vap}}}\left(\frac{1}{T_2} - \frac{1}{T_1}\right) \tag{8.30}$$

Step 5 *Calculate ΔZ^{vap}*
Applying the definition of Z gives

$$Z^V = \frac{PV}{RT} = \frac{(10^6 \, \text{Pa})\left(0.1944 \, \dfrac{\text{m}^3}{\text{kg}}\right)\left(18.015 \, \dfrac{\text{g}}{\text{mol}}\right)\left(\dfrac{1 \, \text{kg}}{1000 \, \text{g}}\right)}{\left(8.314 \, \dfrac{\text{Pa} \cdot \text{m}^3}{\text{mol} \cdot \text{K}}\right)[(179.88 + 273.15) \, \text{K}]\left(\dfrac{1 \, \text{m}}{100 \, \text{cm}}\right)^3} = 0.9298 \quad (8.31)$$

$$Z^L = \frac{PV}{RT} = \frac{(10^6 \, \text{Pa})\left(0.001127 \, \dfrac{\text{m}^3}{\text{kg}}\right)\left(18.015 \, \dfrac{\text{g}}{\text{mol}}\right)\left(\dfrac{1 \, \text{kg}}{1000 \, \text{g}}\right)}{\left(8.314 \, \dfrac{\text{Pa} \cdot \text{m}^3}{\text{mol} \cdot \text{K}}\right)[(179.88 + 273.15) \, \text{K}]\left(\dfrac{1 \, \text{m}}{100 \, \text{cm}}\right)^3} = 0.00539 \quad (8.32)$$

$$\Delta Z^{vap} = 0.9298 - 0.00539 = 0.9244 \quad (8.33)$$

Step 6 *Apply known values to Equation 8.30*
We use the known data point at T_1, P_1 to find the vapor pressure at $T_2 = 200°C$:

$$\ln\left(\frac{P_2^{sat}}{1 \, \text{MPa}}\right) \quad (8.34)$$

$$= \frac{-\left(2014.59 \, \dfrac{\text{kJ}}{\text{kg}}\right)\left(\dfrac{18.015 \, \text{g}}{\text{mol}}\right)\left(\dfrac{1 \, \text{kg}}{1000 \, \text{g}}\right)}{\left(8.314 \, \dfrac{\text{J}}{\text{mol} \cdot \text{K}}\right)(0.9244)\left(\dfrac{1 \, \text{kJ}}{1000 \, \text{J}}\right)}\left[\frac{1}{(200 + 273.15) \, \text{K}} - \frac{1}{(179.88 + 273.15) \, \text{K}}\right]$$

$$\boxed{P_2^{sat} = 1.558 \, \text{MPa}}$$

Vapor pressures for other temperatures are computed in an analogous manner and the results are summarized in Table 8-2.

TABLE 8-2 Comparison to data of vapor pressures for water as predicted using Equation 8.30.

Temperature (°C)	Calculated P^{sat} (MPa)	P^{sat} from Steam Tables (MPa)
200	1.558	1.555
250	4.044	3.976
300	8.887	8.588
350	17.21	16.53

SOLUTION B: *Compare model results to data from steam tables.*
Table 8-2 reveals that the predictions become less accurate as the temperature difference $T_2 - T_1$ becomes larger, but even at $T_2 = 350°C$, the error is only $\approx 4\%$. This is a good result considering we were basing our calculation on a single data point and extrapolating over a large temperature interval of $\approx 160°C$. By contrast, using the Clausius-Clapeyron equation and the same known data point yields an estimate of $P^{sat} = 13.88$ MPa at $T_2 = 350°C$, which is an error of 16%.

FOOD FOR THOUGHT 8-4

Based on these values of Z, how well would you expect the Clausius-Clapeyron equation to work if we used it for this example?

Example 8-2 illustrates a strategy for modeling vapor–liquid equilibrium at high-pressure conditions. The assumption that $\Delta \underline{H}^{\text{vap}}/\Delta \underline{Z}^{\text{vap}}$ is a constant proved to work well for H_2O over the range of temperatures and pressures examined in the example. This assumption allowed us to derive Equation 8.30, as

$$\ln \left(\frac{P_2^{\text{sat}}}{P_1^{\text{sat}}}\right) = \frac{-\Delta \underline{H}^{\text{vap}}}{R \Delta \underline{Z}^{\text{vap}}}\left(\frac{1}{T_2} - \frac{1}{T_1}\right)$$

In Example 8-2, solution of Equation 8.30 for an unknown vapor pressure (P_2^{sat}) required a known vapor pressure (P_1^{sat}). It was noted in Section 7.2 that the critical point and acentric factor have been tabulated for many compounds. We can make use of these by introducing them into Equation 8.30. Start by specifying "state 1" (P_1 and T_1) as equal to the critical point:

$$\ln \left(\frac{P_2^{\text{sat}}}{P_c}\right) = \frac{-\Delta \underline{H}^{\text{vap}}}{R \Delta \underline{Z}^{\text{vap}}}\left(\frac{1}{T_2} - \frac{1}{T_c}\right) \tag{8.35}$$

And express state 2 again in terms of reduced properties. By definition, $P = P_r P_c$ and $T = T_r T_c$.

$$\ln \left(\frac{P_r^{\text{sat}} P_c}{P_c}\right) = \frac{-\Delta \underline{H}^{\text{vap}}}{R \Delta \underline{Z}^{\text{vap}}}\left(\frac{1}{T_r T_c} - \frac{1}{T_c}\right)$$

$$\ln \left(P_r^{\text{sat}}\right) = \frac{-\Delta \underline{H}^{\text{vap}}}{\left(R T_c \Delta \underline{Z}^{\text{vap}}\right)}\left(\frac{1}{T_r} - 1\right) \tag{8.36}$$

Or converting from natural logarithms into common (base 10) logarithms:

$$\frac{\log_{10}(P_r^{\text{sat}})}{\log_{10} e} = \frac{-\Delta \underline{H}^{\text{vap}}}{\left(R T_c \Delta \underline{Z}^{\text{vap}}\right)}\left(\frac{1}{T_r} - 1\right)$$

$$\log_{10}(P_r^{\text{sat}}) = \frac{-(\log_{10} e) \Delta \underline{H}^{\text{vap}}}{R T_c \Delta \underline{Z}^{\text{vap}}}\left(\frac{1}{T_r} - 1\right) \tag{8.37}$$

The acentric factor is, in effect, a known value of P_r^{sat}, since it is defined from the vapor pressure at a reduced temperature $T_r = 0.7$ as

$$\omega = -1 - \left. \log_{10}(P_r^{\text{sat}})\right|_{T_r = 0.7}$$

Introducing the acentric factor into Equation 8.37:

$$-(\omega + 1) = \frac{-(\log_{10} e) \Delta \underline{H}^{\text{vap}}}{\left(R T_c \Delta \underline{Z}^{\text{vap}}\right)}\left(\frac{1}{0.7} - 1\right) \tag{8.38}$$

which can be rearranged as

$$\frac{(1 + \omega)}{\left(\frac{1}{0.7} - 1\right)} = \frac{(\log_{10} e) \Delta \underline{H}^{\text{vap}}}{\left(R T_c \Delta \underline{Z}^{\text{vap}}\right)} \tag{8.39}$$

The premise of this modeling approach is that the dimensionless quantity on the right-hand side is approximately constant, as $\Delta \underline{H}^{\text{vap}}$ and $\Delta \underline{Z}^{\text{vap}}$ both approach 0 at the critical point. In effect, Equation 8.39 uses the acentric factor to tell us this constant is approximately equal to:

$$\frac{(\log_{10} e) \Delta \underline{H}^{\text{vap}}}{R T_c \Delta \underline{Z}^{\text{vap}}} \approx \frac{7}{3}(1 + \omega) \tag{8.40}$$

The critical point can be used in Equation 8.30 because it does represent a vapor pressure; it is the point at the end of the vapor pressure curve.

The subscripts "1" and "2" are no longer necessary in Equation 8.36. "State 1" specifies as the critical point and "state 2" represents the vapor pressure at any other temperature.

This step uses the logarithmic property:

$$\log_B A = \frac{\log_x A}{\log_x B}$$

The acentric factor was introduced in Section 7.2.5.

T_c is different for every compound, but can be treated as a constant when we are modeling one specific compound, so it is considered a constant in Equations 8.38 and 8.39.

Introducing this approximation into Equation 8.37 produces what Elliot and Lira (Lira, 1999) term the **shortcut vapor pressure equation**.

$$\log_{10} P_r^{sat} = \frac{7}{3}(\omega + 1)\left(1 - \frac{1}{T_r}\right) \tag{8.41}$$

The acentric factor is based upon the reduced vapor pressure (P^{sat}/P_c) at a reduced temperature of 0.7. The fact that the vapor pressure of water happens to be a round number (1.0 MPa) at $T_r = 0.7$ was convenient in Example 8-2 but has no special significance.

The known data point in Example 8-2 ($P^{sat} = 1.0$ MPa at $T = 179.88°C$) was actually chosen because for water, the temperature of $T = 179.88°C = 453.03$ K corresponds to $T_r = 0.7$ ($T_r = T/T_c = 453.03$ K/647.1 K $= 0.700$). Thus the acentric factor for water is:

$$\omega = -1 - \log_{10}\left(\frac{1.0\,\text{MPa}}{22.06\,\text{MPa}}\right)\Bigg|_{T_r=0.7} = 0.344 \tag{8.42}$$

Using this value of the acentric factor in Equation 8.41 produces the same estimates of water vapor pressure that are shown in Table 8-2.

The shortcut vapor pressure equation is valid at high pressures, and is therefore a useful complement to the Clausius-Clapeyron equation, which is valid at ideal gas conditions. When using the shortcut equation, always keep in mind that it is based upon the assumption that $\Delta \underline{H}^{vap}/\Delta \underline{Z}^{vap}$ is a constant, which becomes more valid as one gets closer to the critical point. Elliott and Lira (Lira, 1999) recommend that the *shortcut equation be applied in the range $0.5 < T_r < 1.0$.*

The complementary roles of the Clausius-Clapeyron equation and the shortcut equation are demonstrated for water in Figure 8-4. Here, the vapor pressure data for water from the steam tables are compared to predictions made using (1) the Clausius-Clapeyron equation with the vapor pressure and $\Delta \underline{H}^{vap}$ at $T = 20°C$ as the basis for predicting all other vapor pressures and (2) the shortcut equation, using the acentric factor value $\omega = 0.344$. Figure 8-4a shows excellent agreement between the Clausius-Clapeyron equation and the real data for temperatures up to $T = 50°C$, Figure 8-4b reveals that the predictions are identical at $T \sim 95°C$, and the shortcut equation prediction is better at temperatures above this. Figure 8-4c demonstrates the excellent agreement between the shortcut equation and the real data at elevated temperatures.

8.2.6 The Antoine Equation

FOOD FOR THOUGHT 8-5

Throughout the book, we have frequently reminded the reader that T must be expressed on an absolute scale in thermodynamic equations. Now suddenly we say that in the Antoine equation, temperature must be expressed in degrees Celsius. Why the apparent contradiction?

The two previous sections introduce two equations that can be used to estimate vapor pressure for a pure compound. A third is the Antoine equation, which should be familiar to many readers:

$$\log_{10}(P^{sat}) = A - \frac{B}{T + C} \tag{8.43}$$

This equation has been shown to be useful for fitting vapor pressure vs. temperature data for a wide variety of compounds. Values for the constants A, B, and C have been tabulated and are given for a number of common compounds in Appendix E. Note that the constants are listed with a recommended temperature range over which they can be applied.

While the form of the Antoine equation is somewhat similar to that of Equation 8.26 or Equation 8.41, the equation does not have a rigorous theoretical basis comparable to that of the Clapeyron equation. The Antoine equation constants are best

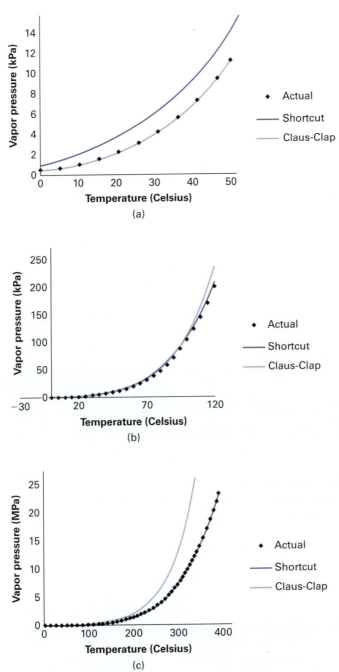

FIGURE 8-4 Vapor pressure of water vs. temperature:
Comparisons of the predictions from Clausius-Clapeyron equation
and the shortcut equation.

regarded as empirical. Because they are derived from real data, they will produce excellent estimates of vapor pressure within the recommended temperature range. However, one should not extrapolate beyond the temperature range over which the constants have been validated, as there is no theoretical reason to assume vapor pressure will follow Equation 8.43 at all temperatures.

The data in Appendix E relate temperature expressed in degrees Celsius to pressure in mm Hg, but be careful when using Antoine constants obtained from other sources—the units used for pressure and temperature may differ. Also, when using published Antoine constants, take care to check whether the logarithm term is expressed in base 10 (as in Equation 8.43) or base e (ln). Either can be used as a basis for the equation, but different values of constant A will be the result.

| EXAMPLE 8-3 | **CALCULATING VAPOR PRESSURE WITH THE ANTOINE EQUATION** |

You are designing a process in which toluene is used as a solvent. As part of your safety analysis, you need to ensure that toluene will not boil at process conditions. Use the Antoine equation to find

A. The vapor pressure at $T = 100°C$.

B. The normal boiling point.

SOLUTION A:

Step 1 *Look up data*
From Appendix E, for toluene, the Antoine constants are $A = 6.95464$, $B = 1344.8$, and $C = 219.48$.

Step 2 *Solve Antoine equation*
We calculate the vapor pressure at $T = 100°C$, with T expressed in Celsius:

$$\log_{10}(P^{\text{sat}}) = 6.95464 - \frac{1344.8}{(100°C) + 219.48}$$

$$P^{\text{sat}} = \textbf{556.3 mm Hg} \tag{8.44}$$

SOLUTION B:

Step 3 *Back-solve Antoine equation*
Here we wish to know the normal boiling point; in other words, the boiling temperature that corresponds to a pressure of 1 atm, or 760 mm Hg:

$$\log_{10}(760 \text{ mm Hg}) = 6.95464 - \frac{1344.8}{T + 219.48}$$

$$T = \textbf{110.6°C} \tag{8.45}$$

556.3 mm Hg is equivalent to 74.1 kPa.

8.3 Fugacity and Its Use in Modeling Phase Equilibrium

Section 8.2 demonstrated some problem-solving strategies for estimating the pressures and temperatures at which phase equilibriums (primarily vapor–liquid equilibrium) occur. All of these depended, to varying degrees, on the availability of known data points to use as a reference and were only valid under specific circumstances. These strategies were all derived from the equilibrium criterion, which for vapor–liquid equilibrium is $\underline{G}^L = \underline{G}^V$. In Chapter 6, we saw that changes in molar enthalpy ($\underline{H}$) and molar entropy ($\underline{S}$) can be modeled using an equation of state. These are the state properties that contribute to the molar Gibbs energy ($\underline{G} = \underline{H} - T\underline{S}$), so it is also possible to compute changes in $\underline{G}$ from an equation of state. In fact, one can predict the entire vapor–liquid equilibrium curve using no information beyond an equation of state that describes the vapor and liquid phases.

The approaches described in this section will probably look more mathematically rigorous than the Clausius-Clapeyron equation or the shortcut equation, and in a sense, they are. Those equations were derived using simplifying approximations that are only reasonable under specific conditions, while the mathematics presented in this section are completely general. Keep in mind, however, that an equation of state IS an approximation. When we start with an equation of state, even if we make no further assumptions or approximations in building our model, our answers are only accurate to the extent that the equation of state accurately represents the chemical we are modeling at the conditions of interest.

8.3.1 Calculating Changes in Gibbs Energy

The phase equilibrium criterion for pure compounds states that, for phases in equilibrium, the molar Gibbs energy of each phase must be identical. For vapor–liquid equilibrium, specifically, this is written as $\underline{G}^L = \underline{G}^V$. Vapor–liquid equilibrium is singled out at this point because it is the phase equilibrium of most frequent interest to chemical engineers, and as shown in Section 7.2, it is a routine practice to model the liquid and vapor phases with a single equation of state. If we can compute $\underline{G}$ at any temperature and pressure for either the liquid or the vapor phase from an equation of state, then we can identify the temperatures and pressures where $\underline{G}^L = \underline{G}^V$. This sounds like a very plausible problem-solving strategy, because we already have a mathematical relationship among $\underline{G}$, T, and P. The fundamental property relationship first introduced as Equation 6.46 gives

$$d\underline{G} = \underline{V}\,dP - \underline{S}\,dT$$

However, a practical difficulty arises when working with $\underline{G}$ (at least directly). This will be illustrated by considering an isothermal change in $\underline{G}$.

CALCULATING GIBBS ENERGY ON AN ISOTHERMAL PATH	EXAMPLE 8-4

In specifying the molar Gibbs free energy of a compound, the liquid phase at $T = 300$ K and $P = 100$ kPa is being used as the reference state. Therefore $\underline{G}^L = 0$ at $T = 300$ K and $P = 100$ kPa. It has been estimated that for pure vapor at $T = 300$ K and $P = 100$ kPa, $\underline{G}^V = 500$ J/mol. The molar volume of the liquid at the reference state is $\underline{V}^L = 125$ cm³/mol.

A. Estimate the liquid phase $\underline{G}$ at $T = 300$ K and $P = 1.0$ kPa, relative to this reference state.

B. Estimate the vapor phase $\underline{G}$ at $T = 300$ K and $P = 1.0$ kPa, relative to this reference state.

C. Estimate the vapor pressure at $T = 300$ K.

The lower value of $\underline{G}$ corresponds to the more stable phase. Therefore, this compound is in the liquid phase at $T = 300$ K and $P = 100$ kPa ($0 < 500$ J/mol)

SOLUTION:

Step 1 *Mathematically relate $\underline{G}$ to P*
In parts A and B of this problem, we know $\underline{G}$ at $T = 300$ K and $P = 100$ kPa, and we need to determine $\underline{G}$ at the same temperature and a lower pressure. If following the methodical problem-solving strategy that was recommended in Section 6.2, we would start by writing an exact differential that relates the intensive property we wish to model, which is the change in Gibbs energy ($d\underline{G}$), to the two intensive properties about which we know the most (in this case, T and P). Such an equation would look like

$$d\underline{G} = \left(\frac{\partial \underline{G}}{\partial P}\right)_T dP + \left(\frac{\partial \underline{G}}{\partial T}\right)_P dT \qquad (8.46)$$

Writing Equation 8.46 isn't a necessary step in this problem—you can start with the fundamental property relationship directly. But an exact differential is always a good place to start if you're not sure where to start.

We then look to apply the mathematical tools learned in Chapter 6 to re-express $(\partial \underline{G}/\partial P)_T$ and $(\partial \underline{G}/\partial T)_P$ in terms of temperature and pressure. In any problem involving $\underline{U}$, $\underline{H}$, $\underline{A}$, or $\underline{G}$, it makes sense to consider using the relevant fundamental property relation in some way. In this case, there are no mathematical manipulations to do; we need only recognize that the fundamental property relationship for $\underline{G}$ as

$$d\underline{G} = \underline{V}dP - \underline{S}dT$$

is exactly the same form as Equation 8.46; it relates $\underline{G}$ directly to T and P.

All three parts of this problem involve an isothermal path ($\int \underline{S}\,dT = 0$), so the equation for $d\underline{G}$ simplifies to

$$d\underline{G} = \underline{V}dP \tag{8.47}$$

The only way Equations 6.46 and 8.46 can both be true is if $(\partial \underline{G}/\partial P)_T = \underline{V}$ and $(\partial \underline{G}/\partial T)_P = -\underline{S}$.

SOLUTION A:

Step 2 *Integrate for liquid phase*
In order to integrate Equation 8.47, we must express $\underline{V}$ as a mathematical function of P. If we had an equation of state that describes the liquid phase or a value or expression for the isothermal compressibility, we could do this without making any assumptions. As it is, we will assume $\underline{V}$ is a constant for the liquid phase. The molar Gibbs energy of the liquid at $T = 300$ K and $P = 100$ kPa, which we call $\underline{G}_1^L$, is our reference state, and it is 0 by definition. $\underline{G}_2^L$ is the Gibbs energy of the liquid at $T = 300$ K and $P = 1.0$ kPa and is what we are trying to determine.

The values of P^{sat} do not depend upon what reference state is chosen. Changing the reference state changes the values of G^L and G^V, but the outcome $\underline{G}^L > \underline{G}^V$, $\underline{G}^L < \underline{G}^V$ or $\underline{G}^L = \underline{G}^V$ will be the same for any reference state.

$$\int_{\underline{G}_1^L = 0}^{\underline{G}_2^L = \underline{G}_2^L} d\underline{G} = \int_{P_1 = 100\,\text{kPa}}^{P_2 = 1.0\,\text{kPa}} \underline{V}^L \, dP \tag{8.48}$$

$$\underline{G}_2^L - \underline{G}_1^L = \underline{V}^L(P_2 - P_1)$$

$$\underline{G}_2^L = \left(125\,\frac{\text{cm}^3}{\text{mol}}\right)(1\,\text{kPa} - 100\,\text{kPa})\left(\frac{1\,\text{m}^3}{10^6\,\text{cm}^3}\right)\left(\frac{10^3\,\text{Pa}}{\text{kPa}}\right) = -\textbf{12.4}\,\frac{\textbf{J}}{\textbf{mol}} \tag{8.49}$$

SOLUTION B:

Step 3 *Integrate for vapor phase*
The molar Gibbs energy of vapor at $T = 300$ K and $P = 100$ kPa, which we call $\underline{G}_1^V$, is known relative to our reference state; $\underline{G}_1^V = 500$ J/mol. $\underline{G}_2^V$ represents the molar Gibbs energy at $T = 300$ K and $P = 1.0$ kPa, which is our unknown. Since we are examining a vapor at pressures below atmospheric, we will use the ideal gas law to relate $\underline{V}$ to P.

$$\int_{\underline{G}_1^V = 500\,\text{J/mol}}^{\underline{G}_2^V = \underline{G}_2^V} d\underline{G} = \int_{P_1 = 100\,\text{kPa}}^{P_2 = 1.0\,\text{kPa}} \underline{V}^V \, dP \tag{8.50}$$

$$\int_{\underline{G}_1^V = 500\,\text{J/mol}}^{\underline{G}_2^V = \underline{G}_2^V} d\underline{G} = \int_{P_1 = 100\,\text{kPa}}^{P_2 = 1.0\,\text{kPa}} \frac{RT}{P} \, dP$$

$$\underline{G}_2^V - \underline{G}_1^V = RT \ln\left(\frac{P_2}{P_1}\right)$$

$$\underline{G}_2^V = 500\,\frac{\text{J}}{\text{mol}} + \left(8.314\,\frac{\text{J}}{\text{mol} \cdot \text{K}}\right)(300\,\text{K}) \ln\left(\frac{1.0\,\text{kPa}}{100\,\text{kPa}}\right) \tag{8.51}$$

$$= -11{,}000\,\frac{\text{J}}{\text{mol}}$$

The answer is a negative number that is large in magnitude. Equation 8.51 includes ln (P_2/P_1), which means that $\underline{G}^V$ will approach negative infinity as P_2 approaches 0.

SOLUTION C:

Step 4 *Find pressure at which $\underline{G}^L = \underline{G}^V$*
Notice that for $T = 300$ K, $\underline{G}^L < \underline{G}^V$ at $P = 100$ kPa and $\underline{G}^V < \underline{G}^L$ at $P = 1.0$ kPa; this means the compound is a vapor at $P = 1.0$ kPa and a liquid at $P = 100$ kPa. The vapor pressure is the unique pressure at which $\underline{G}^L = \underline{G}^V$. In Equations 8.49 and 8.51, we essentially derived expressions for $\underline{G}^L$ and $\underline{G}^V$ at $T = 300$ K. In parts A and B, P_2 was known; here we can use these same expressions for $\underline{G}^L$ and $\underline{G}^V$ and express P_2 as an unknown:

$$\underline{G}_2^L = \underline{G}_2^V \tag{8.52}$$

$$\left(125 \frac{\text{cm}^3}{\text{mol}}\right)(P_2 - 100\,\text{kPa})\left(\frac{1\,\text{m}^3}{10^6\,\text{cm}^3}\right)\left(\frac{10^3\,\text{Pa}}{\text{kPa}}\right) = 500 \frac{\text{J}}{\text{mol}} + \left(8.314 \frac{\text{J}}{\text{mol}\cdot\text{K}}\right)(300\,\text{K})\ln\left(\frac{P_2}{100\,\text{kPa}}\right)$$

$$\boldsymbol{P_2 = 81.76 \text{ kPa}}$$

Thus, if Equation 8.51 is a correct formula for $\underline{G}_2^V$ and Equation 8.49 is a correct formula for $\underline{G}_2^L$, the vapor pressure of the compound at $T = 300$ K is $P^{\text{sat}} = 81.76$ kPa. However, Figure 8-5 helps to reveal some practical challenges with using molar Gibbs energy for the calculation. For the vapor phase, $\underline{G}$ is extremely sensitive to pressure and in fact goes to $-\infty$ as P goes to 0, as noted in step 2. By contrast, the liquid phase $\underline{G}$ is comparatively insensitive to pressure. The vapor pressure is reported to four significant figures, at $P^{\text{sat}} = 81.76$ kPa. Considering the way the given information is reported ($\underline{V}^L = 125$ cm³/mol, $\underline{G}^V = 500$ J/mol at $P = 100$ kPa), it isn't realistic to claim our answer is correct to four significant figures—we might more reasonably say $P^{\text{sat}} = 82$ kPa or $P^{\text{sat}} = 81.8$ kPa. However, $\underline{G}^V$ is so sensitive to pressure that if we plugged $P = 82$ kPa or $P = 81.8$ kPa into Equation 8.52, we wouldn't necessarily recognize these as "correct" answers; the molar Gibbs energies of the two phases look significantly different, as shown in Table 8-3.

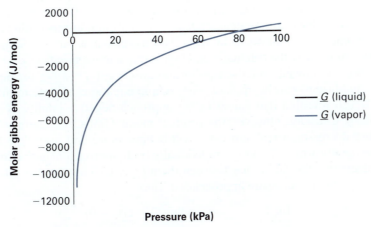

FIGURE 8-5 Molar Gibbs energy vs. pressure at $T = 300$ K for the compound examined in Example 8-4.

TABLE 8-3 Values of $\underline{G}^L$ and $\underline{G}^V$ obtained from Equations 8.49 and 8.51, for some values of pressure that are identical to two significant figures.

Pressure (kPa)	$\underline{G}^L$ from Equation 8.49 (J/mol)	$\underline{G}^V$ from Equation 8.51 (J/mol)
81.75	−2.281	−2.592
81.76	−2.280	−2.287
81.80	−2.275	−1.067
82.00	−2.250	5.024

Example 8-4 illustrates that it is possible to apply the equilibrium criterion $\underline{G}^L = \underline{G}^V$ to estimate vapor pressure from simple models: We assumed ideal gas behavior for the vapor and incompressible fluid behavior for the liquid. We could just as readily produce estimates of vapor pressure using more complex models; any equation of state could be used to relate $\underline{V}$ to P in steps 1 and 2.

However, the example also illustrates that $\underline{G}$ is an awkward mathematical function to use. For the vapor phase, Table 8-3 illustrates how seemingly insignificant changes to P can result in significant changes in $\underline{G}$. Furthermore, we showed mathematically that $\underline{G}^V$ approaches negative infinity as pressure goes to 0. This aspect of $\underline{G}$ will become a much more significant concern when we begin to model mixtures. While a chemical process being carried out at an *absolute* pressure of $P = 1$ kPa might be quite rare, it is not at all unusual for a vapor phase mixture to contain one or more compounds that have a *partial* pressure on the order of $P = 1$ kPa or less.

Section 8.3.2 illustrates how we can use the residual molar Gibbs free energy—rather than the molar Gibbs free energy itself—as the basis for phase equilibrium calculations.

8.3.2 Mathematical Definition of Fugacity

Residual functions were first introduced in Section 6.3. They compare real properties of a compound to the properties of a hypothetical ideal gas at the same temperature and pressure.

This section introduces fugacity, which is a property that, for pure compounds, is readily computed through the residual molar Gibbs free energy. Conceptually, fugacity can be used as the basis for building models for equilibrium just as readily as Gibbs energy, and practically it allows us to avoid the problems illustrated in Section 8.3.1. Initially, we discuss why the residual molar Gibbs free energy, specifically, is chosen.

Section 8.3.1 showed how the equilibrium criterion ($\underline{G}^L = \underline{G}^V$ for liquid and vapor in equilibrium) can be applied using simple equations of state to model $\underline{G}^L$ and $\underline{G}^V$. It also revealed that values of $\underline{G}^V$ approach negative infinity as pressure approaches 0. Consider, however, the residual molar Gibbs free energy: $\underline{G} - \underline{G}^{ig}$. As pressure decreases, a real substance will behave more and more like an ideal gas. Consequently, while $\underline{G}$ and $\underline{G}^{ig}$ individually both approach negative infinity as pressure decreases, the difference between them ($\underline{G} - \underline{G}^{ig}$) approaches 0 (as do all residual properties) as pressure approaches 0. Thus,

$$\lim_{P \to 0} \underline{G} = -\infty \qquad \lim_{P \to 0}(\underline{G} - \underline{G}^{ig}) = 0 \qquad (8.53)$$

Phases in equilibrium, which have the same $\underline{G}$, are at the same temperature and pressure as each other, so they must also have the same $\underline{G}^{ig}$. Thus, an equilibrium model built upon the residual molar Gibbs free energy is conceptually not different than one built on the molar Gibbs free energy itself.

The **fugacity** (f) of a pure substance is defined through

$$dG = RTd(\ln f)$$ (8.54)

This definition is chosen to mirror Equation 8.50, which was derived for an ideal gas undergoing an isothermal process:

$$d\underline{G}^{ig} = RT\frac{dP}{P} = RTd(\ln P)$$ (8.55)

Subtracting Equation 8.55 from Equation 8.54 produces an expression for the residual molar Gibbs free energy, as

$$d(\underline{G} - \underline{G}^{ig}) = RT[d(\ln f) - d(\ln P)]$$ (8.56)

When integrated and rearranged, this produces the equation for fugacity that we will find most useful for computing the fugacity of pure compounds:

$$\ln\left(\frac{f}{P}\right) = \frac{\underline{G} - \underline{G}^{ig}}{RT} = \frac{\underline{G}^{R}}{RT}$$ (8.57)

First, we note that *fugacity (f) is a state property*. Because it is derived entirely from state properties, it is also a state property. Second, we note that Equation 8.57 can be applied to a substance in any phase—vapor, liquid, or solid. We will use superscripts to distinguish phases from each other.

- f^{V} represents the fugacity of a pure vapor phase.
- f^{L} represents the fugacity of a pure liquid phase.
- f^{S} represents the fugacity of a pure solid phase.
- f^{scf}, which is the fugacity of a supercritical fluid, can be calculated but will not be of practical significance in this chapter. Our goal is to model phase equilibrium for a pure compound. A supercritical fluid is, by definition, above the temperature and pressure at which phase equilibrium can occur.

Equation 8.53 and the discussion around it explain our interest in using the molar Gibbs free energy residual function in some form, but why *this* definition specifically? Fugacity as defined here has units of pressure, and in some special cases it simplifies to properties with which we're well acquainted.

- We can see immediately from Equation 8.57 that under ideal gas conditions, the right-hand side is 0 and $f^{V} = P$ (where f^{V} represents the fugacity of a pure vapor).
- We will learn in Section 8.3.3 that under some very specific, but common, conditions, the fugacity of a pure liquid (f^{L}) is well approximated by the vapor pressure P^{sat}.

You can use these special cases as starting points to build an intuitive feel for fugacity. For example, imagine a vapor at $P = 200$ kPa. If we calculate $\underline{G}^{R} = -1500$ J/mol, we might wonder if this is a "big" or "small" value. Whereas, if we calculate $f^{V} = 199$ kPa, we readily recognize that $f^{V} \approx P$ and interpret this as evidence that the vapor departs only slightly from ideal gas behavior.

More significantly, we will see, starting in Chapter 10, that fugacity is a convenient property for use in the modeling of mixtures.

Because fugacity is related directly to the residual molar Gibbs free energy (through Equation 8.57), the equilibrium criterion ($\underline{G}^{L} = \underline{G}^{V}$ for VLE, etc.) also can be expressed in terms of fugacity.

- $f^{V} = f^{L}$ at liquid–vapor equilibrium conditions.
- $f^{V} = f^{S}$ at solid–vapor equilibrium conditions.

The word "fugacity" is derived from the Latin "fugere," which means "to flee." You might think of the vapor pressure of a compound as a measure of its ability to "escape" into the vapor phase.

Fugacity was introduced in 1908 by Gilbert Norton Lewis (Lewis, 1908), after whom Lewis dot structures and Lewis acids and bases are named (Olson, 1998).

- $f^L = f^S$ at solid–liquid equilibrium conditions.
- The triple point of a compound is the one temperature and pressure at which $f^V = f^L = f^S$.

Expressions for $H -$ H^{ig} and $S - S^{ig}$ were first presented in Section 6.3.

An expression for the residual molar Gibbs energy itself can be derived from the definition $\underline{G} = \underline{H} - T\underline{S}$, which applies to an ideal gas, just as it does for Gibbs energy in general:

$$\underline{G} = \underline{H} - T\underline{S} \quad \text{and} \quad \underline{G}^{ig} = \underline{H}^{ig} - T\underline{S}^{ig} \tag{8.58}$$

$$\underline{G} - \underline{G}^{ig} = (\underline{H} - T\underline{S}) - (\underline{H}^{ig} - T\underline{S}^{ig})$$

$$\underline{G} - \underline{G}^{ig} = (\underline{H} - \underline{H}^{ig}) - T(\underline{S} - \underline{S}^{ig}) \tag{8.59}$$

Thus, we can combine previously derived expressions for the enthalpy and entropy residual functions (Equations 6.130 and 6.131) with Equation 8.57 to relate fugacity to measurable properties:

$$\ln\left(\frac{f}{P}\right) = \frac{\underline{G} - \underline{G}^{ig}}{RT} = \frac{(\underline{H} - \underline{H}^{ig})}{RT} - \frac{T(\underline{S} - \underline{S}^{ig})}{RT}$$

$$\ln\left(\frac{f}{P}\right) = \frac{1}{RT}\int_{T=T,P=0}^{T=T,P=P}\left[\underline{V} - T\left(\frac{\partial \underline{V}}{\partial T}\right)_P\right]dP + \frac{T}{RT}\int_{T=T,P=0}^{T=T,P=P}\left[\left(\frac{\partial \underline{V}}{\partial T}\right)_P - \frac{R}{P}\right]dP$$

Recall that temperature is a constant in the evaluation of a residual function, so T (and R) can be moved inside or outside the integral as convenient.

When we combine the integrals, the $(\partial \underline{V}/\partial T)_P$ terms cancel each other and we are left with

$$\ln\left(\frac{f}{P}\right) = \frac{1}{RT}\int_{T=T,P=0}^{T=T,P=P}\left[\underline{V} - \frac{RT}{P}\right]dP \tag{8.60}$$

The dimensionless ratio (f/P) which appears in Equation 8.60 is called the **fugacity coefficient** and is given the symbol ϕ:

$$\phi = \frac{f}{P} \tag{8.61}$$

In Chapter 10, we will find the fugacity coefficient a convenient property to use in modeling mixtures.

In this chapter, our primary use for fugacity is to locate equilibrium conditions for pure compounds. It doesn't much matter whether we express calculations in terms the fugacity (f) or fugacity coefficient (φ); the criterion for vapor–liquid equilibrium can be expressed as either $f^L = f^V$ or $\phi^L = \phi^V$.

Equation 8.60 is conveniently expressed in terms of the compressibility factor Z:

Here we are applying the definition $Z = P\underline{V}/RT$.

$$\ln\left(\frac{f}{P}\right) = \ln \phi = \int_{T=T,P=0}^{T=T,P=P}\left[\frac{\underline{V}}{RT} - \frac{1}{P}\right]dP$$

Recall from Chapter 6 that residuals can be computed by integrating with respect to either dP or $d\underline{V}$. This is significant because it allows us to work with equations of state in either the form $P = f(T,\underline{V})$ or $\underline{V} = f(P,T)$.

$$\ln\left(\frac{f}{P}\right) = \ln \phi = \int_{T=T,P=0}^{T=T,P=P}\left[\frac{Z - 1}{P}\right]dP \tag{8.62}$$

If we start with the alternative ($d\underline{V}$) expressions for the residual molar enthalpy and entropy and perform another derivation analogous to Equations 8.54 through 8.60, the counterpart to Equation 8.62 is

$$\ln\left(\frac{f}{P}\right) = \ln \phi = (Z - 1) - \ln Z + \int_{T=T, \underline{V}=\infty}^{T=T, \underline{V}=\underline{V}}\frac{(1 - Z)}{\underline{V}}d\underline{V} \tag{8.63}$$

Estimation of a vapor pressure using these expressions for fugacity is illustrated in Example 8-5.

ESTIMATING A VAPOR PRESSURE FOR FREON USING THE VAN DER WAALS EOS	**EXAMPLE 8-5**

The boiler of a refrigerator is designed to operate at $T = 263.71$ K. Assuming Freon® 22 in the liquid and vapor phases can be described by the van der Waals equation with the parameters $a = 4.888 \times 10^6$ Pa m⁶/kmol² and $b = 0.09988$ m³/kmol, compute the vapor pressure of Freon® 22 at $T = 263.71$ K.

SOLUTION:

Step 1 *Apply VLE criterion; consider degrees of freedom*
At the vapor pressure, saturated liquid and saturated vapor can exist in equilibrium, and the equilibrium criteria that describes this condition are $P^L = P^V$, $T^L = T^V$, and $f^L = f^V$. According to the Gibbs phase rule, when there are two phases and a single component, there is only one degree of freedom ($F = 2 - \pi + C = 2 - 2 + 1$). That one degree of freedom is specified in this problem by the temperature $T^L = T^V = 263.71$ K. Thus, all other intensive properties of the liquid and vapor phases—including the pressure $P^{sat} = P^V = P^L$—have fixed values. The constraint $f^L = f^V$ can be used to find this unknown pressure if we can relate f^L and f^V to T and P.

Step 2 *Apply van der Waals equation to the general expression for fugacity*
In calculating fugacity from an equation of state, we can always start with either Equations 8.62 or 8.63. The van der Waals equation, when written in terms of the compressibility Z, is

> Section 7.2.7 gives the van der Waals, Soave, and Peng-Robinson equations in the forms $P = P(T, \underline{V})$ and $Z = Z(T, \underline{V})$.

$$Z = \frac{\underline{V}}{\underline{V} - b} - \frac{a}{\underline{V}RT}$$

This is readily integrated with respect to $d\underline{V}$, so we use Equation 8.63

$$\ln\left(\frac{f}{P}\right) = (Z - 1) - \ln(Z) + \int_{T=T, \underline{V}=\infty}^{T=T, \underline{V}=\underline{V}} \frac{(1 - Z)}{\underline{V}} d\underline{V}$$

> **PITFALL PREVENTION**
>
> The terms $(Z - 1)$ and $-\ln(Z)$ are outside the integral. Be careful to avoid accidentally moving them inside the integral.

Focusing first on the integral, introducing the expression for Z gives

$$\ln\left(\frac{f}{P}\right) = (Z - 1) - \ln Z + \int_{T=T, \underline{V}=\infty}^{T=T, \underline{V}=\underline{V}} \left(\frac{1 - \dfrac{\underline{V}}{\underline{V} - b} + \dfrac{a}{\underline{V}RT}}{\underline{V}}\right) d\underline{V} \quad (8.64)$$

$$\ln\left(\frac{f}{P}\right) = (Z - 1) - \ln Z + \int_{T=T, \underline{V}=\infty}^{T=T, \underline{V}=\underline{V}} \left(\frac{1}{\underline{V}} - \frac{1}{\underline{V} - b} + \frac{a}{\underline{V}^2 RT}\right) d\underline{V}$$

which integrates to

$$\ln\left(\frac{f}{P}\right) = (Z - 1) - \ln Z + \left[\ln(\underline{V}) - \ln(\underline{V} - b) - \left(\frac{a}{\underline{V}RT}\right)\right]\Bigg|_{\underline{V}=\infty}^{\underline{V}=\underline{V}}$$

$$\ln\left(\frac{f}{P}\right) = (Z - 1) - \ln Z + \left[\ln\left(\frac{\underline{V}}{\underline{V} - b}\right) - \left(\frac{a}{\underline{V}RT}\right)\right]\Bigg|_{\underline{V}=\infty}^{\underline{V}=\underline{V}} \quad (8.65)$$

> This step uses the property of logarithms: $\ln x - \ln y = \ln\frac{x}{y}$. The individual terms $\ln(\underline{V})$ and $\ln(\underline{V} - b)$ are both problematic at the limit $\underline{V} = \infty$, but $\underline{V}/(\underline{V} - b)$ is 1 when $\underline{V} = \infty$.

At the lower limit ($\underline{V} = \infty$) of integration the integrand is 0 ($\ln 1 = 0$; $1/\infty = 0$) so the final expression is

$$\ln\left(\frac{f}{P}\right) = (Z - 1) - \ln Z + \ln\left(\frac{\underline{V}}{\underline{V} - b}\right) - \left(\frac{a}{\underline{V}RT}\right) \quad (8.66)$$

Up to this point we have assumed nothing other than that the vapor and liquid phases are described by the van der Waals EOS.

Section 7.2.3 discussed solving cubic equations of state for V.

Step 3 *Relate liquid and vapor fugacity to measurable properties*

For a particular temperature and pressure, the van der Waals EOS will produce three solutions for $\underline{V}$—each of which will yield a different value of Z. For example, when $T = 263.71$ and $P = 68.94$ kPa, the three solutions for $\underline{V}$ are $\underline{V} = 0.1394$ m³/kmol, $\underline{V} = 0.3563$ m³/kmol, and $\underline{V} = 31.42$ m³/kmol.

Physically, we interpret the smallest of these as $\underline{V}^L$, the largest as $\underline{V}^V$, and the middle value as physically unrealistic, as discussed in Section 7.2.4. Using Equation 8.66, the corresponding fugacity for both liquid and vapor can be found, as summarized in Table 8-4. We can simply introduce the $\underline{V}^L$ and $\underline{Z}^L$ values into Equation 8.66 when we are calculating the liquid fugacity (f^L) and introduce $\underline{V}^V$ and $\underline{Z}^V$ into Equation 8.66 when calculating the vapor fugacity (f^V). This may look odd—fugacity is defined from a Gibbs energy residual function, which is the difference between the actual $\underline{G}$ and $\underline{G}^{ig}$. Is it really valid to compare a *liquid* property to an ideal gas property in this way? This question is examined further after the example.

Table 8-4 summarizes the fugacity calculations at $P = 68.95$ kPa and $T = 263.71$ K. Using a spreadsheet, calculating f for many different temperatures and pressures requires little more effort than calculating f for a single T and P.

TABLE 8-4 Computed liquid and vapor fugacity for Freon® at $T = 68.94$ kPa and $P = 263.15$ K.

	V (m³/kmol)	Z	ln (f/P)	f (kPa)
Liquid	0.1394	0.004571	2.13	605.49
Vapor	31.42	0.9876	−0.0124	68.10
Non-Physical	5.71	0.0112	2.45	116

In Table 8-4, the lower value of fugacity (f) corresponds to the lower value of molar Gibbs energy ($\underline{G}$), and as outlined in Section 8.2.2, the lower $\underline{G}$ corresponds to the more stable phase. Thus, at 68.94 kPa, the vapor phase is more stable, so the vapor pressure must be higher than 68.94 kPa. We can complete similar calculations to those summarized in Table 8-4 for any pressure. Table 8-5 summarizes the results for a range of pressures.

TABLE 8-5 Summary of vapor and liquid fugacity values for $T = 263.71$ K.

T (K)	P (kPa)	$\underline{V}^L$ (m³/kmol)	$\underline{V}^V$ (m³/kmol)	Z^L	Z^V	f^L (kPa)	f^V (kPa)	Most Stable Phase
263.71	68.94	0.1394	31.42	.004571	0.9876	605.50	68.09	Vapor
263.71	344.73	0.1389	5.95	0.02176	0.935	615.97	320.67	Vapor
263.71	689.47	0.1384	2.73	0.04365	0.8581	629.70	605.91	Vapor
263.71	725.33	0.1384	2.57	0.04573	0.8494	631.01	631.01	Vapor/Liquid
263.71	758.42	0.1383	2.43	.04782	0.8409	632.38	655.76	Liquid

Section 8.2.1 provides a general discussion of why the solid, liquid, and vapor phases exist and what kinds of conditions favor each.

Thus, the vapor pressure is approximately **725.33 kPa**. We can continue iterating until f^L and f^V agree to as many significant figures as desired. However, the van der Waals a and b were both given to three significant figures, so we cannot expect that requiring f^L and f^V to agree to 5 or 6 significant figures will give us a more accurate answer than this one.

It is instructive to compare the data at 689.47 kPa to the data at 725.33 kPa in Table 8-5.

- The two pressures differ by ~5%: 725.33 vs. 689.47 kPa.

- The vapor phase fugacities differ by ~4%: 631.07 vs. 605.91 kPa.

- Contrast these observations with Table 8-3, in which changes to the third or fourth significant figure of the pressure affected the first significant figure of the molar Gibbs energy. This illustrates why fugacity is a more practical and convenient property to use in calculations than $\underline{G}$.

Example 8-5 illustrates how we can apply the equilibrium criterion ($f^V = f^L$ for VLE) in combination with an equation of state to generate an estimate of vapor pressure from an equation of state. The example showed the calculation at 263.71 K, but the same process could be applied to any temperature, producing a complete vapor pressure curve, like that shown in Figure 8-1. The accuracy of the estimate is dependent upon the suitability of the equation of state. For example, Figure 8-6 illustrates that the van der Waals equation provides poor estimates of the vapor pressure for the normal alkanes, while the Peng-Robinson equation predicts vapor pressure quite accurately. A derivation analogous to that shown in step 2 of Example 8-5 can be carried out for the Peng-Robinson equation, or any other equation of state. For the Peng-Robinson equation, the outcome is the expression for fugacity:

$$\ln\left(\frac{f}{P}\right) = (Z - 1) - \ln(Z - B) - \frac{A}{2B\sqrt{2}} \times \ln\left[\frac{Z + (1 + \sqrt{2})B}{Z + (1 - \sqrt{2})B}\right] \qquad (8.67)$$

where A and B are dimensionless groups that include the Peng-Robinson parameters a and b:

$$A = \frac{aP}{R^2 T^2} \qquad (8.68)$$

$$B = \frac{bP}{RT} \qquad (8.69)$$

These expressions are used in Example 8-7. Recall two primary differences between the Peng-Robinson and van der Waals equations.

- The adjustable parameters a and b are both constants in the van der Waals equation, but in the Peng-Robinson equation, a is a function of temperature.

- The acentric factor is used in computing a and b for the Peng-Robinson equation.

Since the Peng-Robinson equation incorporates the acentric factor ω, which is in effect a point on the vapor pressure curve, one would expect it to produce more accurate estimates of the vapor pressure. Figure 8-6 demonstrates this is indeed the case for the normal alkanes.

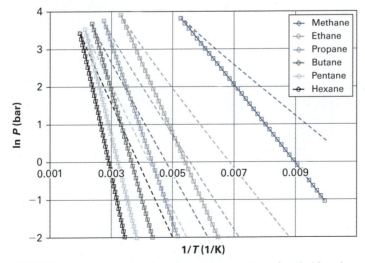

FIGURE 8-6 Comparison between vapor pressure data (symbols) and predictions obtained from the van der Waals (dashed lines) and Peng-Robinson (solid lines) equations, for six normal alkanes.

In practice, equations of state aren't commonly used to describe solids, but conceptually one could use an appropriate EOS to compute the fugacity of a solid.

Example 8-5 raises two other points that are worth further consideration.

1. The example illustrates that fugacity can be estimated for something that doesn't actually exist at equilibrium, such as a pure liquid at a pressure that is below the vapor pressure or a pure vapor at a pressure that is above the vapor pressure.

2. The example illustrates that the equation for calculating f from an EOS is the same—whether we are modeling a vapor or liquid.

The remainder of this section elaborates on these points.

Looking back at Example 8-5, we computed a value of fugacity for liquid Freon® at many conditions, such as $f^L = 605.50$ kPa at $T = 263.71$ K and $P = 68.94$ kPa. This calculation was a step in determining that the vapor pressure is about 723.95 kPa; meaning that Freon® at $T = 263.71$ K and $P = 68.94$ kPa exists only in the vapor phase. Thus, the value $f^L = 605.50$ kPa applies to something that doesn't actually exist, namely pure liquid Freon® at $T = 263.71$ K and $P = 68.94$ kPa. A crucial word in the last sentence is "pure." Everything we have studied up to this point has been a pure compound, but in Chapter 9 we will begin examining mixtures. Consider H_2O: We know from personal experience that H_2O can exist in the vapor phase at ~300 K and atmospheric pressure (as we have seen puddles of water evaporate at around this temperature and have felt the effects of humidity in the air). However, we also know that *pure* H_2O is a liquid at $T = 300$ K and $P = 101.325$ kPa atmospheric pressure; water vapor at these conditions exists only as a component of a mixture. We will learn starting in Chapter 9 that pure component properties provide a foundation for the modeling of mixtures. Thus, even though at equilibrium there is no such thing as pure water vapor at $T = 300$ K and $P = 101.325$ kPa, a numerical estimate for the fugacity of pure water vapor *if* it existed at $T = 300$ K and $P = 101.325$ kPa has utility in modeling water as a component of a vapor phase mixture at 300 K and $P = 101.325$ kPa. Consequently, in this chapter, we seek to develop strategies that allow us to estimate the fugacity of a pure compound in any phase at any temperature and pressure.

This brings us to residual functions, which compare a real compound to a hypothetical ideal gas that exists at the same temperature and pressure. The idea of comparing a liquid to an ideal gas might seem less natural than comparing a vapor to an ideal gas, and it might seem surprising that both can be done using exactly the same Equations 8.62 and 8.63. Nonetheless, the math is valid. Consider Figure 8-7, which shows an isotherm on a P vs. $\underline{V}$ diagram, as it would be modeled using a cubic equation of state. When we calculate fugacity using Equation 8.62, we have

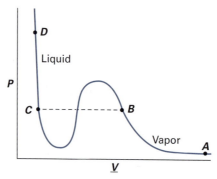

FIGURE 8-7 Path by which liquid fugacity is calculated from a cubic EOS. A = ideal gas condition, B = saturated vapor, C = saturated liquid, and D = high-pressure liquid.

$$\ln\left(\frac{f}{P}\right) = (Z-1) - \ln Z + \int_{T=T,\underline{V}=\infty}^{T=T,\underline{V}=\underline{V}} \frac{(1-Z)}{\underline{V}} d\underline{V}$$

We integrate from $\underline{V} = \infty$ to the actual volume $\underline{V}$ at the temperature and pressure of interest. Consider the possibilities:

- The curve AB on Figure 8-7 represents vapor conditions, and Equation 8.62 can be applied to calculate the vapor phase fugacity at any point on this curve.
- Point B represents saturated vapor, and point C represents saturated liquid. The part of the isotherm between B and C does not represent stable phases.
- If we wished to compute the fugacity of a compressed liquid, represented by point D, we would apply Equation 8.62 from $\underline{V} = \infty$ to point D. It might seem counter-intuitive that this process includes integrating over a curve (BC) that doesn't represent anything physically real. However, the fugacity at points B and C is identical—in fact, Example 8-5 illustrated that this equality is precisely how we can recognize the vapor pressure. Thus, $f_B = f_C$, which means that the "non-physical" portion of the curve does not actually affect our final answer.

Thus, conceptually, an equation of state provides a workable method of computing fugacity for any fluid, whether liquid, vapor, or supercritical. The next section, however, presents an alternative that is frequently applied to liquids and solids.

8.3.3 Poynting Method of Estimating Liquid and Solid Fugacity

Section 8.3.2 demonstrated that the fugacity of a pure compound—whether it is in the solid, liquid, vapor, or supercritical state—always can be estimated from Equation 8.62 or 8.63, IF an equation of state is available that describes the compound in the state of interest. Throughout this book, the fugacity for a vapor phase (or supercritical fluid) will be determined precisely this way. In practice, however, reliable and accurate equations of state are less routinely available for liquids and solids than for vapors. Consequently, this section describes an alternative approach—the Poynting correction—that is commonly used to estimate liquid or solid fugacity at elevated pressures.

John Henry Poynting was a professor of physics at Mason College, which during his tenure became Birmingham University (Porter, 2000).

ESTIMATING THE FUGACITY OF COMPRESSED WATER	EXAMPLE 8-6

The vapor pressure of water at $T = 298.15$ K is $P^{sat} = 3.17$ kPa, and the molar volume of saturated liquid at this temperature is $\underline{V} = 18.054 \times 10^{-6}$ m³/mol. *Using only this information*, estimate the fugacity of liquid water at $T = 298.15$ K and the following pressures: $P = 0.1$ MPa, $P = 1$ MPa, $P = 10$ MPa, and $P = 100$ MPa.

SOLUTION:

Step 1 *Recognize fugacity is known at the saturation point*
If we had an equation of state that we knew accurately modeled liquid water, we could apply Equation 8.62 or 8.63 to find the fugacity. Lacking the information needed for this straightforward solution approach, we begin by asking ourselves: Do we know the fugacity of liquid water at *any* point?

We learned in Section 8.3.2 that at saturation conditions, $f^V = f^L$. Thus, at $P = 3.17$ kPa and $T = 298.15$ K, $f^V = f^L$; if we can determine one of f^V or f^L, we will also know the other. The given information does not include an equation of state describing water vapor, but at low pressure, we can assume water vapor behaves as an ideal gas. For an ideal gas, $f^V = P$. Thus,

$$f^L = f^V \approx P^{sat} = 3.17 \text{ kPa} \quad \text{at } T = 298.15 \text{ K and } P = 3.17 \text{ kPa} \quad (8.70)$$

With fugacity known at one point, we can use the techniques learned in Chapter 6 to quantify the effect on fugacity when the pressure is changed from $P = 3.17$ kPa to the pressures of interest in this example ($P = 0.1, 1, 10,$ and 100 MPa), as illustrated in Figure 8-8.

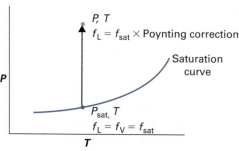

FIGURE 8-8 Illustration of the path used to calculate f_L.

Step 2 *Quantify effect of pressure on molar Gibbs free energy residual function*

Fugacity is defined from the residual molar Gibbs free energy, so let's consider the relationship between Gibbs free energy and pressure, beginning with the fundamental property relationship for $\underline{G}$ as

This fundamental property relationship was first introduced as Equation 6.46 in Section 6.2.2.

$$d\underline{G} = \underline{V}\,dP - \underline{S}\,dT$$

Here, f is known at $T_1 = 298.15$ K and $P_1 = 3.17$ kPa. We wish to quantify the change in $\underline{G}$ when the state is changed from this known point to $T_2 = 298.15$ K, $P_2 = 0.1$ MPa. This is an isothermal path, so the dT term is zero.

$$d\underline{G} = \underline{V}\,dP \tag{8.71}$$

Equation 8.71 can be applied to a pressure change for an ideal gas, just as it can be applied to the actual compound we are modeling. Thus,

Equation 8.71 is valid for any isothermal path or process.

$$d\underline{G}^{ig} = \underline{V}^{ig}\,dP \tag{8.72}$$

and combining Equations 8.71 and 8.72 to quantify the change in the molar Gibbs free energy residual function gives

$$d(\underline{G} - \underline{G}^{ig}) = (\underline{V} - \underline{V}^{ig})\,dP \tag{8.73}$$

But $\underline{V}^{ig}$ can be related to pressure through the ideal gas law as

$$d(\underline{G} - \underline{G}^{ig}) = \left(\underline{V} - \frac{RT}{P}\right)dP \tag{8.74}$$

Step 3 *Relate changes in fugacity to changes in Gibbs energy*

The definition of fugacity as given in Equation 8.57 is

$$\ln\left(\frac{f}{P}\right) = \frac{\underline{G} - \underline{G}^{ig}}{RT}$$

Here we apply $d(xy) = x\,dy + y\,dx$, where $x = (\underline{G} - \underline{G}^{ig})$ and $y = 1/T$. R is a constant.

Differentiating this using the chain rule gives

$$d\left[\ln\left(\frac{f}{P}\right)\right] = \frac{d(\underline{G} - \underline{G}^{ig})}{RT} + \frac{(\underline{G} - \underline{G}^{ig})}{R}\,d\left(\frac{1}{T}\right) \tag{8.75}$$

But here, we are looking at a constant temperature path ($T_1 = T_2 = 298.15$ K), so $1/T$ is constant ($d(1/T) = 0$), and the differential simplifies to

$$d\left[\ln\left(\frac{f}{P}\right)\right] = \frac{d(\underline{G} - \underline{G}^{ig})}{RT} \tag{8.76}$$

Substituting Equation 8.74 into Equation 8.76 gives

$$d\left[\ln\left(\frac{f}{P}\right)\right] = \frac{\left(\underline{V} - \frac{RT}{P}\right)dP}{RT} \tag{8.77}$$

or

$$d\left[\ln\left(\frac{f}{P}\right)\right] = \left(\frac{\underline{V}}{RT} - \frac{1}{P}\right)dP \tag{8.78}$$

Step 4 *Integrate*

R is a constant, and T can be treated as a constant on this isothermal path. Consequently, the only thing preventing us from integrating Equation 8.78 is that we have no mathematical relationship between $\underline{V}$ and P. For most liquids, $\underline{V}$ is reasonably approximated as constant (except close to the critical point) and is only known here at one pressure, so we will assume $\underline{V}$ is constant. The integration is

$$\int_{P_1}^{P_2} d\left[\ln\left(\frac{f}{P}\right)\right] = \int_{P_1}^{P_2}\left(\frac{\underline{V}}{RT} - \frac{1}{P}\right)dP \tag{8.79}$$

$$\ln\left(\frac{f}{P}\right)\Big|_{P_1}^{P_2} = \frac{\underline{V}P}{RT} - \ln(P)\Big|_{P_1}^{P_2}$$

$$\ln\left(\frac{f_2}{P_2}\right) - \ln\left(\frac{f_1}{P_1}\right) = \frac{\underline{V}(P_2 - P_1)}{RT} - \ln\left(\frac{P_2}{P_1}\right)$$

$$\ln\left(\frac{f_2}{f_1}\right) - \ln\left(\frac{P_2}{P_1}\right) = \frac{\underline{V}(P_2 - P_1)}{RT} - \ln\left(\frac{P_2}{P_1}\right)$$

This simplifies to

$$\frac{f_2}{f_1} = \exp\left\{\frac{\underline{V}(P_2 - P_1)}{RT}\right\} \tag{8.80}$$

> This step uses the property $-\ln(x) = \ln(1/x)$

where the exponential term is called the Poynting correction factor.

Step 5 *Solve*

We are trying to determine the fugacity (f_2) corresponding to $P_2 = 0.1$ MPa with the known data point $f_1 = 3.17$ kPa at $P_1 = 3.17$ kPa:

$$\frac{f_2}{3.17\text{ kPa}} = \exp\left\{\frac{\left(18.054 \times 10^{-6}\,\frac{\text{m}^3}{\text{mol}}\right)(100\text{ kPa} - 3.17\text{ kPa})}{\left(8.314\,\frac{\text{Pa m}^3}{\text{mol K}}\right)(298.15\text{ K})} \times \frac{(1000\text{ Pa})}{(1\text{ kPa})}\right\} \tag{8.81}$$

$$\frac{f_2}{3.17\text{ kPa}} = e^{0.000705} \tag{8.82}$$

> f_1 in this equation represents the fugacity at the vapor pressure $P^{\text{sat}} = 3.17$ kPa, and is therefore the same, regardless of the value of P_2.

In this case, the Poynting correction factor is essentially equal to unity (1.000705) and $f_2 = 3.17$ kPa. This is typical; the Poynting correction factor only becomes significant at elevated pressures; as summarized in Table 8-6. The solution approach of the other cases ($P = 1$ MPa, 10 MPa, etc.) is identical to this one, except the value of P_2 is different.

TABLE 8-6 Summary of results of Example 8-6 with estimates of liquid water fugacity at $T = 298.15$ K.

Actual Pressure (P_2)	Vapor Pressure (P_1)	$\exp\left\{\dfrac{V(P_2 - P_1)}{RT}\right\}$	Fugacity of Liquid Water (f_L)
0.1 MPa	3.17 kPa	1.0007	3.17 kPa
1 MPa	3.17 kPa	1.0073	3.19 kPa
10 MPa	3.17 kPa	1.0755	3.41 kPa
100 MPa	3.17 kPa	2.072	6.57 kPa

Example 8-6 illustrates the Poynting method for quantifying the effect of pressure changes on fugacity. While the example examined a liquid, the only assumption used in deriving Equation 8.80 was that the molar volume $\underline{V}$ was a constant—an approximation that logically is applied to solids as well as liquids.

Conceptually, the equation can be applied to any two pressures P_1 and P_2 for

$$\frac{f_2}{f_1} = \exp\left\{\frac{\underline{V}(P_2 - P_1)}{RT}\right\}$$

Practically speaking, the fugacity is most often known at the vapor pressure, because at the vapor pressure, $f^L = f^V$ for vapor–liquid equilibrium and $f^S = f^V$ for solid–vapor equilibrium. Consequently, the Poynting equation is often written as

$$f = f^{sat}\exp\left\{\frac{\underline{V}(P - P^{sat})}{RT}\right\} \tag{8.83}$$

where

f is the fugacity of the liquid or solid at the actual temperature and pressure of interest, T and P.

$\underline{V}$ is the molar volume of the liquid or solid and is assumed constant.

P^{sat} is the vapor pressure of the liquid or solid at the temperature of interest T.

f^{sat} is the fugacity of saturated vapor at the temperature T and the vapor pressure P^{sat}.

Common simplifications that occur when using the Poynting correction include:

- When P (actual pressure) is low, ideal gas behavior is assumed and the Poynting correction factor is negligible, as illustrated in Example 8-6, so $f = f^{sat} = P^{sat}$.
- Even when the actual pressure (P) is high, if P^{sat} is low enough to assume ideal gas behavior, the simplification $f^{sat} = P^{sat}$ still applies.

At higher vapor pressures, f^{sat} can be determined from an equation of state that describes the vapor phase. We close this chapter with an example that illustrates this process.

| **FUGACITY OF COMPRESSED LIQUID WATER AT HIGHER TEMPERATURES** | **EXAMPLE 8-7** |

The vapor pressure of water at $T = 473.15\,\text{K}$ ($200\,°\text{C}$) is $P^{\text{sat}} = 1.555\,\text{MPa}$, and the molar volume of saturated liquid at this temperature is $\underline{V} = 20.85 \times 10^{-6}\,\text{m}^3/\text{mol}$. Estimate the fugacity of liquid water at $T = 473.15\,\text{K}$ and the following pressures: $P = 0.1\,\text{MPa}$, $P = 10\,\text{MPa}$, and $P = 100\,\text{MPa}$.

> The values of P^{sat} and $\underline{V}$ given here are consistent with the data in the steam tables.

SOLUTION:

Step 1 *Consider approaches, examine known and unknown information*
The Poynting equation (Equation 8.83) was derived in Example 8-6.

$$f = f^{\text{sat}} \exp\left\{\frac{\underline{V}(P - P^{\text{sat}})}{RT}\right\}$$

The derivation won't be repeated, but we note that this is the very situation for which the Poynting equation is most useful. We know the vapor pressure at our temperature of interest ($T = 473.15\,\text{K}$), but we are trying to determine fugacity of a solid or liquid at a different pressure and this temperature.

If we assume $\underline{V}$ is constant with respect to pressure, then the entire exponential term can be considered known: T, $\underline{V}$, and P^{sat} are given, R is a constant, and P is any one of the three pressures at which we are asked to compute the fugacity. However, f^{sat} is not known. In Example 8-6 we applied an ideal gas assumption to saturated vapor. Here we cannot do that because saturated vapor has a pressure of more than 1.5 MPa. We must, therefore, estimate f^{sat} using an equation of state.

We will begin by solving the Peng-Robinson equation for saturated water vapor.

Step 2 *Determine Peng-Robinson EOS parameters for water*
According to Appendix C, water has $T_c = 647.3\,\text{K}$, $P_c = 22.12\,\text{MPa}$ and $\omega = 0.344$. This is everything we need to know to calculate parameters a and b for the Peng-Robinson equation, using the equations introduced in Section 7.2.5. The parameter b is given by

$$b = 0.07780\,R\,\frac{T_c}{P_c}$$

> The parameter a is a function of temperature. But this problem involves only one temperature and does not require taking any derivatives with respect to T, so once we've found the value of a for our temperature $T = 200°\text{C}$, we can use it like a constant.

$$b = 0.07780\left(8.314\,\frac{\text{Pa}\cdot\text{m}^3}{\text{mol}\cdot\text{K}}\right)\frac{(647.3\,\text{K})\times(1\,\text{MPa}/10^6\,\text{Pa})}{(22.12\,\text{MPa})} = 18.93 \times 10^{-6}\,\frac{\text{m}^3}{\text{mol}} \qquad (8.84)$$

Finding the parameter a requires several steps, as described in Section 7.2.5. First we compute κ as

$$\kappa = 0.37464 + 1.54226\omega - 0.269932\omega^2$$

$$\kappa = 0.37464 + 1.54226(0.344) - 0.269932(0.344)^2 = 0.8732 \qquad (8.85)$$

Our temperature $T = 473.15\,\text{K}$ can be expressed as a reduced temperature:

$$T_r = \frac{T}{T_c} = \frac{473.15\,\text{K}}{647.3\,\text{K}} = 0.7310 \qquad (8.86)$$

And κ and T_r are used to determine parameter α as

$$\alpha = [1 + \kappa(1 - T_r^{0.5})]^2$$

$$\alpha = [1 + (0.8732)(1 - \sqrt{0.7310})]^2 = 1.269 \qquad (8.87)$$

The parameter a_c is a function of the critical temperature and pressure:

$$a_c = 0.45724\,\frac{R^2 T_c^2}{P_c}$$

$$a_c = 0.45724 \frac{\left(8.314 \dfrac{\text{Pa} \cdot \text{m}^3}{\text{mol} \cdot \text{K}}\right)^2 (647.3\,\text{K})^2 \times \left(\dfrac{1\,\text{MPa}}{10^6\,\text{Pa}}\right)}{22.12\,\text{MPa}} = 0.5987\,\frac{\text{Pa} \cdot \text{m}^6}{\text{mol}^2} \qquad (8.88)$$

Finally, a is determined from α and a_c to be

$$a = a_c \alpha = \left(0.5987\,\frac{\text{Pa} \cdot \text{m}^6}{\text{mol}^2}\right)(1.269) = 0.7599\,\frac{\text{Pa} \cdot \text{m}^6}{\text{mol}^2} \qquad (8.89)$$

This form of the Peng-Robinson equation was first introduced in Section 7.2.2.

Step 3 *Solve Peng-Robinson equation for water vapor at $T = 473.15$ K and $P = 1.555$ MPa*
The Peng-Robinson equation, in the $P = P(\underline{V},T)$ form, is

$$P = \frac{RT}{\underline{V} - b} - \frac{a}{\underline{V}(\underline{V} + b) + b(\underline{V} - b)}$$

For $T = 473.15$ K, $P = 1.555$ MPa, and the values of a and b computed in step 2, the three solutions are

The interpretation of the three solutions for $\underline{V}$ from a cubic equation of state is discussed in Section 7.2.4.

$$\underline{V} = 0.2535 \times 10^{-4}\,\frac{\text{m}^3}{\text{mol}},\ 1.402 \times 10^{-4}\,\frac{\text{m}^3}{\text{mol}},\ \text{and } 23.45 \times 10^{-4}\,\frac{\text{m}^3}{\text{mol}}$$

The largest value of $\underline{V}$ is the one that corresponds to the vapor phase, and this is the value we will use when we compute f^{sat}, which is the fugacity of saturated vapor at $T = 473.15$ K. However, because $P = 1.555$ MPa is the vapor pressure at $T = 473.15$ K, water can exist in either the liquid or vapor phase at this temperature and pressure, and the smallest solution for $\underline{V}$ corresponds to the liquid phase. Here it is worth comparing the estimates from the Peng-Robinson equation to the data in the steam tables, as shown in Table 8-7.

The calculations in Table 8-7 are carried out for the vapor pressure value, 1.555 MPa, that appears in the steam tables. If the Peng-Robinson EOS is used to estimate the vapor pressure, as in Example 8-5, the result is 1.561 MPa.

TABLE 8-7 Comparison of molar volumes for saturated liquid water and water vapor at 200°C.

State	$\hat{V}$(m³/kg) from Steam Tables	$\underline{V}$ (m³/mol) from Steam Tables	$\underline{V}$ (m³/mol) from Peng-Robinson Equation	% Difference
Saturated Liquid	0.001157	20.85×10^{-6}	25.35×10^{-6}	21.6%
Saturated Vapor	0.12721	22.93×10^{-4}	23.45×10^{-4}	2.3%

The Peng-Robinson equation predicts the vapor phase molar volume within $\approx 2\%$, but the relative accuracy is significantly worse for the saturated liquid molar volume. This result illustrates why we are going to use the Poynting method for estimating liquid fugacity, rather than attempting to apply the Peng-Robinson equation directly to the liquid phase at $T = 473.15$ K and $P = 0.1$, 10, or 1000 MPa.

Step 4 *Compute f^V at $T = 473.15$ K and $P = 1.55$ MPa*
Equation 8.67 gives the fugacity of a compound as described by the Peng-Robinson equation:

$$\ln\left(\frac{f}{P}\right) = (Z - 1) - \ln(Z - B) - \frac{A}{2B\sqrt{2}} \times \ln\left[\frac{Z + (1 + \sqrt{2})B}{Z + (1 - \sqrt{2})B}\right]$$

in which A and B are dimensionless groups given by Equations 8.68 and 8.69:

$$A = \frac{aP}{R^2T^2}$$

$$B = \frac{bP}{RT}$$

In step 4, $P = 1.555$ MPa, because we are calculating the fugacity of *saturated* vapor. In step 5, $P = 0.1, 10$ or 100 MPa, because those are the actual pressures at which we wish to compute the liquid fugacity.

Equation 8.67 is valid for any fluid that is described by the Peng-Robinson equation of state. In this example, we are using it to find the fugacity of saturated water vapor at $T = 473.15$ K. To apply Equation 8.67, we need to determine A, B, and Z at $T = 473.15$ K and $P = 1.555$ MPa. A and B can be found immediately, as

$$A = \frac{aP}{R^2T^2} = \frac{\left(0.7599 \frac{\text{Pa} \cdot \text{m}^6}{\text{mol}^2}\right) 1.555 \text{ MPa} \times \frac{(10^6 \text{ Pa})}{(1 \text{ MPa})}}{\left(8.314 \frac{\text{Pa} \cdot \text{m}^3}{\text{mol} \cdot \text{K}}\right)^2 (473.15 \text{ K})^2} = 0.07636 \quad (8.90)$$

$$B = \frac{bP}{RT} = \frac{\left(18.93 \times 10^{-6} \frac{\text{m}^3}{\text{mol}}\right) 1.555 \text{ MPa} \times \frac{(10^6 \text{ Pa})}{(1 \text{ MPa})}}{\left(8.314 \frac{\text{Pa} \cdot \text{m}^3}{\text{mol} \cdot \text{K}}\right)(473.15 \text{ K})} = 0.00748 \quad (8.91)$$

The value of Z for saturated vapor is determined using the largest of the three values of molar volume ($\underline{V}$) found in step 3:

$$Z = \frac{P\underline{V}}{RT} = \frac{\left(1.555 \text{ MPa} \times 10^6 \frac{\text{Pa}}{1\text{MPa}}\right)\left(23.45 \times 10^{-4} \frac{\text{m}^3}{\text{mol}}\right)}{\left(8.314 \frac{\text{Pa} \cdot \text{m}^3}{\text{mol} \cdot \text{K}}\right)(473.15 \text{ K})} = 0.9270 \quad (8.92)$$

Introducing these results into Equation 8.67 gives us the fugacity of the saturated vapor:

$$\ln\left(\frac{f^{\text{sat}}}{P}\right) = (Z - 1) - \ln(Z - B) - \left(\frac{A}{2B\sqrt{2}}\right)\ln\left[\frac{Z + (1 + \sqrt{2})B}{Z + (1 - \sqrt{2})B}\right]$$

$$\ln\left(\frac{f^{\text{sat}}}{P}\right) = (0.9270 - 1) - \ln(0.9270 - 0.00748)$$

$$-\left(\frac{0.07636}{2(0.00748)\sqrt{2}}\right)\ln\left[\frac{0.9270 + (1 + \sqrt{2})0.00748}{0.9270 + (1 - \sqrt{2})0.00748}\right] = -0.0708$$

$$f^{\text{sat}} = (0.9316)P = (0.9316)(1.555 \text{ MPa}) = 1.449 \text{ MPa} \quad (8.93)$$

Step 5 *Use Poynting equation to find liquid fugacity*
The result of step 4 is our best estimate of the fugacity of saturated water vapor at $T = 473.15$ K, which is equal to the fugacity of saturated liquid water at $T = 473.15$ K for liquid–vapor equilibrium, $f^V = f^L = f^{\text{sat}}$. We can now use the Poynting equation

$$f = f^{\text{sat}} \exp\left\{\frac{\underline{V}(P - P^{\text{sat}})}{RT}\right\}$$

to find the fugacity of liquid water at $T = 473.15$ K and other pressures. Beginning with $P = 100$ MPa:

The steam tables give $\hat{V} = 0.001083 \frac{\text{m}^3}{\text{kg}}$ at $T = 473.15$ K and $P = 100$ MPa, and $\hat{V} = 0.001157 \frac{\text{m}^3}{\text{kg}}$ for saturated liquid at $T = 200°C$, so approximating $\hat{V}$ as constant is accurate within ~6% in this case.

$$f = (1.449 \text{ MPa}) \exp \left\{ \frac{\left(20.85 \times 10^{-6}\frac{\text{m}^3}{\text{mol}} \right)(100 \text{ MPa} - 1.555 \text{ MPa})\left(10^6 \frac{\text{Pa}}{1 \text{ MPa}} \right)}{\left(8.314 \frac{\text{Pa} \cdot \text{m}^3}{\text{mol}} \right)(473.15 \text{ K})} \right\} \quad (8.94)$$

$$f = (1.449 \text{ MPa}) (1.685) = \textbf{2.441 MPa}$$

Thus our estimate of the fugacity of liquid water at $T = 473.15$ K and $P = 100$ MPa is $f = 2.441$ MPa. By comparison, if we apply Equation 8.67 at $P = 100$ MPa and $T = 473.15$ K directly, we obtain $P = 2.692$ MPa.

Using the same method, we obtain $f = \textbf{1.515 MPa}$ at $P = 10$ MPa and $f = \textbf{1.438 MPa}$ at $P = 0.1$ MPa. Note that $P = 0.1$ MPa is below the vapor pressure. Pure liquid water does not exist at $P = 0.1$ MPa and $T = 473.15$ K; water would be in the vapor phase at these conditions. However, we nonetheless have an estimate of the fugacity that pure liquid water would have if it existed at $P = 0.1$ MPa and $T = 473.15$ K, which could be useful in modeling a liquid mixture that contains water.

Equation 8.67 is used to calculate fugacity for any T and P. Equation 8.67 assumes the fluid is accurately modeled by the Peng-Robinson EOS.

8.4 SUMMARY OF CHAPTER EIGHT

- Phase transitions, and the temperatures and pressures at which they occur, are of central importance to the design and analysis of many engineering processes.

- For a pure compound occurring in multiple phases in equilibrium, the molar Gibbs energy in each of the equilibrium phases are identical.

- The *Clapeyron equation* can be used to model any phase equilibrium (vapor–liquid, vapor–solid, or liquid–solid). It gives the slope of the P^{sat} vs. T curve for phases in equilibrium.

- Given the enthalpy of vaporization and vapor pressure at one temperature, the *Clausius-Clapeyron* equation can be used to estimate vapor pressures at other temperatures. The Clausius-Clapeyron equation is valid at ideal gas conditions over small temperature intervals.

- Given the critical point and the acentric factor, the *shortcut equation* can be used to estimate vapor pressures or boiling points in the range $0.5 < T_r < 1.0$.

- The *Antoine equation* is an empirical equation that is frequently used to model the relationship between vapor pressure and temperature.

- The molar Gibbs energy of a vapor phase approaches negative infinity as P approaches zero.

- The residual molar Gibbs energy approaches zero as P approaches zero.

- Fugacity (f) is expressed in the same units as pressure, and is defined as a function of the residual molar Gibbs energy.

- The phase equilibrium criterion can be expressed in terms of fugacity:
  - $f^V = f^L$ at liquid–vapor equilibrium conditions
  - $f^V = f^S$ at solid–vapor equilibrium conditions
  - $f^L = f^S$ at solid–liquid equilibrium conditions
  - The triple point of a compound is the one temperature and pressure at which $f^V = f^L = f^S$

- The fugacity of a pure substance in any phase at any T and P can be determined from an equation of state that accurately describes the substance in that phase.
- The fugacity of a liquid or solid at elevated pressure can be estimated using the Poynting equation.
- The fugacity of an ideal gas is equal to the pressure $f = P$.
- The fugacity of a liquid or solid at low pressure is approximately equal to the vapor pressure, $f \sim P^{sat}$.

8.5 EXERCISES

8-1. Estimate the vapor pressure of benzene at temperatures of $T = 0, 50, 100$ and $150°C$, using the following methods.
 A. The Antoine equation
 B. The Clausius-Clapeyron equation with $\Delta H^{vap} = 33.9$ kJ/mol and a vapor pressure of 24.2 mm of Hg at $T = 0°C$
 C. The shortcut equation

8-2. Estimate the vapor pressure of n-pentane at temperatures of $T = 273.15, 323.15, 373.15$, and 423.15 K, using the following methods.
 A. The Antoine equation
 B. The Clausius-Clapeyron equation, with $\Delta H^{vap} = 25.8$ kJ/mol at the normal boiling point of $T = 309.15$ K
 C. The shortcut equation

8-3. Estimate the boiling points of n-hexane at pressures of $P = 10, 50, 100$ and 500 kPa, using the following methods.
 A. The Antoine equation
 B. The Clausius-Clapeyron equation, with $\Delta H^{vap} = 28.85$ kJ/mol at the normal boiling point of $T = 341.95$ K
 C. The shortcut equation

8-4. Estimate the boiling points of toluene at pressures of $P = 0.01, 0.05, 0.1$, and 0.5 MPa:
 A. The Antoine equation
 B. The Clausius-Clapeyron equation, with $\Delta H^{vap} = 33.2$ kJ/mol at the normal boiling point of $T = 383.85$ K
 C. The shortcut equation

8-5. Use the van der Waals EOS to estimate the fugacity of propane at each of the following conditions.
 A. $T = 200$ K, $P = 0.05$ MPa (vapor)
 B. $T = 300$ K, $P = 0.1$ MPa (vapor)
 C. $T = 400$ K, $P = 0.5$ MPa (gas)
 D. $T = 500$ K, $P = 2.5$ MPa (gas)

8-6. Use the Peng-Robinson EOS to estimate the fugacity of butane at each of the following conditions (find both f^L and f^V if possible).
 A. $T = 300$ K, $P = 0.05$ MPa
 B. $T = 350$ K, $P = 0.5$ MPa
 C. $T = 400$ K, $P = 0.1$ MPa
 D. $T = 500$ K, $P = 1$ MPa

8-7. Use the Peng-Robinson EOS to estimate the fugacity of 1-propanol at each of the following conditions (find both f^L and f^V if possible).
 A. $T = 300$ K, $P = 0.1$ MPa
 B. $T = 300$ K, $P = 0.5$ MPa
 C. $T = 400$ K, $P = 0.3$ MPa
 D. $T = 500$ K, $P = 1$ MPa

8-8. A compound has a liquid molar volume of 100 cm³/ mol and a vapor pressure of 0.03 MPa at $T = 300$ K. Estimate the fugacity of this compound in the liquid phase at $T = 300$ K and each of the following pressures: 0.1, 1, 10, and 100 MPa.

8-9. A compound has a solid molar volume of 8.5×10^{-5} m³/mol and at a pressure of 250 kPa, it sublimates at $T = 275$ K. Estimate the fugacity of this compound in the solid phase at $T = 275$ K and each of the following pressures: 0.1, 1, 10, and 100 MPa.

8.6 PROBLEMS

8-10. Estimate the vapor pressure of ethanol at temperatures of $T = 0, 50, 100$ and $150°C$, using the following methods.
 A. The Antoine equation
 B. The Clausius-Clapeyron equation with $\Delta H^{vap} = 42.0$ kJ/mol at the normal boiling point of $T = 351.5$ K
 C. The shortcut equation
 D. For each temperature, identify the estimate you consider the most accurate, and explain why.

8-11. Using only information in the appendices of this book, estimate the boiling points of 1-propanol at pressures of $P = 0.01, 0.05, 0.1$, and 0.5 MPa, using the following methods.
 A. The Antoine equation
 B. The Clausius-Clapeyron equation
 C. The shortcut equation
 D. For each pressure, identify the estimate you consider the most accurate, and explain why.

8-12. Four vapor pressure data points—two representing solid–vapor equilibrium and two representing liquid–vapor equilibrium—are available for a compound:

Temperature (°C)	Pressure (kPa)	Phases
0	8	Solid–vapor
25	10	Solid–vapor
50	11.5	Liquid–vapor
75	12.7	Liquid–vapor

The molar volume of the compound is $\underline{V}^S = 8 \times 10^{-5}$ m³/mol at 25°C and 101.325 kPa, and $\underline{V}^L = 9 \times 10^{-5}$ m³/mol at 50°C and 101.325 kPa.

A. Give your best estimate of the triple point pressure and temperature for this compound.

B. Give your best estimate of pressure at which solid–liquid equilibrium occurs at $T = 50°C$.

C. Give your best estimate of the temperature at which liquid–vapor equilibrium occurs at $P = 15$ kPa.

8-13. Four vapor pressure data points—two representing solid–vapor equilibrium and two representing liquid–vapor equilibrium—are available for a compound:

Temperature (K)	Pressure (kPa)	Phases
260.93	50.66	Solid–vapor
263.71	55.73	Solid–vapor
266.48	59.78	Liquid–vapor
269.26	63.83	Liquid–vapor

The molar volume of the compound is $\underline{V}^S = 1 \times 10^{-4}$ m³/mol at 288.15 K and 101.325 kPa, and $\underline{V}^L = 1.15 \times 10^{-4}$ m³/mol at 283.15 K and 101.325 kPa.

A. Give your best estimate of the triple point pressure and temperature for this compound.

B. Give your best estimate of pressure at which solid–liquid equilibrium occurs at $T = 269.26$ K.

C. Give your best estimate of the normal boiling point.

8-14. Compare Problems 8-12 and 8-13. You presumably approached both in exactly the same way. Are you confident in the accuracy of your answers in both cases, or are some of them more questionable than others? Explain.

8-15. Ten moles of a solid is placed in a piston-cylinder device, and its pressure is maintained constant at $P = 70.93$ kPa throughout the following process.

■ The solid is heated to 320.15 K, at which temperature it melts.

■ It takes 55,000 J of heat to melt the solid completely.

■ The liquid is heated to 348.15 K, at which temperature it boils.

■ It takes 429,000 J of heat to boil the liquid completely.

■ The volume is 570 cm³ when melting begins, 630 cm³ when melting ends, and 640 cm³ when boiling begins.

Give your best estimate of each of the following properties of this compound.

A. The coefficient of thermal expansion in the liquid phase

B. The normal boiling point

C. The normal melting point

D. The triple point

8-16. Ten moles of a gas is placed in a piston-cylinder device, initially at $P = 10$ kPa, and its temperature is maintained constant temperature $T = 323.15$ K throughout the following process:

■ The gas is compressed to $P = 67$ kPa, at which pressure condensation begins.

■ At the end of the condensation, the volume is 0.82 dm³. 490 kJ of heat was removed during the condensation.

■ The liquid is compressed to $P = 1.322$ MPa and $V = 0.80$ dm³, at which point fusion begins.

■ At the end of the conversion of liquid into solid, the volume is 0.73 dm³. 23 kJ of heat was removed during the fusion.

Give your best estimate of each of the following properties of this compound.

A. The isothermal compressibility of the liquid phase

B. The normal boiling point

C. The normal melting point

D. The triple point

8-17. For each of the following compounds, make a table and graph that compare the vapor pressure vs. temperature predictions obtained from the Antoine and shortcut equations over the entire region $0.5 < T_r < 1.0$. Identify regions in which the disagreement is significant and comment on which you expect is more accurate.

A. Ethanol

B. Acetone

C. Benzene

D. n-Pentane

8-18. This problem involves the same compound that was examined in Problems 6-14 through 6-17, which in the vapor phase was described by the EOS:

$$\underline{V} = \frac{RT}{P} + aTP^2$$

with $a = 0.3 \times 10^{-16}$ m³/mol Pa² K.

This compound has vapor pressures of $P^{sat} = 50$ kPa at $T = 323$ K and $P = 500$ kPa at $T = 373$ K. Saturated liquid at both $T = 323$ K and $T = 373$ K has $\underline{V} \approx 1.25 \times 10^{-4}$ m³/mol. Estimate each of the following.

A. The fugacity in the vapor phase at $T = 323$ K and $P = 10$ kPa.

B. The fugacity in the liquid phase at $T = 323$ K and $P = 10$ MPa.

C. The fugacity in the vapor phase at $T = 373$ K and $P = 300$ kPa.

D. The fugacity in the liquid phase at $T = 373$ K and $P = 10$ MPa.

8-19. This problem concerns the gas studied in problem 6-18, which is known to follow the EOS:

$$\underline{V} = RT/P + aP^2$$

where $a = 0.01 \ 10^{-13}$ m³/Pa² mol.

A. Find a general equation for the fugacity of this compound as a function of T and P.

B. Find the fugacity of this compound at $T = 500$ K and $P = 0.5$ MPa.

8-20. Use the Peng-Robinson EOS to estimate each of the following properties of 1-propanol.

A. The vapor pressure at $T = 373$ K.

B. The fugacity of saturated vapor at $T = 373$ K.

C. The fugacity of saturated liquid at $T = 373$ K.

D. The fugacity of compressed liquid at $T = 373$ K and $P = 50$ MPa.

E. The normal boiling point.

8-21. Use the Peng-Robinson EOS to estimate each of the following properties of propylene.

A. The vapor pressure at $T = 25°C$.

B. The fugacity at the critical point.

C. The fugacity of saturated liquid at $T = 25°C$.

D. The fugacity of compressed liquid at $T = 25°C$ and $P = 50$ MPa.

E. The normal boiling point.

8-22. Complete the following for liquid water at $T = 373.15$ K.

A. Plot f^L vs. P, using the Peng-Robinson equation of state to calculate all values of f^L.

B. Plot f^L vs. P, using the Poynting method to calculate all values of f^L. Explain and justify your approach for determining f^{sat} for this calculation.

C. Re-do parts A and B for liquid water at $T = 473.15$ K. Discuss any differences in your approach for the two temperatures.

8-23. For each of the following compounds, give your best estimate of the vapor pressure at $T = 400$ K, and your best estimate of the fugacity in the liquid phase at $T = 400$ K and $P = 10$ MPa.

A. Benzene

B. n-hexane

C. Phenol

8-24. A compound has a critical point of $T = 800$ K and $P = 4.5$ MPa, and an acentric factor of $\omega = 0.3$. Estimate of the vapor pressure at each of the temperatures 300 K, 500 K, and 700 K, using the following methods.

A. The shortcut equation

B. The Peng-Robinson equation of state

8-25. You are designing a process that involves mixtures of the compounds n-butane, n-hexane, 1,3-butadiene, and benzene. You are considering whether distillation will be an effective means of separating such mixtures. As a first step, you have decided to determine the boiling points of each of the four compounds at each of the following pressures: $P = 50$ kPa, $P = 100$ kPa, $P = 300$ kPa, and $P = 500$ kPa. Thus, you are determining 16 unknown boiling points—four for each of the four compounds.

A. Give your best estimate of each of the 16 boiling points, using only information contained in the Appendices of this book.

B. Give your best estimate of each of the 16 boiling points, using the best research resources you have available to you.

C. Comment on your answers to parts A and B. How many of your answers were significantly different?

D. Compare what you now consider your best estimates of the boiling points for the four compounds at the four pressures. Does the range of boiling points get larger, smaller, or stay roughly the same as pressure increases?

E. What, if anything, do you learn from the outcomes of parts A through D that might be useful in designing a separation process for these four compounds?

8-26. You are designing a fermentation process that produces mixtures of water, acetone, ethanol, and 1-butanol. You are considering whether distillation will be an effective means of separating these compounds. As a first step, you have decided to determine the vapor pressures of each of the four compounds at each of the following temperatures: $T = 20, 50, 100,$ and $200°C$. Thus, you are determining 16 vapor pressures—four for each of the four compounds.

A. Give your best estimate of each of the 16 vapor pressures, using only information contained in Appendices of this book.

B. Give your best estimate of each of the 16 vapor pressures, using the best research resources you have available to you.

C. Comment on your answers to parts A and B. How many of your answers were significantly different?

D. Using what you now consider the best estimates of vapor pressure, find the ratios of the vapor pressures of the four compounds at each temperature. Do these ratios get smaller or larger as temperature increases?

E. What, if anything, do you learn from the outcomes of parts A through D that might be useful in designing a separation process for these four compounds?

8.7 GLOSSARY OF SYMBOLS

$\underline{A}$	molar Helmholtz energy	
a	parameter in van der Waals, Soave, or Peng-Robinson EOS	
b	parameter in van der Waals, Soave, or Peng-Robinson EOS	
C	number of chemical compounds	
C_P	constant pressure heat capacity	
C_P^*	constant pressure heat capacity for ideal gas	
C_V	constant volume heat capacity	
C_V^*	constant volume heat capacity for ideal gas	
C.O.P.	coefficient of performance	
F	degrees of freedom	
f	fugacity	
f^{sat}	the fugacity of a liquid or vapor that is at its vapor pressure	
G	Gibbs free energy	
$\underline{G}$	molar Gibbs free energy	
$\underline{G}^{ig}$	molar Gibbs free energy for an ideal gas state	
$\underline{G}^R$	residual molar Gibbs free energy	
H	enthalpy	

$\underline{H}$	molar enthalpy
$\underline{H}^{ig}$	molar enthalpy for an ideal gas state
n	number of moles
P	pressure
P_c	critical pressure
P_r	reduced pressure
P^{sat}	vapor pressure
Q	heat
R	gas constant
S	entropy
$\hat{S}$	specific entropy
$\underline{S}$	molar entropy
$\underline{S}^{ig}$	molar entropy for an ideal gas state
T	temperature
T_c	critical temperature
T_r	reduced temperature
U	internal energy
$\underline{U}$	molar internal energy
$\underline{U}^{ig}$	molar internal energy for an ideal gas state
V	volume
$\underline{V}$	molar volume

$\underline{V}_c$	molar volume at critical point
$\underline{V}_r$	reduced molar volume
W	work
$\dot{W}$	power
W_S	shaft work
Z	compressibility factor
$\Delta\underline{G}_{vap}$	change in molar Gibbs energy from vaporization
$\Delta\underline{H}_{fus}$	change in molar enthalpy from fusion
$\Delta\underline{H}_{sub}$	change in molar enthalpy from sublimation
$\Delta\underline{H}_{vap}$	change in molar enthalpy from vaporization
$\Delta\underline{S}_{vap}$	change in molar entropy from vaporization
$\Delta\underline{U}_{vap}$	change in molar internal energy from vaporization
$\Delta\underline{V}_{vap}$	change in molar volume from vaporization
$\Delta\underline{Z}_{vap}$	change in compressibility from vaporization
μ	chemical potential
π	number of phases
ϕ	fugacity coefficient

8.8 REFERENCES

Benedick, R. E., *Ozone Diplomacy*, Harvard University Press, Cambridge, MA, 1991.

Lewis, G. "The Osmotic Pressure of Concentrated Solutions, and the Laws of the Perfect Solution," *Journal of the American Chemical Society*, 1908.

Lira, J. R., Elliott C. T., *Introductory Chemical Engineering Thermodynamics*, Prentice-Hall, 1999.

Olson, R. (editor), *Biographical Encyclopedia of Scientists*, Marshall Cavendish, Tarrytown, NY, 1998.

Porter, R., Ogilvie, M. (editors), *The Biographical Dictionary of Scientists*, 3rd edition, Oxford University Press, 2000.

An Introduction to Mixtures 9

LEARNING OBJECTIVES

This chapter is intended to help you learn how to:

- Understand that your intuition, in many cases, might be best modeled by what is called the *ideal solution*
- Quantify the differences between an ideal solution and a real solution (quantified by what is known as an *excess property*)
- Understand that the entropy and Gibbs free energy of a mixture behave differently than the internal energy, enthalpy, and volume (and why)
- Calculate a property of mixing for a system
- Understand the utility of a thermodynamic construction called a *partial molar property* that describes how a component in the mixture behaves at a given temperature, pressure, and composition
- Calculate the partial molar property of a system from either an expression for the solution property or using a figure

The first eight chapters of this book have dealt exclusively with pure components and their properties, be they in a solid, liquid, vapor, or supercritical state. However, one signature area of the chemical engineering profession is chemical engineers' special training for when systems are made up of more than one component, which can be broadly termed mixtures. Indeed, the existence of chemical engineering as a profession is deeply rooted in the ability of these professionals to understand chemical reactions and the processing of what leaves a chemical reactor, which is almost always a mixture.

As we show in the motivational example for this chapter, your intuition of how the properties of mixtures behave is often counter to reality. Thus, it is easy to fool yourself into making a wrong prediction on how a system might behave. Therefore, the aim of this chapter is to show you how and why your intuition in predicting mixture properties will work in some cases and will not work in others. We will also introduce some of the framework for studying the thermodynamics of mixtures that we will build upon in later chapters. ■

9.1 MOTIVATIONAL EXAMPLE: Mixing Chemicals—Intuition

In Chapter 14, we will discuss an important outcome of chemical thermodynamics, namely the ability to determine the equilibrium composition of a chemical reaction. When you mix certain chemicals at certain conditions and if they can overcome the

A typical example of an exothermic reaction is the burning of a hydrocarbon. For example, methane gas will burn in the presence of oxygen to yield carbon dioxide and water and give off 890 kJ of heat per mole of methane burned.

activation energy barrier to reaction, they will react and either release heat or absorb heat, depending on the chemicals used and the temperature at which you run the reaction. Instant heating or cooling packs are examples where you may have experienced the difference between reactions that give off heat (heating pack) or absorb heat (cooling pack). In both of those instances, bonds break on the reactant chemicals and different bonds form on the product chemicals. The relative energies of the products to the reactants determine whether the chemical reaction is an exothermic reaction (gives off heat to its surroundings) or an endothermic reaction (absorbs heat from its surroundings).

However, what if you mix two chemicals and they *do not* react? Since there is no breaking or forming of chemical bonds, is this a rather benign process? Let us try a thought experiment.

You have 10 moles of a liquid (substance *A*) in a beaker at room temperature (293.15 K), and you have looked up the substance's molar enthalpy at this temperature: 25 J/mol.

You also have 10 moles of another liquid (substance *B*) in a different beaker at room temperature (293.15 K). This substance has a molar enthalpy of 75 J/mol at this temperature.

We use the term "equimolar" often throughout the next several chapters, which means the number of moles (or mole fraction) of the components in the mixture are the same. Thus, for a two-component (or binary) mixture, the mole fraction is 0.5 or 50%.

Next, you pour the substance in the first beaker (containing *A*) into the substance in the second beaker (containing *B*). You know there is no chemical reaction between these two substances. What is the molar enthalpy of the resulting mixture?

You might intuit that this is like a weighted average—similar to how you would calculate your grade in, say, a thermodynamics course based on your test scores. Each substance (or exam) contributes to the overall mixture molar enthalpy (or exam average) based on its pure component molar enthalpy (or individual exam score) and the percentage of the substance in the whole (or percentage contribution).

Thus, substance *A* has a mole fraction of 0.5 since x_A = 10 moles /(10 moles + 10 moles). Likewise, the same for substance *B* that also has a mole fraction of x_B = 0.5. Finally, the mixture molar enthalpy becomes

$$\underline{H} = x_A \underline{H}_A + x_B \underline{H}_B = 0.5\left(25\frac{J}{mol}\right) + 0.5\left(75\frac{J}{mol}\right) = 50 \text{ J/mol} \qquad (9.1)$$

n-hexane

n-heptane

1-hexanol

Cyclohexane

This is the intuitive answer. But is it right?

Let's explore this question more by comparing four scenarios where we mix the same number of moles of two pure substances together. Note that this mixture is traditionally called an *equimolar* mixture. But before we get started, let's define a quantity that measures the *difference* between the actual mixture molar enthalpy and our "intuitive" value as calculated by Equation 9.1. This term, which we'll give the symbol H^E, will help us quantify the difference among the four scenarios for an easier comparison.

Scenario 1: We mix 0.5 moles of *n*-hexane with 0.5 moles of *n*-hexane at 298.15 K.

Scenario 2: We mix 0.5 moles of *n*-hexane with 0.5 moles of *n*-heptane at 298.15 K.

Scenario 3: We mix 0.5 moles of *n*-hexane with 0.5 moles of 1-hexanol at 298.15 K.

Scenario 4: We mix 0.5 moles of *n*-hexane with 0.5 moles of cyclohexane at 298.15 K.

The result of this mixing is described in Table 9-1.

Scenario 1 provides a good reference for the analysis, since you are mixing the *same* substance with itself. Your intuition would tell you that the weighted average formula would have to work in this case and it does. Let's look at Scenario 2, where

TABLE 9-1 Four mixing scenarios and the values for $\underline{H}^E$.

Scenario	System	$\underline{H}^E$(J/mol)
1	*n*-hexane + *n*-hexane	0
2[a]	*n*-hexane + *n*-heptane	0.46
3[b]	*n*-hexane + 1-hexanol	460
4[c]	*n*-hexane + cyclohexane	216

Based on data from F. Kimura, G. Benson and C. Halpin, Fluid Phase Equilibria, 11, 245–250 (1983) AND C. Christensen, J. Gmehling, R. Rasmussen, U. Weidlich, Heats of Mixing Data Collection, Vol. III, DECHEMA, Frankfurt, 1984; Smith and Robinson, J Chem Eng Data, 15, 391 (1970).

we mix two compounds together that are similar, but not the same. The $\underline{H}^E$ value is very small, especially when compared with Scenarios 3 and 4.

Let's compare Scenario 2 with Scenario 3. Both have *n*-hexane mixed with another chemical of approximately the same size as *n*-hexane. However, 1-hexanol is a polar substance while *n*-heptane is not. The intermolecular interactions between the nonpolar *n*-hexane and the non-polar *n*-heptane will be different than that of the non-polar*n*-hexane with the polar 1-hexanol. The 1-hexanol will induce a dipole on the *n*-hexaneresulting in dipole-induced dipole interactions between the two molecules. Such intermolecular interactions manifest themselves in an $\underline{H}^E$ value much different from the intuitive result, as seen in Table 9-1. Comparing Scenarios 1, 2, and 3, we see that, if molecules have the same (Scenario 1) or similar (Scenario 2) intermolecular interactions, our intuition at estimating the molar enthalpy of the mixture will work well. However, if molecules do not have similar intermolecular interactions (Scenario 3), our intuition fails.

Now let's compare Scenario 2 with Scenario 4. In each case, we have *n*-hexane interacting with a non-polar substance: *n*-heptane in Scenario 2 and cyclohexane in Scenario 4. Thus, we will only have non-polar interactions between the substances. What's different then? It is the size of the molecules. While *n*-hexane and *n*-heptane are of similar size (both are about the same length and in a chain), *n*-hexane and cyclohexane are of different sizes (one a chain structure and the other a ringed structure). This size (here, shape) difference affects orientation benefits for the pure components that are disrupted when the shapes of the two components are different. So, *n*-hexane has some orientation ordering when by itself that gets disrupted when mixed with cyclohexane. This manifests itself in the relatively large $\underline{H}^E$ value seen in Table 9-1.

Going back to our thought question that opened up this chapter, is the molar enthalpy of the mixture equal to 50 J/mol? By now, your answer should be: It depends. Namely, it depends on whether substances *A* and *B* have similar sizes/shapes and similar interactions. When they do, our intuitive approach will work well and a good estimate is 50 J/mol. We call mixtures of compounds that have similar sizes and similar intermolecular interactions **ideal solutions**. Note that this is NOT to be confused with an ideal gas. For an ideal gas, the molecules have negligible volume (compared to the system volume) and do not interact at all. For an ideal solution, on the other hand, the molecules most certainly do have appreciable volume (compared to the system) and do interact. However, they interact in such a way that the mixture shows essentially the same macroscopic behavior as the individual pure components, because individual molecules in the mixture will have essentially the same intermolecular interactions that they would have if the system was a pure compound.

A personification of this idea is exemplified through the use of chemical engineering students, like many of you reading this book. Consider a group of chemical engineering students interacting among themselves at, say, a student chapter of

"Non-polar" means that the substance does not have a permanent dipole moment.

Section 7.4.4 introduced three types of intermolecular interactions: dipole-dipole, dipole-induced dipole, and London dispersion forces.

If you substitute benzene for cyclohexane in Scenario 4, the differences are more dramatic. $\underline{H}^E = 883$ J/mol for that equimolar mixture at 25°C. (Smith and Robinson, 1970).

Benzene

a professional organization meeting. Compare that with a group of chemical engineering students interacting with a group of mechanical engineering students at the same professional organization meeting. While the interactions between the two groups might not be exactly the same, the likelihood is that the conversation will turn to a math course, a design project, or even thermodynamics. However, contrast both of these situations with a group of chemical engineering students interacting with a group of art history students. The interactions (i.e., conversations) that occur between the chemical engineering students and students like themselves (other chemical engineering students or those mechanical engineering students) will likely be much different than the interactions between the chemical engineering students and the art history students. The bigger the difference, the larger the magnitude of the $\underline{H}^E$ value.

> The conversation *should* turn to thermodynamics at some point, right?

Why is it important to concern yourself with whether mixtures behave according to an ideal solution or not? What we describe in this introductory thermodynamics course is normally limited to binary mixtures. However, just consider an extremely important problem that chemical engineers face, namely determining the phase behavior of a petroleum reservoir that needs to be modeled. Often, engineers are not exactly sure of the contents of the petroleum reservoir, and those components can range from substances as small as methane to those large enough to contain 200 carbon atoms—and many compounds in between. Additionally, not only do the engineers not know exactly what the components are, they don't know how much of each component is contained in the reservoir as well. Clearly, if one wants to try and extract this mixture from the ground, it is important to estimate the properties of the mixture in order to size the equipment needed to safely and successfully perform this function. Such modeling and estimation work is the job of the chemical engineer.

FOOD FOR THOUGHT 9-1

Do chemical engineers need to know the composition of *all* the components in the petroleum reservoir?

Working with a system where you don't know the number of compounds, the structures of all of the compounds, and the amount of each compound is quite challenging—but is it a hopeless problem? No, since engineers have developed (and continue to develop) strategies to make useful estimates of mixture properties. As such, a reasonable approach is to start with the assumptions of the ideal solution (namely similar size and similar interactions) and then develop models that account for the interaction and size differences between the unlike molecules. Basically, the ideal solution part becomes the known (since we can accurately determine properties for ideal solutions—your intuitive approach) and the modeling focuses on how the system deviates from the ideal solution. Accordingly, it is important to understand the key features and aspects of the ideal solution first. This is presented in Section 9.2.

9.2 Ideal Solutions

> The subscripts on the variables indicate the component in the mixture, so $\underline{H}_1$ means the "molar enthalpy of pure component 1"

Recall in Section 9.1 that our intuition equation for the molar enthalpy of a solution was based on a weighted average of the pure component values. We list this again for a binary mixture through

$$\underline{H} = x_1\underline{H}_1 + x_2\underline{H}_2 \tag{9.2}$$

What does the molar enthalpy look like for an ideal solution as a function of composition? Substituting $x_1 = 1 - x_2$ into Equation 9.2 yields

$$\underline{H} = x_1\underline{H}_1 + (1 - x_1)\underline{H}_2 \tag{9.3}$$

$$\underline{H} = x_1(\underline{H}_1 - \underline{H}_2) + \underline{H}_2 \tag{9.4}$$

If one plots $\underline{H}$ as a function of x_1 from Equation 9.4, this is the equation of a line ($y = mx + b$) where the slope is ($\underline{H}_1 - \underline{H}_2$) and the y-intercept is $\underline{H}_2$. However, you will not often see a plot of molar enthalpy vs. concentration for a mixture, since molar enthalpy is referenced (i.e., set equal to zero) for a pure component at a particular temperature, pressure, and phase. Since where one decides to set one's reference for a particular pure component is arbitrary, it is convenient to place the reference for the pure component at the same temperature and pressure where you have calculated the solution molar enthalpy.

For example, say we were interested in calculating the molar enthalpy of a liquid mixture of ethanol and benzene at 318 K and 100 kPa. If we set $\underline{H}_1 = \underline{H}_2 = 0$ at 318 K and 100 kPa, our ideal solution molar enthalpy is (via Equation 9.4) equal to 0 for *all* values of x_1 and our plot of $\underline{H}$ vs. x_1 for the mixture is a line with zero slope, as seen in Figure 9-1 (the solid blue line).

Based on data from Williamson, A. G. and Scott, R.L., "Heats of mixing of non-electrolyte solutions. I. Ethanol + benzene and methanol + benzene", JPCA, 64, 440-442 (1960).

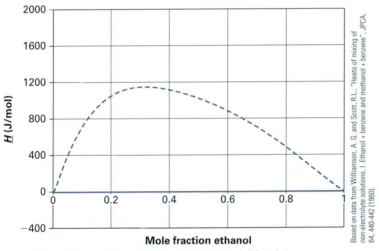

FIGURE 9-1 The molar enthalpy of the ethanol (1) + benzene (2) mixture at 318 K and 100 kPa.

Of course, a mixture of ethanol and benzene does not behave as an ideal solution and the actual molar enthalpy of this mixture is plotted as a function of composition at 318 K and 100 kPa in Figure 9-1 as well (the dashed blue curve). Recall that we defined our metric earlier ($\underline{H}^E$), which was the difference between the solution molar enthalpy of the real solution minus that of the molar enthalpy of the solution if you described it as an ideal solution. Let's give this quantity ($\underline{H}^E$) a name. It is called the **excess molar enthalpy** of the mixture. True to its name, it describes the "excess" molar enthalpy above and beyond the result you would obtain if you calculated it from the ideal solution model (your intuition). Graphically, $\underline{H}^E$ easily is seen as just the difference between the curve and the line in Figure 9-1. You'll notice a few things about $\underline{H}^E$. First, it changes as a function of the composition of the solution. Second, its value at the endpoints (that represent pure solutions of either ethanol or benzene) is zero.

Think about Scenario 1 from Section 9.1. The "perfect" ideal solution is not a solution at all, but it is one where the two components are the *same* substance. Thus, at the endpoints of Figure 9-1, $\underline{H}^E$ would be zero. Near both endpoints of Figure 9-1, the solution is well approximated by the ideal solution. Why is that? Near the endpoints, one of the components is highly concentrated, while the other is dilute. At a molecular level, even if the components are very different, the highly concentrated component does not "know" the other component "is around." Recall again the chemical

---

In $y = mx + b$, y is the dependent variable (its value *depends* on what x is), x is the independent variable, m is the slope, and b is the y-intercept.

The reference requires specification of the *phase*, since the molar enthalpy of liquid water at 373.15 K and 101.325 kPa is different than the molar enthalpy of water vapor at 373.15 K and 101.325 kPa (with the difference being the latent heat).

Ethanol

PITFALL PREVENTION

Often the words "mixture" and "solution" are used interchangeably. All solutions (where one substance is dissolved in another) are mixtures, but not all mixtures are solutions. You can have a mixture of different socks in your dresser drawer, but it isn't a solution. Also, a solution refers to a single phase, while a mixture refers to the system overall. In short, it is safe to use the word "mixture" when describing multi-component systems, but the word "solution" implies something more specific.

In this example, the art history students won't even know that other art history students are around either, but they know they are speaking with chemical engineering students. We discuss the situation from the art history student's perspective later in this chapter.

engineering student/art history student example. If there are a lot of chemical engineering students in a room with only a few art history students, most of the chemical engineering students won't even know the art history students are there and behave as though the room is, basically, all chemical engineering students.

> In general, one can define an excess property not just for the molar enthalpy but for any solution property. Thus,
>
> $$\underline{M}^E = \underline{M} - \underline{M}^{ID} \tag{9.5}$$
>
> where we have now introduced the superscript ID to signify the **ideal solution**. Note that in addition to the molar enthalpy $(\underline{H})$ this relationship can be written for $\underline{V}$, $\underline{U}$, $\underline{S}$, and $\underline{G}$.

There is a subtlety here that might work against our intuition that needs a more detailed explanation. Recall for the ideal solution molar enthalpy, our intuition allowed for the relationship:

$$\underline{H}^{ID} = x_1 \underline{H}_1 + x_2 \underline{H}_2 \tag{9.6}$$

Using Equation 9.6, $\underline{H}^E$ becomes

$$\underline{H}^E = \underline{H} - \underline{H}^{ID} = \underline{H} - (x_1 \underline{H}_1 + x_2 \underline{H}_2) \tag{9.7}$$

Likewise, one can reason that

We will work almost exclusively with binary mixtures when developing the theories and models in solution thermodynamics covered in this first course in chemical thermodynamics.

$$\underline{U}^{ID} = x_1 \underline{U}_1 + x_2 \underline{U}_2 \tag{9.8}$$

and

$$\underline{V}^{ID} = x_1 \underline{V}_1 + x_2 \underline{V}_2 \tag{9.9}$$

Applying Equation 9.5 to Equations 9.8 and 9.9 we arrive at

$$\underline{U}^E = \underline{U} - \underline{U}^{ID} = \underline{U} - (x_1 \underline{U}_1 + x_2 \underline{U}_2) \tag{9.10}$$

and

$$\underline{V}^E = \underline{V} - \underline{V}^{ID} = \underline{V} - (x_1 \underline{V}_1 + x_2 \underline{V}_2) \tag{9.11}$$

FOOD FOR THOUGHT 9-2

What do you think might be the expression for the ideal solution molar enthalpy for a *three-component* (ternary) mixture?

What about entropy? Let's revisit the same thought experiment from the beginning of this chapter.

You have 10 moles of a liquid (substance A) in a beaker at room temperature (293.15 K), and you have looked up the substance's molar entropy at this temperature: 25 J/mol-K.

You also have 10 moles of another liquid (substance B) in a different beaker at room temperature (293.15 K). This substance has a molar entropy of 75 J/mol-K at room temperature.

Next, you pour the substance in the first beaker (containing A) into the substance in the second beaker (containing B). You know there is no chemical reaction between these two substances. What is the molar entropy of the resulting mixture?

While your intuition might be to say 50 J/mol-K using the same process as before, even in an ideal solution, this would *not* be the correct answer for molar entropy. Your intuition would be partially correct, as each component will contribute its pure component molar entropy value weighted by its mole fraction. But there would be something else that is missing that requires a more detailed discussion. Recall that entropy was, in Section 4.3.3, linked to spontaneity—any spontaneous process leads to an increase in entropy. When 10 moles of *n*-hexane are poured into a beaker containing 10 moles of *n*-heptane, they mix together spontaneously. However, intuitively, you would never imagine a spontaneous process in which the mixture separated back into 10 moles of pure *n*-hexane and 10 moles of pure *n*-heptane. Even though *n*-hexane and *n*-heptane form an ideal solution, the molar entropy of this solution is higher than what you would calculate from your intuition.

Let's further illustrate this point by going down to the level of individual molecules. Consider a system that has a blue molecule that can be in either of two available locations. Both of the distributions are shown in figure (a) in the margin.

(a)

Next, you have another system that has one black molecule that can be in either of the two available locations (different than the blue system). Those two distributions are shown in figure (b) in the margin.

(b)

If you mix these two systems together, the blue and the black, how many ways are there to put one blue and one black molecule in the (now) four available locations? Your first inclination (or intuition) might be that there are four ways to put one black and one blue molecule in the four available locations, since the blue system alone had two ways and the black system alone had two ways, and now you are putting them together. However, your intuition would fail you in this case.

As you can see there are 12 ways that you can put one blue and one black molecule in four locations as given in figure (c) in the margin. Using the terminology introduced in Chapter 4, these are 12 *microstates*. The *macrostate* is described as *the state containing one black and one blue molecule* and there are 12 different microstates that correspond to this macrostate.

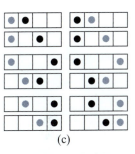
(c)

As introduced in Chapter 4, the fundamental equation in the microscopic view of entropy is

$$S = k_b \ln \Omega \tag{9.12}$$

This is Equation 4.36.

where k_b is Boltzmann's constant and Ω is the number of distributions (the number of microstates corresponding to the system's macrostate).

For our simple example, consider a calculation of the entropy change that occurred on mixing. We calculate the entropy of the first system that had the one blue particle with two available locations.

$$S_{\text{Blue}} = k_b \ln (2) \tag{9.13}$$

Likewise, the second system that had one black particle with two available locations gives an entropy value of

$$S_{\text{Black}} = k_b \ln (2) \tag{9.14}$$

If the mixture entropy was given by our intuition, we would have only four distributions.

So,

$$S_{\text{Mix,intuition}} = k_b \ln (4) \tag{9.15}$$

Thus, our entropy of mixing in this case would be

$$\Delta S = k_b \ln (4) - (k_b \ln (2) + k_b \ln (2)) \tag{9.16}$$

$$\Delta S = k_b \ln (4) - (2 \, k_b \ln (2)) = k_b \ln (4) - k_b \ln (2)^2 = 0 \tag{9.17}$$

This result would be the same as that for the change in mixing of the enthalpy, internal energy, and volume for the system modeled as an ideal solution, namely zero. However, as we already know, the number of distributions for our mixture is not equal to four but 12. Therefore, when we mixed these two simple systems together, the entropy increased, and this is true *even if there are no intermolecular interactions between the black and blue molecules at all.*

While the theory required to apply Equation 9.12 directly to real mixtures is beyond the scope of this introductory book, our thought experiment involving two molecules and four locations shows that mixing inherently leads to an increase in Ω; therefore, an increase is entropy. This simple example demonstrates that the entropy of mixing for an ideal solution is different than that for enthalpy, internal energy, or volume; it is positive, not zero.

Probability theory provides the actual expression for the entropy due to this mixing, which results in an ideal solution molar entropy that is not the weighted sum of the pure component molar entropy, but is equal to

$$\underline{S}^{ID} = x_1 \underline{S}_1 + x_2 \underline{S}_2 - Rx_1 \ln(x_1) - Rx_2 \ln(x_2) \tag{9.18}$$

The last two terms to the right of the equal sign reflect the additional, or configurational, entropy. Note that each logarithmic term is negative (since the mole fraction must be between 0 and 1, exclusive of the endpoints), which results in a positive effect for each term, since they are multiplied by a negative value. The net result is an increase in molar entropy as a result of mixing for ideal solutions.

Applying Equation 9.5, the excess molar entropy becomes

$$\underline{S}^{E} = \underline{S} - \underline{S}^{ID} = \underline{S} - [x_1 \underline{S}_1 + x_2 \underline{S}_2 - Rx_1 \ln(x_1) - Rx_2 \ln(x_2)] \tag{9.19}$$

Since $\underline{G} = \underline{H} - T\underline{S}$, we can arrive at the **ideal solution molar Gibbs free energy** as

$$\underline{G}^{ID} = \underline{H}^{ID} - T\underline{S}^{ID} \tag{9.20}$$

$$\underline{G}^{ID} = (x_1 \underline{H}_1 + x_2 \underline{H}_2) - T[x_1 \underline{S}_1 + x_2 \underline{S}_2 - Rx_1 \ln(x_1) - Rx_2 \ln(x_2)] \tag{9.21}$$

$$\underline{G}^{ID} = (x_1 \underline{G}_1 + x_2 \underline{G}_2) + (RTx_1 \ln(x_1) + RTx_2 \ln(x_2) \tag{9.22}$$

Likewise, the **excess molar Gibbs free energy** is

$$\underline{G}^{E} = \underline{G} - \underline{G}^{ID} = \underline{G} - [x_1 \underline{G}_1 + x_2 \underline{G}_2 + RTx_1 \ln(x_1) + RTx_2 \ln(x_2)] \tag{9.23}$$

9.3 Properties of Mixing

Our example from Section 9.1 presented a scenario in which two pure substances were mixed together. This example allowed the exploration of the ideal solution and excess properties. However, let us return again to that mixing example. In that example, we were interested in the change in the enthalpy of the solution as a result of the mixing process. A reasonable definition for this "change of mixing" would be to take what you finish with and subtract out what you started with. What is left over would be your **change of mixing of the molar enthalpy** due to this process. This is written as

$$\Delta \underline{H} = \underline{H} - (x_1 \underline{H}_1 + x_2 \underline{H}_2) \tag{9.24}$$

In general, we can write this molar change of mixing for any thermodynamic property as

$$\Delta \underline{M} = \underline{M} - (x_1 \underline{M}_1 + x_2 \underline{M}_2) \tag{9.25}$$

At this point, you may interject with "*Didn't we do this already? Isn't this an excess property?*" Not exactly. Here, the definition of change of mixing is that you subtract the pure component contributions (weighted by their mole fraction) from the mixture property. For an excess property, you subtract the ideal solution from the real mixture property. There is a subtlety here that requires emphasizing.

For the molar enthalpy of mixing, we have Equation 9.24, as

$$\Delta \underline{H} = \underline{H} - (x_1 \underline{H}_1 + x_2 \underline{H}_2)$$

For the excess molar enthalpy of mixing, we have Equation 9.7, which was written as

$$\underline{H}^E = \underline{H} - \underline{H}^{ID} = \underline{H} - (x_1 \underline{H}_1 + x_2 \underline{H}_2)$$

It is clear from Equations 9.7 and 9.24 that the *molar enthalpy of mixing and the excess molar enthalpy are the same* ($\Delta \underline{H} = \underline{H}^E$), because the ideal solution molar enthalpy is written as $x_1 \underline{H}_1 + x_2 \underline{H}_2$. This is also the case for the molar internal energy and the molar volume. However, this is NOT the case for molar entropy (or the molar Gibbs free energy, for that matter), since the ideal solution for both of those properties are not as simple as the average of the pure component properties, weighted by their mole fraction. For clarity, Table 9-2 contains the molar mixing and molar excess terms for enthalpy, internal energy, volume, entropy, and Gibbs free energy.

> We use the symbol M to denote any thermodynamic property (H, U, V, etc.).

TABLE 9-2 Ideal solution, molar mixing property and excess property.

	Ideal Solution ($\underline{M}^{ID}$)	Mixing Property for Ideal Solution ($\Delta \underline{M}^{ID}$)	Excess Property ($\underline{M}^E$)
Enthalpy	$x_1 \underline{H}_1 + x_2 \underline{H}_2$	0	$\underline{H} - (x_1 \underline{H}_1 + x_2 \underline{H}_2)$
Internal Energy	$x_1 \underline{U}_1 + x_2 \underline{U}_2$	0	$\underline{U} - (x_1 \underline{U}_1 + x_2 \underline{U}_2)$
Volume	$x_1 \underline{V}_1 + x_2 \underline{V}_2$	0	$\underline{V} - (x_1 \underline{V}_1 + x_2 \underline{V}_2)$
Entropy	$x_1 \underline{S}_1 + x_2 \underline{S}_2 - Rx_1 \ln(x_1) - Rx_2 \ln(x_2)$	$-Rx_1 \ln(x_1)$ $-Rx_2 \ln(x_2)$	$\underline{S} - [x_1 \underline{S}_1 + x_2 \underline{S}_2 - Rx_1 \ln(x_1) - Rx_2 \ln(x_2)]$
Gibbs free energy	$x_1 \underline{G}_1 + x_2 \underline{G}_2 + RTx_1 \ln(x_1)$ $+ RTx_2 \ln(x_2)$	$RTx_1 \ln(x_1)$ $+ RTx_2 \ln(x_2)$	$\underline{G} - [x_1 \underline{G}_1 + x_2 \underline{G}_2 + RTx_1 \ln(x_1)$ $+ RTx_2 \ln(x_2)]$

Let's explore a mixing property in Example 9-1.

ADDING VOLUMES | **EXAMPLE 9-1**

You have 1 dm³ of acetone and would like to make a 50% by mass mixture of acetone (1) and *n*-heptane (2). What volume of *n*-heptane would you have to add to accomplish this? What is the final volume of the mixture? Assume this process occurs at room temperature (25°C) and 101.325 kPa.

SOLUTION:

Step 1 *Draw a picture and label your diagram*
Here, m_1 is the mass of the acetone, m_2 is the mass of the *n*-heptane, and m_3 is the mass of the final mixture. The subscripts indicate the total amount of each stream entering or exiting the mixing process and not a flow rate.

Step 2 *A degree of freedom analysis and material balances*
From a material balance standpoint, you have three unknowns in the problem: the masses of all three streams, m_1, m_2, and m_3. There are two independent material balances you can

> When working with binary mixtures, it is common practice to identify one of the compounds as "1" and the other as "2" (in parentheses). This allows the use of subscripts to unambiguously identify which component is being referenced when using symbols for mole fraction, pure component enthalpy, etc.

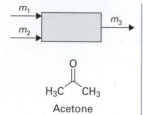

H₃C   CH₃

Acetone

write since this is a two-component system. Also, you will have an equation that uses the known specific volume and volume of stream 1 to obtain m_1. Thus, you have three equations and three unknowns in order to solve the problem.

Step 3 *Obtaining the masses for each stream*
To obtain the mass of stream 1, we need to know the specific volume of the acetone. From Appendix C, we know the molar volume is 73.94 cm³/mol. Since the molecular weight of acetone is 58.08 g/mol, the specific volume is

$$\hat{V} = \frac{73.94 \text{ cm}^3/\text{mol}}{58.08 \text{ g/mol}} = 1.273 \frac{\text{cm}^3}{\text{g}} = 1.273 \frac{\text{cm}^3}{\text{g}}$$

Thus,

$$m_1 = \frac{V}{\hat{V}} = \frac{1000 \text{ cm}^3}{1.273 \text{ cm}^3/\text{g}} = 785.546 \text{ g}$$

Using the process specification on the final mass fraction of the mixture, 50%, a material balance on acetone yields

$$(785.546 \text{ g})(1.0) + m_2(0) = m_3(0.5)$$

$$m_3 = 1571.092 \text{ g}$$

Finally, the overall material balance will yield

$$m_1 + m_2 = m_3$$

$$785.546 \text{ g} + m_2 = 1571.092 \text{ g}$$

$$m_2 = 785.546 \text{ g}$$

Step 4 *Finding the volume of n-heptane needed*
We can use the specific volume of *n*-heptane, 1.472 cm³/g (calculated from values in Appendix C, just like acetone) to obtain the volume required.

$$V = m_2\hat{V} = (785.546 \text{ g})\left(\frac{1.472 \text{ cm}^3}{\text{g}}\right) = \textbf{1156.324 cm}^3$$

Step 5 *Finding the volume of the 50% by mass mixture*
If we start with 1000 cm³ of acetone and add 1156.324 cm³ of *n*-heptane to it, we will end up with a mixture that contains 2156.324 cm³. This answers the question posed in the problem statement. *Or does it?*

Step 6 *Checking our answer*
As it turns out, someone took our 50% by mass mixture of acetone + *n*-heptane, and measured the specific volume. The reading was 1.414 cm³/g (Marino et al., 2001). Thus, we can check our answer from step 5 using this value.

$$V = m_3\hat{V} = (1571.092 \text{ g})\left(\frac{1.414 \text{ cm}^3}{\text{g}}\right) = \textbf{2221.524 cm}^3$$

What happened? Since we assume the data measurement for the specific volume was correct, *why was the answer from step 5, 2156.324 cm³, wrong?*

   Let's take an extreme case. Take a beaker and fill it with marbles up to the point on the beaker where it reads 500 cm³. Then take 10 cm³ of water and add it to the beaker. The resulting marble + water mixture still has a volume of 500 cm³. The water filled in the interstitial areas between the marbles. At a microscopic level, this happens as well. Small voids exist in systems (owing to steric effects or hydrogen bonding, for example) where other, perhaps smaller, molecules can fit and, thus, the volumes of the two systems are then not additive. Of course, as always, the overall mass is conserved.

When the system behaves as we expect (our intuition—when additive volumes hold), we have an ideal solution. The error from step 5 is where we added the volumes of the two streams together to obtain the third stream volume. This is incorrect, since our system does *not* behave as an ideal solution and there is a non-zero volume of mixing. Step 6 is correct, since we obtained the volume using the *actual* specific volume of the mixture at the concentration of interest.

In Figure 9-2, we provide molar volume of mixing data as a function of composition for a variety of mixtures. The solid blue line is the 1-propanol (1) + 1-heptene (2) system at 298.15 K (Andrzej et al., 2010). The solid black line is the 1-propanol (1) + 1-hexene (2) system at 298.15 K (Andrzej et al., 2010). The dashed blue line is the acetone (1) + methanol (2) system at 288.15 K (Murakami et al., 1964). The dashed black line is the methanol (1) + ethanol (2) system at 298.15 K (Albuquerque et al., 1996).

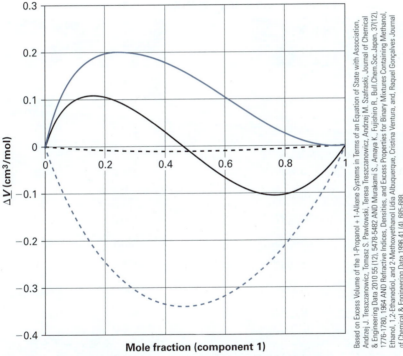

Based on Excess Volume of the 1-Propanol + 1-Alkene Systems in Terms of an Equation of State with Association, Andrzej J. Treszczanowicz, Tomasz S. Pawlowski, Teresa Treszczanowicz, Andrzej M. Szafraski, Journal of Chemical & Engineering Data 2010 55 (12), 5478-5482 AND Murakami S., Amaya K., Fujishiro R., Bull.Chem.Soc.Japan, 37(12), 1776-1780, 1964 AND Refractive Indices, Densities, and Excess Properties for Binary Mixtures Containing Methanol, Ethanol, 1,2-Ethanediol, and 2-Methoxyethanol Lidia Albuquerque, Cristina Ventura, and, Raquel Gonçalves Journal of Chemical & Engineering Data 1996 41 (4), 685-688.

FIGURE 9-2 Molar volume of mixing for different binary mixtures.

As shown in Figure 9-2, the mixture that best adheres to the "similar size, similar interactions" assumption of the ideal solution is the methanol + ethanol system. Accordingly, it has the smallest molar volume of mixing (black dashed line). On the other hand, 1-propanol and 1-heptene, a strongly polar (hydrogen bonding) substance in 1-propanol with a slightly polar hydrocarbon in 1-heptene, do not have similar size or interactions and, thus, have a large positive molar volume of mixing (solid blue line). The 1-propanol and 1-hexene system shows both positive and negative molar volumes of mixing, depending on the concentration. Hydrogen bonding opportunities between acetone and methanol are exploited when they are in a mixture together, resulting in a negative molar volume of mixing.

The terms "heat of mixing" and "enthalpy of mixing" are often used interchangeably.

In addition to the molar volume of mixing, the molar enthalpy of mixing takes an important role in the field of chemical engineering. This is because the mixing of two chemicals will likely result in the absorption or the evolution of heat. (When heat is not evolved or absorbed, we have an ideal solution!)

In this next example, we will use molar enthalpy of mixing data to make an enthalpy vs. concentration curve.

EXAMPLE 9-2	MOLAR ENTHALPY OF MIXING

In a process at your plant, you are mixing two liquids: benzene (1) and 2-propanol (2). You would like to create a molar enthalpy vs. composition diagram for this mixture and have found molar enthalpy of mixing data available in Table 9-3.

TABLE 9-3 Molar enthalpy of mixing for the benzene (1) + 2-propanol (2) system at 298.15 K and 308.15 K.

$T = 298.15$ K		$T = 308.15$ K	
$\Delta \underline{H}$ [J/mol]	x_1	$\Delta \underline{H}$ [J/mol]	x_1
0	0	0	0
232.2	0.0632	92.2	0.0213
518.8	0.1451	284.5	0.0668
789.9	0.2346	511.4	0.1237
986.2	0.3129	735.8	0.1856
1110.4	0.3734	925.2	0.2440
1186.6	0.4192	1100.5	0.3069
1234.3	0.4509	1225.4	0.3606
1281.1	0.4933	1319.5	0.4099
1319.2	0.5511	1408.0	0.4790
1328.4	0.6180	1454.9	0.5349
1287.4	0.6941	1473.4	0.6036
1201.6	0.7614	1431.3	0.6869
1071.1	0.8222	1294.3	0.7788
906.3	0.8773	1064.0	0.8615
687.4	0.9290	766.5	0.9245
418.4	0.9669	436.8	0.9664
161.1	0.9889	153.8	0.9893
0	1	0	1

Based on data from Nagata I., Asano H., Fujiwara K., Fluid Phase Equilib., 1., 211-217, 1977-1978.

SOLUTION:

Step 1 *From heat of mixing to the solution property*
Recall Equation 9.24:

$$\Delta \underline{H} = \underline{H} - (x_1 \underline{H}_1 + x_2 \underline{H}_2)$$

We can write this in terms of $\underline{H}$ (since that is what we want):

$$\underline{H} = \Delta \underline{H} + (x_1 \underline{H}_1 + x_2 \underline{H}_2)$$

Since we have data for $\Delta \underline{H}$, all we need to do is find $\underline{H}_1$ and $\underline{H}_2$ in order to create the desired figure.

Step 2 *Finding $\underline{H}_1$ and $\underline{H}_2$*
Some of the discussion surrounding Figure 9-1 concerned where to place the reference, or zero value, for the enthalpy of the pure components. In this problem, all we have molar enthalpy changes (Table 9-3) and the molar enthalpy of the pure components is not specified (or referenced). Thus, we are able to choose the reference (or zero value) wherever we want. A reasonable choice is to set $\underline{H}_1$ and $\underline{H}_2$ equal to zero at $T = 298.15$ K.

What is the value of $\underline{H}_1$ and $\underline{H}_2$ at 308.15 K? Does enthalpy change with temperature? We know it does, and we need to calculate the change in molar enthalpy going from 298.15 K to 308.15 K for each of the pure components. However, we've done this before using the heat capacity. For example, we can write this for component 1 as

$$\underline{H}_1(T) = \underline{H}_1(T_{\text{Ref}}) + \int_{T_{\text{Ref}}}^{T} C_{P,1}\, dT$$

Here we have already fixed our reference at 298.15 K and set $\underline{H}_1(T_{\text{Ref}} = 298.15\,\text{K}) = 0$ at that temperature. Thus, the first term on the right-hand side is zero.

The heat capacity, or C_P, is a function of temperature, but the temperature change is small (10 K), so we make the assumption that it is constant over this range. Therefore,

$$\underline{H}_1(T) = 0 + C_{P,1}(308.15\,\text{K} - 298.15\,\text{K})$$

Likewise, since we set $\underline{H}_2(T_{\text{Ref}} = 298.15\,\text{K}) = 0$

$$\underline{H}_2(T) = 0 + C_{P,2}(308.15\,\text{K} - 298.15\,\text{K})$$

We can obtain the liquid heat capacity for both substances in Appendix D at 298.15 K as

$$C_{P,1} = 135.69 \ \text{J/mol·K}$$

$$C_{P,2} = 155.0 \ \text{J/mol·K}$$

Thus, using our reference values, we find that $\underline{H}_1(308.15\,\text{K}) = 1356.9$ J/mol and $\underline{H}_2(308.15\ \text{K}) = 1547.5$ J/mol.

Step 3 *Plotting $\underline{H}$ at both temperatures*
We can calculate $\underline{H}$ as a function of composition at both temperatures using Equation 9.24

$$\underline{H} = \Delta \underline{H} + (x_1 \underline{H}_1 + x_2 \underline{H}_2)$$

For $T = 298.15$ K, the term in the parentheses is zero, because the pure component molar enthalpies were set equal to zero at 298.15 K. Therefore, the molar enthalpy of the solution is just the molar enthalpy of mixing. For $T = 308.15$ K, we must include the pure component molar enthalpies (which are now not equal to zero) calculated in step 2. We plot both of these in Figure 9-3.

H₃C ⟍ ⟋ CH₃
  CH
   |
  OH
2-propanol

Pressure is not given in this problem, but it is reasonable to assume 100 kPa (or not consider pressure at all, since molar enthalpy is a weak function of pressure).

PITFALL PREVENTION

We can't set the reference equal to 0 for the same substance at two different temperatures.

PITFALL PREVENTION

Once again, the pure component molar enthalpy of benzene at 308.15 K is 1356.9 J/mol *referenced to 0 at 298.15 K*. If you look up the molar enthalpy of benzene in the NIST Webbook at 308.15 K and 100 kPa, it is −6360 J/mol because it uses a difference reference. However, this is not an issue since any *changes* in molar enthalpy will be the same regardless of the reference used.

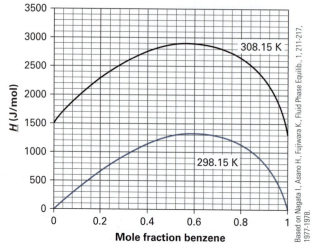

Based on Nagata I., Asano H., Fujiwara K., Fluid Phase Equilib., 1, 211-217, 1977-1978.

FIGURE 9-3 Molar enthalpy for the benzene + 2-propanol system at two temperatures (blue = 298.15 K and black = 308.15 K).

A molar enthalpy versus concentration diagram, just like the one in Figure 9-3, provides an opportunity to explore "heat of mixing" problems. We do just that in Example 9-3.

EXAMPLE 9-3

USING A MOLAR ENTHALPY DIAGRAM FOR A SOLUTION

In a mixing unit at your plant, you mix 4 moles/s of an equimolar mixture of benzene (1) + 2-propanol (2) at 298.15 K with 8 moles/s of pure benzene at 298.15 K. If you desire to keep the resulting mixture at 298.15 K, what is the heat load on this system? Are you adding or removing heat?

SOLUTION:

Step 1 *Draw a picture and label your diagram*
We will have to find the concentration of the resulting mixture, so let's find that first.

Here, $\dot{n}_A$ is the molar flow rate of the equimolar mixture (stream A), $\dot{n}_B$ is the molar flow rate of the benzene (stream B), and $\dot{n}_C$ is the molar flow rate of the final mixture (stream C). The concentrations of each stream are provided as well.

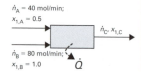

$\dot{n}_A = 40$ mol/min;
$x_{1,A} = 0.5$

$\dot{n}_C, x_{1,C}$

$\dot{n}_B = 80$ mol/min;
$x_{1,B} = 1.0$ $\dot{Q}$

Step 2 *A degree of freedom analysis*
From an unknown standpoint, we have three unknowns ($\dot{n}_C, x_{1,C}$ and $\dot{Q}$). Our three equations are comprised of the two material balances for our two-component mixture, plus the first law of thermodynamics for an open system.

Step 3 *Solving for stream C*
Adding the molar flow rates of the two streams entering the mixer (the overall material balance), we find $\dot{n}_C = 12$ mol/s. Next, a 2-propanol material balance will yield (there is no 2-propanol in stream B)

$$\dot{n}_C(1 - x_{1,C}) = \dot{n}_A(1 - x_{1,A})$$

$$\left(12 \frac{\text{mol}}{\text{s}}\right)(1 - x_{1,C}) = \left(4 \frac{\text{mol}}{\text{s}}\right)(1 - 0.5)$$

$$x_{1,C} = 0.83$$

Step 4 *First law for an open system*
Recall our first law equation for an open system:

$$\frac{d}{dt}\left\{M\left(\hat{U} + \frac{v^2}{2} + gh\right)\right\} = \sum_{j=1}^{j=J}\left\{\dot{m}_{j,in}\left(\hat{H}_j + \frac{v_j^2}{2} + gh_j\right)\right\}$$

$$-\sum_{k=1}^{k=K}\left\{\dot{m}_{k,out}\left(\hat{H}_k + \frac{v_k^2}{2} + gh_k\right)\right\} + \dot{W}_S + \dot{W}_{EC} + \dot{Q}$$

If we draw our system boundary around the mixing unit, assume steady state, no expansion/contraction or shaft work and use a molar basis, we arrive at

$$0 = \left(\dot{n}_B \underline{H}_B + \dot{n}_A \underline{H}_A\right) - \dot{n}_C \underline{H}_C + \dot{Q}$$

Here, recall that the subscripts for the molar enthalpy are not for the pure components, but for the stream. Thus, stream "C" is what is flowing "out" of the system in the first law of thermodynamics, while streams "A" and "B" are flowing "in" to the system in the first law of thermodynamics. To obtain these values, we use Figure 9-3 at 298.15 K.

$$\underline{H}_A(x_{1,A} = 0.5) = 1280 \text{ J/mol}$$

$$\underline{H}_B(x_{1,B} = 1.0) = 0 \text{ J/mol}$$

$$\underline{H}_C(x_{1,C} = 0.83) = 1050 \text{ J/mol}$$

Obviously, reading from a graph will introduce an error in your calculations, so if you can functionalize the curve (using a curve-fitting function), that can reduce some of the human error. At any rate, when plugging these values into the equation above, we arrive at

$$0 = \left(4\frac{\text{mol}}{\text{s}}\right)\left(1280\frac{\text{J}}{\text{mol}}\right) - \left(12\frac{\text{mol}}{\text{s}}\right)\left(1050\frac{\text{J}}{\text{mol}}\right) + \dot{Q}$$

$$\dot{Q} = 7480\frac{\text{J}}{\text{s}}$$

Since this is a positive value, we are adding heat to our system to keep the temperature at 298.15 K.

9.4 Mathematical Framework for Solutions

Up to this point, we have discussed some of the differences between a real solution and the ideal solution, namely how ideal solutions have similar size and similar interactions among their components. Additionally, we examined the data (through tables and figures) showing how this non-ideality manifests itself, such as in a non-zero molar volume of mixing. The reality is, however, we could just do experiments every time we ever wanted to determine a particular property of a mixture at a particular concentration, temperature, and pressure. Of course, this isn't practical. Experiments are costly, time-consuming, and sometimes dangerous.

At the end of Section 9.1, we described a scenario where one would need to model a system. There are many other situations when a reasonably accurate model can help a chemical engineer make a process decision in an efficient manner. You may have already used process simulation software in another course to evaluate a chemical process in a particular manner. The size of a piece of chemical equipment for a unit operation (say, a distillation column) is determined by the thermodynamic information input in order to achieve the desired goal. It is very helpful to have this thermodynamic information available at various states (temperatures, pressures, and compositions) in order to evaluate many alternative processes, and this is facilitated by the useful estimates from thermodynamic models.

We start our mathematical framework for solutions by playing off our intuition. For example, we know that the molar enthalpy of the ideal solution is given conveniently as the sum of the pure component properties, weighted by the mole fractions of each of the components.

$$\underline{H}^{ID} = x_1 \underline{H}_1 + x_2 \underline{H}_2 \tag{9.26}$$

We know that only in special cases (similar sized molecules with similar interactions) does Equation 9.26 hold in reality. However, the simplicity and convenience of the relationship written in that way is intuitive. What if we could define a relationship that *looks* like Equation 9.26, but instead of using the pure component properties, we use a sort of "pseudo" pure component that would make the equation true for *real* solutions? This would look like

$$\underline{M} = x_1 \overline{M}_1 + x_2 \overline{M}_2 \tag{9.27}$$

Equation 9.27 is sometimes referred to as the "summability" relationship.

Here (once again), $\overline{M}_1$ and $\overline{M}_2$ are those pseudo pure components, which are quantities that make the relationship true for all non-ideal (i.e., real) solutions. Before we discuss more about $\overline{M}_1$ and $\overline{M}_2$, let us explore their required functionality.

Recall our non-ideal mixture of ethanol and benzene from Figure 9-1. Using the data from Figure 9-1, we provide Table 9-4, which lists the molar enthalpy for the

TABLE 9-4 Molar enthalpy for the ethanol (1) + benzene (2) system at 318 K and 100 kPa.

If one was simply solving each row in Table 9-4 for $\overline{H}_1$ and $\overline{H}_2$ using Equation 9.27, there would be an infinite number of solutions (1 equation and 2 unknowns every row). The values given in Table 9-4, however, represent a unique set of solutions that satisfy Equation 9.28.

Mole Fraction Ethanol	$\underline{H}$ (J/mol)	$\overline{H}_1$ (J/mol)	$\overline{H}_2$ (J/mol)
0	0	10206.0	0
0.168	1000.0	3198.6	556.0
0.264	1125.5	1598.3	955.9
0.346	1142.2	941.2	1248.6
0.398	1112.9	698.3	1387.1
0.482	1041.8	466.8	1576.9
0.548	970.7	351.9	1720.9
0.549	958.1	339.0	1711.8
0.551	970.7	351.9	1731.3
0.571	937.2	312.5	1768.7
0.572	949.8	324.9	1784.9
0.588	891.2	264.5	1785.6
0.629	820.1	196.2	1877.7
0.758	569.0	34.4	2243.6
0.887	292.9	6.6	2540.4
1	0	0.0	2131.0

Based on data from Williamson, A. G. and Scott, R.L., "Heats of mixing of non-electrolyte solutions. I. Ethanol + benzene and methanol + benzene", JPCA, 64, 440-442 (1960).

ethanol (1) + benzene (2) system at different compositions as well as the values of $\overline{H}_1$ and $\overline{H}_2$ that make Equation 9.27 true.

There are a few things to notice about $\overline{H}_1$ and $\overline{H}_2$ just by looking at the entries in Table 9-4. First, $\overline{H}_1$ and $\overline{H}_2$ are functions of composition. So clearly, these variables ($\overline{H}_1$ and $\overline{H}_2$) are properties of the mixture. Second, $\overline{H}_1 = 0$ when $x_1 = 1$, or (in other words) when the system is pure ethanol, this mixture property vanishes. Likewise, $\overline{H}_2$ vanishes to zero when $x_2 = 1$ or (in other words) when the system is pure benzene. This makes sense, because we do not have a mixture anymore (just a pure component) when either $x_1 = 1$ or $x_2 = 1$.

You may want to ask "*Why doesn't $\overline{H}_1$ go to zero also when $x_1 = 0$? Why is it equal to 10206 J/mol?*" That value represents what is called an **infinite dilution** value. Certainly when $x_1 = 0$, $\overline{H}_1$ has no meaning, since there is no ethanol in the mixture. However, the *limit* of $\overline{H}_1$ when x_1 goes to 0 does not have to be zero and reflects the value of $\overline{H}_1$ as the system gets exceedingly dilute with respect to ethanol.

> The quantity $\overline{M}_i$ is called a **partial molar property**. Thus, $\overline{H}_i$ would be the partial molar enthalpy of component i. In general, a partial molar property is, as you might expect from the name, *a partial derivative of the solution property with respect to changes in the number of moles of that component, keeping the temperature, pressure, and molar amounts of the rest of the components constant.* We define the partial molar property "*M*" of component "*i*" as
>
> $$\overline{M}_i = \left(\frac{\partial (nM)}{\partial n_i} \right)_{T,P,n_{j(j \neq i)}} \tag{9.28}$$
>
> Here, *M* can be molar enthalpy, molar internal energy, molar volume, etc., while *n* is the number of moles in the system.

In the definition of the partial derivative, we use the notation $n_{j(j \neq i)}$. This means that we keep the number of moles of all of the substances in the mixture constant when evaluating this partial derivative—except substance *i*.

Looking at the definition from Equation 9.28 for a partial molar property, we can imagine an experiment that would give us an estimate for the partial molar property. Say you are interested in obtaining the partial molar volume of substance *A* in a binary mixture of *A* and *B*. If we keep our system at constant temperature and constant pressure and then add an exceedingly small amount of component *A* (while keeping the amount of *B* the same) to the *A* + *B* mixture, we would measure the change in the molar volume of the system as a result of adding the small amount of component *A*. This would give us an estimate of the partial molar volume of component *A* in the mixture at this temperature, pressure and composition. In essence, a partial molar property is a response function that describes how a particular mixture property changes (i.e., responds) when you make a small change in the amount of one of its components, while keeping temperature, pressure, and the amount of the other components constant.

Although we've provided both the definition of the partial molar property (Equation 9.28) as well as the relationship that used the partial molar properties to determine the mixture properties (Equation 9.27), we have not provided a reason why Equation 9.27 is true. It certainly isn't an obvious result. Since we make use of partial molar properties throughout the next few chapters, it is illustrative to present how the definition of partial molar property will lead to Equation 9.27. Additionally, using just knowledge of differentials, we can arrive at an additional constraint that is also a powerful tool in solution thermodynamics.

Consider the molar property ($\underline{M}$) of a binary mixture for a given phase. Recall the Gibbs phase rule that was defined in Chapter 2 as

$$F = C - \pi + 2 \tag{2.2}$$

Since there are two components ($C = 2$) and one phase ($\pi = 1$), we see that we can specify three intensive variables (F). It is reasonable to identify T, P, and x_1 as the three intensive variables, so we can write $\underline{M} = \underline{M}(T, P, x_1)$.

However, since we are looking to get from the definition of a partial molar property (Equation 9.28) to the solution relationship of Equation 9.27, we would like to have $n\underline{M}$, as in Equation 9.28. Thus, we multiply n by $\underline{M}$ in Equation 9.27. However, the functionality is now no longer in terms of the mole fraction, but in terms of the number of moles of the components. This allows us to expose the impact of changing the moles of one substance while keeping the moles of the other substances constant. Note that if we changed *mole fraction* of a component instead, this could be arrived at by changing the number of moles of *any* component (per the definition of mole fraction), thus not isolating the effect of a single component. This functionality is as follows.

$$n\underline{M} = f(T, P, n_1, n_2) \tag{9.29}$$

In order to explore how $n\underline{M}$ changes with the four independent variables above, we can write the total differential of this term as

$$d(n\underline{M}) = \left(\frac{\partial [n\underline{M}]}{\partial T}\right)_{P,n_1,n_2} dT + \left(\frac{\partial [n\underline{M}]}{\partial P}\right)_{T,n_1,n_2} dP + \left(\frac{\partial [n\underline{M}]}{\partial n_1}\right)_{P,T,n_2} dn_1 \tag{9.30}$$

$$+ \left(\frac{\partial [n\underline{M}]}{\partial n_2}\right)_{P,T,n_1} dn_2$$

The last two partial derivatives in the above equation are, in fact, partial molar properties (Equation 9.28), so we can make this substitution directly. Thus,

$$d(n\underline{M}) = \left(\frac{\partial [n\underline{M}]}{\partial T}\right)_{P,n_1,n_2} dT + \left(\frac{\partial [n\underline{M}]}{\partial P}\right)_{T,n_1,n_2} dP + \overline{M}_1\, dn_1 + \overline{M}_2\, dn_2 \tag{9.31}$$

Let's make another simplification at this point. In the first two terms, we are holding n_1 and n_2 constant. Since $n = n_1 + n_2$, then n must also be constant. Thus, we can pull n out of those partial derivatives, which yields

$$d(n\underline{M}) = n\left(\frac{\partial [\underline{M}]}{\partial T}\right)_{P,n_1,n_2} dT + n\left(\frac{\partial [\underline{M}]}{\partial P}\right)_{T,n_1,n_2} dP + \overline{M}_1\, dn_1 + \overline{M}_2\, dn_2 \tag{9.32}$$

Also, since it is much more convenient to work in terms of mole fractions than number of moles of each of the species, we can modify Equation 9.32 in the following ways. First, if you keep n_1 and n_2 constant, this implies that x_1 and x_2 must also be constant, since $x_1 = n_1/(n_1 + n_2)$. Therefore, we can change n_1 and n_2 to x_1 and x_2 in the partial derivatives of the first two terms. Thus,

$$d(n\underline{M}) = n\left(\frac{\partial [\underline{M}]}{\partial T}\right)_{P,x_1,x_2} dT + n\left(\frac{\partial [\underline{M}]}{\partial P}\right)_{T,x_1,x_2} dP + \overline{M}_1\, dn_1 + \overline{M}_2\, dn_2 \tag{9.33}$$

Second, we can expand the left-hand side to yield

$$d(n\underline{M}) = n(d\underline{M}) + \underline{M}(dn) \tag{9.34}$$

Third, we can rewrite dn_1 and dn_2 in terms of mole fractions as

$$dn_1 = d(nx_1) = n(dx_1) + x_1\, dn \tag{9.35}$$

$$dn_2 = d(nx_2) = n(dx_2) + x_2\, dn \tag{9.36}$$

FOOD FOR THOUGHT 9-3

$\underline{M}$ was previously listed as a function of three variables (T, P, x_1). But $n\underline{M}$ is shown as a function of four variables. Did we do something to "add" a degree of freedom?

It should be clear what the differential would look like for a 3-component mixture and generalize to an "*n*-component" mixture

Putting Equations 9.34, 9.35, and 9.36 into Equation 9.33 yields

$$n(d\underline{M}) + \underline{M}(dn) = n\left(\frac{\partial[M]}{\partial T}\right)_{P,x_1,x_2} dT + n\left(\frac{\partial[M]}{\partial P}\right)_{T,x_1,x_2} dP \tag{9.37}$$

$$+ \overline{M}_1(n(dx_1) + x_1\,dn) + \overline{M}_2(n(dx_2) + x_2\,dn)$$

The detailed steps are presented so that you can see how the relationships are arrived at in this text. We encourage readers to do the math on their own (and use these steps as a check or a guide).

Grouping terms multiplying 'dn' and those multiplying 'n'

$$(\underline{M} - \overline{M}_1x_1 - \overline{M}_2x_2)\,dn + \left[d\underline{M} - \left(\frac{\partial[M]}{\partial T}\right)_{P,x_1,x_2} dT - \left(\frac{\partial[M]}{\partial P}\right)_{T,x_1,x_2} dP \right. \tag{9.38}$$

$$\left. - \overline{M}_1\,dx_1 - \overline{M}_2\,dx_2\right]n = 0$$

Writing Equation 9.38 explicitly in terms of one of the variables (n) as well as the change in one of the variables (dn) allows us to equate *both* of the terms multiplying dn and n to zero. This is because both n and dn, as variables in Equation 9.38, are arbitrary. As such, both can take any value and still Equation 9.38 must hold. Note that there is a mathematical solution where $n = 0$ and $dn = 0$, but a system where $n = 0$ has no practical meaning.

First, by setting the term multiplying dn equal to zero in Equation 9.38, we arrive at

$$\underline{M} = x_1\overline{M}_1 + x_2\overline{M}_2 \tag{9.27}$$

This equation is exactly what we had before (Equation 9.27), and we have confirmation of the relationship we surmised earlier in this section. That is, the solution property of any solution is just the sum of the partial molar properties weighted by their mole fraction.

Second, by setting the term multiplying n equal to zero in Equation 9.38, we also arrive at

$$d\underline{M} - \left(\frac{\partial[M]}{\partial T}\right)_{P,x_1,x_2} dT - \left(\frac{\partial[M]}{\partial P}\right)_{T,x_1,x_2} dP - \overline{M}_1\,dx_1 - \overline{M}_2\,dx_2 = 0 \tag{9.39}$$

The $d\underline{M}$ in Equation 9.39 can be expressed by taking the derivative of Equation 9.27:

$$\underline{M} = x_1\overline{M}_1 + x_2\overline{M}_2 \tag{9.27}$$

$$d\underline{M} = \overline{M}_1\,dx_1 + x_1\,d\overline{M}_1 + \overline{M}_2\,dx_2 + x_2\,d\overline{M}_2 \tag{9.40}$$

Plugging Equation 9.40 into Equation 9.39 yields

$$\overline{M}_1\,dx_1 + x_1\,d\overline{M}_1 + \overline{M}_2\,dx_2 + x_2\,d\overline{M}_2 - \left(\frac{\partial[M]}{\partial T}\right)_{P,x_1,x_2} dT \tag{9.41}$$

$$- \left(\frac{\partial[M]}{\partial P}\right)_{T,x_1,x_2} dP - \overline{M}_1\,dx_1 - \overline{M}_2\,dx_2 = 0$$

Simplifying, we arrive at

$$x_1\,d\overline{M}_1 + x_2\,d\overline{M}_2 - \left(\frac{\partial[M]}{\partial T}\right)_{P,x_1,x_2} dT - \left(\frac{\partial[M]}{\partial P}\right)_{T,x_1,x_2} dP = 0 \tag{9.42}$$

Equation 9.42 is a constraint equation in that it expresses a required relationship on how a thermodynamic property will change as a function of temperature, pressure, and mole fraction. This important relationship is called the Gibbs-Duhem equation.

While the Gibbs-Duhem equation is general for a single phase, it finds utility in vapor–liquid phase equilibrium calculations, which is a common and important situation in chemical engineering. In particular, quantitative information on vapor–liquid equilibrium is required to size separation equipment, such as distillation columns. In order to have accurate information to assess various separation schemes, phase equilibrium data is required. Thus, researchers perform smaller-scale experiments to determine this phase equilibrium. The Gibbs-Duhem equation allows them, in certain situations, to evaluate the accuracy of their experimental measurements.

Previously, we identified an expression for the properties of a solution in terms of its pure components through our intuition. We labeled this as the ideal solution, and for molar volume, molar internal energy, and molar enthalpy, it confirmed our intuition in the case of molecules with similar sizes and similar interactions. For molar entropy and molar Gibbs free energy, we had to add an extra "mixing" term whose manifestation was explained in Section 9.2. For real solutions, we used a similar approach to that of the ideal solution, but replaced the pure component property with a mixture property called a partial molar property. Next, using the definition of the partial molar property, we showed how this relationship holds for real solutions. In closing the loop between ideal solutions and partial molar properties, we can now determine the partial molar properties for ideal solutions.

It is direct to look at, say, the molar enthalpy and examine the relationships between the ideal solution and the partial molar property.

$$\underline{H}^{ID} = x_1\underline{H}_1 + x_2\underline{H}_2 \tag{9.6}$$

$$\underline{H} = x_1\overline{H}_1 + x_2\overline{H}_2 \tag{9.43}$$

It is clear by looking at Equations 9.6 and 9.43 that the partial molar enthalpy for a component in a mixture, when modeled as an ideal solution, is just the pure component property. One can also obtain this from the definition of the partial molar property applied to the molar enthalpy as

$$\overline{H}_1 = \left(\frac{\partial(n\underline{H})}{\partial n_1}\right)_{T,P,n_2} \tag{9.44}$$

$$\overline{H}_1^{ID} = \left(\frac{\partial(n\underline{H}^{ID})}{\partial n_1}\right)_{T,P,n_2} = \left(\frac{\partial([n_1 + n_2](x_1\underline{H}_1 + x_2\underline{H}_2))}{\partial n_1}\right)_{T,P,n_2} \tag{9.45}$$

$$\overline{H}_1^{ID} = \left(\frac{\partial([n_1 + n_2]\left(\frac{n_1}{n_1 + n_2}\underline{H}_1 + \frac{n_2}{n_1 + n_2}\underline{H}_2\right))}{\partial n_1}\right)_{T,P,n_2} \tag{9.46}$$

$$\overline{H}_1^{ID} = \left(\frac{\partial(n_1\underline{H}_1 + n_2\underline{H}_2)}{\partial n_1}\right)_{T,P,n_2} \tag{9.47}$$

$$\overline{H}_1^{ID} = \underline{H}_1 \tag{9.48}$$

One can use the same approach to find that $\overline{V}_1^{ID} = \underline{V}_1$ and $\overline{U}_1^{ID} = \underline{U}_1$. As you might expect by now, this is not the case for the molar entropy and molar Gibbs free energy. Here, $\overline{S}_1^{ID} = \underline{S}_1 - R\ln(x_1)$ and $\overline{G}_1^{ID} = \underline{G}_1 + RT\ln(x_1)$.

Let's explore the partial molar property in a little more detail, following an approach used elsewhere (Smith, 2005). Consider again Equation 9.27.

$$\underline{M} = x_1\overline{M}_1 + x_2\overline{M}_2 \tag{9.27}$$

Taking the differential, we find

$$d\underline{M} = d[x_1\overline{M}_1 + x_2\overline{M}_2] \tag{9.49}$$

Expanding with the product rule, we find

$$d\underline{M} = x_1 d\overline{M}_1 + \overline{M}_1 dx_1 + x_2 d\overline{M}_2 + \overline{M}_2 dx_2 \tag{9.50}$$

Grouping terms differently, we have

$$d\underline{M} = x_1 d\overline{M}_1 + x_2 d\overline{M}_2 + \overline{M}_1 dx_1 + \overline{M}_2 dx_2 \tag{9.51}$$

The sum of the first two terms on the right-hand side of Equation 9.51 is just the Gibbs-Duhem equation (9.42), written at constant temperature and pressure. Therefore, those first two terms will sum to zero and we can eliminate their sum from Equation 9.51.

$$d\underline{M} = \overline{M}_1 dx_1 + \overline{M}_2 dx_2 \tag{9.52}$$

Now, if we write x_2 as $1 - x_1$, we have

$$d\underline{M} = \overline{M}_1 dx_1 + \overline{M}_2 d(1 - x_1) = \overline{M}_1 dx_1 - \overline{M}_2 dx_1 \tag{9.53}$$

If we factor and divide out by dx_1, we have

$$\frac{d\underline{M}}{dx_1} = \overline{M}_1 - \overline{M}_2 \tag{9.54}$$

Solving for $\overline{M}_1$ from Equation 9.54 and eliminating $\overline{M}_2$ by Equation 9.27, we have

$$\overline{M}_1 = \overline{M}_2 + \frac{d\underline{M}}{dx_1} = \frac{\underline{M} - x_1\overline{M}_1}{x_2} + \frac{d\underline{M}}{dx_1} \tag{9.55}$$

Multiplying through by x_2 in Equation 9.55 and getting $\overline{M}_1$ by itself on the left-hand side will yield

> For simplicity, the partial molar properties of both components can be written in terms of derivatives with respect to x_1, so that is how it is presented here.

$$x_2\overline{M}_1 + x_1\overline{M}_1 = \underline{M} + x_2\frac{d\underline{M}}{dx_1} \tag{9.56}$$

$$\overline{M}_1 = \underline{M} + x_2\frac{d\underline{M}}{dx_1} \tag{9.57}$$

In a similar way, we can start with Equation 9.54, solve for $\overline{M}_2$, and eliminate $\overline{M}_1$, to find that

> Here, "functionality" means that we have a relationship for $\underline{M}$ in terms of the composition (since we need to take the derivative with respect to composition).

$$\overline{M}_2 = \underline{M} - x_1\frac{d\underline{M}}{dx_1} \tag{9.58}$$

Equations 9.57 and 9.58 provide useful expressions to calculate the partial molar property if the functionality of the solution property ($\underline{M}$) is known. In the following examples, we practice using partial molar properties.

PARTIAL MOLAR QUANTITIES: INTRODUCTORY CALCULATIONS	EXAMPLE 9-4

Suppose that the liquid molar enthalpy for a binary mixture of benzene (1) and cyclohexane (2) has been fit to the following functional form at 298.15 K and 100 kPa (Abello, 1973). Thus,

$$\underline{H} = -3399 + 4516x_1 - 3215x_1^2$$

The symbol [=] means 'has units of'

Cyclohexane

Benzene

where $\underline{H}$ [=] J/mol.

A. What is the partial molar enthalpy of benzene in this mixture when it is equimolar at 298.15 K and 100 kPa?

B. What is the partial molar enthalpy of cyclohexane in this mixture when it is equimolar at 298.15 K and 100 kPa?

C. What is the *pure* component molar enthalpy of cyclohexane at 298.15 K and 100 kPa?

SOLUTION:

Step 1 *Identify what the problem is asking*

Right before this example problem, we derived a formula for the partial molar property if an expression for the solution property is known. For parts (A) and (B) of this question, this is precisely what is being asked.

Step 2 *Write down the needed formulas*

Using Equations 9.57 and 9.58 written for the molar enthalpy, we have

$$\overline{H}_1 = \underline{H} + x_2 \frac{d\underline{H}}{dx_1}$$

$$\overline{H}_2 = \underline{H} - x_1 \frac{d\underline{H}}{dx_1}$$

Step 3 *Obtaining the partial molar expressions*

The functionality of the solution molar enthalpy is given in terms of a quadratic in the composition. Thus, it is simple to take the derivative of the quadratic expression as follows:

$$\underline{H} = -3399 + 4516x_1 - 3215x_1^2$$

yielding

$$\frac{d\underline{H}}{dx_1} = 4516 - 6430x_1$$

Once the expression for the derivative is known, we can plug this into the partial molar enthalpy relationships listed in step 2 to obtain

$$\overline{H}_1 = 1117 - 6430x_1 + 3215x_1^2$$

$$\overline{H}_2 = -3399 + 3215x_1^2$$

Step 4 *Plugging in the numbers*

Evaluating the expressions in step 3 for $x_1 = 0.5$ provides the required answers for parts (A) and (B).

$$\overline{H}_1(x_1 = 0.5) = -1294.25 \text{ J/mol}$$

$$\overline{H}_2(x_1 = 0.5) = -2595.25 \text{ J/mol}$$

Note that even though the partial molar enthalpies are being evaluated at the same composition, their values are different.

Step 5 *Finding the pure component molar enthalpy for cyclohexane at this temperature and pressure*

Remember that a binary mixture where one of the components has a mole fraction of zero or one, is not a mixture at all, but a pure component. Thus, the mixture expression given at the beginning of this problem will allow you to calculate the pure component molar enthalpy for cyclohexane. Since cyclohexane is component 2, this implies that $x_1 = 0$

when the mixture is pure cyclohexane. We can plug this in directly into the mixture expression to give the molar enthalpy of pure cyclohexane as **−3399 J/mol** at 298.15 K and 100 kPa.

In this next example, rather than utilize a given solution relationship to calculate the partial molar properties, you will first find your own solution relationship (from the experimental data) and then calculate the partial molar properties.

PARTIAL MOLAR QUANTITIES: DEVELOPING YOUR OWN SOLUTION EXPRESSION	EXAMPLE 9-5

As an undergraduate chemical engineering student, you are involved in a summer research project that requires you to measure the molar volume of a methanol (1) + water (2) binary system at 298.15 K along an isobar. You measure the following data, which is reported in Table 9-5 as

TABLE 9-5 Molar volume for the methanol (1) + water (2) system at 298.15 K and 100 kPa.

x_1	Molar Volume (cm³/mol)
0	18.070
0.0545	19.130
0.1016	20.045
0.1532	21.054
0.2042	22.065
0.2516	23.022
0.2960	23.936
0.3438	24.939
0.3937	26.010
0.4396	27.016
0.5057	28.502
0.5416	29.326
0.5904	30.464
0.6450	31.760
0.6946	32.958
0.7391	34.048
0.7855	35.199
0.8388	36.540
0.8797	37.583
0.9284	38.843
1	40.732

A. Arce, A. Blanco, A. Soto and I. Vidal.: Densities, Refractive Indices, and Excess Molar Volumes of the Based on data from Ternary Systems Water + Methanol + 1-Octanol and Water + Ethanol + 1-Octanol and Their Binary Mixtures at 298.15 K. J.Chem.Eng.Data 38 (1993) 336-340.

A month later, after you have finished the project and have left the facility, you get a message from your research advisor asking you to *estimate* (as accurately as you can) the molar volume of the methanol + water mixture with 75% by mole methanol at the same *T* and *P* of your experiment. What molar volume will you report? What are the partial molar volumes of each component at this state?

SOLUTION:

Step 1 *Identify what the problem is asking*
This seems similar to Example 9-4, except that we are not given an expression for the solution property. However, we have experimental data and can make our own expression.

Step 2 *Plotting the data and obtaining a best fit model*
Once you plot the data, a very reasonable model would again be a quadratic function. Using any of a number of curve fitting techniques, one can arrive at the following relationship:

$$\underline{V} = 18.102 + 18.491x_1 + 4.1506x_1^2$$

where $\underline{V}$ [=] cm³/mol.
This function and the experimental data are plotted in Figure 9-4.

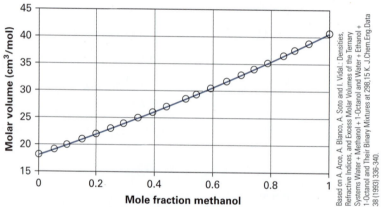

Based on A. Arce, A. Blanco, A. Soto and I. Vidal.: Densities, Refractive Indices, and Excess Molar Volumes of the Ternary Systems Water + Methanol + 1-Octanol and Water + Ethanol + 1-Octanol and Their Binary Mixtures at 298,15 K. J.Chem.Eng.Data 38 (1993) 336–340.

FIGURE 9-4 Fitting the experimental molar volume of the methanol + water mixture to a functional form at 298.15 K and 100 kPa. The experimental data are given as empty circles, while the function is the solid curve.

Step 3 *Estimating the molar volume of the 75% methanol mixture*
The model generated in step 2 provides a relationship for the molar volume of this mixture as a function of composition. We use this model to estimate the molar volume at the required composition, $x_1 = 0.75$. The result is $\underline{V} = \textbf{34.305 cm}^3\textbf{/mol}$.

Step 4 *Obtaining the partial molar expressions*
To obtain the partial molar volumes, we use the same expressions as in the previous example (Equations 9.57 and 9.58), except the molar volume is substituted for the molar enthalpy. Thus,

$$\overline{V}_1 = \underline{V} + x_2 \frac{d\underline{V}}{dx_1}$$

$$\overline{V}_2 = \underline{V} - x_1 \frac{d\underline{V}}{dx_1}$$

We take the derivative of the expression for the molar volume of the solution

$$\underline{V} = 18.102 + 18.491x_1 + 4.1506x_1^2$$

$$\frac{d\underline{V}}{dx_1} = 18.491 + 8.3012x_1$$

Once the expression for the derivative is known, we can plug this into the partial molar volume expressions to obtain the needed relationships:

$$\overline{V}_1 = 36.593 + 8.3012x_1 - 4.1506x_1^2$$

$$\overline{V}_2 = 18.102 - 4.1506x_1^2$$

Step 5 *Plugging in the numbers*
Evaluating the expressions in step 4 for $x_1 = 0.75$ provides the required answers.

$$\overline{V}_1(x_1 = 0.75) = \textbf{40.484 cm}^3\textbf{/mol}$$

$$\overline{V}_2(x_1 = 0.75) = \textbf{16.767 cm}^3\textbf{/mol}$$

Once again, note that even though the partial molar volumes are being evaluated at the same composition, their values are different.

Partial molar properties are sometimes best viewed graphically. For example, let's revisit Figure 9-1, which provided the molar enthalpy as a function of composition for the ethanol (1) + benzene (2) system.

The slope of the tangent line to the curve at any point represents the derivative at that point. Thus, if we wanted to know the derivative of any point on Figure 9-1, which is $d\underline{H}/dx_1$, we would find the slope of the tangent line at the point. If we extend the tangent line so that it intersects the vertical axis (where $x_1 = 0$), we will now have two points to evaluate this slope. We will call the point where the tangent line intersects the vertical axis B, so its coordinates are $(B, 0)$. Thus,

$$\frac{d\underline{H}}{dx_1} = \frac{\underline{H}(x_1) - \underline{H}(x_1 = 0)}{x_1 - 0} = \frac{\underline{H}(x_1) - B}{x_1} \tag{9.59}$$

We can rearrange the expression to give

$$B = \underline{H}(x_1) - x_1\frac{d\underline{H}}{dx_1} \tag{9.60}$$

Comparing this to Equation 9.58 written for the molar enthalpy

$$\overline{H}_2 = \underline{H} - x_1\frac{d\underline{H}}{dx_1} \tag{9.61}$$

allows us to see that the $\underline{H}$-axis intercept is, in fact, the partial molar enthalpy of component 2 at the composition of interest. In a similar way, one can extend the tangent line through the other vertical axis (where $x_1 = 1$) and find that this intercept is the partial molar enthalpy of component 1 at that composition $(\overline{H}_1)$. This is visualized in Figure 9-5.

By inspection of Figure 9-5, it can be seen that the partial molar enthalpy of component 1, evaluated at $x_1 = 0.6$, is about 250 J/mol, while the partial molar enthalpy of component 2 at this same concentration is about 1850 J/mol. Also, at

These partial molar enthalpy estimates compare favorably to the actual values from Table 9-4.

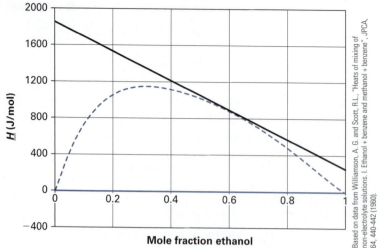

Based on data from Williamson, A. G. and Scott, R.L., "Heats of mixing of non-electrolyte solutions. I. Ethanol + benzene and methanol + benzene". JPCA, 64, 440-442 (1960).

FIGURE 9-5 The molar enthalpy of the ethanol (1) + benzene (2) mixture at 318 K and 100 kPa. The tangent line is drawn at $x_1 = 0.6$.

around $x_1 = 0.3$, the slope of the tangent line is flat, which means that this line will intersect both the $\underline{H}$-axis at $x_1 = 0$ and the $\underline{H}$-axis at $x_1 = 1$ at about the same value, just below 1200 J/mol.

9.5 Ideal Gas Mixtures

While the term "solution" (as in ideal solution) is often associated with the liquid phase, it does not preclude the solid or vapor phases. Indeed, an ideal solution can be used to describe solids and vapors as well as liquids. A discussion of the solid phase, as well as ideal solutions in the solid phase, is presented later in this text.

What about an ideal solution of *vapors or gases*? Nothing prevents the use of the ideal solution approach to describe a mixture of vapors or gases. Indeed, in a limited window (where the ideal solution is valid—molecules have similar size and similar interactions—and at elevated pressures), there is some utility to the ideal solution model for vapors or gases. However, it is not the best reference system to use for vapors and gases—the ideal gas is a better reference and there are approaches (such as what we've already learned for pure components) that take into account deviations from the ideal gas model.

> As we show in Chapter 13, systems that behave according to the ideal solution model are quite rare in the solid phase.

Consider Figure 9-6, which qualitatively shows particles in a box representative of a liquid and vapor phase. Here, the particles are of different types so there is a mixture in each box.

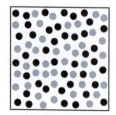
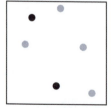

Liquid mixture   Vapor mixture

FIGURE 9-6 Cartoon representations of liquid and vapor mixtures.

The particles are close to each other in the liquid, while they are much further apart in the vapor phase. Since the particles are very far apart in the vapor, whether the particles have similar interactions (such as *n*-hexane and *n*-heptane) or much different interactions (such as *n*-hexane and 1-hexanol) or are of different sizes (such as *n*-hexane and cyclohexane) becomes, somewhat, immaterial. The particles are, in general, so far apart in the vapor phase that their sizes and intermolecular interactions do not normally play a significant role. So the question becomes: do we want to use an ideal solution to describe our gas or vapor mixture, or use a different reference system that, while more restrictive, is normally readily satisfied by gases and vapors? The choice is the latter and, as expected, the modeling of gas mixtures regularly starts with the ideal gas (a concept we introduced in Chapter 2 for pure components).

The ideal gas for mixtures is basically a more restrictive version of the ideal solution, since instead of all of the molecules in the system behaving as though they have similar size and similar interactions, the ideal gas particles have no size nor any interactions. The formalism of the ideal gas mixture begins with the definition of the partial pressure of a component in a mixture. As you likely recall from a general chemistry course, the partial pressure of component *i* in an ideal gas mixture is

> Here we use y_i to describe the mole fraction of component *i* in the vapor phase, which is the popular convention.

$$P_i = y_i P^{ig} = \frac{y_i RT}{\underline{V}} \qquad (9.62)$$

This approach allows us to model the contribution of the components of an ideal gas mixture to the properties of the mixture through their contribution to the partial pressure (since the sum of the partial pressures of all of the components in a mixture will be the pressure of the mixture). Therefore, we can define the partial molar property of a substance in an ideal gas state as

$$\overline{M}_i^{ig}(T, P, y) = \underline{M}_i^{ig}(T, P_i) \qquad (9.63)$$

> You might recall Dalton's Law of partial pressures from a chemistry course, which states that you can add up the partial pressures of the components of an ideal gas mixture to equal the system pressure.

How does this look for our thermodynamic variables of $\underline{H}$, $\underline{U}$, $\underline{V}$, $\underline{S}$, and $\underline{G}$? For the molar enthalpy and molar internal energy, the relationship is straightforward, since both of those quantities are independent of pressure when modeled as an ideal gas. Thus, it doesn't matter what pressure (P_i, P, or any other pressure) you use to evaluate the molar enthalpy and molar internal energy when modeled as an ideal gas.

$$\overline{H}_i^{ig}(T, P) = \underline{H}_i^{ig}(T, P_i) = \underline{H}_i^{ig}(T, P) \qquad (9.64)$$

$$\overline{U}_i^{ig}(T, P) = \underline{U}_i^{ig}(T, P_i) = \underline{U}_i^{ig}(T, P) \qquad (9.65)$$

For the molar entropy and molar Gibbs free energy, it *does* matter what pressure you use since both the molar entropy and molar Gibbs free energy are functions of pressure for an ideal gas. For example,

$$\overline{S}_i^{ig}(T, P) = \underline{S}_i^{ig}(T, P_i) \neq \underline{S}_i^{ig}(T, P) \qquad (9.66)$$

We can express the impact of pressure on the molar entropy by using Equation 9.67, which provides the needed differential, written for the ideal gas as

$$d\underline{S}_i^{ig} = \int \frac{C_{P,i}^*}{T} dT - R \int \frac{dP}{P} \qquad (9.67)$$

$$\int_{\underline{S}_i(T, P_i)}^{\underline{S}_i(T, P)} d\underline{S}_i^{IG} = \int_T^T \frac{C_{P,i}^*}{T} dT - R \int_{P_i}^P \frac{dP}{P} \tag{9.68}$$

$$\underline{S}_i^{ig}(T, P) - \underline{S}_i^{ig}(T, P_i) = -R \ln\left(\frac{P}{P_i}\right) = -R \ln\left(\frac{P}{y_i P}\right) = R \ln(y_i) \tag{9.69}$$

Rewriting this, we have

$$\underline{S}_i^{ig}(T, P_i) = \underline{S}_i^{ig}(T, P) - R \ln(y_i) \tag{9.70}$$

Thus, we can update Equation 9.66 to read as

$$\overline{S}_i^{ig}(T, P) = \underline{S}_i^{ig}(T, P_i) = \underline{S}_i^{ig}(T, P) - R \ln(y_i) \tag{9.71}$$

You can follow the same strategy to obtain the needed relationship for the partial molar Gibbs free energy in the ideal gas state:

$$\overline{G}_i^{ig}(T, P) = \underline{G}_i^{ig}(T, P_i) = \underline{G}_i^{ig}(T, P) + RT \ln(y_i) \tag{9.72}$$

What we have done up to this point is write the partial molar enthalpy, internal energy, entropy, and Gibbs free energy of component i in an ideal gas mixture in terms of their pure component, ideal gas properties (plus the effect of pressure, where needed, in terms of the composition). However, we have not yet presented the partial molar volume of an ideal gas. We explore this separately because Equation 9.63 does NOT hold for the partial molar volume of an ideal gas based on how the partial pressure is defined. Specifically, consider Equation 9.28 as now written for the molar volume of an ideal gas mixture:

$$\overline{V}_i^{ig} = \left(\frac{\partial(n\underline{V}^{ig})}{\partial n_i}\right)_{T,P,n_{j\neq i}} \tag{9.73}$$

We know what the molar volume is for an ideal gas, namely

$$\underline{V}^{ig} = \frac{RT}{P} \tag{9.74}$$

Thus, Equation 9.73 becomes

$$\overline{V}_i^{ig} = \left(\frac{\partial(nRT/P)}{\partial n_i}\right)_{T,P,n_{j\neq i}} = \frac{RT}{P}\left(\frac{\partial(n)}{\partial n_i}\right)_{T,P,n_{j\neq i}} = \frac{RT}{P} \tag{9.75}$$

Therefore, by inspection of Equations 9.74 and 9.75, we have

$$\overline{V}_i^{ig}(T, P) = \underline{V}^{ig}(T, P) \tag{9.76}$$

This result says that the partial molar volume of component i, when modeled as an ideal gas mixture, is equal to the molar volume of the ideal gas mixture at the same temperature and pressure. Likewise, we know that the molar volume of *any* ideal gas (mixture or pure component) is only a function of the temperature and pressure. Thus,

$$\underline{V}^{ig}(T, P) = \underline{V}_i^{ig}(T, P) \tag{9.77}$$

Therefore, if we use Equation 9.63 written for the molar volume, we would have

$$\overline{V}_i^{ig}(T, P) = \underline{V}_i^{ig}(T, P_i) \tag{9.78}$$

Remember, Equation 9.78 is NOT true.

Since the partial molar volume and the pure component molar volume must be the same (from Equations 9.74 and 9.75) when modeled as an ideal gas, and they are both functions of temperature and pressure, they CANNOT be equivalent (as implied by Equation 9.78) when they are at different pressures (here, P and P_i). Accordingly, Equation 9.78 is NOT true.

The development of the partial molar quantities for ideal gas mixtures allows us to use Equation 9.27, but now written for the ideal gas mixture as

$$\underline{M}^{ig} = y_1 \overline{M}_1^{ig} + y_2 \overline{M}_2^{ig} \tag{9.79}$$

> This is the summability relationship

Therefore, molar enthalpy and molar internal energy are easily written at a given temperature (the pressure has no impact here). Thus,

$$\underline{H}^{ig} = y_1 \overline{H}_1^{ig} + y_2 \overline{H}_2^{ig} = y_1 \underline{H}_1^{ig} + y_2 \underline{H}_2^{ig} \tag{9.80}$$

$$\underline{U}^{ig} = y_1 \overline{U}_1^{ig} + y_2 \overline{U}_2^{ig} = y_1 \underline{U}_1^{ig} + y_2 \underline{U}_2^{ig} \tag{9.81}$$

The molar entropy and molar Gibbs free energy are a little different, as they have the effect of pressure included. Note that for clarity, the functionality of the molar entropy and molar Gibbs free energy is given explicitly in the square brackets. Thus,

$$\underline{S}^{ig}[T, P, y] = y_1 \overline{S}_1^{ig}[T, P, y] + y_2 \overline{S}_2^{ig}[T, P, y] \tag{9.82}$$
$$= y_1 (\underline{S}_1^{ig}[T, P] - R \ln y_1) + y_2 (\underline{S}_2^{ig}[T, P] - R \ln y_2)$$

$$\underline{G}^{ig}[T, P, y] = y_1 \overline{G}_1^{ig}[T, P, y] + y_2 \overline{G}_2^{ig}[T, P, y] \tag{9.83}$$
$$= y_1 (\underline{G}_1^{ig}[T, P] + RT \ln y_1) + y_2 (\underline{G}_2^{ig}[T, P] + RT \ln y_2)$$

Finally, we can write the molar volume of the ideal gas in terms of the molar volumes of the pure components (which are exactly the same as the mixture molar volumes). So

$$\underline{V}^{ig}[T, P] = y_1 \overline{V}_1^{ig}[T, P] + y_2 \overline{V}_2^{ig}[T, P] = y_1 \underline{V}_1^{ig}[T, P] + y_2 \underline{V}_2^{ig}[T, P] \tag{9.84}$$

If Equation 9.84 is multiplied through by the number of moles in the system, the terms on the right-hand side are now extensive volumes. This expression is another way of stating Amagat's law of additive volumes for ideal gases.

We can define the change of mixing properties in the ideal gas state, borrowing from Equation 9.25, to be

$$\Delta \underline{M}^{ig} = \underline{M}^{ig} - (y_1 \underline{M}_1^{ig} + y_2 \underline{M}_2^{ig}) \tag{9.85}$$

Using Equations 9.80, 9.81, and 9.84, we have

$$\Delta \underline{H}^{ig} = \Delta \underline{U}^{ig} = \Delta \underline{V}^{ig} = 0 \tag{9.86}$$

We can use Equations 9.82 and 9.83 to arrive at

$$\Delta \underline{S}^{ig} = -Ry_1 \ln y_1 - Ry_2 \ln y_2 \tag{9.87}$$
$$\Delta \underline{G}^{ig} = RT y_1 \ln y_1 + RT y_2 \ln y_2 \tag{9.88}$$

Now that we have a definition for the various thermodynamic quantities as calculated from an ideal gas mixture standpoint, the concept of a residual property can be introduced in a manner similar to that from Chapter 6.

In particular, a **residual property** for the mixture is defined as

$$\underline{M}^R = \underline{M} - \underline{M}^{ig} \tag{9.89}$$

This quantity is valid for the molar enthalpy, molar internal energy, molar volume, molar entropy, and molar Gibbs free energy.

EXAMPLE 9-6	IDEAL GAS ENTROPY OF MIXING

What is the entropy change when you mix one mole of butane gas with one mole of propane gas, modeled as an ideal gas mixture? Assume the pressure and temperature of the system are constant during the process.

SOLUTION:

Step 1 *Identify what the problem is asking*
We are looking for the ideal gas entropy change of mixing, which is precisely what is in Equation 9.87.

Step 2 *Utilize Equation 9.87*
Since we have one mole of each substance, we have an equimolar mixture, so $y_1 = y_2 = 0.5$.

$$\Delta \underline{S}^{ig} = -Ry_1 \ln y_1 - Ry_2 \ln y_2 \qquad (9.87)$$

$$\Delta \underline{S}^{ig} = -8.314 \frac{J}{mol \cdot K}(0.5)\ln(0.5) - 8.314 \frac{J}{mol \cdot K}(0.5)\ln(0.5)$$

$$\Delta \underline{S}^{ig} = 5.76 \frac{J}{mol \cdot K}$$

Step 3 *Convert to an extensive value, per problem statement*
The problem provided extensive values (moles) and asked for the entropy change, not the molar entropy change. Thus, we need to multiply our answer by the number of moles in the system.

$$\Delta S^{ig} = (2 \, mol)\left(5.76 \frac{J}{mol \cdot K}\right) = 11.52 \, J/K$$

Chapter 9 has provided an introduction to mixtures, contrasting your intuitive approach to present when and why your intuition is valid for mixtures, and when it fails. We have started to provide the framework for working with the thermodynamics of mixtures, namely the partial molar properties. Since much of solution thermodynamics deviates from our intuition, this data is often gathered through experimentation, presented in both graphical and tabular form. In the next chapter we explore some of the rich phase behavior observed when vapors are in equilibrium with liquids. We also continue with our intuitive modeling approach, owing to its simplicity and utility for some systems.

9.6 SUMMARY OF CHAPTER NINE

- Our intuition in calculating mixture properties from pure components will work for enthalpy, internal energy, and volume—provided the components behave as an **ideal solution**, meaning they are of similar type (similar interactions and similar size/shape).

- We can quantify deviations from an ideal solution model by an **excess** property.

- Even when a system behaves as an ideal solution, it will have non-zero **mixing properties** for entropy and Gibbs free energy.

- **Partial molar properties** measure how a system's properties will change with a small change in the amount of one of the components, keeping the temperature, pressure, and amount of the rest of the components in the mixture constant.

- A **mixture property** can be calculated as the sum of the partial molar properties of its components, weighted by the mole fraction of each component.

9.7 EXERCISES

9-1. What is the entropy change when you mix 1 mole of 1-propanol with 4 moles of ethanol at 298.15 K and 101.325 kPa? Assume an ideal solution.

9-2. A two-component, two-phase mixture has how many degrees of freedom, according to Gibbs phase rule?

9-3. Using the data for the benzene (1) + 2-propanol (2) system in Table 9-3, fit the experimental data to an *appropriate* polynomial equation whose independent variable is the mole fraction of 2-propanol. Work with only the 298.15 K data set.

9-4. Consider two atoms: one red and the other blue. There are five locations to place those two atoms. How many microstates exist for this system?

9-5. The NIST WebBook provides very good estimates of thermodynamic properties for dozens of compounds at a variety of states (http://webbook.nist.gov/chemistry/fluid/). Using the NIST WebBook, find the liquid molar enthalpy, entropy, internal energy, and Gibbs free energy for the following substances, all at their normal boiling point:
A. water
B. propane
C. R-125

9-6. The density of an equal-mass water + ethanol mixture is 913 kg/m³ at 293.15 K. If the density of water is 998 kg/m³ and ethanol is 789 kg/m³ with both at 293.15 K, does this equal-mass mixture possess a positive or negative excess volume at 293.15 K?

9-7. Using the NIST Webbook, estimate the following properties of an equimolar *n*-pentane + *n*-hexane mixture at 298.15 K and 101.325 kPa, assuming an ideal solution: molar enthalpy, molar internal energy, molar Gibbs free energy and molar entropy.

9-8. You mix two pure components together (A and B) at the same temperature and pressure, and the mixing process is exothermic.
A. If you desire to do the mixing in an adiabatic manner, will the temperature of the system increase, decrease or stay the same?

B. How would your answer change if you treated the system as an ideal solution?

9-9. You are in a mentoring program where you help your former secondary school by answering questions from students taking Advanced Chemistry. A student emails you the following question: "*I don't understand the different between an ideal gas and an ideal solution. I also don't understand excess or residual properties. Aren't they the same? Please help!*" How would you respond to this student? Remember your audience.

9-10. Consider the following term:

$$\overline{A}_i = \left(\frac{\partial (nA)}{\partial (n_i)} \right)_{V,T,n_j \neq i}$$

Would you call $\overline{A}_i$ the *partial molar Helmholtz free energy*? Why or why not?

9-11. Experimental data is available that shows the excess molar volume of acetone with three different normal alkanes at 298.15 K: *n*-hexane, *n*-heptane, and *n*-octane. Each of the three mixtures shows positive excess molar volumes across the entire composition range. Consider their values in Table E9-11 for an equimolar mixture.

TABLE E9-11 Excess molar volumes for three *n*-alkane + acetone equimolar systems at 298.15 K.

System	$\underline{V}^E$ (cm³/mol)
n-hexane + acetone	1.08
n-heptane + acetone	1.14
n-octane + acetone	1.18

Based on data from Marino, G. et al., "Temperature Dependence of Binary Mixing Properties for Acetone, Methanol, and Linear Aliphatic Alkanes (C6–C8)," J. Chem. Eng. Data, 2001, 46, 728, (2010).

From an interactions perspective, describe the trend in the data in Table E9-11.

9.8 PROBLEMS

9-12. Equation 9.88 provides the expression for the molar Gibbs free energy of mixing for an ideal gas mixture. Starting with the relationship between the molar Gibbs free energy and the molar entropy, derive Equation 9.88.

9-13. You are a chemical engineer working at Regional Chemical Company. The upper management has done some preliminary cost estimates and is looking to manufacture mixtures of chloroform and methanol. The end use of this mixture would be for

lipid extraction in the biological field. Your boss, thinking about the hydroxyl group on the methanol and the three chlorines on the chloroform, opines that the mixture is probably very non-ideal. Since you are standing next to her, she turns to you and asks, "*Can you plot for me the partial molar enthalpies of this mixture as a function of composition at 323.15 K?*" Enthusiastically, you say "*yes*" and wander off back to your office.

After a half hour of searching, you come up with enthalpy of mixing data for this mixture at 323.15 K, presented in Table P9-13.

TABLE P9-13 Enthalpy of mixing data for the methanol (1) + chloroform (2) system at 323.15 K.

$\Delta \underline{H}_{mix}$ (J/mol)	x_1
0	0
−96.8	0.0256
−194.4	0.0559
−283.4	0.0930
−297.4	0.0974
−364.6	0.1374
−369.5	0.1447
−400.3	0.1755
−407.4	0.2014
−409.9	0.2175
−394.8	0.2568
−369.1	0.2860
−364.6	0.2921
−328.9	0.3184
−316.2	0.3272
−287.2	0.3456
−284.3	0.3468
−255.7	0.3622
−237.5	0.3740
−208.5	0.3870
−189.4	0.3963
−90.1	0.4431
−87.9	0.4433
31.2	0.4953
51.5	0.5053
194.3	0.5655
241.0	0.5908
390.3	0.6563
524.5	0.7331
587.2	0.7855
573.3	0.7909
590.9	0.8563
523.9	0.9132
489.1	0.9224
334.0	0.9578
186.3	0.9797
33.9	0.9967
0	1

Based on data from van Ness H.C.: I. Acetone-Chloroform-Methanol at 50 °C. J.Chem.Eng.Data 20 (1975) 403-405

Please provide the graphs that the boss has asked you for. Put both curves on the same graph.

9-14. The molar enthalpy of a mixture of hydrogen fluoride (1) and water (2) at 293.15 K is given as (Tyner, 1949)

$$\underline{H} = -1850 - 28240\,x_1 + 26800\,x_1^2$$

where $\underline{H}\,[=]\,$J/mol

Calculate the partial molar enthalpy of water in this mixture at the following compositions.
A. $x_1 = 0.4$
B. $x_1 = 0$

9-15. Figure 9-3 describes the molar enthalpy for the benzene (1) + 2-propanol (2) system at two temperatures: 298.15 K and 308.15 K. Answer the following questions related to this figure.
A. A flow process mixes 10 moles per second of benzene with 10 moles per second of 2-propanol at 308.15 K. Would this system give off heat, absorb heat, or stay at the same temperature? Please explain.
B. If you assume the benzene + 2-propanol mixture behaved as an ideal solution, how would your answer from part (A) change?

9-16. Using the NIST Webbook, if one looks up the molar enthalpy of pure benzene at 308.15 K and 100 kPa, the reported value is −6359.6 J/mol. Figure 9-3 has this same property as being equal to 1356.9 J/mol (via Example 9-2). Whose value, if any, is in error? Please explain your answer.

9-17. In a flow process, you mix 0.2 kg/s of benzene at 308.15 K with 0.3 kg/s of 2-propanol at 298.15 K. Using Figure 9-3, answer the following;
A. What is the concentration of the resulting mixture?
B. Estimate the heat effect (added/removed) of this mixing if it is done such that the resulting temperature is 298.15 K.
C. If the mixing were done in an adiabatic fashion, estimate the resulting solution temperature.

9-18. The molar volume for a mixture of methanol (1) + water (2) at 298.15 K and 100 kPa is given as

$$\underline{V} = 18.102 + 18.491\,x_1 + 4.1506\,x_1^2$$

where $\underline{V}$ is in units of cm³/mol.
A. What is the partial molar volume of methanol at this T and P when $x_1 = 0.4$?
B. What is the pure component molar volume of methanol at this T and P?
C. What is the pure component molar volume of water at this T and P?
D. Let's assume you have an equimolar mixture of water and methanol. What will be the molar volume of this mixture according to the functional form given in the problem? What will be the

Ortho-xylene

Para-xylene

molar volume of this mixture if you assumed this mixture behaved as an ideal solution?

9-19. Consider the following three liquid mixtures.
A. Water + n-propane
B. n-hexane + benzene
C. Ortho-xylene + para-xylene
Which mixtures, if any, do you expect to behave as an ideal solution and why? For those that you do not expect to behave as an ideal liquid solution, explain why.

9-20. Using the information from Table 9-3, plot the partial molar enthalpies of both components at both temperatures on the same curve as a function of composition. You may have already completed some of the needed work in Exercise 9-3.

9-21. Using tabulated experimental data from the literature for either the excess molar volume or excess molar enthalpy of a system of your choice, provide the following information.
A. A plot of the excess property (either molar volume or molar enthalpy) as a function of composition.
B. A functional description of the excess property using the composition as an independent variable.
C. A plot of the partial molar property (either volume or enthalpy) as a function of composition.
D. Does the system behave according to the ideal solution model? If not, please explain why.

For the next three problems, use Figure P9-22 for the sulfuric acid + water system (Smith et al., 2005).

9-22. At 26.7°C, we mix 10 kg of sulfuric acid with 20 kg of water. What is the resulting heat of mixing for this process? Is the heat liberated or absorbed? Use Figure P9-22.

9-23. A mass of 500 kg of 40 wt% sulfuric acid solution at 60°C is diluted with 200 kg of pure water at 37.8°C. What is the concentration of the resulting solution? What is the heat effect (liberated or absorbed) of this mixing (report an extensive number) if the mixing is done such that the resulting solution is at 37.8°C? If the mixing was done adiabatically, what would be the resulting solution temperature? Use Figure P9-22.

9-24. Estimate the partial molar enthalpy of sulfuric acid at 60°C at the following two compositions using Figure P9-22.
A. 30% by wt sulfuric acid
B. 80% by wt sulfuric acid

9-25. You have 0.1 kg of water at 298.15 K in a container that holds exactly 0.2 dm³. What mass of methanol do you need to add to the system such that the container is filled without overflowing? See the solution of Example 9.5 for additional details.

9-26. Consider the density of the mixture 1,2-dicholoroethane (1) + chlorobenzene (2) at 298.15 K. Use Table P9-26 to answer the following questions.

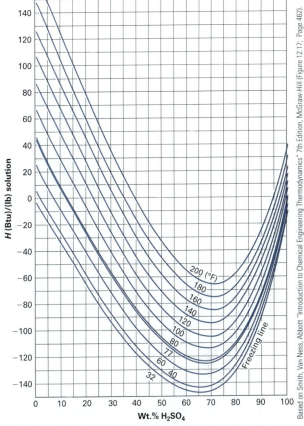

FIGURE P9-22 Solution specific enthalpy of the sulfuric acid + water system at various temperatures.

Based on Smith, Van Ness, Abbott. "Introduction to Chemical Engineering Thermodynamics." 7th Edition, McGraw-Hill (Figure 12.17, Page 462).

TABLE P9-26 Mixture density of the 1,2-dichloroethane (1) + chlorobenzene (2) system at 298.15 K.

x_1 (mole fraction)	ρ (kg/m³)
0	1100.40
0.1022	1111.56
0.2015	1123.05
0.3013	1135.29
0.4039	1148.64
0.4999	1161.91
0.5977	1176.36
0.6986	1192.24
0.8000	1209.30
0.8990	1227.12
1.0	1246.70

Based on data from Thermodynamic Properties of the Binary Mixtures of 1,2-Dichloroethane with Chlorobenzene and Bromobenzene from (298.15 to 313.15) K. Bangqing Ni, Liyan Su, Haijun Wang *, and Haigang Qiu J. Chem. Eng. Data, 2010, 55 (10), pp 4541-4545

A. Does this mixture behave as an ideal solution?
B. If you mixed 0.1 dm³ of both components together at 298.15 K, will you have more or less

than 0.2 dm³? What percentage different from the ideal solution value is the experimental value?

9-27. You have a tank with a partition in it. On one side of the tank, you load in 0.112 kg of nitrogen gas (348.15 K and 300 kPa). On the other side you load in 2.5 moles of argon gas at 348.15 K and 300 kPa. You remove the partition and the gases mix in an adiabatic fashion. Calculate the change in enthalpy and change in entropy of this system if you assume the gases to behave according to the ideal gas model.

9-28. You have a tank with a partition in it. On one side of the tank, you load in 0.112 kg of nitrogen gas (348.15 K and 3 MPa). On the other side, you load in 2.5 moles of argon gas at 403.15 K and 2 MPa. You remove the partition and the gases mix in an adiabatic fashion. Calculate the change in enthalpy and change in entropy of this system if you assume the gases to behave according to the ideal gas model.

9-29. Experimental data for the excess molar volume of 1-propanol (1) + 1-hexene (2) system at 298.15 K is provided in Table P9-29. If the pure component densities at 298.15 K are 799.65 m³/kg for 1-propanol and 668.28 kg/m³ for 1-hexene, do the following.
  A. Plot the solution molar volume as a function of composition of 1-propanol.
  B. Fit the solution molar volume to an appropriate functional form.
  C. Determine the partial molar volume for both substances when the system is equimolar.

9-30. Chemical engineers need to find data (either from published papers or through modeling techniques) for the systems they are to analyze. Sometimes it is a challenge to extract the information needed from a published journal paper. To that end, consider the following article from the *Journal of Chemical and*

TABLE P9-29 Excess molar volume of 1-propanol (1) + 1-hexene (2) system at 298.15 K (Treszczanowicz, 2010).

x_1	$\underline{V}^E$ (cm³/mol)
0.0559	0.0747
0.1058	0.0994
0.1992	0.1002
0.1507	0.1068
0.2544	0.0863
0.3009	0.0722
0.3520	0.0528
0.3964	0.0317
0.5009	−0.0220
0.5518	−0.0520
0.6016	−0.0697
0.7038	−0.0951
0.8088	−0.1026
0.8497	−0.0950
0.8990	−0.0709
0.9490	−0.0383

Engineering Data (W. Fan et al., *J. Chem. Eng. Data,* **53**, 1836 (2008)) on the excess molar volume of the methanol + methyl methacrylate system at a variety of temperatures. The paper provides a variety of correlations for the excess molar volume at different compositions and temperatures. Using Equations 3 and 4 in that journal article, estimate the molar volume of the methanol + methyl methacrylate system at 295.45 K and 42% (by mole) methanol.

9.9 GLOSSARY OF SYMBOLS

C	number of components in a system	$\underline{G}_A$	Gibbs free energy of pure component A	$\overline{G}_A^{ID}$	partial molar Gibbs free energy of component A in a mixture modeled as an ideal solution
$C_{P,1}$	constant pressure molar heat capacity of component 1	$\underline{G}$	molar Gibbs free energy of the mixture		
C_P^*	constant pressure heat capacity for ideal gas	$\underline{G}^{ID}$	molar Gibbs free energy of the mixture modeled as an ideal solution	$\overline{G}_A^{ig}$	partial molar Gibbs free energy of component A in a mixture modeled as an ideal gas
F	degrees of freedom				
G^E	excess Gibbs free energy	$\Delta\underline{G}$	molar Gibbs free energy of mixing	$\underline{G}^R$	residual molar Gibbs free energy of the mixture
$\underline{G}^E$	excess molar Gibbs free energy of the mixture	$\Delta\underline{G}^{ig}$	molar Gibbs free energy of mixing modeled as an ideal gas	$\underline{H}^E$	excess molar enthalpy of the mixture
$\underline{G}_A$	molar Gibbs free energy of pure component A	$\overline{G}_A$	partial molar Gibbs free energy of component A in a mixture	$\underline{H}_A$	molar enthalpy of pure component A

$\underline{H}$	molar enthalpy of the mixture
$\underline{H}^{\text{ID}}$	molar enthalpy of the mixture modeled as an ideal solution
$\Delta\underline{H}$	molar enthalpy of mixing
$\Delta\underline{H}^{ig}$	molar enthalpy of mixing modeled as an ideal gas
$\overline{H}_A$	partial molar enthalpy of component A in a mixture
$\overline{H}_A^{\text{ID}}$	partial molar enthalpy of component A in a mixture modeled as an ideal solution
$\overline{H}_A^{ig}$	partial molar enthalpy of component A in a mixture modeled as an ideal gas
$\underline{H}^{\text{R}}$	residual molar enthalpy of the mixture
k_b	Boltzmann's constant
m	mass
$\underline{M}$	an arbitrary thermodynamic property (normally $\underline{U}, \underline{H}, \underline{V}, \underline{G},$ or $\underline{S}$)
n	moles
$\dot{n}_1$	molar flow rate of component 1
P	pressure
P_i	partial pressure of component i
$\dot{Q}$	heat flow rate
R	universal gas constant
S^{E}	excess entropy
$\underline{S}^{\text{E}}$	excess molar entropy of the mixture
$\underline{S}_A$	molar entropy of pure component A
S_A	entropy of pure component A
$\underline{S}$	molar entropy of the mixture

$\underline{S}^{\text{ID}}$	molar entropy of the mixture modeled as an ideal solution
$\Delta\underline{S}$	molar entropy of mixing
$\Delta\underline{S}^{ig}$	molar entropy of mixing modeled as an ideal gas
$\overline{S}_A$	partial molar entropy of component A in a mixture
$\overline{S}_A^{\text{ID}}$	partial molar entropy of component A in a mixture modeled as an ideal solution
$\overline{S}_A^{ig}$	partial molar entropy of component A in a mixture modeled as an ideal gas
$\underline{S}^{\text{R}}$	residual molar entropy of the mixture
T	temperature
U^{E}	excess internal energy
$\underline{U}^{\text{E}}$	excess molar internal energy of the mixture
$\underline{U}_A$	molar internal energy of pure component A
U_A	molar internal energy of pure component A
$\underline{U}$	molar internal energy of the mixture
$\underline{U}^{\text{ID}}$	molar internal energy of the mixture modeled as an ideal solution
$\Delta\underline{U}$	molar internal energy of mixing
$\Delta\underline{U}^{ig}$	molar internal energy of mixing modeled as an ideal gas
$\overline{U}_A$	partial molar internal energy of component A in a mixture
$\overline{U}_A^{\text{ID}}$	partial molar internal energy of component A in a mixture modeled as an ideal solution

$\overline{U}_A^{ig}$	partial molar internal energy of component A in a mixture modeled as an ideal gas
U^{R}	residual molar internal energy of the mixture
$\hat{V}$	specific volume
V^{E}	excess volume
$\underline{V}^{\text{E}}$	excess molar volume of the mixture
$\underline{V}_A$	molar volume of pure component A
V_A	volume of pure component A
$\underline{V}$	molar volume of the mixture
$\underline{V}^{\text{ID}}$	molar volume of the mixture as an ideal solution
$\Delta\underline{V}$	molar volume of mixing
$\Delta\underline{V}^{ig}$	molar volume of mixing as an ideal gas
$\overline{V}_A$	partial molar volume of component A in a mixture
$\overline{V}_A^{\text{ID}}$	partial molar volume of component A in a mixture modeled as an ideal solution
$\overline{V}_A^{ig}$	partial molar volume of component A in a mixture modeled as an ideal gas
V^{R}	residual molar volume of the mixture
x_A	mole fraction of component A (normally in the liquid phase)
y_A	mole fraction of component A (normally in the vapor phase)
Ω	microstates
π	number of phases in equilibrium

9.10 REFERENCES

Abello, L., "Excess Heats of Binary Systems Containing Benzene, Hydrocarbons, and Chloroform or Methylchloroform." *J. Chim. Phys. Phys-Chim. Biol.*, 70, 1355 (1973).

Albuquerque, L., Ventura, C., Gonçalves, R. "Refractive Indices, Densities, and Excess Properties for Binary Mixtures Containing Methanol, Ethanol, 1, 2-Ethanol, and 2-Methoxyethanol." *J. Chem. Eng. Data,* 41, 685 (1996).

Arenosa, R. L., Menduiña, C., Tardajos, G, Diaz Peña, M. "Excess Enthalpies at 298.15 K of Binary Mixtures of Cyclohexane with *n*-Alkanes" *J. Chem. Thermodyn.* 11, 159 (1979).

Arce, A., Blanco, A., Soto, A., Vidal, I. "Densities, Refractive Indices, and Excess Molar Volumes of the Ternary Systems Water + Methanol + 1-Octanol and Water + Ethanol + 1-Octanol and their Binary Mixtures at 298.15 K." *J. Chem. Eng. Data*, 38, 336 (1993).

Bangqing Ni, K., Wang, L. S. H., Haigang Q. "Thermodynamic Properties of the Binary Mixtures of 1,2-Dichloroethane with Chlorobenzene and Bromobenzene from (298.15 to 313.15)K." *J. Chem. Eng. Data*, 55, 4541 (2010).

Christensen, C., Gmehling, J., Rasmussen, R., Weidlich, U. Heats of Mixing Data Collection, Vol. III, DECHEMA, Frankfurt, 1984.

Fan, W. et al., "Excess Molar Volume and Viscosity Deviation for the Methanol + Methyl Methacrylate Binary System at T = (283.15 to 333.15) K." *J. Chem. Eng. Data*, 53, 1836 (2008).

Kimura, F., Benson, G., Halpin, C. "Excess enthalpies of Binary Mixtures of n-heptane with hexane Isomers." *Fluid Phase Equilibria*, 11, 245 (1983).

Marino, G. et al., "Temperature Dependence of Binary Mixing Properties for Acetone, Methanol, and Linear Aliphatic Alkanes (C_6–C_8)." *J. Chem. Eng. Data*, 46, 728 (2001).

Murakami, S., Amaya, K., Fujishiro, R. "Heats of Mixing for Binary Mixtures. The Energy of Hydrogen Bonding Between Alcohol and Ketone Molecules." *Bull. Chem. Soc. Japan*, 37, 1776 (1964).

Nagata, I., Asano, H., Fujiwara, K. "Excess Enthalpies for Systems of 2-Propanol-Benzene-Methylcyclohexane." *Fluid Phase Equil.*, 1, 211 (1977–1978).

Redrawn from Smith, J. M., Van Ness, H. C., Abbott, M. M. "Introduction to Chemical Engineering Thermodynamics" 7th Edition, McGraw-Hill, Figure 12.17; Page 462, 2005.

Smith, V. C., Robinson, Jr., R. L. "Vapor–Liquid Equilibrium at 25 deg. in the Binary Mixtures Formed by Hexane, Benzene, and Ethanol." *J. Chem. Eng. Data*, 15, 391 (1970).

Treszczanowicz, A. J., Pawlowski, T. S., Treszczanowicz, T.,Szafraski, A. M. "Excess Volume of the 1-Propanol + 1-Alkene Systems in Terms of an Equation of State with Association." *J. Chem. Eng. Data*, 55, 5478 (2010).

Tyner, M. "Enthalpy Concentration Diagram for Hydrogen Fluoride Water System at one Atmosphere." *Chem. Eng. Prog.*, 45, 49 (1949).

Van Ness, H. C. et al., "Excess Thermodynamic Functions for Ternary Systems. I. Acetone-Chloroform-Methanol at 50. deg." *J. Chem. Eng. Data*, 20, 403 (1975).

Williamson, A. G., Scott, R. L. "Heats of Mixing of Non-Electrolyte Solutions. I. Ethanol + Benzene and Methanol + Benzene." *J. Phys. Chem.*, 64, 440 (1960).

Vapor–Liquid Equilibrium

10

LEARNING OBJECTIVES

This chapter is intended to help you learn how to:

- Understand that liquid and vapor phases of a mixture in equilibrium will almost never have the same composition (and when they do, it is called an azeotrope)

- Read mixture data tables and phase diagrams for vapor–liquid equilibrium and the difference between a *Txy* and *Pxy* plot

- Know when the simplest model for vapor–liquid equilibrium of mixtures, called Raoult's Law, will work and when it will not

- Identify when a mixture of two substances will be simpler to separate using distillation and when it will be more challenging to separate

- Understand that in some special circumstances, you can lower the pressure of a mixture and increase the amount of liquid in the system, which runs counter to our intuition

- Solve combined material balance and phase equilibrium problems, called "flash problems", for multi-component mixtures that are used to separate the components in a mixture.

Owing to both their size and importance, distillation columns are key features of chemical plants across the world. Distillation columns are normally the very tall, cylindrical-shaped columns that you will see in chemical plants performing distillation, whose unit operation separates the components of a mixture based on their boiling point. As you might expect, information on how mixtures separate into vapor and liquid phases becomes the key piece of information associated with distillation. In this chapter, we introduce how this data is presented (in both tabular and graphical form) as well as what these tables and graphs can tell us. We also begin to introduce a simple approach for modeling vapor–liquid equilibrium of mixtures based on what we have learned in Chapter 9, namely the ideal solution and the ideal gas. ■

10.1 Motivational Example

When you boil a pot of water at 101.325 kPa, there is liquid H_2O in the pot and vapor H_2O leaving the pot, all happening at 100°C. Instead of boiling water (a single substance, H_2O) alone, suppose we had an equimolar liquid mixture of water and another compound, say ethanol, in the pot at 101.325 kPa. What temperature would the mixture boil at? Would it be at 100°C (the normal boiling point of water)? Would it be 78°C (the normal boiling point of ethanol)? Would it be around 89°C, which is a weighted average of the normal boiling points based on composition (our "intuitive" approach from Chapter 9)? Also, would the vapor composition leaving the pot be equimolar as well?

Experiment tells us that an equimolar liquid mixture of water and ethanol will boil at around 80°C at 101.325 kPa, while the vapor phase composition will not be equimolar (like the liquid), but will be 65% by mole ethanol. These results work against our intuitive approach, but also provide a tremendous possibility. Just by boiling a liquid mixture at one composition, we can obtain a different composition in the vapor phase. So in the example above, our vapor composition will be enriched in ethanol, in effect "separating" some of the ethanol from the water. This is the basis for one of the most important and visible unit operations in chemical engineering: the distillation column. Obviously, there is a great need to understand why our intuitive approach does not work, as well as develop a way to predict and/or model what these results will be for any vapor–liquid mixture of interest in the chemical process industry.

In order to understand the experimental results described in the previous paragraph, we have to go back to the phenomena of evaporation and boiling to consider the nature of the process at a molecular level. Consider a pure, compressed liquid in a fixed volume container. The container is inside a heat bath that controls the temperature and keeps it constant. The liquid fills all of the space in the container. If left alone, will the liquid evaporate? No, because this is a closed system and there is no space for the vapor to fill. However, say that container is connected to a second container by a divider and the contents of the second container are evacuated (Figure 10-1), so the pressure of the second container is 0 kPa. When the divider is removed, what happens? The liquid will start to expand into the evacuated space and a portion of the liquid will start to vaporize because any vapor bubbles that attempt to form from the surface of the liquid will not be stopped (there is no pressure, thus no force to stop them) above the liquid at that point. Both the expansion of the liquid and the vaporization of the liquid will continue until the point at which some of the vapor starts to condense. A dynamic equilibrium will be established where the rate of vaporization will equal the rate of condensation. The pressure exerted by this vapor above the surface of this liquid is now balanced by the pressure exerted in the liquid phase. This pressure is called the vapor pressure, which is the unique pressure at which liquid and vapor can exist in equilibrium, and it will be a function of the temperature of the system.

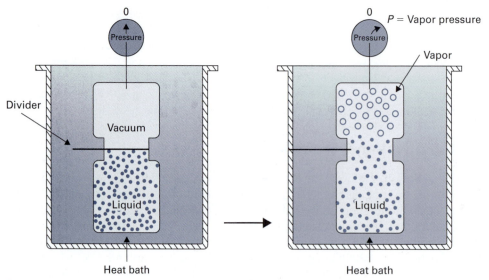

FIGURE 10-1 A liquid expanding and vaporizing into the evacuated space.

What if we did a different experiment? The setup is initially the same, but now you have a mixture of two liquids instead of one pure compound in the first container. The first container is again connected to a second container (whose contents are evacuated) by a divider. We remove the divider and what happens? Is it the same as before? Yes and no. Yes, there is expansion and vaporization and, ultimately, a dynamic equilibrium is established as before. But what is the final pressure in the system? Since there are now two liquids, is the final pressure just the sum of the vapor pressures of both liquids? No, it isn't. Vapor pressure of a *pure* substance is the pressure above the surface of the *pure* liquid. However, when you have a mixture, more than one liquid is present at the surface. Thus, for a fixed surface area (as in this problem), there will be less of each component at the liquid–vapor interface than if they were there by themselves. Therefore, the partial pressure that each component will exert will be different than if it were a pure component. How much less? Well, we've seen before that when we want to calculate a mixture property, we need to use the mole fractions to account for the percentage of each material in a mixture. The molar enthalpy of a mixture of two components (a and b) was introduced in Equation 9.1 as

$$\underline{H} = x_a \underline{H}_a + x_b \underline{H}_b \tag{9.1}$$

So the question becomes, "Is the pressure over the liquid mixture simply a sum of the vapor pressures of the pure components (weighted by their mole fractions), like this?"

$$P = x_a P_a^{sat} + x_b P_b^{sat} \tag{10.1}$$

This mirrors our "intuitive" approach from the previous chapter. The answer we provided there is the same as it is now. If the system behaves as an ideal solution, this would be the case. If not, then (as we discovered from other thermophysical properties in Chapter 9) we'll have to account for those non-ideal cases as well.

Recall again Chapter 9, where we discussed the variety of behavior that can occur in the liquid, relative to the ideal solution. For example, we saw in Figure 9-2 that there were situations where we had both positive and negative excess molar volumes, for the *same* system, as a function of composition. The liquid phase interactions that occur in the mixture impact the "combined" vapor pressure that this mixture possesses. Because of this variety in behavior as well as the importance of vapor–liquid equilibrium in separating chemicals, it is essential for chemical engineers to be comfortable extracting the proper information for these systems from the data, provided in both tabular and graphical format. We describe this in the Section 10.2.

Remember, Equation 9.1 is valid only for an ideal solution (our intuitive approach).

10.2 Raoult's Law and the Presentation of Data

All the way back in Chapter 2, the Gibbs phase rule was introduced to provide a constraint on the number of intensive variables that can be specified for a system, with knowledge of the number of components and equilibrium phases.

$$F = C - \pi + 2 \tag{2.2}$$

This was also discussed in Chapter 9. Previously, we asked the question, "What is the boiling point of water?" From a Gibbs phase rule analysis, we find that F (the degrees of freedom) will equal 1 for water at its boiling point ($C = 1$; $\pi = 2$). Thus, there is no single, unique answer to the question about the boiling point of water—we need to specify another intensive variable, since $F = 1$. So a proper question could be "What is the boiling point of water at 101.325 kPa?" The answer is 100°C. If we specified a different pressure, the answer would be a different temperature. In fact, if we

Figure 8.1 shows a plot of the vapor pressure as a function of temperature.

wanted to be efficient about things, we could prepare a table that lists the boiling point of water as a function of pressure. Here, pressure would play the role of the independent variable while the temperature would be the dependent variable. This table is, of course, part of the steam tables included in Appendix A.

Now that we have moved into the realm of mixtures, how does this question and answer change? Specifically, what if we asked the question, "*What is the boiling point of an ethanol + water mixture?*" Once again, we employ Gibbs phase rule and find that, now, $F = 2$ ($C = 2$; $\pi = 2$). Thus, we need to fix two intensive variables. We could ask the question this way: "What is the boiling point of an ethanol–water mixture at $P = 101.325$ kPa and a liquid mole fraction of 50% ethanol?" The answer is around 80°C. (Recall that we answered this question in the second paragraph of Section 10.1.) If we change the value of either or both of our two intensive variables (the system pressure and the mole fraction of ethanol in the liquid phase), the boiling point changes. Say, like before, we wanted to be efficient in presenting our answer. In this case, rather than having a table that showed how the boiling point changed with different pressures (as in the pure component example in the previous paragraph), we need to fix one of our intensive variables (e.g. pressure) and vary the other intensive variable (e.g. liquid mole fraction) to see the effect on the boiling point. Additionally, we could do a similar presentation of this data—not in tabular form—but in graphical form. In the rest of this section, we introduce the presentation of vapor–liquid equilibrium data through the use of tables and graphs.

Let's consider a mixture of two commercial refrigerants, R-134a (1), which is 1,1,1,2-tetrafluoroethane (CH_2FCF_3), and R-245fa (2), which is 1,1,1,3,3-pentafluoropropane ($C_3H_3F_5$). We would like to efficiently present the vapor–liquid equilibrium for this mixture. However, per the previous discussion, we have two degrees of freedom and, thus, need to keep one of our variables constant in order to present the data in tabular form. Normally, data of this type is presented by keeping either the temperature or pressure constant and showing how the other variables change as we take the mixture across all ranges of liquid composition (from pure R-134a to pure R-245fa and the compositions in between). To this end, in Table 10-1, we show how the system pressure will change at constant temperature (293.15 K) as we vary the composition of the mixture. The endpoints typically provide the pure components for both species in the mixture. When $x_1 = 0$ (the first row of Table 10-1), we have pure R-245fa ($x_2 = 1$), and what is displayed in the row is the vapor pressure for R-245fa at a temperature of 293.15 K. Likewise, the fifth row of Table 10-1 (when $x_1 = 1$) is for pure R-134a, and the pressure displayed is the vapor pressure at 293.15 K. The three entries in the middle are for various compositions of the mixture—all at 293.15 K. Note that, as mentioned earlier in this chapter, the vapor phase composition is almost never the same as the liquid phase composition. For completeness (though normally not shown), we provide the values for x_2 and y_2 in Table 10-1, realizing that $x_2 = 1 - x_1$ and $y_2 = 1 - y_1$.

The standard practice is to denote the liquid phase composition with x, the vapor phase composition with y and, where applicable, the solid phase composition with z.

R-134a is the more volatile substance (its vapor pressure is higher at 293.15 K compared to R-245fa). This is also seen by its increased mole fraction in the vapor phase, which is relative to its composition in the liquid phase (at the same temperature and pressure).

F and F structure diagram (R-134a):

```
    F  F
    |  |
F—C—C—H
    |  |
    F  H
```
R-134a

```
    F  H  H
    |  |  |
F—C—C—C—F
    |  |  |
    F  H  F
```
R-245fa

TABLE 10-1 Vapor–liquid equilibrium data for the mixture R-134a (1) + R-245fa (2) at 293.15 K.

Temperature (K)	Pressure (kPa)	x_1	x_2	y_1	y_2
293.15	123	0	1	0	1
293.15	202	0.18	0.82	0.48	0.52
293.15	315	0.45	0.55	0.76	0.24
293.15	439	0.72	0.28	0.91	0.09
293.15	571	1	0	1	0

In addition to fixing the temperature and varying the liquid phase mole fraction, one could also fix the pressure and vary the liquid phase mole fraction. Accordingly, a typical data table would look like Table 10-2, which shows the vapor–liquid equilibrium for benzene (1) (C_6H_6) and toluene (2) ($C_6H_5CH_3$) at 101.3 kPa.

TABLE 10-2 Vapor–liquid equilibrium data for the mixture benzene (1) + toluene (2) at 101.3 kPa.

Pressure (kPa)	Temperature (K)	x_1	x_2	y_1	y_2
101.3	383.85	0	1	0	1
101.3	371.55	0.30	0.70	0.51	0.49
101.3	365.75	0.49	0.51	0.70	0.30
101.3	358.25	0.77	0.23	0.89	0.11
101.3	353.35	1	0	1	0

Based on data from H. Schuberth, J. Prakt. Chem., 6, 129 (1958).

For Table 10-2 here the first and last rows denote the vapor pressure of the pure components, toluene (top row) and benzene (bottom row). The interior rows represent the composition of the liquid that is in equilibrium with the vapor at that temperature and pressure.

In addition to tabular form, vapor–liquid equilibrium data is often displayed in graphical form. For example, in Table 10-1 we presented the data for the R-134a (1) + R-245fa (2) mixture, where we have made the temperature constant (T = 293.15 K) while varying the liquid–phase composition. The resulting pressures and vapor-phase compositions were provided in Table 10-1. To present this same data in graphical form, we normally denote what is being fixed somewhere on the graph or in the figure caption (here, the temperature) and plot the system pressure as a function of the liquid and vapor phase mole fraction. This is normally called a "*Pxy*" graph and is presented in Figure 10-2.

You have other choices you could make for the two intensive variables that you fix. For example, you can specify the pressure and the vapor phase mole fraction (according to Gibbs phase rule) and then vary the vapor phase mole fraction to measure the values of temperature and liquid phase mole fraction. However, this is much more difficult to do experimentally, so data is not regularly reported in this way.

Benzene

Toluene

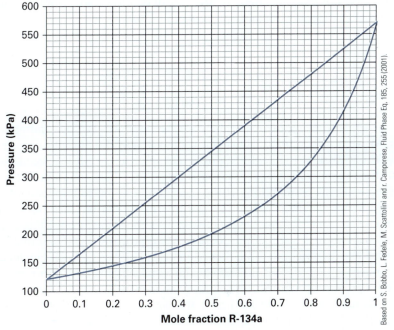

Based on S. Bobbo, L. Fedele, M. Scattolini and r. Camporese, Fluid Phase Eq, 185, 255 (2001).

FIGURE 10-2 *Pxy* plot for the R-134a (1) + R-245fa (2) system at 293.15 K.

You wouldn't normally show all of the minor grid lines on these type of plots, but we do here to aid in reading the compositions and pressures.

FOOD FOR THOUGHT 10-2

What do you think the letters after the numbers represent in the refrigerant numbering system?

Once again, Figure 10-2 describes the system where the temperature is equal to 293.15 K, with the curves themselves including the information in Table 10-1, which is when the vapor is in equilibrium with the liquid at that temperature and pressure. Note that the area above the top curve is for the mixture as a liquid alone (a single liquid phase with no vapor phase), while the area below the bottom curve is for the mixture as a vapor alone (a single vapor phase with no liquid). Let's describe some of the key features of Figure 10-2 in more detail, including what it means to be "in between" the curves.

The vertical axis on the left (where $x_1 = 0$) describes the pure component R-245fa. The point you see at $P = 123$ kPa (which is also the first row of Table 10-1) is the vapor pressure for R-245fa at 293.15 K. Recall from Chapter 2, this is the pressure at which pure R-245fa can exist in the liquid and vapor phases at equilibrium. If we go to a pressure higher than the vapor pressure for pure R-245fa, say 200 kPa, we now have only one equilibrium phase—a liquid. At this point (200 kPa), if we begin to add a small amount of R-134a to our system (isothermally), we will move horizontally to the right on Figure 10-2 and we still have one equilibrium phase—a liquid. However, as we continue to add more R-134a to our system, we will continue to move horizontally to the right on Figure 10-2, and eventually run into a curve at $x_1 = 0.17$. This is called the **bubble-point curve**, and it describes the composition at which the first bubble of vapor is in equilibrium with the liquid.

Let's go backward for a moment. We said that the vertical axis at $x_1 = 0$ describes pure R-245fa—specifically, the point where the two curves meet (at $x_1 = y_1 = 0$) is the vapor pressure at 293.15 K (here, 123 kPa). Likewise, the vertical axis at $x_1 = 1$ will describe pure R-134a—specifically, the point where the two curves meet again (now at $x_2 = y_2 = 0$) is the vapor pressure at 293.15 K (here, 571 kPa, which is the last row of Table 10-1). Let's call on our intuition again as we did in Chapter 9. Say we wanted to estimate this "vapor pressure" for the mixture at a composition of $x_1 = 0.17$ (now called a "bubble-point pressure"). Could we just do as we did at the beginning of Chapter 9 for the molar enthalpy and take a weighted average of the vapor pressures of the pure components? Recall that we used a similar approach at the outset of this chapter. At any rate, using this intuition, we again find

$$P = x_1 P_1^{sat} + x_2 P_2^{sat} \tag{10.2}$$

$$P = 0.17(571 \text{ kPa}) + 0.83(123 \text{ kPa})$$

$$P = 199 \text{ kPa}$$

Since we are comparing the experimental value (200 kPa) to our "intuitive" value (199 kPa), we conclude that this worked very well. Why did our intuition work here, but it did not work at the beginning of this chapter in the Motivational Example? Recall that the model for our intuition is called the "ideal solution." For systems, to best validate the assumptions of the ideal solution, the components of the mixture need to have similar sizes and similar intermolecular interactions. What about R-134a and R-245fa? R-134a has an ethane backbone that is substituted with four fluorine and two hydrogen atoms. R-245fa has a propane backbone that is substituted with five fluorine and three hydrogen atoms. So, we have two molecules that have somewhat similar sizes and similar intermolecular interaction types. Thus, it turns out that this system behaves similar to an ideal solution.

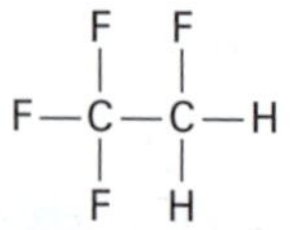

Let us take an idea from earlier in the chapter in an attempt to explore this point further. Recall that the vapor pressure for a pure component is the pressure that exists above a pure liquid in a closed system. We described how the surface of

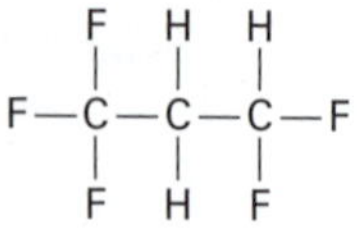

Sidebar (left margin):

$x_1 = \dfrac{n_1}{(n_1 + n_2)}$. As we add more n_1, the value of x_1 increases.

What would be the bubble-point pressure, using our intuition, for a three-component mixture?

$P = x_1 P_1^{sat} + x_2 P_2^{sat} + x_3 P_3^{sat}$

F F
| |
F—C—C—H
| |
F H

R-134a

F H H
| | |
F—C—C—C—F
| | |
F H F

R-245fa

the liquid was important in this process. In a mixture, however, each component is not present alone at the surface; thus, each cannot exert its full vapor pressure. The vapor pressure each component can exert will be a function of its surface coverage (composition) as well as how the different components interact with each other. That we were able to use the ideal solution assumption here allowed us to simply use the weighted average of the vapor pressures of the pure components.

Since our ideal solution model provided a good estimate of the bubble-point pressure for this mixture at $x_1 = 0.17$, can it also help us estimate the composition of the vapor that is in equilibrium with this liquid mixture? Recall that we said the bubble point was the point at which the first bubble of vapor appears. What is the composition of this vapor? Is it also $y_1 = 0.17$, or is it at something different? *Should* it be something different? Well, the liquid interacted as an ideal solution and the system is at 200 kPa. This is a moderately low pressure, at which many (though not all) systems can reasonably be modeled as an ideal gas. Here we will examine the vapor phase by assuming ideal gas behavior, and then test the accuracy of our final answer.

Recall that the partial pressures of an ideal gas mixture sum to give the total pressure of the system, and since we know the total pressure of the system from our model is 199 kPa, we can then determine the composition. But what is the partial pressure of each component? Since our liquid is well modeled as an ideal solution, each component contributes to the bubble-point pressure of the mixture based on the percentage of its own molecules at the surface, multiplied by its own vapor pressure. Thus, the partial pressure of the substance is just its pure component vapor pressure times its mole fraction in the liquid (we use the liquid composition since the surface is formed in the liquid phase). Mathematically, this becomes $p_1 = x_1 P_1^{sat}$, where P_1^{sat} is the vapor pressure of component 1. Since we are treating the vapor as an ideal gas, the partial pressure of component 1 becomes $p_1 = Py_1$. Equating both partial pressures for component 1 yields

$$Py_1 = x_1 P_1^{sat} \qquad (10.3)$$

Finally, we can solve for vapor phase mole fraction, y_1, as

$$y_1 = \frac{x_1 P_1^{sat}}{P} \qquad (10.4)$$

Plugging in the numbers for our problem, we arrive at

$$y_1 = \frac{0.17(571 \text{ kPa})}{199 \text{ kPa}} = 0.49$$

So if we assume that the R-134a + R-245fa mixture in the vapor phase behaves as an ideal gas, the mole fraction of this vapor that is in equilibrium with the liquid at a liquid composition of $x_1 = 0.17$ and a temperature of 293 K is $y_1 = 0.49$. Is this correct? To find our answer, we return to the experimental data presented in Figure 10-2. Here, to find the vapor phase composition in equilibrium with the liquid phase, just move horizontally from the bubble-point curve (at 200 kPa and $x_1 = 0.17$) to the point where you intersect the next curve. That point, at around $y_1 = 0.49$, is called the **dew point**, and the curve itself is the dew-point curve. The dew point is where the system is basically all vapor and the first drop of liquid (or "dew") appears. Getting back to our problem at hand, when we compare the ideal gas model results of $y_1 = 0.49$ (calculated above) to the actual value of around $y_1 = 0.49$, this is a very good estimate.

The assumption of an ideal solution for the liquid, combined with the assumption of an ideal gas for the vapor, is a very quick and popular way to estimate vapor–liquid equilibrium for mixtures.

Not all systems can be effectively modeled as an ideal gas at 200 kPa. Hydrogen fluoride, for example, will not behave as an ideal gas at 200 kPa (or 100 kPa, for that matter). Much lower pressures are required for hydrogen fluoride to behave as an ideal gas owing to the extensive hydrogen bonding in this system.

Saturated water vapor has a compressibility factor of 0.975 at 200 kPa, meaning it is reasonably modeled as an ideal gas at this state.

We move horizontally between the liquid (bubble point) and vapor (dew point) curves to identify the phases in equilibrium because the temperature is constant when moving horizontally, and constant temperature is one of the requirements of phase equilibrium.

Raoult's Law is based on the work of François-Marie Raoult, and this relationship is attributed to his contribution from 1882. Raoult also investigated freezing point depressions of solvents when a small amount of solute is introduced.

These combined assumptions are called **Raoult's Law** and can be written for each component individually as

$$y_1 = \frac{x_1 P_1^{sat}}{P} \tag{10.4}$$

$$y_2 = \frac{x_2 P_2^{sat}}{P} \tag{10.5}$$

If one adds these equations together, the sum becomes

$$y_1 + y_2 = 1 = \frac{x_1 P_1^{sat}}{P} + \frac{x_2 P_2^{sat}}{P} \tag{10.6}$$

Solving for the system pressure yields

$$P = x_1 P_1^{sat} + x_2 P_2^{sat} \tag{10.2}$$

$$P = x_1 P_1^{sat} + (1 - x_1) P_2^{sat} \tag{10.7}$$

$$P = x_1 (P_1^{sat} - P_2^{sat}) + P_2^{sat} \tag{10.8}$$

Recall that this is at constant temperature, so the vapor pressures (which are functions of temperature) are fixed.

Equation 10.8 is that of a line if one plots P as a function of x_1, where the slope is $P_1^{sat} - P_2^{sat}$ and the P-axis intercept is P_2^{sat}. This is an important result, since the Raoult's Law P vs. x_1 curve (which is the bubble-point curve, by definition) must be a line. That the R-134a + R-245fa system would be well-approximated by Raoult's Law is seen in Figure 10-2, which has a bubble-point curve that looks (by eye) like a line.

Next, let's consider the following situation. You have a mixture made up of 3 moles of R-134a and 7 moles of R-245fa. The system is at 293.15 K and 500 kPa. What is the equilibrium phase of this system? The bubble-point pressure for this composition ($x_1 = 0.3$) is about 255 kPa and our system pressure (500 kPa) is above the bubble-point pressure; thus the system is a liquid. This is analogous to the situation where our pure component system is at a pressure above its vapor pressure at a given temperature.

Let's consider the same system as in the previous paragraph, but now we lower the system pressure from 500 kPa to the 150 kPa. What is the equilibrium phase of our mixture at this state? When we look at Figure 10-2, we find that we are at a pressure below the dew-point curve with the dew-point pressure being about 160 kPa for this composition ($y_1 = 0.3$). Since our system pressure (150 kPa) is lower than the dew-point pressure (160 kPa), the system is a vapor. This is analogous to the situation where our pure component system is at a pressure below its vapor pressure at a given temperature.

Finally, let's consider the same system as in the previous two paragraphs, but now we raise the system pressure from 150 kPa to 200 kPa. What is the equilibrium phase of our mixture at this state? When we look at Figure 10-2, we find that now we are above the dew-point curve, but below the bubble-point curve. We are somewhere between the bubble-point pressure and the dew-point pressure at our system composition (30% by mole of R-134a). When we are between the bubble- and dew-point curves (inclusive of the curves themselves), we have two equilibrium phases: a liquid and a vapor. What is the composition of each phase? We just move horizontally (both directions) from our mixture composition (30% by mole of R-134a) and 200 kPa until we intersect both the bubble-point and dew-point curves. Where we intersect the bubble-point curve ($x_1 = 0.17$) is the composition of the liquid phase, and where we intersect the dew-point curve ($y_1 = 0.49$) is the composition of the vapor phase. Both of those points are at the same pressure and same temperature, which are required by phase equilibrium.

Just as there is a graphical representation of constant temperature data, there is the same for constant pressure data. In Figure 10-3, we show the data of Table 10-2 in a graphical format that now plots the temperature as a function of the equilibrium liquid and vapor compositions, called a *Txy* plot. Like the *Pxy* plot, the *Txy* plot has the pure component information as the vertical lines at $x_1 = 0$ (pure toluene) and $x_1 = 1$ (pure benzene). For toluene, the boiling point at 101.3 kPa is a little over 110°C as seen on Figure 10-3 (at $x_1 = 0$).

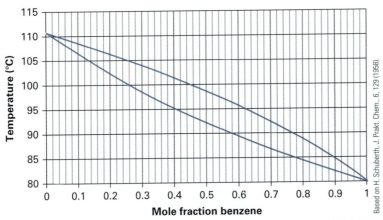

Based on H. Schuberth, J. Prakt. Chem., 6, 129 (1958).

FIGURE 10-3 *Txy* plot for the benzene (1) + toluene (2) system at 101.3 kPa.

Just as before, Figure 10-3 spans the entire composition range, but now it is the pressure that is fixed (instead of the temperature) and the pressure is equal to 101.3 kPa. The vertical axis on the left (where $x_1 = 0$) describes the pure component toluene. The point you see at around $T = 110°C$ (which is the first row of Table 10-2) is the temperature for toluene when the vapor pressure is 101.3 kPa. If we go to a temperature higher than this saturation temperature for pure toluene, say 115°C, we now have only one equilibrium phase—a vapor. Likewise, if we go to a temperature below the saturation temperature, say 100°C, we now have only one equilibrium phase—a liquid.

Let's begin with pure toluene ($x_1 = 0$) at 100°C and 101.3 kPa (which is a liquid, according to Figure 10-3) and start to add some benzene while maintaining a constant temperature. We represent the increasing benzene concentration by moving along horizontally on Figure 10-3 until we intersect the curve at $x_1 = 0.26$. Since we had all liquid, this intersection with the curve brings the arrival of the first bubble of vapor that is in equilibrium with the liquid, so this curve (as before) is called the bubble-point curve. To find the composition of this vapor in equilibrium with the liquid, once again we move horizontally from the bubble-point curve (at 100°C and $x_1 = 0.26$) to the point where we intersect the next curve. That point, at $y_1 = 0.45$, is called the dew point, and the whole curve is called the dew-point curve, as before.

Unlike the *Pxy* curve (for which each component has a fixed vapor pressure since the temperature is constant), here the vapor pressure for the pure component will change with temperature. Therefore, the conceptual analysis done previously for *Pxy* plots involving the ideal solution and the ideal gas (e.g. a line for the bubble-point curve implies a system that behaves according to Raoult's Law) is not as accessible for a *Txy* plot. However, we can still attempt to validate our intuition about how a benzene + toluene mixture would behave through use of this model. Since benzene (C_6H_6) and toluene ($C_6H_5CH_3$) are similar substances with similar sizes and interactions and our system is at a low pressure, we would expect that this system is well modeled by Raoult's Law. Let's examine this conjecture.

PITFALL PREVENTION

Don't memorize whether it is a liquid or vapor that is above or below a *Pxy* or *Txy* plot. You already know that higher temperature (or lower pressure) means a vapor, so use that as your guide.

Benzene

CH_3

Toluene

EXAMPLE 10-1	**RAOULT'S LAW**

What is the system temperature and vapor phase mole fraction for a mixture of benzene (1) + toluene (2) that is 26% by mole benzene in the liquid phase and at a system pressure of 101.3 kPa if the system is modeled according to Raoult's Law? How does this result compare to the experimental value?

SOLUTION:

Step 1 *Write out our equations and unknowns*

Recall the expression for Raoult's Law written for each component is given by

$$Py_1 = x_1 P_1^{sat} \tag{10.3}$$

$$Py_2 = x_2 P_2^{sat} \tag{10.9}$$

Here, x_1 and x_2 are known (since $x_2 = 1 - x_1$) and the pressure (P) is fixed. The unknowns in the problem are y_1, y_2, P_1^{sat}, and P_2^{sat}. Additionally, since our goal is to find the temperature (T), we have to explicitly expose where it occurs in our equations, namely through the vapor pressure equation. Thus, we have five unknowns (T, y_1, y_2, P_1^{sat}, and P_2^{sat}), but only two equations (the Raoult's Law expressions for both components). Therefore, we need three additional independent relationships among our unknown variables.

The first additional equation is simply $y_2 = 1 - y_1$. The other two additional equations come from the Antoine equation, which provides the vapor pressure as a function of temperature.

Recall that the Antoine equation, introduced in Section 8.2.6, is written as

$$\log_{10} P^{sat} = A - B/(T + C)$$

where P is in units of mm Hg and T is in Celsius. The values for A, B, and C are provided in Appendix E and repeated here.

Substance	A	B	C
Benzene	6.90565	1211.033	220.790
Toluene	6.95464	1344.800	219.48

Since we have one Antoine equation expression for benzene and another for toluene, we now have five equations and five unknowns. Next, we can solve the problem.

Step 2 *Solve the problem*

To review, there are five equations in five unknowns. The five equations are as follows:

$$\log_{10} P_1^{sat} = 6.90565 - 1211.033/(T + 220.70)$$

$$\log_{10} P_2^{sat} = 6.95464 - 1344.800/(T + 219.48)$$

$$(760 \text{ mm Hg})y_1 = (0.26)P_1^{sat}$$

$$(760 \text{ mm Hg})y_2 = (1 - 0.26)P_2^{sat}$$

$$y_2 = (1 - y_1)$$

The five unknowns are T, y_1, y_2, P_1^{sat} and P_2^{sat}. Here, we have already made the change in pressure units to match the vapor pressure from the Antoine equation (mm Hg).

There are many resources available for you to solve five non-linear equations and five unknowns—be it Excel, some other program, or even your calculator. The solution to the problem for the relevant unknowns is

$$y_1 = 0.46 \qquad P_1^{sat} = 1344.0 \text{ mm Hg (179.2 kPa)}$$

$$y_2 = 0.54 \qquad P_2^{sat} = 554.8 \text{ mm Hg (74.0 kPa)}$$

$$T = 99.9°C$$

Recall that the experimental values were (if we estimate from the graph) $y_1 = 0.45$, $y_2 = 0.55$, and $T = 100°C$. Thus, the conjecture that the benzene + toluene system forms an ideal solution in the liquid and an ideal gas in the vapor is well justified based on the comparison of the model results to the experiment.

Now that we have learned how to read both Pxy and Txy plots, we present a variety of different systems in order to demonstrate the diversity of behavior that can occur when systems do not behave ideally as with R134a + R-245fa or benzene + toluene. In Figure 10-4, we show the system n-pentane (1) (C_5H_{12}) + methanol (2) (CH_3OH) at 422.6 K. Here we have a hydrocarbon and an alcohol. They not only have different sizes, but have much different types of self-interactions. The hydrocarbon is non-polar and would have only induced dipole-type interactions. On the other hand, methanol is a polar compound that forms hydrogen bonds. Clearly, this is a system that would not be well modeled by the ideal solution assumption. Thus, we would anticipate that Raoult's Law would be a poor approximation to model the system behavior (not to mention the fact that the system is at a high pressure, making the ideal gas portion of Raoult's Law not appealing as well). The black lines in Figure 10-4, which are model predictions for the bubble- and-dew point curves for this system, clearly show that Raoult's Law does not work for this system. The actual phase behavior is shown with the blue curves and is read in the same way as in Figure 10-2. The top curve is the bubble-point curve, while the bottom curve is the dew-point curve. However, unlike Figure 10-2, this curve has a point where the liquid and vapor compositions

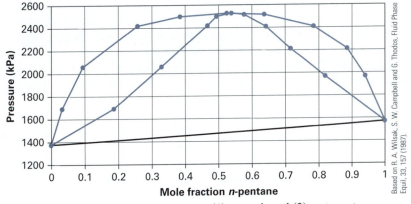

FIGURE 10-4 Pxy plot for the n-pentane (1) + methanol (2) system at 422.6 K. Raoult's Law is the black curve (both bubble- and dew-point curves are shown) in the figure.

The experimental data are given as symbols, while the blue curves are just helpful guides for the eye here.

Based on R. A. Wilsak, S. W. Campbell and G. Thodos, Fluid Phase Equil, 33, 157 (1987).

Azeotropes are not uncommon. Indeed, they are so common that systems that do NOT exhibit an azeotrope have their own name—zeotropic systems. Note that systems can be azeotropic at one set of conditions and zeotropic at another set of conditions. This is because the azeotropic composition changes with temperature or pressure.

are the same, namely at $x_1 = y_1 = 0.53$. This point is called the **azeotrope** and systems that show an azeotrope are called azeotropic systems. The knowledge of the location of an azeotrope is important when designing distillation columns for separating two components via their boiling point. A typical chemical engineering curriculum includes at least one course that includes a discussion of separations, and in that course you learn that the azeotropic composition provides a limit on the purity of a separation, since both the liquid and vapor phases have the same composition. As such, chemical engineers strategize different approaches to move past this constraint, such as adding a third chemical to the system, changing the system temperature, or even trying a different separation approach (such as one based on freezing point instead of boiling point, for example).

One major failing of Raoult's Law for this system is the fact that the model does not produce an azeotrope, while experimentally an azeotrope is observed. Additionally, the system pressure is at around 2500 kPa at the azeotropic composition while the Raoult's Law predicted pressure is just below 1500 kPa. Systems that exhibit a pressure above that predicted by Raoult's Law are said to have "**positive deviations**" from Raoult's Law, meaning the pressure is greater (or positive) relative to what Raoult's Law would predict. Note that, as we show later, some systems will show negative deviations from Raoult's Law, meaning that the system pressure is less than that predicted by Raoult's Law.

Methanol will hydrogen bond with itself in a mixture with n-pentane. The O-atom in methanol can not hydrogen bond with the H-atom in n-pentane since the C-atoms in n-pentane are not electronegative enough to draw enough electron density from the H-atom it is bonded to. On the other hand, the O-atom (which is more electronegative than a C-atom) in the alcohol will draw enough electron density from the H-atom to which it is bonded, making the H-atom on the alcohol available for hydrogen bonding (with another alcohol molecule).

Mechanistically, why should a mixture that does not adhere to the ideal solution assumption yield a pressure above that of the ideal-solution predicted pressure? In order to explore this issue in more detail, we need to return to the concept of vapor pressure. We know from our previous discussion of vapor pressure that a substance with a high vapor pressure is volatile, meaning it will readily vaporize to the vapor phase from the liquid phase. This, of course, has to do with the molecular level interactions in the liquid phase (and, specifically, at the surface). When a mixture behaves as an ideal solution, the similar size and interactions of the molecules involved mean that a snapshot (of both the interactions and the sizes) for the mixture would be similar to that of either pure component. A good example of an ideal solution, as we discussed previously, is the benzene + toluene mixture. However, when you have a solution that is not ideal, the situation changes. Relative to either pure component, the overall intermolecular interactions in the mixture normally will be less advantageous as the unlike interactions between the two different molecules are not as energetically beneficial as those between like substances. This decrease in attractive interactions (relative to the pure components) results in a system that is more volatile. A good example of a non-ideal solution behaving in this manner is the n-pentane + methanol system from Figure 10-4. Therefore, for a given composition, the system pressure is higher than what the ideal solution would predict and, thus, there are positive deviations from Raoult's Law.

On the other hand, there are some systems that will form hydrogen bonds between the different components in a mixture but do not as individual pure components. This additional increase in attractive interactions (relative to the pure components alone) results in a system that is less volatile. Accordingly, the system will exhibit negative deviations from Raoult's Law. One such system is acetone $(CH_3)_2 CO$ + chloroform $(CHCl_3)$, as demonstrated in Figure 10-5. Here, neither pure component can hydrogen bond, but they will participate in hydrogen bonding interactions with each other. Since this is a somewhat unusual set of conditions, negative deviations from Raoult's Law are far less common than positive deviations.

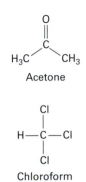

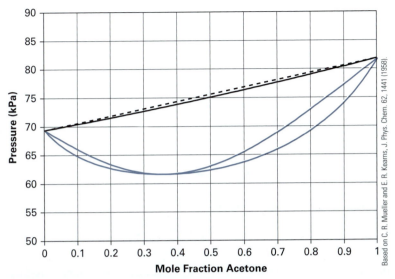

FIGURE 10-5 *Pxy* plot for the acetone (1) + chloroform (2) system at 50°C. Raoult's Law is the dashed curve in the figure.

In Figures 10-4 and 10-5, the azeotrope for both systems appeared near the middle of the composition region on the *Pxy* diagram. However, this certainly need not be the case, in general. For example, the well-studied water (1) + ethanol (2) system exhibits positive deviations from Raoult's Law and has an azeotropic composition that is almost all ethanol, as seen in Figure 10-6. Indeed, this system becomes zeotropic (without an azeotrope) as the temperature is lowered.

The three electronegative chlorines attached to the carbon in chloroform will help draw electron density from the H-atom, making it available for hydrogen bonding with the acetone (which has an O-atom available for hydrogen bonding). By themselves, they can't hydrogen bond, but together they will form a hydrogen bond (or "cross associate").

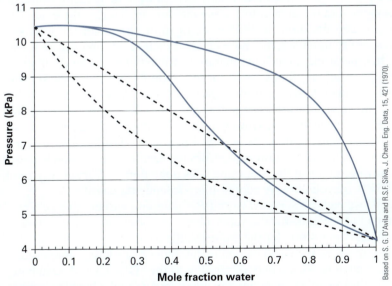

FIGURE 10-6 *Pxy* plot for the water (1) + ethanol (2) system at 30°C. The dashed curves are Raoult's Law.

10.2.1 Distribution Coefficients, Relative Volatility, and *xy* Diagrams

Now that we have introduced both tabular and graphical representations of data for vapor–liquid equilibrium of binary mixtures, we take an aside to mention additional

Another term for the K-factor is called the "distribution ratio."

ways of expressing this data. In particular, it is common to use the terms "**partition coefficient**," "**distribution coefficient**," or "**K-factor**" to describe the ratio of the mole fraction of a component in one phase relative to another phase. For vapor–liquid equilibrium, we define the K-factor as:

$$K_i = \frac{y_i}{x_i} \tag{10.10}$$

For example, from Table 10-2 we find that the K-factor for benzene at 101.3 kPa and 371.55 K is:

$$K_b = \frac{y_b}{x_b} = \frac{0.51}{0.30} = 1.7$$

While for toluene at the same state it is:

$$K_t = \frac{y_t}{x_t} = \frac{0.49}{0.70} = 0.7$$

For Raoult's Law, the K-factor simply reduces to:

$$K_i = \frac{y_i}{x_i} = \frac{P_i^{sat}}{P}$$

Since benzene is more volatile (relative to toluene), it will have a K-factor larger than 1 (it appears preferentially in the vapor), while toluene will have a K-factor less than 1.

Another common term used for vapor–liquid equilibrium is called the relative volatility and it provides the ratio of the K-factors for a mixture. For example, in the benzene + toluene system, we find that at 101.3 kPa and 371.55 K, the **relative volatility** ($\alpha_{b,t}$) is

$$\alpha_{b,t} = \frac{K_b}{K_t} = \frac{1.7}{0.7} = 2.43 \tag{10.11}$$

For Raoult's Law, the relative volatility simply reduces to:

$$a_{i,j} = \frac{K_i}{K_j} = \frac{P_i^{sat}}{P_j^{sat}}$$

The relative volatility, which is normally written with the K-factor of the more volatile component in the numerator, is one measure to describe the ease of separation of two components through distillation. In short, the larger the relative volatility between two components, the more efficient their separation will be through distillation. In general, more efficient separations will require smaller distillation columns, less energy usage, etc. For example, for the ethanol + water system shown in Figure 10-6, we find the relative volatility at 303.15 K and 7 kPa equal to $a_{e,w} = 7.36$, which means that (in general) the ethanol + water system is easier to separate than the benzene + toluene system. However, this is a state-dependent observation since the ethanol + water system (as observed in Figure 10-6) has an azeotrope, making separation as the composition approaches the azeotrope exceedingly difficult (since the relative volatility at the azeotrope is, by definition, 1). Very clearly, the K-factors and the relative volatility are composition dependent quantities.

While we have presented Txy and Pxy diagrams earlier in this chapter, it is not uncommon to see plots for the vapor phase composition plotted as function of the liquid phase composition for the more volatile component in a mixture. This ratio is, of course, the K-factor for that component. Indeed, the large vapor-equilibrium database for mixtures, commonly referred to as the "DECHEMA Database," graphically provides these xy diagrams for many mixtures (J. Gmehling and U. Onken, 1977). While Figure 10-3 provides the Txy diagram for the benzene (1) + toluene (2) system at 101.3 kPa, we show the xy diagram for this system at the same pressure in Figure 10-7.

For the benzene + toluene system at this state, the xy diagram is typical for a system obeying Raoult's Law. A 45-degree line, where $y = x$, is typically shown in a xy diagram and is often used as a quick guide to estimate the distillation column size

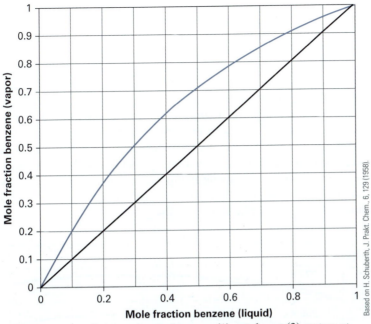

FIGURE 10-7 *xy* diagram for the benzene (1) + toluene (2) system at 101.3 kPa.

needed to effect a desired separation, though that is beyond the scope of this class. Also, where an azeotrope is present, the 45-degree line bisects the equilibrium line at the azetropic composition. An example of this is shown in Figure 10-8 for the *n*-pentane + methanol system at 422.6 K, where the azetropic composition is readily observed at about 55% *n*-pentane.

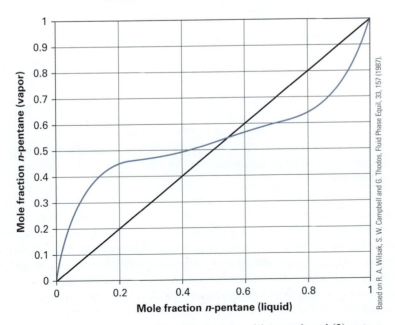

This is the same system as in Figure 10.4 at the same temperature.

FIGURE 10-8 A *xy* diagram for the *n*-pentane (1) + methanol (2) system at 422.6 K. Note the azeotrope at about 55% *n*-pentane.

10.3 Mixture Critical Points

F
|
F—C—H
|
F

Trifluoromethane

F
|
F———Cl
|
F

Chlorotrifluoromethane

It is often convenient to present vapor–liquid equilibrium data for a system at multiple states on the same graph. This allows for a quick examination of the effect of a particular state change (temperature or pressure) on the phase equilibrium that exists. For example, Figure 10-9 shows the trifluoromethane (1) (CHF_3) and chlorotrifluoromethane (2) ($CClF_3$) system at four temperatures that span a range of about 75 K. On the vertical axes (both left and right), you can see how the vapor pressures of the pure components change with the temperature. Because of this large change in vapor pressures, the data is often best presented in a semi-logarithmic plot, where the pressure axis is logarithmic. Note that this system does exhibit an azeotrope at all four temperatures, and the azeotropic composition migrates from 55% by mole trifluoromethane at 199 K to 65% by mole trifluoromethane at 273 K. The azeotropic composition is denoted as a filled circle on the graph.

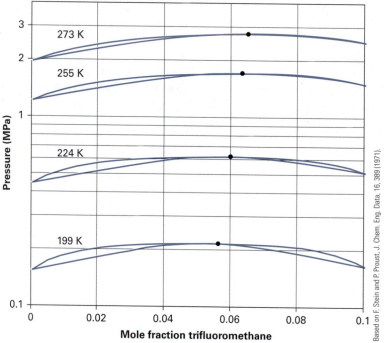

FIGURE 10-9 *Pxy* plot for the trifluoromethane (1) + chlorotrifluoromethane (2) system at four different temperatures. The filled circles are the azeotropic compositions.

F
|
F—C—F
|
F

Carbon tetrafluoride

The second component in the mixture is chlorotrifluoromethane and has one chlorine and three fluorine atoms bonded to a single carbon atom. What would be the impact on our mixture if we replaced that one chlorine atom with a fluorine atom, resulting in tetrafluoromethane, which is more commonly known as carbon tetrafluoride? Figure 10-10 shows the trifluoromethane (1) + tetrafluoromethane (2) system.

This simple swapping of two halogens (chlorine for fluorine) impacts the phase equilibrium in a significant way. Rather than being able to look at both vertical axes to find the pure component vapor pressure, we see that the left axis (the one corresponding to the pure tetrafluoromethane) "ends" somewhere between 225 K and 255 K. This is because the critical temperature of carbon tetrafluoride is 227 K and, thus, this will impact the "critical point" for the mixture. Note that, by way of

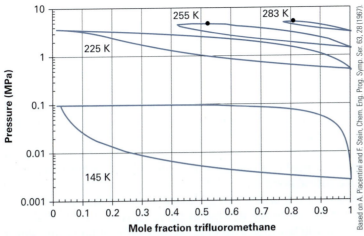

FIGURE 10-10 *Pxy* plot for the trifluoromethane (1) + tetrafluoromethane (2) system at four different temperatures. The "•" denotes the mixture critical points.

comparison, the critical temperature for chlorotrifluoromethane is 302 K, while the critical point for trifluoromethane is 299 K. Thus, this issue of a critical point for a mixture is not observed for the states shown in Figure 10-9.

As can be seen in Figure 10-10, the critical point for the mixture is different at two different temperatures (255 K and 283 K). At each of those critical temperatures, a different critical pressure and critical composition is realized as well (as shown with an "•" on the curves). On a graph such as this, if above the critical point, the maximum pressure always will be the critical pressure, since that is the point where the tie lines (that describe the composition of the liquid in equilibrium with the vapor) converge to a single point. Thus, to the right of the "•" will be the liquid phase, and to the left of the "•" will be the vapor phase. Since the dew-point curve "wraps around" here, we have two dew-point pressures associated with one temperature. Such a situation results in an unusual, yet useful, phenomenon called **retrograde condensation**. Let's take a moment to describe retrograde condensation through a qualitative description, as given in Figure 10-11.

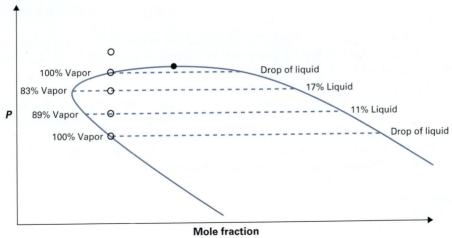

FIGURE 10-11 A qualitative description of a *Pxy* plot for a system showing retrograde condensation.

Figure 10-11 is basically an enlarged version of the portion of the *Pxy* curve near the critical point, similar to what you might see in Figure 10-10 at *T* = 255 K or *T* = 283 K. It is drawn this way to be representative to help the reader understand the phenomenon.

In Figure 10-11, the "•" is the critical point of some mixture at a given temperature, with the dew point and bubble point curves appearing to the left and right of the critical point, respectively. Let's follow the five empty black circles from the top point to the bottom point, assuming they are all at the same overall composition and same temperature.

The top-most circle depicts a mixture that is above the critical point. The phase of the mixture is neither liquid nor vapor but is a super-critical mixture (as it is above the mixture critical point).

If we lower the pressure a little, we will arrive at the second empty black circle from the top. It sits on the dew-point curve (it is to the left of the "•") and, thus, is made up of 100% vapor with only the first drop of liquid present. In our experience up to this point, if we were to drop the pressure even more from the dew point, we would end up with a superheated vapor (such as in the *T* = 225 K curve in Figure 10-10; remember that the bottom curve is the dew-point curve for a *Pxy* plot). Here, the situation is different, since the dew point curve "wraps around" as a result of the mixture having a critical point at this state. When we lower the pressure even more (the middle, or third, empty black circle), we find that, in fact, we *increase* the amount of liquid (from the first drop of liquid at the dew point to around 17% here). Thus, lowering the pressure resulted in an *increase* in liquid, which is a counter-intuitive result. This phenomenon is known as retrograde condensation since we are condensing the vapor to form liquid by *lowering* the pressure. Ultimately, as we decrease the pressure even more, we now decrease the amount of liquid present (from 17% to 11% for the fourth empty black circle). Finally, when we decrease the pressure enough, we make it back to the dew-point curve where we have 100% vapor and a drop of liquid. The phenomenon of retrograde condensation is important in the processing of hydrocarbon mixtures, especially in their transport in pipelines.

As seen in Figure 10-10, two mixture critical points are shown, which are functions of the temperature and pressure of the system. The functionality of the three variables (composition, temperature, and pressure) is best displayed in a pressure vs. temperature (*P-T*) plot for the mixture at different specified compositions. For example, Figure 10-12 shows the pressure versus temperature plot for a mixture of ethylene (1) and *n*-heptane (2).

The left-most curve in Figure 10-12 is just the vapor pressure for pure ethylene. As you would expect, it terminates at its critical point. Likewise, the right-most curve provides the vapor pressure for pure *n*-heptane and it terminates at its critical point. What is in between, however, is the *P-T* plots for this mixture at various compositions. Let's follow the second curve from the left that shows the *P-T* information for a mixture at 78% ethylene and 22% *n*-heptane. The curve goes up to a maximum at about 10.3 MPa and 125°C and then comes back down again. Let's pick a pressure on this curve, say 2 MPa. We see that for 2 MPa, there are two intersections with this curve: one at around −20°C and the other at around 145°C. The right-most intersection (the one that occurs at a *higher* temperature) is the vapor, whereas the left-most intersection is the liquid. The point near the top (though not necessarily at the top) is the critical point for this composition (and is where two phases cease to coexist and one "super-critical" phase exists).

An understanding of the behavior of hydrocarbons (including retrograde condensation) is important for chemical engineers, as many work in the petroleum industry, where hydrocarbon processing plays a major role.

Ethylene is such an important industrial compound that it is much more often referred to using this name—as opposed to its official IUPAC name of "ethene."

Ethene (ethylene)

n-heptane

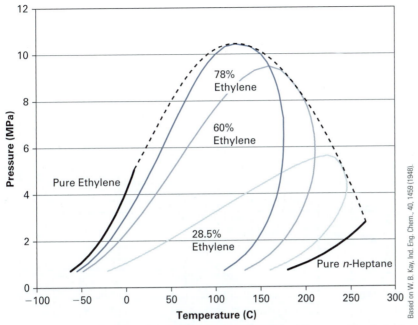

FIGURE 10-12 A *P-T* plot for the ethylene (1) + *n*-heptane (2) system. The dashed curve is the locus of critical points for the mixture.

One can present multiple *P-T* plots for different compositions and denote where the critical point occurs (we show three mixture compositions in Figure 10-12). Then we can connect these critical points as a function of composition, and this denotes the locus of critical point (that starts and finishes at the critical points of the pure components).

10.4 Lever Rule and the Flash Problem

Recall the previous discussion defining bubble point and dew point. For the bubble point, we had an equilibrium mixture that was all liquid—except for the first bubble of vapor. And, as such, our phase point was on the bubble-point curve. Likewise, the dew point described a system that was an equilibrium mixture of all vapor—except for the first drop of liquid (or dew). Between the bubble and dew point, a variety of *amounts* (mass or moles) of each equilibrium phase can be realized. At the bubble point, the system is approximately 100% liquid and the single bubble of vapor. At the dew point, it is approximately 100% vapor and the single drop of liquid. Depending on where the equilibrium point appears on the line connecting the phases (called the tie line), this will tell us the amount of each phase present. This can be solved two ways: (1) using a lever rule and (2) performing a material balance.

The **lever rule** is a simple geometric argument that exploits the linearity of the relationship between composition and mass (or moles). If the composition on a *Pxy* or *Txy* plot is given as a mole fraction, the resulting lever rule is for moles. Alternatively, if the composition is given in terms of a mass fraction, the resulting lever rule is for mass. Since the lever rule works the same way for a *Pxy* or a *Txy* plot, we describe it for the *Pxy* plot in the following example.

While there is a person named Lever in the chemical process industry (William Hesketh Lever founded Lever Brothers, now Unilever, which makes many brands you will likely recognize), the name "lever rule" is referencing the simple machine with the fulcrum as the location of the overall composition of the mixture that splits into two phases.

| EXAMPLE 10-2 | **LEVER RULE** |

If we mix 3 moles of R-134a (1) and 7 moles of R-245fa (2) together at 293 K and 200 kPa, what are the composition and amounts of the phase(s) present at equilibrium? Use the lever rule.

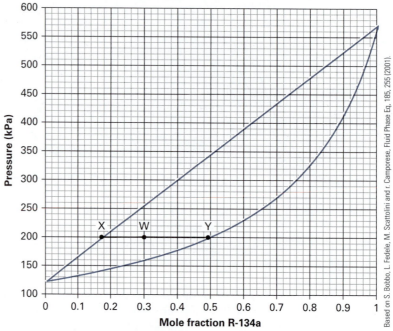

FIGURE 10-2 *Pxy* plot for the R-134a (1) + R-245fa (2) system at 293.15K (Repeated).

SOLUTION:

Step 1 *Where is the point on the mixture phase diagram? Liquid, vapor, or both?*

Fortunately, we have a phase diagram for this system (Figure 10-2) that encompasses the state (T and P) relevant to this problem. When we mix the pure compounds together, the composition of the mixture is 30% by mole of R-134a. To avoid confusion with the nomenclature we've used to denote mole fraction of liquids (x) and vapors (y), we can call this overall composition "w." Therefore, the overall composition is $w_1 = 0.3$.

When we look at Figure 10-2, it is for a fixed temperature (293.15 K), and we can find on the figure where the system is both at 200 kPa and $w_1 = 0.3$. We see that it is in between the bubble- and dew-point curves, so we have both phases (liquid and vapor) present. Recall that if the point describing our system is above the bubble-point curve at this composition and temperature, such as at 300 kPa, the answer to the problem will be very easy: We would have 10 moles of liquid with a composition $x_1 = 0.3$. Likewise, if we are below the bubble-point curve at this composition and temperature, such as at 100 kPa, once again the answer will be very easy: We would have 10 moles of vapor with a composition of $y_1 = 0.3$. However, both of these situations are not the case here, and we need to determine the composition and amount of the liquid and vapor pressure.

Step 2 *Application of the lever rule*

We know that no single equilibrium phase exists in our problem for this mixture at the temperature, pressure, and composition we have assigned, but that our system splits into two phases. Specifically, this mixture will split into a liquid phase at $x_1 = 0.17$ and a vapor phase at $y_1 = 0.49$ (per Figure 10-2). The lever rule will help us find the amount (here, the number of moles) of each phase present.

Let X represent the point where the tie line intersects the left-most curve (bubble for a Pxy plot; dew for a Txy plot). In our problem, this is at $x_1 = 0.17$ for the $P = 200$ kPa tie line. Let Y represent the point where the tie line intersects the right-most curve (dew for a Pxy plot; bubble for a Txy plot). In our problem, this is at $y_1 = 0.49$ for the $P = 200$ kPa tie line. Let W represent the composition of the overall mixture, which is at $w_1 = 0.3$ for the $P = 200$ kPa tie line.

The amount of the left-most phase (the liquid phase for a Pxy plot; vapor for a Txy plot) is given as a ratio of the length of the line segment $\overline{WY}$ to that of the length of the line segment $\overline{XY}$, multiplied by the overall system amount (N). Thus,

$$\text{Liquid amount} = N(\overline{WY}/\overline{XY}) \tag{10.12}$$

Likewise, the vapor amount (for a Pxy plot) is given as

$$\text{Vapor amount} = N(\overline{XW}/\overline{XY}) \tag{10.13}$$

Alternatively, the vapor amount can be determined by subtracting the liquid amount from the overall system amount through a simple material balance.

Step 3 *Using the lever rule to obtain the amounts of each equilibrium phase*
Implementing Equations 10.12 and 10.13 for the distances relevant in Figure 10-2, we arrive at

$$\overline{XW} = 10.3 \text{ mm}$$
$$\overline{XY} = 26.2 \text{ mm}$$
$$\overline{WY} = 15.9 \text{ mm}$$

Therefore, the amount of liquid becomes

$$\text{Liquid amount} = 10 \text{ moles} \left(\frac{15.9 \text{ mm}}{26.2 \text{ mm}}\right) = \textbf{6.1 moles}$$

The amount of vapor is

$$\text{Vapor amount} = 10 \text{ moles} \left(\frac{10.3 \text{ mm}}{26.2 \text{ mm}}\right) = \textbf{3.9 moles}$$

By the way, you might ask yourself, why is this called the lever rule, since you are using the length of the right side of the "lever" to get the amount for the liquid phase, the composition of which is given on the left side of the lever? And you would be correct as this technique, used here, is sometimes called the *inverse* lever rule. Whatever it is called, it is a simple, quick, and intuitive way to estimate the amounts of two equilibrium phases. However, what if we wanted to do a more precise calculation? We certainly can, and that falls under the heading of a "flash problem."

A flash distillation unit (Figure 10-13) takes a liquid mixture and lowers the pressure so that the mixture is now between the bubble- and dew-point curves. On entering the vessel, a portion of the liquid mixture "flashes" to the vapor phase (at a different composition). Thus, this approach can be used as a technique to separate two components in the liquid (since the more volatile liquid will appear in a higher concentration in the vapor). This is precisely what was done in the previous problem, where we took our liquid mixture of R-134a and R-245fa and lowered the pressure.

While many people understand and use the lever rule in an intuitive way, it can also be derived and proven mathematically, as shown in the following "flash problem."

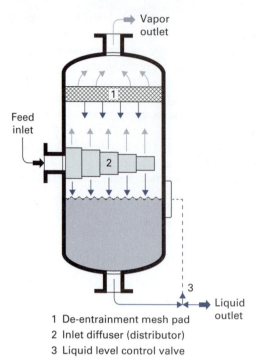

FIGURE 10-13 A flash evaporator (or flash drum).

Practically, the flash vessel itself is just an empty chamber with a large diameter relative to the inlet pipe diameter. This sudden expansion upon entering the vessel from the inlet pipe is responsible for the pressure drop that (hopefully) places the liquid into the two-phase region. Flash vessels can be adiabatic (if insulated) or isothermal (if provided with a mechanism to transfer heat, such as heating jacket). Here, we consider the same basic problem as Example 10-2 but now solve it as a flash problem.

EXAMPLE 10-3	FLASH PROBLEM USING EXPERIMENTAL DATA

In problems involving material and/or energy balances, the first step has normally been to define a system and write the balance equations. It wouldn't be wrong to start this problem that way, but we'd immediately realize we need to know *how many exit streams* there are. For example, if no flash occurs, there is only one stream (liquid) leaving. Here step 1 demonstrates that our balance equations must account for two exit streams.

A liquid mixture of R-134a (1) + R-245fa (2) contains 30% by mole R-134a. The mixture is at 600 kPa and 293.15 K and is fed at a rate of 10 mol/s into a steady-state flash distillation unit operating at 200 kPa. What is the resulting composition and molar flow rate of the equilibrium streams exiting the flash distillation unit?

SOLUTION:

Step 1 *Does the system flash at all? Does it flash partially? Does it flash fully?*
From Figure 10-2, the system is a liquid at 600 kPa and 293 K. We know from Example 10-2 that, at 200 kPa and 293.15 K, the system is indeed in the two-phase region. Thus, we know that it will flash partially, since 200 kPa is between the bubble (255 kPa) and dew point (160 kPa) pressures at this temperature and composition, per Figure 10-2. Next, we can move to find the amount and composition of the liquid that is in equilibrium with the vapor.

Step 2 *Draw a picture*
The composition of the inlet stream can be described as $w_1 = 0.3$. A diagram of the system and known information is shown in Figure 10-14.

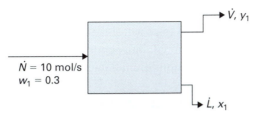

FIGURE 10-14 A diagram of the flash process.

Conventionally, we will use L to represent the liquid stream (such as an amount or flow rate) and V to represent the vapor stream (such as an amount or flow rate). For flow rates, we use a "dot" above the letter to denote it as a flow rate.

Step 3 *Write material balances and perform a degree of freedom analysis*
Since this is a binary system, we will be able to write two independent material balances. Let us write these for the overall material balance and for component 1.

$$\text{Overall Balance:} \qquad \dot{N} = \dot{L} + \dot{V}$$

$$\text{Component 1:} \qquad w_1\dot{N} = x_1\dot{L} + y_1\dot{V}$$

What are the unknowns in the problem? There are four: $\dot{L}$, $\dot{V}$, y_1, and x_1. With only two equations and four unknowns, our problem is underspecified. However, in this case, we do have the Pxy plot (Figure 10-2) to provide us with two additional process specifications, namely y_1 and x_1. As before, $x_1 = 0.17$ and $y_1 = 0.49$. We now have zero degrees of freedom and can proceed to solve the problem.

Step 4 *Solve the problem*
Performing the simple algebra on the two equations yields the following relationship for $\dot{L}$, as

$$\dot{L} = \dot{N}\frac{(y_1 - w_1)}{(y_1 - x_1)} \qquad (10.14)$$

Note that this relationship looks very familiar and is, in fact, the *same result* as the lever rule, as

$$\text{Liquid amount} = N(\overline{WY}/\overline{XY}) \qquad (10.12)$$

except that in the lever rule you had to *measure* line segments, and here we are using the values of the endpoints of the line segments to determine, effectively, the length of the line segment. Thus, the lever rule *is* the material balance equation.

Plugging in the composition values yields

$$\dot{L} = \left(10\ \frac{\text{mol}}{\text{s}}\right)\left[\frac{(0.49 - 0.30)}{(0.49 - 0.17)}\right] = \textbf{5.94 mol/s}$$

Likewise, $\dot{V}$ is given as $\dot{N} - \dot{L}$ from an overall material balance. Thus, $\dot{V} = \textbf{4.06 mol/s}$.

The reason why the answer for $\dot{L}$ and $\dot{V}$ is different than the answer from Example 10-2—although we are solving the same problem—is associated with how accurately you can measure something with a ruler.

What if we have a problem where we do *not* have experimental data for x_1 and y_1, nor do we know the bubble point and dew point information for the system at the "feed" composition? At this point, we will still have two equations (the two material balances as demonstrated in Example 10-3), but now we will have four unknowns (assuming the system partially flashes). Therefore, we need two additional, independent equations that relate the unknown variables in the problems. This is precisely where modeling the relationship between the equilibrium liquid and vapor phases is required. To illustrate this point, we will solve a problem using a model (Raoult's Law) to provide our two additional, independent equations (the relationship between the compositions of the equilibrium liquid and vapor phases).

| EXAMPLE 10-4 | **FLASH PROBLEM USING A MODEL FOR THE LIQUID AND VAPOR PHASES** |

Ten moles per second of an equimolar liquid mixture of water (1) + ethanol (2) at 30°C and 11 kPa is flashed to 7.0 kPa. What is the resulting composition and molar flow rate of the equilibrium streams exiting the flash distillation unit? Model this mixture using Raoult's Law.

SOLUTION:

Step 1 *Does the system flash at all? Does it flash partially? Does it flash fully?*
Here our goal is to determine the bubble-point pressure and dew-point pressure for our mixture at 30°C. If the bubble-point pressure is less than the system pressure (7.0 kPa), we still have an equimolar liquid mixture, just at a lower pressure. If the dew-point pressure is higher than the system pressure (7.0 kPa), we have flashed all of our mixture to a vapor and now have an equimolar vapor mixture. Both scenarios do not result in any separation, which is the point of the distillation.

Recall that Raoult's Law for both components is given as

$$Py_1 = x_1 P_1^{sat} \tag{10.3}$$

$$Py_2 = x_2 P_2^{sat} \tag{10.9}$$

We will use these relationships to estimate the bubble-point and dew-point pressures at our mixture composition (equimolar) and 30°C. The first step is to determine the vapor pressures of the pure components. We can obtain the vapor pressures for water (1) and ethanol (2) at 30°C using the Antoine equation, with parameters provided in Appendix E. Plugging in the values, we arrive at

$$P_1^{sat}(30°C) = 4.25 \text{ kPa}$$

$$P_2^{sat}(30°C) = 10.56 \text{ kPa}$$

PITFALL PREVENTION

Remember, the left-hand side would be
$Py_1 + Py_2 = P(y_1 + y_2)$
$= P$

As mentioned previously, we can add the Raoult's law expressions to obtain the bubble-point pressure relationship as

$$P = x_1 P_1^{sat} + x_2 P_2^{sat} \tag{10.2}$$

Plugging in numbers for our problem, we arrive at

$$P = (0.5)(4.25 \text{ kPa}) + (0.5)(10.56 \text{ kPa})$$

$$P = 7.41 \text{ kPa}$$

Since the bubble-point pressure is above that of the system pressure (7.0 kPa), we know that we do not have all liquid. Next, let's determine the dew-point pressure to see if we have just a vapor or both liquid and vapor.

Once again, we use Raoult's Law to model our system, but now we use it to calculate the dew-point pressure. However, Raoult's Law is given in terms of the liquid composition, and we need the vapor composition to estimate the dew-point pressure. Thus, we have to go back to the definition of Raoult's Law in order to expose the vapor composition (y) to obtain the dew-point pressure.

Raoult's Law is given as

$$Py_1 = x_1 P_1^{sat} \tag{10.3}$$

$$Py_2 = x_2 P_2^{sat} \tag{10.9}$$

Rewriting this pair of Raoult's Law equations, we have

$$\frac{Py_1}{P_1^{sat}} = x_1$$

$$\frac{Py_2}{P_2^{sat}} = x_2$$

Adding these two expressions together gives us a right-hand side equal to one (since $x_1 + x_2 = 1$). Thus,

$$\frac{Py_1}{P_1^{sat}} + \frac{Py_2}{P_2^{sat}} = 1$$

Solving for P, we obtain the expression we are looking for, which is the dew-point pressure expression from Raoult's Law.

$$P = \frac{1}{\left(\dfrac{y_1}{P_1^{sat}} + \dfrac{y_2}{P_2^{sat}}\right)} \tag{10.15}$$

Remember, this is for the dew point pressure at the feed composition.

Now we substitute the feed composition for the dew-point composition to obtain the dew-point pressure:

$$P = \frac{1}{\left(\dfrac{0.5}{4.25 \text{ kPa}} + \dfrac{0.5}{10.56 \text{ kPa}}\right)} = 6.06 \text{ kPa}$$

Since the system pressure (7.0 kPa) is between the bubble (7.41 kPa) and dew point (6.06 kPa) pressure, our liquid system will partially flash, allowing a separation to occur.

Step 2 *Draw a picture*
Like the previous example problem, we can draw a diagram and label the variables. This is given in Figure 10-15.

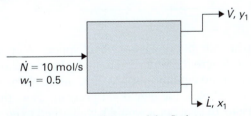

FIGURE 10-15 A diagram of the flash process.

Step 3 *Write material balances and perform a degree of freedom analysis*
Since this is a binary system, we will be able to write two independent material balances. Let's write this for both the overall material balance and the material balance of component 1.

Overall Balance: $\dot{N} = \dot{L} + \dot{V}$

Component 1: $w_1 \dot{N} = x_1 \dot{L} + y_1 \dot{V}$

What are the unknowns in the problem? There are four: $\dot{L}$, $\dot{V}$, y_1, and x_1. With only two equations and four unknowns, our problem is underspecified. However, we do have another pair of equations that relate the unknown variables in the problem, namely the Raoult's Law expressions for components 1 and 2.

The problem now has four equations and four unknowns and is properly specified.

Step 4 *Solve the problem*

This four-equation, four-unknown system is solved easily using the Solver feature in MS Excel (or any equation solver, including many calculators).

To reiterate, the four equations are

$$10\,\frac{\text{mol}}{\text{s}} = \dot{L} + \dot{V}$$

$$(0.5)\left(10\,\frac{\text{mol}}{\text{s}}\right) = x_1\dot{L} + y_1\dot{V}$$

$$(7\text{ kPa})y_1 = x_1(4.25\text{ kPa})$$

$$(7\text{ kPa})(1 - y_1) = (1 - x_1)(10.56\text{ kPa})$$

and four unknowns: $\dot{L}$, $\dot{V}$, y_1 and x_1.

The results are as follows:

$$\dot{L} = 7.11\,\frac{\text{mol}}{\text{s}}$$

$$\dot{V} = 2.89\,\frac{\text{mol}}{\text{s}}$$

$$x_1 = 0.56$$

$$y_1 = 0.34$$

To recap, we have taken 10 mol/s of an equimolar liquid mixture of water (1) and ethanol (2) at 30°C and 11 kPa and flashed it to 7.0 kPa. A liquid stream exited the flash distillation unit at 7.11 mol/s with a composition of 56% (by mole) water, while a vapor stream also exited the flash distillation unit at 2.89 mol/s with a composition of 34% (by mole) water vapor. Thus, we have somewhat separated our equimolar liquid mixture into a liquid portion that has an enhanced water composition and a vapor portion that has an enhanced ethanol composition. The more volatile component (ethanol) is concentrated in vapor phase.

When using models or theories to help describe how systems behave, engineers have to make sure that they are validating the assumptions inherent for those models or theories. A quick and obvious way to demonstrate to others that you don't know what you are doing is by using models or theories outside of their assumptions (such as the ideal gas law to describe a liquid).

In reality, the recap in Example 10-4 is *not* what happens. Why? Because reality is depicted in Figure 10-6, where the water + ethanol system is shown as solid lines (the experimental vapor–liquid equilibrium). At 30°C and 7.0 kPa, an equimolar mixture is already at a pressure below the dew-point pressure. Thus, the system would have flashed from 11 kPa to 7.0 kPa, but it would have flashed to *all* vapor, resulting in no separation effect.

As is illustrated through Example 10-4, models relating the composition of the liquid and vapor phases are needed to solve flash problems when no experimental data exists. However, just because a model provides an answer, it doesn't make it the right answer. The chemical engineer is tasked with identifying what models to use and when to use them in order to make reasonable estimations for analyzing processes. Here, we already knew that the water + ethanol system would not behave as an ideal solution. Thus, using Raoult's Law as a model for this system was, from the start, not a reasonable strategy, and this was validated in comparing the model results to that of experiment.

Since all we know about vapor–liquid models for mixtures is Raoult's Law (and this is only valid for a relatively small subset systems), how do we develop better correlative and predictive approaches? We will do this in Chapter 11, but we will have to start out at the beginning and introduce the formality of equilibrium between the liquid and vapor phases for mixtures. Such an approach will provide the framework for developing the modeling approaches used by chemical engineers for mixture phase equilibrium (for liquids–vapors and other types of phase equilibrium).

Before we turn our attention to the more advanced modeling techniques in the next chapter, we solve a vapor–liquid equilibrium problem using Raoult's Law where the system is *not* at constant temperature—but at constant pressure. Such an approach will allow for a contrast in the calculations involving constant temperature versus constant pressure systems.

CALCULATING THE *TXY* CURVE FOR A SYSTEM | **EXAMPLE 10-5**

Does an equimolar mixture of methanol (1) + acetone (2) exist as one or two phases at 101.325 kPa and 334 K? Plot the entire *Txy* curve to find out. Use Raoult's Law to estimate the phase coexistence.

SOLUTION:

Step 1 *Make a selection of the variables to specify*
The problem asks us to create the entire *Txy* curve, so the implication is that we have a fixed pressure system (101.325 kPa). When we consider Gibbs phase rule, we have a two-component system and two equilibrium phases (since we are creating a *Txy* curve). This leaves us with two independent variables we can specify. We have already fixed one ($P = 101.325$ kPa), so we have to decide on the other. As we have seen in Example 10-4, it is a little more straightforward to fix the liquid phase composition than the vapor phase composition—but either will work, and both must give you the same *Txy* curve. Thus, we will fix the liquid composition for our calculations.

Step 2 *Develop a strategy to create the Txy curve*
Since we want to create the entire *Txy* curve at $P = 101.325$ kPa, we have to start with a liquid phase mole fraction of $x_1 = 0$ and go to $x_1 = 1$. This requires us to vary the liquid phase composition from 0 to 1. We have a choice in how we vary the liquid phase mole fraction. We can take a small step size in liquid phase mole fraction (say, 0.001), but that would require us to solve for the system temperature and vapor phase composition about 1000 times. Alternatively, we could take very large steps in the liquid phase mole fraction (say, 0.25), but that really wouldn't provide us with enough detail to adequately present the *Txy* curve (although we would only have to solve the problem three times: $x_1 = 0.25$, $x_1 = 0.50$, and $x_1 = 0.75$). A reasonable step size that will provide adequate detail for the *Txy* curve and not require us to solve the problem nearly 1000 times is a step size for the liquid phase mole fraction of 0.05.

We wouldn't solve the problem at $x_1 = 0$ or $x_1 = 1$, since we already know the temperature and vapor phase compositions at the endpoints (which are the pure components).

Step 3 *Solving the problem at $x_1 = 0.05$*
Unlike Example 10.4, we are not solving a flash problem here, since we are not interested in the *amounts* of the components in the liquid and vapor phases. We are only concerned with the compositions. Thus, we do not have any material balances to consider, nor do we have those unknowns. The only unknowns in the problem (once we specify the pressure and liquid phase mole fraction) are the temperature (T) and the vapor phase mole

We show the vapor pressure of the pure components explicitly here as a function of temperature to remind the reader that the vapor pressure will change with temperature (since temperature is not fixed in this problem).

fraction of component 1 (y_1). For these two unknowns, we need two equations and those are given by Raoult's Law:

$$Py_1 = x_1 P_1^{sat}(T) \tag{10.3}$$

$$Py_2 = x_2 P_2^{sat}(T) \tag{10.9}$$

Once again, we can add these equations together to obtain

$$P = x_1 P_1^{sat}(T) + x_2 P_2^{sat}(T) \tag{10.2}$$

If we fixed the vapor phase composition instead of the liquid phase composition, we would use Equation 10.15 instead of Equation 10.2

Since we are fixing x_1, we know x_2. However, since we don't know the temperature, we cannot calculate the vapor pressure directly (since we need to know the temperature). Thus, the phase equilibrium problem where we fix the pressure (as in this example) is more challenging, since we have temperature as an unknown (and cannot calculate the vapor pressure directly).

More precisely, if we provide the vapor pressure using Antoine's equation, we can see directly where the issue is:

$$P = x_1 P_1^{sat}(T) + (1 - x_1)P_2^{sat}(T)$$

$$P = x_1 10^{\left(A_1 - \frac{B_1}{T + C_1}\right)} + (1 - x_1)10^{\left(A_2 - \frac{B_2}{T + C_2}\right)}$$

If we know the Antoine coefficients for our two substances (and we do, from the Appendix E in the back of the text), the pressure (101.325 kPa), and the liquid phase mole fraction (here, $x_1 = 0.05$), we can solve for T using any of a number of root-finding techniques (Solver from Excel, your calculator, etc.). When you do this, you find that $T = 329.68$ K.

In solving the Antoine equation, you actually obtain T in Celcius. We have converted back to Kelvin here.

To find the vapor phase mole fraction (y_1) when $x_1 = 0.05$, we employ Raoult's Law for component 1:

$$y_1 = (x_1 P_1^{sat}(T))/P \tag{10.4}$$

Since we have already solved for T at this point in the problem, y_1 can be obtained directly and is equal to 0.036.

Step 4 *Solving the problem at $x_1 = 0.10, 0.15, \dots, 0.90, 0.95$*
Once we solve the problem at $x_1 = 0.05$, we step to $x_1 = 0.10$ and solve the problem all over again, obtaining new values of the temperature and the vapor phase mole fraction. We repeat this process of stepping in the liquid phase mole fraction until we have spanned the entire composition region, recording the values of the temperature and vapor phase mole fraction at each step.

One can write a computer code or a macro in Excel, for example, to do this iterative process more efficiently.

Step 5 *Plotting the results*
We plot T vs. x_1 and T vs. y_1, which give us our dew-point and bubble-point curves, shown in Figure 10-16 as the solid black lines. Note again that the bubble-point curve is not a straight line for Raoult's Law on a Txy plot, since the vapor pressure changes with temperature.

To answer the question in the problem, Raoult's Law *does* predict a two-phase system for an equimolar mixture of methanol (1) + acetone (2) at 101.325 kPa and 334 K, as can be seen in Figure 10-16. That point on the diagram is between the dew-point and bubble-point curves, so Raoult's Law predicts the existence of a liquid phase (of about 47% by mole methanol) in equilibrium with a vapor phase (of about 56% by mole methanol).

What is also shown on Figure 10-16 is the experimental data for this system. Clearly, this is a system that does *not* behave according to Raoult's Law, as the system is not an ideal solution. Acetone is a ketone that does not hydrogen bond with itself, while methanol is an alcohol that does hydrogen bond with itself. Also, acetone can form hydrogen bonds with the methanol in the mixture.

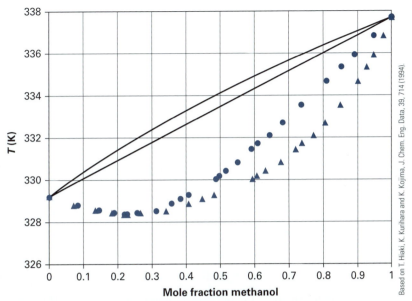

Mole fraction methanol

FIGURE 10-16 A *Txy* plot of the methanol (1) + acetone (2) system at 101.325 kPa. The Raoult's Law predictions are given as the solid curves, while the experimental data are provided in the symbols.

Once again, this problem motivates the need to account for the non-idealities in the liquid phase in order to make the modeling of liquid–vapor phase equilibrium for mixtures more useful and predictive. We discuss this in Chapter 11.

10.4.1 Ternary Systems

Up to this point in the book, we have not discussed mixtures beyond binary systems. Of course, mixtures with three, four, five, etc. components are commonplace, especially when leaving a chemical reactor, for instance, where normally the products (and some reactants) are in the effluent of the reactor. So while the approaches developed in this book for mixtures are easily extended to a system of n components, we demonstrate the flash problem for a ternary system to expose this more clearly for the reader.

FLASH PROBLEM FOR TERNARY SYSTEMS | **EXAMPLE 10-6**

Benzene (1), toluene (2), and ethylbenzene (3) are by-products of a reaction. 10 moles/s of this liquid mixture is flashed from 200 mm Hg (26.7 kPa) and 50°C to 100 mm Hg (13.3 kPa). If the mole fraction of benzene and toluene are both 40% (with the balance ethylbenzene) entering the flash distillation unit, what is the resulting composition and molar flow rate of the equilibrium streams exiting the unit? Model this mixture using Raoult's Law.

SOLUTION:

Step 1 *Does the system flash at all? Does it flash partially? Does it flash fully?*
As before, we want to determine if the mixture will flash partially. If it doesn't flash, all we have done is lower the pressure of a liquid mixture. If it flashes fully, then all we have done is transform a liquid mixture (at a particular composition) to a vapor mixture

Benzene

Toluene

Ethylbenzene

> Now the left-hand side would be
> $$Py_1 + Py_2 + Py_3 =$$
> $$P(y_1 + y_2 + y_3) = P$$

(at the same composition). Thus, our first step is to determine the bubble- and dew-point pressures of this mixture at 50°C at the specified composition.

Raoult's Law is a reasonable approach here since each component possesses a benzene ring with only one and two methyl groups added for toluene and ethylbenzene, respectively. Raoult's Law for each of the three components is as follows:

$$Py_1 = x_1 P_1^{sat} \tag{10.3}$$

$$Py_2 = x_2 P_2^{sat} \tag{10.9}$$

$$Py_3 = x_3 P_3^{sat} \tag{10.16}$$

As before, we can use Appendix E to find our Antoine coefficients to estimate the vapor pressure of each of the pure components at 50°C. Plugging in the values, we arrive at

$$P_1^{sat}(50°C) = 271.29 \text{ mm Hg (36.2 kPa)}$$

$$P_2^{sat}(50°C) = 92.11 \text{ mm Hg (12.3 kPa)}$$

$$P_2^{sat}(50°C) = 35.16 \text{ mm Hg (4.7 kPa)}$$

For the ternary mixture, the bubble-point pressure looks (as you would expect) just like the binary mixture for Raoult's Law:

$$P = x_1 P_1^{sat} + x_2 P_2^{sat} + x_3 P_3^{sat} \tag{10.17}$$

Plugging in numbers for our problem, we arrive at:

$$P = (0.4)(271.29 \text{ mm Hg}) + (0.4)(92.11 \text{ mm Hg}) + (0.2)(35.16 \text{ mm Hg})$$

$$P = 152.39 \text{ mm Hg (20.3 kPa)}$$

Since the bubble point pressure is above that of the system pressure (100 mm Hg), we know that we do not have all liquid. We now need to determine the dew-point pressure to determine if we have some liquid or no liquid at all.

For the ternary mixture, we now have:

$$\frac{Py_1}{P_1^{sat}} = x_1$$

$$\frac{Py_2}{P_2^{sat}} = x_2$$

$$\frac{Py_3}{P_3^{sat}} = x_3$$

Adding together, the right-hand side (as before with the binary) will sum to 1, leaving:

$$\frac{Py_1}{P_1^{sat}} + \frac{Py_2}{P_2^{sat}} + \frac{Py_3}{P_3^{sat}} = 1$$

Solving for P, we obtain the expression we are looking for, which is the dew-point pressure expression from Raoult's Law for the ternary system.

$$P = \frac{1}{\left(\dfrac{y_1}{P_1^{sat}} + \dfrac{y_2}{P_2^{sat}} + \dfrac{y_3}{P_3^{sat}} \right)} \tag{10.18}$$

Now we substitute the feed composition for the dew-point composition to obtain the dew-point pressure:

$$P = \frac{1}{\left(\dfrac{0.4}{271.29 \text{ mm Hg}} + \dfrac{0.4}{92.11 \text{ mm Hg}} + \dfrac{0.2}{35.16 \text{ mm Hg}}\right)} = 86.92 \text{ mm Hg (11.6 kPa)}$$

Since the system pressure (100 mm Hg) is between the bubble (152.39 mm Hg) and dew point (86.92 mm Hg) pressure, our liquid system will partially flash, allowing a separation to occur.

Step 2 *Draw a picture*
We can draw a picture and identify the variables through Figure 10.17.

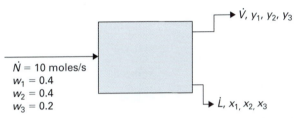

FIGURE 10-17 A diagram of the flash process for the ternary system.

Step 3 *Write material balances and perform a degree of freedom analysis*
For our ternary system, we will be able to write three independent material balances. Let's write this for the overall moles, the moles of component 1 and component 2.

$$\text{Overall Balance:} \qquad \dot{N} = \dot{L} + \dot{V}$$

$$\text{Component 1:} \qquad w_1 \dot{N} = x_1 \dot{L} + y_1 \dot{V}$$

$$\text{Component 2:} \qquad w_2 \dot{N} = x_2 \dot{L} + y_2 \dot{V}$$

What are the unknowns in the problem? There are six: $\dot{L}$, $\dot{V}$, y_1, y_2, x_1 and x_2. With only three equations and six unknowns, our problem is underspecified. However, as before, we have the relationship between the composition of the liquid and vapor phases of each of the components in equilibrium (or, in other words, Raoult's Law). Since we have three Raoult's Law relationships, we have six equations and six unknowns, so the problem is properly specified.

Step 4 *Solve the problem*
To reiterate, the six equations are

$$10 \frac{\text{moles}}{\text{s}} = \dot{L} + \dot{V}$$

$$(0.4)\left(10 \frac{\text{moles}}{\text{s}}\right) = x_1 \dot{L} + y_1 \dot{V}$$

$$(0.4)\left(10 \frac{\text{moles}}{\text{s}}\right) = x_2 \dot{L} + y_2 \dot{V}$$

The reader can treat this as eight equations in eight unknowns, with x_3 and y_3 being two additional unknowns and $x_1 + x_2 + x_3 = 1$ and $y_1 + y_2 + y_3 = 1$ being two additional equations. But, as before, it is easy to write x_3 as $1 - x_1 - x_2$ and y_3 as $1 - y_1 - y_2$.

$$(100 \text{ mm Hg})y_1 = x_1(271.29 \text{ mm Hg})$$

$$(100 \text{ mm Hg})y_2 = x_2(92.11 \text{ mm Hg})$$

$$(100 \text{ mm Hg})(1 - y_1 - y_2) = (1 - x_1 - x_2)(35.16 \text{ mm Hg})$$

and six unknowns: $\dot{L}$, $\dot{V}$, y_1, y_2, x_1 and x_2. This six-equation, six-unknown system is readily solved using the Solver feature in MS Excel (or any equation solver, including many calculators).

The results are as follows:

$$\dot{L} = 2.248 \, \frac{\text{mol}}{\text{s}}$$

$$\dot{V} = 7.752 \, \frac{\text{mol}}{\text{s}}$$

$$x_1 = 0.172$$

$$x_2 = 0.426$$

$$y_1 = 0.466$$

$$y_2 = 0.392$$

And, since the mole fractions in a phase have to sum to 1, $x_3 = 0.402$ and $y_3 = 0.142$

As before, we have somewhat separated our liquid mixture, concentrating the component that boils last (has the smallest vapor pressure—is least volatile), ethylbenzene, into the liquid phase.

160 mm Hg is 21.3 kPa.

As a computational aid for the reader, what would the solution "look like" if we were to select a flash pressure above the bubble-point pressure or below the dew-point pressure? For example, if we solve the above problem for a system pressure of 160 mm Hg (above the bubble-point pressure), the answers are as follows:

$$\dot{L} = 112.16 \, \frac{\text{mol}}{\text{min}}$$

$$\dot{V} = -12.16 \, \frac{\text{mol}}{\text{min}}$$

$$x_1 = 0.437$$

$$x_2 = 0.380$$

$$y_1 = 0.741$$

$$y_2 = 0.219$$

Though the answers satisfy the equations, they are nonsense. The liquid flow rate must be 100 moles/min since the mixture didn't flash. Likewise the composition leaving the flash distillation must be the feed composition since the liquid didn't flash. Thus, if you solve your flash problem and find that either your liquid or vapor flow rates are negative numbers, you likely have an error somewhere in your conceptual understanding of the physical situation (such as being above/below the calculated bubble/dew points).

10.5 SUMMARY OF CHAPTER TEN

■ The data for the vapor–liquid equilibrium of mixtures can be presented either in tabular form or graphical form. The most typical way to present information is to keep either the temperature or the pressure constant and show how the liquid and vapor phase compositions vary. In graphical format, if temperature is kept constant, we call that a *Pxy* plot; if pressure is kept constant, we call that a *Txy* plot.

■ Many mixtures possess **azeotropes**, which means the liquid composition is the same as the vapor composition somewhere in the composition range between the pure component endpoints. Knowledge of the presence of azeotropes is important for separations, since distillation will only be able to separate the components up to the azeotropic composition. Azeotropic compositions for a mixture will change if the temperature or pressure of the system is changed.

■ **Raoult's Law** is a modeling approach for vapor–liquid equilibrium of mixtures where the liquid is treated as an ideal solution, while the vapor is treated as an ideal gas. Raoult's Law is only valid for systems that best conform to the assumptions of ideal solution and ideal gas.

■ On a *Pxy* plot, you can tell quickly if a mixture behaves according to Raoult's Law if the bubble-point curve is a line.

■ **Positive deviations from Raoult's Law** occur for mixtures when the components of the mixture interact more strongly with themselves than with the other component in the mixture. This is normally the situation, although some mixtures show negative deviations from Raoult's Law when the interactions between the components are very beneficial (relative to their self-interactions). The acetone + chloroform mixture is one such system that shows **negative deviations from Raoult's Law**.

■ Numerically, vapor-liquid equilibrium for mixtures is often presented in terms of "**K-factors.**" The ratio of these K-factors is called the **relative volatility** of the system, $\alpha_{i,j}$, and is used to help characterize the difficulty to separate the components in the mixture through distillation. The larger the relative volatility, the easier (normally) it is to separate.

■ Like pure components, mixtures have critical points above which the mixture exists not as a liquid or vapor but as a supercritical fluid. Unlike a pure component, which has a single critical point, a mixture will have multiple critical points that vary with composition.

■ **Retrograde condensation** is a phenomenon for some systems where a mixture in vapor–liquid equilibrium is expanded (i.e., the pressure is lowered), yet the system forms a greater percentage of liquid, which is counterintuitive.

■ The **lever rule** is a graphical approach that allows you to estimate the amount of the two phases in vapor–liquid equilibrium. It is not as accurate as solving the problem using a material balance, however.

■ **Tie lines** connect equilibrium states on a *Txy* or *Pxy* plot.

■ **Flash distillation** is a process where the pressure of a liquid mixture is lowered below the bubble-point pressure but above the dew-point pressure, so that the mixture "flashes" into two phases: a liquid and a vapor. This approach can be used to separate the components in a mixture.

■ Modeling of vapor–liquid equilibrium for mixtures is more challenging for constant pressure systems as compared to constant temperature systems, since the

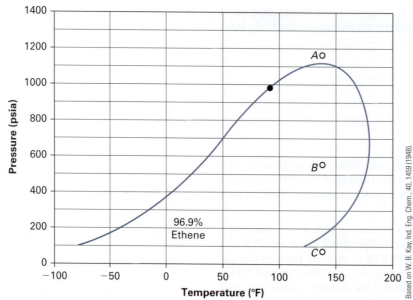

FIGURE P10-30 *P-T* diagram for the ethene + *n*-heptane system at 96.9% by mole ethene.

10-30. Figure P10-30 is a *P-T* diagram for the ethene + *n*-heptane system at 96.9% by mole ethene (Kay, 1948). Please answer the following questions based on this figure. Note that the "•" in the figure describes the critical point for the mixture at this composition.

A. What is the curve to the left of the critical point called?

B. What is the curve to the right of the critical point called?

C. What is the percentage of liquid and vapor at point A?

D. What is the percentage of liquid and vapor at point C?

E. Would you expect point B to be 100% liquid, 100% vapor, or something in between?

F. Does the figure imply retrograde condensation? Please explain.

10.8 GLOSSARY OF SYMBOLS

A, B, C	Antoine equation coefficients	M	an arbitrary thermodynamic property (normally U, H, V, G, or S)	x_A	mole fraction of component A (normally in the liquid phase)
C	# of components in a system			y_A	mole fraction of component A (normally in the vapor phase)
F	degrees of freedom	n	moles		
$\underline{H}_A$	molar enthalpy of pure component A	P	pressure		
		P^{sat}	vapor pressure	π	number of phases in equilibrium
$\underline{H}$	molar enthalpy of the mixture	R	universal gas constant		
$\dot{L}$	molar flow rate of the liquid	T	temperature	$\alpha_{i,j}$	relative volatility
m	mass	$\dot{V}$	molar flow rate of the vapor	K_i	K-factor (distribution coefficient) of component i.

10.9 REFERENCES

Amer, H. H., Paxton, R. R., Van Winkle, M. "Methanol-Ethanol-Acetone." *Ind. Eng. Chem.*, 48, 142 (1956).

Benson, G. C. "Molar Excess Gibbs Free Energies of Benzene-m-Xylene Mixtures," *Can. J. Chem.* 47 (1969) 539–542.

Bobbo, S., Fedele, L., Scattolini, M., Camporese, R. "Isothermal VLE measurements for the Binary Mixtures HFC-134a + HFC-245fa and HC-600 + HFC-245fa." *Fluid Phase Equil.*, 185, 255 (2001).

D'Avila, S. G., Silva, R. S. F. "Isothermal Vapor-Liquid Equilibrium Data by Total Pressure Method. Systems acetaldehyde-ethanol, acetaldehyde-water, and ethanol-water." *J. Chem. Eng. Data*, 15, 421 (1970).

Geier, K., Bittrich, J. "Zur Thermodynamik der Flüssig-Dampf-Gleichgewichte der binären Systeme *n*-Hexan/Benzol, Cyclohexan/Benzol und Tetrahydrofuran mit *n*-hexan, cyclohexan, benzol und dimethylformamid". *Z. Phys. Chem. (Leipzig),* 206, 705 (1979).

Gmehling, J., Onken, U. "Vapor-Liquid Equilibrium Data Collection," DECHEMA, 1977.

Hiaki, T., Kurihara, K., Kojima, K. "Vapor-Liquid Equilibria for Acetone + Chloroform + Methanol and Constituent Binary Systems at 101.3 kPa." *J. Chem. Eng. Data*, 39, 714 (1994).

Kay, W. B. "Liquid-Vapor Equilibrium Relations in Binary Systems. Ethylene-Heptane System." *Ind. Eng. Chem.*, 40, 1459 (1948).

Kirss, H., Kudryavtseva, L. S., Eizen, O. G. "Vapor-Liquid Equilibrium in Ternary Systems hexene-1-hexane-octane, Behzol-heptene-1-heptane, 1-heptene-heptane and toluene in the Corresponding Binary Systems at the Temperature 55°C." *Eesti NSV Tead. Akad. Toim. Keem. Geol.* 24, 15 (1975).

Mueller, C. R., Kearns, E. R. "Thermodynamic Studies of the System: Acetone + Chloroform." *J. Phys. Chem.*, 62, 1441 (1958).

Piacentini, A., Stein, F. "An Experimental and Correlative Study of the Vapor-Liquid Equilibria of the Tetrafluoromethane System." *Chem. Eng. Prog. Symp. Ser.* 63, 28 (1967).

Schuberth, H. "Phasengleichgewichtsmessungen. II. Vorausberechnung von Gleichgewichtsdaten dampfförmig-flussig idealer binrer Systeme und Prüfung derselben mittels einer neuen Gleichgewichtsapparatur." *J. Prakt. Chem.*, 6, 129 (1958).

Sinor, J. E., Weber, J. H. "Vapor-Liquid Equilibria at Atmospheric Pressure. Systems Containing Ethyl Alcohol, *n*-Hexane, Benzene, and Methylcyclopentane." *J. Chem. Eng. Data*, 5, 243 (1960).

Stein, F., Proust, P. "Vapor-Liquid Equilibriums of the trifluoromethane-trifluorochloromethane System." *J. Chem. Eng. Data*, 16, 389 (1971).

Wilsak, R. A., Campbell, S. W., Thodos, G. "Vapor-Liquid Equilibrium Measurements for the *n*-pentane-methanol System at 372.7, 397.7 and 422.6 K." *Fluid Phase Equil.*, 33, 157 (1987).

Theories and Models for Vapor–Liquid Equilibrium of Mixtures: Modified Raoult's Law Approaches

11

LEARNING OBJECTIVES

This chapter is intended to help you learn how to:

- Use the activity coefficient to correlate, predict, and/or quantify deviations to ideal solution behavior using excess molar Gibbs free energy models
- Understand the chemical equilibrium condition for mixtures
- Use Henry's Law to model the phase behavior of gases that have limited solubility in liquids
- Use modified Raoult's Law to model a variety of systems, through both correlative and predictive approaches
- Relate the natural logarithm of the activity coefficient to the excess molar Gibbs free energy
- Use the Gibbs-Duhem equation to examine the thermodynamic consistency of experimental data

In Chapter 10, we laid the groundwork for modeling the vapor–liquid equilibrium of mixtures, namely the assumption of ideal gas and ideal solution; in other words, Raoult's Law. However, as we saw in Chapter 10, the assumptions of Raoult's Law are quite limiting and, thus, this model cannot be used in practice outside of only a small subset of mixtures. In Chapter 11 we focus on techniques that treat liquids not as ideal solutions, but as real solutions. Such modeling approaches are generally denoted as modified Raoult's Law. ∎

11.1 Motivational Example

Bioethanol is ethanol created from a biomass source. During the process, the cellulose from the biomass is extracted, and enzymes (or other processes) are used to convert the cellulose into simple sugars. Yeast is then employed to ferment the sugars into ethanol. Water is used in the process as a wash agent. Accordingly, there is an important need to remove the water from the bioethanol produced and, as such, knowledge of the phase equilibrium for the ethanol + water mixture is important.

While methane (CH_4), ethane (C_2H_6), and propane (C_3H_8) each have only one isomer, decane ($C_{10}H_{22}$) has 75 isomers. $C_{20}H_{42}$ has over 350,000 isomers. All of this from just two atoms (C and H). You can imagine how the number of isomers grows very large very quickly when you add nitrogen and oxygen.

Refer to Figure 10-6, which presents the vapor–liquid equilibrium for the water + ethanol system at 303.15 K. Recall that the solid curve is the experimental data, while the dashed line is the model predictions (Raoult's Law). Clearly, the model does not describe this system accurately. Ultimately, when we size separation equipment in a plant (such as in the bioethanol separation example), we would need experimental data. Thus, what is the main point of the modeling if we will need experimental data anyway?

The answer to this question is rooted in combinatorics. There are estimates that around 10^{60} compounds can be created for biological systems, which is an unfathomable number (Dobson, 2004). How can we even contemplate the number of binary, ternary, etc. mixtures that can be created? Surely, we cannot experimentally evaluate *all* of the properties of *all* of the mixtures at *all* of the conditions (temperature, pressure, and composition) of interest. However, if we had a way to estimate some of the properties of some of the mixtures at a variety of conditions, we would be able to make more-informed process-related decisions up front, rather than having to perform costly (time and money) experiments—at least initially. The way we can do this is through modeling. Therefore, we are motivated to make models that are not limited to specific conditions (ideal solution, for example), and this is the focus of the chapter. In order to do this, we need to go backward and create a foundation for the basic fundamentals of mixture phase equilibrium so we can build our modeling approaches to more accurately describe what is experimentally observed. In short, we need to address the "why" of phase equilibrium for mixtures, as we did in Section 8.2.1 when discussing phase equilibrium for pure components.

Chemical plants (and their various process lines) are such a big monetary investment that experimental data must be used to provide the most accurate information in the final plant or process designs.

11.2 Phase Equilibrium for Mixtures

Let's return to Chapter 1, where we identified three driving forces for changes in a system: mechanical, thermal, and chemical. In Chapter 8, we established that, for a pure compound to exist in two phases at equilibrium, these driving forces must all be zero. Thus, the two phases must have identical temperature (no thermal driving force), identical pressure (no mechanical driving force), and identical molar Gibbs free energy (no chemical driving force). Do these constraints also hold for mixtures? Let's consider a thought experiment.

In Figure 11-1, we have a closed system that encases two open systems (α and β). Each open system contains two components (black and blue). Recall that an open system means that mass and energy can flow across the system boundaries (represented by the dashed enclosures). If we say that the two open systems are in phase equilibrium, it implies the temperatures of both phases (α and β) are the same. Additionally, it implies that the pressures of both phases (α and β) are the same. We know from Chapter 8 that, for a *pure* component in phase equilibrium, the Gibbs free energy of the phases is the same.

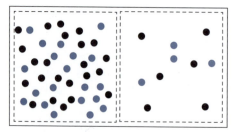

FIGURE 11-1 Two open systems (α and β) that are encased in a closed system.

The expression for the Gibbs free energy of a two-component mixture can be written as

$$n\underline{G} = f(T, P, n_1, n_2) \tag{11.1}$$

$$d(n\underline{G}) = \left[\frac{\partial(n\underline{G})}{\partial(T)}\right]_{P,n_1,n_2} dT + \left[\frac{\partial(n\underline{G})}{\partial(P)}\right]_{T,n_1,n_2} dP + \left[\frac{\partial(n\underline{G})}{\partial(n_1)}\right]_{T,P,n_2} dn_1$$

$$+ \left[\frac{\partial(n\underline{G})}{\partial(n_2)}\right]_{T,P,n_1} dn_2 \tag{11.2}$$

From Section 8.4, we know

$$n\underline{V} = \left[\frac{\partial(n\underline{G})}{\partial(P)}\right]_T$$

$$n\underline{S} = -\left[\frac{\partial(n\underline{G})}{\partial(T)}\right]_P$$

These relationships look almost exactly the same as those in the first two terms of Equation 11.2. However, since the number of moles is being kept constant in the first two terms of Equation 11.2, they are mathematically equivalent to the equations from Section 8.4. Therefore, we can make the direct substitution and Equation 11.2 simplifies to

$$d(n\underline{G}) = (-n\underline{S})\, dT + (n\underline{V})\, dP + \left[\frac{\partial(n\underline{G})}{\partial(n_1)}\right]_{T,P,n_2} dn_1 + \left[\frac{\partial(n\underline{G})}{\partial(n_2)}\right]_{T,P,n_1} dn_2 \tag{11.3}$$

This expression is considerably more compact, but let us explore those last two terms. The first thing to notice is that it looks like it is a partial molar quantity. Recall that we defined partial molar quantities through Equation 9.28 as

$$\overline{M}_i = \left(\frac{\partial(nM)}{\partial n_i}\right)_{T,P,n_{j(j\neq i)}} \tag{9.28}$$

Thus, the last two terms in Equation 11.3 are the partial molar Gibbs free energy. This partial derivative, however, has an additional name called the **chemical potential** (μ_i). The chemical potential of a pure substance was introduced in Chapter 8. For pure components, we saw that the chemical potential was *equal* to the molar Gibbs free energy. For mixtures, however, this is *not* the case, and the chemical potential of a component in a mixture is equal to the partial molar Gibbs free energy of that component (not the molar Gibbs free energy of the solution).

$$\overline{G}_i = \left(\frac{\partial(n\underline{G})}{\partial n_i}\right)_{T,P,n_{j(j\neq i)}} = \mu_i \tag{11.4}$$

More specifically, for our two-component system, we have

$$\mu_1 = \overline{G}_1 = \left[\frac{\partial(n\underline{G})}{\partial(n_1)}\right]_{T,P,n_2} \tag{11.5}$$

$$\mu_2 = \overline{G}_2 = \left[\frac{\partial(n\underline{G})}{\partial(n_2)}\right]_{T,P,n_1} \tag{11.6}$$

For a pure compound, in Equation 11.5, $n = n_1$, and the equation simplifies to

$$\mu_1 = \overline{G}_1 = \underline{G}_1$$

which is equivalent to Equation 8.8.

As with many aspects of thermodynamics, the concept of chemical potential was introduced by J. W. Gibbs, the first person to earn a Ph.D. in Engineering in the U.S. (1863, Yale University).

We can now make the substitution back into Equation 11.3 to yield

$$d(n\underline{G}) = (-n\underline{S})\, dT + (n\underline{V})\, dP + \mu_1 dn_1 + \mu_2 dn_2 \tag{11.7}$$

Returning to Figure 11.1, if we say our overall system is at equilibrium, we can now write Equation 11.7 for each of our three systems (overall and the two open systems— α and β). Here we already know that the temperature and pressure of the two open systems (α and β) are the same so we do not need superscripts for T and P.

Overall (closed):

$$d(n\underline{G}) = (-n\underline{S})\, dT + (n\underline{V})\, dP + \mu_1 dn_1 + \mu_2 dn_2 \tag{11.8}$$

α (open):

$$d(n\underline{G}^\alpha) = (-n\underline{S}^\alpha)\, dT + (n\underline{V}^\alpha)\, dP + \mu_1^\alpha dn_1^\alpha + \mu_2^\alpha dn_2^\alpha \tag{11.9}$$

β (open):

$$d(n\underline{G}^\beta) = (-n\underline{S}^\beta)\, dT + (n\underline{V}^\beta)\, dP + \mu_1^\beta dn_1^\beta + \mu_2^\beta dn_2^\beta \tag{11.10}$$

For the closed system, no mass can leave the system boundaries. Thus, n_1 and n_2 are constant within this closed system and, thus, dn_1 and dn_2 are both zero. This simplifies the closed expression to:

Overall (closed):

$$d(n\underline{G}) = (-n\underline{S})\, dT + (n\underline{V})\, dP \tag{11.11}$$

Any change in the Gibbs free energy of either open system—α or β—would be part of a change for the overall system. Thus,

$$d(n\underline{G}) = d(n\underline{G}^\alpha) + d(n\underline{G}^\beta) \tag{11.12}$$

Inserting Equations 11.9, 11.10, and 11.11 into Equation 11.12 yields

$$\begin{aligned}(-n\underline{S})\, dT + (n\underline{V})\, dP &= (-n\underline{S}^\alpha)\, dT + (n\underline{V}^\alpha)\, dP + \mu_1^\alpha dn_1^\alpha + \mu_2^\alpha dn_2^\alpha \\ &\quad + (-n\underline{S}^\beta)\, dT + (n\underline{V}^\beta)\, dP + \mu_1^\beta dn_1^\beta + \mu_2^\beta dn_2^\beta\end{aligned} \tag{11.13}$$

The same additive relationship from Equation 11.12 can be made for the molar entropy and molar volume:

$$d(n\underline{S}) = d(n\underline{S}^\alpha) + d(n\underline{S}^\beta) \tag{11.14}$$

$$d(n\underline{V}) = d(n\underline{V}^\alpha) + d(n\underline{V}^\beta) \tag{11.15}$$

This reduces the above expression to

$$0 = \mu_1^\alpha dn_1^\alpha + \mu_2^\alpha dn_2^\alpha + \mu_1^\beta dn_1^\beta + \mu_2^\beta dn_2^\beta \tag{11.16}$$

We can simplify Equation 11.16 even more by noticing the following relationship. If, say, one mole of component 1 leaves phase β and enters phase α, the change in the number of moles of component 1 in phase β (which will be negative) is met by an equivalent change in the number of moles of component 1 in phase α (which will be positive). From a mathematical standpoint, this is expressed as

> If I give you one dollar, my bank account decreases by one dollar, and your bank account increases by one dollar.

$$dn_1^\alpha = -dn_1^\beta \tag{11.17}$$

The same discussion follows for component 2, and Equation 11.16 is simplified to

$$0 = \mu_1^\alpha dn_1^\alpha + \mu_2^\alpha dn_2^\alpha - \mu_1^\beta dn_1^\alpha - \mu_2^\beta dn_2^\alpha \tag{11.18}$$

Grouping terms yields

$$0 = (\mu_1^\alpha - \mu_1^\beta)\,dn_1^\alpha + (\mu_2^\alpha - \mu_2^\beta)\,dn_2^\alpha \tag{11.19}$$

From the most general standpoint, there are only two ways to satisfy this equation. First, we could make dn_1^α and dn_2^α equal to zero. This certainly is a mathematical solution to Equation 11.19, but it would exclude the possibility of mass transfer across the system boundaries that connect phases α and β. From an experimental observation standpoint, as well as from what we presented in Chapter 8 regarding phase equilibrium, we know this solution is not what happens in reality. The other general result to make Equation 11.19 true is to force the terms in the parentheses to each be zero (while allowing dn_1^α and dn_2^α to be general).

This yields

$$(\mu_1^\alpha - \mu_1^\beta) = 0 \tag{11.20}$$

$$(\mu_2^\alpha - \mu_2^\beta) = 0 \tag{11.21}$$

or

$$\mu_1^\alpha = \mu_1^\beta \tag{11.22}$$

$$\mu_2^\alpha = \mu_2^\beta \tag{11.23}$$

Equations 11.22 and 11.23 form a powerful result and state that *the chemical potential of a species has the same value in any phases that are in equilibrium in a given system.* When a system has a chemical potential difference in one or more of its species, it is not in equilibrium, and this creates the driving force for mass transfer as a system equilibrates (in the same way as temperature is the driving force for thermal equilibrium and pressure is the driving force for mechanical equilibrium). This equality of chemical potentials for the individual species is called the **chemical equilibrium condition.**

> Recall the discussion on dynamic equilibrium presented in Chapter 8.

> Stated generally, for phases in equilibrium, we have
> $$\mu_A^\alpha = \mu_A^\beta$$
> for any component A, regardless of how many other components are present and whether α and β represent solid, liquid, or vapor phases.

> The phase equilibrium criterion is derived here for a two-component system, but generalizes to any type of phase equilibrium and any number of components.

11.3 Fugacity in Mixtures

It is very easy to get lost in the maze of overbars, underlines, subscripts, superscripts, etc. that are pervasive in solution thermodynamics. Thus, the presentation of fugacity for mixtures in this section will be done in a methodical fashion for enhanced clarity.

In Chapter 8, the fugacity of a pure component was introduced as an alternative to the Gibbs free energy as

$$d\underline{G} \equiv RT\,d\ln f \tag{8.53}$$

In a similar way, we can define the **fugacity of a component in a mixture** via Equation 11.24 as

$$d\mu_i \equiv RT\,d\ln \hat{f}_i \tag{11.24}$$

The phrase "fugacity of component *i* in a mixture" is a bit long-winded. Therefore, we will refer to this term as the "**mixture fugacity of component *i*.**"

PITFALL PREVENTION

While there are lists of variables and symbols in the back of each chapter, you might want to make a special list on your own that defines overbars, underlines, carats, etc. Then you can refer to it quickly while learning the material and reading the book. The best strategy, if you get confused, is to consider the context in which the symbol is being used.

Note the symbol for the mixture fugacity of component i has a carat ($\hat{}$) over it to distinguish it from the pure component fugacity. Also, we do not use an overbar here, since the mixture fugacity of component i, though a mixture property, is NOT a partial molar property.

There are two things that will be presented in this subsection related to the mixture fugacity of component i. First, we will set the stage for the modeling that will be done in the rest of this chapter via the mixture fugacity of component i. Second, we will rewrite the phase equilibrium requirements in terms of the mixture fugacity of component i, just as we did with the pure component fugacity in Chapter 8. To start, we return to the definition of the mixture fugacity of component i as given in Equation 11.24. We can integrate this expression from the pure component ($x_i = 1$) to the mixture composition at the same temperature and pressure as

$$\int_{\mu_i(T,P)}^{\mu_i(T,P,x)} d\mu_i = \int_{\ln f_i(T,P)}^{\ln \hat{f}_i(T,P,x)} RT d \ln \hat{f}_i \tag{11.25}$$

Performing the integration yields

$$\mu_i(T,P,x) - \mu_i(T,P) = RT \ln \hat{f}_i(T,P,x) - RT \ln f_i(T,P) \tag{11.26}$$

Since the symbol for chemical potential is the same whether it is a pure component or mixture, we can avoid some confusion by substituting the pure component excess molar Gibbs free energy, $\underline{G}_i$, for the pure component chemical potential. Thus,

$$\mu_i(T,P,x) = \underline{G}_i + RT \ln \left[\frac{\hat{f}_i(T,P,x)}{f_i(T,P)} \right] = \overline{G}_i \tag{11.27}$$

Note that, per Equation 11.4, the chemical potential of component i in the mixture is also the partial molar Gibbs free energy of that component, and this is indicated in Equation 11.27.

With Equation 11.27, we can explore two of the important models for our vapor–liquid equilibrium modeling, namely the ideal solution for the liquid and the ideal gas for the vapor.

If we wanted to know the chemical potential of species i in a mixture modeled by an ideal solution, we would have

$$\mu_i^{ID}(T,P,x) = \underline{G}_i + RT \ln \left[\frac{\hat{f}_i^{ID}(T,P,x)}{f_i(T,P)} \right] \tag{11.28}$$

In order to simplify Equation 11.28, we have to go back to Chapter 9 and revisit the expression for the molar Gibbs free energy of an ideal solution. Recall that Equation 9.22 gave

$$\underline{G}^{ID} = (x_1 \underline{G}_1 + x_2 \underline{G}_2) + [RTx_1 \ln(x_1) + RTx_2 \ln(x_2)] \tag{9.22}$$

We also know (from Equation 9.27, where $\underline{M} = \underline{G}^{ID}$) that

$$\underline{G}^{ID} = x_1 \overline{G}_1^{ID} + x_2 \overline{G}_2^{ID} \tag{11.29}$$

Comparing Equations 9.22 and 11.29, we can see that

$$\overline{G}_1^{ID} = \underline{G}_1 + RT \ln(x_1) \tag{11.30}$$

$$\overline{G}_2^{ID} = \underline{G}_2 + RT \ln(x_2) \tag{11.31}$$

or in general,

$$\overline{G}_i^{ID} = \underline{G}_i + RT \ln(x_i) \tag{11.32}$$

Equation 11.32 is *exactly* Equation 11.28, since the ideal solution chemical potential of component i is the same as the ideal solution partial molar Gibbs free energy of component i. Thus, just by comparing terms between Equations 11.28 and 11.32, we see that

$$\frac{\hat{f}_i^{ID}(T,P,x)}{f_i(T,P)} = x_i \tag{11.33}$$

This relationship is regularly written as

$$\hat{f}_i^{ID}(T,P,x) = x_i f_i(T,P) \tag{11.34}$$

and is known as the **Lewis-Randall rule**. It says that the mixture fugacity of component i modeled as an ideal solution is equal to its pure component fugacity at the mixture temperature and pressure multiplied by its mole fraction.

Plugging this relationship back into Equation 11.28 yields

$$\mu_i^{ID}(T,P,x) = \underline{G}_i + RT \ln\left[\frac{x_i f_i(T,P)}{f_i(T,P)}\right] \tag{11.35}$$

$$\mu_i^{ID}(T,P,x) = \underline{G}_i + RT \ln x_i \tag{11.36}$$

Equation 11.36 is not necessarily a new relationship, since we knew about this (as the partial molar Gibbs free energy) from the information in Chapter 9. What we did learn, however, was that the mixture fugacity of component i as modeled as an ideal solution is equal to the pure component fugacity multiplied by its mole fraction (or in other words, the Lewis-Randall rule).

Moving on to the ideal gas, we can follow the same approach that we used for the ideal solution, but now applied to the ideal gas.

$$\mu_i^{ig}(T,P,y) = \underline{G}_i^{ig} + RT \ln\left[\frac{\hat{f}_i^{ig}(T,P,y)}{f_i^{ig}(T,P)}\right] \tag{11.37}$$

Let's act more directly now and obtain the partial molar Gibbs free energy of component i modeled as an ideal gas. This was done previously in Chapter 9 using

$$\overline{G}_i^{ig} = \underline{G}_i^{ig} + RT \ln(y_i) \tag{9.78}$$

Since the ideal gas chemical potential of component i is the same as the ideal gas partial molar Gibbs free energy of component i, we can equate the Equations 9.78 and 11.37 and, by inspection, determine that

$$\frac{\hat{f}_i^{ig}(T,P,y)}{f_i^{ig}(T,P)} = y_i \tag{11.38}$$

From Chapter 8, we know that the pure component fugacity in an ideal gas is exactly the system pressure; thus, we can write the mixture fugacity of component i when modeled as an ideal gas by

$$\hat{f}_i^{ig}(T,P,y) = Py_i \tag{11.39}$$

Plugging Equation 11.39 back into 11.37 yields

$$\mu_i^{ig}(T,P,y) = \underline{G}_i^{ig} + RT \ln\left[\frac{Py_i}{P}\right] \tag{11.40}$$

$$\mu_i^{ig}(T,P,y) = \underline{G}_i^{ig} + RT \ln y_i \tag{11.41}$$

The "Lewis" referenced here is Gilbert Lewis, who was nominated for the Nobel Prize 35 times, but never won.

Merle Randall worked as a researcher under Gilbert Lewis at both MIT and UC-Berkeley. The pair combined to author an influential book on thermodynamics in the 1920s.

Even though we are using the molar Gibbs free energy of pure component i on the right-hand side of Equation 11.37, we need the superscript "*ig*" to denote that this is to be evaluated in the ideal gas state (as opposed to its actual state).

FOOD FOR THOUGHT 11-1

How and why is Equation 11.38 different than Equation 11.33?

To recap so far, we have identified how we can write the chemical potential of component i in a mixture when modeled as both an ideal solution and an ideal gas. Additionally, we identified how we can write the mixture fugacity of component i in both the ideal solution (Equation 11.34) and the ideal gas (Equation 11.39) states. Our next step involves identifying the relationships that will help us in modeling systems when they deviate from ideal solution and ideal gas behavior.

To capture the deviation from ideal solution, we use excess properties. Therefore, we can calculate the excess chemical potential of component i in a mixture using Equations 11.27 and 11.36. Thus,

$$\mu_i^E(T,P,x) = \mu_i(T,P,x) - \mu_i^{ID}(T,P,x)$$

$$= \underline{G}_i + RT \ln\left[\frac{\hat{f}_i(T,P,x)}{f_i(T,P)}\right] - (\underline{G}_i + RT \ln x_i) \tag{11.42}$$

Gilbert Lewis introduced the concept of the activity coefficient in 1901.

$$\mu_i^E(T,P,x) = RT \ln\left[\frac{\hat{f}_i(T,P,x)}{x_i f_i(T,P)}\right] \equiv RT \ln \gamma_i \tag{11.43}$$

Within the logarithm, we have defined the **activity coefficient** of component i, γ_i as the ratio of the mixture fugacity of component i to its ideal solution value (i.e., the Lewis-Randall rule). So, the activity coefficient captures the deviation from ideal solution.

In other words,

$$\hat{f}_i(T,P,x) = \hat{f}_i^{ID} \gamma_i = x_i f_i(T,P) \gamma_i \tag{11.44}$$

As you might be able to tell from Equations 11.43 or 11.44, when your system behaves as an ideal solution, γ_i has to equal 1.

Now we move to the vapor phase and look to capture the deviation from ideal gas. To do this, we use residual properties. Therefore, we can calculate the residual chemical potential of component i in a mixture using Equations 11.27 and 11.41. Thus,

$$\mu_i^R(T,P,y) = \mu_i(T,P,y) - \mu_i^{ig}(T,P,y)$$

$$= \underline{G}_i + RT \ln\left[\frac{\hat{f}_i(T,P,y)}{f_i(T,P)}\right] - \left(\underline{G}_i^{ig} + RT \ln y_i\right) \tag{11.45}$$

$$\mu_i^R(T,P,y) = \left(\underline{G}_i - \underline{G}_i^{ig}\right) + RT \ln\left[\frac{\hat{f}_i(T,P,y)}{y_i f_i(T,P)}\right] \tag{11.46}$$

Unlike with the excess chemical potential, the pure component molar Gibbs free energies in Equation 11.46 do *not* cancel out. We have to account for this pure component deviation from the ideal gas state. However, we have already done this in Chapter 8, and the equation we need is Equation 8.56.

$$\left(\underline{G}_i - \underline{G}_i^{ig}\right) = RT \ln\left[\frac{f_i(T,P)}{P}\right] \tag{8.56}$$

Thus,

$$\mu_i^R(T,P,y) = RT \ln\left[\frac{f_i(T,P)}{P}\right] + RT \ln\left[\frac{\hat{f}_i(T,P,y)}{y_i f_i(T,P)}\right] \tag{11.47}$$

Combining the logarithmic terms yields

$$\mu_i^R(T, P, y) = RT \ln\left[\frac{\hat{f}_i(T, P, y)}{Py_i}\right] = RT \ln[\hat{\varphi}_i] \qquad (11.48)$$

Within the logarithm, we have defined the **mixture fugacity coefficient** of component i, $\hat{\varphi}_i$ as the ratio of the mixture fugacity of component i to its ideal gas value. We use a carat here above the symbol to distinguish it from the pure component fugacity coefficient introduced in Chapter 8.

As we can see, the mixture fugacity coefficient captures the deviation from ideal gas behavior. So,

$$\hat{f}_i(T, P, y) = \hat{f}_i^{ig}\hat{\varphi}_i = y_i P \hat{\varphi}_i \qquad (11.49)$$

Before we leave this section and move on to using the activity coefficient and fugacity coefficients for modeling vapor–liquid equilibrium, we need to rewrite the chemical equilibrium requirement in terms of fugacity.

Our chemical equilibrium condition for phase equilibrium in a mixture is

$$\mu_i^\alpha = \mu_i^\beta \qquad (11.22)$$

So, we can make the direct substitution using Equation 11.27 as

$$\underline{G}_i^\alpha + RT \ln\left[\frac{\hat{f}_i(T,P,x)^\alpha}{f_i(T,P)^\alpha}\right] = \underline{G}_i^\beta + RT \ln\left[\frac{\hat{f}_i(T,P,x)^\beta}{f_i(T,P)^\beta}\right] \qquad (11.50)$$

Here, the temperature and pressure are the same in both phases (α and β), since equality of temperatures and pressures are also a requirement of phase equilibrium. Also, since the pure components ($\underline{G}_i$ and f_i) are evaluated at the same temperature and pressure, they will be the same. This reduces to

$$RT \ln \hat{f}_i(T,P,x)^\alpha = RT \ln \hat{f}_i(T,P,x)^\beta \qquad (11.51)$$

or more succinctly,

$$\hat{f}_i(T,P,x)^\alpha = \hat{f}_i(T,P,x)^\beta \qquad (11.52)$$

This tells us that, similar to the pure component result, the mixture chemical equilibrium condition is equivalent to that for the mixture fugacity condition. Note that this result is *not* a new piece of information to be used "in addition to" the chemical equilibrium condition. It is the *same result*—written in a different way.

When we introduced the fugacity during Chapter 8 for pure components, we also introduced the fugacity coefficient. We found that the equivalence of fugacity coefficients was another way of writing the chemical equilibrium constraint for a pure component. However, this is *not* the case for mixtures and stems from how the fugacity coefficient in a mixture is defined. Specifically, using Equation 11.52 and substituting Equation 11.49, we have

$$\hat{f}_i(T,P,y)^\alpha = \hat{f}_i(T,P,y)^\beta \qquad (11.53)$$

$$y_i^\alpha P \hat{\varphi}_i^\alpha = y_i^\beta P \hat{\varphi}_i^\beta \qquad (11.54)$$

Since the pressure is the same in both phases at equilibrium, we arrive at

$$y_i^\alpha \hat{\varphi}_i^\alpha = y_i^\beta \hat{\varphi}_i^\beta \qquad (11.55)$$

Remember, the mixture fugacity of component i is in the ideal gas state, so the denominator in Equation 11.48 is Equation 11.39

As you might be able to tell from Equations 11.48 or 11.49, $\hat{\varphi}_i$ must equal 1 when your system behaves as an ideal gas.

When the term "mixture fugacity" is used, it does NOT mean the fugacity of the mixture. That latter term "fugacity of the mixture" has no meaning in how the fugacity is defined. Mixture fugacity implies the "mixture fugacity of component i." This is true for the mixture fugacity coefficient as well.

In Equation 11.52, α and β can represent equilibrium solid, liquid, or vapor phases, in any combination.

x or y does not imply a particular phase (liquid or vapor) here and is arbitrary

Therefore, the chemical equilibrium condition includes the mole fraction when using the mixture fugacity coefficient. And, as we saw in Chapter 9, the mole fractions of the different phases are almost always *not* equal to each other in equilibrium (except at an azeotrope).

Now that we have the phase equilibrium condition for mixtures written in terms of the fugacity as well as the framework to characterize the deviation from the ideal solution and ideal gas behavior, we can begin our modeling in earnest.

11.4 Gamma-Phi Modeling

In modeling the phase equilibrium of mixtures, we will always start with the formal phase equilibrium condition, namely that a component in a mixture will have the same mixture fugacity in any of the phases in which it appears (Equation 11.52). Written for the specific case of a liquid mixture in equilibrium with a vapor mixture, we have:

We use superscripts "L" and "V" to distinguish the different phases in equilibrium—here, liquid and vapor.

$$\hat{f}_i^{\mathrm{L}} = \hat{f}_i^{\mathrm{V}} \tag{11.56}$$

To describe the mixture fugacity of component i in the liquid phase, we use an approach that uses the ideal solution as the reference. Thus, by Equation 11.44, we have:

$$\hat{f}_i^{\mathrm{L}}(T,P,x) = x_i f_i(T,P)\gamma_i \tag{11.44}$$

where the deviations from ideal solution behavior are captured by the activity coefficient, γ_i (gamma).

Likewise, to describe the mixture fugacity of component i in the vapor phase, we use an approach that uses the ideal gas as the reference. Recall

$$\hat{f}_i^{\mathrm{V}}(T,P,y) = y_i P \hat{\varphi}_i \tag{11.49}$$

While it is not necessary to show the functionality for f_i, we do so here just to emphasize that it is for the pure component and not a function of the composition of the mixture.

in which the deviation from ideal gas behavior is captured by the mixture fugacity coefficient of component i, $\hat{\varphi}_i$ (phi).

Using Equation 11.44 for the left-hand-side of Equation 11.56 and Equation 11.49 for the right-hand-side of Equation 11.56, we have:

$$x_i f_i(T,P)\gamma_i = y_i P \hat{\varphi}_i \tag{11.57}$$

This is the "gamma-phi" equation and is a popular approach to modeling the vapor–liquid equilibrium behavior for mixtures. The next steps in this chapter are to develop models for how to best (or most easily) describe the liquid and vapor phases using Equation 11.57.

11.4.1 Raoult's Law Revisited

We don't refer to the activity coefficient of component i as the "*mixture activity coefficient of component i*," since activity coefficient is a concept only defined for mixtures.

The simplest model for vapor–liquid phase equilibrium of mixtures (and one we've introduced before) is Raoult's Law. Using this approach, we model the liquid as an ideal solution and the vapor as an ideal gas. We have obtained these mixture fugacities in the previous section and provide them here again. For the ideal solution, we have again

$$\hat{f}_i^{\mathrm{L}} = \hat{f}_i^{\mathrm{ID}} = x_i f_i \tag{11.34}$$

which means that the activity coefficient of component i is equal to 1 when modeled as an ideal solution.

Likewise, the mixture fugacity for a component in a vapor mixture modeled as an ideal gas becomes

$$\hat{f}_i^{\mathrm{V}} = \hat{f}_i^{\mathrm{ig}} = y_i P \tag{11.39}$$

which means that the mixture fugacity coefficient of component i is equal to 1 when modeled as an ideal gas.

Plugging these results back into Equation 11.56 yields

$$x_i f_i(T, P) = y_i P \tag{11.58}$$

As in Chapter 8, we can calculate the fugacity of pure component i in the liquid phase at the mixture temperature and the *vapor pressure* of the pure component at that temperature, and then adjust to the mixture pressure by using the Poynting correction. When we do this, we have

$$x_i f_i^{\text{sat}}(T, P_i^{\text{sat}}) \exp\left\{ \frac{\underline{V}_i(P - P_i^{\text{sat}})}{RT} \right\} = y_i P \tag{11.59}$$

Since the pure component fugacity is equal to the pressure times the fugacity coefficient, we can write this as

$$x_i P_i^{\text{sat}} \varphi_i^{\text{sat}}(T, P_i^{\text{sat}}) \exp\left\{ \frac{\underline{V}_i(P - P_i^{\text{sat}})}{RT} \right\} = y_i P \tag{11.60}$$

> Be sensitive to all of the steps and assumptions going from the general relationship (the equality of mixture fugacities of component i in the liquid and vapor phase) to Raoult's Law.

But as we saw in Example 8.4, the fugacity of a liquid has to be at a greatly elevated pressure to be a value that is appreciably different than the vapor pressure at that temperature. Since we are already assuming an ideal gas for the vapor, we are already assuming a low pressure. Therefore, the fugacity coefficient will be very close to one, as will the Poynting correction. Therefore, their product can be assumed as

$$\varphi_i^{\text{sat}} \exp\left\{ \frac{\underline{V}_i(P - P_i^{\text{sat}})}{RT} \right\} \sim 1 \tag{11.61}$$

and Equation 11.60 simplifies to the Raoult's Law relationship we have described and already used in Chapter 10. Thus,

$$x_i P_i^{\text{sat}} = y_i P \tag{11.62}$$

> Equation 11.62 is a general version of Equation 10.3 and Equation 10.4

Recall that Raoult's Law is valid only for systems that have similar sizes and similar intermolecular interactions. We saw that a benzene + toluene system behaved according to Raoult's Law, while the ethanol + water system did not. However, there is a certain condition where most every system at low to moderate pressures will have a component behave according to Raoult's Law—regardless of whether the molecules have similar sizes or similar intermolecular interactions— and that is near the pure-component endpoints.

11.4.2 Henry's Law

Consider Figure 11-2, which is a mixture of blue and black molecules. As is observed, the system is concentrated in the blue molecules and dilute in the black molecules. By and large, many of the blue molecules do not have a nearest neighbor that is a black molecule. In essence, many of the blue molecules are not influenced at all by the presence of the black molecules, since there are so few of them. When this is the situation, the blue molecules are basically interacting almost exclusively with molecules that have "similar size and similar intermolecular interactions"—the other blue molecules. The mixture fugacity of the blue molecules in this system is not influenced from the presence of the black molecules (by a good approximation) and is linearly proportional to its amount present (with the proportionality constant the pure component fugacity). In other words, the blue molecules are well-modeled by the ideal solution at this state. A cartoon plot of

> William Henry, an English chemist, did many experiments to demonstrate the proportionality now known as Henry's Law. He published this work in 1803.

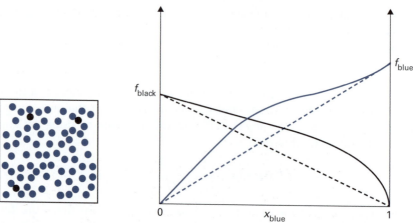

FIGURE 11-2 A cartoon representation showing a mixture of blue and black molecules (concentrated in blue) as well as the mixture fugacity of the blue and black molecules as a function of composition. The dashed lines indicate the Lewis-Randall rule.

the mixture fugacity of the blue molecules in solution as a function of composition is given in Figure 11-2. Thus, when the concentration of the blue molecules is high, the Lewis-Randall rule holds for the blue molecules.

What about on the other end of the spectrum? What about the black molecules? Previously, we mentioned that the amount of black molecules were such that most blue molecules were not surrounded by any black molecules—in essence, not experiencing the influence of any other black molecules in a significant way. Likewise, when the concentration of black molecules is low, a black molecule will almost never be surrounded by another black molecule and not feel the influence of any black molecules. It is as if each black molecule was in its own sea of blue molecules. Therefore, the impact of each black molecule is not influenced (by a good approximation) by the rest of the black molecules, and its mixture fugacity is linearly proportional to its concentration. However, the proportionality constant is *not* the pure component fugacity of a black molecule in this case, but some term that encompasses how a single black molecule interacts with the sea of blue molecules.

> Notice in Figure 11-2 that the Henry's Law constant (the slope) for a blue molecule in a sea of black molecules is *not* the same as a black molecule in a sea of blue molecules—nor should it be. A few art history majors interacting with a sea of chemical engineering students would be different than a few chemical engineering students interacting with a sea of art history majors.

This proportionality constant is called Henry's constant (H) and is both unique to the type of molecules in the mixture and the system temperature. We then write the mixture fugacity of the black molecules by **Henry's Law**:

$$\hat{f}_{black} = x_{black}H_{black,blue} \qquad (11.63)$$

While we have already spent some time describing and calculating Raoult's Law results in the previous chapter, this is our first introduction to Henry's Law. So while Henry's Law is certainly applicable in the dilute region for mixtures we have already encountered, the approach is most often reserved to describe gases dissolving sparingly in liquids. Henry's Law tells us that a gas (recall that a gas is above its critical temperature, but at low enough pressures to resemble a vapor) in contact and in equilibrium with a liquid will have some small amount of gas dissolved in that liquid. It is best to think of it as a "gas dissolved in a liquid" rather than *condensing* into the liquid phase, since the gas (by definition) cannot condense. Our daily existence is surrounded by gases and not vapors (oxygen, nitrogen, carbon dioxide, etc.). Thus, Henry's law provides a good model to describe, say, how much of these gases are in liquids, such as water. Indeed, when you put water in a pot on your stove to boil, those tiny bubbles that leave the liquid as you are heating (but well below the boiling temperature) are the dissolved gases in the water leaving the liquid phase. This common observation implies that the concentration of, say, oxygen in water will be less at higher temperatures than lower temperatures. Let's confirm this through an example problem.

HENRY'S LAW	EXAMPLE 11-1

Estimate the mole fraction of oxygen in water at 25°C and at 75°C and 101.325 kPa.

SOLUTION:

Step 1 *Phase equilibrium for Henry's Law (the dilute component—oxygen)*
When we were deriving Raoult's Law, we started with the equality of fugacities of the components in solution. Rather than assume an ideal solution for this system, we write Henry's Law for the dilute species here (and reasonably assume oxygen gas and water vapor comprise an ideal gas at this pressure).

$$x_o H_{o,w} = y_o P \tag{11.64}$$

The Henry's Law constants for oxygen in water at 25°C and 75°C are given here (Sander, 1999).

T (°C)	$H_{o,w}$ (MPa)
25	4310.21
75	8877.53

Based on data from Rolf Sander (sander@mpch-mainz.mpg.de)

Examining the previous equation, we have two unknowns, x_o and y_o, but only one equation (Henry's law for the oxygen in water). We need a second equation and that will come from the other component in our system, water.

Step 2 *Phase equilibrium for Raoult's Law (the concentrated component—water)*
Water is the concentrated component in the liquid phase and, thus, Raoult's Law will be used to describe the phase equilibrium for water using Equation 11.62:

$$x_w P_w^{sat} = y_w P$$

But is our system in equilibrium? Recall that the term "relative humidity" states how much water vapor the air contains at a certain temperature, divided by the maximum amount of water vapor that air *could* contain at that temperature. For the illustrative purposes of this problem, let's assume that the air contains that maximum amount (the relative humidity is 100%). Thus, we can use the previous equation in this instance.

Step 3 *Solving the combined problem*
We obtain the vapor pressure of water at the two temperatures of interest from the steam tables in Appendix A for the data here.

T (°C)	P^{sat} (kPa)
25	3.17
75	38.59

The Henry's Law relationship is defined a few different ways in the literature (it is more than 200 years old), so the units may be a different depending on the definition. Be sensitive to that when using a Henry's Law constant—look to how it is being defined.

Remember, since the liquid phase is almost pure water, it is reasonable to model the water as an ideal solution.

We use the term "air" here, though we have only considered oxygen for the purposes of this problem.

Now we have two unknowns (x_o, y_o) in the two equilibrium equations (Henry's Law for oxygen and Raoult's Law for water). Note that we have already made the substitution that $x_w = 1 - x_o$ and $y_w = 1 - y_o$.

Solving the two equations in two unknowns at both temperatures yields the following results.

	$T = 25°C$	$T = 75°C$
x_o	2.2×10^{-5}	7.1×10^{-6}
y_o	0.97	0.62

Clearly the amount of the oxygen dissolved in the water is exceedingly small and decreases as a function of temperature. Thus, when you heat the water on your stove, you are decreasing the amount of oxygen that is dissolved in the water and this is what you will observe during this process.

Note that this is a contrived example. We would have to include nitrogen in the air to obtain a better estimate of the dissolved amounts of both gases. We leave this for an end-of-chapter problem.

> You may not think about the dissolved oxygen in water, but the fish in your fish tank care! The minimum mole fraction required for a typical fish at 25°C is about 4×10^{-6}, so there is enough oxygen for the fish to survive.

More formally, the Henry's Law constant is really a limiting value of the ratio of the mixture fugacity of component i to the mole fraction of component i. Through Equation 11.63 we can write this as

$$H_i = \lim_{x_i \to 0}\left[\frac{\hat{f}_i(T,P,x)}{x_i}\right] = \lim_{x_i \to 0}\left[\frac{x_i f_i(T,P)\gamma_i}{x_i}\right] = \lim_{x_i \to 0}[f_i(T,P)\gamma_i]$$
$$= f_i(T,P)\lim_{x_i \to 0}[\gamma_i] \tag{11.65}$$

Inside the limit, we find the activity coefficient of component i as the composition of i becomes increasingly dilute. We call this value the **infinite dilution activity coefficient** of component i. Thus,

$$H_i = f_i(T,P)\lim_{x_i \to 0}[\gamma_i] = f_i(T,P)\gamma_i^{\infty} \tag{11.66}$$

where γ_i^{∞} is the infinite dilution activity coefficient of component i. With knowledge of the infinite dilution activity coefficient, the Henry's Law constant can be obtained.

EXAMPLE 11-2	**HENRY'S LAW USING INFINITE DILUTION ACTIVITY COEFFICIENTS**

One strategy to lower the concentration of certain substances in gas mixtures is to put the gas in contact with a liquid, since the gases will be absorbed in different amounts depending on the type of gas. Such a process unit in the chemical process industry is called an absorber.

In an intermediate calculation to size an absorber, you need to determine the partial pressure of benzene in the vapor phase, in equilibrium with a liquid composed of 99.9 mol% water and 0.1 mol% benzene. You know the infinite dilution activity coefficient for benzene in water at this temperature is 2500.

SOLUTION:

Step 1 *Write out the phase equilibrium*
For benzene, we can write a Henry's Law relationship to expose the partial pressure of benzene using Equation 11.64:

$$x_b H_{b,w} = y_b P = P_b$$

Step 2 *Calculate the Henry's Law constant*

It was previously shown in Equation 11.66 that

$$H_{b,w} = f_b(T,P)\gamma_b^{\infty}$$

So now we need to determine the pure-component fugacity of benzene. Recall from Chapter 8 that

$$f_b(T,P) = P_b^{sat}\varphi_b^{sat}\left(T,P_b^{sat}\right)\exp\left\{\frac{V_b\left(P - P_b^{sat}\right)}{RT}\right\}$$

Using the same simplifications as before (assume low pressure–Equation 11.61), we find that

$$f_b(T,P) \approx P_b^{sat}$$

From the Antoine equation using the parameters in Appendix E, we find that the vapor pressure for benzene at 293 K is 10.13 kPa. Thus,

$$H_{b,w} = (10.13\,\text{kPa})(2500) = 25331\,\text{kPa}$$

Step 3 *Calculate the partial pressure*

Finally, we can calculate the partial pressure by multiplying the Henry's Law constant by the mole fraction to obtain:

$$\boldsymbol{P_b = x_b H_{b,w} = (0.001)(25331\,\text{kPa}) = 25.3\,\text{kPa}}$$

> From the literature (Sander, 1999), the Henry's Law constant for benzene in water at this temperature is 25,635 kPa, so the estimate in this problem is consistent with the data.

11.5 Modified Raoult's Law

In the previous section, we explored both Raoult's Law, which describes the vapor–liquid equilibrium for an ideal solution in equilibrium with an ideal gas, and Henry's Law for dissolved gases in liquids. The next step on our journey in phase equilibrium modeling of mixtures is to examine the much more common scenario: vapor–liquid equilibrium involving non-ideal solutions and (eventually) non-ideal gases. Since the deviations from ideal solution have a much bigger impact on modeling the phase behavior than the deviations from ideal gas, we will focus mainly on strategies to account for non-idealities in the liquid phase.

Let's start back at the beginning, namely the equality of mixture fugacities for components in the phases where they appear in equilibrium (here, liquid and vapor).

$$\hat{f}_i^{L} = \hat{f}_i^{V} \tag{11.56}$$

For the liquid, we can use Equation 11.44, and that allows us to write the mixture fugacity of component i in terms of the pure component fugacity, composition, and the activity coefficient.

$$\hat{f}_i^{L}(T,P,x) = x_i f_i(T,P)\gamma_i \tag{11.44}$$

Remember, the activity coefficient of species i is the factor (γ_i) that accounts for the deviations to ideal solution behavior. Recall that when $\gamma_i = 1$, we recover an ideal solution. Thus, how far γ_i deviates from 1 tells you how non-ideal the ith component is in that liquid mixture. If we rewrite the pure component fugacity as before and assume an ideal gas for the vapor phase ($\hat{\varphi}_i = 1$), we have

$$x_i\gamma_i f_i^{sat}\exp\left\{\frac{V_i\left(P - P_i^{sat}\right)}{RT}\right\} = y_i P \tag{11.67}$$

> Even though we define only the liquid in terms of the activity coefficient, nothing stops us from using this for the vapor phase (other than the fact that the ideal gas provides a better reference for the vapor).

or using $f_i^{\text{sat}} = \varphi_i^{\text{sat}} P_i^{\text{sat}}$ we have

$$x_i \gamma_i P_i^{\text{sat}} \varphi_i^{\text{sat}} \exp\left\{ \frac{\underline{V}_i \left(P - P_i^{\text{sat}} \right)}{RT} \right\} = y_i P \tag{11.68}$$

Once again, since we have assumed ideal gas for the vapor, we will be at a fairly low pressure. Thus, Equation 11.61 holds as

$$\varphi_i^{\text{sat}} \exp\left\{ \frac{\underline{V}_i \left(P - P_i^{\text{sat}} \right)}{RT} \right\} \sim 1 \tag{11.61}$$

And we can simplify Equation 11.68 to

$$\gamma_i x_i P_i^{\text{sat}} = y_i P \tag{11.69}$$

Equation 11.69 is commonly known as **modified Raoult's Law**.

Much of the modeling of solution thermodynamics is centered around obtaining good expressions and estimates for the activity coefficient. Therefore, we need to discuss the activity coefficient in more detail.

Recall that Equation 11.43 provided us with the definition of the activity coefficient and its relationship to the excess chemical potential.

$$\mu_i^{\text{E}}(T,P,x) = RT \ln\left[\frac{\hat{f}_i(T,P,x)}{x_i f_i(T,P)} \right] \equiv RT \ln \gamma_i \tag{11.43}$$

We know that the chemical potential of species i in a mixture is also the partial molar Gibbs free energy of species i, so the excess chemical potential is the excess partial molar Gibbs free energy.

$$\mu_i^{\text{E}}(T, P, x) = RT \ln \gamma_i = \overline{G}_i^{\text{E}} \tag{11.70}$$

Recall from Equation 11.4 that we can write the chemical potential as

$$\mu_i = \overline{G}_i = \left(\frac{\partial (n\underline{G})}{\partial n_i} \right)_{T,P,n_{j(j \neq i)}} \tag{11.4}$$

Therefore, we can write the same for the excess properties as

$$\mu_i^{\text{E}} = \overline{G}_i^{\text{E}} = \left(\frac{\partial (n\underline{G}^{\text{E}})}{\partial n_i} \right)_{T,P,n_{j(j \neq i)}} \tag{11.71}$$

Since the partial derivative is evaluated at constant T, we change nothing by dividing by RT (the benefit we will see in a moment). We have

$$\frac{\mu_i^{\text{E}}}{RT} = \frac{\overline{G}_i^{\text{E}}}{RT} = \left(\frac{\partial (n\underline{G}^{\text{E}}/RT)}{\partial n_i} \right)_{T,P,n_{j(j \neq i)}} \tag{11.72}$$

Using Equation 11.43, we see that this means that the natural logarithm of the activity coefficient is, in fact, a partial molar property of the excess molar Gibbs free energy divided by RT. Thus,

$$\ln \gamma_i = \frac{\mu_i^{\text{E}}}{RT} = \frac{\overline{G}_i^{\text{E}}}{RT} = \left(\frac{\partial (n\underline{G}^{\text{E}}/RT)}{\partial n_i} \right)_{T,P,n_{j(j \neq i)}} \tag{11.73}$$

That the natural logarithm of the activity coefficient is a partial molar property of the excess molar Gibbs free energy is an important result in two ways: (1) We want to model real solutions using the activity coefficient and (2) there are useful relationships that apply to partial molar properties (that were derived in Chapter 9). Therefore, we exploit the fact that $\ln[\gamma_i]$ is a partial molar property in modeling approaches.

Since $\ln[\gamma_i]$ is a partial molar property of the excess molar Gibbs free energy, it allows us to use Equation 9.27 (the "summability" relationship) for partial molar properties as

$$\underline{G}^E/RT = x_1\left(\frac{\overline{G}_1^E}{RT}\right) + x_2\left(\frac{\overline{G}_2^E}{RT}\right) = x_1\ln[\gamma_1] + x_2\ln[\gamma_2] \qquad (11.74)$$

Practically, modelers develop approaches to characterize the excess molar Gibbs free energy for real solutions. From these models and the relationship between the activity coefficient and the excess molar Gibbs free energy, they obtain the activity coefficient for use in modeling vapor–liquid equilibrium with approaches such as modified Raoult's Law (Equation 11.69). The next section introduces some excess molar Gibbs free energy models and the application of the models.

PITFALL PREVENTION

Remember, this relationship,

$\underline{G}^E/RT$
$= x_1\ln[\gamma_1] + x_2\ln[\gamma_2]$

is general and is *not* based on a model.

11.6 Excess Molar Gibbs Free Energy Models: An Introduction

In developing excess molar Gibbs free energy models, we have some built-in constraints that the models must satisfy. In particular, when our binary liquid mixture becomes very concentrated in one of the substances (i.e., $x_1 \to 1$ or $x_2 \to 1$), we know that we approach an ideal solution. More specifically, when $x_1 = 1$ and $x_2 = 1$, $\underline{G}^E/RT$ must be zero, since we have a pure component at both of those points (and certainly pure components do a great job of validating the assumptions of an ideal solution—similar size and similar intermolecular interactions!).

What would be the simplest expression that would satisfy these endpoints? The answer is a line that is equal to zero at all compositions, not just the endpoints (i.e., pure components), would be the simplest expression for $\underline{G}^E/RT$. Indeed, this is provided in Figure 11-3. However, we already know this model—it is the ideal solution, since the excess molar Gibbs free energy is zero at all compositions for the ideal solution.

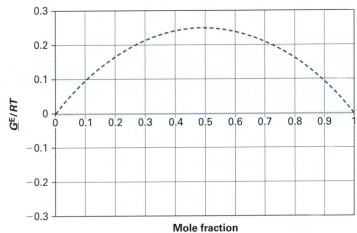

FIGURE 11-3 The excess molar Gibbs free energy versus mole fraction for an ideal solution (solid line) and for the 1-parameter Margules equation, $A = 1$ (dashed line).

So the simplest expression ($\underline{G}^E/RT = 0$) does not help us model real solutions. What is the next simplest model that must satisfy those endpoint constraints? This would be a parabola with a line of symmetry at $x_1 = 0.5$ and this is also shown in Figure 11-3.

The expression for this parabola is

$$\frac{\underline{G}^E}{RT} = Ax_1x_2 \tag{11.75}$$

This simplest model for real solutions is known as the 1-parameter **Margules equation.**

In Chapter 9, we discussed two useful relationships for partial molar properties. Namely, they are used in summability relationships (Equation 9.27) and in the Gibbs-Duhem equation (Equation 9.42). For the former, we showed in Equation 11.74 how we can write a summability relationship for $\underline{G}^E/RT$ in terms of its partial molar property (namely, the natural logarithm of the activity coefficient, $\ln[\gamma_1]$). Let's explore the latter relationship for the Gibbs-Duhem equation.

Equation 9.48 is the Gibbs-Duhem equation written for a solution property $\underline{M}$ and its partial molar property. Recall that the Gibbs-Duhem equation provides a constraint on how different thermodynamic variables can change. Thus,

$$x_1 d\overline{M}_1 + x_2 d\overline{M}_2 - \left(\frac{\partial[\underline{M}]}{\partial T}\right)_{P,x_1,x_2} dT - \left(\frac{\partial[\underline{M}]}{\partial P}\right)_{T,x_1,x_2} dP = 0 \tag{9.48}$$

We can write this equation where $\underline{G}^E/RT$ is our solution property ($\underline{M}$) and $\ln[\gamma_1]$ is its partial molar property. If we consider this when evaluating Pxy data, we know our temperature is constant. Also, as we will show later in this chapter, $\underline{G}^E/RT$ does not change much with the pressure. Thus, Equation 9.48 reduces to

$$x_1 d\ln\gamma_1 + x_2 d\ln\gamma_2 = 0 \tag{11.76}$$

where $\ln[\gamma_1]$ and $\ln[\gamma_1]$ are the partial molar properties being considered here.

Equation 11.76 is a constraint that provides a relationship between activity coefficients that *must* be observed. Do the activity coefficients from the 1-parameter Margules equation model obey this constraint? In order to test this, we need to first obtain the activity coefficient expressions from the excess molar Gibbs free energy model. Recall our relationship between the activity coefficient and the excess molar Gibbs free energy, as given by

$$\ln\gamma_i = \left(\frac{\partial(n\underline{G}^E/RT)}{\partial n_i}\right)_{T,P,n_{j(j\neq i)}} \tag{11.73}$$

For a binary mixture, this is written as

$$\ln\gamma_1 = \left(\frac{\partial(n\underline{G}^E/RT)}{\partial n_1}\right)_{T,P,n_2} \tag{11.77}$$

$$\ln\gamma_2 = \left(\frac{\partial(n\underline{G}^E/RT)}{\partial n_2}\right)_{T,P,n_1} \tag{11.78}$$

Let's first determine the $\ln[\gamma_1]$ expression for the 1-parameter Margules equation. Here we have to make sure to expose the dependency on the number of moles in the mole fraction, so we make the appropriate substitutions. Thus,

$$\ln \gamma_1 = \frac{\partial}{\partial n_1}[n(Ax_1x_2)]_{T,P,n_2} = \frac{\partial}{\partial n_1}\left[\frac{(n_1 + n_2)A(n_1)(n_2)}{(n_1 + n_2)(n_1 + n_2)}\right]_{T,P,n_2} \qquad (11.79)$$

Remember,
$$n = n_1 + n_2;$$
$$x_1 = \frac{n_1}{n_1 + n_2};$$
$$x_2 = \frac{n_2}{n_1 + n_2}$$

$$\ln \gamma_1 = \frac{\partial}{\partial n_1}\left[\frac{A(n_1)(n_2)}{(n_1 + n_2)}\right]_{T,P,n_2} \qquad (11.80)$$

Pulling out the constant A and n_2 from the derivative gives

$$\ln \gamma_1 = An_2 \frac{\partial}{\partial n_1}\left[\frac{(n_1)}{(n_1 + n_2)}\right]_{T,P,n_2}$$

$$\ln \gamma_1 = An_2\left[\frac{(n_1 + n_2)(1) - (n_1)(1)}{(n_1 + n_2)^2}\right]$$

$$\ln \gamma_1 = \frac{An_2}{(n_1 + n_2)}\left[\frac{(n_1 + n_2)(1) - (n_1)(1)}{(n_1 + n_2)}\right]$$

$$\ln \gamma_1 = Ax_2[1 - x_1]$$

$$\ln \gamma_1 = Ax_2^2 \qquad (11.81)$$

Likewise, if you go through the same exercise for $\ln[\gamma_2]$, you will arrive at

$$\ln \gamma_2 = Ax_1^2 \qquad (11.82)$$

You should go through the exercise to derive Equation 11.82 to help you understand how we arrived at Equation 11.81.

So, does the 1-parameter Margules equation satisfy the Gibbs-Duhem equation as given by Equation 11.74?

$$x_1d \ln \gamma_1 + x_2d \ln \gamma_2 = 0 \qquad (11.76)$$

Here, we can evaluate this differential relationship for small changes in dx_1. Thus,

$$x_1\frac{d \ln \gamma_1}{dx_1} + x_2\frac{d \ln \gamma_2}{dx_1} = 0 \qquad (11.83)$$

Plugging in the activity coefficient expressions obtained for the 1-parameter Margules Equation (11.81 and 11.82), we arrive at

$$x_1\frac{d(Ax_2^2)}{dx_1} + x_2\frac{d(Ax_1^2)}{dx_1} = 0 \qquad (11.84)$$

$$x_1\frac{d(A[1 - x_1]^2)}{dx_1} + x_2\frac{d(Ax_1^2)}{dx_1} = 0$$

$$x_1[2A(1 - x_1)(-1)] + x_2 2Ax_1 = 0$$

$$- x_1[2Ax_2] + x_2 2Ax_1 = 0$$

$$0 = 0 \qquad (11.85)$$

Thus, we have shown that the 1-parameter Margules equation does, indeed, satisfy the Gibbs-Duhem equation as required. If one develops an excess molar Gibbs free energy model that does not satisfy the Gibbs-Duhem equation, it is not thermodynamically correct and, therefore, not a self-consistent model.

In the course of demonstrating that the 1-parameter Margules equation does satisfy the Gibbs-Duhem equation, we determined the activity coefficient expressions from this model (Equations 11.81 and 11.82). If we plot those expressions as a function of mole fraction (Figure 11-4), we see that the activity coefficients are symmetric about $x_1 = 0.5$. Therefore, the use of the 1-parameter Margules equation is limiting in that, while it takes into account the deviations from ideal solution, it does so in a way that does not distinguish between the components. Thus, a system concentrated in one of the components A and dilute in the other component B will interact the same as a system concentrated in B and dilute in A, according to the 1-parameter Margules equation.

> Recall that we are not saying the 1-parameter Margules equation treats ALL of the interactions between all the molecules the same—that would be the ideal solution.

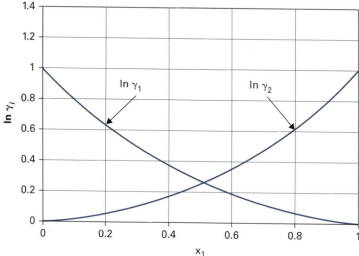

> As is required, the activity coefficients of both pure components are equal to one (and the natural logarithm is zero).

FIGURE 11-4 The natural logarithm of the activity coefficients vs. mole fraction for the 1-parameter Margules equation ($A = 1$).

To recap, at this point we have introduced the simplest excess molar Gibbs free energy model that can describe a non-ideal solution, the 1-parameter Margules equation. We have obtained the activity coefficient expressions for this model and have shown that the model does satisfy the Gibbs-Duhem equation. We also plotted the symmetric behavior of the activity coefficients from this model as a function of composition. However, rather than provide a list of different excess molar Gibbs free energy models (and there are several in use) we are going to focus on the utility of excess molar Gibbs free energy modeling approaches in solving problems. We will introduce the excess molar Gibbs free energy models in the context of the problems. Note that, once we develop the utility of a model, the same approach can be used for other excess molar Gibbs free energy models.

11.7 Excess Molar Gibbs Free Energy Models: Usage

We have discussed, somewhat, the importance of obtaining the activity coefficient to effectively model liquid mixtures that do not behave according to the ideal solution (which is the most common situation). It would be very useful to have models that can *predict* the activity coefficient for mixtures from individual pure-component properties. While there are approaches like this (and we will discuss a few later in this chapter), they have restrictions on their predictive ability. Therefore, the most useful strategy comes from measuring a few vapor–liquid equilibrium data points experimentally, determining the parameters of the excess molar Gibbs free energy model that best describe that experimental data, and then predicting the behavior for the mixture over the entire composition range and/or extrapolating the parameters for use in the prediction of phase behavior at different temperatures. Alternatively, if we have many data points for a mixture across the composition region, we can effectively reduce the data to an equation. Thus, the term "data reduction" applies in this context (where a large table of data can be *reduced* to a single mathematical expression).

USING THE 1-PARAMETER MARGULES EQUATION ON A SINGLE EXPERIMENTAL DATA POINT	**EXAMPLE 11-3**

A single vapor–liquid equilibrium point for the water (1) + ethanol (2) system is experimentally measured at 30°C. The experiment provides the following information: $x_1 = 0.30$, $y_1 = 0.23$, and $P = 10.1$ kPa. Use this information to estimate the system pressure and vapor-phase mole fraction when $x_1 = 0.8$.

SOLUTION:

Step 1 *Gibbs phase rule*
According to the Gibbs phase rule, we require two intensive variables to be specified for our 2-component, 2-phase system. We have specified the liquid phase mole fraction ($x_1 = 0.8$) and the temperature (30°C), so our system is properly specified. And while this is a somewhat contrived example, it is good practice to think about the Gibbs phase rule in such a context when starting *any* phase equilibrium problem.

Step 2 *Model choice*
In all phase equilibrium problems, we start with the equality of mixtures fugacities of the components in solution. Here, we have a system at low pressure, so the ideal gas law is a good model for the vapor phase. However, we already know from the definition of the ideal solution that water and ethanol do *not* behave as an ideal solution. Thus, modified Raoult's Law is the approach.

$$\gamma_i x_i P_i^{sat} = y_i P \qquad (11.69)$$

We need to find a way to estimate the activity coefficient at $x_1 = 0.8$ in order to solve the problem. Our strategy, then, is to use the known data at $x_1 = 0.3$ to obtain the parameters for an excess molar Gibbs free energy model. Then we can use this model at $x_1 = 0.8$ to estimate the activity coefficient and, ultimately, estimate the vapor phase composition and system pressure.

Step 3 *Calculating the activity coefficient from experimental data*
Using Equation 11.69, we can obtain the activity coefficients for water and ethanol at the known compositions.

$$\gamma_1 = \frac{y_1 P}{x_1 P_1^{sat}} \qquad (11.86)$$

$$\gamma_2 = \frac{y_2 P}{x_2 P_2^{sat}} \qquad (11.87)$$

We can find the vapor pressure for the pure components from the Antoine equation at 30°C using the parameters in Appendix E. At this temperature, $P_1^{sat}(30°C) = 4.20$ kPa and $P_2^{sat}(30°C) = 10.42$ kPa.

Water

Ethanol

Plugging the numbers into Equations 11.86 and 11.87, we obtain

$$\gamma_1 = \frac{(0.23)(10.1\,\mathrm{kPa})}{(0.3)(4.20\,\mathrm{kPa})} = 1.844$$

$$\gamma_2 = \frac{(0.77)(10.1\,\mathrm{kPa})}{(0.7)(10.42\,\mathrm{kPa})} = 1.066$$

To reiterate, the values above are the *experimental* activity coefficients at $x_1 = 0.3$. If this were an ideal solution, the values would be equal to one. That they are greater than one indicates that this solution shows positive deviations from Raoult's Law, as do most solutions.

Step 4 *Excess molar Gibbs free energy model selection and parameterization*
Since we have only learned one excess molar Gibbs free energy model up to this point, we will choose our 1-parameter Margules equation. This is

$$\frac{G^{\mathrm{E}}}{RT} = Ax_1x_2 \tag{11.75}$$

So how do we determine the A value from our experimental data? Recall Equation 11.74, which provides us with a summability relationship to use for the excess molar Gibbs free energy:

$$\frac{G^{\mathrm{E}}}{RT} = x_1 \ln[\gamma_1] + x_2 \ln[\gamma_2] \tag{11.74}$$

Thus, our experimental value for $\underline{G}^{\mathrm{E}}/RT$ is

$$\frac{G^{\mathrm{E}}}{RT} = (0.3)\ln[1.844] + (0.7)\ln[1.066]$$

$$\frac{G^{\mathrm{E}}}{RT} = 0.228$$

Now we can solve for the parameter A from our 1-parameter Margules equation:

$$0.228 = A(0.3)(0.7)$$

$$A = 1.086$$

Note that solving for the A parameter directly here was only feasible because there was only one experimental data point to use. If there were multiple experimental data points, but one parameter, we would need to develop an objective function and minimize its value in order to determine the "best fit" parameter that would describe our data best over the range of data provided. We do that in the next example.

Step 5 *Predicting the activity coefficients at the new point*
Now that we know A, we can use it to estimate the activity coefficients at $x_1 = 0.8$.

Recall that we already obtained the activity coefficient expressions for the 1-parameter Margules equation when evaluating it for compliance to the Gibbs-Duhem equation (Equations 11.81 and 11.82). Thus,

$$\ln \gamma_1 = Ax_2^2 \tag{11.81}$$

$$\ln \gamma_2 = Ax_1^2 \tag{11.82}$$

So, for $x_1 = 0.8$ we find

$$\ln \gamma_1 = 1.086(1 - 0.8)^2$$

$$\gamma_1 = 1.044$$

$$\ln \gamma_2 = 1.086(0.8)^2$$

$$\gamma_2 = 2.004$$

Step 6 *Predicting the vapor phase composition and system pressure*

Revisiting our modified Raoult's Law expression (Equation 11.69) written for components 1 and 2 gives

$$\gamma_1 x_1 P_1^{sat} = y_1 P$$

$$\gamma_2(1 - x_1)P_2^{sat} = (1 - y_1)P$$

We see that we have two equations and two unknowns (y_1 and P). These can be rearranged to solve consecutively as

$$y_1 = \frac{\gamma_1 x_1 P_1^{sat}}{\gamma_1 x_1 P_1^{sat} + \gamma_2 x_2 P_2^{sat}} \tag{11.88}$$

$$P = \gamma_1 x_1 P_1^{sat} + \gamma_2 x_2 P_2^{sat} \tag{11.89}$$

These are readily solved, resulting in

$$y_1 = \textbf{0.46}$$

$$P = \textbf{7.69\,kPa}$$

The answer we have solved for is the modeling result. However, we know experimentally that $y_1 = 0.44$ and $P = 8.23$ kPa while, for comparison, the Raoult's Law values are $y_1 = 0.62$ and $P = 5.45$ kPa. Thus, using a real solution model to estimate the VLE for this system was beneficial. However, the pressure was off by about 7% and there are better excess molar Gibbs free energy models to use that will result in more accurate predictions.

A modification of the 1-parameter Margules equation is the **2-parameter Margules equation**, given as

$$\frac{G^E}{RT} = x_1 x_2 (A_{21}x_1 + A_{12}x_2) \tag{11.90}$$

Using Equations 11.77 and 11.78, one can obtain the activity coefficient expressions for the 2-parameter Margules equation in a manner similar to that for the 1-parameter Margules equation. Thus,

$$\ln \gamma_1 = x_2^2(A_{12} + 2[A_{21} - A_{12}]x_1) \tag{11.91}$$

$$\ln \gamma_2 = x_1^2(A_{21} + 2[A_{12} - A_{21}]x_2) \tag{11.92}$$

Unlike the 1-parameter Margules equation, the 2-parameter Margules equation is not necessarily symmetric. In this next example, we will compare both models.

EXAMPLE 11-4	DATA REDUCTION USING THE 1- AND 2-PARAMETER MARGULES EQUATION

di-isopropyl ether

1-propanol

Perform a reduction of the data in Table 11-1 for the di-isopropyl ether (1) + 1-propanol (2) system at 303.15 K using both the 1-parameter and 2-parameter Margules equations.

TABLE 11-1 Experimental Pxy data for the di-isopropyl ether (1) + 1-propanol (2) system at 303.15 K.

P (kPa)	x_1	y_1	P (kPa)	x_1	y_1
3.77	0	0	19.51	0.5296	0.8774
5.05	0.0199	0.2671	20.23	0.5902	0.8890
6.15	0.0399	0.4090	20.71	0.6505	0.8974
7.22	0.0601	0.5061	21.35	0.7101	0.9093
8.29	0.0799	0.5783	21.92	0.7685	0.9209
10.60	0.1192	0.6847	22.62	0.8300	0.9372
12.16	0.1694	0.7346	23.20	0.8803	0.9521
14.07	0.2294	0.7822	23.59	0.9179	0.9637
15.62	0.2891	0.8133	23.80	0.9397	0.9709
16.81	0.3495	0.8343	23.99	0.9581	0.9785
17.91	0.4090	0.8524	24.19	0.9804	0.9885
18.77	0.4708	0.8659	24.36	1	1

Based on data from I. Hwang et al., J. Chem. Eng. Data, 52, 2503 (2007).

SOLUTION:

Step 1 *Obtain the experimental activity coefficients and the experimental excess molar Gibbs free energy*

We can obtain the experimental activity coefficients and the excess molar Gibbs free energy in a manner similar to the Example 11-2 by using the data in Table 11-1 and

$$\gamma_1 = \frac{y_1 P}{x_1 P_1^{\text{sat}}} \tag{11.86}$$

$$\gamma_2 = \frac{(1 - y_1)P}{(1 - x_1)P_2^{\text{sat}}} \tag{11.87}$$

$$\frac{G^{\text{E}}}{RT} = x_1 \ln[\gamma_1] + x_2 \ln[\gamma_2] \tag{11.74}$$

The resulting data is given in Table 11-2. At this point it is important to note a few things. First, the activity coefficient for component 1 when $x_1 = 1$ does not exist (since component 1 is not present at this point). Likewise, the activity coefficient for component 2 when $x_1 = 0$ also does not exist. Thus, those entries are not included in Table 11-2.

Second, by definition, the excess molar Gibbs free energy when $x_1 = 1$ or $x_2 = 1$ is zero. Thus, we can put those endpoint entries into the table by hand.

Third, when a substance becomes very dilute in a mixture, its activity coefficient approaches a limiting value called the **infinite dilution activity coefficient** and is

TABLE 11-2 Experimental activity coefficient data for the di-isopropyl (1) + 1-propanol (2) system at 303.15 K.

P (kPa)	x_1	y_1	γ_1	γ_2	$\underline{G}^E/RT$
3.77	0	0	—	1	0
5.05	0.0199	0.2671	2.782	1.001	0.0213
6.15	0.0399	0.4090	2.587	1.003	0.0408
7.22	0.0601	0.5061	2.499	1.005	0.0597
8.29	0.0799	0.5783	2.464	1.006	0.0776
10.60	0.1192	0.6847	2.498	1.005	0.1135
12.16	0.1694	0.7346	2.165	1.030	0.1554
14.07	0.2294	0.7822	1.970	1.053	0.1953
15.62	0.2891	0.8133	1.804	1.087	0.2299
16.81	0.3495	0.8343	1.648	1.134	0.2564
17.91	0.4090	0.8524	1.533	1.185	0.2751
18.77	0.4708	0.8659	1.417	1.260	0.2864
19.51	0.5296	0.8774	1.327	1.346	0.2896
20.23	0.5902	0.8890	1.251	1.452	0.2850
20.71	0.6505	0.8974	1.173	1.611	0.2705
21.35	0.7101	0.9093	1.123	1.770	0.2479
21.92	0.7685	0.9209	1.078	1.984	0.2163
22.62	0.8300	0.9372	1.049	2.214	0.1748
23.20	0.8803	0.9521	1.030	2.460	0.1338
23.59	0.9179	0.9637	1.017	2.764	0.0989
23.80	0.9397	0.9709	1.009	3.047	0.0756
23.99	0.9581	0.9785	1.006	3.259	0.0552
24.19	0.9804	0.9885	1.001	3.750	0.0269
24.36	1	1	1	—	0

denoted as γ_i^∞. We introduced this earlier in the chapter when working with Henry's Law. Recall that γ_i^∞ changes depending on which component is dilute and which component is concentrated. For example, in this problem, when 1-propanol is dilute in di-isopropyl ether, its infinite dilution activity coefficient would seem to be a little above 3.75, according to Table 11-2. On the other hand, when di-isopropyl ether is dilute in 1-propanol, its infinite dilution activity coefficient is likely a little above 2.78. Just by looking at the data, we can tell that the 1-parameter Margules equation will have trouble modeling this system, since it predicts a symmetric system (and the same infinite dilution activity coefficient for each compound). At any rate, we have the experimental data we need for modeling and the next step is to find the model parameters.

FOOD FOR THOUGHT 11-2

If there is no value for the activity coefficients of 1 when $x_1 = 0$ or of 2 when $x_2 = 0$, how is this different than the infinite dilution activity coefficient discussed previously?

The infinite dilution activity coefficients for both components in a binary mixture, according to the 1-parameter Margules equation, are the same and are easily calculated by taking the limit as the composition of the component of interest goes to zero (or, equivalently, while the other goes to 1).

$$\ln \gamma_1^\infty = \lim_{x_2 \to 1} [\ln \gamma_1]$$
$$= \lim_{x_2 \to 1} [Ax_2^2] = A$$
$$\ln \gamma_2^\infty = \lim_{x_1 \to 1} [\ln \gamma_2]$$
$$= \lim_{x_1 \to 1} [Ax_1^2] = A$$

In Section 11.4, we discussed the Henry's Law constants and how they are different for the different components in the same mixture. The same discussion applies here for the activity coefficients at infinite dilution.

Another objective function you could use is to minimize the difference between the experimental pressure and the model-predicted pressure. Both approaches are used in practice.

We use only 22 points for the modeling, since the endpoints would provide a denominator of 11.93 that is divided by zero.

Step 2 *Defining the objective function*

In Example 11-2 we had a single data point. Thus, we used that one data point to solve for the A parameter of the 1-parameter Margules equation. Here, we have 24 data points and we need to find the best value for A (1-parameter Margules equation) and the best value for A_{12} and A_{21} (2-parameter Margules equation) across all of the data. Therefore, we need to define an objective function that, once minimized, will provide us with the best values for the parameters.

One approach is to try to minimize the deviation between the predicted value of G^E/RT from the model and the experimental value for the 24 points. There are a few procedural issues that one must be aware of when performing minimizations of this kind. First, if you have a program that does minimization (such as Solver in MS Excel), it will try to minimize your objective function—all the way to negative infinity, if possible! That's certainly not what you want. Therefore, you need to define your objective function such that only positive terms exist. Second, it is important to make the terms in your objective function have the same order of magnitude (which is just good practice). This way, if fitting data whose values change across orders of magnitude, the minimization process isn't biased toward the larger values (whose actual deviations from the model will be a much smaller percentage than the smaller values).

A reasonable objective function, therefore, looks like

$$\text{OBJ} = \frac{1}{22} \sum_{i=1}^{22} \left[\frac{\left(\dfrac{G^E}{RT}\right)_i^{\text{Model}} - \left(\dfrac{G^E}{RT}\right)_i^{\text{Expt.}}}{\left(\dfrac{G^E}{RT}\right)_i^{\text{Expt.}}} \right]^2 \tag{11.93}$$

Here, OBJ will always be positive, and the terms are scaled so that they are the same order of magnitude. Note that the endpoints ($x_1 = 0$ and $x_1 = 1$) have been removed since the denominator of the objective function for both of those points will be zero and, thus, the objective function will become undefined.

If we minimize the objective function (OBJ) to find the best values for both models, we arrive at the following result, presented in Table 11-3.

TABLE 11-3 Model parameters for the di-isopropyl (1) + 1-propanol (2) system at 303.15 K.

Model	A	A_{12}	A_{21}	OBJ
1-parameter Margules	1.165			0.0653
2-parameter Margules		1.041	1.317	0.0065

Notice how the value A_{12} and A_{21} bracket the value of A.

Note that the 2-parameter Margules equation fits the data better than the 1-parameter Margules equation, as evidenced by its smaller OBJ value (an order of magnitude smaller). You might have anticipated this result since the model contains a second parameter and, thus, more flexibility to fit the experimental data.

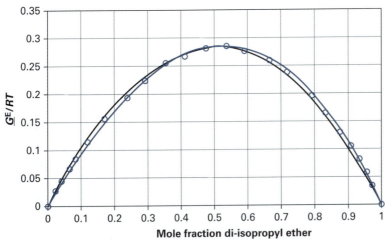

FIGURE 11-5 The excess molar Gibbs free energy for the di-isopropyl ether (1) + 1-propanol (2) system at 303.15 K. The experimental data is given by empty circles while the models are given by the solid lines: black for the 1-parameter Margules equation and blue for the 2-parameter Margules equation.

The infinite dilution activity coefficients for both components in a binary mixture, according to the 2-parameter Margules equation, are *different*.

$$\ln \gamma_1^{\infty} = \lim_{x_2 \to 1}[\ln \gamma_1]$$
$$= A_{12}$$

$$\ln \gamma_2^{\infty} = \lim_{x_1 \to 1}[\ln \gamma_2]$$
$$= A_{21}$$

These are obtained through observation of Equations 11.91 and 11.92.

We can plot the results as well to compare the two models relative to the experimental data in Figure 11-5.

Likewise, we can plot the activity coefficients as well for both models relative to the experimental data in Figure 11-6.

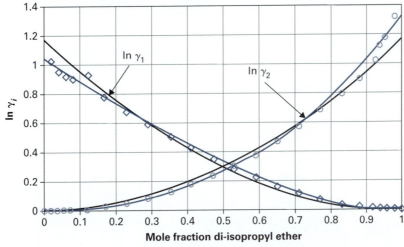

FIGURE 11-6 The natural logarithm of the activity coefficients for the di-isopropyl ether (1) + 1-propanol (2) system at 303.15 K. The experimental data is given by empty circles while the models are given by the solid lines: black for the 1-parameter Margules equation and blue for the 2-parameter Margules equation.

Note that while the black curves (1-parameter Margules equation) predict the same infinite dilution activity coefficient for both substances, the blue curves (2-parameter Margules equation) do not and better reflect what is observed experimentally.

In the next example, we introduce another excess molar Gibbs free energy model that can be used in order to take fitted parameters at one temperature and apply them to another temperature.

| EXAMPLE 11-5 | **PARAMETER DETERMINATION USING THE WILSON EQUATION** |

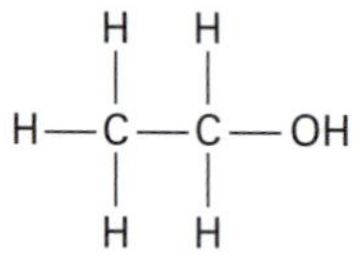

You are part of a team evaluating the separation of a mixture of ethanol (1) + butyl methyl ether (2). This was a project your company explored a few years earlier. In that project, team members measured the VLE for this system at 323.15 K. However, based on some additional operating constraints, you now need the VLE for this system at 338.15 K to provide an estimate on the size of the distillation column required. Rather than perform the experiment again at 338.15 K, you decide to use the data at 323.15 K to estimate the VLE at 338.15 K. The data at 323.15 K for this system is given in Table 11-4.

Ethanol

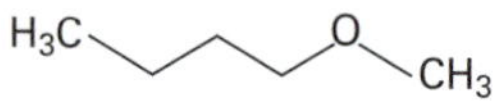

Butyl methyl ether

Butyl methyl ether is also known as 1-methoxybutane.

TABLE 11-4 Experimental Pxy data for the ethanol (1) + butyl methyl ether (2) system at 323.15 K.

x_1	P (kPa)	y_1
0	50.69	0
0.0316	52.37	0.0592
0.1067	53.95	0.1597
0.1899	55.83	0.2426
0.3393	56.73	0.2990
0.4303	56.10	0.3317
0.6225	53.75	0.4030
0.6649	52.74	0.4334
0.7352	50.52	0.4902
0.8124	47.49	0.5284
0.8525	45.52	0.6058
0.9091	41.41	0.6677
0.9542	36.75	0.7751
0.9820	32.49	0.8956
1	29.34	1

Based on data from T. Hofman, A. Sporzynski and A. Goldon, J. Chem. Eng. Data, 45, 169 (2000).

SOLUTION:

Step 1 *Make a plan*

Up to this point, we have not discussed the temperature dependence of the excess molar Gibbs free energy model parameters. The Margules equation has no provision for parameter temperature dependence built in, but other modeling approaches do. One popular excess molar Gibbs free energy model that has a temperature-dependent element for the parameters is called the Wilson equation. Therefore, in this example, we will fit the parameters for the Wilson equation at the temperature where we have experimental data (323.15 K) and use those parameters to predict the mixture phase behavior at 338.15 K.

Step 2 *Obtain the experimental activity coefficients and excess molar Gibbs free energy at 323.15 K*

As in the previous examples, we can obtain the experimental activity coefficients and the excess molar Gibbs free energy in a manner similar to Example 11-4 by using the following expressions and data in Table 11-4. Thus,

$$\gamma_1 = \frac{y_1 P}{x_1 P_1^{sat}} \tag{11.86}$$

$$\gamma_2 = \frac{(1 - y_1)P}{(1 - x_1)P_2^{sat}} \tag{11.87}$$

$$\frac{G^E}{RT} = x_1 \ln[\gamma_1] + x_2 \ln[\gamma_2] \tag{11.74}$$

The results are shown in Table 11-5.

TABLE 11-5 Experimental activity coefficient data for the ethanol (1) + butyl methyl ether (2) system at 323.15 K.

x_1	P (kPa)	y_1	γ_1	γ_2	G^E/RT
0	50.69	0		1	0
0.0316	52.37	0.0592	3.3924	1.0204	0.0582
0.1067	53.95	0.1597	2.7839	1.0182	0.1254
0.1899	55.83	0.2426	2.4537	1.0479	0.2084
0.3393	56.73	0.2990	1.7176	1.2089	0.3089
0.4303	56.10	0.3317	1.4847	1.3222	0.3292
0.6225	53.75	0.4030	1.1929	1.7093	0.3122
0.6649	52.74	0.4334	1.1778	1.7939	0.3046
0.7352	50.52	0.4902	1.1529	1.9583	0.2826
0.8124	47.49	0.5284	1.0566	2.4052	0.2094
0.8525	45.52	0.6058	1.1053	2.4545	0.2178
0.9091	41.41	0.6677	1.0385	3.0582	0.1360
0.9542	36.75	0.7751	1.0183	3.6548	0.0767
0.9820	32.49	0.8956	1.0101	3.8290	0.0340
1	29.34	1	1		0

Step 3 *The Wilson equation*

The Wilson equation is given as

Grant Wilson, a Ph.D. physical chemist who graduated from MIT, proposed this excess molar Gibbs free energy model in 1964.

$$\frac{G^E}{RT} = -x_1 \ln(x_1 + \Lambda_{12}x_2) - x_2 \ln(x_2 + \Lambda_{21}x_1) \tag{11.94}$$

The activity coefficients for this model are given using the same approach as before (using Equations 11.77 and 11.78), to obtain

$$\ln \gamma_1 = -\ln(x_1 + \Lambda_{12}x_2) + x_2\left(\frac{\Lambda_{12}}{x_1 + \Lambda_{12}x_2} - \frac{\Lambda_{21}}{x_2 + \Lambda_{21}x_1}\right) \tag{11.95}$$

$$\ln \gamma_2 = -\ln(x_2 + \Lambda_{21}x_1) - x_1\left(\frac{\Lambda_{12}}{x_1 + \Lambda_{12}x_2} - \frac{\Lambda_{21}}{x_2 + \Lambda_{21}x_1}\right) \tag{11.96}$$

The Wilson equation is based on local composition theory, which states that while the overall composition of a mixture is fixed, the local composition throughout the system is not necessarily homogeneous.

The temperature dependence of the parameters Λ_{12} and Λ_{21} is given as

$$\Lambda_{12} = \frac{\underline{V}_2}{\underline{V}_1} \exp\left(\frac{-\alpha_{12}}{RT}\right) \tag{11.97}$$

$$\Lambda_{21} = \frac{\underline{V}_1}{\underline{V}_2} \exp\left(\frac{-\alpha_{21}}{RT}\right) \tag{11.98}$$

Here, $\underline{V}$ is the liquid molar volume of the pure component at the mixture temperature, while α_{12} and α_{21} are composition-independent parameters that describe how the interactions between the unlike components differ from the like components. Our fitting approach involves finding the best-fit parameters for Λ_{12} and Λ_{21} and then solving for the values of α_{12} and α_{21} at that temperature. For other temperatures, we just use Equations 11.97 and 11.98 to find Λ_{12} and Λ_{21} at the new temperature of interest, using the temperature-independent parameters, α_{12} and α_{21}.

Step 4 *Parameter fitting at 323.15 K*
Our objective function, OBJ, for the data set in Table 11-5 becomes using Equation 11.93:

$$\text{OBJ} = \frac{1}{13}\sum_{i=1}^{13}\left[\frac{\left(\dfrac{G^E}{RT}\right)_i^{\text{Model}} - \left(\dfrac{G^E}{RT}\right)_i^{\text{Expt.}}}{\left(\dfrac{G^E}{RT}\right)_i^{\text{Expt.}}}\right]^2$$

Thus, at 323.15 K we fit the data using the objective function and obtain the best fit values for Λ_{12} and Λ_{21}, shown in Table 11-6.

TABLE 11-6 Model parameters for the ethanol (1) + butyl methyl ether (2) system at 323.15 K and 338.15 K.

Model	T (K)	Λ_{12}	Λ_{21}	OBJ
Wilson equation	323.15	0.5045	0.3031	0.145
Wilson equation	338.15	0.5367	0.3097	

Step 5 *Obtaining the Wilson parameters at 338.15 K*
Now that we have found the values for Λ_{12} and Λ_{21} at 323.15 K, we need to find the values for α_{12} and α_{21} to obtain the new values for Λ_{12} and Λ_{21} at 338.15 K. For that, we will need the pure-component molar volumes at both 323.15 K and 338.15 K. While many values of the molar volume are found in the Appendix C (including ethanol), butyl methyl ether is not. In these situations, a sound strategy is to measure the molar volume yourself, using the same chemicals that were used to calculate the VLE. For the purposes of this example, assume that they were measured at 298.15 K (they were—Hofman et al., 2000). We can make the reasonable assumption that the molar volumes of liquids do not change appreciably with temperature, especially far from the critical point. Helping even more is the fact that we need the ratio, so whatever change occurs for the density of one component is likely met with a similar change in the other component. Thus, this ratio would be fairly constant as a function of temperature.

$$\underline{V}_1 = 58.67 \text{ cm}^3/\text{mol}$$

$$\underline{V}_2 = 119.25 \text{ cm}^3/\text{mol}$$

Accordingly, α_{12}/R and α_{21}/R are given as

$$\alpha_{12}/R = 450.2789 \text{ K}$$

$$\alpha_{21}/R = 156.5022 \text{ K}$$

Plugging this into Equations 11.97 and 11.98 we obtain the new values for Λ_{12} and Λ_{21} at 338.15 K, which are listed in Table 11-6.

Step 6 *Predicting the VLE at 338.15 K*

Using the parameter values at 338.15 K, we can determine the activity coefficients for both components from the Wilson equation using Equations 11.95 and 11.96. Since we are interested in predicting the *Pxy* data for this system at this temperature, we use the following equations to obtain the vapor-phase mole fraction and the system pressure.

$$y_1 = \frac{\gamma_1 x_1 P_1^{sat}}{\gamma_1 x_1 P_1^{sat} + \gamma_2 x_2 P_2^{sat}} \tag{11.88}$$

$$P = \gamma_1 x_1 P_1^{sat} + \gamma_2 x_2 P_2^{sat} \tag{11.89}$$

We can plot the model results in Figure 11-7.

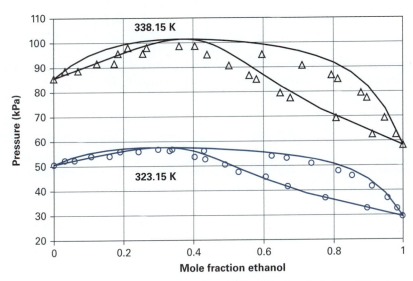

FIGURE 11-7 A *Pxy* diagram of the ethanol (1) + butyl methyl ether (2) system at two different temperatures. The empty symbols are experimental data.

To provide perspective, we show the experimental values for both *T* = 323.15 K and 338.15 K relative to the model correlations (323.15 K) and model predictions (338.15K). Here, the benefit of the model predictions at 338.15 K provides an estimate of the VLE that can be used during process analysis. However, the actual sizing of the equipment would require experimental data.

11.7.1 Temperature and Pressure Dependence of the Activity Coefficient

Up to this point, we have not discussed the effect of temperature and pressure on the activity coefficients themselves. All of our work so far has been at constant temperature and the implication (not stated) was that the pressure effects on the excess molar Gibbs free energy (and, in turn, the activity coefficients) did not strongly impact the results. Indeed, none of the models were explicit in the pressure. Before moving further, let's explore this notion in this subsection as it also sets the stage for something we will need in a later chapter.

If we wanted to determine how temperature and pressure impact the excess molar Gibbs free energy, we need to return to Chapter 9, which describes the fundamental property relationship on how $\underline{M}$ changes with the system variables (Equation 9.36):

$$d(n\underline{M}) = \left(\frac{\partial[n\underline{M}]}{\partial T}\right)_{P,n_1,n_2} dT + \left(\frac{\partial[n\underline{M}]}{\partial P}\right)_{T,n_1,n_2} dP + \left(\frac{\partial[n\underline{M}]}{\partial n_1}\right)_{P,T,n_2} dn_1$$
$$+ \left(\frac{\partial[n\underline{M}]}{\partial n_2}\right)_{P,T,n_1} dn_2 \tag{9.36}$$

Plugging in $\underline{G}^E/RT$ for $\underline{M}$ yields

$$d\left(\frac{n\underline{G}^E}{RT}\right) = \frac{\left(\partial\left[\dfrac{n\underline{G}^E}{RT}\right]\right)}{\partial T}\Bigg|_{P,n_1,n_2} dT + \frac{\left(\partial\left[\dfrac{n\underline{G}^E}{RT}\right]\right)}{\partial P}\Bigg|_{T,n_1,n_2} dP$$
$$+ \frac{\left(\partial\left[\dfrac{n\underline{G}^E}{RT}\right]\right)}{\partial n_1}\Bigg|_{P,T,n_2} dn_1 + \frac{\left(\partial\left[\dfrac{n\underline{G}^E}{RT}\right]\right)}{\partial n_2}\Bigg|_{P,T,n_1} dn_2 \tag{11.99}$$

We can make a few simplifications here. First, those last two partial derivatives are the definitions of the activity coefficient (see Equation 11.73).

$$d\left(\frac{n\underline{G}^E}{RT}\right) = \frac{\left(\partial\left[\dfrac{n\underline{G}^E}{RT}\right]\right)}{\partial T}\Bigg|_{P,n_1,n_2} dT + \frac{\left(\partial\left[\dfrac{n\underline{G}^E}{RT}\right]\right)}{\partial P}\Bigg|_{T,n_1,n_2} dP + \ln\gamma_1 dn_1 + \ln\gamma_2 dn_2 \tag{11.100}$$

Second, we can pull some variables out of the derivatives as they are constants:

$$d\left(\frac{n\underline{G}^E}{RT}\right) = \left(\frac{n}{R}\right)\frac{\left(\partial\left[\dfrac{\underline{G}^E}{T}\right]\right)}{\partial T}\Bigg|_{P,n_1,n_2} dT + \left(\frac{n}{RT}\right)\left(\frac{\partial[\underline{G}^E]}{\partial P}\right)_{T,n_1,n_2} dP + \ln\gamma_1 dn_1 + \ln\gamma_2 dn_2 \tag{11.101}$$

Third, we know from Section 6.2.2 that

$$\underline{V} = \left(\frac{\partial\underline{G}}{\partial P}\right)_T$$

Therefore,

$$\underline{V}^E = \left(\frac{\partial\underline{G}^E}{\partial P}\right)_T \tag{11.102}$$

And we can make the substitution as

$$d\left(\frac{n\underline{G}^E}{RT}\right) = \left(\frac{n}{R}\right)\frac{\left(\partial\left[\dfrac{\underline{G}^E}{T}\right]\right)}{\partial T}\Bigg|_{P,n_1,n_2} dT + \left(\frac{n}{RT}\right)\underline{V}^E dP + \ln\gamma_1 dn_1 + \ln\gamma_2 dn_2 \tag{11.103}$$

The first term on the right-hand side of Equation 11.103 looks like it is related to the excess molar entropy since

$$\underline{S} = \left(\frac{\partial\underline{G}}{\partial T}\right)_P$$

Thus,

$$\underline{S}^E = \left(\frac{\partial \underline{G}^E}{\partial T}\right)_P \tag{11.104}$$

But it is not this simple since the temperature is *inside* the derivative (it isn't constant). However, we can expose that derivative into something that might be more recognizable using the quotient rule. Thus,

$$\frac{\left(\partial\left[\dfrac{\underline{G}^E}{T}\right]\right)}{\partial T}_{P,n_1,n_2} = \frac{T\left(\dfrac{\partial[\underline{G}^E]}{\partial T}\right)_{P,n_1,n_2} - (1)\underline{G}^E}{T^2} \tag{11.105}$$

Using Equation 11.104, we have

$$\frac{\left(\partial\left[\dfrac{\underline{G}^E}{T}\right]\right)}{\partial T}_{P,n_1,n_2} = \frac{-T\underline{S}^E - \underline{G}^E}{T^2} \tag{11.106}$$

Since we know that $\underline{H} = \underline{G} + T\underline{S}$, then $\underline{H}^E = \underline{G}^E + T\underline{S}^E$. So

$$\frac{\left(\partial\left[\dfrac{\underline{G}^E}{T}\right]\right)}{\partial T}_{P,n_1,n_2} = \frac{-\underline{H}^E}{T^2} \tag{11.107}$$

Now we can plug Equation 11.107 into Equation 11.103 to obtain

$$d\left(\frac{n\underline{G}^E}{RT}\right) = -\left(\frac{n\underline{H}^E}{RT^2}\right)dT + \left(\frac{n\underline{V}^E}{RT}\right)dP + \ln(\gamma_1)\,dn_1 + \ln(\gamma_2)\,dn_2 \tag{11.108}$$

Equation 11.108 is an expression we can use to answer our question about the impact of temperature and pressure on the excess molar Gibbs free energy.

To explore the effect of temperature, we divide by a differential temperature at constant pressure and number of moles of each component to obtain the required partial derivative.

$$\frac{\partial}{\partial T}\left(\frac{n\underline{G}^E}{RT}\right)_{P,n_1,n_2} = -\left(\frac{n\underline{H}^E}{RT^2}\right)\left(\frac{\partial T}{\partial T}\right)_{P,n_1,n_2} + \left(\frac{n\underline{V}^E}{RT}\right)\left(\frac{\partial P}{\partial T}\right)_{P,n_1,n_2}$$
$$+ \ln(\gamma_1)\left(\frac{\partial n_1}{\partial T}\right)_{P,n_1,n_2} + \ln(\gamma_2)\left(\frac{\partial n_2}{\partial T}\right)_{P,n_1,n_2} \tag{11.109}$$

The last three terms on the right hand side of Equation 11.109 are all zero. Do you see why?

This simplifies to

$$\frac{\partial}{\partial T}\left(\frac{n\underline{G}^E}{RT}\right)_{P,n_1,n_2} = -\left(\frac{n\underline{H}^E}{RT^2}\right) \tag{11.110}$$

or dividing out the number of moles, this yields

$$\frac{\partial}{\partial T}\left(\frac{\underline{G}^E}{RT}\right)_{P,n_1,n_2} = -\left(\frac{\underline{H}^E}{RT^2}\right) \tag{11.111}$$

Similarly for pressure, we divide Equation 11.108 by a differential pressure at constant temperature and number of moles of each component to obtain

$$\frac{\partial}{\partial P}\left(\frac{n\underline{G}^E}{RT}\right)_{T,n_1,n_2} = -\left(\frac{n\underline{H}^E}{RT^2}\right)\left(\frac{\partial T}{\partial P}\right)_{T,n_1,n_2} + \left(\frac{n\underline{V}^E}{RT}\right)\left(\frac{\partial P}{\partial P}\right)_{T,n_1,n_2}$$
$$+ \ln(\gamma_1)\left(\frac{\partial n_1}{\partial P}\right)_{T,n_1,n_2} + \ln(\gamma_2)\left(\frac{\partial n_2}{\partial P}\right)_{T,n_1,n_2} \tag{11.112}$$

which simplifies to

$$\frac{\partial}{\partial P}\left(\frac{n\underline{G}^E}{RT}\right)_{T,n_1,n_2} = \frac{n\underline{V}^E}{RT} \tag{11.113}$$

or dividing out the number of moles yields

$$\frac{\partial}{\partial P}\left(\frac{\underline{G}^E}{RT}\right)_{T,n_1,n_2} = \frac{\underline{V}^E}{RT} \tag{11.114}$$

Equations 11.111 and 11.114 allow us to examine the impact of temperature and pressure, respectively, on the excess molar Gibbs free energy and, consequently, the activity coefficients. For example, consider the system tetrahydrofuran (1) + 2,2,2-trifluoroethanol (2) at 298.15 K. The measured excess molar enthalpy for an equimolar mixture is about -2.0 kJ/mol, while the measured excess molar volume for an equimolar mixture is about 1 cm³/mol (Perez et al., 2003). Plugging these values into Equations 11.111 and 11.114 (with proper choice of the universal gas constant) yields

$$\frac{\partial}{\partial T}\left(\frac{\underline{G}^E}{RT}\right)_{P,n_1,n_2} = \frac{2.7 \times 10^{-3}}{K}$$

$$\frac{\partial}{\partial P}\left(\frac{\underline{G}^E}{RT}\right)_{T,n_1,n_2} = \frac{4.03 \times 10^{-10}}{Pa} = \frac{4.03 \times 10^{-7}}{kPa}$$

Tetrahydrofuran

2,2,2-trifluoroethanol

For the derivative with respect to temperature, we see that there is a change on the order of 10^{-3} per change in 1 Kelvin. For the derivative with respect to pressure, we see that there is a change on the order of 10^{-7} for every change in 1 kPa. Thus, the impact of a change in temperature far exceeds that of pressure for the excess molar Gibbs free energy. Therefore, models for the excess molar Gibbs free energy will not normally include the effect of pressure and, accordingly, activity coefficients are often assumed to be independent of pressure.

11.7.2 Excess Molar Gibbs Free Energy Models and the Flash Problem

In Chapter 10 we solved the flash problem using both experimental data and a thermodynamic model (Raoult's Law). In this subsection, we show the solution of the flash problem using modified Raoult's Law.

EXAMPLE 11-6	THE FLASH PROBLEM USING MODIFIED RAOULT'S LAW

10 moles/s of an equimolar mixture of methyl ethyl ketone (1) and toluene (2) are flashed from 760 mm Hg (101.3 kPa) to 460 mm Hg (61.3 kPa) at 75°C. You know, experimentally, that the infinite dilution activity coefficient of methyl ethyl ketone in toluene is 1.44 at this temperature. Likewise, you know that the experimental value for the infinite dilution activity coefficient of toluene in methyl ethyl ketone is 1.33 at this temperature. Solve for the composition and amount of each equilibrium phase as a result of the flash if

(a) The liquid is treated as an ideal solution

(b) The liquid is treated as modeled by the 2-parameter Margules equation

SOLUTION:

Step 1 *Does the system flash at all? Does it flash partially? Does it flash fully? Raoult's Law model*

Let's determine if the system will flash at all, flash partially or flash fully. If the system does not flash or flashes fully, the material balance problem is trivial. Thus, the first step is to calculate the bubble point pressure for the system at this temperature. Let's do this for part (a) by modeling the liquid according to the ideal solution model. Since the system pressure is low, the vapor is sufficiently modeled by the ideal gas. Thus, we are using Raoult's Law.

Recall that Raoult's Law for both components is given as follows:

$$Py_1 = x_1 P_1^{sat} \qquad (10.3)$$

$$Py_2 = x_2 P_2^{sat} \qquad (10.9)$$

And the bubble-point pressure is given by

$$P = x_1 P_1^{sat} + x_2 P_2^{sat} \qquad (10.2)$$

Thus, we need the vapor pressure of both components at 75°C. Using Appendix E, we find that only the parameters for toluene are listed. With those parameters, we have:

$$P_2^{sat}(75°C) = 244.3 \text{ mm Hg } (32.6 \text{ kPa})$$

What about methyl ethyl ketone? There are a lot of resources available to determine the vapor pressure of methyl ethyl ketone (or 2-butanone, which is its IUPAC name) at a temperature of interest. The NIST Webbook provides Antoine equation parameters for many substances, including methyl ethyl ketone. However, the user must take precaution when using Antoine equation parameters since the parameter values are specific to the functional form of the equation used and the units specified. For methyl ethyl ketone, the NIST WebBook provides parameter values of:

$$A = 3.9894$$

$$B = 1150.207$$

$$C = -63.904$$

If you use those parameters for the Antoine equation in Appendix E of *this text*, you will find the vapor pressure of methyl ethyl ketone at 75°C is around 10^{-100} mm Hg. The actual expression for this set of parameters is

$$\log_{10}P = A - \frac{B}{T + C}$$

Here, T is in Kelvin and P is in bar.

Using the proper formula and converting to our pressure unit of interest for this problem, we find:

$$P_1^{sat}(75°C) = 657.6 \text{ mm Hg } (87.7 \text{ kPa})$$

Accordingly, our bubble point pressure becomes:

$$P = (0.5)(657.6 \text{ mm Hg}) + (0.5)(244.3 \text{ mm Hg})$$

$$P = 450.9 \text{ mm Hg } (60.1 \text{ kPa})$$

CH₃

Toluene

H₃C — O — CH₃

Methyl ethyl ketone

The bubble point pressure (450.9 mm Hg) is below the system pressure (460 mm Hg), thus the system will not flash and no separation will occur. The material balance is trivial as we still have an equimolar liquid mixture of methyl ethyl ketone and toluene, and part (a) of the problem is solved.

Step 2 *Does the system flash at all? Does it flash partially? Does it flash fully? Modified Raoult's Law model*

For part (b), we need to do the same thing again: determine if the system will flash at all and, if so, will it flash partially or fully. The bubble-point pressure, according to modified Raoult's Law, is

$$P = \gamma_1 x_1 P_1^{sat} + \gamma_2 x_2 P_2^{sat} \tag{11.89}$$

Thus, we need to determine the activity coefficients of the two components to determine the bubble-point pressure.

In the problem statement we are given a useful piece of experimental information that will allow us to define the parameters for our excess molar Gibbs free energy model. Recall that, for the 2-parameter Margules equation, the infinite dilution activity coefficients are as follows:

$$\ln \gamma_1^\infty = \lim_{x_2 \to 1}[\ln \gamma_1] = A_{12}$$

$$\ln \gamma_2^\infty = \lim_{x_1 \to 1}[\ln \gamma_2] = A_{21}$$

Thus,

$$A_{12} = \ln \gamma_1^\infty = \ln(1.44) = 0.3646$$

$$A_{21} = \ln \gamma_2^\infty = \ln(1.33) = 0.2852$$

Now that we have defined the parameters of the model, we can use Equations 11.91 and 11.92 to determine the activity coefficients at the composition of interest (here, equimolar). We find that:

$$\gamma_1(x_1 = 0.5) = 1.074$$

$$\gamma_2(x_1 = 0.5) = 1.095$$

Plugging that into the bubble-point expression for modified Raoult's Law

$$P = (1.074)(0.5)(657.6 \text{ mm Hg}) + (1.095)(0.5)(244.3 \text{ mm Hg})$$

$$P = 486.9 \text{ mm Hg } (64.9 \text{ kPa})$$

Thus, the bubble-point pressure (486.9 mm Hg) is above the system pressure (460 mm Hg), so it will flash. But will it flash partially or fully? To answer that question, we need to determine the dew-point pressure.

For modified Raoult's Law, we have

$$\frac{Py_1}{\gamma_1 P_1^{sat}} = x_1 \tag{11.115}$$

$$\frac{Py_2}{\gamma_2 P_2^{sat}} = x_2 \tag{11.116}$$

Adding these two expressions together gives us a right-hand side equal to one (since $x_1 + x_2 = 1$) and, thus,

$$\frac{Py_1}{\gamma_1 P_1^{sat}} + \frac{Py_2}{\gamma_2 P_2^{sat}} = 1 \tag{11.117}$$

Solving for P, we obtain the expression for the dew-point pressure from modified Raoult's Law.

$$P = \frac{1}{\left(\dfrac{y_1}{\gamma_1 P_1^{\text{sat}}} + \dfrac{y_2}{\gamma_2 P_2^{\text{sat}}}\right)} \qquad (11.118)$$

One very important point to note here is that we are substituting the feed composition for the dew-point (i.e. vapor-phase) composition. In other words, we are setting $y_1 = 0.5$. However, the activity coefficients in Equation 11.118 are evaluated at a liquid-phase composition that is in equilibrium with the vapor-phase composition ($y_1 = 0.5$). Thus, the liquid-phase compositions are unknown and, accordingly, the activity coefficients are unknown (since they are functions of the liquid-phase composition).

The strategy to address this is to treat the dew-point pressure (P) and the liquid-phase composition (x_1) as unknown variables. In addition to Equation 11.118, we need a second equation relating those two unknown variables — Equation 11.115.

$$\frac{P y_1}{\gamma_1 P_1^{\text{sat}}} = x_1 \qquad (11.115)$$

Solving Equations 11.115 and 11.118 for the two unknowns, P and x_1, using a non-linear equation solver (such as the Solver function in Microsoft Excel), we arrive at a dew-point pressure of 380.39 mm Hg (with $x_1 = 0.239$).

Thus, since the system pressure (460 mm Hg) is below the bubble-point pressure (486.9 mm Hg) and above the dew-point pressure (380.39 mm Hg), the system will partially flash. Therefore, we need to move to the next stage of the problem: the flash.

> 380.39 mm Hg is 50.7 kPa.

Step 3 *Draw a picture*
We can draw a diagram of the flash process and label the variables. This is given in Figure 11-8.

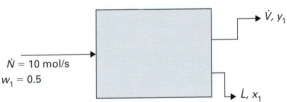

$\dot{N} = 10$ mol/s
$w_1 = 0.5$

$\dot{V}, y_1$

$\dot{L}, x_1$

> In calculating the bubble-point pressure, $\gamma_1(x_1 = 0.5) = 1.074$ and $\gamma_2(x_1 = 0.5) = 1.095$. For the dew-point pressure, we find $\gamma_1(x_1 = 0.239) = 1.208$ and $\gamma_2(x_1 = 0.239) = 1.024$.

FIGURE 11-8 A diagram of the material balance for the flash process.

Step 4 *Write material balances and perform a degree of freedom analysis*
As before when solving the flash problem in Chapter 10 for Raoult's Law, we will be able to write two independent material balances. Let's write this for the overall moles and for moles of component 1.

Overall Balance: $\qquad\qquad \dot{N} = \dot{L} + \dot{V}$

Component 1: $\qquad\qquad w_1 \dot{N} = x_1 \dot{L} + y_1 \dot{V}$

What are the unknowns in the problem? There are four: $\dot{L}, \dot{V}, y_1$, and x_1. With only two equations and four unknowns, our problem is underspecified. However, we do have another pair of equations that relate the unknown variables in the problem: the modified Raoult's Law expressions for components 1 and 2.

The problem now has four equations and four unknowns, and is now properly specified. Note that while the activity coefficient values are not known, they are funtions of another unknown (x_1) and, thus, do not add any additional unknowns to the problem.

>
> **PITFALL PREVENTION**
>
> You **cannot** use the values for the activity coefficients from step 2 of this problem here since your liquid phase compositions are *unknown* (and, thus, the activity coefficient values are unknown, since they are functions of the composition).

Step 5 *Solve the problem*

To reiterate, the four equations are

$$10 \frac{\text{mol}}{\text{s}} = \dot{L} + \dot{V}$$

$$(0.5)\left(10 \frac{\text{mol}}{\text{s}}\right) = x_1 \dot{L} + y_1 \dot{V}$$

$$(460 \text{ mm Hg})y_1 = x_1(657.6 \text{ mm Hg}) \exp\left[(1 - x_1)^2(A_{12} + 2[A_{21} - A_{12}]x_1)\right]$$

$$(460 \text{ mm Hg})(1 - y_1) = (1 - x_1)(244.3 \text{ mm Hg}) \exp\left[(x_1^2)(A_{21} + 2[A_{12} - A_{21}](1 - x_1))\right]$$

and four unknowns: $\dot{L}, \dot{V}, y_1$ and x_1. This four non-linear equation, four-unknown system is readily solved using the Solver feature in Microsoft Excel (or any equation solver, including many calculators).

The results are as follows.

$$\dot{L} = 7.092 \frac{\text{mol}}{\text{s}}$$

$$\dot{V} = 2.908 \frac{\text{mol}}{\text{s}}$$

$$x_1 = 0.428$$

$$y_1 = 0.675$$

To recap, we have taken 10 mol/s of an equimolar liquid mixture of methyl ethyl ketone (1) and toluene (2) at 75°C and 760 mm Hg (101.3 kPa) and flashed it to 460 mm Hg (61.3 kPa). A liquid stream exited the flash distillation unit at 7.092 mol/s with a composition of 42.8% (by mole) methyl ethyl ketone, while a vapor stream also exited the flash distillation unit at 2.908 mol/s with a composition of 67.5% (by mole) methyl ethyl ketone. The substance that boils first (methyl ethyl ketone) appears preferentially in the vapor phase. This is the result from the modified Raoult's Law approach that incorporates the non-idealities in the liquid phase.

At this point, we have presented three different excess molar Gibbs free energy models (1- and 2-parameter Margules equations and the Wilson equation) and have demonstrated the impact of temperature and pressure on the activity coefficients. We have also shown how these models are regularly used as both data reduction tools and to predict VLE data at one state given experimental VLE data at another state. Finally, we have employed an excess molar Gibbs free energy model in a flash calculation and compared it to the ideal solution approach. In the next section, we introduce predictive approaches to obtain the needed activity coefficients from pure-component data.

11.8 Predictive Excess Molar Gibbs Free Energy Models

Up to this point, we have used experimental data for the mixture to fit excess molar Gibbs free energy model parameters. While we have used these fitted mixture parameters to predict behavior at different compositions (Example 11-3) or at different temperatures (Example 11-5), we always required mixture data to start. What if we didn't have experimental information on the mixture? This has, somewhat, been an ever present goal of mixture thermodynamics. Namely, how can we use pure component information to predict mixture properties? In this next section, we will review some of these approaches to provide the reader with an introduction to the topic.

11.8.1 Van Laar Equation and Regular Solution Theory

Recall that the definition of an excess property is the mixture property minus that if the mixture was modeled by an ideal solution, as provided by Equation 9.5. This equation, applied to molar volume, molar internal energy and molar entropy, is as follows:

$$\underline{V}^E = \underline{V} - \underline{V}^{ID} \tag{9.11}$$

$$\underline{U}^E = \underline{U} - \underline{U}^{ID} \tag{9.10}$$

$$\underline{S}^E = \underline{S} - \underline{S}^{ID} \tag{9.19}$$

Since $\underline{G} = \underline{U} + P\underline{V} - T\underline{S}$, we can use the previous three equations to show that

$$\underline{G}^E = \underline{U}^E + P\underline{V}^E - T\underline{S}^E \tag{11.119}$$

Johannes van Laar, a Dutch chemist, made two main assumptions when considering Equation 11.119. First, he assumed that the excess molar volume and excess molar entropy of the system were zero, which is known as "regular solution theory." Second, he assumed that both the pure components and the mixture would behave according to the van der Waals equation of state. The latter assumption was, perhaps, not surprising as van Laar was a Ph.D. student of van der Waals.

Joel Hildebrand, an American chemist working at UC-Berkeley, introduced the term "regular solution" in 1927.

The first assumption of van Laar isolates the excess internal energy as the key to calculating the excess molar Gibbs free energy (since $\underline{V}^E$ and $\underline{S}^E$ are set to zero). Van Laar's strategy was to create a hypothetical process with three steps. The first step is to decompress both pure liquids to an ideal gas in an isothermal manner, then use the van der Waals equation of state to calculate this internal energy change. The second step is to mix these ideal gases in an isothermal manner to form a mixture. Recall that the internal energy of mixing for an ideal gas is zero, so this step contributes nothing to the calculation. Finally, he would recompress this ideal gas mixture to a liquid at the pressure of interest (again in an isothermal manner) and calculate this internal energy change via the van der Waals equation of state.

From a calculation standpoint, the excess molar internal energy is equivalent to the internal energy of mixing.

$$\Delta \underline{U} = \underline{U}^E = \underline{U} - (x_1 \underline{U}_1 + x_2 \underline{U}_2) \tag{9.25}$$

For an ideal gas, we can write this mixing term using Equation 9.86 as

$$\Delta \underline{U}^{ig} = \underline{U}^{ig} - \left(x_1 \underline{U}_1^{ig} + x_2 \underline{U}_2^{ig}\right) = 0 \tag{11.120}$$

Since Equation 11.120 is equal to zero, we can subtract it from Equation 9.25 and still have the excess molar internal energy. Thus,

$$\Delta \underline{U} = \underline{U} - (x_1 \underline{U}_1 + x_2 \underline{U}_2) - \left[\underline{U}^{ig} - \left(x_1 \underline{U}_1^{ig} + x_2 \underline{U}_2^{ig}\right)\right] \tag{11.121}$$

Rearranging, we get

$$\Delta \underline{U} = (\underline{U} - \underline{U}^{ig}) - x_1\left(\underline{U}_1 - \underline{U}_1^{ig}\right) - x_2\left(\underline{U}_2 - \underline{U}_2^{ig}\right) \tag{11.122}$$

This is a key step and shows the power of thermodynamics and emphasizes the utility of state functions.

Equation 11.122 describes what van Laar proposed, namely to calculate the molar internal energies he needed referenced back to the ideal gas state, and then describe those residual values through the van der Waals equation. In other words, each of the three residuals in Equation 11.122 is described by the van der Waals equation, the former for the mixture and the latter two for the pure components.

While we have calculated pure-component parameters for the van der Waals equation in Chapter 8, we have yet to calculate mixture parameters. In order to do this, we introduce the **van der Waals one-fluid mixing rules** here.

For the a and b parameters in the van der Waals equation, we use a quadratic mixing rule and denote that these are mixture parameters with a subscript m for clarity.

$$a_m = x_1x_1a_{11} + x_1x_2a_{12} + x_2x_1a_{21} + x_2x_2a_{22} \tag{11.123}$$

$$b_m = x_1x_1b_{11} + x_1x_2b_{12} + x_2x_1b_{21} + x_2x_2b_{22} \tag{11.124}$$

The two subscripts on the a and b parameters indicate the components involved. For those with the same subscripts, the pure component is implied. Thus, $a_{11} = a_1, b_{11} = b_1$, $a_{22} = a_2, b_{22} = b_2$.

The parameters where the subscripts are not the same are called **cross-parameters**, and they are defined as follows. For the cross-parameter on a, the normal approach is a geometric-mean combining rule:

$$a_{12} = a_{21} = \sqrt{a_{11}a_{22}} = \sqrt{a_1a_2} \tag{11.125}$$

For b, the cross-parameter is defined using an arithmetic-mean combining rule as

$$b_{12} = b_{21} = \left(\frac{b_{11} + b_{22}}{2}\right) = \left(\frac{b_1 + b_2}{2}\right) \tag{11.126}$$

If you plug Equation 11.126 into Equation 11.124, the mixing rule for b_m will simplify to

$$b_m = x_1b_1 + x_2b_2 \tag{11.127}$$

If one follows the strategy described above in calculating the excess molar internal energy via Equation 11.122 through use of the van der Waals equation of state, the result is an excess molar Gibbs free energy model that is

$$\frac{G^{\mathrm{E}}}{RT} = x_1x_2\frac{L_{12}L_{21}}{L_{12}x_1 + L_{21}x_2} \tag{11.128}$$

where the activity coefficients, as obtained via Equation 11.73, are

$$\ln[\gamma_1] = L_{12}\left(1 + \frac{L_{12}x_1}{L_{21}x_2}\right)^{-2} \tag{11.129}$$

$$\ln[\gamma_2] = L_{21}\left(1 + \frac{L_{21}x_2}{L_{12}x_1}\right)^{-2} \tag{11.130}$$

Here, L_{12} and L_{21} are given in terms of the van der Waals parameters as

$$L_{12} = \frac{b_1}{RT}\left(\frac{\sqrt{a_1}}{b_1} - \frac{\sqrt{a_2}}{b_2}\right)^2 \tag{11.131}$$

and

$$L_{21} = \frac{b_2}{RT}\left(\frac{\sqrt{a_1}}{b_1} - \frac{\sqrt{a_2}}{b_2}\right)^2 \tag{11.132}$$

The derivation of Equation 11.127 using Equations 11.126 and 11.124 are part of an end-of-chapter exercise.

George Scatchard was an American physical chemist who also introduced the term "excess" to denote differences between real and ideal solutions.

As described in Chapter 7, the van der Waals equation of state is not very accurate from a predictive standpoint. Thus, George Scatchard and Joel Hildebrand (independently) made modifications to the van Laar approach that utilizes experimental pure component data instead of the van der Waals parameters. In particular, they use the experimental molar internal energy of vaporization (normally calculated at 298.15 K) as well as the molar volumes of the pure liquids in order to estimate both the mixture internal energy of vaporization and mixture molar volume. The resulting form of the excess molar Gibbs free energy expression using the Scatchard-Hildebrand approach is the same as that from van Laar. However, the parameters (called L_{12} and L_{21} in the van Laar equation) are defined in a different manner and are denoted as M_{12} and M_{21} here.

$$\frac{G^E}{RT} = x_1 x_2 \frac{M_{12}M_{21}}{M_{12}x_1 + M_{21}x_2} \tag{11.133}$$

$$\ln[\gamma_1] = M_{12}\left(1 + \frac{M_{12}x_1}{M_{21}x_2}\right)^{-2} \tag{11.134}$$

$$\ln[\gamma_2] = M_{21}\left(1 + \frac{M_{21}x_2}{M_{12}x_1}\right)^{-2} \tag{11.135}$$

where

$$M_{12} = \frac{V_1}{RT}(\delta_1 - \delta_2)^2 \tag{11.136}$$

$$M_{21} = \frac{V_2}{RT}(\delta_1 - \delta_2)^2 \tag{11.137}$$

Here, δ_1 and δ_2 are known as **solubility parameters** and are given as

$$\delta_1 = \sqrt{\frac{\Delta U_1^{vap}}{V_1}} \tag{11.138}$$

$$\delta_2 = \sqrt{\frac{\Delta U_2^{vap}}{V_2}} \tag{11.139}$$

The δ terms are often known as Hildebrand solubility parameters.

As evidenced by the definitions for M_{12} and M_{21}, the solubility parameters from this modeling approach provide a quick estimate of a mixture's deviation from ideal solution. When the solubility parameters of two compounds are the same, the mixture is predicted to behave as an ideal solution. The larger the difference of the solubility parameters from each other, the more non-ideal the mixture is predicted to behave. Not surprisingly, however, systems that best validate the assumptions inherent in the approach, namely that excess molar volume and excess molar entropy are zero, are where predictive results provide the most benefit. This is normally for non-polar or slightly polar systems with small excess molar volumes. We demonstrate this in Example 11-7.

| REGULAR SOLUTION THEORY, WHERE IT WORKS | EXAMPLE 11-7 |

Estimate the vapor-liquid equilibrium for the *n*-pentane (1) + benzene (2) system at 313.15 K using the van Laar equation.

SOLUTION:

Step 1 *Make a plan*
The first step for VLE problems is to decide the modeling approach needed. Should we try Raoult's Law? From the liquid side, the two molecules, while non-polar, are not similar in size and shape (pentane is a chain and benzene is a ringed structure). Thus, we will need an excess molar Gibbs free energy model for the liquid (and, as provided in the problem statement, we should use the van Laar model). What about the vapor side? Do we need to account for vapor phase non-idealities? A quick check for this would be to look at the vapor pressure of the pure components at the temperature of interest.

n-pentane

Benzene

The Antoine equation can help us find a good estimate of the vapor pressure of both compounds at 313.15 K. Using the Antoine parameters for the two components (found in Appendix E), we find

$$P_1^{\text{sat}}(313.15 \text{ K}) = 119 \text{ kPa}$$

$$P_2^{\text{sat}}(313.15 \text{ K}) = 24 \text{ kPa}$$

Unless this mixture produces very large, positive deviations from Raoult's Law, the pressure of the mixture is not likely to go well above 119 kPa. And since these are two non-polar compounds, the ideal gas is a reasonable assumption in the absence of experimental data. Thus, assuming an ideal gas for the vapor phase is quite reasonable here. Accordingly, our modeling approach will be modified Raoult's Law:

$$\gamma_i x_i P_i^{\text{sat}} = y_i P \tag{11.69}$$

The next step would be to find the activity coefficients so we can predict the bubble and dew point curves for the mixture at this temperature, per the problem statement. We will use the van Laar equation to find our activity coefficients. As we have seen above, the van Laar approach from a *predictive* standpoint means using pure component information to predict the activity coefficients. We can do this in two ways: (1) using the van der Waals parameters and (2) using the solubility parameters from Scatchard-Hildebrand.

Step 2 *Van Laar with VDW parameters*

Recall that we need to predict the activity coefficients and those are a function of L_{12} and L_{21}.

$$\ln[\gamma_1] = L_{12}\left(1 + \frac{L_{12}x_1}{L_{21}x_2}\right)^{-2} \tag{11.129}$$

$$\ln[\gamma_2] = L_{21}\left(1 + \frac{L_{21}x_2}{L_{12}x_1}\right)^{-2} \tag{11.130}$$

L_{12} and L_{21} are given in terms of the van der Waals parameters, a and b, whose values are obtained from the critical point of the pure components.

$$L_{12} = \frac{b_1}{RT}\left(\frac{\sqrt{a_1}}{b_1} - \frac{\sqrt{a_2}}{b_2}\right)^2 \tag{11.131}$$

$$L_{21} = \frac{b_2}{RT}\left(\frac{\sqrt{a_1}}{b_1} - \frac{\sqrt{a_2}}{b_2}\right)^2 \tag{11.132}$$

Recall from Chapter 7, we have

$$a = \frac{27R^2T_c^2}{64P_c}$$

$$b = \frac{RT_c}{8P_c}$$

The critical point information for our two compounds from Appendix C is

$$T_{c,1} = 469.7 \text{ K}$$

$$P_{c,1} = 3.37 \text{ MPa}$$

$$T_{c,2} = 562.05 \text{ K}$$

$$P_{c,2} = 4.895 \text{ MPa}$$

You will often see "van der Waals" abbreviated by "VDW"

Plugging in the critical point information yields the following values for a and b:

$$a_1 = 1.909 \text{ m}^6 \text{ Pa/mol}^2$$
$$b_1 = 144.853 \times 10^{-6} \text{ m}^3/\text{mol}$$
$$a_2 = 1.882 \text{ m}^6 \text{ Pa/mol}^2$$
$$b_2 = 119.333 \times 10^{-6} \text{ m}^3/\text{mol}$$

This, in turn, using Equations 11.131 and 11.132 at $T = 313.15$ K, yields the following set of parameters:

$$L_{12} = 0.2132$$
$$L_{21} = 0.1756$$

Once we have our parameters, we can plug them into Equations 11.129 and 11.130 to obtain the predicted activity coefficients at any composition of interest. Recall the predicted pressure and vapor phase composition are obtained using modified Raoult's Law:

$$y_1 = \frac{\gamma_1 x_1 P_1^{sat}}{\gamma_1 x_1 P_1^{sat} + \gamma_2 x_2 P_2^{sat}} \tag{11.88}$$

$$P = \gamma_1 x_1 P_1^{sat} + \gamma_2 x_2 P_2^{sat} \tag{11.89}$$

Varying x_1 from 0 to 1 allows us to plot both the bubble-point and dew-point pressure, which are the solid lines in Figure 11-9. With the actual experimental data for this system plotted on the same graph, one can conclude that the prediction using the van Laar approach with the predicted parameters from the van der Waals equation provides a reasonable estimate of the phase equilibrium. However, can we do better with the use of the solubility parameters as described previously? Let's examine this approach.

FOOD FOR THOUGHT 11-3

What do the values of L_{12} and L_{21} have to be for a system that behaves according to the ideal solution?

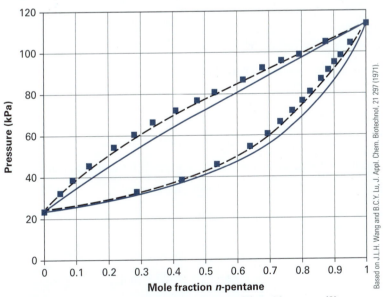

Based on J.L.H. Wang and B.C.Y. Lu, J. Appl. Chem. Biotechnol, 21 297 (1971).

FIGURE 11-9 A *Pxy* diagram of the *n*-pentane (1) and benzene (2) system at 313.15 K. The symbols are experimental data. The solid line is the prediction from the van Laar equation using the VDW-defined parameters. The dashed line is the prediction from the van Laar equation using the Scatchard-Hildebrand defined parameters.

Step 3 *Van Laar with Scatchard-Hildebrand parameters*

Within the activity coefficient expressions for the van Laar equation, M_{12} and M_{21} play the role of L_{12} and L_{21}. It is how those parameters are calculated that is different. According to the Scatchard-Hildebrand approach, the parameters are defined in terms of the solubility parameters, δ_1 and δ_2, of the pure components, as well as the pure-component molar volumes. The pure component molar volume is straight-forward and one typically uses the liquid molar volume at 298.15 K, which is given in Appendix C.

$$\underline{V}_1 = 1.1522 \times 10^{-4}\, \text{m}^3/\text{mol}$$

$$\underline{V}_2 = 8.941 \times 10^{-5}\, \text{m}^3/\text{mol}$$

The solubility parameters are defined as

$$\delta_1 = \sqrt{\frac{\Delta \underline{U}_1^{\text{vap}}}{\underline{V}_1}} \tag{11.138}$$

$$\delta_2 = \sqrt{\frac{\Delta \underline{U}_2^{\text{vap}}}{\underline{V}_2}} \tag{11.139}$$

The molar internal energy change due to vaporization of the liquid is a function of temperature. In order to provide a standard point to calculate this value, 298.15 K is normally chosen. Additionally, some simplifications are imposed on the calculation. First, the molar internal energy of vaporization is written in terms of the molar enthalpy of vaporization, as

$$\Delta \underline{U}^{\text{vap}} = \Delta \underline{H}^{\text{vap}} - \Delta(P\underline{V}) \tag{11.140}$$

Since pressure does not change during vaporization, we can pull that out of the "delta" term.

$$\Delta \underline{U}^{\text{vap}} = \Delta \underline{H}^{\text{vap}} - P\Delta(\underline{V}) \tag{11.141}$$

or

$$\Delta \underline{U}^{\text{vap}} = \Delta \underline{H}^{\text{vap}} - P(\underline{V}^{\text{vap}} - \underline{V}^{\text{liq}}) \tag{11.142}$$

We can make the assumption here that $\underline{V}^{\text{vap}} \gg \underline{V}^{\text{liq}}$, thus,

$$\underline{V}^{\text{vap}} - \underline{V}^{\text{liq}} \approx \underline{V}^{\text{vap}} \tag{11.143}$$

Therefore,

$$\Delta \underline{U}^{\text{vap}} = \Delta \underline{H}^{\text{vap}} - P\underline{V}^{\text{vap}} \tag{11.144}$$

> We made a similar assumption about the saturated liquid and vapor molar volumes when deriving the Clausius-Clapeyron equation.

We conclude with the assumption that the vapor behaves as an ideal gas. Thus,

$$\Delta \underline{U}^{\text{vap}} = \Delta \underline{H}^{\text{vap}} - RT \tag{11.145}$$

Values for the molar enthalpy of vaporization at 25°C are readily tabulated and taken from Appendix C.

$$\Delta \underline{H}_1^{\text{vap}} = 25790 \text{ J/mol}$$

$$\Delta \underline{H}_2^{\text{vap}} = 30720 \text{ J/mol}$$

> Traditionally, the units of the solubility parameter (δ) are given as $\text{cal}^{1/2}\, \text{cm}^{-3/2}$. They are presented in this example in the proper SI base units for calculation clarity.

Thus, plugging in the molar enthalpy of vaporization and the molar volume (with $T = 313.15$ K) yields solubility parameters equal to

$$\delta_1 = 14185.79 \text{ Pa}^{1/2}$$

$$\delta_2 = 17726.90 \text{ Pa}^{1/2}$$

Once the solubility parameters have been obtained, the values for the van Laar parameters, M_{12} and M_{21}, can be calculated. Note that in this portion of the problem, R is chosen to be 8.314 J/mol-K in order to make M_{12} and M_{21} dimensionless.

$$M_{12} = \frac{V_1}{RT}(\delta_1 - \delta_2)^2 \tag{11.136}$$

$$M_{21} = \frac{V_2}{RT}(\delta_1 - \delta_2)^2 \tag{11.137}$$

Plugging in these values into Equation 11.136 and 11.137 yields

$$M_{12} = .5549$$

$$M_{21} = .4306$$

Once the parameters are obtained, the problem proceeds along the same path as in step 2 in order to obtain the activity coefficients and, ultimately, the bubble-point curve and dew-point curve. Those predictions are given as the dashed lines in Figure 11-9.

> You use Equations 11.134 and 11.135 to obtain the needed activity coefficients.

As can be seen in Figure 11-9, the parameters obtained from the Scatchard-Hildebrand approach provide a very good match with the experimental data. The improved estimate of the van Laar parameters (relative to the use of the VDW parameters) by using the experimental vaporization energies proved predictive in this instance. Note two things, however. First, the two components in the system are non-polar and, as mentioned previously, the assumptions of regular solution are more likely to be met by these types of systems. Second, the experimental excess molar volume for this system is, at a maximum, around 0.15 cm³/mol (Mahl et al., 1971), meaning the assumption of a zero excess molar volume for this system is reasonable. In this next example, we show a limitation of regular solution theory.

| **REGULAR SOLUTION THEORY, WHERE IT DOES NOT WORK** | **EXAMPLE 11-8** |

Estimate the vapor–liquid equilibrium for the hexafluorobenzene (1) + cyclohexane (2) system at 323.15 K using the van Laar equation.

SOLUTION:

Step 1 *Make a plan*
As in Example 11-7, we decide the modeling approach needed. Here, while both components are non-polar, the presence of the six fluorine atoms is certainly cause for concern if considering an ideal solution (since fluorine is the most electronegative element). Accordingly, we choose to use an excess molar Gibbs free energy model for the liquid (and the van Laar model at that, per the problem statement). For the vapor phase, looking at the vapor pressures at 323.15 K from the Antoine equation (with parameters found in Appendix E) yields

$$P_1^{\text{sat}}(323.15 \text{ K}) = 34.1 \text{ kPa}$$

$$P_2^{\text{sat}}(323.15 \text{ K}) = 36.3 \text{ kPa}$$

Hexafluorobenzene

Cyclohexane

Thus, an ideal gas seems appropriate here, and our modeling approach will use modified Raoult's Law:

$$\gamma_i x_i P_i^{sat} = y_i P \tag{11.69}$$

As in the previous problem, we predict the activity coefficients two ways: (1) using the van der Waals parameters and (2) using the solubility parameters from Scatchard-Hildebrand.

Step 2 *Van Laar with VDW parameters*
Since the steps here are the same as in Example 11-7, we can go right to providing the pure-component critical temperature and critical pressure to obtain L_{12} and L_{21}.

The critical point information for our two compounds are obtained from Appendix C

$$T_{c,1} = 516.73 \text{ K}$$

$$P_{c,1} = 3.275 \text{ MPa}$$

$$T_{c,2} = 553.5 \text{ K}$$

$$P_{c,2} = 4.073 \text{ MPa}$$

Plugging in the critical point information yields the following values for a and b:

$$a_1 = 2.738 \times \text{ m}^6 \text{ Pa/mol}^2$$

$$b_1 = 163.979 \times 10^{-6} \text{ m}^3/\text{mol}$$

$$a_2 = 2.194 \times \text{ m}^6 \text{ Pa/mol}^2$$

$$b_2 = 141.234 \times 10^{-6} \text{ m}^3/\text{mol}$$

Using Equations 11.131 and 11.132 at $T = 323.15$ K, yields the following set of parameters.

$$L_{12} = 0.0716$$

$$L_{21} = 0.0617$$

Once we have our parameters, we can plug them into Equations 11.129 and 11.130 to obtain the predicted activity coefficients at any composition of interest. We show this in Figure 11-10 as the small dashed lines. Note that the small values of the parameters (relative to the previous problem) indicate a system that is predicted to have only small deviations from ideal solution. However, comparing the predictions to the experimental data (symbols) shows that the system is, in fact, not ideal and possesses an azeotrope.

Step 3 *Van Laar with Scatchard-Hildebrand parameters*
In the previous problem there was improvement in the van Laar model predictions when the parameters were calculated using the Scatchard-Hildebrand approach, relative to those obtained from the VDW parameters. Accordingly, we try that approach again in this step.

In order to calculate the van Laar parameters using the Scatchard-Hildebrand approach, we need the molar volumes and the molar enthalpy of vaporization for the pure components. These are calculated in the same way as in Example 11-7 both at 298.15 K.

$$\underline{V}_1 = 1.154 \times 10^{-4} \text{ m}^3/\text{mol}$$

$$\underline{V}_2 = 1.087 \times 10^{-4} \text{ m}^3/\text{mol}$$

$$\Delta \underline{H}_1^{vap} = 31660 \text{ J/mol}$$

$$\Delta \underline{H}_2^{vap} = 29970 \text{ J/mol}$$

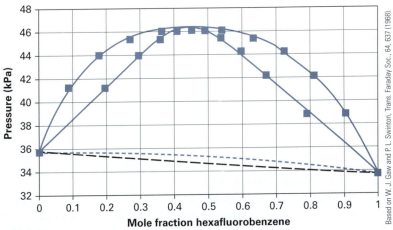

Based on W. J. Gaw and P. L. Swinton, Trans. Faraday Soc., 64, 637 (1968).

FIGURE 11-10 A *Pxy* diagram of the hexafluorobenzene (1) + cyclohexane (2) system at 323.15 K. The symbols are experimental data. The small blue dashed lines are predictions from the van Laar equation using the VDW-defined parameters, while the larger black dashed lines are predictions from the van Laar equation using the Scatchard-Hildebrand defined parameters. The solid lines are the correlations from the van Laar model. All models show both the bubble-point and dew-point curves.

Thus, plugging in the molar enthalpy of vaporization and the molar volume (with $T = 323.15$ K) yields solubility parameters equal to

$$\delta_1 = 15844.08 \ \text{Pa}^{1/2}$$

$$\delta_2 = 15839.23 \ \text{Pa}^{1/2}$$

Using these solubility parameters into the expressions for the van Laar parameters (Equations 11.136 and 11.137) yields

$$M_{12} = 1.008 \times 10^{-6}$$

$$M_{21} = 9.503 \times 10^{-7}$$

Just through observation of the values of the parameters, ideal solution is predicted from this model for this system. Indeed, when one calculates the activity coefficients using Equations 11.134 and 11.135 and plots the bubble-point and dew-point curves using Equations 11.88 and 11.89 in Figure 11-10 (long, dashed line), an ideal solution is realized.

What went wrong here? Why were the predictions of the van Laar parameters using both the VDW approach and the Scatchard-Hildebrand approach so poor? It lies in the assumptions of using the van Laar model in a predictive way, namely regular solution. In the previous example, the *n*-pentane + benzene system exhibited small deviations from the assumptions of regular solution as evidenced by the small maximum value for the excess molar volume (around 0.15 cm³/mol). On the other hand, the maximum excess molar volume for the hexafluorobenzene + cyclohexane system is about 2.5 cm³/mol, more than 15 times larger than the *n*-pentane + benzene system (Battino, 1971). Thus, the assumptions of the regular solution have not been met by this system. Therefore, caution must be taken when using a regular solution model for predicting the phase behavior of systems, even non-polar systems. Note that if one treats the van Laar equation as a *correlative* model (as we have done with other excess molar Gibbs free energy models several times in this chapter), it works very well and this is shown as the solid line in Figure 11-10 ($L_{12} = 1.289$; $L_{21} = 1.038$).

While the van Laar approach allows for prediction of mixture properties, it is not the only activity coefficient model that can be predictive. An approach called the *universal functional activity coefficient model (UNIFAC)* is available that uses group contributions to obtain activity coefficient values for systems. If you recall, we discussed group contributions for obtaining pure component properties in Chapter 7. Part of the UNIFAC approach involves breaking molecules into functional groups and those contribute on an individual basis and in combination with other functional groups. The former part is similar to what we've discussed previously, as here the groups provide a contribution to the activity coefficient (through parameters) depending on how often they appear in an individual molecule. Additionally, how the individual groups in the system as a whole interact with other functional groups is also included in the UNIFAC approach. Here, as you might imagine, one requires both a large table that contains regressed values of how two groups interact and an assumption that two groups interact the same way regardless of where they appear. Thus, for example, when a compound containing an -OH group is mixed with a compound containing a -CH_2- group, the interaction between these groups will provide the same contribution to the mixture properties, regardless of the specific compounds where they appear. The interaction between an -OH group and a -CH_2- group is assumed to be the same, regardless of what the -OH group and the -CH_2- group are attached to. At any rate, the UNIFAC approach has proven to be more predictive than van Laar—but certainly at a greater computational expense.

The excess molar Gibbs free energy models mentioned in this chapter are just a few of the most popular and common ones. Others exist, such as the non-random two liquid (NRTL) model.

The expressions defining UNIFAC are somewhat lengthy and the by-hand approach to determine their parameters is also lengthy and tedious. If you use UNIFAC, you are very likely going to be doing this by selecting the UNIFAC model from a pull-down menu in a process simulator. Thus, we do not show the model and calculation within this text (Fredenslund et al., 1979). It is noted that another correlative (not predictive) excess molar Gibbs free energy model sounds like UNIFAC, namely the *universal quasi-chemical model (UNIQUAC)*. Indeed, the UNIFAC model shares a common functional form with UNIQUAC. But, like other correlative models, UNIQUAC is not predictive and is used (like the Margules equation and Wilson equation) to correlate experimental data. Once again, the UNIQUAC approach is lengthy and, thus, we have excluded it from this introductory text on chemical thermodynamics (Abrams and Prausnitz, 1975).

11.9 Thermodynamic Consistency

Consider two streams entering a process unit at steady state. Without measuring the mass flow rate of the output stream of the process unit, you can calculate that value by simply adding the incoming mass flow rates of the two streams (a material balance). Of course, you can measure the output stream to "close the material balance" (as it is called) to verify your assumption of steady state. What if (within reasonable error) you find that your measurement of the third stream is not what has been predicted by your simple material balance? This would cause you to investigate whether there is a measurement error with any of the streams, a leak somewhere (if you are losing material), perhaps material is accumulating in the process unit that you haven't accounted for, etc. This use of a simple material balance (compared to what is measured) has helped you identify a problem.

In this chapter, we have done a lot of work with experimental data and have assumed the experimental data is correct, since it was measured experimentally. Is there an analog to the material balance that we can use here to check whether we might have concern about the accuracy of our experimental data? We do, and it has been introduced earlier this chapter: the Gibbs-Duhem equation.

The Gibbs-Duhem equation is a relationship among the thermodynamic variables in our system and can be used in two ways (analogous to the material balance). First, if you measure all but one of the independent thermodynamic variables in the system, the Gibbs-Duhem equation can be used to "solve" for the one variable that is not measured. This is similar to the material balance example above, where we measure the two mass flow rates of the input streams and solve for the mass flow rate of the output stream. Relative to the modeling of vapor–liquid equilibrium, the Gibbs-Duhem equation, for example, can be used to solve for the vapor-phase composition in a binary mixture without having to measure it experimentally.

The second way the Gibbs-Duhem equation can be used (and the way we will use it here) is to measure all of the independent thermodynamic variables of our system and then test to see if they make the Gibbs-Duhem equation true (within some reasonable experimental error). If the data have large deviations from the Gibbs-Duhem equation, this will cause the user to have an increasing level of skepticism as to the accuracy and utility of the data. (The end of Example 11-10 gives heuristics for quantifying what is meant by "large" deviations.)

We have already used the Gibbs-Duhem equation in this chapter to check whether the 1-parameter Margules equation satisfied this required consistency. Indeed, we started with

$$x_1 d\overline{M}_1 + x_2 d\overline{M}_2 - \left(\frac{\partial [M]}{\partial T}\right)_{P,x_1,x_2} dT - \left(\frac{\partial [M]}{\partial P}\right)_{T,x_1,x_2} dP = 0 \qquad (9.48)$$

Once again, for example purposes, we assume we are evaluating Pxy data and that means the temperature is constant. Also, as we have shown in Section 11.7.1, activity coefficients are only slight functions of pressure. Thus, Equation 9.48 reduces, given these assumptions, to

One can evaluate the thermodynamic consistency of constant *pressure* data as well (*Txy* data), but the resulting version of the Gibbs-Duhem equation becomes more challenging to work with.

$$x_1 d\ln \gamma_1 + x_2 d\ln \gamma_2 = 0 \qquad (11.76)$$

Once again, we can evaluate this differential relationship for small changes in dx_1. Thus,

$$x_1 \frac{d\ln \gamma_1}{dx_1} + x_2 \frac{d\ln \gamma_2}{dx_1} = 0 \qquad (11.83)$$

This is a convenient form of the Gibbs-Duhem equation, since it contains relationships between the activity coefficients.

11.9.1 Integral (Area) Test

Let's assume one has experimental thermodynamic data for a binary mixture at constant temperature, which means that Pxy data exist at a given temperature. From this data, one can solve for the experimental activity coefficients as has been done a few times in this chapter. Additionally, as has also been done in this chapter, experimental excess molar Gibbs free energies can be calculated from these experimental activity coefficients. More precisely, we use the summability relationship (Equation 11.74) to obtain

$$\frac{G^{E*}}{RT} = x_1 \ln[\gamma_1]^* + x_2 \ln[\gamma_2]^* \qquad (11.146)$$

In order to discriminate between the experimental and model values in what follows, we assign an asterisk to the experimental values.

If we take the derivative of both sides with respect to x_1 and, of course, write $x_2 = 1 - x_1$, we end up with

$$\frac{G^{E^*}}{RT} = x_1 \ln[\gamma_1]^* + (1 - x_1)\ln[\gamma_2]^* \tag{11.147}$$

$$\frac{d}{dx_1}\left(\frac{G^E}{RT}\right)^* = \frac{d}{dx_1}(x_1 \ln[\gamma_1]^*) + \frac{d}{dx_1}((1 - x_1)\ln[\gamma_2]^*) \tag{11.148}$$

$$\frac{d}{dx_1}\left(\frac{G^E}{RT}\right)^* = x_1\frac{d}{dx_1}(\ln[\gamma_1]^*) + \ln[\gamma_1]^*\frac{d(x_1)}{dx_1} + (1 - x_1)\frac{d}{dx_1}(\ln[\gamma_2]^*) + \ln[\gamma_2]^*\frac{d(1 - x_1)}{dx_1} \tag{11.149}$$

$$\frac{d}{dx_1}\left(\frac{G^E}{RT}\right)^* = x_1\frac{d}{dx_1}(\ln[\gamma_1]^*) + \ln[\gamma_1]^*(1) + (1 - x_1)\frac{d}{dx_1}(\ln[\gamma_2]^*) + \ln[\gamma_2]^*(-1) \tag{11.150}$$

Grouping terms, this can be written as

$$\frac{d}{dx_1}\left(\frac{G^E}{RT}\right)^* = x_1\frac{d}{dx_1}(\ln[\gamma_1]^*) + x_2\frac{d}{dx_1}(\ln[\gamma_2]^*) + \ln\left[\frac{\gamma_1}{\gamma_2}\right]^* \tag{11.151}$$

What we see is that the first two terms on the right-hand side of Equation 11.151 are, in fact, the Gibbs-Duhem equation. Accordingly, if the experimental data is consistent with the Gibbs-Duhem equation, those terms must sum to zero. This implies that

$$\frac{d}{dx_1}\left(\frac{G^E}{RT}\right)^* = \ln\left[\frac{\gamma_1}{\gamma_2}\right]^* \tag{11.152}$$

or alternatively,

$$d\left(\frac{G^E}{RT}\right)^* = \ln\left[\frac{\gamma_1}{\gamma_2}\right]^* dx_1 \tag{11.153}$$

To make use of this relationship, we can integrate both sides from $x_1 = 0$ to $x_1 = 1$, as

$$\int_{x_1=0}^{x_1=1} d\left(\frac{G^E}{RT}\right)^* = \int_{x_1=0}^{x_1=1} \ln\left[\frac{\gamma_1}{\gamma_2}\right]^* dx_1 \tag{11.154}$$

$$\frac{G^{E^*}}{RT}(x_1 = 1) - \frac{G^{E^*}}{RT}(x_1 = 0) = \int_{x_1=0}^{x_1=1} \ln\left[\frac{\gamma_1}{\gamma_2}\right]^* dx_1 \tag{11.155}$$

Each of the terms on the left-hand side are equal to zero by themselves, since they define the pure component endpoints ($x_1 = 0$ and $x_1 = 1$), both of which have an excess molar Gibbs free energy equal to zero. Thus, we are left with

$$0 = \int_{x_1=0}^{x_1=1} \ln\left[\frac{\gamma_1}{\gamma_2}\right]^* dx_1 \tag{11.156}$$

What Equation 11.156 says is that, if we plot the natural logarithm of the ratio of the activity coefficients and obtain the area under the curve from $x_1 = 0$ to $x_1 = 1$, this area should be zero if the experimental data are to satisfy the Gibbs-Duhem equation. Let's explore the thermodynamic consistency of a data set previously used in this chapter for its adherence to the integral test, namely the system from Example 11-4.

DETERMINING THE THERMODYNAMIC CONSISTENCY OF A SYSTEM: INTEGRAL TEST	EXAMPLE 11-9

Using the experimental data from Example 11-4 for the di-isopropyl ether (1) + 1-propanol (2) system at 303.15 K, determine if the data set passes the integral test.

SOLUTION:

Step 1 *Finding the ratio of the natural logarithm of the activity coefficients*
From Table 11-2 provided in Example 11-4, we add a column that provides the natural logarithm of the ratio of the experimental activity coefficients, and this is Table 11-7.

TABLE 11-7 Experimental data, including the activity coefficient, for the di-isopropyl (1) + 1-propanol (2) system at 303.15 K.

P (kPa)	x_1	y_1	γ_1	γ_2	$\ln(\gamma_1/\gamma_2)$
3.77	0	0		1	
5.05	0.0199	0.2671	2.782	1.001	1.0222
6.15	0.0399	0.4090	2.587	1.003	0.9475
7.22	0.0601	0.5061	2.499	1.005	0.9109
8.29	0.0799	0.5783	2.464	1.006	0.8958
10.60	0.1192	0.6847	2.498	1.005	0.9105
12.16	0.1694	0.7346	2.165	1.030	0.7429
14.07	0.2294	0.7822	1.970	1.053	0.6264
15.62	0.2891	0.8133	1.804	1.087	0.5066
16.81	0.3495	0.8343	1.648	1.134	0.3738
17.91	0.4090	0.8524	1.533	1.185	0.2575
18.77	0.4708	0.8659	1.417	1.260	0.1174
19.51	0.5296	0.8774	1.327	1.346	−0.0142
20.23	0.5902	0.8890	1.251	1.452	−0.1490
20.71	0.6505	0.8974	1.173	1.611	−0.3173
21.35	0.7101	0.9093	1.123	1.770	−0.4550
21.92	0.7685	0.9209	1.078	1.984	−0.6100
22.62	0.8300	0.9372	1.049	2.214	−0.7470
23.20	0.8803	0.9521	1.030	2.460	−0.8706
23.59	0.9179	0.9637	1.017	2.764	−0.9998
23.80	0.9397	0.9709	1.009	3.047	−1.1052
23.99	0.9581	0.9785	1.006	3.259	−1.1754
24.19	0.9804	0.9885	1.001	3.750	−1.3208
24.36	1	1	1		

Step 2 *Plot the relevant data*

Since we are interested in evaluating the area under the curve for the integrand of Equation 11.156, we need to plot ln (γ_1/γ_2) vs. x_1. This is plotted in Figure 11-11.

$$0 = \int_{x_1=0}^{x_1=1} \ln \left[\frac{\gamma_1}{\gamma_2}\right]^* dx_1 \qquad (11.156)$$

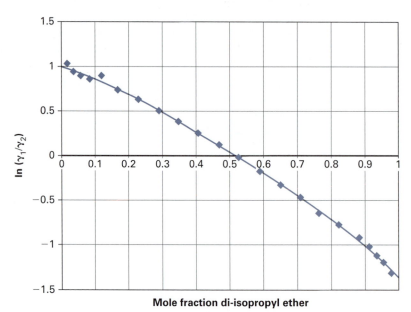

FIGURE 11-11 A plot of the natural logarithm of the ratio of the activity coefficients for the di-isopropyl ether (1) + 1-propanol (2) system at 303.15 K. A fitting function is provided on the plot as a blue curve, whose functional form is ln $(\gamma_1/\gamma_2) = -4.118733x_1^4 + 7.634390x_1^3 - 5.121398x_1^2 - 0.764514x_1^1 + 1.004767$.

Step 3 *Evaluate the area under the curve*

In order to determine the area under the curve, which is needed to evaluate how well this data set satisfies Equation 11.156, we have some options. We could use a numerical technique such as the trapezoidal rule or Simpson's rule to numerically evaluate the integral. On the other hand, fitting the data to a polynomial and then performing the integration analytically is another strategy. Note in the latter case, a fitting polynomial can be extrapolated to estimate the contribution from $x_1 = 0$ to the first point ($x_1 = 0.0199$) as well as from the last point ($x_1 = .9804$) to $x_1 = 1$. Accordingly, we have fit the data to a fourth-degree polynomial (with x_1 as the independent variable) whose function is found in the caption of Figure 11-11.

The fitting function allows us to solve for a value of x_1 where the function ln (γ_1/γ_2) goes from positive to negative. Here, that value is $x_1 = .5176$. Thus, we can integrate our fitting function from $x_1 = 0$ to $x_1 = 0.5176$ to estimate the area above the curve. Likewise, we can integrate our fitting function from $x_1 = 0.5176$ to $x_1 = 1$ to estimate the area below the curve. Accordingly, our results are as follows:

$$\text{Area above curve} = 0.28732$$

$$\text{Area below curve} = -0.28709$$

While these numbers are close in magnitude, they do not sum to zero. This brings up another question: What kind of deviation is considered reasonable? One heuristic is to take the absolute values of the difference in the area above and below the curve to the absolute values of their sums. If this result is within 2%, the data set can be said to pass the integral test.

For the areas obtained in this problem, we find

$$\frac{|\text{Area above curve}| - |\text{Area below curve}|}{|\text{Area above curve}| + |\text{Area below curve}|} \tag{11.157}$$

$$\frac{|0.28732| - |-0.28709|}{|0.28732| + |-0.28709|} = 0.00040 = 0.04\%$$

Thus, we conclude that this data set has passed the integral test.

There is a limitation, however, with the integral test, and it requires some explanation.

Let's consider the term that we are evaluating, namely the natural logarithm of the ratio of the activity coefficients.

$$\ln \left[\frac{\gamma_1}{\gamma_2} \right]^*$$

Recall that these activity coefficients were calculated through

$$\ln \left[\frac{\left(\dfrac{y_1 P}{x_1 P_1^{sat}} \right)}{\left(\dfrac{y_2 P}{x_2 P_2^{sat}} \right)} \right]^*$$

Note that the experimentally-measured system pressure appears in *both* the numerator and the denominator. Thus, it is cancelled out of this ratio, leading to

$$\ln \left[\frac{y_1 x_2 P_2^{sat}}{y_2 x_1 P_1^{sat}} \right]^*$$

This is a problematic result, since one can have poorly measured experimental system pressures, yet those values would have no impact on the assessment of the quality of the experimental data through the integral test. Indeed, it has been shown that a data set can pass this integral test and still be thermodynamically inconsistent. Thus, the integral test is now known as a necessary—but not sufficient—test of thermodynamic consistency for a data set. It can be used quickly to determine if a data set fails the integral test. If it does fail the integral test, you can conclude that the data set is of poor quality. If it passes the integral test, you still do not know if the data set is thermodynamically consistent.

So if the integral test is only a *necessary* condition for thermodynamic consistency, is there another way to examine the data using the Gibbs-Duhem equation that gives us more insight as to the quality of the experimentally measured data? Fortunately there is and it is called the **direct test**.

11.9.2 Direct Test

The direct test is a different, more robust, application of the Gibbs-Duhem equation in the evaluation of experimental data. The thought behind the direct test is as follows: If we have an excess molar Gibbs free energy model and fit that model to experimental data, how well do the model activity coefficients compare with the experimental activity coefficients? If the root mean square of the difference is too large, you do not have good confidence in that data set.

In particular, we start with

$$\frac{d}{dx_1}\left(\frac{G^E}{RT}\right)^* = x_1\frac{d}{dx_1}(\ln[\gamma_1]^*) + x_2\frac{d}{dx_1}(\ln[\gamma_2]^*) + \ln\left[\frac{\gamma_1}{\gamma_2}\right]^* \tag{11.158}$$

but now write the same equation for the model (it will not have the "asterisk"):

$$\frac{d}{dx_1}\left(\frac{G^E}{RT}\right) = x_1\frac{d}{dx_1}(\ln[\gamma_1]) + x_2\frac{d}{dx_1}(\ln[\gamma_2]) + \ln\left[\frac{\gamma_1}{\gamma_2}\right] \tag{11.159}$$

If we subtract Equation 11.158 from 11.159, we obtain

$$\begin{aligned}
\frac{d}{dx_1}\left(\frac{G^E}{RT}\right) &- \frac{d}{dx_1}\left(\frac{G^E}{RT}\right)^* \\
&= x_1\frac{d}{dx_1}\left(\ln[\gamma_1]\right) + x_2\frac{d}{dx_1}\left(\ln[\gamma_2]\right) + \ln\left[\frac{\gamma_1}{\gamma_2}\right] \\
&- \left(x_1\frac{d}{dx_1}\left(\ln[\gamma_1]^*\right) + x_2\frac{d}{dx_1}\left(\ln[\gamma_2]^*\right) + \ln\left[\frac{\gamma_1}{\gamma_2}\right]^*\right)
\end{aligned} \tag{11.160}$$

We already know that the first two terms on the right-hand side of Equation 11.160 make up the Gibbs-Duhem equation and, thus, their sum is zero. We also know that the first two terms in the parenthesis *should* be zero as well (if the data is to be consistent), since that is also the Gibbs-Duhem equation. Thus, we are left with

$$\frac{d}{dx_1}\left(\frac{G^E}{RT}\right) - \frac{d}{dx_1}\left(\frac{G^E}{RT}\right)^* = \ln\left[\frac{\gamma_1}{\gamma_2}\right] - \ln\left[\frac{\gamma_1}{\gamma_2}\right]^* \tag{11.161}$$

While we have previously used the term "residual" to specifically describe a system property above that of the ideal gas value, residual has a broader meaning. Here, residual is used as the difference between the model value and the experimental value. Thus, we can write the above expression in terms of residuals, δ = model − experimental, as

> The symbol δ has been used for both solubility parameter and residual in this chapter alone. You need to recognize the context of the expression to understand its meaning as used.

$$\delta\left[\frac{d}{dx_1}\left(\frac{G^E}{RT}\right)\right] = \delta\left[\ln\left[\frac{\gamma_1}{\gamma_2}\right]\right] \tag{11.162}$$

Recall that the objective function we use to correlate the experimental data to an excess molar Gibbs free energy model is written in terms of the excess molar Gibbs free energy (as we have done throughout this chapter), which means that we are trying to make the difference between G^E/RT and G^{E*}/RT as small as possible (hence its use in the objective function). Thus, residuals in G^E/RT will scatter about zero (as it the purpose of minimizing the objective function). Accordingly, residuals in the derivative of G^E/RT, which is what we have on the left-hand side of Equation 11.162, will be basically zero. Therefore, the residuals on the right-hand side (in terms of the natural logarithm of the ratio of the activity coefficients), *should* also be close to zero. When those values deviate greatly from zero, our confidence in a particular data set wanes. Of course, all of this is dependent on a good excess molar Gibbs free energy model found to correlate the experimental data. We show the application of the direct test to our data from the previous problem.

| DETERMINING THE THERMODYNAMIC CONSISTENCY OF A SYSTEM: DIRECT TEST | EXAMPLE 11-10 |

Using the experimental data from Example 11-4 for the di-isopropyl ether (1) + 1-propanol (2) system at 303.15 K, determine if the data set passes the direct test.

SOLUTION:

Step 1 *Obtaining an excess molar Gibbs free energy model for this system*
We did this already in Example 11-4; twice in fact, since we used both the 1- and 2-parameter Margules equation. This requisite information is found in Table 11-2, and we refer the reader back to that table.

Let's focus on the 2-parameter Margules equation, since that provided a fit that was an order of magnitude better (i.e. smaller discrepancy between the data and the model) than that of the 1-parameter Margules equation. As a reminder, this is given as

$$\frac{G^{\mathrm{E}}}{RT} = x_1 x_2 (A_{21} x_1 + A_{12} x_2) \tag{11.90}$$

Recall, the objective function we used was

$$\mathrm{OBJ} = \frac{1}{22} \sum_{i=1}^{22} \left[\frac{\left(\dfrac{G^{\mathrm{E}}}{RT}\right)_i^{\mathrm{Model}} - \left(\dfrac{G^{\mathrm{E}}}{RT}\right)_i^{\mathrm{Expt.}}}{\left(\dfrac{G^{\mathrm{E}}}{RT}\right)_i^{\mathrm{Expt.}}} \right]^2 \tag{11.93}$$

The 2-parameter model that resulted was

$$\frac{G^{\mathrm{E}}}{RT} = x_1 x_2 (1.317 x_1 + 1.041 x_2)$$

Step 2 *Plotting the residuals of the excess molar Gibbs free energy*
As an illustration that the residuals in the excess molar Gibbs free energy do, in fact, fall about zero, they are plotted in Figure 11-12. Note that all of the residuals are scaled by the experimental value of G^{E}/RT at that point, as this was the approach used in the objective function.

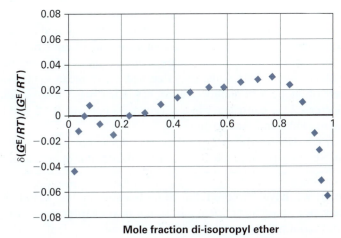

FIGURE 11-12 A plot of scaled residuals of the excess molar Gibbs free energy for the di-isopropyl ether (1) + 1-propanol (2) system at 303.15 K using a 2-parameter Margules equation.

Step 3 *Calculating and plotting the residuals of the natural logarithm of the ratio of the activity coefficients*

Once we have an excess molar Gibbs free energy model, we can impose the direct test on the data set by calculating the residuals in the natural logarithm of the ratios of the activity coefficients. Specifically, we are looking to calculate the following at each point:

$$\delta\left[\ln\left[\frac{\gamma_1}{\gamma_2}\right]\right] \tag{11.163}$$

Since we already have the experimental activity coefficients for this system in Table 11-2 and the $\underline{G}^E/RT$ model parameters in Table 11-3, it is a quick spreadsheet calculation to find the residuals needed. These are plotted in Figure 11-13 as a function of the composition.

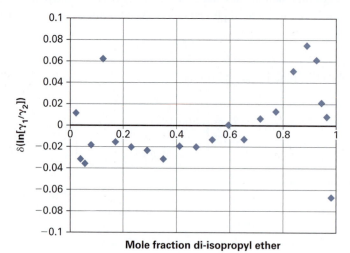

Mole fraction di-isopropyl ether

FIGURE 11-13 A plot of the residuals in the natural logarithm of the ratios of the activity coefficients for the di-isopropyl ether (1) + 1-propanol (2) system at 303.15 K using a 2-parameter Margules equation.

Hendrick "Hank" van Ness was a chemical engineer (Ph.D. from Yale University) who worked for many years at RPI.

While the data does scatter about zero, the values at the higher compositions are a little more suspect than those at the lower compositions. However, in order to evaluate the data set overall, we need a heuristic to use. Van Ness suggested a scale that ranks the data on a *consistency index* based on the root mean square value of $\delta \ln(\gamma_1/\gamma_2)$ (Van Ness, 1995). Data that had a root mean square value less than 0.025 were of the highest quality (consistency index = 1), while data that had a root mean square value greater than 0.225 were of the poorest quality (consistency index = 10). Every additional 0.025 in the RMS increased the consistency index. Thus, data between 0.025 and 0.050 would have a consistency index of 2, and so on.

For the current data set, the root mean square value of $\delta\ln(\gamma_1/\gamma_2)$ is 0.035. Thus, this data has a van Ness consistency index of 2, which indicates it is of high quality.

In this chapter, we have explored the non-idealities in the liquid phase, while modeling the vapor phase as an ideal gas. In the next chapter, we incorporate non-idealities in the vapor phase using both the ideal gas and the ideal solution as a reference. Additionally, for the former, we model the vapor liquid equilibrium of mixtures using equations of state that can describe both the liquid and vapor phase, much like we did for pure components in Chapter 8.

11.10 SUMMARY OF CHAPTER ELEVEN

- Modeling for mixture vapor–liquid equilibrium is required because the acquisition of experimental data is time consuming and relatively expensive. Modeling allows for the evaluation of many systems at various states as a way to estimate the size of chemical plant equipment.

- The **chemical potential** for component i in a mixture is the partial molar Gibbs free energy of that component.

- For phase equilibrium, a component in a mixture will have the same chemical potential in all of the equilibrium phases. This is called the **chemical equilibrium condition**.

- The **mixture fugacity of a component** was introduced as an alternative to the chemical potential. The chemical equilibrium condition can also be written in terms of the mixture fugacity of the components.

- The **Lewis-Randall rule** stated that the mixture fugacity of component *i* in a mixture modeled by an ideal solution was just the pure component fugacity at the mixture temperature and pressure multiplied by the mole fraction of component *i* in the mixture.

- The **activity coefficient of component *i*** in a mixture was defined as the mixture fugacity of component *i* divided by the mixture fugacity of component *i* if modeled by an ideal solution. Thus, when the activity coefficient of component *i* in a mixture was equal to 1, that component in the mixture behaved according to the ideal solution.

- The **mixture fugacity coefficient of component *i*** was defined as the mixture fugacity of component *i* divided by the mixture fugacity of component *i* if modeled by an ideal gas mixture. Thus, when the mixture fugacity of component *i* was equal to 1, that component in the mixture behaved according to the ideal gas model.

- **Gamma-phi modeling** $(\gamma_i\text{-}\hat{\varphi}_i)$ refers to an approach to model the vapor–liquid equilibrium of mixtures. The "gamma" part is for the liquid (the activity coefficient) is used, while the "phi" part is for the vapor (the mixture fugacity coefficient) is used.

- **Henry's Law** is an approach normally used to model sparingly soluble gases in liquids.

- **Raoult's Law**, while valid for liquids modeled as ideal solutions and vapors modeled as ideal gases, can be used when modeling the component in *any* mixture whose composition is very close to 1.

- **Modified Raoult's Law** is a popular modeling approach for the vapor–liquid equilibrium of mixtures. Here, the deviations from ideal solution in the liquid are modeled through an activity coefficient while the vapor is assumed to behave as an ideal gas.

- The natural logarithm of the activity coefficient of a component in a mixture is a partial molar property of the excess molar Gibbs free energy, divided by RT.

- The natural logarithm of the mixture fugacity coefficient of a component in a mixture is a partial molar property of the residual molar Gibbs free energy, divided by RT.

- The simplest model for real (non-ideal) solutions is the **1-parameter Margules equation**. This model is for the excess molar Gibbs free energy.

- The **2-parameter Margules equation** (used for the excess molar Gibbs free energy) allows for additional flexibility in the modeling of real solutions.

- **Data reduction** refers to reducing a table of data to a single mathematical expression. This is a common practice for the excess molar Gibbs free energy.

- The **infinite dilution activity coefficient** of component i in j, describes the interaction between a single molecule of i in a sea of j molecules.

- The **Wilson equation** is another excess molar Gibbs free energy model that contains parameters that are temperature dependent.

- The activity coefficients are weak functions of the pressure and are much stronger functions of the temperature.

- Predictive excess molar Gibbs free energy models exist, such as the **van Laar equation** and **UNIFAC**, that rely upon pure component information to predict the activity coefficients for the mixtures.

- The quality of experimental vapor–liquid equilibrium data can be evaluated using the Gibbs-Duhem equation.

- The **integral test** provides a necessary, but not sufficient, method of evaluating the thermodynamic consistency of experimental data (using the Gibbs-Duhem equation).

- The **direct test** provides a sufficient method of evaluating the thermodynamic consistency of experimental data (using the Gibbs-Duhem equation).

11.11 EXERCISES

11-1. The derivation of the expression for the natural logarithm of the activity coefficient of component "1" for a binary mixture modeled by the 1-parameter Margules equation is given in Section 11.6. Please derive the expression for component "2."

11-2. Derive the expression for the natural logarithm of the activity coefficient for component "1" in a binary mixture modeled by the 2-parameter Margules equation.

11-3. Given an equimolar binary mixture, calculate the activity coefficients for both components from the 1-parameter Margules equation whose parameter value is $A = 1.5$. What would be the excess molar Gibbs free for this system?

11-4. Given an equimolar binary mixture, calculate the activity coefficients for both components from the 2-parameter Margules equation whose parameter values are $A_{12} = 0.5$ and $A_{21} = 1.5$. What would be the excess molar Gibbs free for this system?

11-5. Estimate the system pressure and vapor-phase composition for a binary mixture ($x_1 = 0.2$) modeled by the 1-parameter Margules equation ($A = 0.832$) with the following properties: $P_1^{sat} = 10$ kPa and $P_2^{sat} = 110$ kPa. What would be the excess molar Gibbs free for this system?

11-6. You calculate the Wilson equation parameters for the ethanol (1) + 1-propanol (2) system at 298.15 K and find they are $\Lambda_{12} = 0.7$ and $\Lambda_{21} = 1.1$. Estimate the value of the parameters at 323.15 K?

11-7. Calculate the van Laar parameters, L_{12} and L_{21}, for the benzene (1) + toluene (2) system at 298.15 K through use of the van der Waals equation of state. What would be the value for the activity coefficients of an equimolar mixture of benzene + toluene at 298.15 K?

11-8. Calculate the van Laar parameters using the Scatchard-Hildebrand approach, M_{12} and M_{21}, for the benzene (1) + toluene (2) system at 298.15 K. What would be the value for the activity coefficients of an equimolar mixture of benzene + toluene at 298.15 K?

11-9. Explain why the integral test for thermodynamic consistency is only a necessary condition, while the direct test is a sufficient condition.

11-10. An experiment on the vapor-liquid equilibrium for the methanol (1) + dimethyl carbonate (2) system at 337.35 K provides the following information (S. Yunhai et al., 2005):

Dimethyl carbonate

- $x_1 = 0.0$, $y_1 = 0.0$ and $P = 41.02$ kPa.
- $x_1 = 0.20$, $y_1 = 0.51$ and $P = 68.23$ kPa.
- $x_1 = 1.0$, $y_1 = 1.0$ and $P = 99.91$ kPa.

Use this information to estimate the system pressure and vapor-phase mole fraction when $x_1 = 0.8$. Use the 1-parameter Margules equation.

11-11. Consider the following experimental data in Table E11-11 for the ethanol (1) + 2, 2, 4-trimethyl pentane (2) system at 333.15 K.

2,2,4 trimethyl pentane

TABLE E11-11 Vapor–liquid equilibrium data for the ethanol (1) + 2,2,4-trimethyl pentane system at 333.15 K.

P (kPa)	x_1	y_1	γ_1	γ_2
41.54	0.0196	0.3315	12.528	1.002
62.52	0.2904	0.5764	2.687	1.280
63.77	0.5303	0.6066	1.542	1.869
62.98	0.7705	0.6428	1.131	3.351
50.24	0.9799	0.9206	1.001	8.328

Based on data from Oshea, S. J. and Stokes, R. H., "Activity coefficients and excel partial molar enthalpies for ethanol + hexane from 283 to 318 K," J. Chem. Thermodyn., 18, 691–696, 1986.

Considering the values of the activity coefficients, you know that this system will not behave as an ideal solution. However, would the 1-parameter Margules equation be a reasonable model to account for the liquid phase non-idealities? Why or why not? Explain your answer based on the information provided in the table.

11-12. Which system provided here, if any, would be best modeled by an ideal solution? If any of the solutions are non-ideal, discuss whether the Scatchard-Hildebrand approach would be appropriate to model the non-idealities. Explain your answer.

(i) ethane + n-decane
(ii) water + 1-butanol
(iii) benzene + toluene

11.12 PROBLEMS

11-13. A binary liquid containing mostly component 2 is in equilibrium with a vapor phase containing both components 1 and 2. The pressure of this two-phase system is 100 kPa; the temperature is 298.15 K. Estimate x_1 and y_1 if the Henry's Law constant is equal to 20 MPa and the vapor pressure of component 2 at 298.15 K is 10 kPa.

11-14. Provide an estimate of the composition of N_2 dissolved in water at 298 K. Assume $A = 6.52$ for the 1-parameter Margules equation for this system and the partial pressure of the Nitrogen is 4.509 MPa at this temperature.

11-15. Resolve Example Problem 11.1, but now include the nitrogen gas into the system. Note the Henry's Law constants for nitrogen in water in Table P11-15:

TABLE P11-15 Henry's Law constants for N_2 in water.

T (°C)	$H_{N,w}$ (MPa)
25	8816.14
75	16490.47

Based on data from R. Sander, Compilation of Henry's Law Constants for Inorganic and Organic Species of Potential Importance in Environmental Chemistry, 1999. sander@mpch-mainz.mpg.de

11-16. A liquid mixture of 20% by mole benzene and the rest 2-propanol is in equilibrium with its vapor at 69°C. Estimate the equilibrium pressure and the composition of the vapor phase. Assume the 1-parameter Margules equation holds. At this

CH3—CH—CH3
|
OH

2-propanol

temperature, the vapor pressure of the benzene is 70.66 kPa and that of the 2-propanol is 57.86 kPa. Note that the system does form an azeotrope at this temperature whose composition is 61% by mole benzene (at a pressure of 91.32 kPa) (Storonkin and Morachevsky, 1956). If you make an assumption, please *justify* your assumption based on the results.

11-17. You are interested in finding the pressure at which the first bubble of vapor will form from a liquid mixture of methanol (1) and 2-methyl-1-propanol (2) (49% by mole methanol) at 50°C. The Margules parameters for this mixture are $A_{12} = 0.2565$ and $A_{21} = 0.2404$ (Gmehling and Onken, 1977). Note that the vapor pressure of methanol is 56.26 kPa and that for 2-methyl-1-propanol is 7.47 kPa at 323.15 K. Find the pressure and vapor-phase composition.

2-methyl-1-propanol

11-18. For a chloroform (1) + ethanol (2) system at 328.15 K, a Margules equation has been written as $G^E/RT = (0.5579x_1 + 1.5254x_2)x_1x_2$ (Gmehling and Onken, 1977). Calculate the pressure and vapor-phase composition of the system when $x_1 = .25$. What would be the predicted pressure and vapor-phase composition if you treated the liquid as an *ideal solution*?

Chloroform

11-19. You are interested in finding the pressure at which the first bubble of vapor will form from a liquid mixture of ethanol (1) and benzene (2) (49% by mole ethanol) at 313 K. The Margules parameters for this mixture are $A_{12} = 2.173$ and $A_{21} = 1.539$, while the Wilson parameters for this mixture are $a_{12}/R = 653.13$ K and $a_{21}/R = 66.16$ K. Find the pressure and vapor-phase composition using three ways.
A. The 2-parameter Margules equation
B. The Wilson equation
C. Ideal solution

11-20. The azeotrope of a binary mixture, being an important point on a mixture phase diagram, is often used for parameter estimation. To that end, use the azeotropic information for the acetone (1) + cyclohexane (2) system at 308.15 K to determine the A parameter of the 1-parameter Margules equation. Then plot the Pxy predictions from this model and compare it to the experimental data in Table P11-20. Note that you will need to plot the experimental data first to estimate the location of the azeotrope.

Acetone

Cyclohexane

TABLE P11-20 Vapor-liquid equilibrium of acetone (1) + cyclohexane (2) at 308.15 K.

P [kPa]	x_1	y_1
19.625	0	0
37.877	0.098	0.482
45.476	0.198	0.601
47.969	0.283	0.635
49.489	0.387	0.665
50.316	0.489	0.686
50.969	0.598	0.709
51.302	0.700	0.732
51.409	0.771	0.754
50.196	0.895	0.841
45.863	1	1

Based on data from Marinichev A.N., Susarev M.P., J. Appl. Chem. USSR, 38(2), 371-375, 1965.

11-21. Consider the experimental data in Table P11-21 for the ethyl acetate (1) + cyclohexane (2) system at 293.15 K. Please fit this system to the 2-parameter Margules equation and plot the Pxy curve for the system (along with the

Ethyl acetate

experimental data and Raoult's Law predictions). Also, plot the natural logarithm of activity coefficients and the excess molar Gibbs free energy (divided by RT) both on the same curve.

Cyclohexane

TABLE P11-21 Vapor-liquid equilibrium of ethyl acetate (1) + cyclohexane (2) at 293.15 K.

P [kPa]	x_1	y_1
10.386	0	0
11.132	0.025	0.075
11.666	0.050	0.148
12.199	0.075	0.197
12.959	0.144	0.286
13.732	0.218	0.350
13.919	0.274	0.390
14.066	0.347	0.425
14.132	0.406	0.442
14.132	0.494	0.476
14.052	0.575	0.505
13.719	0.690	0.556
12.946	0.787	0.619
12.026	0.887	0.730
10.986	0.948	0.840
9.786	1	1

Based on data from Slavin A.A., Abramzon A.A., J. Appl. Chem. USSR, 50(4), 739-743, 1977.

11-22. In the sizing of separation equipment, you need to know the vapor–liquid equilibrium for the benzene (1) + acetonitrile (2) system. You have data for this system at 293.15 K, but not at your desired temperature, which is 318.15 K. You also know this system has an azeotrope at 293.15 K at about 12.7 kPa and 53% benzene.
A. Use the Wilson equation to fit all of the experimental data in Table P11-22 at 293.15 K.
B. Use the Wilson parameters determined from part (A) to predict the system behavior at 318.15 K.
C. Now that you have predicted the behavior (part (B)), how well does your prediction compare with the experimental data for this azeotrope at 318.15 K, which is $P = 37.1$ kPa and $x_1 = y_1 = 0.52$?

Acetonitrile

TABLE P11-22 Vapor-liquid equilibrum of benzene (1) + acetonitrile (2) at 293.15 K.

P [kPa]	x_1	y_1
9.799	0.018	0.054
10.106	0.033	0.096
10.599	0.063	0.161
10.639	0.064	0.162
11.106	0.098	0.221
11.386	0.128	0.256
11.719	0.168	0.300
11.852	0.187	0.321
11.932	0.198	0.330
12.186	0.243	0.370
12.226	0.256	0.376
12.359	0.284	0.395
12.439	0.310	0.412
12.466	0.320	0.417
12.612	0.368	0.442
12.666	0.424	0.473
12.679	0.435	0.479
12.679	0.446	0.484
12.692	0.446	0.486
12.719	0.470	0.498
12.692	0.481	0.501
12.746	0.511	0.518
12.732	0.512	0.519
12.746	0.535	0.532
12.719	0.572	0.551
12.719	0.607	0.571
12.666	0.648	0.593
12.639	0.657	0.597
12.612	0.679	0.612
12.519	0.710	0.632
12.359	0.753	0.662
12.186	0.793	0.695
12.119	0.815	0.709
12.012	0.841	0.734
11.866	0.857	0.757
11.706	0.874	0.767
11.386	0.906	0.808
11.252	0.920	0.829

Based on data from Werner G., Schuberth H., J. Prakt. Chem. Ser. 4, 31(5-6), 225-239, 1966.

P [kPa]	x_1	y_1
11.106	0.933	0.849
11.026	0.939	0.860
10.879	0.953	0.883
10.692	0.966	0.906
10.506	0.975	0.930
10.319	0.983	0.950
10.186	0.991	0.972

11-23. You are interested in evaluating how well you can predict the phase-behavior of a system using an excess molar Gibbs free energy model. Here, use the van Laar predictions (from the VDW-defined parameters) relative to fitting the data (in Table P11-23) with the van Laar model for the methanol (1) + water (2) system at 328.15 K. What is your conclusion based on these results?

TABLE P11-23 Vapor-liquid equilibrium of methanol (1) + water (2) at 328.15 K.

P [kPa]	x_1	y_1
30.710	0.1587	0.5660
33.368	0.1980	0.6154
39.038	0.2961	0.7019
40.215	0.3171	0.7182
40.835	0.3339	0.7280
42.143	0.3610	0.7432
42.760	0.3733	0.7504
44.237	0.4137	0.7693
45.184	0.4346	0.7799
46.743	0.4753	0.7947
48.181	0.5078	0.8079
50.500	0.5669	0.8315
51.100	0.5897	0.8401
51.974	0.6030	0.8495
52.227	0.6091	0.8502
52.664	0.6217	0.8557
54.245	0.6600	0.8709
54.956	0.6781	0.8770
55.980	0.7032	0.8866
59.208	0.7808	0.9183

Based on data from Kurihara, K., Minoura, T., Takeda, K., and Kojima, K., "Isothermal Vapor-Liquid Equilibria for Methanol + Ethanol + Water, Methanol + Water, and Ethanol + Water," J.Chem. Eng. Data, 40, 679-684, 1995.

11-24. Compare the van Laar predictions if using VDW-defined parameters relative to those from Scatchard-Hildebrand in order to calculate the *Pxy* diagram of

Benzene

the benzene (1) + m-xylene (2) system at 310.15 K Plot the results from both approaches as well as the experimental data from Table P11-24.

TABLE P11-24 The vapor-liquid equilibrium of benzene (1) + m-xylene (2) at 310.15 K.

P [kPa]	x_1	y_1
2.944	0.037	0.296
3.040	0.042	0.314
3.609	0.072	0.446
3.925	0.085	0.501
4.877	0.137	0.619
7.341	0.268	0.782
9.821	0.394	0.866
11.423	0.474	0.900
12.634	0.535	0.916
14.852	0.651	0.948
16.679	0.745	0.967
18.065	0.819	0.977
20.092	0.928	0.991

Based on data from Boublik T., Benson G.C., Can. J. Chem., 47, 539-542, 1969.

11-25. You desire to flash 20 moles/min of a liquid mixture containing 20% by mole chloroform (1) and 80% by mole ethanol (2) at 332.4 K in a flash distillation unit operating at 70 kPa. You do not know a lot about this mixture at this temperature, other than the fact it exhibits an azeotrope ($x_1 = y_1 = 0.84$ at 101.3 kPa; Chen, 1995). Since this system forms an azeotrope, you know not to treat it as an ideal solution. Model the system using the 1-parameter Margules equation and estimate whether the mixture flashes at 332.4 K and 70 kPa. If it does flash, calculate the resulting flowrate and composition of the stream(s) leaving the flash distillation unit.

11-26. Your company needs to evaluate the separation of an equimolar mixture of ethanol (1) + n-hexane (2) at 318.15 K. While looking at some company notebooks, you find data for this system at 318.15 K as follows:

TABLE P11-26 Vapor-liquid equilibrium of ethanol (1) + n-hexane (2) at 318.15 K.

P [kPa]	x_1	y_1
45.75	0	0
48.251	0.00484	0.0682
25.256	0.9947	0.9107
23.08	1	1

Based on data from O'Shea, S. J. and Stokes, R. H., "Activity coefficients and excel partial molar enthalpies for ethanol 1 hexane from 283 to 318 K," J. Chem. Thermodyn., 18, 691–696, 1986.

You show this to your boss and she mentions "*Oh yeah . . . I remember when this was done. It was to determine infinite dilution values quickly in order to parameterize a model. I don't think we ever did that, however. You should do this to see if we can flash the mixture at 65 kPa.*"

Since you follow the instructions of your boss, use the data in Table P11-26 to parameterize the 2-parameter Margules equation and determine if the mixture will flash at 65 kPa and 318.15 K. If it does flash, determine the amount and composition of the stream(s) exiting the unit, if the feed enters at 50 mol/min.

11-27. In a process you need to evaluate the flash separation of an equimolar mixture of 1,3-dioxolane (1) + n-heptane (2) at 343.15 K. However, you can only find experimental data at 313.15 K. From your thermodynamics class, you remember that the Wilson equation can be used to predict behavior at one temperature if you parameterize your system at another temperature. This seems to be the perfect opportunity to use the Wilson equation.

1,3-dioxolane

n-heptane

A. Using the data in Table P11-27, parameterize the system using the Wilson equation at 40°C.

B. Determine the parameters for this system at 343.15 K.

C. Run a flash calculation at 343.15 K and 90 kPa on an equimolar feed of 10 moles/s. Determine the amount and composition of the stream(s) leaving the flash distillation unit.

D. After completing parts (A, B, and C), you find experimental data for this system at 343.15 K. You notice the following: $P^{\text{bubble}} (x_1 = 0.5) = 88.66$ kPa; $P^{\text{dew}} (y_1 = 0.5) = 68.93$ kPa. Discuss the results from part (C) in light of this information.

TABLE P11-27 Vapor-liquid equilibrium of 1,3-dioxolane (1) + n-heptane (2) at 40°C.

P [kPa]	x_1	y_1
12.24	0	0
19.75	0.0748	0.4022
23.20	0.1348	0.5268
26.09	0.2315	0.6017
26.62	0.2426	0.6090

P [kPa]	x_1	y_1
27.88	0.3060	0.6415
28.16	0.3400	0.6537
28.96	0.4007	0.6652
29.16	0.4400	0.6738
29.51	0.4961	0.6837
29.72	0.5209	0.6970
29.92	0.5633	0.7040
30.13	0.6266	0.7188
30.12	0.6320	0.7203
30.25	0.6961	0.7325
30.29	0.7275	0.7501
30.33	0.7603	0.7580
30.29	0.8169	0.7788
29.94	0.8977	0.8264
29.27	0.9373	0.8668
27.50	0.9857	0.9589
26.89	0.9951	0.9838
26.56	1	1

Based on data from Wu, H. S. and Sandler, S. Y., "Vapor–Liquid equilibrium of 1,3-dioxolane systems," J. Chem. Eng. Data, 34, 209–213, 1989.

11-28. You desire to flash separate 10 mol/s of an equimolar liquid mixture of 1-butene (1) + n-heptane (2) at 100 kPa and 273.15 K. If your flash distillation unit operates at 50 kPa and 273.15 K, what is the flow rate and composition of the stream(s) exiting the flash distillation unit? Use the van Laar equation with van der Waals predicted parameters to account for the deviations of this mixture from the ideal solution model.

1-butene

11-29. In a process analysis application, you are working with the di-n-propyl ether (1) and 2-propanol (2) system at 25°C. You think you have an error in the spreadsheet you have been working with, but you can't seem to find the problem. After a while, you begin to wonder if the data you are using is thermodynamically consistent. Not that you are suspicious of the data, but you've checked everything else by this point.

Di-n-propyl ether

CH_3—CH—CH_3
|
OH

2-propanol

For the data presented in Table P11-29, examine the thermodynamic consistency using both the integral test and direct test.

TABLE P11-29 Vapor-liquid equilibrium of di-n-propyl ether (1) + 2-propanol (2) at 298.15 K.

P [kPa]	x_1	y_1
2.878	0	0
3.110	0.0637	0.1379
3.290	0.1377	0.2033
3.308	0.1751	0.2249
3.312	0.2096	0.2409
3.306	0.2922	0.2699
3.316	0.3570	0.2858
3.248	0.4682	0.3163
3.225	0.5165	0.3287
3.142	0.6011	0.3559
2.798	0.7656	0.4459
2.604	0.8334	0.5041
2.142	0.9207	0.6583
1.519	1	1

Based on data from Garriga, R., Sanchez, F., Perez, P., and Gracia, M., "Isothermal vapor-liquid equilibrium at eight temperatures and excess functions at 298.15 K of di-n-propylether with 1-propanol or 2-propanol," Fluid Phase Equil, 138, 131–144, 1997.

11-30. You have found a set of thermodynamic data in the literature for benzene (1) + 2-propanol (2) at 298.15 K that you need to evaluate for a particular process. However, you want to verify the quality of the data prior to using it. To that end, please evaluate the thermodynamic consistency of the data set in Table P11-30.

TABLE P11-30 Vapor-liquid equilibrium of benzene (1) + 2-propanol (2) at 298.15 K.

P(kPa)	x_1	y_1
5.866	0	0
8.853	0.076	0.365
11.199	0.164	0.530
13.306	0.300	0.635
14.106	0.479	0.712
14.452	0.638	0.745
14.532	0.854	0.795
13.932	0.941	0.877
12.586	1	1

Based on data from Olsen A.L., Washburn E.R., J.Phys.Chem., 41, 457-462, 1937.

11.13 GLOSSARY OF SYMBOLS

α_{12}, α_{21} temperature-independent parameters of the Wilson equation

a_{12}, a_{21} cross-parameters for equations of state

a_1, a_2 equation of state parameter

a_m equation of state parameter for the mixture

A parameter of the 1-parameter Margules equation

A_{12}, A_{21} parameters of the 2-parameter Margules equation

b_{12}, b_{21} cross-parameters for equations of state

b_1, b_2 equation of state parameter

b_m equation of state parameter for the mixture

f_i fugacity of pure component i

f_i^{ig} fugacity of pure component i modeled as an ideal gas

f_i^{sat} fugacity of pure component i at saturation conditions

$\hat{f}_i$ mixture fugacity of component i

$\hat{f}_i^{ID}$ mixture fugacity of component i modeled as an ideal solution

$\hat{f}_i^{ig}$ mixture fugacity of component i modeled as an ideal gas mixture

$\underline{G}^E$ excess molar Gibbs free energy of the mixture

$\underline{G}_i$ molar Gibbs free energy of pure component i

$\underline{G}_i^{ig}$ molar Gibbs free energy of pure component i modeled as an ideal gas

$\underline{G}$ molar Gibbs free energy of the mixture

$\underline{G}^{ID}$ molar Gibbs free energy of the mixture modeled as an ideal solution

$\overline{G}_i$ partial molar Gibbs free energy of component i in a mixture

$\overline{G}_i^{ID}$ partial molar Gibbs free energy of component i in a mixture modeled as an ideal solution

$\overline{G}_i^E$ partial molar excess Gibbs free energy of component i in a mixture

$\overline{G}_i^{ig}$ partial molar Gibbs free energy of component i in a mixture modeled as an ideal gas mixture

$\underline{H}^E$ excess molar enthalpy of the mixture

$\underline{H}$ molar enthalpy of the mixture

$\Delta \underline{H}_i^{vap}$ molar enthalpy of vaporization of component i

L_{12}, L_{21} parameters of the van Laar equation using the VDW definitions

M an arbitrary thermodynamic property (normally $U, H, V, G,$ or S)

M_{12}, M_{21} parameters of the van Laar equation using the Scatchard-Hildebrand approach

n moles

OBJ   objective function value

P pressure

P_c critical pressure

P_i^{sat} vapor pressure of component i

R universal gas constant

$\underline{S}^E$ excess molar entropy of the mixture

$\underline{S}$ molar entropy of the mixture

$\underline{S}^{ID}$ molar entropy of the mixture as an ideal solution

T temperature

T_c critical temperature

$\underline{U}^E$ excess molar internal energy of the mixture

$\underline{U}_A$ molar internal energy of pure component A

$\underline{U}_A^{ig}$ molar internal energy of pure component A as an ideal gas

$\underline{U}$ molar internal energy of the mixture

$\underline{U}^{ID}$ molar internal energy of the mixture modeled as an ideal solution

$\Delta \underline{U}$ molar internal energy of mixing

$\Delta \underline{U}_i^{vap}$ molar internal energy of vaporization of component i

$\underline{V}^E$ excess molar volume of the mixture

$\underline{V}_i$ molar volume of pure component i

$\underline{V}$ molar volume of the mixture

$\underline{V}^{ID}$ molar volume of the mixture modeled as an ideal solution

x_A mole fraction of component A (normally in the liquid phase)

y_A mole fraction of component A (normally in the vapor phase)

δ_i solubility parameter of component i

δ residual

$\hat{\varphi}_i$ mixture fugacity coefficient of component i

$\hat{\varphi}_i^{sat}$ mixture fugacity coefficient of component i at saturation conditions

γ_i activity coefficient of component i

γ_i^{∞} infinite dilution activity coefficient of component i

$\Lambda_{12}, \Lambda_{21}$ temperature-dependent parameters of the Wilson equation

| μ_i | chemical potential of component i in the mixture | μ_i^{ID} | chemical potential of component i modeled as an ideal solution | μ_i^{R} | residual chemical potential of component i |
| μ_i^{E} | excess chemical potential of component i | μ_i^{ig} | chemical potential of component i modeled as an ideal gas mixture | | |

11.14 REFERENCES

Abrams, D. S., Prausnitz, J. M. "Statistical Thermodynamics of Liquid Mixtures: A New Expression for the Excess Gibbs Energy of Partly of Completely Miscible Systems." *AIChE J*, 21, 116 (1975).

Battino, R. "Volume Changes on Mixing for Binary Mixtures of Liquids." *Chem. A. Fredenslund J. Gmehling P. RasmussenReviews*, 71, 5 (1971).

Bredig, G., Bayer, R., "The Steam Pressures of the Binary System Methyl Alcohol–Water." *Z. Phys. Chem. (Leipzig)*, 130(1), 1–14 (1927).

Boublik, T., Benson G. C., "Molar Excess Gibbs Free Energies of Benzene – m-Xylene Mixtures." *Can. J. Chem.*, 47, 539–542 (1969).

Chen, G. H., Wang, Q., Ma, Z., Yan, X. H., Han, S. J. "Phase equilibria at superatmospheric pressure Water", for systems containing halohydrocarbon, aromatic hydrocarbon, and alcohol," *J.Chem.Eng.Data*, 40, 361–366 (1995).

Dobson, C. M. "Chemical Space and Biology." *Nature*, 432, 824 (2004).

Fredenslund, A., Gmehling J., Rasmussen, P. "Vapor-liquid equilibria using UNIFAC: a group contribution method", Elsevier Scientific, New York, 1979.

Garriga, R., Sanchez, F., Perez, P., Gracia, M. "Isothermal vapor-liquid equilibrium at eight temperatures and excess functions at 298.15 K of di-*n*-propyl ether with 1-propanol or 2-propanol," *Fluid Phase Equil*, 138, 131–144 (1997).

Gaw, W. J., Swinton, P. L. "Thermodynamic Properties of Binary Systems Containing Hexafluorobenzene. Part 3. Excess Gibbs Free Energy of the System Hexafluorobenzene + Cyclohexane." *Trans. Faraday Soc.*, 64, 637 (1968).

Gmehling, J., Onken, U. "Vapor-Liquid Equilibrium Data Collection," Vol 1, Part 2a, DECHEMA, 1977.

Gritsenko, N. N., "Determination of polarizability and the radius of acetonitrile and dimethylacetamide molecules," *Zh. Fiz. Khim.* 54, 198–199 (1980).

Hofman, T., Sporzynski, A., Goldon, A. "Vapor-Liquid Equilibria in Ethanol + (Butyl Methyl Ether or Dipropyl Ether) Systems at 308.15, 323.15, and 338.15 K." *J. Chem. Eng. Data*, 45, 169 (2000).

Hwang et al., "Isothermal VLE and V^E at 303.15 K for the Binary and Ternary Mixtures of Di-isopropyl Ether (DIPE) + 1-Propanol + 2,2,4-Trimethylpentane." *J. Chem. Eng. Data*, 52, 2503 (2007).

Kurihara, K., Minoura, T., Takeda, K., Kojima, K. "Isothermal Vapor-Liquid Equilibria for Methanol + Ethanol + Water, Methanol + Water, and Ethanol + Water," *J.Chem. Eng. Data*, 40, 679–684 (1995).

Mahl, B., et al., "Thermodynamics of Binary Mixtures. The Effect of Substituents in Aromatics on their Excess Volumes of Mixing with *n*-pentane." *J. Chem. Thermodynamics*, 3, 363 (1971).

Marinichev, A. N., Susarev M. P. "Investigation of the liquid-vapor equilibrium in the systems acetone- methanol and acetone- cyclohexane at 35, 45, 55.deg. and 760 mm Hg pressure" *J. Appl. Chem. USSR*, 38(2), 378–383 (1965).

Oh, J. H, Hwang, I. C., Park, S. J. "Isothermal vapor-liquid equilibrium at 333.15 K and excess molar volumes and refractive indices at 298.15 K for the mixtures of dimethyl carbonate, ethanol and 2, 2, 4-trimethylpentane", *Fluid Phase Equil*, 276, 142–149 (2009).

Olsen, A. L., Washburn, E. R. "The Vapor Pressure of Binary Solutions of Isopropyl Alcohol and Benzene at 25°C." *J. Phys. Chem.*, 41, 457–462 (1937).

Oshea, S. J., Stokes, R. H. "Activity coefficients and excel partial molar enthalpies for ethanol + hexane from 283 to 318 K," *J. Chem. Thermodyn.*, 18, 691–696 (1986).

Perez, E. et al., "Excess Enthalpy, Excess Volume, Viscosity Deviation, and Speed of Sound Deviation for the mixture Tetrahydrofuran + 2,2,2-Trifluoroethanol at (283.15, 298.15, and 313.15) K." *J. Chem. Eng. Data*, 48, 1306 (2003).

Sander, R., Compilation of Henry's Law Constants for Inorganic and Organic Species of Potential Importance in Environmental Chemistry, 1999. sander@mpch-mainz. mpg.de

Slavin, A. A., Abramzon A. A. *J. Appl. Chem. USSR*, 50(4), 739–743 (1977).

Storonkin, V., Morachevskii, A. G., *Zh. Fiz. Khim*, 30, 1297–1307 (1956).

Van Ness, H. C. "Thermodynamics in the Treatment of Vapor/Liquid Equilibrium (VLE) Data." *Pure & Appl. Chem.,* 67, 859 (1995).

Wang, J. L. H., Lu, B. C. Y. "Vapour-Liquid Equilibrium Data for the *n*-Pentane-Benzene System." *J. Appl. Chem. Biotechnol,* 21, 297 (1971).

Werner, G., Schuberth, H. "Das Phasengleichgewicht Flüssig-Flüssig des Systems Benzol/*n*-Heptan/Acetonitrill sonic die Phasegleichgewichte dampfforming-flussing der entsprechenden binären systeme bei 20, 0°C," *J. Prakt. Chem.*, 4, 225 (1966).

Wu, H. S., Sandler, S. Y. " Vapor–Liquid equilibrium of 1, 3-dioxolane systems," *J. Chem. Eng. Data*, 34, 209–213 (1989).

Yunhai, S., Honglai, L., Wang, K., Xiao, W., Ying, H. "Measurements of isothermal vapor-liquid equilibrium of binary methanol/dimethyl carbonates system under pressure", *Fluid Phase Equil*, 234, 1–10 (2005).

Theories and Models for Vapor–Liquid Equilibrium of Mixtures: Using Equations of State

12

LEARNING OBJECTIVES

This chapter is intended to help you learn how to:

- Use an equation of state in a gamma-phi modeling approach for mixtures
- Use an equation of state in a phi-phi modeling approach for mixtures
- Calculate an appropriate binary interaction parameter to correlate an equation of state to experimental mixture data
- Model vapor–liquid equilibrium for ideal solutions at elevated pressures

---

In Chapter 11, we examined the modeling of vapor–liquid equilibrium for mixtures using modified Raoult's Law. A limitation of that approach was the assumption of an ideal gas for the vapor phase—an assumption best met at lower pressures (normally a few hundred kPa, at most). In this chapter, we formally account for the deviations from ideal gas behavior in the modeling of vapor–liquid equilibrium, which is readily addressed through the use of equations of state. Additionally, since equations of state can be used for both the liquid and vapor phases, we present the formal modeling of vapor–liquid equilibrium using an equation of state for both the liquid and vapor phases. ■

12.1 Motivational Example

Much of what has been presented about mixtures thus far has focused on lower pressures. However, mixture properties change with pressure, and this provides an opportunity for the chemical engineer to design more improved, efficient processes at elevated pressures. Additionally, sometimes conditions are fixed at high pressure, such as the removal of carbon dioxide from natural gas sources, which can be up to 70 times greater than atmospheric pressure (Parrish and Kidnay, 2006).

For a pure component, the vapor pressure does not increase linearly in temperature, but in an exponential way (see Figure 8-1). When components are in a mixture, we see that the resulting system pressure is normally greater than our intuitive approach (positive deviations from Raoult's Law result in a greater pressure than what ideal solution would predict). It is clear, then, that system pressures will grow quickly when we increase system temperatures. When the system pressure is increased, the ideal gas model becomes an increasingly poor assumption. To that end, we need to account for the non-idealities in the vapor phase.

Accounting for the deviations from ideal gas behavior in the vapor phase requires that we develop models that provide us with mixture fugacity coefficients for the various components that are not equal to one (as in the ideal gas approach). This is the domain of real gas equations of state, as was explored in Chapter 7. In what follows in this chapter, we show the use of an equation of state to account for deviations from ideal gas behavior. However, as we saw in Chapter 7, equations of state can be used for both the liquid and vapor phases. We extend this treatment to mixtures in this chapter as well.

12.2 Deviations from the Ideal Gas Model for the Vapor Phase

When modeling vapor–liquid equilibrium for mixtures, we start at the same place we have started before: Equation 11.56. Recall that this equation requires that the mixture fugacity of component i are the same in both the liquid and vapor phases.

$$\hat{f}_i^L = \hat{f}_i^V \qquad (11.56)$$

We start with Equation 11.44 for the liquid:

$$\hat{f}_i^L(T,P,x) = x_i f_i(T,P)\gamma_i \qquad (11.44)$$

As before, we write out the fugacity of the pure component, f_i, in terms of the vapor pressure of component i, the fugacity coefficient of the pure component at saturation and the Poynting correction. This yields the left-hand side of Equation 11.68, which is

$$\hat{f}_i^L = x_i \gamma_i P_i^{sat} \varphi_i^{sat} \exp\left\{\frac{V_i(P - P_i^{sat})}{RT}\right\} \qquad (12.1)$$

On the vapor side, we use Equation 11.49, which provides the mixture fugacity of component i in the vapor phase in terms of the mixture fugacity coefficient of component i. Thus,

$$\hat{f}_i^V = y_i P\hat{\varphi}_i \qquad (11.49)$$

Equating these two mixture fugacities, as required by equilibrium (Equation 11.56), we have

$$x_i \gamma_i P_i^{sat} \varphi_i^{sat} \exp\left\{\frac{V_i(P - P_i^{sat})}{RT}\right\} = y_i P\hat{\varphi}_i \qquad (12.2)$$

This equation is general, and we have introduced it before. It is called the "gamma-phi" approach and is essentially Equation 11.57. The "gamma" part requires a model for the activity coefficients of the components in the liquid phase, while the "phi" part requires a model for the mixture fugacity coefficients of the components in the vapor phase. We have developed the formalism for the activity coefficient (gamma) part in Chapter 11; we develop the formalism for the fugacity coefficient (phi) part in the next subsection.

12.2.1 Mixture Fugacity Coefficients

In Chapter 11, we showed how the natural logarithm of the activity coefficient is a partial molar property of the excess molar Gibbs free energy (divided by RT). We will show something similar here for the natural logarithm of the mixture fugacity coefficient and its relationship to the residual molar Gibbs free energy. Indeed, the approaches in Chapter 11 and this section will mirror each other to emphasize their similarities.

Recall that Equation 11.48 provided us with the definition of the mixture fugacity coefficient of species i and its relationship to the residual chemical potential.

$$\mu_i^R(T,P,y) = RT \ln \left[\frac{\hat{f}_i(T,P,y)}{Py_i}\right] = RT \ln[\hat{\varphi}_i] \tag{11.48}$$

We know that the chemical potential of species i in a mixture is also the partial molar Gibbs free energy of species i, so the residual chemical potential is the residual partial molar Gibbs free energy. Thus,

$$\mu_i^R(T,P,y) = RT \ln[\hat{\varphi}_i] = \overline{G}_i^R \tag{12.3}$$

Recall from Equation 11.4 that we can write the chemical potential as

$$\mu_i = \overline{G}_i = \left(\frac{\partial(nG)}{\partial n_i}\right)_{T,P,n_{j(j \neq i)}} \tag{11.4}$$

Therefore, we can write the same for the residual properties:

$$\mu_i^R = \overline{G}_i^R = \left(\frac{\partial(n\underline{G}^R)}{\partial n_i}\right)_{T,P,n_{j(j \neq i)}} \tag{12.4}$$

Since the partial derivative is evaluated at constant T, it is valid to divide through by RT, the benefit of which we will see in a moment. So

$$\frac{\mu_i^R}{RT} = \frac{\overline{G}_i^R}{RT} = \left(\frac{\partial(n\underline{G}^R/RT)}{\partial n_i}\right)_{T,P,n_{j(j \neq i)}} \tag{12.5}$$

Equation 12.5 means that the natural logarithm of the mixture fugacity coefficient of component i is, in fact, a partial molar property of the residual molar Gibbs free energy, divided by RT. Thus,

$$\ln \hat{\varphi}_i = \frac{\mu_i^R}{RT} = \frac{\overline{G}_i^R}{RT} = \left(\frac{\partial(n\underline{G}^R/RT)}{\partial n_i}\right)_{T,P,n_{j(j \neq i)}} \tag{12.6}$$

To complete the comparison, we can use the natural logarithm of the mixture fugacity coefficient of component i in a Gibbs-Duhem type relationship, but the usefulness of this approach is strained, since most of the non-idealities are found in the liquid phase. Likewise, a summability relationship exists, as

> The summability relationship was introduced in Equation 9.27

$$\frac{G^R}{RT} = x_1 \frac{\overline{G}_1^R}{RT} + x_2 \frac{\overline{G}_2^R}{RT} = x_1 \ln[\hat{\varphi}_1] + x_2 \ln[\hat{\varphi}_2] \tag{12.7}$$

From a calculation standpoint, we need to write the mixture fugacity coefficient of component i in terms of thermodynamic variables. Since the definitions of fugacity coefficient differ depending on whether you are working with a mixture or a pure component, we cannot use the pure-component expression for the fugacity coefficient previously presented in Chapter 8. We must derive it again.

Picking out the first and third terms in Equation 12.6 and multiplying by 'RT', we can write Equation 12.8, which provides the relationship between the natural logarithm of the mixture fugacity coefficient of species i and the partial molar residual Gibbs free energy.

$$\overline{G}_i^R = \overline{G}_i - \overline{G}_i^{ig} = RT \ln[\hat{\varphi}_i] \tag{12.8}$$

> We employ this strategy of calculating differences relative to the same reference all the time when we calculate changes in state functions, since those values are provided relative to a reference value (normally set to zero – but it doesn't have to be).

We are interested in calculating this difference at the mixture temperature, mixture pressure, and vapor phase composition. One strategy to calculate this difference is to determine the changes in both terms (the partial molar Gibbs free energy and the partial molar Gibbs free energy in the ideal gas state) relative to the *same* reference. A reasonable reference is at zero pressure since, as we will see, this allows for a simplification.

For a system that has a fixed composition, we can write the differential of the Gibbs free energy as

$$d(n\underline{G}) = (-n\underline{S})dT + (n\underline{V})dP \tag{12.9}$$

However, since we are interested in the partial molar Gibbs free energy, we can re-write Equation 12.9 for that variable as

$$d(n\overline{G}_i) = (-n\overline{S}_i)dT + (n\overline{V}_i)dP \tag{12.10}$$

If we divide by a differential change in pressure at constant temperature and number of moles (i.e. fixed composition), we have

$$\left(\frac{\partial(n\overline{G}_i)}{\partial P}\right)_{T,n} = (-n\overline{S}_i)\left(\frac{\partial T}{\partial P}\right)_{T,n} + (n\overline{V}_i)\left(\frac{\partial P}{\partial P}\right)_{T,n} \tag{12.11}$$

This simplifies to

$$\left(\frac{\partial(n\overline{G}_i)}{\partial P}\right)_{T,n} = (n\overline{V}_i) \tag{12.12}$$

or by dividing out the number of moles, we have

$$\left(\frac{\partial\overline{G}_i}{\partial P}\right)_{T,n} = \overline{V}_i \tag{12.13}$$

We can use Equation 12.13 to determine the effect of a pressure change on the partial molar Gibbs free energy for both the real mixture and the ideal gas mixture, referenced back to zero pressure. For the real system (at fixed composition and temperature), we have

$$d\overline{G}_i = \overline{V}_i\,dP \tag{12.14}$$

Integrating from zero pressure to the system pressure, we have

$$\int_{\overline{G}_i(P=0)}^{\overline{G}_i(P=P)} d\overline{G}_i = \overline{G}_i(P) - \overline{G}_i(P=0) = \int_{P=0}^{P=P}\overline{V}_i\,dP \tag{12.15}$$

or

$$\overline{G}_i(P) = \overline{G}_i(P=0) + \int_{P=0}^{P=P}\overline{V}_i\,dP \tag{12.16}$$

Similarly, for our ideal gas mixture, we have

$$\int_{\overline{G}_i^{ig}(P=0)}^{\overline{G}_i^{ig}(P=P)} d\overline{G}_i^{ig} = \overline{G}_i^{ig}(P) - \overline{G}_i^{ig}(P=0) = \int_{P=0}^{P=P}\overline{V}_i^{ig}\,dP \tag{12.17}$$

or

$$\overline{G}_i^{ig}(P) = \overline{G}_i^{ig}(P=0) + \int_{P=0}^{P=P}\overline{V}_i^{ig}\,dP \tag{12.18}$$

We can now calculate the partial molar residual Gibbs free energy of component i by subtracting Equation 12.18 from Equation 12.16:

$$\overline{G}_i^{R} = \overline{G}_i(P) - \overline{G}_i^{ig}(P) = \overline{G}_i(P=0) + \int_{P=0}^{P=P}\overline{V}_i\,dP - \left[\overline{G}_i^{ig}(P=0)\right.$$
$$\left. + \int_{P=0}^{P=P}\overline{V}_i^{ig}\,dP\right] \tag{12.19}$$

PITFALL PREVENTION ⚠️

We have purposefully not skipped steps here and have been very explicit so that the reader can follow the derivation.

Since the real mixture will behave as an ideal gas mixture in the limit as pressure goes to zero, we can safely eliminate the partial molar residual Gibbs free energy terms evaluated at $P = 0$ for both the real and ideal gas mixtures (they will cancel each other out). This leaves

$$\overline{G}_i^R = \overline{G}_i(P) - \overline{G}_i^{ig}(P) = \int_{P=0}^{P=P} \overline{V}_i \, dP - \int_{P=0}^{P=P} \overline{V}_i^{ig} \, dP \tag{12.20}$$

From Equation 9.76, we know that the partial molar volume of a component in an ideal gas mixture is equal to the molar volume of the mixture. Thus,

$$\overline{V}_i^{ig}(T,P) = \underline{V}^{ig}(T,P) \tag{9.76}$$

The molar volume of an ideal gas mixture is just RT/P, so we can make the substitution directly into Equation 12.20 and group both terms into the integral:

$$\overline{G}_i^R = \overline{G}_i(P) - \overline{G}_i^{ig}(P) = \int_{P=0}^{P=P} \left(\overline{V}_i - \frac{RT}{P} \right) dP \tag{12.21}$$

And since the natural logarithm of the mixture fugacity coefficient of species i is equal to the partial molar Gibbs free energy (divided by RT), by Equation 12.6, we arrive at an expression that allows us to evaluate the mixture fugacity coefficient of species *i*.

$$\ln[\hat{\varphi}_i] = \frac{1}{RT} \int_{P=0}^{P=P} \left[\overline{V}_i - \frac{RT}{P} \right] dP \tag{12.22}$$

By way of comparison, recall that Equation 8.60 gave the expression for the natural logarithm of the *pure component* fugacity coefficient.

$$\ln(\varphi) = \frac{1}{RT} \int_{P=0}^{P=P} \left[\underline{V} - \frac{RT}{P} \right] dP \tag{8.60}$$

Both equations are similar, but for the mixture fugacity coefficient of component i, the partial molar volume is in the integral, while it is just the molar volume for the pure component fugacity coefficient.

Returning to the mixture, while Equation 12.22 can be used directly for relationships that provide $\underline{V} = \underline{V}(T,P)$ for mixtures, equations of state are normally provided in $P = P(T,\underline{V})$ form. Therefore, a coordinate transformation is needed in these instances such that Equation 12.22 is written with the molar volume and not the pressure as the integration variable. That relationship is covered later in this chapter. However, we will show the utility of Equation 12.22 directly for use in determining the mixture fugacity coefficient in gamma-phi modeling.

12.2.2 Incorporating the Mixture Fugacity Coefficient

There are nuances with gamma-phi modeling that we have not had to address up to this point. Let's consider the left-hand side (or liquid side) of Equation 12.2

$$x_i \gamma_i P_i^{sat} \varphi_i^{sat} \exp\left\{ \frac{\left[\underline{V}_i (P - P_i^{sat}) \right]}{RT} \right\} = y_i P \hat{\varphi}_i \tag{12.2}$$

Moving both of the terms with the fugacity coefficient over to the vapor side is a convenience that allows us to use the same thermodynamic model for both.

Recalling how this equation was developed, φ_i^{sat} is the pure component fugacity coefficient of species i evaluated at the mixture temperature and the saturation pressure at that temperature in the *liquid* phase. Since we know that the fugacity coefficient for a pure liquid must equal that for a pure vapor at saturation (that is one statement of chemical equilibrium), we can move the evaluation of this term from the liquid side to the vapor side of the equation (the right-hand side of the equality in Equation 12.2).

Therefore, we can move this term over in the expression to separate our liquid and vapor sides as before yielding our most useful gamma-phi expression.

$$x_i \gamma_i P_i^{\text{sat}} \exp\left\{ \frac{\underline{V}_i(P - P_i^{\text{sat}})}{RT} \right\} = y_i P \left(\frac{\hat{\varphi}_i}{\varphi_i^{\text{sat}}} \right) \tag{12.23}$$

Let's write out the functionality of each of the terms on both sides of this expression.

Liquid:

$$\gamma_i \to \underline{G}^{\text{E}} = \underline{G}^{\text{E}}(T, x)$$

$$P_i^{\text{sat}} = P_i^{\text{sat}}(T)$$

FOOD FOR THOUGHT 12-1

When would assuming that the liquid volume is a constant be a poor assumption?

Note that $\underline{V}_i$ is a function of T and P, but without a model that describes the functionality, it is reasonable to assume the molar volume of the liquid is a constant at most states.

If we fix the liquid-phase mole fraction and temperature, as allowed by the Gibbs phase rule (such as in a bubble-point pressure calculation), we obtain the activity coefficient and vapor pressure directly. If we were to assume an ideal gas for the vapor phase, our expressions for the activity coefficient and vapor pressure would be found directly (as in Example 11.3). In that situation, we used our modeling approach to solve for the system pressure and the vapor-phase composition (our unknown variables directly). Here, however, our vapor phase is NOT an ideal gas; thus, gamma-phi modeling is warranted.

Vapor:

$$\varphi_i^{\text{sat}} = \varphi_i^{\text{sat}}(T, P)$$

$$\hat{\varphi}_i = \hat{\varphi}_i(T, P, y)$$

When the vapor is not an ideal gas, the pressure and vapor-phase mole fraction are the variables needed to solve for the mixture fugacity coefficient of component i. However, we do not know the system pressure or the vapor-phase mole fraction as those are the variables we are looking to solve for. In the next section, we will demonstrate the solution strategy when this is the case through an example.

12.2.3 Gamma-Phi Modeling: Application Example

The best way to understand the steps in gamma-phi modeling is to demonstrate this through an example. However, in order to set up the problem, we need to discuss some items related to the new part of this approach, namely the phi model. Here we will use the virial equation to describe the vapor phase, but this time applied to mixtures.

Recall the virial equation as provided in Equation 7.91:

$$Z = 1 + \frac{B'P}{RT} \tag{7.91}$$

where B' is the second-virial coefficient, but now we have a mixture. How does one calculate a parameter value for a mixture? We approached this in the previous chapter when needing the mixture parameter values for the van der Waals equation of state in the van Laar modeling approach.

What do we know about B'? For a pure component, it describes two-body interactions and it is a function of temperature. Certainly the value of B' for water at 523 K and the value of B' for methane at the same temperature will be different (B'_{H_2O} = -149 cm³/mol; $B'_{CH_4} = 2.16$ cm³/mol). Therefore, a mixture value for B' will need to be a function of the mole fraction of each substance—just as before with the van der Waals one-fluid mixing rules. Statistical mechanics provides an exact definition of this mixing rule for B', and for a binary mixture, it becomes

$$B'_m = y_1 y_1 B'_{11} + y_1 y_2 B'_{12} + y_2 y_1 B'_{21} + y_2 y_2 B'_{22} \qquad (12.24)$$

The mixture value of B', which we call B'_m, uses a mixing rule that is quadratic in the mole fraction and the subscripts denote the component involved. B'_{11} is just the second-virial coefficient for pure component "1." Likewise, B'_{22} is just the second-virial coefficient of pure component "2." B'_{12} is the cross-term and describes how molecule "1" interacts with molecule "2." Note that $B'_{11} = B'_1$ and $B'_{22} = B'_2$.

The calculation of cross-terms is still an area of active research, and several approaches have been proposed for various equations of state. While it would be convenient to calculate the cross-term as a function of the pure component (such as in an arithmetic or geometric mean approach), B'_{12} is often determined using the *same* correlations for the pure components. However, the inputs to those correlations (pure-component critical temperatures and critical pressures) to calculate mixture values use both geometric and arithmetic means. For simplicity and demonstration purposes, we use a simple geometric mean for the cross-term. Thus,

$$B'_{12} = B'_{21} = \sqrt{B'_{11} B'_{22}} \qquad (12.25)$$

> This geometric mean approach was what we followed for the cross-term a in the van der Waals equation of state.

What will the fugacity coefficient look like from our virial equation of state? We need to determine this both for the pure component and the mixture, since φ_i^{sat} is a pure-component term.

First the pure component, whose fugacity coefficient expression is given by

$$\ln(\varphi) = \frac{1}{RT} \int_{P=0}^{P=P} \left[\underline{V} - \frac{RT}{P} \right] dP \qquad (8.60)$$

Our equation of state is the virial coefficient, as given by Equation 7.78

$$Z = \frac{P\underline{V}}{RT} = 1 + \frac{B'P}{RT} \qquad (12.26)$$

> **PITFALL PREVENTION**
>
> Since B'_{11} and B'_{22} are almost always negative, be sure to choose the negative solution to the square root of Equation 12.25.

We rewrite Equation 12.26 such that the molar volume is a function of T and P.

$$\underline{V} = \frac{RT}{P} + B' \qquad (12.27)$$

Plugging Equation 12.27 for the molar volume back into the pure component fugacity expression (Equation 8.60) yields

$$\ln(\phi) = \frac{1}{RT} \int_{P=0}^{P=P} \left[\frac{RT}{P} + B' - \frac{RT}{P} \right] dP = \frac{1}{RT} \int_{P=0}^{P=P} B' \, dP = \frac{B'P}{RT} \qquad (12.28)$$

> We can pull the second virial coefficient (B') out of the integral, since it is not a function of pressure, only temperature.

For the mixture fugacity coefficient of component i, we return to Equation 12.22

$$\ln[\hat{\varphi}_i] = \frac{1}{RT} \int_{P=0}^{P=P} \left[\overline{V}_i - \frac{RT}{P} \right] dP \qquad (12.22)$$

We still use Equation 12.27 for the molar volume

$$\underline{V} = \frac{RT}{P} + B'_m \tag{12.27}$$

but we use the mixture second-virial coefficient. Additionally, we now require the partial molar volume instead of the mixture molar volume:

$$\overline{V}_i = \left(\frac{\partial (n\underline{V})}{\partial n_i}\right)_{T,P,n_{j(j\neq i)}} = \left(\frac{\partial \left(n\left[\frac{RT}{P} + B'_m\right]\right)}{\partial n_i}\right)_{T,P,n_{j(j\neq i)}} \tag{12.29}$$

We use the product rule to go from Equation 12.29 to 12.30.

$$\overline{V}_i = \frac{RT}{P} + \left(\frac{\partial (nB'_m)}{\partial n_i}\right)_{T,P,n_{j(j\neq i)}} \tag{12.30}$$

For the pure component, B' was just a function of temperature. However, the B' for our mixture (B'_m) is a function of temperature plus the composition (through Equation 12.24). We leave the completion of this derivation as an exercise for the reader in an end-of-chapter problem and report just the final expressions here.

$$\ln(\hat{\varphi}_1) = \frac{P}{RT}[B'_{11} + y_2^2(2B'_{12} - B'_{11} - B'_{22})] \tag{12.31}$$

$$\ln(\hat{\varphi}_2) = \frac{P}{RT}[B'_{22} + y_1^2(2B'_{12} - B'_{11} - B'_{22})] \tag{12.32}$$

Now we have all of the expressions we need to incorporate a real gas model for the vapor phase in our gamma-phi modeling approach. We demonstrate its utility in the following example.

EXAMPLE 12-1 | **GAMMA-PHI MODELING (NON-IDEAL SOLUTION AND NON-IDEAL GAS)**

Consider the data set in Table 12-1 for the methanol (1) + dimethyl carbonate (2) system at two different temperatures: 337.35 K and 428.15 K.

TABLE 12-1 Experimental data for the methanol (1) + dimethyl carbonate (2) system at 337.35 K and 428.15 K.

T = 337.35 K			T = 428.15 K		
x_1	y_1	P (kPa)	x_1	y_1	P (kPa)
0	0	41.02	0	0	566.47
0.0069	0.0230	42.11	0.0062	0.0189	576.56
0.0955	0.3241	55.19	0.0930	0.2609	719.04
0.1999	0.5072	68.23	0.1838	0.4327	862.38
0.3033	0.6168	78.79	0.2758	0.5509	996.22
0.3974	0.6857	86.55	0.3757	0.6432	1127.59

Based on data from S. Yunhai et al.. Fluid Phase Equil, 234, 1 (2005)

(continued)

T = 337.35 K		
x_1	y_1	P (kPa)
0.4993	0.7406	93.16
0.5986	0.7823	97.96
0.6982	0.8161	101.23
0.7999	0.8431	102.91
0.8500	0.8517	103.05
0.9001	0.8851	102.67
0.9502	0.9373	101.71
1	1	99.91

T = 428.15 K		
x_1	y_1	P (kPa)
0.4759	0.7145	1245.56
0.5716	0.7705	1345.49
0.6855	0.8267	1445.39
0.7935	0.8735	1515.80
0.9001	0.9201	1556.04
0.9426	0.9456	1563.29
1	1	1565.62

Develop a modeling approach that can properly characterize the experimental data.

SOLUTION:

Step 1 *Selection of models*

As can be seen, the data set at the lower temperature has a maximum pressure a little over 100 kPa. Thus, an ideal gas for the vapor phase would certainly seem reasonable at these conditions. On the other hand, the data set at 428.15 K reaches a maximum pressure of about 1550 kPa, and at this pressure an ideal gas for the vapor phase would be much less reasonable.

To model the lower temperature system, we use modified Raoult's Law with the Wilson equation to account for the liquid phase non-idealities. To model the higher temperature system, we will use gamma-phi modeling (Wilson equation for the liquid phase and the virial equation for the vapor phase).

Step 2 *Modeling the system at 337.35 K*

As mentioned, we use modified Raoult's Law here with the Wilson equation. We obtain the experimental activity coefficients from

$$\gamma_1 = \frac{y_1 P}{x_1 P_1^{\text{sat}}} \tag{11.86}$$

and

$$\gamma_2 = \frac{y_2 P}{x_2 P_2^{\text{sat}}} \tag{11.87}$$

These results are provided in Table 12-2.

H—C—OH with H above and H below

Methanol

H₃C—O—C(=O)—O—CH₃

Dimethyl carbonate

TABLE 12-2 Activity coefficients and fugacity coefficient ratios for the methanol (1) + dimethyl carbonate (2) system at 337.35 K.

x_1	γ_1	γ_2	$\dfrac{\hat{\varphi}_1}{\varphi_1^{\text{sat}}}$	$\dfrac{\hat{\varphi}_2}{\varphi_2^{\text{sat}}}$
0		1	1.0151	1
0.0069	1.4062	1.0099	1.0148	0.9995
0.0955	1.8760	1.0049	1.0112	0.9939

The last two columns of Table 12-2 provide evidence that, indeed, the ideal gas is a reasonable model for the vapor phase, since the ratio of the mixture fugacity of the components relative to their pure component value, is close to 1.

(continued)

TABLE 12-2 *(continued)*

x_1	γ_1	γ_2	$\dfrac{\hat{\varphi}_1}{\varphi_1^{sat}}$	$\dfrac{\hat{\varphi}_2}{\varphi_2^{sat}}$
0.1999	1.7336	1.0236	1.0079	0.9886
0.3033	1.6043	1.0552	1.0053	0.9844
0.3974	1.4950	1.0989	1.0034	0.9814
0.4993	1.3832	1.1746	1.0018	0.9789
0.5986	1.2814	1.2928	1.0006	0.9771
0.6982	1.1843	1.5008	0.9998	0.9760
0.7999	1.0856	1.9632	0.9994	0.9754
0.8500	1.0334	2.4787	0.9994	0.9754
0.9001	1.0105	2.8730	0.9994	0.9758
0.9502	1.0042	3.1157	0.9996	0.9765
1	1		1	0.9776

> We calculate the model (Wilson equation) excess molar Gibbs free energy values through Equation 11.94. We calculate the experimental values through Equation 11.74.

As in Example 11.5, we can find the best fit parameters by minimizing the objective function as

$$\text{OBJ} = \frac{1}{12}\sum_{i=1}^{12}\left[\frac{\left(\dfrac{\underline{G}^{\text{E}}}{RT}\right)_i^{\text{Model}} - \left(\dfrac{\underline{G}^{\text{E}}}{RT}\right)_i^{\text{Expt.}}}{\left(\dfrac{\underline{G}^{\text{E}}}{RT}\right)_i^{\text{Expt.}}}\right]^2$$

The best fit values for Λ_{12} and Λ_{21} at 337.15 are presented in Table 12-3.

TABLE 12-3 Wilson equation parameters for the methanol (1) + dimethyl carbonate (2) system.

Model	T (K)	Λ_{12}	Λ_{21}	OBJ
Wilson Equation	337.35	0.9181	0.2671	0.027
Wilson Equation (predicted)	428.15	1.0922	0.3024	
Wilson Equation	428.15	1.4255	0.2592	0.018

> **PITFALL PREVENTION**
>
> Remember, we don't include the pure component endpoints when minimizing OBJ.

Once the model parameters are obtained, the pressure and vapor phase mole fraction can be determined from modified Raoult's Law using Equations 11.88 and 11.89, with activity coefficients values obtained from Equations 11.95 and 11.96. That the fit to the experimental data is good is evidenced by Figure 12-1 and, thus, the model (Wilson equation) and the assumption of an ideal gas are reasonable at this pressure.

In Example 11.5, we used the Wilson equation parameters at one temperature to predict the phase behavior at a different temperature. We can do this here (and assume an ideal gas for the vapor) to demonstrate the impact of violating the assumption of an ideal gas.

Step 3 *Modeling the system at 428.15 K using modified Raoult's Law*
When using the temperature-dependent feature of the Wilson equation, we need to know the molar volumes for the pure components, and these are found from Appendix C.

$\underline{V}_1 = 40.73 \text{ cm}^3/\text{mol}$
$\underline{V}_2 = 84.82 \text{ cm}^3/\text{mol}$

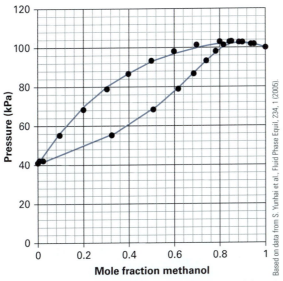

FIGURE 12-1 A *Pxy* diagram of the methanol (1) and
dimethyl carbonate (2) system at 337.35 K. The symbols
are experimental data and the curve is the correlation
from modified Raoult's Law using the Wilson equation.

With these molar volumes, we can solve for the temperature-independent parameters
α_{12} and α_{21} from Equations 11.97 and 11.98. Using those solved values for α_{12} and α_{21} we
can calculate the new Wilson parameters for $T = 428.15$ K, and these are provided in
Table 12-3 (listed as 'predicted').

If we assume ideal gas for the vapor, we can now predict the bubble-point and dew-
point curves for this mixture at 428.15 K, and this is provided in the Figure 12-2. The dashed

We use the liquid
molar volumes from
Appendix C, which
are nominally at
298.15 K. However,
if you can obtain the
liquid molar volumes
at the temperature
of interest (here,
428.15 K), always use
those instead.

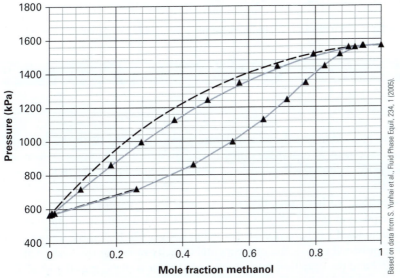

FIGURE 12-2 A *Pxy* diagram of the methanol (1) and dimethyl carbonate
(2) system at 428.15 K. The symbols are experimental data. The dashed curve
is the prediction for the Wilson equation assuming an ideal gas for the vapor.
The solid curve is the correlation using the Wilson equation for the liquid
and the virial equation for the vapor phase. Both approaches provide, basically,
the same result for the dew point pressure curve, so they overlap in the figure.

black lines are the bubble-point and dew-point predictions, while the filled-in triangles are the experimental data. As can be seen, the bubble-point pressure is over-predicted.

Step 4 *Modeling the system at 428.15 K using the gamma-phi approach*

At this point, one might want to argue that the fit using the Wilson parameters at 337.35 K might not work well when extrapolated to 428.15 K. Thus, fitting the parameters at 428.15 K will correlate the data better. However, when we look at the expression to obtain the *experimental* activity coefficients, we have to incorporate the vapor phase non-idealities.

For example, consider the gamma-phi expression

$$x_i \gamma_i P_i^{sat} \exp\left\{\frac{V_i(P - P_i^{sat})}{RT}\right\} = y_i P\left(\frac{\hat{\varphi}_i}{\varphi_i^{sat}}\right) \tag{12.23}$$

Rewritten for the activity coefficient, this expression becomes

$$\gamma_i = \frac{y_i P \hat{\varphi}_i}{x_i P_i^{sat} \varphi_i^{sat} \exp\left\{\dfrac{V_i(P - P_i^{sat})}{RT}\right\}} \tag{12.33}$$

If you read a journal article that reports vapor-liquid equilibria for a binary system, you will often see the authors report experimental activity coefficients. Normally, regardless of the pressure, they will include fugacity coefficient models and the Poynting correction to provide estimates of the most accurate value.

Thus, a model for the vapor phase is needed in order to determine the necessary fugacity coefficients. This is where the vapor phase model (the virial equation) will be employed. Note that for the system pressure and the vapor phase composition, we use the *experimental values* as provided in Table 12-1 above, since we are determining an *experimental value* for the activity coefficient. Equations 12.28, 12.31, and 12.32 can be used to find the necessary fugacity coefficients, both pure component and in the mixture, if the second-virial coefficients are known. The pure-component second-virial coefficients are obtained from the critical temperature, critical pressure, and acentric factor through Equations 7.80 through 7.82. In order to determine the cross-term (B'_{12}), we can use a geometric mean approach (Equation 12.25). The virial coefficients at 428.15K are provided in Table 12-4.

TABLE 12-4 Second-virial coefficients for the methanol (1) + dimethyl carbonate (2) system at 428.15 K.

B'_{11} (cm³/mol)	B'_{22} (cm³/mol)	B'_{12} (cm³/mol)
−320.29	−663.46	−460.98

Also, the pure component molar volumes previously used (with the assumption that they are independent of temperature) can be used for the Poynting correction terms in Equation 12.33; thus, the experimental activity coefficients at 428.15 K can be given in Table 12-5.

Note that we have provided the *ratios* of the experimental fugacity coefficients (mixture:pure) in Table 12-5 as well to show that they deviate from unity; thus, you cannot assume ideal gas behavior at this temperature (428.15 K). By way of comparison, we have provided those same experimental fugacity ratios for this system at 337.35 K in Table 12-2 and the deviation from ideal gas behavior is slight; thus the ideal gas is a reasonable assumption at this lower temperature.

Finally, to correlate the experimental data with the gamma model (Wilson equation) and the phi model (virial equation), it is not as simple as in modified Raoult's Law. Here, the pressure cannot be isolated on the left-hand side of the equation since it is a variable in determining the contribution of the fugacity coefficients. The situation is the same for the vapor phase composition, which is also a variable in determining the

FOOD FOR THOUGHT 12-2

What if you didn't want to include the fugacity coefficient into this analysis (pure or mixture), but wanted to include the Poynting correction? Would this make sense?

TABLE 12-5 Activity coefficients and fugacity coefficient ratios for the methanol (1) + dimethyl carbonate (2) system at 428.15 K.

x_1	γ_1	γ_2	$\dfrac{\hat{\varphi}_1}{\varphi_1^{sat}}$	$\dfrac{\hat{\varphi}_2}{\varphi_2^{sat}}$
0		1	1.1049	1
0.0062	1.2531	1.0027	1.1037	0.9981
0.093	1.4135	1.0026	1.0865	0.9728
0.1838	1.3993	0.9971	1.0705	0.9490
0.2758	1.3513	1.0016	1.0562	0.9279
0.3757	1.2922	1.0193	1.0428	0.9080
0.4759	1.2360	1.0500	1.0310	0.8909
0.5716	1.1861	1.0953	1.0213	0.8769
0.6855	1.1279	1.1890	1.0116	0.8636
0.7935	1.0716	1.3699	1.0049	0.8548
0.9001	1.0171	1.8257	1.0010	0.8508
0.9426	1.0020	2.1731	1.0003	0.8509
1	1		1	0.8530

mixture fugacity coefficient. From a solution standpoint, we could employ an iterative approach (which we demonstrate in this problem) or a root-finding approach (which we demonstrate in Section 12.4). Specifically, if we return to our gamma-phi approach, our equation is

$$x_i \gamma_i P_i^{sat} \exp\left\{\frac{V_i(P - P_i^{sat})}{RT}\right\} = y_i P\left(\frac{\hat{\varphi}_i}{\varphi_i^{sat}}\right) \tag{12.23}$$

We isolate the system pressure and vapor-phase mole fractions and write this for a binary mixture:

$$\frac{x_1 \gamma_1 P_1^{sat}}{\hat{\varphi}_1/\varphi_1^{sat}} \exp\left\{\frac{V_1(P - P_1^{sat})}{RT}\right\} = y_1 P \tag{12.34}$$

$$\frac{x_2 \gamma_2 P_2^{sat}}{\hat{\varphi}_2/\varphi_2^{sat}} \exp\left\{\frac{V_2(P - P_2^{sat})}{RT}\right\} = y_2 P \tag{12.35}$$

Summing this as before, we find

$$\frac{x_1 \gamma_1 P_1^{sat}}{\hat{\varphi}_1/\varphi_1^{sat}} \exp\left\{\frac{V_1(P - P_1^{sat})}{RT}\right\} + \frac{x_2 \gamma_2 P_2^{sat}}{\hat{\varphi}_2/\varphi_2^{sat}} \exp\left\{\frac{V_2(P - P_2^{sat})}{RT}\right\} = (y_1 + y_2)P \tag{12.36}$$

Since $y_1 + y_2 = 1$, we have

$$P = \frac{x_1 \gamma_1 P_1^{sat}}{\hat{\varphi}_1/\varphi_1^{sat}} \exp\left\{\frac{V_1(P - P_1^{sat})}{RT}\right\} + \frac{x_2 \gamma_2 P_2^{sat}}{\hat{\varphi}_2/\varphi_2^{sat}} \exp\left\{\frac{V_2(P - P_2^{sat})}{RT}\right\} \tag{12.37}$$

Likewise, y_1 is determined from Equation 12.34 as

$$y_1 = \frac{x_1 \gamma_1 P_1^{\text{sat}}}{P\hat{\varphi}_1/\varphi_1^{\text{sat}}} \exp\left\{\frac{\underline{V}_1(P - P_1^{\text{sat}})}{RT}\right\} \tag{12.38}$$

The pressure (P) appears on both sides of Equation 12.37 (it is being solved for on the left-hand side and appears on the right-hand side explicitly in the Poynting correction and as an independent variable in calculating fugacity coefficient). Thus, one solution approach is to converge on a value of the pressure that makes equation 12.37 true (i.e.: the left-hand side equal to the right-hand side).

> The Gibbs phase rule says we can specify only two intensive variables for our two-component, two-phase system. Here we specify T and x_1, while solving for P and y_1 from our modeling approach.

While there are a few approaches that will work, one common algorithm is given here.

Step 1 For the experimental activity coefficients in Table 12-5, find the best fit parameters for the excess molar Gibbs free energy model (minimizing an objective function).

Step 2 Set $\hat{\varphi}_1$, φ_1^{sat}, $\hat{\varphi}_2$, and φ_2^{sat} all equal to 1.0.

Step 3 Solve for P (using Equation 12.37) and y_1 (using Equation 12.38) at each x_1 value. Note that y_2 is obtained by $y_2 = 1 - y_1$.

Step 4 With the values of P, y_1, and y_2 from step 3, solve for $\hat{\varphi}_1$, φ_1^{sat}, $\hat{\varphi}_2$, and φ_2^{sat} at each x_1 value via Equations 12.28, 12.31, and 12.32.

Step 5 Using the solved values for $\hat{\varphi}_1$, φ_1^{sat}, $\hat{\varphi}_2$ and φ_2^{sat} from step 4, solve for P, y_1, and y_2 at each x_1 value.

Step 6 Sum up the absolute value of the differences in P from step 5 to that of P from step 3. If those differences are below a certain threshold value (user defined), the pressures have converged. Print out those pressures and vapor phase mole fractions. However, if the differences are above a threshold, you have not converged. Go back to step 4 to get new values for $\hat{\varphi}_1$, φ_1^{sat}, $\hat{\varphi}_2$, and φ_2^{sat} using the values of P, y_1 and y_2 from step 5.

This algorithm should converge quite quickly and is short work for a small computer code, though it would still be workable (though longer) in Microsoft Excel. The threshold value mentioned in step 6 should be a small number that is a fraction of a percent of the representative system pressure. Additionally, and as alluded to earlier this section as "the root-finding approach," one can treat the pair of Equations (12.37 and 12.38) as a system of two non-linear equations in two unknowns (P and y_1). Once again, the solver function in Excel or a small computer code (this time linked to a non-linear equation solver) will provide a solution at each value of x_1.

The algorithm presented is only for a bubble-point pressure calculation, as we specified the temperature and liquid phase mole fraction in the modeling of this problem. Other similar algorithms exist for the other types of calculations required (depending on the variables that are specified): bubble-point temperature, dew-point pressure, and dew-point temperature. These are not provided here, but we refer the user to another text for those algorithms, if needed (Smith et al., 2006).

Returning to the specifics associated with the implementation of the algorithm to our problem, this approach yielded Wilson parameters provided in the last row of Table 12-3. Note that these values are different than those that were extrapolated to 428.15 K from the values at 337.35 K. The output of the algorithm, once converged, provides the model pressure and vapor-phase mole fractions. Therefore, we can now plot the bubble and dew point curves from this approach at 428.15 K, and this is given as the solid blue line in Figure 12-2. As can be seen, the incorporation of the vapor phase non-idealities into the modeling approach results in a much better match with the experimental data.

This is to be expected, as the system pressure was high enough to provide ratios of fugacity coefficients that deviated by almost 15% from ideal gas behavior, as seen in Table 12-5.

12.3 Phi-Phi Modeling

In Chapter 11 and in this chapter (up to this point), we have modeled vapor–liquid equilibrium in mixtures using several approaches. For those mixtures made up of similar compounds at low pressure, we invoked Raoult's Law. When the system components were not similar and the system was at low pressure, we used modified Raoult's Law to model the phase behavior. For modified Raoult's Law, we employed excess molar Gibbs free energy models to generate activity coefficients for the components in the liquid phase. We even showed how to predict the activity coefficients when experimental data was not available (such as through temperature extrapolation using the Wilson equation or in the predictive excess molar Gibbs free energy models, such as the van Laar approach). Finally, we introduced what was called "gamma-phi" modeling for systems whose components are not similar and the pressure was not low (i.e. we could not use the ideal gas assumption for the vapor).

In this section, we move away from the gamma-phi modeling approach for the vapor–liquid equilibrium of mixtures and introduce an approach called "phi-phi" modeling. Here, rather than use an equation of state only for the vapor phase (which requires the mixture fugacity coefficients), we use an equation of state for *both* the liquid AND vapor phases. As we have learned in Chapter 7, some equations of state describe both the liquid and vapor phases in one model (such as the van der Waals or Peng-Robinson equation). Therefore, we will use those for modeling our mixture phase behavior.

FOOD FOR THOUGHT 12-3

Can you think of an equation of state that does *not* describe the liquid and vapor phases in one model?

12.3.1 Equality of Mixture Fugacities

We start where we always start when modeling phase equilibrium: equality of mixture fugacities of the components in the phases in which they appear. Written for vapor-liquid equilibrium, this is

$$\hat{f}_i^{\mathrm{L}}(T,P,x) = \hat{f}_i^{\mathrm{V}}(T,P,y) \qquad (11.56)$$

In what we have done in this chapter and the previous chapter, the next step was to write the mixture fugacity for the liquid in terms of the activity coefficient and the mixture fugacity in the vapor in terms of the mixture fugacity coefficient. However, now we will write *both* phases in terms of the mixture fugacity coefficient:

$$\hat{f}_i^{\mathrm{L}}(T,P,x) = x_i P\hat{\varphi}_i^{\mathrm{L}} \qquad (12.39)$$

$$\hat{f}_i^{\mathrm{V}}(T,P,y) = y_i P\hat{\varphi}_i^{\mathrm{V}} \qquad (12.40)$$

Inserting these two expressions into Equation 11.56 yields

$$x_i P\hat{\varphi}_i^{\mathrm{L}} = y_i P\hat{\varphi}_i^{\mathrm{V}} \qquad (12.41)$$

Since we know the pressure of each phase is the same at equilibrium, we can cancel those out in Equation 12.41 to yield

$$x_i \hat{\varphi}_i^{\mathrm{L}} = y_i \hat{\varphi}_i^{\mathrm{V}} \qquad (12.42)$$

This relationship is not a new result. It is Equation 11.55 from Section 11.3, now written for the liquid and vapor phases. Equation 12.42 requires a model to describe the

You might have wondered about modeling the liquid *and* the vapor phases through an activity coefficient model (gamma-gamma). While there is nothing fundamentally wrong with that approach, real gases are closer to ideal gases than they are to ideal solutions.

liquid and vapor phase fugacity coefficients of the components in the mixture. This is where we utilize an equation of state that has the ability to describe both phases simultaneously, such as a cubic equation of state.

12.3.2 Mixture Fugacity Coefficients when Pressure Is a Dependent Variable

A general expression for the mixture fugacity coefficient of component i was given previously as

$$\ln[\hat{\varphi}_i] = \frac{1}{RT} \int_{P=0}^{P=P} \left[\overline{V}_i - \frac{RT}{P} \right] dP \tag{12.22}$$

Recall that we used this expression to obtain the mixture fugacity coefficient of component i for the virial equation of state. However, for equations of state that are written in the form of $P = P\,(T, \underline{V})$, we need to transform Equation 12.22 into a form where molar volume is the integration variable. This is done through the use of the triple product rule (Equation 6.25) relating the three variables P, $\underline{V}$, and n_i.

The resulting equation is

$$\ln[\hat{\varphi}_i] = \frac{1}{RT} \int_{\underline{V}=\infty}^{\underline{V}=\underline{V}} \left[\frac{RT}{\underline{V}} - N\left(\frac{\partial P}{\partial n_i}\right)_{T,V,n_{j(j\neq i)}} \right] d\underline{V} - \ln(Z) \tag{12.43}$$

Expressions for the fugacity coefficients from various pressure-explicit equations of state are readily evaluated from Equation 12.43.

In the following example problems, we explore the application of phi-phi modeling in a variety of contexts using different cubic equations of state (requiring evaluation of Equation 12.43).

EXAMPLE 12-2	PREDICTING VAPOR–LIQUID EQUILIBRIUM OF MIXTURES USING EQUATIONS OF STATE

H—C—C—C—H (with H atoms above and below)

Propane

H—C—C—C—C—H (with H atoms above and below)

n-butane

You are a student doing a project where you are evaluating process options in the separation of a propane (1) + n-butane (2) system at 323.15 K. There are choices of equations of state that you can use and you need to make a selection. Prior to this, you decide to predict the phase behavior yourself using two of the choices: the van der Waals equation of state and the Peng-Robinson equation of state. Provide the Pxy diagram for this system using both models.

SOLUTION:

Step 1 *Gibbs phase rule and a degree of freedom analysis*

For our two-component, two-phase system, we can specify two intensive variables at equilibrium. While we have a choice, we already know that our cubic equations of state in the problem (van der Waals—VDW; Peng-Robinson—PR) have pressure as a dependent variable. Thus, we can choose temperature and the liquid phase mole fraction (x_1) as our independent variables to specify.

The four equations for our phase equilibrium problem are

$$x_1 \hat{\varphi}_1^L = y_1 \hat{\varphi}_1^V \tag{12.44}$$

$$(1 - x_1)\hat{\varphi}_2^L = (1 - y_1)\hat{\varphi}_2^V \tag{12.45}$$

$$T^L = T^V \tag{12.46}$$

$$P^L = P^V \tag{12.47}$$

Since we are already specifying the temperature (per Gibbs phase rule), Equation 12.46 is not a relationship amongst unknown variables. Thus, we have only three independent equations. How many unknowns do we have? To answer that question, we need to look at our expressions. The first thing we need is our equation of state (it is directly used in Equations 12.47 above and will be used to determine the fugacity coefficients in Equations 12.44 and 12.45, via Equation 12.43).

We show both the van der Waals equation of state and the Peng-Robinson equation of state here in the first part of this problem. We write this for the liquid and vapor phases as

Van der Waals (VDW):

$$P^L = \frac{RT}{\underline{V}^L - b_m^L} - \frac{a_m^L}{(\underline{V}^L)^2} \tag{12.48}$$

$$P^V = \frac{RT}{\underline{V}^V - b_m^V} - \frac{a_m^V}{(\underline{V}^V)^2} \tag{12.49}$$

Peng-Robinson (PR):

$$P^L = \frac{RT}{\underline{V}^L - b_m^L} - \frac{a_m^L}{\underline{V}^L(\underline{V}^L + b_m^L) + b_m^L(\underline{V}^L - b_m^L)} \tag{12.50}$$

$$P^V = \frac{RT}{\underline{V}^V - b_m^V} - \frac{a_m^V}{\underline{V}^V(\underline{V}^V + b_m^V) + b_m^V(\underline{V}^V - b_m^V)} \tag{12.51}$$

We have been very explicit with the superscripts to clearly denote whether the equation is for the liquid or vapor phase, as well as the subscripts to emphasize that the parameters are "mixture parameters." Since we are specifying the temperature, we do not put a superscript on this variable since it is the same in both phases.

The molar volumes of both phases, $\underline{V}^L$ and $\underline{V}^V$, are unknown, just like before when we worked with pure component vapor–liquid equilibrium and equations of state. Thus, we have already identified two of our unknowns.

What about the parameters, b_m^L, b_m^V, a_m^L, and a_m^V? When working with pure components, we could specify those parameters through information known about the critical point (T_c and P_c) for the pure component (as well as the temperature and acentric factor, needed for the Peng-Robinson equation of state). But what about the mixture parameters? This was introduced previously. The first time was when we used mixture parameters for the van der Waals equation using the van Laar approach to obtain activity coefficients in Chapter 11. The second time was when we needed a mixing rule for the second virial coefficient during our gamma-phi problem (Example 12-1).

In determining the equation of state parameters for the mixture, which T_c and P_c do we use? Those for propane or those for butane? Your intuition is probably telling you "both" somehow, and you would be correct. There is also the fact that even if you were to use both components in calculating the mixture parameters (a_m and b_m), how do you account for the *composition* of the mixture, since the mixture parameters for a 99% ethanol + 1% water system should be a different value than for a 1% ethanol + 99% water system? If this sounds like an interesting and complex problem, you would be on target. The subject of how to go from a pure component parameter in an equation of state to a mixture parameter has garnered much research in the thermodynamics community over the decades. It is an unresolved issue and many strategies have been devised, some simple and some quite complex. That it is still an active area of research underscores the fact that the last word on equation of state mixing rules has not yet been written. The user, however, is warned that when you choose an equation of state to use in a process simulator for evaluating a process involving a mixture, your job is not complete. You must also identify the mixing rules that you would like to use to estimate the properties of your mixture.

As we have mentioned and used previously, the most typical and popular mixing rules are of the van der Waals one-fluid type. We use that here because of its simplicity and ubiquity. We repeat Equations 11.123 and 11.124 here.

$$a_m = x_1 x_1 a_{11} + x_1 x_2 a_{12} + x_2 x_1 a_{21} + x_2 x_2 a_{22} \tag{11.123}$$

$$b_m = x_1 x_1 b_{11} + x_1 x_2 b_{12} + x_2 x_1 b_{21} + x_2 x_2 b_{22} \tag{11.124}$$

Chances are the instructor teaching this course or the teaching assistant/ grader helping with this course is doing/ has done research in thermodynamics. You might ask that person about his/her research in thermodynamics to give you an idea as to how that "cutting edge" relates to the fundamentals you are learning in this class.

While we have used x to denote mole fraction (regardless of phase), we know that we will have different mixture parameters (depending on the phase) because of the difference in composition between the phases.

Once again, the cross-parameters are defined for a using a geometric mean combining rule:

$$a_{12} = a_{21} = \sqrt{a_{11} a_{22}} \tag{11.125}$$

while for b, the cross-parameter is defined using an arithmetic mean combining rule:

$$b_{12} = b_{21} = \left(\frac{b_{11} + b_{22}}{2} \right) \tag{11.126}$$

And, as mentioned previously, if you plug Equation 11.126 into Equation 11.124, the mixing rule for b will simplify to

$$b_m = x_1 b_{11} + x_2 b_{22} \tag{11.127}$$

Recall that when the subscripts are the same, we are referencing the pure component. Thus, the mixing rules with the combining rules identified in Equations 11.124 and 11.128 reduce to

Remember, $a_{11} = a_1$ and $b_{11} = b_1$.

$$a_m = x_1^2 a_1 + x_2^2 a_2 + 2 x_1 x_2 \sqrt{a_1 a_2} \tag{12.52}$$

$$b_m = x_1 b_1 + x_2 b_2 \tag{12.53}$$

Returning back to the task at hand (identifying unknowns in our degree of freedom analysis), we have

$$a_m^L = x_1^2 a_1 + x_2^2 a_2 + 2 x_1 x_2 \sqrt{a_1 a_2} \tag{12.54}$$

$$b_m^L = x_1 b_1 + x_2 b_2 \tag{12.55}$$

$$a_m^V = y_1^2 a_1 + y_2^2 a_2 + 2 y_1 y_2 \sqrt{a_1 a_2} \tag{12.56}$$

$$b_m^V = y_1 b_1 + y_2 b_2 \tag{12.57}$$

Remember, $x_2 = 1 - x_1$ and $y_2 = 1 - y_1$; thus x_2 and y_2 are not new unknowns.

Since the pure-component parameters (a_1, b_1, a_2, and b_2) are a function of their critical point (and ω and T for PR) and considering we are also fixing x_1, these mixing rules introduce only one additional unknown into our system: y_1. Thus, overall, we have three equations and three unknowns (the molar volume of the liquid, the molar volume of the vapor, and the vapor phase composition for component 1). Since the determination of the expressions for the mixture fugacity coefficient involve just mathematical applications and does not introduce new variables (as we will show next), we can solve this problem as it contains the same number of unknowns as equations that relate the unknowns.

Step 2 *Mixture fugacity coefficients for the van der Waals and Peng-Robinson equations of state*

The next step is to plug in Equation 12.48 (or 12.49) and Equation 12.50 (or 12.51) into Equation 12.43 to obtain an expression for the mixture fugacity coefficient of species i of the equation of state of interest (VDW and PR). The result is, of course, dependent on both the mixing rule and the combining rule used. The results are as follows.

VDW (Mixture):

$$\ln(\hat{\varphi}_i) = \ln\left(\frac{\underline{V}}{\underline{V} - b_m}\right) + \left(\frac{b_i}{\underline{V} - b_m}\right) - \frac{2}{\underline{V}RT}(x_1 a_{i1} + x_2 a_{i2}) - \ln(Z_m) \qquad (12.58)$$

VDW (Pure):

$$\ln\left(\frac{f}{P}\right) = \ln(\varphi) = (Z - 1) - \ln Z + \ln\left(\frac{\underline{V}}{\underline{V} - b}\right) - \left(\frac{a}{\underline{V}RT}\right) \qquad (8.66)$$

PR (Mixture):

$$\ln[\hat{\varphi}_i] = \frac{b_i}{b_m}(Z_m - 1) - \ln\left(Z_m - \frac{b_m P}{RT}\right) - \frac{a_m}{2\sqrt{2}\,b_m RT}$$

$$\left[\frac{2(x_1 a_{i1} + x_2 a_{i2})}{a_m} - \frac{b_i}{b_m}\right] \ln\left[\frac{Z_m + (1 + \sqrt{2})\dfrac{b_m P}{RT}}{Z_m + (1 - \sqrt{2})\dfrac{b_m P}{RT}}\right] \qquad (12.59)$$

PR (Pure):

$$\ln\left(\frac{f}{P}\right) = \ln(\varphi) = (Z - 1) - \ln(Z - B) - \frac{A}{2B\sqrt{2}} \times \ln\left[\frac{Z + (1 + \sqrt{2})B}{Z + (1 - \sqrt{2})B}\right] \qquad (8.67)$$

Here, $Z_m = P\underline{V}/(RT)$.

Note that we provide the pure component fugacity coefficients, though not needed in this problem, for perspective so that the reader can see how these expressions change when going from a pure component to the mixture (and how the mixture expressions reduce to the pure component expressions as the mole fraction of the component goes to 1).

Step 3 *Predicting the phase behavior*

Recall that we have three equations (Equations 12.44, 12.45, and 12.47) and three unknowns: the liquid molar volume ($\underline{V}^L$), the vapor molar volume ($\underline{V}^V$), and the vapor-phase mole fraction (y_1). We can solve these three equations and three unknowns in a variety of ways. As mentioned before, Microsoft Excel has a "solver" function that allows you to cast one of your equations as an objective function to be "solved," while the other two equations become constraints. The three variables are the three that need to be changed while using the root-finding algorithm inside of Microsoft Excel. Of course, the user can solve these equations with their own source code and a call to a root-finding algorithm. Another approach that can be used to "solve" these three equations and three unknowns is to treat the root-finding problem as a minimization problem. For example, when we fix temperature and liquid phase mole fraction (a bubble-point pressure calculation), our equations are 12.44, 12.45, and 12.47. Instead of finding the roots of the equations, we can minimize the absolute value (or squared value) of the difference between the left- and right-hand

PITFALL PREVENTION

Equations 12.58 and 12.59 have pressure on the right-hand side (either explicit or through Z_m). This is just a way to reduce the size of the equation written in the text. While doing the actual calculation, you will insert the equation of state written explicitly for P (Equations 12.48–12.51).

sides for each of the three equations, while varying the independent variables (here, once again, $\underline{V}^L$, $\underline{V}^v$ and y_1). As we make the differences in each of the individual equations (12.44, 12.45, and 12.47) as small as possible, we approach the roots of the set of equations. In some situations, this approach will provide results almost exactly the same as treating the system as a root-finding problem, though it can be less useful in other situations. It may also be used to provide reasonable initial guesses for the actual roots when initiating a root-finding algorithm.

For the van der Waals equation, we need the critical temperature and critical pressure of our pure components to obtain the pure component a and b parameters (a_1, b_1, a_2, and b_2), while we require the acentric factor for the Peng-Robinson equation of state. This pure-component data is readily accessible in Appendix C. From this information, we can predict the Pxy behavior from both equations of state and present the results in Figure 12-3.

Two things are clear from the examination of Figure 12-3. First, the Peng-Robinson equation of state predicts the phase behavior of this system very well. Second, the van der Waals equation of state predicts the vapor pressure of the pure components very poorly (the system pressure at $x_1 = 0$ and $x_1 = 1$). Therefore, it has no chance to accurately predict the mixture behavior.

> While the pressure isn't one of the three unknown variables in the problem, once we solve for the system molar volume (either liquid or vapor) at the temperature and liquid phase mole fraction of interest, we plug the molar volume back into the equation of state to obtain the pressure. We can use either molar volume at this point since both must give us the same pressure!

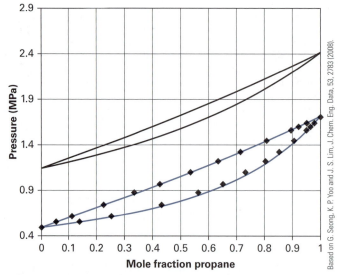

FIGURE 12-3 A Pxy diagram of the propane (1) and n-butane (2) system at 323.15 K. The symbols are experimental data. The blue curve is the prediction from the Peng-Robinson equation of state, while the black curve is the prediction from the van der Waals equation of state.

Based on G. Seong, K. P. Yoo and J. S. Lim, J. Chem. Eng. Data, 53, 2783 (2008).

The issue of the van der Waals equation of state not predicting accurate vapor pressures for n-alkanes when solving the phase equilibrium problem is a known problem, and we already encountered this in Figure 8-6. The Peng-Robinson equation of state has a temperature-dependent a parameter, while the van der Waals equation of state a parameter is constant. Additionally, the Peng-Robinson equation of state uses an additional fitting parameter (the acentric factor, ω) that improves the correlative ability of the model (relative to van der Waals, which does not use ω).

PREDICTING THE BUBBLE-POINT PRESSURE AND VAPOR PHASE COMPOSITION USING THE PENG-ROBINSON EQUATION OF STATE	EXAMPLE 12-3

A single vapor–liquid equilibrium point for the water (1) + ethanol (2) system is experimentally measured at 303.15 K. The experiment provides the following information: $x_1 = 0.30$, $y_1 = 0.23$, and $P = 10.1$ kPa.

Use this information to estimate the system pressure and vapor-phase mole fraction when $x_1 = 0.8$ using the Peng-Robinson equation of state.

SOLUTION:

Step 1 *Gibbs phase rule and a degree of freedom analysis*
Like in Example 12-2, we have a two-phase, two-component system, so we can specify two intensive variables at equilibrium (according to the Gibbs phase rule). From the problem statement, we are fixing the liquid phase composition and the temperature in order to determine the system pressure and vapor-phase mole fraction.

Step 2 *Peng-Robinson pressure and mixture fugacity coefficient*
As in Example 12-2, we have four equations and four unknowns. However, since we are fixing temperature, this reduces to three equations in three unknowns, as in Example 12-2. We use Equations 12.50, 12.51 and 12.59 to provide the needed pressure and fugacity coefficient expressions for the Peng-Robinson equation of state.

Step 3 *Predicting the phase behavior*
We need the pure-component information to determine the pure-component equation of state parameters in order to obtain the mixture parameters. This is readily available in Appendix C.

With this information, we can calculate the mixture parameters for the PR equation of state. Finally, we are able to solve our three equations (12.44, 12.45 and 12.47) for our three unknowns at $x_1 = 0.3$ and $T = 303.15$ K. Doing this, we arrive at the following result to the problem of interest: $\underline{V}^L = 50.53$ cm³/mol; $\underline{V}^V = 184200$ cm³/mol; $y_1 = 0.40$. This yields a system pressure of 13.6 kPa. Recall that the experimental data for the system was $P = 10.1$ kPa and $y_1 = 0.23$ (per Example 11-3).

Note that the approach in this problem has been purely predictive at this point in that we only used pure-component information (T_C, P_C, and ω) to predict mixture behavior. We can contrast this with the approach we used in Example 11-3 to solve this problem, specifically our use of a modified Raoult's Law approach (1-parameter Margules equation for the liquid phase and an ideal gas for the vapor). Recall for the 1-parameter Margules equation, we *fit* the known experimental data for the *mixture* to obtain our Margules parameter (A). At this point you might ask "*is there a way we can also use known experimental data for the mixture to aid our phi-phi approach?*" We describe that in the next step.

Step 4 *Correlating the phase behavior*
Recall the current situation. Our phi-phi modeling approach using an equation of state (PR) for both phases has resulted in a prediction of the bubble-point pressure and vapor-phase composition that is much different than the experimental value. How can we adjust our model to better correlate the known experimental data? The traditional route is to introduce a **binary interaction parameter** (k_{12}) on one or more of the cross-parameters. This is typically done for the cross-parameter for a as

$$a_{12} = \sqrt{a_1 a_2}\,(1 - k_{12}) = a_{21} \qquad (12.60)$$

Here, k_{12} is a parameter that is adjusted to best match the experimental data for the mixture. When $k_{12} = 0$, the model is purely predictive, and we end up with the result from step 3. We show a few values of k_{12} and the resulting bubble-point pressures and vapor phase compositions, relative to the experimental value, in Table 12-6.

PITFALL PREVENTION

Some binary interaction parameters are defined differently than described in Equation 12.60. Be sensitive to how they are defined if using a published binary interaction parameter for a model.

TABLE 12-6 Correlating the binary interaction parameter for the water (1) + ethanol (2) system.

	P(kPa)	y_1		
Expt.	10.1	0.23		
k_{12}	P(kPa)	y_1	V^L(cm³/mol)	V^V(cm³/mol)
0.0	13.6	0.40	50.53	184200
0.1	22.3	0.61	50.87	112500
−0.1	9.9	0.22	50.22	254500
−0.09	10.1	0.24	50.25	248000

It is clear that a negative value for the k_{12} parameter results in values that are closer to experiment. We settle on a value of $k_{12} = -0.09$, which is the value for the binary interaction parameter that best matches both the experimental system pressure and the vapor-phase mole fraction. Do note that the k_{12} value is both a function of composition and temperature, and its non-zero value for our system indicates a modeling shortcoming of the Peng-Robinson approach with the mixing rules as defined.

Step 5 *Predicting the system behavior using the fitted k_{12} value*
The final step in the problem is to use the k_{12} value from $x_1 = 0.3$ to help predict the phase behavior at $x_1 = 0.8$. When we solve the phase equilibrium problem at $x_1 = 0.8$ and $T = 30°C$ (with $k_{12} = -0.09$) we arrive at a value given in Table 12-7.

TABLE 12-7 Prediction of the phase behavior for the water (1) + ethanol (2) system at $T = 30°C$ and $x_1 = 0.8$.

	P(kPa)	y_1		
Expt.	8.2	0.44		
k_{12}	P(kPa)	y_1	V^L(cm³/mol)	V^V(cm³/mol)
−0.09	9.3	0.38	29.52	269500
0	21.1	0.20	29.69	118900

Note that the value for the pressure and vapor-phase composition using $k_{12} = -0.09$, while not a perfect match to experiment, is a much better estimate than if we predicted the property at this composition (i.e. used $k_{12} = 0$), as is also indicated in Table 12-7. By way of comparison, the predicted value at this composition from the excess molar Gibbs free energy modeling approach (as in Example 11-3) is $P = 7.69$ kPa and $y_1 = 0.46$. Thus, both approaches (the Peng-Robinson EOS and the 1-parameter Margules equation) are reasonably accurate.

In the next problem, we revisit Example 11-5, where we produced the entire phase equilibrium curve at one temperature given experimental data at another temperature. As in the previous problem, it gives us an opportunity to compare this approach with that using an excess molar Gibbs free energy model.

MIXTURE PREDICTIONS USING THE PENG-ROBINSON EQUATION OF STATE	EXAMPLE 12-4

You are part of a team evaluating the separation of a mixture of ethanol (1) + butyl methyl ether (2). This was a project your company explored a few years earlier. In that project, team members measured the VLE for this system at 323.15 K. However, based on some additional operating constraints, you now need the VLE for this system at 338.15 K to provide an estimate on the size of the distillation column required. Rather than perform the experiment again at 338.15 K, you decide to use the data at 323.15 K to estimate the VLE at 338.15 K.

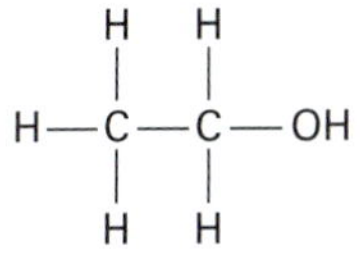

The data at 323.15 K for this system is given in Table 12-8.

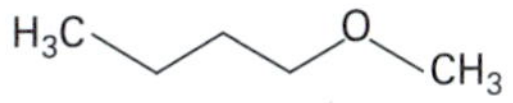

TABLE 12-8 Experimental Pxy data for the ethanol (1) + butyl methyl ether (2) system at 323.15 K.

x_1	P (kPa)	y_1
0	50.69	0
0.0316	52.37	0.0592
0.1067	53.95	0.1597
0.1899	55.83	0.2426
0.3393	56.73	0.2990
0.4303	56.10	0.3317
0.6225	53.75	0.4030
0.6649	52.74	0.4334
0.7352	50.52	0.4902
0.8124	47.49	0.5284
0.8525	45.52	0.6058
0.9091	41.41	0.6677
0.9542	36.75	0.7751
0.9820	32.49	0.8956
1	29.34	1

Based on data from T. Hofman, A. Sporzynski and A. Goldon, J. Chem. Eng. Data, 45, 169 (2000).

Ethanol

Butyl methyl ether

SOLUTION:

Step 1 *Make a plan*
The approach is similar to the previous problem. We will use the Peng-Robinson equation of state to model the system, but we will fit a binary interaction parameter for use over the entire data set. Once we determine this parameter value, we will use that value to predict the behavior at a new temperature.

Step 2 *Pure-component values*
We use the VDW 1-fluid mixing rule again for the mixture parameters for the PR equation of state. This, in turn, requires knowledge of the critical temperature, pressure, and acentric factor to obtain the pure component parameters. This data is provided in Appendix C.

Step 3 *Determining the best-fit k_{12} value at 323.15 K*
In Example 11-5, we created an objective function for the excess molar Gibbs free energy and fit the experimental data to the Wilson equation, minimizing the model parameters. In this problem, however, we have a choice of experimental data to fit to: the system

pressure and/or the vapor phase mole fraction. Typically, the system pressure is used as the variable of choice, though many examples in the literature exist where both variables are used. For this example, we use both approaches for illustration purpose.

Our options are

$$\text{OBJ\_}P = \frac{1}{13} \sum_{i=1}^{13} \left[\frac{(P)_i^{\text{Model}} - (P)_i^{\text{Expt.}}}{(P)_i^{\text{Expt.}}} \right]^2 \qquad (12.61)$$

$$\text{OBJ\_}y_1 = \frac{1}{13} \sum_{i=1}^{13} \left[\frac{(y_1)_i^{\text{Model}} - (y_1)_i^{\text{Expt.}}}{(y_1)_i^{\text{Expt.}}} \right]^2 \qquad (12.62)$$

Equation 12.61 considers only the pressure, while Equation 12.62 considers only the vapor-phase mole fraction. The selection of the objective function to use is user-dependent and ultimately depends on whether it is more important to predict the pressure or the vapor-phase mole fraction with more accuracy. If they are of equal importance, one could define an objective function that contains both variables. We can explore the impacts of our choices in the rest of the problem.

Using various choices for the binary interaction parameter (k_{12}), we determine the minimum for the two objective functions used (Equations 12.61 and 12.62). This is presented in Table 12-9.

TABLE 12-9 Binary interaction parameters for pressure-defined and vapor-phase mole fraction defined objective functions. Optimal values are in bold face italics.

k_{12}	OBJ_P	OBJ_y_1
0	3.77×10^{-3}	3.07×10^{-2}
0.010	1.84×10^{-3}	2.19×10^{-2}
0.012	1.72×10^{-3}	2.05×10^{-2}
0.013	*1.70×10^{-3}*	1.98×10^{-2}
0.014	1.71×10^{-3}	1.92×10^{-2}
0.015	1.75×10^{-3}	1.86×10^{-2}
0.020	2.36×10^{-3}	1.61×10^{-2}
0.025	3.79×10^{-3}	1.45×10^{-2}
0.030	6.15×10^{-3}	1.39×10^{-2}
0.031	6.74×10^{-3}	*1.39×10^{-2}*
0.032	7.39×10^{-3}	1.39×10^{-2}
0.035	9.49×10^{-3}	1.43×10^{-2}

As can be seen in Table 12-9, a different value of the binary interaction parameter produces the minimum in the objective function values defined for the pressure and the vapor-phase mole fraction. We provide the *Pxy* plot for the optimal value (identified in boldface italics in Table 12-9) for both objectives functions, relative to the experimental data, in Figure 12-4. Note two things. First, the pure-component vapor-pressure from the Peng-Robinson equation of state using the pure-component parameters as defined through the critical point is a little off for ethanol. Second, while both correlations are an improvement over the predicted value ($k_{12} = 0$), they are both not as accurate as the Wilson equation (excess molar Gibbs free energy model approach), as seen previously in Figure 11.7.

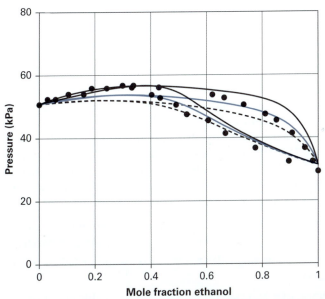

FIGURE 12-4 A Pxy diagram of the ethanol (1) and methyl butyl ether (2) system at 323.15 K. The symbols are experimental data (Hofman et al., 2000). The two solid curves are the correlation for the Peng-Robinson equation of state using a binary interaction parameter optimized with respect to the pressure ($k_{12} = 0.013$; black solid curve) and vapor-phase composition ($k_{12} = 0.031$; blue solid curve). The dashed black curve is the Peng-Robinson equation of state prediction ($k_{12} = 0.0$).

Step 4 *Predicting the phase behavior at 338.15 K*
The final part of the problem involves taking the k_{12} values fit at 323.15 K to predict the phase behavior at 338.15 K using the Peng-Robinson equation of state. We provide that in Figure 12-5.

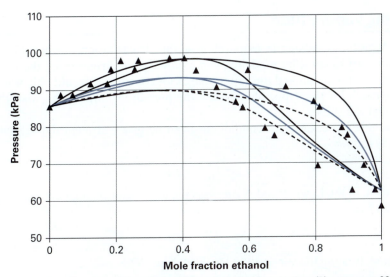

FIGURE 12-5 A Pxy diagram of the ethanol (1) and methyl butyl ether (2) system at 338.15 K. The symbols are experimental data (Hofman et al., 2000). The two solid curves are the predictions for the Peng-Robinson equation of state using a binary interaction parameter optimized with respect to the pressure ($k_{12} = 0.013$; black solid curve) and vapor-phase composition ($k_{12} = 0.031$; blue solid curve) at 323.15 K. The black dashed curve is the Peng-Robinson equation of state prediction ($k_{12} = 0.0$).

Once again, the predictions using the binary interaction parameter optimized at the lower temperature are better than predictions without any binary interaction parameter (the dashed curve). However, the predictions are not as accurate as that realized in Example 11-5 and shown in Figure 11.7.

The reasons that both the correlations at 323.15 K and the subsequent predictions at 338.15 K using the Peng-Robinson equation of state are not accurate are varied. The main reason stems from the fact that the Peng-Robinson equation of state was designed for non-polar substances (its utility for non-polar mixtures is demonstrated in Example 12-2), while the system in Example 12-4 has two polar compounds and has hydrogen-bonding interactions. We do have choices, though, if we would like to use an equation of state approach (phi-phi modeling) for mixtures that contain polar compounds. Example approaches currently in use in this field of research include:

■ A change in how the *a* parameter in the Peng-Robinson equation of state is calculated by bringing in an additional, compound-specific parameter.

■ A change in the mixing rule that uses excess molar Gibbs free energy modeling information.

■ The addition of a hydrogen-bonding term to a cubic equation of state.

■ Equations of state that are not cubic, but are based on theories from statistical mechanics.

Any and all of these approaches have been used to improve the accuracy, both from a correlative and predictive standpoint, when working with equations of state for the vapor–liquid equilibrium of mixtures. Those approaches are beyond the scope of this introductory text, however. But the general concept of phi-phi modeling and parameter optimization is basically the same, whether you are working with the Peng-Robinson equation of state shown in this example or something much more advanced.

12.4 Ideal Solution for the Vapor Phase

For systems that validate the assumptions of the ideal solution, it is possible (though not common) to consider modeling vapor–liquid equilibrium using the ideal solution for *both* the liquid and the vapor phases. Of course, one would only want to use this approach for the vapor phase when the pressure is large. Otherwise, the ideal gas model will suffice.

The mixture fugacity of component i can be given through the fugacity coefficient as follows:

$$\hat{f}_i^V = y_i P \hat{\varphi}_i \tag{11.49}$$

We know that the Lewis-Randall rule says the ideal solution fugacity for component i is given by:

$$\hat{f}_i^{ID} = y_i f_i \tag{11.34}$$

Therefore, treating the vapor as an ideal solution yields:

$$\hat{f}_i^{ID} = y_i f_i = y_i P \hat{\varphi}_i^{ID} \tag{12.62}$$

This implies that

$$\hat{\varphi}_i^{ID} = \frac{f_i}{P} \tag{12.63}$$

We already know from Chapter 8 that f_i/P is equal to the pure component fugacity coefficient. Thus,

$$\hat{\varphi}_i^{ID} = \frac{f_i}{P} = \varphi_i \qquad (12.64)$$

Therefore, the mixture fugacity of component i for the vapor becomes

$$\hat{f}_i^{V(ID)} = y_i P \varphi_i \qquad (12.65)$$

For the liquid side, we are also treating the system as an ideal solution, so the Lewis-Randall rule provides:

$$\hat{f}_i^{L(ID)} = x_i f_i \qquad (12.66)$$

Unlike our efforts in Chapters 10 and 11, we are at an elevated pressure and, thus, must include the impact of pressure on the liquid phase fugacity through the Poynting correction. Accordingly, we have

$$\hat{f}_i^{L(ID)} = x_i P_i^{sat} \varphi_i^{sat} \exp\left\{\frac{V_i(P - P_i^{sat})}{RT}\right\} \qquad (12.67)$$

Setting both Equations 12.65 and 12.67 equal to each other, as required for phase equilibrium, yields:

$$x_i P_i^{sat} \varphi_i^{sat} \exp\left\{\frac{V_i(P - P_i^{sat})}{RT}\right\} = y_i P \varphi_i \qquad (12.68)$$

Or, written as before, where we move the saturation fugacity coefficient of pure component i to the right-hand side, we have:

$$x_i P_i^{sat} \exp\left\{\frac{V_i(P - P_i^{sat})}{RT}\right\} = y_i P \frac{\varphi_i}{\varphi_i^{sat}} \qquad (12.69)$$

Equation 12.69, while looking much like the gamma-phi modeling approach in Equation 12.2, is different in two important ways: (1) the activity coefficient is set equal to 1 here since the liquid is an ideal solution and (2) the mixture fugacity coefficient of component i in Equation 12.2 is now just the pure component fugacity coefficient of component i, since the vapor is an ideal solution. We demonstrate the utility of this approach through Example 12-5.

VAPOR–LIQUID EQUILIBRIUM FOR AN IDEAL SOLUTION FOR BOTH THE LIQUID AND VAPOR PHASES	**EXAMPLE 12-5**

You are interested in modeling the vapor-liquid equilibrium of the propane (1) + n-pentane (2) system at 360.93 K. You know that the vapor pressure of propane at this temperature is 3.6334 MPa while the vapor pressure of the n-pentane is 0.4464 MPa. Since the components in the mixture are relatively similar in size and interactions, you want to use the ideal solution model. However, the mixture pressure will be well above that needed to comfortably use the ideal gas model. Thus, Raoult's Law is not the preferred approach. Create a *Pxy* plot showing the behavior of this system if modeled as an ideal solution for both the liquid and the vapor phases. Compare your model results with both the Raoult's Law predictions and the experimental data given in Table 12-10.

H H H
| | |
H—C—C—C—H
| | |
H H H

Propane

H H H H H
| | | | |
H—C—C—C—C—C—H
| | | | |
H H H H H

n-pentane

TABLE 12-10 Vapor-liquid equilibrium of propane (1) + n-pentane (2) at 360.93 K.

P (kPa)	x_1	y_1
446.4	0	0
551.6	0.040	0.193
689.5	0.090	0.355
861.8	0.150	0.486
1034.2	0.213	0.583
1378.9	0.339	0.716
1723.7	0.463	0.802
2068.4	0.578	0.861
2413.1	0.685	0.906
2757.9	0.783	0.940
3102.6	0.876	0.968
3447.3	0.962	0.991
3633.4	1	1

Based on data from Sage B. H. and Lacey, W. N. "Phase Equilibria in Hydrocarbon Systems." Industrial and Engineering Chemistry, 32, 992–996, 1940.

SOLUTION:

Step 1 *Make a plan*

The plan is to first model the system using Raoult's Law and then model the system by treating the vapor phase as an ideal solution (as opposed to the ideal gas of Raoult's Law).

Recall that the assumption of ideal gas for Raoult's Law impacts not only the vapor phase (through the use of the ideal gas assumption), but the liquid phase as well. Setting the mixture fugacity coefficient of component *i* equal to 1 in the vapor phase for Raoult's Law implies that this value is equal to 1 at any pressure, so that the fugacity coefficient of component *i* in the liquid at the saturation pressure (φ_i^{sat}) is equal to 1 as well. If not, the expressions for the pressure from the model will not return the pure-component vapor pressure at the pure-component endpoints.

On the other hand, the elevated pressure, just as in the example to begin this chapter, implies that the Poynting correction for the liquid (to correct the pure-component liquid fugacity) is not necessarily negligible. Thus, it can (and should) be included in the modeling for the system.

To demonstrate the impact of the Poynting correction, it is useful to show the model predictions with and without the Poynting correction when making the ideal gas assumption for the vapor.

Step 2 *Raoult's Law, no Poynting correction*

We have modeled many systems with Raoult's Law before. The equations are as follows:

$$P = x_1 P_1^{sat} + (1 - x_1) P_2^{sat} \tag{10.7}$$

$$y_1 = \frac{x_1 P_1^{sat}}{P} \tag{10.4}$$

Remember, the *P* in Equation 10.4 is not the experimental pressure, but the pressure from the model (Equation 10.7).

The vapor pressures are not from the Antoine equation this time, but from the data table given in the problem statement (Table 12-10). Such an approach keeps the modeling work self-consistent if there are any discrepancies between the Antoine equation vapor pressure correlations and the measured experimental values. Here, $P_1^{sat} = 3633.4$ kPa and $P_2^{sat} = 446.4$ kPa.

Solving Equations 10.4 and 10.7 for P and y_1 at a given T, incrementing in x_1, provides the results in Figure 12-6. As is observed, the model overpredicts the bubble-point pressure and underpredicts the dew-point pressure.

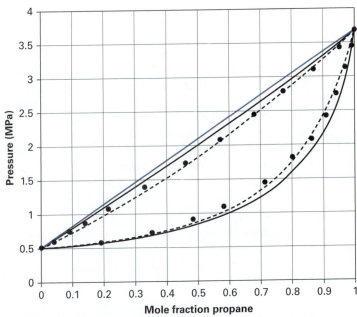

FIGURE 12-6 A Pxy diagram of the propane (1) and n-pentane (2) system at 360.93 K. The symbols are experimental data (Sage and Lacey, 1940). The blue curve is the prediction using Raoult's Law. The black curve is the prediction using Raoult's Law with the Poynting correction. The black dashed curve is the prediction using the ideal solution model for the liquid and vapor phase. The solid black and blue curves overlap for the dew-point predictions.

Step 3 *Raoult's Law, including the Poynting correction*
Here, unlike Raoult's Law, the variables we are solving for (P and y_1) are not all given explicitly. Specifically, the pressure is both on the left-hand side of equation 12.70 and inside the exponential for the Poynting correction.

$$P = x_1 P_1^{\text{sat}} \exp\left\{\frac{\underline{V}_1(P - P_1^{\text{sat}})}{RT}\right\} + (1 - x_1)P_2^{\text{sat}} \exp\left\{\frac{\underline{V}_2(P - P_2^{\text{sat}})}{RT}\right\} \tag{12.70}$$

$$y_1 = \frac{x_1 P_1^{\text{sat}} \exp\left\{\frac{\underline{V}_1(P - P_1^{\text{sat}})}{RT}\right\}}{x_1 P_1^{\text{sat}} \exp\left\{\frac{\underline{V}_1(P - P_1^{\text{sat}})}{RT}\right\} + (1 - x_1)P_2^{\text{sat}} \exp\left\{\frac{\underline{V}_2(P - P_2^{\text{sat}})}{RT}\right\}} \tag{12.71}$$

Additionally, we need to know the pure-component molar volumes. These can be estimated through Appendix C.

$$\underline{V}_1 = 74.87 \text{ cm}^3/\text{mol}$$
$$\underline{V}_2 = 115.22 \text{ cm}^3/\text{mol}$$

We now have two equations (12.70 and 12.71) and two unknowns (P and y_1), so we can solve this using an iterative technique or, more directly, through a non-linear equation solving algorithm (such as the Solver function in MS Excel). This is done at increments of x_1

from $x_1 = 0$ to $x_1 = 1$ at a fixed temperature to predict the bubble-point and dew-point pressures, which are presented in Figure 12-6. As can be seen, there is some improvement in predicting the bubble-point pressure when including the Poynting correction to the liquid fugacity, but little impact is made on the dew-point pressure predictions. The latter is understandable, as the dew-point curve describes the vapor phase.

Using the Poynting correction in this way does not provide all of the pressure impact on the liquid phase fugacity. Indeed, since we see from Figure 12-6 that the Poynting correction does impact the model predictions, this implies that the pressure is an important consideration for both phases. Recall that the impact of pressure on the liquid phase is NOT just in the Poynting correction, but is also in the fugacity coefficient of pure component i at saturation. The strategy employed earlier in the chapter was to move this term to the "vapor side" of the gamma-phi approach in order to use the same equation of state to describe both the deviations from ideal gas behavior for the vapor and the fugacity coefficient of the pure components at saturation. Therefore, the next step, which treats the vapor not as an ideal gas but as an ideal solution, should impact the liquid phase as well.

Step 4 *Ideal solution for the vapor phase*
Recall that the ideal solution for the vapor phase implies that the mixture fugacity coefficient of component i is equal to the pure component fugacity coefficient at the mixture temperature and pressure. Thus, including the fugacity coefficient at saturation for the pure component yields the following model for the pressure and the vapor phase composition:

$$P = \frac{x_1 P_1^{sat}}{\varphi_1/\varphi_1^{sat}} \exp\left\{\frac{V_1(P - P_1^{sat})}{RT}\right\} + \frac{(1 - x_1)P_2^{sat}}{\varphi_2/\varphi_2^{sat}} \exp\left\{\frac{V_2(P - P_2^{sat})}{RT}\right\} \tag{12.72}$$

$$y_1 = \frac{\dfrac{x_1 P_1^{sat}}{\varphi_1/\varphi_1^{sat}} \exp\left\{\dfrac{V_1(P - P_1^{sat})}{RT}\right\}}{\dfrac{x_1 P_1^{sat}}{\varphi_1/\varphi_1^{sat}} \exp\left\{\dfrac{V_1(P - P_1^{sat})}{RT}\right\} + \dfrac{(1 - x_1)P_2^{sat}}{\varphi_2/\varphi_2^{sat}} \exp\left\{\dfrac{V_2(P - P_2^{sat})}{RT}\right\}} \tag{12.73}$$

Unlike the previous two approaches in this problem so far, we now need a fugacity coefficient model for the system. This is the domain of equations of state and we have a few options. One approach is to use the Peng-Robinson equation of state; in this approach we would not need the mixture fugacity coefficient expression, but the pure-component fugacity coefficient expression, since all of the fugacity coefficients in Equations 12.72 and 12.73 are for the pure components. However, using cubic equations of state is problematic since we are attempting to calculate the pure component fugacity for the vapor at conditions where the less volatile component is not a vapor, but a liquid. For example, we know that the saturation pressure for n-pentane is 0.4464 MPa at the temperature of interest for the problem (360.93 K). Thus, the liquid and vapor phase fugacity coefficients for n-pentane are the same at this temperature and pressure (as is required for pure-component phase equilibrium). But the system pressure in this problem will be quite large (greater than 3.6 MPa, the saturation pressure of propane at 360.93 K) and the Peng-Robinson equation will only provide a single real root (liquid) at this temperature and pressure for n-pentane. Indeed, when one increases the pressure past 1.3 MPa and 360.93 K, only one single real root exists for n-pentane and it is the liquid. Thus, one cannot obtain the necessary estimates of the pure component fugacity coefficients in this way.

A more useful approach comes from using the virial equation of state as the model to obtain the needed fugacity coefficients. The virial equation of state is only for the vapor-phase and, thus, will provide vapor-like predictions at pressures where the pure compound would (from an equilibrium point of view) exist as only a liquid. We have seen use of these hypothetical states before when creating thermodynamic paths (such as in the approach van Laar used to develop his excess molar Gibbs free energy expression). From Equation 12.28, the fugacity coefficient for pure component i is given as:

$$\varphi_i = \exp\left[\frac{B'_i P}{RT}\right] \tag{12.74}$$

We can obtain the second-virial coefficients using the critical temperature, critical pressure, and acentric factor (Appendix C) via Equations 7.80 – 7.82. Thus, we have

$$B'_1(360.93\,\text{K}) = -263.19\,\frac{\text{cm}^3}{\text{mol}}$$

$$B'_2(360.93\,\text{K}) = -760.49\,\frac{\text{cm}^3}{\text{mol}}$$

The fugacity coefficients at saturation are evaluated at the system temperature and the saturation pressure of each pure component and, thus, can be calculated from Equation 12.74:

$$\varphi_1^{\text{sat}} = 0.727$$

$$\varphi_2^{\text{sat}} = 0.893$$

Clearly the fugacity coefficient of the first component at saturation deviates more from the ideal gas value than the second component since it is evaluated at a pressure much greater than the second component. This is because the vapor pressure for the first component is much larger than that of the second component.

At this point, we have two equations (12.72 and 12.73) in terms of two unknowns: P and y_1. These pair of equations can be solved at the temperature of interest (360.93 K) and at increments of the liquid phase mole fraction (x_1) from $x_1 = 0$ to $x_1 = 1$. The result is plotted on Figure 12-6.

As is readily observed, accounting for the pressure using the fugacity coefficients has improved the predictions of both the bubble-point and dew-point curves. The simplicity of using the ideal solution model for the vapor (where the mixture fugacity coefficient of component i is modeled as its pure component value at the mixture temperature and pressure) has utility even for systems that would not behave as an ideal solution in the liquid phase. Indeed, with the particles of a vapor or gas so far apart (relative to the liquid phase), accounting for size and interaction effects through a non-ideal gas approach is important for elevated pressures (even when the sizes of the molecules and their interactions are not necessarily similar). An end-of-chapter problem explores the use of the ideal solution model for a vapor that is far from an ideal solution in the liquid phase.

> While pure n-pentane exists as a liquid at $P > 0.4464$ MPa at this temperature, here we estimate the fugacity coefficient that pure n-pentane *would* have if it existed in the vapor phase, because it is needed for modeling how n-pentane behaves in a VLE mixture.

> The vapor pressure of propane is 3.6334 MPa, while that of n-pentane is 0.4464 MPa.

> Equation 12.72 is not a function of y_1, so you are really solving one equation in one unknown (P). Once you find the pressure, you can calculate the vapor-phase mole fraction (y_1) directly from Equation 12.73.

At this point, over the last two chapters we have provided an overview of the various modeling options for vapor–liquid equilibrium of mixtures, through Raoult's Law, modified Raoult's Law, gamma-phi approaches (excess molar Gibbs free energy model for the liquid and an equation of state for the vapor), and phi-phi approaches (equation of state for both phases). Additionally, we showed that for ideal solutions at elevated pressures, we can use the ideal solution to model our system for both phases. However, vapor–liquid equilibrium is not the only mixture phase equilibrium of interest to chemical engineers. Liquid–liquid equilibrium and solid–liquid equilibrium are important for separations as well. Both of these phase equilibria are presented in the next chapter in terms of how the experimental information is presented and modeling approaches.

12.5 SUMMARY OF CHAPTER TWELVE

- **Gamma-phi** modeling requires an excess molar Gibbs free energy model to obtain the activity coefficient (gamma) and an equation of state to obtain the mixture fugacity coefficient (phi). The phi part of the approach is required when pressures are high enough where the ideal gas model is not a good description of the vapor phase.

- Equations of state can be used to predict mixture phase behavior based on pure component properties.

- **Mixing rules** describe how parameters for a mixture are obtained from an equation of state. They are normally functions of the pure component parameters and the composition of the mixture.

- **Combining rules** describe how the pure component parameters "combine" to make a cross-parameter (how component 1 interacts with component 2).

- **Van der Waals one-fluid mixing rules** are common and are quadratic in terms of the mole fraction. Combining rules for cubic equations of state regularly treat the a cross-parameter as a geometric mean and the b cross-parameter as an arithmetic mean.

- A **binary interaction parameter (k_{12})** is normally used to modify the a cross-parameter in an equation of state such that it better matches the experimental data for the mixture.

- If the pure component properties aren't correlated well from an equation of state, the mixture properties will not be predicted accurately.

- Many modifications exist in modeling mixtures through equations of state, to include modifications to mixing rules as well as statistical-mechanics based equation of state approaches.

- Ideal solutions can be used to model the vapor phase at elevated pressures.

12.6 EXERCISES

12-1. Consider an equimolar mixture of n-butane (1) + 1-butanol (2) at 323.15 K. Calculate the second-virial coefficient of the mixture using Equation 12.24.

n-butane   1-butanol

12-2. Derive Equation 12.31, which is the expression for the mixture fugacity coefficient of component 1 in a binary mixture described by the virial equation.

12-3. Calculate the mixture fugacity coefficient for both components of a *n*-butane (1) + propane (2) mixture at 323.15 K using the virial equation. Provide the result at five compositions:
  A. $x_1 = 0$
  B. $x_1 = 0.25$
  C. $x_1 = 0.5$
  D. $x_1 = 0.75$
  E. $x_1 = 1.0$

12-4. For an equimolar acetone (1) + methyl ethyl ketone (2) mixture, calculate the van der Waals and Peng-Robinson equation of state parameters at 308.15 K.

Acetone   Methyl ethyl ketone

12-5. For an equimolar mixture of *n*-hexane (1) + benzene (2) at 423.15 K and 500 kPa, estimate the molar volume of the mixture three ways:
  A. Ideal gas law
  B. van der Waals equation of state
  C. Peng-Robinson equation of state

12-6. For an equimolar mixture of water (1) + chloroform (2) mixture at 473.15 K and 500 kPa, estimate the molar volume of the mixture three ways:
  A. Ideal gas law
  B. van der Waals equation of state
  C. Peng-Robinson equation of state

Chloroform

12-7. For a binary mixture you desire to produce a *Txy* diagram from an equation of state. List the equations needed and the unknown variables.

12-8. Starting with Equation 11.124 and assuming the combining rule in Equation 11.126, derive Equation 11.127.

12-9. Consider an equimolar *n*-butane (1) + *n*-pentane (2) mixture at 125°C and 400 kPa. Calculate the fugacity coefficient for *n*-butane in this mixture in the following ways:
  A. Assume the system behaves as an ideal gas.
  B. Assume the system behaves according to the Peng-Robinson equation of state.
  C. Assume the system behaves according to the Peng-Robinson equation of state, but your system is modeled as an ideal solution.

12-10. The binary interaction parameters for the Peng-Robinson equation of state are reported (Moshfeghian et al., 1992) for the following systems:

  ◼ Ethane + isobutane: $k_{12} = -0.0102$
  ◼ Trifluorochloromethane + *n*-butane: $k_{12} = 0.0735$
  ◼ Ammonia + water: $k_{12} = -0.2694$

Isobutane   Trifluorochloromethane   Ammonia

12-11. You need to determine the binary interaction parameter (k_{12}) for the Peng-Robinson equation of state for the tetrafluoromethane + trifluorochloromethane system at 250 K and 225 K. The literature lists the following values (Abu-Eishan, 1991):

Tetrafluoromethane

Temperature (K)	k_{12}
233	0.0481
243	0.0374
253	0.0350

What values will you report? Which values will you have the most confidence in and why?

Does the magnitude of the k_{12} values make sense to you? Please explain.

12-12. You are interested in predicting the phase behavior of a binary mixture that contains an azeotrope. An accurate prediction of the azeotrope is important since it provides a distillation limit for separation. You have experimental *Pxy* data for the system at one temperature, but not for the temperature of interest. Thus, your plan is to correlate the experimental *Pxy* data for your system using an equation of state and a k_{12} value at one temperature, and predict the behavior at the temperature of interest. Since, as you know, you have a choice of objective functions to use to obtain the "best fit" k_{12} value, describe an approach to make this decision.

12.7 PROBLEMS

12-13. An equimolar mixture of methane and propane is discharged from a compressor at 5500 kPa and 363.15 K at a rate of 1.4 kg/s. If the velocity in the discharge line is not to exceed 30 m/s, what is the minimum diameter of the discharge line?

12-14. Consider the propane (1) + *n*-butane (2) system at 50°C. Using a gamma-phi modeling approach, predict the *Pxy* diagram for the system using the van Laar equation and the virial equation. Compare the predicted values with the experimental data provided (on the same plot) in Table P12-14. Also add the Raoult's Law predictions and comment on the utility of the gamma-phi modeling approach for this system at this state.

TABLE P12-14 Vapor-liquid equilibrium of propane (1) + *n*-butane (2) at 50°C.

P (MPa)	x_1	y_1
0.493	0	0
0.557	0.054	0.139
0.618	0.112	0.254

P (MPa)	x_1	y_1
0.743	0.226	0.432
0.878	0.334	0.563
0.971	0.426	0.651
1.101	0.535	0.732
1.222	0.634	0.804
1.324	0.714	0.853
1.448	0.808	0.906
1.564	0.895	0.950
1.602	0.922	0.964
1.643	0.951	0.978
1.709	1	1

12-15. Consider the 1,1,1- trifluoroethane [R-143a] (1) + *n*-butane (2) system at 50°C. Using a gamma-phi

1,1,1-trifluoroethane

modeling approach, calculate the *Pxy* diagram for the system using the 2-parameter Margules equation and the virial equation. Compare the predicted values with the experimental data provided in Table P12-15 (on the same plot). Also, please provide two *additional* modeling approaches to the plot.

1. Model the system using modified Raoult's Law (ideal gas for the vapor). This includes calculating the activity coefficients assuming an ideal gas for the vapor (as in Chapter 11).

2. Model the system using Raoult's Law.

TABLE P12-15 Vapor-liquid equilibrium for R-143a (1) + *n*-butane (2) at 50°C.

P(MPa)	x_1	y_1
0.494	0	0
0.690	0.042	0.254
0.854	0.085	0.396
1.320	0.255	0.629
1.503	0.355	0.696
1.697	0.482	0.747
1.783	0.546	0.773
1.868	0.614	0.798
2.014	0.733	0.844
2.079	0.789	0.870
2.164	0.863	0.907
2.202	0.900	0.929
2.243	0.939	0.954
2.298	1	1

Based on data from G. Seong, J. S. Lim and H-S. Byun, J. Chem. Eng. Data, 53, 1470 (2008b).

12-16. Predict the *Pxy* behavior for a mixture of propane (1) + isobutane (2) at 303.15 K using the Peng-Robinson equation of state. Compare the predicted values with experimental data as given in Table P12-16.

CH₃
|
H₃C   CH₃
Isobutane

12-17. Predict the *Pxy* behavior for a mixture of pentafluoroethane [R-125] (1) + isobutane (2) at 303.15 K using the Peng-Robinson equation of state. Compare the predicted values with experimental data given in Table P12-17. How would you suggest improving the modeling results from the Peng-Robinson equation relative to the experimental data?

F   F
| |
F—C—C—H
| |
F   F
Pentafluoroethane

TABLE P12-16 Vapor-liquid equilibrium of propane (1) + isobutane (2) at 303.15 K.

x_1	y_1	P(MPa)
0	0	0.4052
0.119	0.226	0.4770
0.220	0.385	0.5407
0.332	0.517	0.6086
0.441	0.625	0.6802
0.535	0.716	0.7449
0.621	0.780	0.8098
0.706	0.840	0.8682
0.802	0.895	0.9372
0.900	0.949	1.0063
1	1	1.0784

Based on data from J. S. Lim, et al., J. Chem. Eng. Data, 2004, 49, 192–198.

TABLE P12-17 Vapor-liquid equilibrium of R-125 (1) + isobutane (2) at 303.15 K.

P(MPa)	x_1	y_1
0.4070	0	0
0.6900	0.0816	0.3998
0.7376	0.0977	0.4474
0.8492	0.1492	0.5277
0.9714	0.2085	0.5977
1.0858	0.2889	0.6515
1.2240	0.4115	0.7125
1.3412	0.5371	0.7611
1.4530	0.6894	0.8200
1.5372	0.8510	0.8920
1.5652	0.9260	0.9384
1.5700	1	1

Based on data from B-G Lee, et al., J. Chem. Eng. Data, 2000, 45, 760–763.

12-18. Consider the pentafluorethane [R-125] (1) + isobutane (2) system at 303.15 K. Using a gamma-phi modeling approach, calculate the *Pxy* diagram for the system using the 2-parameter Margules equation and the virial equation. Compare the predicted values with the experimental data provided in Table P12-18 (on the same plot). Also please provide two *additional* modeling approaches to the plot.

1. Model the system using modified Raoult's Law (ideal gas for the vapor). This includes calculating the activity coefficients assuming an ideal gas for the vapor (as in Chapter 11).

2. Model the system using Raoult's Law.

TABLE P12-18 Vapor-liquid equilibrium of R-125 (1) + isobutane (2) at 303.15 K.

P (MPa)	x_1	y_1
0.4070	0	0
0.6900	0.0816	0.3998
0.7376	0.0977	0.4474
0.8492	0.1492	0.5277
0.9714	0.2085	0.5977
1.0858	0.2889	0.6515
1.2240	0.4115	0.7125
1.3412	0.5371	0.7611
1.4530	0.6894	0.8200
1.5372	0.8510	0.8920
1.5652	0.9260	0.9384
1.5700	1	1

Based on data from B-G Lee, et al., J. Chem. Eng. Data, 2000, 45, 760 –763.

12-19. You work in a developing nation for a large chemical company. Your division works on refrigerants and foam-blowing agents. You have need to correlate a set of data for the trifluoromethane (1) + trifluorochloromethane (2) system at 273.1 K. You know the following about your system:

Compound	T_c (K)	P_c (MPa)	ω	$\underline{V}$ (cm³/mol)
Trifluoromethane	299.07	4.836	0.2654	67.07
Trifluorochloromethane	302.01	3.870	0.1703	93.45

Trifluoromethane

Trifluorochloromethane

Using a gamma-phi modeling approach, calculate the Pxy diagram for the system using the Wilson equation and the virial equation. Compare the predicted values with the experimental data provided (on the same plot) in Table P12-19. Also, please provide two *additional* modeling approaches to the plot.

1. Model the system using modified Raoult's Law (ideal gas for the vapor). This includes calculating the activity coefficients assuming an ideal gas for the vapor (as in Chapter 11).

2. Model the system using Raoult's Law.

TABLE P12-19 Vapor-liquid equilibrium of trifluoromethane (1) + trifluorochloromethane (2) at 273.1 K.

P (MPa)	x_1	y_1
1.966	0	0
2.135	0.0615	0.0979
2.279	0.1240	0.1890
2.422	0.2021	0.2783
2.528	0.2768	0.3480
2.624	0.3650	0.4360
2.693	0.4620	0.5060
2.732	0.5380	0.5640
2.748	0.5640	0.5860
2.752	0.5820	0.5980
2.761	0.6520	0.6520
2.750	0.7371	0.7240
2.722	0.8062	0.7829
2.682	0.8598	0.8355
2.597	0.9378	0.9194
2.498	1	1

Based on data from Stein F.P. and Proust, P. C. "Vapor-liquid equilibria of the trifluoromethane-trifluorochloromethane system." J. Chem. Eng. Data, 16, 389-393, 1971.

12-20. Predict the Txy behavior for a mixture of n-hexane (1) + p-xylene (2) at 101.33 kPa using the Peng-Robinson equation of state. Compare the predictions to the experimental data, as given in Table P12-20. Is the Peng-Robinson equation of state a reasonable model for this system at this state? Please explain.

n-hexane

p-xylene

TABLE P12-20 Vapor-liquid equilibrium of n-hexane (1) + p-xylene (2) at 101.33 kPa.

T [K]	x_1	y_1
409.25	0.011	0.069
403.15	0.042	0.225
391.45	0.123	0.487
385.25	0.165	0.587
382.95	0.192	0.625
369.45	0.339	0.785
364.45	0.409	0.831

Based on data from S. Saito.: Asahi-Garasu-Kogyo-Gijutsu-Shoreikai-Kenkyu-Hokoku 15, 397 (1969)

(continued)

359.75	0.497	0.867
356.05	0.578	0.898
351.35	0.701	0.941
347.95	0.797	0.966
345.75	0.868	0.978
343.15	0.953	0.993

12-21. Predict the Txy behavior for a mixture of ethanol (1) + 1-butanol (2) at 101.33 kPa using the Peng-Robinson equation of state Compare the predictions to the experimental data given in Table P12-21. Is the Peng-Robinson equation of state a reasonable model for this system at this state? Please explain.

Ethanol   1-butanol

TABLE P12-21 Vapor-liquid equilibrium of ethanol (1) + 1-butanol (2) at 101.33 kPa.

T [K]	x_1	y_1
390.75	0	0
388.15	0.0345	0.1250
385.65	0.0685	0.2285
383.15	0.1055	0.3270
380.65	0.1450	0.4160
378.15	0.1880	0.4960
373.15	0.2840	0.6345
368.15	0.3990	0.7495
363.15	0.5365	0.8430
360.65	0.6160	0.8830
358.15	0.7030	0.9169
355.65	0.7995	0.9508
353.15	0.9080	0.9798
351.45	1	1

Based on data from L. Gay : Chim.Ind.Genie Chim. 18, 187 (1927).

12-22. Predict the Txy behavior for a mixture of acetone (1) + 1-hexene (2) at 101.33 kPa using the Peng-Robinson equation of state. Compare the predictions to the experimental data given in Table P12-22. Determine an optimal binary interaction value by defining an objective function in terms of the temperature (it can be called OBJ_T).

Is the Peng-Robinson equation of state a reasonable model for this system at this state? Please explain.

Acetone   1-hexene

TABLE P12-22 Vapor-liquid equilibrium of acetone(1) + 1-hexene (2) at 101.33 kPa.

T [K]	x_1	y_1
336.75	0	0
331.05	0.060	0.184
327.85	0.116	0.299
326.35	0.184	0.365
324.65	0.287	0.441
323.85	0.387	0.500
323.35	0.497	0.555
323.25	0.593	0.605
323.25	0.605	0.605
323.45	0.698	0.658
324.15	0.805	0.730
325.95	0.908	0.838
329.55	1	1

Based on data from S. K. Ogorodnikov, V. B. Kogan and M. S. Nemtsov, J. Appl. Chem. USSR 34, 313 (1961).

12-23. Predict the Pxy behavior for a mixture of 1,3-butadiene (1) + n-hexane (2) at 413.15 K using the Peng-Robinson equation of state. Compare the predictions to the experimental data given in Table P12-23. Using the OBJ_P objective function, calculate an optimal binary interaction parameter. Is the Peng-Robinson equation of state a reasonable model for this system at this state? Please explain.

1,3-butadiene

TABLE P12-23 Vapor-liquid equilibrium of 1,3-butadiene (1) + n-hexane (2) at 413.15 K.

P [kPa]	x_1	y_1
613.016	0	0
709.275	0.038	0.156
810.600	0.078	0.288
911.925	0.114	0.388
1215.900	0.222	0.622

Based on data from M.S. Rozhnov, Khim. Prom. 43, 288 (1967).

P [kPa]	x_1	y_1
1519.875	0.330	0.740
1823.850	0.436	0.812
2127.825	0.542	0.868
2431.800	0.644	0.910
2735.775	0.742	0.942
3039.750	0.838	0.974
3343.725	0.936	0.994
3537.256	1	1

12-24. Predict the *Pxy* behavior for a mixture of diethyl ether (1) + methanol (2) at 303.15 K using the Peng-Robinson equation of state. Compare the predictions to the experimental data given in Table P12-24. Using the OBJ_P objective function, calculate an optimal binary interaction parameter. Is the Peng-Robinson equation of state a reasonable model for this system at this state? Please explain.

Diethyl ether

TABLE P12-24 Vapor-liquid equilibrium of diethyl ether (1) + methanol (2) at 303.15K.

P [kPa]	x_1	y_1
25.035	0.0121	0.1296
39.243	0.0774	0.4647
52.562	0.1640	0.6235
61.302	0.2401	0.6982
70.381	0.3719	0.7539
74.754	0.4661	0.7867
77.540	0.5409	0.8089
83.686	0.7724	0.8688
84.820	0.8252	0.8840
86.620	0.9767	0.9764

Based on data from J. Gmehling, U. Onken and H. W. Schulte, J.Chem.Eng.Data 25, 29 (1980).

12-25. Predict the *Pxy* behavior for a mixture of cyclohexane (1) + 1-butanol (2) at 383.15 K using the Peng-Robinson equation of state. Compare the predictions to the experimental data given in Table P12-25. Determine a "best fit" binary interaction parameter (k_{12}) that best matches the equation of state pressures to the experimental pressures. Plot those new predictions on the same curve. Comment on your results.

Cyclohexane

1-butanol

TABLE P12-25 Vapor-liquid equilibrium of cyclohexane (1) + 1-butanol (2) at 383.15K.

P [kPa]	x_1	y_1
77.754	0	0
121.107	0.1	0.4117
153.897	0.2	0.5779
178.521	0.3	0.6563
196.896	0.4	0.7017
210.448	0.5	0.7522
220.123	0.6	0.7792
226.387	0.7	0.7984
229.212	0.8	0.8261
228.093	0.9	0.8750
222.040	1	1

Based on data from R. S. Ramalho and J. Delmas. J.Chem.Eng.Data 13, 161 (1968).

12-26. Predict the *Pxy* behavior for a mixture of acetone (1) + 2-propanol (2) at 328.15 K using the Peng-Robinson equation of state. Compare the predictions to the experimental data given in Table P12-26. Determine a "best fit" binary interaction parameter that best matches the equation of state pressures to the experimental pressures. Plot those new predictions on the same curve. Comment on your results.

Acetone

2-propanol

TABLE P12-26 Vapor-liquid equilibrium of acetone (1) + 2-propanol (2) at 328.15 K.

P [kPa]	x_1	y_1
34.393	0.0237	0.1166
39.930	0.0642	0.2777
44.208	0.0971	0.3625
52.026	0.1591	0.4762
59.050	0.2353	0.5722

Based on data from D. C. Freshwater and K. A. Pike, J.Chem.Eng.Data 12, 179 (1967).

(continued)

60.848	0.2687	0.6024
69.000	0.3879	0.6995
71.177	0.4314	0.7284
75.068	0.5234	0.7655
80.287	0.6084	0.8098
85.038	0.7216	0.8617
85.942	0.7338	0.8729
91.220	0.8569	0.9240
94.308	0.9214	0.9629

12-27. You are interested in the location of the azeotrope for the acetonitrile (1) + benzene (2) system at 346.85 K. However, you only have that information for this system at 318.15 K. At that state (318.15 K) the azeotropic pressure is 37.197 kPa, while the azeotrope is located at $x_1 = y_1 = 0.53$ (Palmer and Smith, 1972). Use this information to predict the azeotropic pressure and composition at the temperature of interest (346.85 K). [Note: P at the azeotrope (346.85 K) = 101.325 kPa; $x_1 = y_1 = 0.44$ (Lecat, 1946)] Solve the problem using the Peng-Robinson equation of state.

Acetonitrile

Benzene

12-28. In Example 12-2 in this chapter, you evaluated two equation of state approaches for the prediction of the phase behavior for the propane (1) + n-butane (2) system at 323.15 K. Once you completed this chapter, being an inquisitive student, you wondered if this system could also be modeled by treating the liquid phase as an ideal solution and the vapor phase as an ideal solution (since the system pressure was over 1500 kPa). Explore your curiosity by providing the following modeling approaches for this system on the same plot:

A. Raoult's Law (ideal solution for the liquid; ideal gas for the vapor)

B. Ideal Solution—Ideal Solution (ideal solution for the liquid, but include the impact of pressure; ideal solution for the vapor—virial equation is your model).

Compare your results with the experimental data plotted as symbols.

TABLE P12-28 Vapor-liquid equilibrium of propane (1) + n-butane (2) at 338.15K.

P (MPa)	x_1	y_1
0.493	0	0
0.557	0.054	0.139
0.618	0.112	0.254
0.743	0.226	0.432
0.878	0.334	0.563
0.971	0.426	0.651
1.101	0.535	0.732
1.222	0.634	0.804
1.324	0.714	0.853
1.448	0.808	0.906
1.564	0.895	0.950
1.602	0.922	0.964
1.643	0.951	0.978
1.709	1	1

12-29. In Problem 12-18 in this section, you used a gamma-phi modeling approach for the pentafluoroethane [R-125] (1) + isobutane (2) system at 303.15 K. There (if you solved that problem), you realized the benefit of incorporating a gamma-phi approach (i.e., treating the vapor phase as a real gas rather than an ideal gas) as compared to using modified Raoult's Law.

In this problem repeat the gamma-phi modeling, but treat the vapor-phase as an ideal solution. Here, you are not including the composition effects on the fugacity coefficient, but modeling it as a pure component at the mixture temperature and pressure. Plot both results (the full gamma-phi approach from Problem 12-18 and the current approach) as well as the experimental data (as symbols). Additionally, report the following information in tabular form:

■ The experimental activity coefficients for both approaches

■ The ratio of the mixture fugacity coefficient of component i to the saturation fugacity coefficient of component i

What can you conclude about the ideal solution approach to the vapor phase in the context of this problem?

TABLE P12-29 Vapor-liquid equilibrium of R-125 (1) + isobutane (2) at 303.15 K.

P (MPa)	x_1	y_1
0.4070	0	0
0.6900	0.0816	0.3998
0.7376	0.0977	0.4474
0.8492	0.1492	0.5277
0.9714	0.2085	0.5977
1.0858	0.2889	0.6515
1.2240	0.4115	0.7125
1.3412	0.5371	0.7611
1.4530	0.6894	0.8200

Based on data from B-G Lee, et al., J. Chem. Eng. Data, 2000, 45, 760 - 763.

12-30. Use a γ-φ approach to model the vapor-liquid equilibrium of an ethyne [acetylene] (1) + 1, 1 difluoro ethane {R-152a] (2) system at 303.2 K. Treat the

H—C≡C—H   H—C(F)(F)—CH₃
Ethyne   1,1 difluoroethane

liquid using the 2-parameter Margules equation and the vapor as an ideal solution (described by the virial equation). Report the following:

■ Raoult's Law predictions
■ Modified Raoult's Law predictions (ideal gas for the vapor phase)
■ γ-φ modeling results
■ Experimental data as symbols given in Table P12-30

TABLE P12-30 Vapor-liquid equilibrium of ethyne (1) + R-152a (2) at 303.2K.

P (kPa)	x_1	y_1
701	0	0
780	0.0380	0.1572
850	0.1283	0.3849
1175	0.2509	0.5984
1775	0.4243	0.7603
2500	0.5685	0.8370
3249	0.6956	0.8905
3920	0.8014	0.9378
4720	0.9093	0.9627
5476	1	1

Based on data from Lim, J. S., Lee, Y.W., Kim J. D. and Lee, Y.Y. "Vapor-liquid equilibria for 1,1-difluoroethane + acetylene and 1,1-difluoroethane + 1,1-dichloroethane at 303.2 K and 323.2 K." J. Chem. Eng. Data, 41, 1168-1170, 1996.

12.8 GLOSSARY OF SYMBOLS

a_{12}, a_{21} cross-parameters for equations of state

a_1, a_2 equation of state parameter

a_m equation of state parameter for the mixture

b_{12}, b_{21} cross-parameters for equations of state

b_1, b_2 equation of state parameter

b_m equation of state parameter for the mixture

B' second-virial coefficient

B'_m second-virial coefficient of the mixture

B'_{ij} second-virial coefficient of component i interacting with component j

f_i fugacity of pure component i

$\hat{f}_i$ mixture fugacity of component i

$\underline{G}^E$ excess molar Gibbs free energy of the mixture

$\underline{G}$ molar Gibbs free energy of the mixture

$\overline{G}_i$ partial molar Gibbs free energy of component i in a mixture

$\overline{G}_i^R$ partial molar residual Gibbs free energy of component i in a mixture

$\overline{G}_i^{ig}$ partial molar Gibbs free energy of component i in a mixture modeled as an ideal gas mixture

$\underline{G}^R$ residual molar Gibbs free energy of the mixture

k_{12} binary interaction parameter

n moles

OBJ   objective function value

P pressure

P_c critical pressure

P_i^{sat} vapor pressure of component i

R universal gas constant

$\underline{S}$ molar entropy of the mixture

$\overline{S}_i$ partial molar entropy of component i in a mixture

T temperature

T_c critical temperature

$\underline{V}_i$ molar volume of pure component i

$\underline{V}^{ig}$ molar volume of the mixture modeled as an ideal gas

$\overline{V}_i$ partial molar volume of component i in a mixture

$\overline{V}_i^{ig}$ partial molar volume of component i in a mixture modeled as an ideal gas

x_A mole fraction of component A (normally in the liquid phase)

y_A mole fraction of component A (normally in the vapor phase)

Z compressibility factor

$\hat{\varphi}_i$ mixture fugacity coefficient of component i

φ_i^{sat} fugacity coefficient of component i at saturation conditions

$\hat{\varphi}_i^{ID}$ mixture fugacity coefficient of component i modeled as an ideal solution

γ_i activity coefficient of component i

$\Lambda_{12}, \Lambda_{21}$ temperature-dependent parameters of the Wilson equation

μ_i chemical potential of component i in the mixture

μ_i^R residual chemical potential of component i

ω acentric factor

12.9 REFERENCES

Abu-Eishan, S. I. "Calculation of vapour–liquid equilibrium data for binary chlorofluorocarbon mixtures using the Peng-Robinson equation of state," *Fluid Phase Equil*, 62, 41–52 (1991).

Freshwater, D. C., Pike, K. A. "Vapor-Liquid Equilibrium Data for Systems of Acetone-Methanol-Isopropanol." *J. Chem. Eng. Data*, 12, 179 (1967).

Gay, L. "Deuxième mémoire (Second Submission)" *Chim. Ind. Genie Chim.*, 18, 187 (1927).

Gmehling, J., Onken, U., Schulte, H. W. "Vapor- Liquid Equilibriums for the Binary Systems Diethyl Ether-Halothane (1, 1, 1-trifluoro-2-bromo-chloroethane), Halothane-Methanol, and Diethyl Ether-Methanol." *J. Chem. Eng. Data*, 25, 29 (1980).

Lee, B-G. et al., "Vapor-Liquid Equilibria for Isobutane + Pentafluoroethane (HFC-125) at 293.15 to 313.15 K and + 1,1,1,2,3,3,3-Heptafluoropropane (HFC-227ea) at 303.115 to 323.15 K." *J. Chem. Eng. Data*, 45, 760–763 (2000).

Lecat, M. "Some Orthobaric Azeotropes." *C. R. Hebd. Seances Acad. Sci.*, 222, 733 (1946).

Lim, J. S. et al., "Measurement of Vapor-Liquid Equilibria for the Binary Mixture of Propane (R-290) + Isobutane (R-600a)." *J. Chem. Eng. Data*, 49, 192–198 (2004).

Lim, J. S., Lee, Y. W., Kim J. D., Lee, Y. Y. "Vapor-liquid equilibria for 1,1-difluoroethane + acetylene and 1,1-difluoroethane + 1,1-dichloroethane at 303.2 K and 323.2 K." *J. Chem. Eng. Data*, 41, 1168–1170 (1996).

Moshfeghian, M., Shariat, A., Maddox, R. N. "Prediction of refrigerant thermodynamic properties by equations of state: vapor liquid equilibrium behavior of binary mixtures," *Fluid Phase Equil*, 80, 33–44 (1992).

Ogorodinkov, S. K., Kogan, V. B., Nemtsov, M. S. "Vapor-liquid equilibria in binary systems formed by hydrocarbons and acetone," *J. Appl. Chem. USSR*, 34, 313 (1961).

Parrish, W. R., Kidnay, A. J., *Gas Treating in Fundamentals of Natural Gas Processing*, CRC Press, Boca Raton, 2006.

Palmer, D. A., Smith, B. D. "Thermodynamic Excess Property Measurements for Acetonitrile-Benzene-*n*-Heptane System at 45.deg." *J. Chem. Eng. Data*, 17, 71 (1972).

Ramalho R. S., Delmas, J. "Isothermal and Isobaric Vapor-Liquid Equilibrium data and Excess Free Energies by the Total Pressure Method. Systems: 2,2,4-trimethylpentane-toluene, cyclohexane-1-butanol, and ethanol-*n*-heptane." *J. Chem. Eng. Data*, 13, 161 (1968).

Rozhnov, M. S. "Phase and Volume Ratios in Systems Divinyl Hydrocarbons." *Khim. Prom.*, 43, 288 (1967).

Sage B. H., Lacey, W. N. "Phase Equilibria in Hydrocarbon Systems," *Ind. Eng. Chem. Res*, 32, 992–996 (1940).

Saito S., "Separation of Hydrocarbons." *Asahi-Garasu-Kogyo-Gijutsu-Shoreikai-Kenkyu-Hokuku*, 15, 397 (1969).

Seong, G., Yoo, K. P., Lim, J. S. "Vapor-Liquid Equilibria for Propane (R290) + *n*-Butane (R600) at Various Tempratures." *J. Chem. Eng. Data*, 53, 2783 (2008a).

Seong, G., Lim, J. S., Byun H-S. "Vapor-Liquid Equilibria for the Binary System of 1, 1, 1-Tetrifluoroethane (HFC-143a) + Butane (R600) at Various Temprature." *J. Chem. Eng. Data*, 53, 1470 (2008b).

Smith, J. M., Van Ness, H. C., Abbott, M. M. *Introduction to Chemical Engineering Thermodynamics*, 7th Edition, McGraw-Hill, 2006.

Stein F. P., Proust, P. C. "Vapor-liquid equilibria of the trifluoromethane-trifluorochloromethane system," *J. Chem. Eng. Data*, 16, 389–393 (1971).

Yunhai, S. et al., "Measurements of Isothermal Vapor–Liquid Equilibrium of Binary Methanol/Dimethyl Carbonate System Under Pressure." *Fluid Phase Equil*, 234, 1 (2005).

Liquid–Liquid, Vapor–Liquid–Liquid, and Solid–Liquid Equilibrium

13

LEARNING OBJECTIVES

This chapter is intended to help you learn how to:

- Understand why liquid–liquid equilibrium occurs
- Through modeling, calculate the compositions where liquid–liquid, vapor-liquid-liquid, and solid–liquid equilibrium occur

In Chapters 9 through 12, we have focused exclusively on vapor–liquid equilibrium, owing to its primary importance in separations within the chemical process industry. However, other types of phase equilibrium exist for mixtures and are utilized (or accounted for) by chemical engineers. In this chapter, liquid–liquid, vapor–liquid–liquid, and solid–liquid equilibrium are presented. ■

13.1 Motivational Example

You have two liquids, methanol and cyclohexane, both at 310 K. You mix 10 moles of methanol and 90 moles of cyclohexane together; you call that mixture *A*. You mix 50 moles each of methanol and cyclohexane together; you call that mixture *B*. Finally, you mix 90 moles of methanol and 10 moles of cyclohexane together; you call that mixture *C*.

We have examined an exercise like this in Chapter 9, where we introduced non-ideal mixtures and how (and why) they would deviate from our intuitive approach when trying to calculate the properties of these newly formed mixtures. However, in the current example, something different happens. While mixtures *A* and *C* (upon inspection) are in one liquid phase, mixture *B* forms two liquid phases. Why does this happen? How do chemical engineers exploit this behavior, and can we model it?

With regards to liquid–liquid equilibrium, the chemical process industry often makes use of liquid–liquid extraction, where a liquid mixture of two substances (say an equimolar mixture of acetone and water), when exposed to a liquid containing a third substance (such as methyl isobutyl ketone), will form two liquid phases (one liquid phase that is rich in the acetone and the other liquid phase that is rich in the water). This liquid–liquid extraction provides a route for separation that is an alternative to distillation and is very important in the pharmaceutical industry, where heat-sensitive materials often limit the high temperatures required for distillation.

Other types of phase equilibrium exist, including solid–liquid equilibrium, and are exploited for separations as well. In this chapter, we discuss liquid–liquid equilibrium, vapor–liquid–liquid equilibrium, and solid–liquid equilibrium.

13.2 Liquid–Liquid Equilibrium

Returning to our motivational example, why should it be that an equimolar mixture of methanol and cyclohexane forms two liquid phases at 310 K, while the other two mixtures we created (10% methanol/90% cyclohexane and 90% methanol/10% cyclohexane) form a single phase? That answer is fundamental to thermodynamics and equilibrium. Specifically, a system at constant temperature and constant pressure is at equilibrium if it exists in the state that minimizes its Gibbs free energy. This is why H_2O is a liquid at room temperature and atmospheric pressure rather than a vapor or solid. Following that logic forward, an equimolar mixture of methanol and cyclohexane forms two liquid phases at 310 K, since that is the state that results in an overall lowering of the system Gibbs free energy (relative to keeping the system in a single, liquid phase).

Before we go further with the details on the "why," let's explore how these phase diagrams look for liquid–liquid equilibrium. In particular, Figure 13-1 shows the methanol + cyclohexane system at 100 kPa. The symbols are experimental data points, while the line is a guide for the eye. There are several things to notice about liquid–liquid equilibrium from Figure 13-1. First, the mixtures inside the dome are

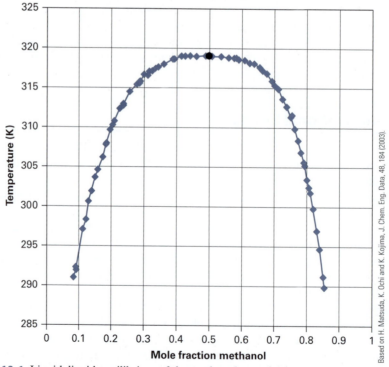

Based on H. Matsuda, K. Ochi and K. Kojima, J. Chem. Eng. Data, 48, 184 (2003).

FIGURE 13-1 Liquid–liquid equilibrium of the methanol + cyclohexane system at 100 kPa. Inside the dome, called the miscibility gap, the system is not stable and will split into two liquid phases, whose composition is given by the blue curve. Outside of the curve, the mixture exists as a single liquid phase. The experimental data is provided as symbols, with the line as a guide to the eye. The UCST is the black circle at the top of the dome.

not stable, and they split into two liquid phases. This is very similar to when we showed a mixture between the bubble-point and dew-point curves. In that situation, the mixture did not exist as a single phase (liquid *or* vapor) but existed as two equilibrium phases (liquid *and* vapor). Here, the mixture does not exist as a single phase inside the dome but as two equilibrium phases (liquid *and* liquid). Just as with vapor–liquid equilibrium, we draw a tie line horizontally to find the composition of the phases that exist at equilibrium.

Returning to our motivational example, we had an equimolar mixture of methanol + cyclohexane at 310 K. If we look at Figure 13-1, we see that it is inside the dome and, indeed, it will split into two liquid phases:

■ Liquid Phase 1: 20% methanol/80% cyclohexane
■ Liquid Phase 2: 76% methanol/24% cyclohexane

What about the two other mixtures, *A* and *C*? Mixture *A* (10% methanol + 90% cyclohexane) is to the left of the dome in Figure 13-1 at 310 K; thus it exists as one liquid phase. Likewise, mixture *C* (90% methanol + 10% cyclohexane) is to the right of the dome in Figure 13-1 at 310 K, and it will exist as one liquid phase as well.

What if we take our equimolar mixture of methanol and cyclohexane at 310 K (which is in two phases) and heat it such that the temperature of system is now 320 K instead of 310 K? As is observed in Figure 13-1, this state is now above the dome and, thus, it is once again a single liquid phase.

The black circle in Figure 13-1 at around 318 K and 50% (by mole) methanol is the point where two phases cease to exist and a single phase emerges. Recall that we saw this when examining vapor–liquid equilibrium curves (both pure components and in mixtures). There, the vapor and liquid phases came together to form a single (supercritical) phase. We called that the "critical point." In this situation, two liquid phases come together at this point to form a single liquid phase, which is also called a "critical point." However, for liquid–liquid equilibrium, this is given a special name: critical solution temperature. Note that since it occurs at the top or "upper" part of the dome, it is called an **upper critical solution temperature (UCST)**. Additional detail is provided in Figure 13-2.

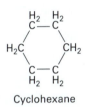

Methanol

Cyclohexane

PITFALL PREVENTION

Though we use the analogy of a "critical point" for the UCST as compared to a VLE critical point, you would *not* refer to states above the UCST as "supercritical." That terminology is reserved for states above the liquid–vapor supercritical point.

The upper critical solution temperature is sometimes called the *"upper consolute solution temperature."*

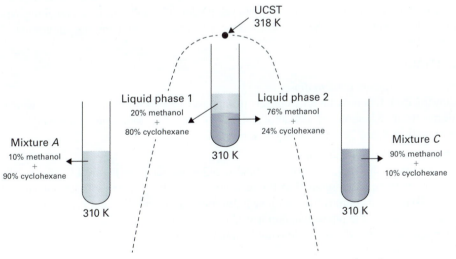

FIGURE 13-2 A single liquid phase and two liquid phases for the methanol + cyclohexane system at 100 kPa. Note that when two liquid phases exist, the denser phase (here, liquid phase 2) is on the bottom.

The language of liquid–liquid equilibrium (LLE) often uses the word **miscibility**. In this context, a miscible liquid mixture forms a single liquid phase. At states where the liquid is not completely miscible, a **miscibility gap** exists, and the liquids are termed **immiscible** in this region. So the region inside the dome for the methanol + cyclohexane system in Figure 13-1 is the miscibility gap for this system. In that region, the system is immiscible.

In these next two examples, we practice reading liquid–liquid phase equilibrium diagrams and perform material balances to determine the amount of each phase.

EXAMPLE 13-1	READING LIQUID-LIQUID PHASE DIAGRAMS (NO PHASE SPLIT)

Here, "stable" means that the liquid remains in one phase and does not split into two phases.

You mix 80 g of methanol (1) with 20 g of cyclohexane (2) at 295 K. Is the resulting system stable? If not, what is the composition (in mass fraction) and amount (in grams) of the resulting phases?

SOLUTION:

Step 1 *Convert from mass to moles*
One measures out real quantities in mass, so this problem requires conversion of mass to moles in order to use Figure 13-1, which provides the relevant data in terms of mole fraction. The molecular weight of methanol is 32.04 g/mol, while the molecular weight of cyclohexane is 84.16 g/mol. Converting from mass to moles, via their respective molecular weights, results in 2.497 moles of methanol and 0.238 moles of cyclohexane. The composition of this mixture in terms of moles is then 91% methanol.

Step 2 *Using the phase diagram*
For a mixture with a mole fraction $x_1 = 0.91$ at 295 K, we see that the mixture falls outside of the two-phase region. Thus, the resulting mixture is stable and one liquid phase exists. The mass of that one phase is 100 g and the composition is 80% (by mass) methanol and 20% (by mass) cyclohexane.

EXAMPLE 13-2	READING LIQUID–LIQUID PHASE DIAGRAMS (PHASE SPLIT)

You mix 60 g of methanol (1) with 40 g of cyclohexane (2) at 295 K. Is the resulting system stable? If not, what is the composition (in mass fraction) and amount (in grams) of the resulting phases?

SOLUTION:

Step 1 *Convert from mass to moles*
Following from Example 13-1, we convert from mass to moles using the molecular weight. This results in 1.873 moles of methanol and 0.475 moles of cyclohexane, which is 80% (by mole) methanol and 20% (by mole) cyclohexane.

Step 2 *Using the phase diagram*
For a mixture with a mole fraction $x_1 = 0.80$ at 295 K, we see from Figure 13-1 that it falls inside of the two-phase region; thus, the mixture is *not* stable and will split into two liquid phases. To find the composition of those liquid phases that are in equilibrium, we move horizontally along the phase diagram just like we did before with the bubble-point and dew-point curves for vapor–liquid equilibrium. For clarity, we'll call the first phase α and the second phase β.

$$\text{Phase } \alpha: \ x_1^\alpha = 0.10; x_2^\alpha = 0.90$$

$$\text{Phase } \beta: \ x_1^\beta = 0.84; x_2^\beta = 0.16$$

Step 3 *Mass balance*

Now that we have determined the composition of the existing phases by mole, we need to determine the amount of each phase. Here we have two choices: the lever rule or a material balance. Recall that the lever rule was a technique in which we used the overall composition and the resulting equilibrium phase compositions to determine the amounts of each phase in a graphical manner. Indeed, we can use the lever rule as a quick way to qualitatively determine that phase β will contain most of the 100 g of the original mixture. However, let's use the material balance here, as it is both simple to use and more accurate than the lever rule.

> As we saw previously, the lever rule *is* the material balance, just written in a graphical way.

$$\text{Overall Material Balance: } N = L^\alpha + L^\beta$$

Here, N indicates the overall number of moles, while L^α and L^β represent the moles in phases α and β, respectively.

$$\text{Methanol Material Balance: } x_1 N = x_1^\alpha L^\alpha + x_1^\beta L^\beta$$

If we plug in the known values, we have two equations in terms of two unknowns (L^α, L^β) for

$$(1.873 \text{ moles} + 0.475 \text{ moles}) = L^\alpha + L^\beta$$

$$(0.8)(1.873 \text{ moles} + 0.475 \text{ moles}) = (0.1)L^\alpha + (0.84)L^\beta$$

Solving these two equations in two unknowns yields

$$L^\alpha = 0.135 \text{ moles}$$
$$L^\beta = 2.213 \text{ moles}$$

Step 4 *Conversion back to mass*

We can transform this back to mass (both amounts and compositions) via the molecular weights.

PHASE α

$$\text{Methanol: } 0.1(0.135 \text{ moles})(32.04 \text{ g/mol}) = 0.433 \text{ grams}$$

$$\text{Cyclohexane: } 0.9(0.135 \text{ moles})(84.16 \text{ g/mol}) = 10.225 \text{ grams}$$

Using these masses, we find the composition about 4.1% by mass methanol and 95.9% by mass cyclohexane. Additionally, the total mass of phase α is 10.658 grams.

PHASE β

$$\text{Methanol: } 0.84(2.213 \text{ moles}) (32.04 \text{ g/mol}) = 59.560 \text{ grams}$$

$$\text{Cyclohexane: } 0.16(2.213 \text{ moles}) (84.16 \text{ g/mol}) = 29.799 \text{ grams}$$

Using these masses, we find the composition is about 66.7% by mass methanol and 33.3% by mass cyclohexane. Additionally, the total mass of phase β is 89.359 grams.

> If you sum the mass of both phases, you get a total that is slightly more than what you started with: 100 g. This is due to round-off error in performing the intermediate calculations.

13.2.1 Impact of Pressure on Liquid–Liquid Equilibrium

Let's revisit Figure 13-1 again, which shows the LLE for the methanol + cyclohexane system at 100 kPa. This figure might well represent the LLE for this same system at 200 kPa. The impact of pressure on LLE for a particular system is quite small and, thus, it is often omitted when presenting LLE data. As we saw in Chapter 11, the impact of pressure on the activity coefficients for liquid mixtures is much smaller than that of temperature. By extension, the Gibbs free energy for liquids will be a weak function of pressure. Since the Gibbs free energy is the important quantity for determining

LLE, it is reasonable to expect that LLE would be a weak function of the pressure, but (as observed in Figure 13-1) it is a strong function of temperature.

Note that when we use the phrase "weak function of pressure" in the previous paragraph, it is not the same as if we said "not a function of pressure." LLE *is* a function of pressure, just not as strong as the temperature. For example, the UCST for the methanol + cyclohexane system is at around 318 K at 101.325 kPa, as seen in Figure 13-1. However, at 100 MPa, the UCST is around 340 K. It is just that 100 MPa is well outside any typical operating system pressure, so considering a UCST as independent of pressure is normally a reasonable approximation for most practical purposes.

13.2.2 LLE—Components of the Gibbs Free Energy

Liquid–liquid equilibrium occurs because the mixture can lower its overall Gibbs free energy by separating into two liquid phases rather than staying in one phase. Consider Table 13-1, which shows the various components of the molar Gibbs free energy for pure methanol at 310 K and 101.3 kPa.

TABLE 13-1 Thermodynamic properties of methanol at 310 K and 101.3 kPa.

$\underline{G}$ (J/mol)	$\underline{U}$ (J/mol)	$-T\underline{S}$ (J/mol)	$P\underline{V}$ (J/mol)
-103	-2408	2301	4

Remember,
$$\underline{G} = \underline{U} - T\underline{S} + P\underline{V}$$

It is clear that the largest impact on the molar Gibbs free energy occurs through the molar internal energy term ($\underline{U}$) and the molar entropy term ($-T\underline{S}$). Consider the methanol + cyclohexane system presented in Figure 13-1. Methanol will form hydrogen bonds with itself, while cyclohexane is a non-polar hydrocarbon. As is regularly the case in phase equilibrium, there is a balance between the internal energy term and the entropic term, whose balance is mediated by the temperature. The more hydrogen bonding in the system, the more negative the internal energy term and the more negative the molar Gibbs free energy.

The addition of cyclohexane to methanol interrupts the energetically-beneficial hydrogen bonding that can occur between the methanol molecules. The system is then left with a "dilemma." It can split into two phases—one richer in the methanol—so than the energetic hydrogen bonding between the methanol molecules can be increased. However, the cost of splitting one liquid phase into two liquid phases is a lowering in the entropy. Indeed, this "de-mixing" of one phase into two phases (with each of the two phases concentrated in one of the components) lowers entropy. Here, once again, is where the temperature plays an important role. When the temperature is high (relatively speaking), the entropy loss associated with splitting into two phases is not overcome by the benefit from the hydrogen bonding to the internal energy term. Thus, the system will remain in one liquid phase (as seen in Figure 12-1). On the other hand, when the temperature is lowered (at certain compositions), the entropy term plays an increasingly smaller role (remember, it is $-T\underline{S}$), and the benefit to splitting one liquid phase into two liquid phases for the sake of hydrogen bonding results in the lowest Gibbs free energy for the system. Thus, the system will split into two liquid phases (as seen in Figure 12-1). It is important to note that the temperature where this occurs is a function of the composition of the system. While we called the maximum temperature associated with this phase split the upper critical solution temperature (UCST), at some compositions, the temperature needs to go increasingly lower for this energetic benefit to phase splitting to overcome the

For systems that have similar intermolecular interactions and reasonably similar sizes/shapes, you will not have liquid-liquid phase splitting at any temperature. This supports the rule of thumb that says "like dissolves like".

entropy cost. In fact, if this system is either highly concentrated or highly dilute in methanol, there is not a temperature low enough that will result in a phase split.

13.3 Various Types of LLE

What has been described in the previous section for the methanol + cyclohexane system as it relates to LLE is fairly standard. However, it is possible, though less regularly observed, for liquid–liquid equilibrium to occur for mixtures where a **lower critical solution temperature**, or LCST, is realized. For example, the propylene glycol n-propyl ether + water system will show an LCST at high water concentrations, as seen in Figure 13-3.

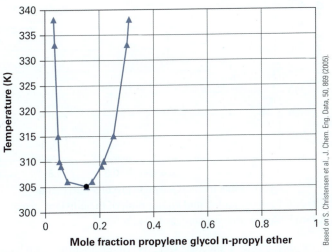

FIGURE 13-3 Liquid-liquid equilibrium of the propylene glycol
n-propyl ether + water system at 100 kPa.

The behavior seen in this glycol ether + water system in Figure 13-3 can be explained in terms of the hydrogen bonding. In particular, while propylene glycol n-propyl ether does not hydrogen bond with itself, its oxygen atom on the −OH group is an electron donor, which can hydrogen bond with an electron acceptor (the hydrogen atoms on the water molecules).

This cross-species hydrogen bonding plays an important role at the lower temperatures where the thermal energy contribution to the internal energy becomes less positive, resulting in a mixture that is completely miscible at all compositions below a certain temperature (the LCST). Above the LCST and in the dilute ether region of Figure 13-3, we see the system split into two liquid phases. The reason for this, once again, has to do with hydrogen bonding, but this time it is the hydrogen bonding for the water molecules with themselves. In the concentrated glycol ether regions above the LCST, one liquid phase exists as the energetic benefits to the system from the hydrogen bonding in the smaller number of water molecules do not overcome the entropic loss to the system as a result of splitting into two phases. However, at higher water concentrations (the dilute glycol ether region), larger numbers of energetically-favorable hydrogen bonds occur between the water molecules, and this system benefit becomes large (i.e.: more negative) enough to overcome the entropic loss of phase splitting. Accordingly, the system is not stable in this region and will split into two liquid phases.

Note that the entropic benefits to the system staying in one phase will increase with increasing temperature (as seen in Figure 13-1, which provides the UCST for the

FOOD FOR THOUGHT 13-2

Some polymer + water systems possess a UCST and an LCST, but it does not result in closed-loop phase behavior. What do you think that phase diagram would look like?

methanol + cyclohexane system). Thus, it can happen where some glycol ether + water systems will possess *both* a LCST (for the reasons mentioned for the propylene glycol *n*-propyl ether + water system) as well as an UCST. Indeed, this is observed in the ethylene glycol *n*-butyl ether + water system. This type of liquid–liquid equilibrium for these systems is referred to as a **closed-loop miscibility gap**. This is shown in Figure 13-4.

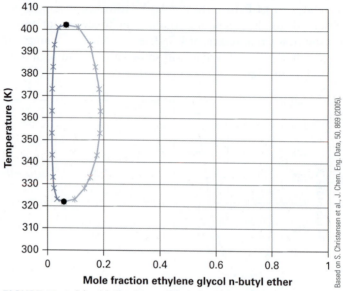

Ethylene glycol
n-butyl ether

FIGURE 13-4 Liquid–liquid equilibrium of the ethylene glycol *n*-butyl ether + water system at 100 kPa.

Note that the disparity in molecular weights (water: 18 g/mol; ethylene glycol *n*-butyl ether: 118 g/mol) of the molecules in this system allows for the data to be presented in a way that is easier to read. As such, something like Figure 13-4 would often be displayed where the mass fraction or mass percent is used, and this is shown in Figure 13-5. Both

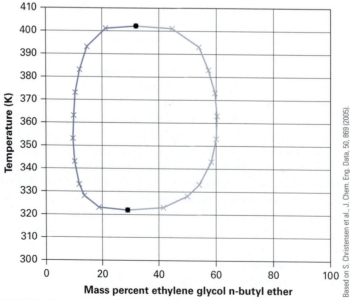

FIGURE 13-5 Liquid–liquid equilibrium of the ethylene glycol *n*-butyl ether + water system at 100 kPa. Data are presented as mass percent.

Figures 13-4 and 13-5 contain the same data and carry the same information, but the latter is presented in a way that uses a more appropriate variable for this system.

13.4 Miscibility Gaps from a ΔG of Mixing Perspective

Earlier in this chapter, we discussed the reason why a system would split into two liquid phases as opposed to staying in a single, liquid phase. In particular, the system will find the state (either one or two phases) that minimizes the overall Gibbs free energy of the system. Examining this from a graphical perspective provides useful insights, as well as allows for the development of a rule based on this graphical interpretation, and we provide this approach next.

If one mixes two liquids together at constant temperature and pressure, the system will go toward equilibrium as a result of the mixing process. Two scenarios are available: (1) the system keeps the overall Gibbs free energy the same or (2) the overall Gibbs free energy of the system is lowered. In the former situation, one could envision pouring a less dense liquid into a denser liquid and having the less dense liquid sit on the top of the denser liquid with absolutely no mixing. This "totally immiscible" situation would result in no change in the Gibbs free energy of the system as a result of this "mixing" process (which is scenario 1 above). This creates an upper bound on the process and, thus, all other observed mixing situations must result in a lowering of the overall system Gibbs free energy. We can make a mathematical relationship out of the description above.

The total Gibbs free energy of the system must be less than or equal to the Gibbs free energy of the individual components. We can write this as

$$G = n\underline{G} \le n_1 \underline{G}_1 + n_2 \underline{G}_2 \tag{13.1}$$

Dividing by the total number of moles, n, yields

$$\underline{G} \le \frac{(n_1 \underline{G}_1 + n_2 \underline{G}_2)}{n} \tag{13.2}$$

We can also write this in terms of the mole fraction as

$$\underline{G} \le x_1 \underline{G}_1 + x_2 \underline{G}_2 \tag{13.3}$$

or

$$\underline{G} - (x_1 \underline{G}_1 + x_2 \underline{G}_2) \le 0 \tag{13.4}$$

Recall that the definition of $\Delta \underline{G}_{mix}$ is

$$\Delta \underline{G}_{mix} = \underline{G} - (x_1 \underline{G}_1 + x_2 \underline{G}_2) \tag{13.5}$$

Thus, we can write Equation 13.4 as simply

$$\Delta \underline{G}_{mix} \le 0 \tag{13.6}$$

What does a graph of $\Delta \underline{G}_{mix}$ as a function of mole fraction look like? To help with this, let's employ an excess molar Gibbs free energy model. Recall that back in Chapter 9, we wrote

$$\underline{G}^E = \underline{G} - \underline{G}^{ID} = \underline{G} - (x_1 \underline{G}_1 + x_2 \underline{G}_2 + RTx_1 \ln(x_1) + RTx_2 \ln(x_2)) \tag{9.23}$$

which we can rewrite as

$$\underline{G}^E = \underline{G} - (x_1 \underline{G}_1 + x_2 \underline{G}_2) - RTx_1 \ln(x_1) - RTx_2 \ln(x_2)$$

or

$$\underline{G}^E + RTx_1 \ln(x_1) + RTx_2 \ln(x_2) = \underline{G} - (x_1 \underline{G}_1 + x_2 \underline{G}_2)$$

Thus, we can write $\Delta \underline{G}_{mix}$ in terms of the excess molar Gibbs free energy, since

$$\Delta \underline{G}_{mix} = \underline{G} - (x_1 \underline{G}_1 + x_2 \underline{G}_2) \tag{13.5}$$

and using Equation 9.23 gives

$$\Delta \underline{G}_{mix} = \underline{G}^E + RTx_1 \ln(x_1) + RTx_2 \ln(x_2) \tag{13.7}$$

Now that we have the molar Gibbs free energy of mixing in terms of the excess molar Gibbs free energy, we can use a model (such as the 2-parameter Margules equation):

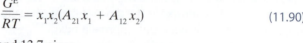

$$\frac{\underline{G}^E}{RT} = x_1 x_2 (A_{21} x_1 + A_{12} x_2) \tag{11.90}$$

Thus, using Equations 11.90 and 13.7 gives

$$\frac{\Delta \underline{G}_{mix}}{RT} = x_1 x_2 (A_{21} x_1 + A_{12} x_2) + x_1 \ln(x_1) + x_2 \ln(x_2) \tag{13.8}$$

If we choose $A_{12} = 3$ and $A_{21} = 2$, we arrive at Figure 13-6. This provides the molar Gibbs free energy of mixing for the system described by these Margules parameters as a function of composition.

We are actually plotting the molar Gibbs free energy of mixing divided by RT, but since we are interested in the process at a single temperature, the RT term is constant. Thus, the shape of the curve will not change at all when dividing by RT.

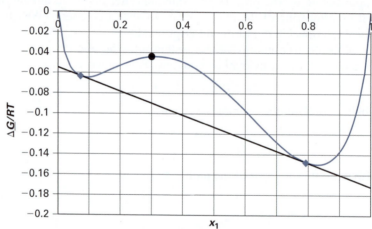

FIGURE 13-6 The molar Gibbs free energy of mixing for a system described by the 2-parameter Margules equation ($A_{12} = 3$ and $A_{21} = 2$).

Let's consider what Figure 13-6 is telling us. Basically, if we make mixtures of various compositions from pure components 1 and 2, the blue curve tells us the molar Gibbs free energy change as a result of this mixing at a particular composition. For example, if we mix 30 moles of substance (1) and 70 moles of substance (2), we will arrive with a mixture described on Figure 13-6 by the black circle. Clearly, compared to the pure components, this state will result in a lowering of the molar Gibbs free energy of the system since $\Delta \underline{G}_{mix}$ is negative. However, does the system exist as a single liquid phase or two liquid phases? To answer that, we first need to consider what possibilities are available to the system.

Consider a tangent line to the $\Delta \underline{G}_{mix}$ curve that touches the curve in two places, as shown in Figure 13-6. Imagine you had two systems: one of the systems had a composition of one of the diamonds and the other had a composition of the other diamond. It is clear that both systems have a $\Delta \underline{G}_{mix}$ lower than the mixture indicated by the black circle. Thus, if this one liquid phase (composition 30% of component 1) could split into the two liquid phases identified by the two diamonds, the resulting system would have an overall *lower* Gibbs free energy since the $\Delta \underline{G}_{mix}$ would have a larger negative value. Therefore, the system is not stable and would split into two liquid phases. Indeed, the ability to draw a tangent line to this curve that can touch the curve at *two compositions* provides an opportunity for phase splitting between those compositions.

The compositions at those two equilibrium liquid phases is something that can be estimated graphically or solved analytically; the latter approach is left for a end of chapter problem. For this system, the tangent line hits the curve at two points, as can be seen in Figure 13-6:

$$x_1 = 0.072$$

and

$$x_1 = 0.792$$

Note that these points are *not* the local minima seen in the curve, which occur at $x_1 = 0.092$ and $x_1 = 0.827$.

From a graphical perspective, if a $\Delta \underline{G}_{mix}$ curve possesses a tangent line that can touch the $\Delta \underline{G}_{mix}$ curve twice, it must change concavity from concave up to concave down and back to concave up. If it does not, then the curve will not possess this double tangency. Thus, it will not phase split into two liquid phases at any composition. This observation allows us to write a graphical rule associated with mixture stability. Namely, if a plot of the molar Gibbs free energy of mixing for the mixture is always concave up across the composition range, then it is stable (i.e., won't split into two phases).

> Mathematically, this is written as
>
> If $\left(\dfrac{\partial^2 \Delta \underline{G}_{mix}}{\partial x_1^2} \right)_{T,P} > 0$, then the mixture is stable (13.9)

For example, if our two-component system discussed above was at a different (normally higher) temperature, the Margules parameters would likely change. Thus, if those Margules equation parameters at this new temperature were $A_{12} = 1$ and $A_{21} = 2$, we would arrive at Figure 13-7. As can be seen, there is no double tangency

If you are already thinking that your analytical solution approach will involve equality of mixture fugacities of the components, you would be correct!

Since T is constant when evaluating the derivative, Equation 13.9 can be written equivalently as:

If $\left(\dfrac{\partial^2 [\Delta \underline{G}_{mix}/RT]}{\partial x_1^2} \right)_{T,P} > 0,$

then the mixture is stable.

FOOD FOR THOUGHT 13-3

Why did we write "normally higher" associated with the temperature when discussing that a liquid mixture would go from not stable to stable?

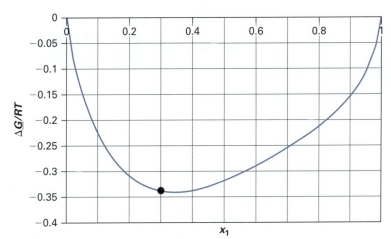

FIGURE 13-7 The molar Gibbs free energy of mixing for a system described by the 2-parameter Margules equation ($A_{12} = 1$ and $A_{21} = 2$).

here and the curve is always concave up. Thus, this mixture is stable (does not phase split) and is miscible across all compositions at this temperature. This is not a surprising result at all, considering we have already discussed the concepts of a UCST and LCST. Here, the new parameters at this new temperature (that does not show a liquid–liquid phase split) would be above the UCST, while the parameters used before would describe the same mixture—but at a temperature below the UCST, for example.

The existence of a concavity change in a plot of $\Delta \underline{G}_{\text{mix}}$ vs. composition is an easily visible way to determine the presence of a miscibility gap for a system. From a modeling standpoint, however, it is more convenient to use relationships derived in terms of the excess molar Gibbs free energy, since much of liquid modeling is done in terms of the excess molar Gibbs free energy. In the next section, we will do just that.

13.5 Stability Criterion for Liquid Mixtures

Our goal is to start with Equation 13.9

$$\text{If } \left(\frac{\partial^2 \Delta \underline{G}_{\text{mix}}}{\partial x_1^2} \right)_{T,P} > 0, \text{then the mixture is stable} \tag{13.9}$$

and derive this relationship in terms of the excess molar Gibbs free energy instead.

Earlier in this chapter, we reminded the reader of the relationship between the excess molar Gibbs free energy and $\Delta \underline{G}_{\text{mix}}$:

$$\Delta \underline{G}_{\text{mix}} = \underline{G}^{\text{E}} + RTx_1 \ln(x_1) + RTx_2 \ln(x_2) \tag{13.7}$$

Our goal is to obtain the second derivative of $\Delta \underline{G}_{\text{mix}}$, with respect to x_1 (at constant T and P), so we divide out by RT and take the first derivative (making sure to replace x_2 with $1 - x_1$) as

$$\left(\frac{\partial [\Delta \underline{G}_{\text{mix}}/RT]}{\partial x_1} \right)_{T,P} = \left(\frac{\partial [\underline{G}^{\text{E}}/RT]}{\partial x_1} \right)_{T,P} + \left[\frac{x_1}{x_1} + \ln(x_1) + \frac{-(1 - x_1)}{1 - x_1} - \ln(1 - x_1) \right] \tag{13.10}$$

$$\left(\frac{\partial [\Delta \underline{G}_{\text{mix}}/RT]}{\partial x_1} \right)_{T,P} = \left(\frac{\partial [\underline{G}^{\text{E}}/RT]}{\partial x_1} \right)_{T,P} + [\ln(x_1) - \ln(1 - x_1)] \tag{13.11}$$

Moving to the second derivative,

$$\left(\frac{\partial^2 [\Delta \underline{G}_{\text{mix}}/RT]}{\partial x_1^2} \right)_{T,P} = \left(\frac{\partial^2 [\underline{G}^{\text{E}}/RT]}{\partial x_1^2} \right)_{T,P} + \left(\frac{1}{x_1} + \frac{1}{1 - x_1} \right) \tag{13.12}$$

$$\left(\frac{\partial^2 [\Delta \underline{G}_{\text{mix}}/RT]}{\partial x_1^2} \right)_{T,P} = \left(\frac{\partial^2 [\underline{G}^{\text{E}}/RT]}{\partial x_1^2} \right)_{T,P} + \left(\frac{1}{x_1} + \frac{1}{x_2} \right) \tag{13.13}$$

$$\left(\frac{\partial^2 [\Delta \underline{G}_{\text{mix}}/RT]}{\partial x_1^2} \right)_{T,P} = \left(\frac{\partial^2 [\underline{G}^{\text{E}}/RT]}{\partial x_1^2} \right)_{T,P} + \left(\frac{1}{x_1 x_2} \right) \tag{13.14}$$

Therefore, the stability criterion becomes:

$$\text{If} \left(\frac{\partial^2 [\underline{G}^E/RT]}{\partial x_1^2}\right)_{T,P} + \left(\frac{1}{x_1 x_2}\right) > 0, \text{ then the mixture is stable} \qquad (13.15)$$

Before we use this criterion, we note that one can arrive at another version of the above stability criterion in terms of the activity coefficient. There are situations where this expression is more easily used and, thus, we provide it here as

$$\text{If} \left(\frac{\partial [\ln(\gamma_1)]}{\partial x_1}\right)_{T,P} > -\left(\frac{1}{x_1}\right), \text{ then the mixture is stable} \qquad (13.16)$$

The derivation of this relationship is left as an end-of-chapter problem.

FOOD FOR THOUGHT 13-4

Equation 13.16 can be written equivalently as

$$\left(\frac{\partial [\ln(x_1 \gamma_1)]}{\partial x_1}\right)_{T,P} > 0.$$

Can you derive it?

| **USING THE LIQUID PHASE STABILITY CRITERIA** | **EXAMPLE 13-3** |

Can the following excess molar Gibbs free energy models predict a liquid–liquid phase split?

A. Ideal solution

B. 1-Parameter Margules equation

SOLUTION:

Step 1 *Ideal solution—your intuition*
Your intuition should be telling you, "*How can an ideal solution predict phase splitting for the liquid, since the model assumes similar size and similar molecular interactions . . . and it is the dissimilar size and/or dissimilar interactions that give rise to liquid–liquid phase splitting?*"

Your intuition would be correct here, but let's go through the formality of the approach so we can use this to evaluate part B of this problem.

Step 2 *Ideal solution—excess molar Gibbs free energy stability criterion*
Let's use the excess molar Gibbs free energy stability criterion for evaluation of the ideal solution. The excess molar Gibbs free energy expression for the ideal solution is simply zero.

$$\frac{\underline{G}^E}{RT} = 0 \qquad (13.17)$$

The relevant version of the stability criterion is given in Equation 13.15

$$\text{If} \left(\frac{\partial^2 [\underline{G}^E/RT]}{\partial x_1^2}\right)_{T,P} + \left(\frac{1}{x_1 x_2}\right) > 0, \text{ then the mixture is stable} \qquad (13.15)$$

The second derivative of 0 with respect to x_1 is still 0. Thus, the mixture is stable if the left-hand side is positive (i.e., greater than zero). The range of values that x_1 and x_2 can take in a mixture are between 0 and 1 (exclusive of the endpoints). Therefore, the left-hand side can never be negative or zero for the ideal solution. Thus, the ideal solution model will always produce a stable liquid phase. Our intuition is confirmed.

Step 3 *1-parameter Margules equation*
We've already shown that the 2-parameter Margules equation allows for liquid–liquid phase splitting (see Figure 13-6). But what about the 1-parameter Margules equation?

$$\frac{\underline{G}^E}{RT} = Ax_1 x_2 \qquad (11.75)$$

You are taking the second derivative of a function which is quadratic in x_1. Thus, you should expect the second derivative to include only constants.

We have to take the second derivative of this expression, which is

$$\left(\frac{\partial^2[\underline{G}^E/RT]}{\partial x_1^2}\right)_{T,P} = \left(\frac{\partial^2[Ax_1x_2]}{\partial x_1^2}\right)_{T,P} = A\left(\frac{\partial^2[x_1(1-x_1)]}{\partial x_1^2}\right)_{T,P}$$

$$= A\left(\frac{\partial^2[x_1 - x_1^2]}{\partial x_1^2}\right)_{T,P} = A(0-2) = -2A$$

(13.18)

So the criterion for the 1-parameter Margules equation becomes

$$\text{If} - 2A > -\left(\frac{1}{x_1x_2}\right), \text{ then the mixture is stable}$$

(13.19)

We can solve Equation 13.19 for A and arrive at the relationship

$$\text{If } A < \left(\frac{1}{2x_1x_2}\right), \text{ then the mixture is stable}$$

(13.20)

So, when the value of A for a particular system does not make this relationship true (for a particular composition), the mixture is not stable and splits into two phases. This leads to the question: Is there a value of A such that the mixture will *always* be stable for all compositions? Since the relationship is a "less-than" relationship, we want to know the smallest value that the right-hand side can take. If A is smaller than that value from $x_1 = 0$ to $x_1 = 1$, then the mixture will always be stable. While there are a few ways to do this, the simplest would be to plot the right-hand side, $1/(2x_1x_2)$, as a function of x_1, which is provided in Figure 13-8.

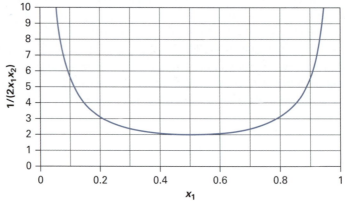

FIGURE 13-8 Evaluating the range of stability for the 1-parameter Margules equation. If $A < 2$, the mixture is stable for all compositions.

As can be seen in Figure 13-8, if A is less than 2, the mixture will be stable for all compositions. If the Margules parameter is greater than 2, then the mixture will have a region of liquid–liquid instability (or a miscibility gap).

We can apply the stability criterion (in any of its forms) to other excess molar Gibbs free energy models discussed in Chapter 11. It is noted that the Wilson equation does not allow for phase splitting, though modified versions of the Wilson equation are available that do allow that do allow for the prediction/correlation of miscibility gaps.

13.6 Modeling Liquid–Liquid Equilibrium

Now that we have shown what LLE looks like on phase diagrams and how it can be viewed from a stability standpoint, our next approach is to model this phase behavior. We start at the beginning: equality of mixture fugacities of the components.

$$\hat{f}_i^\alpha = \hat{f}_i^\beta \tag{11.52}$$

Since this is two liquid phases, one approach is to rewrite the mixture fugacity of component i through the use of the activity coefficient, as we did several times in Chapter 11 (see Equation 11.44). However, this time it is written for both phases as

$$x_i^\alpha f_i^\alpha \gamma_i^\alpha = x_i^\beta f_i^\beta \gamma_i^\beta \tag{13.21}$$

Since f_i is the *pure* component fugacity evaluated at the mixture T and P in the liquid phase, this term will be the same value on both sides of the equation. Thus, the relationship simplifies to

$$x_i^\alpha \gamma_i^\alpha = x_i^\beta \gamma_i^\beta \tag{13.22}$$

Written for a binary mixture, we have

$$x_1^\alpha \gamma_1^\alpha = x_1^\beta \gamma_1^\beta \tag{13.23}$$

$$\left(1 - x_1^\alpha\right)\gamma_2^\alpha = \left(1 - x_1^\beta\right)\gamma_2^\beta \tag{13.24}$$

So now that we have a modeling approach using the activity coefficients, what is its utility? One issue to identify right away is that this is not like vapor–liquid modeling using modified Raoult's Law. There, the vapor side was modeled using an ideal gas, which brought the pressure into the modeling in a very explicit way. Indeed, we could write out the pressure expression directly from that vapor–liquid modeling approach. However, here the pressure dependence is tied up very weakly into the activity coefficients and is not explicit in Equation 13.22. With this difference in mind, we explore the use of excess molar Gibbs free energy modeling approaches as they apply to liquid–liquid equilibrium.

For Modified Raoult's Law, the pressure is given as
$$P = x_1 \gamma_1 P_1^{\text{sat}} + x_2 \gamma_2 P_2^{\text{sat}}$$

USING MODEL PARAMETERS TO PREDICT THE EXISTENCE AND LOCATION OF A MISCIBILITY GAP	EXAMPLE 13-4

Consider the 1-butanol (1) + water (2) system. The parameters for this liquid mixture at 333.15 K using the 2-parameter Margules equation are $A_{12} = 2.7576$ and $A_{21} = 1.3064$ (Gmehling and Onken, 1977). Determine whether this system will split into two phases and, if so, what will be the composition of each phase.

1-butanol

SOLUTION:

Step 1 *Write out the modeling expressions for the 2-parameter Margules equation*
We provide Equations 13.23 and 13.24, which are general for LLE. However, we need expressions for the activity coefficient to use these equations.

$$x_1^\alpha \gamma_1^\alpha = x_1^\beta \gamma_1^\beta \tag{13.23}$$

$$\left(1 - x_1^\alpha\right)\gamma_2^\alpha = \left(1 - x_1^\beta\right)\gamma_2^\beta \tag{13.24}$$

Recall we have used the 2-parameter Margules equation a few times before and have provided the activity coefficient expressions through the equations:

$$\ln[\gamma_1] = x_2^2(A_{12} + 2[A_{21} - A_{12}]x_1) \tag{11.91}$$

$$\ln[\gamma_2] = x_1^2(A_{21} + 2[A_{12} - A_{21}]x_2) \tag{11.92}$$

Step 2 *Solving for the equilibrium liquid compositions*

Since the activity coefficient expressions are in terms of the given parameters and the composition of each phase, we have two unknowns (x_1^α and x_1^β) in the two Equations 13.23 and 13.24. This system of two non-linear equations in two unknowns is readily solved through a variety of approaches (such as your calculator or the solver function in MS Excel) to arrive at

$$x_1^\alpha = 0.093$$

$$x_1^\beta = 0.645$$

That solutions are obtained with real compositions (between 0 and 1) indicates the existence of the miscibility gap.

Note that the experimental result for this system at this temperature is

$$x_1^\alpha = 0.016$$

$$x_1^\beta = 0.440$$

So while the model parameters used in the correlation of the VLE data allowed for the identification of the presence of the miscibility gap, they were not very accurate in predicting the composition of the equilibrium liquid phases.

One can do a similar calculation with the van Laar equation ($L_{12} = 3.7739$; $L_{21} = 1.3244$) (Gmehling and Onken, 1977) and improve on the predicted results: $x_1^\alpha = 0.025$ and $x_1^\beta = 0.574$.

EXAMPLE 13-5	DETERMINING MODEL PARAMETERS FROM LLE DATA

Consider the methanol (1) + cyclohexane (2) system as depicted in Figure 13-1. Use the data at 300 K to determine the model parameter for the 1-parameter Margules equation.

SOLUTION:

Step 1 *Obtaining the experimental information from the figure*

From Figure 13-1, we can identify composition of the coexisting phases at 300 K:

$$x_1^\alpha = 0.13$$

$$x_1^\beta = 0.82$$

Step 2 *Using the model equations*

We use Equations 13.23 and 13.24, as in Example 13-4, but now our activity coefficient expressions are given by the 1-parameter Margules equation:

$$x_1^\alpha \gamma_1^\alpha = x_1^\beta \gamma_1^\beta \tag{13.23}$$

$$(1 - x_1^\alpha)\gamma_2^\alpha = (1 - x_1^\beta)\gamma_2^\beta \tag{13.24}$$

$$\ln[\gamma_1] = Ax_2^2 \tag{11.81}$$

$$\ln[\gamma_2] = Ax_1^2 \tag{11.82}$$

Step 3 *Finding the A parameter*

Unlike Example 13-4, where we knew the model parameters and determined the equilibrium compositions, here we know the compositions but do not know the parameter value. Thus, the unknown in the modeling approach is A, found through the activity coefficient expressions.

We have one unknown (A) in the two equations (13.23 and 13.24). Rather than use the equilibrium composition of one of the phases to solve for A at the expense of the equilibrium composition of the other phase, the best course of action is to write an objective function that gives a "best fit" A value. Accordingly, we can use Equations 13.23 and 13.24 to write an objective function that we will minimize to determine the best fit A value. Thus,

$$\text{OBJ} = [x_1^\alpha \gamma_1^\alpha - x_1^\beta \gamma_1^\beta]^2 + [(1 - x_1^\alpha)\gamma_2^\alpha - (1 - x_1^\beta)\gamma_2^\beta]^2$$

Plugging in the 1-parameter Margules equation expressions for the activity coefficients (Equations 11.81 and 11.82) yields

$$\text{OBJ} = \left[x_1^\alpha \exp\left[A(1 - x_1^\alpha)^2\right] - x_1^\beta \exp\left[A(1 - x_1^\beta)^2\right]\right]^2$$
$$+ \left[(1 - x_1^\alpha)\exp\left[A(x_1^\alpha)^2\right] - (1 - x_1^\beta)\exp\left[A(x_1^\beta)^2\right]\right]^2$$

Upon plugging in the known experimental compositions given in the problem statement, we can minimize the objective function to find the best fit value for A, which is $A = \mathbf{2.475}$. Note that this value makes sense since we showed earlier in the chapter that for values of $A > 2$, a miscibility gap will exist for the 1-parameter Margules equation.

When we modeled vapor–liquid equilibrium in Chapters 11 and 12, we used parameter values to help predict the behavior at different states (temperatures or compositions). Indeed, we used one experimental data point to help predict the equilibrium VLE curve at other states. Is it possible to use the same approach for LLE? Not exactly. Recall that those were isothermal systems where only the pressure changed. And since we know that the excess molar Gibbs free energy is a weak function of pressure, we were able to use those parameters at different pressures (but the same temperature) in order to predict a new phase point. For our LLE work, we would be interested in predicting the phase behavior at different temperatures (such as in Figure 13-1). However, like pressure, temperature is not explicit in our modeling approach, and the only way temperature impacts the modeling is through the activity coefficients (and those are parameterized at different temperatures). Accordingly, we will need to estimate how our parameter values change with temperature. We explore this approach in Example 13-6.

USING MODEL PARAMETERS TO PREDICT THE LLE CURVE	**EXAMPLE 13-6**

Using the 1-parameter Margules equation model obtained for LLE for the methanol (1) + cyclohexane (2) system at 300 K from Example 13-5, estimate the LLE curve as a function of temperature.

SOLUTION:

Step 1 *Determining how A changes with temperature*
In order to determine how A changes with temperature, we need to remind ourselves how $\underline{G}^E/RT$ changes with temperature. Recall from Chapter 11 that Equation 11.111 gave us an approach to use:

$$\frac{\partial}{\partial T}\left(\frac{\underline{G}^E}{RT}\right)_{P,n_1,n_2} = -\left(\frac{\underline{H}^E}{RT^2}\right) \tag{11.111}$$

Our $\underline{G}^E/RT$ model is the Margules equation (1-parameter):

$$\frac{\underline{G}^E}{RT} = Ax_1 x_2 \tag{11.75}$$

Plugging Equation 11.75 into 11.111, we have

$$\frac{\partial}{\partial T}(Ax_1x_2)_{P,n_1,n_2} = -\left(\frac{\underline{H}^E}{RT^2}\right) \tag{13.25}$$

$$x_1x_2\frac{\partial}{\partial T}(A)_{P,n_1,n_2} = -\left(\frac{\underline{H}^E}{RT^2}\right) \tag{13.26}$$

We can separate variables to obtain:

$$x_1x_2\,dA = -\left(\frac{\underline{H}^E}{RT^2}\right)dT \tag{13.27}$$

Step 2 *A relationship for the excess molar enthalpy*

As we know from Figure 9-3, in general the excess molar enthalpy is both a function of temperature and a function of composition. What does the excess molar enthalpy look like for the system in our problem? In Figure 13-9, we plot the excess molar enthalpy at two temperatures: one above the UCST and one below the UCST.

Recall that the experimental UCST is 318 K.

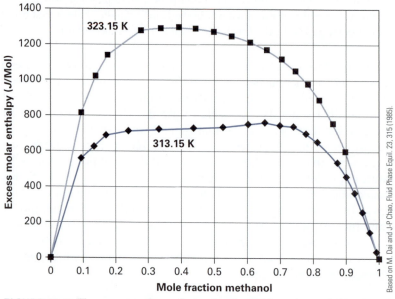

FIGURE 13-9 The excess molar enthalpy for the methanol + cyclohexane system at two temperatures.

Based on M. Dai and J-P Chao, Fluid Phase Equil. 23, 315 (1985).

When the system is in a single liquid phase (323.15 K), we have a parabolic-like shape whose behavior can be fit with a polynomial in terms of the mole fraction. However, when the excess molar enthalpy goes through a miscibility gap (313.15 K), the function has a plateau. Indeed, one can roughly estimate the compositions of the miscibility gap through examination of the excess molar enthalpy.

The miscibility gap at 313.15 K goes from about 25% methanol to 72% methanol (by mole) according to Figure 13-1.

For demonstration/illustration purposes in this problem, let's parameterize the excess molar enthalpy curve above the UCST at 323.15 K, where the data can be estimated with a parabola. Here, our best fit model becomes

$$\underline{H}^E = 5718.65x_1x_2 \tag{13.28}$$

with $\underline{H}^E$ in units of J/mol.

Step 3 *The functional relationship for A = A(T)*

We can plug Equation 13.28 back into Equation 13.27 for

$$x_1x_2\,dA = -\left(\frac{5718.65x_1x_2}{RT^2}\right)dT \tag{13.29}$$

$$dA = -\left(\frac{5718.65}{RT^2}\right)dT \tag{13.30}$$

Integration yields

$$\int_{A(300\,\text{K})}^{A(T)} dA = -\int_{300\,\text{K}}^{T}\left(\frac{5718.65}{RT^2}\right)dT \tag{13.31}$$

We know an estimate for the value of A at 300 K from Example 13-5, which was 2.475.

$$A(T) - A(300\,\text{K}) = -\frac{5718.65}{R}\left[\frac{1}{300\,\text{K}} - \frac{1}{T}\right] \tag{13.32}$$

$$A(T) = 2.475 - \frac{5718.65}{R}\left[\frac{1}{300\,\text{K}} - \frac{1}{T}\right] \tag{13.33}$$

Equation 13.33 was our goal, which was a relationship for how A changed as a function of the temperature.

Step 4 *Predicting the LLE behavior*

Our mixture fugacity expressions are, once again, Equations 13.23 and 13.24. We have three unknowns, A, x_1^α, and x_1^β. However, A is a function of the temperature (Equation 13.33), so we can fix the temperature and solve for x_1^α and x_1^β. Equations 13.23 and 13.24, written for the 1-parameter Margules equation, are as follows:

$$x_1^\alpha \exp\left[A(1 - x_1^\alpha)^2\right] = x_1^\beta \exp\left[A(1 - x_1^\beta)^2\right]$$

$$(1 - x_1^\alpha)\exp\left[A(x_1^\alpha)^2\right] = (1 - x_1^\beta)\exp\left[A(x_1^\beta)^2\right]$$

When we solve these equations at a range of temperatures, we can create the desired LLE predictions. This is given in Figure 13-10.

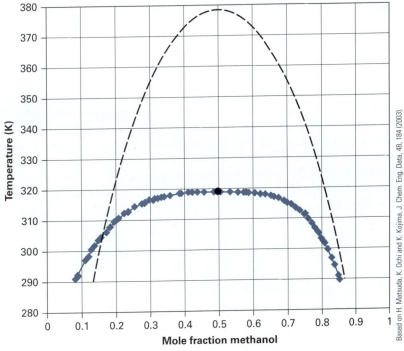

Based on H. Matsuda, K. Ochi and K. Kojima, J. Chem. Eng. Data, 48, 184 (2003).

FIGURE 13-10 The methanol + cyclohexane system experimental data (blue diamonds) and model predictions (black dashed line).

There are a few things to notice about Figure 13-10. First, even at 300 K, the correlation of experimental data using the 1-parameter Margules equation is not exact. Second, the model is symmetric, per the symmetry previously discussed related to the 1-parameter Margules equation. This means that the predicted composition at the UCST is equimolar. Third, the UCST is predicted to be much larger than the experimental value. Improvements on this approach can be realized through better excess molar Gibbs free energy models as well as a more accurate description of the excess molar enthalpy.

There is another way to estimate the LLE for various systems that runs in parallel to what we did previously in Chapter 12. Recall that we employed a phi-phi modeling approach, where an equation of state was used to estimate the mixture fugacity in order to predict vapor–liquid phase behavior. Likewise, we can do something similar, but rather than predict VLE by an equation of state, we can predict *LLE* by an equation of state. However, this is beyond the scope of an introductory thermodynamics textbook. Instead, we will go the other way and explore simplifications for very immiscible systems.

13.6.1 Modeling Liquid–Liquid Equilibrium— Immiscible Systems

In general, activity coefficients less than about 10 will result in totally miscible systems. As the value of the activity coefficients increases, the likelihood for liquid–liquid equilibrium grows (along with the size of the miscibility gap) (Kontogeorgis and Folas, 2009).

To model systems that have miscibility gaps across ranges of composition, we employed excess molar Gibbs free energy models to obtain the activity coefficients. However, for systems where the miscibility gap span most, if not all, of the composition region, simplifications arise. Functionally, these are called *immiscible systems* because each liquid phase is very rich in one of the components. Modeling for these systems starts as before using Equations 13.23 and 13.24 for

$$x_1^\alpha \gamma_1^\alpha = x_1^\beta \gamma_1^\beta \tag{13.23}$$

$$(1 - x_1^\alpha)\gamma_2^\alpha = (1 - x_1^\beta)\gamma_2^\beta \tag{13.24}$$

Consider α to be the phase rich in component 1 and β to the phase rich in component 2. For phase α, the activity coefficient for component 1, γ_1^α can be set to a value of 1 since the phase is almost all component 1. For phase β, the activity coefficient for component 2, γ_2^β can be set to a value of 1 since the phase is almost all component 2.

The activity coefficient of the dilute species can be equated to the infinite dilution activity coefficient of the dilute species in the concentrated species. Thus, the model simplifies to

$$x_1^\alpha = x_1^\beta \gamma_1^\infty \tag{13.34}$$

$$(1 - x_1^\alpha)\gamma_2^\infty = (1 - x_1^\beta) \tag{13.35}$$

With knowledge of the infinite dilution activity coefficients, one can estimate the compositions, or *vice versa*. This is demonstrated in Example 13-7.

IMMISCIBLE LIQUID SYSTEMS	EXAMPLE 13-7

The water (1) + di-ethyl ether (2) system is almost totally immiscible across all composition regions. At 307.3 K, two liquid phases exist. The water-rich phase has a composition of $x_1^\alpha = 0.9889$ while the ether-rich phase has a composition of $x_1^\beta = 0.0495$ (Gomis et al., 2000). Estimate the infinite-dilution activity coefficients for this mixture at this temperature.

Di-ethyl ether

SOLUTION:

Step 1 *List the equations and unknowns*
For this immiscible system we will have two equations and two unknowns (using Equations 13.34 and 13.35).

$$x_1^\alpha = x_1^\beta \gamma_1^\infty \tag{13.34}$$

$$(1 - x_1^\alpha)\gamma_2^\infty = (1 - x_1^\beta) \tag{13.35}$$

The unknowns are γ_1^∞ and γ_2^∞. They can be solved for directly from each equation.

Step 2 *Solving for the infinite-dilution activity coefficients*
Plugging in the known compositions given from the problem statement, we find the following:

$$0.9889 = (0.0495)\gamma_1^\infty$$

$$\gamma_1^\infty = 19.98$$

$$(1 - 0.9889)\gamma_2^\infty = (1 - 0.0495)$$

$$\gamma_2^\infty = 85.63$$

These infinite-dilution activity coefficients are far greater than an ideal solution (which has $\gamma = 1$) as well as those values obtained for our other systems in Chapter 11. For comparison's sake, the value for our ethanol + butyl methyl ether system explored in Example 11-5 had infinite dilution activity coefficients under 4.

13.7 Vapor–Liquid–Liquid Equilibrium (VLLE)

Consider Figure 13-11, that models the 1,1,1,3,3-pentafluropropane (or, more commonly, R-245fa) + *n*-pentane system. We see three *Txy* curves for this system at different pressures (1 MPa, 500 kPa, and 100 kPa). Toward the bottom of Figure 13-11, at the lower temperatures, we see a typical miscibility gap dome indicative of liquid–liquid equilibrium. Unlike the VLE, we don't need to associate a pressure with the LLE since it is only a very weak function of pressure. Indeed, both the 100 kPa and 1 MPa predictions are shown for LLE, but they overlay each other so you only see one dome.

As noticed in Figure 13-11, as the pressure decreases, the temperature associated with the VLE will also decrease. This is a result we know from Chapter 10. However, what if we are interested in the *Txy* curve at 10 kPa? If we project down to lower temperatures associated with this lower pressure, it seems that the *Txy* plot will intersect the LLE dome. We show this on Figure 13-12.

First recall that a *Txy* plot describes a vapor phase (the dew-point curve) in equilibrium with a liquid phase (the bubble-point curve). Additionally, the lower

1,1,1,3,3-pentafluoropropane

n-pentane

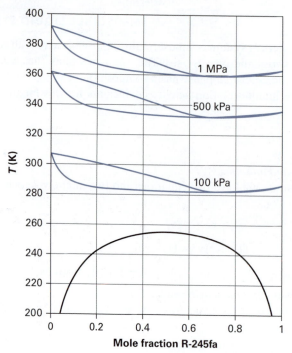

FIGURE 13-11 Representative VLE and LLE for the
R-245fa + *n*-pentane system at multiple pressures.

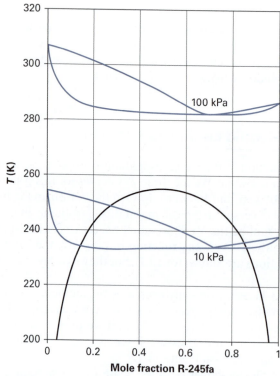

FIGURE 13-12 Representative VLE and LLE for the
R-245fa + *n*-pentane system at 100 kPa and 10 kPa.

portion of the LLE curve is properly represented below the temperature in Figure 13-12 where the LLE curve intersects the bubble-point curve. However, at the intersection of the LLE curve with the bubble-point curve in Figure 13-12, the liquid does not exist as one phase, but as two liquid phases (which are both in equilibrium with a vapor phase). This three-phase equilibrium is called vapor–liquid–liquid equilibrium (VLLE) and occurs at the one unique temperature where the LLE curve intersects the bubble-point curve. We draw a tie line to show this on the plot and remove the part of the LLE above the bubble-point curve. This is shown on Figure 13-13.

It is instructive to use Figure 13-13 to emphasize the phases that are in equilibrium at certain states. We do this through Example 13-8.

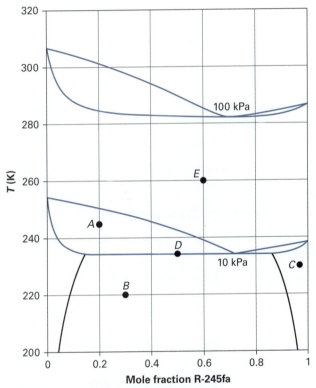

This mixture possesses what is called a heterogenous azeotrope since the miscibility gap encompasses the azeotrope and, thus, the vapor-phase composition for VLLE is at the azeotrope.

FIGURE 13-13 Representative VLE, LLE, and VLLE for the R-245fa + n-pentane system at 100 kPa and 10 kPa.

| READING A VLLE PHASE DIAGRAM | EXAMPLE 13-8 |

Consider the R-245fa (1) + n-pentane (2) system in Figure 13-13. At the different state points (black dots) provided on Figure 13-13, name the equilibrium phases and the compositions of those equilibrium phases. Note that the overall mixture composition is given using the symbol w.

Point A: $T = 245$ K; $P = 10$ kPa; $w_1 = 0.2$

Point B: $T = 220$ K; $P = 10$ kPa; $w_1 = 0.3$

Point C: $T = 228$ K; $P = 10$ kPa; $w_1 = 0.97$

Point D: $T = 234$ K; $P = 10$ kPa; $w_1 = 0.5$

Point E: $T = 260$ K; $P = 10$ kPa; $w_1 = 0.6$

SOLUTION:

Step 1 *Point A*

Point A: $T = 245$ K; $P = 10$ kPa; $w_1 = 0.2$

Point A is below the dew-point curve and above the bubble-point curve. Therefore, it splits into two phases: one liquid phase and one vapor phase.

Liquid phase: $x_1 = 0.02$

Vapor phase: $y_1 = 0.42$

Step 2 *Point B*

Point B: $T = 220$ K; $P = 10$ kPa; $w_1 = 0.3$

Point B is below the bubble point-curve, so it is a liquid. However, it is within the LLE dome, so it will split into two liquid phases.

Liquid phase α: $x_1^\alpha = 0.07$

Liquid phase β : $x_1^\beta = 0.92$

Step 3 *Point C*

Point C: $T = 228$ K; $P = 10$ kPa; $w_1 = 0.97$

Point C is below the bubble-point curve, so it is a liquid. It is also outside the LLE dome, so it is one liquid phase.

Liquid phase: $x_1 = 0.97$

Step 4 *Point D*

Point D: $T = 234$ K; $P = 10$ kPa; $w_1 = 0.5$

Point D is on the VLLE tie line. This three-phase equilibrium exists of two liquid phases and one vapor phase.

Liquid phase α : $x_1^\alpha = 0.14$

Liquid phase β : $x_1^\beta = 0.86$

Vapor phase: $y_1 = 0.71$

Step 5 *Point E*

Point E: $T = 260$ K; $P = 10$ kPa; $w_1 = 0.6$

Point E is above the dew-point curve, so the mixture exists only as a single vapor phase.

Vapor phase: $y_1 = 0.6$

Note that if this problem identified point E as $T = 260$ K; $P = 100$ kPa; $w_1 = 0.6$, we would use the upper curve. This point would now be below the bubble-point curve and, thus, would exist only as a single *liquid* phase.

Actual *amounts* of each equilibrium phase can be determined using a material balance, if given an input amount (in moles or mass).

13.8 Modeling of Vapor–Liquid–Liquid Equilibrium (VLLE)

We start at the same place in modeling phase equilibrium, namely the equality of mixture fugacities of component i. However, now we have three phases in equilibrium instead of two:

L1 means "liquid phase 1" and L2 means "liquid phase 2."

$$\hat{f}_i^{L1} = \hat{f}_i^{L2} = \hat{f}_i^{V} \tag{13.36}$$

For a binary mixture, we have

$$\hat{f}_1^{L1} = \hat{f}_1^{L2} = \hat{f}_1^{V} \tag{13.37}$$

$$\hat{f}_2^{L1} = \hat{f}_2^{L2} = \hat{f}_2^{V} \tag{13.38}$$

Substituting as before, with activity coefficients for the liquid phases (Equation 11.44) and fugacity coefficients for the vapor phase (Equation 11.49), we have:

$$x_1^{L1} f_1^{L1} \gamma_1^{L1} = x_1^{L2} f_1^{L2} \gamma_1^{L2} = \hat{\varphi}_1 y_1 P \tag{13.39}$$

$$x_2^{L1} f_2^{L1} \gamma_2^{L1} = x_2^{L2} f_2^{L2} \gamma_2^{L2} = \hat{\varphi}_2 y_2 P \tag{13.40}$$

There are six equations contained within the collection of Equations 13.39 and 13.40, but only four are independent. The two equations for the phase equilibrium between the two liquid phases are:

$$x_1^{L1} f_1^{L1} \gamma_1^{L1} = x_1^{L2} f_1^{L2} \gamma_1^{L2} \tag{13.39a}$$

$$x_2^{L1} f_2^{L1} \gamma_2^{L1} = x_2^{L2} f_2^{L2} \gamma_2^{L2} \tag{13.40a}$$

As before with our LLE modeling, we eliminate the pure-component fugacities on both sides of these two equations since the pure component fugacities are for the same substance, at the same temperature and pressure.

$$x_1^{L1} \gamma_1^{L1} = x_1^{L2} \gamma_1^{L2} \tag{13.41}$$

$$\left(1 - x_1^{L1}\right) \gamma_2^{L1} = \left(1 - x_1^{L2}\right) \gamma_2^{L2} \tag{13.42}$$

For the VLE, we have a choice of equations for the liquid part among Equations 13.39 and 13.40. One approach that has been suggested in the past is equating the concentrated liquid phase with the vapor phase, per component, and we follow that here (Smith et al., 2006). Assuming L1 is rich in component 1 and L2 is rich in component 2 in Equations 13.39 and 13.40, we arrive at

$$x_1^{L1} f_1^{L1} \gamma_1^{L1} = \hat{\varphi}_1 y_1 P \tag{13.43}$$

$$\left(1 - x_1^{L2}\right) f_2^{L2} \gamma_2^{L2} = \hat{\varphi}_2 (1 - y_1) P \tag{13.44}$$

We model the liquid part of Equations 13.43 and 13.44 using the form from Chapter 11 as

$$x_1^{L1} \gamma_1^{L1} P_1^{\text{sat}} \varphi_1^{\text{sat}} \exp\left\{\frac{V_1(P - P_1^{\text{sat}})}{RT}\right\} = y_1 P \hat{\varphi}_1 \tag{13.45}$$

$$\left(1 - x_1^{L2}\right) \gamma_2^{L2} P_2^{\text{sat}} \varphi_2^{\text{sat}} \exp\left\{\frac{V_2(P - P_2^{\text{sat}})}{RT}\right\} = (1 - y_1) P \hat{\varphi}_2 \tag{13.46}$$

As before when modeling VLE, we can simplify the general form presented by Equations 13.45 and 13.46 by assuming low pressure, which results in a modified Raoult's Law relationship (Equation 11.69) written here as

$$x_1^{L1} \gamma_1^{L1} P_1^{\text{sat}} = y_1 P \tag{13.47}$$

$$\left(1 - x_1^{L2}\right) \gamma_2^{L2} P_2^{\text{sat}} = (1 - y_1) P \tag{13.48}$$

From Equations, 13.41, 13.42, 13.47, and 13.48, we have five unknowns: x_1^{L1}, x_1^{L2}, y_1, T, and P. Note that we do not include the activity coefficients or the vapor pressures as unknowns here, since they are given in terms of other unknowns, namely the liquid phase composition and temperature.

We have three equilibrium phases (vapor, $L1$, and $L2$) for our binary mixture. Thus, according to Gibbs phase rule, we have one degree of freedom. ($F = C - \pi + 2$; here, $C = 2$ and $\pi = 3$. Thus, $F = 1$.)

When we specify that one degree of freedom from our five unknowns, we can solve our four non-linear equations in four unknowns. A typical approach might be to specify the temperature where the VLLE occurs and, thus, use the model to estimate the pressure where the VLLE occurs as well as the composition of the three equilibrium phases. We demonstrate this approach in Example 13-9.

EXAMPLE 13-9	**MODELING VLLE**

Using the 1-parameter Margules equation, estimate the pressure and composition for VLLE at 308 K for the methanol (1) + cyclohexane (2) system.

SOLUTION:

Step 1 *Setting up the equations*
The four equations that describe our VLLE are:

$$x_1^{L1}\gamma_1^{L1} = x_1^{L2}\gamma_1^{L2} \tag{13.41}$$

$$\left(1 - x_1^{L1}\right)\gamma_2^{L1} = \left(1 - x_1^{L2}\right)\gamma_2^{L2} \tag{13.42}$$

$$x_1^{L1}\gamma_1^{L1}P_1^{\text{sat}} = y_1 P \tag{13.47}$$

$$\left(1 - x_1^{L2}\right)\gamma_2^{L2}P_2^{\text{sat}} = (1 - y_1)P \tag{13.48}$$

The Antoine equation provides values for the pressure in mmHg. We have converted to kPa here.

At 308 K, the vapor pressure for methanol is 27.6 kPa, while the vapor pressure of cyclohexane is 20 kPa. These values are obtained from the Antoine equation parameters in Appendix E.

Recall that for the 1-parameter Margules equation, the activity coefficient relationships are

$$\ln[\gamma_1] = A x_2^2 \tag{11.81}$$

$$\ln[\gamma_2] = A x_1^2 \tag{11.82}$$

In Example 13-5 we determined that best fit value for the 1-parameter Margules equation at 300 K was $A = 2.475$. We can use Equation 13.33, which provided $A = A(T)$, to give us an estimate for $A(T = 308$ K$)$. When calculated, this value is 2.415. Accordingly, we now have four equations in four unknowns (x_1^{L1}, x_1^{L2}, P, y_1).

$$x_1^{L1} \exp\left(2.415[1 - x_1^{L1}]^2\right) = x_1^{L2} \exp\left(2.415[1 - x_1^{L2}]^2\right)$$

$$(1 - x_1^{L1}) \exp\left(2.415[x_1^{L1}]^2\right) = (1 - x_1^{L2}) \exp\left(2.415[x_1^{L2}]^2\right)$$

$$x_1^{L1} \exp\left(2.415[1 - x_1^{L1}]^2\right)(27.6 \text{ kPa}) = y_1 P$$

$$(1 - x_1^{L2}) \exp\left(2.415[x_1^{L1}]^2\right)(20 \text{ kPa}) = (1 - y_1)P$$

You can solve this system of equations using your calculator (if it has this ability) or the Solver function in MS Excel. Remember to provide reasonable guesses for the values of the unknowns to help aid in the convergence to a solution.

Step 2 *Solving the problem*
We can readily solve our four equations in four unknowns to obtain the following results:

$$x_1^{L1} = 0.166$$

$$x_1^{L2} = 0.834$$

$$y_1 = 0.579$$

$$P = 42.2 \text{ kPa}$$

An experimental estimate of the system pressure is 43 kPa, while the vapor-phase composition is about 0.57 (Marinchev and Susarev, 1965). Thus, the modeling approach does very well relative to the experimental data.

Note that for liquid systems that are basically immiscible, a simplification occurs to our four VLLE equations. First, we do not need to model the LLE part of the VLLE, since we assume that each liquid phase only contains one of the components. Second, the VLE part, which is

$$x_1^{L1} \gamma_1^{L1} P_1^{\text{sat}} = y_1 P \tag{13.47}$$

$$\left(1 - x_1^{L2}\right) \gamma_2^{L2} P_2^{\text{sat}} = (1 - y_1)P \tag{13.48}$$

will simplify to

$$(1)(1)P_1^{\text{sat}} = y_1 P \tag{13.49}$$

$$(1 - 0)(1)P_2^{\text{sat}} = (1 - y_1)P \tag{13.50}$$

When Equations 13.49 and 13.50 are added together, they become

$$P_1^{\text{sat}} + P_2^{\text{sat}} = P \tag{13.51}$$

with

$$y_1 = \frac{P_1^{\text{sat}}}{P} \tag{13.52}$$

Thus, the system pressure of the VLLE for immiscible liquid systems is just the sum of the vapor pressure of the pure components, while the vapor-phase mole fraction is just the vapor pressure of one of the components divided by the total system pressure.

> If we would have assumed an immiscible liquid in Example 13-9, the system pressure would have been 47.6 kPa with a vapor phase mole fraction (y_1) of 0.42. Such an approach provides a quick estimate of the pressure and vapor-phase composition of VLLE for systems at low pressure with large miscibility gaps.

13.9 Solid–Liquid Equilibrium (SLE)

Let's revisit the methanol + cyclohexane LLE from Figure 13-1 presented earlier in this chapter. You will notice that the curve terminates at about 290 K. This is only because we wanted to focus on the temperatures near the UCST, but LLE exists below 290 K as well—it is just not shown on Figure 13-1. How far below? In other words, when does the curve actually stop? Your intuition is probably telling you that if you go to lower and lower temperatures, eventually the system will freeze. Thus, the LLE curve will terminate at the point where the liquid freezes to a solid. When this occurs, we have solid–liquid equilibrium (or for methanol + cyclohexane, we would have solid–liquid–liquid equilibrium since we have liquid phase-splitting already for this system).

As there is rich phase behavior when LLE meets the solid phase, we leave the details of that discussion to a more advanced treatment in chemical thermodynamics. Thus, in this introductory text, we start more simply: a mixture that is in a single liquid phase that is cooled to its freezing point. Note that solid–liquid equilibrium (SLE) is very important in the chemical process industry as some systems are not well separated by their boiling point (such as certain isomers), but are well separated by their melting point. Thus, distillation is not an option for separation, but crystallization is.

Although crystallization for separation is important in the chemical process industry, metallic systems dominate research in solid–liquid equilibrium because of the importance metals play in our world, as well as their existence in the solid phase at ambient

> The boiling point range for the three xylene (di-methyl benzene) isomers (ortho, meta and para) is less than 10°C, but the melting point range is more than 60°C. Thus, crystallization provides a more attractive separation approach relative to distillation.

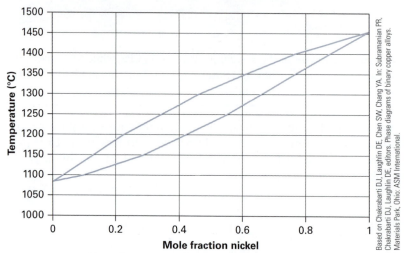

Based on Chakrabarti DJ, Laughlin DE, Chen SW, Chang YA. In: Subramanian PR, Chakrabarti DJ, Laughlin DE, editors. Phase diagrams of binary copper alloys. Materials Park, Ohio: ASM International.

FIGURE 13-14 The solid–liquid equilibrium for the nickel + copper system.

Ni-Cu face-centered cubic lattice structure

temperatures and pressures. Accordingly, we show in Figure 13-14 the solid–liquid equilibrium for the nickel + copper system at a representative pressure. Very much like LLE, the pressure impact on SLE is very weak and, thus, you will see SLE regularly presented without specifying the actual pressure. The SLE for nickel + copper looks similar to the ideal systems presented in Chapter 10 when we discussed VLE. However, this is an *atypical* situation for SLE, owing to the fact that it is very difficult to have a solid phase where the components are mixed as they would be in a liquid or vapor. Solids have unit cells, and atoms/molecules appear at regular intervals on a lattice. Thus, when one makes a solid mixture of nickel and copper, the atoms must order regularly on a lattice. This will only occur in very special situations, such as when the atomic/molecular radii of the atoms/molecules are similar, the pure components have the same crystal structure (meaning they pack the same way on a lattice, such as body-centered cubic, face-centered cubic, etc.) and the components have the same type of interactions. That these conditions are met for the nickel + copper system are evidenced by the phase behavior indentified in Figure 13-14. Systems that behave in this manner are called **substitutional alloys**, since the atoms in the mixture substitute for each other on the lattice sites.

Any state above the top curve is all liquid. The top curve in Figure 13-14 describes an all liquid state with the first bit of solid forming and is called the **liquidus**. Likewise, the bottom curve describes an all solid state with the first drop of liquid forming and is given the name **solidus**. These are the SLE analogs to the dew- and bubble-point curves we discussed previously with VLE. Also, the point at either end of the phase diagram (either pure copper or pure nickel) where the curves meet identifies the melting point for the pure component.

Considering the stringent requirements for a completely miscible solid solution, a much more common situation is when the solid phase has a miscibility gap that encompasses part of or, in many cases, all of the composition region. An example of the former is the copper + silver system as shown in Figure 13-15. Here, only in a very small region (copper-rich and silver-rich) will the solid exist in a single phase. This accounts for the accommodations an otherwise pure solid lattice will make when a small amount of the other component is introduced into the system. After this point, the solid will split into two solid phases with the equilibrium compositions given in a manner similar to that from LLE and VLLE. In many cases, this large miscibility gap will span the entire composition region, resulting in two totally immiscible

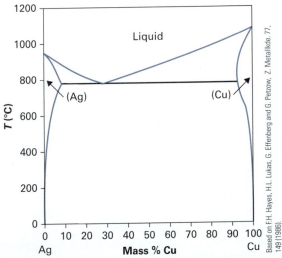

FIGURE 13-15 The solid–liquid equilibrium for the copper + silver system. The eutectic composition is 28.1% by weight copper.

solids. Note also that a three-phase equilibrium exists where a single liquid phase is in equilibrium with two solids phases (or LSSE). This point is known as the **eutectic**.

Since we are showing phase equilibrium among *atomic* species rather than molecular species, compounds can form readily at different atomic compositions, and this gives rise to very rich phase behavior in metallic system for solid–liquid equilibrium. For example, in Figure 13-16 we show the phase behavior for the lead + strontium system. Atomic compositions are shown on the bottom axis while the weight compositions are shown on the top axis. The top curve separates the liquid phase from those many solid phases (with different compounds) that can form.

There are systems where a minimum exists in the solidus curve without the solid splitting into two solid phases. This is not a eutectic point, but would be equivalent to an azeotrope in VLE. The gold + nickel system exhibits this behavior (though it will split into two solid phases at lower temperatures).

FOOD FOR THOUGHT 13-5

The dashed horizontal line in Figure 13-16 represents a transition between the body-centered cubic and face-centered cubic unit cell for the pure solid strontium. What do you think drives this transition?

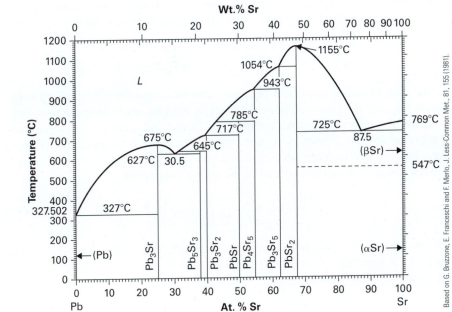

FIGURE 13-16 The solid–liquid equilibrium for the lead + strontium system. Note the many solid compounds that form in the solid phase.

13.10 Modeling Solid–Liquid Equilibrium (SLE)

While modeling a system such as that presented in Figure 13-16 is well beyond the scope of this book, we can make progress on modeling some of the simpler SLE phase behavior. Once again, to model phase equilibrium we start out with equality of mixture fugacities for the components in the system. Thus, Equation 11.52 written for the solid and liquid phases becomes

$$\hat{f}_i^S = \hat{f}_i^L \tag{13.53}$$

As with LLE, we rewrite the mixture fugacity through the use of the activity coefficient:

$$z_i f_i^S \gamma_i^S = x_i f_i^L \gamma_i^L \tag{13.54}$$

Just as it is common practice to denote the liquid phase composition with x and the vapor phase composition with y, it is common practice to denote the composition in the solid phase with z, so no superscript is required.

Unlike LLE, the fugacity of the pure solid is not the same as the pure liquid (since they are different phases), so we cannot eliminate those pure-component fugacities from Equation 13.54, but must account for their differences. We follow a useful approach to evaluate these pure component fugacities by first moving them to the same side of the equal sign, as in Equation 13.55 (Smith et al., 2006).

$$z_i \left(\frac{f_i^S}{f_i^L} \right) \gamma_i^S = x_i \gamma_i^L \tag{13.55}$$

The term in parentheses is the ratio we must evaluate at the mixture T and P. We can write this in a different way, however, by using a multiplicative identity:

$$\frac{f_i^S(T,P)}{f_i^L(T,P)} = \frac{f_i^S(T,P)}{f_i^L(T,P)} \frac{f_i^S(T_m,P)}{f_i^S(T_m,P)} \frac{f_i^L(T_m,P)}{f_i^L(T_m,P)} \tag{13.56}$$

Next we exploit the fact that the pure solid and pure liquid (for component i) must have the same fugacity at their own melting point (as required by phase equilibrium), so $f_i^S(T_m,P) = f_i^L(T_m,P)$. Using this relationship, we can rewrite Equation 13.56 as

$$\frac{f_i^S(T,P)}{f_i^L(T,P)} = \frac{f_i^S(T,P)}{f_i^L(T,P)} \frac{f_i^L(T_m,P)}{f_i^S(T_m,P)} = \frac{f_i^S(T,P)}{f_i^S(T_m,P)} \frac{f_i^L(T_m,P)}{f_i^L(T,P)} \tag{13.57}$$

Thus, this ratio of pure-component fugacities can be determined by evaluating the change in fugacity for pure component i from its melting point to the system temperature (at the system pressure) for both the liquid and solid phases, separately. The relationship that allows us to determine how the pure-component fugacity changes with temperature at constant pressure is a variant of Equation 11.111, written for the residual properties instead of the excess properties.

Recall that the relationship between the fugacity and the residual molar Gibbs free energy was provided in Equation 8.56.

$$\left(\frac{\partial \ln f_i}{\partial T} \right)_P = \frac{\partial}{\partial T} \left[\frac{G_i^R}{RT} \right]_P = -\frac{H_i^R}{RT^2} \tag{13.58}$$

We can use the relationship between the fugacity and the residual molar enthalpy for each phase, separate variables and integrate from the melting point temperature to the temperature of the mixture:

$$\int_{\ln f_i(T_m)}^{\ln f_i(T)} d\ln f_i = -\int_{T_m}^{T} \frac{H_i^R}{RT^2} dT \tag{13.59}$$

$$\ln f_i(T,P) - \ln f_i(T_m,P) = -\int_{T_m}^{T} \frac{\underline{H}_i^R}{RT^2} dT \qquad (13.60)$$

$$\ln \frac{f_i(T,P)}{f_i(T_m,P)} = -\int_{T_m}^{T} \frac{\underline{H}_i^R}{RT^2} dT \qquad (13.61)$$

$$\frac{f_i(T,P)}{f_i(T_m,P)} = \exp\left(\int_{T_m}^{T} \frac{-\underline{H}_i^R}{RT^2} dT\right) \qquad (13.62)$$

Writing Equation 13.62 for both the solid and liquid phases yields

$$\frac{f_i^S(T,P)}{f_i^L(T,P)} = \frac{f_i^S(T,P)}{f_i^S(T_m,P)} \frac{f_i^L(T_m,P)}{f_i^L(T,P)} = \left[\exp\int_{T_m}^{T} \frac{-\underline{H}_i^{R,S}}{RT^2} dT\right]\left[\exp\int_{T}^{T_m} \frac{-\underline{H}_i^{R,L}}{RT^2} dT\right] \qquad (13.63)$$

Swapping the limits of integration for the liquid integral removes the negative sign for the liquid, and we can combine the exponentials, yielding

$$\frac{f_i^S(T,P)}{f_i^L(T,P)} = \left[\exp\int_{T_m}^{T} \frac{-\underline{H}_i^{R,S}}{RT^2} dT\right]\left[\exp\int_{T_m}^{T} \frac{\underline{H}_i^{R,L}}{RT^2} dT\right] \qquad (13.64)$$

$$\frac{f_i^S(T,P)}{f_i^L(T,P)} = \left[\exp\int_{T_m}^{T} \frac{\underline{H}_i^{R,L} - \underline{H}_i^{R,S}}{RT^2} dT\right] \qquad (13.65)$$

Up to this point, Equation 13.65 is exact, as there were no assumptions made in the derivation. The residual molar enthalpies are functions of temperature and this functionality can be provided via the heat capacity. A simplification, however, is to assume that the difference between the residual molar enthalpies of the liquid and solid is a constant (not a function of temperature). This moves those terms out of the integral.

$$\frac{f_i^S(T,P)}{f_i^L(T,P)} = \exp\left[\left(\underline{H}_i^{R,L} - \underline{H}_i^{R,S}\right)\int_{T_m}^{T} \frac{dT}{RT^2}\right] \qquad (13.66)$$

Since residuals (regardless of phase) are referenced back to the ideal gas at the same T and P, the difference between the liquid and solid residual molar enthalpy at the same T and P is just the molar enthalpy of fusion at that temperature. As we have already assumed the difference in the residuals is a constant and independent of temperature, the reasonable place to evaluate this difference is at the normal melting temperature, T_m.

$$\frac{f_i^S(T,P)}{f_i^L(T,P)} = \exp\left[\Delta\underline{H}_i^{Fus}(T_m)\int_{T_m}^{T} \frac{dT}{RT^2}\right] \qquad (13.67)$$

The integration yields the final relationship as

$$\frac{f_i^S(T,P)}{f_i^L(T,P)} = \exp\left[\frac{\Delta\underline{H}_i^{Fus}(T_m)}{RT_m} \frac{(T - T_m)}{T}\right] \qquad (13.68)$$

Smith (Smith et al., 2005) gives this ratio the symbol ψ and we follow that nomenclature here.

$$\psi_i = \exp\left[\frac{\Delta\underline{H}_i^{Fus}(T_m)}{RT_m} \frac{(T - T_m)}{T}\right] \qquad (13.69)$$

Molar enthalpy of fusion data for some compounds are provided in Appendix C.

We can now use Equations 13.68 and 13.69 in our equilibrium relationship for the solid and the liquid (Equation 13.55) to yield

$$z_i \psi_i \gamma_i^S = x_i \gamma_i^L \tag{13.70}$$

13.10.1 Modeling Solid–Liquid Equilibrium (SLE): Simplifications

While Equation 13.70 can be used "as is" to model solid–liquid equilibrium, two simplifications are realized. One simplification can be used to help model the systems of the type in Figure 13-14, while the other simplification can be used to model the systems of the type in Figure 13-15. We present both of those approaches in the following two example problems.

| EXAMPLE 13-10 | MODELING SLE—IDEAL SOLUTIONS |

Model the SLE phase behavior for the copper (1) + nickel (2) system by assuming an ideal solution for both the liquid and solid phases.

SOLUTION:

Step 1 *Assume ideal solution for both the liquid and solid phases*
We start with Equation 13.70 written for a binary mixture:

$$z_1 \psi_1 \gamma_1^S = x_1 \gamma_1^L \tag{13.70a}$$

$$(1 - z_1) \psi_2 \gamma_2^S = (1 - x_1) \gamma_2^L \tag{13.70b}$$

The ideal solution assumption for both phases means that

$$\gamma_1^S = \gamma_1^L = 1$$

$$\gamma_2^S = \gamma_2^L = 1$$

Thus, the phase equilibrium equations reduce to

$$z_1 \psi_1 = x_1 \tag{13.71}$$

$$(1 - z_1) \psi_2 = (1 - x_1) \tag{13.72}$$

According to the Gibbs phase rule, we have a two-component system and two equilibrium phases. Thus, we can specify two variables. It is typical to specify both the temperature and pressure, and use the above two equations to determine the composition of the coexisting phases. Of course, the pressure does not explicitly appear in the modeling approach (it could have appeared in the activity coefficients, but those were set equal to 1), so the pressure choice is arbitrary.

One can rearrange Equations 13.71 and 13.72 to provide expressions for both x_1 and z_1 as functions of ψ_1 and ψ_2. Thus,

$$x_1 = \frac{\psi_1(1 - \psi_2)}{\psi_1 - \psi_2} \tag{13.73}$$

$$z_1 = \frac{(1 - \psi_2)}{\psi_1 - \psi_2} \tag{13.74}$$

Thus, one can vary T (which will change ψ_1 and ψ_2) and plot the compositions (x_1 and z_1) to create the coexistence curve, starting with the melting point of one of the components and ending with the melting point of the other component.

Step 2 *The physical properties of the pure components*
In order to calculate ψ_1 and ψ_2, we require the melting temperature and the heat of fusion for both of the pure components. For nickel, the melting point is 1728 K while the heat of fusion is 17,500 J/mol, while for copper the melting point is 1356 K while the heat of fusion is 13,000 J/mol (Knovel, 2008).

Step 3 *Plotting x_1 and z_1 as a function of temperature*
Once we have the pure-component data from step 2, we can simply plot x_1 and z_1 (using Equations 13.73 and 13.74) as a function of T, as shown in Figure 13-17. We provide the model results in black, while the experimental data for this system are given in blue. The ideal solution assumption gives a reasonable reproduction of the trend of the data, but the system is not an ideal solution. Indeed, experimental data for the infinite-dilution activity coefficients for this system in the solid phase range from about 2.5 to 4.0, depending on the temperature (Oishi et al., 2003).

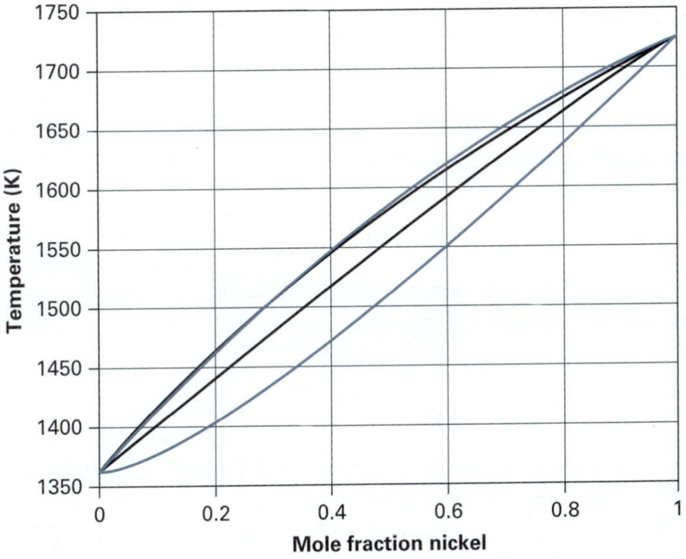

> The terms *heat of fusion* and *heat of melting* are often used instead of molar enthalpy of fusion.

Based on Chakrabarti DJ, Laughlin DE, Chen SW, Chang YA. In: Subramanian PR, Chakrabarti DJ, Laughlin DE, editors. Phase diagrams of binary copper alloys. Materials Park, Ohio: ASM International; 1994.

FIGURE 13-17 The solid–liquid equilibrium for the nickel + copper system. The black curve is the model predictions (ideal solution for the liquid and solid phases), while the blue curve is the experimental data.

In the next example, we describe the other type of simplification, namely where the solid phases are immiscible.

MODELING SLE—IMMISCIBLE SOLID PHASES | **EXAMPLE 13-11**

You are interested in modeling the SLE phase behavior for the hexane-1,6 diamine (1) + naphthalene system by assuming an ideal solution for the liquid phase, but treating the solid phases as immiscible.

(a) Produce a plot of the SLE for this system.
(b) At what temperature and composition does the eutectic occur?
(c) If you cool a liquid system of 60% hexane-1,6 diamine from 330 K to 300 K, describe the equilibrium phases as the temperature is lowered.

SOLUTION:

Step 1 *Assume ideal solution for the liquid, while the solid phases is immiscible*
We start with Equation 13.70 written for a binary mixture:

$$z_1 \psi_1 \gamma_1^S = x_1 \gamma_1^L \tag{13.70a}$$

$$(1 - z_1) \psi_2 \gamma_2^S = (1 - x_1) \gamma_2^L \tag{13.70b}$$

The ideal solution assumption for liquid phase means that

$$\gamma_1^L = \gamma_2^L = 1$$

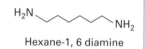

Hexane-1, 6 diamine

Naphthalene

For the solid, two phases form: one with all hexane-1,6 diamine and the other with all naphthalene. This is what is observed experimentally since hexane-1,6 diamine is a molecule whose structure is a hexane with an amine group at both ends, while naphthalene is two benzene rings fused together. What is the implication of the total immiscibility on the modeling of the system? To understand this, we return to the fugacity of component i in the solid solution.

$$\hat{f}_i^S = z_i f_i^S \gamma_i^S \tag{13.75}$$

The fugacity of component i in the solid solution *must* be equal to its pure-component fugacity at that temperature and pressure, since it is not present in the other solid phase. Thus, we have

$$\hat{f}_i^S = z_i f_i^S \gamma_i^S = f_i^S \tag{13.76}$$

In order for Equation 13.76 to be true, $z_i \gamma_i^S = 1$. Therefore, our phase equilibrium relationships (Equations 13.70a and 13.70b) reduce to

$$\psi_1 = x_1 \gamma_1^L \tag{13.77}$$

$$\psi_2 = (1 - x_1) \gamma_2^L \tag{13.78}$$

Note that if an excess molar Gibbs free energy model for the liquid phase was available, the activity coefficients for the liquid phase could be provided (as a function of composition). However, we have assumed an ideal solution for the liquid in this example ($\gamma_1^L = \gamma_2^L = 1$) in order to demonstrate the approach. Therefore, Equations 13.77 and 13.78 reduce to

$$\psi_1 = x_1 \tag{13.79}$$

$$1 - \psi_2 = x_1 \tag{13.80}$$

Step 2 *The physical properties of the pure components*
Once again, we need the melting point and heat of fusion. For hexane-1,6 diamine, the melting point is 311.6 K while the heat of fusion is 39.38 kJ/mol, while for naphthalene, the melting point is 354.7 K and the heat of fusion is 19.55 kJ/mol (Khimeche and Dahmani, 2006).

Step 3 *Do we raise or lower the temperature?*
Before we plot both Equations 13.79 and 13.80, we have to think about whether we are raising or lowering the temperature from the pure component melting points. For the substitutional solid solution shown in Figure 13-14 for copper + nickel mixture, the temperature of the system was bounded between the melting point of the pure components. Here, however, we do not have a substitutional solid solution but, in fact, have an immiscible solid solution. How should the temperature change? To answer that question, we need to consider a simple example.

Consider a pure solid (component A) that is in equilibrium with its liquid. If you add a small amount of another solid (component B), the chemical potential of component A in the liquid will decrease slightly (if assuming an ideal solution for the liquid), since (according to Equation 11.36):

$$\mu_A^{ID}(T,P,x) = \underline{G}_A + RT \ln x_A \qquad (11.36)$$

Here, x_A will now be a number less than one (since we added some B). Thus, $\ln x_A$ will be a negative number. As RT is a positive number, μ_A^{ID} will have to decrease (relative to its pure-component value, $\underline{G}_A$). Thus, for this mixture to be in solid–liquid equilibrium again, the temperature of the system must decrease (in order to lower the chemical potential of pure solid A such that it is equal the chemical potential of component A in the liquid phase). This is precisely the reason why we add salt to ice in order to melt the ice (since the freezing point of the mixture is now lower relative to pure ice.).

In light of this information, when plotting the system in our problem, we have to *lower* the system temperatures (starting from both pure-component melting points) as we plot Equations 13.79 and 13.80.

Step 4 *Plotting x_1 and z_1 as a function of temperature*
Using the pure component data from step 2, we can simply plot x_1 twice on the same figure using Equations 13.79 and 13.80. We will plot both equations as a function of T by lowering the temperature from their respective pure-component melting points.

There will be a temperature at which the curves cross each other. Below that temperature, only the two solid phases will exist (both pure components) and the point at which this occurs is the eutectic. We plot the model predictions in Figure 13-18 relative to the experimental data for this system. As can be seen, the model overpredicts the temperature of the eutectic with a corresponding underprediction of the eutectic composition. Such a result is due to the ideal solution assumption for the liquid phase.

Step 5 *The eutectic temperature and composition*
There are two ways to determine the eutectic temperature and composition. One way is to directly observe the location on Figure 13-18 where the curves cross. That point is the eutectic composition and temperature. For this problem, by inspection of Figure 13-18, an estimate of the result is $T = 305$ K and $x_1 = 0.71$.

If one was not interested in creating the entire SLE curve, but only the location of the eutectic, setting Equations 13.79 and 13.80 equal to each other and solving for T will provide the eutectic temperature. Once that temperature is known, it can be plugged into either Equation 13.79 or 13.80 to find the eutectic composition. For this problem, the precise answer is $T = 305.02$ K and $x_1 = 0.715$.

Step 6 *Cooling a system of $x_1 = 0.6$ from 330 K to 300 K*
Using the model results from Figure 13-18, we see that at the $T = 330$ K and $x_1 = 0.6$, we have one liquid phase. As we cool the system at this composition, we run into the liquidus curve at about 315 K. At this temperature, the first bit of solid is formed that is in equilibrium with a liquid phase of $x_1 = 0.6$. However, the solid phase is pure component 2 (naphthalene) according to Figure 13-18. As the system is cooled further, more solid will form (by inspection using the lever rule), but the composition of the solid is still pure component 2. However, the liquid phase in equilibrium with the solid phase at this temperature now has a higher composition of component 1. Once the system reaches the eutectic temperature 305.02 K, we have three-phase equilibrium. Here we have one liquid phase in equilibrium with two solid phases. The liquid phase has a composition of $x_1 = 0.715$ (our eutectic composition), while the two solid phases are pure naphthalene and pure hexane-1, 6 diamine. Finally, as we cool the system to

You will get a chance to incorporate an activity coefficient model to the liquid phase to explore its impact in an end-of-chapter problem.

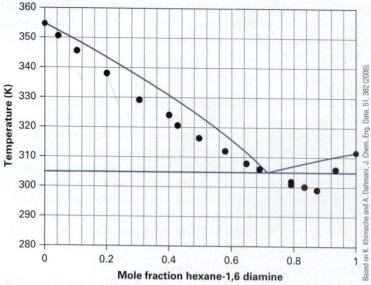

FIGURE 13-18 The solid–liquid equilibrium for the hexane-1,6 diamine + naphthalene system as modeled by an ideal solution for the liquid and an immiscible solid solution. The experimental data are provided as solid circles.

Based on K. Khimeche and A. Dahmani, J. Chem. Eng. Data, 51, 382 (2006).

300 K, only two solid phases remain—pure naphthalene and pure hexane-1,6 diamine. Considering the composition of the initial liquid ($x_1 = 0.6$), for every 6 moles of solid hexane-1,6 diamine present, we would have 4 moles of solid naphthalene, as required by a material balance.

13.11 SUMMARY OF CHAPTER THIRTEEN

- Many mixtures exhibit **liquid–liquid equilibrium (LLE)**, where two liquid phases exist. This occurs since the system can lower its Gibbs free energy by splitting into two phases.

- A balance between entropic and energetic effects determines at what temperature and composition the LLE will occur.

- The **upper critical solution temperature (UCST)** is the temperature at which a liquid mixture possesses a single, stable liquid phase. At temperatures directly below the UCST, the liquid mixture exists in two stable liquid phases of differing compositions. Likewise, the **lower critical solution temperature (LCST)** is the temperature at which a liquid mixture possesses a single, stable liquid phase. At temperatures directly above the LCST, the liquid mixture exists in two, stable liquid phases of differing compositions. Some systems have both a UCST and an LCST.

- The range of compositions over which LLE exists for a system is called a **miscibility gap** and the mixture is called **immiscible** in that region.

- When a mixture does not exhibit LLE, it is termed a **miscible** or **stable** mixture.
- The concavity of the Gibbs free energy of mixing as plotted against composition for a system can be used as a criterion to determine (from modeling approaches) if a mixture is predicted to exhibit LLE.
- LLE is a weak function of the system pressure.
- When the LLE curve encounters the VLE curve, three-phase **(VLLE)** equilibrium exists.
- **Solid–liquid equilibrium (SLE)** is important for separations based on differences in melting point (crystallization), often when separation by boiling point (distillation) is not an option.
- Metallic systems show a rich behavior of solid–liquid equilibrium, especially in the solid phase, where different atomic compositions can give rise to different compounds.
- **Substitutional alloys** are metal mixtures where the different types of atoms in the system can replace each other on the lattice, although this is rare.
- Most solid mixtures will show a miscibility gap, since the requirements for solids (filling a lattice) create a challenge for different molecular sizes and interactions in a mixture. Many of these miscibility gaps run the entire composition range, resulting in two solid phases that are both pure in one of the components. These are **immiscible solid solutions**.
- The **liquidus** curve for SLE is where the system is all liquid with the first bit of solid forming.
- The **solidus** curve for SLE is where the system is all solid with the first drop of liquid forming.
- The **eutectic** is the point where the liquidus curve intersects an immiscible solid mixture and where three-phase equilibrium (liquid-solid-solid) exists.

13.12 EXERCISES

13.1. *N*-formylmorpholine can be used as a solvent in an extraction process for producing high-purity aromatic compounds. To that end, liquid–liquid equilibrium data has been prepared for this compound with a variety of aromatics, including methylcyclopentane. Using the LLE diagram in Figure E13-1 for the methylcyclopentane (1) + *N*-formylmorpholine (2) system, answer the following questions:
  A. For an equimolar mixture at 320 K, what is the composition of the stable phase(s)?
  B. For an equimolar mixture at 420 K, what is the composition of the stable phase(s)?
  C. Estimate the UCST for this system and the composition of the UCST.
  D. Provide the structure for both compounds. By examining the structure, explain why this system would produce a miscibility gap.

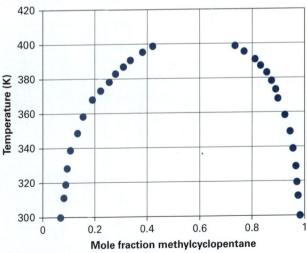

FIGURE E13-1 Liquid-liquid equilibrium for the methylcyclopentane + N-formylmorpholine system.

Answer Exercises 13.2–13.4 based on Figure 13-5.

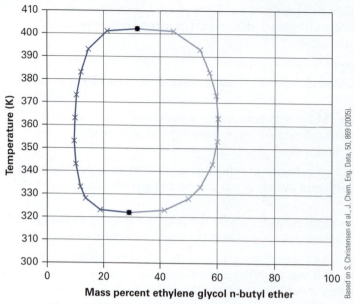

FIGURE 13-5 Liquid-liquid equilibrium of the ethylene glycol n-butyl ether + water system at 100 kPa. Data are presented as mass percent.

13.2. For an ethylene glycol *n*-butyl ether (1) + water (2) system at 360 K with 70% by mass water, determine if the system is one stable liquid phase or two stable liquid phases at equilibrium. If the latter, provide the mass fraction of the co-existing phases.

13.3. For an ethylene glycol *n*-butyl ether (1) + water (2) system at 310 K with 70% by mass water, determine if the system is one stable liquid phase or two stable liquid phases at equilibrium. If the latter, provide the mass fraction of the co-existing phases.

13.4. For an ethylene glycol *n*-butyl ether (1) + water (2) system at 340 K with 80% by mass water, determine if the system is one stable liquid phase or two stable liquid phases at equilibrium. If the latter, provide the mass fraction of the co-existing phases and the amount of each phase.

Answer Exercises 13.5 and 13.6 based on Figure 13-3.

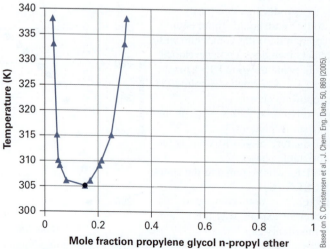

FIGURE 13-3 Liquid–liquid equilibrium of the propylene glycol *n*-propyl ether + water system at 100 kPa.

13.5. For a propylene glycol n-propyl ether (1) + water (2) system at 335 K with 80% by mole water, determine if the system is one stable liquid phase or two stable liquid phases at equilibrium. If the latter, provide the mole fraction of the co-existing phases.

13.6. For a propylene glycol n-propyl ether (1) + water (2) system at 315 K with 20% by mole water, determine if the system is one stable liquid phase or two stable liquid phases at equilibrium. If the latter, provide the mole fraction of the co-existing phases.

13.7. Chopade and co-workers in 2003 reported the vapor–liquid, liquid–liquid, and vapor–liquid–liquid equilibrium for the system 2-methyl-1, 3 dioxolane (1) + water (2) system at 101.3 kPa. They abbreviate the name of the former substance by "2MD."

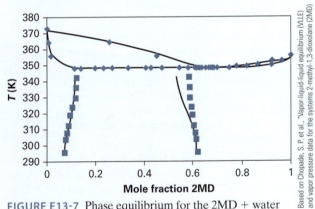

They show the experimental data as symbols and model results (here, UNIQUAC) with the solid curves. Answer the following questions that are based on Figure E13-7:

2-methyl 1,3 dioxolane

FIGURE E13-7 Phase equilibrium for the 2MD + water system at 101.3 kPa.

Based on Chopade, S. P. et al., "Vapor-liquid-liquid equilibrium (VLLE) and vapor pressure data for the systems 2-methyl-1,3-dioxolane (2MD) + water and 2,4-dimethyl-1,3, dioxolane (24 DMD) + water," J. Chem. Eng. Data, 48, 44–47 (2003).

A. What is the normal boiling point of 2MD?
B. At what temperature does the VLLE occur at 101.3 kPa and what are the equilibrium compositions for VLLE?
C. For a 20% by mole 2MD system, what are the equilibrium phases (and their composition) that exist at 101.3 kPa and the following temperatures:
   a. 380 K
   b. 360 K
   c. 340 K

13.8. In 1924, Gilbert Lewis and Merle Randall published *Thermodynamics and the Free Energy of Chemical Substances* (1st Edition, McGraw-Hill Book Co., Inc., New York). On page 225 these authors state, "*As far as is known, all gases are miscible with one another in all proportions.*" In short, they were referencing that no known evidence has been presented to indicate a miscibility gap exists

in the vapor phase of mixtures (in other words, vapor–vapor equilibrium).

With the benefit of about a century of research since that time, what is the current knowledge about the existence of vapor–vapor equilibrium for mixtures? Provides evidence (a brief description) plus a citation (if applicable).

13.9. Will the 2-parameter Margules equation show a miscibility gap for a system described by the following parameters: ($A_{12} = 2.1$ and $A_{21} = 3.2$)?

13.10. In an interstitial alloy, one of the components in the solid mixture does not occupy lattice sites, but is small enough to fit within the spaces (or interstitials) between the lattice sites of the other component. Find an example of an interstitial alloy.

13.11. Calculate the psi (ψ) value of each component in the methanol (1) + cyclohexane (2) system at 170 K.

13.12. Prove that stability criterion given in Equation 13.16,

$$\left(\frac{\partial[\ln(\gamma_1)]}{\partial x_1}\right)_{T,P} > -\left(\frac{1}{x_1}\right) \text{ is true.}$$

13.13. Additional stability criteria exist for liquid-liquid equilibrium. Thus, using other stability criteria presented in the text, prove the stability criterion

$$\left(\frac{\partial \hat{f}_1}{\partial x_1}\right)_{T,P} > 0 \text{ is true.}$$

Answer Exercises 13.14 and 13.15 based on Figure 13-15.

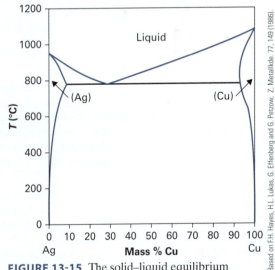

FIGURE 13-15 The solid–liquid equilibrium for the copper + silver system. The eutectic composition is 28.1% by weight copper.

Based on F.H. Hayes, H.L. Lukas, G. Effenberg and G. Petzow, Z. Metallkde. 77, 149 (1986).

13.14. For the copper (1) + silver (2) system, identify the type and number of stable phases at equilibrium and their composition.
   a. 30 wt% copper and 1000 °C
   b. 70 wt% copper and 600 °C
   c. 70 wt% copper and 825 °C

13.15. For the copper (1) + silver (2) system, identify the number of stable phases and their composition at the eutectic.

13.13 PROBLEMS

13.16. Liquid–liquid equilibrium is realized for the system carbon tetrachloride and water at 298.15 K. The aqueous-rich phase contains 0.083 wt% organic and the organic-rich phase contains 0.011 wt% water. Estimate the activity coefficient of the carbon tetrachloride in the aqueous phase and the water in the organic phase.

Cl—C—Cl (with Cl above and Cl below)

Carbon tetrachloride

13.17. You have liquid–liquid equilibrium data for several normal alkanes in water at 25°C, provided in Table P13-17. Based on this data, what would be a reasonable prediction for the amount of n-decane ($C_{10}H_{22}$) in the aqueous phase and the infinite dilution activity coefficient of n-decane in water at 298.15 K?

TABLE P13-17 Liquid-liquid equilibrium data for some normal alkanes in water at 25°C. (Maczynski et al., 2004)

n-alkane	x^{α}_{alkane}	x^{β}_{water}
pentane	1.10×10^{-5}	6.20×10^{-4}
hexane	2.40×10^{-6}	6.10×10^{-4}
heptane	5.30×10^{-7}	6.00×10^{-4}
octane	1.00×10^{-7}	6.10×10^{-4}
nonane	1.90×10^{-8}	6.20×10^{-4}

13.18. Given the 1-parameter Margules equation, plot the molar Gibbs free energy of mixing as a function of x_1 for three temperatures: 325 K, 300.7 K, and 250 K on the same plot. Determine if there is a miscibility gap at each of the three temperatures. For this study, let the product of the Margules parameter (A) and RT be equal to 5000 J/mol.

13.19. In Figure 13-6 it was shown that the double tangency occurred at $x_1 = 0.072$ and $x_1 = 0.792$, while the local minima in Figure 13-6 was at $x_1 = 0.092$ and $x_1 = 0.827$. Since the double tangency and not the local minima defines the coexisting phases, find the values for the double tangency yourself.

13.20. Using the double-tangency method, determine if the following systems (defined by their Margules equation parameter values) exhibit a miscibility gap. If so, identify the composition of the coexisting phases.

A. $A_{12} = 2.5$; $A_{21} = 3.0$
B. $A_{12} = 1.2$; $A_{21} = 3.0$
C. $A_{12} = 3.3$; $A_{21} = 0.3$
D. $A_{12} = 2.0$; $A_{21} = 1.4$

13.21. Does a mixture of water (1) and 1-butanol (2) form a miscibility gap at 365.15 K? If it does, what is the range of compositions over which this miscibility gap exists? Note: You know that the van Laar parameters for this system are as follows: $L_{12} = 1.2739$ and $L_{21} = 3.9771$ (Gmehling and Onken, 1977).

H—C—C—C—C—OH (with H H H H above and H H H H below)

1-butanol

13.22. The infinite-dilution activity coefficients for the 1-butanol (1) + p-xylene (2) mixture at 333.15 K are $\gamma_1^{\infty} = 7.2360$ and $\gamma_2^{\infty} = 4.9720$ (Prasad, 1998). Will the 2-parameter Margules equation predict a miscibility gap for this system at this temperature? If so, what is the composition of the equilibrium phases?

CH_3 (benzene ring) CH_3

p-xylene

13.23. The infinite dilution activity coefficients for the methanol (1) + n-heptane (2) mixture at 303.15 K are $\gamma_1^{\infty} = 84.20$ (Wobst et al., 1992) and $\gamma_2^{\infty} = 35.10$ (Gmehling et al., 1986). You know that this system shows a miscibility gap at this temperature ($x_1^{\alpha} = .167$; $x_1^{\beta} = .884$) (Sorensen and Arlt, 1979). Will the 2-parameter Margules equation predict a miscibility gap for this system? If so, what is the composition of the equilibrium phases and how do they compare with the experimental data?

13.24. When you learned about the solubility parameters (δ) from the Scatchard-Hildebrand approach in Chapter 11, it was discussed that the difference in the values of the solubility parameters between two substances can provide you with a quick way of identifying the degree of deviation from the ideal solution. Thus, a binary mixture whose difference in solubility parameters is small will have smaller deviations from ideal solution behavior. You conjecture that a mixture with a large difference in their solubility parameters might be so non-ideal that the system displays a miscibility gap. You decide to test your conjecture with the methanol (1) + n-hexane (2) system at 300 K. Does your mixture form a miscibility gap when calculated using the solubility parameters? If so, how does your value compare with the experimental value at this temperature $(x_1^\alpha = .288; x_1^\beta = .769)$(Hradetzky and Lempe, 1991)?

13.25. At 10°C, n-pentane (1) + water (2) shows a miscibility gap. The composition of the phases in equilibrium is as follows: $x_1^\alpha = 0.00107$ and $x_2^\beta = 0.0184$, where "α" is for the water-rich phase and "β" is for the organic-rich phase (Sorensen and Arlt, 1979). Estimate the pressure and composition where this mixture would show vapor–liquid–liquid equilibrium at this temprature.

13.26. Estimate the pressure and composition for VLLE for the diethyl ether (1) + water (2) system at 308.15 K. Assume the liquid can be modeled by the 2-parameter Margules equation where $A_{12} = 4.62$ and $A_{21} = 3.35$ (Villamanan et al., 1984). If you treated the liquid as immiscible, how would your results change?

Di-ethyl ether

13.27. You are tasked with determining a good estimate of the pressure and vapor–phase composition for a vapor–liquid–liquid equilibrium system of benzene (1) + water (2) at two temperatures: 303.15 K and 342.35 K. You have the following liquid–liquid equilibrium data for this system at 323.15 K: $x_1^\alpha = .00051$ and $x_1^\beta = .99366$ (Griswold et al., 1950). Please provide those estimates while making reasonable assumptions.

13.28. Produce the SLE phase diagram for the m-chloronitrobenzene (1) + p-chloronitrobenzene (2) system at 101.3 kPa. Treat the liquid phase as an ideal solution and the solid phase as immiscible. On each section of the phase diagram, denote which phases are in equilibrium. Please plot your phase diagram using component 1 as the independent variable. What is the eutectic temperature and composition for this system? If you have an equimolar liquid mixture and cool it until you meet the liquidus line, what is the composition of the solid precipitate and the temperature at which this occurs?

p-chloronitrobenzene

m-chloronitrobenzene

13.29. Produce the SLE phase diagram for the p-xylene (1) + m-xylene system. Treat the liquid phase as an ideal solution and the solid phase as immiscible. Please plot your phase diagram using the p-xylene as the independent variable. What is the eutectic temperature and composition from the model? Compare your work with the experimental data (Jakob et al., 1995). If you were to improve the modeling result

p-xylene

through the use of an excess molar Gibbs free energy model, would the activity coefficients for the system need to be greater than or less than 1 to bring the model closer to the experimental data? Please demonstrate how you arrived at your answer. Note that the pure component data you need is available in the reference in this problem statement.

m-xylene

13.30. Produce the SLE phase diagram for the p-dichlorobenzene (1) + p-dibromobenzene system at 101.3 kPa. You will do the modeling in two ways and answer each part of the question. Some helpful data are provided in Table P13-30a:

p-dichlorobenzene p-dibromobenzene

TABLE P13-30a Relevant pure component data for p-dichlorobenzene and p-dibromobenzene at 101.3 kPa.

	T_m(K)	ΔH^{Fus} (J/mol)	ΔH^{vap} (J/mol)	V (m³/mol)
p-dichlorobenzene[1,3]	327.15	18160	44811	0.00011846
p-dibromobenzene[2,3]	362.15	20040	66203	0.00012

[1] Heat of melting and molar volume data are from Yaws, 2003.
[2] Heat of melting data is from the NIST Webbook (Linstrom and Mallard, 2012). Molar volume data are from Hildebrand, 1919.
[3] Enthalpy of vaporization data are estimated using the Antoine equation from the NIST Webbook and the Clausius-Clapeyron equation.

A. Treat the liquid phase as an ideal solution and the solid phase as immiscible. Please plot your phase diagram using the p-dichlorobenzene as the independent variable. What is the eutectic temperature and composition from the model? Compare your work with the experimental data provided in Table P13-30b.

B. Treat the liquid phase as described by regular solution theory using the Scatchard-Hildebrand approach and the solid phase as immiscible. Please plot your phase diagram using the p-dichlorobenzene as the independent variable. What is the eutectic temperature and composition from this model? Compare your work with the experimental data provided in Table P13-30b.

C. If you have a liquid mixture that is 76% p-dichlorobenzene and cool it until you meet the liquidus line, what is the composition of the solid precipitate and the temperature at which this occurs for both models? How does this compare to the experimental result?

TABLE P13-30b Solid-liquid equilibrium for the p-dichlorobenzene + p-dibromobenzene system at 101.3 kPa.

x_1	T(K)
1	327.15
0.9275	324.15
0.8640	320.65

x_1	T(K)
0.8325	315.65
0.8247	317.35
0.7335	327.35
0.6472	334.95
0.5560	339.25
0.4525	343.25
0.3158	349.15
0.2455	353.65
0.1250	357.15
0	362.15

Singh, N. B. et al., "Solid-Liquid Equilibria for p-Dichlorobenzene + p-Dibromobenzene and p-Dibromobenzene + Resorcinol." J. Chem. Eng. Data, 44, 605 (1999).

13.14 GLOSSARY OF SYMBOLS

A	parameter of the 1-parameter Margules equation
A_{12}, A_{21}	parameters of the 2-parameter Margules equation
C	# of components in a system
F	degrees of freedom
f_i	fugacity of pure component i
$\hat{f}_i$	mixture fugacity of component i
$\underline{G}^E$	excess molar Gibbs free energy of the mixture
G_i	molar Gibbs free energy of pure component i
G_i^R	residual molar Gibbs free energy of pure component i
$\underline{G}$	molar Gibbs free energy of the mixture
G^t	total Gibbs free energy of the mixture (extensive)
$\underline{G}^{ID}$	molar Gibbs free energy of the mixture as an ideal solution

$\Delta\underline{G}$	molar Gibbs free energy of mixing
$\underline{H}^E$	excess molar enthalpy of the mixture
$\Delta\underline{H}_i^{Fus}$	molar enthalpy of Fusion of component i
H_i^R	residual molar enthalpy of pure component i
L_{12}, L_{21}	parameters of the van Laar equation
n	moles
P	pressure
P_i^{sat}	vapor pressure of component i
R	universal gas constant
$\underline{S}$	molar entropy of the mixture
T	temperature
T_m	melting temperature
$\underline{U}$	molar internal energy of the mixture
$\underline{V}$	molar volume of the mixture

x_A	mole fraction of component A (normally in the liquid phase)
y_A	mole fraction of component A (normally in the vapor phase)
z_A	mole fraction of component A (normally in the solid phase)
γ_i	activity coefficient of component i
γ_i^∞	infinite dilution activity coefficient of component i
μ_i^{ID}	chemical potential of component i in an ideal solution
ψ_i	dimensionless ratio of solid to liquid fugacity used in SLE
π	number of phases in equilibrium

13.15 REFERENCES

Bruzzone, G., Franceschi, E., Merlo, F. "On the Sr-Pb System." *J. Less-Common Met.*, 81, 155 (1981).

Chakrabarti, D. J., Laughlin, D. E., Chen, SW, Chang YA. In: Subramanian, PR, Chakrabarti, D. J., Laughlin,

D. E. editors. *Phase diagrams of binary copper alloys.* Materials Park, Ohio: ASM International, 1994.

Chopade, S. P. et al., "Vapor-liquid-liquid equilibrium (VLLE) and vapor pressure data for the systems 2-methyl-1,3-dioxolane (2MD) + water and 2,4-dimethyl-1,3, dioxolane (24 DMD) + water", *J. Chem. Eng. Data*, 48, 44–47 (2003).

Christensen, S. et al., "Mutual Solubility and Lower Critical Solution Temperature for Water + Glycol Ether Systems." *J. Chem. Eng. Data*, 50, 869 (2005).

Dai, M., Chao, J-P. Heats of Mixing of the Partially Miscible Liquid System Cyclohexane + Methanol." *Fluid Phase Equil.* 23, 315 (1985).

Gmehling, J., Onken, U. *Vapor-Liquid Equilibrium Data Collection*, Vol. 1, Part 1, DECHEMA, 1977.

Ghemling, J. et al, *Activity Coefficients at Infinite Dilution* C_1–C_9, DECHEMA, Vol IX, Part 1., 1986.

Gomis, V., Ruiz, F., Asensi, J. C. "The Application of Ultrasound in the Determination of Isobaric Vapour-Liquid-Liquid Equilibrium Data." *Fluid Phase Equil.*, 172, 245 (2000).

Griswold, J., Chew, J-N., Klecka, M. "Pure hydrocarbons from petroleum", *Ind. Eng. Chem.*, 42, 1246–1251 (1950).

Hayes, F. H., Lukas, H. L., Effenberg, G., Petzow, G. "Thermodynamic Calculation of the Al-rich Corner of the Al-Ti-B System." *Z. Metallkde.* 77, 149 (1986).

Hildebrand, J. "Solubility. III. Relative values of internal pressures and their practical application", *JACS*, 41, 1067–1080 (1919).

Hradetzky, G., Lempe, D. "Phase equilibria in binary and higher systems methanol + hydrocarbon(s). Part 1. Experimental determination of liquid-liquid equilibrium data and their representation using the NRTL equation", *Fluid Phase Equil*, 19, 285–301 (1991).

lwakabe, K., Kosuge, H. "Isobaric vapor-liquid-liquid equilibria with a newly developed still", *Fluid Phase Equil*, 192, 171–186 (2001).

Jakob, R., Rose, C., Gmehling, J. "Solid-liquid equilibria in binary mixtures of organic compounds," *Fluid Phase Equil*, 113, 117–126 (1995).

Khimeche, K., Dahmani, A. "Solid-Liquid Equilibria of Naphthalene + Alkanediamine Mixtures." *J. Chem. Eng. Data*, 51, 382 (2006).

Knovel Critical Tablets (2nd Edition). (2008). Knovel. Online version available at: http://www.knovel.com/web/portal/browse/display?_EXT_KNOVEL_DISPLAY_bookid=761&VerticallD=0

Ko, M., Na, S., Kwon, S., Lee, S., Kim, H. "Liquid-Liquid Equilibria for the Binary Systems of *N*-Formylmorpholine with Branched Cycloalkanes", *J. Chem. Eng. Data*, 48, 699–702 (2003).

Kontogeorgis, G. M., Folas, G. K. *Thermodynamic Models for Industrial Applications*, John Wiley & Sons, 2009.

Linstrom, P. J.; Mallard, W. G. Eds., NIST Chemistry WebBook, NIST Standard Reference Database Number 69, National Institute of Standards and Technology, Gaithersburg MD, 20899, http://webbook.nist.gov (retrieved November 4, 2012).

Maczynski, A., Wisniewska-Goclowska, B., Goral, M. "Recommended liquid-liquid equilibrium data. Part 1. Binary alkane-water systems", *J. Phys. Chem. Ref. Data*, 33, 549–577 (2004).

Marinichev, N., Susarev, M. P., *Zh. Prinkl. Khim.* 38, 1619 (1965).

Matsuda, H., Ochi, K., Kojima, K. "Determination and Correlation of LLE and SLE Data for the Methanol + Cyclohexane, Aniline + Heptane, and Phenol + Hexane System." *J. Chem. Eng. Data*, 48, 184 (2003).

Oishi, T., Tagawa, S., Tanegashima, S. "Activity Measurements of Copper in Solid Copper-Nickel Alloys Using Copper-Beta-Alumina." *Materials Transactions*, 44, 1120 (2003).

Prasad, D. "Isothermal phase equilibrium studies on the binary mixtures of n-butanol and n-pentanol with alkylbenzenes", *Phys. Chem. Liq.*, 36, 149–158 (1998).

Singh, N. B. et al., "Solid-Liquid Equilibria for *p*-Dichlorobenzene + *p*-Dibromobenzene and *p*-Dibromobenzene + Resorcinol", *J. Chem. Eng. Data*, 44, 605 (1999).

Smith, J. M., Van Ness, H. C., Abbott, M. M. *Introduction to Chemical Engineering Thermodynamics*, 7th Edition, McGraw Hill, 2006.

Sorensen, J., Arlt, W. *Liquid-Liquid Equilibrium Data Collection: Binary Systems*, DECHEMA, 1979.

Villamanan, M., Allawi, A., Van Ness, H. C. "Vapor/liquid/liquid equilibrium and heats of mixing for diethyl ether/water at 35", *J. Chem. Eng. Data*, 29, 431–435 (1984).

Wobst, M., Hradetzky, G., Bittrich, H. "Measurements of activity coefficients in highly dilute solutions. Part II." *Fluid Phase Equil*, 77, 297–312 (1992).

Yaws, C. *Yaws' Handbook of Thermodynamic and Physical Properties of Chemical Compounds*, Knovel, 2003.

Fundamentals of Chemical Reaction Equilibrium

14

LEARNING OBJECTIVES

This chapter is intended to help you learn how to:

- Apply *mole balances* to systems that involve chemical reactions
- Define the *equilibrium constant* of a reaction, and quantify it as a *function of temperature*
- Combine mole balances with principles of equilibrium to determine the *extent of a reaction*
- Model chemical reaction equilibrium using fugacity models that are appropriate for the physical situation and that account for pressure, phase, and ideal vs. real solution behavior
- Apply equilibrium models to systems in which *multiple reactions* occur simultaneously

Chemical reactions are an integral component of real chemical processes. Recall that the opening of this book included Figure 1-1, a schematic of a typical chemical process. This figure showed raw materials *A* and *B* converted via at least one chemical reaction into a desired product *P*. Since then, while we have explored a variety of thermodynamics principles and discussed their importance in the design and analysis of chemical processes, chemical reactions have barely been mentioned. This shouldn't be taken as a reflection on the importance of the topic— it is essential! But chemical reactions inherently involve multiple compounds, so we can't model them effectively without the "thermodynamics of mixtures" foundation developed in Chapters 9 through 13. This chapter provides a thorough overview of chemical reaction equilibrium.

Using the reaction from Figure 1-1 as an example:

$$A + B \leftrightarrow P$$

When pure *A* and pure *B* are mixed together, what happens? We can use the principle of equilibrium to determine how much *A*, *B*, and *P* will be present—IF the system is allowed to reach equilibrium. The answer can be highly dependent upon temperature and pressure. As engineers, we seek a set of conditions that allow us to maximize production of the desired product *P*.

The *rate* of the reaction is also a crucial consideration in the design and analysis of the chemical process. There are chemical reactions that reach equilibrium so quickly they can be modeled as instantaneous. There are also chemical reactions that are so slow they can be modeled as not occurring at all—even though equilibrium strongly favors the product *P*. Rate, like equilibrium, can be highly dependent upon temperature and pressure. The goal of maximizing rate is sometimes competing with and contradictory to the goal of

Figure 1-1 also shows the undesired by-product *U*, which is not included in this discussion. Side products and by-products are a common complication in chemical processes.

maximizing the extent of the reaction that is reached at equilibrium. Reaction rate, however, is not something we can determine from thermodynamics. Thus our focus in this chapter is on determining the composition of a reacting mixture at equilibrium, but we recognize there is more to the story in modeling and analyzing chemical reactions. A typical chemical engineering curriculum includes a separate course on reaction kinetics and the design of chemical reactors. ■

14.1 MOTIVATIONAL EXAMPLE: Ethylene from Ethane

While the compound $H_2C{=}CH_2$ is commonly called ethylene, the name by IUPAC nomenclature is ethene.

Worldwide, over 100 million tons of the compound ethylene ($H_2C{=}CH_2$) are produced per year, making it the highest volume organic chemical product in the world, and one of the top five chemical products overall. Ethylene is not widely used as an "end product." It is primarily an intermediate; it is produced and then consumed in the synthesis of other chemical products. Most straightforwardly, it is polymerized into different kinds of polyethylene. This section examines the formation of ethylene through the vapor phase chemical reaction:

$$C_2H_6 \leftrightarrow C_2H_4 + H_2$$

In practice ethylene is produced by thermal cracking of mixtures of hydrocarbons. This is one of the many distinct chemical reactions that occur simultaneously in such a process.

Here we revisit a question that was first posed in Chapter 4: If one mole of ethane, 0.5 mole of ethylene, and 0.5 mole of hydrogen are placed in a sealed container, will the reaction run forward or backward? At that time, we simply stated that the answer depends upon T and P; a quantitative solution was not yet possible. Since then we've learned two essential concepts that can be applied.

1. In Chapter 8, we learned that spontaneous processes act to decrease the Gibbs free energy (G) of a system, and that equilibrium is achieved when the Gibbs free energy of the system is minimized.

2. In Chapter 8, we defined the chemical potential μ for a pure compound as the partial molar Gibbs free energy of that compound. We then learned to quantify the chemical potential for components of mixtures, for both ideal solutions and real solutions.

The standard Gibbs free energy of formation, ΔG_f^0, is reviewed in Section 14.3.4.

Throughout this chapter, we will apply these concepts to chemical reactions. We will begin with Example 14-1, in which the reaction is carried out at $T = 25°C$ and $P = 100$ kPa, the commonly used standard state at which ΔG_f^0 values are tabulated.

| EXAMPLE 14-1 | DEHYDROGENATION OF ETHANE TO FORM ETHYLENE |

A sealed container (Figure 14-1) initially contains 1.0 mole of ethane, 0.5 mole of ethylene, and 0.5 mole of hydrogen. The vessel is maintained at a constant $T = 25°C$ and $P = 100$ kPa, as the following reversible reaction occurs:

The principle of equilibrium tells us nothing about how quickly a reaction will progress. Kinetics (rates of chemical reactions) is a vitally important subject in chemical engineering, which is typically treated in a separate course.

$$C_2H_6(g) \leftrightarrow C_2H_4(g) + H_2(g)$$

If the reaction is allowed to progress until the system reaches equilibrium, what is the mole fraction of each of the three gases at the end of the process?

SOLUTION:

Step 1 *Write and simplify mole balances for each compound*
We are asked to find the contents of the reactor at the end of the process—the final values of $y_{C_2H_6}$, $y_{C_2H_4}$, and y_{H_2}. Thus it is logical to begin by writing mole balances for each compound.

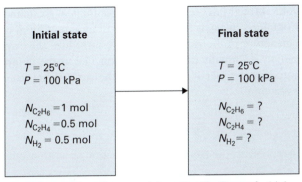

FIGURE 14-1 Reaction vessel described in Example 14-1. The system is closed and the process is isothermal and isobaric.

In this example, no information is provided about the rate of the process, so we will use a time-independent approach to writing balances. Equation 3.31 gave the time-independent mass balance:

$$M_{final} - M_{initial} = \sum_{j=1}^{j=J} m_{j,in} - \sum_{k=1}^{k=K} m_{k,out} \tag{3.31}$$

This expression is for total mass of a system. Throughout the book we have modeled mass as a conserved quantity; it cannot be created or destroyed. However, the number of moles of a compound is not a conserved quantity during a chemical reaction. Therefore a mole balance expression analogous to Equation 3.31 must also account for the generation or consumption of the compound by a chemical reaction:

$$N_{final} - N_{initial} = \sum_{j=1}^{j=J} n_{j,in} - \sum_{k=1}^{k=K} n_{k,out} + N_{gen} \tag{14.1}$$

In Equation 14.1, N_{gen} can be positive or negative. A negative value would represent that the compound is consumed by the chemical reaction.

In the current example, the reaction is occurring in a closed system—no material enters or leaves. Thus, the mole balance for ethylene is

$$N_{C_2H_4,final} - N_{C_2H_4,initial} = N_{C_2H_4,gen} \tag{14.2}$$

Since our goal is to find the final state of the system, this is more conveniently written as

$$N_{C_2H_4,final} = N_{C_2H_4,initial} + N_{C_2H_4,gen} \tag{14.3}$$

An analogous expression can be written for hydrogen and for ethane:

$$N_{H_2,final} = N_{H_2,initial} + N_{H_2,gen} \tag{14.4}$$

$$N_{C_2H_6,final} = N_{C_2H_6,initial} + N_{C_2H_6,gen} \tag{14.5}$$

At first glance it appears we have made no real progress—we've written three equations but introduced three new unknowns ($N_{C_2H_6,gen}$, $N_{C_2H_4,gen}$, and $N_{H_2,gen}$) into the problem, so we are apparently no closer to solving for $N_{C_2H_6,final}$, $N_{C_2H_4,final}$, and $N_{H_2,final}$. However, we can relate $N_{C_2H_6,gen}$, $N_{C_2H_4,gen}$, and $N_{H_2,gen}$ to each other through the stoichiometry of the chemical reaction—when one ethylene molecule is generated, one hydrogen is also generated and one ethane is consumed.

$$N_{C_2H_4,gen} = N_{H_2,gen} \tag{14.6}$$

$$N_{C_2H_4,gen} = -N_{C_2H_6,gen} \tag{14.7}$$

You can think of the extent of reaction, ξ, as a measure of how far the reaction has progressed from a starting point. A negative extent of reaction indicates progress in the "reverse" direction.

Thus, Equations 14.3 through 14.7 represent five equations in six unknowns. In step 3, a sixth equation is obtained through the calculation of an equilibrium constant. First, to simplify the nomenclature in the mass balances, we introduce the *extent of reaction, ξ*. The extent of reaction is formally defined in Section 14.2 as the number of moles of a compound generated by the reaction, divided by the stoichiometric coefficient of that compound in the reaction:

$$\xi = \frac{N_{C_2H_4,gen}}{+1} = \frac{N_{H_2,gen}}{+1} = \frac{N_{C_2H_6,gen}}{-1} \tag{14.8}$$

The −1 for ethane represents the fact that in the reaction as written, ethane is consumed, not generated.

We can introduce ξ to express the N_{gen} terms in Equations 14.3 through 14.5, and also insert the known initial values of the number of moles of each compound.

$$N_{C_2H_4,final} = 0.5 + \xi \tag{14.9}$$

$$N_{H_2,final} = 0.5 + \xi \tag{14.10}$$

$$N_{C_2H_6,final} = 1 - \xi \tag{14.11}$$

Equation 14.12 reveals that this is an example of a reaction in which "total moles" is not a conserved quantity; it increases as the reaction progresses.

Thus if the extent of reaction ξ can be determined, the exact contents of the vessel at equilibrium ("final") can be found using Equations 14.9 through 14.11.

Step 2 *Write expressions for mole fractions of each compound*

We're asked to find the mole fraction of each compound. Equations 14.9 through 14.11 give the final number of moles for each compound, and summing these gives the total number of moles present. Thus,

$$N_{tot,final} = N_{C_2H_4,final} + N_{C_2H_6,final} + N_{H_2,final}$$

$$N_{tot,final} = 2 + \xi \tag{14.12}$$

Which means the mole fraction of each species in the final equilibrium state can be expressed as

In introductory chemistry, you likely learned to use the equilibrium constant K_p or K_c, depending upon whether the quantities of reactants and products were expressed as partial pressures (p) or concentrations (c). In Section 14.3, we will define the equilibrium constant in its most general form.

$$y_{C_2H_4} = \frac{N_{C_2H_4,final}}{N_{tot,final}} = \frac{0.5 + \xi}{2 + \xi} \tag{14.13}$$

$$y_{H_2} = \frac{N_{H_2,final}}{N_{tot,final}} = \frac{0.5 + \xi}{2 + \xi} \tag{14.14}$$

$$y_{C_2H_6} = \frac{N_{C_2H_6,final}}{N_{tot,final}} = \frac{1 - \xi}{2 + \xi} \tag{14.15}$$

Step 3 *Evaluate the equilibrium constant for the reaction*

You likely learned in introductory chemistry courses that the equilibrium constant for a gas phase reaction can be determined from the Gibbs free energy of reaction, using

$$K_P = \exp\left(-\frac{\Delta \underline{G}_T^0}{RT}\right) \tag{14.16}$$

The Gibbs free energy of a reaction is frequently written as in Equation 14.17 (e.g., Zumdahl, 2009).

in which $\Delta \underline{G}_T^0$ is the standard change in Gibbs free energy of the reaction. For this example,

$$\Delta \underline{G}_T^0 = \sum \Delta \underline{G}_{f,products}^0 - \sum \Delta \underline{G}_{f,reactants}^0 \tag{14.17}$$

$$\Delta \underline{G}_T^0 = \Delta \underline{G}_{f,ethylene}^0 + \Delta \underline{G}_{f,hydrogen}^0 - \Delta \underline{G}_{f,ethane}^0$$

From Appendix C, we find the following data is available at $T = 298.15$ K (25°C) and $P = 100$ kPa:

For ethane, $\Delta \underline{G}_f^0 = -31.86$ kJ/mol

For ethylene, $\Delta \underline{G}_f^0 = 68.43$ kJ/mol

For hydrogen, $\Delta \underline{G}_f^0 = 0$ kJ/mol

Section 14.3 formalizes our definition for the standard Gibbs free energy of a reaction.

Step 4 will use this data to find the numerical value of the equilibrium constant, but first we make some qualitative observations regarding the significance of the data.

Aside — the reactant is more stable than the products

The reaction was written as if ethane is being converted into ethylene and hydrogen. However, it's a reversible reaction, and the vessel initially contains one mole of "reactant" (ethane) and one mole of "products" (ethylene and hydrogen), so the reaction could realistically progress in either direction. We learned in Chapter 8 to associate low values of G with stability. The data reveals the "reactant" is more stable than the "products," and so we now expect the reaction to proceed in the "reverse" direction. From the perspective of obtaining numerical answers, this does not matter—the fact that the reverse reaction is favored will be reflected in negative values of $N_{C_2H_4,\text{gen}}$ and $N_{H_2,\text{gen}}$.

The molar Gibbs free energy of formation is defined as the change in $\underline{G}$ that occurs when the compound is formed from its elements at STP. It it zero for hydrogen because hydrogen IS an element.

Offhand, it might appear that the container should contain pure ethane at equilibrium. We learned in Chapter 8 that any spontaneous process will act to minimize G. Since $\Delta \underline{G}_f^0$ is negative for ethane and positive for ethylene, the Gibbs free energy G of the system is decreased when ethylene and hydrogen combine to form ethane. Isn't G therefore minimized when all of the ethylene is converted into ethane?

By the above reasoning, ALL chemical reactions would progress to completion. The error in the above reasoning is that it overlooks the entropy of mixing. The $\Delta \underline{G}_f^0$ values above are for *pure* ethane, ethylene, and hydrogen at the standard state of $T = 298.15$ K and $P = 100$ kPa. While pure ethane has a lower $\Delta \underline{G}_f^0$ than pure ethylene or pure hydrogen, because of the entropy of mixing, a mixture of ethane, ethylene, and hydrogen can have a lower G than that of an equivalent mass of pure ethane. The equilibrium constant accounts for the entropy of mixing, as we will see explicitly in Section 14.3.1.

These values of $\Delta \underline{G}_f^0$ are accurate at $T = 298.15$ K and $P = 100$ kPa specifically. Thus the conclusion that ethane is more stable than the products, ethylene and hydrogen, is true at $T = 298.15$ K and $P = 100$ kPa, but can't be assumed true at all conditions.

Step 4 *Compute numerical value of equilibrium constant*

Using the data listed in step 3, we can evaluate the standard Gibbs free energy of this reaction as

$$\Delta \underline{G}_R^0 = \Delta \underline{G}_{f,\text{ethylene}}^0 + \Delta \underline{G}_{f,\text{hydrogen}}^0 - \Delta \underline{G}_{f,\text{ethane}}^0$$

$$\Delta \underline{G}_R^0 = 68.43 \frac{\text{kJ}}{\text{mol}} + 0 - \left(-31.86 \frac{\text{kJ}}{\text{mol}}\right) = 100.29 \frac{\text{kJ}}{\text{mol}} \tag{14.18}$$

$$\Delta \underline{G}_R^0 = 100{,}290 \frac{\text{J}}{\text{mol}}$$

The extent of reaction ξ can be either positive or negative. Here, we start with one mole of reactant and one mole of products, and the reactant ethane is more stable than the products, so we expect a negative ξ.

Inserting this value of $\Delta \underline{G}_R^0$ into Equation 14.16 allows us to quantify the equilibrium constant as

$$K_P = \exp\left(-\frac{\Delta \underline{G}_R^0}{RT}\right) = \exp\left[-\frac{100{,}290 \frac{\text{J}}{\text{mol}}}{\left(8.314 \frac{\text{J}}{\text{mol} \cdot \text{K}}\right)(298.15 \text{ K})}\right] = 2.69 \times 10^{-18} \tag{14.19}$$

Step 5 *Relate equilibrium constant to mole fraction of each compound*

For a gas phase reaction, the equilibrium constant can be related to the partial pressures of the reactants and products as

$$K_P = \frac{P_{C_2H_4} P_{H_2}}{P_{C_2H_6}} \tag{14.20}$$

Equilibrium constants that are extremely small (or extremely large) are not unusual, but these results are temperature dependent; as temperature increases the products in this reaction will be more favored. Section 14.3 examines quantifying the effect of temperature.

This relationship is only valid under *ideal gas* conditions. In Section 14.3.1, we will establish the theoretical basis for Equation 14.20, derive a model for chemical reaction equilibrium in general, and see that the rigorous model only simplifies to Equation 14.20 if the ethane, hydrogen, and ethylene are modeled as a mixture of ideal gases. In the current example, the pressure is 100 kPa, so it is indeed reasonable to assume ideal gas behavior.

Equation 14.20 has three unknown partial pressures. By definition, the partial pressure of a gas is equal to the mole fraction times the total pressure, so

$$K_P = \frac{\left(y_{C_2H_4}P\right)\left(y_{H_2}P\right)}{\left(y_{C_2H_6}P\right)} \tag{14.21}$$

> It is bad practice to drop or ignore the units on a number, and normally the authors would not do it or recommend it. It is done here only to illustrate a result that is likely to be familiar to the reader from introductory chemistry courses.

Units can cause confusion in using equilibrium constants. In Equation 14.19, it appears K_P should be dimensionless, but in Equation 14.21, the right-hand side does not appear to be dimensionless—it has units of pressure. This apparent discrepancy is explained in Example 14-4, in which we illustrate that Equations 14.20 and 14.21 are only valid if pressure is expressed in bar. Here $P = 1$ bar (100 kPa), so Equation 14.21 simplifies to:

$$K_P = \frac{\left(y_{C_2H_4}P\right)\left(y_{H_2}P\right)}{\left(y_{C_2H_6}P\right)} = \frac{\left(y_{C_2H_4}\right)\left(y_{H_2}\right)}{\left(y_{C_2H_6}\right)} \tag{14.22}$$

Step 6 *Solve Equation 14.22*

We can introduce the expressions for mole fraction into Equation 14.22:

$$K_P = \frac{y_{C_2H_4}y_{H_2}}{y_{C_2H_6}}$$

$$K_P = \frac{\left(\dfrac{0.5 + \xi}{2 + \xi}\right)\left(\dfrac{0.5 + \xi}{2 + \xi}\right)}{\left(\dfrac{1 - \xi}{2 + \xi}\right)} \tag{14.23}$$

which simplifies to

$$K_P = \left(\frac{0.5 + \xi}{2 + \xi}\right)\left(\frac{0.5 + \xi}{1 - \xi}\right) \tag{14.24}$$

> If you multiply the terms in parentheses together and cross-multiply, you will find that Equation 14.24 is a quadratic polynomial in ξ.

Since the value of K_P is known from step 3, Equation 14.24 is one equation in one unknown, which can be solved numerically or by using the quadratic equation. The result is

$$\xi \sim -0.5 \text{ moles}$$

Notice that $(0.5 + \xi)$ appears on the right-hand side of Equation 14.24, so if $K_P = 0$, then $\xi = -0.5$. While our calculated K_P is not identically zero, it is so small that our calculated ξ only differs from -0.5 if we carry 10 significant digits. Thus $y_{C_2H_6} \sim 1$ and $y_{C_2H_4} = y_{H_2} \sim 0$.

In some applications, even if the mole fraction of a compound is extremely small, it is still important to accurately estimate its value—particularly if the small mole fraction represents an impurity that is highly toxic or otherwise dangerous. In this case, the simplest way to estimate accurately the equilibrium mole fractions of hydrogen and ethylene is to return to Equation 14.22. Recognizing that $y_{C_2H_6} \sim 1$, Equation 14.22 simplifies to

$$K_P = \frac{\left(y_{C_2H_4}\right)\left(y_{H_2}\right)}{\left(y_{C_2H_6}\right)}$$

$$2.69 \times 10^{-18} = \frac{\left(y_{C_2H_4}\right)\left(y_{H_2}\right)}{1} \tag{14.25}$$

We learned from Equations 14.13 and 14.14 that the mole fractions of ethylene and hydrogen must be equal to each other, not only in the equilibrium state but at any point in the process. Thus,

$$2.69 \times 10^{-18} = \frac{\left(y_{C_2H_4}\right)\left(y_{C_2H_4}\right)}{1} \tag{14.26}$$

And $y_{C_2H_4} = y_{H_2} = \mathbf{1.64 \times 10^{-9}}$.

For many applications, however, this last step would not be necessary. If our goal is to produce ethylene, then we certainly know this reaction will not work—at least, not at this T and P.

Problems like Example 14-1 should be familiar from introductory chemistry courses, but chemical engineering requires a more in-depth knowledge of chemical reaction equilibrium than that represented in Example 14-1. We will see that Equation 14.20 is only valid if we model this system as an ideal mixture of real gases. This is a good model at 100 kPa, but Equation 14.20 could produce very misleading results if applied at higher pressures. Other questions and points raised by Example 14-1 include the following.

- Does an equilibrium constant have units or not? Note the apparent dimension of pressure on the right-hand sides of Equations 14.20 and 14.21, while Equation 14.19 implies the equilibrium constant is dimensionless.

- Calculation of equilibrium constants requires knowledge of the Gibbs free energies of formation, which are conventionally tabulated at $P = 100$ kPa and $T = 298.15$ K. How do we account for reactions occurring at other temperatures and pressures?

- How do we model chemical reaction equilibrium in situations where an ideal gas approximation is not realistic?

This chapter addresses these considerations, and includes several examples, some of which involve multiple reactions occurring simultaneously. Our investigation of chemical reaction equilibrium begins in Section 14.3. First, recognizing that we could not have solved Example 14-1 without accurate mole balances for each reactant and product, we examine reaction stoichiometry in Section 14.2.

14.2 Chemical Reaction Stoichiometry

While the primary focus of this chapter is chemical reaction equilibrium, the motivational example illustrated that, in order to apply a reaction equilibrium model in a useful way, we first needed to write mole balances for each species and account for reaction stoichiometry. In the motivational example, we examined the chemical reaction:

$$C_2H_6(g) \leftrightarrow C_2H_4(g) + H_2(g)$$

The example illustrated how to relate the "generation" terms in each species mole balance to each other (see step 2 in Example 14-1). However, this step was simplified in that the reaction had a 1:1:1 stoichiometry. Slightly more complex is a reaction like:

$$N_2(g) + 3H_2(g) \leftrightarrow 2NH_3(g)$$

in which the number of moles of nitrogen consumed is half the number of moles of ammonia generated:

$$\frac{N_{NH_3,\text{gen}}}{2} = -N_{N_2,\text{gen}} \tag{14.27}$$

FOOD FOR THOUGHT 14-1

How would Equation 14.27 be affected if the reaction were instead written:

$$\frac{1}{2}N_2(g) + \frac{3}{2}H_2(g)$$
$$\leftrightarrow NH_3(g)$$

The negative sign reflects the stoichiometry of the reaction, in which, when ammonia is generated, hydrogen is consumed ("generation" is negative). As we learned in the motivational example, the reaction may progress in the reverse of the direction written, which is simply reflected in negative "generation" of products and positive "generation" of reactants. The relationship between ammonia and hydrogen is

$$\frac{N_{NH_3,gen}}{2} = \frac{N_{H_2,gen}}{-3} \tag{14.28}$$

We define a parameter called the **extent of reaction** (ξ) as the number of moles of a compound generated or consumed by a reaction divided by the stoichiometric coefficient of that compound in the reaction:

$$\xi = \frac{N_{N_2,gen}}{-1} = \frac{N_{H_2,gen}}{-3} = \frac{N_{NH_3,gen}}{2} \tag{14.29}$$

We are using the convention that *stoichiometric coefficients are positive for products and negative for reactants.*

Step 1 of Example 14-1 illustrated how ξ is a useful parameter. We were able to express all of the unknown mole fractions as functions of ξ, such that when we applied an equilibrium constant, ξ was the only unknown in the equation. Section 14.3 details the application of the equilibrium criterion to chemical reactions. In this section, we demonstrate the use of the parameter ξ in both time-independent (Section 14.2.1) and time-dependent (Section 14.2.2) mole balances.

14.2.1 Extent of Reaction and Time-Independent Mole Balances

Chapters 3 and 4 demonstrated that balance equations—whether for material, energy or entropy—could be written in either a time-dependent or a time-independent form. A time-independent mole balance has the form given in Equation 14.1:

$$N_{final} - N_{initial} = \sum_{j=1}^{j=J} n_{j,in} - \sum_{k=1}^{k=K} n_{k,out} + N_{gen}$$

In effect, reaction stoichiometry is used to account for inter-dependencies in the N_{gen} terms of the mole balances for reactants and products. The definition of extent of reaction generalizes as given here.

> When the sum of the stoichiometric coefficients of a reaction is zero, that reaction is a mole-conserving reaction.

For a reaction of the form

$$\nu_A A + \nu_B B \leftrightarrow \nu_C C + \nu_D D$$

the **extent of reaction** ξ is given by

$$\xi = \frac{N_{A,gen}}{\nu_A} = \frac{N_{B,gen}}{\nu_B} = \frac{N_{C,gen}}{\nu_C} = \frac{N_{D,gen}}{\nu_D} \tag{14.30}$$

where the stoichiometric coefficients ν are positive for products ($\nu_C > 0$ and $\nu_D > 0$) and negative for reactants ($\nu_A < 0$ and $\nu_B < 0$).

The extent of reaction can also be expressed in differential form, as

> The differential form in Equation 14.31 is useful in the derivation of the reaction-equilibrium model in Section 14.3.

$$d\xi = \frac{dN_{A,gen}}{\nu_A} = \frac{dN_{B,gen}}{\nu_B} = \frac{dN_{C,gen}}{\nu_C} = \frac{dN_{D,gen}}{\nu_D} \tag{14.31}$$

This definition can be applied to multiple reactions simultaneously. In a system with multiple reactions ξ_1 would quantify the moles of each compound generated by reaction 1 specifically, ξ_2 would refer to the moles generated by reaction 2, etc. This use is illustrated in Example 14-2.

COMBUSTION OF AMMONIA	EXAMPLE 14-2

This sequence of two gas phase reactions can occur in the catalyzed combustion of ammonia:

(R1): $4NH_3 + 5O_2 \leftrightarrow 4NO + 6H_2O$

(R2): $2NO + O_2 \leftrightarrow 2NO_2$

A closed vessel initially contains 5 moles of ammonia and 10 moles of oxygen (Figure 14-2). Assuming these are the only two chemical reactions that occur, write expressions for the mole fractions of each of the species present, in terms of the extents of the two reactions, ξ_1 and ξ_2.

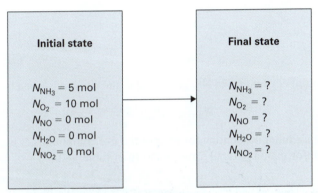

FIGURE 14-2 Ammonia combustion process modeled in Example 14-2.

SOLUTION:

Step 1 *Write mole balances for each species*
Equation 14.1 gave the mole balance in its most general form as

$$N_{final} - N_{initial} = \sum_{j=1}^{j=J} n_{j,in} - \sum_{k=1}^{k=K} n_{j,out} + N_{gen}$$

As in Example 14-1, this is a closed system. With no material entering or leaving, the only way the number of moles of a compound can change is generation (or consumption) by a chemical reaction.

Thus the mole balance for each individual species simplifies to

$$N_{final} = N_{initial} + N_{gen} \tag{14.32}$$

We can write a mole balance of this form for each of the five compounds present in the reaction (though in some cases $N_{initial}$ equals 0). The main thing that distinguishes the five mole balances from each other is the generation terms, which reflect how they participate in the two reactions.

Step 2 *Apply definition of extent of reaction*
Equation 14.30 gives the definition of the extent of reaction, ξ, as the number of moles of a compound generated by a reaction divided by the stoichiometric coefficient of that compound in the reaction. This definition is applicable, regardless of how many chemical reactions are occurring in the system. For example, in R1, the stoichiometric coefficient for ammonia is −4 (the negative value signifying it is a reactant) so applying Equation 14.30, the extent of reaction 1 is

$$\xi_1 = \frac{N_{NH_3,gen}}{-4} \tag{14.33}$$

Some books write "generation" and "consumption" as separate terms in a material balance. Here, they are one term; a positive N_{gen} represents generation and a negative N_{gen} represents consumption.

which can be rearranged as

$$N_{NH_3,gen} = -4\xi_1 \tag{14.34}$$

Similarly, NO is a product in R1, with a stoichiometric coefficient of $+4$, so

$$\xi_1 = \frac{N_{NO,gen,1}}{4}$$

$$N_{NO,gen,1} = 4\xi_1 \tag{14.35}$$

The additional subscript "1" in the generation variable ($N_{NO,gen,1}$) identifies the particular reaction being considered.

But this represents only the generation of NO by the reaction R1. Unlike ammonia, NO also participates in reaction R2 as a reactant with a stoichiometric coefficient of -2:

$$\xi_2 = \frac{N_{NO,gen,2}}{-2}$$

$$N_{NO,gen,2} = -2\xi_2 \tag{14.36}$$

Thus, the total number of moles of NO generated in the process is given by:

$$N_{NO,gen} = 4\xi_1 - 2\xi_2 \tag{14.37}$$

We can conduct a similar accounting for each of the five compounds. It is convenient to summarize the information in table form, as shown in Table 14-1.

TABLE 14-1 Mole balances and reaction stoichiometry for Example 14-2.

	NH_3	O_2	NO	H_2O	NO_2	Total Moles
$N_{initial}$ (moles)	5	10	0	0	0	15
N_{gen} (R1)	$-4\xi_1$	$-5\xi_1$	$4\xi_1$	$6\xi_1$		ξ_1
N_{gen} (R2)		$-\xi_2$	$-2\xi_2$		$2\xi_2$	$-\xi_2$
N_{final} (moles)	$5-4\xi_1$	$10-5\xi_1-\xi_2$	$4\xi_1-2\xi_2$	$6\xi_1$	$2\xi_2$	$15+\xi_1-\xi_2$
y	$\dfrac{5-4\xi_1}{15+\xi_1-\xi_2}$	$\dfrac{10-5\xi_1-\xi_2}{15+\xi_1-\xi_2}$	$\dfrac{4\xi_1-2\xi_2}{15+\xi_1-\xi_2}$	$\dfrac{6\xi_1}{15+\xi_1-\xi_2}$	$\dfrac{2\xi_2}{15+\xi_1-\xi_2}$	

Step 3 *Write expressions for mole fractions of each compound*
We can find the total number of moles in the system by summing the mole balances for the five individual compounds. Recall that for a compound (C) the mole fraction y_c is by definition N_C/N_{total}. Expressions for the mole fraction of each compound are also shown in Table 14-1.

A "batch reactor" is one in which the entire feed mixture is placed in the reactor at the beginning, and then the reactor operates as a closed system for the duration of the process.

Constructing a stoichiometric table analogous to that shown in Table 14-1 is a key step in the solution of typical problems involving chemical reactions. Example 14-2 investigated a closed system—a simple batch reactor. While the batch reactor is a very practical system to study, large-scale chemical processes are most often designed to operate continuously. Consequently, the next section applies the principles of stoichiometry to an open-system reactor operating at steady state.

14.2.2 Extent of Reaction and Time-Dependent Material Balances

Time-dependent material, energy, and entropy balances were introduced in Chapters 3 and 4, and are applicable to problems in which either known information or the desired results are expressed as rates. A time-dependent mole balance takes a form analogous to Equation 3.1:

$$\frac{dN}{dt} = \sum_{k=1}^{k=K} \dot{n}_{k,\text{in}} - \sum_{k=1}^{k=K} \dot{n}_{k,\text{out}} + \dot{N}_{\text{gen}} \tag{14.38}$$

with $\dot{N}_{\text{gen}}$ representing the rate at which the compound is generated or consumed by a reaction. For a reaction of the form

$$\nu_A A + \nu_B B \leftrightarrow \nu_C C + \nu_D D$$

we can modify our definition of the extent of reaction to relate the rates at which individual compounds are generated by a reaction. Thus,

$$\dot{\xi} = \frac{\dot{N}_{A,\text{gen}}}{\nu_A} = \frac{\dot{N}_{B,\text{gen}}}{\nu_B} = \frac{\dot{N}_{C,\text{gen}}}{\nu_C} = \frac{\dot{N}_{D,\text{gen}}}{\nu_D} \tag{14.39}$$

> Remember, reactants have negative stoichiometric coefficients, thus in Equation 14.39; ν_A and ν_B are negative numbers, and ν_C and ν_D are positive numbers.

This definition is applied in Example 14-3.

SYNTHESIS OF MALEIC ANHYDRIDE	EXAMPLE 14-3

Maleic anhydride, which has the chemical formula $C_2H_2(CO)_2O$, can be synthesized through the oxidation of *n*-butane, using the chemical reaction:

$$2C_4H_{10} + 7O_2 \leftrightarrow 2C_2H_2(CO)_2O + 8H_2O$$

10,000 mol/hr of *n*-butane and 50,000 mol/hr of oxygen enter a steady-state chemical reactor (Figure 14-3). Derive expressions that quantify the flow rate (in mol/hr) and mole fraction of each reactant and product in the exit stream. The extent of reaction $\dot{\xi}$ should be the only unknown in the expressions.

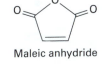

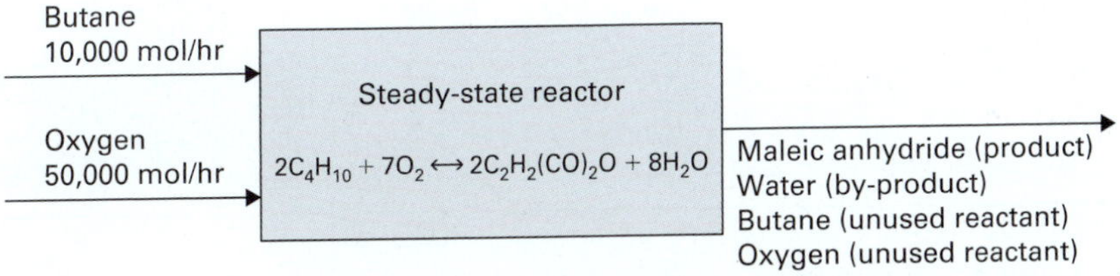

Maleic anhydride

Butane
10,000 mol/hr

Oxygen
50,000 mol/hr

Steady-state reactor

$2C_4H_{10} + 7O_2 \leftrightarrow 2C_2H_2(CO)_2O + 8H_2O$

Maleic anhydride (product)
Water (by-product)
Butane (unused reactant)
Oxygen (unused reactant)

FIGURE 14-3 Synthesis of maleic anhydride as modeled in Example 14-3.

SOLUTION:

Step 1 *Apply and simplify mole balance equation*
The time-dependent mole balance in its most general form is

$$\frac{dN}{dt} = \sum_{j=1}^{j=J} \dot{n}_{j,\text{in}} - \sum_{k=1}^{k=K} \dot{n}_{k,\text{out}} + \dot{N}_{\text{gen}}$$

This is a steady-state process, with only one stream entering and one stream leaving. Consequently the mole balance for this case simplifies to

$$0 = \dot{n}_{\text{in}} - \dot{n}_{\text{out}} + \dot{N}_{\text{gen}} \tag{14.40}$$

which is conveniently rearranged as

$$\dot{n}_{out} = \dot{n}_{in} + \dot{N}_{gen}$$

(14.41)

We can write an equation analogous to Equation 14.41 for either total moles or for any individual chemical compound in the process. For *n*-butane, for example, the flow rate in the inlet stream is given, and the flow rate in the outlet stream is

$$\dot{n}_{butane,out} = 10,000 \frac{mol}{hr} + \dot{N}_{butane,gen}$$

(14.42)

Step 2 *Apply definition of extent of reaction*
The stoichiometric coefficients in this reaction are −2 for *n*-butane, −7 for oxygen, +2 for maleic anhydride, and +8 for water. Applying the definition of extent of reaction in Equation 14.39 gives

$$\dot{\xi} = \frac{\dot{N}_{butane,gen}}{-2} = \frac{\dot{N}_{oxygen,gen}}{-7} = \frac{\dot{N}_{mal.anh.,gen}}{+2} = \frac{\dot{N}_{water,gen}}{+8}$$

(14.43)

Step 3 *Construct stoichiometric table that summarizes mole balances*
When steps 1 and 2 are applied to every compound in the system, the results can be summarized in a table similar to Table 14-1, except instead of N_{final}, the number of moles of each compound at the end of the process, our major outcome is $\dot{n}_{out}$, the flow rate at which each compound leaves the reactor. Since the material leaving the reactor is a single mixed stream, the mole fraction of each compound in this stream can be determined from the compound molar flow rate and the total molar flow rate. The results are summarized in Table 14-2.

TABLE 14-2 Mole balances and reaction stoichiometry for Example 14-3.

	Butane	Oxygen	Maleic Anhydride	Water	Total Moles
$\dot{n}_{in}$ (mol/hr)	10,000	50,000	0	0	60,000
$\dot{N}_{gen}$ (mol/hr)	$-2\dot{\xi}$	$-7\dot{\xi}$	$+2\dot{\xi}$	$+8\dot{\xi}$	$+\dot{\xi}$
$\dot{n}_{out}$ (mol/hr)	$10,000 - 2\dot{\xi}$	$50,000 - 7\dot{\xi}$	$2\dot{\xi}$	$8\dot{\xi}$	$60,000 + \dot{\xi}$
y in outlet stream	$\dfrac{10,000 - 2\dot{\xi}}{60,000 + \dot{\xi}}$	$\dfrac{50,000 - 7\dot{\xi}}{60,000 + \dot{\xi}}$	$\dfrac{2\dot{\xi}}{60,000 + \dot{\xi}}$	$\dfrac{8\dot{\xi}}{60,000 + \dot{\xi}}$	

Examples 14-2 and 14-3 illustrate the application of mole balances to chemical reactions, and demonstrate how the results are conveniently summarized in tabular form. In the next section, we develop strategies for modeling chemical reaction equilibrium. We will see that the mole fraction is a fundamental quantity in these models, so quantifying mole fractions as we did in Tables 14-1 and 14-2 is essential.

14.3 The Equilibrium Criterion Applied to a Chemical Reaction

In this section, we examine the theoretical basis behind the equilibrium constant for a chemical reaction. Most readers are familiar with equilibrium constants from introductory chemistry courses. For example, the freshman chemistry text *Chemical Principles* (Zumdahl, 2009) notes that for a chemical reaction of the form:

$$\nu_A A + \nu_B B \leftrightarrow \nu_C C + \nu_D D$$

the equilibrium state is described by an equation of the form:

$$K = \frac{[C]^{v_c}[D]^{v_D}}{[A]^{v_A}[B]^{v_B}}$$

(14.44)

with $[C]$, $[D]$, $[A]$, and $[B]$ representing appropriate measures of concentration, such as partial pressures for a gas phase process or molarities for a reaction occurring in solution. Equation 14.44 is useful for solving problems like the one posed in Example 14-1, but is only valid for ideal gases and ideal liquid solutions. While Zumdahl acknowledges this limitation, quantitative accounting for departures from ideal solution behavior is beyond the scope of a typical freshman chemistry course. However, in Chapters 9 through 13, we have learned to model deviations from the ideal gas and ideal solution models, largely in the context of phase equilibrium. In this section, we will see how to apply the same principles to reaction equilibrium, devise a model for chemical reaction equilibrium, and demonstrate its equivalence to Equation 14.44 for the ideal solution case.

Like the ideal gas law, Equation 14.44 was actually determined empirically prior to its derivation from theoretical models (Zumdahl, 2009). The importance of the theory explored in this section is that it allows us to identify precisely when an equation like Equation 14.44 should apply, and when a more complex treatment is needed.

14.3.1 The Equilibrium Constant

Our discussions of chemical equilibrium began in Chapter 8, when we established that spontaneous processes progress in a direction that decreases the Gibbs free energy and that a system is at equilibrium when the Gibbs free energy reaches a minimum. The change in Gibbs free energy for a system can be written as

$$dG = -S\,dT + V\,dP + \sum_i \mu_i\,dN_i$$

(14.45)

This equation is valid whether changes in the number of moles of a compound (dn_i) occur because of material entering or leaving the system, or because of chemical reactions occurring inside the system.

Our goal is to develop a way of modeling chemical reactions at specific conditions (e.g., T and P) of interest, so let us assume a chemical reaction occurs in a closed system at constant temperature and pressure, like the system in Example 14-1. For this case, Equation 14.45 simplifies to

$$dG = \sum_i \mu_i\,dN_i$$

(14.46)

Since we are modeling a closed system, changes in the number of moles of a species can only occur by chemical reaction, which means they are related to each other by stoichiometry. While values of dN_i are different in different terms of the summation, we can use Equation 14.31 to relate the change in moles of each compound to the extent of the reaction:

$$dG = \sum_i \mu_i \nu_i\,d\xi$$

(14.47)

and $d\xi$ can be divided out of the summation as

$$\frac{dG}{d\xi} = \sum_i \mu_i \nu_i$$

(14.48)

We know that G is minimized at equilibrium, which means that at equilibrium the derivative of G is zero:

$$\frac{dG}{d\xi} = \sum_i \mu_i \nu_i = 0$$

(14.49)

While we use a constant T constant P process in this derivation, the result is applicable to processes that aren't isothermal or isobaric. The equilibrium state at a specific T and P is independent of the path that led to that state.

We will find the equation more convenient to use if we apply the definition of the mixture fugacity of component i, as established in Section 11.3 (see Equation 11.26):

$$\sum_i \mu_i \nu_i = \sum_i \nu_i \left[\underline{G}_i^0 + RT \ln \left(\frac{\hat{f}_i}{f_i^0} \right) \right] = 0 \qquad (14.50)$$

Recall that Gibbs free energy, like enthalpy and internal energy, is measured relative to a reference state. Here $\underline{G}_i^0$ is the molar Gibbs free energy of pure compound i at the chosen reference state, and f_i^0 is the fugacity of pure compound i at that same reference state. It is instructional to separate the two terms of the summation in Equation 14.50. Thus, at equilibrium:

$$\sum_i \nu_i \underline{G}_i^0 + \sum_i \nu_i \left[RT \ln \left(\frac{\hat{f}_i}{f_i^0} \right) \right] = 0 \qquad (14.51)$$

Here, we recall a point that was raised in the motivational example. The first term in Equation 14.51 represents the Gibbs free energy of pure components—reactants and products. We know that Gibbs free energy is minimized at equilibrium, so if for example the Gibbs free energy of the products is lower than the Gibbs free energy of the reactants, wouldn't the reaction simply progress until at least one of the reactants is completely consumed? The answer is "No," because comparing the Gibbs free energy of pure products to the Gibbs free energy of pure reactants omits the phenomenon of entropy of mixing. The fugacity term in Equation 14.51 incorporates the entropy of mixing, because mixture fugacity is derived from the partial molar Gibbs free energy (through Equations 11.4 and 11.24). Equation 14.51 can be rearranged as

$$-\sum_i \nu_i \underline{G}_i^0 = \sum_i \nu_i \left[RT \ln \left(\frac{\hat{f}_i}{f_i^0} \right) \right] \qquad (14.52)$$

First, consider the term on the left-hand side. We will define this summation as the **standard Gibbs free energy of reaction ($\Delta \underline{G}_T^0$)**:

$$\Delta \underline{G}_T^0 = \sum_i \nu_i \underline{G}_i^0 \qquad (14.53)$$

$\underline{G}_i^0$ represents the Gibbs free energy of compound i at a reference state. Conventionally, the reference state used is $\underline{G} = 0$ for elements at $T = 298.15$ K and $P = 100$ kPa. The change in Gibbs free energy when a compound is formed from elements at $T = 25°C$ and $P = 100$ kPa is called $\Delta \underline{G}_f^0$, and values for some common chemicals are tabulated in Appendix C.

Introducing the standard Gibbs free energy of reaction allows us to re-write Equation 14.52 as

$$-\frac{\Delta \underline{G}_T^0}{RT} = \sum_i \nu_i \left[\ln \left(\frac{\hat{f}_i}{f_i^0} \right) \right] \qquad (14.54)$$

Taking the exponential of both sides yields

$$\exp \left(\frac{-\Delta \underline{G}_T^0}{RT} \right) = \prod_i \left(\frac{\hat{f}_i}{f_i^0} \right)^{\nu_i} \qquad (14.55)$$

The left-hand side is likely familiar from introductory chemistry.

Here we will use Equation 14.56 to define the **equilibrium constant K_T** for a chemical reaction:

$$K_T = \exp\left(\frac{-\Delta \underline{G}_T^0}{RT}\right) = \prod_i \left(\frac{\hat{f}_i}{f_i^0}\right)^{\nu_i} \tag{14.56}$$

where

T is the temperature.

$\Delta \underline{G}_T^0$ is the change in standard molar Gibbs free energy for the reaction at the temperature T.

K_T is the equilibrium constant of the reaction at the temperature T.

$\hat{f}_i$ is the mixture fugacity of component i.

f_i^0 is the fugacity of pure component i in its reference state at temperature T.

ν_i is the stoichiometric coefficient of compound i in the reaction, with the convention that it is negative for reactants and positive for products.

Conceptually, we can apply Equation 14.56 using any reference state we wish. The crucial point is that, for each compound i, f_i^0 on the right-hand side must be computed for the same reference state that is used in determining its standard Gibbs free energy $\underline{G}_i^0$.

In practice, it is conventional to define the reference state for gaseous compounds as *the ideal gas state at $P = 100$ kPa*. This is convenient for two reasons:

1. When the reference state is chosen as an ideal gas state, $f_i^0 = P^0 = 100$ kPa.

2. Values of $\underline{G}_i^0$ are often not available at the temperature of interest. Section 14.3.3 illustrates the calculations used in such a case. Selecting the ideal gas state as the reference state allows us to use published values of the ideal gas heat capacity C_P^* in these calculations.

Meanwhile, the reference state for liquid compounds is most commonly the pure liquid at $P = 100$ kPa.

The simplest case to analyze is one in which the actual reaction conditions are identical to the reference state. The motivational example, Example 14-1, was such a case, and was solved using an approach that is likely familiar from introductory chemistry courses. In Example 14-4, we examine the same reaction. One purpose of this example is to help reconcile any apparent contradictions between the approach introduced in this chapter and what was likely presented in prior courses. The example demonstrates the connection between the generalized reaction equilibrium expression derived in this section:

$$K_T = \exp\left(\frac{-\Delta \underline{G}_T^0}{RT}\right) = \prod_i \left(\frac{\hat{f}_i}{f_i^0}\right)^{\nu_i}$$

And the K_P expression that is likely familiar from introductory chemistry courses:

$$K_P = \frac{P_{C_2H_4} P_{H_2}}{P_{C_2H_6}}$$

PITFALL PREVENTION

K is conventionally called "the equilibrium constant," but this is potentially misleading in that K depends upon temperature. In this book we use the symbol K_T to emphasize that the equilibrium constant is only valid at the specific T at which it is calculated.

The Π operator is analogous to the more frequently used Σ operator. Σ indicates the sum of the terms, and Π indicates the product of the terms.

| EXAMPLE 14-4 | **MOTIVATIONAL EXAMPLE REVISITED** |

The chemical reaction:

$$C_2H_6(g) \leftrightarrow C_2H_4(g) + H_2(g)$$

is carried out at $P = 100$ kPa and $T = 298.15$ K. Derive an equilibrium expression for this reaction in which the only unknowns are the mole fractions of the three gases.

SOLUTION:

Step 1 *Calculate numerical value of equilibrium constant*
The standard Gibbs free energy of the reaction is found using Equation 14.53, and the data in Appendix C. For this reaction, the stoichiometric coefficients are -1 for C_2H_6, $+1$ for C_2H_4 and $+1$ for H_2. Consequently, $\Delta \underline{G}_T^0$ is equal to

$$\Delta \underline{G}_T^0 = \sum_i \nu_i \underline{G}_i^0 = \underline{G}_{C_2H_4}^0 + \underline{G}_{H_2}^0 - \underline{G}_{C_2H_6}^0 \tag{14.57}$$

If you're confused by the way the symbols $\underline{G}_i^0$ and $\Delta \underline{G}_{f,i}^0$ seem to get used interchangeably, see Section 14.3.4 for a discussion of nomenclature and reference states.

Using the ideal gas state at $P = 100$ kPa as the reference state, we can obtain values of $\underline{G}_i^0$ for each compound from Appendix C. The calculations required to determine K_T from this point were in step 3 of the motivational example and are not repeated here.

$$\Delta \underline{G}_{298.15}^0 = 100{,}290 \frac{J}{mol} \quad \text{and} \quad K_{298.15} = 2.69 \times 10^{-18}$$

Step 2 *Apply equilibrium criterion*
The general expression derived for chemical reaction equilibrium is

$$K_T = \exp\left(\frac{-\Delta \underline{G}_T^0}{RT}\right) = \prod_i \left(\frac{\hat{f}_i}{f_i^0}\right)^{\nu_i}$$

The Π operator is analogous to the more frequently used Σ operator. Σ indicates the sum of the terms, and Π indicates the product of the terms.

The numerical value of $K_{298.15}$ is known. Recognizing again that the stoichiometric coefficients are -1 for C_2H_6, $+1$ for C_2H_4, and $+1$ for H_2, the product in the rightmost term becomes

$$K_T = \frac{\left(\dfrac{\hat{f}_{C_2H_4}}{f_{C_2H_4}^0}\right)\left(\dfrac{\hat{f}_{H_2}}{f_{H_2}^0}\right)}{\left(\dfrac{\hat{f}_{C_2H_6}}{f_{C_2H_6}^0}\right)} \tag{14.58}$$

As the reaction is being carried out at $P = 100$ kPa $= 1$ bar we will assume here that ethane, ethylene, and hydrogen form an ideal gas mixture. Consequently, $\hat{f}_i$ for each compound is simply the partial pressure.

$$K_T = \frac{\left(\dfrac{p_{C_2H_4}}{f_{C_2H_4}^0}\right)\left(\dfrac{p_{H_2}}{f_{H_2}^0}\right)}{\left(\dfrac{p_{C_2H_6}}{f_{C_2H_6}^0}\right)} \tag{14.59}$$

However, the reference state is also an ideal gas, which means for all compounds, $f^0 = P^0$. If we *always* use an ideal gas at $P^0 = 1$ bar as our reference state for all compounds, and we *always* express all pressures in bar, then f^0 will always be numerically equal to one. Recognizing this, one could eliminate f^0 from the equilibrium expression completely for

$$K_T = \frac{P_{C_2H_4} P_{H_2}}{P_{C_2H_6}} \tag{14.60}$$

This is now identical to the K_p expression that is familiar to most readers from freshman chemistry and was used in solving Example 14-1. However, we can now recognize that Equation 14.60 is a shortcut that is only valid when the gas is modeled as a mixture of ideal gases and when all partial pressures are expressed in bar. Instead of using this shortcut, we will progress from Equation 14.59 by recognizing that $f^0 = P^0$. Thus,

$$K_T = \frac{\left(\dfrac{P_{C_2H_4}}{P^0}\right)\left(\dfrac{P_{H_2}}{P^0}\right)}{\left(\dfrac{P_{C_2H_6}}{P^0}\right)} \tag{14.61}$$

The partial pressure of a gas is equal to its mole fraction times the total pressure:

$$K_T = \frac{\left(\dfrac{y_{C_2H_4} P}{P^0}\right)\left(\dfrac{y_{H_2} P}{P^0}\right)}{\left(\dfrac{y_{C_2H_6} P}{P^0}\right)} \tag{14.62}$$

And in this example, both the actual pressure P and the reference pressure P^0 are 100 kPa, so they cancel each other. Introducing the numerical value of K_T produces the answer:

$$2.69 \times 10^{-18} = \frac{y_{C_2H_4} y_{H_2}}{y_{C_2H_6}} \tag{14.63}$$

> The equilibrium constant in Equation 14.60 appears to have units because we neglected f^0. Keep in mind that f^0 is not simply equal to 1; it is equal to 1 bar, or 100 kPa.

> Step 6 of Example 14-1 illustrated how we can progress from this equilibrium expression to a quantitative solution for a specific problem.

Example 14-4 asked us to find an equilibrium expression in which the equilibrium mole fractions of the reactant and product gases were the only unknowns, and this is shown in Equation 14.63. This result is identical to what we found in step 6 of Example 14-1. Thus, the reaction equilibrium criterion in Equation 14.56 is consistent with simplified approaches you may have learned previously but doesn't share their limitations. In Example 14-4, we applied Equation 14.56 to a mixture of ideal gases. As this chapter progresses, we will apply it to real gases (see Example 14-6) and to liquids (see Example 14-7).

Example 14-4 represents the simplest possible chemical reaction to analyze, because the reaction was carried out at the same pressure and temperature at which the reference state data was available. The next two sections examine the effects of changes in pressure and temperature.

14.3.2 Accounting for the Effects of Pressure

The previous section derived Equation 14.56, the equilibrium criterion for chemical reactions:

$$K_T = \exp\left(\frac{-\Delta G_T^0}{RT}\right) = \prod_i \left(\frac{\hat{f}_i}{f_i^0}\right)^{\nu_i}$$

The standard Gibbs free energy of the reaction ($\Delta \underline{G}^0_T$) is computed from the Gibbs free energy of individual reactant and product compounds, which are most commonly tabulated at $P = 100$ kPa and $T = 25°C$. The ideal gas state at $P = 100$ kPa is commonly used as a reference state. We can apply Equation 14.56 at pressures other than $P = 100$ kPa. We must simply be mindful of the distinction between the actual pressure P and the reference pressure P^0, and we must be careful to compute the mixture fugacity coefficient $\hat{f}_i$ using an approach that is appropriate for the system and state. Example 14-5 illustrates application of the equilibrium criterion to a low-pressure gaseous system, Example 14-6 to a high-pressure system in which an ideal gas approximation is not reasonable, and Example 14-7 a liquid reaction. These examples also demonstrate that changes in pressure can have a profound influence on the equilibrium conversion attained at a given temperature.

EXAMPLE 14-5	SYNGAS FROM NATURAL GAS

"Syngas" is a gaseous mixture of CO and H_2. The ratio of CO/H_2 varies in different applications.

The steam reforming of natural gas into syngas features the gas phase reaction:

$$CH_4 + H_2O \leftrightarrow CO + 3H_2$$

which has an equilibrium constant $K_{1000} = 26.6$ (Lu, 2010). This reaction is to be carried out in a sealed reaction vessel (Figure 14-4) that has a constant $T = 1000$ K and a constant pressure. The reactor initially contains 5 moles each of methane and water vapor, and no carbon monoxide or hydrogen. For each of the following cases, find the extent of reaction at equilibrium.

A. The reactor has constant $P = 100$ kPa.

B. The reactor has constant $P = 20$ kPa.

C. The reactor has constant $P = 100$ kPa, and initially contains 40 moles of nitrogen in addition to the methane and water vapor.

Initial state (A)	Initial state (B)	Initial state (C)
$T = 1000$ K $P = 100$ kPa	$T = 1000$ K $P = 20$ kPa	$T = 1000$ K $P = 100$ kPa
$N_{CH4} = 5$ mol $N_{H2O} = 5$ mol $N_{CO} = 0$ mol $N_{H2} = 0$ mol	$N_{CH4} = 5$ mol $N_{H2O} = 5$ mol $N_{CO} = 0$ mol $N_{H2} = 0$ mol	$N_{N2} = 40$ mol $N_{CH4} = 5$ mol $N_{H2O} = 5$ mol $N_{CO} = 0$ mol $N_{H2} = 0$ mol

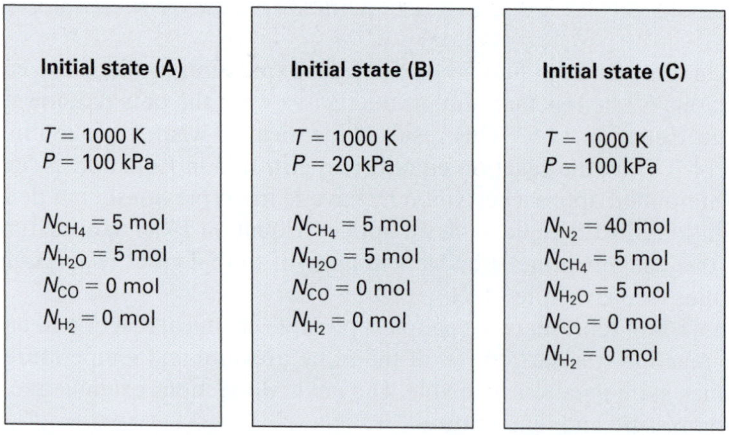

FIGURE 14-4 Syngas formation in three different batch reactors modeled in Example 14-5.

Since P and T are both being held constant throughout this process, the vessel cannot be a constant-volume vessel.

SOLUTION A:

Step 1 *Construct stoichiometric table for reaction*
The vessel initially contains 5 moles each of methane and water vapor, and nothing else. The mole balances for each compound are summarized in Table 14-3.

TABLE 14-3 Stoichiometric table for parts A and B of Example 14-5.

	CH$_4$	H$_2$O	CO	H$_2$	Total Moles
$N_{initial}$ (moles)	5	5	0	0	10
N_{gen} (moles)	$-\xi$	$-\xi$	$+\xi$	$+3\xi_1$	$+2\xi$
N_{final} (moles)	$5-\xi$	$5-\xi$	ξ	3ξ	$10+2\xi$
y	$\dfrac{5-\xi}{10+2\xi}$	$\dfrac{5-\xi}{10+2\xi}$	$\dfrac{\xi}{10+2\xi}$	$\dfrac{3\xi}{10+2\xi}$	

Step 2 *Apply equilibrium criterion*

Equation 14.56 gives the equilibrium criterion for a reaction in its most general form. Applying Equation 14.56 to this reaction gives:

$$K_T = \exp\left(\frac{-\Delta G_T^0}{RT}\right) = \frac{\left(\dfrac{\hat{f}_{CO}}{f_{CO}^0}\right)\left(\dfrac{\hat{f}_{H_2}}{f_{H_2}^0}\right)^3}{\left(\dfrac{\hat{f}_{CH_4}}{f_{CH_4}^0}\right)\left(\dfrac{\hat{f}_{H_2O}}{f_{H_2O}^0}\right)} \tag{14.64}$$

Calculation of the standard Gibbs free energy of the reaction is unnecessary in this case since a numerical value of the equilibrium constant K_T is given at the temperature of interest. At $P = 100$ kPa, an ideal gas assumption is quite reasonable, which means $\hat{f}_i$ for each compound is equal to the partial pressure $y_i P$.

$$K_T = \frac{\left(\dfrac{y_{CO}P}{P^0}\right)\left(\dfrac{y_{H_2}P}{P^0}\right)^3}{\left(\dfrac{y_{CH_4}P}{P^0}\right)\left(\dfrac{y_{H_2O}P}{P^0}\right)} \tag{14.65}$$

In addition, the actual pressure P is identical to the reference pressure P^0. Consequently:

$$K_T = \frac{(y_{CO})(y_{H_2})^3}{(y_{CH_4})(y_{H_2O})} \tag{14.66}$$

Step 3 *Solve for extent of reaction*

Plugging in the known value of K_{1000} and the expressions for mole fraction from Table 14-3 gives one equation in one unknown as

$$26.6 = \frac{\left(\dfrac{\xi}{10+2\xi}\right)\left(\dfrac{3\xi}{10+2\xi}\right)^3}{\left(\dfrac{5-\xi}{10+2\xi}\right)\left(\dfrac{5-\xi}{10+2\xi}\right)} \tag{14.67}$$

This is a fourth-degree equation in ξ, which many modern calculators are capable of solving. Standard tools such as EXCEL, MATHCAD, or POLYMATH can also be used. Solutions occur at $\xi = -7.1, -4.08, +4.08,$ and $+7.1$. Conceptually, negative extents of reaction are possible in modeling chemical reaction equilibrium, as demonstrated in Example 14-1.

FOOD FOR THOUGHT 14-2

Why are the four solutions symmetrical around 0?

In this particular example, however, there are 0 moles of product in the vessel initially, so a negative value of ξ would imply negative final quantities of CO and H_2, which is a physically unrealistic result. Similarly, there are only 5 moles each of methane and water vapor present initially, so if $\xi = 7.1$, then the final number of moles of each reactant is -2.1. The only physically realistic solution is $\xi = \textbf{4.08 moles}$. Plugging this result into the expressions in Table 14-3 reveals that at equilibrium, the vessel contains approximately 67.4 mol% hydrogen, 22.5 mol% carbon monoxide, and 5.1 mol% each of water vapor and methane.

> Realistic mole fractions must fall between 0 and 1. Inspection of the expressions in Table 14-3 reveals that realistic values of ξ must therefore fall between 0 and 5 in this example.

SOLUTION B:

Step 4 *Compare part B to part A*

The chemical reaction and initial numbers of moles for each compound are the same in parts A and B, so the stoichiometric table from step 1 still applies. The pressure is here lower than it was in part A, so the ideal gas model reflected in Equation 14.65 is still valid.

$$K_T = \frac{\left(\dfrac{y_{CO}P}{P^0}\right)\left(\dfrac{y_{H_2}P}{P^0}\right)^3}{\left(\dfrac{y_{CH_4}P}{P^0}\right)\left(\dfrac{y_{H_2O}P}{P^0}\right)}$$

This time, however, the actual pressure P is not equal to the reference pressure P^0, so pressure does not cancel out. It is instructive to combine the pressure terms and rearrange the equation this way:

$$K_T\left(\frac{P^0}{P}\right)^2 = \frac{(y_{CO})(y_{H_2})^3}{(y_{CH_4})(y_{H_2O})} \tag{14.68}$$

> Comparing Equations 14.66 and 14.68 reveals the effect of decreasing the pressure in this case; the $(P^0/P)^2$ term has the same effect as increasing the equilibrium constant by a factor of 25.

Step 5 *Solve for extent of reaction*

Inserting known values into Equation 14.68 gives

$$26.6\left(\frac{100\ kPa}{20\ kPa}\right)^2 = \frac{\left(\dfrac{\xi}{10 + 2\xi}\right)\left(\dfrac{3\xi}{10 + 2\xi}\right)^3}{\left(\dfrac{5 - \xi}{10 + 2\xi}\right)\left(\dfrac{5 - \xi}{10 + 2\xi}\right)} \tag{14.69}$$

$$665 = \frac{\left(\dfrac{\xi}{10 + 2\xi}\right)\left(\dfrac{3\xi}{10 + 2\xi}\right)^3}{\left(\dfrac{5 - \xi}{10 + 2\xi}\right)\left(\dfrac{5 - \xi}{10 + 2\xi}\right)}$$

> Le Chatelier's principle states that, if a reacting system at equilibrium is disturbed, the reaction will act to counteract the disturbance and a new equilibrium will be reached. If more of one compound is added, the chemical reaction will act to consume that compound, and if pressure is increased, the reaction will act to reduce the total moles of gas.

As in part A, this equation has four solutions, but three are physically unrealistic because $\xi < 0$ or $\xi > 5$ produce mole fractions that aren't between 0 and 1. The physically realistic solution is $\xi = \textbf{4.77 moles}$. The equilibrium mixture here contains only 1.2 mol% water vapor and 1.2 mol% methane. Compared to the case examined in part A, the equilibrium has shifted toward the products. Qualitatively, this outcome can be understood in terms of Le Chatelier's principle—when the system pressure is lowered, the equilibrium shifts in the direction that produces more moles of gas. Equation 14.68 illustrates the effect mathematically, because P^2 appears in the denominator of the left-hand side. Reducing P has the same effect as increasing the equilibrium constant, while increasing P has the same effect as decreasing the equilibrium constant. These observations are specific to the current example, and the effects of pressure are discussed more generally after the example.

SOLUTION C:

Step 6 *Compare part C to part A*

Comparing part C to part A, we note the following observations:

- Because $P = 100$ kPa in part C, the ideal gas model is a reasonable approximation; thus Equation 14.65, which was derived in part A, is also valid in part C.

- Because $P = 100$ kPa, the simplification $P = P^0$, which applied in part A but not in part B, is valid in part C. Thus, Equation 14.66 is correct for part C.

- The difference between parts A and C is the presence of nitrogen in part C, which is accounted for in the next step.

Step 7 *Construct stoichiometric table for part C*

Table 14-4 shows the stoichiometric table for part C. It is identical to Table 14-3 with the exception of the presence of 40 moles of nitrogen. Since nitrogen does not participate in the reaction, the number of moles of nitrogen is constant throughout the process, but its presence effectively lowers the mole fractions of all other compounds.

TABLE14-4 Stoichiometric table for part C of Example 14-5.

	CH$_4$	H$_2$O	CO	H$_2$	N$_2$	Total Moles
$N_{initial}$ (moles)	5	5	0	0	40	50
N_{gen} (moles)	$-\xi$	$-\xi$	$+\xi$	$+3\xi_1$	0	$+2\xi$
N_{final} (moles)	$5-\xi$	$5-\xi$	ξ	3ξ	40	$50+2\xi$
y	$\dfrac{5-\xi}{50+2\xi}$	$\dfrac{5-\xi}{50+2\xi}$	$\dfrac{\xi}{50+2\xi}$	$\dfrac{3\xi}{50+2\xi}$	$\dfrac{40}{50+2\xi}$	

Thus, Equation 14.66, for this case, becomes:

$$26.6 = \frac{\left(\dfrac{\xi}{50+2\xi}\right)\left(\dfrac{3\xi}{50+2\xi}\right)^3}{\left(\dfrac{5-\xi}{50+2\xi}\right)\left(\dfrac{5-\xi}{50+2\xi}\right)} \tag{14.70}$$

And the physically realistic solution is $\xi = 4.63$ **moles**. The equilibrium mixture contains 67.5% nitrogen, 23.4% hydrogen, 7.8% CO, and 0.6% each methane and water.

The effect of pressure on equilibrium conversion in this gas phase reaction is apparent in Equation 14.68, in which the equilibrium constant is multiplied by $(P^0/P)^2$. While Equation 14.68 is specific to Example 14-5, the pressure effect can be generalized. The equilibrium criterion in Equation 14.56 is

$$K_T = \prod_i \left(\frac{\hat{f}_i}{f_i^0}\right)^{\nu_i}$$

and can be re-expressed in terms of the fugacity coefficient through Equation 11.49 as

$$K_T = \prod_i \left(\frac{y_i \hat{\phi}_i P}{f_i^0}\right)^{\nu_i} \tag{14.71}$$

If the ideal gas state is used as the reference state, the reference state fugacity is the reference pressure:

$$K_T = \prod_i \left(\frac{y_i \hat{\phi}_i P}{P^0} \right)^{\nu_i} \tag{14.72}$$

The pressure terms can be factored out and moved to the left-hand side.

$$K_T \left(\frac{P^0}{P} \right)^{\Sigma \nu_i} = \prod_i (y_i \hat{\phi}_i)^{\nu_i} \tag{14.73}$$

Equation 14.73 is valid for any gas phase reaction so long as the ideal gas state is used as the reference state.

In the reaction in Example 14-5, there are more moles of gas on the product side than the reactant side. This is reflected mathematically in that $\Sigma \nu_i$, which is the sum of the stoichiometric coefficients, is positive. The equilibrium constant K_T is divided by $P^{\Sigma \nu_i}$, meaning that increasing pressure will shift the equilibrium toward the reactants. If there are more moles of reactant than product, then $\Sigma \nu_i$ is negative, meaning increasing pressure will increase equilibrium conversion. In either case, increasing the pressure shifts the equilibrium toward the side of the reaction with fewer moles of gas.

Le Chatelier's principle is that if an equilibrium system is disturbed, the reaction will shift to counteract the disturbance. If pressure is increased, then shifting the reaction equilibrium towards the side with fewer moles of gas will reduce the pressure.

Example 14-5 also illustrates that compounds which don't participate in the reaction directly can still influence the equilibrium conversion. In parts A and B the vessel contained only reactants and products, while in part C nitrogen was also present. Did the presence of the nitrogen "help" in achieving a higher equilibrium conversion? The answer depends upon how we do our comparison. Comparing part A to part C, we note that in both cases the total pressure was $P = 100$ kPa, but ξ is higher with the nitrogen present. In effect, we might say the addition of nitrogen lowered the mole fraction of all the other compounds, allowing us to obtain some of the benefit of a lower pressure without actually operating at vacuum conditions.

However, we now compare part B to part C. The temperatures are the same. In part B, the vessel initially contains 10 moles of gas, and in part C, the vessel initially contains 5 times as much gas but has 5 times the pressure (100 kPa versus 20 kPa). Thus, assuming ideal gas behavior, both vessels have the same initial volume—the vessel in part B looks exactly like the vessel in part C would if we pretended the nitrogen wasn't there. However, when both vessels are maintained at constant pressures, the vessel in part B progresses to a higher equilibrium conversion. Thus, diluting the mixture with nitrogen isn't *exactly* the same as lowering the pressure, as the comparison between A and B might have implied.

Example 14-5 relied entirely upon an ideal gas model. The next example considers a reaction at $P = 2.5$ MPa, which is a condition at which an ideal gas assumption isn't realistic.

EXAMPLE 14-6	**THERMAL CRACKING OF BUTANE TO FORM PROPYLENE**

Propylene ($CH_3–CH=CH_2$) can be formed by the gas phase thermal cracking of n-butane:

$$C_4H_{10} \leftrightarrow C_3H_6 + CH_4$$

A steady-state reactor (Figure 14-5) is maintained at a constant $T = 500$ K and a constant P. The feed to the reactor is 10 mol/s of n-butane. Assuming the stream leaving the reactor is at equilibrium, find the flow rate of propylene in the stream for each of the following cases.

A. The vessel has $P = 100$ kPa.

B. The vessel has $P = 2.5$ MPa, and the contents are modeled as an ideal gas mixture.

C. The vessel has $P = 2.5$ MPa, and the mixture fugacity of each compound is determined from the Peng-Robinson equation.

The reaction at 500 K has an equilibrium constant $K_{500} = 0.931$.

The motivational example examined ethylene, which is (by mass) the #1 organic chemical product in the world. Propylene, which shares many attributes with ethylene, is also among the top chemical products in annual worldwide sales.

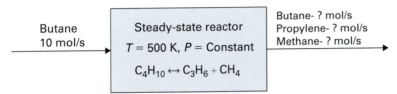

FIGURE 14-5 Thermal cracking of *n*-butane modeled in Example 14-6.

SOLUTION:

Step 1 *Construct a stoichiometric table*
In this case, the inlet stream contains 10 mol/s of pure *n*-butane. The stoichiometric coefficients are −1 for *n*-butane, +1 for propylene, and +1 for methane. The mole balances are summarized in Table 14-5, which is valid for all three parts of the example.

TABLE 14-5 Stoichiometric table for Example 14-6.

	C_4H_{10}	C_3H_6	CH_4	Total Moles
$\dot{n}_{in}$ (mol/s)	10	0	0	10
$\dot{n}_{gen}$ (mol/s)	$-\dot{\xi}$	$+\dot{\xi}$	$+\dot{\xi}$	$+\dot{\xi}$
$\dot{n}_{out}$ (mol/s)	$10-\dot{\xi}$	$+\dot{\xi}$	$+\dot{\xi}$	$10+\dot{\xi}$
y	$\dfrac{10-\dot{\xi}}{10+\dot{\xi}}$	$\dfrac{\dot{\xi}}{10+\dot{\xi}}$	$\dfrac{\dot{\xi}}{10+\dot{\xi}}$	

Anytime you are looking at changing *P* or *T*, you should always consider the possibility of a resulting phase change. Here, 500 K is above the critical point of all three compounds, so we needn't worry about one of them condensing as we increase the pressure—they will be gases throughout the example.

SOLUTION A:

Step 2 *Apply the equilibrium criterion*
Equation 14.73 expresses the equilibrium criterion for any gas phase reaction, so long as the ideal gas state is used as the reference state. As is conventional, we choose here the ideal gas state at $P = 100$ kPa as our reference state, which will allow us to make use of the data in Appendix C.

$$K_T \left(\frac{P^0}{P}\right)^{\Sigma \nu_i} = \prod_i (y_i \hat{\phi}_i)^{\nu_i}$$

In case A, the actual pressure is $P = 100$ kPa. This is significant in two ways:

1. In this case, $P = P^0$; the pressure dependence of Equation 14.73 cancels.

2. The pressure is low enough to assume ideal gas behavior; $\hat{\phi}_i = 1$ for all compounds

Thus, Equation 14.73 for this case simplifies to

$$K_T = \prod_i (y_i)^{\nu_i} \qquad (14.74)$$

Applying the stoichiometric coefficients of -1 for *n*-butane, $+1$ for propylene, and $+1$ for methane gives

$$K_T = \frac{\left(y_{C_3H_6}\right)\left(y_{CH_4}\right)}{\left(y_{C_4H_{10}}\right)} \tag{14.75}$$

Step 3 *Calculate extent of reaction*
Plugging the given value of K_{500} and the mole fraction expressions from Table 14-5 into Equation 14.75 gives

$$K_{500} = \frac{\left(y_{C_3H_6}\right)\left(y_{CH_4}\right)}{\left(y_{C_4H_{10}}\right)}$$

$$0.931 = \frac{\left(\dfrac{\dot{\xi}}{10 + \dot{\xi}}\right)\left(\dfrac{\dot{\xi}}{10 + \dot{\xi}}\right)}{\left(\dfrac{10 - \dot{\xi}}{10 + \dot{\xi}}\right)} \tag{14.76}$$

While chemical reaction equilibrium problems like this often produce complex equations that are best solved numerically, this one does have a simple algebraic solution:

$$0.931 = \frac{\dot{\xi}^2}{(10 + \dot{\xi})(10 - \dot{\xi})}$$

$$0.931 = \frac{\dot{\xi}^2}{100 - \dot{\xi}^2}$$

$$93.1 = 1.931\dot{\xi}^2$$

$$\dot{\xi} = -6.94 \text{ or } +6.94 \text{ mol/s}$$

Table 14-5 reveals that the equilibrium flow rate of propylene is equal to $\dot{\xi}$, so the negative solution is physically unrealistic and 6.94 mol/s of propylene are present in the product stream.

SOLUTION B:

Step 4 *Apply equilibrium criterion using ideal gas model at 2.5 MPa*
As in part A, we begin our evaluation of the equilibrium condition using Equation 14.73:

$$K_T\left(\frac{P^0}{P}\right)^{\Sigma\nu_i} = \prod_i (y_i\hat{\phi}_i)^{\nu_i}$$

This time, $P = 2.5$ MPa while $P^0 = 100$ kPa. However, we are using an ideal gas model. The simplification $\hat{\phi}_i = 1$ is applicable for any ideal gas mixture. Equation 14.73 thus becomes

$$K_T\left(\frac{P^0}{P}\right) = \prod_i (y_i)^{\nu_i} \tag{14.77}$$

$$K_T\left(\frac{P^0}{P}\right) = \frac{\left(y_{C_3H_6}\right)\left(y_{CH_4}\right)}{\left(y_{C_4H_{10}}\right)}$$

The sum of the stoichiometric coefficients is 1 ($+1$ methane, $+1$ propylene, -1 *n*-butane).

PITFALL PREVENTION

P is the actual system pressure, and is thus different in parts A and B. P^0 is the reference state pressure, and is the same throughout the example.

$$(0.931)\left(\frac{100 \text{ kPa}}{2500 \text{ kPa}}\right) = \frac{\left(\dfrac{\dot{\xi}}{10 + \dot{\xi}}\right)\left(\dfrac{\dot{\xi}}{10 + \dot{\xi}}\right)}{\left(\dfrac{10 - \dot{\xi}}{10 + \dot{\xi}}\right)} \qquad (14.78)$$

An algebraic solution analogous to step 4 produces

$$\xi = \textbf{1.89 mol/sec}$$

According to this calculation, the amount of propylene in the equilibrium mixture decreases significantly when the pressure is increased to 2.5 MPa; instead of converting ~70% of the reactant into the desired product, we convert only ~19% at equilibrium. Qualitatively, this result is logical in that increasing the pressure favored the side of the reaction—the reactants—that has fewer moles of gas. However, this calculation is suspect in that we used an ideal gas model at 2.5 MPa. Part C applies a more realistic model.

SOLUTION C:

Step 5 *Apply equilibrium criterion for real gas at 2.5 MPa*
Once again we begin with Equation 14.73.

$$K_T\left(\frac{P^0}{P}\right)^{\Sigma \nu_i} = \prod_i (y_i \hat{\phi}_i)^{\nu_i}$$

Modeling real gases through the use of the ideal solution model was first covered in Section 12.4.

While we are not assuming ideal gas behavior, we can reasonably assume this mixture of three hydrocarbons behaves as an ideal solution at $P = 2.5$ MPa. Consequently, while we are no longer assuming $\hat{\phi}_i = 1$, we are assuming that $\hat{\phi}_i = \phi_i$, which is Equation 12.64. Thus, the equilibrium criterion for this case is

$$K_T\left(\frac{P^0}{P}\right) = \frac{\left(y_{C_3H_6}\phi_{C_3H_6}\right)\left(y_{CH_4}\phi_{CH_4}\right)}{\left(y_{C_4H_{10}}\phi_{C_4H_{10}}\right)} \qquad (14.79)$$

We can relate mole fractions to the extent of reaction as we did in parts A and B, but must now account for the fugacity coefficient of each of these three pure compounds at $T = 500$ K and $P = 2.5$ MPa.

Step 6 *Compute Peng-Robinson parameters for pure n-butane, propylene, and methane*
The critical point and acentric factor for each of the compounds is available in Appendix C. The parameters a and b in the Peng-Robinson equation are determined from Equations 7.48 through 7.52, and the results are summarized in Table 14-6.

TABLE 14-6 Calculation of Peng-Robinson parameters for *n*-butane, propylene and methane at $T = 500$ K.

	T_c (K)	P_c (MPa)	ω	T_r	a (Pa · m⁶/mol²)	b (m³/mol)
Butane	425.2	3.797	0.193	1.18	1.34	72.4×10^{-6}
Propylene	364.8	4.613	0.142	1.37	0.738	51.2×10^{-6}
Methane	190.6	4.604	0.011	2.62	0.143	26.8×10^{-6}

Step 7 *Compute fugacity coefficients for pure n-butane, propylene and methane*

The fugacity coefficient for a pure compound was first defined in Section 8.3.2 as $\varphi = f/P$. Equation 8.67 gave the fugacity coefficient for a fluid described by the Peng-Robinson equation of state as

$$\ln \phi = \ln\left(\frac{f}{P}\right) = (Z - 1) - \ln(Z - B) - \frac{A}{2B\sqrt{2}} \ln\left[\frac{Z + (1 + \sqrt{2})B}{Z + (1 - \sqrt{2})B}\right]$$

$T = 500$ K is above the critical temperature for each of these compounds, so there is only one solution (Z) for each pressure. A and B are found from Equations 8.68 and 8.69, and the results are summarized in Table 14-7.

TABLE 14-7 Calculation of fugacity coefficients for *n*-butane, propylene, and methane at $T = 500$ K and $P = 2.5$ MPa.

	Z	**A**	**B**	**ln ϕ**	**ϕ**
Butane	0.846	0.194	0.0436	−0.152	0.859
Propylene	0.926	0.107	0.0308	−0.0750	0.928
Methane	0.996	0.0207	0.0161	−0.0041	0.996

The departure from the ideal gas model is most prominent in the results for *n*-butane.

Step 8 *Compute equilibrium extent of reaction*

The fugacity coefficients from Table 14-7 can be inserted into Equation 14.79, which can now be solved.

$$K_T\left(\frac{P^0}{P}\right) = \frac{(y_{C_3H_6}\phi_{C_3H_6})(y_{CH_4}\phi_{CH_4})}{(y_{C_4H_{10}}\phi_{C_4H_{10}})}$$

$$(0.931)\left(\frac{100 \text{ kPa}}{2500 \text{ kPa}}\right) = \frac{\left(\dfrac{\xi}{10 + \xi}\right)(0.928)\left(\dfrac{\xi}{10 + \xi}\right)(0.996)}{\left(\dfrac{10 - \xi}{10 + \xi}\right)(0.859)} \qquad (14.80)$$

$$\xi = \textbf{1.83 moles}$$

When we account for non-ideality of the gases using the Peng-Robinson equation, we obtain an answer that is different from the ideal gas estimate by about 5%. It remains clear that the equilibrium conversion was significantly reduced when the pressure increased.

Cost of chemical process equipment is linked directly to size, and the size of a chemical reactor can typically be decreased if the rate of the reaction is increased.

Offhand, Example 14-6 might seem like an impractical example. We established in Example 14-5 that if there are more moles of gas on the product side of a reaction, then the equilibrium conversion is improved by lowering the pressure. Since Example 14-6 features another reaction with more moles of gas on the product side, why would we look at high pressures? One answer is that chemical engineers are very concerned with the *rate* of a chemical reaction. Typically, gas phase chemical reactions proceed faster at higher pressures because inter-molecular collisions are more frequent. Consequently, designers are often faced with a trade-off between the higher equilibrium conversions attainable at low pressures and the faster rate of reaction available at high pressures. The principles of thermodynamics only allow us to quantify equilibrium states; the rate of the chemical reaction (or kinetics) is a subject normally addressed in a separate chemical engineering course.

The chapter up to this point has exclusively featured gas phase examples. This emphasis is deliberate in that chemical processes are typically designed to operate with high reaction rates, which usually requires high temperatures. This in turn often means they are carried out in the gas phase. Consequently, we need to be able to calculate equilibrium constants at temperatures different from (and often well above) the reference temperature. This is covered in Section 14.3.3. First, however, we demonstrate the application of chemical reaction equilibrium to compounds that are in the liquid phase at the commonly used reference state of $T = 298.15$ K and $P = 100$ kPa.

> Many "gas phase" reactions are actually carried out in fixed beds with solid catalysts, but whether or not a catalyst is used has no effect on equilibrium composition, only on the rate.

A LIQUID PHASE ESTERIFICATION REACTION	**EXAMPLE 14-7**

Ethyl acetate can be formed from ethanol and acetic acid, by the liquid phase reaction

$$C_2H_5OH + CH_3COOH \leftrightarrow H_2O + CH_3COOC_2H_5$$

$$\text{ethanol} + \text{acetic acid} \leftrightarrow \text{water} + \text{ethyl acetate}$$

> You can verify this is a liquid phase reaction- If you look up the normal boiling points of all four compounds, you will see they are significantly above 25°C.

A liquid phase reactor (Figure 14-6) with a constant temperature $T = 298.15$ K initially contains 5 moles each of ethanol, acetic acid, water, and ethyl acetate. Find the contents of the reaction at equilibrium if

A. The reaction is carried out at constant $P = 100$ kPa

B. The reaction is carried out at constant $P = 5$ MPa

For the purposes of this problem, assume the mixture can be modeled as an ideal solution.

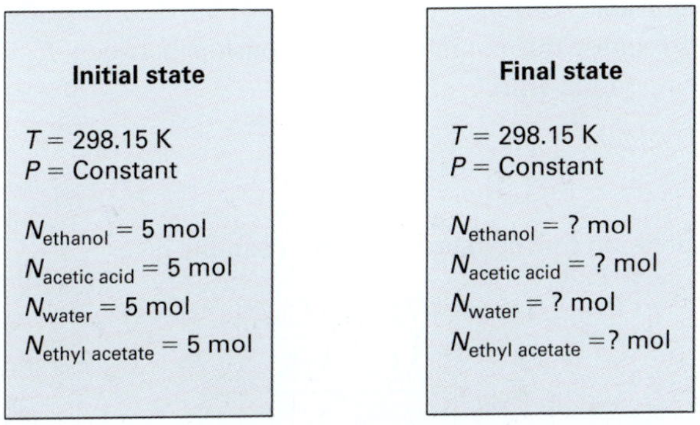

Initial state	Final state
$T = 298.15$ K	$T = 298.15$ K
$P =$ Constant	$P =$ Constant
$N_{ethanol} = 5$ mol	$N_{ethanol} = ?$ mol
$N_{acetic\ acid} = 5$ mol	$N_{acetic\ acid} = ?$ mol
$N_{water} = 5$ mol	$N_{water} = ?$ mol
$N_{ethyl\ acetate} = 5$ mol	$N_{ethyl\ acetate} = ?$ mol

> This is called an esterification reaction because ethyl acetate is an ester. Its IUPAC name is ethyl ethanoate.

FIGURE 14-6 Esterification reaction modeled in Example 14-7.

SOLUTION A:

Step 1 *Construct stoichiometric table*
Here, like in the motivational example, the vessel initially contains both reactants and products, so the extent of reaction ξ could realistically be positive or negative. Stoichiometric coefficients are -1 for ethanol and acetic acid and $+1$ for water and ethyl acetate. The full stoichiometric table is in Table 14-8.

Step 2 *Calculate equilibrium constant at $T = 298.15$ K*
In prior examples, we have used values of $\underline{G}^0$ taken directly from Appendix C-2, but data isn't always available in exactly this form. To illustrate this, literature values of $\Delta \underline{H}_f^0$ and $\underline{S}^0$ for each compound are given in Table 14-9. These values are valid at 298.15 K and $\bar{P} = 100$ kPa. According to Table 14-9, this reaction has $\Delta \underline{H}_{298.15}^0 = -6.17$ kJ/mol and

> The definition of Gibbs free energy, $G = H - TS$, was first introduced in Chapter 4.

$\Delta S^0_{298.15} = +11.49$ J/mol · K. We can apply the definition of Gibbs free energy to find the change in Gibbs free energy for the reaction, $\Delta G^0_{298.15}$:

$$\Delta G^0_{298.15} = \Delta H^0_{298.15} - T\Delta S^0_{298.15} \qquad (14.81)$$

$$\Delta G^0_{298.15} = \left(-6170\frac{\text{J}}{\text{mol}}\right) - (298.15 \text{ K})\left(11.49\frac{\text{J}}{\text{mol} \cdot \text{K}}\right) = -9596\frac{\text{J}}{\text{mol}}$$

TABLE 14-8 Stoichiometric table for esterification reaction.

	C_2H_5OH	CH_3COOH	H_2O	$CH_3COOC_2H_5$	Total Moles
$N_{initial}$ (moles)	5	5	5	5	20
N_{gen} (moles)	$-\xi$	$-\xi$	$+\xi$	$+\xi$	0
N_{final} (moles)	$5-\xi$	$5-\xi$	$5+\xi$	$5+\xi$	20
y	$\dfrac{5-\xi}{20}$	$\dfrac{5-\xi}{20}$	$\dfrac{5+\xi}{20}$	$\dfrac{5+\xi}{20}$	

FOOD FOR THOUGHT 14-3

The ΔH^0_f values in Table 14-9 are changes in enthalpy resulting from when the compound is formed from its elements at 298.15 K and 100 kPa. The S^0 values are calculated on an absolute scale, with a perfect crystal at $T = 0$ K representing $S = 0$. Does this difference in reference state cause a problem when we combine the values in computing ΔG^0_T?

Step 3 *Calculate equilibrium constant at 298.15 K*
Applying the definition of the equilibrium constant in Equation 14.56 gives

$$K_{298.15} = \exp\left(\frac{-\Delta G^0_{298.15}}{RT}\right) = \exp\left[\frac{-\left(-9596\frac{\text{J}}{\text{mol}}\right)}{\left(8.314\frac{\text{J}}{\text{mol} \cdot \text{K}}\right)(298.15 \text{ K})}\right] = 48.0 \quad (14.82)$$

Step 4 *Apply equilibrium criterion*
Equation 14.56 also states the equilibrium relationship between K_T and the mixture fugacities of the compounds. Thus,

$$K_T = \prod_i \left(\frac{\hat{f}_i}{f^0_i}\right)^{\nu_i}$$

Expanding this expression for the reaction in this example gives

$$K_{298.15} = \frac{\left(\dfrac{\hat{f}_{water}}{f^0_{water}}\right)\left(\dfrac{\hat{f}_{ethylacetate}}{f^0_{ethylacetate}}\right)}{\left(\dfrac{\hat{f}_{ethanol}}{f^0_{ethanol}}\right)\left(\dfrac{\hat{f}_{aceticacid}}{f^0_{aceticacid}}\right)} \qquad (14.83)$$

TABLE 14-9 Calculation of $\Delta G^0_{298.15}$ for esterification reaction.

Compound	N	ΔH^0_f (kJ/mol)	S^0 (J/mol · K)	Sources
C_2H_5OH	-1	-276.0	159.86	Afeefy, 2012
CH_3COOH	-1	-483.52	158	Afeefy, 2012; Domalski, 2012
H_2O	$+1$	-285.83	69.95	Chase, 1998
$CH_3COOC_2H_5$	$+1$	-479.86	259.4	Afeefy, 2012
Δ		-6.17	11.49	

How does this equilibrium criterion apply to liquids? In computing the equilibrium constant, we used pure liquids at $P = 100$ kPa as the reference state, because that was the data we had available. Thus, the f^0 for each compound in Equation 14.84 represents the fugacity of each pure liquid at $P = 100$ kPa. As always, $\hat{f}_i$ for each compound represents the mixture fugacity of that compound in the equilibrium mixture. We established in Section 11.3 (specifically Equation 11.44) that the mixture fugacity of a compound could be expressed as $\hat{f}_i = \gamma_i x_i f_i$. Two observations are critical here:

1. We are modeling the liquid as an ideal solution, so the activity coefficient γ for each compound is 1. (This will still be true in part B.)

2. f_i represents the fugacity of pure compound i at the actual temperature and pressure of the mixture. Here, the actual pressure 100 kPa is identical to the reference pressure. Thus f_i and f^0_i are equal to each other; they both represent the fugacity of pure compound i at $P = 100$ kPa and $T = 298.15$ K. This will not be true in part B, when the actual pressure is 5 MPa.

With the simplifications $\gamma_i = 1$ and $f_i = f^0_i$, Equation 14.83 simplifies to

$$K_{298.15} = \frac{(x_{water})(x_{ethylacetate})}{(x_{ethanol})(x_{acetic\ acid})} \tag{14.84}$$

Step 5 *Relate equilibrium constant to mole fractions*
Inserting the known numerical value of the equilibrium constant, and the mole fraction expressions from Table 14-8, into Equation 14.84 gives

$$K_{298.15} = \frac{\left(\dfrac{5+\xi}{20}\right)\left(\dfrac{5+\xi}{20}\right)}{\left(\dfrac{5-\xi}{20}\right)\left(\dfrac{5-\xi}{20}\right)} \tag{14.85}$$

$$48.0 = \frac{(5+\xi)(5+\xi)}{(5-\xi)(5-\xi)}$$

Equation 14.85 can be solved either numerically or using the quadratic equation. The solutions are $\xi = \mathbf{3.74\ moles}$ and $\xi = 6.69$ moles, the latter of which is unrealistic because it produces negative final values for the moles of the two reactants. Plugging the extent of reaction into the expressions in Table 14-8 reveals that the contents of the reactor at equilibrium are **8.74 moles each of water and ethyl acetate and 1.26 moles each of ethanol and acetic acid**.

SOLUTION B:

Step 6 *Compare to part A and evaluate role of pressure*
In part B, the reactor again contains 5 moles of each compound initially, and the reaction is again carried out at $T = 298.15$ K. Here, however, $P = 5.0$ MPa instead of 100 kPa. What is the significance of this change in terms of the model we developed in part A? The stoichiometry summarized in step 1 and Table 14-8 reflects only material balances; it is not affected by pressure. The equilibrium constant is a function of temperature only. Thus, the value of $K_{298.15}$ computed in steps 2 and 3 is again applicable to part B, but we must keep in mind that it was computed using pure liquids at $P^0 = 100$ kPa as a reference state.

Here the example specifies modeling the liquid as an ideal solution. We could, however, model departures from ideality by quantifying γ using any of the models introduced in Chapter 11.

The equilibrium criterion for this reaction is again Equation 14.83:

$$K_{298.15} = \frac{\left(\dfrac{\hat{f}_{water}}{f^0_{water}}\right)\left(\dfrac{\hat{f}_{ethylacetate}}{f^0_{ethylacetate}}\right)}{\left(\dfrac{\hat{f}_{ethanol}}{f^0_{ethanol}}\right)\left(\dfrac{\hat{f}_{aceticacid}}{f^0_{aceticacid}}\right)}$$

The fugacity of a compound in a liquid mixture is modeled as $\hat{f}_i = \gamma_i x_i f_i$:

$$K_{298.15} = \frac{\left(\dfrac{\gamma_{water}\,x_{water}\,f_{water}}{f^0_{water}}\right)\left(\dfrac{\gamma_{ethylacetate}\,x_{ethylacetate}\,f_{ethylacetate}}{f^0_{ethylacetate}}\right)}{\left(\dfrac{\gamma_{ethanol}\,x_{ethanol}\,f_{ethanol}}{f^0_{ethanol}}\right)\left(\dfrac{\gamma_{acetic\,acid}\,x_{acetic\,acid}\,f_{acetic\,acid}}{f^0_{acetic\,acid}}\right)} \qquad (14.86)$$

As in part A, we are using an ideal solution approximation, so γ for each compound is 1.

$$K_{298.15} = \frac{\left(\dfrac{x_{water}\,f_{water}}{f^0_{water}}\right)\left(\dfrac{x_{ethylacetate}\,f_{ethylacetate}}{f^0_{ethylacetate}}\right)}{\left(\dfrac{x_{ethanol}\,f_{ethanol}}{f^0_{ethanol}}\right)\left(\dfrac{x_{acetic\,acid}\,f_{acetic\,acid}}{f^0_{acetic\,acid}}\right)} \qquad (14.87)$$

Here, however, we cannot make the simplification that $f_i = f^0_i$. For example, f_{water} represents the fugacity of pure water at the actual temperature and pressure (5 MPa) of the reaction mixture, while f^0_{water} represents the fugacity of pure water at the actual temperature and the reference pressure (100 kPa). In Chapter 8, we learned to account for the effect of pressure on liquid fugacity using the Poynting method.

Step 7 *Apply Poynting correction to liquid fugacities*
The Poynting correction was first introduced as Equation 8.80, in which it was revealed that, if the molar volume of the liquid (or solid) V was assumed constant with respect to pressure, the fugacities of a liquid at two different pressures (but the same temperature) were related as

$$\frac{f_2}{f_1} = \exp\left\{\frac{V(P_2 - P_1)}{RT}\right\}$$

Notice when we use this method, it isn't necessary to find absolute values of the actual or reference fugacity—only their ratio appears in Equation 14.88, and the Poynting method provides a good estimate of their ratio.

Here, our interest is in relating the fugacity at the actual reaction pressure to the fugacity at the reference pressure. For water

$$\frac{f_{water}}{f^0_{water}} = \exp\left\{\frac{V_{water}(P - P_0)}{RT}\right\} \qquad (14.88)$$

The molar volume of water at 298.15 K and 100 kPa is 18.07×10^{-6} m³/mol; thus Equation 14.88 is solved:

$$\frac{f_{water}}{f^0_{water}} = \exp\left\{\frac{\left(18.07 \times 10^{-6}\,\dfrac{m^3}{mol}\right)(5 - 0.1\,MPa)\left(\dfrac{10^6\,Pa}{1\,MPa}\right)}{\left(8.314\,\dfrac{Pa \cdot m^3}{mol \cdot K}\right)(298.15\,K)}\right\} = 1.036 \qquad (14.89)$$

Analogous expressions can be written for each liquid. The results are summarized in Table 14-10.

Step 8 *Calculate equilibrium extent of reaction*
The results in Table 14-10 are inserted into Equation 14.87 for

$$K_{298.15} = \frac{\left(\dfrac{x_{water} f_{water}}{f^0_{water}}\right)\left(\dfrac{x_{ethylacetate} f_{ethylacetate}}{f^0_{ethylacetate}}\right)}{\left(\dfrac{x_{ethanol} f_{ethanol}}{f^0_{ethanol}}\right)\left(\dfrac{x_{acetic\,acid} f_{acetic\,acid}}{f^0_{acetic\,acid}}\right)}$$

$$48.0 = \frac{(x_{water})(1.036)(x_{ethylacetate})(1.215)}{(x_{ethanol})(1.123)(x_{acetic\,acid})(1.120)} \tag{14.90}$$

$$48.0 = (1.0009)\frac{(x_{water})(x_{ethylacetate})}{(x_{ethanol})(x_{acetic\,acid})}$$

The combined molar volume of the products is almost identical to the combined molar volume of the reactants, as illustrated in Table 14-10. Consequently, while the individual Poynting correction terms are measurably different from 1, a great deal of cancellation occurs when they are multiplied together. Comparing Equation 14.90 to Equation 14.85 reveals that they are identical to at least three significant figures. Thus, the answer to part B is essentially identical to part A: **8.74 moles each of water and ethyl acetate and 1.26 moles each of ethanol and acetic acid.**

TABLE 14-10 Poynting correction applied to fugacity of each compound in esterification reaction.

Compound	$\underline{V}$ (m³/mol) (Perry, 1997)	f/f^0
Water	18.07×10^{-6}	1.036
Ethyl Acetate	98.59×10^{-6}	1.215
Ethanol	58.68×10^{-6}	1.123
Acetic Acid	57.52×10^{-6}	1.120

Throughout the chapter, we have used the ideal gas state at $P^0 = 100$ kPa as a reference state in modeling gas phase reaction equilibrium. Example 14-7 illustrated modeling a liquid phase reaction, using the pure liquid at $P = 100$ kPa as a reference state. The reaction equilibrium criterion, as it applies to liquids, is

$$K_T = \prod_i \left(\frac{\gamma_i x_i f_i}{f^0_i}\right)^{\nu_i} \tag{14.91}$$

Step 7 of Example 14-7 illustrates that if the Poynting method is used to relate f_i to f^0_i, the result is

$$K_T = \prod_i \left\{\gamma_i x_i \exp\left[\frac{\underline{V}_i(P - P_0)}{RT}\right]\right\}^{\nu_i} \tag{14.92}$$

We learned in Chapter 8 that the "Poynting correction factor" for a pure compound is often negligible. Example 14-7 further demonstrates that even if the correction

Recall from Chapter 8 that the Poynting method assumes $\underline{V}$ is constant; an approximation that is reasonable for most liquids and solids but is not valid for liquids near their critical point.

factors of the reactants and the products are significant, they can largely cancel each other out. Consequently, the effect of pressure on reaction equilibrium in the liquid phase is often assumed negligible, in which case Equation 14.92 simplifies to:

$$K_T = \prod_i (\gamma_i x_i)^{\nu_i} \tag{14.93}$$

14.3.3 Accounting for Changes in Temperature

The equilibrium constant, as defined in Equation 14.56, is independent of pressure but is dependent upon temperature. For modeling chemical reaction equilibrium, the simplest case occurs when a measurement of the equilibrium constant at the temperature of interest is available, as in Example 14-5. If $\underline{G}^0$ for each compound is known at the temperature of interest, then $\Delta \underline{G}_T^0$ and the equilibrium constant K_T can be computed directly using Equation 14.56. Often, however, values of $\underline{G}^0$ are only available at a specific reference temperature (conventionally, 298.15 $\overline{K}$) and one must estimate $\Delta \underline{G}_T$ at a different temperature. This section examines a popular simplified approach in Example 14-8, and a more rigorous approach that relies upon accurate estimates of C_P in Example 14-9.

EXAMPLE 14-8	**SHORTCUT ESTIMATION OF T EFFECTS**

The compound chloromethane can be synthesized by the chemical reaction:

$$CH_4 + Cl_2 \leftrightarrow CH_3Cl + HCl$$

This reaction progresses by a free-radical mechanism and typically requires high temperatures.

The reaction is carried out in a closed vessel (Figure 14-7) at $P = 100$ kPa and constant temperature. The reactor initially contains 2 moles of methane and 1 mole of chlorine. Assuming this is the only chemical reaction that occurs, determine the extent of reaction at equilibrium for temperatures of:

A. 298.15 K
B. 800 K

FOOD FOR THOUGHT 14-4

The problem statement specifies a pressure of 100 kPa. How would the solution of the problem be impacted if the pressure was different, or was not specified at all?

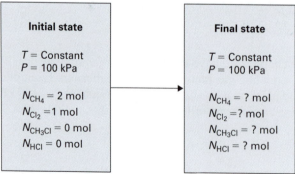

FIGURE 14-7 Synthesis of chloromethane modeled in Example 14-8.

SOLUTION A:

Step 1 *Construct stoichiometric table*
Example 14-5 presented a reaction in which methane and water vapor were consumed in equal quantities, and the reactor initially contained 5 moles of each. Thus, the reactants

were present in a *stoichiometric ratio*. While it is common for feedstocks to enter real chemical processes in a stoichiometric ratio, there are in some cases compelling reasons to do otherwise. The theory we have developed throughout this chapter in no way depends upon a stoichiometric ratio. Here, there is initially twice as much methane as chlorine, but they are consumed at equal rates. The full stoichiometric table is shown in Table 14-11.

One example of a reason to use a feed ratio other than a stoichiometric ratio occurs when one reactant also undergoes an undesired side reaction. The concentration of this reactant is minimized and other reactants are fed in excess to increase the rate of the desired reaction.

TABLE 14-11 Stoichiometric table for Example 14-8.

	CH_4	Cl_2	CH_3Cl	HCl	Total Moles
$N_{initial}$ (moles)	2	1	0	0	3
N_{gen} (moles)	$-\xi$	$-\xi$	$+\xi$	$+\xi$	0
N_{final} (moles)	$2-\xi$	$1-\xi$	ξ	ξ	3
y	$\dfrac{2-\xi}{3}$	$\dfrac{1-\xi}{3}$	$\dfrac{\xi}{3}$	$\dfrac{\xi}{3}$	

Step 2 *Find equilibrium constant at 298.15 K.*
As usual for gas phase reactions, we use the ideal gas state at $P = 100\,kPa$ as the reference state. The standard molar Gibbs free energy of each compound is known at 25°C from Appendix C. Applying Equation 14.53 gives

$$\Delta \underline{G}^0_{298.15} = \sum_i \nu_i \underline{G}^0_i = \Delta \underline{G}^0_{f,CH_3Cl} + \Delta \underline{G}^0_{f,HCl} - \Delta \underline{G}^0_{f,CH_4} - \Delta \underline{G}^0_{f,Cl_2} \qquad (14.94)$$

$$\Delta \underline{G}^0_{298.15} = \left(-62.8855\,\frac{kJ}{mol}\right) + \left(-95.2864\,\frac{kJ}{mol}\right) - \left(-50.45\,\frac{kJ}{mol}\right) - (0)$$

$$\Delta \underline{G}^0_{298.15} = -107.72\,\frac{kJ}{mol}$$

Applying the definition of the equilibrium constant in Equation 14.56 gives

$$K_{298.15} = \exp\left(\frac{-\Delta \underline{G}^0_{298.15}}{RT}\right) = \exp\left[\frac{107{,}720\,\dfrac{J}{mol}}{\left(8.314\,\dfrac{J}{mol \cdot K}\right)(298.15\,K)}\right] = 7.47 \times 10^{18} \quad (14.95)$$

Step 3 *Find equilibrium extent of reaction*
The actual pressure is $P = 100\,kPa$, so we assume ideal gas behavior ($\hat{\phi}_i = 1$) and Equation 14.73 simplifies:

$$K_T\left(\frac{P^0}{P}\right)^{\Sigma \nu_i} = \prod_i (y_i \hat{\phi}_i)^{\nu_i}$$

$$K_{298.15} = \frac{y_{CH_3Cl}\, y_{HCl}}{y_{CH_4}\, y_{Cl_2}} \qquad (14.96)$$

Here the pressure factor on the left-hand side is 1, because $P = P^0$, and even if it were not, the exponent $\Sigma \nu_i = 0$.

Substituting in the mole fraction expressions from the stoichiometric table gives

$$7.47 \times 10^{18} = \frac{\left(\dfrac{\xi}{3}\right)\left(\dfrac{\xi}{3}\right)}{\left(\dfrac{2-\xi}{3}\right)\left(\dfrac{1-\xi}{3}\right)} \qquad (14.97)$$

Because the equilibrium constant is so large, the extent of reaction is $\xi \sim 1$; the limiting reagent chlorine is essentially entirely consumed. However, this result is specific to $T = 298.15$ K. As mentioned in Section 14.3.2, designers are often motivated to increase the rate of a chemical reaction, and increasing temperature is often the most straightforward way to do this. If we increase the temperature to $T = 800$ K, will the products remain strongly favored at equilibrium?

Step 4 *Express mathematical effect of T on equilibrium constant*
The equilibrium constant K_T is found from $(\Delta \underline{G}^0_T / RT)$, and we can only calculate $\Delta \underline{G}^0_T$ directly at 298.15 K. Consequently, our strategy will be to determine how the quantity $(\Delta \underline{G}^0_T / RT)$ is influenced by changes in temperature. We follow the methodology first outlined in Chapter 6—we write a total derivative relating the quantity of interest $(\Delta \underline{G}^0_T / RT)$ to changes in T and P:

$$d\left(\Delta \underline{G}^0_T / RT\right) = \left[\frac{\partial \left(\Delta \underline{G}^0_T / RT\right)}{\partial P}\right]_T dP + \left[\frac{\partial \left(\Delta \underline{G}^0_T / RT\right)}{\partial T}\right]_P dT \qquad (14.98)$$

$\Delta \underline{G}^0_T$ is defined relative to a reference pressure; here $P^0 = 100$ kPa is the reference pressure regardless of the temperature. Consequently, we can treat pressure as a constant and Equation 14.98 simplifies to

$$d\left(\Delta \underline{G}^0_T / RT\right) = \left[\frac{\partial \left(\Delta \underline{G}^0_T / RT\right)}{\partial T}\right]_P dT \qquad (14.99)$$

> This step applies the rule of differentiation: $d(xy) = x\,dy + y\,dx$

$$d\left(\Delta \underline{G}^0_T / RT\right) = \frac{1}{RT}\left(\frac{\partial \Delta \underline{G}^0_T}{\partial T}\right)_P + \Delta \underline{G}^0_T\left(\frac{-1}{RT^2}\right) dT \qquad (14.100)$$

One of the results of Chapter 6 is that $(\partial \underline{G} / \partial T)_P = -\underline{S}$; this comes from the fundamental property relation for $\underline{G}$ (Equation 6.46). We can apply this result to $\Delta \underline{G}^0_T$ as well as to $\underline{G}$, because $\Delta \underline{G}^0_T$ is nothing more than a sum of individual $\underline{G}$ values for individual pure compounds.

$$d\left(\Delta \underline{G}^0_T / RT\right) = \frac{1}{RT}\left(-\Delta \underline{S}^0_T\right) + \Delta \underline{G}^0_T\left(\frac{-1}{RT^2}\right) dT \qquad (14.101)$$

When we introduce the definition $\underline{G} = \underline{H} - T\underline{S}$, a significant simplification occurs:

$$d\left(\Delta \underline{G}^0_T / RT\right) = \frac{1}{RT}\left(-\Delta \underline{S}^0_T\right) + \left(\Delta \underline{H}^0_T - T\Delta \underline{S}^0_T\right)\left(\frac{-1}{RT^2}\right) dT$$

$$\frac{d\left(\Delta \underline{G}^0_T / RT\right)}{dT} = -\left(\frac{\Delta \underline{H}^0_T}{RT^2}\right) \qquad (14.102)$$

Step 5 *Determine standard enthalpy of reaction at reference T*

The quantity $\Delta\underline{H}_T^0$, which appears in Equation 14.102, is the **standard enthalpy of a reaction.** It is analogous to $\Delta\underline{G}_T^0$, which was formally defined in Equation 14.53. Its definition is

$$\Delta\underline{H}_T^0 = \sum_i \nu_i \underline{H}_i^0 \qquad (14.103)$$

Like $\Delta\underline{G}_T^0$, the quantity $\Delta\underline{H}_T^0$ is readily computed using data contained in Appendix C, for the temperature 298.15 K or 25°C. We will refer to the values of the standard Gibbs free energy and standard enthalpy of the reaction at 298.15 K as $\Delta\underline{G}_R^0$ and $\Delta\underline{H}_R^0$. The R stands for "reference" and is used to distinguish the reference temperature 298.15 K from the temperature of interest, 800 K.

Since the reference temperature is 298.15 K, $\Delta\underline{G}_R^0$ is the $\Delta\underline{G}_{298.15}^0$ from step 2. $\Delta\underline{H}_R^0$ is found analogously as

$$\Delta\underline{H}_R^0 = \sum_i \nu_i \underline{H}_i^0 = \Delta\underline{H}_{f,CH_3Cl}^0 + \Delta\underline{H}_{f,HCl}^0 - \Delta\underline{H}_{f,CH_4}^0 - \Delta\underline{H}_{f,Cl_2}^0 \qquad (14.104)$$

$$\Delta\underline{H}_R^0 = \sum_i \nu_i \underline{H}_i^0 = \left(-80.7512\ \frac{kJ}{mol}\right) + \left(-92.3074\ \frac{kJ}{mol}\right)$$

$$- \left(-74.8936\ \frac{kJ}{mol}\right) - (0)$$

$$\Delta\underline{H}_R^0 = -98.165\ \frac{kJ}{mol}$$

Step 6 *Determine relationship between standard enthalpy of reaction and T*

The value of $\underline{H}_i^0$ for each individual compound is known at 298.15 K. The value of each compound at 800 K (or any other temperature) can be quantified through the heat capacity, using the relationship $dH = C_P dT$ which was first introduced in Chapter 2. $\Delta\underline{H}_T^0$ can be computed by summing the contributions of each individual compound for

$$\Delta\underline{H}_T^0 = \Delta\underline{H}_R^0 + \int_{T_R = 298.15\ K}^{T = 800\ K} \sum_i \nu_i C_{P,i}\ dT \qquad (14.105)$$

We will define the summation as ΔC_P, since in effect it represents the difference in heat capacity between the products and reactants:

$$\Delta C_P = \sum_i \nu_i C_{P,i} \qquad (14.106)$$

For this example, it becomes

$$\Delta C_P = C_{P,CH_4} + C_{P,Cl_2} - C_{P,CH_3Cl} - C_{P,HCl} \qquad (14.107)$$

In this case, C_P^* of each compound is known, and is shown in Table 14-12 along with ΔC_P.

Table 14-12 shows the calculation of ΔC_P for this reaction. The A term (which is frequently the most significant term in a calculation involving C_P^*) is very small for ΔC_P, compared to the values of A for the individual components. While you probably wouldn't describe the B, C, D, and E terms in ΔC_P as "small," note how they alternate in sign, leading to a great deal of cancellation when ΔC_P is computed. Figure 14-8 reveals that the total heat capacity of the reactants is in fact very similar to the total heat capacity of the products across a wide range of temperatures. This result highlights a simplifying

Reference books routinely publish tables of ΔH_f^0 and ΔG_f^0 values side by side. This derivation illustrates one reason why. ΔG_T^0 is needed to find the equilibrium constant. But if the temperature of interest is anything other than the reference T, then ΔH_T^0 is needed to find ΔG_T^0.

Recall that C_P^* represents an ideal gas heat capacity. For this system we have already assumed the ideal gas model is valid, so we can use ideal gas values of C_P in equation 14.107.

assumption that is commonly used in modeling chemical reaction equilibrium—the approximation that $\Delta C_P \sim 0$.

TABLE 14-12 Ideal gas heat capacities and ΔC_P for chloromethane synthesis reaction. Heat capacity is modeled using the form $C_P^*/R = A + BT + CT^2 + DT^3 + ET^4$ with T expressed in Kelvin (Poling, 2001).

	N	A	$B \times 10^3$	$C \times 10^5$	$D \times 10^8$	$E \times 10^{11}$
CH_4	−1	4.568	−8.975	3.631	−3.407	1.091
Cl_2	−1	3.0560	5.3708	−0.8098	0.5693	−0.15256
CH_3Cl	+1	3.578	−1.750	3.071	−3.714	1.408
HCl	+1	3.827	−2.936	0.879	−1.031	0.439
$\Sigma C_{P,reactants}$		7.624	−3.6042	2.8212	−2.8377	0.93844
$\Sigma C_{P,products}$		7.405	−4.686	3.95	−4.745	1.847
ΔC_P		−0.219	−1.0818	1.1288	−1.9073	0.90856

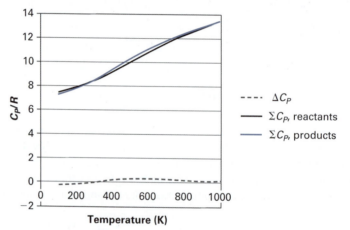

FIGURE 14-8 Heat capacity of reactants and products in chloromethane synthesis reaction.

Step 7 *Apply simplifying assumption to find* $\Delta \underline{G}_T^0$

Equation 14.105 can be used to find $\Delta \underline{H}_T^0$ at any temperature, once $\Delta \underline{H}_R^0$ and ΔC_P are known. However, the approximation that $\Delta C_P \sim 0$ implies that $\Delta \underline{H}_T^0$ is constant; $\Delta \underline{H}_T^0 = \Delta \underline{H}_R^0$. This approximation greatly simplifies Equation 14.102 as

$$\frac{d\left(\dfrac{\Delta \underline{G}_T^0}{RT}\right)}{dT} = -\left(\frac{\Delta \underline{H}_T^0}{RT^2}\right)$$

$$d\left(\frac{\Delta \underline{G}_T^0}{RT}\right) = -\left(\frac{\Delta \underline{H}_T^0}{RT^2}\right) dT \tag{14.108}$$

This can be integrated from the reference temperature (T_R) to the temperature of interest (T):

$$\int_{T_R = 298.15\,K}^{T = 800\,K} d\left(\frac{\Delta \underline{G}_T^0}{RT}\right) = \int_{T_R = 298.15\,K}^{T = 800\,K} -\left(\frac{\Delta \underline{H}_T^0}{RT^2}\right) dT \tag{14.109}$$

$$\frac{\Delta \underline{G}_T^0}{RT} - \frac{\Delta \underline{G}_R^0}{RT_R} = \left(\frac{\Delta \underline{H}_T^0}{R}\right)\left(\frac{1}{T} - \frac{1}{T_R}\right)$$

$$\frac{\Delta \underline{G}_T^0}{RT} = \frac{\Delta \underline{G}_R^0}{RT_R} + \left(\frac{\Delta \underline{H}_T^0}{R}\right)\left(\frac{1}{T} - \frac{1}{T_R}\right) \qquad (14.110)$$

Substituting in known values and multiplying through by R (which eliminates it from the equation) gives

$$\frac{\Delta \underline{G}_{800}^0}{(800\,\mathrm{K})} = \frac{\left(-107.72\dfrac{\mathrm{kJ}}{\mathrm{mol}}\right)}{298.15\,\mathrm{K}} + \left(-98.165\dfrac{\mathrm{kJ}}{\mathrm{mol}}\right)\left(\frac{1}{800\,\mathrm{K}} - \frac{1}{298.15\,\mathrm{K}}\right) \qquad (14.111)$$

$$\Delta \underline{G}_{800}^0 = -123.8\frac{\mathrm{kJ}}{\mathrm{mol}}$$

The standard Gibbs free energy of this reaction is negative, and is actually larger in magnitude at 800 K than at 298.15 K. We can now use this value to determine an equilibrium constant at 800 K.

Step 8 *Determine K_{800}*
Calculation of the equilibrium constant is analogous to Equation 14.95 in part A. Thus,

$$K_{800} = \exp\left(\frac{-\Delta \underline{G}_{800}^0}{RT}\right) = \exp\left[\frac{123{,}800\dfrac{\mathrm{J}}{\mathrm{mol}}}{\left(8.314\dfrac{\mathrm{J}}{\mathrm{mol}\cdot\mathrm{K}}\right)(800\,\mathrm{K})}\right] = 7.47 \times 10^{18} \quad (14.112)$$

$$K_{800} = 1.21 \times 10^8$$

The equilibrium constant is significantly smaller than it was in part A. While $\Delta \underline{G}_T^0$ is larger in magnitude at 800 K than it was at 298 K, it is also being divided by a larger T in the calculation of K_T. Calculation of the equilibrium extent of reaction is identical to Equation 14.97, but using the numerical value of K_{800}:

$$1.21 \times 10^8 = \frac{\left(\dfrac{\xi}{3}\right)\left(\dfrac{\xi}{3}\right)}{\left(\dfrac{2-\xi}{3}\right)\left(\dfrac{1-\xi}{3}\right)} \qquad (14.113)$$

While K_{800} is several orders of magnitude smaller than K_{298}, it is still large enough that ξ is essentially equal to 1. The number of moles of chlorine remaining at equilibrium $(1 - \xi)$ is on the order of $\sim10^{-8}$.

Example 14-8 illustrates an approach to modeling the effect of temperature on reaction equilibrium. The fundamental result is Equation 14.102:

$$\frac{d\left(\Delta \underline{G}_T^0 / RT\right)}{dT} = -\left(\frac{\Delta \underline{H}_T^0}{RT^2}\right)$$

Which is known as the **van 't Hoff equation**. There are no assumptions or approximations involved in the derivation of the van 't Hoff equation, but two different strategies are commonly employed to progress to a solution from this point.

FOOD FOR THOUGHT 14-5

With exothermic reactions, a trade-off often occurs: increasing temperature increases the rate of the reaction but also reduces the equilibrium constant. What is the highest temperature at which this system can operate, and still obtain at least 99.5% consumption of the chlorine?

Jacobus H. van 't Hoff won the Nobel Prize in Chemistry in 1901.

The "shortcut" approach is the one illustrated in Example 14-8, where the standard enthalpy change of the reaction is assumed to be a constant with respect to temperature; $\Delta \underline{H}^0_R$ is calculated at a reference temperature but assumed valid at all temperatures. This is equivalent to assuming that the heat capacity of the reactants is identical to the heat capacity of the products, stated mathematically:

$$\Delta C_P = \sum_i \nu_i C_{P,i} = 0 \tag{14.114}$$

This assumption produces the result in Equation 14.110, which we will call the **short-cut van 't Hoff equation**:

$$\frac{\Delta \underline{G}^0_T}{RT} = \frac{\Delta \underline{G}^0_R}{RT_R} + \left(\frac{\Delta \underline{H}^0_R}{R}\right)\left(\frac{1}{T} - \frac{1}{T_R}\right)$$

Using the shortcut method, one can rapidly estimate K_T at any temperature, needing only values of $\Delta \underline{G}^0_R$ and $\Delta \underline{H}^0_R$ at one reference temperature T_R. Conceptually, T_R can be any temperature at which data is available. In practice, data are most commonly tabulated at $T = 298.15$ K, as in Appendix C.

The more general approach to solving the van 't Hoff equation involves recognizing that $\Delta \underline{H}^0_T$ is a function of temperature, and accounting for this fact through the heat capacity of each compound, using Equation 14.115. Thus,

$$\Delta \underline{H}^0_T = \Delta \underline{H}^0_R + \int_{T_R}^{T} \Delta C_P \, dT \tag{14.115}$$

When the heat capacity is expressed as $C_p = A + BT + CT^2 + DT^3 + ET^4$, as it is in Appendix D, this becomes

$$\Delta \underline{H}^0_T = \Delta \underline{H}^0_R + \Delta A(T - T_R) + \frac{\Delta B}{2}(T^2 - T^2_R) + \frac{\Delta C}{3}(T^3 - T^3_R) \tag{14.116}$$
$$+ \frac{\Delta D}{4}(T^4 - T^4_R) + \frac{\Delta E}{5}(T^5 - T^5_R)$$

With ΔA, ΔB, ΔC, ΔD, and ΔE defined analogously to ΔC_P. The constant $\Delta \underline{H}^0_R$ and the various T_R terms can be combined into a single constant J for

$$\Delta \underline{H}^0_T = J + \Delta A T + \frac{\Delta B}{2} T^2 + \frac{\Delta C}{3} T^3 + \frac{\Delta D}{4} T^4 + \frac{\Delta E}{5} T^5 \tag{14.117}$$

with

$$J = \Delta \underline{H}^0_R - \Delta A T_R - \frac{\Delta B}{2} T^2_R - \frac{\Delta C}{3} T^3_R - \frac{\Delta D}{4} T^4_R - \frac{\Delta E}{5} T^5_R \tag{14.118}$$

When the expression for $\Delta \underline{H}^0_T$ from Equation 14.117 is introduced into Equation 14.102, the resulting integration is

$$\frac{\Delta \underline{G}^0_T}{RT} = \frac{\Delta \underline{G}^0_R}{RT_R} + \frac{J}{R}\left(\frac{1}{T} - \frac{1}{T_R}\right) - \frac{\Delta A}{R} \ln\left(\frac{T}{T_R}\right) - \frac{\Delta B}{2R}(T - T_R) \tag{14.119}$$
$$- \frac{\Delta C}{6R}(T^2 - T^2_R) - \frac{\Delta D}{12R}(T^3 - T^3_R) - \frac{\Delta E}{20R}(T^4 - T^4_R)$$

The use of this expression is demonstrated in the next example.

| ESTIMATION OF *T* EFFECTS USING IDEAL GAS HEAT CAPACITIES | **EXAMPLE 14-9** |

Example 14-6 considered the formation of propylene via the reaction:

$$C_4H_{10} \leftrightarrow C_3H_6 + CH_4$$

Estimate the equilibrium constant of this reaction at

A. 298.15 K
B. 500 K
C. 1000 K

SOLUTION:

Step 1 *Find standard Gibbs free energy and enthalpy of reaction at reference temperature*
In this reaction, the stoichiometric coefficients are −1 for *n*-butane, +1 for propylene, and +1 for methane. Standard enthalpy and Gibbs free energy for this reaction at the reference temperature 298.15 K can be found using data in Appendix C, as summarized in Table 14-13.

$$\Delta \underline{G}^0_R = \sum_i \nu_i \underline{G}^0_i = 62.14 + (-50.45) - (-16.57)\,\frac{kJ}{mol} = 28.26\,\frac{kJ}{mol} \quad (14.120)$$

$$\Delta \underline{H}^0_R = \sum_i \nu_i \underline{H}^0_i = 19.71 + (-74.8936) - 125.79\,\frac{kJ}{mol} = 70.61\,\frac{kJ}{mol} \quad (14.121)$$

TABLE 14-13 Data for calculation of standard enthalpy and Gibbs free energy for reaction.

	$\Delta \underline{H}^0_f$	$\Delta \underline{G}^0_f$
n-butane	−125.79	−16.57
Propylene	19.71	62.14
Methane	−74.8936	−50.45

SOLUTION A:

Step 2 *Calculate* $K_{298.15}$
In part A, the temperature of interest is identical to the reference temperature ($\Delta \underline{G}^0_T = \Delta \underline{G}^0_R$), so K_T is found directly from Equation 14.56 as

$$K_T = \exp\left(\frac{-\Delta \underline{G}^0_T}{RT}\right) = \exp\left[\frac{-28,260\,\dfrac{J}{mol}}{\left(8.314\,\dfrac{J}{mol \cdot K}\right)(298.15\ K)}\right] \quad (14.122)$$

$$K_T = 1.12 \times 10^{-5}$$

Step 3 *Calculate* ΔC_p
The heat capacity data in Appendix D is for the ideal gas state. In computing $\Delta \underline{G}^0_T$ we use the ideal gas state as a reference state, in part so that it is valid to apply values for the ideal gas heat capacity (C^*_p). The calculation of ΔC_p is summarized in Table 14-14.

TABLE 14-14 Ideal gas heat capacities and ΔC_p for propylene synthesis reaction. Heat capacity is modeled using the form $C_p^* = A + BT + CT^2 + DT^3 + ET^4$ with T expressed in Kelvin and C_p^* in J/mol·K.

	υ	A	B	C	D	E
C_4H_{10}	-1	46.12	0.04603	6.699×10^{-4}	-8.789×10^{-7}	3.437×10^{-10}
C_3H_6	$+1$	31.88	0.03237	3.898×10^{-4}	-4.999×10^{-7}	1.898×10^{-10}
CH_4	$+1$	37.98	-0.07462	3.019×10^{-4}	-2.833×10^{-7}	9.071×10^{-11}
ΔC_p		23.74	-0.08828	2.178×10^{-5}	9.569×10^{-8}	-6.319×10^{-11}

Step 4 *Find temperature-dependent expression for $\Delta \underline{G}_T^0$*
Equation 14.119 gives the expression for $\Delta \underline{G}_T^0$ as a function of temperature:

$$\frac{\Delta \underline{G}_T^0}{RT} = \frac{\Delta \underline{G}_R^0}{RT_R} + \frac{J}{R}\left(\frac{1}{T} - \frac{1}{T_R}\right) - \frac{\Delta A}{R}\ln\left(\frac{T}{T_R}\right) - \frac{\Delta B}{2R}(T - T_R) - \frac{\Delta C}{6R}\left(T^2 - T_R^2\right)$$

$$- \frac{\Delta D}{12R}\left(T^3 - T_R^3\right) - \frac{\Delta E}{20R}\left(T^4 - T_R^4\right)$$

Our goal is to progress to the point at which T and $\Delta \underline{G}_T^0$ are the only unknowns. Then we can find $\Delta \underline{G}_T^0$ at any temperature with 500 and 1000 K being the specific temperatures of interest in this example. T_R, the reference temperature, has to be 298.15 K, because that is the temperature at which we have data. $\Delta \underline{G}_R^0$ is known from step 1, and the heat capacity parameters ΔA, ΔB, etc. are known from step 3. This leaves only the calculation of the constant J from Equation 14.118. Thus,

$$J = \Delta \underline{H}_R^0 - \Delta A T_R - \frac{\Delta B}{2}T_R^2 - \frac{\Delta C}{3}T_R^3 - \frac{\Delta D}{4}T_R^4 - \frac{\Delta E}{5}T_R^5$$

$$J = \left(70{,}610 \, \frac{J}{mol}\right) - \left(23.74 \, \frac{J}{mol \cdot K}\right)(298.15K) - \frac{\left(-0.08288 \, \frac{J}{mol \cdot K^2}\right)}{2}(298.15 \, K)^2$$

$$- \frac{\left(2.718 \times 10^{-5} \, \frac{J}{mol \cdot K^3}\right)}{3}(298.15 \, K)^3 - \frac{\left(9.569 \times 10^{-8} \, \frac{J}{mol \cdot K^4}\right)}{4}(298.15 \, K)^4$$

$$- \frac{\left(-6.319 \times 10^{-11} \, \frac{J}{mol \cdot K^5}\right)}{5}(298.15 \, K)^5 = 68{,}575 \, \frac{J}{mol}$$

Step 5 *Find equilibrium constants at (B) $T = 500$ K and (C) $T = 1000$ K*
While Equation 14.119 is long, it is purely algebraic, and everything in it is now a known constant except the temperature T and the standard Gibbs free energy of reaction $\Delta \underline{G}_T^0$. The results are

$$\Delta \underline{G}_{500}^0 = 296.9 \, \frac{J}{mol} \quad \text{and} \quad \Delta \underline{G}_{1000}^0 = -59{,}598 \, \frac{J}{mol}$$

Inserting these into Equation 14.56 gives

$$K_{500} = 0.931 \quad \text{and} \quad K_{1000} = 1298$$

Notice that as the temperature increases, the standard Gibbs free energy decreases and the equilibrium constant increases. Mathematically, this can be understood through the van't Hoff equation. The change in Gibbs free energy is proportional to $-\Delta \underline{H}_T^0$. Because of the negative sign, if the reaction is endothermic ($\Delta \underline{H}_T^0 > 0$) then the standard Gibbs free energy decreases with increasing temperature. On a molecular level, this result can be understood through recognizing that an endothermic reaction requires an input of energy in order to progress in the forward direction. Thus the products are more favored as temperature increases, as the required energy input is more and more readily available.

This numerical value of K_{500} was used in Example 14-6. The result was that less than 20% of the initial n-butane was converted to propylene and methane at $T = 500$ K and $P = 2.5$ MPa. An analogous calculation using K_{1000} reveals that ~99% of the initial n-butane is converted to propylene and methane at $T = 1000$ K and $P = 2.5$ MPa.

Example 14-9 illustrated a strategy for modeling the effect of temperature on chemical reaction equilibrium. The shortcut method in Example 14-8 relied upon the assumption that $\Delta C_p = 0$ or, in words, that the heat capacity of the reactants is equal to the heat capacity of the products. The rigorous method in Example 14-9 does not rely on any such simplifying assumptions or approximations, although of course the results are only as accurate as the values of C_p upon which they are based.

Example 14-9 also demonstrates that for an endothermic reaction, the equilibrium constant increases as temperature increases. It has been mentioned previously that in many cases the goal of maximizing equilibrium conversion of reactants into products is in conflict with the goal of maximizing the rate of the reaction, and the designer is required to make a trade-off between these two goals. Example 14-9 illustrates a case in which increasing temperature will increase the equilibrium conversion in addition to increasing the reaction rate. Thus, if the goal is to make propylene, there is no trade-off in this case. A calculation like the one shown in Example 14-9 might tell the designer that the reaction should be carried out at the highest temperature that is possible, within the confines of what is safe and practical.

14.3.4 Reference States and Nomenclature

In this chapter, we use the symbol $\underline{G}^0$ to represent the molar Gibbs free energy of a compound at a standard state (see Equations 14.53 and 14.57), but in carrying out the calculations, we use $\Delta \underline{G}_f^0$ (see Equation 14.94). This section is intended to clarify possible confusion on this usage—why use two different symbols for the same thing? The answer is they aren't precisely the same thing.

- $\underline{G}^0$ represents the standard molar Gibbs free energy at a reference state. Conceptually it could be *any* reference state.

- $\Delta \underline{G}_f^0$ is the standard molar Gibbs free energy of formation; the change in Gibbs free energy when the compound is formed from elements at $T = 298.15$ K and $P = 100$ kPa. It is the same property, but we use this symbol to indicate that we have chosen a *specific* reference state in which $\underline{H} = \underline{G} = 0$ for all elements at $T = 25°C$ and $P = 100$ kPa.

Some books use slightly different definitions of the standard state; for example, some books use $P = 1$ atm instead of $P = 100$ kPa as the standard pressure.

Similarly, in an energy balance (Equations 3.34 or 3.35), we use the symbol $\underline{H}$ to represent molar enthalpy of a stream as it enters or leaves a system. In many instances, $\underline{H}$ can either be set equal to $\Delta \underline{H}_f^0$ (if the stream is at the standard temperature and pressure) or $\Delta \underline{H}_f^0$ can be used as a starting point in computing $\underline{H}$. But if we use $\Delta \underline{H}_f^0$, then we must use "$\underline{H} = 0$ for elements at $T = 25°C$ and $P = 100$ kPa" as the reference state throughout the energy balance.

14.4 Multiple Reaction Equilibrium

Every example we considered in Section 14.3 involved a single chemical reaction, but in a real chemical process reactor, it is quite common to have multiple chemical reactions occurring simultaneously. When multiple reactions are present, each is modeled individually as illustrated in Section 14.3. An equilibrium constant is calculated for each reaction, and the mixture fugacities of each participating compound are quantified using a method appropriate for the relevant phase, pressure, and temperature. The equilibrium condition of the system is one in which the equilibrium criterion:

$$K_T = \exp\left(\frac{-\Delta G^0_T}{RT}\right) = \prod_i \left(\frac{\hat{f}_i}{f^0_i}\right)^{\nu_i}$$

is simultaneously satisfied for *every* reaction. We illustrate multiple-reaction equilibrium through an expansion of Example 14-8.

EXAMPLE 14-10	SEQUENTIAL CHLORINATION OF METHANE

Chloroform (CHCl₃) and carbon tetrachloride (CCl₄) are common names that predate IUPAC nomenclature. The IUPAC names are trichloromethane and tetrachloromethane.

Example 14-8 illustrated the chlorination of methane to form chloromethane:

(R1): $CH_4 + Cl_2 \leftrightarrow CH_3Cl + HCl$

Example 14-8 also revealed that at equilibrium, the products were strongly favored. However, the substitution of chlorine for hydrogen can continue in subsequent reactions, forming dichloromethane, chloroform, and carbon tetrachloride:

(R2): $CH_3Cl + Cl_2 \leftrightarrow CH_2Cl_2 + HCl$

(R3): $CH_2Cl_2 + Cl_2 \leftrightarrow CHCl_3 + HCl$

(R4): $CHCl_3 + Cl_2 \leftrightarrow CCl_4 + HCl$

A steady-state reactor is maintained at $P = 100$ kPa and $T = 1000°C$. Assuming the product stream leaving the reactor is at equilibrium, find its composition if the reactor feed consists of the following.

A.   1 mol/s of methane and 4 mol/s of chlorine

B.   1 mol/s of methane and 1 mol/s of chlorine

Use the shortcut method to account for the effects of temperature.

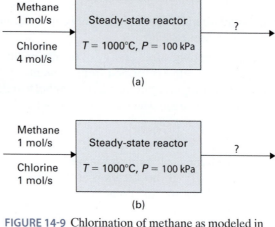

FIGURE 14-9 Chlorination of methane as modeled in Example 14-10.

SOLUTION:

Step 1 *Construct stoichiometric table*

As introduced in Example 14-2, when we are analyzing a system of multiple reactions, we can define a separate "extent of reaction" for each. In this case, we are using time-dependent mole balances, so the extents of reaction will be expressed as rates: $\dot{\xi}_1, \dot{\xi}_2, \dot{\xi}_3,$ and $\dot{\xi}_4$. The definition in Equation 14.38 is applied to each reactant and product in each reaction, and the individual "generation" terms are summed to determine the total generation for a compound. Chloroform, for example, has a stoichiometric coefficient of $+1$ in reaction R3 and -1 in reaction R4, so the generation of chloroform can be quantified as $+\dot{\xi}_3 - \dot{\xi}_4$. Stoichiometric tables for both cases A and B are given in Table 14-15 and Table 14-16.

TABLE 14-15 Stoichiometric table for sequential chlorination of methane, part A.

	CH_4	Cl_2	CH_3Cl	HCl	CH_2Cl_2	$CHCl_3$	CCl_4	Total
In (mol/s)	1	4						5
Gen, R1 (mol/s)	$-\dot{\xi}_1$	$-\dot{\xi}_1$	$+\dot{\xi}_1$	$+\dot{\xi}_1$				0
Gen, R2 (mol/s)		$-\dot{\xi}_2$	$-\dot{\xi}_2$	$+\dot{\xi}_2$	$+\dot{\xi}_2$			0
Gen, R3 (mol/s)		$-\dot{\xi}_3$		$+\dot{\xi}_3$	$-\dot{\xi}_3$	$+\dot{\xi}_3$		0
Gen, R4 (mol/s)		$-\dot{\xi}_4$		$+\dot{\xi}_4$		$-\dot{\xi}_4$	$+\dot{\xi}_4$	0
Out (mol/s)	$1-\dot{\xi}_1$	$4-\dot{\xi}_1-\dot{\xi}_2-\dot{\xi}_3-\dot{\xi}_4$	$\dot{\xi}_1-\dot{\xi}_2$	$\dot{\xi}_1+\dot{\xi}_2+\dot{\xi}_3+\dot{\xi}_4$	$\dot{\xi}_2-\dot{\xi}_3$	$\dot{\xi}_3-\dot{\xi}_4$	$\dot{\xi}_4$	5
y	$\dfrac{1-\dot{\xi}_1}{5}$	$\dfrac{4-\dot{\xi}_1-\dot{\xi}_2-\dot{\xi}_3-\dot{\xi}_4}{5}$	$\dfrac{\dot{\xi}_1-\dot{\xi}_2}{5}$	$\dfrac{\dot{\xi}_1+\dot{\xi}_2+\dot{\xi}_3+\dot{\xi}_4}{5}$	$\dfrac{\dot{\xi}_2-\dot{\xi}_3}{5}$	$\dfrac{\dot{\xi}_3-\dot{\xi}_4}{5}$	$\dfrac{\dot{\xi}_4}{5}$	

TABLE 14-16 Stoichiometric table for sequential chlorination of methane, part B.

	CH_4	Cl_2	CH_3Cl	HCl	CH_2Cl_2	$CHCl_3$	CCl_4	Total
In (mol/s)	1	1						2
Gen, R1 (mol/s)	$-\dot{\xi}_1$	$-\dot{\xi}_1$	$+\dot{\xi}_1$	$+\dot{\xi}_1$				0
Gen, R2 (mol/s)		$-\dot{\xi}_2$	$-\dot{\xi}_2$	$+\dot{\xi}_2$	$+\dot{\xi}_2$			0
Gen, R3 (mol/s)		$-\dot{\xi}_3$		$+\dot{\xi}_3$	$-\dot{\xi}_3$	$+\dot{\xi}_3$		0
Gen, R4 (mol/s)		$-\dot{\xi}_4$		$+\dot{\xi}_4$		$-\dot{\xi}_4$	$+\dot{\xi}_4$	0
Out (mol/s)	$1-\dot{\xi}_1$	$1-\dot{\xi}_1-\dot{\xi}_2-\dot{\xi}_3-\dot{\xi}_4$	$\dot{\xi}_1-\dot{\xi}_2$	$\dot{\xi}_1+\dot{\xi}_2+\dot{\xi}_3+\dot{\xi}_4$	$\dot{\xi}_2-\dot{\xi}_3$	$\dot{\xi}_3-\dot{\xi}_4$	$\dot{\xi}_4$	2
y	$\dfrac{1-\dot{\xi}_1}{2}$	$\dfrac{1-\dot{\xi}_1-\dot{\xi}_2-\dot{\xi}_3-\dot{\xi}_4}{2}$	$\dfrac{\dot{\xi}_1-\dot{\xi}_2}{2}$	$\dfrac{\dot{\xi}_1+\dot{\xi}_2+\dot{\xi}_3+\dot{\xi}_4}{2}$	$\dfrac{\dot{\xi}_2-\dot{\xi}_3}{2}$	$\dfrac{\dot{\xi}_3-\dot{\xi}_4}{2}$	$\dfrac{\dot{\xi}_4}{2}$	

The only difference between case A and case B is the entering flow rate of chlorine. Note that since all of the reactions have $\Sigma \nu = 0$, the total numbers of moles of gas entering and leaving the reactor are identical, regardless of the extents of the reactions.

Step 2 *Calculate standard Gibbs free energy and standard enthalpy of each reaction*
Table 14-17 shows the calculation of $\Delta \underline{G}_T^0$ and $\Delta \underline{H}_T^0$ for each reaction from data in Appendix C.

TABLE 14-17 Summary of contributions of each compound to the standard Gibbs free energy and enthalpy for reactions R1 through R4.

	R1		R2		R3		R4	
	$\Delta \underline{H}_R^0$ (kJ/mol)	$\Delta \underline{G}_R^0$ (kJ/mol)	$\Delta \underline{H}_R^0$ (kJ/mol)	$\Delta \underline{G}_R^0$ (kJ/mol)	$\Delta \underline{H}_R^0$ (kJ/mol)	$\Delta \underline{G}_R^0$ (kJ/mol)	$\Delta \underline{H}_R^0$ (kJ/mol)	$\Delta \underline{G}_R^0$ (kJ/mol)
CH_4	$-(-74.894)$	$-(-50.45)$						
Cl_2	$-(0)$	$-(0)$	$-(0)$	$-(0)$	$-(0)$	$-(0)$	$-(0)$	$-(0)$
HCl	-92.307	-95.296	-92.307	-95.296	-92.307	-95.296	-92.307	-95.296
CH_3Cl	-80.751	-62.886	$-(-80.751)$	$-(-62.886)$				
CH_2Cl_2			-95.395	-68.869	$-(-95.395)$	$-(-68.869)$		
$CHCl_3$					-103.345	-68.534	$-(-103.345)$	$-(-68.534)$
CCl_4							-95.8136	-60.6261
Σ	-98.165	-107.722	-106.951	-101.27	-100.257	-94.9516	-84.776	-87.3786

Equation 14.110 was first derived in the context of Example 14-8. It incorporates the assumption that $\Delta \underline{H}_T^0$ is constant with respect to T; thus $\Delta \underline{H}_T^0 = \Delta \underline{H}_R^0$

Step 3 *Estimate rate constants at T = 1000 °C for each reaction*
The bottom row of Table 14-17 shows the change in Gibbs free energy and the change in enthalpy of each reaction at the reference temperature 298.15 K; thus these are $\Delta \underline{H}_R^0$ and $\Delta \underline{G}_R^0$. To estimate the rate constant K_{1273} we begin by applying Equation 14.110, the shortcut van 't Hoff equation:

$$\frac{\Delta \underline{G}_T^0}{RT} = \frac{\Delta \underline{G}_R^0}{RT_R} + \left(\frac{\Delta \underline{H}_R^0}{R}\right)\left(\frac{1}{T} - \frac{1}{T_R}\right)$$

For reaction R1, for example,

$$\frac{\Delta \underline{G}_T^0}{RT} = \frac{\left(-107{,}222 \, \dfrac{J}{mol}\right)}{\left(8.314 \, \dfrac{J}{mol \cdot K}\right)(298.15 \, K)} + \left(\frac{-98.165 \, \dfrac{J}{mol}}{8.314 \, \dfrac{J}{mol \cdot K}}\right)\left(\frac{1}{1273.15 \, K} - \frac{1}{298.15 \, K}\right)$$

$$\frac{\Delta \underline{G}_T^0}{RT} = -13.13 \tag{14.123}$$

Applying Equation 14.56 gives

$$K_{1273.15} = \exp(-13.13) = 5.035 \times 10^5 \tag{14.124}$$

Equilibrium constants for the other reactions are estimated in an analogous manner and summarized in Table 14-18.

TABLE 14-18 Equilibrium constants for sequential chlorination of methane at 1000°C.

	$\Delta \underline{G}^0_R$ (kJ/mol)	$\Delta \underline{H}^0_R$ (kJ/mol)	$\dfrac{\Delta \underline{G}^0_T}{RT}$	$K_{1273.15}$
R1	−107.722	−98.165	−13.129	5.035×10^5
R2	−101.27	−106.951	−7.812	2470
R3	−94.9516	−100.257	−7.331	1527
R4	−87.3786	−84.776	−9.059	8596

Step 4 *Apply the equilibrium criterion to each reaction*
The equilibrium criterion in its most general form is given in Equation 14.56 as

$$K_T = \prod_i \left(\frac{\hat{f}_i}{f_i^0} \right)^{\nu_i}$$

The process is at $P = 100$ kPa, so we assume ideal gas behavior and the equilibrium criterion simplifies to

$$K_T = \prod_i \left(\frac{y_i P}{P^0} \right)^{\nu_i} \tag{14.125}$$

which, because the system pressure and reference pressure are both 100 kPa, further simplifies to

$$K_T = \prod_i (y_i)^{\nu_i} \tag{14.126}$$

An equation in the form of Equation 14.126 is written for each of the four reactions:

$$K_{R1,1273} = \frac{y_{HCl} y_{CH_3Cl}}{y_{CH_4} y_{Cl_2}} \tag{14.127}$$

$$K_{R2,1273} = \frac{y_{HCl} y_{CH_2Cl_2}}{y_{CH_3Cl} y_{Cl_2}} \tag{14.128}$$

$$K_{R3,1273} = \frac{y_{HCl} y_{CHCl_3}}{y_{CH_2Cl_2} y_{Cl_2}} \tag{14.129}$$

$$K_{R4,1273} = \frac{y_{HCl} y_{CCl_4}}{y_{CHCl_3} y_{Cl_2}} \tag{14.130}$$

Step 5 *Solve for equilibrium composition*
When the mole fraction expressions from Table 14-15 and the equilibrium constants from Table 14-18 are inserted into Equations 14.127 through 14.130, the result is

$$5.035 \times 10^5 = \frac{\left(\dfrac{\dot{\xi}_1 + \dot{\xi}_2 + \dot{\xi}_3 + \dot{\xi}_4}{5} \right)\left(\dfrac{\dot{\xi}_1 - \dot{\xi}_2}{5} \right)}{\left(\dfrac{1 - \dot{\xi}_1}{5} \right)\left(\dfrac{4 - \dot{\xi}_1 - \dot{\xi}_2 - \dot{\xi}_3 - \dot{\xi}_4}{5} \right)} \tag{14.131}$$

$$2470 = \frac{\left(\dfrac{\dot{\xi}_1 + \dot{\xi}_2 + \dot{\xi}_3 + \dot{\xi}_4}{5}\right)\left(\dfrac{\dot{\xi}_2 - \dot{\xi}_3}{5}\right)}{\left(\dfrac{\dot{\xi}_1 - \dot{\xi}_2}{5}\right)\left(\dfrac{4 - \dot{\xi}_1 - \dot{\xi}_2 - \dot{\xi}_3 - \dot{\xi}_4}{5}\right)} \tag{14.132}$$

$$1527 = \frac{\left(\dfrac{\dot{\xi}_1 + \dot{\xi}_2 + \dot{\xi}_3 + \dot{\xi}_4}{5}\right)\left(\dfrac{\dot{\xi}_3 - \dot{\xi}_4}{5}\right)}{\left(\dfrac{\dot{\xi}_2 - \dot{\xi}_3}{5}\right)\left(\dfrac{4 - \dot{\xi}_1 - \dot{\xi}_2 - \dot{\xi}_3 - \dot{\xi}_4}{5}\right)} \tag{14.133}$$

$$8596 = \frac{\left(\dfrac{\dot{\xi}_1 + \dot{\xi}_2 + \dot{\xi}_3 + \dot{\xi}_4}{5}\right)\left(\dfrac{\dot{\xi}_4}{5}\right)}{\left(\dfrac{\dot{\xi}_3 - \dot{\xi}_4}{5}\right)\left(\dfrac{4 - \dot{\xi}_1 - \dot{\xi}_2 - \dot{\xi}_3 - \dot{\xi}_4}{5}\right)} \tag{14.134}$$

Analogous equations can be written for part B, using the mole fraction expressions in Table 14-16.

At this point, we have finished the "thermodynamics" portion of the problem—what remains is purely math. Equations 14.131 through 14.134 represent a system of four algebraic equations in four unknowns, which can be solved. Once the extent of each reaction is known, the mole fractions are computed. In practice, a typical non-linear equation solver likely wouldn't converge on the solution of Equations 14.131 through 14.134 without good initial guesses. The solutions for parts A and B are shown in Table 14-19.

TABLE 14-19 Equilibrium compositions for sequential chlorination of methane at $T = 1000°C$.

Compound	Mole Fraction (part A)	Mole Fraction (part B)
CH_4	9.3×10^{-9}	0.0324
Cl_2	4.75×10^{-3}	1.35×10^{-5}
CH_3Cl	2.8×10^{-5}	0.4372
HCl	0.7952	0.5000
CH_2Cl_2	4.12×10^{-4}	0.0290
$CHCl_3$	3.85×10^{-3}	1.19×10^{-3}
CCl_4	0.1957	2.76×10^{-4}

Example 14-10 illustrates a reaction system in which four chlorination reactions occur sequentially. Because all four equilibrium constants are quite large, we might recognize that the products in all four reactions are strongly favored, and intuitively expect that the chlorination progresses almost to completion. This is exactly what happens in part A, when the system initially contains four moles of chlorine for each mole of methane; about 98% of the original methane is converted into carbon tetrachloride at equilibrium. Part B, however, illustrates that we can achieve a high equilibrium concentration of methyl chloride by limiting the amount of chlorine in the system. In part B, 93.5% of the methane is converted into methyl chloride, and while the equilibrium constant of R2 strongly favors the products, <7% of this methyl chloride actually reacts further. This can be understood by recognizing that there isn't enough chlorine to convert all of the methane into carbon tetrachloride, and while all four reactions have large equilibrium constants, R1 has the largest.

14.5 SUMMARY OF CHAPTER FOURTEEN

▪ The *extent of a reaction*, ξ, is defined as the moles of a compound generated (or consumed) by a reaction divided by the stoichiometric coefficient of the compound, and can be expressed on either an absolute $\left(\xi = \dfrac{N_{i,gen}}{\nu_i}\right)$ or rate $\left(\dot{\xi} = \dfrac{\dot{N}_{i,gen}}{\nu_i}\right)$ basis.

▪ The *equilibrium constant* for a reaction is a function of temperature, and its value is given by

$$K_T = \exp\left(\frac{-\Delta \underline{G}_T^0}{RT}\right)$$

▪ The equilibrium composition of a reacting system is described by

$$K_T = \prod_i \left(\frac{\hat{f}_i}{f_i^0}\right)^{\nu_i}$$

One can model chemical reaction equilibrium by combining this equilibrium expression with reaction stoichiometry.

▪ The equilibrium constant can be applied to chemical reactions occurring in liquid or vapor phases, at high or low pressure, and as long as fugacity is quantified using a model appropriate for system conditions.

▪ The equilibrium constant can be calculated at temperatures other than the reference temperature using the van 't Hoff equation, which has both a shortcut and a rigorous form.

▪ One can model equilibrium for a system in which multiple reactions occur simultaneously by applying the equilibrium constant constraint to each reaction individually.

14.6 EXERCISES

14-1. Determine the equilibrium constant at $T = 298.15$ K for each of the following reactions:

A. Formation of ethanol from ethylene and water:

$$C_2H_4(g) + H_2O(g) \leftrightarrow C_2H_5OH(g)$$

B. Dehydrogenation of cyclohexane to form benzene:

$$C_6H_{12}(g) \leftrightarrow C_6H_6(g) + 3H_2(g)$$

C. Combustion of methane:

$$CH_4(g) + 2O_2(g) \leftrightarrow CO_2(g) + 2H_2O(g)$$

14-2. For each of the three reactions listed in Exercise 14-1, use the shortcut van 't Hoff equation to compute the equilibrium constant at $T = 500$ K.

14-3. Determine the equilibrium constant at $T = 298.15$ K for each of the following reactions.

A. $2NO(g) + O_2(g) \leftrightarrow N_2O_4(g)$
B. $3O_2(g) \leftrightarrow 2O_3(g)$
C. $2SO_2(g) + O_2(g) \leftrightarrow 2SO_3(g)$

14-4. For each of the three reactions listed in Exercise 14-3, use the shortcut van 't Hoff equation to compute the equilibrium constant at $T = 600$ K.

reactions have $\Delta C_P = 0$ and that $A, B, P,$ and U form ideal solutions.

The feed entering a steady-state reactor is 1000 mol/hr each of compounds A and B. The reactor is at a uniform pressure of 100 kPa.

A. Determine the equilibrium composition of the exit stream at 300 K.

B. Determine the equilibrium composition of the exit stream at 600 K.

C. The second reaction has a larger equilibrium constant than the first, yet there is, at equilibrium, more P than U. Why?

TABLE P14-18

Compound	H°_{300} (kJ/mol)	G°_{300} (kJ/mol)
A	40	30
B	40	25
P	95	70
U	200	150

14-19. Thermal decomposition of propane can progress by two different gas phase reaction pathways:

(R1): $C_3H_8 \leftrightarrow C_2H_4 + CH_4$

(R2): $C_3H_8 \leftrightarrow C_3H_6 + H_2$

10 mol/s of propane enter a reactor in which R1 and R2 occur simultaneously. The exiting stream is at equilibrium. Find the composition of the exiting stream if the reactor is at

A. $T = 500$ K and $P = 100$ kPa.

B. $T = 1000$ K and $P = 100$ kPa.

C. $T = 1000$ K and $P = 500$ kPa.

14-20. Thermal decomposition of n-butane can progress by three different gas-phase reaction pathways, forming ethylene, propylene, or 1-butene:

(R1): $C_4H_{10} \leftrightarrow C_2H_4 + C_2H_6$

(R2): $C_4H_{10} \leftrightarrow C_3H_6 + CH_4$

(R3): $C_4H_{10} \leftrightarrow C_4H_8 + H_2$

1000 mol/hr of n-butane enter a reactor in which these three reactions occur simultaneously. The exiting stream is at equilibrium, with $P = 100$ kPa. Use the rigorous method of accounting for the effect of temperature on equilibrium constants.

A. Find the composition of the exiting stream if the reactor is at $T = 500°C$.

B. Find the composition of the exiting stream if the reactor is at $T = 1000°C$.

C. What temperature leads to the maximum yield of propylene?

14-21. Ethanol can be converted into either ethylene or acetaldehyde, by the following pair of reactions:

$$C_2H_5OH \leftrightarrow C_2H_4 + H_2O$$

$$C_2H_5OH \leftrightarrow CH_3CHO + H_2$$

But ethylene and acetaldehyde can also be converted into butadiene:

$$C_2H_4 + CH_3CHO \leftrightarrow C_4H_6 + H_2$$

A reactor initially contains 10 moles of pure ethanol. Assuming all three of these reactions—and no others—occur, find the equilibrium composition of the reactor for the following.

A. The reactor is at $T = 500$ K and $P = 100$ kPa.

B. The reactor is at $T = 1000$ K and $P = 100$ kPa.

C. The reactor is at $T = 500$ K and $P = 500$ kPa.

State any assumptions you make, and indicate the source of any data not obtained from the Appendices.

14.8 GLOSSARY OF SYMBOLS

A	parameter used in computing fugacity from Peng-Robinson EOS	f	fugacity	K_P	equilibrium constant for a gas phase reaction
a	parameter in van der Waals, Soave, or Peng-Robinson EOS	f^0	fugacity at standard state	K_T	equilibrium constant for reaction at temperature T
		$\hat{f}$	fugacity of a compound in a mixture		
B	parameter used in computing fugacity from Peng-Robinson EOS			M	mass of system
		G	Gibbs free energy	m	mass added to or removed from system
		$\underline{G}$	molar Gibbs free energy		
b	parameter in van der Waals, Soave, or Peng-Robinson EOS	$\underline{G}^0$	molar Gibbs free energy at reference state	$\dot{m}$	mass flow rate
C_P	constant pressure heat capacity	H	enthalpy	N	number of moles in system
C_P^*	constant pressure heat capacity for ideal gas	$\underline{H}$	molar enthalpy	N_{gen}	number of moles generated by reaction
		$\underline{H}^0$	molar enthalpy at reference state	$\dot{N}_{gen}$	rate at which moles are generated by reaction

n	number of moles added to or removed from system	$\underline{V}$	molar volume	$\Delta\underline{H}_f^0$	molar enthalpy of formation at standard state
$\dot{n}$	molar flow rate	x	liquid phase mole fraction	$\Delta\underline{H}_R^0$	change in standard molar enthalpy for reaction at reference temperature
P	pressure	y	vapor phase mole fraction		
P^0	reference pressure	ΔC_P	difference in heat capacity between reactants and products		
p	partial pressure			$\Delta\underline{H}_T^0$	change in standard molar enthalpy for reaction at temperature T
R	gas constant	$\Delta\underline{G}_f^0$	molar Gibbs free energy of formation at standard state		
S	entropy			γ	activity coefficient
$\underline{S}$	molar entropy	$\Delta\underline{G}_R^0$	change in standard molar Gibbs free energy for reaction at reference temperature	ν	stoichiometric coefficient
$\underline{S}^0$	molar entropy at standard state			ξ	extent of reaction
T	temperature	$\Delta\underline{G}_T^0$	change in standard molar Gibbs free energy for reaction at temperature T	φ	fugacity coefficient of a pure compound
T_R	reference temperature			$\hat{\varphi}$	fugacity coefficient of a component in a mixture
V	volume				

14.9 REFERENCES

Afeefy, H. Y., Liebman, J. F., Stein, S. E. "Neutral Thermochemical Data" in *NIST Chemistry WebBook, NIST Standard Reference Database Number 69*, Eds. P. J. Linstrom and W. G. Mallard, National Institute of Standards and Technology, Gaithersburg MD, 20899, http://webbook.nist.gov (retrieved April 21, 2012).

Chase, M. W., Jr. "NIST-JANAF Themochemical Tables, Fourth Edition," *J. Phys. Chem. Ref. Data, Monograph 9*, 1998, 1–1951.

Domalski, E. S., Hearing, E. D. "Condensed Phase Heat Capacity Data" in *NIST Chemistry WebBook, NIST Standard Reference Database Number 69*, Eds. P. J. Linstrom and W. G. Mallard, National Institute of Standards and Technology, Gaithersburg MD, 20899, http://webbook.nist.gov (retrieved April 21, 2012).

Perry, R. H., Green, D. W. editors, *Perry's Chemical Engineers' Handbook*, 7th ed., McGraw-Hill, 1997.

Poling, B. E., Prausnitz, J. M., O'Connell, J. P. *The Properties of Gases and Liquids*. 5th edition, McGraw-Hill, New York, 2001.

Song, C., Lu, K., Subramani, V. *Hydrogen and Syngas Production and Purification Technologies*, John Wiley & Sons, 2010.

Zumdahl, Stephen S. *Chemical Principles*, 6th ed., Houghton Mifflin Company, New York, NY, 2009.

Synthesis of Thermodynamic Principles

15

Throughout this book, we have developed a number of tools for solving a wide range of engineering problems. Examples include energy balances, entropy balances, VLE models for both ideal and real solutions, and models for chemical reaction equilibrium. A possible danger is that newly learned concepts and problem-solving strategies may become understood only in a disconnected or compartmentalized way. It was easy to recognize "apply an energy balance" as a necessary step for the solution of the problems at the end of Chapter 3, because Chapter 3 was about energy balances. Unfortunately, we cannot expect real-world engineering challenges to be presented with labels like "energy balance problem," "gamma-phi modeling problem," or "multi-reaction equilibrium problem." Nor can we expect real-world engineering applications to fall neatly into one—and only one—such category. Consider, for example, the chapter we just completed. There were no energy balances in Chapter 14, but in practice, conservation of energy and chemical reaction equilibrium are essential principles that are applied in the design of real chemical reactors. As another example, there was no estimation of vapor pressures in Chapter 14; it was simply presented as "known" that each reaction was carried out in either the liquid or vapor phase. In designing real chemical processes, however, the engineer must always be mindful of the possibility that any change in condition (T or P) can lead to a change in phase, and the principles introduced in Chapter 8 are directly relevant for evaluating this.

Consequently, we close the book with a chapter that examines some more complex examples in which a variety of thermodynamic principles are relevant. ■

15.1 MOTIVATIONAL EXAMPLE: Reactive Distillation

The authors have tried to use meaningful examples throughout this book, but inevitably the primary focus of examples (and end-of-chapter problems) is on the "new" material that is being presented. Here we examine applications that involve

FOOD FOR THOUGHT 15-1

The definition of equilibrium constant used by Bessling et al. is a little different from the one presented in Chapter 14. Can you guess what the symbol K_x means, exactly?

a wider array of these principles applied in conjunction. Consider the chemical reaction:

$$CH_3COOH + CH_3OH \leftrightarrow CH_3COOCH_3 + H_2O$$

Acetic acid + methanol $\leftrightarrow$ methyl acetate + water

Methyl acetate is the desired product in this reaction. Like the analogous ethyl acetate synthesis reaction that was examined in Example 14-7, the conversion of this reaction is limited by equilibrium; Bessling et al. (Bessling, 1998) report $K_x = 5.2$ as a typical value for the equilibrium constant. This means that, if acetic acid and methanol are fed to a reactor in a stoichiometric ratio, approximately 70% of the reactants are converted into product at equilibrium. A logical first idea is to send the product stream to a distillation train and recycle the unused reactants, as in Figure 15-1.

The mixture leaving the reactor has four components, and separating it requires more than one unit operation. A series of separation steps is often called a "separation train," or if they are all distillation columns, a "distillation train."

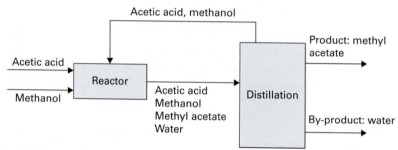

FIGURE 15-1 Schematic outline of a methyl acetate production process.

Unfortunately, separating this particular mixture of four compounds via distillation is problematic. Binary mixtures of methyl acetate and methanol form an azeotrope, as do binary mixtures of methyl acetate and water. As a result, it is difficult to obtain a pure methyl acetate stream through conventional distillation. It is reported (Bessling, 1998) that a conventional process like the one in Figure 15-1 requires two reactors, eight distillation columns, and a feed that uses a large excess of one reactant to overcome the equilibrium limitations on both the reaction and the separation. However, high conversions and high product purity can be obtained with a much smaller number of unit operations—and using a stoichiometric feed—through *reactive distillation*.

A conventional distillation column is a counter-current cascade of trays, as shown in Figure 15-2. On each tray, liquid mixes with the vapor moving upward from the tray below, and a portion of the more volatile component (or components) boils as a portion of the less volatile component(s) condense, approaching vapor–liquid equilibrium conditions. In principle, **reactive distillation** is a simple extension of this—the column has functionally the same design, but in addition to evaporation and condensation occurring in each stage, a chemical reaction also occurs. The methyl acetate synthesis reaction has a negligible rate unless a catalyst is used, so in this case, the column can have "reactive" and "non-reactive" sections. In the "reactive" section, the liquid is in contact with a catalyst, and any stage with no catalyst will function like a conventional distillation tray.

The potential advantages of reactive distillation are significant. By separating the products from the reactants *as they are formed*, one can achieve conversions much higher than would be obtained in a simple, equilibrium-limited reactor.

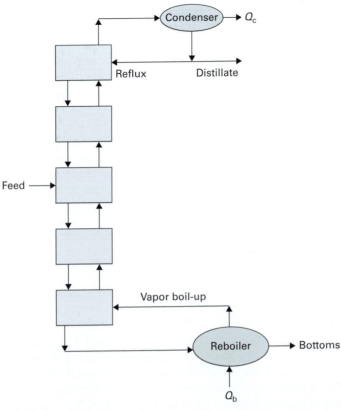

FIGURE 15-2 Schematic of a conventional distillation column.

However, design of a complete reactive distillation process is quite complex. Developing optimal design strategies for reactive distillation is an area of active research, and the methyl acetate reaction is frequently used as a test system (Huss, 2003).

Reliable models are valuable tools for design, so consider now the complexity of developing an accurate model for reactive distillation. If we defined one "reactive" stage as the system, the following are true:

- There are at least two distinct entering streams—more for any stage at which feed enters—and two exiting streams.

- Each stream is a mixture of several components—four, in the case of the methyl acetate process.

- The compositions of the liquid and vapor streams exiting a stage can be related to each other by vapor–liquid equilibrium, but in addition, the composition of the liquid must be consistent with the reaction equilibrium constant.

- Chemical reactions involve a $\Delta \underline{H}_T^0$, which means they affect the temperature of the system, and an energy balance can be used to quantify this. But the equilibrium constant is itself a function of temperature; thus, the progression of the reaction determines the temperature, but the temperature determines the progression of the reaction. (This kind of interaction between T and extent of reaction is explored in Example 15-2.)

- Evaporation and condensation also involve a $\Delta \underline{H}$, so these processes also influence temperature. But here again, physical properties like vapor pressure, which are fundamental in determining vapor–liquid equilibrium compositions, are themselves functions of temperature.

Clearly, building a model for reactive distillation would be quite complex. But all of the points on the previous page can be addressed quantitatively using the principles covered throughout this book—material balances, energy balances, solution models for computing the physical properties of mixtures, phase equilibrium models, and reaction equilibrium models. We've done all of these things—but not in a single problem. In a sense, this chapter contains no "new" information. What it does is explore how the various principles covered in the first 14 chapters can be synthesized in the solution of more complex problems.

An additional aspect of the challenge in modeling reactive distillation is that the bullet list above assumes the phase changes and reaction all progress to equilibrium, which they may not. As a result, a true model of the process would apply a full breadth of thermodynamics principles in combination with principles of kinetics that are outside the scope of this book. Thus, this motivational example embodies not only the synthesis of thermodynamics principles—which is the topic of this chapter—but also the synthesis of thermodynamics with other subjects within a typical chemical engineering curriculum.

15.2 Energy Balances on Chemical Reactors

One of the points made in Section 15.1 concerns energy and chemical reactions. Endothermic reactions require energy, which means energy must be added in order to maintain the reaction at a desired temperature. Exothermic reactions produce energy, which can mean that reactions must be carefully designed and controlled to avoid temperatures climbing to unsafe levels. In either case, energy balances are essential components for building models of the process. We begin by revisiting the esterification reaction that was originally introduced in Example 14-7.

EXAMPLE 15-1	**ENERGY BALANCE APPLIED TO A CHEMICAL REACTOR**

Ethyl acetate is synthesized from ethanol and acetic acid by the liquid phase esterification reaction shown in Figure 15-3.

$$C_2H_5OH + CH_3COOH \leftrightarrow H_2O + CH_3COOC_2H_5$$

Ethanol + acetic acid $\leftrightarrow$ water + ethyl acetate

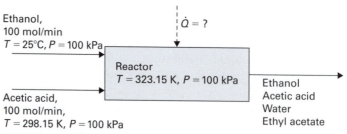

The ideal solution assumption is used in Examples 15-1 and 15-2 for illustrative purposes, but this mixture doesn't meet the "similar size, similar molecular interactions" criterion for an ideal solution. This is discussed further immediately after Example 15-2.

FIGURE 15-3 Esterification process modeled in Example 15-1.

Two separate feeds containing 100 mol/min each of ethanol and acetic acid enter a steady-state reactor at $T = 298.15$ K and $P = 100$ kPa. The reactor is maintained at $P = 100$ kPa and is isothermal at $T = 323.15$ K, and large enough that the reaction can be modeled as progressing to equilibrium. Assume the liquid phase can be modeled as an ideal solution.

A. Determine the composition of the stream leaving the reactor.

B. Determine the rate at which heat must be added to or removed from the reactor in order to maintain a constant temperature of $T = 323.15$ K.

SOLUTION A:

Step 1 *Construct stoichiometric table*

Part A asks us to determine the composition of the exiting stream. In order to determine this, we need to know how far the reaction progresses. We can apply the definition of extent of reaction in Equation 14.38 to relate the flow rate of each compound in the exiting stream to the extent of reaction $\dot{\xi}$, as summarized in Table 15-1. Note that the entering flow rates of ethanol and acetic acid are given, and the stoichiometric coefficients are -1 for ethanol and acetic acid and $+1$ for ethyl acetate and water.

TABLE 15-1 Mole balances and reaction stoichiometry for Example 15-1.

	Ethanol	Acetic Acid	Ethyl Acetate	Water	Total Flow Rate
$\dot{n}_{in}$ (mol/min)	100	100	0	0	200
$\dot{N}_{gen}$ (mol/min)	$-\dot{\xi}$	$-\dot{\xi}$	$+\dot{\xi}$	$+\dot{\xi}$	0
$\dot{n}_{out}$ (mol/min)	$100 - \dot{\xi}$	$100 - \dot{\xi}$	$\dot{\xi}$	$\dot{\xi}$	200
x in outlet stream	$\dfrac{100 - \dot{\xi}}{200}$	$\dfrac{100 - \dot{\xi}}{200}$	$\dfrac{\dot{\xi}}{200}$	$\dfrac{\dot{\xi}}{200}$	

Step 2 *Apply equilibrium criterion for chemical reactions*

The chemical reaction equilibrium criterion in its most general form was given in Equation 14.56:

$$K_T = \exp\left(\frac{-\Delta G_T^0}{RT}\right) = \prod_i \left(\frac{\hat{f}_i}{f_i^0}\right)^{\nu_i}$$

First we evaluate the right-hand side. Applying the known stoichiometric coefficients for each of the four compounds in this reaction, and recognizing that $\hat{f}_i = \gamma_i x_i f_i$ gives

$$K_T = \frac{\left(\dfrac{\gamma_{H_2O} x_{H_2O} f_{H_2O}}{f_{H_2O}^0}\right)\left(\dfrac{\gamma_{CH_3COOC_2H_5} x_{CH_3COOC_2H_5} f_{CH_3COOC_2H_5}}{f_{CH_3COOC_2H_5}^0}\right)}{\left(\dfrac{\gamma_{C_2H_5OH} x_{C_2H_5OH} f_{C_2H_5OH}}{f_{C_2H_5OH}^0}\right)\left(\dfrac{\gamma_{CH_3COOH} x_{CH_3COOH} f_{CH_3COOH}}{f_{CH_3COOH}^0}\right)} \tag{15.1}$$

Recall from Chapter 14 that f_i^0 is the fugacity of pure liquid i at the reference pressure while f_i is the fugacity of pure liquid i at the actual conditions of the equilibrium mixture. As usual the data we have available uses $P^0 = 100$ kPa, so we define this as the reference state, which means $P^0 = P$ and $f^0 = f$.

$$K_T = \frac{\left(\gamma_{H_2O} x_{H_2O}\right)\left(\gamma_{CH_3COOC_2H_5} x_{CH_3COOC_2H_5}\right)}{\left(\gamma_{C_2H_5OH} x_{C_2H_5OH}\right)\left(\gamma_{CH_3COOH} x_{CH_3COOH}\right)} \tag{15.2}$$

The mole fraction for each compound can be related to the extent of reaction through the expressions in the bottom row of Table 15-1. We are assuming that the liquid phase forms an ideal solution, so $\gamma = 1$ for all four compounds. Consequently, Equation 15.2 simplifies to

$$K_{\mathrm{T}} = \frac{\left(\dfrac{\dot{\xi}}{200}\right)\left(\dfrac{\dot{\xi}}{200}\right)}{\left(\dfrac{100 - \dot{\xi}}{200}\right)\left(\dfrac{100 - \dot{\xi}}{200}\right)} \tag{15.3}$$

Equation 15.3 can be solved for ξ once we find a numerical value for K_{T}. This requires us to find $\Delta \underline{G}^0_{\mathrm{T}}$, the standard change in Gibbs free energy for the reaction at the actual temperature of 323.15 K.

Step 3 *Collect data*
Example 14-7 gives the changes in molar enthalpy and molar Gibbs free energy for this reaction at a reference state of $T = 25°C$ and $P = 100$ kPa: $\Delta \underline{H}^0_{298.15} = -6.17$ kJ/mol and $\Delta \underline{G}^0_{298.15} = -9.596$ kJ/mol.

Step 4 *Determine equilibrium constant*
Here the actual temperature is $T = 323.15$ K; only 25 K higher than the reference temperature. Over this relatively small temperature range it is reasonable to use the shortcut van 't Hoff equation to find $\Delta \underline{G}_{\mathrm{T}}^{0}$:

$$\frac{\Delta \underline{G}^0_{\mathrm{T}}}{RT} = \frac{\Delta \underline{G}^0_{\mathrm{R}}}{RT_{\mathrm{R}}} + \left(\frac{\Delta \underline{H}^0_{\mathrm{R}}}{R}\right)\left(\frac{1}{T} - \frac{1}{T_{\mathrm{R}}}\right)$$

$$\frac{\Delta \underline{G}^0_{\mathrm{T}}}{RT} = \frac{\left(-9596\,\dfrac{\mathrm{J}}{\mathrm{mol}}\right)}{\left(8.314\dfrac{\mathrm{J}}{\mathrm{mol}\cdot\mathrm{K}}\right)(298.15\,\mathrm{K})} + \frac{\left(-6170\,\dfrac{\mathrm{J}}{\mathrm{mol}}\right)}{8.314\dfrac{\mathrm{J}}{\mathrm{mol}\cdot\mathrm{K}}}\left(\frac{1}{323.15\,\mathrm{K}} - \frac{1}{298.15\,\mathrm{K}}\right) \tag{15.4}$$

$$\frac{\Delta \underline{G}^0_{\mathrm{T}}}{RT} = -3.871 + (-742.1\,\mathrm{K})\left(-0.00026\,\frac{1}{\mathrm{K}}\right) = -3.679$$

From the equilibrium criterion in Equation 14.56, we have

$$K_{\mathrm{T}} = \exp\left(\frac{-\Delta \underline{G}^0_{\mathrm{T}}}{RT}\right) = \exp(3.679) = 39.6 \tag{15.5}$$

Step 5 *Determine composition of outlet stream*
When this value of the equilibrium constant is substituted into Equation 15.3, the extent of reaction $\dot{\xi}$ is the only unknown in the equation. The solutions are $\dot{\xi} = 86.3$ and $\dot{\xi} = 118.9$ mol/min. The physically realistic solution is 86.3 mol/min; the other solution produces negative molar flow rates for ethanol and acetic acid. Consequently the outlet stream contains **86.3 mol/min each of ethyl acetate and water**, and **13.7 mol/min each of ethanol and acetic acid**.

SOLUTION B:

Step 6 *Apply and simplify energy balance*
Part B asks us to find the rate at which heat ($\dot{Q}$) is added to the reactor. This can be determined using an energy balance. The energy balance in its most general form is Equation 3.34:

$$\frac{d}{dt}\left\{M\left(\hat{U} + \frac{v^2}{2} + gh\right)\right\} = \sum_{j=1}^{j=J}\left\{\dot{m}_{j,\text{in}}\left(\hat{H}_j + \frac{v_j^2}{2} + gh_j\right)\right\} - \sum_{k=1}^{k=K}\left\{\dot{m}_{k,\text{out}}\left(\hat{H}_k + \frac{v_k^2}{2} + gh_k\right)\right\}$$
$$+ \dot{W}_{\mathrm{S}} + \dot{W}_{\mathrm{EC}} + \dot{Q}$$

If we define the reactor as the system, then two simplifications occur immediately because the system is at steady state: the accumulation term (left-hand side) is 0, and the volume of the system is not changing, so $\dot{W}_{EC}$ is also 0.

$$0 = \sum_{j=1}^{j=J}\left\{\dot{m}_{j,\text{in}}\left(\hat{H}_j + \frac{v_j^2}{2} + gh_j\right)\right\} - \sum_{k=1}^{k=K}\left\{\dot{m}_{k,\text{out}}\left(\hat{H}_k + \frac{v_k^2}{2} + gh_k\right)\right\} + \dot{W}_S + \dot{Q} \qquad (15.6)$$

We learned in Chapter 3 that potential and kinetic energy terms are frequently negligible in standard chemical process equipment, and the problem statement gives us no information that would make such an assumption seem unreasonable, so we will assume these terms are negligible here. Similarly, the problem statement does not mention a mixer or any other feature of the reactor that would add shaft work, so assuming this term is negligible as well, we are left with

$$0 = \sum_{j=1}^{j=J}\dot{m}_{j,\text{in}}(\hat{H}_j) - \sum_{k=1}^{k=K}\dot{m}_{k,\text{out}}(\hat{H}_k) + \dot{Q} \qquad (15.7)$$

But all of the given information is on a molar basis, so it is convenient to express the equation in terms of molar properties. We also solve the equation for $\dot{Q}$, since that is what we wish to find. Thus,

$$\dot{Q} = -\sum_{j=1}^{j=J}\dot{n}_{j,\text{in}}(\underline{H}_j) + \sum_{k=1}^{k=K}\dot{n}_{k,\text{out}}(\underline{H}_k) \qquad (15.8)$$

Step 7 *Identify relevant data*

There are two entering streams, one composed of pure acetic acid and one of pure ethanol. There is a single outlet stream.

$$\dot{Q} = -\dot{n}_{\text{ethanol,in}}(\underline{H}_{\text{ethanol,in}}) - \dot{n}_{\text{acet.acid,in}}(\underline{H}_{\text{acet.acid,in}}) + \dot{n}_{\text{out}}(\underline{H}_{\text{out}}) \qquad (15.9)$$

The flow rates of the inlet streams are known, and they are at the exact conditions ($T = 298.15$ K, $P = 100$ kPa) as the reference state for the data in Appendix C, so the molar enthalpies can also be considered known.

The exiting stream is more complex in that it is a single stream, with a total molar flow rate of 200 mol/min, but $\underline{H}$ for this stream is the molar enthalpy of a mixture of four compounds. Because we are modeling the exiting liquid as an ideal solution, there is no "enthalpy of mixing." The molar enthalpy of the stream is the sum of the molar enthalpies of the individual components, weighted by their mole fractions, as first introduced (for a two-component mixture) in Equation 9.1:

$$\underline{H} = x_a\underline{H}_a + x_b\underline{H}_b$$

Applying this ideal-solution model to the four-component mixture in the outlet stream:

$$\underline{H}_{\text{out}} = x_{\text{ethanol}}\underline{H}_{\text{ethanol}} + x_{\text{acet.acid}}\underline{H}_{\text{acet.acid}} + x_{\text{water}}\underline{H}_{\text{water}} + x_{\text{ethylacet}}\underline{H}_{\text{ethylacet}} \qquad (15.10)$$

Here the stream temperature is 323.15 K, but data for $\underline{H}$ is only directly available at the reference temperature of $T = 298.15$ K.

Step 8 *Compute H for each compound at 323.15 K*

The enthalpy of formation of each compound is known at 25°C and 100 kPa (Appendix C-2). Since the actual pressure is also 100 kPa, we can use the constant pressure heat capacity to relate the unknown molar enthalpy at 323.15 K to the known value at 25°C.

$$d\underline{H} = C_P\,dT$$

$$\underline{H}_2 - \underline{H}_1 = \int_{T_1 = 298.15\text{ K}}^{T_2 = 323.15\text{ K}} C_P\,dT \qquad (15.11)$$

With $\underline{H}_1$ representing the enthalpy of formation at the reference state of $T = 298.15$ K and $P = 100$ kPa, and $\underline{H}_2$ representing the molar enthalpy at $T = 323.15$ K. Here we will model C_P of each liquid as a constant.

The values in Table 15-2 can be used in Equation 15.10 to find the molar enthalpy of the exiting stream:

$$\underline{H}_{out} = x_{ethanol}\underline{H}_{ethanol} + x_{acet.acid}\underline{H}_{acet.acid} + x_{water}\underline{H}_{water} + x_{ethylacet}\underline{H}_{ethylacet}$$

$$\underline{H}_{out} = (0.0685)(-273.2 \text{ kJ/mol}) + (0.0685)(-480.4 \text{ kJ/mol}) \tag{15.12}$$

$$+ (0.4315)(-283.9 \text{ kJ/mol}) + (0.4315)(-475.6 \text{ kJ/mol})$$

$$\underline{H}_{out} = -379.4 \frac{\text{kJ}}{\text{mol}}$$

TABLE 15-2 Calculation of molar enthalpy for each liquid at 50°C.

Compound	ΔH_f° (kJ/mol) ($\underline{H}_1$ in Eqn. 15.11)	C_P (J/mol·K) (Source: Domalski, 2012)	H at $T = 50°C$ (kJ/mol) ($\underline{H}_2$ in Eqn. 15.11)
Ethanol	−276.0	112.4	−273.2
Acetic Acid	−483.52	123.1	−480.4
Water	−285.83	75.4	−283.9
Ethyl Acetate	−479.86	168.9	−475.6

Now that this value is known, we can solve Equation 15.9 for $\dot{Q}$ as

$$\dot{Q} = -\dot{n}_{ethanol,in}(\underline{H}_{ethanol,in}) - \dot{n}_{scet.acid,in}(\underline{H}_{acet.acid,in}) + \dot{n}_{out}(\underline{H}_{out})$$

$$\dot{Q} = -\left(100\frac{\text{mol}}{\text{min}}\right)\left(-276.0\frac{\text{kJ}}{\text{mol}}\right) - \left(100\frac{\text{mol}}{\text{min}}\right)\left(-483.5\frac{\text{kJ}}{\text{mol}}\right) \tag{15.13}$$

$$+ \left(200\frac{\text{mol}}{\text{min}}\right)\left(-379.4\frac{\text{kJ}}{\text{mol}}\right) = 75.3\frac{\text{kJ}}{\text{min}}$$

Example 15-1 shows the answer to a straightforward and practical engineering question: How much heat must be added to, or removed from, a reactor to maintain it at the desired temperature? While Example 15-1 contains both an energy balance and a chemical reaction, it doesn't represent a particularly demanding "synthesis" of these principles. Part A was essentially a "chemical reaction equilibrium" problem typical of those we saw in Chapter 14, and part B was essentially an "energy balance" problem similar to those we saw in Chapter 3 (though in Chapter 3 we focused on pure compounds, not mixtures). Example 15-2 revisits the process of ethyl acetate synthesis, but this time, there is a deeper interdependence between the energy balance and the reaction equilibrium calculations. We will find that we cannot solve either one without the other.

| AN ADIABATIC REACTOR | EXAMPLE 15-2 |

Ethyl acetate is synthesized from ethanol and acetic acid by the liquid phase esterification reaction:

$$C_2H_5OH + CH_3COOH \leftrightarrow H_2O + CH_3COOC_2H_5$$

Ethanol + acetic acid $\leftrightarrow$ water + ethyl acetate

Two separate feeds containing 100 mol/min each of ethanol and acetic acid enter a steady-state reactor at $T = 323.15$ K and $P = 100$ kPa. The reactor is adiabatic and is large enough that the reaction can be modeled as progressing to equilibrium. Assume the liquid phase can be modeled as an ideal solution at $P = 100$ kPa.

Determine the temperature and composition of the stream leaving the reactor.

SOLUTION:

Step 1 *Construct a stoichiometric table*
The mass balance equations for this process are identical to those in step 1 of Example 15-1; the chemical reaction is the same, and the entering streams have the same molar flow rates. Consequently, the stoichiometric table for this example is identical to that given in Table 15-1, though we will see that the extent of reaction $\dot{\xi}$ is not the same.

Step 2 *Relate extent of reaction to equilibrium constant*
This step, too, is identical to Example 15-1. Since we are modeling the liquid phase as an ideal solution, the equilibrium criterion for this reaction simplifies to

$$K_T = \frac{(x_{H_2O})(x_{CH_3COOC_2H_5})}{(x_{C_2H_5OH})(x_{CH_3COOH})} \quad (15.14)$$

Using the mole fraction expressions in Table 15-1, Equation 15.14 becomes:

$$K_T = \frac{\left(\frac{\dot{\xi}}{200}\right)\left(\frac{\dot{\xi}}{200}\right)}{\left(\frac{100 - \dot{\xi}}{200}\right)\left(\frac{100 - \dot{\xi}}{200}\right)} \quad (15.15)$$

Step 3 *Quantify relationship between equilibrium constant and temperature*
In Example 15-1, we used the shortcut van 't Hoff equation to model the relationship between $\Delta \underline{G}_T^0$ and temperature. The result from step 4 was

$$\frac{\Delta \underline{G}_T^0}{RT} = \frac{\Delta \underline{G}_R^0}{RT_R} + \left(\frac{\Delta H_R^0}{R}\right)\left(\frac{1}{T} - \frac{1}{T_R}\right)$$

$$\frac{\Delta G_T^0}{RT} = \frac{\left(-9596\,\frac{J}{mol}\right)}{\left(8.314\,\frac{J}{mol \cdot K}\right)(298.15\,K)} + \frac{\left(-6170\,\frac{J}{mol}\right)}{\left(8.314\,\frac{J}{mol \cdot K}\right)}\left(\frac{1}{T} - \frac{1}{298.15\,K}\right) \quad (15.16)$$

Section 14.3.3 illustrated two methods of quantifying the relationship between equilibrium constant and temperature: the "shortcut" method and the "rigorous" method.

However, in Example 15-1, the temperature of the reactor was known, so ΔG_T^0 could be computed directly. Here T is unknown and the equation cannot be solved immediately. The fundamental definition of the equilibrium constant is

$$K_T = \exp\left(\frac{-\Delta G_T^0}{RT}\right) \quad (14.56)$$

Step 4 *Consider degrees of freedom*
Up to here, we have applied the same principles of chemical reaction equilibrium that we learned in Chapter 14, and have been needed for every problem involving a chemical

reaction. Now, however, we have three equations (Equations 15.15, 15.16, and 14.56) in four unknowns: T, K_T, ξ, and $\Delta \underline{G}_T^0$. Thus, we need one more relationship to find an answer, and this will be the energy balance.

Step 5 *Apply and simplify the energy balance*

The energy balance, in its most general form, is Equation 3.34:

$$\frac{d}{dt}\left\{M\left(\hat{U} + \frac{v^2}{2} + gh\right)\right\} = \sum_{j=1}^{j=J}\left\{\dot{m}_{j,\text{in}}\left(\hat{H}_j + \frac{v_j^2}{2} + gh_j\right)\right\} - \sum_{k=1}^{k=K}\left\{\dot{m}_{k,\text{out}}\left(\hat{H}_k + \frac{v_k^2}{2} + gh_k\right)\right\}$$
$$+ \dot{W}_S + \dot{W}_{EC} + \dot{Q}$$

This is a steady-state process, so the accumulation term is zero, and there is no expansion or contraction of the system. As in Example 15-1, we assume shaft work and potential and kinetic energy terms are zero. In this case, however, the reactor is adiabatic; $\dot{Q}$ is also 0. Thus, the only energy entering or leaving the system is the enthalpy of the inlet and outlet streams.

$$0 = \sum_{j=1}^{j=J}\dot{m}_{j,\text{in}}(\hat{H}_j) - \sum_{k=1}^{k=K}\dot{m}_{k,\text{out}}(\hat{H}_k) \tag{15.17}$$

The given information is on a molar basis, so we re-express the equation in terms of molar properties.

$$0 = \sum_{j=1}^{j=J}\dot{n}_{j,\text{in}}(\underline{H}_j) - \sum_{k=1}^{k=K}\dot{n}_{k,\text{out}}(\underline{H}_k) \tag{15.18}$$

Step 6 *Relate enthalpy to temperature for each stream*

There are two entering streams, consisting of known flow rates of pure ethanol and acetic acid at $T = 323.15$ K and $P = 100$ kPa. The molar enthalpies of ethanol and acetic acid at these conditions were computed in step 8 of Example 15-1, and are equal to $\underline{H}_{\text{ethanol,in}} = -273.2$ kJ/mol and $\underline{H}_{\text{acet.acid,in}} = -480.4$ kJ/mol.

$$\left(100\frac{\text{mol}}{\text{min}}\right)\left(-273.2\frac{\text{kJ}}{\text{mol}}\right) + \left(100\frac{\text{mol}}{\text{min}}\right)\left(-480.4\frac{\text{kJ}}{\text{mol}}\right) = \sum_{k=1}^{k=K}\dot{n}_{k,\text{out}}(\underline{H}_k) \tag{15.19}$$

On the right-hand side of Equation 15.19 there is only one exiting stream. While the extent of reaction is unknown, Table 15-1 shows that the total flow rate of this stream has to be 200 mol/min, regardless of the extent of reaction. Thus

$$\left(100\frac{\text{mol}}{\text{min}}\right)\left(-273.2\frac{\text{kJ}}{\text{mol}}\right) + \left(100\frac{\text{mol}}{\text{min}}\right)\left(-480.4\frac{\text{kJ}}{\text{mol}}\right) = \left(200\frac{\text{mol}}{\text{min}}\right)(\underline{H}_{\text{out}}) \tag{15.20}$$

$$\underline{H}_{\text{out}} = -376.8\frac{\text{kJ}}{\text{mol}}$$

This result provides us with a numerical value for the molar enthalpy of the exiting stream. However, remember that the purpose of writing the energy balance was that we needed an additional relationship between T, K_T, ξ, and $\Delta \underline{G}_T^0$. For $\underline{H}_{\text{out}}$ to be useful, we must connect it to these variables. As in Example 15-1, we are modeling the reactor effluent as an ideal solution, so the molar enthalpy of the exiting stream is a simple weighted average of the molar enthalpies of the four components.

$$\underline{H}_{\text{out}} = x_{\text{ethanol}}\underline{H}_{\text{out,ethanol}} + x_{\text{acet.acid}}\underline{H}_{\text{out,acet.acid}} + x_{\text{water}}\underline{H}_{\text{out,water}} \tag{15.21}$$
$$+ x_{\text{eth.acet.}}\underline{H}_{\text{out,eth.acet.}}$$

Temperature is one of our four unknowns, and energy balances are often useful for relating temperature to other variables.

By definition $H = m\hat{H} = n\underline{H}$

The mole fraction of each compound is known (see Table 15-1) as a function of $\dot{\xi}$. The molar enthalpy of each compound is known (see Table 15-2) at $T = 298.15$ K, as is the heat capacity. We can relate molar enthalpy of each component (i) of the outlet stream to temperature by integrating the equation $d\underline{H} = C_P dT$ from the reference temperature of $T = 298.15$ K to the outlet temperature of the reactor T_{out}:

$$\underline{H}_{out,i} - \Delta \underline{H}^0_{f,i} = \int_{T_{ref}=298.15 \text{ K}}^{T_{out}} C_{P,i} \, dT \tag{15.22}$$

Because we are modeling C_P as constant, this is simply

$$\underline{H}_{out,i} = \Delta \underline{H}^0_{f,i} + C_{P,i}(T_{out} - 298.15 \text{ K}) \tag{15.23}$$

When this is done for all four compounds, the results are summarized in Table 15-3

Step 7 *Formulate a solution strategy*
Inserting the information from Tables 15-1 and 15-3 into Equation 15.21 leaves us with four equations and four unknowns, but there isn't a compact algebraic solution—they are not simple linear equations. We could use a non-linear equation solver, or we could use an iterative solution procedure, such as:

1. Guess a temperature, T.
2. Calculate $\Delta \underline{G}^0_T/RT$ using Equation 15.16.
3. Calculate the equilibrium constant K_T using Equation 14.56.
4. Calculate the extent of reaction $\dot{\xi}$ using Equation 15.15.
5. With T and $\dot{\xi}$ known, determine $\underline{H}_{out}$ using Equation 15.21 and see if it is equal to -376.8 kJ/mol. If not, go back to step 1 and try a different temperature.

TABLE 15-3 Molar enthalpy as a function of temperature for each compound in Example 15-2.

Compound	ΔH^0_f (kJ/mol)	C_p (J/mol · K)	$\underline{H}_{out}$ (kJ/mol) as Function of T (K)
Ethanol	−276.0	112.4	$-276.0 + 0.1124(T - 298.15 \text{ K})$
Acetic Acid	−483.52	123.1	$-483.5 + 0.1231(T - 298.15 \text{ K})$
Water	−285.83	75.4	$-285.8 + 0.0754(T - 298.15 \text{ K})$
Ethyl Acetate	−479.86	168.9	$-479.9 + 0.1689(T - 298.15 \text{ K})$

This procedure is readily implemented in a spreadsheet. The result is

$$T = 368.85 \text{ K}$$
$$\Delta \underline{G}^0_T/RT = -3.540$$
$$K_T = 34.48$$
$$\dot{\xi} = 85.4 \text{ mol/min}$$
$$\underline{H}_{out} = -376.8 \text{ kJ/mol}$$

This value of extent of reaction means that the reactor produces 85.4 mol/min each of ethyl acetate and water. The exiting stream also contains 14.6 mol/min each of unreacted ethanol and acetic acid.

Example 15-2 illustrates a situation that is more complex than the reaction examples we have considered previously, because the temperature of the reactor is neither constant nor known (except at the inlet, where it is known to be 323.15 K). The exothermic reaction releases energy, which increases the temperature of the liquid as it progresses through the reactor, and this influences the equilibrium constant. Consequently the equilibrium conversion is a bit lower that it was in Example 15-1 when the reactor was isothermal at $T = 323.15$ K.

Despite this complexity, Example 15-2 is actually simplified in several important respects.

- Heat capacity was modeled as a constant, but it is in principle more accurate to model it as a function of temperature.
- The shortcut van 't Hoff equation was used to model the temperature dependence of the equilibrium constant, but the rigorous approach illustrated in Example 14-9 is in principle more accurate.

In this particular case, these two "shortcuts" can both be defended in that the temperature changes are comparatively small. One simplifying assumption that is harder to justify, however, is modeling the liquid as an ideal solution. In this case we cannot realistically say that the four compounds are all chemically and structurally "similar" to each other. Process simulation software is frequently used to implement more complex, real-solution modeling approaches. As an example, this problem was solved using ASPEN software and the "PSRK" option for computing physical properties. In this approach:

- The Soave equation of state, introduced in Section 7.2.2, is used to estimate all mixture properties.
- The UNIFAC method, which was introduced briefly in Section 11.8.1, is used to estimate activity coefficients.

PSRK stands for "Predictive Soave-Redlich-Kwong." The equation that was introduced as the Soave equation in Chapter 7 is also sometimes called the Redlich-Kwong-Soave, or Soave-Redlich-Kwong, equation.

The result was an outlet temperature of $T = 354.15$ K, and an outlet stream containing 80.3 mol/min each of water and ethyl acetate, and 19.7 mol/min each of ethanol and acetic acid.

15.3 Simultaneous Reaction and Phase Equilibrium

In some applications, chemical reactions are carried out in multi-phase systems. There are various reasons why this might be done. In some cases, the chemical reaction itself is inherently a multi-phase reaction. For example, the familiar process of metal rusting is a reaction between a solid and oxygen gas. Even when this is not the case, there are scenarios in which a second phase has a positive impact on the yield of desired products. We begin with a comparatively simple example that illustrates the effect a second phase can have on chemical reaction equilibrium.

| **EXAMPLE 15-3** | **NITROGEN DIOXIDE** |

At ambient conditions, nitrogen dioxide and dinitrogen tetraoxide form a vapor phase equilibrium mixture with each other:

$$2NO_2(g) \leftrightarrow N_2O_4(g)$$

A. 10 moles of nitrogen dioxide are placed in a reactor that is maintained at $T = 25°C$ and $P = 100$ kPa. Find the equilibrium composition of the vapor phase.

B. 10 moles of nitrogen dioxide and 50 moles of water are placed in a reactor that is maintained at $T = 25°C$ and $P = 100$ kPa. Find the equilibrium compositions of the liquid and vapor phases.

Assume that no chemical reaction occurs in the liquid phase. The Henry's Law constants at this temperature are 469.5 MPa for NO_2 in water and 4.0 MPa for N_2O_4 in water (Shwartz, 1981).

SOLUTION A:

The solution of part A is analogous to several reaction equilibrium examples in Chapter 14.

Step 1 *Construct stoichiometric table*

The stoichiometric table for part A is shown in Table 15-4.

TABLE 15-4 Stoichiometric table for Example 15-3, part A.

	NO_2	N_2O_4	Total Moles
$N_{initial}$ (moles)	10	0	10
N_{gen} (moles)	-2ξ	$+\xi$	$-\xi$
N_{final} (moles)	$10 - 2\xi$	$+\xi$	$10 - \xi$
y	$\dfrac{10 - 2\xi}{10 - \xi}$	$\dfrac{\xi}{10 - \xi}$	

Step 2 *Compute equilibrium constant for reaction*

From Appendix C-2, $\Delta\underline{G}_f^0 = 51.3$ kJ/mol for NO_2 and $\Delta\underline{G}_f^0 = 97.82$ kJ/mol for N_2O_4. Applying Equations 14.53 and 14.56, $\Delta\underline{G}_R^0 = -4.78$ kJ/mol and the equilibrium constant is $K_{298} = 6.88$.

Step 3 *Apply Equation 14.56 to relate equilibrium constant to mole fractions*

Here the reference and the actual pressure are both 100 kPa ($P^0 = P$). Assuming ideal gas behavior at $P = 100$ kPa, the fugacity of each compound is equal to its partial pressure, and Equation 14.56 simplifies to

$$K_T = \frac{y_{N_2O_4}}{y_{NO_2}^2} \tag{15.24}$$

$$6.88 = \frac{\dfrac{\xi}{10 - \xi}}{\left(\dfrac{10 - 2\xi}{10 - \xi}\right)^2}$$

$$\xi = 4.06 \text{ moles}$$

Thus there are **4.06 moles of N_2O_4 and 1.87 moles of NO_2** vapor at equilibrium.

SOLUTION B:

Step 4 *Consider degrees of freedom*

Now we add water to the mix. At 25°C, the vapor pressure is only 3.17 kPa (Appendix A), so we expect the water to be primarily, but not exclusively, in the liquid phase. Both gases are soluble in water, and the solubility can be quantified using Henry's Law. Thus, we expect that each of the three compounds will occur in both the liquid and the vapor phases. This is turn means there are six total unknowns needed to describe the contents of the reactor at equilibrium: $N_{H_2O}^L, N_{H_2O}^V, N_{NO_2}^L, N_{NO_2}^V, N_{N_2O_4}^L$, and $N_{N_2O_4}^V$. We must identify six independent equations. We begin with material balances.

FOOD FOR THOUGHT 15-3

Why aren't we making a stoichiometric table for part B, and why aren't we using the extent of reaction in part B?

In this example, the reference temperature and the actual temperature are both 25°C, so $\Delta\underline{G}_T^0 = \Delta\underline{G}_R^0$.

Step 5 *Write material balances*

Water does not participate in the chemical reaction, so we can view the number of moles of water as a conserved quantity; 50 moles are placed in the reactor, but we don't know how they will distribute between the two phases:

$$50\,\text{moles} = N^{\text{L}}_{\text{H}_2\text{O}} + N^{\text{V}}_{\text{H}_2\text{O}} \tag{15.25}$$

The reactor initially contains 10 moles of nitrogen dioxide. While the total number of moles of NO_2 is changed by the reaction, the number of moles of nitrogen atoms is a conserved quantity:

$$10\,\text{moles} = N^{\text{L}}_{\text{NO}_2} + N^{\text{V}}_{\text{NO}_2} + 2N^{\text{L}}_{\text{N}_2\text{O}_4} + 2N^{\text{V}}_{\text{N}_2\text{O}_4} \tag{15.26}$$

We now have two independent equations. Next we can account for the fact that the chemical reaction is at equilibrium.

Step 6 *Apply known equilibrium constant*

The temperature is again $T = 25°C$, so the equilibrium constant computed in step 2 is again valid. Assuming ideal gas behavior, the fundamental equation that models equilibrium for the vapor phase reaction is the same as it was in step 3:

$$K_{\text{T}} = \frac{y_{\text{N}_2\text{O}_4}}{y^2_{\text{NO}_2}}$$

However, the expressions for the mole fractions of NO_2 and N_2O_4 in Table 15-4 are no longer valid, as the table does not account for the possibility of material leaving the vapor phase entirely. For the moment, we simply apply the definition of mole fraction:

$$K_{\text{T}} = \frac{\left(\dfrac{N^{\text{V}}_{\text{N}_2\text{O}_4}}{N^{\text{V}}_{\text{N}_2\text{O}_4} + N^{\text{V}}_{\text{NO}_2} + N^{\text{V}}_{\text{H}_2\text{O}}}\right)}{\left(\dfrac{N^{\text{V}}_{\text{NO}_2}}{N^{\text{V}}_{\text{N}_2\text{O}_4} + N^{\text{V}}_{\text{NO}_2} + N^{\text{V}}_{\text{H}_2\text{O}}}\right)^2} \tag{15.27}$$

Since K_{T} is known, we now have three independent equations for our original six unknowns. We now account for vapor–liquid equilibrium.

Step 7 *Apply Raoult's Law for water*

Because the total pressure is 100 kPa, it is quite reasonable to model the vapor phase as an ideal gas, so we would like to use Raoult's Law. Raoult's Law also assumes the liquid is an ideal solution, which is, on the whole, not a reasonable assumption—we might say the nitrogen oxides are "similar" compounds to each other, but not similar to water. However, our expectation is that the liquid phase is going to be primarily water with a small amount of the two gases dissolved in the water. As discussed in Section 11.4.2, we can use an ideal solution model for the primary component of a dilute solution. Thus we can apply Raoult's Law *to water only* and, at the end, we should verify our assumption that the liquid phase is nearly pure water. Applying Raoult's Law:

$$x_{\text{H}_2\text{O}} P^{\text{sat}}_{\text{H}_2\text{O}} = y_{\text{H}_2\text{O}} P$$

The total pressure and vapor pressure of water are known, and the mole fractions can be expressed in terms of our original six unknowns as

$$\left(\frac{N^{\text{L}}_{\text{H}_2\text{O}}}{N^{\text{L}}_{\text{N}_2\text{O}_4} + N^{\text{L}}_{\text{NO}_2} + N^{\text{L}}_{\text{H}_2\text{O}}}\right)(3.17\,\text{kPa}) = \left(\frac{N^{\text{V}}_{\text{H}_2\text{O}}}{N^{\text{V}}_{\text{N}_2\text{O}_4} + N^{\text{V}}_{\text{NO}_2} + N^{\text{V}}_{\text{H}_2\text{O}}}\right)(100\,\text{kPa}) \tag{15.28}$$

This gives us four independent equations. We cannot use Raoult's Law to model VLE for the other two compounds; for these we use Henry's Law.

Step 8 *Apply Henry's Law for NO_2 and N_2O_4*

Example 11-1 demonstrated that when the vapor phase is modeled as an ideal gas, Henry's Law can be used to describe vapor-liquid equilibrium for a gas dissolved in the liquid phase:

$$x_i H_i = y_i P$$

where x_i is the mole fraction in the liquid phase and $y_i P$ is the partial pressure of the gas in the vapor phase, with the total pressure P known to equal 100 kPa. Applying this expression to NO_2 and N_2O_4:

$$\left(\frac{N^L_{NO_2}}{N^L_{N_2O_4} + N^L_{NO_2} + N^L_{H_2O}}\right) H_{NO_2} = \left(\frac{N^V_{NO_2}}{N^V_{N_2O_4} + N^V_{NO_2} + N^V_{H_2O}}\right)(100 \text{ kPa}) \qquad (15.29)$$

$$\left(\frac{N^L_{N_2O_4}}{N^L_{N_2O_4} + N^L_{NO_2} + N^L_{H_2O}}\right) H_{N_2O_4} = \left(\frac{N^V_{N_2O_4}}{N^V_{N_2O_4} + N^V_{NO_2} + N^V_{H_2O}}\right)(100 \text{ kPa}) \qquad (15.30)$$

The values of the Henry's Law constants (H_i) were given, so these are two equations in which the only unknowns are our original six unknowns. We now have six independent equations: two material balances (Equations 15.25 and 15.26), one equation that models the VLE for each of the three species (Equations 15.28, 15.29 and 15.30), and one that accounts for equilibrium in the vapor-phase chemical reaction (Equation 15.27).

Step 9 *Solve equations*

This system of six equations was solved using the POLYMATH non-linear equation solver. The results are summarized:

Compound	N^L (moles)	N^V (moles)	Total Moles
H_2O	49.82	0.18	50.0
NO_2	0.0033	1.75	1.75
N_2O_4	0.39	3.73	4.12

More N_2O_4 is formed in part B than part A. This is because N_2O_4 is more soluble in water, so in effect it is being "removed" from the vapor phase, shifting the equilibrium of the vapor phase reaction to the right.

Note that the liquid phase is >99 mol% water, so our decision to apply an ideal solution model to the water in the liquid phase appears reasonable.

Example 15-3 illustrated one form that a multi-phase reaction can take: the reaction occurs in the vapor phase, but there is also a liquid present in which some or all of the gases are soluble. Another possibility is that the reaction occurs in the liquid phase, but the system is above the bubble-point temperature, so a vapor phase is also present. Example 15-4 examines such a system.

| EXAMPLE 15-4 | **ISOMERIZATION OF XYLENE** |

The three isomers of xylene are *o*-xylene, *m*-xylene, and *p*-xylene (standing for "ortho," "meta," and "para") shown in Figure 15-4.

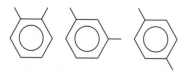

o-xylene   m-xylene   p-xylene

FIGURE 15-4 The isomers of xylene.

Inter-conversions between these isomers can be modeled through these two chemical reactions:

$$(R1): o\text{-xylene} \leftrightarrow m\text{-xylene}$$

$$(R2): m\text{-xylene} \leftrightarrow p\text{-xylene}$$

FOOD FOR THOUGHT 15-5

Throughout this book, we've used $P^0 = 100$ kPa as a standard reference state, but $P^0 = 1$ atm is also logical and some books use it. How different would you expect values (e.g., $\Delta \underline{H}_f^0$) on these two different reference states to be?

A. 10 moles of xylene are placed in a vessel that is maintained at $T = 373.15$ K and $P = 101.325$ kPa (1 atm). Find the amounts of *o*-, *p*-, and *m*-xylene that are present at equilibrium.

B. 10 moles of xylene are placed in a vessel that is maintained at $T = 373.15$ K and has a volume of 0.6 m³. The vessel contains both liquid and vapor. The liquid phase is exposed to a catalyst but the vapor is not; thus we can assume reactions R1 and R2 attain equilibrium in the liquid phase but do not occur at all in the vapor phase. Find the contents of the reactor at equilibrium.

We will use the following data, which uses $T = 25°C$ and $P = 1$ atm, rather than $P = 100$ kPa, as a reference state.

	$\Delta \underline{H}_f^0$ (kJ/mol)[a]	$\Delta \underline{G}_f^0$ (kJ/mol)[a]	P^{sat} (atm)[b]	$\underline{V}^L$ (m³/mol)[b]
o-xylene	19.08	122.05	0.261	1.2125×10^{-4}
m-xylene	17.32	118.89	0.307	1.2347×10^{-4}
p-xylene	18.03	121.48	0.316	1.2393×10^{-4}

[a]Poling, Prausnitz and O'Connell, 2001.
[b]Brown and Stein, 2012.

Based on data from Brown, R. L. and Stein, S. E., "Boiling Point Data" in NIST Chemistry WebBook, NIST Standard Reference Database Number 69, Eds. Linstrom, P. J. and Mallard, W. G., National Institute of Standards and Technology, Gaithersburg MD, 20899, http://webbook.nist.gov, (retrieved May 27, 2012).

SOLUTION A:

Step 1 *Construct stoichiometric table*
In this case, making the stoichiometric table (Table 15-5) is complicated by the fact that the problem statement says that 10 moles of xylene are initially present—it doesn't say which isomer(s) comprise the initial 10 moles. However, the equilibrium state will be the same for any initial distribution of isomers. We will assume the entire 10 moles is *o*-xylene initially, since it is the reactant in R1 the way we've written the reactions.

You can verify that the equilibrium state is the same for any initial distribution of isomers by re-writing Table 15-5 with different values of N_{init}.

TABLE 15-5 Mole balances and reaction stoichiometry for Example 15-4.

	o-Xylene	m-Xylene	p-Xylene	Total Moles
N_{init} (mol)	10	0	0	10
N_{gen} (mol)	$-\xi_1$	$+\xi_1$	0	10
N_{final} (mol)	0	$-\xi_2$	$+\xi_2$	10
X	$\dfrac{10 - \xi_1}{10}$	$\dfrac{\xi_1 - \xi_2}{10}$	$\dfrac{\xi_2}{10}$	

Step 2 *Compute equilibrium constants*
Here we use the shortcut method (Equation 14.110) of accounting for the effect of temperature. The calculations are summarized in the following data.

	ΔH^0_R (kJ/mol)	ΔG^0_R (kJ/mol)	T_R (K)	T (K)	K_{373}
R1	-1.76	-3.16	298.15	373.15	3.10
R2	0.71	2.59	298.15	373.15	0.373

Step 3 *Apply equilibrium criterion*
Since the three isomers of xylene are three chemically and structurally similar compounds, we will assume ideal solution behavior. Since the reference pressure and the actual pressure are identical, the equilibrium criteria (Equation 14.56) for these reactions simplify to:

$$K_{373,\text{R1}} = \frac{x_m}{x_o} \qquad K_{373,\text{R2}} = \frac{x_p}{x_m} \tag{15.31}$$

Plugging the mole fraction expressions from Table 15-5 and the known numerical values of K_T into these expressions gives a system of two equations in two unknowns. The solution is $\xi_1 = 8.10$ moles and $\xi_2 = 2.20$ moles. Thus at equilibrium the liquid contains **1.90 moles of o-xylene, 5.90 moles of m-xylene, and 2.20 moles of p-xylene**.

While p-xylene is not the most favored isomer at equilibrium, it is the most volatile of the three, as the given vapor pressures indicate. How will the addition of a vapor phase influence the outcome?

SOLUTION B:

Step 4 *Consider degrees of freedom*
Here there are again 10 total moles of xylene, but now we need to account for the presence of two phases. Thus, there are six unknowns, which are the numbers of moles of each isomer in each of the two phases. The overall mass balance provides one equation:

$$N^L_O + N^L_M + N^L_P + N^V_O + N^V_M + N^V_P = 10 \text{ moles} \tag{15.32}$$

Thus, we need to establish five additional independent equations.

Step 5 *Apply reaction equilibrium criterion*

Here, unlike part A, the system pressure is not equal to the reference pressure ($P^0 = 1$ atm). In fact, we do not know the system pressure. However, as illustrated in Example 14-7, liquid fugacity is not very sensitive to pressure, and the effects of pressure on the reactant and product fugacities often largely cancel each other out. Consequently, we will use the common simplifying assumption that the effect of Poynting correction is negligible (Equation 14.93).

The equilibrium expressions given in Equation 15.31 are again valid in part B. This time, however, we cannot use the mole fraction expressions from Table 15-4; the stoichiometric table does not account for the possibility of evaporation. For now, we will simply express equilibrium in terms of the six unknowns we identified in step 4:

$$K_{373,R1} = \frac{x_m}{x_o} \qquad K_{373,R2} = \frac{x_p}{x_m}$$

$$K_{373,R1} = \frac{N_m^L}{N_o^L} \qquad K_{373,R2} = \frac{N_p^L}{N_m^L} \qquad (15.33)$$

We now have three independent equations relating to six unknowns.

Step 6 *Apply VLE criterion*

We have already decided to model the liquid phase as an ideal solution ($\gamma = 1$). If we now assume the vapor phase behaves as an ideal gas, we can use Raoult's Law to model the VLE. We will do this for now and revisit the assumption at the end of the solution.

Raoult's Law is

$$x_i P_i^{\,\text{sat}} = y_i P$$

And it can be applied to each of the three isomers of xylene:

$$\left(\frac{N_o^L}{N_o^L + N_m^L + N_p^L}\right) P_o^{\,\text{sat}} = \left(\frac{N_o^V}{N_o^V + N_m^V + N_p^V}\right) P \qquad (15.34)$$

$$\left(\frac{N_m^L}{N_o^L + N_m^L + N_p^L}\right) P_m^{\,\text{sat}} = \left(\frac{N_m^V}{N_o^V + N_m^V + N_p^V}\right) P \qquad (15.35)$$

$$\left(\frac{N_p^L}{N_o^L + N_m^L + N_p^L}\right) P_p^{\,\text{sat}} = \left(\frac{N_p^V}{N_o^V + N_m^V + N_p^V}\right) P \qquad (15.36)$$

We now have six independent equations, but have introduced a seventh unknown; the total pressure of the system. We can quantify the pressure through the known total volume.

Step 7 *Quantify volume of liquid and vapor phases*

The total volume is known, but the volumes of the liquid and vapor phases are not. According to the given information, the molar volumes of all three compounds are less than 1.25×10^{-4} m³/mol. This means that even if all 10 moles are in the liquid phase, the liquid volume would be less than 1.25×10^{-4} m³, which is ~0.2% of the total volume of the vessel. Consequently, we will consider the liquid volume negligible and use the ideal gas law to relate the vapor volume (0.6 m³) to the unknown system pressure.

$$P = \frac{NRT}{V} = \frac{\left(N_o^V + N_m^V + N_p^V\right)\left(8.205746 \times 10^{-5}\,\dfrac{\text{m}^3 \cdot \text{atm}}{\text{mol} \cdot \text{K}}\right)(373.15\ \text{K})}{0.6\ \text{m}^3} \qquad (15.37)$$

Step 8 *Solve the equations*

This system of seven nonlinear equations was solved using POLYMATH software. The results are given in the following data.

	Liquid Moles	Vapor Moles	Total Moles
o-Xylene	0.78	0.97	1.75
m-Xylene	2.43	3.55	5.98
p-Xylene	0.91	1.36	2.27
Total Moles	4.12	5.88	10

In addition, the total pressure is **0.300 atm**. The decision to model the vapor phase as an ideal gas thus appears quite valid.

In examining part B of Example 15-4, we see that at equilibrium the xylene is almost evenly distributed between the liquid (5.1 moles) and vapor (4.9 moles) phases. This is very dependent upon the volume of the container. If the container is made larger, more xylene will evaporate to fill the space available. The p-xylene is the most volatile of the three isomers, as measured by vapor pressure at $T = 373.15$ K. Because of this volatility, the total production of p-xylene increased in part B (when a vapor phase was present) compared to part A (when the process is carried out entirely in the liquid phase). The differences in volatility among the three isomers of xylene are small, so the increase in yield of p-xylene resulting from the addition of a vapor phase is slight. In principle, however, this example illustrates how reactive distillation, which was introduced in the motivational example, can lead to higher yields of desired products than the equilibrium constraint would seem to allow. Furthermore, we now see how we could potentially build a model of a reactive distillation process; each individual stage can be modeled similarly to the system in Example 15-4.

15.4 A Complete Chemical Process

The examples throughout this book have examined a variety of physical systems—including a rocket, a tree, and a sculpture—but the vast majority of the examples have been directly related to the chemical process industry. Most of these examples have focused on individual unit operations—a compressor, a reactor, a separation process, etc. The motivational example in Chapter 5 discussed combining these unit operations into designs for complete chemical processes, using ammonia synthesis as an example. In Chapter 5, we were able to discuss the challenges surrounding the design of this process in a qualitative way, but did not yet possess the tools needed for a quantitative analysis. We close the chapter by revisiting that motivational example. We will start with a simple "single-pass" reaction system that illustrates some of the challenges associated with this process.

| EXAMPLE 15-5 | **SINGLE-PASS SYNTHESIS OF AMMONIA** |

30 mol/s of hydrogen gas and 15 mol/s of air, each compressed to 2.5 MPa, enter a steady-state reactor as shown in Figure 15-5, where the nitrogen in the air reacts with the hydrogen to form ammonia:

$$N_2 + 3H_2 \leftrightarrow 2NH_3$$

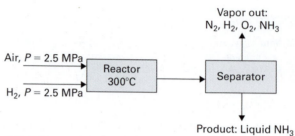

FIGURE 15-5 Schematic of ammonia synthesis process described in Example 15-5.

The stream leaving the reactor is at equilibrium at a temperature of 300°C (573.15 K). It is then cooled (though it remains at 2.5 MPa) and sent to a vessel in which the liquid and vapor are separated from each other. The liquid ammonia is the desired product. Find the flow rate of liquid product and the composition and flow rate of the vapor stream for the following.

A. The streams leaving the separator are in VLE at 25°C.

B. The streams leaving the separator are in VLE at 0°C.

Model air as 79 mol% nitrogen and 21 mol% oxygen, model all gas/vapor phases as ideal, and assume the liquid product is pure ammonia (in other words, nitrogen, oxygen, and hydrogen are modeled as gases that are insoluble in ammonia).

There is no catalyst in the separator, so it can be assumed the reaction occurs in the reactor only.

SOLUTION:

Our unknowns describe the streams leaving the separator, but we begin with no information about what's going into the separator. The most straightforward way to solve this problem is to determine what is coming out of the reactor using principles of reaction equilibrium, then move on to the separator.

Step 1 *Construct stoichiometric table*
In Chapter 14, we learned to summarize mole balances around a reactor using a stoichiometric table, as in Table 15-6.

Step 2 *Calculate $\Delta \underline{H}_R^0$ and $\Delta \underline{G}_R^0$*
From Appendix C, ammonia has $\Delta \underline{G}_f^0 = -16.45$ kJ/mol and $\Delta \underline{H}_f^0 = -46.11$ kJ/mol at the standard state of 100 kPa and 25°C. **Nitrogen and hydrogen are both elements, and have $\Delta \underline{G}_f^0 = \Delta \underline{H}_f^0 = 0$.** Consequently, the reaction has $\Delta \underline{G}_R^0 = -32.90$ kJ/mol and $\Delta \underline{H}_R^0 = -92.22$ kJ/mol.

Step 3 *Calculate equilibrium constant*
We can estimate the equilibrium constant at 300°C (573.15 K) using the shortcut van 't Hoff Equation 14.110 for

PITFALL PREVENTION !

Don't forget the stoichiometric coefficient of +2 for ammonia when computing $\Delta \underline{G}_R^0$ and $\Delta \underline{H}_R^0$.

$$\frac{\Delta \underline{G}_T^0}{RT} = \frac{\Delta \underline{G}_R^0}{RT_R} + \left(\frac{\Delta \underline{H}_R^0}{R}\right)\left(\frac{1}{T} - \frac{1}{T_R}\right)$$

$$\frac{\Delta \underline{G}_T^0}{RT} = \frac{\left(-32{,}900\,\dfrac{J}{mol}\right)}{\left(8.314\,\dfrac{J}{mol \cdot K}\right)(298.15\,K)} + \left(\frac{-92{,}220\,\dfrac{J}{mol}}{8.314\,\dfrac{J}{mol \cdot K}}\right)\left(\frac{1}{573.15\,K} - \frac{1}{298.15\,K}\right) \quad (15.38)$$

$$\frac{\Delta \underline{G}_T^0}{RT} = 4.578$$

which, when inserted into Equation 14.56, gives an equilibrium constant of $K_{573.15} = 0.01028$.

TABLE 15-6 Stoichiometric table for ammonia synthesis process modeled in Example 15-5.

	Nitrogen	Oxygen	Hydrogen	Ammonia	Total
$\dot{n}_{in}$ (mol/sec)	$0.79\left(15\dfrac{mol}{s}\right) = 11.85\dfrac{mol}{s}$	$0.21\left(15\dfrac{mol}{s}\right) = 3.15\dfrac{mol}{s}$	$30\dfrac{mol}{s}$	0	$45\dfrac{mol}{s}$
$\dot{N}_{gen}$ (mol/sec)	$-\dot{\xi}$	0	$-3\dot{\xi}$	$+2\dot{\xi}$	$-2\dot{\xi}$
$\dot{n}_{out}$ (mol/sec)	$11.85\dfrac{mol}{s} - \dot{\xi}$	$3.15\dfrac{mol}{s}$	$30\dfrac{mol}{s} - 3\dot{\xi}$	$+2\dot{\xi}$	$45\dfrac{mol}{s} - 2\dot{\xi}$
y	$\dfrac{11.85\dfrac{mol}{s} - \dot{\xi}}{45\dfrac{mol}{s} - 2\dot{\xi}}$	$\dfrac{3.15\dfrac{mol}{s}}{45\dfrac{mol}{s} - 2\dot{\xi}}$	$\dfrac{30\dfrac{mol}{s} - 3\dot{\xi}}{45\dfrac{mol}{s} - 2\dot{\xi}}$	$\dfrac{+2\dot{\xi}}{45\dfrac{mol}{s} - 2\dot{\xi}}$	

Step 4 *Apply equilibrium criterion*

We can apply Equation 14.73, which describes vapor phase equilibrium, to

$$K_T\left(\frac{P^0}{P}\right)^{\Sigma \nu_i} = \prod_i (y_i \hat{\phi}_i)^{\nu_i}$$

Here the sum of the stoichiometric coefficients is -2 (-3 for hydrogen, -1 for nitrogen, and $+2$ for ammonia), the reference pressure is 100 kPa, and the actual pressure is 2.5 MPa. Since we are modeling this as an ideal gas, the mixture fugacity coefficient for all compounds is 1, and the equation simplifies to

$$K_T\left(\frac{P^0}{P}\right)^{\Sigma \nu_i} = \frac{y_{NH_3}^2}{y_{N_2} y_{H_2}^3}$$

$$(0.01028)\left(\frac{100\,kPa}{2500\,kPa}\right)^{-2} = \frac{\left(\dfrac{2\dot{\xi}}{45\dfrac{mol}{sec} - 2\dot{\xi}}\right)^2}{\left(\dfrac{11.85\dfrac{mol}{sec} - \dot{\xi}}{45\dfrac{mol}{sec} - 2\dot{\xi}}\right)\left(\dfrac{30\dfrac{mol}{sec} - 3\dot{\xi}}{45\dfrac{mol}{sec} - 2\dot{\xi}}\right)^3} \quad (15.39)$$

The result is $\dot{\xi} = $ **5.178 mol/s.**

Step 5 *Consider knowns and unknowns in separator*

Plugging the extent of reaction computed in step 4 into the expressions determined in step 1 allows us to find the exact composition of the stream exiting the reactor: 10.356 mol/s of ammonia, 14.467 mol/s of hydrogen, 6.672 mol/s of nitrogen, and 3.150 mol/s of oxygen. This is the only stream entering the separator.

There are five unknowns — the flow rate of the liquid ammonia product, which we'll call $\dot{n}_{prod}$, and the flow rates of each of the four components in the vapor outlet stream, which we'll call $\dot{n}^V_{NH_3}$, $\dot{n}^V_{N_2}$, $\dot{n}^V_{H_2}$ and $\dot{n}^V_{O_2}$.

Since no reactions occur in the separator, we can treat the number of moles of each compound as a conserved quantity, $\dot{n}_{in} = \dot{n}_{out}$. There is no nitrogen, hydrogen, or oxygen in the liquid stream. For these three compounds, all of the material that enters from the reactor leaves through the vapor outlet, so $\dot{n}^V_{N_2} = 6.672$ mol/s, $\dot{n}^V_{H_2} = 14.467$ mol/s, and $\dot{n}^V_{O_2} = 3.150$ mol/s.

There is one stream entering the separator and two streams exiting the separator that contain ammonia, so the mole balance is

$$10.356 \frac{mol}{s} = \dot{n}^V_{NH_3} + \dot{n}_{prod} \tag{15.40}$$

So we are left with two unknowns and one equation. We obtain the second equation by accounting for the fact that the vapor and liquid streams are at equilibrium.

Step 6 *Apply vapor–liquid equilibrium*

Raoult's Law is based upon two assumptions: the liquid phase is an ideal solution and the vapor phase is an ideal gas. Here, we've specifically been told to model the vapor phase as an ideal gas and that the liquid phase is pure ammonia — a pure compound is the most ideal solution possible. So we apply Raoult's Law for ammonia

$$x_{NH_3} P^{sat}_{NH_3} = y_{NH_3} P$$

where $x = 1$, and y_{NH_3} can be expressed in terms of the unknown number of moles of ammonia in the vapor phase.

$$(1) P^{sat}_{NH_3} = \left(\frac{\dot{n}^V_{NH_3}}{\dot{n}^V_{NH_3} + \dot{n}^V_{N_2} + \dot{n}^V_{H_2} + \dot{n}^V_{O_2}} \right) (2.5 \, MPa) \tag{15.41}$$

$$(1) P^{sat}_{NH_3} = \left(\frac{\dot{n}^V_{NH_3}}{\dot{n}^V_{NH_3} + 24.288 \frac{mol}{s}} \right) (2.5 \, MPa)$$

Step 7 *Solve for ammonia flow rates*

The only difference between case A and case B is the temperature of the separator. Therefore, the vapor pressure of ammonia is different. If the vapor pressure is known, Equation 15.41 can be solved for $\dot{n}^V_{NH_3}$, and Equation 15.40 can be solved for $\dot{n}_{prod}$.

The relevant vapor pressures are $P^{sat} = 0.994$ MPa at $T = 25°C$ and $P^{sat} = 0.423$ MPa at $0°C$ (Brown and Stein, 2012). The solutions are given here.

	A	B
$P^{sat}_{NH_3}$	0.994 MPa	0.423 MPa
$\dot{n}^V_{NH_3}$	16.03 mol/s	4.946 mol/s
$\dot{n}_{prod}$	−5.674 mol/s	5.410 mol/s

Air actually contains N_2, O_2, and other trace gases; argon is the third most prominent component. Here, argon and oxygen would both act as inerts. If we included argon in the model, it would behave essentially the same as the oxygen.

Examining the solution to Example 15-5, the answers obtained for part A are physically unrealistic, with a negative molar flow rate of product. In our calculations, we assumed that there would be a vapor phase and a liquid phase (in equilibrium) leaving the separator. In reality, at 25°C, the reactor is not producing enough ammonia to obtain a partial pressure of 0.994 MPa of ammonia. Thus, no material will condense. We are making over 10 mol/s of ammonia in the reactor, but not producing any pure liquid product. With the separator at 0°C (part B), there is still a significant amount of ammonia in the vapor phase, but there is a product stream, so this is at least a viable process. This is a simple example of the sort of issue designers are faced with constantly. We changed a single parameter—a temperature—by the relatively modest increment of 25 degrees, and it had a profound effect on the operation of the process.

Example 15-5 also illustrates that only about half of the entering nitrogen and hydrogen are converted into ammonia at this reaction pressure and temperature, so one way to improve the process is to recycle the unused reactants. However, if we simply send the "Vapor out" stream of Figure 15-5 back to the recycle stream, this creates a problem because of the presence of oxygen. If oxygen enters the system, but there is no vapor stream leaving, oxygen will accumulate in the system—there can be no steady state. Consequently, in Example 15-6 we implement a solution that was mentioned during the motivational example in Chapter 5—we "purge" a portion of the recycle stream from the system.

> A purge stream for inerts is a common feature of chemical processes, to which you may have been introduced in a course on material balances.

| COMPLETE PROCESS FOR SYNTHESIS OF AMMONIA | **EXAMPLE 15-6** |

Ammonia is synthesized in Figure 15-6 by the gas phase chemical reaction:

$$N_2 + 3H_2 \leftrightarrow 2NH_3$$

> Oxygen is treated as an "inert" in this problem because it does not participate in the chemical reaction examined here. But it is certainly not an "inert" compound in general. The presence of oxygen in a system that also contains flammable/explosive compounds is a severe hazard.

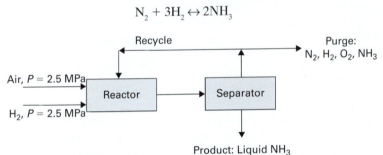

FIGURE 15-6 Schematic of ammonia synthesis process.

The steady-state process is to be designed as follows.

- 30 mol/s of hydrogen gas and 15 mol/s of air, both compressed to 2.5 MPa, enter the reactor.

- A recycle stream, also at 2.5 MPa, enters the reactor.

- The stream leaving the reactor is at $T = 300°C$ and $P = 2.5$ MPa, and it is cooled to 0°C and $P = 2.5$ MPa. This condenses much of the ammonia into liquid.

- The liquid and vapor are separated. The liquid ammonia is removed as product.

- 5% of the vapor stream is purged from the process, and the remainder is recycled. The purge and recycle streams have identical compositions.

Determine the molar flow rate of product, and the composition and flow rate of the purge stream, for each of the following cases.

A. The stream leaving the reactor is at equilibrium.

B. The stream leaving the reactor is not at equilibrium. Five percent of the hydrogen entering the reactor gets converted into ammonia.

Model air as 79 mol% nitrogen and 21 mol% oxygen, and assume the liquid product is pure ammonia (in other words, nitrogen, oxygen, and hydrogen are modeled as gases that are insoluble in ammonia).

SOLUTION:

Step 1 *Identify unknowns and count degrees of freedom*

The product stream is known to be pure ammonia, but its flow rate $\dot{n}_{prod}$ is unknown.

The purge stream has unknown flow rate ($\dot{n}_{pur}$) and an unknown composition. Since the stream can contain all four of the compounds in the process, we can consider these four more unknowns: $y_{N_2,pur}$, $y_{O_2,pur}$, $y_{NH_3,pur}$, and $y_{H_2,pur}$.

We have identified six unknowns. Naturally, the mole fractions must sum to one:

$$y_{O_2,pur} + y_{N_2,pur} + y_{H_2,pur} + y_{NH_3,pur} = 1 \qquad (15.42)$$

We need to identify five more independent equations. In the previous example, we analyzed the reactor, and once we knew exactly what was coming out of the reactor, we moved on to the next unit. Here, the presence of the recycle stream means we can't use this approach—we don't know exactly what's entering the reactor. Instead, we need a more holistic approach to modeling the system.

Step 2 *Material balances on entire system*

The known and unknown information all concerns streams that enter and leave the process, so it makes sense to draw a control volume around the entire process and write material balances for this system.

Since this is a steady-state process, a mass balance has the form

$$0 = \sum \dot{n}_{in} - \sum \dot{n}_{out}$$

Oxygen only enters the system through the air inlet and only leaves the system through the purge, so

$$0 = (0.21)\left(15\frac{mol}{sec}\right) - y_{O_2,pur}\dot{n}_{pur} \qquad (15.43)$$

We cannot write a similar balance equation for the compounds N_2, H_2, or NH_3, because they participate in the chemical reaction; the number of moles of these compounds is not a conserved quantity. However, we can write balance equations for the nitrogen and hydrogen *atoms*.

The nitrogen atom balance is

$$0 = (0.79)(2)\left(15\frac{mol}{sec}\right) - 2y_{N_2,pur}\dot{n}_{pur} - y_{NH_3,pur}\dot{n}_{pur} - \dot{n}_{prod} \qquad (15.44)$$

and the hydrogen atom balance

$$0 = 2\left(30\frac{mol}{sec}\right) - 2y_{H_2,pur}\dot{n}_{pur} - 3y_{NH_3,pur}\dot{n}_{pur} - 3\dot{n}_{prod} \qquad (15.45)$$

The product stream is modeled as pure ammonia,

$$x_{NH_3,prod} = 1$$

We have now identified four independent equations for six unknowns. What else do we know? In the separation step, all nitrogen, hydrogen, and oxygen remain in the vapor phase, but ammonia distributes between the vapor and liquid.

Step 3 *Apply VLE model for ammonia*

It would be convenient if all of the ammonia condensed when the reactor effluent was cooled to 0°C, but we saw in the previous example that ammonia has a very significant vapor pressure at this temperature. We will again account for the vapor–liquid equilibrium with Raoult's Law. Note that the purge and recycle streams have identical composition; this is normal if the purge is accomplished by a simple split of the stream. Thus,

$y_{NH_3,pur}$, which is one of our six unknowns, is identical to the mole fraction of ammonia in the recycle stream and the mole fraction of ammonia in the vapor stream leaving the separator.

For ammonia, Raoult's Law is written as

$$x_{NH_3,prod} P^{sat}_{NH_3} = y_{NH_3,pur} P \tag{15.46}$$

Here, the liquid phase is pure ammonia, the total pressure is 2.5 MPa, and the vapor pressure at 0°C is 0.423 MPa.

$$(1)(0.423 \, \text{MPa}) = y_{NH_3,pur}(2.5 \, \text{MPa}) \tag{15.47}$$

$$y_{NH_3,pur} = \mathbf{0.169}$$

We've now solved for one of our six unknowns, but have only four equations relating the other five. The information we have not used is the fact that the stream leaving the reactor is at equilibrium.

Step 4 *Construct stoichiometric table*
The stoichiometric table in Table 15-7 is similar to the one in Example 15-5, but the molar flow rate entering, for each component, has an additional term representing the recycle stream.

TABLE 15-7 Stoichiometric table for reactor in Example 15-6.

	Nitrogen	Oxygen	Hydrogen	Ammonia	Total Moles
$\dot{n}_{in}$ (mol/s)	$0.79\left(15\dfrac{\text{mol}}{\text{s}}\right)$ $+ y_{N_2,rec}\dot{n}_{rec}$	$0.21\left(15\dfrac{\text{mol}}{\text{s}}\right)$ $+ y_{O_2,rec}\dot{n}_{rec}$	$30\dfrac{\text{mol}}{\text{s}}$ $+ y_{H_2,rec}\dot{n}_{rec}$	$y_{NH_3,rec}\dot{n}_{rec}$	$45\dfrac{\text{mol}}{\text{s}} + \dot{n}_{rec}$
$\dot{N}_{gen}$ (mol/s)	$-\dot{\xi}$	0	$-3\dot{\xi}$	$+2\dot{\xi}$	$-2\dot{\xi}$
$\dot{n}_{out}$ (mol/s)	$0.79\left(15\dfrac{\text{mol}}{\text{s}}\right)$ $+ y_{N_2,rec}\dot{n}_{rec}$ $-\dot{\xi}$	$0.21\left(15\dfrac{\text{mol}}{\text{s}}\right)$ $+ y_{O_2,rec}\dot{n}_{rec}$	$30\dfrac{\text{mol}}{\text{s}}$ $+y_{H_2,rec}\dot{n}_{rec}$ $-3\dot{\xi}$	$y_{NH_3,rec}\dot{n}_{rec}$ $+2\dot{\xi}$	$45\dfrac{\text{mol}}{\text{s}}$ $+ \dot{n}_{rec} - 2\dot{\xi}$
y	$\dfrac{0.79\left(15\frac{\text{mol}}{\text{s}}\right) + y_{N_2,rec}\dot{n}_{rec} - \dot{\xi}}{45\frac{\text{mol}}{\text{s}} + \dot{n}_{rec} - 2\dot{\xi}}$	$\dfrac{0.21\left(15\frac{\text{mol}}{\text{s}}\right) + y_{O_2,rec}\dot{n}_{rec}}{45\frac{\text{mol}}{\text{s}} + \dot{n}_{rec} - 2\dot{\xi}}$	$\dfrac{30\frac{\text{mol}}{\text{s}} + y_{H_2,rec}\dot{n}_{rec} - 3\dot{\xi}}{45\frac{\text{mol}}{\text{s}} + \dot{n}_{rec} - 2\dot{\xi}}$	$\dfrac{y_{NH_3,rec}\dot{n}_{rec} + 2\dot{\xi}}{45\frac{\text{mol}}{\text{s}} + \dot{n}_{rec} - 2\dot{\xi}}$	

Step 5 *Apply equilibrium criterion*
In the previous example, we determined that the equilibrium state for this reaction could be modeled using

$$K_T\left(\frac{P^0}{P}\right)^{\Sigma \nu_i} = \frac{y^2_{NH_3}}{y_{N_2} y^3_{H_2}}$$

Here the reactor is at the same temperature and pressure as it was in Example 15-5, so again $P = 2.5$ MPa, $K_T = 0.01028$, and $P^0 = 100$ kPa. But the mole fraction expressions are different from those in Example 15-5; they are obtained from Table 15-7.

$$(0.01028)\left(\frac{0.1\ \text{MPa}}{2.5\ \text{MPa}}\right)^{-2} = \frac{\left(\dfrac{y_{NH_3,rec}\dot{n}_{rec} + 2\dot{\xi}}{45\dfrac{\text{mol}}{\text{sec}} + \dot{n}_{rec} - 2\dot{\xi}}\right)^{2}}{\left(\dfrac{0.79\left(15\dfrac{\text{mol}}{\text{sec}}\right) + y_{N_2,rec}\dot{n}_{rec} - \dot{\xi}}{45\dfrac{\text{mol}}{\text{sec}} + \dot{n}_{rec} - 2\dot{\xi}}\right)\left(\dfrac{30\dfrac{\text{mol}}{\text{sec}} + y_{H_2,rec}\dot{n}_{rec} - 3\dot{\xi}}{45\dfrac{\text{mol}}{\text{sec}} + \dot{n}_{rec} - 2\dot{\xi}}\right)^{3}}$$

$$(15.48)$$

Offhand, this doesn't look particularly useful. We've written an equation that introduces a number of apparently new unknowns into the problem, including the flow rate and composition of the recycle stream. However, the recycle stream has the same composition as the purge stream—and the composition of the purge stream is exactly what we've been trying to find. Also, we know that 5% of the vapor leaving the separator is purged and the rest is recycled. Thus,

$$\frac{\dot{n}_{rec}}{\dot{n}_{pur}} = \frac{95}{5} = 19 \qquad (15.49)$$

Introducing these facts into Equation 15.48 gives

$$(0.01028)\left(\frac{1}{25}\right)^{-2} = \frac{\left(\dfrac{19 y_{NH_3,pur}\dot{n}_{pur} + 2\dot{\xi}}{45\dfrac{\text{mol}}{\text{sec}} + 19\dot{n}_{pur} - 2\dot{\xi}}\right)^{2}}{\left[\dfrac{0.79\left(15\dfrac{\text{mol}}{\text{sec}}\right) + 19 y_{N_2,pur}\dot{n}_{pur} - \dot{\xi}}{45\dfrac{\text{mol}}{\text{sec}} + 19\dot{n}_{pur} - 2\dot{\xi}}\right]\left[\dfrac{30\dfrac{\text{mol}}{\text{sec}} + 19 y_{H_2,pur}\dot{n}_{pur} - 3\dot{\xi}}{45\dfrac{\text{mol}}{\text{sec}} + 19\dot{n}_{pur} - 2\dot{\xi}}\right]^{3}}$$

$$(15.50)$$

Thus, the only "new" unknown in Equation 15.50 is the extent of reaction $\dot{\xi}$. We have added one new equation and one new unknown to the problem—so what have we gained? What we've gained is that we now have an expression for the amount of ammonia entering the separator in terms of $\dot{\xi}$. We can now finish off the problem with a mole balance for ammonia in the separator, similar to Equation 15.40 in Example 15-5.

Step 6 *Apply mole balance to separator*
Because no reaction occurs in the separator, we can consider moles of ammonia a conserved quantity in the separator. For the system as drawn in Figure 15-7, ammonia enters the system in the reactor outlet stream, and exits through three streams: the purge, the recycle, and the liquid.

$$\dot{n}_{NH_3,rout} = \dot{n}_{NH_3,rec} + \dot{n}_{prod} + \dot{n}_{NH_3,pur}$$

Equation 15.52 could also have been derived using an "ammonia balance" for the entire process as the system.

$$y_{NH_3,rec}\dot{n}_{rec} + 2\dot{\xi} = y_{NH_3,rec}\dot{n}_{rec} + \dot{n}_{prod} + y_{NH_3,pur}\dot{n}_{pur} \qquad (15.51)$$

The ammonia in the recycle stream can be subtracted from both sides, yielding

$$2\dot{\xi} = \dot{n}_{prod} + y_{NH_3,pur}\dot{n}_{pur} \qquad (15.52)$$

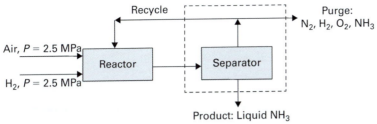

FIGURE 15-7 System analyzed in step 6 of Example 15-6.

Step 7 *Solve case A*

We now have, in total, seven unknowns (four mole fractions, purge and product flow rates, and the extent of reaction) and seven equations—Equations 15.42 through 45, 15.47, 15.50, and 15.52. The solution was determined using POLYMATH nonlinear equation solver. The result is

$$\dot{\xi} = 8.26 \, \frac{mol}{s}$$

Liquid ammonia product flow rate $(\dot{n}_{prod})$ = **14.088 mol/s**

Purge flow rate $(\dot{n}_{pur})$ = **14.434 mol/s**, with composition **36.1% H$_2$, 24.8% N$_2$, 16.9% NH$_3$, and 22.2% O$_2$.**

Step 8 *Compare part A to part B*

The difference between parts A and B is that in part B we are now considering a scenario in which the stream exiting the reactor is not at equilibrium. Reviewing the seven equations we used to solve part A, we see six of them still apply directly to part B.

- Mole fractions for a stream must always sum to 1 (Equation 15.42).

- In step 2, we wrote three material balance equations for the entire system. Both cases have the same known feed streams, and the same unknowns required to describe the exit streams. Thus, while the solutions to the material balances will be different, the balance equations themselves are the same in part B (Equations 15.43, 15.44, and 15.45).

- Similarly, the balance equation for ammonia in the separator (Equation 15.52) is the same in both cases. Ammonia enters and exits through the same streams, although the quantities will be different.

- The separator is again at 0°C, so the mole fraction of ammonia, as determined by its vapor pressure, is again $y_{NH_3,pur}$ = 0.169.

However, the seventh equation (Equation 15.50), which models equilibrium in the reactor, is no longer valid. We need a new approach to modeling the reactor.

Step 9 *Material balances on reactor*

We are accustomed to summarizing the material balances in a reactor using a stoichiometric table, as in Table 15-7. This table still applies to case B—the reactor inlet streams and reaction stoichiometry have not changed. What we cannot do is use an equilibrium model to find $\dot{\xi}$. Instead, the given information states that the hydrogen consumed ($-3\dot{\xi}$) is equal to 5% of the hydrogen entering the reactor:

$$-3\dot{\xi} = 0.05 \left(30 \, \frac{mol}{s} + y_{H_2,rec} \dot{n}_{rec} \right) \tag{15.53}$$

We can relate the properties of the recycle stream to the properties of the purge stream as in step 5:

$$-3\dot{\xi} = 0.05\left(30 \frac{\text{mol}}{\text{sec}} + 19 y_{H_2,\text{pur}} \dot{n}_{\text{pur}}\right) \tag{15.54}$$

Step 10 *Solve case B*
We now have seven equations in seven unknowns for the situation described in part B. The solution is now:

Liquid ammonia product flow rate $(\dot{n}_{\text{prod}})$ = **5.27 mol/s**

Purge flow rate $(\dot{n}_{\text{pur}})$ = **29.47 mol/s**, with composition **49.6% H₂, 22.8% N₂, 16.9% NH₃,** and **10.7% O₂.**

Comparing part B of Example 15-5 to part A of Example 15-6, we see that the addition of the recycle stream nearly triples the amount of pure ammonia product compared to the simple "one pass" reaction system. What might not be obvious from the calculations shown here is that this improvement would come with some cost. The recycle stream means larger flow rates circulating throughout the process, which means larger and more expensive equipment.

Unfortunately, it's often not practical to design reactors in which a reversible reaction progresses all the way to equilibrium. Part B of Example 15-6 examines the consequences of a lower reaction conversion on the production of liquid ammonia. It also illustrates a simple way of modeling a nonequilibrium reactor.

There are several thoughts we might have for improving the process as described in Example 15-6 part B, and thermodynamics has a clear role in assessing the impact of each on the process.

■ We could lower the vapor pressure of ammonia by further decreasing the temperature in the separator, but this would require a more expensive cooling process. It is instructive to remember the link between the temperature of the cold reservoir and coefficient of performance for a refrigerator from Chapter 5.

■ We could lower the mole fraction of ammonia in the vapor phase by further increasing the pressure throughout the system, but this would require more work in the compressors. We learned to model series of compressors in Chapter 5, but we only considered pure components at that time; enthalpy and entropy of mixing would also be relevant in modeling compression of a mixture.

■ We could increase the fractional conversion of hydrogen and nitrogen into ammonia that is achieved in the reactor—getting closer to the equilibrium condition described in part A. The most straightforward way to increase conversion in a reactor is often to design a larger and more expensive reactor with more catalyst. While principles of catalysis and reaction kinetics are outside the sphere of introductory thermodynamics, the principles of chemical reaction equilibrium do still represent a limit on the conversion that can be attained.

■ We could increase the rate of the reaction by increasing the temperature of the reactor. This would involve adding more heat. Because this is an exothermic reaction, increasing temperature would also lower the equilibrium constant of the reaction. Both of these effects can be quantified using thermodynamics.

Some design variants on the ammonia synthesis example are considered in Problems 15-3 and 15-4.

15.5 SUMMARY OF CHAPTER FIFTEEN

- Applications covered in this book include material, energy and entropy balances, phase equilibrium, and chemical reaction equilibrium. All of these can require models of the interrelationships between physical properties of pure compounds and mixtures.

- Real engineering challenges routinely involve more than one of these applications simultaneously.

- The examples in this chapter illustrate that energy conservation, reaction equilibrium, and phase equilibrium are inter-dependent—often in ways that may not be intuitively obvious.

- More complex (and realistic) engineering applications often involve larger sets of equations and unknowns that cannot be solved analytically.

15.6 PROBLEMS

15-1. This problem revisits Example 15-3, part B. Re-do the problem three times, changing the amount of water to the following.
  A. 25 moles
  B. 100 moles
  C. 500 moles
  D. Use graphs or tables to compare the results of parts A through C to the original problem that used 50 moles of water. Give interpretations based on physical phenomena for any trends that are evident.

15-2. This problem revisits Example 15-4. Re-do the problem three times, changing the volume of the system to the following.
  A. 0.3 m³
  B. 1 m³
  C. 1.5 m³
  D. Use graphs or tables to compare the results of parts A through C to the original problem that used a volume of 0.6 m³. Give interpretations based on physical phenomena for any trends that are evident.

15-3. This problem considers the ammonia synthesis process described in Example 15-5, part B. The fresh air and hydrogen each enter the process at $T = 298.15$ K and $P = 100$ kPa, and need to be compressed to $P = 2.5$ MPa. This is done in two compressions stages, with the gas leaving the first compressor cooled to $T = 298.15$ K and $P = 500$ kPa before it enters the second stage. Assume all compressors have $\eta = 0.8$.
  A. Find the rate at which work is added to each compressor, and the rate at which heat is removed in the interstage cooler for the hydrogen.

B. Find the rate at work is added to each compressor, and the rate at which heat is removed in the interstage cooler, for the air.
C. Find the rate at which heat must be added to (or removed from) the reactor in order to maintain its constant temperature of 300°C.
Use an ideal gas model for all streams that are at pressures below 1 MPa, and model streams above 1 MPa as ideal solutions of real gases. State and explain any other assumptions that you make.

15-4. This problem re-examines Example 15-5 by considering the effect of changing the pressure.
  A. Re-do Example 15-5. Use $P = 4.9$ MPa for all pressures throughout the system, assume all mixtures are ideal solutions of real gases, and assume all other specifications are the same as in Example 15-5, part B.
  B. Use relevant system parameters to quantify the advantages and disadvantages of a system pressure $P = 4.9$ MPa compared to $P = 2.5$ MPa.

15-5. A vessel contains two compartments separated by a partition. One compartment contains 1 kg of pure liquid water at $T = 323.15$ K and $P = 100$ kPa. The other compartment contains 0.1 kg of pure carbon dioxide at $T = 323.15$ K and $P = 100$ kPa. The partition is removed, and the contents of the vessel are permitted to come to equilibrium.
  A. The vessel is rigid and is maintained at a constant temperature of 323.15 K. Determine the final pressure, the contents of the liquid and vapor phases, and the amount of heat added to (or removed from) the system.
  B. The vessel is rigid and adiabatic. Determine the final temperature and pressure and the contents of the liquid and vapor phases.

State any assumptions you make.

15-6. Acetone can be converted into ketene and methane by a vapor-phase reaction:

$$CH_3COCH_3 \leftrightarrow CH_2CO + CH_4$$

Assume $\Delta C_P = 0$ for this process.

A. Since Appendix C contains no data for ketene, use the Joback method to estimate ΔH_f^0 and ΔG_f^0 for this compound.

B. Plot the equilibrium constant of this reaction, vs. temperature, from $T = 25°C$ to $T = 800°C$.

C. A popular Chemical Reaction Engineering text (Fogler, 1992) contains an example in which this reaction is carried out at 1035 K. Based on part B, does 1035 K seem like a well-chosen temperature?

D. If a reactor has constant $T = 1035$ K and $P = 100$ kPa, and is filled with pure acetone at the beginning, what fraction of the acetone is converted into products at equilibrium?

E. 1000 mol/min of pure acetone enters a steady-state, adiabatic flow reactor at $T = 1035$ K and $P = 100$ kPa, and the stream exiting the reactor is at $P = 100$ kPa and at equilibrium. What is the temperature and composition of the stream leaving the reactor?

15-7. Example 14-6 examined the thermal decomposition of butane forming propylene:

$$C_4H_{10} \leftrightarrow C_3H_6 + CH_4$$

But a competing thermal decomposition reaction forming ethylene and ethane is also possible:

$$C_4H_{10} \leftrightarrow C_2H_4 + C_2H_6$$

and ethane can react further as

$$C_2H_6 \leftrightarrow C_2H_4 + H_2$$

Assume these are the only three reactions that occur. Use the shortcut van 't Hoff method to calculate rate constants.

A. 10 mol/s of butane enter a reactor at $T = 298$ K and $P = 100$ kPa. The exiting stream is at $T = 1000$ K and $P = 100$ kPa and is at equilibrium. At what rate is heat added to the reactor, and what is the composition of the exiting stream?

B. 10 moles/s of butane and 10 mol/s of nitrogen enter a reactor at $T = 298$ K and $P = 100$ kPa. The exiting stream is at $T = 1000$ K and $P = 100$ kPa and is at equilibrium. At what rate is heat added to the reactor, and what is the composition of the exiting stream?

C. 10 mol/s of butane enter a reactor at $T = 1000$ K and $P = 100$ kPa. The reactor is isobaric, adiabatic, and large enough for the contents to reach equilibrium. Find the temperature and composition of the outlet stream.

D. 10 moles/s of butane and 10 mol/s of nitrogen enter a reactor at $T = 1000$ K and $P = 100$ kPa. The reactor is isobaric, adiabatic, and large enough for the contents to reach equilibrium. Find the temperature and composition of the outlet stream.

15-8. This example revisits the pair of reactions in Example 14-2.

$$4NH_3 + 5O_2 \leftrightarrow 4NO + 6H_2O$$
$$2NO + O_2 \leftrightarrow 2NO_2$$

100 mol/min of ammonia and 150 mol/min of oxygen enter an isobaric steady state reactor at $T = 800$ K.

A. If the reactions progress to equilibrium at $P = 100$ kPa and $T = 800$ K, what is the composition of the exiting stream, and at what rate is heat added to or removed from the reactor?

B. Repeat part A for a reactor pressure of 300 kPa. Assume ideal gas behavior at this pressure.

15-9. You have 1.5 moles of pure water and 1 mole of CO, both at 298.15 K and 100 kPa. You want to mix them together to make carbon dioxide by the following *gas phase* reaction at 500 K:

$$CO + H_2O \rightarrow CO_2 + H_2$$

Once the reaction is at equilibrium, you want to heat up the resulting mixture to 750 K. This heating is done quickly enough that it can be assumed no reaction occurs; the equilibrium composition at 500 K is still the composition at 750 K. The entire process is to be carried out at 100 kPa.

What is the TOTAL heat load required for this process (from 298.15 K to 750 K)? Will you be adding heat overall or removing heat?

15-10. An equimolar mixture of carbon monoxide and water vapor enters a steady state gas phase reactor at $T = 500$ K and $P = 100$ kPa, where it undergoes the same reaction examined in the previous problem. The reactor is adiabatic and the exit stream leaves the reactor at $P = 100$ kPa.

A. What is the temperature of the exit stream if 25% of the entering carbon monoxide is converted to carbon dioxide?

B. What is the temperature of the exit stream if 50% of the entering carbon monoxide is converted to carbon dioxide?

C. What is the temperature and composition of the exit stream if the reaction progresses to equilibrium?

15-11. This problem expands upon Example 15-4. A reaction vessel is rigid and has a volume of 0.5 m³ and initially contains 10 moles of o-xylene. The liquid phase is exposed to catalyst that facilitates isomerization reactions between the three isomers of xylene, but it is realistic to assume no reactions occur in the vapor phase. The liquid phase can be modeled as an ideal solution, and the vapor phase can be modeled as an ideal gas.

A. The contents of the vessel are allowed to reach equilibrium at 373.15 K. Find the pressure, and the contents of the liquid, and the vapor phase at equilibrium.

B. The contents of the vessel are cooled from the 373.15 K equilibrium state to 348.15 K, and a new equilibrium state is established at 348.15 K. Find the new pressure and the contents of the liquid and vapor phase. Also determine the amount of heat that was removed.

C. The vessel is at the equilibrium state described in part B when 5 kJ of heat are added to the vessel and the vessel is again allowed to reach equilibrium. Find the new pressure, temperature, and contents of the liquid and vapor phases.

15-12. The Diels-Alder reaction is a reaction mechanism in which a diene (a compound with two double bonds) and an alkene (a compound with a double bond) combine to form a ring. Perhaps the simplest example is the combination of 1,3-butadiene and ethylene to form cyclohexene:

$$C_2H_4 + C_4H_6 \leftrightarrow C_6H_{10}$$

5 mol/s each of ethylene and 1,3-butadiene enter a reactor. The reaction is carried out in the vapor phase. Model the reaction mixture as either an ideal gas or an ideal mixture of real gases, as appropriate for the conditions of the reactor.

A. The entering ethylene and butadiene are each at $T = 25°C$ and $P = 100$ kPa, but the reactor is maintained at $T = 300°C$ and $P = 100$ kPa. If the exiting mixture is at equilibrium, find the composition of the exiting stream and the rate at which heat is added to the reactor.

B. The entering ethylene and butadiene are each at $T = 300°C$ and $P = 500$ kPa, and the reactor is also maintained at $T = 300°C$ and $P = 500$ kPa. If the exiting mixture is at equilibrium, find the composition of the exiting stream and the rate at which heat is added to the reactor.

C. The entering ethylene and butadiene are each at $T = 300°C$ and $P = 100$ kPa, and the reactor is adiabatic and maintained at $P = 100$ kPa. Find the composition and temperature of the exiting stream.

15.7 GLOSSARY OF SYMBOLS

Symbol	Definition	Symbol	Definition	Symbol	Definition
C_P	constant pressure heat capacity	N	number of moles in system	x	liquid phase mole fraction
C_P^*	constant pressure heat capacity for ideal gas	N_{gen}	number of moles generated by reaction	y	vapor phase mole fraction
f	fugacity	$\dot{N}_{gen}$	rate at which moles are generated by reaction	$\Delta \underline{G}_f^0$	molar Gibbs energy of formation at standard state
f^0	fugacity at standard state			$\Delta \underline{G}_R^0$	change in standard molar Gibbs energy for reaction at reference temperature
$\hat{f}$	mixture fugacity of a compound	n	number of moles added to or removed from system		
$\underline{G}^0$	molar Gibbs free energy at reference state	$\dot{n}$	molar flow rate	$\Delta \underline{G}_T^0$	change in standard molar Gibbs energy for reaction at temperature T
$\underline{H}$	molar enthalpy	P	pressure		
$\underline{H}^0$	molar enthalpy at reference state	P^0	reference pressure	$\Delta \underline{H}_f^0$	molar enthalpy of formation at standard state
$\hat{H}$	specific enthalpy	p	partial pressure		
H	Henry's Law constant	P^{sat}	vapor pressure	$\Delta \underline{H}_R^0$	change in standard molar enthalpy for reaction at reference temperature
K_T	equilibrium constant for reaction at temperature T	$\dot{Q}$	rate of heat transfer		
		R	gas constant		
K_x	equilibrium constant for a reaction expressed as a function of liquid mole fractions only	T	temperature	$\Delta \underline{H}_T^0$	change in standard molar enthalpy for reaction at temperature T
		T_R	reference temperature		
		V	volume		
m	mass added to or removed from system	$\underline{V}$	molar volume	ν	stoichiometric coefficient
		W_{EC}	work of expansion/contraction	γ	activity coefficient
				ξ	extent of reaction
$\dot{m}$	mass flow rate	W_S	shaft work	$\dot{\xi}$	extent of reaction, rate basis

15.8 REFERENCES

Bessling, B., Longing, J., Ohligshlager, A., Schembecker G., Sundmacher, K. "Investigations on the Synthesis of Methyl Acetate in a Heterogeneous Reactive Distillation Process," *Chemical Engineering Technology*, 21, 5 (1998).

Brown, R. L., Stein, S. E. "Boiling Point Data" in *NIST Chemistry WebBook, NIST Standard Reference Database Number 69*, Eds. Linstrom, P. J. and Mallard, W. G., National Institute of Standards and Technology, Gaithersburg MD, 20899, *http://webbook.nist.gov* (retrieved May 27, 2012).

Domalski, E. S., Hearing, E. D. "Condensed Phase Heat Capacity Data" in *NIST Chemistry WebBook, NIST Standard Reference Database Number 69*, Eds. Linstrom, P. J. and Mallard, W. G., National Institute of Standards and Technology, Gaithersburg MD, 20899, *http://webbook.nist.gov* (retrieved April 21, 2012).

Fogler, H. S. *Elements of Chemical Reaction Engineering*, 2nd ed. Prentice Hall, Upper Saddle River, NJ, 1992.

Huss, R., Chen, F., Malone, M., Doherty, M. "Reactive distillation for methyl acetate production," *Computers & Chemical Engineering*, 27, 2003.

Poling, B. E., Prausnitz, J. M., O'Connell, J. P. *The Properties of Gases and Liquids*. McGraw-Hill, New York, 2001.

Schwartz, S. E., White, W. H. "Solubility equilibria of the nitrogen oxides and oxyacids in dilute aqueous solution," Pfafflin J. R. and Ziegler, E. N. editors, *Advances in Environmental Science and Engineering*, volume 4, pages 1–45. Gordon and Breach Science Publishers, NY, 1981.

Appendix A: Steam Tables

A-1 Saturated Steam-Pressure Increments

A-2 Saturated Steam-Temperature Increments

A-3 Superheated Steam

A-4 Compressed Liquid

These steam tables were generated using FLUIDCAL, a set of computer programs provided by the International Association for the Properties of Water and Steam (IAPWS). FLUIDCAL was also used to generate the data in Table 2-2, Table 7-1, Figures 4-21, 5-4, 5-5, and 7-3 through 7-5, and Examples 1-5, 8-1, and 8-2—all of which present properties of liquid water and/or steam.

The authors gratefully acknowledge Professor Wolfgang Wagner and the IAPWS for making FLUIDCAL available for the production of this book.

The FLUIDCAL programs are used to compute the properties of water and steam according to the IAPWS-95 formulation—a description of which was available at time of writing at the following location: http://www.iapws.org/relguide/IAPWS95-Rev.pdf

A-1 Saturated Steam-Pressure Increments

T (Celsius)	P (kPa)	$\hat{V}^L$ (m³/kg)	$\hat{V}^V$ (m³/kg)	$\hat{U}^L$ (kJ/kg)	$\hat{U}^V$ kJ/kg	H^L (kJ/kg)	H^V (kJ/kg)	$\hat{S}^L$ (kJ/kg · K)	$\hat{S}^V$ (kJ/kg · K)
6.97	1	0.001000	129.18	29.3	2384.5	29.3	2513.7	0.1059	8.9749
17.49	2	0.001001	66.987	73.4	2398.9	73.4	2532.9	0.2606	8.7226
24.08	3	0.001003	45.653	101.0	2407.9	101.0	2544.8	0.3543	8.5764
28.96	4	0.001004	34.791	121.4	2414.5	121.4	2553.7	0.4224	8.4734
32.88	5	0.001005	28.185	137.7	2419.8	137.7	2560.7	0.4762	8.3938
36.16	6	0.001006	23.733	151.5	2424.2	151.5	2566.6	0.5208	8.3290
39.00	7	0.001008	20.524	163.3	2428.0	163.4	2571.7	0.5590	8.2745
41.51	8	0.001008	18.099	173.8	2431.4	173.8	2576.2	0.5925	8.2273
43.76	9	0.001009	16.199	183.2	2434.4	183.3	2580.2	0.6223	8.1858
45.81	10	0.001010	14.670	191.8	2437.2	191.8	2583.9	0.6492	8.1488
53.97	15	0.001014	10.020	225.9	2448.0	225.9	2598.3	0.7549	8.0071
60.06	20	0.001017	7.6480	251.4	2456.0	251.4	2608.9	0.8320	7.9072
64.96	25	0.001020	6.2032	271.9	2462.4	272.0	2617.4	0.8932	7.8302
69.10	30	0.001022	5.2284	289.2	2467.7	289.3	2624.5	0.9441	7.7675
72.68	35	0.001024	4.5251	304.3	2472.3	304.3	2630.7	0.9877	7.7146
75.86	40	0.001026	3.9930	317.6	2476.3	317.6	2636.1	1.0261	7.6690
78.71	45	0.001028	3.5759	329.6	2479.9	329.6	2640.9	1.0603	7.6288
81.32	50	0.001030	3.2400	340.5	2483.2	340.5	2645.2	1.0912	7.5930
85.93	60	0.001033	2.7317	359.8	2489.0	359.9	2652.9	1.1454	7.5311
89.93	70	0.001036	2.3648	376.7	2493.9	376.8	2659.4	1.1921	7.4790
93.48	80	0.001039	2.0871	391.6	2498.2	391.7	2665.2	1.2330	7.4339
96.69	90	0.001041	1.8694	405.1	2502.1	405.2	2670.3	1.2696	7.3943
99.61	100	0.001043	1.6939	417.4	2505.6	417.5	2674.9	1.3028	7.3588
102.29	110	0.001045	1.5495	428.7	2508.7	428.8	2679.2	1.3330	7.3269
104.78	120	0.001047	1.4284	439.2	2511.7	439.4	2683.1	1.3609	7.2977
107.11	130	0.001049	1.3253	449.1	2514.4	449.2	2686.6	1.3868	7.2709
109.29	140	0.001051	1.2366	458.3	2516.9	458.4	2690.0	1.4110	7.2461
111.35	150	0.001053	1.1593	467.0	2519.2	467.1	2693.1	1.4337	7.2230

T (Celsius)	P (MPa)	$\hat{V}^L$ (m³/kg)	$\hat{V}^V$ (m³/kg)	$\hat{U}^L$ (kJ/kg)	$\hat{U}^V$ kJ/kg)	$\hat{H}^L$ (kJ/kg)	$\hat{H}^V$ (kJ/kg)	$\hat{S}^L$ (kJ/kg·K)	$\hat{S}^V$ (kJ/kg·K)
113.30	0.16	0.001054	1.0914	475.2	2521.4	475.4	2696.0	1.4551	7.2014
115.15	0.17	0.001056	1.0312	483.0	2523.5	483.2	2698.8	1.4753	7.1812
116.91	0.18	0.001058	0.9775	490.5	2525.5	490.7	2701.4	1.4945	7.1621
118.60	0.19	0.001059	0.9292	497.6	2527.3	497.9	2703.9	1.5127	7.1440
120.21	0.2	0.001061	0.8857	504.5	2529.1	504.7	2706.2	1.5302	7.1269
127.41	0.25	0.001067	0.7187	535.1	2536.8	535.3	2716.5	1.6072	7.0524
133.52	0.3	0.001073	0.6058	561.1	2543.2	561.4	2724.9	1.6717	6.9916
138.86	0.35	0.001079	0.5242	583.9	2548.5	584.3	2732.0	1.7274	6.9401
143.61	0.4	0.001084	0.4624	604.2	2553.1	604.7	2738.1	1.7765	6.8955
147.90	0.45	0.001088	0.4139	622.6	2557.1	623.1	2743.4	1.8205	6.8560
151.83	0.5	0.001093	0.3748	639.5	2560.7	640.1	2748.1	1.8604	6.8207
155.46	0.55	0.001097	0.3426	655.2	2563.9	655.8	2752.3	1.8970	6.7886
158.83	0.6	0.001101	0.3156	669.7	2566.8	670.4	2756.1	1.9308	6.7592
161.98	0.65	0.001104	0.2926	683.4	2569.4	684.1	2759.6	1.9623	6.7322
164.95	0.7	0.001108	0.2728	696.2	2571.8	697.0	2762.8	1.9918	6.7071
167.75	0.75	0.001111	0.2555	708.4	2574.0	709.2	2765.6	2.0195	6.6836
170.41	0.8	0.001115	0.2403	720.0	2576.0	720.9	2768.3	2.0457	6.6616
172.94	0.85	0.001118	0.2269	731.0	2577.9	732.0	2770.8	2.0705	6.6409
175.35	0.9	0.001121	0.2149	741.6	2579.6	742.6	2773.0	2.0940	6.6213
177.66	0.95	0.001124	0.2041	751.7	2581.2	752.7	2775.1	2.1165	6.6027
179.88	1	0.001127	0.1944	761.4	2582.7	762.5	2777.1	2.1381	6.5850
184.06	1.1	0.001133	0.1774	779.8	2585.5	781.0	2780.6	2.1785	6.5520
187.96	1.2	0.001139	0.1633	797.0	2587.8	798.3	2783.7	2.2159	6.5217
191.60	1.3	0.001144	0.1512	813.1	2589.9	814.6	2786.5	2.2508	6.4936
195.04	1.4	0.001149	0.1408	828.4	2591.8	830.0	2788.8	2.2835	6.4675
198.29	1.5	0.001154	0.1317	842.8	2593.4	844.6	2791.0	2.3143	6.4430
201.37	1.6	0.001159	0.1237	856.6	2594.8	858.5	2792.8	2.3435	6.4199
204.31	1.7	0.001163	0.1167	869.8	2596.1	871.7	2794.5	2.3711	6.3981
207.11	1.8	0.001168	0.1104	882.4	2597.2	884.5	2795.9	2.3975	6.3775
209.80	1.9	0.001172	0.1047	894.5	2598.2	896.7	2797.2	2.4227	6.3578

(Continued)

T (Celsius)	P (MPa)	$\hat{V}^L$ (m³/kg)	$\hat{V}^V$ (m³/kg)	$\hat{U}^L$ (kJ/kg)	$\hat{U}^V$ (kJ/kg)	$\hat{H}^L$ (kJ/kg)	$\hat{H}^V$ (kJ/kg)	$\hat{S}^L$ (kJ/kg·K)	$\hat{S}^V$ (kJ/kg·K)
212.38	2	0.001177	0.099585	906.1	2599.1	908.5	2798.3	2.4468	6.3390
223.95	2.5	0.001197	0.079949	958.9	2602.1	961.9	2801.9	2.5543	6.2558
233.85	3	0.001217	0.066664	1004.7	2603.2	1008.3	2803.2	2.6455	6.1856
242.56	3.5	0.001235	0.057058	1045.5	2602.9	1049.8	2802.6	2.7254	6.1243
250.35	4	0.001253	0.049776	1082.5	2601.7	1087.5	2800.8	2.7968	6.0696
257.44	4.5	0.001270	0.044059	1116.5	2599.7	1122.2	2797.9	2.8615	6.0197
263.94	5	0.001286	0.039446	1148.2	2597.0	1154.6	2794.2	2.9210	5.9737
275.58	6	0.001319	0.032448	1206.0	2589.9	1213.9	2784.6	3.0278	5.8901
285.83	7	0.001352	0.027378	1258.2	2581.0	1267.7	2772.6	3.1224	5.8148
295.01	8	0.001385	0.023526	1306.2	2570.5	1317.3	2758.7	3.2081	5.7450
303.34	9	0.001418	0.020490	1351.1	2558.5	1363.9	2742.9	3.2870	5.6791
311.00	10	0.001453	0.018030	1393.5	2545.2	1408.1	2725.5	3.3606	5.6160
318.08	11	0.001489	0.015990	1434.1	2530.5	1450.4	2706.3	3.4303	5.5545
324.68	12	0.001526	0.014264	1473.1	2514.3	1491.5	2685.4	3.4967	5.4939
330.85	13	0.001566	0.012780	1511.1	2496.5	1531.5	2662.7	3.5608	5.4336
336.67	14	0.001610	0.011485	1548.4	2477.1	1571.0	2637.9	3.6232	5.3727
342.16	15	0.001657	0.010338	1585.3	2455.6	1610.2	2610.7	3.6846	5.3106
347.35	16	0.001709	0.009309	1622.3	2431.8	1649.7	2580.8	3.7457	5.2463
352.29	17	0.001769	0.008371	1659.9	2405.2	1690.0	2547.5	3.8077	5.1787
356.99	18	0.001840	0.007502	1699.0	2374.8	1732.1	2509.8	3.8718	5.1061
361.47	19	0.001927	0.006677	1740.5	2339.1	1777.2	2466.0	3.9401	5.0256
365.75	20	0.002040	0.005865	1786.4	2295.0	1827.2	2412.3	4.0156	4.9314
369.83	21	0.002206	0.004996	1841.2	2233.7	1887.6	2338.6	4.1064	4.8079
373.95	22.064	0.003106	0.003106	2015.7	2015.7	2084.3	2084.3	4.4070	4.4070

A-2 Saturated Steam-Temperature Increments

T (Celsius)	P (kPa)	$\hat{V}^L$ (m³/kg)	$\hat{V}^V$ (m³/kg)	$\hat{U}^L$ (kJ/kg)	$\hat{U}^V$ (kJ/kg)	H^L (kJ/kg)	H^V (kJ/kg)	$\hat{S}^L$ (kJ/kg·K)	$\hat{S}^V$ (kJ/kg·K)
0	0.61	0.001000	206.13	−0.04	2374.9	−0.04	2500.9	−0.0002	9.1558
5	0.87	0.001000	147.01	21.0	2381.8	21.0	2510.1	0.0763	9.0248
10	1.23	0.001000	106.30	42.0	2388.6	42.0	2519.2	0.1511	8.8998
15	1.76	0.001001	77.875	63.0	2395.5	63.0	2528.3	0.2245	8.7803
20	2.34	0.001002	57.757	83.9	2402.3	83.9	2537.4	0.2965	8.6660
25	3.17	0.001003	43.337	104.8	2409.1	104.8	2546.5	0.3672	8.5566
30	4.25	0.001004	32.878	125.7	2415.9	125.7	2555.5	0.4368	8.4520
35	5.63	0.001006	25.205	146.6	2422.7	146.6	2564.5	0.5051	8.3517
40	7.38	0.001008	19.515	167.5	2429.4	167.5	2573.5	0.5724	8.2555
45	9.59	0.001010	15.252	188.4	2436.1	188.4	2582.4	0.6386	8.1633
50	12.35	0.001012	12.027	209.3	2442.7	209.3	2591.3	0.7038	8.0748
55	15.76	0.001015	9.5643	230.2	2449.3	230.3	2600.1	0.7680	7.9898
60	19.95	0.001017	7.6672	251.2	2455.9	251.2	2608.8	0.8313	7.9081
65	25.04	0.001020	6.1935	272.1	2462.4	272.1	2617.5	0.8937	7.8296
70	31.2	0.001023	5.0395	293.0	2468.9	293.1	2626.1	0.9551	7.7540
75	38.6	0.001026	4.1289	314.0	2475.2	314.0	2634.6	1.0158	7.6812
80	47.41	0.001029	3.4052	335.0	2481.6	335.0	2643.0	1.0756	7.6111
85	57.87	0.001032	2.8258	356.0	2487.8	356.0	2651.3	1.1346	7.5434
90	70.18	0.001036	2.3591	377.0	2494.0	377.0	2659.5	1.1929	7.4781
95	84.61	0.001040	1.9806	398.0	2500.0	398.1	2667.6	1.2504	7.4151
100	101.42	0.001043	1.6718	419.1	2506.0	419.2	2675.6	1.3072	7.3541
105	120.9	0.001047	1.4184	440.1	2511.9	440.3	2683.4	1.3633	7.2952
110	143.38	0.001052	1.2093	461.3	2517.7	461.4	2691.1	1.4188	7.2381
115	169.18	0.001056	1.0358	482.4	2523.3	482.6	2698.6	1.4737	7.1828
120	198.67	0.001060	0.891212	503.6	2528.9	503.8	2705.9	1.5279	7.1291
125	232.24	0.001065	0.770026	524.8	2534.3	525.1	2713.1	1.5816	7.0770
130	270.28	0.001070	0.668004	546.1	2539.5	546.4	2720.1	1.6346	7.0264
135	313.23	0.001075	0.581729	567.4	2544.7	567.7	2726.9	1.6872	6.9772

(Continued)

T (Celsius)	P (MPa)	$\hat{V}^L$ (m³/kg)	$\hat{V}^V$ (m³/kg)	$\hat{U}^L$ (kJ/kg)	$\hat{U}^V$ (kJ/kg)	$\hat{H}^L$ (kJ/kg)	$\hat{H}^V$ (kJ/kg)	$\hat{S}^L$ (kJ/kg·K)	$\hat{S}^V$ (kJ/kg·K)
140	0.36154	0.001080	0.508454	588.8	2549.6	589.2	2733.4	1.7392	6.9293
145	0.41568	0.001085	0.445962	610.2	2554.4	610.6	2739.8	1.7907	6.8826
150	0.47616	0.001091	0.392453	631.7	2559.1	632.2	2745.9	1.8418	6.8371
155	0.5435	0.001096	0.346460	653.2	2563.5	653.8	2751.8	1.8924	6.7926
160	0.61823	0.001102	0.306782	674.8	2567.8	675.5	2757.4	1.9426	6.7491
165	0.70093	0.001108	0.272431	696.5	2571.9	697.2	2762.8	1.9923	6.7066
170	0.79219	0.001114	0.242589	718.2	2575.7	719.1	2767.9	2.0417	6.6650
175	0.8926	0.001121	0.216581	740.0	2579.4	741.0	2772.7	2.0906	6.6241
180	1.00281	0.001127	0.193842	761.9	2582.8	763.1	2777.2	2.1392	6.5840
185	1.12346	0.001134	0.173901	783.9	2586.0	785.2	2781.4	2.1875	6.5447
190	1.25524	0.001141	0.156362	806.0	2589.0	807.4	2785.3	2.2355	6.5059
195	1.39882	0.001149	0.140892	828.2	2591.7	829.8	2788.8	2.2832	6.4678
200	1.55493	0.001157	0.127210	850.5	2594.2	852.3	2792.0	2.3305	6.4302
205	1.7243	0.001164	0.115078	872.9	2596.4	874.9	2794.8	2.3777	6.3930
210	1.90767	0.001173	0.104292	895.4	2598.3	897.6	2797.3	2.4245	6.3563
215	2.10584	0.001181	0.094679	918.0	2599.9	920.5	2799.3	2.4712	6.3200
220	2.31959	0.001190	0.086092	940.8	2601.2	943.6	2800.9	2.5177	6.2840
225	2.54972	0.001199	0.078403	963.7	2602.2	966.8	2802.1	2.5640	6.2483
230	2.79709	0.001209	0.071503	986.8	2602.9	990.2	2802.9	2.6101	6.2128
235	3.06253	0.001219	0.065298	1010.0	2603.2	1013.8	2803.2	2.6561	6.1775
240	3.34693	0.001229	0.059705	1033.4	2603.1	1037.6	2803.0	2.7020	6.1423
245	3.65117	0.001240	0.054654	1057.0	2602.7	1061.5	2802.2	2.7478	6.1072
250	3.97617	0.001252	0.050083	1080.8	2601.8	1085.8	2800.9	2.7935	6.0721
255	4.32289	0.001264	0.045938	1104.8	2600.5	1110.2	2799.1	2.8392	6.0369
260	4.69226	0.001276	0.042173	1129.0	2598.7	1135.0	2796.6	2.8849	6.0016
265	5.08529	0.001289	0.038746	1153.4	2596.5	1160.0	2793.5	2.9307	5.9661
270	5.50299	0.001303	0.035621	1178.1	2593.7	1185.3	2789.7	2.9765	5.9304
275	5.94639	0.001318	0.032766	1203.1	2590.3	1210.9	2785.2	3.0224	5.8944
280	6.41658	0.001333	0.030153	1228.3	2586.4	1236.9	2779.9	3.0685	5.8579
285	6.91466	0.001349	0.027756	1253.9	2581.8	1263.2	2773.7	3.1147	5.8209
290	7.44178	0.001366	0.025555	1279.9	2576.5	1290.0	2766.7	3.1612	5.7834

T (Celsius)	P (MPa)	V̂ᴸ (m³/kg)	V̂ᵛ (m³/kg)	Ûᴸ (kJ/kg)	Ûᵛ (kJ/kg)	Ĥᴸ (kJ/kg)	Ĥᵛ (kJ/kg)	Ŝᴸ (kJ/kg·K)	Ŝᵛ (kJ/kg·K)
295	7.99911	0.001385	0.023529	1306.2	2570.5	1317.3	2758.7	3.2080	5.7451
300	8.5879	0.001404	0.021660	1332.9	2563.6	1345.0	2749.6	3.2552	5.7059
305	9.20943	0.001425	0.019933	1360.2	2555.9	1373.3	2739.4	3.3028	5.6657
310	9.86505	0.001448	0.018335	1387.9	2547.1	1402.2	2727.9	3.3510	5.6244
315	10.55617	0.001472	0.016851	1416.3	2537.2	1431.8	2715.1	3.3998	5.5816
320	11.28429	0.001499	0.015471	1445.3	2526.0	1462.2	2700.6	3.4494	5.5372
325	12.05101	0.001528	0.014183	1475.1	2513.4	1493.5	2684.3	3.5000	5.4908
330	12.85805	0.001561	0.012979	1505.8	2499.2	1525.9	2666.0	3.5518	5.4422
335	13.70726	0.001597	0.011847	1537.6	2483.0	1559.5	2645.4	3.6050	5.3906
340	14.60068	0.001638	0.010781	1570.6	2464.4	1594.5	2621.8	3.6601	5.3356
345	15.54055	0.001685	0.009769	1605.3	2443.1	1631.5	2594.9	3.7176	5.2762
350	16.52942	0.001740	0.008802	1642.1	2418.1	1670.9	2563.6	3.7784	5.2110
355	17.57012	0.001808	0.007868	1682.0	2388.4	1713.7	2526.6	3.8439	5.1380
360	18.66601	0.001895	0.006949	1726.3	2351.8	1761.7	2481.5	3.9167	5.0536
365	19.82136	0.002017	0.006012	1777.8	2303.8	1817.8	2422.9	4.0014	4.9497
370	21.04356	0.002215	0.004954	1844.1	2230.3	1890.7	2334.5	4.1112	4.8012
373.95	22.064	0.003106	0.003106	2015.7	2015.7	2084.3	2084.3	4.4070	4.4070

A-3 Superheated Steam

	$P = 10$ kPa					$P = 50$ kPa				
T (°C)	$\hat{V}$ (m³/kg)	$\hat{U}$ (kJ/kg)	$\hat{H}$ (kJ/kg)	$\hat{S}$ (kJ/kg·K)	T (°C)	$\hat{V}$ (m³/kg)	$\hat{U}$ (kJ/kg)	$\hat{H}$ (kJ/kg)	$\hat{S}$ (kJ/kg·K)
45.806	14.670	2437.2	2583.9	8.1488	81.317	3.240	2483.2	2645.2	7.5930
50	14.867	2443.3	2592.0	8.1741	50				
100	17.196	2515.5	2687.5	8.4489	100	3.419	2511.5	2682.4	7.6953
150	19.513	2587.9	2783.0	8.6892	150	3.890	2585.7	2780.2	7.9413
200	21.826	2661.3	2879.6	8.9049	200	4.356	2660.0	2877.8	8.1592
250	24.136	2736.1	2977.4	9.1015	250	4.821	2735.1	2976.1	8.3568
300	26.446	2812.3	3076.7	9.2827	300	5.284	2811.6	3075.8	8.5386
350	28.755	2890.0	3177.5	9.4513	350	5.747	2889.4	3176.8	8.7076
400	31.063	2969.3	3279.9	9.6094	400	6.209	2968.9	3279.3	8.8659
450	33.371	3050.3	3384.0	9.7584	450	6.672	3049.9	3383.5	9.0151
500	35.680	3132.9	3489.7	9.8998	500	7.134	3132.6	3489.3	9.1566
550	37.988	3217.2	3597.1	10.0344	550	7.596	3217.0	3596.8	9.2913
600	40.296	3303.3	3706.3	10.1631	600	8.058	3303.1	3706.0	9.4201
650	42.603	3391.2	3817.2	10.2866	650	8.519	3391.0	3816.9	9.5436
700	44.911	3480.8	3929.9	10.4055	700	8.981	3480.6	3929.7	9.6625
750	47.219	3572.2	4044.4	10.5202	750	9.443	3572.0	4044.2	9.7773
800	49.527	3665.3	4160.6	10.6311	800	9.905	3665.2	4160.4	9.8882
850	51.835	3760.3	4278.6	10.7386	850	10.366	3760.1	4278.5	9.9957
900	54.142	3856.9	4398.3	10.8429	900	10.828	3856.8	4398.2	10.1000
950	56.450	3955.2	4519.7	10.9442	950	11.290	3955.1	4519.6	10.2014
1000	58.758	4055.2	4642.8	11.0428	1000	11.751	4055.1	4642.7	10.3000

P = 100 kPa

T (°C)	V̂ (m³/kg)	Û (kJ/kg)	Ĥ (kJ/kg)	Ŝ (kJ/kg·K)
99.606	1.694	2505.6	2674.9	7.3588
50				
100	1.696	2506.2	2675.8	7.3610
150	1.937	2582.9	2776.6	7.6148
200	2.172	2658.2	2875.5	7.8356
250	2.406	2733.9	2974.5	8.0346
300	2.639	2810.6	3074.5	8.2172
350	2.871	2888.7	3175.8	8.3866
400	3.103	2968.3	3278.6	8.5452
450	3.334	3049.4	3382.8	8.6946
500	3.566	3132.2	3488.7	8.8361
550	3.797	3216.6	3596.3	8.9709
600	4.028	3302.8	3705.6	9.0998
650	4.259	3390.7	3816.6	9.2234
700	4.490	3480.4	3929.4	9.3424
750	4.721	3571.8	4043.9	9.4572
800	4.952	3665.0	4160.2	9.5681
850	5.183	3760.0	4278.2	9.6757
900	5.414	3856.6	4398.0	9.7800
950	5.645	3955.0	4519.5	9.8813
1000	5.875	4055.0	4642.6	9.9800

P = 200 kPa

T (°C)	V̂ (m³/kg)	Û (kJ/kg)	Ĥ (kJ/kg)	Ŝ (kJ/kg·K)
120.21	0.886	2529.1	2706.2	7.1269
50				
100				
150	0.960	2577.1	2769.1	7.2810
200	1.080	2654.6	2870.7	7.5081
250	1.199	2731.4	2971.2	7.7100
300	1.316	2808.8	3072.1	7.8941
350	1.433	2887.3	3173.9	8.0644
400	1.549	2967.1	3277.0	8.2236
450	1.665	3048.5	3381.6	8.3734
500	1.781	3131.4	3487.7	8.5152
550	1.897	3215.9	3595.4	8.6502
600	2.013	3302.2	3704.8	8.7792
650	2.129	3390.2	3815.9	8.9030
700	2.244	3479.9	3928.8	9.0220
750	2.360	3571.4	4043.4	9.1369
800	2.475	3664.7	4159.8	9.2479
850	2.591	3759.6	4277.8	9.3555
900	2.707	3856.3	4397.6	9.4598
950	2.822	3954.7	4519.1	9.5612
1000	2.938	4054.8	4642.3	9.6599

	P = 300 kPa					P = 400 kPa			
T (°C)	V̂ (m³/kg)	Û (kJ/kg)	Ĥ (kJ/kg)	Ŝ (kJ/kg·K)	T (°C)	V̂ (m³/kg)	Û (kJ/kg)	Ĥ (kJ/kg)	Ŝ (kJ/kg·K)
133.522	0.606	2543.2	2724.9	6.9916	143.608	0.462	2553.1	2738.1	6.8955
50					50				
100					100				
150	0.634	2571.0	2761.2	7.0791	150	0.471	2564.4	2752.8	6.9306
200	0.716	2651.0	2865.9	7.3131	200	0.534	2647.2	2860.9	7.1723
250	0.796	2728.9	2967.9	7.5180	250	0.595	2726.4	2964.5	7.3804
300	0.875	2807.0	3069.6	7.7037	300	0.655	2805.1	3067.1	7.5677
350	0.954	2885.9	3172.0	7.8750	350	0.714	2884.4	3170.0	7.7399
400	1.032	2966.0	3275.5	8.0347	400	0.773	2964.9	3273.9	7.9002
450	1.109	3047.5	3380.3	8.1849	450	0.831	3046.6	3379.0	8.0508
500	1.187	3130.6	3486.6	8.3270	500	0.889	3129.8	3485.5	8.1933
550	1.264	3215.3	3594.5	8.4623	550	0.948	3214.6	3593.6	8.3287
600	1.341	3301.6	3704.0	8.5914	600	1.006	3301.0	3703.2	8.4580
650	1.419	3389.7	3815.3	8.7153	650	1.064	3389.1	3814.6	8.5820
700	1.496	3479.5	3928.2	8.8344	700	1.122	3479.0	3927.6	8.7012
750	1.573	3571.0	4042.9	8.9494	750	1.179	3570.6	4042.4	8.8162
800	1.650	3664.3	4159.3	9.0604	800	1.237	3663.9	4158.8	8.9273
850	1.727	3759.3	4277.4	9.1680	850	1.295	3759.0	4277.0	9.0350
900	1.804	3856.0	4397.3	9.2724	900	1.353	3855.7	4396.9	9.1394
950	1.881	3954.4	4518.8	9.3739	950	1.411	3954.2	4518.5	9.2409
1000	1.958	4054.5	4642.0	9.4726	1000	1.469	4054.3	4641.7	9.3396

P = 500 kPa

T (°C)	V̂ (m³/kg)	Û (kJ/kg)	Ĥ (kJ/kg)	Ŝ (kJ/kg · K)
151.831	0.375	2560.7	2748.1	6.8207
50				
100				
150				
200	0.425	2643.3	2855.8	7.0610
250	0.474	2723.8	2961.0	7.2724
300	0.523	2803.2	3064.6	7.4614
350	0.570	2883.0	3168.1	7.6346
400	0.617	2963.7	3272.4	7.7955
450	0.664	3045.6	3377.7	7.9465
500	0.711	3129.0	3484.5	8.0892
550	0.758	3213.9	3592.7	8.2249
600	0.804	3300.4	3702.5	8.3543
650	0.851	3388.6	3813.9	8.4784
700	0.897	3478.5	3927.0	8.5977
750	0.943	3570.2	4041.8	8.7128
800	0.990	3663.6	4158.4	8.8240
850	1.036	3758.6	4276.6	8.9317
900	1.082	3855.4	4396.6	9.0362
950	1.129	3953.9	4518.2	9.1377
1000	1.175	4054.0	4641.4	9.2364

P = 600 kPa

T (°C)	V̂ (m³/kg)	Û (kJ/kg)	Ĥ (kJ/kg)	Ŝ (kJ/kg · K)
158.826	0.316	2566.8	2756.1	6.7592
50				
100				
150				
200	0.352	2639.3	2850.6	6.9683
250	0.394	2721.2	2957.6	7.1832
300	0.434	2801.4	3062.0	7.3740
350	0.474	2881.6	3166.1	7.5481
400	0.514	2962.5	3270.8	7.7097
450	0.553	3044.7	3376.5	7.8611
500	0.592	3128.2	3483.4	8.0041
550	0.631	3213.2	3591.8	8.1399
600	0.670	3299.8	3701.7	8.2695
650	0.709	3388.1	3813.2	8.3937
700	0.747	3478.1	3926.4	8.5131
750	0.786	3569.8	4041.3	8.6283
800	0.825	3663.2	4157.9	8.7395
850	0.863	3758.3	4276.2	8.8472
900	0.902	3855.1	4396.2	8.9518
950	0.940	3953.6	4517.8	9.0533
1000	0.979	4053.7	4641.1	9.1521

P = 800 kPa

T (°C)	V̂ (m³/kg)	Û (kJ/kg)	Ĥ (kJ/kg)	Ŝ (kJ/kg·K)
170.406	0.240	2576.0	2768.3	6.6616
50				
100				
150				
200	0.261	2631.0	2839.7	6.8176
250	0.293	2715.9	2950.4	7.0401
300	0.324	2797.5	3056.9	7.2345
350	0.354	2878.6	3162.2	7.4106
400	0.384	2960.2	3267.6	7.5734
450	0.414	3042.8	3373.9	7.7257
500	0.443	3126.6	3481.3	7.8692
550	0.473	3211.9	3590.0	8.0054
600	0.502	3298.7	3700.1	8.1354
650	0.531	3387.1	3811.9	8.2598
700	0.560	3477.2	3925.3	8.3794
750	0.589	3569.0	4040.3	8.4947
800	0.618	3662.4	4157.0	8.6061
850	0.647	3757.6	4275.4	8.7139
900	0.676	3854.5	4395.5	8.8185
950	0.705	3953.1	4517.2	8.9201
1000	0.734	4053.2	4640.5	9.0189

P = 1 MPa

T (°C)	V̂ (m³/kg)	Û (kJ/kg)	Ĥ (kJ/kg)	Ŝ (kJ/kg·K)
179.878	0.194	2582.7	2777.1	6.5850
50				
100				
150				
200	0.206	2622.2	2828.3	6.6955
250	0.233	2710.4	2943.1	6.9265
300	0.258	2793.6	3051.6	7.1246
350	0.283	2875.7	3158.2	7.3029
400	0.307	2957.9	3264.5	7.4669
450	0.330	3040.9	3371.3	7.6200
500	0.354	3125.0	3479.1	7.7641
550	0.378	3210.5	3588.1	7.9008
600	0.401	3297.5	3698.6	8.0310
650	0.424	3386.0	3810.5	8.1557
700	0.448	3476.2	3924.1	8.2755
750	0.471	3568.1	4039.3	8.3909
800	0.494	3661.7	4156.1	8.5024
850	0.518	3757.0	4274.6	8.6103
900	0.541	3853.9	4394.8	8.7150
950	0.564	3952.5	4516.5	8.8166
1000	0.587	4052.7	4639.9	8.9155

		P = 1.2 MPa						P = 1.4 MPa			
T (°C)	V̂ (m³/kg)	Û (kJ/kg)	Ĥ (kJ/kg)	Ŝ (kJ/kg·K)		T (°C)	V̂ (m³/kg)	Û (kJ/kg)	Ĥ (kJ/kg)	Ŝ (kJ/kg·K)
187.957	0.163	2587.8	2783.7	6.5217		195.039	0.141	2591.8	2788.8	6.4675
50						50				
100						100				
150						150				
200	0.169	2612.9	2816.1	6.5909		200	0.143	2602.7	2803.0	6.4975
250	0.192	2704.7	2935.6	6.8313		250	0.164	2698.9	2927.9	6.7488
300	0.214	2789.7	3046.3	7.0335		300	0.182	2785.7	3040.9	6.9552
350	0.235	2872.7	3154.2	7.2139		350	0.200	2869.7	3150.1	7.1379
400	0.255	2955.5	3261.3	7.3793		400	0.218	2953.1	3258.1	7.3046
450	0.275	3038.9	3368.7	7.5332		450	0.235	3037.0	3366.1	7.4594
500	0.295	3123.4	3476.9	7.6779		500	0.252	3121.8	3474.8	7.6047
550	0.314	3209.1	3586.3	7.8150		550	0.269	3207.7	3584.5	7.7422
600	0.334	3296.3	3697.0	7.9455		600	0.286	3295.1	3695.4	7.8730
650	0.353	3385.0	3809.2	8.0704		650	0.303	3384.0	3807.8	7.9982
700	0.373	3475.3	3922.9	8.1904		700	0.320	3474.4	3921.7	8.1183
750	0.392	3567.3	4038.2	8.3060		750	0.336	3566.5	4037.2	8.2340
800	0.412	3661.0	4155.2	8.4176		800	0.353	3660.2	4154.3	8.3457
850	0.431	3756.3	4273.8	8.5256		850	0.370	3755.6	4273.0	8.4538
900	0.451	3853.3	4394.0	8.6303		900	0.386	3852.7	4393.3	8.5587
950	0.470	3952.0	4515.9	8.7320		950	0.403	3951.4	4515.2	8.6604
1000	0.489	4052.2	4639.4	8.8310		1000	0.419	4051.7	4638.8	8.7594

P = 1.6 MPa				
T (°C)	V̂ (m³/kg)	Û (kJ/kg)	Ĥ (kJ/kg)	Ŝ (kJ/kg·K)
201.37	0.124	2594.8	2792.8	6.4199
50				
100				
150				
200				
250	0.142	2692.9	2919.9	6.6753
300	0.159	2781.6	3035.4	6.8863
350	0.175	2866.6	3146.0	7.0713
400	0.190	2950.7	3254.9	7.2394
450	0.205	3035.0	3363.5	7.3950
500	0.220	3120.1	3472.6	7.5409
550	0.235	3206.3	3582.6	7.6788
600	0.250	3293.9	3693.9	7.8100
650	0.265	3382.9	3806.5	7.9354
700	0.279	3473.5	3920.5	8.0557
750	0.294	3565.7	4036.1	8.1716
800	0.309	3659.5	4153.3	8.2834
850	0.323	3755.0	4272.2	8.3916
900	0.338	3852.1	4392.6	8.4965
950	0.352	3950.9	4514.6	8.5984
1000	0.367	4051.2	4638.2	8.6974

P = 1.8 MPa				
T (°C)	V̂ (m³/kg)	Û (kJ/kg)	Ĥ (kJ/kg)	Ŝ (kJ/kg·K)
207.112	0.110	2597.2	2795.9	6.3775
50				
100				
150				
200				
250	0.125	2686.7	2911.7	6.6087
300	0.140	2777.4	3029.9	6.8246
350	0.155	2863.6	3141.8	7.0120
400	0.168	2948.3	3251.6	7.1814
450	0.182	3033.1	3360.9	7.3380
500	0.196	3118.5	3470.4	7.4845
550	0.209	3205.0	3580.8	7.6228
600	0.222	3292.7	3692.3	7.7543
650	0.235	3381.9	3805.1	7.8799
700	0.248	3472.6	3919.4	8.0004
750	0.261	3564.9	4035.1	8.1164
800	0.274	3658.8	4152.4	8.2284
850	0.287	3754.3	4271.3	8.3367
900	0.300	3851.5	4391.9	8.4416
950	0.313	3950.3	4514.0	8.5435
1000	0.326	4050.7	4637.6	8.6426

		P = 2.0 MPa					P = 2.5 MPa			
T (°C)	V̂ (m³/kg)	Û (kJ/kg)	Ĥ (kJ/kg)	Ŝ (kJ/kg · K)	T (°C)	V̂ (m³/kg)	Û (kJ/kg)	Ĥ (kJ/kg)	Ŝ (kJ/kg · K)
212.377	0.100	2599.1	2798.3	6.3390	223.95	0.0799	2602.1	2801.9	6.2558
50					50				
100					100				
150					150				
200					200				
250	0.111	2680.2	2903.2	6.5475	250	0.0871	2663.3	2880.9	6.4107
300	0.126	2773.2	3024.2	6.7684	300	0.0989	2762.2	3009.6	6.6459
350	0.139	2860.5	3137.7	6.9583	350	0.1098	2852.5	3127.0	6.8424
400	0.151	2945.9	3248.3	7.1292	400	0.1201	2939.8	3240.1	7.0170
450	0.164	3031.1	3358.2	7.2866	450	0.1302	3026.2	3351.6	7.1767
500	0.176	3116.9	3468.2	7.4337	500	0.1400	3112.8	3462.7	7.3254
550	0.188	3203.6	3579.0	7.5725	550	0.1497	3200.1	3574.3	7.4653
600	0.200	3291.5	3690.7	7.7043	600	0.1593	3288.5	3686.8	7.5979
650	0.211	3380.8	3803.8	7.8302	650	0.1689	3378.2	3800.4	7.7243
700	0.223	3471.6	3918.2	7.9509	700	0.1783	3469.3	3915.2	7.8455
750	0.235	3564.0	4034.1	8.0670	750	0.1878	3562.0	4031.5	7.9620
800	0.247	3658.0	4151.5	8.1790	800	0.1972	3656.2	4149.2	8.0743
850	0.258	3753.6	4270.5	8.2874	850	0.2066	3752.0	4268.5	8.1830
900	0.270	3850.9	4391.1	8.3925	900	0.2160	3849.4	4389.3	8.2882
950	0.282	3949.8	4513.3	8.4945	950	0.2253	3948.4	4511.7	8.3904
1000	0.293	4050.2	4637.0	8.5936	1000	0.2347	4048.9	4635.6	8.4896

P = 3.0 MPa

T (°C)	V̂ (m³/kg)	Û (kJ/kg)	Ĥ (kJ/kg)	Ŝ (kJ/kg · K)
233.853	0.0667	2603.2	2803.2	6.1856
50				
100				
150				
200				
250	0.0706	2644.7	2856.5	6.2893
300	0.0812	2750.8	2994.3	6.5412
350	0.0906	2844.4	3116.1	6.7449
400	0.0994	2933.5	3231.7	6.9234
450	0.1079	3021.2	3344.8	7.0856
500	0.1162	3108.6	3457.2	7.2359
550	0.1244	3196.6	3569.7	7.3768
600	0.1324	3285.5	3682.8	7.5103
650	0.1405	3375.6	3796.9	7.6373
700	0.1484	3467.0	3912.2	7.7590
750	0.1563	3559.9	4028.9	7.8758
800	0.1642	3654.3	4146.9	7.9885
850	0.1720	3750.3	4266.5	8.0973
900	0.1799	3847.9	4387.5	8.2028
950	0.1877	3947.0	4510.1	8.3051
1000	0.1955	4047.7	4634.1	8.4045

P = 3.5 MPa

T (°C)	V̂ (m³/kg)	Û (kJ/kg)	Ĥ (kJ/kg)	Ŝ (kJ/kg · K)
242.557	0.0571	2602.9	2802.6	6.1243
50				
100				
150				
200				
250	0.0588	2624.0	2829.7	6.1764
300	0.0685	2738.8	2978.4	6.4484
350	0.0768	2836.0	3104.8	6.6601
400	0.0846	2927.2	3223.2	6.8427
450	0.0920	3016.1	3338.0	7.0074
500	0.0992	3104.5	3451.6	7.1593
550	0.1063	3193.1	3565.0	7.3014
600	0.1133	3282.5	3678.9	7.4356
650	0.1202	3372.9	3793.5	7.5633
700	0.1270	3464.7	3909.3	7.6854
750	0.1338	3557.9	4026.3	7.8027
800	0.1406	3652.5	4144.6	7.9156
850	0.1474	3748.6	4264.4	8.0247
900	0.1541	3846.4	4385.7	8.1303
950	0.1608	3945.6	4508.4	8.2328
1000	0.1675	4046.4	4632.7	8.3324

		P = 4.0 MPa						P = 4.5 MPa			
T (°C)	V̂ (m³/kg)	Û (kJ/kg)	Ĥ (kJ/kg)	Ŝ (kJ/kg·K)		T (°C)	V̂ (m³/kg)	Û (kJ/kg)	Ĥ (kJ/kg)	Ŝ (kJ/kg·K)
250.354	0.0498	2601.7	2800.8	6.0696		257.437	0.0441	2599.7	2797.9	6.0197
50						50				
100						100				
150						150				
200						200				
250						250				
300	0.0589	2726.2	2961.7	6.3639		300	0.0514	2713.0	2944.2	6.2854
350	0.0665	2827.4	3093.3	6.5842		350	0.0584	2818.6	3081.5	6.5153
400	0.0734	2920.7	3214.5	6.7714		400	0.0648	2914.2	3205.7	6.7070
450	0.0800	3011.0	3331.2	6.9386		450	0.0708	3005.8	3324.2	6.8770
500	0.0864	3100.3	3446.0	7.0922		500	0.0765	3096.0	3440.4	7.0323
550	0.0927	3189.5	3560.3	7.2355		550	0.0821	3186.0	3555.6	7.1767
600	0.0989	3279.4	3674.9	7.3705		600	0.0877	3276.4	3670.9	7.3127
650	0.1049	3370.3	3790.1	7.4988		650	0.0931	3367.7	3786.6	7.4416
700	0.1110	3462.4	3906.3	7.6214		700	0.0985	3460.0	3903.3	7.5646
750	0.1170	3555.8	4023.6	7.7390		750	0.1038	3553.7	4021.0	7.6826
800	0.1229	3650.6	4142.3	7.8523		800	0.1092	3648.8	4140.0	7.7962
850	0.1289	3747.0	4262.4	7.9616		850	0.1145	3745.3	4260.3	7.9057
900	0.1348	3844.8	4383.9	8.0674		900	0.1197	3843.3	4382.1	8.0118
950	0.1406	3944.2	4506.8	8.1701		950	0.1250	3942.8	4505.2	8.1146
1000	0.1465	4045.1	4631.2	8.2697		1000	0.1302	4043.9	4629.8	8.2144

	$P = 5.0$ MPa					$P = 6.0$ MPa				
T (°C)	$\hat{V}$ (m³/kg)	$\hat{U}$ (kJ/kg)	$\hat{H}$ (kJ/kg)	$\hat{S}$ (kJ/kg·K)	T (°C)	$\hat{V}$ (m³/kg)	$\hat{U}$ (kJ/kg)	$\hat{H}$ (kJ/kg)	$\hat{S}$ (kJ/kg·K)
263.941	0.0394	2597.0	2794.2	5.9737	275.585	0.0324	2589.9	2784.6	5.8901
50					50				
100					100				
150					150				
200					200				
250					250				
300	0.0453	2699.0	2925.7	6.2110	300	0.0362	2668.4	2885.5	6.0703
350	0.0520	2809.5	3069.3	6.4516	350	0.0423	2790.4	3043.9	6.3357
400	0.0578	2907.5	3196.7	6.6483	400	0.0474	2893.7	3178.2	6.5432
450	0.0633	3000.6	3317.2	6.8210	450	0.0522	2989.9	3302.9	6.7219
500	0.0686	3091.7	3434.7	6.9781	500	0.0567	3083.1	3423.1	6.8826
550	0.0737	3182.4	3550.9	7.1237	550	0.0610	3175.2	3541.3	7.0307
600	0.0787	3273.3	3666.8	7.2605	600	0.0653	3267.2	3658.7	7.1693
650	0.0836	3365.0	3783.2	7.3901	650	0.0694	3359.6	3776.2	7.3001
700	0.0885	3457.7	3900.3	7.5136	700	0.0735	3453.0	3894.3	7.4246
750	0.0934	3551.6	4018.4	7.6320	750	0.0776	3547.5	4013.2	7.5438
800	0.0982	3646.9	4137.7	7.7458	800	0.0816	3643.2	4133.1	7.6582
850	0.1029	3743.6	4258.3	7.8556	850	0.0857	3740.3	4254.2	7.7685
900	0.1077	3841.8	4380.2	7.9618	900	0.0896	3838.8	4376.6	7.8751
950	0.1124	3941.5	4503.6	8.0648	950	0.0936	3938.7	4500.3	7.9784
1000	0.1171	4042.6	4628.3	8.1648	1000	0.0976	4040.1	4625.4	8.0786

P = 7.0 MPa

T (°C)	V̂ (m³/kg)	Û (kJ/kg)	Ĥ (kJ/kg)	Ŝ (kJ/kg·K)
285.829	0.0274	2581.0	2772.6	5.8148
50				
100				
150				
200				
250				
300	0.0295	2633.5	2839.9	5.9337
350	0.0353	2770.1	3016.9	6.2304
400	0.0400	2879.5	3159.2	6.4502
450	0.0442	2979.0	3288.3	6.6353
500	0.0482	3074.3	3411.4	6.8000
550	0.0520	3167.9	3531.6	6.9506
600	0.0557	3260.9	3650.6	7.0910
650	0.0593	3354.3	3769.3	7.2231
700	0.0629	3448.3	3888.2	7.3486
750	0.0664	3543.3	4007.9	7.4685
800	0.0699	3639.5	4128.4	7.5836
850	0.0733	3736.9	4250.1	7.6944
900	0.0767	3835.7	4373.0	7.8014
950	0.0802	3935.9	4497.1	7.9050
1000	0.0836	4037.5	4622.5	8.0055

P = 8.0 MPa

T (°C)	V̂ (m³/kg)	Û (kJ/kg)	Ĥ (kJ/kg)	Ŝ (kJ/kg·K)
295.008	0.0235	2570.5	2758.7	5.7450
50				
100				
150				
200				
250				
300	0.0243	2592.3	2786.5	5.7937
350	0.0300	2748.3	2988.1	6.1321
400	0.0343	2864.6	3139.4	6.3658
450	0.0382	2967.8	3273.3	6.5579
500	0.0418	3065.4	3399.5	6.7266
550	0.0452	3160.5	3521.8	6.8799
600	0.0485	3254.7	3642.4	7.0221
650	0.0517	3348.9	3762.3	7.1556
700	0.0548	3443.6	3882.2	7.2821
750	0.0579	3539.1	4002.6	7.4028
800	0.0610	3635.7	4123.8	7.5184
850	0.0641	3733.5	4246.0	7.6297
900	0.0671	3832.6	4369.3	7.7371
950	0.0701	3933.1	4493.8	7.8411
1000	0.0731	4035.0	4619.6	7.9419

| | P = 9 MPa | | | | | P = 10 MPa | | | |
T (°C)	V̂ (m³/kg)	Û (kJ/kg)	Ĥ (kJ/kg)	Ŝ (kJ/kg·K)	T (°C)	V̂ (m³/kg)	Û (kJ/kg)	Ĥ (kJ/kg)	Ŝ (kJ/kg·K)
303.345	0.0205	2558.5	2742.9	5.6791	310.997	0.0180	2545.2	2725.5	5.616
50					50				
100					100				
150					150				
200					200				
250					250				
300					300				
350	0.0258	2724.9	2957.3	6.0380	350	0.0224	2699.6	2924.0	5.9459
400	0.0300	2849.2	3118.8	6.2876	400	0.0264	2833.1	3097.4	6.2141
450	0.0335	2956.3	3258.0	6.4872	450	0.0298	2944.5	3242.3	6.4219
500	0.0368	3056.3	3387.4	6.6603	500	0.0328	3047.0	3375.1	6.5995
550	0.0399	3153.0	3512.0	6.8164	550	0.0357	3145.4	3502.0	6.7585
600	0.0429	3248.4	3634.1	6.9605	600	0.0384	3242.0	3625.8	6.9045
650	0.0458	3343.4	3755.2	7.0953	650	0.0410	3337.9	3748.1	7.0408
700	0.0486	3438.8	3876.1	7.2229	700	0.0436	3434.0	3870.0	7.1693
750	0.0514	3534.9	3997.3	7.3443	750	0.0461	3530.7	3992.0	7.2916
800	0.0541	3632.0	4119.1	7.4606	800	0.0486	3628.2	4114.5	7.4085
850	0.0569	3730.2	4241.9	7.5724	850	0.0511	3726.8	4237.8	7.5207
900	0.0596	3829.6	4365.7	7.6802	900	0.0535	3826.5	4362.0	7.6290
950	0.0622	3930.3	4490.6	7.7844	950	0.0560	3927.5	4487.3	7.7335
1000	0.0649	4032.4	4616.7	7.8855	1000	0.0584	4029.9	4613.8	7.8349

P = 15 MPa

T (°C)	V̂ (m³/kg)	Û (kJ/kg)	Ĥ (kJ/kg)	Ŝ (kJ/kg · K)
342.155	0.0103	2455.6	2610.7	5.3106
50				
100				
150				
200				
250				
300				
350	0.0115	2520.9	2693.1	5.4437
400	0.0157	2740.6	2975.7	5.8819
450	0.0185	2880.7	3157.9	6.1434
500	0.0208	2998.4	3310.8	6.3480
550	0.0229	3106.2	3450.4	6.5230
600	0.0249	3209.3	3583.1	6.6796
650	0.0268	3310.1	3712.1	6.8233
700	0.0286	3409.8	3839.1	6.9572
750	0.0304	3509.4	3965.2	7.0836
800	0.0321	3609.2	4091.1	7.2037
850	0.0338	3709.8	4217.1	7.3185
900	0.0355	3811.2	4343.7	7.4288
950	0.0372	3913.6	4471.0	7.5350
1000	0.0388	4017.1	4599.2	7.6378

P = 20 MPa

T (°C)	V̂ (m³/kg)	Û (kJ/kg)	Ĥ (kJ/kg)	Ŝ (kJ/kg · K)
365.749	0.00587	2295.0	2412.3	4.9314
50				
100				
150				
200				
250				
300				
350				
400	0.00995	2617.9	2816.9	5.5525
450	0.01272	2807.2	3061.7	5.9043
500	0.01479	2945.3	3241.2	6.1446
550	0.01657	3064.7	3396.1	6.3389
600	0.01818	3175.3	3539.0	6.5075
650	0.01969	3281.4	3675.3	6.6593
700	0.02113	3385.1	3807.8	6.7990
750	0.02252	3487.7	3938.1	6.9297
800	0.02387	3590.1	4067.5	7.0531
850	0.02519	3692.6	4196.4	7.1705
900	0.02648	3795.7	4325.4	7.2829
950	0.02776	3899.5	4454.7	7.3909
1000	0.02902	4004.3	4584.7	7.4950

| | $P = 25$ MPa | | | | | $P = 30$ MPa | | | |
T (°C)	$\hat{V}$ (m³/kg)	$\hat{U}$ (kJ/kg)	$\hat{H}$ (kJ/kg)	$\hat{S}$ (kJ/kg·K)	T (°C)	$\hat{V}$ (m³/kg)	$\hat{U}$ (kJ/kg)	$\hat{H}$ (kJ/kg)	$\hat{S}$ (kJ/kg·K)
50					50				
100					100				
150					150				
200					200				
250					250				
300					300				
350					350				
400	0.00600	2428.5	2578.6	5.1400	400	0.00280	2068.9	2152.8	4.4757
450	0.00918	2721.2	2950.6	5.6759	450	0.00674	2618.9	2821.0	5.4421
500	0.01114	2887.3	3165.9	5.9642	500	0.00869	2824.0	3084.7	5.7956
550	0.01274	3020.8	3339.2	6.1816	550	0.01018	2974.5	3279.7	6.0402
600	0.01414	3140.0	3493.5	6.3637	600	0.01144	3103.4	3446.7	6.2373
650	0.01543	3251.9	3637.7	6.5242	650	0.01259	3221.7	3599.4	6.4074
700	0.01664	3359.9	3776.0	6.6702	700	0.01365	3334.3	3743.9	6.5598
750	0.01780	3465.8	3910.9	6.8054	750	0.01466	3443.6	3883.4	6.6997
800	0.01892	3570.7	4043.8	6.9322	800	0.01563	3551.2	4020.0	6.8300
850	0.02001	3675.4	4175.6	7.0523	850	0.01656	3658.0	4154.9	6.9529
900	0.02108	3780.2	4307.1	7.1668	900	0.01747	3764.6	4288.8	7.0695
950	0.02212	3885.5	4438.5	7.2765	950	0.01836	3871.4	4422.3	7.1810
1000	0.02315	3991.5	4570.2	7.3820	1000	0.01924	3978.6	4555.8	7.2880

P = 40 MPa

T (°C)	V̂ (m³/kg)	Û (kJ/kg)	Ĥ (kJ/kg)	Ŝ (kJ/kg·K)
50				
100				
150				
200				
250				
300				
350				
400	0.00191	1854.9	1931.4	4.1145
450	0.00369	2364.2	2511.8	4.9448
500	0.00562	2681.6	2906.5	5.4744
550	0.00698	2875.0	3154.4	5.7857
600	0.00809	3026.8	3350.4	6.0170
650	0.00905	3159.5	3521.6	6.2078
700	0.00993	3282.0	3679.1	6.3740
750	0.01075	3398.6	3828.4	6.5236
800	0.01152	3511.8	3972.6	6.6612
850	0.01226	3623.1	4113.6	6.7896
900	0.01298	3733.3	4252.5	6.9106
950	0.01368	3843.1	4390.2	7.0256
1000	0.01436	3952.9	4527.3	7.1355

P = 50 MPa

T (°C)	V̂ (m³/kg)	Û (kJ/kg)	Ĥ (kJ/kg)	Ŝ (kJ/kg·K)
50				
100				
150				
200				
250				
300				
350				
400	0.00173	1787.8	1874.4	4.0029
450	0.00249	2160.3	2284.7	4.5896
500	0.00389	2528.1	2722.6	5.1762
550	0.00512	2769.5	3025.3	5.5563
600	0.00611	2947.1	3252.5	5.8245
650	0.00696	3095.6	3443.4	6.0373
700	0.00772	3228.7	3614.6	6.2178
750	0.00842	3353.1	3773.9	6.3775
800	0.00907	3472.2	3925.8	6.5225
850	0.00970	3588.0	4072.9	6.6565
900	0.01030	3702.0	4216.8	6.7819
950	0.01088	3814.9	4358.7	6.9004
1000	0.01144	3927.3	4499.4	7.0131

P = 60 MPa

T (°C)	$\hat{V}$ (m³/kg)	$\hat{U}$ (kJ/kg)	$\hat{H}$ (kJ/kg)	$\hat{S}$ (kJ/kg·K)
50				
100				
150				
200				
250				
300				
350				
400	0.00163	1745.2	1843.2	3.9317
450	0.00209	2055.1	2180.2	4.4140
500	0.00295	2393.2	2570.3	4.9356
550	0.00396	2664.5	2901.9	5.3517
600	0.00483	2866.8	3156.8	5.6527
650	0.00559	3031.3	3366.7	5.8867
700	0.00626	3175.4	3551.3	6.0814
750	0.00688	3307.6	3720.5	6.2510
800	0.00746	3432.6	3880.0	6.4033
850	0.00800	3553.2	4033.1	6.5428
900	0.00852	3670.9	4182.0	6.6725
950	0.00902	3786.9	4328.1	6.7944
1000	0.00950	3901.9	4472.2	6.9099

P = 80 MPa

T (°C)	$\hat{V}$ (m³/kg)	$\hat{U}$ (kJ/kg)	$\hat{H}$ (kJ/kg)	$\hat{S}$ (kJ/kg·K)
50				
100				
150				
200				
250				
300				
350				
400	0.00152	1687.5	1808.8	3.8340
450	0.00177	1945.9	2087.8	4.2335
500	0.00219	2222.4	2397.4	4.6473
550	0.00276	2489.1	2709.9	5.0391
600	0.00338	2717.4	2988.1	5.3674
650	0.00398	2907.5	3225.5	5.6321
700	0.00452	3071.4	3432.7	5.8507
750	0.00501	3218.6	3619.7	6.0382
800	0.00548	3355.1	3793.3	6.2038
850	0.00591	3484.8	3957.7	6.3537
900	0.00633	3609.9	4115.9	6.4915
950	0.00672	3732.0	4269.8	6.6199
1000	0.00711	3852.1	4420.5	6.7407

		$P = 100$ MPa			
T (°C)	$\hat{V}$ (m³/kg)	$\hat{U}$ (kJ/kg)	$\hat{H}$ (kJ/kg)	$\hat{S}$ (kJ/kg·K)	
50					
100					
150					
200					
250					
300					
350					
400	0.00144	1646.8	1791.1	3.7639	
450	0.00163	1881.9	2044.7	4.1271	
500	0.00189	2126.9	2316.2	4.4900	
550	0.00225	2371.0	2595.9	4.8405	
600	0.00267	2597.9	2865.1	5.1581	
650	0.00312	2798.9	3110.5	5.4315	
700	0.00355	2976.1	3330.7	5.6639	
750	0.00395	3135.2	3530.5	5.8642	
800	0.00434	3281.7	3715.3	6.0406	
850	0.00470	3419.5	3889.3	6.1991	
900	0.00504	3551.4	4055.6	6.3440	
950	0.00537	3679.1	4216.3	6.4782	
1000	0.00569	3804.0	4373.0	6.6038	

A-4 Compressed Liquid

	P = 500 kPa					P = 1.0 MPa			
T (°C)	V̂ (m³/kg)	Û (kJ/kg)	Ĥ (kJ/kg)	Ŝ (kJ/kg·K)	T (°C)	V̂ (m³/kg)	Û (kJ/kg)	Ĥ (kJ/kg)	Ŝ (kJ/kg·K)
151.831	0.001093	639.5	640.1	1.8604	179.878	0.001127	761.4	762.5	2.1381
0	0.001000	0.0	0.5	-0.0001	0	0.001000	0.0	1.0	-0.0001
20	0.001002	83.9	84.4	0.2964	20	0.001001	83.9	84.9	0.2963
40	0.001008	167.5	168.0	0.5722	40	0.001007	167.4	168.4	0.5720
60	0.001017	251.1	251.6	0.8310	60	0.001017	251.0	252.0	0.8308
80	0.001029	334.9	335.4	1.0753	80	0.001029	334.7	335.8	1.0750
100	0.001043	418.9	419.5	1.3069	100	0.001043	418.8	419.8	1.3065
120	0.001060	503.5	504.0	1.5276	120	0.001060	503.3	504.4	1.5272
140	0.001080	588.7	589.3	1.7391	140	0.001079	588.5	589.6	1.7386
160					160	0.001102	674.6	675.7	1.9421
180					180				
200					200				
220					220				
240					240				
260					260				
280					280				
300					300				
320					320				
340					340				
360					360				

P = 5 MPa

T (°C)	V̂ (m³/kg)	Û (kJ/kg)	Ĥ (kJ/kg)	Ŝ (kJ/kg·K)
263.941	0.001286	1148.2	1154.6	2.9210
0	0.000998	0.0	5.0	0.0001
20	0.001000	83.6	88.6	0.2954
40	0.001006	166.9	172.0	0.5705
60	0.001015	250.3	255.4	0.8287
80	0.001027	333.8	339.0	1.0723
100	0.001041	417.6	422.8	1.3034
120	0.001058	501.9	507.2	1.5236
140	0.001077	586.8	592.2	1.7344
160	0.001099	672.5	678.0	1.9374
180	0.001124	759.5	765.1	2.1338
200	0.001153	847.9	853.7	2.3251
220	0.001187	938.4	944.3	2.5127
240	0.001227	1031.6	1037.7	2.6983
260	0.001275	1128.5	1134.9	2.8841
280				
300				
320				
340				
360				

P = 10 MPa

T (°C)	V̂ (m³/kg)	Û (kJ/kg)	Ĥ (kJ/kg)	Ŝ (kJ/kg·K)
310.997	0.001453	1393.5	1408.1	3.3606
0	0.000995	0.1	10.1	0.0003
20	0.000997	83.3	93.3	0.2943
40	0.001003	166.3	176.4	0.5685
60	0.001013	249.4	259.6	0.8260
80	0.001024	332.7	342.9	1.0691
100	0.001038	416.2	426.6	1.2996
120	0.001055	500.2	510.7	1.5191
140	0.001074	584.7	595.5	1.7293
160	0.001095	670.1	681.0	1.9315
180	0.001120	756.5	767.7	2.1271
200	0.001148	844.3	855.8	2.3174
220	0.001181	934.0	945.8	2.5037
240	0.001219	1026.1	1038.3	2.6876
260	0.001265	1121.6	1134.3	2.8710
280	0.001323	1221.8	1235.0	3.0565
300	0.001398	1329.4	1343.3	3.2488
320				
340				
360				

P = 15 MPa

T (°C)	V̂ (m³/kg)	Û (kJ/kg)	Ĥ (kJ/kg)	Ŝ (kJ/kg·K)
342.155	0.001657	1585.3	1610.2	3.6846
0	0.000993	0.2	15.1	0.0004
20	0.000995	83.0	97.9	0.2932
40	0.001001	165.7	180.8	0.5666
60	0.001011	248.6	263.7	0.8234
80	0.001022	331.6	346.9	1.0659
100	0.001036	414.8	430.4	1.2958
120	0.001052	498.5	514.3	1.5148
140	0.001071	582.7	598.7	1.7243
160	0.001092	667.6	684.0	1.9259
180	0.001116	753.6	770.3	2.1206
200	0.001144	840.8	858.0	2.3100
220	0.001175	929.8	947.4	2.4951
240	0.001212	1021.0	1039.2	2.6774
260	0.001256	1115.1	1134.0	2.8586
280	0.001310	1213.4	1233.0	3.0409
300	0.001378	1317.6	1338.3	3.2279
320	0.001473	1431.9	1454.0	3.4263
340	0.001631	1567.9	1592.4	3.6555
360				

P = 20 MPa

T (°C)	V̂ (m³/kg)	Û (kJ/kg)	Ĥ (kJ/kg)	Ŝ (kJ/kg·K)
365.749	0.002040	1786.4	1827.2	4.0156
0	0.000990	0.2	20.0	0.0005
20	0.000993	82.7	102.6	0.2921
40	0.000999	165.2	185.2	0.5646
60	0.001008	247.8	267.9	0.8208
80	0.001020	330.5	350.9	1.0627
100	0.001034	413.5	434.2	1.2920
120	0.001050	496.8	517.9	1.5105
140	0.001068	580.7	602.1	1.7194
160	0.001089	665.3	687.0	1.9203
180	0.001112	750.8	773.0	2.1143
200	0.001139	837.5	860.3	2.3027
220	0.001170	925.8	949.2	2.4867
240	0.001205	1016.1	1040.2	2.6676
260	0.001247	1109.0	1134.0	2.8469
280	0.001298	1205.5	1231.5	3.0265
300	0.001361	1307.1	1334.4	3.2091
320	0.001445	1416.6	1445.5	3.3996
340	0.001569	1540.2	1571.6	3.6086
360	0.001825	1703.6	1740.1	3.8787

P = 25 MPa

T (°C)	V̂ (m³/kg)	Û (kJ/kg)	Ĥ (kJ/kg)	Ŝ (kJ/kg·K)
0	0.000988	0.3	25.0	0.0004
20	0.000991	82.4	107.2	0.2909
40	0.000997	164.6	189.5	0.5627
60	0.001006	246.9	272.1	0.8182
80	0.001018	329.4	354.9	1.0595
100	0.001031	412.2	438.0	1.2883
120	0.001047	495.2	521.4	1.5062
140	0.001065	578.8	605.4	1.7146
160	0.001085	663.0	690.1	1.9148
180	0.001108	748.0	775.8	2.1081
200	0.001135	834.2	862.6	2.2956
220	0.001164	921.9	951.0	2.4786
240	0.001199	1011.4	1041.3	2.6582
260	0.001239	1103.2	1134.2	2.8357
280	0.001287	1198.3	1230.5	3.0129
300	0.001346	1297.6	1331.3	3.1919
320	0.001421	1403.4	1438.9	3.3764
340	0.001526	1519.4	1557.5	3.5731
360	0.001697	1656.2	1698.6	3.7993

P = 50 MPa

T (°C)	V̂ (m³/kg)	Û (kJ/kg)	Ĥ (kJ/kg)	Ŝ (kJ/kg·K)
0	0.000977	0.3	49.1	−0.0010
20	0.000980	80.9	130.0	0.2845
40	0.000987	161.9	211.3	0.5528
60	0.000996	243.1	292.9	0.8054
80	0.001007	324.4	374.8	1.0442
100	0.001020	405.9	456.9	1.2705
120	0.001035	487.7	539.4	1.4859
140	0.001052	569.8	622.4	1.6916
160	0.001070	652.3	705.8	1.8889
180	0.001091	735.5	790.1	2.0790
200	0.001115	819.4	875.2	2.2628
220	0.001141	904.4	961.4	2.4414
240	0.001171	990.6	1049.1	2.6156
260	0.001204	1078.2	1138.4	2.7864
280	0.001243	1167.7	1229.9	2.9547
300	0.001288	1259.6	1324.0	3.1218
320	0.001341	1354.3	1421.4	3.2888
340	0.001405	1452.9	1523.1	3.4575
360	0.001485	1556.5	1630.7	3.6301

		$P = 100$ MPa		
T (°C)	$\hat{V}$ (m³/kg)	$\hat{U}$ (kJ/kg)	$\hat{H}$ (kJ/kg)	$\hat{S}$ (kJ/kg · K)
0	0.000957	−0.3	95.4	−0.0085
20	0.000962	78.0	174.2	0.2699
40	0.000969	157.0	253.9	0.5328
60	0.000978	236.2	334.0	0.7809
80	0.000988	315.6	414.5	1.0153
100	0.001000	395.1	495.1	1.2375
120	0.001014	474.6	576.0	1.4487
140	0.001028	554.4	657.2	1.6501
160	0.001045	634.3	738.8	1.8429
180	0.001063	714.5	820.8	2.0280
200	0.001083	795.1	903.4	2.2064
220	0.001104	876.3	986.7	2.3788
240	0.001128	958.0	1070.8	2.5459
260	0.001154	1040.3	1155.8	2.7084
280	0.001183	1123.5	1241.8	2.8669
300	0.001215	1207.6	1329.1	3.0219
320	0.001250	1292.8	1417.8	3.1740
340	0.001290	1379.1	1508.2	3.3238
360	0.001335	1466.8	1600.3	3.4717

Appendix B: Mathematical Techniques

B.1 Linear Interpolation

Appendix A contains the steam tables, which provide physical properties of water as functions of temperature and pressure. The quantities in the steam tables (specific volume, specific internal energy, specific enthalpy, and specific entropy) are not linear functions of temperature and pressure for water or any other compound. However, the increments between data points published in the steam tables are small enough that it is acceptably accurate (and standard practice) to use linear interpolation between published data points.

Thus, for example, if you wish to know the specific enthalpy $\hat{H}$ at a temperature T which falls between T_1 and T_2, this could be estimated as

$$\hat{H} = \hat{H}_1 + (\hat{H}_2 - \hat{H}_1)\left(\frac{T - T_1}{T_2 - T_1}\right)$$

with $\hat{H}_1$ and $\hat{H}_2$ representing the known values of the specific enthalpy at T_1 and T_2.

This assumes the pressure of interest is explicitly given in the steam tables. An analogous equation can be used if data is available at the temperature of interest, but the pressure of interest P falls between two known pressures P_1 and P_2.

$$\hat{H} = \hat{H}_1 + (\hat{H}_2 - \hat{H}_1)\left(\frac{P - P_1}{P_2 - P_1}\right)$$

> Linear interpolation is just the point-slope formula for a line you learned many years ago, with $\hat{H}$ playing the role of "y" and T playing the role of "x"

B.2 Solution of Cubic Equations

Cubic equations can be solved analytically. The solution as presented here is adapted from (Kaw, 2011), which shows the derivations in detail. The cubic equation is expressed in the form

$$ax^3 + bx^2 + cx + d = 0$$

where a, b, c, and d are real numbers.

A series of three substitutions is used to convert the cubic equation into a quadratic equation. The substitutions are

$$x = y - \frac{b}{3a}$$

$$y = z + \frac{s}{z}$$

$$w = z^3$$

733

The end result of these substitutions is

$$w^2 + fw - \frac{e^3}{27} = 0$$

with e and f given by

$$e = \frac{1}{a}\left(c - \frac{b^2}{3a}\right)$$

$$f = \frac{1}{a}\left(d + \frac{2b^3}{27a^2} - \frac{bc}{3a}\right)$$

And s, which appears in one of the substitutions, is equal to

$$s = \frac{-e}{3}$$

Since a, b, c, and d are all real numbers, this means e, f, and s are also real numbers. Consequently, the equation can be solved for w using the quadratic formula:

$$w = \frac{-f \pm \sqrt{f^2 + \frac{4e^3}{27}}}{2}$$

Thus, there are two solutions for w. Finding the solutions for the original variable x requires working backward through the three substitutions that were used to transform the original cubic equation into the quadratic equation. First solve for z using

$$w = z^3$$

There are in general two solutions for w, each of which has three corresponding solutions for z. If w is a real number, for example, then it will yield one real and two imaginary roots for z. We might be inclined to say "This is an equation of state, we are calculating the molar volume, an imaginary solution has no relevance." However, it is our original variable (x) that represents a solution to the equation of state, and imaginary values of z can produce real values of x, so we cannot eliminate imaginary solutions at this point.

Next y is found by solving

$$y = z + \frac{s}{z}$$

Each value of z will give one solution for y. However, it can be proved (Kaw, 2011) that the six values of y will come in three pairs of identical solutions. Thus, there will be three *distinct* values of y, and the three roots of the original equation are found from these using

$$x = y - \frac{b}{3a}$$

Kaw (2011) gives two worked examples showing this solution procedure.[1]

---

[1]Kaw, A. " Solution of Cubic Equations" http://numericalmethods.eng.usf.edu (Holistic Numerical Methods: Transforming Numerical Methods Education for the STEM Undergraduate), 2011, accessed May 31, 2012, available at http://numericalmethods.eng.usf.edu/mws/gen/03nle/mws_gen_nle_bck_exactcubic.pdf

Appendix C: Physical Properties

C.1 Critical Point, Enthalpy of Phase Change, and Liquid Molar Volume

Values of enthalpy of vaporization ($\Delta \underline{H}^{\text{vap}}$) and enthalpy of fusion ($\Delta \underline{H}^{\text{fus}}$) in this table are valid at **atmospheric pressure**. The liquid molar volume ($\underline{V}^{\text{L}}$) is measured at the temperature (T) indicated in the rightmost column.

Name	Formula	T_c (K)	P_c (MPa)	ω	$\Delta \underline{H}^{\text{vap}}$ (kJ/mol)	$\Delta \underline{H}^{\text{fus}}$ (kJ/mol)	$\underline{V}^{\text{L}}$ (cm³/mol)	T (K) for $\underline{V}^{\text{L}}$
Argon	Ar	150.86	4.898	−0.002	6.43		29.10	90.00
Tetrachloromethane (carbon tetrachloride)	CCl_4	556.3	4.557	0.194[a]	29.82	3.28	97.07	298.15
Tetrafluoromethane	CF_4	227.51	3.745	0.177			56.41	153.15
Chlorodifloromethane (R-22)	$CHClF_2$	369.28	4.986	0.221	20.22	4.12	59.08	213.15
Trichloromethane (chloroform)	$CHCl_3$	536.50	5.500	0.216[a]	29.24	8.80	80.68	298.15
Methane	CH_4	190.56	4.599	0.011	8.17	0.94	35.54	90.68
Methanol	CH_3OH	512.64	8.097	0.565	35.21	3.18	40.73	298.15
Carbon monoxide	CO	132.85	3.494	0.045	6.04	0.84	34.88	81.00
Carbon dioxide	CO_2	304.12	7.374	0.225		9.02		
Pentafluoroethane	C_2HF_5	339.17	3.615	0.305			98.61	293.48
Ethyne (acetylene)	C_2H_2	308.30	6.114	0.189		21.28	43.47	203.15
1,1,1-Trifluoroethane	$C_2H_3F_3$	346.30	3.792	0.259	18.99	6.19	75.38	245.00
Acetonitrile	C_2H_3N	545.5[a]	4.833[a]	0.353[a]				
Ethene (ethylene)	C_2H_4	282.34	5.041	0.087	13.53	3.35	51.07	183.15
1,2-Dichloroethane	$C_2H_4Cl_2$	523.00	5.100		28.85	7.87	84.73	298.15
Ethane	C_2H_6	305.32	4.872	0.099	14.70	2.86	46.15	90.36
Ethanol	C_2H_6O	513.92	6.148	0.649	38.56	5.01	58.68	298.15
Propene (propylene)	C_3H_6	364.90	4.600	0.142	18.42	3.00		
Propanone (acetone)	C_3H_6O	508.10	4.700	0.307	29.10	5.69	73.94	298.15
Dimethylcarbonate	$C_3H_6O_3$	557.00	4.800	0.336	37.7		84.82	298.15
Propane	C_3H_8	369.83	4.248	0.152	19.04	3.53	74.87	233.15
1-Propanol	C_3H_8O	536.78	5.175	0.629	41.44	5.20	75.14	298.15
2-Propanol	C_3H_8O	508.30	4.762	0.665	39.85	5.38	76.92	298.15

(Continued)

Name	Formula	T_c (K)	P_c (MPa)	ω	ΔH^{vap} (kJ/mol)	ΔH^{fus} (kJ/mol)	V^L (cm³/mol)	T (K) for V^L
1,3-Butadiene	C_4H_6	425.00	4.320	0.195	22.47	7.98	88.04	298.15
1-Butene	C_4H_8	419.50	4.020	0.194	22.07	3.96	95.34	298.15
Butanone (methyl ethyl ketone)	C_4H_8O	536.80	4.210	0.322	31.30	8.44	90.13	298.15
n-Butane	C_4H_{10}	425.12	3.796	0.200	22.44	4.66	100.48	298.15
2-Methylpropane (isobutane)	C_4H_{10}	407.85	3.640	0.186	21.30	4.61	104.36	298.15
1-Butanol	$C_4H_{10}O$	563.05	4.423	0.590	43.29	9.28	91.96	298.15
Diethyl ether	$C_4H_{10}O$	466.70	3.640	0.281	26.52	7.27	104.75	298.15
n-Pentane	C_5H_{12}	469.70	3.370	0.252	25.79	8.40	115.22	298.15
Butyl methyl ether	$C_5H_{12}O$	512.8[b]	3.371[c]	0.316[d]				
Hexafluorobenzene	C_6F_6	516.73	3.275	0.396	31.66			
Chlorobenzene	C_6H_5Cl	632.40	4.520	0.251	35.19	9.61	102.22	298.15
Benzene	C_6H_6	562.05	4.895	0.210	30.72	9.95	89.41	298.15
Phenol	C_6H_6O	694.25	6.130	0.442	46.18	11.29	87.87	298.15
Cyclohexane	C_6H_{12}	553.50	4.073	0.211	29.97	2.63	108.75	298.15
1-Hexene	C_6H_{12}	504.00	3.143	0.281	28.28	7.52	125.90	298.15
n-Hexane	C_6H_{14}	507.60	3.025	0.300	28.85	13.07	131.59	298.15
Toluene	C_7H_8	591.75	4.108	0.264	33.18	6.85	106.87	298.15
n-Heptane	C_7H_{16}	540.20	2.740	0.350	31.77	14.03	147.47	298.15
Ethylbenzene	C_8H_{10}	617.15	3.609	0.304	35.57	9.18	123.08	298.15
o-Xylene	C_8H_{10}	630.30	3.732	0.312	36.24	13.60	121.25	298.15
m-Xylene	C_8H_{10}	617.00	3.541	0.327	35.66	11.57	123.47	298.15
p-Xylene	C_8H_{10}	616.20	3.511	0.322	35.67	16.81	123.93	298.15
n-Octane	C_8H_{18}	568.70	2.490	0.399	34.41	20.65	163.53	298.15
n-Nonane	C_9H_{20}	594.60	2.290	0.445	36.91	15.50	179.70	298.15
Napthalene	$C_{10}H_8$	748.40	4.050	0.304	43.40	19.12	129.13	333.15
n-Decane	$C_{10}H_{22}$	617.70	2.110	0.490	38.75	28.78	195.95	298.15
Chlorine	Cl_2	417.00	7.700	0.069[a]	20.41		45.36	239.00
Hydrogen	H_2	33.25	1.297	−0.216	0.89	0.12	29.39	20.00
Water	H_2O	647.14	22.064	0.344	40.66	6.01	18.07	298.15
Nitrogen	N_2	126.2	3.398	0.037	5.58	0.72	34.84	78.00
Ammonia	NH_3	406.6[a]	11.27[a]	0.252[a]			25.0[e]	240[e]
Oxygen	O_2	154.58	5.043	0.022[a]	6.82	0.44	27.85	90.00

Based on data from B. E. Poling, J. M. Prausnitz, J. P. O'Connell, The Properties of Gases and Liquids. 5th edition, McGraw-Hill, New York, 2001, except as noted.

[a]Elliott, J. R. and Lira, C. T., Introductory Chemical Engineering Thermodynamics, Prentice Hall, Upper Saddle River, NJ, 1999.

[b]Majer, V.; Svoboda, V., Enthalpies of Vaporization of Organic Compounds: A Critical Review and Data Compilation, Blackwell Scientific Publications, Oxford, 1985, 300.

[c]Ambrose, D.; Broderick, B.E.; Townsend, R., The Critical Temperatures and Pressures of Thirty Organic Compounds, J. Appl. Chem. Biotechnol., 1974, 24, 359

[d]Chemical Engineering Research Information Center, retrieved from http://www.cheric.org/kdb/kdb/hcprop/showprop.php?cmpid=1005 on December 10, 2012.

[e]Perry, R. H., Green, D. W. Perry's Chemical Engineers Handbook, 7th ed., McGraw-Hill, 1997.

C.2 Standard Enthalpy and Gibbs Energy of Formation

The data in this table is at a standard state of $T = 25°C$ (298.15 K) and $P = 100$ kPa.

Name	Formula	$\Delta \underline{H}_f$ (kJ/mol)	$\Delta \underline{G}_f$ (kJ/mol)	Source
Diamond	C (s)	1.90	2.90	1
Graphite	C (s)	0	0	1
Carbon tetrachloride	CCl_4 (g)	−95.7	−60.63	1
Carbon monoxide	CO	−110.53	−137.16	3
Carbon dioxide	CO_2	−393.51	−394.38	3
Methanol	CH_3OH (g)	−201.0	−162.3	1
	CH_3OH (l)	−239.2	−166.6	1
Methane	CH_4 (g)	−74.6	−50.5	1
Ethyne (acetylene)	C_2H_2 (g)	227.4	209.9	1
Ethene (ethylene)	C_2H_4 (g)	52.4	68.4	1
Acetaldehyde	C_2H_4O (g)	−166.47	−133.39	2
Acetic acid	$C_2H_4O_2$ (g)	−432.2	−374.2	1
	$C_2H_4O_2$ (l)	−484.3	−389.9	1
Ethane	C_2H_6 (g)	−84.0	−32.0	1
Ethanol	C_2H_6O (g)	−234.8	−167.9	1
	C_2H_6O (g) (l)	−277.6	−174.8	1
Propene (propylene)	C_3H_6 (g)	20.43	62.76	2
Acetone	C_3H_6O (g)	−217.1	−152.7	1
	C_3H_6O (l)	−248.4	−155.7	1
Propane	C_3H_8 (g)	−103.8	−23.4	1
1,3-Butadiene	C_4H_6 (l)	85.41	149.68	2
	C_4H_6 (g)	110.24	150.77	2
1-Butene	C_4H_8 (g)	−0.54	70.24	3
n-Butane	C_4H_{10} (l)	−147.75	−15.07	2
	C_4H_{10} (g)	−126.23	−17.17	2
Isobutane	C_4H_{10} (l)	−158.55	−21.98	2
	C_4H_{10} (g)	−134.61	−20.89	2
n-Pentane	C_5H_{12} (l)	−173.33	−9.46	2
	C_5H_{12} (g)	−146.54	−8.37	2
Benzene	C_6H_6 (g)	82.9	129.7	1
	C_6H_6 (l)	49.1	124.5	1

(*Continued*)

Name	Formula	$\Delta \underline{H}_f$ (kJ/mol)	$\Delta \underline{G}_f$ (kJ/mol)	Source
Cyclohexane	C_6H_{12} (g)	−123.22	31.78	2
	C_6H_{12} (l)	−156.34	26.89	2
n-Hexane	C_6H_{14} (l)	−198.96	−4.35	2
	C_6H_{14} (g)	−167.30	−0.25	2
Toluene	C_7H_8 (l)	12.02	113.84	2
	C_7H_8 (g)	50.03	122.09	2
n-Heptane	C_7H_{16} (l)	−224.54	1.00	2
	C_7H_{16} (g)	−187.90	8.00	2
o-Xylene	C_8H_{10} (l)	19	122.22	3
m-Xylene	C_8H_{10} (l)	17.24	119	3
p-Xylene	C_8H_{10} (l)	17.95	121.26	3
n-Octane	C_8H_{18} (l)	−250.12	6.49	2
	C_8H_{18} (g)	−208.59	16.41	2
n-Nonane	C_9H_{20} (l)	−275.66	11.76	2
	C_9H_{20} (g)	−229.19	24.83	2
n-Decane	$C_{10}H_{22}$ (l)	−301.24	17.25	2
	$C_{10}H_{22}$ (g)	−249.83	33.24	2
Hydrogen	H_2 (g)	0	0	1
Water	H_2O (g)	−241.826	−228.6	1
	H_2O (l)	−285.830	−237.2	1
Hydrogen sulfide	H_2S (g)	−20.6	−33.4	1
Nitrogen	N_2 (g)	0	0	1
Ammonia	NH_3	−46.11	−16.45	2
Nitric oxide	NO (g)	91.3	87.6	1
Nitrous oxide	N_2O (g)	82.05	104.2	1
Nitrogen dioxide	NO_2 (g)	33.2	51.3	1
Dinitrogen tetroxide	N_2O_4 (g)	11.1	97.82	1
	N_2O_4 (l)	−19.5	97.40	1
Oxygen	O_2 (g)	0	0	1
Ozone	O_3 (g)	142.7	163.2	1
Sulfur dioxide	SO_2 (g)	−296.81	−300.1	1
Sulfur trioxide	SO_3 (g)	−395.77	−371.02	2

[1] Laidler, Meiser and Sanctuary, Physical Chemistry, 4th ed., CENGAGE Learning, 2003.
[2] Korestski, Engineering and Chemical Thermodynamics, John Wiley and Sons, 2004.
[3] Elliot and Lira, Introductory Chemical Engineering Thermodynamics, Prentice Hall, Upper Saddle River, NJ, 1999.

Appendix D: Heat Capacity

D.1 Ideal Gas Heat Capacity

The coefficients in the table below are used to compute the ideal gas heat capacity of a compound as a function of temperature, through

$$\frac{C_P^*}{R} = A + BT + CT^2 + DT^3 + ET^4$$

where T is expressed in Kelvin. The coefficients were obtained from Poling (2001), which includes a much more extensive list of compounds.

Name	Formula	A	B × 10³	C × 10⁵	D × 10⁸	E × 10¹¹	T range (K)
Chlorodifluoromethane (R-22)	$CHClF_2$	3.164	10.422	1.179	−2.650	1.222	50–1000
Methane	CH_4	4.568	−8.975	3.631	−3.407	1.091	50–1000
Methanol	CH_4O	4.714	−6.986	4.211	−4.443	1.535	50–1000
Carbon monoxide	CO	3.912	−3.913	1.182	−1.302	0.515	50–1000
Carbon dioxide	CO_2	3.259	1.356	1.502	−2.374	1.056	50–1000
Ethylene (Ethene)	C_2H_4	4.221	−8.782	5.795	−6.792	2.511	50–1000
Ethane	C_2H_6	4.178	−4.427	5.660	−6.651	2.487	50–1000
Ethanol	C_2H_6O	4.396	0.628	5.546	−7.024	2.685	50–1000
Propylene	C_3H_6	3.834	3.893	4.688	−6.013	2.283	50–1000
Propanone (Acetone)	C_3H_6O	5.126	1.511	5.731	−7.177	2.728	200–1000
Propane	C_3H_8	3.847	5.131	6.011	−7.893	3.079	50–1000
1-Propanol	C_3H_8O	4.712	6.565	6.310	−8.341	3.216	50–1000
1,3-Butadiene	C_4H_6	3.607	5.085	8.253	−12.371	5.321	50–1000
1-Butene	C_4H_8	4.389	7.984	6.143	−8.197	3.165	50–1000
n-Butane	C_4H_{10}	5.547	5.536	8.057	−10.571	4.134	200–1000
Benzene	C_6H_6	3.551	−6.184	14.365	−19.807	8.234	50–1000
Toluene	C_7H_8	3.866	3.558	13.356	−18.659	7.690	50–1000
Chlorine	Cl_2	3.0560	5.3708	−0.8098	0.5693	−0.15256	50–1000
Fluorine	F_2	3.347	0.467	0.526	−0.794	0.330	50–1000
Hydrogen	H_2	2.883	3.681	−0.772	0.692	−0.213	50–1000
Water	H_2O	4.395	−4.186	1.405	−1.564	0.632	50–1000
Ammonia	NH_3	4.238	−4.215	2.041	−2.126	0.761	50–1000
Nitrogen	N_2	3.539	−0.261	0.007	0.157	−0.099	50–1000
Oxygen	O_2	3.630	−1.794	0.658	−0.601	0.179	50–1000

Based on data from Poling, B., Prausnitz, J. and O'Connell, J., The Properties of Gases and Liquids, 5th ed., McGraw-Hill, New York, NY, 2001.

D.2 Liquid Heat Capacities

The coefficients in the table below are used to compute the liquid heat capacity of a compound as a function of temperature, through the equation:

$$\frac{C_P}{R} = A + BT + CT^2 + DT^3$$

in which T is expressed in degrees Kelvin.

The "C_P/R" column gives the numerical value for the heat capacity at one specific temperature (T^*), and can be used when C_P is modeled as constant (normally only valid over small temperature intervals).

Compound	Structure	C_P/R (T^*)	T^* (°C)	A	$B \times 10^3$	$C \times 10^6$	$D \times 10^9$	T Range (°C)
Ammonia	NH_3	9.0049	−33.43	−16.4918	266.7154	−951.0848	1180.0655	−77.4 to 100
Water	H_2O	9.0717	25.00	6.1152	25.6276	−75.9393	78.0258	0 to 350
Methanol	CH_3OH	9.8094	25.00	13.5152	−52.0968	133.7651	−2.7234	−97.6 to 220
Carbon dioxide	CO_2	10.1943	−30.00	−427.4549	5638.8614	−24263.0507	34845.3288	−56.5 to 20
Ethanol	C_2H_5OH	12.8706	25.00	−7.7598	211.9910	−839.6255	1210.6486	−114.1 to 180
Propane	C_3H_8	11.8130	−42.10	7.3808	51.7499	−296.4745	669.2867	−187.7 to 80
1-Propanol	C_3H_7OH	17.2489	25.00	−8.3551	259.4291	−1034.9324	1508.5199	−126.2 to 200
Cyclopentane	C_5H_{10}	14.4727	25.00	−2.8635	146.2611	−464.2539	562.6786	−93.88 to 220
1-Butanol	C_4H_9OH	20.9018	25.00	−28.3008	483.8032	−1720.3550	2185.5074	−89.3 to 200
Benzene	C_6H_6	16.1264	25.00	−58.2161	607.7120	−1717.7887	1733.1190	5.53 to 250
Cyclohexane	C_6H_{12}	18.2231	25.00	−54.3817	567.1109	−1486.6013	1366.7414	6.55 to 260
Toluene	C_7H_8	18.3271	25.00	−6.7745	212.5565	−624.1298	660.7615	−95 to 310
Ethyl benzene	$C_6H_5C_2H_5$	22.3473	25.00	2.6400	144.8482	−385.2288	408.6845	−95 to 320
Chloroform	$CHCl_3$	13.5258	25.00	−5.4995	189.1846	−639.2266	744.9634	−63.2 to 250
Carbon tetrachloride	CCl_4	15.4914	25.00	−0.9506	159.3088	−544.9631	666.4961	−22.9 to 260

Based on data from J. W. Miller, Jr., G. R. Schoor and C. L. Yaws, "Heat Capacity of Liquids," Chemical Engineering, 83, 129 (1976).

Appendix E: Antoine Coefficients

The constants in the table below are used to model the relationship between vapor pressure and temperature, through

$$\log_{10} P^{\text{sat}} = A - \frac{B}{T + C}$$

where P^{sat} is expressed in mm Hg and T is expressed in degrees Celsius. Note that 1 mm Hg = 133.3 Pa = 0.1333 kPa.

Name	Structure	A	B	C	T Range
Acetic acid	CH_3COOH	7.38782	1533.313	22.309	Liquid
Acetone	$(CH_3)_2CO$	7.11714	1210.595	229.664	Liquid
Acetonitrile	CH_3CN	7.11988	1314.4	230	Liquid
Acetylene	C_2H_2	9.1402	1232.6	280.9	−130 to −83
		7.0999	711.0	253.4	−82 to −72
Benzene	C_6H_6	9.1064	1885.9	244.2	−12 to 3
		6.90565	1211.033	220.790	8 to 103
Benzyl alcohol	$C_6H_5CH_2OH$	7.19817	1632.593	172.790	122 to 205
Biphenyl	$(C_6H_5)_2$	7.24541	1998.725	202.733	69 to 271
Bromochloromethane	CH_2BrCl	6.49606	942.267	192.587	16 to 68
1,3-Butadiene	C_4H_6	7.03555	998.106	245.233	−80 to −62
		6.84999	930.546	238.854	−58 to 15
n-butane	C_4H_{10}	6.80896	935.86	238.73	−77 to 19
1-butanol	C_4H_9OH	7.47680	1362.39	178.77	15 to 131
2-butanol	C_4H_9OH	7.47431	1314.19	186.55	25 to 120
1-Butene	C_4H_8	6.79290	908.80	238.54	−82 to 13
Chloroethane	C_2H_5Cl	6.98647	1030.01	238.61	−56 to 12.2
Chloroform	$CHCl_3$	6.4934	929.44	196.03	−35 to 61
Chloromethane	CH_3Cl	7.09349	948.58	249.34	−75 to −5
Cyclohexane	C_6H_{12}	6.84130	1201.53	222.65	20 to 81
Diethyl ketone	$C_5H_{10}O$	6.85791	1216.3	204	
Ethanol	C_2H_5OH	8.32109	1718.10	237.52	−2 to 100

(Continued)

Name	Structure	A	B	C	T Range
Ethyl acetate	$CH_3COOCH_2CH_3$	7.10179	1244.95	217.88	15 to 76
Ethyl benzene	C_8H_{10}	6.95719	1424.255	213.21	26 to 164
n-Heptane	C_7H_{16}	6.89677	1264.90	216.54	−2 to 124
n-Hexane	C_6H_{14}	6.87601	1171.17	224.41	−25 to 92
Hexfluorobenzene	C_6F_6	7.03295	1227.98	215.49	5 to 114
Methanol	CH_3OH	7.89750	1474.08	229.13	−14 to 65
		7.97328	1515.14	232.85	64 to 110
2-Methyl-1-propanol	$(CH_3)_2CHCH_2OH$	7.32705	1248.48	172.92	20 to 115
n-Octane	C_8H_{18}	6.91868	1351.99	209.15	19 to 152
n-Pentane	C_5H_{12}	6.85296	1064.84	233.01	−50 to 58
Phenol	C_6H_5OH	7.1330	1516.79	174.95	107 to 182
Propane	C_3H_8	6.80338	804.00	247.04	−108 to −25
1-Propanol	C_3H_7OH	7.84767	1499.21	204.64	2 to 120
2-Propanol	C_3H_7OH	8.11778	1580.92	219.61	0 to 101
Toluene	$C_6H_5CH_3$	6.95464	1344.800	219.48	6 to 137
Water*	H_2O	8.01195	1698.785	231.04	5 to 150
o-Xylene	C_8H_{10}	6.99891	1474.679	213.69	32 to 172
m-Xylene	C_8H_{10}	7.00908	1462.266	215.11	28 to 166
p-Xylene	C_8H_{10}	6.99052	1453.430	215.31	27 to 166

Based on data from James G. Speight, Lange's Handbook of Chemistry, 16th ed., McGraw-Hill, 2005, except as noted.
*Fit to data from steam tables

Appendix F: Thermodynamic Diagrams

Appendix F contains thermodynamic diagrams for the following compounds.

- Nitrogen
- Methane
- R-422A (also known as R-422a or ISCEON® 79)
- Freon® 22 (also known as R-22 or chlorodifluoromethane)

Thermodynamic Properties of Nitrogen

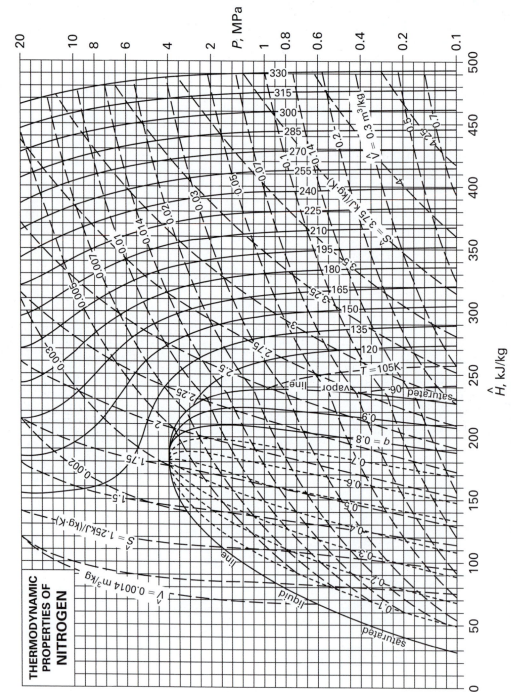

Pressure-Enthalpy Diagram for Methane

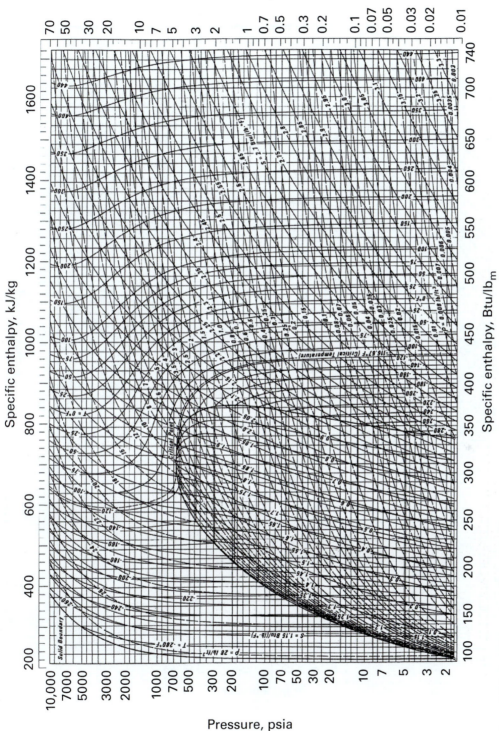

Based on NIST: Thermal Physics Division, Boulder, CO., as seen in Lira, J.R. Elliott and C. T., Introductory Chemical Engineering Thermodynamics, Prentice-Hall, 1999.

Thermodynamic Properties of R-422A (also known as R-422a or ISCEON® 79)

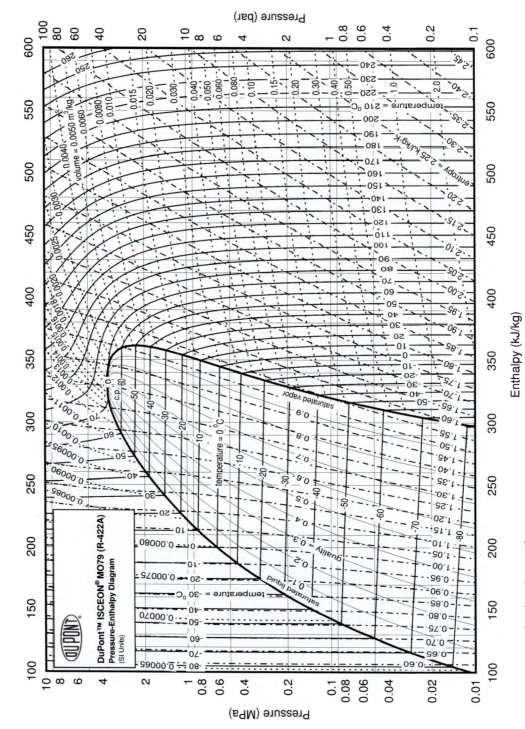

Thermodynamic Properties of HCFC-22 (also known as R-22 or chlorodifluoromethane)

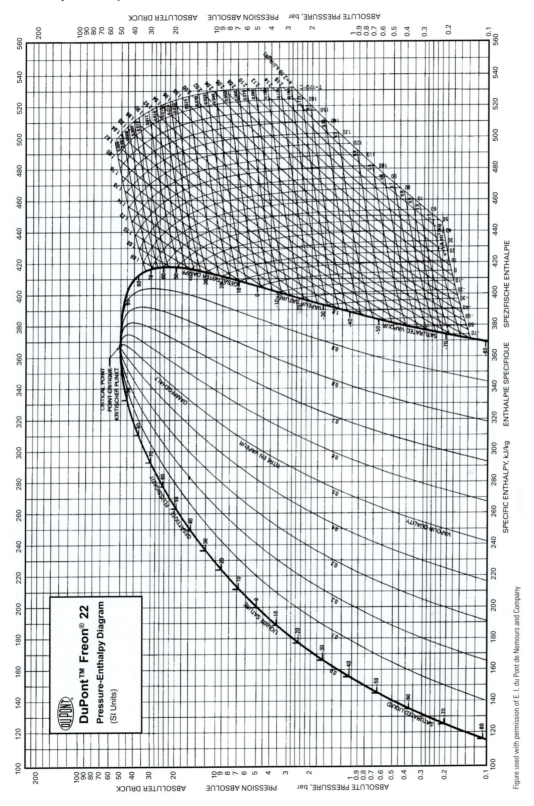

DuPont™ Freon® 22
Pressure-Enthalpy Diagram
(SI Units)

Figure used with permission of E. I. du Pont de Nemours and Company

Appendix G: The Joback Group Additivity Method

The 11 properties that can be estimated using the Joback method are summarized in Table G-1. The symbols are consistent with the nomenclature of this book, and in some cases differ from those published by Joback and Reid (Reid, 1987).

TABLE G-1: Properties that can be estimated using Joback method.

Property	Symbol	Units	Equation
Boiling point at $P = 1$ atm (101.325 kPa)	T_b	K	$T_b = 198.2 + \Sigma$
Freezing point at $P = 1$ atm (101.325 kPa)	T_f	K	$T_f = 122.5 + \Sigma$
Critical temperature	T_c	K	$T_c = T_b[0.584 + 0.965\Sigma - (\Sigma)^2]^{-1}$
Critical pressure	P_c	Bar	$P_c = (0.113 + 0.0032n_a - \Sigma)^{-2}$
Critical molar volume	$\underline{V}_c$	cm³/mol	$\underline{V}_c = 17.5 + \Sigma$
Enthalpy of formation, ideal gas state at 298 K	$\Delta \underline{H}^0_{f,298}$	kJ/mol	$\Delta \underline{H}^0_{f,298} = 68.29 + \Sigma$
Gibbs energy of formation, ideal gas state at 298 K	$\Delta \underline{G}^0_{f,298}$	kJ/mol	$\Delta \underline{G}^0_{f,298} = 53.88 + \Sigma$
Ideal gas heat capacity as a function of temperature	C_p^*	J/mol·K	$C_p^* = \Sigma(a) - 37.93 + [\Sigma(b) + 0.210]T + [\Sigma(c) - 3.91 \times 10^{-4}]T^2 + [\Sigma(d) + 2.06 \times 10^{-7}]T^3$
Enthalpy of vaporization at T_b	$\Delta \underline{H}_{vap}$	kJ/mol	$\Delta \underline{H}_{vap} = 15.30 + \Sigma$
Enthalpy of fusion	$\Delta \underline{H}_{fus}$	kJ/mol	$\Delta \underline{H}_{fus} = -0.88 + \Sigma$
Liquid viscosity as a function of temperature	η_L	N·s/m²	$\eta_L = MW \times \exp\left\{\dfrac{[\Sigma(\eta_a) - 597.82]}{T} + \Sigma(\eta_b) - 11.202\right\}$

■ The symbol Σ signifies the sum of the group contributions for the property of interest. Group contributions are obtained from Table G-2. If a group occurs multiple times in the compound, the group contribution must be multiplied by the number of occurrences.

■ The ideal gas heat capacity and liquid viscosity are functions of temperatures, so these formulas each require multiple parameters and have more than one Σ term.

■ In the formula for P_c, the symbol n_a signifies the number of atoms in the molecule.

■ In the formula for η_L, MW stands for molecular weight of the whole compound.

TABLE G-2: Group contributions to properties, for use with the Joback method (Reid, 1987).

	T_c	P_c	V_c
Non-ring increments			
—CH_3	0.0141	−0.0012	65
—CH_2—	0.0189	0	56
>CH—	0.0164	0.0020	41
>C<	0.0067	0.0043	27
=CH_2	0.0113	−0.0028	56
=CH—	0.0129	−0.0006	46
=C<	0.0117	0.0011	38
=C=	0.0026	0.0028	36
=CH	0.0027	−0.0008	46
≡C—	0.0020	0.0016	37
Ring increments			
—CH_2—	0.0100	0.0025	48
>CH—	0.0122	0.0004	38
>C<	0.0042	0.0061	27
=CH—	0.0082	0.0011	41
=C<	0.0143	0.0008	32
Halogen increments			
—F	0.0111	−0.0057	27
—Cl	0.0105	−0.0049	58
—Br	0.0133	0.0057	71
—I	0.0068	−0.0034	97
Oxygen increments			
—OH (alcohol)	0.0741	0.0112	28
—OH (phenol)	0.0240	0.0184	−25
—O— (nonring)	0.0168	0.0015	18
—O— (ring)	0.0098	0.0048	13
>C=O (nonring)	0.0380	0.0031	62
>C=O (ring)	0.0284	0.0028	55
O=CH— (aldehyde)	0.0379	0.0030	82
—COOH (acid)	0.0791	0.0077	89
—COO— (ester)	0.0481	0.0005	82
=O (expect as above)	0.0143	0.0101	36
Nitrogen increments			
—NH_2	0.0243	0.0109	38
>NH (nonring)	0.0295	0.0077	35
>NH (ring)	0.0130	0.0114	29
>N— (nonring)	0.0169	0.0074	9
—N= (nonring)	0.0255	−0.0099	—
—N= (ring)	0.0085	0.0076	34
=NH	—	—	—
—CN	0.0496	−0.0101	91
—NO_2	0.0437	0.0064	91
Sulfur increments			
—SH	0.0031	0.0084	63
—S— (nonring)	0.0119	0.0049	54
—S— (ring)	0.0019	0.0051	38

TABLE G-2 (*continued*)

	T_b	T_f	$\Delta \underline{H}°_{f,298}$	$\Delta \underline{G}°_{f,298}$
Non-ring increments				
—CH$_3$	23.58	−5.10	−76.45	−43.96
—CH$_2$—	22.88	11.27	−20.64	8.42
>CH—	21.74	12.64	29.89	58.36
>C<	18.25	46.43	82.23	116.02
=CH$_2$	18.18	−4.32	−9.63	3.77
=CH—	24.96	8.73	37.97	48.53
=C<	24.14	11.14	83.99	92.36
=C=	26.15	17.78	142.14	136.70
≡CH	9.20	−11.18	79.30	77.71
≡C—	27.38	64.32	115.51	109.82
Ring increments				
—CH$_2$—	27.15	7.75	−26.80	−3.68
>CH—	21.78	19.88	8.67	40.99
>C<	21.32	60.15	79.72	87.88
=CH—	26.73	8.13	2.09	11.30
=C<	31.01	37.02	46.43	54.05
Halogen increments				
—F	−0.03	−15.78	−251.92	−247.19
—Cl	38.13	13.55	−71.55	−64.31
—Br	66.86	43.43	−29.48	−38.06
—I	93.84	41.69	21.06	5.74
Oxygen increments				
—OH (alcohol)	92.88	44.45	−208.04	−189.20
—OH (phenol)	76.34	82.83	−221.65	−197.37
—O— (nonring)	22.42	22.23	−132.22	−105.00
—O— (ring)	31.22	23.05	−138.16	−98.22
>C=O (nonring)	76.75	61.20	−133.22	−120.50
>C=O (ring)	94.97	75.97	−164.50	−126.27
O=CH— (aldehyde)	72.24	36.90	−162.03	−143.48
—COOH (acid)	169.09	155.50	−426.72	−387.87
—COO— (ester)	81.10	53.60	−337.92	−301.95
=O (except as above)	−10.50	2.08	−247.61	−250.83
Nitrogen increments				
—NH$_2$	73.23	66.89	−22.02	14.07
>NH (nonring)	50.17	52.66	53.47	89.39
>NH (ring)	52.82	101.51	31.65	75.61
>N— (nonring)	11.74	48.84	123.34	163.16
—N= (nonring)	74.60	—	23.61	—
—N= (ring)	57.55	68.40	55.52	79.93
=NH	83.08	68.91	93.70	119.66
—CH	125.66	59.89	88.43	89.22
—NO$_2$	152.54	127.24	−66.57	−16.83
Sulfur increments				
—SH	63.56	20.09	−17.33	−22.99
—S— (nonring)	68.78	34.40	41.87	33.12
—S— (ring)	52.10	79.93	39.10	27.76

TABLE G-2 (*continued*)

	Ideal gas heat capacity			
	(a)	**(b)**	**(c)**	**(d)**
Non-ring increment				
—CH$_3$	1.95×10^1	-8.08×10^{-3}	1.53×10^{-4}	-9.67×10^{-8}
—CH$_2$—	-9.09×10^{-1}	9.50×10^{-2}	-544×10^{-5}	1.19×10^{-8}
>CH—	-2.30×10^1	2.04×10^{-1}	-2.65×10^{-4}	1.20×10^{-7}
>C<	-6.62×10^1	4.27×10^{-1}	-6.41×10^{-4}	3.01×10^{-7}
=CH$_2$	2.36×10^1	-3.81×10^{-2}	1.72×10^{-4}	-1.03×10^{-7}
=CH—	-8.00	1.05×10^{-1}	-9.63×10^{-5}	3.56×10^{-8}
=C<	-2.81×10^1	2.08×10^{-1}	-3.06×10^{-4}	1.46×10^{-7}
=C=	2.74×10^1	-5.57×10^{-2}	1.01×10^{-4}	-5.02×10^{-8}
≡CH	2.45×10^1	-2.71×10^{-2}	1.11×10^{-4}	-6.78×10^{-8}
=C—	7.87	2.01×10^{-2}	-8.33×10^{-6}	1.39×10^{-9}
Ring increments				
—CH$_2$—	-6.03	8.54×10^{-2}	-8.00×10^{-6}	-1.80×10^{-8}
>CH—	-2.05×10^1	1.62×10^{-1}	-1.60×10^{-4}	6.24×10^{-8}
>C<	-9.09×10^1	5.57×10^{-1}	-9.00×10^{-4}	4.69×10^{-7}
=CH—	-2.14	5.74×10^{-2}	-1.64×10^{-6}	-1.59×10^{-8}
=C<	-8.25	1.01×10^{-1}	-1.42×10^{-4}	6.78×10^{-8}
Halogen increments				
—F	2.65×10^1	-9.13×10^{-2}	1.91×10^{-4}	-1.03×10^{-7}
—Cl	3.33×10^1	-9.63×10^{-2}	1.87×10^{-4}	-9.96×10^{-8}
—Br	2.86×10^1	-6.49×10^{-2}	1.36×10^{-4}	-7.45×10^{-8}
—I	3.21×10^1	-6.41×10^{-2}	1.26×10^{-4}	-6.87×10^{-8}
Oxygen increments				
—OH (alcohol)	2.57×10^1	-6.91×10^{-2}	1.77×10^{-4}	-9.88×10^{-8}
—OH (phenol)	-2.81	1.11×10^{-1}	-1.16×10^{-4}	4.94×10^{-8}
—O— (nonring)	2.55×10^1	-6.32×10^{-2}	1.11×10^{-4}	-5.48×10^{-8}
—O— (ring)	1.22×10^1	-1.26×10^{-2}	6.03×10^{-5}	-3.86×10^{-8}
>C=O (nonring)	6.45	6.70×10^{-2}	-3.57×10^{-5}	2.86×10^{-9}
>C=O (ring)	3.04×10^1	-8.29×10^{-2}	2.36×10^{-4}	-1.31×10^{-7}
O=CH— (aldehyde)	3.09×10^1	-3.36×10^{-2}	1.60×10^{-4}	-9.88×10^{-8}
—COOH (acid)	2.41×10^1	4.27×10^{-2}	8.04×10^{-5}	-6.87×10^{-8}
—COO— (ester)	2.45×10^1	4.02×10^{-2}	4.02×10^{-5}	-4.52×10^{-8}
=O (excepts as above)	6.82	1.96×10^{-2}	1.27×10^{-5}	-1.78×10^{-8}
Nitrogen increments				
—NH$_2$	2.69×10^1	-4.12×10^{-2}	1.64×10^{-4}	-9.76×10^{-8}
>NH (nonring)	-1.21	7.62×10^{-2}	-4.86×10^{-5}	1.05×10^{-8}
>NH (ring)	1.18×10^1	-2.03×10^{-2}	1.07×10^{-4}	-6.28×10^{-8}
>N— (nonring)	-3.11×10^1	2.27×10^{-1}	-3.20×10^{-4}	1.46×10^{-7}
—N= (nonring)	—	—	—	—
—N= (ring)	8.83	-3.84×10^{-3}	4.35×10^{-5}	-2.60×10^{-8}
=NH	5.69	-4.12×10^{-3}	1.28×10^{-4}	-8.88×10^{-8}
—CN	3.65×10^1	-7.33×10^{-2}	1.84×10^{-4}	-1.03×10^{-7}
—NO$_2$	2.59×10^1	-3.74×10^{-3}	1.29×10^{-4}	-8.83×10^{-8}
Sulfur increments				
—SH	3.53×10^1	-7.58×10^{-2}	1.85×10^{-4}	-1.03×10^{-7}
—S— (nonring)	1.96×10^1	-5.61×10^{-3}	4.02×10^{-5}	-2.76×10^{-8}
—S— (ring)	1.67×10^1	4.81×10^{-3}	2.77×10^{-5}	-2.11×10^{-8}

TABLE G-2 (*continued*)

	$\Delta\underline{H}_{ub}$	$\Delta\underline{H}_f$	Liquid viscosity (η_A)	(η_b)
Non-ring increments				
—CH₃	2.373	0.908	548.29	−1.719
—CH₂—	2.226	2.590	94.16	−0.199
>CH—	1.691	0.749	−322.15	1.187
>C<	0.636	−1.460	−573.56	2.307
=CH₂	1.724	−0.473	495.01	−1.539
=CH—	2.205	2.691	82.28	−0.242
=C<	2.138	3.063	—	—
=C=	2.661	4.720	—	—
≡CH	1.155	2.322	—	—
=C—	3.302	4.151	—	—
Ring increments				
—CH₂—	2.398	0.490	307.53	−0.798
>CH—	1.942	3.243	−394.29	1.251
>C<	0.644	−1.373	—	—
=CH—	2.544	1.101	259.65	−0.702
=C<	3.059	2.394	−245.74	0.912
Halogen increments				
—F	−0.670	1.398	—	—
—Cl	4.532	2.515	625.45	−1.814
—Br	6.582	3.603	738.91	−2.038
—I	9.520	2.724	809.55	−2.224
Oxygen increments				
—OH (alcohol)	16.826	2.406	2173.72	−5.057
—OH (phenol)	12.499	4.490	3018.17	−7.314
—O— (nonring)	2.410	1.188	122.09	−0.386
—O— (ring)	4.682	5.879	440.24	−0.953
>O=O (nonring)	8.972	4.189	340.35	−0.350
>C=O (ring)	6.645	—	—	—
O=CH— (aldehyde)	9.093	3.197	740.92	−1.713
—COOH (acid)	19.537	11.051	1317.23	−2.578
—COO— (ester)	9.633	6.959	483.88	−0.966
=O (except as above)	5.909	3.624	675.24	−1.340
Nitrogen increments				
—NH₂	10.788	3.515	—	—
>NH (nonring)	6.436	5.009	—	—
>NH (ring)	6.930	7.490	—	—
>H— (nonring)	1.896	4.703	—	—
—N= (nonring)	3.335	—	—	—
—N= (ring)	6.528	3.649	—	—
=NH	12.169	—	—	—
—CH	12.851	2.414	—	—
—NO₂	16.738	9.679	—	—
Sulfur increments				
—SH	6.884	2.360	—	—
—S— (nonring)	6.817	4.130	—	—
—S— (ring)	5.984	1.557	—	—

K.G. Joback, R.C. Reid, "Estimation of pure-component properties from group-contributions," Chemical Engineering Communications, Volume 57, Issue 1-6, 1987, pp. 233-243.

Index

Mass	$1 \text{ kg} = 1000 \text{ g} = 2.20462 \text{ lb}_m$
	$1 \text{ lb}_m = 453.6 \text{ g} = 16 \text{ oz.}$
	$1 \text{ ton} = 2000 \text{ lb}_m = 907.2 \text{ kg}$
	$1 \text{ metric ton} = 1000 \text{ kg}$
Length	$1 \text{ m} = 100 \text{ cm} = 1000 \text{ mm} = 39.370 \text{ in} = 3.2808 \text{ ft} = 1.0936 \text{ yd}$
	$1 \text{ in} = 2.540 \text{ cm}$
	$1 \text{ ft} = 12 \text{ in} = 30.480 \text{ cm} = 0.3048 \text{ m}$
	$1 \text{ yd} = 3 \text{ ft}$
Volume	$1 \text{ m}^3 = 1000 \text{ L} = 10^6 \text{ cm}^3 = 61{,}023.4 \text{ in}^3 = 35.3134 \text{ ft}^3$
	$1 \text{ ft}^3 = 1728 \text{ in}^3 = 0.028318 \text{ m}^3 = 28.318 \text{ L}$
Pressure	$1 \text{ Pa} = 1 \text{ N/m}^2$
	$1 \text{ atm} = 1.01325 \times 10^5 \text{ Pa} = 101.325 \text{ kPa} = 1.01325 \text{ bar} = 14.696 \text{ psia} = 760 \text{ mm Hg}$
	$1 \text{ bar} = 100 \text{ kPa} = 0.1 \text{ MPa} = 10^5 \text{ Pa} = 0.98692 \text{ atm} = 14.504 \text{ psia} = 750.06 \text{ mm Hg}$
Force	$1 \text{ N} = 1 \text{ kg m/s}^2 = 0.22481 \text{ lb}_f$
	$1 \text{ lb}_f = 4.44322 \text{ N} = 32.174 \text{ lb}_m \text{ ft/s}^2$
Energy	$1 \text{ J} = 1 \text{ Nm} = 0.7376 \text{ ft} \cdot \text{lb}_f = 9.486 \times 10^{-4} \text{ BTU} = 0.23901 \text{ cal}$
	$1 \text{ ft} \cdot \text{lb}_f = 1.3558 \text{ J}$
Power	$1 \text{ W} = 1 \text{ J/s} = 9.486 \times 10^{-4} \text{ BTU/s} = 1.341 \times 10^{-3} \text{ hp}$
Values of the Gas Constant, R	$8.3143 \text{ J/(mol K)} = 83.14 \text{ bar cm}^3/\text{(mol K)} = 8.314 \text{ Pa m}^3/\text{(mol K)}$
	$82.056 \text{ cm}^3 \text{ atm/(mol K)} = 0.082056 \text{ L atm/(mol K)}$
	$10.731 \text{ ft}^3 \text{ psia/(lbmol R)}$
	$1.987 \text{ BTU/(lbmol R)}$
	$6.1324 \text{ ft} \cdot \text{lb}_f/\text{mol} \cdot \text{K}$
Gravitational constant (sea level)	$g = 9.8066 \text{ m/s}^2 = 32.174 \text{ ft/s}^2$

PRINCIPAL UNITS USED IN MECHANICS

Quantity	International System (SI)			U.S. Customary System (USCS)		
	Unit	Symbol	Formula	Unit	Symbol	Formula
Acceleration (angular)	radian per second squared		rad/s^2	radian per second squared		rad/s^2
Acceleration (linear)	meter per second squared		m/s^2	foot per second squared		ft/s^2
Area	square meter		m^2	square foot		ft^2
Density (mass) (Specific mass)	kilogram per cubic meter		kg/m^3	slug per cubic foot		$slug/ft^3$
Density (weight) (Specific weight)	newton per cubic meter		N/m^3	pound per cubic foot	pcf	lb/ft^3
Energy; work	joule	J	$N{\cdot}m$	foot-pound		ft-lb
Force	newton	N	$kg{\cdot}m/s^2$	pound	lb	(base unit)
Force per unit length (Intensity of force)	newton per meter		N/m	pound per foot		lb/ft
Frequency	hertz	Hz	s^{-1}	hertz	Hz	s^{-1}
Length	meter	m	(base unit)	foot	ft	(base unit)
Mass	kilogram	kg	(base unit)	slug		$lb{\text-}s^2/ft$
Moment of a force; torque	newton meter		$N{\cdot}m$	pound-foot		lb-ft
Moment of inertia (area)	meter to fourth power		m^4	inch to fourth power		$in.^4$
Moment of inertia (mass)	kilogram meter squared		$kg{\cdot}m^2$	slug foot squared		$slug{\text-}ft^2$
Power	watt	W	J/s $(N{\cdot}m/s)$	foot-pound per second		ft-lb/s
Pressure	pascal	Pa	N/m^2	pound per square foot	psf	lb/ft^2
Section modulus	meter to third power		m^3	inch to third power		$in.^3$
Stress	pascal	Pa	N/m^2	pound per square inch	psi	$lb/in.^2$
Time	second	s	(base unit)	second	s	(base unit)
Velocity (angular)	radian per second		rad/s	radian per second		rad/s
Velocity (linear)	meter per second		m/s	foot per second	fps	ft/s
Volume (liquids)	liter	L	$10^{-3}\ m^3$	gallon	gal.	$231\ in.^3$
Volume (solids)	cubic meter		m^3	cubic foot	cf	ft^3